Humvee HMMV M998 series Technical Manual Unit, Direct Support And General Support Maintenance Repair Parts and Special Tools List Volume 2 For

M998
M1038
M1097A1
M966
M1036
M1045
M1046
M1043
M1044A1
M1037
M1042
M996
M997
M1035A2

ARMY TM 9-2320-280-24P-2
AIR FORCE TO 36A12-1A-3044-2
MARINE CORPS TM 2320-24P/8B

March 2001 w/changes from Jul 2004
edited by
Brian Greul

The M998 HMMV (High Mobility Multipurpose Wheeled Vehicle) was introduced in 1983 to replace the ubiquitous M151 commonly called a Jeep. The HMMV will be replaced by the JLTV with the first fieldings beginning in 2019 for the US Military. This manual is a reprint of the official manual.

Fonts in this reprint may appear differently due to substitutions for unavailable fonts at the time of printing. Every effort has been made to faithfully reproduce the document while cleaning up the pages to make them usable to you the reader.

Black marks in the margins indicate the Change Order Content.

Should you have suggestions or feedback on ways to improve this book please send email to Books@OcotilloPress.com

If you would like to order a copy of this book as a 3 ring punched looseleaf print please contact Books@OcotilloPress.com

---

Edited 2021 Ocotillo Press
ISBN 978-1-954285-09-5

No rights reserved. This content of this book is in the public domain as it is a work of the US Government. It is reproduced by the publisher as a convenience to enthusiasts and others who may wish to own a quality copy of it. It has been adjusted to accomodate the binding process.

Printed in the United States of America

Ocotillo Press
Houston, TX 77017
Books@OcotilloPress.com

Disclaimer: The user of this book is responsible for following safe and lawful practices at all times. The publisher assumes no responsibility for the use of the content of this book. The publisher has made an effort to ensure that the text is complete and properly typeset, however omissions, errors, and other issues may exist that the publisher is unaware of.

Removals: The publisher has removed the DA2028 Recommended Changes to Technical Publications form from the back of this manual along with the mailng label for such forms that were printed in the original publication.

VOLUME NO. 2	ARMY TM 9-2320-280-24P-2 AIR FORCE TO 36A12-1A-3044-2 MARINE CORPS TM 2320-24P/8B (SUPERSEDES TM 9-2320-280-24P-2, 31 January 1996)

TECHNICAL MANUAL
UNIT, DIRECT SUPPORT AND GENERAL SUPPORT MAINTENANCE REPAIR PARTS AND SPECIAL TOOLS LIST
FOR

TRUCK, UTILITY: CARGO/TROOP CARRIER, 1-1/4 TON, 4X4, M998 (2320-01-107-7155) (EIC: BBD); M998A1 (2320-01-371-9577) (EIC: BBN);

TRUCK, UTILITY: CARGO/TROOP CARRIER, 1-1/4 TON, 4X4, W/WINCH, M1038 (2320-01-107-7156) (EIC: BBE); M1038A1 (2320-01-371-9578) (EIC: BBP);

TRUCK, UTILITY: HEAVY VARIANT, 4X4, M1097 (2320-01-346-9317) (EIC: BBM); M1097A1 (2320-01-371-9583) (EIC: BBU); M1097A2 (2320-01-380-8604) (EIC: BB6); M1123 (2320-01-455-9593) (EIC: B6G);

TRUCK, UTILITY: TOW CARRIER, ARMORED, 1-1/4 TON, 4X4, M966 (2320-01-107-7153) (EIC: BBC); M966A1 (2320-01-372-3932) (EIC: BBX); M1121 (2320-01-456-1282) (EIC: B6H);

TRUCK, UTILITY: TOW CARRIER, W/SUPPLEMENTAL ARMOR, 1-1/4 TON, 4X4, M1045 (2320-01-146-7191); M1045A1 (2320-01-371-9580) (EIC: BBR); M1045A2 (2320-01-380-8229) (EIC: BB5);

TRUCK, UTILITY: TOW CARRIER, W/SUPPLEMENTAL ARMOR, 1-1/4 TON, 4X4, W/WINCH, M1046 (2320-01-146-7188);

M1046A1 (2320-01-371-9582) (EIC: BBT);

TRUCK, UTILITY: ARMAMENT CARRIER, ARMORED, 1-1/4 TON, 4X4, M1025 (2320-01-128-9551) (EIC: BBF); M1025A1 (2320-01-371-9584) (EIC: BBV); M1025A2 (2320-01-380-8233) (EIC: BB3);

TRUCK, UTILITY: ARMAMENT CARRIER, ARMORED, 1-1/4 TON, 4X4, W/WINCH, M1026 (2320-01-128-9552) (EIC: BBG);

M1026A1 (2320-01-371-9579) (EIC: BBQ);

TRUCK, UTILITY: ARMAMENT CARRIER, W/SUPPLEMENTAL ARMOR, 1-1/4 TON, 4X4, M1043 (2320-01-146-7190);

M1043A1 (2320-01-372-3933) (EIC: BBY); M1043A2 (2320-01-380-8213) (EIC: BB4);

TRUCK, UTILITY: ARMAMENT CARRIER, W/SUPPLEMENTAL ARMOR, 1-1/4 TON, 4X4, W/WINCH, M1044 (2320-01-146-7189);

M1044A1 (2320-01-371-9581) (EIC: BBS);

TRUCK, UTILITY: S250 SHELTER CARRIER, 4X4, M1037 (2320-01-146-7193) (EIC: BBK);

TRUCK, UTILITY: S250 SHELTER CARRIER, 4X4, W/WINCH, M1042 (2320-01-146-7187);

TRUCK, AMBULANCE, 2-LITTER, ARMORED, 4X4, M996 (2310-01-111-2275) (EIC: BBB); M996A1 (2310-01-372-3935) (EIC: BB2);

TRUCK, AMBULANCE, 4-LITTER, ARMORED, 4X4, M997 (2310-01-111-2274) (EIC: BBA); M997A1 (2310-01-372-3934) (EIC: BBZ); M997A2 (2310-01-380-8225) (EIC: BB8);

TRUCK, AMBULANCE, 2-LITTER, SOFT TOP, 4X4, M1035 (2310-01-146-7194); M1035A1 (2310-01-371-9585) (EIC: BBW);

M1035A2 (2310-01-380-8290) (EIC: BB9).

HEADQUARTERS, DEPARTMENTS OF THE ARMY, THE AIR FORCE, AND MARINE CORPS

MARCH 2001

ARMY TM 9-2320-280-24P-2 AIR FORCE TO 36A12-1A-3044-2 MARINE CORPS TM 2320-24P/8B

CHANGE HEADQUARTERS, DEPARTMENTS

NO. 1 AIR FORCE, AND MARINE CORPS

OF THE ARMY, THE

WASHINGTON, D.C., 15 July 2004

TECHNICAL MANUAL

VOLUME 2 OF 2
UNIT, DIRECT SUPPORT AND GENERAL SUPPORT MAINTENANCE
REPAIR PARTS AND SPECIAL TOOLS LIST
FOR

TRUCK, UTILITY: CARGO/TROOP CARRIER, 1-1/4 TON, 4X4, M998 (2320-01-107-7155) (EIC: BBD); M998A1 (2320-01-371-9577) (EIC: BBN);

TRUCK, UTILITY: CARGO/TROOP CARRIER, 1-1/4 TON, 4X4, W/WINCH, M1038 (2320-01-107-7156) (EIC: BBE); M1038A1 (2320-01-371-9578) (EIC: BBP);

TRUCK, UTILITY: HEAVY VARIANT, 4X4, M1097 (2320-01-346-9317) (EIC: BBM); M1097A1 (2320-01-371-9583) (EIC: BBU); M1097A2 (2320-01-380-8604) (EIC: BB6); M1123 (2320-01-455-9593) (EIC: B6G);

TRUCK, UTILITY: TOW CARRIER, ARMORED, 1-1/4 TON, 4X4, M966 (2320-01-107-7153) (EIC: BBC); M966A1 (2320-01-372-3932) (EIC: BBX); M1121 (2320-01-456-1282) (EIC: B6H);

TRUCK, UTILITY: TOW CARRIER, W/SUPPLEMENTAL ARMOR, 1-1/4 TON, 4X4, M1045 (2320-01-146-7191); M1045A1 (2320-01-371-9580) (EIC: BBR); M1045A2 (2320-01-380-8229) (EIC: BB5);

TRUCK, UTILITY: TOW CARRIER, W/SUPPLEMENTAL ARMOR, 1-1/4 TON, 4X4, W/WINCH, M1046 (2320-01-146-7188);

M1046A1 (2320-01-371-9582) (EIC: BBT);

TRUCK, UTILITY: ARMAMENT CARRIER, ARMORED, 1-1/4 TON, 4X4, M1025 (2320-01-128-9551) (EIC: BBF); M1025A1 (2320-01-371-9584) (EIC: BBV); M1025A2 (2320-01-380-8233) (EIC: BB3);

TRUCK, UTILITY: ARMAMENT CARRIER, ARMORED, 1-1/4 TON, 4X4, W/WINCH, M1026 (2320-01-128-9552) (EIC: BBG);

M1026A1 (2320-01-371-9579) (EIC: BBQ);

TRUCK, UTILITY: ARMAMENT CARRIER, W/SUPPLEMENTAL ARMOR, 1-1/4 TON, 4X4, M1043 (2320-01-146-7190);

M1043A1 (2320-01-372-3933) (EIC: BBY); M1043A2 (2320-01-380-8213) (EIC: BB4);

TRUCK, UTILITY: ARMAMENT CARRIER, W/SUPPLEMENTAL ARMOR, 1-1/4 TON, 4X4, W/WINCH, M1044 (2320-01-146-7189);

M1044A1 (2320-01-371-9581) (EIC: BBS);

TRUCK, UTILITY: S250 SHELTER CARRIER, 4X4, M1037 (2320-01-146-7193) (EIC: BBK);

TRUCK, UTILITY: S250 SHELTER CARRIER, 4X4, W/WINCH, M1042 (2320-01-146-7187);

TRUCK, AMBULANCE, 2-LITTER, ARMORED, 4X4, M996 (2310-01-111-2275) (EIC: BBB); M996A1 (2310-01-372-3935) (EIC: BB2);

TRUCK, AMBULANCE, 4-LITTER, ARMORED, 4X4, M997 (2310-01-111-2274) (EIC: BBA); M997A1 (2310-01-372-3934) (EIC: BBZ); M997A2 (2310-01-380-8225) (EIC: BB8);

TRUCK, AMBULANCE, 2-LITTER, SOFT TOP, 4X4, M1035 (2310-01-146-7194); M1035A1 (2310-01-371-9585) (EIC: BBW); M1035A2 (2310-01-380-8290) (EIC: BB9).

TM 9-2320-280-24P-2, 30 March 2001, is changed as follows:

1. Remove old pages and insert new pages as indicated below.

2. New or changed material is indicated by a vertical bar in the margin of the page and an asterisk (*) next to changed item number.

Remove pages	Insert pages	Remove pages	Insert pages
A/(B Blank)	A and B	429-2	429-1 and 429-2
i through 329-2	i through 329-2	430-2	430-1 and 430-2
330-1 and Figure 331	330-1 and Figure 331	Figure 433	432-1 and Figure 433
331-3 and 331-4	331-3 and 331-4	Figure 438 through 438-2	Figure 438 through 438-2
Figure 333 through Figure 335	Figure 333 through Figure 335	439-1 and 439-2	439-1 and 439-2
Figure 338 through 338-2	Figure 338 through 338-2	440-1 and 440-2	440-1 and 440-2
339-1 and 339-2	339-1 and 339-2	442-2	441-1 and 442-2
342-1 and Figure 343	342-1 and Figure 343	457-3 through Figure 458	457-3 through Figure 458
347-1 through 348-2	347-1 through 348-2	458-5 and Figure 459	458-5 and Figure 459
353-1 and Figure 354	353-1 and Figure 354	461-1 and 461-2	461-1 and 461-2
354-3 and 354-4	354-3 and 354-4	465-1 through Figure 468	465-1 through Figure 468
355-3 and Figure 356	355-3 and Figure 356	471-1 and 471-2	471-1 and 471-2
365-1 and 365-2	365-1 and 365-2	474-1 through 474-4	474-1 through 474-4
366-3 and 366-4	366-3 and 366-4	475-3 and 475-4	475-3 and 475-4
Figure 368	Figure 368	480-3 and 480-4	480-3 and 480-4
368-5 and Figure 369	368-5 through Figure 369	481-3 through 482-2	481-3 through 482-2
372-3 through Figure 373	372-3 through Figure 373	485-1 and 485-2	485-1 and 485-2
375-1 and 375-2	375-1 and 375-2	487A-1 through 487D-2	None
384-1 through Figure 386	384-1 through Figure 386	Figure 488 through Figure 489	Figure 488 through Figure 489
388-1 and Figure 389	388-1 and Figure 389	490-1 and Figure 491	490-1 through Figure 491
396-1 through Figure 398	396-1 through Figure 398	Kits-1 through Figure 500	Kits-1 through Figure 500
401-1 and 401-2	401-1 and 401-2	505-1 and 505-2	505-1 and 505-2
402-1 and 402-2	402-1 and 402-2	509-1 through Figure 511	509-1 through Figure 511
403-1 through 412-2	403-1 through 412-2	I-1 through I-314	I-1 through I-345
413-1 through 414-2	413-1 through 414-2	DA Form 2028-2	DA Form 2028
415-1 and 415-2	415-1 and 415-2		
416-1 and 416-2	416-1 and 416-2		
417-1 and 417-2	417-1 and 417-2		
418-1 through Figure 424	418-1 through Figure 424		
428-1 and 428-2	428-1 and 428-2		
429-1 and			

3. File this change sheet in front of the publication for reference purposes.

By Order of the Secretary of the Army:

PETER J. SCHOOMAKER
General, United States Army

Official:

JOEL B. HUDSON
*Administrative Assistant to the
Secretary of the Army*

By Order of the Secretary of the Air Force:

RONALD R. FOGLEMAN
*General, United States Air Force
Chief of Staff*

Official:

HENRY VICCELLIO, JR.
*General, United States Air Force
Commander, Air Force Materiel Command*

Distribution:

    To be distributed in accordance with the initial distribution number (IDN) 381008, requirements for TM 9-2320-280-24P-2.

TM 9-2320-280-24P-2 C01

LIST OF EFFECTIVE PAGES

Date of issue of volume 2 is:

Original . . . 0 . . . . . 30 March 2001
Change . . . 1 . .   15 July 2004

TOTAL NUMBER OF PAGES IN VOLUME 2 IS 1023. CONSISTING OF THE FOLLOWING:

Page No. . . . .*Change No. Page No. . . . .*Change No. Page No. . . . .*Change No.

Page No.	*Change No.	Page No.	*Change No.	Page No.	*Change No.
VOLUME 2		Figure 369 - 372-2	0	416-1 - 416-2	1
A	1	372-3 - 372-5	1	416-3 - Figure 417	0
B Added	1	Figure 373 - Figure 375	0	417-1	1
i - vi	1	375-1	1	417-2 - Figure 418	0
vii Added	1	375-2 - Figure 384	0	418-1	1
Figure 329	0	384-1	1	Figure 419	0
329-1 - 329-2	1	Figure 385	0	419-1	1
Figure 330 - 330-1	0	385-1	1	Figure 420	0
Figure 331	1	Figure 386 - Figure 388	0	420-1	1
331-1 - 331-3	0	388-1	1	Figure 421	0
331-4	1	Figure 389 - 396-1	0	421-1	1
Figure 332 - 332-2	0	Figure 397	1	Figure 422	0
Figure 333 - 333-1	1	397-1	0	422-1	1
333-2 Added	1	397-2 - 397-6	1	Figure 423	0
Figure 334 - 334-1	1	Figure 397A - 397E-3 Added	1	423-1	1
Figure 335 - 337-2	0	Figure 398 - Figure 401	0	Figure 424 - Figure 428	0
Figure 338 - 338-2	1	401-1	1	428-1 - 428-2	1
Figure 339	0	401-2 - Figure 402	0	428-3 - Figure 429	0
339-1 - 339-2	1	402-1	1	429-1	1
Figure 340 - Figure 342	0	402-2 - Figure 403	0	429-2 - Figure 430	0
342-1	1	403-1	1	430-1 - 430-2	1
Figure 343 - 347-1	0	403-2	0	430-3 - Figure 432	0
Figure 348 - 348-2	1	Figure 404 - 408-5 Deleted	1	432-1	1
Figure 349 - Figure 353	0	Figure 409	0	Figure 433 - 437-2	0
353-1	1	409-1 - 410-3	1	Figure 438 - 438-2	1
Figure 354 - 354-3	0	410-4 Added	1	Figure 439	0
354-4	1	Figure 411	0	439-1 - 439-2	1
Figure 355 - 355-2	0	411-1	1	Figure 440	0
355-3	1	411-2 Added	1	440-1	1
Figure 356 - 365-1	0	Figure 412 - 412-1	0	440-2 - Figure 441	0
365-2	1	412-2	1	441-1	1
365-3 - 366-2	0	Figure 413	0	Figure 442	0
366-3	1	413-1	1	442-1	1
366-4 - 367-2	0	Figure 414	0	442-2 - 457-2	0
Figure 368	1	414-1 - 414-2	1	457-3 - 457-4	1
368-1 - 368-4	0	414-3 - Figure 415	0	457-5	0
368-5	1	415-1	1	Figure 458	1
368-6 Added	1	415-2 - Figure 416	0	458-1 - 458-4	0

*Zero in this column indicates original page.

A

TM 9-2320-280-24P-2 C01

LIST OF EFFECTIVE PAGES (Contd)

Page No. . . . . *Change No.

458-5 1
Figure 459 - Figure 461 0
461-1 1
461-2 - Figure 465 0
465-1 . . . . .1 Figure
466 - 467-1 Deleted . . .1
Figure 468 - Figure 471 0
471-1 1
471-2 - Figure 474 0
474-1 - 474-3 1
474-4 - 475-3 0
475-4 1
Figure 476 - 480-2 0
480-3 1
480-4 - 481-3 0
Figure 482 1
482-1 0
482-2 1
Figure 483 - Figure 485 0
485-1 1
485-2 - 487-2 . . . . .0 Figure
487A - 487D-2 Added . .1
Figure 488 - 488-1 1
488-2 Deleted 1
Figure 489 - Figure 490 0
490-1 . . . . .1 Figure
490A - 490A-1 Added . .1
Figure 491 1
491-1 - 498-1 0
Kits-1 0
Kits-2 - Kits-21 1
Bulk-1 - Bulk-3 1
Figure 499 0
499-1 1
Figure 500 - 505-1 0
505-2 1
Figure 506 - 509-1 0
Figure 510 - 510-1 1
Figure 511 - 512-1 0
I-1 - I-345 1

*Zero in this column indicates original page.

B

*ARMY TM 9-2320-280-24P-2 AIR
FORCE TO 36A12-1A-3044-2 MARINE
CORPS TM 2320-24P/8B

TECHNICAL MANUAL HEADQUARTERS, DEPARTMENTS
NO. 9-2320-280-24P-2 OF THE ARMY, THE
NO. 2320-24P/8B AIR FORCE, AND MARINE CORPS

TECHNICAL ORDER WASHINGTON, D.C. *30 March 2001*
NO. 36A12-1A-3044-2

TECHNICAL MANUAL

VOLUME 2 OF 2
UNIT, DIRECT SUPPORT AND GENERAL SUPPORT MAINTENANCE
REPAIR PARTS AND SPECIAL TOOLS LIST
FOR

TRUCK, UTILITY: CARGO/TROOP CARRIER, 1-1/4 TON, 4X4, M998 (2320-01-107-7155) (EIC: BBD);
M998A1 (2320-01-371-9577) (EIC: BBN);

TRUCK, UTILITY: CARGO/TROOP CARRIER, 1-1/4 TON, 4X4, W/WINCH, M1038 (2320-01-107-7156) (EIC: BBE);
M1038A1 (2320-01-371-9578) (EIC: BBP);

TRUCK, UTILITY: HEAVY VARIANT, 4X4, M1097 (2320-01-346-9317) (EIC: BBM);
M1097A1 (2320-01-371-9583) (EIC: BBU); M1097A2 (2320-01-380-8604) (EIC: BB6);
M1123 (2320-01-455-9593) (EIC: B6G);

TRUCK, UTILITY: TOW CARRIER, ARMORED, 1-1/4 TON, 4X4, M966 (2320-01-107-7153) (EIC: BBC);
M966A1 (2320-01-372-3932) (EIC: BBX); M1121 (2320-01-456-1282) (EIC: B6H);

TRUCK, UTILITY: TOW CARRIER, W/SUPPLEMENTAL ARMOR, 1-1/4 TON, 4X4, M1045 (2320-01-146-7191);
M1045A1 (2320-01-371-9580) (EIC: BBR); M1045A2 (2320-01-380-8229) (EIC: BB5);

TRUCK, UTILITY: TOW CARRIER, W/SUPPLEMENTAL ARMOR, 1-1/4 TON, 4X4, W/WINCH, M1046 (2320-01-146-7188);
M1046A1 (2320-01-371-9582) (EIC: BBT);

TRUCK, UTILITY: ARMAMENT CARRIER, ARMORED, 1-1/4 TON, 4X4, M1025
(2320-01-128-9551) (EIC: BBF); M1025A1 (2320-01-371-9584) (EIC: BBV); M1025A2 (2320-01-380-8233) (EIC: BB3);

TRUCK, UTILITY: ARMAMENT CARRIER, ARMORED, 1-1/4 TON, 4X4, W/WINCH, M1026 (2320-01-128-9552) (EIC: BBG);

M1026A1 (2320-01-371-9579) (EIC: BBQ);

TRUCK, UTILITY: ARMAMENT CARRIER, W/SUPPLEMENTAL ARMOR, 1-1/4 TON, 4X4, M1043 (2320-01-146-7190);
M1043A1 (2320-01-372-3933) (EIC: BBY); M1043A2 (2320-01-380-8213) (EIC: BB4);

TRUCK, UTILITY: ARMAMENT CARRIER, W/SUPPLEMENTAL ARMOR, 1-1/4 TON, 4X4, W/WINCH, M1044 (2320-01-146-7189);

M1044A1 (2320-01-371-9581) (EIC: BBS);

TRUCK, UTILITY: S250 SHELTER CARRIER, 4X4, M1037 (2320-01-146-7193) (EIC: BBK);

TRUCK, UTILITY: S250 SHELTER CARRIER, 4X4, W/WINCH, M1042 (2320-01-146-7187);

TRUCK, AMBULANCE, 2-LITTER, ARMORED, 4X4, M996 (2310-01-111-2275) (EIC: BBB);
M996A1 (2310-01-372-3935) (EIC: BB2);

TRUCK, AMBULANCE, 4-LITTER, ARMORED, 4X4, M997 (2310-01-111-2274) (EIC: BBA); M997A1
(2310-01-372-3934) (EIC: BBZ); M997A2 (2310-01-380-8225) (EIC: BB8);

TRUCK, AMBULANCE, 2-LITTER, SOFT TOP, 4X4, M1035 (2310-01-146-7194); M1035A1 (2310-01-371-9585) (EIC: BBW);
M1035A2 (2310-01-380-8290) (EIC: BB9).

Current as of 1 January 2001

REPORTING ERRORS AND RECOMMENDING IMPROVEMENTS

(Army). You can help improve this publication. If you find any mistakes or if you know of a way to improve the procedures, please let us know. Submit your DA Form 2028 (Recommended Changes to Publication and Blank Forms), through the Internet, on the Army Electronic Product Support (AEPS) website. The Internet address is http://aeps.ria.army.mil. If you need a password, scroll down and click on "ACCESS REQUEST FORM". The DA Form 2028 is located in the ONLINE FORMS PROCESSING section of the AEPS. Fill out the form and click on SUBMIT. Using this form on the AEPS will enable us to respond quicker to your comments and better manage the DA Form 2028 program. You may also mail, fax or E-mail your letter or DA Form 2028 direct to: AMSTA-LC-CI/Tech Pubs, TACOM-RI, 1 Rock Island Arsenal, Rock Island, IL 61299-7630. The email address is TACOM-TECH-PUBS@ria.army.mil. The fax number is DSN 793-0726 or Commercial (309) 782-0726. (Marine Corps) Submit NAVMC Form 10772 to the Commanding General, Marine Corps Logistic Base (Code 850), Albany, GA 31704-5000.

Approved for public release; distribution is unlimited.

* This manual supersedes TM 9-2320-280-24P-2, dated 31 January 1996.

i   Change 1

TABLE OF CONTENTS
VOLUME 2 OF 2

	Page	Illus Figure
Group 22 Body, chassis, and hull accessory items (Cont'd)		
2202 Split rearview mirror and mounting hardware	329-1	329
2202 Reflectors	330-1	330
2202 Diverter assembly, defroster, and heater ducting	331-1	331
2202 A2 diverter assembly, defroster, heater ducting, and radio brackets	332-1	332
2202 Heater assembly, water valve, and controls	333-1	333
2202 Heater fuel lines, filter to tank, M996, M996A1, M997, M997A1, and M997A2 ambulance	334-1	334
2202 Fuel lines, personnel heater to fuel filter, M996 and M996A1 ambulance	335-1	335
2202 Fuel lines, personnel heater to fuel filter, M997, M997A1, and M997A2 ambulance	336-1	336
2202 Personnel heater assembly and attaching hardware, M996 and M996A1 ambulance	337-1	337
2202 Antenna and intercom cables and mounting hardware, M997, M997A1, and M997A2 ambulance	338-1	338
2202 Antenna and intercom cables and mounting hardware, M996 and M996A1 ambulance	339-1	339
2202 Antenna mount and mounting hardware, M996 and M996A1 ambulance	340-1	340
2202 Antenna mount and mounting hardware, M997, M997A1, and M997A2 ambulance	341-1	341
2202 Heater blower assembly, M996 and M996A1 ambulance	342-1	342
2202 Heater, filter, and attaching hardware, M997, M997A1, and M997A2 ambulance	343-1	343
2202 Air intake filter, M996 and M996A1 ambulance	344-1	344
2202 Heater assembly, multifuel, B-model, M996, M996A1, M997, M997A1, and M997A2 ambulance	345-1	345
2202 Heater grille, exhaust pipe, and mounting hardware, M996, and M996A1 ambulance	346-1	346
2202 Heater grille, exhaust pipe, and mounting hardware, M997, M997A1, and M997A2 ambulance	347-1	347
2202 Heater assembly, multifuel, A-model, M996, M996A1, M997, M997A1, and M997A2 ambulance	348-1	348
2202 Heater component bracket assembly, B-model, M996, M996A1, M997, M997A1, and M997A2 ambulance	349-1	349
2202 Burner assembly, B-model, M996, M996A1, M997, M997A1, and M997A2 ambulance	350-1	350
2202 Burner assembly, A-model, M997, M997A1, and M997A2 ambulance	351-1	351
2202 Sun visor assembly and mounting hardware	352-1	352
2202 Ambulance spreader assembly	353-1	353
2210 Data plates, dash panel	354-1	354
2210 Data plates, battery cable and slave receptacle	355-1	355
2210 Data plates, surge tank, fan shroud, and cook stove	356-1	356
2210 Data plates, generator, transfer case, and engine container	357-1	357
2210 Data plates, winch	358-1	358
2210 Data plates, TOW carrier	359-1	359
2210 Data plates, TOW carrier	360-1	360
2210 Data plates, sights, traversing unit, night vision goggles, and radiac meter	361-1	361

TABLE OF CONTENTS (Cont'd)

	Page	Illus Figure
2210 Data plates and identification tags, M996, M996A1, M997, M997A1, and M997A2 ambulance	362-1	362
Group 33 Special purpose kits	363-1	363
3301 Engine shipping container assembly	363-1	363
3303 Arctic kit, fuel injection pump, engine gaskets, and CDR tube	364-1	364
3303 Arctic kit, pan cover and heater tube	365-1	365
3303 Arctic kit, control box, gas cap, fuel pump, and fuel lines	366-1	366
3303 Arctic kit, control box assembly	367-1	367
3303 Arctic kit, heater assembly, plenum assembly, and related components	368-1	368
3303 Arctic kit, battery box components, insulation, and control box connector	369-1	369
3303 Arctic kit, diverter box, defroster, heater ducting, and radio rack bracket	370-1	370
3303 Arctic kit, heater duct assembly	371-1	371
3303 Arctic kit, diverter box, ducting, insulation, and radio mounting brackets	372-1	372
3303 Arctic kit, two man crew, bow, and mounting components	373-1	373
3303 Arctic kit, two man crew curtain fasteners	374-1	374
3303 Arctic kit, two man crew, top curtain, rear curtain, and door assemblies	375-1	375
3303 Arctic kit, four man crew, top enclosure, bows, and mounting components	376-1	376
3303 Arctic kit, door hinges, and curtain channels	377-1	377
3303 Arctic kit, four man crew, top curtain, rear curtain, and door assemblies	378-1	378
3303 Swingfire heater kit and related parts	379-1	379
3303 Swingfire heater kit, circuit breaker and related parts	380-1	380
3303 Swingfire heater kit, harness assemblies	381-1	381
3303 Two door seat and floor covers	382-1	382
3303 Cargo floor cover and harness assembly	383-1	383
3303 Winterization kit, fuel pump and attaching hardware	384-1	384
3303 Fuel pump tube assembly	385-1	385
3303 Winterization kit, fuel line and jumper cable	386-1	386
3303 Winterization kit, heater and exhaust diverter assembly	387-1	387
3303 Winterization kit, control box, filter, and cover assembly	388-1	388
3303 Seat support assemblies and mounting hardware	389-1	389
3303 Step assembly, support assembly, and mounting hardware	390-1	390
3303 Top plate support and attaching hardware	391-1	391
3303 Fuel can bracket, antenna mounting bracket, and composite light housing mounting hardware	392-1	392
3303 Mounting plates and ventilator exhaust ducting	393-1	393
3303 Screw caps and winterization instruction plate	394-1	394
3303 Top cover assembly and mounting hardware	395-1	395
3303 Wiring harness channel and hardware	396-1	396
3303 End closures, door assembly, and mounting hardware	397-1	397
3303 Engine and crew compartment heater and related parts	397A-1	397A
3303 Engine and crew compartment heater support bracket, heat shield, and mounting hardware	397B-1	397B
3303 Engine and crew compartment heater switch, fuel lines, and mounting hardware	397C-1	397C
3303 Engine and crew compartment heater fuel supply and return lines, muffler, and mounting hardware	397D-1	397D
3303 Engine and crew compartment heater fuel pump, lines, and mounting hardware	397E-1	397E

TABLE OF CONTENTS (Cont'd)

Description	Page	Illus Figure
3305 Deep water fording kit, intake and exhaust systems	398-1	398
3305 Deep water fording kit, selector valve, sensor cup, and vent lines	399-1	399
3305 Deep water fording kit, hydro-boost vent line, engine and transmission dipstick, and tube assemblies	400-1	400
3305 Supplemental deep water fording kit	401-1	401
3307 Troop seat kit, safety strap, and mounting hardware	402-1	402
3307 Troop seat assembly, left and right	403-1	403
Deleted	404-1	404
Deleted	405-1	405
Deleted	406-1	406
Deleted	407-1	407
Deleted	408-1	408
3307 Communications kit, radio rack assembly, and mounting hardware, AN/GRC-160 and TSEC/KY-57	409-1	409
3307 Communications kit, radio cables, AN/GRC-160, AN/VRC-46, and TSEC/KY-57	410-1	410
3307 Communications kit, radio rack assembly, and mounting hardware, AN/GRC-160 and TSEC/KY-57, TOW carrier	411-1	411
3307 Communications kit, radio cables, AN/GRC-160, AN/VRC-46, and TSEC/KY-57, TOW carrier	412-1	412
3307 Communications kit, radio rack assembly, AN/VRC-47 and (2) TSEC/KY-57 assemblies	413-1	413
3307 Communications kit, radio cables, AN/VRC-47 and (2) TSEC/KY-57 assemblies	414-1	414
3307 Communications kit, radio rack assembly, AN/VRC-49 and (2) TSEC/KY-57 assemblies	415-1	415
3307 Communications kit, radio cables, AN/VRC-49 and (2) TSEC/KY-57 assemblies	416-1	416
3307 Communications kit, radio rack assembly	417-1	417
3307 Communications kit, radio rack assembly mounting hardware, AN/VRC-46	418-1	418
3307 Communications kit, radio rack assembly mounting hardware, (2) AN/VRC-46	419-1	419
3307 Communications kit, radio rack assembly mounting hardware, (3) AN/VRC-46	420-1	420
3307 Communications kit, radio rack assembly mounting hardware, AN/VRC-47	421-1	421
3307 Communications kit, radio rack assembly mounting hardware, AN/VRC-48	422-1	422
3307 Communications kit, radio rack assembly mounting hardware, AN/VRC-49	423-1	423
3307 Communications kit, radio cable, main power	424-1	424
3307 Communications kit, radio cables, AN/VRC-47 and AN/VRC-48	425-1	425
3307 Communications kit, radio cables, AN/VRC-46, (2) AN/VRC-46, (3) AN/VRC-46, AN/VRC-47, AN/VRC-48, and AN/VRC-49	426-1	426
3307 Communications kit, radio cables, (2) AN/VRC-46, (3) AN/VRC-47, and AN/VRC-49	427-1	427
3307 Communications kit, radio cables, (3) AN/VRC-46	428-1	428
3307 81 MM mortar kit, container assembly, and stowage rack	429-1	429
3307 81 MM mortar kit, equipment rack assembly, barrel stowage bracket, and bipod stowage tray assembly	430-1	430
3307 Traversing bar kit	431-1	431
3307 Rearview mirror mounting kit	432-1	432

TABLE OF CONTENTS (Cont'd)

	Page	Figure
3307 Cargo barrier extension kit	433-1	433
3307 Slave receptacle kit	434-1	434
3307 Slave receptacle with power cable assembly kit	435-1	435
3307 Power steering oil cooler kit	436-1	436
3307 Brushguard kit (Basic and A1 series)	437-1	437
3307 Kit, S250 shelter mounting components	438-1	438
3307 Kit, S250 shelter tailgate mounting components	439-1	439
3307 Kit, camouflage rack and stowage rack assembly	440-1	440
3307 Kit, prime mover, lt. howitzer, 105MM L119, front bumper	441-1	441
3307 Kit, prime mover, lt. howitzer, 105MM L119, rear bumper	442-1	442
3307 Kit, prime mover, lt. howitzer, 105MM L119, winch	443-1	443
3307 Kit, prime mover, lt. howitzer, 105MM L119, winch control and storage box	444-1	444
3307 Kit, prime mover, lt. howitzer, 105MM L119, trailer receptacle and body wiring harness	445-1	445
3307 Kit, prime mover, lt. howitzer, 105MM L119, ammo rack and sight boxes stowage	446-1	446
3307 Kit, prime mover, lt. howitzer, 105MM L119, telephone, remote, and tripod stowage	447-1	447
3307 Kit, prime mover, lt. howitzer, 105MM L119, fuel can, water can, section chest, and spade stowage	448-1	448
3307 Kit, prime mover, lt. howitzer, 105MM L119, G.D.U. and G.D.U. battery stowage	449-1	449
3307 Kit, prime mover, lt. howitzer, 105MM L119, aiming post and cable reel stowage	450-1	450
3307 Kit, prime mover, lt. howitzer, 105MM L119, pioneer tool stowage tray hardware and parking brake hardware	451-1	451
3307 Kit, towed vulcan system	452-1	452
3307 Kit, pintle extension	453-1	453
3307 Kit, rear seat assembly	454-1	454
3307 Kit, man portable air-defense system (MANPADS)	455-1	455
3307 Kit, 9,000 lb winch	456-1	456
3307 Kit, 9,000 lb winch	457-1	457
3307 9,000 lb winch assembly	458-1	458
3307 Modification kit, sun visor	459-1	459
3307 Modification kit, windshield retainer	460-1	460
3307 Left rearview mirror relocation kit	461-1	461
3307 Brushguard kit	462-1	462
3307 Underbody protection kit with and without winch	463-1	463
3307 Underbody protection kit, transfercase, fuel tank shields, and sway bar link nuts	464-1	464
3307 Cover, access cargo floor modification kit	465-1	465
Deleted	466-1	466
3307 R-12 to FR12 refrigerant conversion retrofit kit	467-1	467
3307 Hand throttle control kit	468-1	468
3307 10,500 lb hydraulic winch accessory kit	469-1	469
3307 10,500 lb hydraulic winch winch accessory kit	470-1	470
3307 10,500 lb hydraulic winch accessory kit	471-1	471
3307 100 AMP dual voltage alternator and regulator mounting hardware kit	472-1	472
3307 100/200 AMP dual voltage alternator and regulator mounting hardware kit	473-1	473
3307 100/200 AMP dual voltage alternator wiring harness	474-1	474
3307 200 AMP dual voltage alternator and regulator mounting hardware kit	475-1	475

TABLE OF CONTENTS (Cont'd)

	Page	Illus Figure
3307 400 AMP alternator kit	476-1	476
3307 Power steering pump and hoses-400 AMP alternator kit	477-1	477
3307 Speedometer cable assembly-400 AMP alternator kit	478-1	478
3307 Dipstick and dummy connector plug-400 AMP alternator kit	479-1	479
3307 400 AMP alternator wiring harness cable assemblies	480-1	480
3307 Buss assembly-400 AMP alternator kit	481-1	481
3307 Tachometer parts kit	482-1	482
3307 Tachometer parts kit	483-1	483
3307 Tachometer parts kit	484-1	484
3307 Precision Lightweight Global Positioning System Receiver (PLGR)	485-1	485
3307 Kit, intercom bracket	486-1	486
3307 Kit, rear hatch support	487-1	487
3307 Flexible Brake Line Kit	487A-1	487A
3307 Flexible Brake Line Kit	487B-1	487B
3307 Flexible Brake Line Kit	487C-1	487C
3307 Flexible Brake Line Kit	487D-1	487D
Group 47 Gauges, weighing, and measuring devices	488-1	488
4701 Speedometer, speedometer drive, and related parts	488-1	488
4702 Air restriction indicator assembly	489-1	489
Group 52 Refrigeration, air conditioner, heater, and air conditioning components	490-1	490
5203 Air conditioner compressor and mounting hardware, M997 and M997A1 ambulance	490-1	490
5203 A/C compressor, idler pulley, tensioner, and related parts (Serial Number 196901 and Above)	490A-1	490A
5203 Air conditioner compressor assembly, M997 and M997A1 ambulance	491-1	491
5217 Air conditioner lines and fittings, M997, M997A1, and M997A2 ambulance	492-1	492
5217 Air conditioner condenser and evaporator lines, M997, M997A1, and M997A2 ambulance	493-1	493
5230 Heater and air conditioning condenser, M997, M997A1, and M997A2 ambulance	494-1	494
5230 Heater and air conditioning condenser, M997, M997A1, and M997A2 ambulance	495-1	495
5241 Air conditioning evaporator and mounting hardware, M997, M997A1, and M997A2 ambulance	496-1	496
5243 Evaporator blower motor assembly covers, M997, M997A1, and M997A2 ambulance	497-1	497
5243 Evaporator blower motor assembly, M997, M997A1, and M997A2 ambulance	498-1	498
Group 94 Kits	Kits-1	
9401 Repair parts kits	Kits-1	
Group 95 Bulk material	Bulk-1	
9501 Bulk	Bulk-1	
Section III SPECIAL TOOLS LIST	499-1	
Group 26 Tools and test equipment	499-1	499
2604 Special tools, drive belts and transmission	499-1	499
2604 Special tools, transmission	500-1	500
2604 Special tools, transfer case	501-1	501
2604 Special tools, axle differential	502-1	502
2604 Special tools, geared hub and steering	503-1	503
2604 Special tools, steering pump, body, and speedometer drive	504-1	504
2604 Special tools, multipurpose	505-1	505

TABLE OF CON-

	Page	Figure
2604 Special tools, air conditioner compressor	506-1	506
2604 Special tools, air conditioner compressor	507-1	507
2604 Special tools, air conditioner compressor	508-1	508
2604 Special tools, air conditioner compressor tool kit	509-1	509
2604 Special tools, transmission, transfer case, and engine A2	510-1	510
2604 Special tools, transmission, transfer case, and engine A2	511-1	511
2604 Pinion setting kit	512-1	512
Section IV CROSS-REFERENCE INDEXES	I-1	
National Stock Number Index	I-1	
Part Number Index	I-114	

Section II. TM 9-2320-280-24P-2

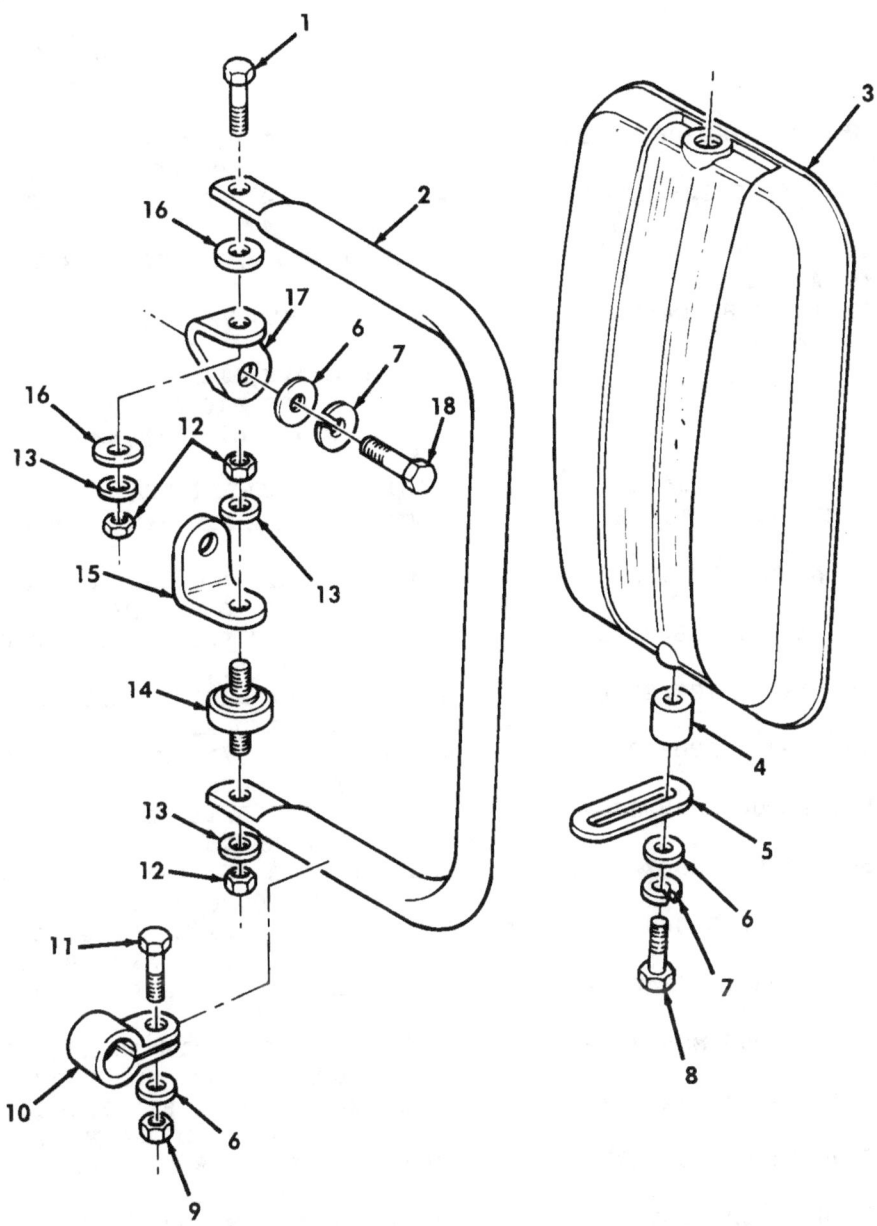

Figure 329. Split Rearview Mirror and Mounting Hardware.

SECTION II   TM9-2320-280-24P C01

(1) ITEM NO	(2) SMR CODE	(3) NSN	(4) CAGEC	(5) PART NUMBER	(6) DESCRIPTION AND USABLE ON CODES (UOC)	(7) QTY

GROUP 2202 ACCESSORY ITEMS

FIG. 329 SPLIT REARVIEW MIRROR AND MOUNTING HARDWARE

1	PAOZZ	5305005432419	80204	B1821BH038C113N	SCREW,CAP,HEXAGON H 3/8-16 X 1-1/8, USBL EFF 100,000 AND ABOVE..........	1
2	PAOZZ	2540013141188	19207	12342131	ARM,REARVIEW MIRROR USBL EFF 100,000 AND ABOVE...................	1
* 3	PAOZZ	2540013141189	70082	0462	MIRROR HEAD,VEHICUL GREEN R.H. USBL EFF 100,000 AND ABOVE..........	1
* 3	PAOZZ	2540013141190	70082	0461	MIRROR HEAD,VEHICUL GREEN L.H. USBL EFF 100,000 AND ABOVE..........	1

UOC:AVY,A11,A13,A14,A15,A20,A24,A25,
A26,A27,B16,B17,B18,HVY,H11,H13,H14,
H15,H16,H17,H18,H20,H21,H24,H25,H26,
H27,H28,MMM

* 3	PAOZZ	2540014873616	70082	0462A	MIRROR ASSEMBLY,REA TAN R.H........	1
* 3	PAOZZ	2540014873626	70082	1305A	MIRROR ASSEMBLY,REA TAN L.H.,LOWER MOUNT ONLY................	1
* 3	PAOZZ	2540014108794	70082	1351	MIRROR HEAD,VEHICUL GREEN L.H, LOWER MOUNT ONLY.............	1
4	PAOZZ	5365013153595	19207	12342139	SPACER,SLEEVE USBL EFF 100,000 AND ABOVE..........................	2
5	PAOZZ	2540013142101	19207	12342138	ARM,REARVIEW MIRROR USBL EFF 100,000 AND ABOVE...................	2
6	PAOZZ	5310013893455	79410	120393	WASHER,LOCK 5/16,USBL EFF 100,000 AND ABOVE....................	6
7	PAOZZ	5310004079566	96906	MS35338-45	WASHER,LOCK 5/16,USBL EFF 100,000 AND ABOVE....................	4
8	PAOZZ	5306002258503	96906	MS90725-39	BOLT,MACHINE 5/16-18 X 1-3/4, USBL EFF 100,000......................	2
9	PAOZZ	5310008140673	96906	MS51943-33	NUT,SELF-LOCKING,HE 5/16-18,USBL EFF 100,000 AND ABOVE.............	2
10	PAOZZ	5340013153611	19207	12342132	CLAMP,LOOP USBL EFF 100,000 AND ABOVE..........................	2
11	PAOZZ	5306002258499	96906	MS90725-34	BOLT,MACHINE 5/16-18 X 1.00 USBL EFF 100,000 AND ABOVE.............	2
12	PAOZZ	5310011269404	24617	9422277	NUT,SELF-LOCKING,HE 3/8-16 USBL EFF 100,000 AND ABOVE.............	6
13	PAOZZ	5310000806004	96906	MS27183-14	WASHER,FLAT 3/8,USBL EFF 100,000 AND ABOVE.......................	6
14	PAOZZ	5307013162986	70082	6185	STUD,LOCKED IN USBL EFF 100,000 AND ABOVE.......................	2
15	PAOZZ	5340014086460	19207	12343063	BRACKET,ANGLE LOWER L.H........... UOC:BVY,B15,B20,B24,B25,C17,NNN	1
15	PFOZZ	5340013141956	19207	12342137	BRACKET,ANGLE LOWER L.H. USBL EFF 100,000 AND ABOVE...................	1

UOC:AVY,A11,A13,A14,A15,A20,A24,A25,
A26,A27,B16,B17,B18,HVY,H11,H13,H14,
H15,H16,H17,H18,H20,H21,H24,H25,H26,
H27,H28,MMM

| 15 | PAOZZ | 5340013162959 | 19207 | 12342136 | BRACKET,ANGLE   LOWER R.H. USBL EFF | 1 |

329-1

SECTION II   TM9-2320-280-24P C01
 (1)   (2)   (3)   (4)   (5)   (6)
(7) ITEM   SMR   PART
     NO   CODE   NSN   CAGEC   NUMBER   DESCRIPTION AND USABLE ON CODES(UOC)
QTY

                                                  100,000 AND ABOVE...................
16 PAOZZ 5310013133562 96906 MS51859-20   WASHER,FLAT 3/8,USBL EFF 100,000   4
AND ABOVE...........................
     17 PAOZZ 5340014086456 19207 12343062   BRACKET,ANGLE UPPER L.H............
1   UOC:BVY,B15,B20,B24,B25,C17,NNN   17
PAOZZ 5340013141955 19207 12342135   BRACKET,ANGLE UPPER L.H.,USBL EFF   1
100,000 AND ABOVE....................   UOC
:AVY,A11,A13,A14,A15,A20,A24,A25,
                                               A26,A27,B16,B17,B18,HVY,H11,H13,H14,
                                               H15,H16,H17,H18,H20,H21,H24,H25,H26,
H27,H28,MMM
     17 PAOZZ 5340013161507 19207 12342134   BRACKET,ANGLE UPPER R.H. USBL 100,
                                                  000 AND ABOVE......................
1
18 PAOZZ 5306002264827 80204 B1821BH031C100N BOLT,MACHINE 5/16-18 X 1.00,USBL   2
EFF 100,000 AND ABOVE...............

Section II.

TM 9-2320-280-24P-2

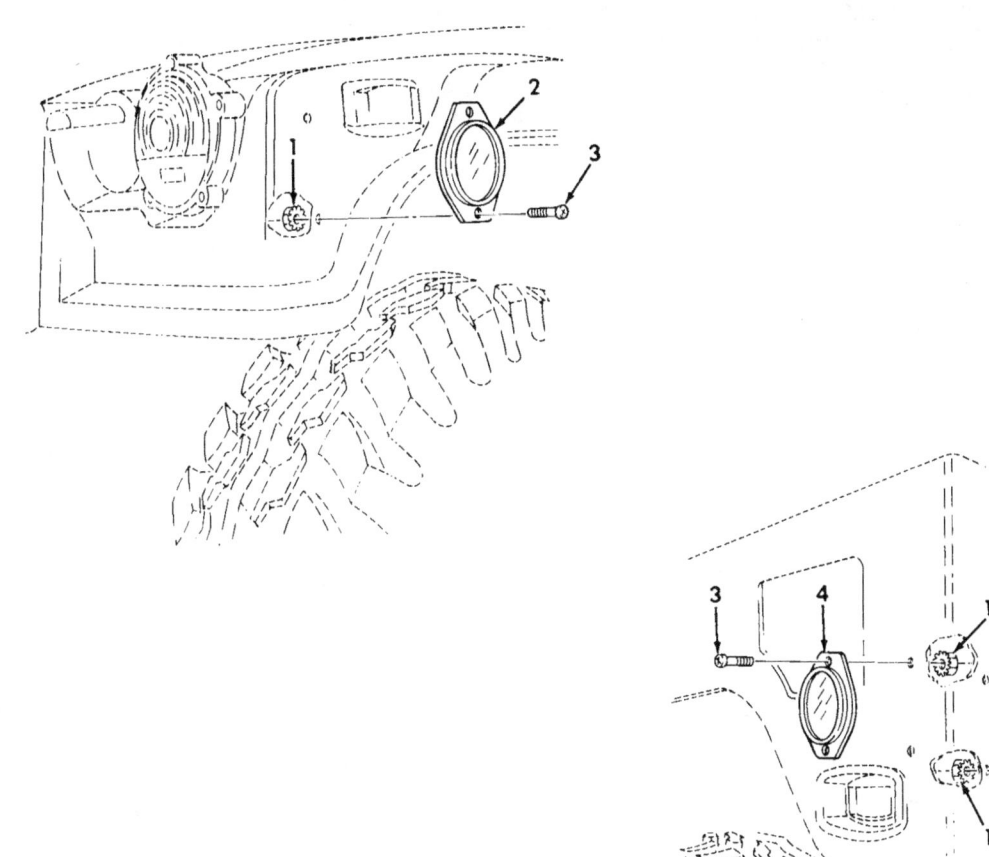

Figure 330. Reflectors.

SECTION II   TM9-2320-280-24P

(1) NO	(2) CODE	(3) NSN	(4) CAGEC	(5) PART NUMBER	(7) ITEM DESCRIPTION AND USABLE ON CODES(UOC)	SMR QTY

GROUP 2202 ACCESSORY ITEMS

FIG. 330 REFLECTORS

1	PAOZZ	5310012522999	34623	5593033	NUT,SELF-LOCKING,AS 1/4-20..........	12
2	PAOZZ	9905002023639	96906	MS35387-2	REFLECTOR,INDICATIN AMBER..........	2
3	PAOZZ	5305009881724	96906	MS35206-280	SCREW,MACHINE 1/4-20 X 5/8..........	12
4	PAOZZ	9905009772727	80753	49RED	REFLECTOR,INDICATIN RED............	4

END OF FIGURE

Section II. TM 9-2320-280-24P-2 C01

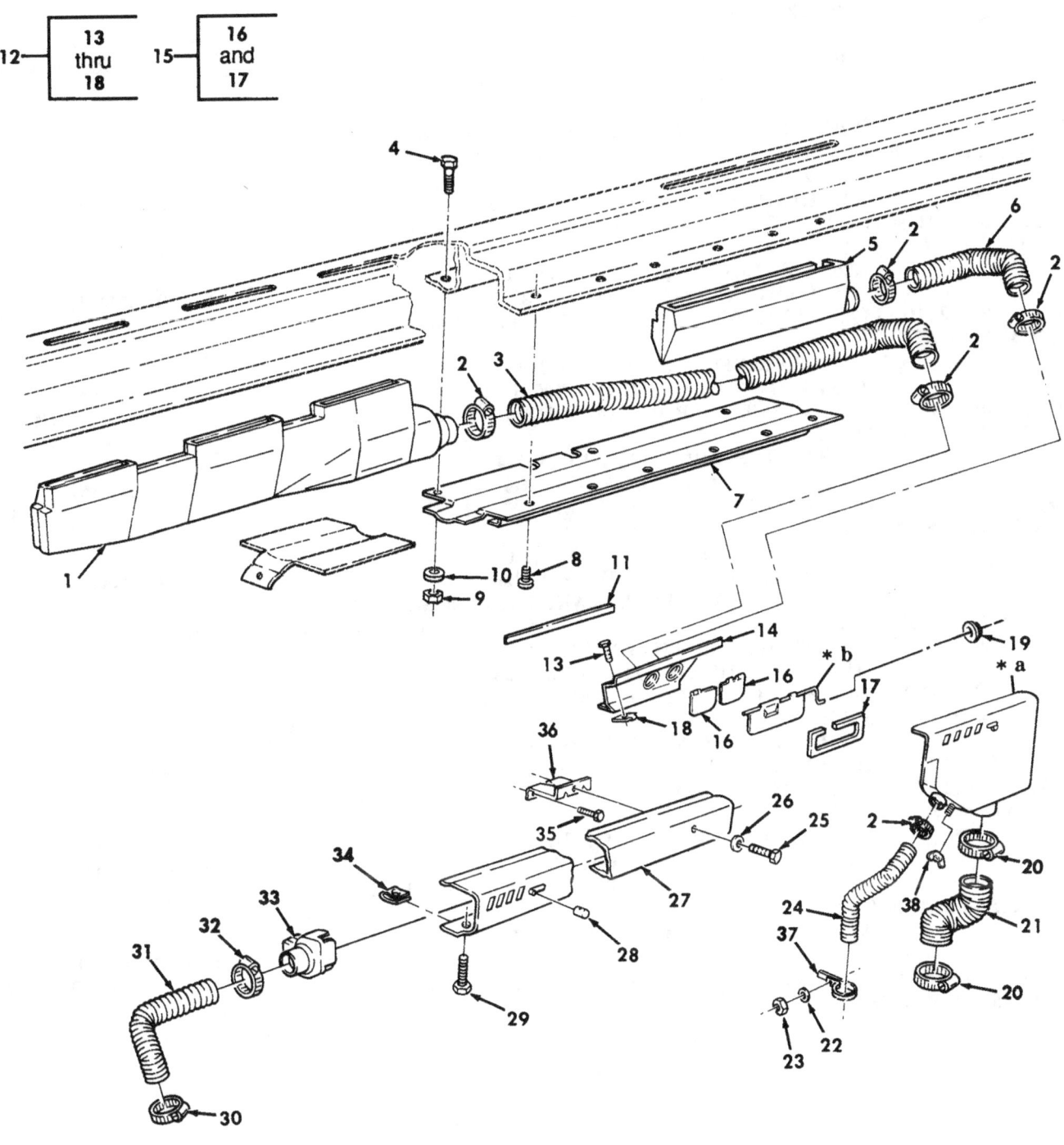

* a PART OF ITEM 12
* b PART OF ITEM 15

Figure 331. Diverter Assembly, Defroster, and Heater Ducting.

SECTION II   TM9-2320-280-24P

(1) NO	(2) SMR CODE	(3) NSN	(4) CAGEC	(5) PART NUMBER	(7) ITEM DESCRIPTION AND USABLE ON CODES(UOC)	QTY

GROUP 2202 ACCESSORY ITEMS

FIG. 331 DIVERTER ASSEMBLY, DEFROSTER, AND HEATER DUCTING

1 PFOZZ 2910011845499 19207 12339289 NOZZLE,DEFROSTER L.H............... 1
UOC:AVY,A11,A13,A14,A15,A20,A24,A25,A26,A27,B16,B17,B18,HVY,H11,H13,H14,H15,H16,H17,H18,H20,H21,H24,H25,H26,H27,H28,MMM

2 PAOZZ 4730010921904 66295 SIZE83 CLAMP,HOSE........................ 5
UOC:AVY,A11,A13,A14,A15,A20,A24,A25,A26,A27,B16,B17,B18,HVY,H11,H13,H14,H15,H16,H17,H18,H20,H21,H24,H25,H26,H27,H28,MMM

3 PAOZZ 4720014481655 19207 12339265-1 HOSE,AIR DUCT...................... 1
UOC:A11,A13,A14,A15,A20,A24,A25,A26,A27,B16,B17,B18,HVY,H11,H13,H14,H15,H16,H17,H18,H20,H21,H24,H25,H26,H27,H28,MMM

4 PAOZZ 5306012707332 34623 5594178 BOLT,MACHINE 1/4-20 X 7/8.......... 3
UOC:AVY,A11,A13,A14,A15,A20,A24,A25,A26,A27,B16,B17,B18,HVY,H11,H13,H14,H15,H16,H17,H18,H20,H21,H24,H25,H26,H27,H28,MMM

5 PFOZZ 2910011845498 19207 12339292 NOZZLE,DEFROSTER R.H............... 1
UOC:AVY,A11,A13,A14,A15,A20,A24,A25,A26,A27,B16,B17,B18,HVY,H11,H13,H14,H15,H16,H17,H18,H20,H21,H24,H25,H26,H27,H28,MMM

6 MOOZZ 19207 12339265-2 DUCT,FLEX MAKE FROM HOSE,P/N AD778-08B79,7 INCHES LONG................. 1
UOC:AVY,A11,A13,A14,A15,A20,A24,A25,A26,A27,B16,B17,B18,HVY,H11,H13,H14,H15,H16,H17,H18,H20,H21,H24,H25,H26,H27,H28,MMM

7 PFOZZ 2510013070152 34623 5593615 PANEL,BODY,VEHICULA................. 1
UOC:AVY,A11,A13,A14,A15,A20,A24,A25,A26,A27,B16,B17,B18,HVY,H11,H13,H14,H15,H16,H17,H18,H20,H21,H24,H25,H26,H27,H28,MMM

8 PAOZZ 5305012553588 7X677 164830 SCREW,TAPPING 3/8-16 X 1/2......... 6
UOC:AVY,A11,A13,A14,A15,A20,A24,A25,A26,A27,B16,B17,B18,HVY,H11,H13,H14,H15,H16,H17,H18,H20,H21,H24,H25,H26,H27,H28,MMM

9 PAOZZ 5310000614650 96906 MS51943-31 NUT,SELF-LOCKING,HE 1/4-20......... 3
UOC:AVY,A11,A13,A14,A15,A20,A24,A25,A26,A27,B16,B17,B18,HVY,H11,H13,H14,H15,H16,H17,H18,H20,H21,H24,H25,H26,H27,H28,MMM

10 PAOZZ 5310011023270 24617 2436161 WASHER,FLAT 1/4.................... 3
UOC:AVY,A11,A13,A14,A15,A20,A24,A25,

331-1

SECTION II   TM9-2320-280-24P

(1) NO	(2) CODE	(3) NSN	(4) CAGEC	(5) NUMBER	(6) DESCRIPTION AND USABLE ON CODES(UOC)	(7) ITEM SMR PART QTY
11	PAOZZ	5330012097817	19207	12339228-7	SEAL,RUBBER STRIP.................. 1	A26,A27,B16,B17,B18,HVY,H11,H13,H14, H15,H16,H17,H18,H20,H21,H24,H25,H26, H27,H28,MMM UOC:AV,A11,A13,A14,A15,A20,A24,A25,
12	PFOZZ	2510011845497	19207	12339291	HOUSING ASSEMBLY,HE................. 1	A26,A27,B16,B17,B18,HVY,H11,H13,H14, H15,H16,H17,H18,H20,H21,H24,H25,H26, H27,H28,MMM UOC:AVY,A11,A13,A14,A15,A20,A24,A25,
13	PAOZZ	5305011878757	7X677	274707	.SCREW,MACHINE #10-24 X 1/2........ 5	A26,A27,B16,B17,B18,HVY,H11,H13,H14, H15,H16,H17,H18,H20,H21,H24,H25,H26, H27,H28,MMM UOC:AVY,A11,A13,A14,A15,A20,A24,A25,
14	PFOZZ	2540012101325	19207	12339283	.PANEL ASSEMBLY,DIVE................ 1	A26,A27,B16,B17,B18,HVY,H11,H13,H14, H15,H16,H17,H18,H20,H21,H24,H25,H26, H27,H28,MMM UOC:AVY,A11,A13,A14,A15,A20,A24,A25,
15	PAOZZ	2510012102748	19207	12339273	.DOOR,METAL,SWINGING................ 1	A26,A27,B16,B17,B18,HVY,H11,H13,H14, H15,H16,H17,H18,H20,H21,H24,H25,H26, H27,H28,MMM UOC AVY,A11,A13,A14,A15,A20,A24,A25,
16	PAOZZ	5330012097842	19207	12339267	..SEAL,NONMETALLIC SP.............. 2	A26,A27,B16,B17,B18,HVY,H11,H13,H14, H15,H16,H17,H18,H20,H21,H24,H25,H26, H27,H28,MMM UOC:AVY,A11,A13,A14,A15,A20,A24,A25,
17	PFOZZ	5330011858709	34623	12339255	..GASKET........................ 1	A26,A27,B16,B17,B18,HVY,H11,H13,H14, H15,H16,H17,H18,H20,H21,H24,H25,H26, H27,H28,MMM UOC:AVY,A11,A13,A14,A15,A20,A24,A25,
18	PAOZZ	5310012155356	78553	C8117-1024	.NUT,SHEET SPRING.................. 5	A26,A27,B16,B17,B18,HVY,H11,H13,H14, H15,H16,H17,H18,H20,H21,H24,H25,H26, H27,H28,MMM UOC:AVY,A11,A13,A14,A15,A20,A24,A25,
19	PAOZZ	5340013327599	19207	12340258	BUTTON,PLUG...................... 1	A26,A27,B16,B17,B18,HVY,H11,H13,H14, H15,H16,H17,H18,H20,H21,H24,H25,H26, H27,H28,MMM UOC:AVY,A11,A13,A14,A15,A20,A24,A25,
20	PAOZZ	4730009086292	96906	MS35842-14	CLAMP,HOSE....................... 2	A26,A27,B16,B17,B18,HVY,H11,H13,H14, H15,H16,H17,H18,H20,H21,H24,H25,H26, H27,H28,MMM UOC:AVY,A11,A13,A14,A15,A20,A24,A25,

SECTION II TM9-2320-280-24P

(1) NO	(2) CODE	(3) NSN	(4) CAGEC	(5) PART NUMBER	(7) ITEM DESCRIPTION AND USABLE ON CODES(UOC)	SMR QTY

21 PAOZZ 4720012795150 19207 12339265-8 DUCT,FLEX........................... 1
UOC:AVY,A11,A13,A14,A15,A20,A24,A25,
A26,A27,B16,B17,B18,HVY,H11,H13,H14,
H15,H16,H17,H18,H20,H21,H24,H25,H26,
H27,H28,MMM

22 PAOZZ 5310000145850 96906 MS27183-42 WASHER,FLAT #10................... 1
UOC:AVY,A11,A13,A14,A15,A20,A24,A25,
A26,A27,B16,B17,B18,HVY,H11,H13,H14,
H15,H16,H17,H18,H20,H21,H24,H25,H26,
H27,H28,MMM

23 PAOZZ 5310012538928 34623 5592995 NUT,PLAIN,EXTENDED #10-24.......... 1
UOC:AVY,A11,A13,A14,A15,A20,A24,A25,
A26,A27,B16,B17,B18,HVY,H11,H13,H14,
H15,H16,H17,H18,H20,H21,H24,H25,H26,
H27,H28,MMM

24 PAOZZ 4720012716985 19207 12339265-9 HOSE,AIR DUCT...................... 1
UOC:AVY,A11,A13,A14,A15,A20,A24,A25,
A26,A27,B16,B17,B18,HVY,H11,H13,H14,
H15,H16,H17,H18,H20,H21,H24,H25,H26,
H27,H28,MMM

25 PAOZZ 5305000680502 96906 MS90725-6 SCREW,CAP,HEXAGON H 1/4-20 X 3/4... 2
UOC:AVY,A11,A13,A14,A15,A20,A24,A25,
A26,A27,B16,B17,B18,HVY,H11,H13,H14,
H15,H16,H17,H18,H20,H21,H24,H25,H26,
H27,H28,MMM

26 PAOZZ 5310008094058 96906 MS27183-10 WASHER,FLAT 1/4.................... 2
UOC:AVY,A11,A13,A14,A15,A20,A24,A25,
A26,A27,B16,B17,B18,HVY,H11,H13,H14,
H15,H16,H17,H18,H20,H21,H24,H25,H26,
H27,H28,MMM

27 PFOZZ 2540012125815 19207 12339281 VENTILATOR,AIR CIRC................ 1
UOC:AVY,A11,A13,A14,A15,A20,A24,A25,
A26,A27,B16,B17,B18,HVY,H11,H13,H14,
H15,H16,H17,H18,H20,H21,H24,H25,H26,
H27,H28,MMM

28 PAOZZ 5355012473593 34623 5596249 KNOB............................... 2
UOC:AVY,A11,A13,A14,A15,A20,A24,A25,
A26,A27,B16,B17,B18,HVY,H11,H13,H14,
H15,H16,H17,H18,H20,H21,H24,H25,H26,
H27,H28,MMM

29 PAOZZ 5305004324205 96906 MS51861-49 SCREW,TAPPING #10-16 X 1.00........ 9
UOC:AVY,A11,A13,A14,A15,A20,A24,A25,
A26,A27,B16,B17,B18,HVY,H11,H13,H14,
H15,H16,H17,H18,H20,H21,H24,H25,H26,
H27,H28,MMM

30 PAOZZ 4730002782523 19207 11608950-6 CLAMP,HOSE......................... 1
UOC:AVY,A11,A13,A14,A15,A20,A24,A25,
A26,A27,B16,B17,B18,HVY,H11,H13,H14,
H15,H16,H17,H18,H20,H21,H24,H25,H26,
H27,H28,MMM

31 PAOZZ 4720012795149 34623 12339265-3 HOSE,AIR DUCT...................... 1
UOC:AVY,A11,A13,A14,A15,A20,A24,A25,
A26,A27,B16,B17,B18,HVY,H11,H13,H14,

SECTION II   TM9-2320-280-24P C01

(1)	(2)	(3)	(4)	(5)	(6)	(7)
ITEM NO	SMR CODE	NSN	CAGEC	PART NUMBER	DESCRIPTION AND USABLE ON CODES (UOC)	QTY
32	PAOZZ	4730012550925	34623	5593309	CLAMP,HOSE.......................... UOC:AVY,A11,A13,A14,A15,A20,A24,A25, A26,A27,B16,B17,B18,HVY,H11,H13,H14, H15,H16,H17,H18,H20,H21,H24,H25,H26, H27,H28,MMM	1
33	PFOZZ	2590011868704	19207	12339252	BRACKET,VEHICULAR C................. UOC:AVY,A11,A13,A14,A15,A20,A24,A25, A26,A27,B16,B17,B18,HVY,H11,H13,H14, H15,H16,H17,H18,H20,H21,H24,H25,H26, H27,H28,MMM	1
34	PAOZZ	5310011947066	72800	C7685-10A	NUT,SHEET SPRING #10-16............ UOC:AVY,A11,A13,A14,A15,A20,A24,A25, A26,A27,B16,B17,B18,HVY,H11,H13,H14, H15,H16,H17,H18,H20,H21,H24,H25,H26, H27,H28,MMM	9
35	PAOZZ	5305000680508	80204	B1821BH025C075N	SCREW,CAP,HEXAGON H 1/4-20 X 3/4... UOC:AVY,A11,A13,A14,A15,A20,A24,A25, A26,A27,B16,B17,B18,HVY,H11,H13,H14, H15,H16,H17,H18,H20,H21,H24,H25,H26, H27,H28,MMM	4
* 36	PFOZZ	5340014764374	19207	12446765	BRACKET,DOUBLE ANGL................. UOC:AVY,A11,A13,A14,A15,A20,A24,A25, A26,A27,B16,B17,B18,HVY,H11,H13,H14, H15,H16,H17,H18,H20,H21,H24,H25,H26, H27,H28,MMM	2
* 37	PAOZA	5340014809600	92878	30232	CLAMP,LOOP........................... UOC:AVY,A11,A13,A14,A15,A20,A24,A25, A26,A27,B16,B17,B18,HVY,H11,H13,H14, H15,H16,H17,H18,H20,H21,H24,H25,H26, H27,H28,MMM	1
* 38	PFOZZ	5310011061144	96906	MS35425-68	WING NUT............................. UOC:AVY,A11,A13,A14,A15,A20,A24,A25, A26,A27,B16,B17,B18,HVY,H11,H13,H14, H15,H16,H17,H18,H20,H21,H24,H25,H26, H27,H28,MMM	1

END OF FIGURE

Section II. TM 9-2320-280-24P-2

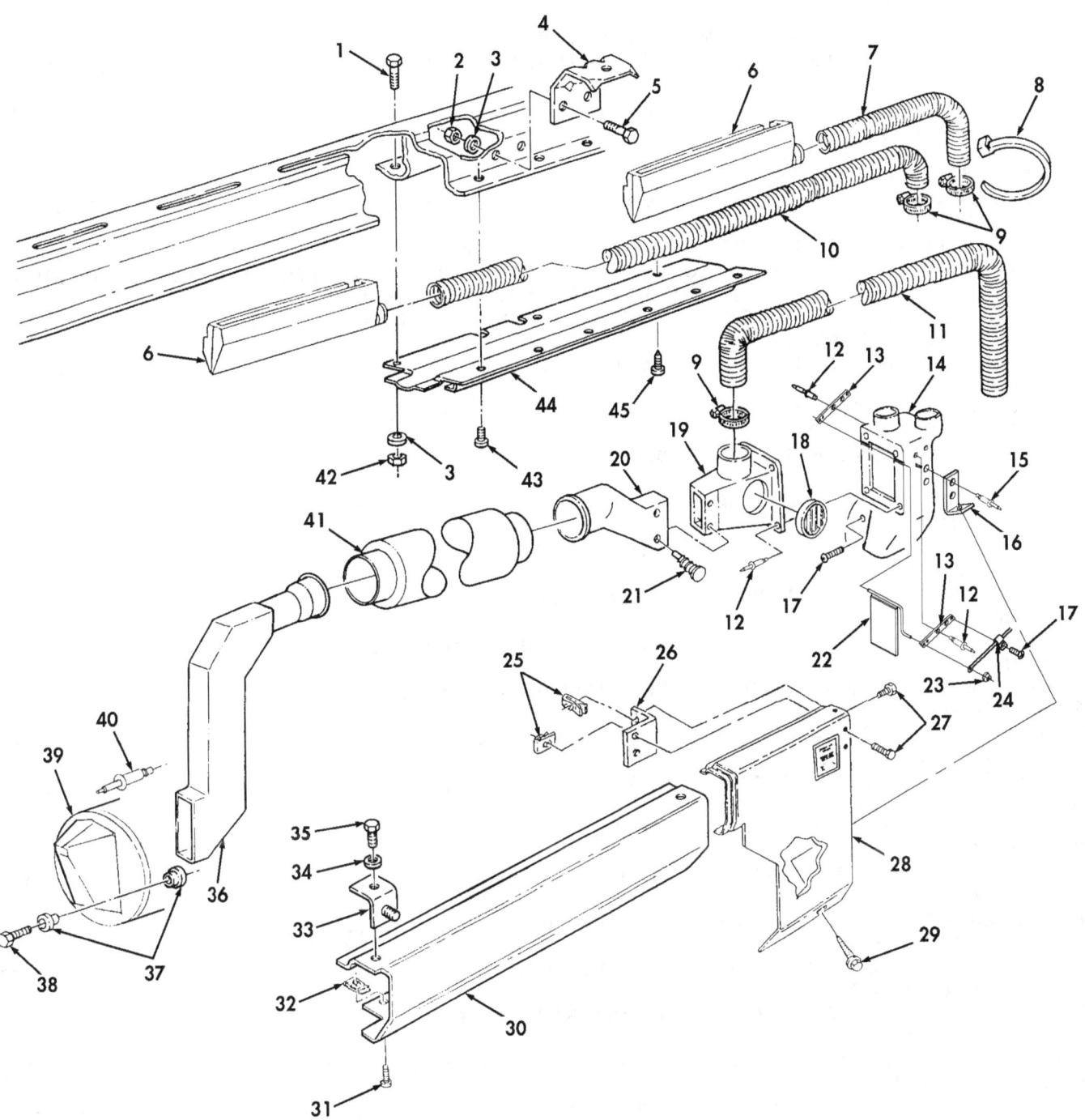

Figure 332. A2 Diverter Assembly, Defroster, Heater Ducting, and Radio Brackets.

SECTION II   TM9-2320-280-24P

(1) NO	(2) SMR CODE	(3) NSN	(4) CAGEC	(5) PART NUMBER	(7) DESCRIPTION AND USABLE ON CODES(UOC)	ITEM QTY

GROUP 2202 ACCESSORY ITEMS

FIG. 332 A2 DIVERTER ASSEMBLY, DEFROSTER, HEATER DUCTING, AND RADIO BRACKETS

1	PAOZZ	5306012707332	34623	5594178	BOLT,MACHINE 1/4-20 X 7/8......... 3 UOC:BVY,B15,B20,B24,B25,C17,NNN
2	PAOZZ	5310014326727	24617	9419471	NUT,SELF-LOCKING,HE 1/4............ 4 UOC:BVY,B15,B20,B24,B25,C17,NNN
3	PAOZZ	5310011023270	24617	2436161	WASHER,FLAT 1/4................... 7 UOC:BVY,B15,B20,B24,B25,C17,NNN
4	PFOZZ	5340014764374	19207	12446765	BRACKET,DOUBLE ANGL.............. 2 UOC:BVY,B15,B20,B24,B25,C17,NNN
5	PAOZZ	5305000680508	80204	B1821BH025C075N	SCREW,CAP,HEXAGON H 1/4-20 X 3/4... 4 UOC:BVY,B15,B20,B24,B25,C17,NNN
6	PAOZZ	2910011845498	34623	12339292-1	NOZZLE,DEFROSTER R.H. AND L.H...... 2 UOC:BVY,B15,B20,B24,B25,C17,NNN
7	PAOZZ	4720014749148	19207	12447010-1	HOSE,AIRDUCT........................ 1 UOC:BVY,B15,B20,B24,B25,C17,NNN
8	PAOZZ	5975010345871	96906	MS3367-7-0	STRAP,TIEDOWN,ELECT................ 1 UOC:BVY,B15,B20,B24,B25,C17,NNN
9	PAOZZ	4730010921904	66295	SIZE 83	CLAMP,HOSE......................... 3 UOC:BVY,B15,B20,B24,B25,C17,NNN
10	PAOZZ	4720014455690	19207	12447010-3	HOSE,AIR DUCT...................... 1 UOC:BVY,B15,B20,B24,B25,C17,NNN
11	PAOZZ	4720014446433	34623	12447010-2	HOSE,AIR DUCT...................... 1 UOC:BVY,B15,B20,B24,B25,C17,NNN
12	PAOZZ	5320012751998	11815	BAPKTR-64	RIVET,BLIND........................ 8 UOC:BVY,B15,B20,B24,B25,C17,NNN
13	PAOZZ	2590014559123	19207	12460233	BRACKET,VEHICULAR C................. 2 UOC:BVY,B15,B20,B24,B25,C17,NNN
14	PAOZZ	2540014472236	34623	12447046	DUCT ASSEMBLY,AIR C................. 1 UOC:BVY,B15,B20,B24,B25,C17,NNN
15	PAOZZ	5320000835009	81349	M24243/1-A403	RIVET,BLIND........................ 2 UOC:BVY,B15,B20,B24,B25,C17,NNN
16	PAOZZ	5340014497352	19207	12447132	BRACKET,ANGLE..................... 1 UOC:BVY,B15,B20,B24,B25,C17,NNN
17	PAOZZ	5305012135024	86403	9414722	SCREW............................. 2 UOC:BVY,B15,B20,B24,B25,C17,NNN
18	PAOZZ	2540014559308	19207	12460434	VENTILATOR,AIR CIRC................. 1 UOC:BVY,B15,B20,B24,B25,C17,NNN
19	PFOZZ	2540014446615	34623	12447130	DUCT ASSEMBLY,AIR C................. 1 UOC:BVY,B15,B20,B24,B25,C17,NNN
20	PAOZZ	2540014445340	34623	12447050	HOUSING,AIR INTAKE,................ 1 UOC:BVY,B15,B20,B24,B25,C17,NNN
21	PAOZZ	5325011973460	02768	354-280308-00-00	FASTENER,SPRING TEN................ 2 78 UOC:BVY,B15,B20,B24,B25,C17,NNN
22	PAOZZ	2540014446655	34623	12447048	DUCT ASSEMBLY,AIR C................. 1 UOC:BVY,B15,B20,B24,B25,C17,NNN
23	PAOZZ	5340013327599	19207	12340258	BUTTON,PLUG....................... 1 UOC:BVY,B15,B20,B24,B25,C17,NNN
24	PAOZZ	5340009907610	96906	MS21333-66	CLAMP,LOOP........................ 1

SECTION II TM9-2320-280-24P

(1) NO	(2) CODE	(3) NSN	(4) CAGEC	(5) PART NUMBER	(7) ITEM DESCRIPTION AND USABLE ON CODES(UOC)	QTY

UOC:BVY,B15,B20,B24,B25,C17,NNN

| 25 | PAOZZ | 5310012155356 | 19207 | 12339694 | NUT,SHEET SPRING | 4 |

UOC:BVY,B15,B20,B24,B25,C17,NNN

| 26 | PAOZZ | 5340011897554 | 19207 | 12339261 | BRACKET,ANGLE | 1 |

UOC:BVY,B15,B20,B24,B25,C17,NNN

| 27 | PAOZZ | 5305011878757 | 7X677 | 274707 | SCREW,MACHINE 24 X 1/2 | 4 |

UOC:BVY,B15,B20,B24,B25,C17,NNN

| 28 | PAOZZ | 5340014702069 | 19207 | 12447013 | COVER,ACCESS | 1 |

UOC:BVY,B15,B20,B24,B25,C17,NNN

| 29 | PAOZZ | 5305012596322 | 34623 | 12342499-1 | SCREW,MACHINE #10 32 X 1/2 | 1 |

UOC:BVY,B15,B20,B24,B25,C17,NNN

| 30 | PAOZZ | 2540014446642 | 34623 | 12447012 | DUCT ASSEMBLY,AIR C | 1 |

UOC:BVY,B15,B20,B24,B25,C17,NNN

| 31 | PAOZZ | 5305014479227 | 24617 | 9414726 | SCREW,TAPPING | 10 |

UOC:BVY,B15,B20,B24,B25,C17,NNN

| 32 | PAOZZ | 5310014483219 | 19207 | 12339426 | NUT,SHEET SPRING | 10 |

UOC:BVY,B15,B20,B24,B25,C17,NNN

| 33 | PAOZZ | 2590014398268 | 34623 | 12446763 | BRACKET,VEHICULAR C | 2 |

UOC:BVY,B15,B20,B24,B25,C17,NNN

| 34 | PAOZZ | 5310000806004 | 96906 | MS27183-14 | WASHER,FLAT | 2 |

UOC:BVY,B15,B20,B24,B25,C17,NNN

| 35 | PAOZZ | 5305005434372 | 80204 | B1821BH038C075N | SCREW,CAP,HEXAGON H 3/8-16 X 3/4 | 2 |

UOC:BVY,B15,B20,B24,B25,C17,NNN

| 36 | PAOZZ | 2540014748562 | 19207 | 12447047 | HOUSING,HEATING DIR | 1 |

UOC:BVY,B15,B20,B24,B25,C17,NNN

| 37 | PAOZZ | 5365014696343 | 19207 | 12460385 | SPACER,SLEEVE | 6 |

UOC:BVY,B15,B20,B24,B25,C17,NNN

| 38 | PAOZZ | 5305014614396 | 3M915 | 9425170 | SCREW,TAPPING #10-24 X 1.00 | 3 |

UOC:BVY,B15,B20,B24,B25,C17,NNN

| 39 | PAOZZ | 2540014633097 | 34623 | 12460386 | DUCT ASSEMBLY,AIR C | 1 |

UOC:BVY,B15,B20,B24,B25,C17,NNN

| 40 | PAOZZ | 5320012542283 | 11815 | BALM-6BP-14 | RIVET | 1 |

UOC:BVY,B15,B20,B24,B25,C17,NNN

| 41 | PAOZZ | 4720014455705 | 19207 | 12447028 | HOSE,AIR DUCT | 1 |

UOC:BVY,B15,B20,B24,B25,C17,NNN

| 42 | PAOZZ | 5310000614650 | 96906 | MS51943-31 | NUT,SELF-LOCKING,HE 1/4-20 | 3 |

UOC:BVY,B15,B20,B24,B25,C17,NNN

| 43 | PAOZZ | 5305012553588 | 7X677 | 164830 | SCREW,TAPPING 3/8-16 X 1/2 | 6 |

UOC:BVY,B15,B20,B24,B25,C17,NNN

| 44 | PFOZZ | 2510013070152 | 19207 | 12339201-1 | PANEL,BODY,VEHICULA | 1 |

UOC:BVY,B15,B20,B24,B25,C17,NNN

| 45 | PAOZZ | 5305012982436 | 96906 | MS51861-22C | SCREW,TAPPING 6-20 X 1/4 | 2 |

UOC:BVY,B15,B20,B24,B25,C17,NNN

END OF FIGURE

Section II. TM 9-2320-280-24P-2 C01

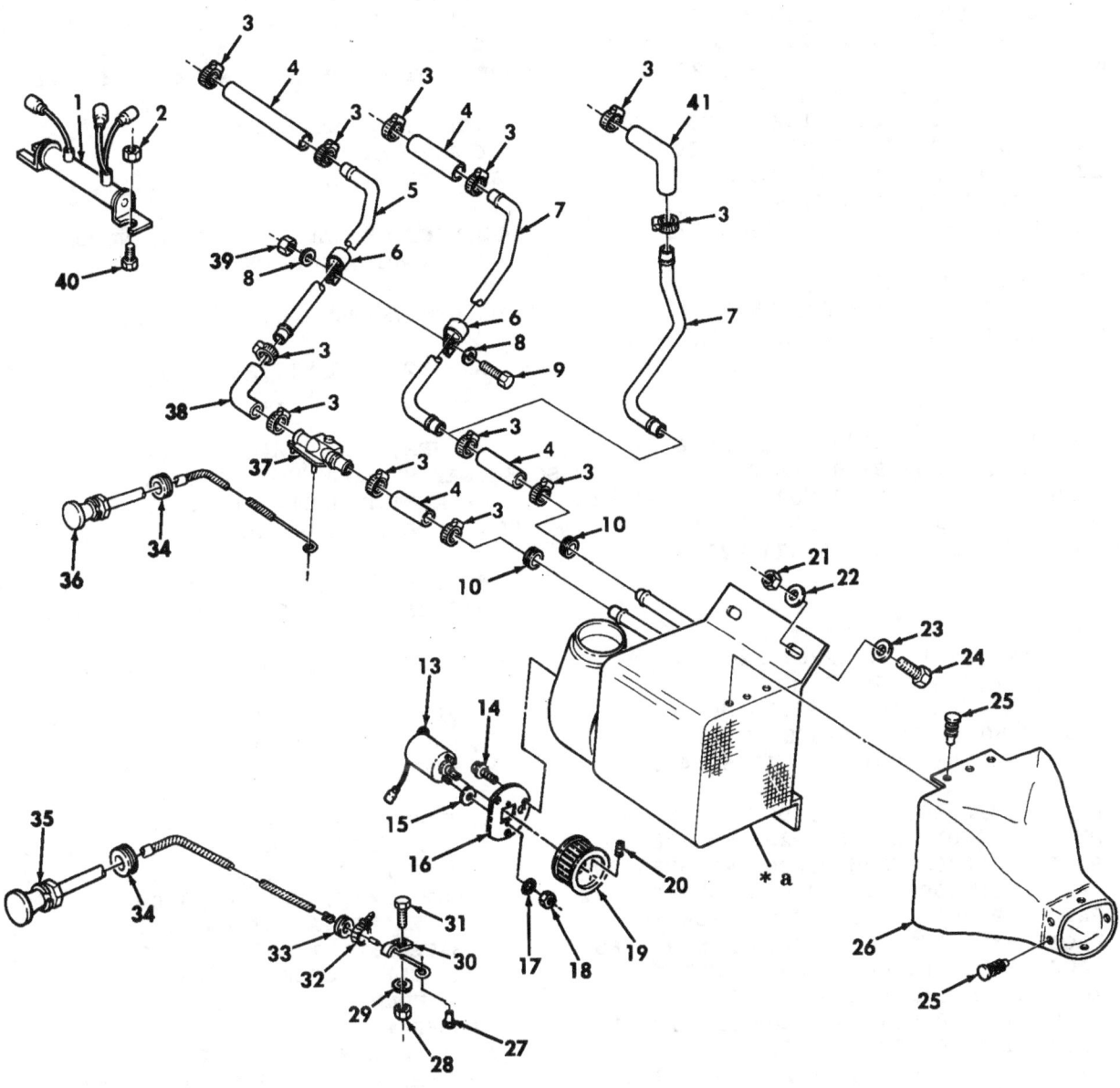

* a PART OF ITEM 11

Figure 333. Heater Assembly, Water Valve, and Controls.

SECTION II   TM9-2320-280-24P C01
  (1)   (2)   (3)   (4)   (5)   (6)
(7)   ITEM   SMR   PART
  NO   CODE   NSN   CAGEC   NUMBER   DESCRIPTION AND USABLE ON CODES(UOC)
QTY

GROUP 2202 ACCESSORY ITEMS

FIG. 333 HEATER ASSEMBLY,WATER
VALVE,AND CONTROLS

(7)	ITEM NO	SMR CODE	NSN	CAGEC	PART NUMBER	DESCRIPTION AND USABLE ON CODES (UOC)	QTY
	1	PAOZZ	5905011839636	19207	12338468	RESISTOR,FIXED,WIRE..................	1
	2	PAOZZ	5310012510711	34623	5593038	NUT,PLAIN,HEXAGON 1/4-20..........	2
	3	PAOZZ	4730010887798	76599	481655-305	CLAMP,HOSE.........................	10
*	4	MFOZZ		19207	12339293-1	TUBE,BENT,NONMETALL MAKE FROM HOSE, P/N MS521301A203R,2.5 INCHES LONG...	4
	5	PAOZZ	4710012103504	19207	12339263	TUBE,BENT,METALLIC..................	1
	6	PFOZZ	5340012143615	21450	137197	CLAMP,LOOP..........................	2
*	7	PAOZZ	4720014732352	34623	RCSK18121	TUBE,BENT,METALLIC.................. UOC:BVY,B15,B20,B24,B25,C17,NNN	1
*	7	PAOZZ	4710014886076	34623	12469479	TUBE,BENT,METALLIC (SERIAL NUMBER 196901 AND ABOVE)................... UOC:BVY,B15,B20,B24,B25,C17,NNN	1
*	7	PAOZZ	4710012103503	34623	5590284	TUBE,BENT,METALLIC.................. UOC:AVY,A11,A13,A14,A15,A20,A24,A25, A26,A27,B16,B17,B18,HVY,H11,H13,H14, H15,H16,H17,H18,H20,H21,H24,H25,H26, H27,H28,MMM	1
	8	PAOZZ	5310011023270	24617	2436161	WASHER,FLAT 1/4....................	2
	9	PAOZZ	5305000680502	96906	MS90725-6	SCREW,CAP,HEXAGON H 1/4-20 X 3/4...	1
	10	PAOZZ	5325002811557	70485	AN931-10-14	GROMMET,NONMETALLIC.................	2
*	11	PFOOO	2540011907079	92878	30250	HEATER,VEHICULAR,CO................. UOC:AVY,A11,A13,A14,A15,A20,A24,A25, A26,A27,B16,B17,B18,HVY,H11,H13,H14, H15,H16,H17,H18,H20,H21,H24,H25,H26, H27,H28,MMM	1
*	11	PFOOO	2540014743217	19207	12447007	HEATER,VEHICULAR,CO................. UOC:BVY,B15,B20,B24,B25,C17,NNN	1
	12	PAOOO	6105012116635	92878	30266	.MOTOR,DIRECT CURREN...............	1
*	13	PAOZZ	6105014635260	92878	30042	..MOTOR,DIRECT CURREN..............	1
	14	PAOZZ	5305013809163	24617	9414313	..SCREW,ASSEMBLED WAS 1/4-20 X 11/32.....	2
	15	PAOZZ	5310000145850	96906	MS27183-42	..WASHER,FLAT #10.................	2
	16	PFOZZ	5340013812248	19207	12342809	..PLATE,MOUNTING...................	1
	17	PAOZZ	5310005765752	96906	MS35333-39	..WASHER,LOCK #10.................	2
	18	PAOZZ	5310009349751	96906	MS35650-302	..NUT,PLAIN,HEXAGON #10-32.......	2
	19	PAOZZ	2930013859108	19207	12342808	..IMPELLER,FAN,AXIAL...............	1
	20	PAOZZ	5305007245812	80205	MS51964-65	..SETSCREW 1/4-28 X .312..........	1
	21	PAOZZ	5310008140673	96906	MS51943-33	NUT,SELF-LOCKING,HE 5/16-18........	4
*	22	PAOZZ	5310011191024	24617	2436162	WASHER,FLAT 5/16...................	4
	23	PAOZZ	5310000814219	96906	MS27183-12	WASHER,FLAT 5/16	4
	24	PAOZZ	5306002258499	96906	MS90725-34	BOLT,MACHINE 5/16-18 X 3/4.........	4
	25	PFOZZ	5325012097843	19207	12339402	GROMMET,NONMETALLIC 1/4 X 5/8......	16
	26	PAOZZ	4720012126403	19207	12340029	DUCT,AIR INTAKE....................	1
	27	PAOZZ	5310011981723	19207	12339998-14	NUT,PLAIN,BLIND RIV................	1
	28	PAOZZ	5310009349747	96906	MS35649-262	NUT,PLAIN,HEXAGON #6-32...........	1
	29	PAOZZ	5310000454007	96906	MS35338-41	WASHER,LOCK #6....................	1
	30	PAOZZ	5340012155478	75272	CWV-0407	CLAMP,LOOP.........................	1
	31	PAOZZ	5305002271543	96906	MS51849-33	SCREW,MACHINE #6-32 X .62.........	1

SECTION II TM9-2320-280-24P C01

(7)	(1) ITEM NO	(2) SMR CODE	(3) NSN	(4) CAGEC	(5) PART NUMBER	(6) DESCRIPTION AND USABLE ON CODES (UOC)	QTY
*	32	PAOZZ	5975010345871	96906	MS3367-7-0	STRAP,TIEDOWN,ELECT................	1
	33	PAOZZ	5325002496345	70485	2758	GROMMET,NONMETALLIC.................	1
	34	PAOZZ	5325002919366	96906	MS35489-11	GROMMET,NONMETALLIC.................	2
	35	PAOZZ	2590011946990	19207	12340531-3	CONTROL ASSEMBLY,PU DEFROSTER......	1
	36	PAOZZ	2590012124956	19207	12340531-4	CONTROL ASSEMBLY,PU HEATER.........	1
	37	PAOZZ	4820011892107	58261	RTCO-2619	VALVE,STOP-CHECK....................	1
	38	PAOZZ	4720011892218	19207	12339251	HOSE,PREFORMED......................	1
	39	PAOZZ	5310000614650	96906	MS51943-31	NUT,SELF-LOCKING,HE 1/4-20.........	1
	40	PAOZZ	5305012685680	34623	5593030	SCREW,ASSEMBLED WAS 1/4-20 X 5/8...	2
*	41	PAOZZ	4730014886163	34623	12469491	ELBOW,HOSE (SERIAL NUMBER 196901 AND ABOVE)............................	1

333-2

Section II. TM 9-2320-280-24P-2 C01

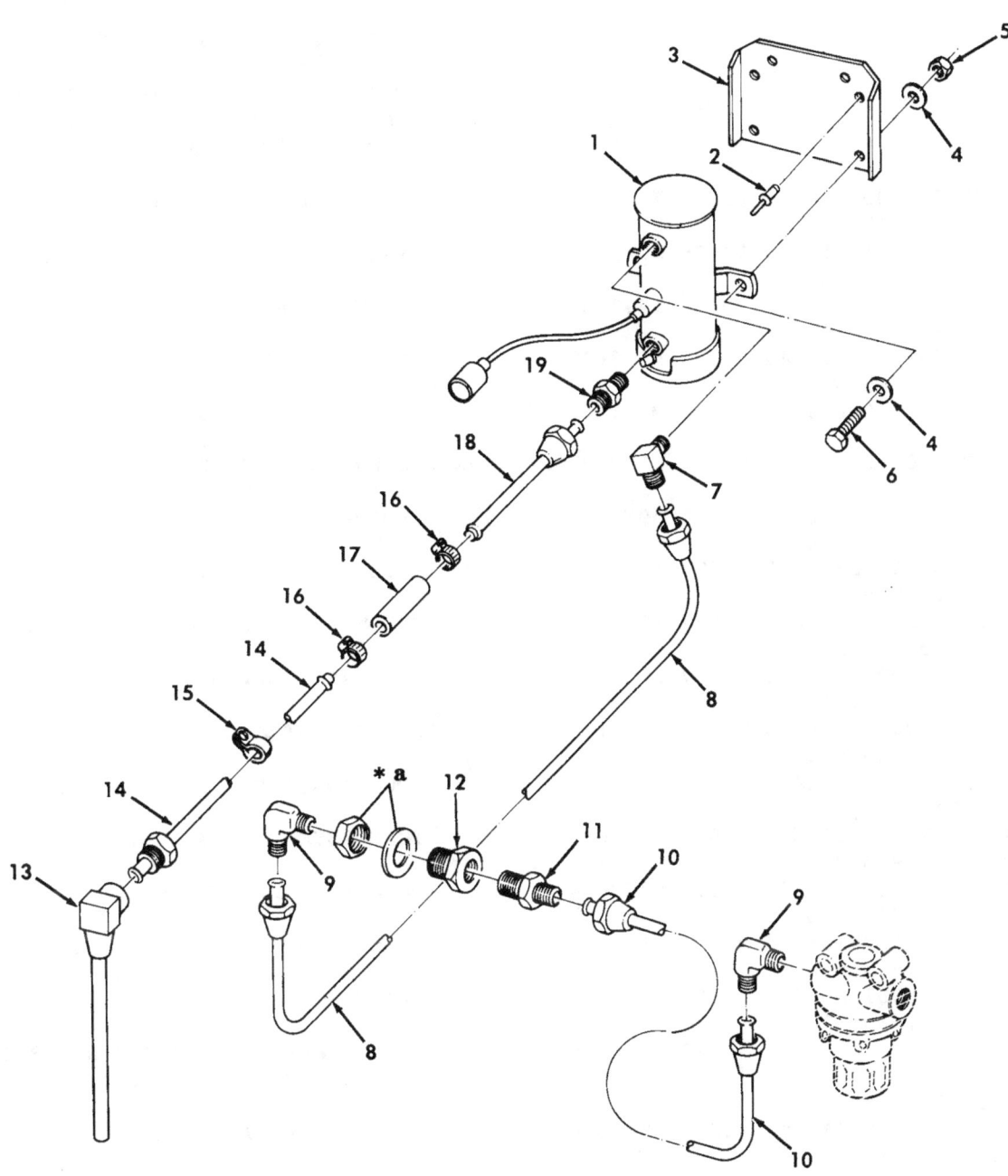

* a PART OF ITEM 12

Figure 334. Heater Fuel Lines, Filter to Tank, M996, M996A1, M997, M997A1, and M997A2 Ambulance.

SECTION II   TM9-2320-280-24P C01

(1) ITEM NO	(2) SMR CODE	(3) NSN	(4) CAGEC	(5) PART NUMBER	(6) DESCRIPTION AND USABLE ON CODES (UOC)	(7) QTY
					GROUP 2202 ACCESSORY ITEMS	
					FIG. 334 HEATER FUEL LINES, FILTER TO TANK, M996, M996A1, M997, M997A1, AND M997A2 AMBULANCE	
1	PAOZZ	2910000730165	19207	7748814	PUMP,FUEL,ELECTRICA................. UOC:A15,B15,B16,H15,H16	1
2	PAOZZ	5320012716357	9K475	MGLP-B6-4	RIVET,BLIND 3/16................... UOC:A15,B15,B16,H15,H16	4
3	PFOZZ	5340012043903	19207	12338559	BRACKET,DOUBLE ANGL................ UOC:A15,B15,B16,H15,H16	1
4	PAOZZ	5310012510748	34623	5593691	WASHER,FLAT 1/4................... UOC:A15,B15,B16,H15,H16	4
5	PAOZZ	5310000614650	96906	MS51943-31	NUT,SELF-LOCKING,HE 1/4 X 20....... UOC:A15,B15,B16,H15,H16	2
6	PAOZZ	5305002253843	80204	B1821BH025C100N	SCREW,CAP,HEXAGON H 1/4-20 X 1.00.. UOC:A15,B15,B16,H15,H16	2
7	PAOZZ	4730002546227	79470	49X6X2	ELBOW,PIPE TO TUBE................. UOC:A15,B15,B16,H15,H16	1
8	PAOZZ	4710012653231	34623	5591614	TUBE ASSEMBLY,METAL................ UOC:A15,B15,B16,H15,H16	1
9	PAOZZ	4730002546211	30327	E14A	ELBOW,PIPE TO TUBE 90 DEGREE....... UOC:A15,B15,B16,H15,H16	2
10	PAOZZ	4710012653275	19207	12341440	TUBE ASSEMBLY,METAL................ UOC:A15,B15,B16,H15,H16	1
11	PAOZZ	4730008036266	72582	8924145	ADAPTER,STRAIGHT,PI................ UOC:A15,B15,B16,H15,H16	1
12	PAOZZ	4730008975497	79470	W21204	COUPLING,PIPE...................... UOC:A15,B15,B16,H15,H16	1
13	PAOZZ	4710012030615	34623	5578694	TUBE ASSEMBLY,METAL................ UOC:A15,B15,B16,H15,H16	1
14	PAOZZ	4710012030607	34623	5588694	TUBE ASSEMBLY,METAL................ UOC:A15,B15,B16,H15,H16	1
15	PAOZZ	5340009936207	96906	MS21333-99	CLAMP,LOOP......................... UOC:A15,B15,B16,H15,H16	1
16	PAOZZ	4730011188278	76599	MMF4-SS	CLAMP,HOSE......................... UOC:A15,B15,B16,H15,H16	2
* 17	MOOZZ		34623	5588698	HOSE MAKE FROM HOSE,P/N 9438315, 2.80 INCHES LONG.................. UOC:A15,B15,B16,H15,H16	1
18	PAOZZ	4710012030608	34623	5588696	TUBE ASSEMBLY,METAL................ UOC:A15,B15,B16,H15,H16	1
19	PAOZZ	4730002660535	81343	5-2 010102B	ADAPTER,STRAIGHT,PI................ UOC:A15,B15,B16,H15,H16	1

END OF FIGURE

Section II. TM 9-2320-280-24P-2

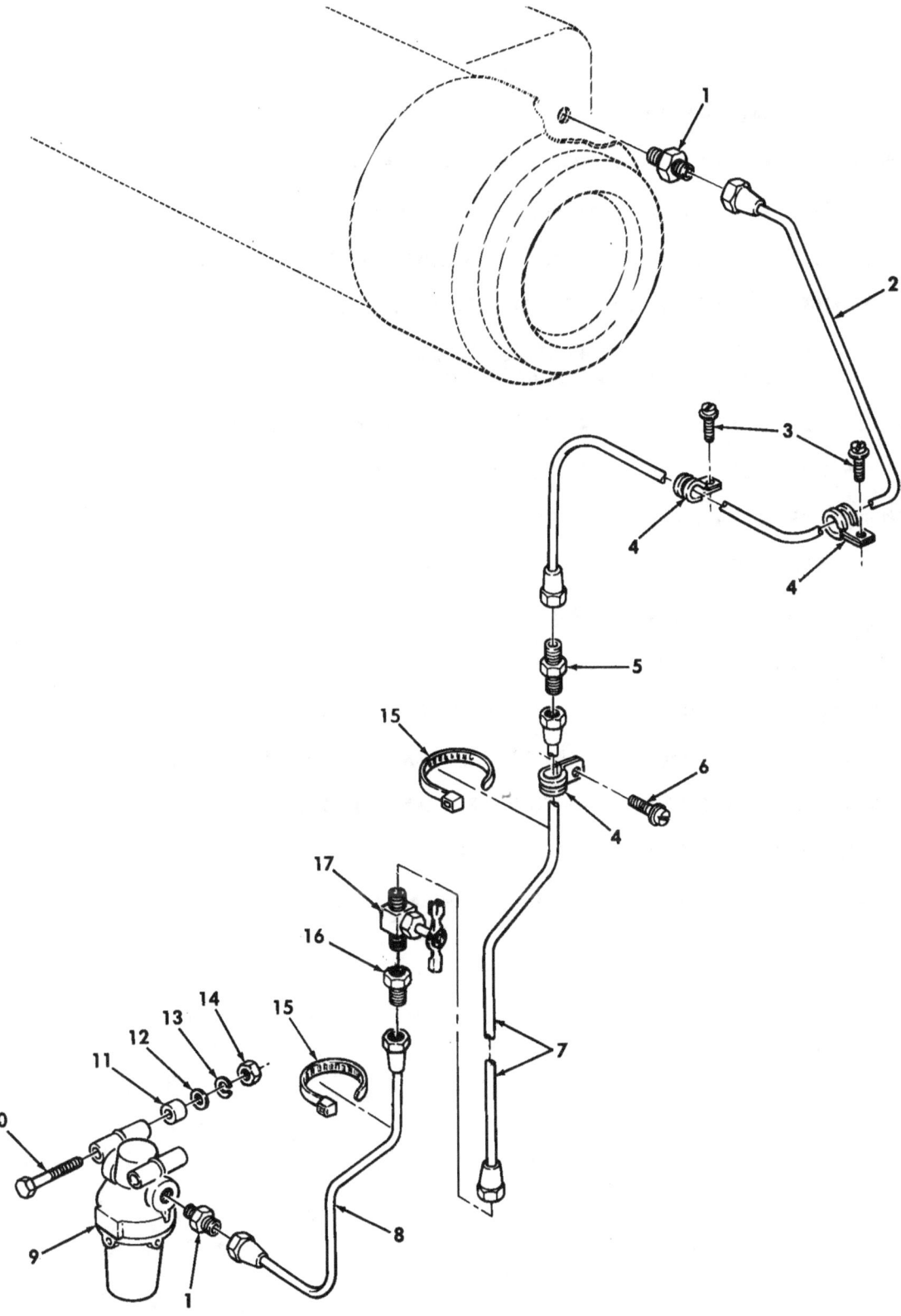

Figure 335. Fuel Lines, Personnel Heater to Fuel Filter, M996 and M996A1 Ambulance.

SECTION II   TM9-2320-280-24P

(1) NO	(2) SMR CODE	(3) NSN	(4) CAGEC	(5) PART NUMBER	(7) ITEM DESCRIPTION AND USABLE ON CODES(UOC)	QTY

GROUP 2202 ACCESSORY ITEMS

FIG. 335 FUEL LINES, PERSONNEL HEATER TO FUEL FILTER, M996 AND M996A1 AMBULANCE

| 1 | PAOZZ | 4730004396021 | 81343 | 6-2 010102B | ADAPTER,STRAIGHT,PI............... | 2 |
UOC:B16,H16
| 2 | PAOZZ | 4710012653228 | 34623 | 5598266 | TUBE ASSEMBLY,METAL............... | 1 |
UOC:B16,H16
| 3 | PAOZZ | 5305012596322 | 34623 | 12342499-1 | SCREW,MACHINE #10-32 X .50........ | 2 |
UOC:B16,H16
| 4 | PAOZZ | 5340007647051 | 96906 | MS21333-69 | CLAMP,LOOP......................... | 3 |
UOC:B16,H16
| 5 | PAOZZ | 4730009498694 | 41947 | A-327 | NIPPLE,TUBE....................... | 1 | UOC:B16,H16
| 6 | PAOZZ | 5305012046502 | 78189 | 112-1008800-003 | SCREW #10-32 X .62............... | 1 |
UOC:B16,H16
| 7 | PAOZZ | 4710012653229 | 34623 | 12341420 | TUBE ASSEMBLY,METAL................ | 1 |
UOC:B16,H16
| 8 | PAOZZ | 4710012653228 | 34623 | 5598266 | TUBE ASSEMBLY,METAL................ | 1 |
UOC:B16,H16
| 9 | PAOZZ | 2910000253493 | 96906 | MS51085-1 | FILTER,FLUID..................... | 1 |
UOC:B16,H16
| 10 | PAOZZ | 5305012705435 | 34623 | 5593421 | SCREW,CAP,HEXAGON H #10-24 X 2.00.. | 2 |
UOC:B16,H16
| 11 | PFOZZ | 5365012727504 | 34623 | 5596814 | SPACER,SLEEVE..................... | 2 |
UOC:B16,H16
| 12 | PAOZZ | 5310012538953 | 34623 | 5593242 | WASHER,FLAT #10................... | 2 |
UOC:B16,H16
| 13 | PAOZZ | 5310000453296 | 96906 | MS35338-43 | WASHER,LOCK #10................... | 2 |
UOC:B16,H16
| 14 | PAOZZ | 5310009349758 | 96906 | MS35649-202 | NUT,PLAIN,HEXAGON #10-24.......... | 2 |
UOC:B16,H16
| 15 | PAOZZ | 5975009856630 | 96906 | MS3367-3-0 | STRAP,TIEDOWN,ELECT................ | 2 |
UOC:B16,H16
| 16 | PAOZZ | 4730002704606 | 81343 | 6-4 010103B | ADAPTER,STRAIGHT,PI................ | 1 |
UOC:B16,H16
| 17 | PAOZZ | 4820002875627 | 19207 | 12341793 | COCK,SHUTOFF,SCREW................ | 1 |
UOC:B16,H16

END OF FIGURE

Section II. TM 9-2320-280-24P-2

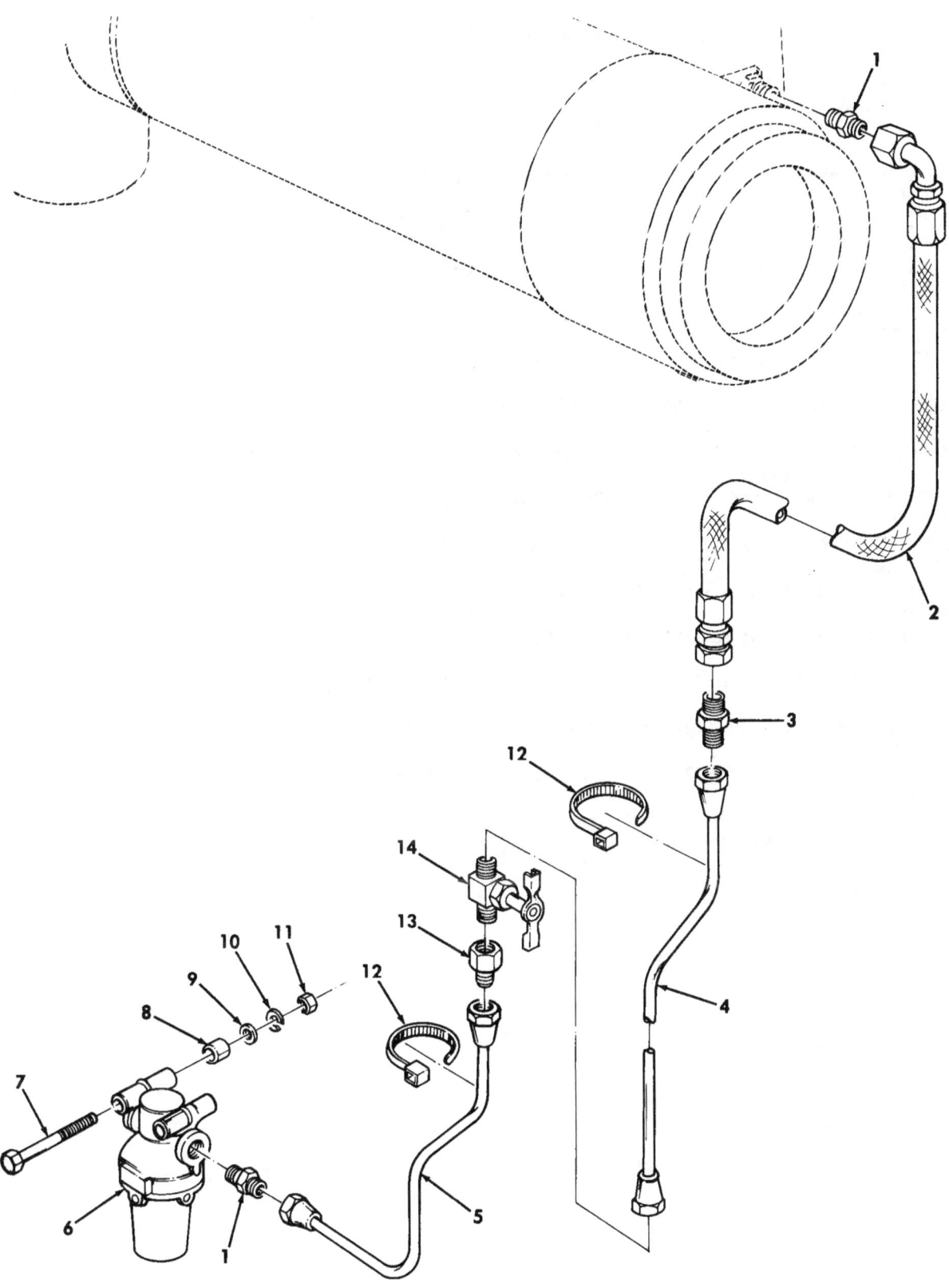

Figure 336. Fuel Lines, Personnel Heater to Fuel Filter, M997, M997A1, and M997A2 Ambulance.

SECTION II   TM9-2320-280-24P

(1) NO	(2) SMR CODE	(3) NSN	(4) CAGEC	(5) PART NUMBER	(7) ITEM DESCRIPTION AND USABLE ON CODES(UOC)	QTY

GROUP 2202 ACCESSORY ITEMS

FIG. 336 FUEL LINES,PERSONNEL HEATER TO FUEL FILTER,M997,M997A1,AND M997A2 AMBULANCE

1 PAOZZ 4730012699530 24617 190652 ADAPTER,STRAIGHT,PI................ 2
UOC:A15,B15,H15
2 PAOZZ 4720012653213 24234 403475 HOSE ASSEMBLY,METAL................. 1
UOC:A15,B15,H15
3 PAOZZ 4730009498694 41947 A-327 NIPPLE,TUBE 3/8..................... 1
UOC:A15,B15,H15
4 PAOZZ 4710012653229 34623 12341420 TUBE ASSEMBLY,METAL................. 1
UOC:A15,B15,H15
5 PAOZZ 4710012653228 34623 5598266 TUBE ASSEMBLY,METAL................. 1
UOC:A15,B15,H15
6 PAOZZ 2910000253493 96906 MS51085-1 FILTER,FLUID........................ 1
UOC:A15,B15,H15
7 PAOZZ 5305012705435 34623 5593421 SCREW,CAP,HEXAGON H #10-24 X 2.00.. 2
UOC:A15,B15,H15
8 PAOZZ 5365012727504 34623 5596814 SPACER,SLEEVE....................... 2
UOC:A15,B15,H15
9 PAOZZ 5310012538953 34623 5593242 WASHER,FLAT #10..................... 2
UOC:A15,B15,H15
10 PAOZZ 5310000453296 96906 MS35338-43 WASHER,LOCK #10..................... 2
UOC:A15,B15,H15
11 PAOZZ 5310009349758 96906 MS35649-202 NUT,PLAIN,HEXAGON #10-24........... 2
UOC:A15,B15,H15
12 PAOZZ 5975009856630 96906 MS3367-3-0 STRAP,TIEDOWN,ELECT................. 4
UOC:A15,B15,H15
13 PAOZZ 4730002704606 81343 6-4 010103B ADAPTER,STRAIGHT,PI................ 1
UOC:A15,B15,H15
14 PAOZZ 4820002875627 34623 5598272 COCK,SHUTOFF,SCREW................. 1
UOC:A15,B15,H15

END OF FIGURE

Section II.

TM 9-2320-280-24P-2

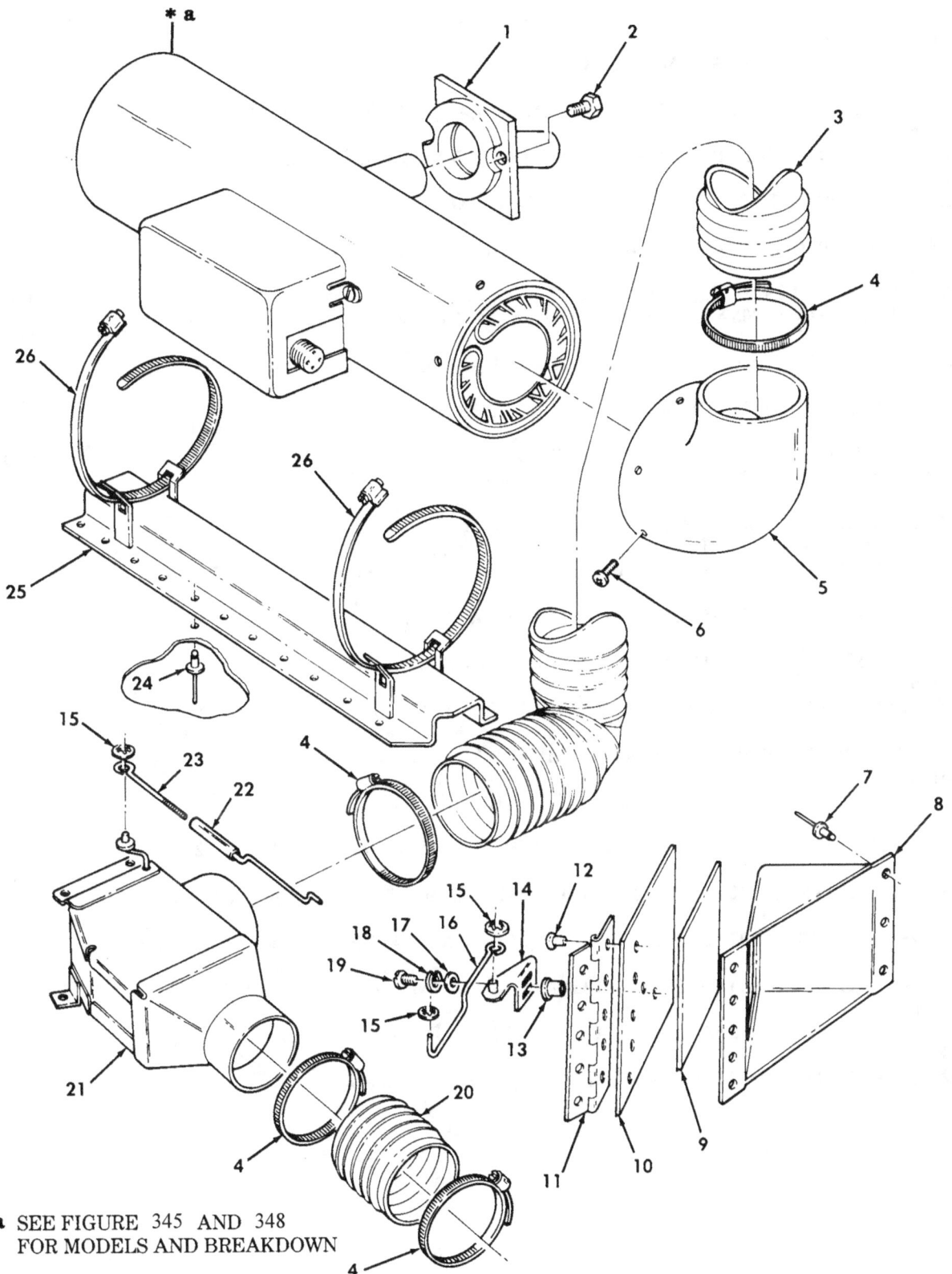

*a SEE FIGURE 345 AND 348 FOR MODELS AND BREAKDOWN

Figure 337. Personnel Heater Assembly and Attaching Hardware, M996 and M996A1 Ambulance.

SECTION II   TM9-2320-280-24P

(1) NO	(2) SMR CODE	(3) NSN	(4) CAGEC	(5) PART NUMBER	(7) ITEM DESCRIPTION AND USABLE ON CODES(UOC)	QTY

GROUP 2202 ACCESSORY ITEMS

FIG. 337 PERSONNEL HEATER ASSEMBLY AND ATTACHING HARDWARE, M996 AND M996A1 AMBULANCE

No	SMR	NSN	CAGEC	Part Number	Description	Qty	UOC
1	PFOZZ	2990012696625	34623	5598002	SHIELD ASSEMBLY,HEA	1	B16,H16
2	PAOZZ	5305005434709	28839	10-35936-7	SCREW,ASSEMBLED WAS	2	B16,H16
3	XBOZZ		19207	12339265-12	TUBING,NONMETALLIC	1	B16,H16
4	PAOZZ	4730012969318	19207	11608950-18	CLAMP,HOSE	4	B16,H16
5	PFOZZ	4730012975111	19207	12341883	ELBOW,PIPE	1	B16,H16
6	PAOZZ	5305009846195	96906	MS35206-247	SCREW,MACHINE	3	B16,H16
7	PAOZZ	5320002758344	90030	AD45BS	RIVET,BLIND 1/8	8	B16,H16
8	PFOZZ	2940012905014	34623	12341662	DEFLECTOR,AIR CLEAN	1	B16,H16
9	PAOZZ	5330012902709	19207	12341657	GASKET	1	B16,H16
10	PFOZZ	5340012893231	34623	5598253	COVER,ACCESS	1	B16,H16
11	PFOZZ	5340012902761	34623	5598256-C	HINGE,BUTT	1	B16,H16
12	PAOZZ	5320008990981	54402	AD42BS	RIVET,BLIND	5	B16,H16
13	PAOZZ	5310010099785	96906	MS27130-20	NUT,PLAIN,BLIND RIV #10-32	2	B16,H16
14	PAOZZ	5340012902263	34623	5598250-B	BRACKET,ANGLE	1	B16,H16
15	PAOZZ	5310013270387	19207	12341800	WASHER,LOCK	3	B16,H16
16	PFOZZ	3040012906758	34623	5598247	CONNECTING LINK,RIG	1	B16,H16
17	PAOZZ	5310000145850	96906	MS27183-42	WASHER,FLAT #10	2	B16,H16
18	PAOZZ	5310000453296	96906	MS35338-43	WASHER,LOCK #10	2	B16,H16
19	PAOZZ	5305009897435	96906	MS35207-264	SCREW,MACHINE #10-32 X .625	2	B16,H16
20	PFOZZ	4720012998469	16632	FLX400132057600	HOSE,AIR DUCT 4.50 INCHES LONG	1	B16,H16
21	PFOZZ	2540012703776	19207	12341227	DIVERTER BOX,HEATER	1	B16,H16
22	PFOZZ	3040012906757	34623	5598309	CONNECTING LINK,RIG	1	B16,H16
23	PFOZZ	3040012906760	34623	5598310	CONNECTING LINK,RIG	1	B16,H16
24	PAOZZ	5320012544251	34623	5593050	RIVET,BLIND	22	

SECTION II   TM9-2320-280-24P

(1) NO	(2) SMR CODE	(3) NSN	(4) CAGEC	(5) PART NUMBER	(6) DESCRIPTION AND USABLE ON CODES(UOC)	(7) QTY
					UOC:B16,H16	
25	PFOZZ	5365012805875	34623	5597035	SPACER,PLATE	1
					UOC:B16,H16	
26	PAOZZ	4730009086294	76599	96HSS	CLAMP,HOSE	2
					UOC:B16,H16	

END OF FIGURE

Section II. TM 9-2320-280-24P-2 C01

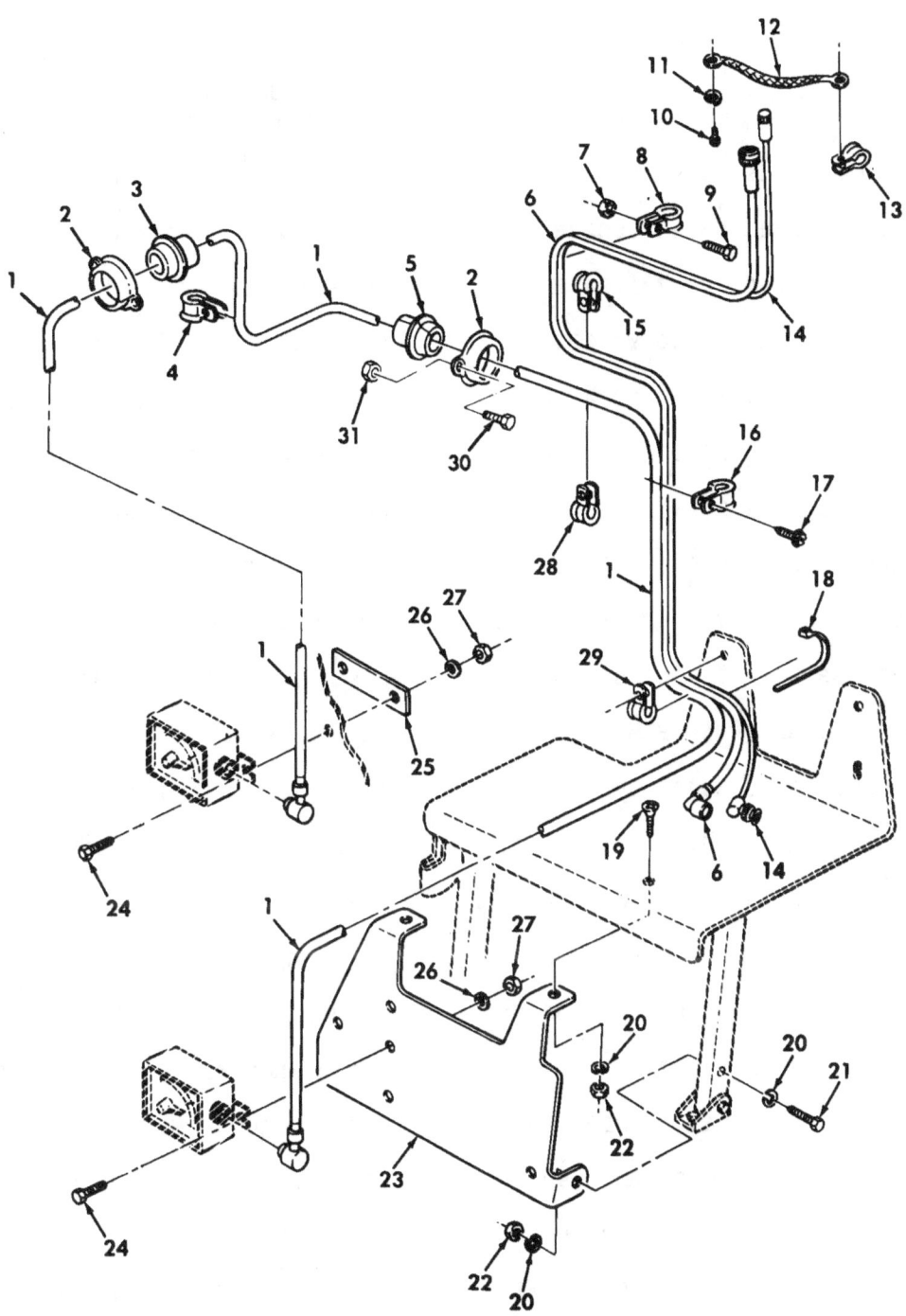

Figure 338. Antenna and Intercom Cables and Mounting Hardware, M997, M997A1, and M997A2 Ambulance.

```
            SECTION II             TM9-2320-280-24P C01
      (1)    (2)     (3)            (4)         (5)                    (6)
(7)   ITEM          SMR                                 PART
      NO    CODE    NSN             CAGEC       NUMBER      DESCRIPTION AND USABLE ON CODES(UOC)
QTY
```

GROUP 2202 ACCESSORY ITEMS

FIG. 338 ANTENNA AND INTERCOM CABLES AND MOUNTING HARDWARE,M997,M997A1, AND M997A2 AMBULANCE

```
         1  PAOZZ  6150012691849  19207  12338100-2        CABLE ASSEMBLY,SPEC..................  1
UOC:A15,B15,H15
         2  PFOZZ  2590004543620  34623  809223            RETAINER,ASSEMBLY....................  2
UOC:A15,B15,H15
         3  PAOZZ  5365012734690  19207  12341231-1        BUSHING,NONMETALLIC..................  1
UOC:A15,B15,H15
         4  PAOZZ  5340009588457  96906  MS21333-78        CLAMP,LOOP...........................  4
UOC:A15,B15,H15
         5  PAOZZ  5365012744674  19207  12341848          BUSHING,NONMETALLIC..................  1
UOC:A15,B15,H15
         6  PAOZZ  6150012691848  19207  12338089-2        CABLE ASSEMBLY,SPEC..................  1
UOC:A15,B15,H15
     *   7  PAOZZ  5310011520598  24617  271172            NUT,SELF-LOCKING,HE  #10-32.........  1
UOC:A15,B15,H15
         8  PAOZZ  5340000572904  96906  MS21333-71        CLAMP,LOOP...........................  1
UOC:A15,B15,H15
         9  PAOZZ  5305009897435  96906  MS35207-264       SCREW,MACHINE    #10-32 X .62.......  1
UOC:A15,B15,H15
        10  PAOZZ  5305009897434  96906  MS35207-263       SCREW,MACHINE    #10-32 X .50.......  1
UOC:A15,B15,H15
        11  PAOZZ  5310000453296  96906  MS35338-43        WASHER,LOCK   #10....................  1
UOC:A15,B15,H15
        12  PAOZZ  5999012720018  34623  5598121           STRIP,ELECTRICAL GR..................  1
UOC:A15,B15,H15
        13  PAOZZ  5340009891771  96906  MS21333-123       CLAMP,LOOP...........................  1
UOC:A15,B15,H15
        14  PAOZZ  5995012699525  19207  12338086-2        CABLE ASSEMBLY,RADI..................  1
UOC:A15,B15,H15
        15  PAOZZ  5340009848540  96906  MS21333-102       CLAMP,LOOP...........................  1
UOC:A15,B15,H15
        16  PAOZZ  5340000502740  96906  MS21333-75        CLAMP,LOOP...........................  3
UOC:A15,B15,H15
        17  PAOZZ  5305012596322  34623  12342499-1        SCREW,MACHINE    #10-32 X 1/2.......  3
UOC:A15,B15,H15
        18  PAOZZ  5975009856630  96906  MS3367-3-0        STRAP,TIEDOWN,ELECT..................  2
UOC:A15,B15,H15
        19  PAOZZ  5305009585246  80205  MS35190-289       SCREW,MACHINE    1/4-20 X .75.......  2
UOC:A15,B15,H15
        20  PAOZZ  5310012510748  34623  5593691           WASHER,FLAT   1/4....................  6
                                                           UOC:A15,B15,H15
        21  PAOZZ  5305000712237  96906  MS90725-14        SCREW,CAP,HEXAGON H  1/4-20 X 2.00..  2
                                                           UOC:A15,B15,H15
        22  PAOZZ  5310000614650  96906  MS51943-31        NUT,SELF-LOCKING,HE  1/4-20.........  4
UOC:A15,B15,H15
        23  PAOZZ  5340012708489  19207  12340879          BRACKET,MOUNTING.....................  1
UOC:A15,B15,H15
        24  PAOZZ  5306002264835  80204  B1821BH031C250N   BOLT,MACHINE    5/16-18 X 2.50......  4
```

338-1

SECTION II   TM9-2320-280-24P C01
  (1)   (2)   (3)   (4)   (5)   (6)
 (7)   ITEM   SMR   PART
       NO   CODE   NSN   CAGEC   NUMBER   DESCRIPTION AND USABLE ON CODES(UOC)
  QTY

                                                 UOC:A15,B15,H15

   25 PAOZZ 5365012805875 34623 5597035   SPACER,PLATE........................ 1
UOC:A15,B15,H15
   26 PAOZZ 5310011191024 24617 2436162   WASHER,FLAT 5/16................... 2
UOC:A15,B15,H15
   27 PAOZZ 5310008140673 96906 MS51943-33   NUT,SELF-LOCKING,HE 5/16-18........ 2
UOC:A15,B15,H15
   28 PAOZZ 5340009226300 96906 MS21333-77   CLAMP,LOOP.......................... 3
UOC:A15,B15,H15
   29 PAOZZ 5340008091494 96906 MS21333-105   CLAMP,LOOP.......................... 2
UOC:A15,B15,H15
   30 PAOZZ 5305000680502 96906 MS90725-6   SCREW,CAP,HEXAGON................... 2
UOC:A15,B15,H15
   31 PAOZZ 5310011520598 24617 271172   NUT,SELF-LOCKING.................... 2
UOC:A15,B15,H15

Section II. TM 9-2320-280-24P-2

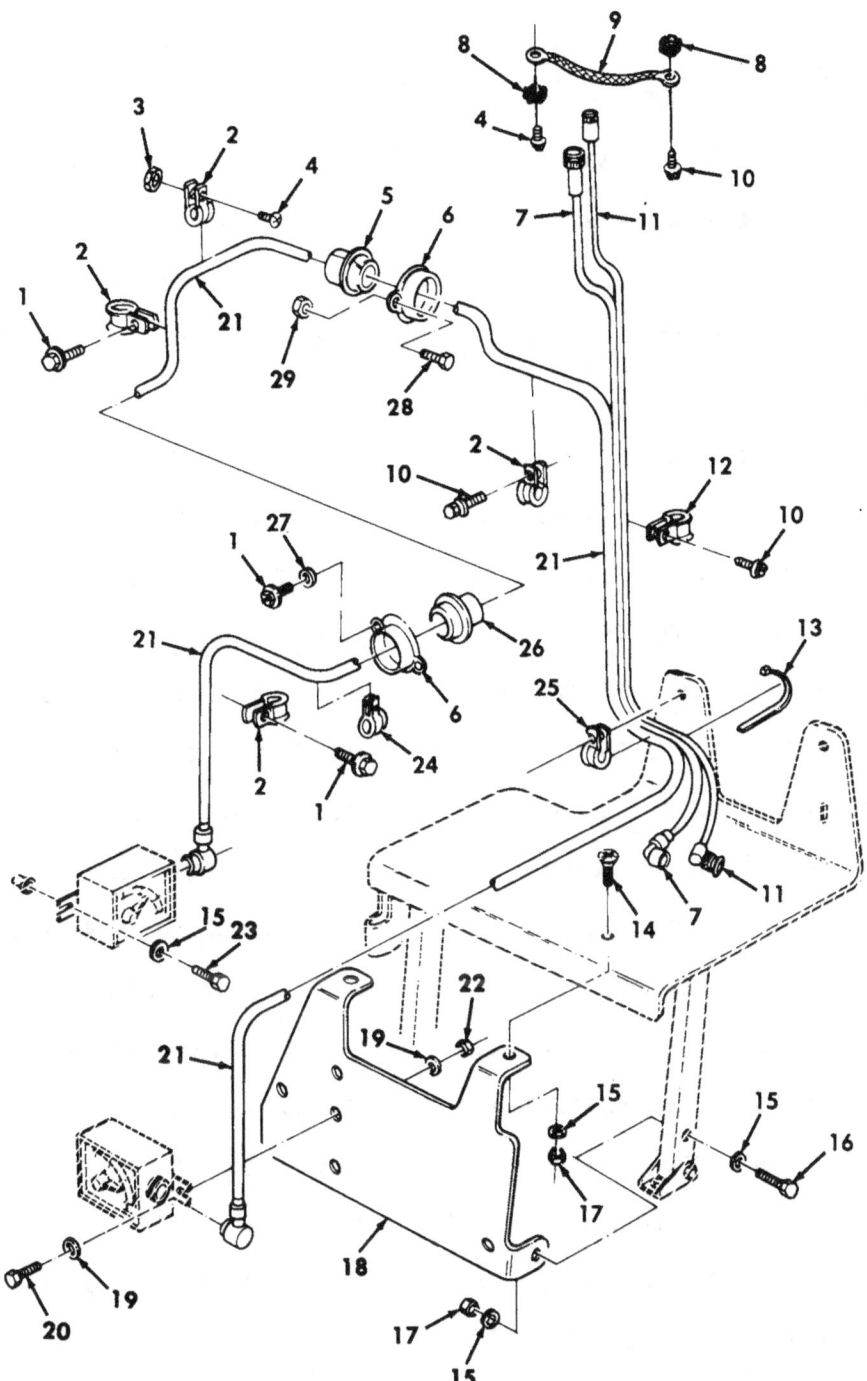

Figure 339. Antenna and Intercom Cables and Mounting Hardware, M996 and M996A1 Ambulance.

```
         SECTION II        TM9-2320-280-24P C01
    (1)     (2)    (3)          (4)       (5)                    (6)
(7)   ITEM   SMR                          PART
      NO    CODE   NSN         CAGEC     NUMBER    DESCRIPTION AND USABLE ON CODES(UOC)
QTY
```

GROUP 2202 ACCESSORY ITEMS

FIG. 339 ANTENNA AND INTERCOM CABLES

AND MOUNTING HARDWARE,M996 AND

M996A1 AMBULANCE

```
         1 PAOZZ 5305012729042 34623 5593409      SCREW,CAP,HEXAGON H   1/4-20 X 1.25..    6
UOC:B16,H16
         2 PAOZZ 5340000572904 96906 MS21333-71   CLAMP,LOOP.........................     8
UOC:B16,H16
         3 PAOZZ 5310012708189 24617 271168       NUT PLAIN,ASSEMBLED  #10-32........      2
UOC:B16,H16
         4 PAOZZ 5305000593659 96906 MS51958-63   SCREW,MACHINE  #10-32 X .50........      3
UOC:B16,H16
         5 PAOZZ 5325012055378 19207 12338084     GROMMET.............................     1
UOC:B16,H16
         6 PFOZZ 2590004543620 34623 809223       RETAINER,ASSEMBLY...................     2
UOC:B15,H15
         7 PAOZZ 6150012691848 19207 12338089-2   CABLE ASSEMBLY,SPEC.................     1
UOC:B16,H16
         8 PAOZZ 5310000809786 96906 MS45904-60   WASHER,LOCK  #10....................     2
UOC:B16,H16
         9 PAOZZ 5999012720018 34623 5598121      STRIP,ELECTRICAL GR.................     1
UOC:B16,H15,H16
        10 PAOZZ 5305012596322 34623 12342499-1   SCREW,MACHINE  #10-32 X 3/4.........     6
UOC:B16,H16
        11 PAOZZ 5995012699525 34623 5589012-B    CABLE ASSEMBLY,RADI.................     1
UOC:B16,H16
        12 PAOZZ 5340000502740 96906 MS21333-75   CLAMP,LOOP..........................     3
UOC:B16,H16
*       13 PAOZZ 5975010345871 96906 MS3367-7-0   STRAP,TIEDOWN,ELECT.................     2
UOC:B16,H16
        14 PAOZZ 5305007195021 96906 MS51959-81   SCREW,MACHINE  1/4-20 X .75.........     2
UOC:B16,H16
        15 PAOZZ 5310012510748 34623 5593691      WASHER,FLAT  1/4....................     8
UOC:B16,H16
        16 PAOZZ 5305000712237 96906 MS90725-14   SCREW,CAP,HEXAGON H   1/4-20 X 2.00..    2
UOC:B16,H16
        17 PAOZZ 5310000614650 96906 MS51943-31   NUT,SELF-LOCKING,HE  1/4-20.........     4
UOC:B16,H16
        18 PFOZZ 5340012708489 19207 12340879     BRACKET,MOUNTING....................     1
UOC:B16,H16
*       19 PAOZZ 5310011191024 24617 2436162      WASHER,FLAT  5/16...................     4
UOC:B16,H16
        20 PAOZZ 5306002258499 96906 MS90725-34   BOLT,MACHINE  5/16-18 X 1.00........     2
UOC:B16,H16
        21 PAOZZ 6150012691839 19207 12338100-1   CABLE ASSEMBLY,SPEC.................     1
UOC:B16,H16
        22 PAOZZ 5310008140673 96906 MS51943-33   NUT,SELF-LOCKING,HE  5/16-18........     2
UOC:B16,H16
        23 PAOZZ 5305012729034 34623 G0045696     SCREW,ASSEMBLED WAS  1/4-20 X 1.25..     2
UOC:B16,H16
        24 PAOZZ 5340009848540 96906 MS21333-102  CLAMP,LOOP..........................     1
```

SECTION II   TM9-2320-280-24P C01
 (1)   (2)   (3)   (4)   (5)   (6)
(7) ITEM   SMR   PART
    NO   CODE   NSN   CAGEC   NUMBER   DESCRIPTION AND USABLE ON CODES (UOC)
TY

                                                 UOC:B16,H16
    25 PAOZZ 5340008091494 96906 MS21333-105   CLAMP,LOOP........................ 2
UOC:B16,H16
    26 PAOZZ 5365012743573 19207 12341231-3   BUSHING,NONMETALLIC................ 1
UOC:B16,H16
    27 PAOZZ 5310000145850 96906 MS27183-42   WASHER,FLAT #10................... 2
UOC:B16,H16
    28 PAOZZ 5305000680502 96906 MS90725-6   SCREW,CAP,HEXAGON H 1/4-20 X 0.75.. 2
UOC:B16,H16
    29 PAOZZ 5310011520598 24617 271172   NUT,SELF-LOCKING,AS 1/4-20........ 2
UOC:B16,H16

339-2

Section II. TM 9-2320-280-24P-2

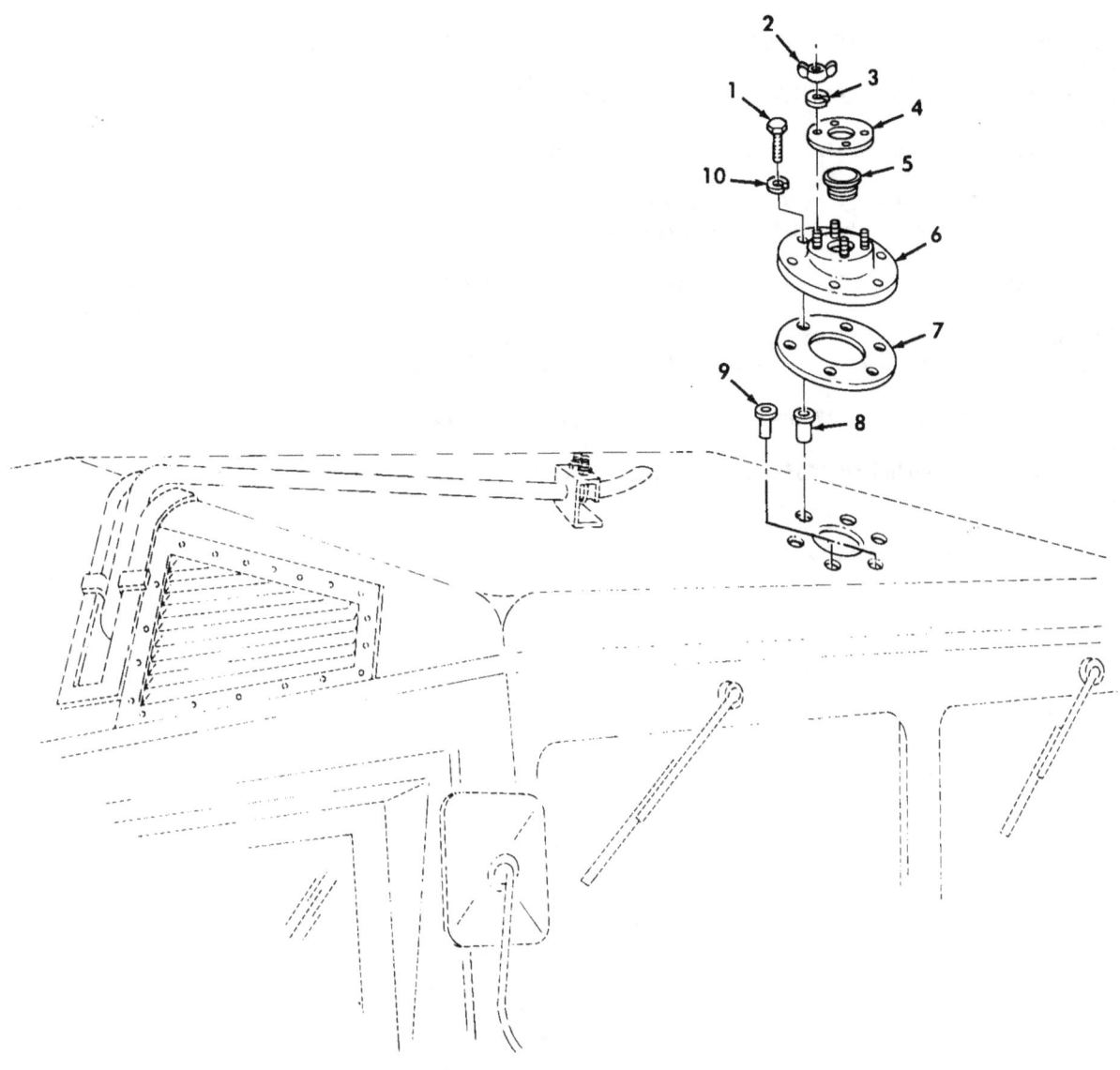

Figure 340. Antenna Mount and Mounting Hardware, M996 and M996A1 Ambulance.

SECTION II TM9-2320-280-24P

(1) NO	(2) SMR CODE	(3) NSN	(4) CAGEC	(5) PART NUMBER	(7) ITEM DESCRIPTION AND USABLE ON CODES(UOC)	QTY

GROUP 2202 ACCESSORY ITEMS

FIG. 340 ANTENNA MOUNT AND MOUNTING HARDWARE, M996 AND M996A1 AMBULANCE

1 PAOZZ 5306002264828 80204 B1821BH031C113N BOLT,MACHINE 5/16-18 X 1.125....... 6 UOC:B16,H16
2 PAOZZ 5310010882490 96906 MS35425-74 NUT,PLAIN,WING...................... 4 UOC:B16,H16
3 PAOZZ 5310006379541 96906 MS35338-46 WASHER,LOCK........................ 4 UOC:B16,H16
4 PAOZZ 5330012708315 19207 12341876 GASKET............................. 1 UOC:B16,H16
5 PAOZZ 5340011940887 99017 BPF-3-1/2 PLUG,PROTECTIVE,DUS................ 1 UOC:B16,H16
6 PAOZZ 5985012698272 19207 12341466 BASE,ANTENNA SUPPOR................ 1 UOC:B16,H16
7 PAOZZ 5330012988127 34623 5597985 GASKET............................. 1 UOC:B16,H16
8 PAOZZ 5310011016046 96906 MS27130-A50 NUT,PLAIN,BLIND RIV................ 4 UOC:B16,H16
9 PAOZZ 5310010465382 96906 MS27130-A43 NUT,PLAIN,BLIND RIV................ 2 UOC:B16,H16
10 PAOZZ 5310004079566 96906 MS35338-45 WASHER,LOCK 5/16................... 6 UOC:B16,H16

END OF FIGURE

Section II. TM 9-2320-280-24P-2

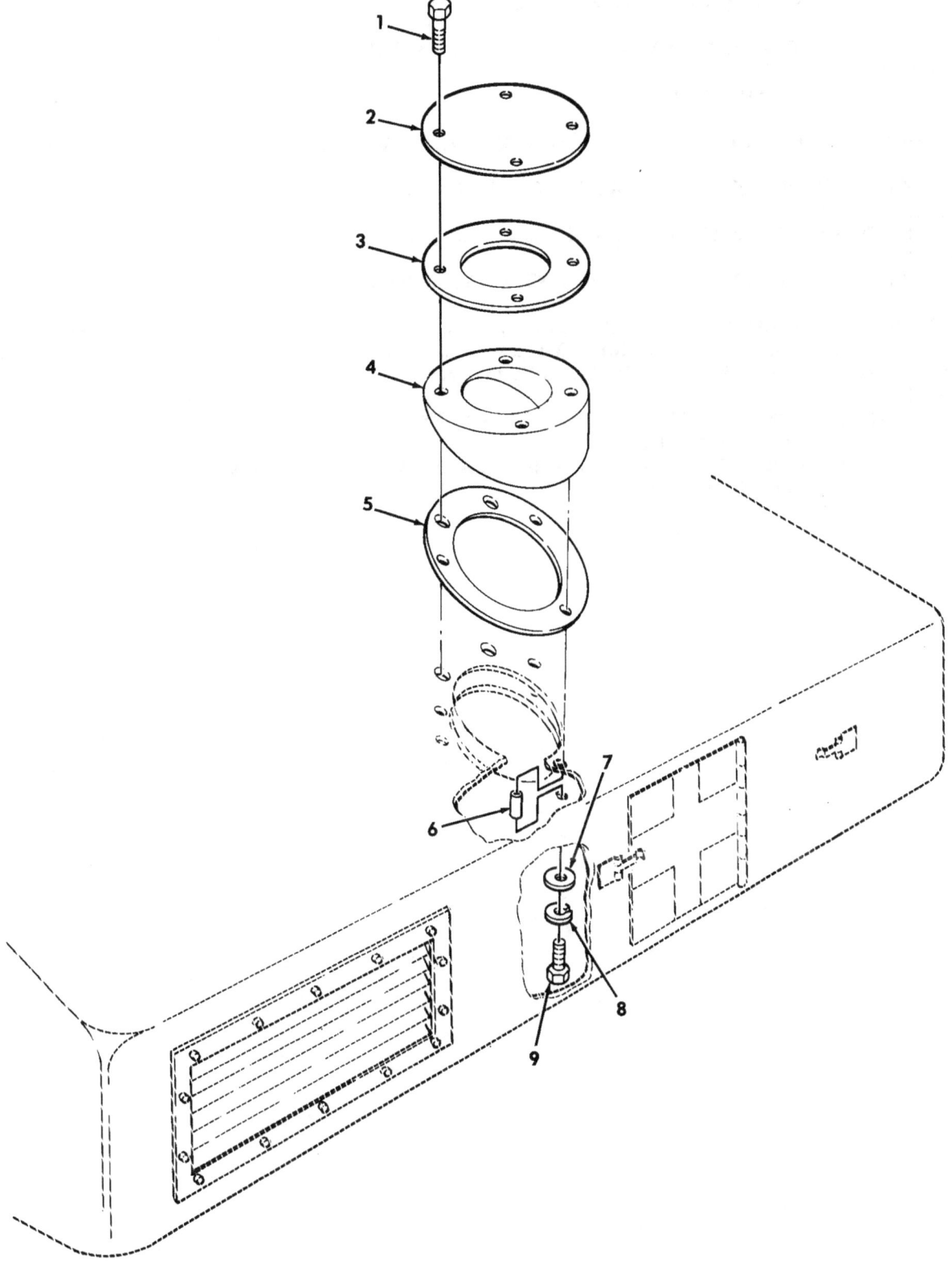

Figure 341. Antenna Mount and Mounting Hardware, M997, M997A1, and M997A2 Ambulance.

SECTION II   TM9-2320-280-24P

(1) NO	(2) SMR CODE	(3) NSN	(4) CAGEC	(5) PART NUMBER	(7) DESCRIPTION AND USABLE ON CODES(UOC)	ITEM QTY

GROUP 2202 ACCESSORY ITEMS

FIG. 341 ANTENNA MOUNT AND MOUNTING HARDWARE,M997,M997A1,AND M997A2 AMBULANCE

```
 1 PAOZZ 5305005432419 80204 B1821BH038C113N  SCREW,CAP,HEXAGON H 3/8-16 X 1.125.  4
UOC:A15,B15,H15
 2 PAOZZ 5985012698271 34623 5598650          SUPPORT,ANTENNA.....................  1
UOC:A15,B15,H15
 3 PAOZZ 5330012708315 19207 12341876         GASKET..............................  1
UOC:A15,B15,H15
 4 PAOZZ 5985012698274 19207 12341108         BASE,ANTENNA SUPPOR.................  1
UOC:A15,B15,H15
 5 PAOZZ 5330012820913 34623 5589970          GASKET..............................  1    UOC:A15,B15,H15
 6 XBOZZ                34623 5934310         SPACER,FRAME........................  3    UOC:A15,B15,H15
 7 PAOZZ 5310014124013 24617 2436163          WASHER,FLAT 3/8.....................  3
UOC:A15,B15,H15
 8 PAOZZ 5310002617340 96906 MS35338-8        WASHER,LOCK 3/8.....................  3
          UOC:A15,B15,H15
 9 PAOZZ 5305006388920 80204 B1821BH038C225N  SCREW,CAP,HEXAGON H 3/8-16 X 2.250.  3
          UOC:A15,B15,H15
```

END OF FIGURE

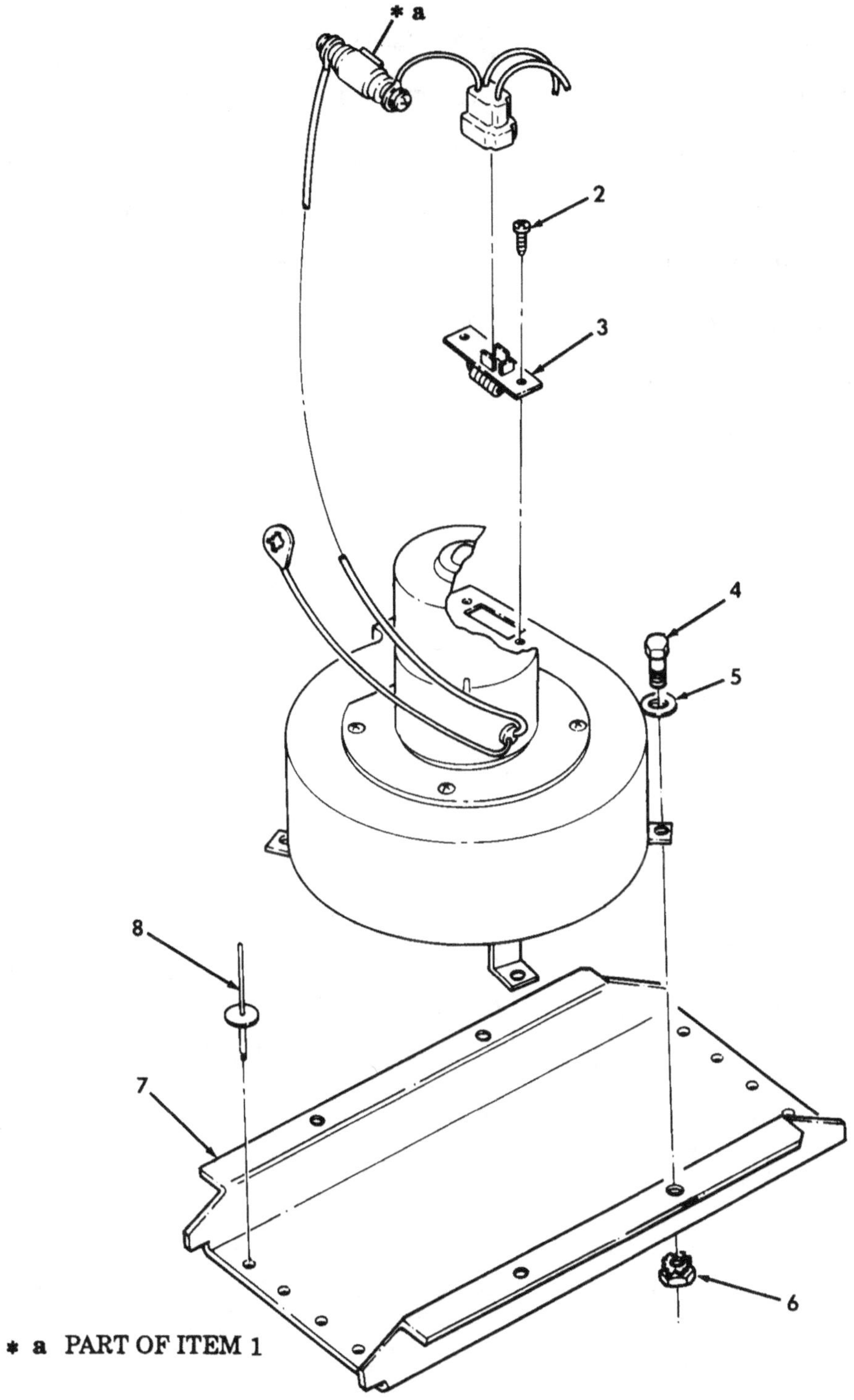

Section II. TM 9-2320-280-24P-2

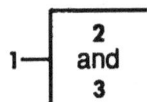

* a PART OF ITEM 1

Figure 342. Heater Blower Assembly, M996 and M996A1 Ambulance.

SECTION II TM9-2320-280-24P C01

(1) ITEM NO	(2) SMR CODE	(3) NSN	(4) CAGEC	(5) PART NUMBER	(6) DESCRIPTION AND USABLE ON CODES (UOC)	(7) QTY
					GROUP 2202 ACCESSORY ITEMS	
					FIG. 342 HEATER BLOWER ASSEMBLY, M996 AND M996A1 AMBULANCE	
1	PAOZZ	4140012653151	19207	12341610	FAN,CENTRIFUGAL..................... UOC:B16,H16	1
2	PAOZZ	5305012705419	24234	200335	.SCREW,TAPPING,THREA................ UOC:B16,H16	2
3	PAOZZ	5905012700966	21002	HV033596	.RESISTOR,FIXED,FILM................ UOC:B16,H16	1
4	PAOZZ	5305000680502	96906	MS90725-6	SCREW,CAP,HEXAGON H 1/4-20 X .75... UOC:B16,H16	3
5	PAOZZ	5310005825965	96906	MS35338-44	WASHER,LOCK 1/4.................... UOC:B16,H16	3
* 6	PAOZZ	5310011981722	34623	5982526	NUT,PLAIN,BLIND RIV................. UOC:B16,H16	3
7	PFOZZ	5340012893244	19207	12341661	BRACKET,MOUNTING.................... UOC:B16,H16	1
8	PAOZZ	5320012645978	19207	12339355-2	RIVET,BLIND 3/16................... UOC:B16,H16	8

Section II. TM 9-2320-280-24P-2

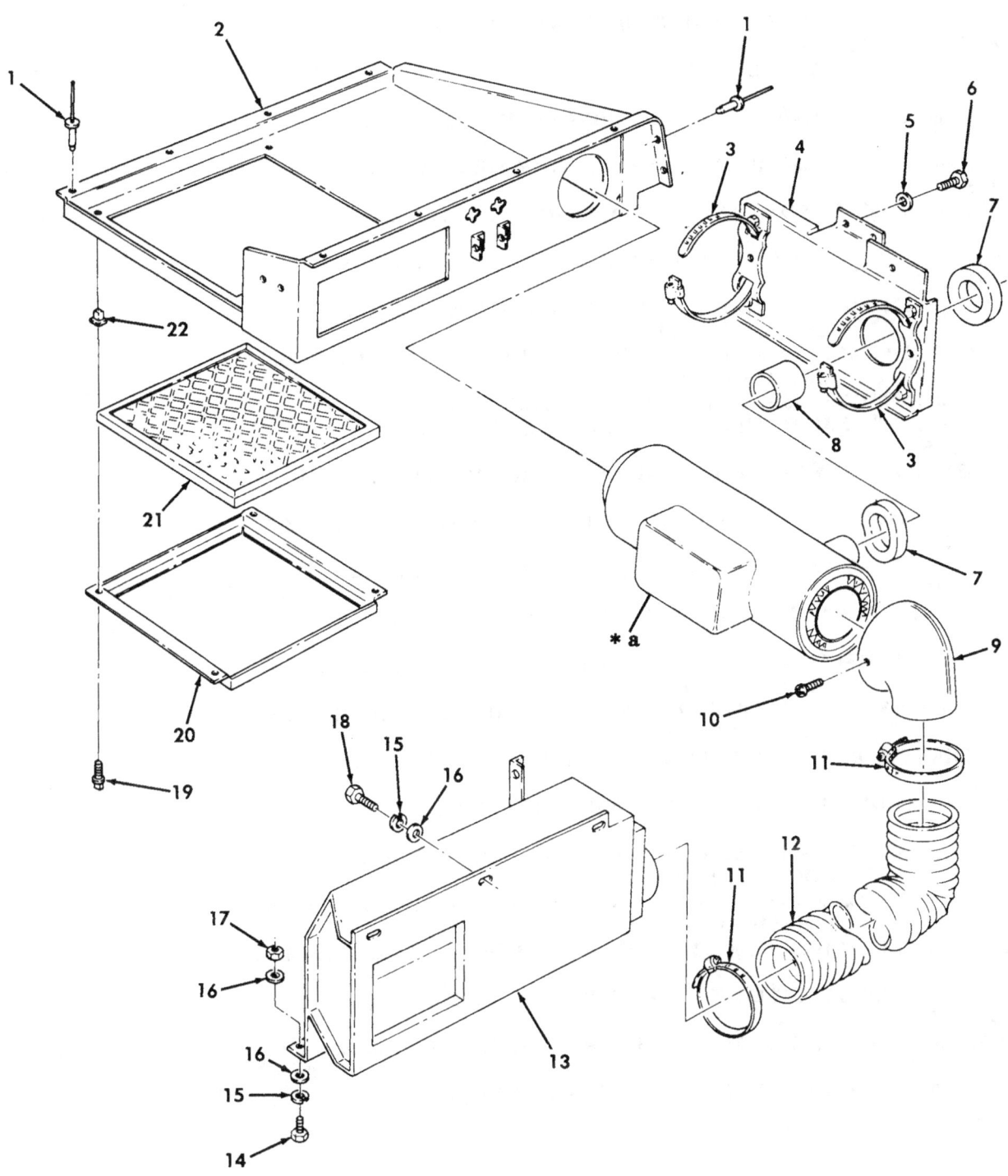

* a SEE FIGURE 345 AND 348
     FOR MODELS AND BREAKDOWN

Figure 343. Heater, Filter, and Attaching Hardware, M997, M997A1, and M997A2 Ambulance.

SECTION II   TM9-2320-280-24P

(1) NO	(2) SMR CODE	(3) NSN	(4) CAGEC	(5) PART NUMBER	(7) ITEM DESCRIPTION AND USABLE ON CODES(UOC)	QTY

GROUP 2202 ACCESSORY ITEMS

FIG. 343 HEATER,FILTER,AND ATTACHING HARDWARE,M997,M997A1,AND M997A2 AMBULANCE

```
 1 PAOZZ 5320012716357 9K475 MGLP-B6-4      RIVET,BLIND.............. 10
UOC:A15,B15,H15
 2 PFOZZ 2540012911043 15434 410912         VENTILATOR,AIR CIRC............. 1
UOC:A15,B15,H15
 3 PAOZZ 4730009086294 76599 96HSS          CLAMP,HOSE................ 2
UOC:A15,B15,H15
 4 XBOZZ                34623 5597900       BRACKET,MOUNTING............. 1
UOC:A15,B15,H15
 5 PAOZZ 5310002748710 24234 202751         WASHER,LOCK #10.............. 2
           UOC:A15,B15,H15
 6 PAOZZ 5305012967762 24234 201686         SCREW,CAP,HEXAGON H 1/4-20 X .75... 2
           UOC:A15,B15,H15
 7 PFOZZ 5330012708317 34623 5741471        GASKET..................... 2       UOC:A15,B15,H15
 8 PFOZZ 2510012730572 34623 5741525        COLLAR,FRAME.............. 1
UOC:A15,B15,H15
 9 PAOZZ 4520011926073 99688 586230         ADAPTER,AIR CONDITI............ 1
UOC:A15,B15,H15
10 PAOZZ 5305009846195 96906 MS35206-247    SCREW,MACHINE............... 3
UOC:A15,B15,H15
11 PAOZZ 4730012737660 7Z588 63060          CLAMP,HOSE................ 2
UOC:A15,B15,H15
12 PFOZZ 4720012653238 24234 319032         HOSE,NONMETALLIC............. 1
UOC:A15,B15,H15
13 PFOZZ 2540012969358 24234 410922         VENTILATOR,AIR CIRC............. 1
           UOC:A15,B15,H15
14 PAOZZ 5305000680502 96906 MS90725-6      SCREW,CAP,HEXAGON H 1/4-20 X 3/4... 4
           UOC:A15,B15,H15
15 PAOZZ 5310005825965 96906 MS35338-44     WASHER,LOCK 1/4................ 7
UOC:A15,B15,H15
16 PAOZZ 5310008094058 96906 MS27183-10     WASHER,FLAT 1/4.............. 11
UOC:A15,B15,H15
17 PAOZZ 5310000614650 96906 MS51943-31     NUT,SELF-LOCKING,HE 1/4-20......... 4
           UOC:A15,B15,H15
18 PAOZZ 5305002253843 80204 B1821BH025C100N SCREW,CAP,HEXAGON H 1/4-20 X 1.00.. 3
           UOC:A15,B15,H15
19 PAOZZ 5305012645874 19207 12340792       SCREW,TAPPING #10 X 1/2............ 4
UOC:A15,B15,H15
20 PFOZZ 2940012852942 24234 513447         FILTER BODY,FLUID............... 1
UOC:A15,B15,H15
21 PAOZZ 2540012653266 24234 407234         FILTER ELEMENT,FLUI............ 1
UOC:A15,B15,H15
22 PAOZZ 5310012705394 34623 5741475        NUT,EXPANSION................ 4
UOC:A15,B15,H15
```

END OF FIGURE

Section II. TM 9-2320-280-24P-2

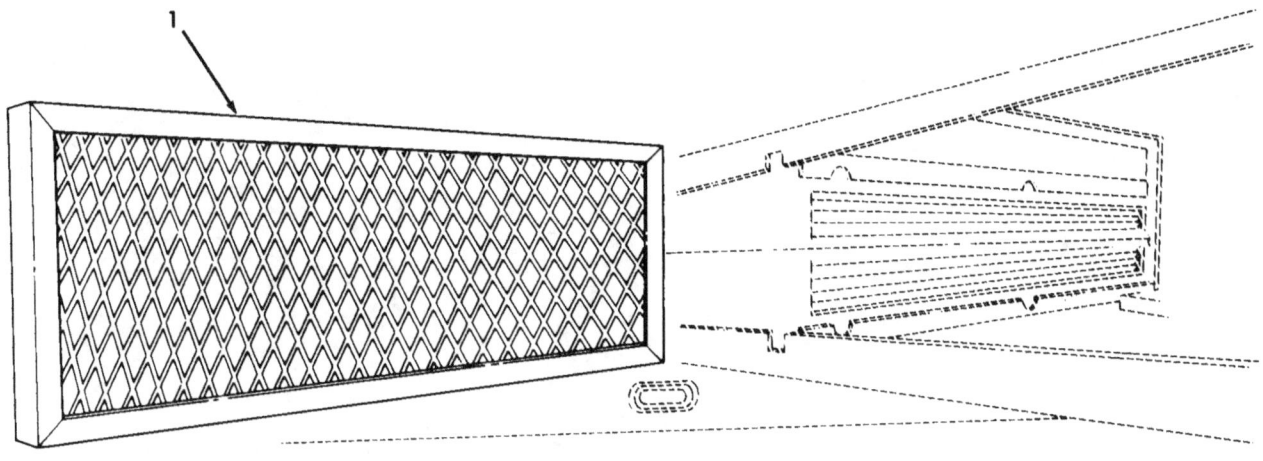

Figure 344. Air Intake Filter, M996 and M996A1 Ambulance.

SECTION II TM9-2320-280-24P

(1) ITEM NO	(2) SMR CODE	(3) NSN	(4) CAGEC	(5) PART NUMBER	(6) DESCRIPTION AND USABLE ON CODES(UOC)	(7) QTY
					GROUP 2202 ACCESSORY ITEMS	
					FIG. 344 AIR INTAKE FILTER, M996 AND M996A1 AMBULANCE	
1	PAOZZ	4130012658691	19207	12341286	FILTER ELEMENT, AIR.................. UOC:B16,H16	1
					END OF FIGURE	

344-1

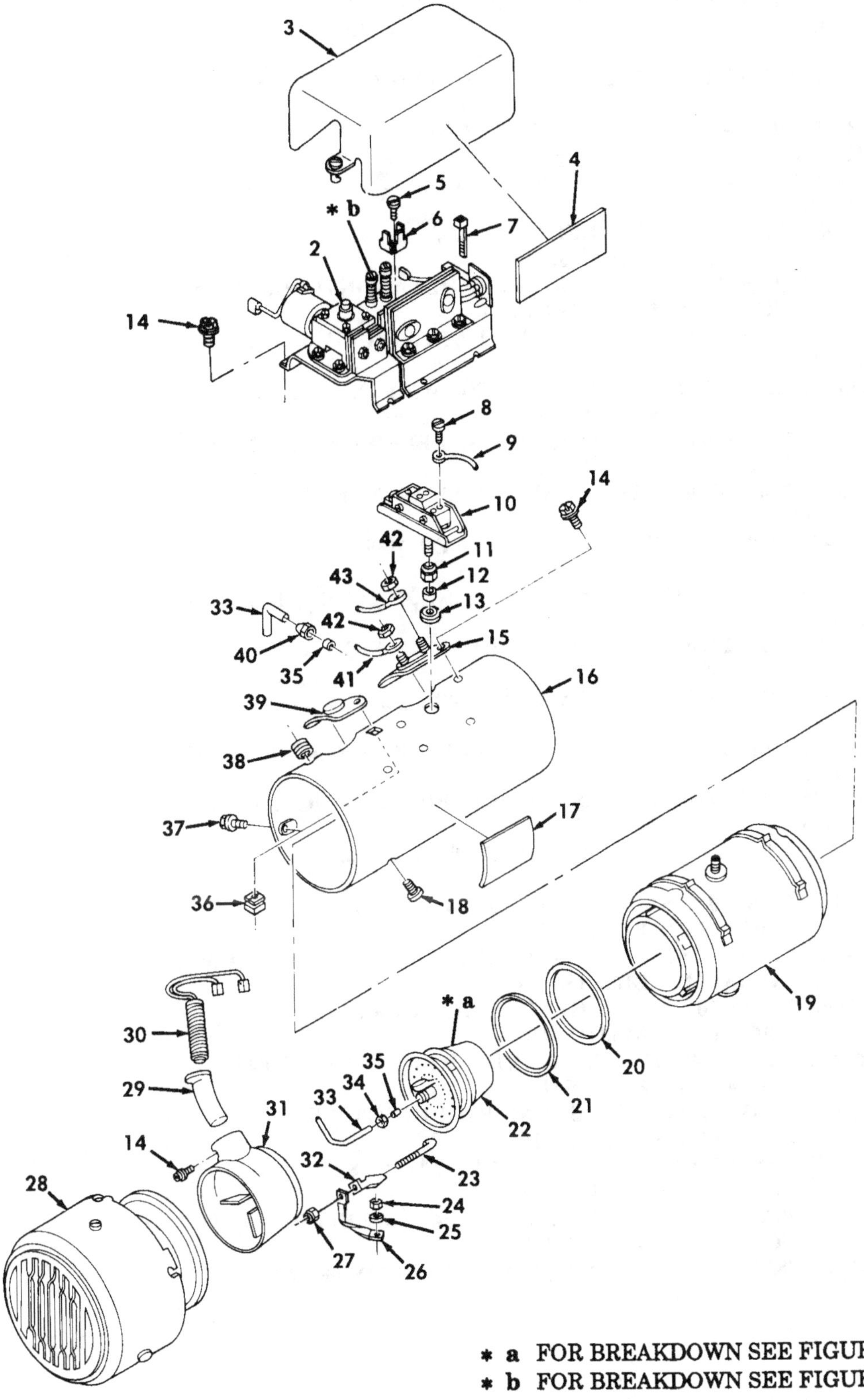

Section II. TM 9-2320-280-24P-2

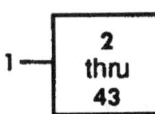

* a FOR BREAKDOWN SEE FIGURE
* b FOR BREAKDOWN SEE FIGURE

Figure 345. Heater Assembly, Multifuel, B=Model, M996, M996A1, M997, M997A1, and M997A2 Ambulance.

SECTION II　　TM9-2320-280-24P

(1) NO	(2) SMR CODE	(3) NSN	(4) CAGEC	(5) PART NUMBER	(6) DESCRIPTION AND USABLE ON CODES(UOC)	(7) QTY

GROUP 2202 ACCESSORY ITEMS

FIG. 345 HEATER ASSEMBLY,MULTIFUEL, B-MODEL,M996, M996A1,M997,M997A1,AND M997A2 AMBULANCE

```
 1 PAOFF 2540011943323 81349 MIL-PRF-62550/3  HEATER,VEHICULAR,CO................ 1
UOC:A15,B15,B16,H15,H16
 2 XBFFF                78385 G706064         .BRACKET,COMPONENT H................ 1
UOC:A15,B15,B16,H15,H16
 3 PFFZZ 5340012906331 78385 G706033          .COVER,ACCESS...................... 1
UOC:A15,B15,B16,H15,H16
 4 PFFZZ 7690012837938 78385 706025           .DECAL............................. 1
                UOC:A15,B15,B16,H15,H16
 5 PAFZZ 5305012906290 78385 706074           .SCREW,CAP,HEXAGON H  #8-32 X 3/8... 2
                UOC:A15,B15,B16,H15,H16
 6 PFFZZ 5940012926907 78385 706070           .TERMINAL,STATIONARY................ 1
UOC:A15,B15,B16,H15,H16
 7 PAFZZ 5975011280390 06383 PLT2S            .STRAP,TIEDOWN,ELECT................ 4
                UOC:A15,B15,B16,H15,H16
 8 PAFZZ 5305004562582 57733 170677           .SCREW,ASSEMBLED WAS  #6-32 X 1/4... 5
                UOC:A15,B15,B16,H15,H16
 9 PFFZZ 6150012879918 78385 G706052-25       .CABLE ASSEMBLY,SPEC................ 1
UOC:A15,B15,B16,H15,H16
10 PAFZZ 5930003455455 19207 11663057         .SWITCH,THERMOSTATIC................ 1
UOC:A15,B15,B16,H15,H16
11 PAFZZ 5975012886594 78385 705267           .NUT,COUPLING,ELECTR  1/4........... 1
UOC:A15,B15,B16,H15,H16
12 PAFZZ 4730011937390 10988 222-652          .SLEEVE,COMPRESSION,................ 1
UOC:A15,B15,B16,H15,H16
13 PAFZZ 5310012878726 78385 705930           .WASHER,FLAT....................... 1
UOC:A15,B15,B16,H15,H16
14 PAFZZ 5305004035130 24617 423531           .SCREW,MACHINE  #8-32 X 1/4......... 12
UOC:A15,B15,B16,H15,H16
15 PAFZZ 5930006795925 19207 10948233         .SWITCH,THERMOSTATIC................ 1
UOC:A15,B15,B16,H15,H16
16 XAFZZ                78385 G706034         .HOUSING ASSEMBLY................... 1
UOC:A15,B15,B16,H15,H16
17 PFOZZ 9905012837937 78385 706015           .PLATE,IDENTIFICATIO................ 1
                UOC:A15,B15,B16,H15,H16
18 PAFZZ 5305005762335 57733 487357           .SCREW,ASSEMBLED WAS  #8-32 X 3/8... 1
                UOC:A15,B15,B16,H15,H16
19 PFFZZ 2540011650465 78385 G704232          .HEAT EXCHANGER.................... 1
                UOC:A15,B15,B16,H15,H16
20 PAFZZ 5330000890978 19207 10948235         .GASKET PART OF KIT P/N 5704052..... 1
                UOC:A15,B15,B16,H15,H16
21 PAFZZ 5331000890998 78385 718768-23        .O-RING PART OF KIT P/N 5704052..... 1
                UOC:A15,B15,B16,H15,H16
22 PFFZZ 4520012847099 78385 G706055          .BURNER ASSEMBLY,SPA............... 1
                UOC:A15,B15,B16,H15,H16
23 PAFZZ 5306013517742 78385 703547           .BOLT,HOOK PART OF KIT P/N 5704052.. 4
                UOC:A15,B15,B16,H15,H16
24 PAFZZ 5310003337341 78385 487283           .NUT,PLAIN,HEXAGON #8-32 PART OF    4
```

345-1

SECTION II   TM9-2320-280-24P

(1) NO	(2) SMR CODE	(3) NSN	(4) CAGEC	(5) PART NUMBER	(7) ITEM DESCRIPTION AND USABLE ON CODES(UOC)	QTY
				KIT P/N 5704052.................... UOC:A15,B15,B16,H15,H16		
25	PAFZZ	5310000610004	57733	475005	.WASHER,LOCK #8................... UOC:A15,B15,B16,H15,H16	1
26	PFFZZ	6150011654667	78385	G704373	.LEAD,ELECTRICAL.................. UOC:A15,B15,B16,H15,H16	1
27	PAFZZ	5310012876543	78385	706131	.NUT,SELF-LOCKING,HE #8-32 PART OF KIT P/N 5704052.................... UOC:A15,B15,B16,H15,H16	4
28	PAFZZ	2540012898328	78385	G706014	.BLOWER ASSEMBLY,HEA................ UOC:A15,B15,B16,H15,H16	1
29	PFFZZ	2540011658176	78385	G704288-1	.IGNITER TUBE AND BR............... UOC:A15,B15,B16,H15,H16	1
30	PAOZZ	4520002175782	16236	CS-4520-SV-0705	.IGNITER,SPARK,FUEL................ UOC:A15,B15,B16,H15,H16	1
31	PFFZZ	2540011658175	78385	G704177	.HOUSING,BAFFLE,HEAT............... UOC:A15,B15,B16,H15,H16	1
32	PFFZZ	5340012365101	78385	703546	.CLAMP,RIM CLENCHING PART OF KIT P/N 5704052............................. UOC:A15,B15,B16,H15,H16	
33	PFFZZ	4710011632805	78385	704363	.TUBE ASSEMBLY,METAL............... UOC:A15,B15,B16,H15,H16	1
34	PFFZZ	4730000433750	21450	120487	.NUT,TUBE COUPLING................. UOC:A15,B15,B16,H15,H16	1
35	PFFZZ	4730007017737	78385	476624	.SLEEVE,COMPRESSION,............... UOC:A15,B15,B16,H15,H16	2
36	PFFZZ	2540012900715	78385	706041	.RECEPTACLE,HEATER H............... UOC:A15,B15,B16,H15,H16	2
37	PAFZZ	5305000188370	78385	489142	.SCREW,MACHINE #8-32 X 3/8......... UOC:A15,B15,B16,H15,H16	4
38	PAFZZ	5325005432902	78385	488993	.GROMMET,NONMETALLIC............... UOC:A15,B15,B16,H15,H16	1
39	PFOZZ	2540012898329	78385	G706035	.HATCH COVER,HEATER................ UOC:A15,B15,B16,H15,H16	1
40	PFFZZ	4730011531871	79470	1611X3	.LOCKNUT,TUBE FITTIN............... UOC:A15,B15,B16,H15,H16	1
41	PFFZZ	6150012914809	78385	G706052-9	.CABLE ASSEMBLY,SPEC............... UOC:A15,B15,B16,H15,H16	1
42	PAFZZ	5310000637360	83385	841-6-32	.NUT,PLAIN,ASSEMBLED #6-32......... UOC:A15,B15,B16,H15,H16	2
43	PFFZZ	6150012879917	78385	G706052-21	.CABLE ASSEMBLY,SPEC............... UOC:A15,B15,B16,H15,H16	1

END OF FIGURE

Section II. TM 9-2320-280-24P-2

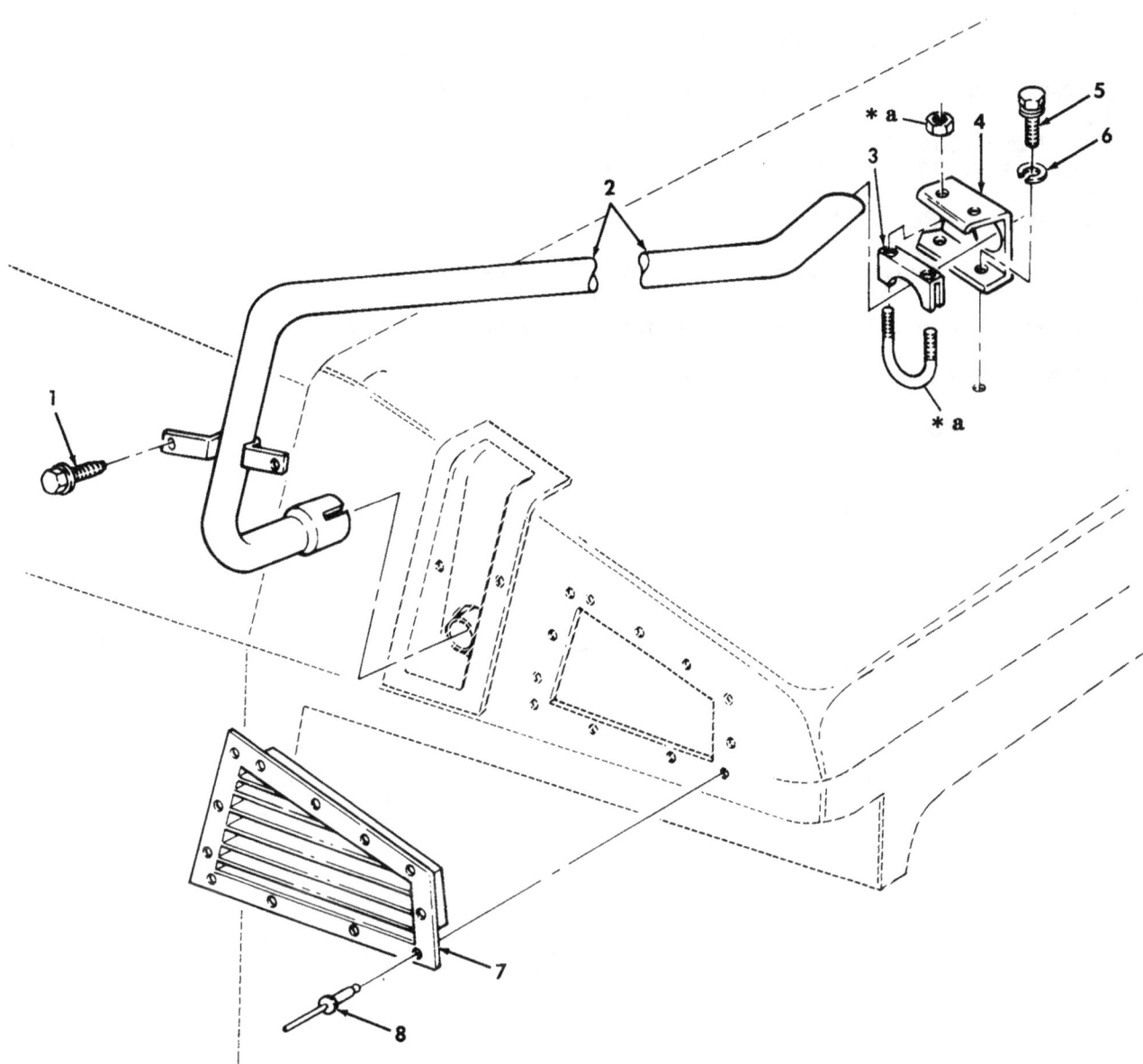

* a PART OF ITEM 3

Figure 346. Heater Grille, Exhaust Pipe, and Mounting Hardware, M996 and M996A1 Ambulance.

SECTION II   TM9-2320-280-24P

(1) NO	(2) SMR CODE	(3) NSN	(4) CAGEC	(5) PART NUMBER	(7) ITEM DESCRIPTION AND USABLE ON CODES(UOC)	QTY

GROUP 2202 ACCESSORY ITEMS

FIG. 346 HEATER GRILLE, EXHAUST PIPE, AND MOUNTING HARDWARE, M996 AND M996A1 AMBULANCE

| 1 | PAOZZ | 5306002264828 | 80204 | B1821BH031C113N | BOLT,MACHINE 5/16-18 X 1.125....... | 2 |

UOC:B16,H16

| 2 | PFOZZ | 2990012653297 | 19207 | 12341271 | PIPE,EXHAUST........................ | 1 |

UOC:B16,H16

| 3 | PAOZZ | 5340013330162 | 58536 | A-A-52406-4SA | CLAMP,LOOP......................... | 1 |

UOC:B16,H16

| 4 | PFOZZ | 5340012663843 | 19207 | 12341229 | BRACKET,DOUBLE ANGL................ | 1 |

UOC:B16,H16

| 5 | PAOZZ | 5305013615353 | 24617 | 9414241 | SCREW,TAPPING #10-32 X .50......... | 4 |

UOC:B16,H16

| 6 | PAOZZ | 5310000809786 | 96906 | MS45904-60 | WASHER,LOCK 5/16................... | 2 |

UOC:B16,H16

| 7 | PFOZZ | 5340012704415 | 19207 | 12341552 | GRILLE,METAL....................... | 1 |

UOC:B16,H16

| 8 | PAOZZ | 5320012711834 | 34623 | 5575940 | RIVET,SOLID 3/16................... | 12 |

UOC:B16,H16

END OF FIGURE

Section II. TM 9-2320-280-24P-2

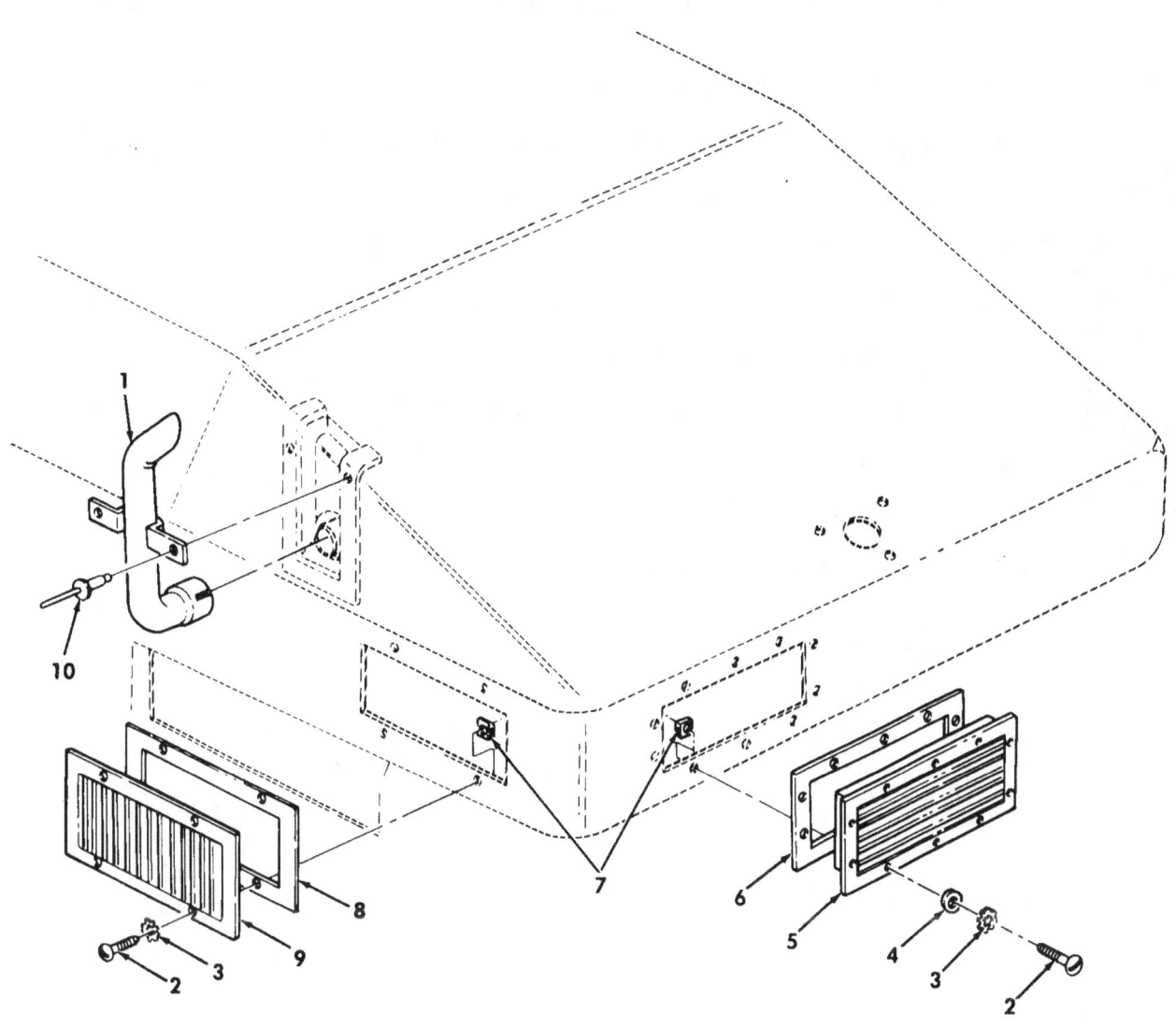

Figure 347. Heater Grille, Exhaust Pipe, and Mounting Hardware, M997, M997A1, and M997A2, Ambulance.

SECTION II   TM9-2320-280-24P

(1) NO	(2) SMR CODE	(3) NSN	(4) CAGEC	(5) PART NUMBER	(7) ITEM DESCRIPTION AND USABLE ON CODES(UOC)	QTY

GROUP 2202 ACCESSORY ITEMS

FIG. 347 HEATER GRILLE, EXHAUST PIPE, AND MOUNTING HARDWARE, M997, M997A1, AND M997A2 AMBULANCE

| 1 | PAOZZ | 2990012653298 | 24234 | 513462 | PIPE,EXHAUST.................... | 1 |

UOC:A15,B15,H15

| 2 | PAOZZ | 5305012554606 | 34623 | 5592900 | SCREW,MACHINE #10-24 X 1/2......... | 14 |

UOC:A15,B15,H15

| 3 | PAOZZ | 5310000809786 | 96906 | MS45904-60 | WASHER,LOCK #10.................... | 14 |

UOC:A15,B15,H15

| 4 | PAOZZ | 5310000145850 | 96906 | MS27183-42 | WASHER,FLAT #10.................... | 10 |

UOC:A15,B15,H15

| 5 | PAOZZ | 2510012707919 | 19207 | 12341533 | GRILLE,RADIATOR,VEH................ | 1 |

UOC:A15,B15,H15

| 6 | PAOZZ | 5330012727472 | 19207 | 12341514 | GASKET............................. | 1 |

UOC:A15,B15,H15

| 7 | PAOZZ | 5310012155356 | 78553 | C8117-1024 | NUT,SHEET SPRING #10-24............ | 14 |

UOC:A15,B15,H15

| 8 | PAOZZ | 5330012727471 | 19207 | 12341487 | GASKET............................. | 1 |

UOC:A15,B15,H15

| 9 | PAOZZ | 2540012712839 | 19207 | 12341088 | VENTILATOR,AIR CIRC................ | 1 |

UOC:A15,B15,H15

| 10 | PAOZZ | 5320012716357 | 9K475 | MGLP-B6-4 | RIVET,BLIND 3/16................... | 2 |

UOC:A15,B15,H15

END OF FIGURE

Section II. TM 9-2320-280-24P-2 C01

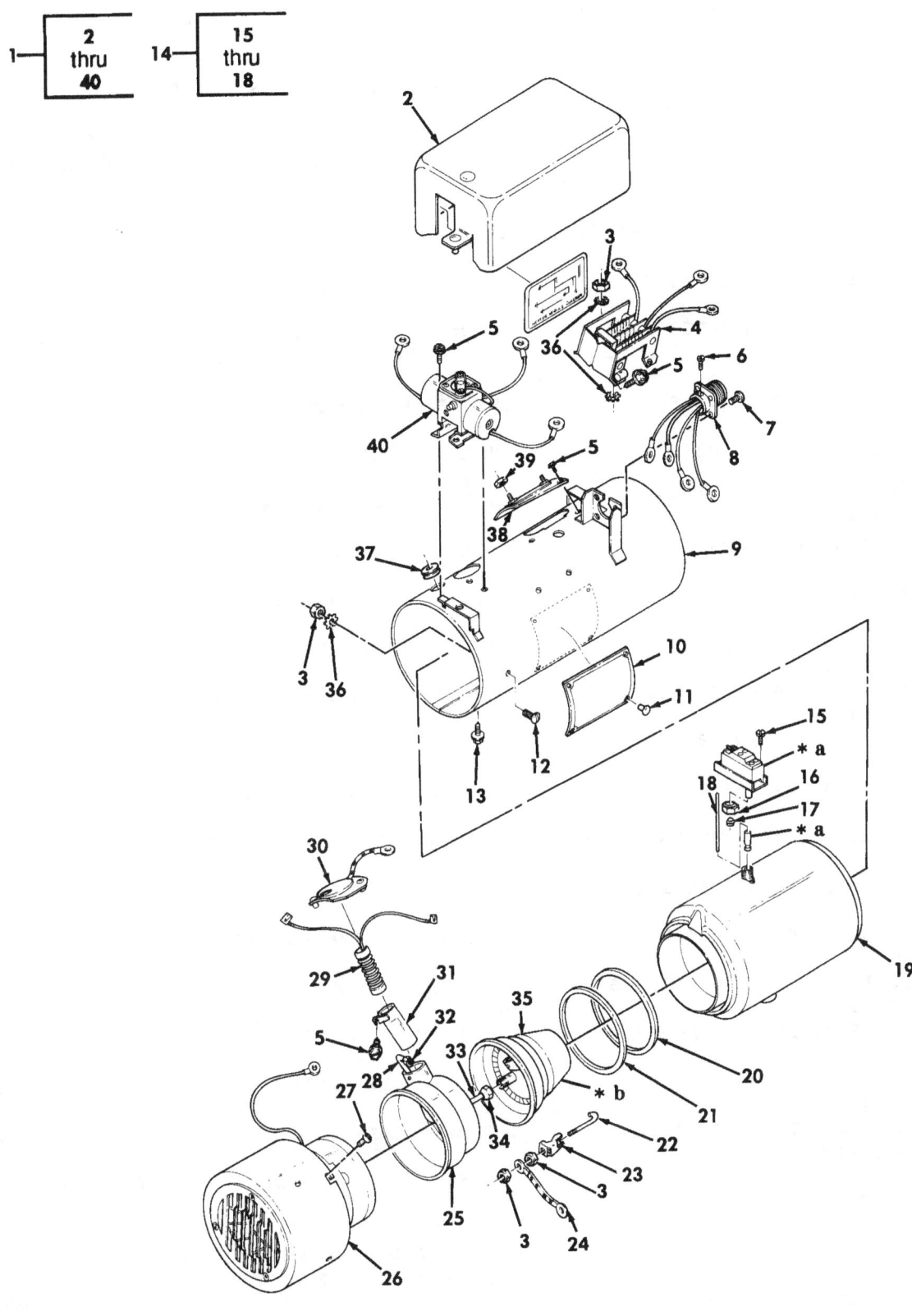

* a PART OF ITEM 14
* b SEE FIGURE 351 FOR BREAKDOWN

Figure 348. Heater Assembly, Multifuel, A-Model, M996, M996A1, M997, M997A1, and M997A2 Ambulance.

SECTION II   TM9-2320-280-24P C01

```
 (7)   (1)    (2)    (3)         (4)        (5)                    (6)
       ITEM   SMR                           PART
 QTY   NO     CODE   NSN         CAGEC      NUMBER     DESCRIPTION AND USABLE ON CODES(UOC)
```

GROUP 2202 ACCESSORY ITEMS

FIG. 348 HEATER ASSEMBLY,MULTIFU-
  A-MODEL,M996,M996A1,M997,M997A1,A
EL,   M997A2 AMBULANCE
ND

```
*  1 PAOFF 2540011943323 81349 MIL-PRF-62550/3  HEATER,VEHICULAR,CO................  1
UOC:A15,B15,B16,H15,H16
*  2 PFFZZ 5935011638981 78385 G704183          .COVER,ELECTRICAL CO...............  1
UOC:A15,B15,B16,H15,H16
*  3 KFFZZ                 78385 487283         .NUT,PLAIN,HEXAGON PART OF KIT P/N
4                                                  5704052 PART OF KIT P/N G704424.....
UOC:A15,B15,B16,H15,H16
*  4 PFFZZ 5905002517145 19207 11663061         .RESISTOR ASSEMBLY.................  1
UOC:A15,B15,B16,H15,H16
*  5 PAFZZ 5305004035130 24617 423531           .SCREW,MACHINE    #8-32 X 1/4.........  3
UOC:A15,B15,B16,H15,H16
*  6 PAFZZ 5305004562582 57733 170677           .SCREW,ASSEMBLED WAS...............  1
UOC:A15,B15,B16,H15,H16
*  7 PAFZZ 5305001353032 78385 488558           .SCREW,ASSEMBLED WAS    #6-32 X 3/8...  4
UOC:A15,B15,B16,H15,H16
*  8 PAFZZ 5935011638987 78385 G704234          .CONNECTOR,RECEPTACL...............  1
UOC:A15,B15,B16,H15,H16
*  9 XAFZZ                 78385 G704213        .HOUSING ASSEMBLY,HE...............  1
UOC:A15,B15,B16,H15,H16
* 10 PFOZZ 9905010174748 78385 704501           ..PLATE,IDENTIFICATIO..............  1
UOC:A15,B15,B16,H15,H16
* 11 PAOZZ 5320008011548 57733 488755           ..RIVET,BLIND......................  4
UOC:A15,B15,B16,H15,H16
* 12 PAFZZ 5305005762335 57733 487357           ..SCREW,ASSEMBLED WAS    #8-32 X 3/8...  1
UOC:A15,B15,B16,H15,H16
* 13 PAFZZ 5305004035130 24617 423531           ..SCREW,MACHINE    #8-32 X 1/4.........  9
UOC:A15,B15,B16,H15,H16
* 14 PAFZZ 5930003455455 19207 11663057         ..SWITCH,THERMOSTATIC..............  1
UOC:A15,B15,B16,H15,H16
* 15 PAFZZ 5305004562582 57733 170677           ..SCREW,ASSEMBLED WAS..............  5
UOC:A15,B15,B16,H15,H16
* 16 PAFZZ 4730000114627 81343 4-060110B        ..NUT,TUBE COUPLING    1/4............  1
UOC:A15,B15,B16,H15,H16
* 17 PFFZZ 4730011937390 10988 222-652          ..SLEEVE,COMPRESSION,..............  1
UOC:A15,B15,B16,H15,H16
* 18 PFFZZ 2540002165722 78385 704225           ..ROD,NONEXPANSIVE.................  1
UOC:A15,B15,B16,H15,H16
* 19 PFFZZ 2540011650465 78385 G704232          .HEAT EXCHANGER....................  1
UOC:A15,B15,B16,H15,H16
* 20 KFFZZ                 78385 702903         .GASKET PART OF KIT P/N 5704052 PART  1
OF KIT P/N G704424 PART OF KIT P/N                                                    G7060
55..............................
              UOC:A15,B15,B16,H15,H16
* 21 KFFZZ                 78385 718768-23      .O-RING PART OF KIT P/N 5704052 PART  1
OF KIT P/N G704424 PART OF KIT P/N                                                    G7060
55..............................
              UOC:A15,B15,B16,H15,H16
```

348-1

SECTION II   TM9-2320-280-24P C01

(1) ITEM NO	(2) SMR CODE	(3) NSN	(4) CAGEC	(5) PART NUMBER	(6) DESCRIPTION AND USABLE ON CODES (UOC)	(7) QTY
22	KFFZZ		78385	703547	.BOLT,HOOK PART OF KIT P/N 5704052 PART OF KIT P/N G704424 PART OF KIT P/N G706055............... UOC:A15,B15,B16,H15,H16	2
23	KFFZZ		78385	703546	.CLAMP,RIM CLENCHING PART OF KIT P/N 5704052 PART OF KIT P/N G704424 PART OF KIT P/N G706055................. UOC:A15,B15,B16,H15,H16	4
24	PFFZZ	6150011654667	78385	G704373	.LEAD,ELECTRICAL.................... UOC:A15,B15,B16,H15,H16	4
25	PFFZZ	2540011658175	78385	G704177	.HOUSING,BAFFLE,HEAT................ UOC:A15,B15,B16,H15,H16	1
26	PAFZZ	2540010081501	78385	G704554	.BLOWER ASSEMBLY.................... UOC:A15,B15,B16,H15,H16	1
27	PAFZZ	5305000188370	78385	489142	.SCREW,MACHINE #8-3/8 X 1.00....... UOC:A15,B15,B16,H15,H16	4
28	PFFZZ	4710011632805	78385	704363	.TUBE ASSEMBLY,METAL................ UOC:A15,B15,B16,H15,H16	4
29	PAOZZ	4520002175782	16236	CS-4520-SV-0705	.IGNITER,SPARK,FUEL................. UOC:A15,B15,B16,H15,H16	1
30	PFFZZ	5340011650745	78385	G704293	.COVER,ACCESS....................... UOC:A15,B15,B16,H15,H16	1
31	PFFZZ	2540011658176	78385	G704288-1	.IGNITER TUBE AND BR................ UOC:A15,B15,B16,H15,H16	1
32	PAFZZ	4730011531871	79470	1611X3	.LOCKNUT,TUBE FITTIN................ UOC:A15,B15,B16,H15,H16	1
33	PFFZZ	4730007017737	78385	476624	.SLEEVE,COMPRESSION,................ UOC:A15,B15,B16,H15,H16	1
34	PAFZZ	4730000433750	21450	120487	.NUT,TUBE COUPLING.................. UOC:A15,B15,B16,H15,H16	1
35	KFFFZ		78385	G706037	.BURNER ASSEMBLY PART OF KIT P/N G704424 PART OF KIT P/N G706055..... UOC:A15,B15,B16,H15,H16	1
36	PAFZZ	5310000610004	57733	475005	.WASHER,LOCK........................ UOC:A15,B15,B16,H15,H16	3
37	PAFZZ	5325005432902	78385	488993	.GROMMET,NONMETALLIC................ UOC:A15,B15,B16,H15,H16	1
38	PAFZZ	5930006795925	19207	10948233	.SWITCH,THERMOSTATIC................ UOC:A15,B15,B16,H15,H16	1
39	PAFZZ	5310000637360	83385	841-6-32	.NUT,PLAIN,ASSEMBLED #6-32......... UOC:A15,B15,B16,H15,H16	2
40	PAFHH	4810002481635	19207	11663058	.VALVE ASSEMBLY,FUEL................ UOC:A15,B15,B16,H15,H16	1

END OF FIGURE

Section II.

TM 9-2320-280-24P-2

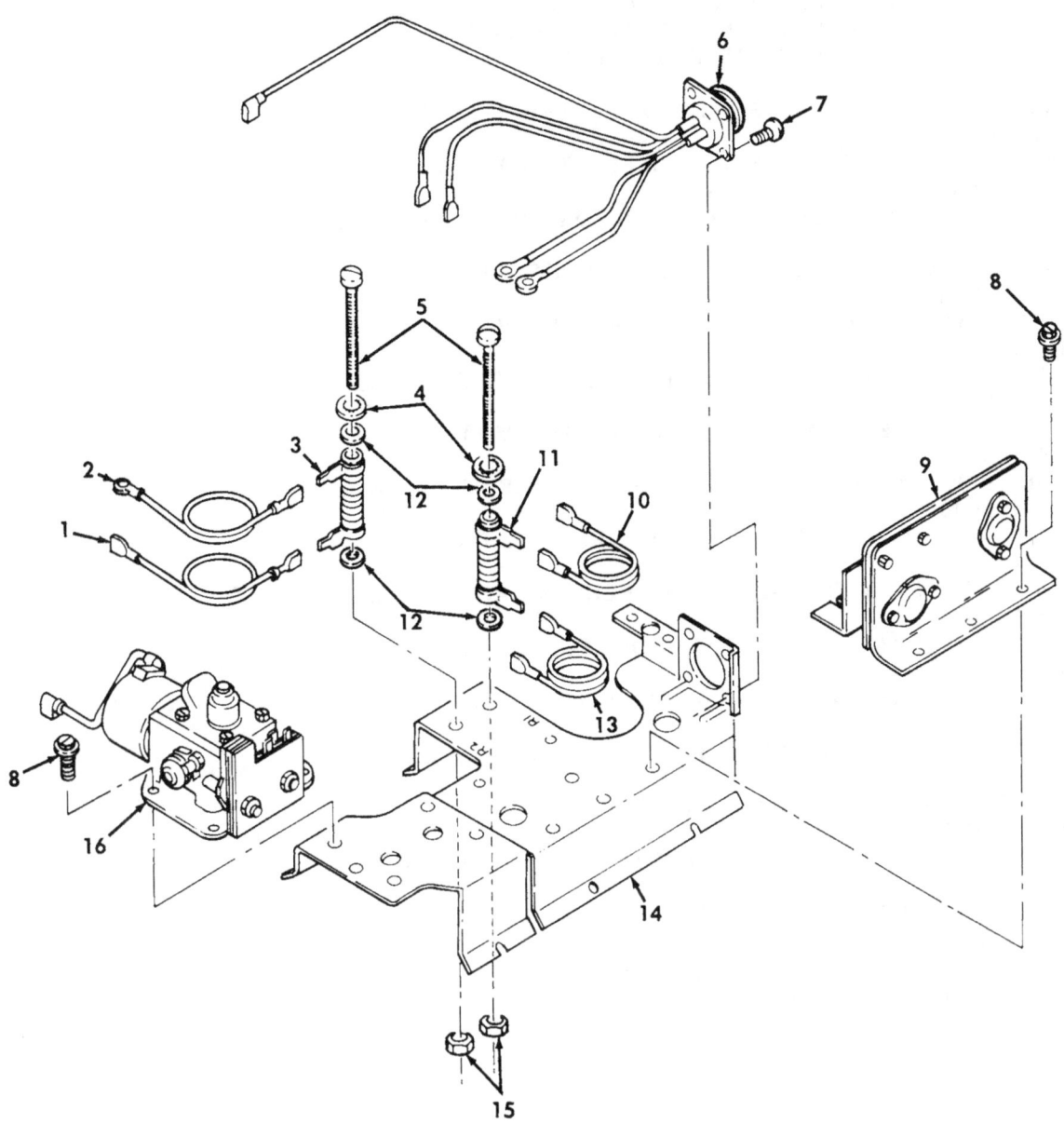

Figure 349. Heater Component Bracket Assembly, B-Model, M996, M996A1, M997, M997A1, and M997A2 Ambulance.

SECTION II TM9-2320-280-24P

(1) NO	(2) SMR CODE	(3) NSN	(4) CAGEC	(5) PART NUMBER	(7) ITEM DESCRIPTION AND USABLE ON CODES(UOC)	QTY

GROUP 2202 ACCESSORY ITEMS

FIG. 349 HEATER COMPONENT BRACKET ASSEMBLY,B-MODEL,M996,M996A1,M997,M997A1,AND M997A2 AMBULANCE

No	SMR	NSN	CAGEC	Part Number	Description	Qty
1	PFFZZ	6150012879925	78385	G706052-13	LEAD,ELECTRICAL	1
					UOC:A15,B15,B16,H15,H16	
2	PFFZZ	6150012879924	78385	G706052-15	LEAD,ELECTRICAL	1
					UOC:A15,B15,B16,H15,H16	
3	PAFZZ	5905012874255	78385	706053	RESISTOR,FIXED,WIRE	1
					UOC:A15,B15,B16,H15,H16	
4	PAFZZ	5310005765752	96906	MS35333-39	WASHER,LOCK #10	2
					UOC:A15,B15,B16,H15,H16	
5	PAFZZ	5305012881130	78385	703611	SCREW,MACHINE #10-32 X 2-1/2	2
					UOC:A15,B15,B16,H15,H16	
6	PAFZZ	6150012934074	78385	G706026	LEAD ASSEMBLY,ELECT	1
					UOC:A15,B15,B16,H15,H16	
7	PAFZZ	5305012881129	78385	705211	SCREW,MACHINE #5-40 X 5/16	4
					UOC:A15,B15,B16,H15,H16	
8	PAFZZ	5305001159406	96906	MS51849-53	SCREW,MACHINE #8-32 X 3/8	7
					UOC:A15,B15,B16,H15,H16	
9	PAFZZ	6110012853902	78385	G706057	VOLTAGE REGULATOR G	1
					UOC:A15,B15,B16,H15,H16	
10	PFFZZ	6150012879923	78385	G706052-19	LEAD,ELECTRICAL	1
					UOC:A15,B15,B16,H15,H16	
11	PAFZZ	5905012874256	78385	706045	RESISTOR,FIXED,WIRE	1
					UOC:A15,B15,B16,H15,H16	
12	PAFZZ	5310012876557	78385	706050GV	WASHER,FLAT 5/16	4
					UOC:A15,B15,B16,H15,H16	
13	PFFZZ	6150012879926	78385	G706052-17	LEAD,ELECTRICAL	1
					UOC:A15,B15,B16,H15,H16	
14	XAFZZ		78385	G706024	COMPONENT BRACKET	1
					UOC:A15,B15,B16,H15,H16	
15	PAFZZ	5310007890398	78385	487370	NUT,PLAIN,ASSEMBLED #10-32	2
					UOC:A15,B15,B16,H15,H16	
16	PAFZZ	2540012536112	78385	G706016	FUEL VALVE	1
					UOC:A15,B15,B16,H15,H16	

END OF FIGURE

Section II. TM 9-2320-280-24P-2

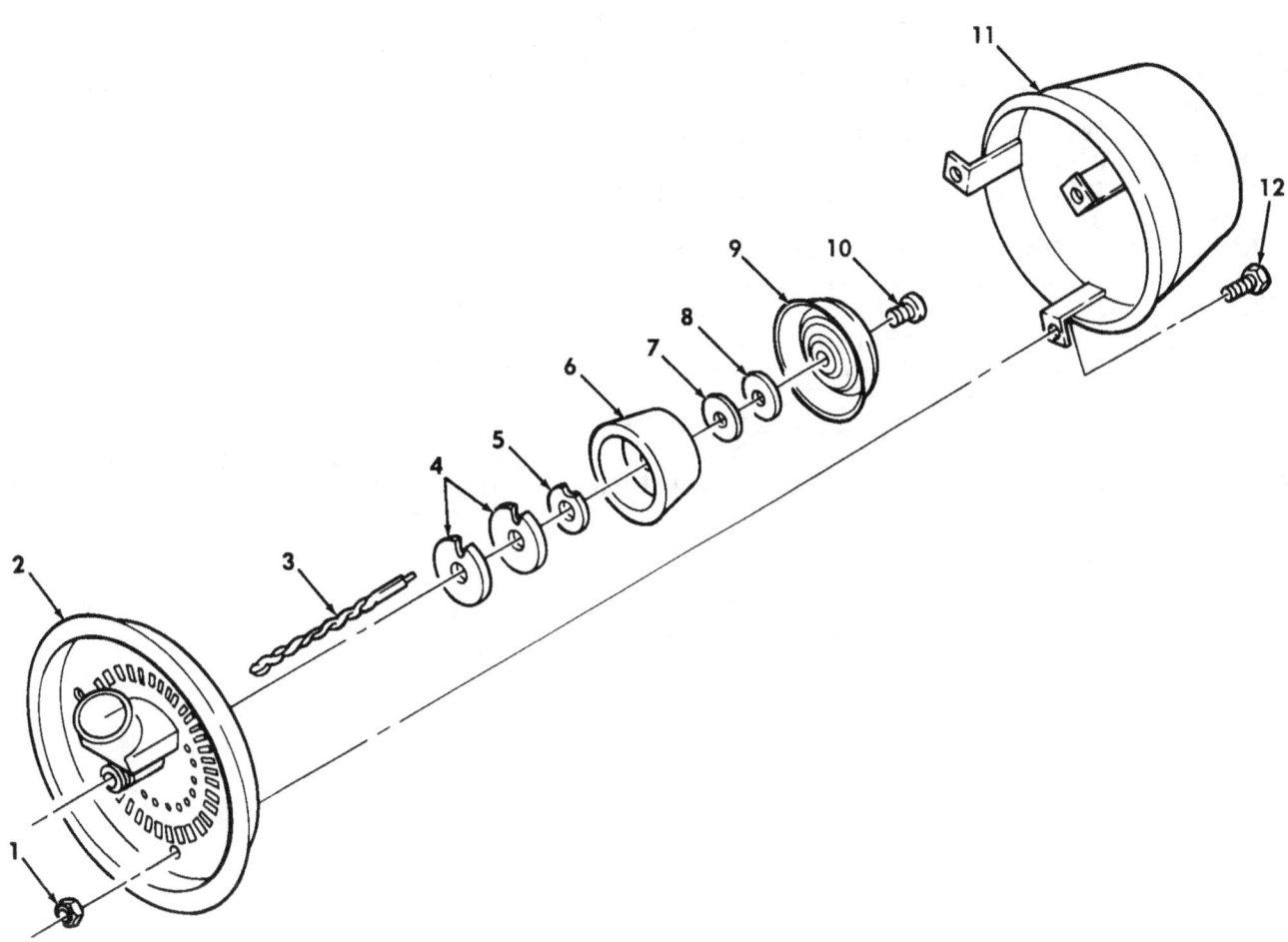

Figure 350. Burner Assembly, B-Model, M996, M996A1, M997, M997A1, and M997A2 Ambulance.

SECTION II   TM9-2320-280-24P

(1) NO	(2) SMR CODE	(3) NSN	(4) CAGEC	(5) PART NUMBER	(6) DESCRIPTION AND USABLE ON CODES(UOC)	(7) QTY

GROUP 2202 ACCESSORY ITEMS

FIG. 350 BURNER ASSEMBLY,B-MODEL, M996,M996A1,M997,M997A1,AND M997A2   AMBULANCE

1	KFFZZ		78385	705587	NUT,HEXAGON #8-32 PART OF KIT P/N 5704052.................... UOC:A15,B15,B16,H15,H16	3
2	PFFZZ	2540011650814	78385	G704284	HEADER PLATE ASSEMB PART OF KIT P/N 5704052..................... UOC:A15,B15,B16,H15,H16	1
3	KFFZZ		78385	705944	WICK PART OF KIT P/N 5704052........ UOC:A15,B15,B16,H15,H16	1
4	KFFZZ		78385	706062	WASHER PART OF KIT P/N 5704052...... UOC:A15,B15,B16,H15,H16	2
5	KFFZZ		78385	706049	WASHER PART OF KIT P/N 5704052...... UOC:A15,B15,B16,H15,H16	1
6	KFFZZ		78385	706063	VAPORIZER PART OF KIT P/N 5704052... UOC:A15,B15,B16,H15,H16	1
7	KFFZZ		78385	704678	WASHER,FLAT PART OF KIT P/N 5704052. UOC:A15,B15,B16,H15,H16	1
8	KFFZZ		78385	704190	WASHER PART OF KIT P/N 5704052...... UOC:A15,B15,B16,H15,H16	1
9	PFFZZ	2540011689482	78385	704181	SHIELD,FUEL VAPOR PART OF KIT P/N 5704052..................... UOC:A15,B15,B16,H15,H16	1
10	PFFZZ	5305010978178	78385	704206	SCREW,MACHINE #8-32 X 1/4 PART OF KIT P/N 5704052..................... UOC:A15,B15,B16,H15,H16	1
11	XAFZZ		78385	G706039	CUP................................ UOC:A15,B15,B16,H15,H16	1
12	KFFZZ		78385	705032	SCREW #8-32 X 3/8 PART OF KIT P/N 5704052.......................... UOC:A15,B15,B16,H15,H16	3

END OF FIGURE

Section II.

TM 9-2320-280-24P-2

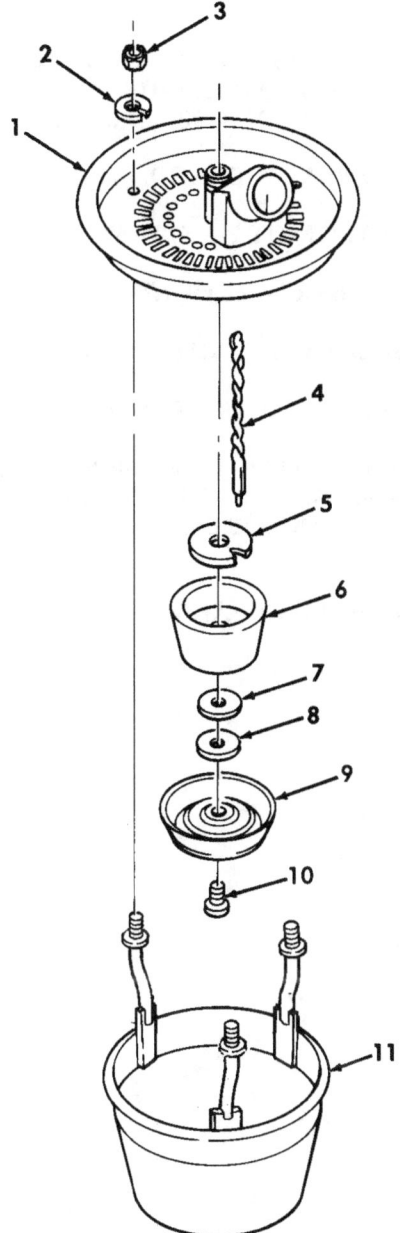

Figure 351. Burner Assembly, A-Model, M997, M997A1, and M997A2 Ambulance.

SECTION II TM9-2320-280-24P

(1) NO	(2) SMR CODE	(3) NSN	(4) CAGEC	(5) PART NUMBER	(7) DESCRIPTION AND USABLE ON CODES(UOC)	ITEM QTY

GROUP 2202 ACCESSORY ITEMS

FIG. 351 BURNER ASSEMBLY, A-MODEL, M997, M997A1, AND M997A2 AMBULANCE

1 PFFZZ 2540011650814 78385 G704284 HEADER PLATE ASSEMB PART OF KIT P/N 5704052............................. 1
  UOC:A15,B15,H15

2 PFFZZ 5310010638522 78385 705237 WASHER,LOCK #8.................... 3
UOC:A15,B15,H15

3 PFFZZ 5310010573098 78385 705068 NUT,PLAIN,HEXAGON #8-32............ 3
UOC:A15,B15,H15

4 KFFZZ 78385 705944 WICK PART OF KIT P/N 5704052........ 1
  UOC:A15,B15,H15

5 KFFZZ 78385 704191 WASHER,FIBER PART OF KIT P/N 5704052 1
  UOC:A15,B15,H15

6 KFFZZ 78385 704192 VAPORIZER PART OF KIT P/N 5704052... 1
  UOC:A15,B15,H15

7 KFFZZ 78385 704678 WASHER,FLAT PART OF KIT P/N 5704052. 1
  UOC:A15,B15,H15

8 KFFZZ 78385 704190 WASHER PART OF KIT P/N 5704052...... 1
UOC:A15,B15,H15

9 PFFZZ 2540011689482 78385 704181 SHIELD,FUEL VAPOR PART OF KIT P/N 5704052............................. 1
  UOC:A15,B15,H15

10 PFFZZ 5305010978178 78385 704206 SCREW,MACHINE #8-32 X 1/4 PART OF KIT P/N 5704052.................... 1
  UOC:A15,B15,H15

11 PFFZZ 2540011689481 78385 6706133 BURNER CUP AND STUD................ 1
UOC:A15,B15,H15

END OF FIGURE

Section II. TM 9-2320-280-24P-2

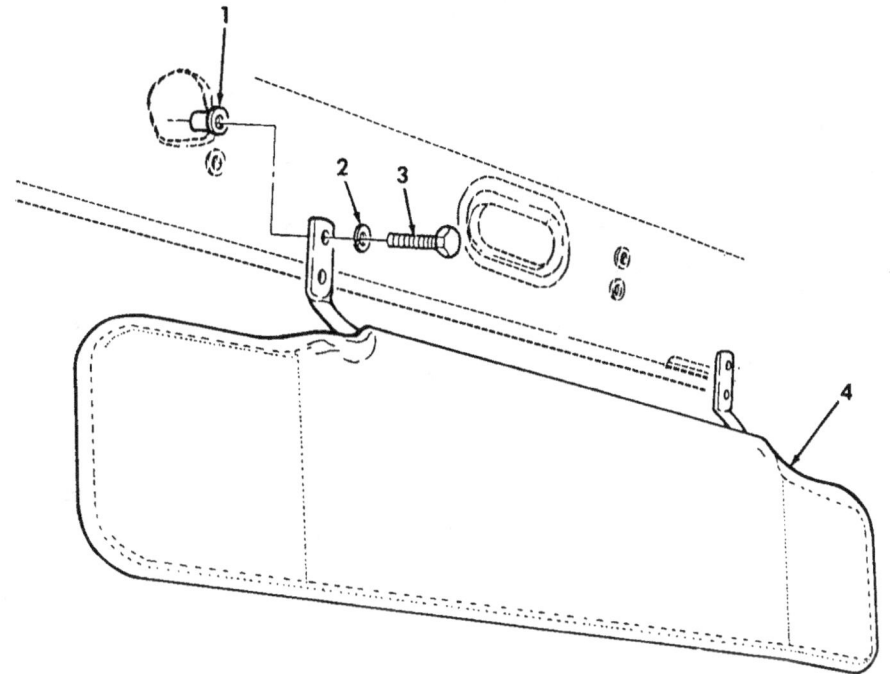

Figure 352. Sun Visor Assembly and Mounting Hardware.

SECTION II   TM9-2320-280-24P

(1) NO	(2) SMR CODE	(3) NSN	(4) CAGEC	(5) PART NUMBER	(6) DESCRIPTION AND USABLE ON CODES(UOC)	(7) QTY

GROUP 2202 ACCESSORY ITEMS

FIG. 352 SUN VISOR ASSEMBLY AND MOUNTING HARDWARE

1	PAOZZ	5310014113422	78276	ALS4-420-165	NUT,PLAIN,BLIND RIV............... UOC:B-VY,B15,B20,B24,B25,C17,NNN	8
2	PAOZZ	5310006379541	96906	MS35338-46	WASHER,LOCK 3/8................. UOC:B-VY,B15,B20,B24,B25,C17,NNN	8
3	PAOZZ	5305002253843	80204	B1821BH025C100N	SCREW,CAP,HEXAGON H 1/4-20 X 1.00.. UOC:A15,BVY,B20,B24,B25,C17,NNN	8
4	PAOZZ	2540014108793	34623	12446819	VISOR,SUN,VEHICLE.................. UOC:B-VY,B15,B20,B24,B25,C17,NNN	2

END OF FIGURE

Section II. TM 9-2320-280-24P-2

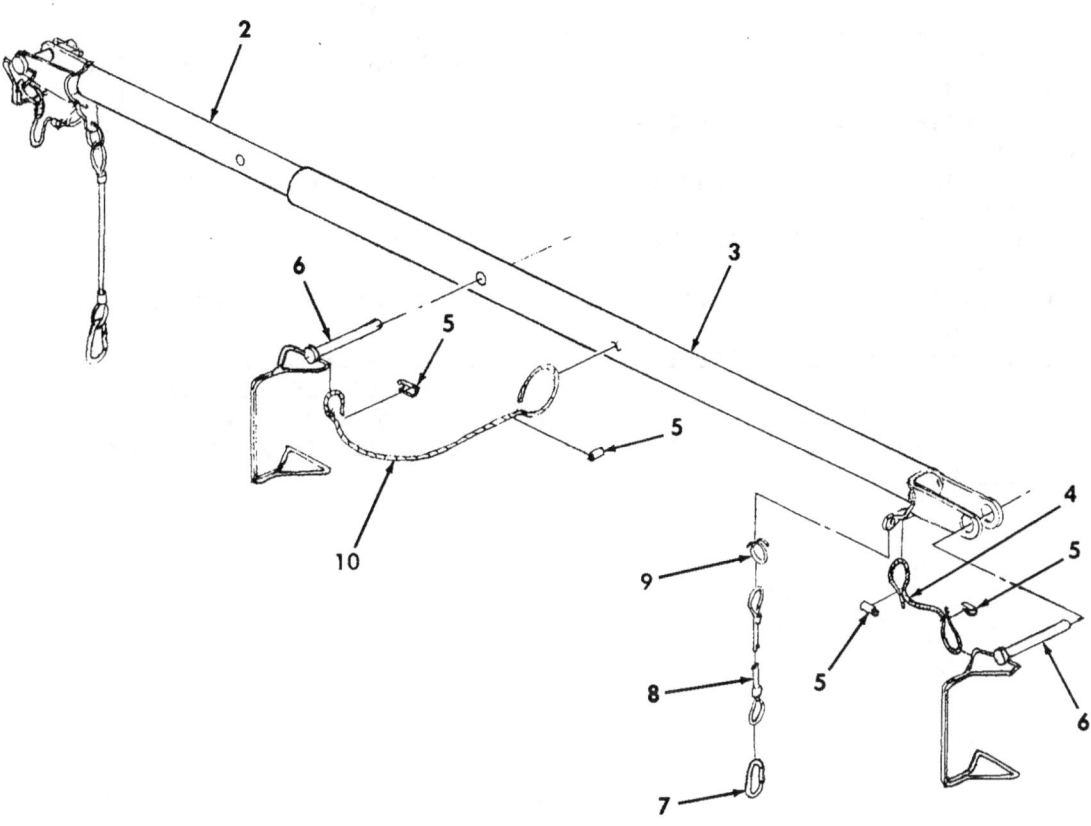

Figure 353. Ambulance Spreader Assembly.

SECTION II   TM9-2320-280-24P C01

(1) ITEM NO	(2) SMR CODE	(3) NSN	(4) CAGEC	(5) PART NUMBER	(6) DESCRIPTION AND USABLE ON CODES (UOC)	(7) QTY

GROUP 2202 ACCESSORY ITEMS

FIG. 353 AMBULANCE SPREADER ASSEMBLY

(1)	(2)	(3)	(4)	(5)	(6)	(7)
1	PAOZZ	4910013138839	19207	12342102	SPREADER,SLING.................... UOC:A15,B15,B16,H15,H16	1
2	XAOZZ		19207	12342097	.EXTENSION BAR ASSY................ UOC:A15,B15,B16,H15,H16	1
3	XAOZZ		19207	12342099	.EXTENSION BAR ASSY................ UOC:A15,B15,B16,H15,H16	1
4	MOOZZ		19207	12342102-3	.CABLE MAKE FROM ROPE,WIRE P/N 8930T31, 10 INCHES LONG............ UOC:A15,B15,B16,H15,H16	2
* 5	PAOZZ	4030011248201	96906	MS51844-82	.SWAGING SLEEVE,WIRE................ UOC:A15,B15,B16,H15,H16	6
6	PAOZZ	5315014065019	19207	12342121	.PIN,STRAIGHT,HEADED................ UOC:A15,B15,B16,H15,H16	3
7	PAOZZ	5340011251682	39428	3933T22	.SNAP HOOK......................... UOC:A15,B15,B16,H15,H16	2
8	PAOZZ	4010014066963	19207	12342141	.WIRE ROPE ASSEMBLY,................ UOC:A15,B15,B16,H15,H16	2
9	PAOZZ	4010014109099	19207	12342102-4	.LINK,CHAIN,LAP.................... UOC:A15,B15,B16,H15,H16	10
	MOOZZ		19207	12342102-2	.CABLE MAKE FROM ROPE,WIRE P/N 8930T31, 20 INCHES LONG............ UOC:A15,B15,B16,H15,H16	1

353-1

Section II.

TM 9-2320-280-24P-2

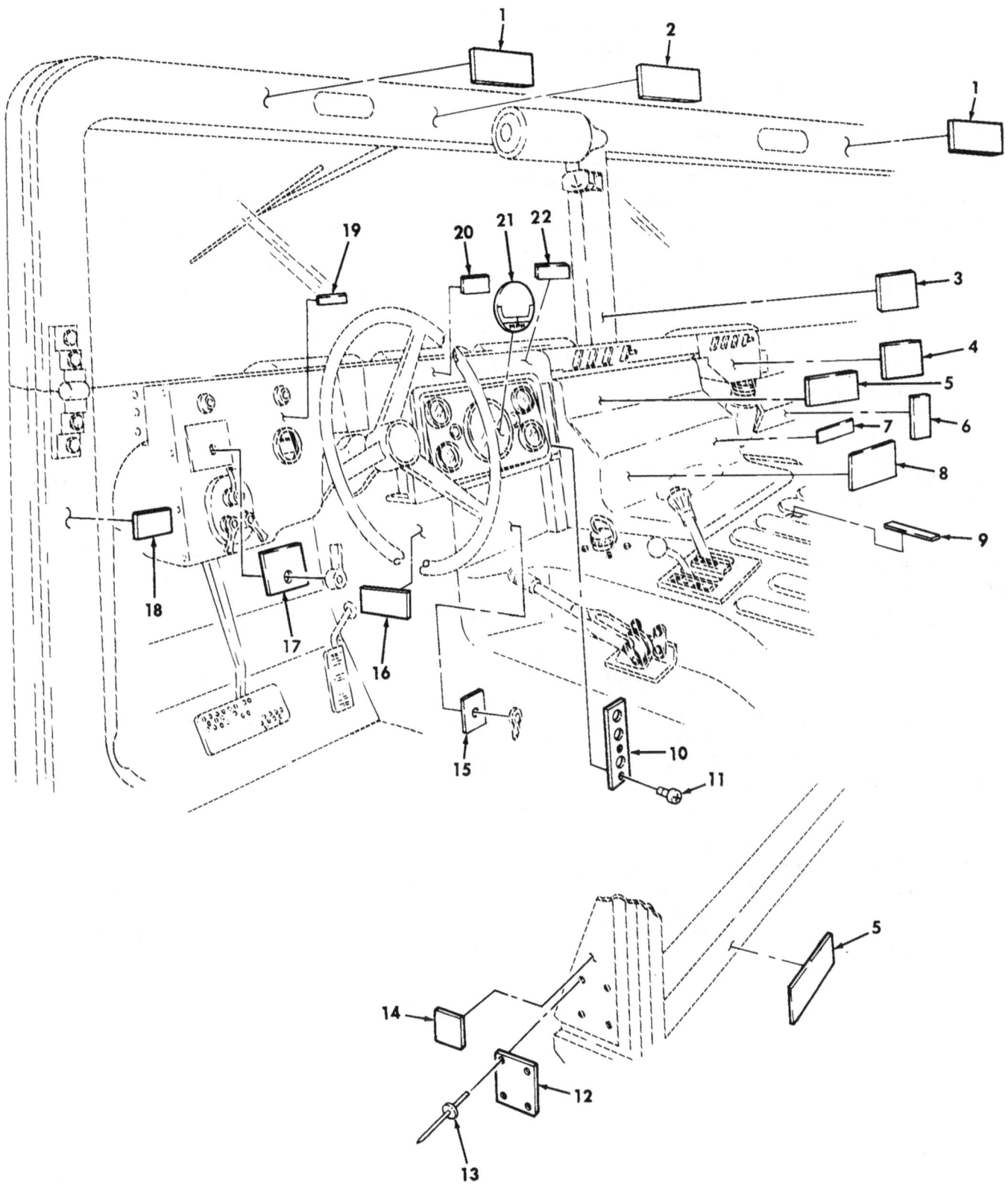

Figure 354. Data Plates, Dash Panel.

SECTION II TM9-2320-280-24P

(1) NO	(2) SMR CODE	(3) NSN	(4) CAGEC	(5) PART NUMBER	(7) ITEM DESCRIPTION AND USABLE ON CODES(UOC)	QTY

GROUP 2210 DATA PLATES AND INSTRUCTION HOLDERS

FIG. 354 DATA PLATES,DASH PANEL

1 PAOZZ 7690012040077 19207 12339053 DECAL............................... 2 UOC:A11,A15,A24,A25,A26,A27,B15,B16,B17,B18,B24,B25,C17,H11,H15,H16,H17,H18,H24,H25,H26,H27,MMM

2 PAOZZ 9905001982728 19207 11643398 PLATE,INSTRUCTION CAUTION,HIGH INTENSITY NOISE.................... 1

3 PAOZZ 7690013158539 19207 12342165 MARKER,IDENTIFICATI VEHICLE BREAK- IN SERVICE......................... 1

4 PFOZZ 7690014766089 19207 12447011 DECAL HEATER AIR CONTROL........... 1 UOC:BVY,B15,B20,B24,B25,C17,NNN

4 PFOZZ 9905012039994 19207 12339100 PLATE,INSTRUCTION HEATER AIR CONTROL............................ 1 UOC:AVY,A11,A13,A14,A15,A20,A24,A25,A26,A27,B16,B17,B18,HVY,H11,H13,H14,H15,H16,H17,H18,H20,H21,H24,H25,H26,H27,H28,MMM

5 PAOZZ 9905011934065 19207 12339106 PLATE,INSTRUCTION SEAT BELT RETRACTOR,USBL EFF 1 THRU 99,999.... 4 UOC:H11,H13,H14,H15,H16,H17,H18,H20,H21,H24,H25,H26,H27,H28,MMM

6 PAOZZ 7690012912974 34623 5597689 DECAL TIEDOWN DATA................. 1 UOC:AVY,A11,A13,A14,A15,A20,A24,A25,A26,A27,B16,B17,B18,HVY,H11,H13,H14,H15,H16,H17,H18,H20,H21,H24,H25,H26,H27,H28,MMM

6 PFOZZ 9905014338554 19207 12460205 PLATE,INSTRUCTION TIEDOWN DATA..... 1 UOC:BVY,B15,B20,B24,B25,C17,NNN

7 PAOZZ 9905012487657 19207 12339111 PLATE,INSTRUCTION SERVICE DATA..... 1 UOC:AVY,A11,A13,A14,A15,A20,A24,A25,A26,A27,B16,B17,B18,HVY,H11,H13,H14,H15,H16,H17,H18,H20,H21,H24,H25,H26,H27,H28,MMM

7 PAOZZ 7690014766101 19207 12460157 DECAL SERVICE DATA................. 1 UOC:BVY,B15,B20,B24,B25,C17,NNN

8 PFOZZ 9905012245860 19207 12339080 PLATE,INSTRUCTION OPERATING INSTRUCTION......................... 1 UOC:HVY,H11,H13,H14,H15,H16,H17,H18,H20,H21,H24,H25,H26,H27,H28,MMM

8 PFOZZ 9905013933794 19207 12343058 PLATE,INSTRUCTION OPERATING INSTRUCTIONS........................ 1 UOC:AVY,A11,A13,A14,A15,A20,A24,A25,A26,A27,B16,B17,B18,

8 PFOZZ 7690014766507 19207 12460156 DECAL OPERATING INSTRUCTION........ 1 UOC:BVY,B15,B20,B24,B25,C17,NNN

9 PAOZZ 9905012047776 19207 12340910 PLATE,IDENTIFICATIO FIRE EXTINGUISHER....................... 1

10 PFOZZ 7690014766510 19207 12447003 DECAL............................. 1 UOC:BVY,B15,B20,B24,B25,C17,NNN

354-1

SECTION II   TM9-2320-280-24P

(1) NO	(2) SMR CODE	(3) NSN	(4) CAGEC	(5) PART NUMBER	(6) DESCRIPTION AND USABLE ON CODES(UOC)	(7) QTY
10	PFOZZ	9905012039992	19207	12338470	PLATE,INSTRUCTION................... UOC:AV A11,A13,A14,A15,A20,A24,A25, A26,A27,B16,B17,B18,HVY,H11,H13,H14, H15,H16,H17,H18,H20,H21,H24,H25,H26, H27,H28,MMM	1
11	PAOZZ	5305009846191	96906	MS35206-243	SCREW,MACHINE #8-32 X 3/8..........	2
12	PFOZZ	9905013647342	19207	12342652	PLATE,IDENTIFICATIO NAME PLATE M1097.............................. UOC:HVY	1
12	PAOZZ	9905011853129	19207	12340868	PLATE,IDENTIFICATIO NAME PLATE M998 UOC:H13	1
12	PAOZZ	9905011853131	19207	12339085	PLATE,IDENTIFICATIO NAME PLATE M996 UOC:H16	1
12	PFOZZ	9905011853132	19207	12339077	PLATE,IDENTIFICATIO NAME PLATE M997 UOC:H15	1
12	PFOZZ	9905011853134	19207	12339082	PLATE,IDENTIFICATIO NAME PLATE M966 UOC:H11	1
12	PAOZZ	9905011863253	19207	12339110	PLATE,IDENTIFICATIO................. UOC:H17	1
12	PFOZZ	9905011863255	19207	12339090	PLATE,IDENTIFICATIO NAME PLATE M1026.................. UOC:H18	1
12	PFOZZ	9905011857972	19207	12339095	PLATE,IDENTIFICATIO NAME PLATE M1035.............................. UOC:H20	1
12	PFOZZ	9905011863259	19207	12339104	PLATE,IDENTIFICATIO NAME PLATE M1037.............................. UOC:H21	1
12	PFOZZ	9905011908425	19207	12340867	PLATE,IDENTIFICATIO NAME PLATE M1038.............................. UOC:H14	1
12	PFOZZ	9905011908427	19207	12341812	PLATE,IDENTIFICATIO NAME PLATE M1042.............................. UOC:H28	1
12	PFOZZ	9905011853138	19207	12339076	PLATE,IDENTIFICATIO NAME PLATE, M1045.............................. UOC:H24	1
12	PFOZZ	9905011853139	19207	12339099	PLATE,IDENTIFICATIO NAME PLATE M1043.............................. UOC:H25	1
12	PFOZZ	9905011857973	19207	12339091	PLATE,IDENTIFICATIO NAME PLATE M1044.............................. UOC:H26	1
12	PFOZZ	9905011857975	19207	12339096	PLATE,IDENTIFICATIO NAME PLATE M1046.............................. UOC:H27	1
12	PFOZZ	9905013852639	19207	12343020	PLATE,IDENTIFICATIO NAME PLATE M1097A1.................. UOC:AVY	1
12	PFOZZ	9905013872761	19207	12342986	PLATE,IDENTIFICATIO NAME PLATE M966A1.............................. UOC:A11	1
12	PFOZZ	9905013845311	19207	12342981	PLATE,IDENTIFICATIO NAME PLATE	1

354-2

SECTION II   TM9-2320-280-24P C01

(1) ITEM NO	(2) SMR CODE	(3) NSN	(4) CAGEC	(5) PART NUMBER	(6) DESCRIPTION AND USABLE ON CODES (UOC)	(7) QTY
					M998A1.............................. UOC:A13	
12	PFOZZ	9905013872762	19207	12343044	PLATE,IDENTIFICATIO NAME PLATE M1038A1.............................. UOC:A14	1
12	PFOZZ	9905013871145	19207	12343010	PLATE,IDENTIFICATIO NAME PLATE M997A1............................... UOC:A15	1
12	PFOZZ	9905013935622	19207	12343002	PLATE,IDENTIFICATIO NAME PLATE M1035A1............................. UOC:A20	1
12	PFOZZ	9905013872752	19207	12342995	PLATE,IDENTIFICATIO NAME PLATE M1045A1.............................	1
12	PFOZZ	9905013939357	19207	12342997	PLATE,IDENTIFICATIO NAME PLATE M1043A1............................. UOC:A25	1
12	PFOZZ	9905013935623	19207	12343050	PLATE,IDENTIFICATIO NAME PLATE M1044A1............................. UOC:A26	1
12	PFOZZ	9905013871146	19207	12343046	PLATE,IDENTIFICATIO NAME PLATE M1046A1............................. UOC:A27	1
12	PFOZZ	9905013872746	19207	12343023	PLATE,IDENTIFICATIO NAME PLATE M996A1.............................. UOC:B16	1
12	PFOZZ	9905013976974	19207	12342992	PLATE,IDENTIFICATIO NAME PLATE M1025A1............................. UOC:B17	1
12	PFOZZ	9905013852633	19207	12343041	PLATE,IDENTIFICATIO NAME PLATE M1026A1............................. UOC:B18	1
12	PFOZZ	7690014588254	19207	12446780	DECAL NAME PLATE M1097A2........... UOC:BVY	1
12	PFOZZ	7690014505481	19207	12446798	DECAL NAME PLATE M997A2............ UOC:B15	1
12	PFOZZ	9905014465770	19207	12446792	PLATE,IDENTIFICATIO NAME PLATE M1035A2.............................	1
12	PFOZZ	9905014489786	19207	12446789	PLATE,IDENTIFICATIO NAME PLATE M1045A2............................. UOC:B20	1
12	PFOZZ	9905014477799	19207	12446786	PLATE,IDENTIFICATIO NAME PLATE M1043A2............................. UOC:B24	1
12	PFOZZ	9905014465769	19207	12446783	PLATE,IDENTIFICATIO NAME PLATE M1025A2............................. UOC:B25	1
					UOC:C17	
12	PFOZZ	7690014762855	34623	EX5061	DECAL NAME PLATE M1123............. UOC:NNN	1
12	PFOZZ	7690014814908	19207	12469101	DECAL................................ UOC:MMM	1
13	PAOZZ	5320000835009	81349	M24243/1-A403	RIVET,BLIND 1/8 DIA.,.126-.187 GRIP	4
14	PFOZZ	7690013828471	19207	12343054	DECAL DRIVER SEAT ADJUSTMENT.......	1

354-3

SECTION II   TM9-2320-280-24P C01
  (1)   (2)   (3)   (4)   (5)   (6)
 ITEM   SMR   PART
  NO   CODE   NSN   CAGEC   NUMBER   DESCRIPTION AND USABLE ON CODES(UOC)   QTY

```
                                                  UOC:AVY,A11,A13,A14,A15,A20,A24,A25,
                                                  A26,A27,BVY,B15,B16,B17,B18,B20,B24,
                                                  B25,C17,NNN
   14   PAOZZ  9905012039995  19207  12339101     PLATE,INSTRUCTION  DRIVER SEAT              1
                                                    ADJUSTMENT.........................
                                                  UOC:HVY,H11,H13,H14,H15,H16,H17
                                                     H18,H20,H21,H24,H25,H26,H27,H28,MMM
   15   PFOZZ  9905012058635  19207  12339109     PLATE,IDENTIFICATIO  DEEP WATER             1
                                                    WORDING.............................
                                                  UOC:AVY,A11,A13,A14,A15,A20,A24,A25,
                                                  A26,A27,BVY,B15,B16,B17,B18,B20,B24,
                                                  B25,C17,HVY,H11,H13,H14,H15,H16,H17,
                                                  H18,H20,H21,H24,H25,H26,H27,H28,MMM,
                                                  NNN
   16   PAOZZ  7690014762842  34623  EX5079       DECAL  BARCODE M1123................       1
                                                  UOC:NNN
   17   PAOZZ  9905012039996  19207  12338475     PLATE,INSTRUCTION  IGNITION.........        1
                                                  UOC:AVY,A11,A13,A20,A24,BVY,B15,B20,
                                                  B24,B25,C17,HVY,H11,H13,H14,H15,H16,
                                                  H17,H18,H20,H21,H24,H25,H26,H27,H28,
                                                  MMM,NNN
   18   PAOZZ  7690012040076  19207  12339063     DECAL  RIFLE,M16....................
                                                  UOC:A11,A24,A25,A26,A27,H11,H24,H25,
                                                     H26,H27,MMM
 * 19   PFOZZ  9905012491612  19207  12340027     PLATE,INSTRUCTION  NEUTRAL START....        1
                                                  UOC:AVY,A11,A13,A14,A15,A20,A24,A25,
                                                  A26,A27,B16,B17,B18,HVY,H11,H13,H14,
                                                  HVY,H11,H13,H14,H15,H16,H17,H18,H20,
                                                  H21,H24,H25,H26,H27,H28,MMM
   20   PAOZZ  9905011857977  19207  12339105     PLATE,IN-
                                                    STRUCTION  STEERING WHEEL  LOCKING
                                                    DEVICE......................          1
   21   PAOZZ  7690012564908  19207  12340652     DECAL  SPEEDOMETER..................       1
   22   PAOZZ  7690012564909  19207  12340917     DECAL  HAND THROTTLE CONTROL........       1

                                        END OF FIGURE
```

Section II.

TM 9-2320-280-24P-2

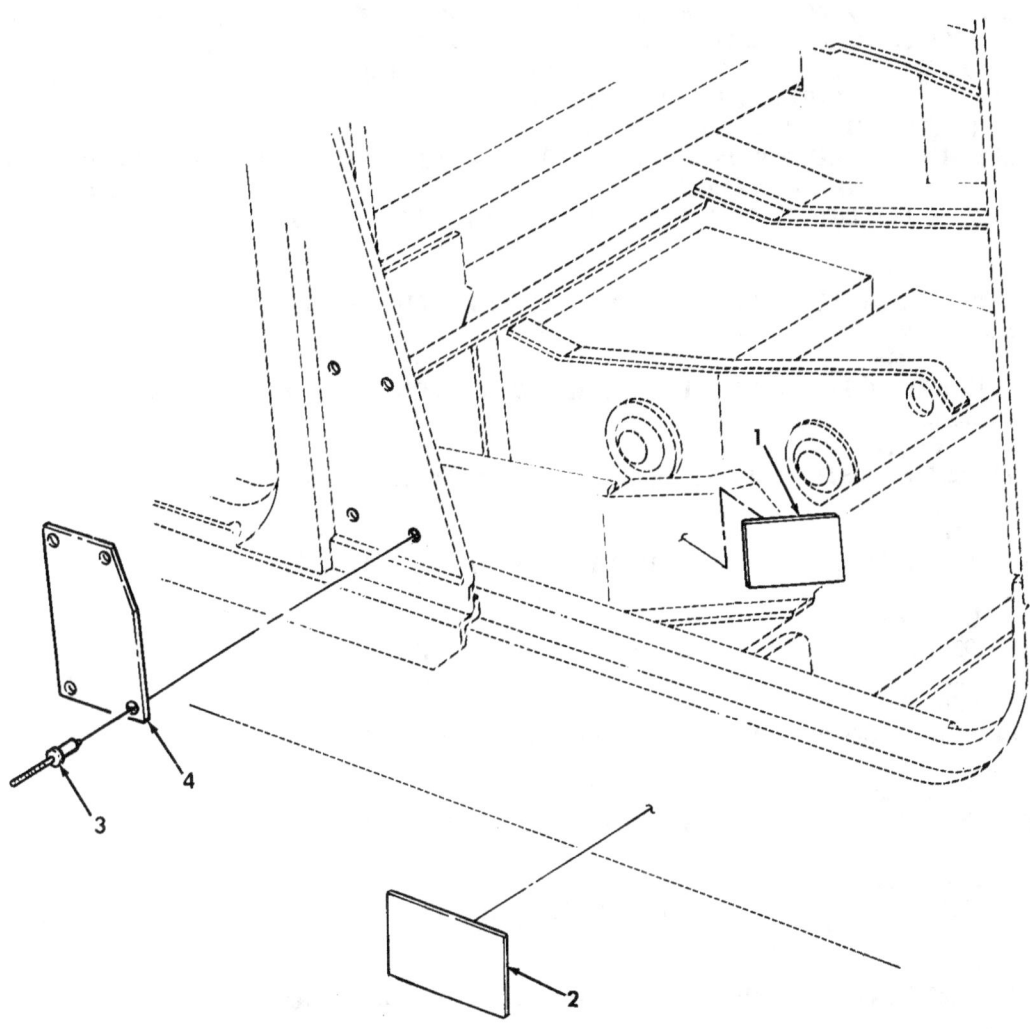

Figure 355. Data Plates, Battery Cable and Slave Receptacle.

SECTION II   TM9-2320-280-24P

(1) NO	(2) SMR CODE	(3) NSN	(4) CAGEC	(5) PART NUMBER	(6) DESCRIPTION AND USABLE ON CODES(UOC)	(7) QTY

GROUP 2210 DATA PLATES AND INSTRUCTION HOLDERS

FIG. 355 DATA PLATES, BATTERY CABLE AND SLAVE RECEPTACLE

1	PAOZZ	9905011853143	19207	12339107	PLATE,INSTRUCTION CONNECTING BATTERY CABLES...................	1
2	PFOZZ	9905011879468	19207	12339087	PLATE,INSTRUCTION SLAVE RECEPTACLE.	1
3	PAOZZ	5320000835009	81349	M24243/1-A403	RIVET,BLIND 1/8 DIA.,.126-.187 GRIP,USBL EFF 100,000 AND ABOVE.....	4
3	PAOZZ	5320010195694	81349	M24243/1A402	RIVET,BLIND 1/8 DIA.,.063-.125 GRIP,USBL EFF 1 THRU 99,999......... UOC:H11,H13,H14,H15,H16,H17,H18,H20, H21,H24,H25,H26,H27,H28,MMM	4
4	PAOZZ	9905013622014	19207	12342653	PLATE,IDENTIFICATIO WEIGHT AND DIMENSIONAL DATA................... UOC:HVY	1
4	PFOZZ	9905012489544	19207	12339900	PLATE,IDENTIFICATIO WEIGHT AND DIMENSIONAL DATA................... UOC:H13	1
4	PFOZZ	9905012481111	19207	12340653	PLATE,IDENTIFICATIO WEIGHT AND DIMENSIONAL DATA................... UOC:H16	1
4	PFOZZ	9905012481112	19207	12340322	PLATE,IDENTIFICATIO WEIGHT AND DIMENSIONAL DATA................... UOC:H11	1
4	PFOZZ	9905012489543	19207	12340321	PLATE,IDENTIFICATIO WEIGHT AND DIMENSIONAL DATA................... UOC:H17	1
4	PFOZZ	9905011857970	19207	12340323	PLATE,IDENTIFICATIO WEIGHT AND DIMENSIONAL DATA................... UOC:H15	1
4	PFOZZ	9905012481113	19207	12340318	PLATE,IDENTIFICATIO WEIGHT AND DIMENSIONAL DATA................... UOC:H18	1
4	PFOZZ	9905012481114	19207	12340320	PLATE,IDENTIFICATIO WEIGHT AND DIMENSIONAL DATA................... UOC:H20	1
4	PFOZZ	9905012489545	19207	12339987	PLATE,IDENTIFICATIO WEIGHT AND DIMENSIONAL DATA................... UOC:H21	1
4	PFOZZ	9905012489546	19207	12339901	PLATE,IDENTIFICATIO WEIGHT AND DIMENSIONAL DATA................... UOC:H14	1
4	PFOZZ	9905012489547	19207	12341836	PLATE,IDENTIFICATIO WEIGHT AND DIMENSIONAL DATA................... UOC:H28	1
4	PFOZZ	9905012489548	19207	12340324	PLATE,IDENTIFICATIO WEIGHT AND DIMENSIONAL DATA................... UOC:H25	1
4	PFOZZ	9905012489549	19207	12340325	PLATE,IDENTIFICATIO WEIGHT AND DIMENSIONAL DATA...................	1

SECTION II TM9-2320-280-24P

(1) ITEM NO	(2) SMR CODE	(3) NSN	(4) CAGEC	(5) PART NUMBER	(6) DESCRIPTION AND USABLE ON CODES(UOC)	(7) QTY
4	PFOZZ	9905012489552	19207	12340326	PLATE,INSTRUCTION WEIGHT AND DIMENSIONAL DATA.................... UOC:H26	1
4	PFOZZ	9905012481115	19207	12340327	PLATE,IDENTIFICATIO WEIGHT AND DIMENSIONAL DATA.................... UOC:H24	1
4	PFOZZ	9905013933795	19207	12342987	PLATE,IDENTIFICATIO WEIGHT AND DIMENSIONAL DATA.................... UOC:H27	1
4	PFOZZ	9905013925795	19207	12342980	PLATE,IDENTIFICATIO WEIGHT AND DIMENSIONAL DATA.................... UOC:A11	1
4	PFOZZ	9905013925794	19207	12343043	PLATE,INSTRUCTION WEIGHT AND DIMENSIONAL DATA.................... UOC:A13	1
4	PFOZZ	9905013925797	19207	12343011	PLATE,IDENTIFICATIO WEIGHT AND DIMENSIONAL DATA.................... UOC:A14	1
4	PFOZZ	9905013925798	19207	12343001	PLATE,IDENTIFICATIO WEIGHT AND DIMENSIONAL DATA.................... UOC:A15	1
4	PFOZZ	9905013925796	19207	12342993	PLATE,IDENTIFICATIO WEIGHT AND DIMENSIONAL DATA.................... UOC:A20	1
4	PFOZZ	9905013931830	19207	12342999	PLATE,IDENTIFICATIO IDENTIFICATION, WEIGHT AND DIMENSIONAL DATA......... UOC:A24	1
4	PFOZZ	9905013925796	19207	12343049	PLATE,IDENTIFICATIO WEIGHT AND DIMENSIONAL DATA.................... UOC:A25	1
4	PFOZZ	9905013931833	19207	12343048	PLATE,IDENTIFICATIO WEIGHT AND DIMENSIONAL DATA.................... UOC:A26	1
4	PFOZZ	9905013937128	19207	12343018	PLATE,IDENTIFICATIO WEIGHT AND DIMENSIONAL DATA.................... UOC:A27	1
4	PFOZZ	9905013925799	19207	12343025	PLATE,IDENTIFICATIO WEIGHT AND DIMENSIONAL DATA.................... UOC:AVY	1
4	PFOZZ	9905013931834	19207	12342991	PLATE,IDENTIFICATIO WEIGHT AND DIMENSIONAL DATA.................... UOC:B16	1
4	PFOZZ	9905013925800	19207	12343040	PLATE,IDENTIFICATIO WEIGHT AND DIMENSIONAL DATA.................... UOC:B17	1
4	PFOZZ	9905014489783	19207	12446782	PLATE,INSTRUCTION WEIGHT AND DIMENSIONAL DATA.................... UOC:B18	1
4	PFOZZ	9905014490476	19207	12446801	PLATE,INSTRUCTION WEIGHT AND DIMENSIONAL DATA.................... UOC:BVY	1
4	PFOZZ	9905014465771	19207	12446794	PLATE,INSTRUCTION WEIGHT AND UOC:B15	1

SECTION II   TM9-2320-280-24P C01
   (1)   (2)   (3)   (4)   (5)   (6)
(7)   ITEM   SMR   PART
NO   CODE   NSN   CAGEC   NUMBER   DESCRIPTION AND USABLE ON CODES(UOC)
QTY

                                                     DIMENSIONAL DATA....................
                                                     UOC:B20
       4 PFOZZ 9905014465768 19207 12446791   PLATE,IDENTIFICATIO WEIGHT AND   1
DIMENSIONAL DATA....................
                                                     UOC:B24
       4 PFOZZ 9905014489784 19207 12446788   PLATE,INSTRUCTION WEIGHT AND   1
DIMENSIONAL DATA....................
                                                     UOC:B25
       4 PFOZZ 9905014466187 19207 12446785   PLATE,INSTRUCTION WEIGHT AND   1
DIMENSIONAL DATA....................
                                                     UOC:C17
* 4 PFOZZ 7690014814906 19207 12469100   DECAL..............................   1
UOC:MMM
* 4 PFOZZ 9905014860051 19207 12469323   PLATE,INSTRUCTION...................   1
UOC:NNN

355-3

Section II. TM 9-2320-280-24P-2

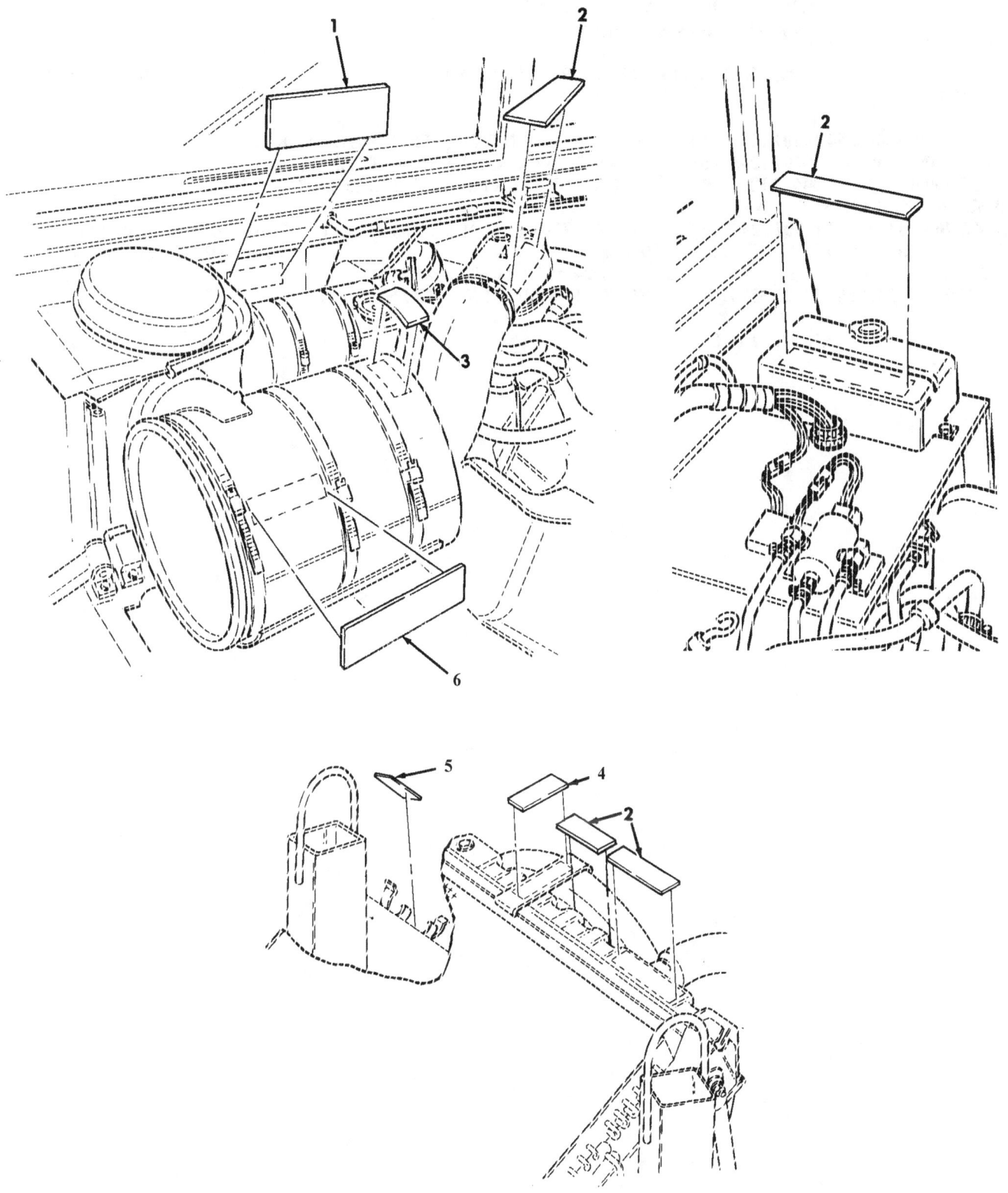

Figure 356. Data Plates, Surge Tank, Fan Shroud, and Cook Stove.

SECTION II TM9-2320-280-24P

(1) NO	(2) SMR CODE	(3) NSN	(4) CAGEC	(5) PART NUMBER	(7) ITEM DESCRIPTION AND USABLE ON CODES(UOC)	QTY

GROUP 2210 DATA PLATES AND INSTRUCTION HOLDERS

FIG. 356 DATA PLATES, SURGE TANK, FAN SHROUD, AND COOK STOVE

1	PAOZZ	7690012090864	19207	12339065	MARKER,IDENTIFICATI SURGE TANK	1
2	PAOZZ	9905012489550	19207	12339055	PLATE,INSTRUCTION NO STEP	4
3	PAOZZ	7690014734550	34623	EX4319	DECAL SERPENTINE BELT UOC:BVY,B15,B20,B24,B25,C17,NNN	1
4	PAOZZ	7690011975500	19207	12339064	DECAL FAN SHROUD	2
5	PAOZZ	7690011885144	19207	12340909	MARKER,IDENTIFICATI GASOLINE BURNER STOVE STOWAGE UOC:A11,A24,A27,H11,H24,H27,MMM	1
6	PAOZZ	9905012489551	34623	5590177	PLATE,INSTRUCTION NO STEP	1

END OF FIGURE

Section II.

TM 9-2320-280-24P-2

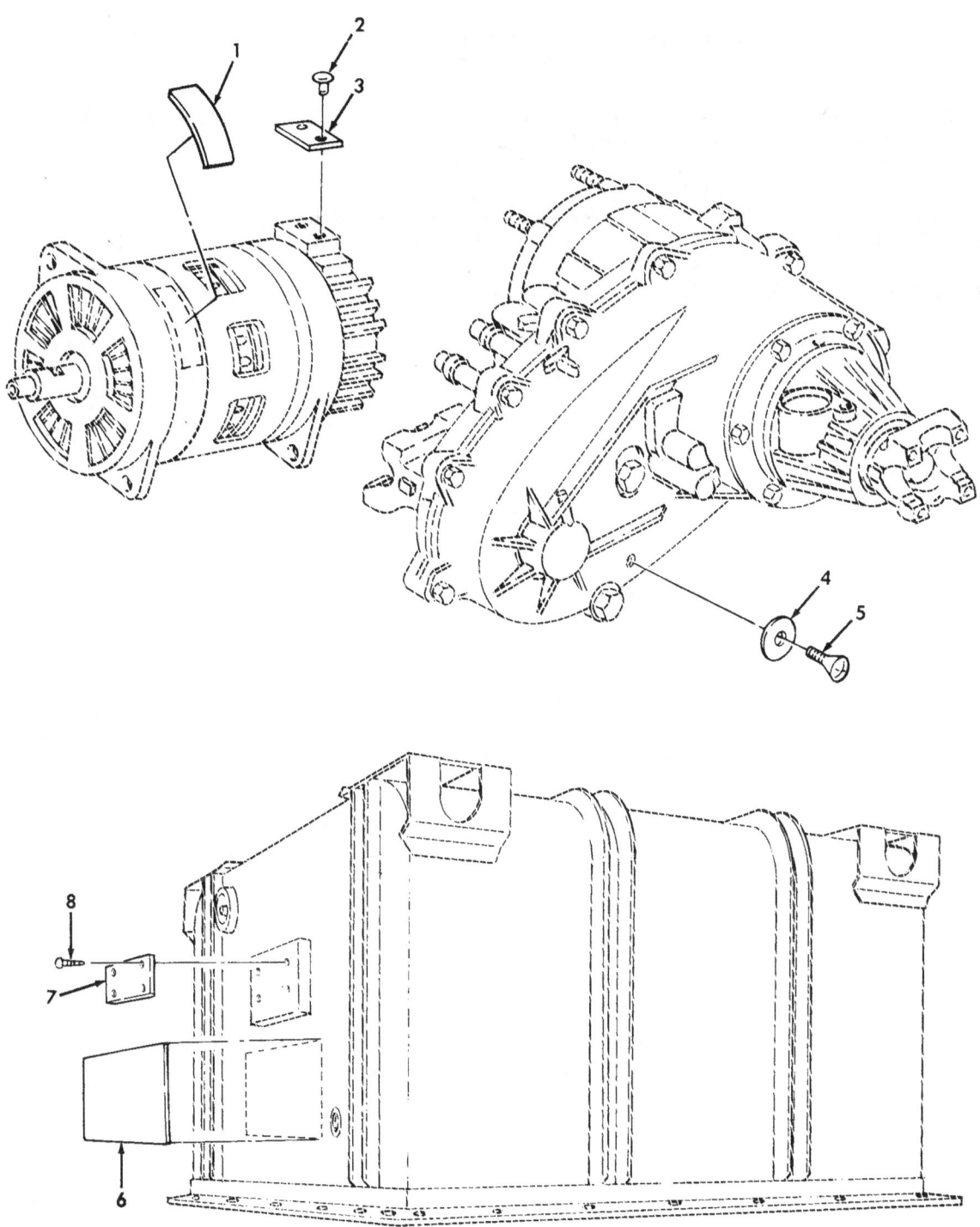

Figure 357. Data Plates, Generator, Transfer Case, and Engine Container.

SECTION II TM9-2320-280-24P

(1) NO	(2) SMR CODE	(3) NSN	(4) CAGEC	(5) PART NUMBER	(7) DESCRIPTION AND USABLE ON CODES(UOC)	ITEM QTY

GROUP 2210 DATA PLATES AND INSTRUCTION HOLDERS

FIG. 357 DATA PLATES,GENERATOR, TRANSFER CASE,AND ENGINE CONTAINER

1 PAOZZ 7690011975500 19207 12339064 DECAL FAN SHROUD.................. 1
2 PAOZZ 5320009572500 96906 MS20604B4W1 RIVET,BLIND 1/8 DIA.,.015-.062 GRIP 2
3 PFOZZ 9905011819456 19207 11630585 PLATE,INSTRUCTION.................. 1
4 PFOZZ 9905011587981 34623 5741083 PLATE,IDENTIFICATIO TRANSFER CASE.. 1
5 PAOZZ 5305012117415 34623 5741084 SCREW,CAP,HEXAGON H #10-32 X 1/4... 1
6 PAOZZ 7690012818341 19207 12338072 DECAL ENGINE CONTAINER............ 1
7 PAOZZ 9905009012942 19207 7973326 PLATE,IDENTIFICATIO ENGINE CONTAINER......................... 1
8 PAOZZ 5305002535627 80205 MS21318-48 SCREW,DRIVE #10-8 X 5/8............ 4

END OF FIGURE

357-1

Section II.

TM 9-2320-280-24P-2

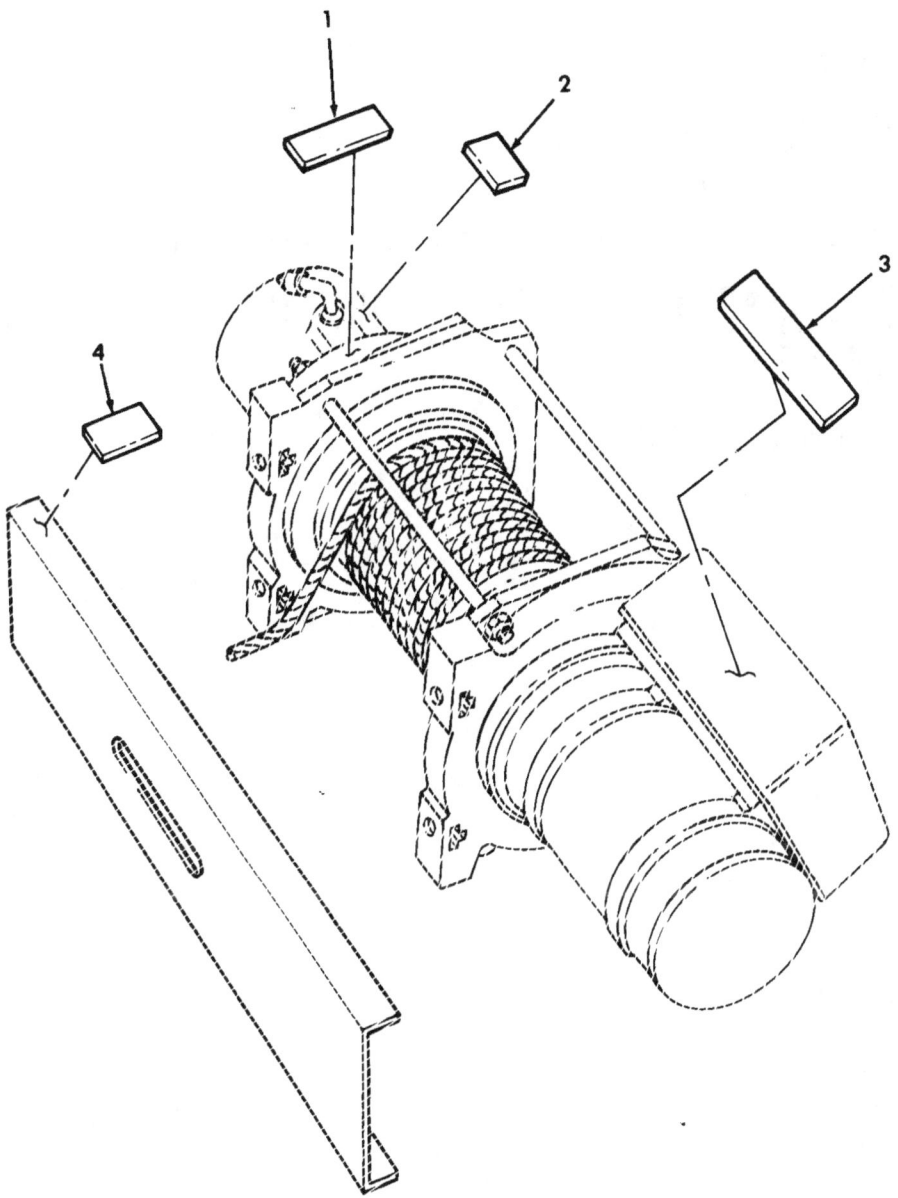

Figure 358. Data Plates, Winch.

SECTION II   TM9-2320-280-24P

(1) NO	(2) SMR CODE	(3) NSN	(4) CAGEC	(5) PART NUMBER	(6) DESCRIPTION AND USABLE ON CODES(UOC)	(7) QTY

GROUP 2210 DATA PLATES AND INSTRUCTION HOLDERS

FIG. 358 DATA PLATES, WINCH

1 PAOZZ 7690011853208 27647 15657   MARKER,IDENTIFICATI................. 1   UOC:A14,A26,A27,B18,H14,H18,H26,H27,H28

2 PFOZZ 9905011853127 27647 15873   PLATE,INSTRUCTION................... 1   UOC:A14,A26,A27,B18,H14,H18,H26,H27,H28

3 PAOZZ 9905011853207 27647 13698   PLATE,INSTRUCTION................... 1   UOC:A14,A26,A27,B18,H14,H18,H26,H27,H28

4 PAOZZ 7690012646536 19207 12340923   DECAL WINCH OPERATION, USBL EFF 1 THRU 99,999......................... 1   UOC:H14,H18,H26,H27,H28

4 PAOZZ 7690013502094 19207 12342439   MARKER,IDENTIFICATI WARNING,WINCH OPERATION USBL EFF 100,000 AND ABOVE   1   UOC:A14,A26,A27,B18,H14,H18,H26,H27,H28

END OF FIGURE

358-1

Section II. TM 9-2320-280-24P-2

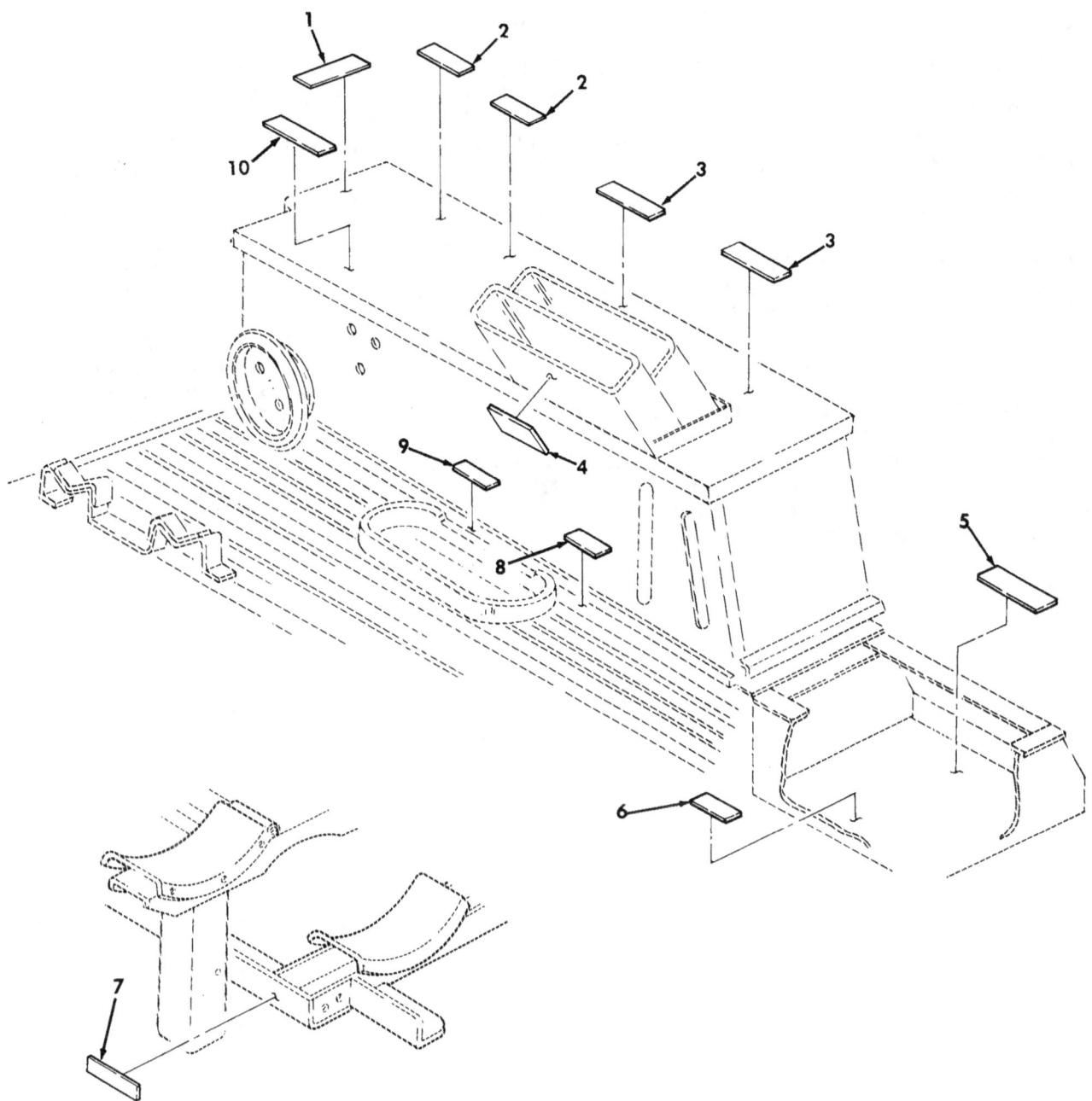

Figure 359. Data Plates, TOW Carrier.

SECTION II   TM9-2320-280-24P

(1) NO	(2) SMR CODE	(3) NSN	(4) CAGEC	(5) PART NUMBER	(7) ITEM DESCRIPTION AND USABLE ON CODES(UOC)	QTY

GROUP 2210 DATA PLATES AND INSTRUCTION HOLDERS

FIG. 359 DATA PLATES, TOW CARRIER

1	PAOZZ	7690011911313	19207	12340898	MARKER,IDENTIFICATI TELEPHONE STOWAGE............................. UOC:A11,A24,A27,B24,H11,H24,H27,MMM	1
2	PAOZZ	7690011893740	19207	12340899	MARKER,IDENTIFICATI COMBAT RATIONS STOWAGE............................. UOC:A11,A24,A27,B24,H11,H24,H27,MMM	2
3	PAOZZ	7690011926371	19207	12340901	MARKER,IDENTIFICATI M-16 AMMUNITION STOWAGE................. UOC:A11,A24,A27,B24,H11,H24,H27,MMM	2
4	PAOZZ	7690011918793	19207	12340890	MARKER,IDENTIFICATI MGS BATTERY STOWAGE............................ UOC:A11,A24,A27,B24,H11,H24,H27,MMM	1
5	PAOZZ	7690011908501	19207	12340904	MARKER,IDENTIFICATI CBR OVERGARMENTS STOWAGE................ UOC:A11,A24,A27,B24,H11,H24,H27,MMM	1
6	PAOZZ	7690011911312	19207	12340908	MARKER,IDENTIFICATI CHEMICAL MASK STOWAGE................. UOC:A11,A24,A27,B24,H11,H24,H27,MMM	1
7	PAOZZ	7690011916467	19207	12339764	MARKER,IDENTIFICATI NOSE END MISSILE RACK....... .... UOC:A11,A24,A27,B24,H11,H24,H27,MMM	1
8	PAOZZ	7690011911316	19207	12340903	MARKER,IDENTIFICATI WATER CAN STOWAGE............................ UOC:A11,A24,A27,B24,H11,H24,H27,MMM	1
9	PAOZZ	7690011911315	19207	12340902	MARKER,IDENTIFICATI FUEL CAN....... UOC:A11,A24,A27,B24,H11,H24,H27,MMM	1
10	PAOZZ	7690011911314	19207	12340900	MARKER,IDENTIFICATI CABLE REEL STOWAGE........................ UOC:A11,A24,A27,B24,H11,H24,H27,MMM	1

END OF FIGURE

Section II.

TM 9-2320-280-24P-2

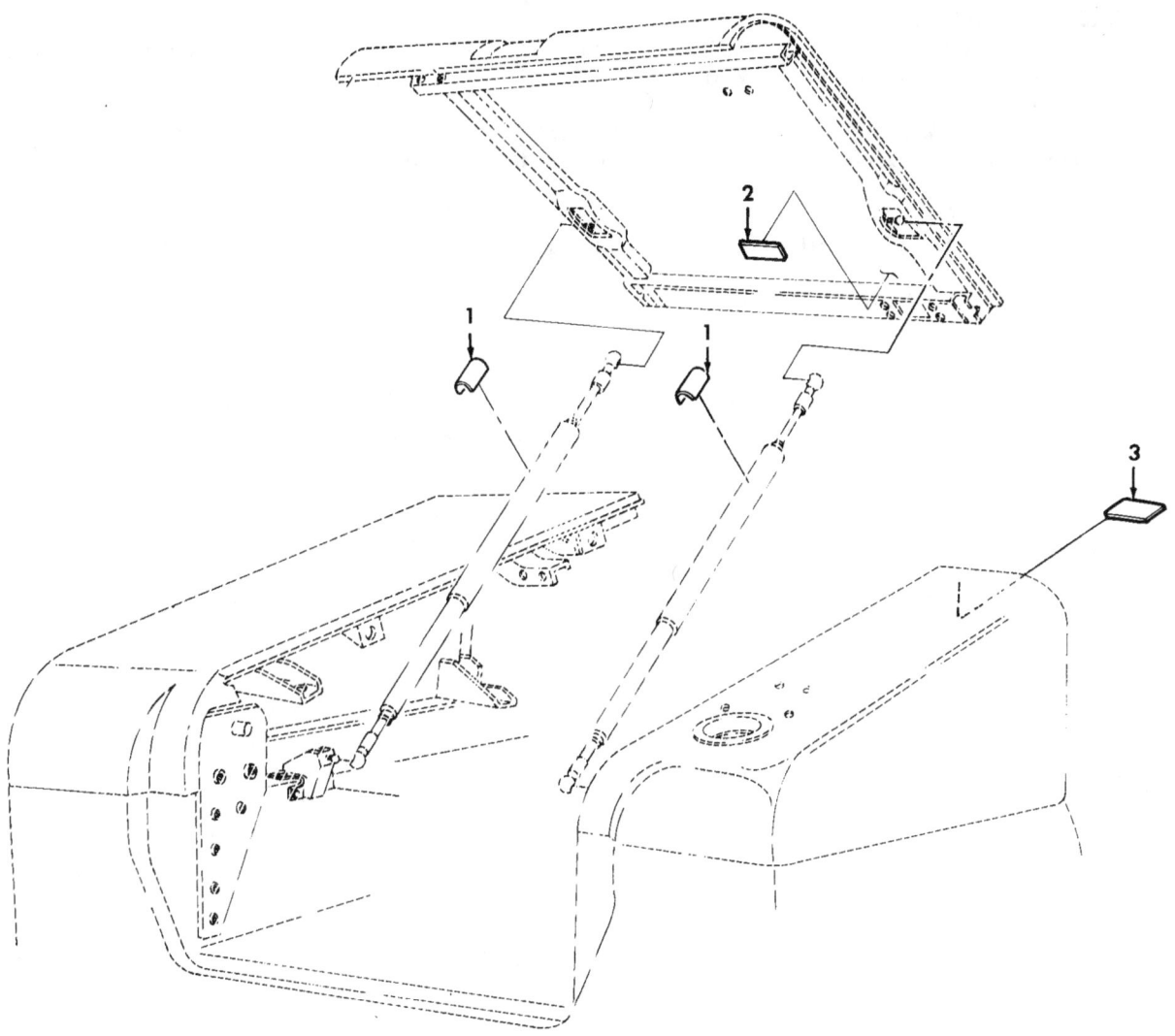

Figure 360. Data Plates, TOW Carrier.

SECTION II   TM9-2320-280-24P

(1) NO	(2) SMR CODE	(3) NSN	(4) CAGEC	(5) PART NUMBER	(6) DESCRIPTION AND USABLE ON CODES(UOC)	(7) QTY

GROUP 2210 DATA PLATES AND INSTRUCTION HOLDERS

FIG. 360 DATA PLATES, TOW CARRIER

1 PAOZZ 7690012047785 19207 12340814   DECAL NO HAND HOLD................. 2   UOC:A11,A24,A25,A26,A27,B17,B18,B24,
B25,C17,H11,H17,H18,H24,H25,H26,H27,MMM

2 PAOZZ 7690012047847 19207 12340852   MARKER,IDENTIFICATI................. 1   UOC:A11,A24,A27,B24,H11,H13,H14,H15,
H16,H17,H18,H20,H21,H24,H25,H26,H27,MMM

3 PAOZZ 7690011908501 19207 12340904   MARKER,IDENTIFICATI CBR OVER GARMENTS........................... 2
UOC:A11,A24,A25,A26,A27,B17,B18,B24,
B25,C17,H11,H17,H18,H24,H25,H26,H27,MMM

END OF FIGURE

360-1

Section II. TM 9-2320-280-24P-2

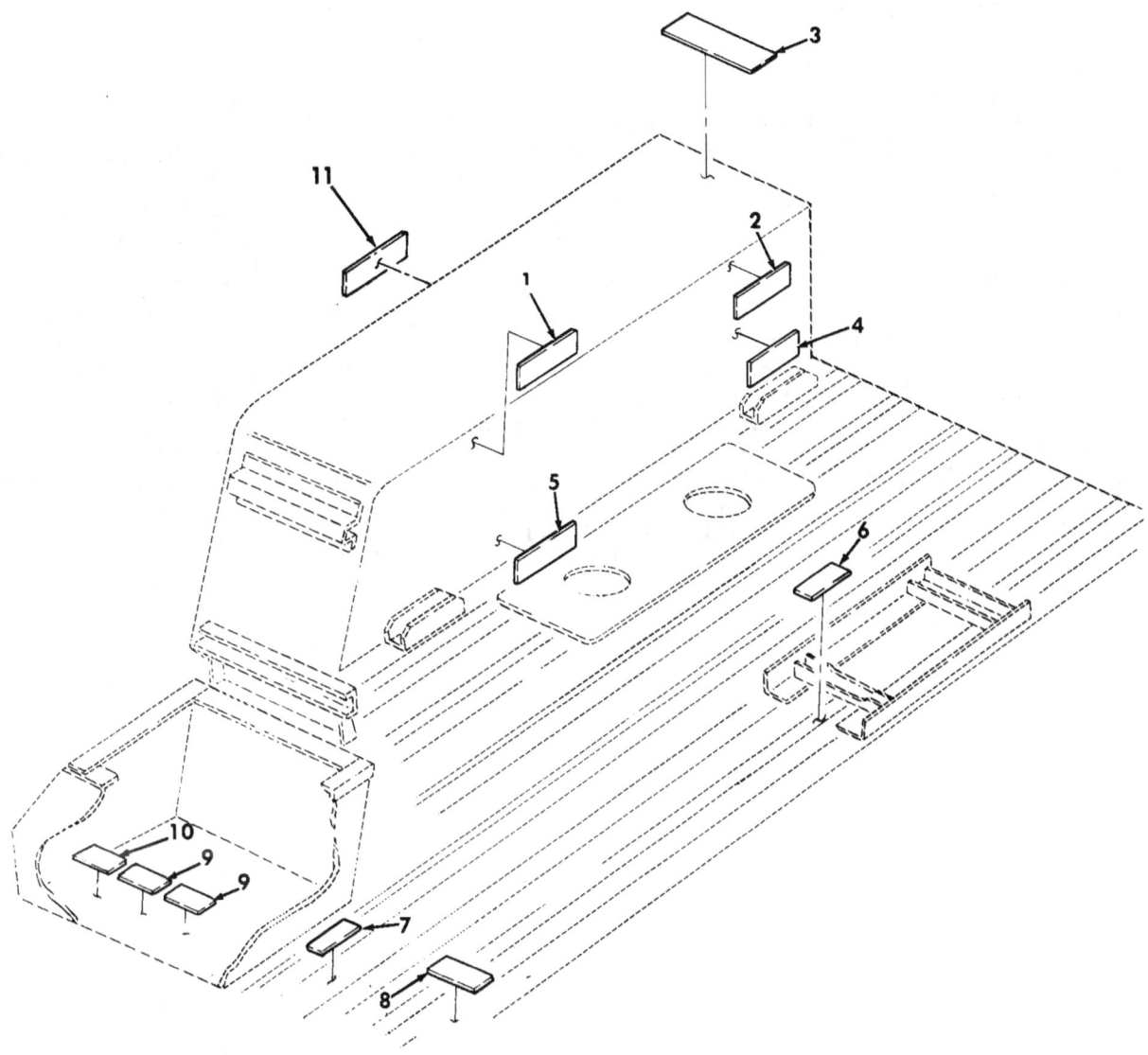

Figure 361. Data Plates, Sights, Traversing Unit, Night Vision Goggles, and Radiac Meter.

SECTION II   TM9-2320-280-24P

(1) NO	(2) SMR CODE	(3) NSN	(4) CAGEC	(5) PART NUMBER	(6) DESCRIPTION AND USABLE ON CODES(UOC)	(7) QTY

GROUP 2210 DATA PLATES AND INSTRUCTION HOLDERS

FIG. 361 DATA PLATES,SIGHTS,TRAVERSING UNIT,NIGHT VISION GOGGLES,AND RADIAC METER

1	PAOZZ	7690011893738	19207	12340894	MARKER,IDENTIFICATI NIGHT SIGHT STOWAGE............................. UOC:A11,A24,A27,B24,H11,H24,H27,MMM	1
2	PAOZZ	7690011893737	19207	12340891	MARKER,IDENTIFICATI BORESIGHT COLLIMATOR STOWAGE.................. UOC:A11,A24,A27,B24,H11,H24,H27,MMM	1
3	PAOZZ	7690011893740	19207	12340899	MARKER,IDENTIFICATI COMBAT RATIONS. UOC:A11,A24,A27,B24,H11,H24,H27,MMM	1
4	PAOZZ	7690012030161	19207	12340896	DECAL NIGHT SIGHT BATTERY STOWAGE.. UOC:A11,A24,A27,B24,H11,H24,H27,MMM	1
5	PAOZZ	7690011893739	19207	12340897	MARKER,IDENTIFICATI NIGHT SIGHT COOLANT STOWAGE.................... UOC:A11,A24,A27,B24,H11,H24,H27,MMM	1
6	PAOZZ	7690011893736	19207	12340893	MARKER,IDENTIFICATI DAYSIGHT STOWAGE............................. UOC:A11,A24,A27,B24,H11,H24,H27,MMM	1
7	PAOZZ	7690012047849	19207	12340895	MARKER,IDENTIFICATI TRAVERSING UNIT UOC:A11,A24,A27,B24,H11,H24,H27,MMM	1
8	PAOZZ	7690012047850	19207	12340892	MARKER,IDENTIFICATI MISSILE GUIDANCE SYSTEM (MGS)............... UOC:A11,A24,A27,B24,H11,H24,H27,MMM	1
9	PAOZZ	7690011926370	19207	12340906	MARKER,IDENTIFICATI NIGHT VISION GOGGLES STOWAGE.................... UOC:A11,A24,A27,B24,H11,H24,H27,MMM	2
10	PAOZZ	7690011926369	19207	12340905	MARKER,IDENTIFICATI RADIAC METER STOWAGE........................... UOC:A11,A24,A27,B24,H11,H24,H27,MMM	1
11	PFOZZ	7690013589391	19207	12342625	MARKER,IDENTIFICATI TIRE PRESSURE, 42 PSI............................. UOC:HVY	2
11	PAOZZ	7690014450456	19207	12446734	MARKER,IDENTIFICATI TIRE PRESSURE, 40 PSI............................. UOC:AVY,BVY	1

END OF FIGURE

361-1

Section II. TM 9-2320-280-24P-2

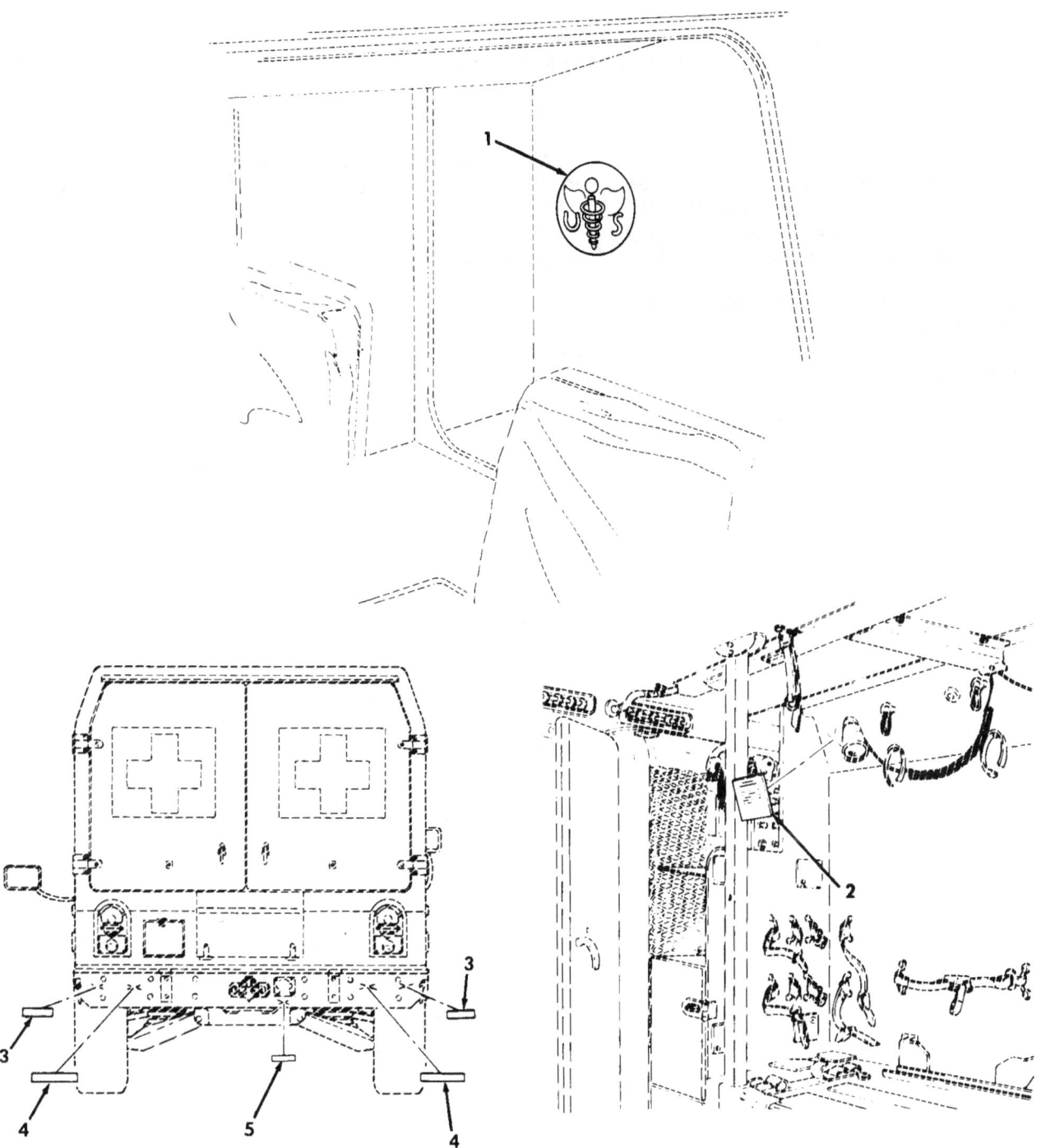

Figure 362. Data Plates and Identification Tags, M996, M996A1, M997, M997A1, and M997A2 Ambulance.

SECTION II   TM9-2320-280-24P

(1) NO	(2) SMR CODE	(3) NSN	(4) CAGEC	(5) PART NUMBER	(6) DESCRIPTION AND USABLE ON CODES(UOC)	(7) QTY

GROUP 2210 DATA PLATES AND INSTRUCTION HOLDERS

FIG. 362 DATA PLATES AND IDENTIFICATION TAGS, M996, M996A1, M997, M997A1, AND M997A2 AMBULANCE

| 1 | PAOZZ | 7690001385788 | 19207 | 11644801 | DECAL ARMY MEDICAL SERVICE INSIGNIA | 1 |

UOC:A15,B15,B16,H15,H16

| 2 | PAOZZ | 9905012663913 | 19207 | 12341757 | PLATE,INSTRUCTION HEATER AND VENT OPERATING INSTRUCTIONS.............. | 1 |

UOC:A15,B15,B16,H15,H16

| 3 | PAOZZ | 7690012651133 | 19207 | 12339057 | DECAL SLING........................ | 2 |

UOC:A15,B15,B16,H15,H16

| 4 | PAOZZ | 7690012651134 | 19207 | 12339059 | DECAL TIEDOWN...................... | 2 |

UOC:A15,B15,B16,H15,H16

| 5 | PAOZZ | 7690012651135 | 19207 | 12339060 | DECAL 24 VOLTS..................... | 1 |

UOC:A15,B15,B16,H15,H16

END OF FIGURE

Section II. TM 9-2320-280-24P-2

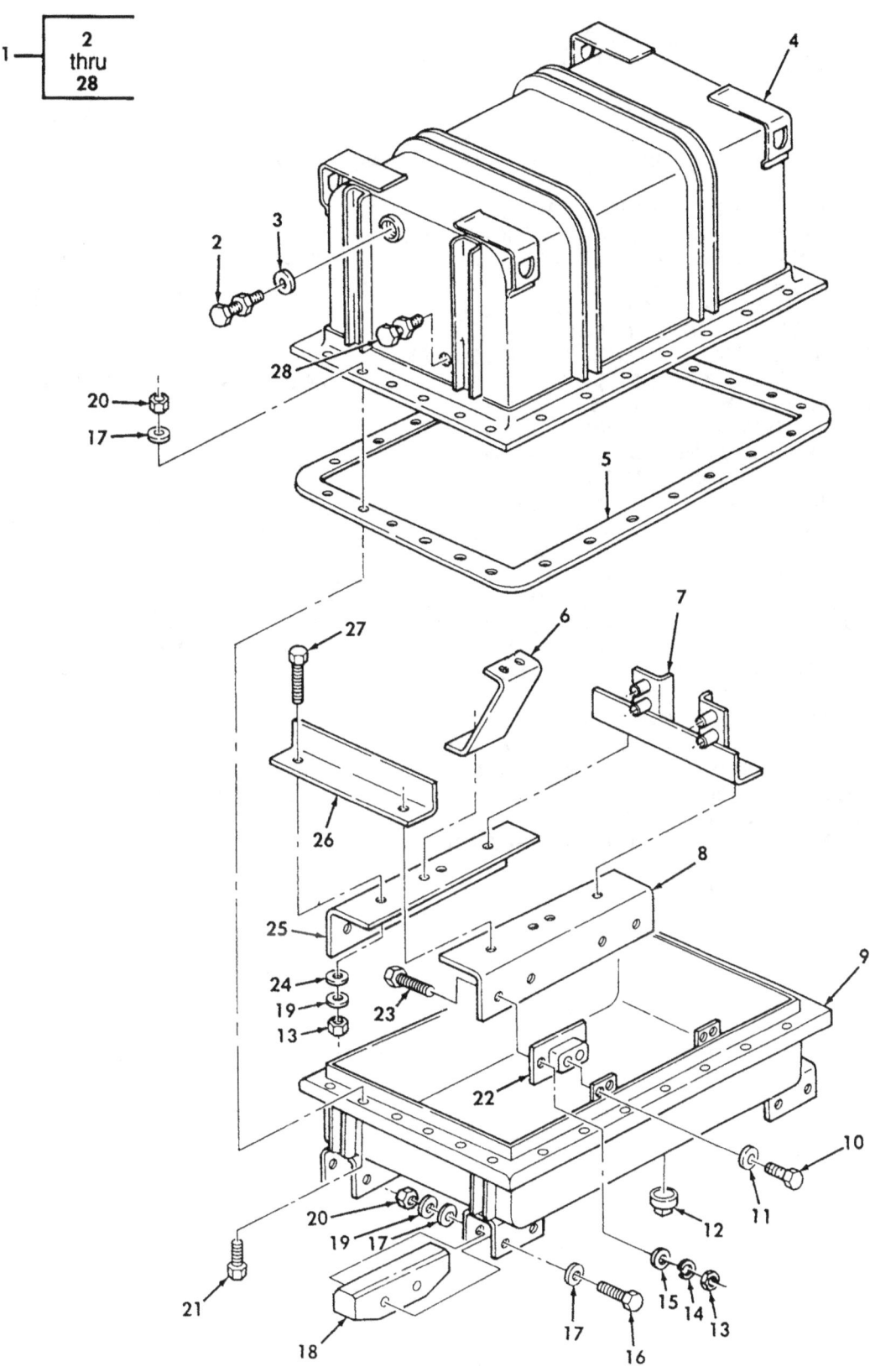

Figure 363. Engine Shipping Container Assembly.

SECTION II   TM9-2320-280-24P

(1) NO	(2) SMR CODE	(3) NSN	(4) CAGEC	(5) PART NUMBER	(7) DESCRIPTION AND USABLE ON CODES(UOC)	QTY

(6) ITEM

GROUP 33 SPECIAL PURPOSE KITS
3301 REUSABLE SHIPPING CONTAINER

FIG. 363 ENGINE SHIPPING CONTAINER ASSEMBLY

```
 1 PFFFF 8145012313747 19207 12338064      SHIPPING AND STORAG CONTAINER,    1
ENGINE..................
 2 PAFZZ 2520012681051 19207 12338078      .BREATHER......................... 1
 3 PAFZZ 4820010052994 01347 330-10-10R    .VALVE,PRESSURE EQUA.............. 1
 4 XAFZZ           19207 12338066          .CONTAINER,UPPER.................. 1
 5 PAFZZ 5330012646537 19207 12338073      .RUBBER ROUND SECTIO CONTAINER    1
                                            ASSEMBLY.......................
 6 PFFZZ 5340012646544 19207 12338068      .BRACKET,DOUBLE ANGL ENGINE SUPPORT 2
 7 PFFZZ 5340012713059 19207 12338067      .BRACKET,MOUNTING ENGINE SUPPORT... 1
 8 PFFZZ 5340012646540 19207 12338074      .BRACKET,ANGLE LEFT................ 1
 9 XAFZZ           19207 12338065          .CONTAINER,LOWER................... 1
10 PAFZZ 5305002693233 80204 B1821BH-038F063N  .SCREW,CAP,HEXAGON H 3/8-24 X 5/8.. 8
11 PAFZZ 5310006379541 96906 MS35338-46    .WASHER,LOCK 3/8................... 8
12 PAFZZ 4730022212140 96906 MS20913-6S    .PLUG,PIPE 3/4-14.................. 1
13 PAFZZ 5310008807744 96906 MS51967-5     .NUT,PLAIN,HEXAGON 5/16-18......... 8
14 PAFZZ 5310004079566 96906 MS35338-45    .WASHER,LOCK 5/16.................. 8
15 PAFZZ 5310000814219 96906 MS27183-12    .WASHER,FLAT 11/32................. 8
16 PAFZZ 5305000712083 80204 B1821BH050C500N  .SCREW,CAP,HEXAGON H 1/2-13 X 5.00. 8
17 PAFZZ 5310008095998 96906 MS27183-18    .WASHER,FLAT 17/32................. 42
18 PAFZZ 5340012653676 19207 12338071      .RUNNER,METAL...................... 4
19 PAFZZ 5310005845272 96906 MS35338-48    .WASHER,LOCK 1/2................... 16
20 PAFZZ 5310007680318 21439 01857-007     .NUT,PLAIN,HEXAGON 1/2-13.......... 42
21 PAFZZ 5305007320511 80204 B1821BH050C113N  .SCREW,CAP,HEXAGON H 1/2-13 X 1-1/8 26
22 PFFZZ 5342012646543 19207 12338070      .MOUNT,RESILIENT................... 4
23 PAFZZ 5306002264828 80204 B1821BH031C113N  .BOLT,MACHINE 5/16-18 X 1-1/8...... 8
24 PAFZZ 5310008093079 96906 MS27183-19    .WASHER,FLAT 9/16.................. 8
25 PFFZZ 5340012646541 19207 12338075      .BRACKET,ANGLE..................... 1
26 PFFZZ 5340012653674 19207 12338076      .BRACKET,ANGLE..................... 1
27 PAFZZ 5305000712069 80204 B1821BH050C150N  .SCREW,CAP,HEXAGON H 1/2-13 X 1-1/2 8
28 PAFZZ 6685006181822 00334 SK2155        .INDICATOR,HUMIDITY................ 1
```

END OF FIGURE

Section II. TM 9-2320-280-24P-2

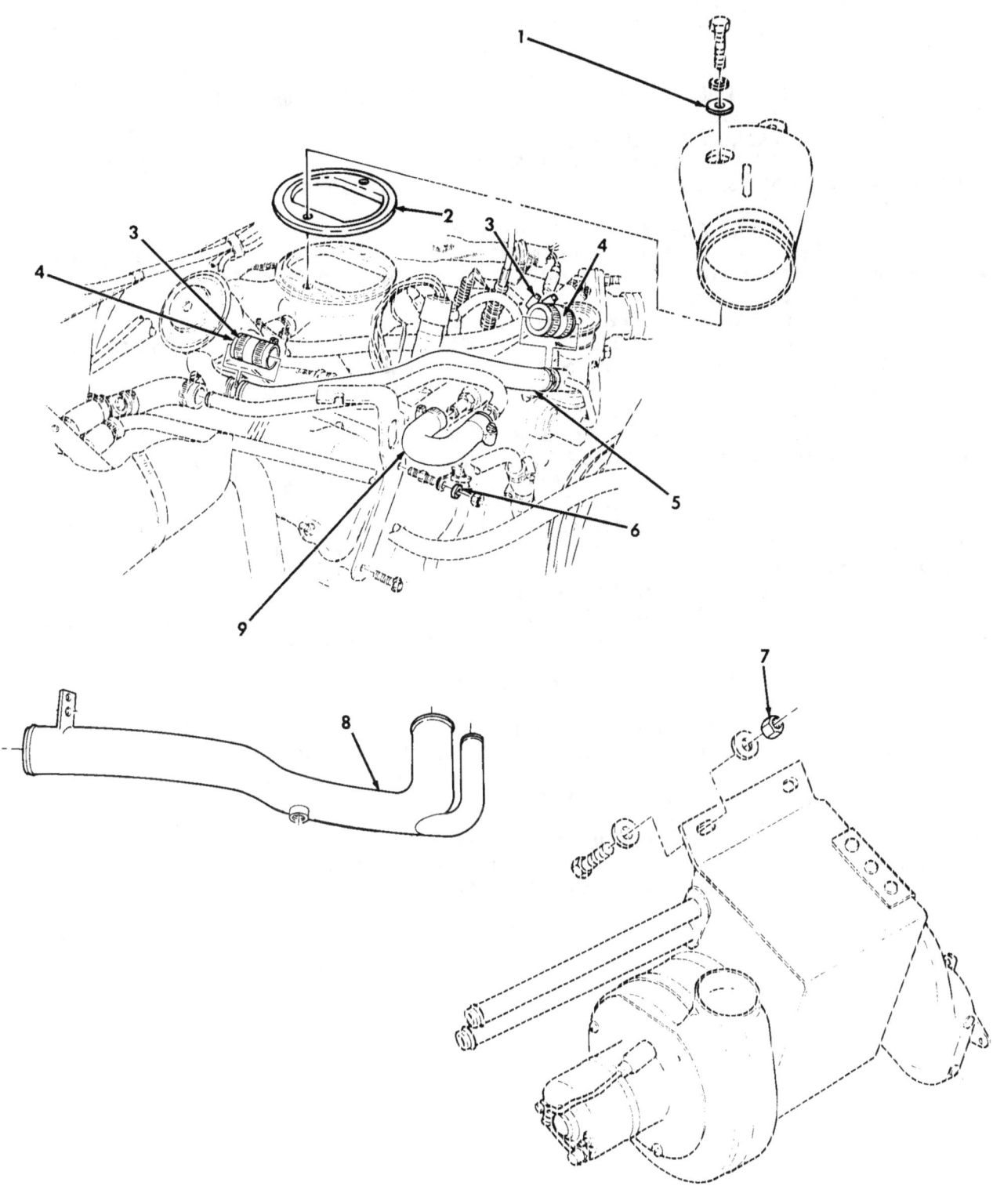

Figure 364. Arctic Kit, Fuel Injection Pump, Engine Gaskets, and CDR Tube.

SECTION II   TM9-2320-280-24P

(1) NO	(2) SMR CODE	(3) NSN	(4) CAGEC	(5) PART NUMBER	(7) ITEM DESCRIPTION AND USABLE ON CODES(UOC)	QTY

GROUP 3303 WINTERIZATION KITS

FIG. 364 ARCTIC KIT,FUEL INJECTION PUMP,ENGINE GASKETS,AND CDR TUBE

1	PAOZZ	5310011857214	19207	12339052	WASHER,FLAT PART OF KIT P/N 5705698. UOC:AVY,A11,A13,A14,A15,A20,A24,A25, A26,A27,B16,B17,B18,HVY,H11,H13,H14, H15,H16,H17,H18,H20,H21,H24,H25,H26, H27,H28,MMM	2
2	PAOZZ	5330012461822	19207	12338382	GASKET PART OF KIT P/N 5705698...... UOC:AVY,A11,A13,A14,A15,A20,A24,A25, A26,A27,B16,B17,B18,HVY,H11,H13,H14, H15,H16,H17,H18,H20,H21,H24,H25,H26, H27,H28,MMM	1
3	PAOZZ	4730009083193	66295	WWD48-58H	CLAMP,HOSE PART OF KIT P/N 5705698.. UOC:AVY,A11,A13,A14,A15,A20,A24,A25, A26,A27,B16,B17,B18,HVY,H11,H13,H14, H15,H16,H17,H18,H20,H21,H24,H25,H26, H27,H28,MMM	2
4	PAOZZ	4720012111998	19207	12340492	HOSE,NONMETALLIC PART OF KIT P/N 5705698............... UOC:AVY,A11,A13,A14,A15,A20,A24,A25, A26,A27,B16,B17,B18,HVY,H11,H13,H14, H15,H16,H17,H18,H20,H21,H24,H25,H26, H27,H28,MMM	2
5	PFOZZ	4710012572649	19207	12340498	TUBE ASSEMBLY,METAL PART OF KIT P/N 5705698............... UOC:AVY,A11,A13,A14,A15,A20,A24,A25, A26,A27,B16,B17,B18,HVY,H11,H13,H14, H15,H16,H17,H18,H20,H21,H24,H25,H26, H27,H28,MMM	1
6	PAOZZ	5310012067306	7X677	11500207	WASHER,LOCK PART OF KIT P/N 5705698. UOC:AVY,A11,A13,A14,A15,A20,A24,A25, A26,A27,B16,B17,B18,HVY,H11,H13,H14, H15,H16,H17,H18,H20,H21,H24,H25,H26, H27,H28,MMM	1
7	PAOZZ	5310008140673	96906	MS51943-33	NUT,SELF-LOCKING,HE 5/16-18 PART OF KIT P/N 5705698.................... UOC:AVY,A11,A13,A14,A15,A20,A24,A25, A26,A27,B16,B17,B18,HVY,H11,H13,H14, H15,H16,H17,H18,H20,H21,H24,H25,H26, H27,H28,MMM	1
8	PAOZZ	4710011880028	34623	5578604	TUBE,BENT,METALLIC PART OF KIT P/N 5705698........................... UOC:AVY,A11,A13,A14,A15,A20,A24,A25, A26,A27,B16,B17,B18,HVY,H11,H13,H14, H15,H16,H17,H18,H20,H21,H24,H25,H26, H27,H28,MMM	1
9	PFOZZ	4720012572655	19207	12340493	HOSE,PREFORMED PART OF KIT P/N 5705698............................. UOC:AVY,A11,A13,A14,A15,A20,A24,A25, A26,A27,B16,B17,B18,HVY,H11,H13,H14,	1

(1) NO	(2) SMR CODE	(3) NSN	(4) CAGEC	(5) PART NUMBER	(6) DESCRIPTION AND USABLE ON CODES(UOC)	(7) QTY
					SECTION II TM9-2320-280-24P ITEM	

H15,H16,H17,H18,H20,H21,H24,H25,H26,H27,H28,MMM

END OF FIGURE

Section II. TM 9-2320-280-24P-2

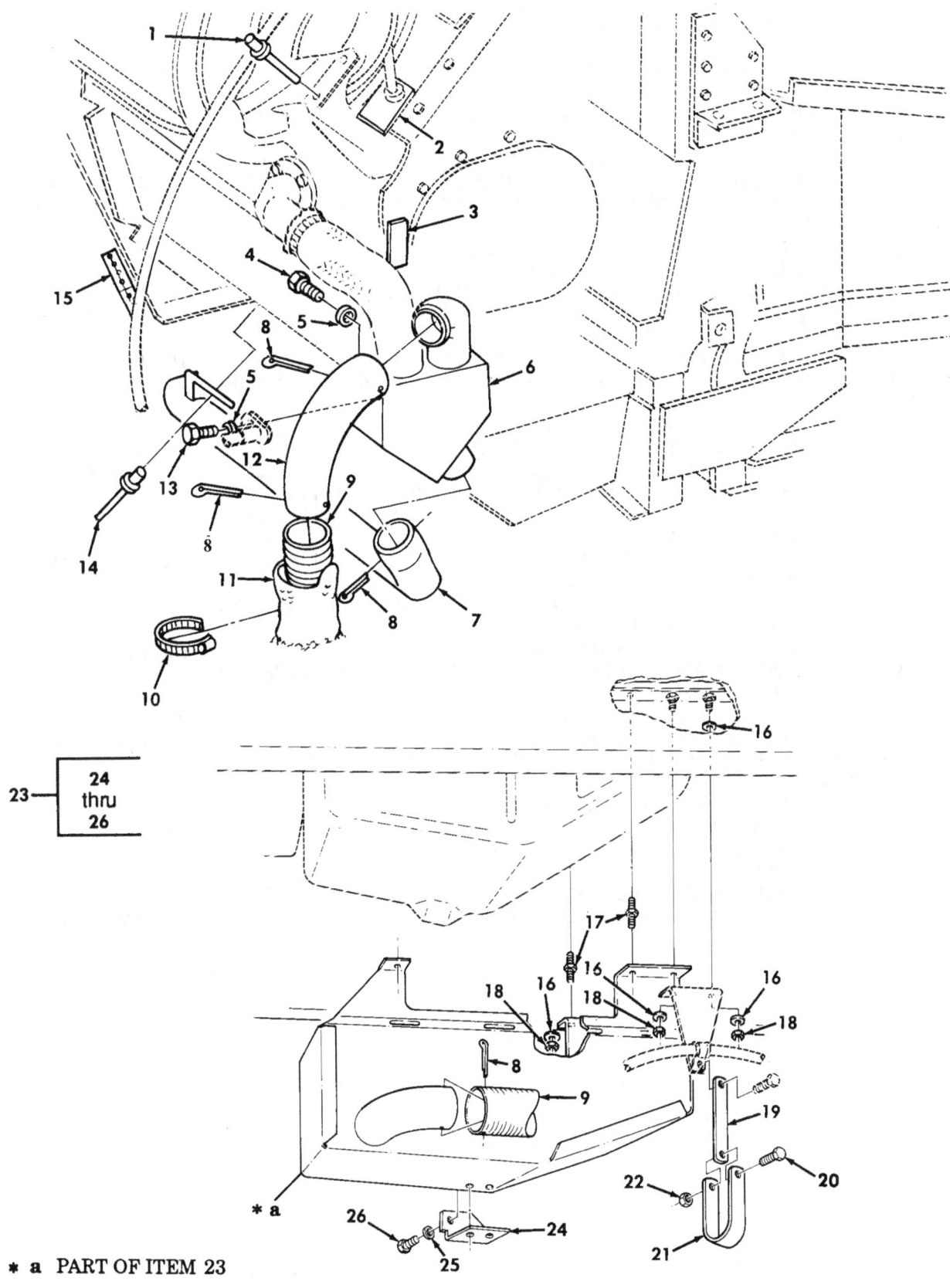

* a PART OF ITEM 23

Figure 365. Arctic Kit, Pan Cover and Heater Tube.

SECTION II   TM9-2320-280-24P

(1) NO	(2) SMR CODE	(3) NSN	(4) CAGEC	(5) PART NUMBER	(6) DESCRIPTION AND USABLE ON CODES(UOC)	(7) QTY

GROUP 3303 WINTERIZATION KITS

FIG. 365 ARCTIC KIT, PAN COVER AND HEATER TUBE

1 PAOZZ 5320011357319 11815 CR-213-6-3   RIVET,BLIND PART OF KIT P/N 5705698. 2
UOC:AVY,A11,A13,A14,A15,A20,A24,A25,
A26,A27,B16,B17,B18,HVY,H11,H13,H14,
H15,H16,H17,H18,H20,H21,H24,H25,H26,
H27,H28,MMM

2 KFFZZ   19207 12340722   TEMPLATE PART OF KIT P/N 5705698.... 1   UOC:AVY,A11,A13,A14,A15,A20,A24,A25,
A26,A27,B16,B17,B18,HVY,H11,H13,H14,
H15,H16,H17,H18,H20,H21,H24,H25,H26,
H27,H28,MMM

3 KFFZZ   19207 12342103   TEMPLATE PART OF KIT P/N 5705698.... 2   UOC:AVY,A11,A13,A14,A15,A20,A24,A25,
A26,A27,B16,B17,B18,HVY,H11,H13,H14,
H15,H16,H17,H18,H20,H21,H24,H25,H26,
H27,H28,MMM

4 PAOZZ 5305000680508 80204 B1821BH025C075N SCREW,CAP,HEXAGON H 1/4-20 X 3/4 2
PART OF KIT P/N 5705698.............
UOC:AVY,A11,A13,A14,A15,A20,A24,A25,
A26,A27,B16,B17,B18,HVY,H11,H13,H14,
H15,H16,H17,H18,H20,H21,H24,H25,H26,
H27,H28,MMM

5 PAOZZ 5310005825965 96906 MS35338-44   WASHER,LOCK 1/4 PART OF KIT P/N   2
5705698.....................   UOC:AVY,A11,A13,A14,A15,A20,A24,A25,
A26,A27,B16,B17,B18,HVY,H11,H13,H14,
H15,H16,H17,H18,H20,H21,H24,H25,H26,
H27,H28,MMM

6 PAOZZ 2540011975449 19207 12342082   DIVERTER ASSEMBLY,E PART OF KIT P/N 1
5705698.............................   UOC:AVY,A11,A13,A14,A15,A20,A24,A25,
A26,A27,B16,B17,B18,HVY,H11,H13,H14,
H15,H16,H17,H18,H20,H21,H24,H25,H26,
H27,H28,MMM

7 PFOZZ 2990013232562 19207 12342094   PIPE,EXHAUST PART OF KIT P/N 5705698 1
UOC:AVY,A11,A13,A14,A15,A20,A24,A25,
A26,A27,B16,B17,B18,HVY,H11,H13,H14,
H15,H16,H17,H18,H20,H21,H24,H25,H26,
H27,H28,MMM

8 PAOZZ 5315002857161 80205 MS24665-377   PIN,COTTER PART OF KIT P/N 5705698.. 4
UOC:AVY,A11,A13,A14,A15,A20,A24,A25,
A26,A27,B16,B17,B18,HVY,H11,H13,H14,
H15,H16,H17,H18,H20,H21,H24,H25,H26,
H27,H28,MMM

9 PAOZZ 2990013949670 34623 12340587-3   PIPE,EXHAUST PART OF KIT P/N 5705698 1
UOC:AVY,A11,A13,A14,A15,A20,A24,A25,
A26,A27,B16,B17,B18,HVY,H11,H13,H14,
H15,H16,H17,H18,H20,H21,H24,H25,H26,
H27,H28,MMM

SECTION II TM9-2320-280-24P C01

(1)	(2)	(3)	(4)	(5)	(6)	(7)
ITEM NO	SMR CODE	NSN	CAGEC	PART NUMBER	DESCRIPTION AND USABLE ON CODES (UOC)	QTY
10	PAOZZ	4730009083193	66295	WWD48-58H	CLAMP,HOSE PART OF KIT P/N 5705698.. UOC:AVY,A11,A13,A14,A15,A20,A24,A25, A26,A27,B16,B17,B18,HVY,H11,H13,H14, H15,H16,H17,H18,H20,H21,H24,H25,H26, H27,H28,MMM	2
11	PAOZZ	2990013229880	19207	12342087-2	CONNECTOR,EXHAUST P PART OF KIT P/N 5705698............................ UOC: AVY,A11,A13,A14,A15,A20,A24,A25, A26,A27,B16,B17,B18,HVY,H11,H13,H14, H15,H16,H17,H18,H20,H21,H24,H25,H26, H27,H28,MMM	1
12	PAOZZ	2990013943751	19207	12342089	PIPE,EXHAUST PART OF KIT P/N 5705698 UOC:AVY,A11,A13,A14,A15,A20,A24,A25, A26,A27,B16,B17,B18,HVY,H11,H13,H14, H15,H16,H17,H18,H20,H21,H24,H25,H26, H27,H28,MMM	1
13	PAOZZ	5305000712506	80204	B1821BH025C050N	SCREW,CAP,HEXAGON H 1/4-20 X 1/2 PART OF KIT P/N 5705698............. UOC:AVY,A11,A13,A14,A15,A20,A24,A25, A26,A27,B16,B17,B18,HVY,H11,H13,H14, H15,H16,H17,H18,H20,H21,H24,H25,H26, H27,H28,MMM	1
14	PAOZZ	5320012716357	19207	12339355-1	RIVET,BLIND PART OF KIT P/N 5705698. UOC:AVY,A11,A13,A14,A15,A20,A24,A25, A26,A27,B16,B17,B18,HVY,H11,H13,H14, H15,H16,H17,H18,H20,H21,H24,H25,H26, H27,H28,MMM	4
15	KFFZZ		19207	12340695	TEMPLATE PART OF KIT P/N 5705698.... UOC:AVY,A11,A13,A14,A15,A20,A24,A25, A26,A27,B16,B17,B18,HVY,H11,H13,H14, H15,H16,H17,H18,H20,H21,H24,H25,H26, H27,H28,MMM	1
16	PAOZZ	5310008094058	96906	MS27183-10	WASHER,FLAT 1/4 PART OF KIT P/N 5705698............................ UOC: AVY,A11,A13,A14,A15,A20,A24,A25, A26,A27,B16,B17,B18,HVY,H11,H13,H14, H15,H16,H17,H18,H20,H21,H24,H25,H26, H27,H28,MMM	10
17	PAOZZ	5307011505992	7X677	14066307	STUD,SHOULDERED PART OF KIT P/N 5705698............................ UOC: AVY,A11,A13,A14,A15,A20,A24,A25, A26,A27,B16,B17,B18,HVY,H11,H13,H14, H15,H16,H17,H18,H20,H21,H24,H25,H26, H27,H28,MMM	7
* 18	PAOZZ	5310012065479	7X677	11516075	NUT,SELF-LOCKING,HE M6-1.0MM PART OF KIT P/N 5705698................... UOC:AVY,A11,A13,A14,A15,A20,A24,A25, A26,A27,B16,B17,B18,HVY,H11,H13,H14, H15,H16,H17,H18,H20,H21,H24,H25,H26, H27,H28,MMM	10
19	PAOZZ	5342013262583	19207	12342085	ANCHOR STRAP PART OF KIT P/N 5705698 UOC:AVY,A11,A13,A14,A15,A20,A24,A25, A26,A27,B16,B17,B18,HVY,H11,H13,H14,	1

SECTION II   TM9-2320-280-24P

(1) NO	(2) SMR CODE	(3) NSN	(4) CAGEC	(5) PART NUMBER	(6) DESCRIPTION AND USABLE ON CODES(UOC)	(7) QTY
20	PAOZZ	5305001804966	96906	MS51849-64	SCREW,MACHINE #10-32 X 1/2 PART OF KIT P/N 5705698..................... UOC:AVY,A11,A13,A14,A15,A20,A24,A25, A26,A27,B16,B17,B18,HVY,H11,H13,H14, H15,H16,H17,H18,H20,H21,H24,H25,H26, H27,H28,MMM	1
21	PAOZZ	5340009883186	96906	MS21333-19	CLAMP,LOOP PART OF KIT P/N 5705698.. UOC:AVY,A11,A13,A14,A15,A20,A24,A25, A26,A27,B16,B17,B18,HVY,H11,H13,H14, H15,H16,H17,H18,H20,H21,H24,H25,H26, H27,H28,MMM	1
22	PAOZZ	5310000617326	96906	MS21045-3	NUT,SELF-LOCKING,HE 3/8-32 PART OF KIT P/N 5705698..................... UOC:AVY,A11,A13,A14,A15,A20,A24,A25, A26,A27,B16,B17,B18,HVY,H11,H13,H14, H15,H16,H17,H18,H20,H21,H24,H25,H26, H27,H28,MMM	1
23	PFOOO	2540011975460	19207	12342088	SHRUOD,ARCTIC PART OF KIT P/N 5705698............................ UOC:AVY,A11,A13,A14,A15,A20,A24,A25, A26,A27,B16,B17,B18,HVY,H11,H13,H14, H15,H16,H17,H18,H20,H21,H24,H25,H26, H27,H28,MMM	1
24	PAOZZ	5340013951244	19207	12340474	.BRACKET,ANGLE...................... UOC:AVY,A11,A13,A14,A15,A20,A24,A25, A26,A27,B16,B17,B18,HVY,H11,H13,H14, H15,H16,H17,H18,H20,H21,H24,H25,H26, H27,H28,MMM	1
25	PAOZZ	5310008094058	96906	MS27183-10	.WASHER,FLAT 1/4................... UOC:AVY,A11,A13,A14,A15,A20,A24,A25, A26,A27,B16,B17,B18,HVY,H11,H13,H14, H15,H16,H17,H18,H20,H21,H24,H25,H26, H27,H28,MMM	4
26	PAOZZ	5305000712506	80204	B1821BH025C050N	.SCREW,CAP,HEXAGON H 1/4-20 X 1/2. UOC:AVY,A11,A13,A14,A15,A20,A24,A25, A26,A27,B16,B17,B18,HVY,H11,H13,H14, H15,H16,H17,H18,H20,H21,H24,H25,H26, H27,H28,MMM	4

END OF FIGURE

Section II. TM 9-2320-280-24P-2

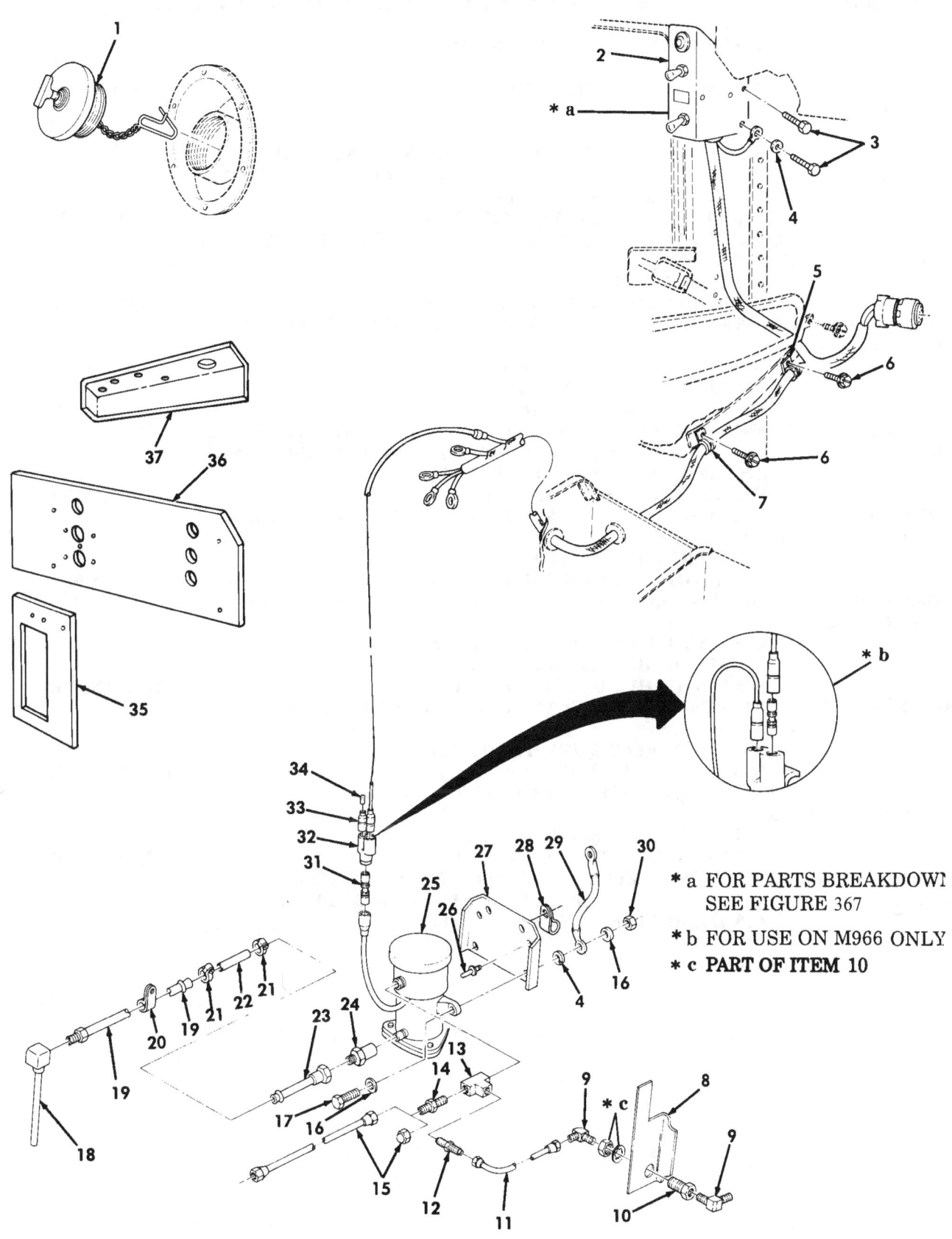

* a FOR PARTS BREAKDOWN SEE FIGURE 367
* b FOR USE ON M966 ONLY
* c PART OF ITEM 10

Figure 366. Arctic Kit, Control Box, Gas Cap, Fuel Pump, and Fuel Lines.

SECTION II   TM9-2320-280-24P

(1) NO	(2) SMR CODE	(3) NSN	(4) CAGEC	(5) PART NUMBER	(7) ITEM DESCRIPTION AND USABLE ON CODES(UOC)	QTY

GROUP 3303 WINTERIZATION KITS

FIG. 366 ARCTIC KIT, CONTROL BOX, GAS CAP, FUEL PUMP, AND FUEL LINES

1	PAOZZ	5342012541498	19207	12338558-2	CAP,FILLER OPENING PART OF KIT P/N 5705698 UOC:AVY,A11,A13,A14,A15,A20,A24,A25, A26,A27,B16,B17,B18,HVY,H11,H13,H14, H15,H16,H17,H18,H20,H21,H24,H25,H26, H27,H28,MMM	1
2	PFOOO	2540011992396	19207	12340436	CONTROL BOX PART OF KIT P/N 5705698. UOC:AVY,A11,A13,A14,A15,A20,A24,A25, A26,A27,B16,B17,B18,HVY,H11,H13,H14, H15,H16,H17,H18,H20,H21,H24,H25,H26, H27,H28,MMM	1
3	PAOZZ	5305010846067	96906	MS51871-1	SCREW,TAPPING 1/4-20 X 1/2 PART OF KIT P/N 5705698 UOC:AVY,A11,A13,A14,A15,A20,A24,A25, A26,A27,B16,B17,B18,HVY,H11,H13,H14, H15,H16,H17,H18,H20,H21,H24,H25,H26, H27,H28,MMM	2
4	PAOZZ	5310005501130	96906	MS35333-40	WASHER,LOCK 1/4 PART OF KIT P/N 5705698 UOC:AVY,A11,A13,A14,A15,A20,A24,A25, A26,A27,B16,B17,B18,HVY,H11,H13,H14, H15,H16,H17,H18,H20,H21,H24,H25,H26, H27,H28,MMM	2
5	PAOZZ	5340000913790	96906	MS21333-72	CLAMP,LOOP PART OF KIT P/N 5705698.. UOC:AVY,A11,A13,A14,A15,A20,A24,A25, A26,A27,B16,B17,B18,HVY,H11,H13,H14, H15,H16,H17,H18,H20,H21,H24,H25,H26, H27,H28,MMM	1
6	PAOZZ	5305014339248	24617	9414295	SCREW,MACHINE #10-32 X 1/2 PART OF KIT P/N 5705698 UOC:AVY,A11,A13,A14,A15,A20,A24,A25, A26,A27,B16,B17,B18,HVY,H11,H13,H14, H15,H16,H17,H18,H20,H21,H24,H25,H26, H27,H28,MMM	2
7	PAOZZ	5340000572904	96906	MS21333-71	CLAMP,LOOP PART OF KIT P/N 5705698.. UOC:AVY,A11,A13,A14,A15,A20,A24,A25, A26,A27,B16,B17,B18,HVY,H11,H13,H14, H15,H16,H17,H18,H20,H21,H24,H25,H26, H27,H28,MMM	1
8	KFFZZ		19207	12340702	TEMPLATE PART OF KIT P/N 5705698.... UOC:AVY,A11,A13,A14,A15,A20,A24,A25, A26,A27,B16,B17,B18,HVY,H11,H13,H14, H15,H16,H17,H18,H20,H21,H24,H25,H26, H27,H28,MMM	1
9	PAOZZ	4730014565446	24617	15670396	ELBOW,PIPE TO TUBE PART OF KIT P/N 5705698 UOC:AVY,A11,A13,A14,A15,A20,A24,A25, A26,A27,B16,B17,B18,HVY,H11,H13,H14,	2

SECTION II   TM9-2320-280-24P

(1) NO	(2) SMR CODE	(3) NSN	(4) CAGEC	(5) PART NUMBER	(7) DESCRIPTION AND USABLE ON CODES(UOC)	ITEM QTY
10	PAOZZ	4730008975497	79470	W21204	COUPLING,PIPE PART OF KIT P/N 5698............. UOC:AVY,A11,A13,A14,A15,A20,A24,A25, A26,A27,B16,B17,B18,HVY,H11,H13,H14, H15,H16,H17,H18,H20,H21,H24,H25,H26, H27,H28,MMM	1 57
11	PAOZZ	4710013953982	19207	12342605	TUBE ASSEMBLY,METAL ARCTIC FUEL INE, PART OF KIT P/N 5705698...... UOC:AVY,A11,A13,A14,A15,A20,A24,A25, A26,A27,B16,B17,B18,HVY,H11,H13,H14, H15,H16,H17,H18,H20,H21,H24,H25,H26, H27,H28,MMM	1
12	PAOZZ	4730002660536	24617	110200	ADAPTER,STRAIGHT,PI ARCTIC FUEL UMP PART OF KIT P/N 5705698........ UOC:AVY,A11,A13,A14,A15,A20,A24,A25, A26,A27,B16,B17,B18,HVY,H11,H13,H14, H15,H16,H17,H18,H20,H21,H24,H25,H26, H27,H28,M-TM	1
13	PAOZZ	4730004713102	81343	4-4-4 130438B	TEE,PIPE PART OF KIT P/N 5705698.... UOC:AVY,A11,A13,A14,A15,A20,A24,A25, A26,A27,B16,B17,B18,HVY,H11,H13,H14, H15,H16,H17,H18,H20,H21,H24,H25,H26, H27,H28,MMM	1
14	PAOZZ	4730002660538	06178	1295	ADAPTER,STRAIGHT,PI PART OF KIT P/N 5705698.......................... UOC:A15,B16,H15,H16	1
14	PAOZZ	4730002660536	24617	110200	ADAPTER,STRAIGHT,PI PART OF KIT P/N 5705698........ UOC:AVY,A11,A13,A14,A20,A24,A25,A26, A27,B17,B18,HVY,H11,H13,H14,H17, H18, H20,H21,H24,H25,H26,H27,H28,MMM	1
15	PAOZZ	4710013946169	19207	12342606	TUBE ASSEMBLY,METAL PART OF KIT P/N 5705698........................... UOC:A15,B16,H15,H16	1
15	PAOZZ	4730002608285	81343	5 010112B(N5)	CAP,TUBE PART OF KIT P/N 5705698.... UOC:AVY,A11,A13,A14,A20,A24,A25,A26, A27,B17,B18,HVY,H11,H13,H14,H17,H18, H20,H21,H24,H25,H26,H27,H28,MMM	1
16	PAFZZ	5310008094058	96906	MS27183-10	WASHER,FLAT 1/4.................... UOC:AVY,A11,A13,A14,A15,A20,A24,A25, A26,A27,B16,B17,B18,HVY,H11,H13,H14, H15,H16,H17,H18,H20,H21,H24,H25,H26, H27,H28,MMM	4
17	PAOZZ	5305002253842	80204	B1821BH025C100N	SCREW,CAP,HEXAGON H 1/4-20 X 1.00 PART OF KIT P/N 5705698............. UOC:AVY,A11,A13,A14,A15,A20,A24,A25, A26,A27,B16,B17,B18,HVY,H11,H13,H14, H15,H16,H17,H18,H20,H21,H24,H25,H26, H27,H28,MMM	2
18	PAOZZ	4710013592956	19207	12338548-2	TUBE ASSEMBLY,METAL PART OF KIT P/N 5705698............................	1

SECTION II   TM9-2320-280-24P C01
  (1)   (2)   (3)   (4)   (5)   (6)
(7)   ITEM SMR   PART
  NO   CODE NSN   CAGEC   NUMBER   DESCRIPTION AND USABLE ON CODES(UOC)
QTY

                                                UOC:AVY,A11,A13,A14,A15,A20,A24,A25,
                                                A26,A27,B16,B17,B18,HVY,H11,H13,H14,
                                                H15,H16,H17,H18,H20,H21,H24,H25,H26,
H27,H28,MMM   19 PAOZZ 4710012030607 34623 5588694   TUBE ASSEMBLY,METAL PART OF
KIT P/N   1   5705698....................
....   UOC:AVY,A11,A13,A14,A15,A20,A2
4,A25,
                                                A26,A27,B16,B17,B18,HVY,H11,H13,H14,
                                                H15,H16,H17,H18,H20,H21,H24,H25,H26,
H27,H28,MMM
    20 PAOZZ 5340009936207 96906 MS21333-99   CLAMP,LOOP PART OF KIT P/N 5705698..   1
UOC:AVY,A11,A13,A14,A15,A20,A24,A25,

                                                A26,A27,B16,B17,B18,HVY,H11,H13,H14,
                                                H15,H16,H17,H18,H20,H21,H24,H25,H26,
H27,H28,MMM
    21 PAOZZ 4730000243971 7Z588 6706   CLAMP,HOSE PART OF KIT P/N 5705698..   2
UOC:AVY,A11,A13,A14,A15,A20,A24,A25,

                                                A26,A27,B16,B17,B18,HVY,H11,H13,H14,
                                                H15,H16,H17,H18,H20,H21,H24,H25,H26,
H27,H28,MMM
  * 22 MOOZZ   19207 12338553-1   HOSE,NONMETALLIC MAKE FROM HOSE,P/N   1
9438315,2.80 INCHES LONG PART OF KIT   P/N
5705698........................

                                                UOC:AVY,A11,A13,A14,A15,A20,A24,A25,
                                                A26,A27,B16,B17,B18,HVY,H11,H13,H14,
                                                H15,H16,H17,H18,H20,H21,H24,H25,H26,
H27,H28,MMM
    23 PAOZZ 4710012030608 19207 12338564   TUBE ASSEMBLY,METAL PART OF KIT P/N   1
5705698............................   UOC
:AVY,A11,A13,A14,A15,A20,A24,A25,

                                                A26,A27,B16,B17,B18,HVY,H11,H13,H14,
                                                H15,H16,H17,H18,H20,H21,H24,H25,H26,
H27,H28,MMM
    24 PAOZZ 4730002776347 20969 246X5X4   ADAPTER,STRAIGHT,PI PART OF KIT P/N   1
5705698............................   UOC
:AVY,A11,A13,A14,A15,A20,A24,A25,

                                                A26,A27,B16,B17,B18,HVY,H11,H13,H14,
                                                H15,H16,H17,H18,H20,H21,H24,H25,H26,
H27,H28,MMM
    25 PAOZZ 2910009309367 53711 2590174   PUMP,FUEL,ELECTRICA PART OF KIT P/N   1
5705698............................   UOC
:AVY,A11,A13,A14,A15,A20,A24,A25,

                                                A26,A27,B16,B17,B18,HVY,H11,H13,H14,
                                                H15,H16,H17,H18,H20,H21,H24,H25,H26,
H27,H28,MMM
    26 PAFZZ 5320011357319 80205 NAS9301BNS-6-03 RIVET,BLIND PART OF KIT P/N 5705698. 4
UOC:AVY,A11,A13,A14,A15,A20,A24,A25,

                                                A26,A27,B16,B17,B18,HVY,H11,H13,H14,
                                                H15,H16,H17,H18,H20,H21,H24,H25,H26,
H27,H28,MMM
    27 PFFZZ 5340012043903 19207 12338559   BRACKET,DOUBLE ANGL PART OF KIT P/N   1
5705698............................   UOC
:AVY,A11,A13,A14,A15,A20,A24,A25,

SECTION II   TM9-2320-280-24P

(1) NO	(2) SMR CODE	(3) NSN	(4) CAGEC	(5) PART NUMBER	(6) DESCRIPTION AND USABLE ON CODES(UOC)	(7) QTY
28	PAOZZ	5340002827537	96906	MS21333-41	CLAMP,LOOP PART OF KIT P/N 5705698.. UOC:AVY,A11,A13,A14,A15,A20,A24,A25, A26,A27,B16,B17,B18,HVY,H11,H13,H14, H15,H16,H17,H18,H20,H21,H24,H25,H26, H27,H28,MMM	1
29	PFOZZ	6150012608000	19207	12340489	LEAD,ELECTRICAL PART OF KIT P/N 5705698............. UOC:AVY,A11,A13,A14,A15,A20,A24,A25, A26,A27,B16,B17,B18,HVY,H11,H13,H14, H15,H16,H17,H18,H20,H21,H24,H25,H26, H27,H28,MMM	1
30	PAOZZ	5310000614650	96906	MS51943-31	NUT,SELF-LOCKING,HE 1/4-20......... UOC:AVY,A11,A13,A14,A15,A20,A24,A25, A26,A27,B16,B17,B18,HVY,H11,H13,H14, H15,H16,H17,H18,H20,H21,H24,H25,H26, H27,H28,MMM	2
31	PAOZZ	5935008074109	19207	8741492	ADAPTER,CONNECTOR PART OF KIT P/N 5698............. UOC:AVY,A11,A13,A14,A15,A20,A24,A25, A26,A27,B16,B17,B18,HVY,H11,H13,H14, H15,H16,H17,H18,H20,H21,H24,H25,H26, H27,H28,MMM	1
32	PAOZZ	5935009006281	96906	MS27147-1	ADAPTER,CONNECTOR PART OF KIT P/N 705698............. UOC:AVY,A11,A13,A14,A15,A20,A24,A25, A26,A27,B16,B17,B18,HVY,H11,H13,H14, H15,H16,H17,H18,H20,H21,H24,H25,H26, H27,H28,MMM	1
33	PAOZZ	5935008338561	19207	8338561	SHELL,ELECTRICAL CO PART OF KIT P/N 705698............. UOC:AVY,A11,A13,A14,A15,A20,A24,A25, A26,A27,B16,B17,B18,HVY,H11,H13,H14, H15,H16,H17,H18,H20,H21,H24,H25,H26, H27,H28,MMM	1
34	PAOZZ	5935002140904	19207	7982907	DUMMY CONNECTOR,PLU PART OF KIT P/N 705698............. UOC:AVY,A11,A13,A14,A15,A20,A24,A25, A26,A27,B16,B17,B18,HVY,H11,H13,H14, H15,H16,H17,H18,H20,H21,H24,H25,H26, H27,H28,MMM	1
35	KFFZZ		19207	12340703	TEMPLATE PART OF KIT P/N 5705698.... UOC:AVY,A11,A13,A14,A15,A20,A24,A25, A26,A27,B16,B17,B18,HVY,H11,H13,H14, H15,H16,H17,H18,H20,H21,H24,H25,H26, H27,H28,MMM	1
36	KFFZZ		19207	12340704	TEMPLATE PART OF KIT P/N 5705698.... UOC:AVY,A11,A13,A14,A15,A20,A24,A25, A26,A27,B16,B17,B18,HVY,H11,H13,H14, H15,H16,H17,H18,H20,H21,H24,H25,H26, H27,H28,MMM	1

SECTION II   TM9-2320-280-24P

(1) NO	(2) CODE	(3) NSN	(4) CAGEC	(5) NUMBER	(6) DESCRIPTION AND USABLE ON CODES(UOC)	(7) ITEM QTY	SMR	PART
37	KFFZZ		34623	12340696	TEMPLATE,FUEL PUMP PART OF KIT P/N 698............................ UOC:AVY,A11,A13,A14,A15,A20,A24,A25, A26,A27,B16,B17,B18,HVY,H11,H13,H14, H15,H16,H17,H18,H20,H21,H24,H25,H26, H27,H28,MMM	1		5705

END OF FIGURE

Section II. TM 9-2320-280-24P-2

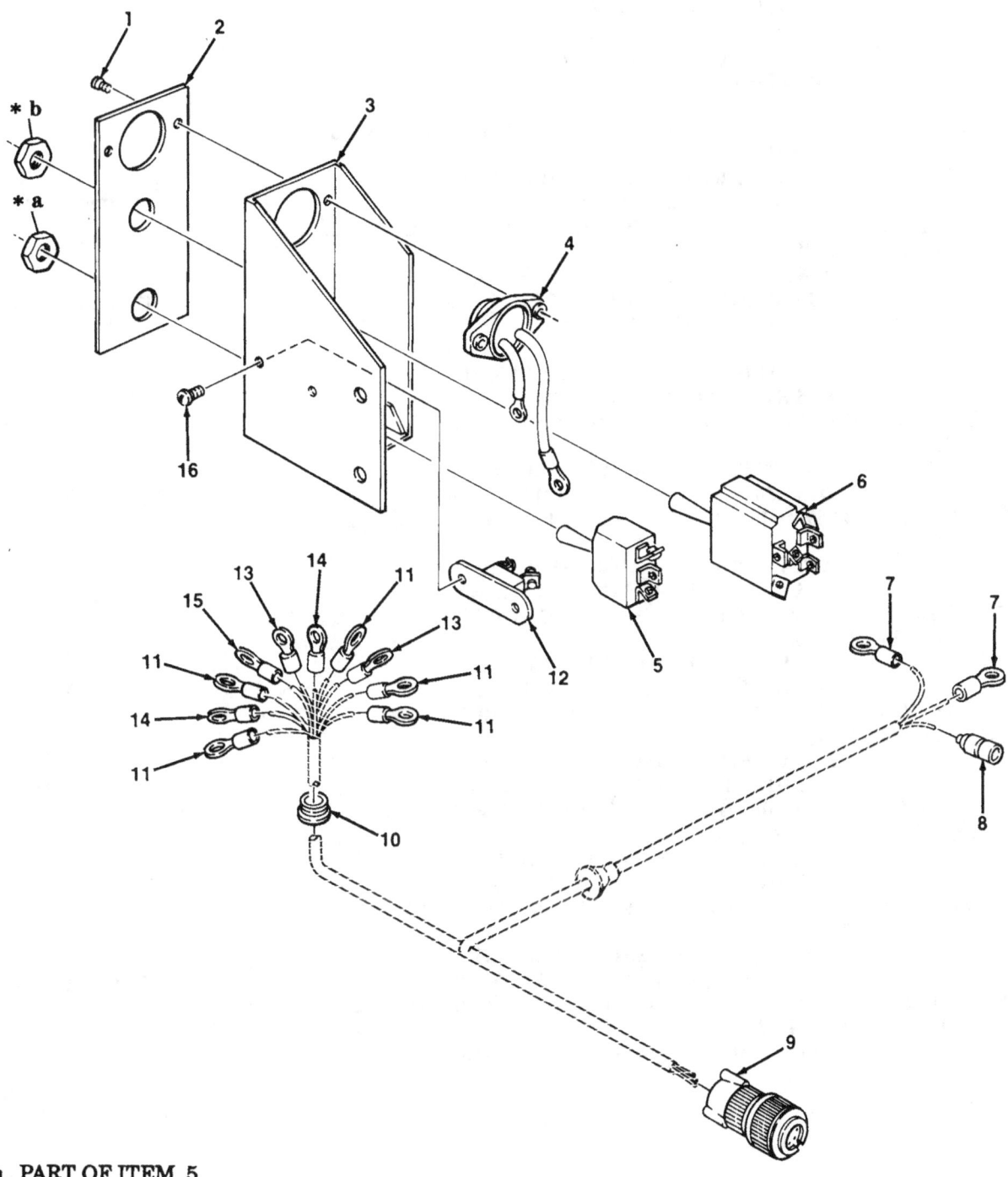

* a PART OF ITEM 5
* b PART OF ITEM 6

Figure 367. Arctic Kit, Control Box Assembly.

SECTION II TM9-2320-280-24P

(1) NO	(2) SMR CODE	(3) NSN	(4) CAGEC	(5) PART NUMBER	(6) DESCRIPTION AND USABLE ON CODES(UOC)	(7) QTY

GROUP 3303 WINTERIZATION KITS

FIG. 367 ARCTIC KIT, CONTROL BOX ASSEMBLY

| 1 | PAOZZ | 5305011389540 | 96906 | MS51863-33 | SCREW,TAPPING #8-32 X 3/8 | 2 |

UOC: AVY,A11,A13,A14,A15,A20,A24,A25, A26,A27,B16,B17,B18,HVY,H11,H13,H14, H15,H16,H17,H18,H20,H21,H24,H25,H26, H27,H28,MMM

| 2 | PFOZZ | 9905011992371 | 19207 | 12340442 | PLATE,IDENTIFICATIO | 1 |

UOC: AVY,A11,A13,A14,A15,A20,A24,A25, A26,A27,B16,B17,B18,HVY,H11,H13,H14, H15,H16,H17,H18,H20,H21,H24,H25,H26, H27,H28,MMM

| 3 | XAOZZ | | 34623 | 5590544 | CONTROL BOX | 1 |

UOC: AVY,A11,A13,A14,A15,A20,A24,A25, A26,A27,B16,B17,B18,HVY,H11,H13,H14, H15,H16,H17,H18,H20,H21,H24,H25,H26, H27,H28,MMM

| 4 | PAOZZ | 6210012032101 | 19207 | 12340466 | LIGHT,INDICATOR | 1 |

UOC: AVY,A11,A13,A14,A15,A20,A24,A25, A26,A27,B16,B17,B18,HVY,H11,H13,H14, H15,H16,H17,H18,H20,H21,H24,H25,H26, H27,H28,MMM

| 5 | PAOZZ | 5930006551514 | 96906 | MS35058-22 | SWITCH,TOGGLE | 1 |

UOC: AVY,A11,A13,A14,A15,A20,A24,A25, A26,A27,B16,B17,B18,HVY,H11,H13,H14, H15,H16,H17,H18,H20,H21,H24,H25,H26, H27,H28,MMM

| 6 | PAOZZ | 5930006157897 | 96906 | MS35059-31 | SWITCH,TOGGLE | 1 |

UOC: AVY,A11,A13,A14,A15,A20,A24,A25, A26,A27,B16,B17,B18,HVY,H11,H13,H14, H15,H16,H17,H18,H20,H21,H24,H25,H26, H27,H28,MMM

| 7 | PAOZZ | 5940001139821 | 96906 | MS20659-166 | TERMINAL,LUG | 2 |

UOC: AVY,A11,A13,A14,A15,A20,A24,A25, A26,A27,B16,B17,B18,HVY,H11,H13,H14, H15,H16,H17,H18,H20,H21,H24,H25,H26, H27,H28,MMM

| 8 | PAOZA | 5935001677775 | 96906 | MS27144-1 | CONNECTOR,PLUG,ELEC | 1 |

UOC: AVY,A11,A13,A14,A15,A20,A24,A25, A26,A27,B16,B17,B18,HVY,H11,H13,H14, H15,H16,H17,H18,H20,H21,H24,H25,H26, H27,H28,MMM

| 9 | PAOZZ | 5935010355139 | 96906 | MS3456W18-11S | CONNECTOR,PLUG,ELEC | 1 |

UOC: AVY,A11,A13,A14,A15,A20,A24,A25, A26,A27,B16,B17,B18,HVY,H11,H13,H14, H15,H16,H17,H18,H20,H21,H24,H25,H26, H27,H28,MMM

| 10 | PAOZZ | 5325002811557 | 79497 | G1895 | GROMMET,NONMETALLIC | 1 |

UOC: AVY,A11,A13,A14,A15,A20,A24,A25, A26,A27,B16,B17,B18,HVY,H11,H13,H14,

SECTION II TM9-2320-280-24P

(1) NO	(2) CODE	(3) NSN	(4) CAGEC	(5) PART NUMBER	(6) DESCRIPTION AND USABLE ON CODES(UOC)	(7) ITEM SMR QTY

11 PAOZZ 5940002835280 96906 MS25036-106 TERMINAL,LUG................ 5
UOC:AV,A11,A13,A14,A15,A20,A24,A25,
A26,A27,B16,B17,B18,HVY,H11,H13,H14,
H15,H16,H17,H18,H20,H21,H24,H25,H26,
H27,H28,MMM

12 PAOZZ 5925013731034 82647 PSA-25 CIRCUIT BREAKER................ 1
UOC:AV,A11,A13,A14,A15,A20,A24,A25,
A26,A27,B16,B17,B18,HVY,H11,H13,H14,
H15,H16,H17,H18,H20,H21,H24,H25,H26,
H27,H28,MMM

13 PAOZZ 5940002048990 96906 MS25036-111 TERMINAL,LUG................ 2
UOC:AV,A11,A13,A14,A15,A20,A24,A25,
A26,A27,B16,B17,B18,HVY,H11,H13,H14,
H15,H16,H17,H18,H20,H21,H24,H25,H26,
H27,H28,MMM

14 PAOZZ 5940001434775 96906 MS25036-156 TERMINAL,LUG................ 2
UOC:AV,A11,A13,A14,A15,A20,A24,A25,
A26,A27,B16,B17,B18,HVY,H11,H13,H14,
H15,H16,H17,H18,H20,H21,H24,H25,H26,
H27,H28,MMM

15 PAOZZ 5940001434774 96906 MS25036-153 TERMINAL,LUG................ 1
UOC:AV,A11,A13,A14,A15,A20,A24,A25,
A26,A27,B16,B17,B18,HVY,H11,H13,H14,
H15,H16,H17,H18,H20,H21,H24,H25,H26,
H27,H28,MMM

16 PAOZZ 5305009844988 96906 MS35206-228 SCREW,MACHINE #6-32 X 3/8.......... 2
UOC:AVY,A11,A13,A14,A15,A20,A24,A25,
A26,A27,B16,B17,B18,HVY,H11,H13,H14,
H15,H16,H17,H18,H20,H21,H24,H25,H26,
H27,H28,MMM

END OF FIGURE

Section II. TM 9-2320-280-24P-2 C02

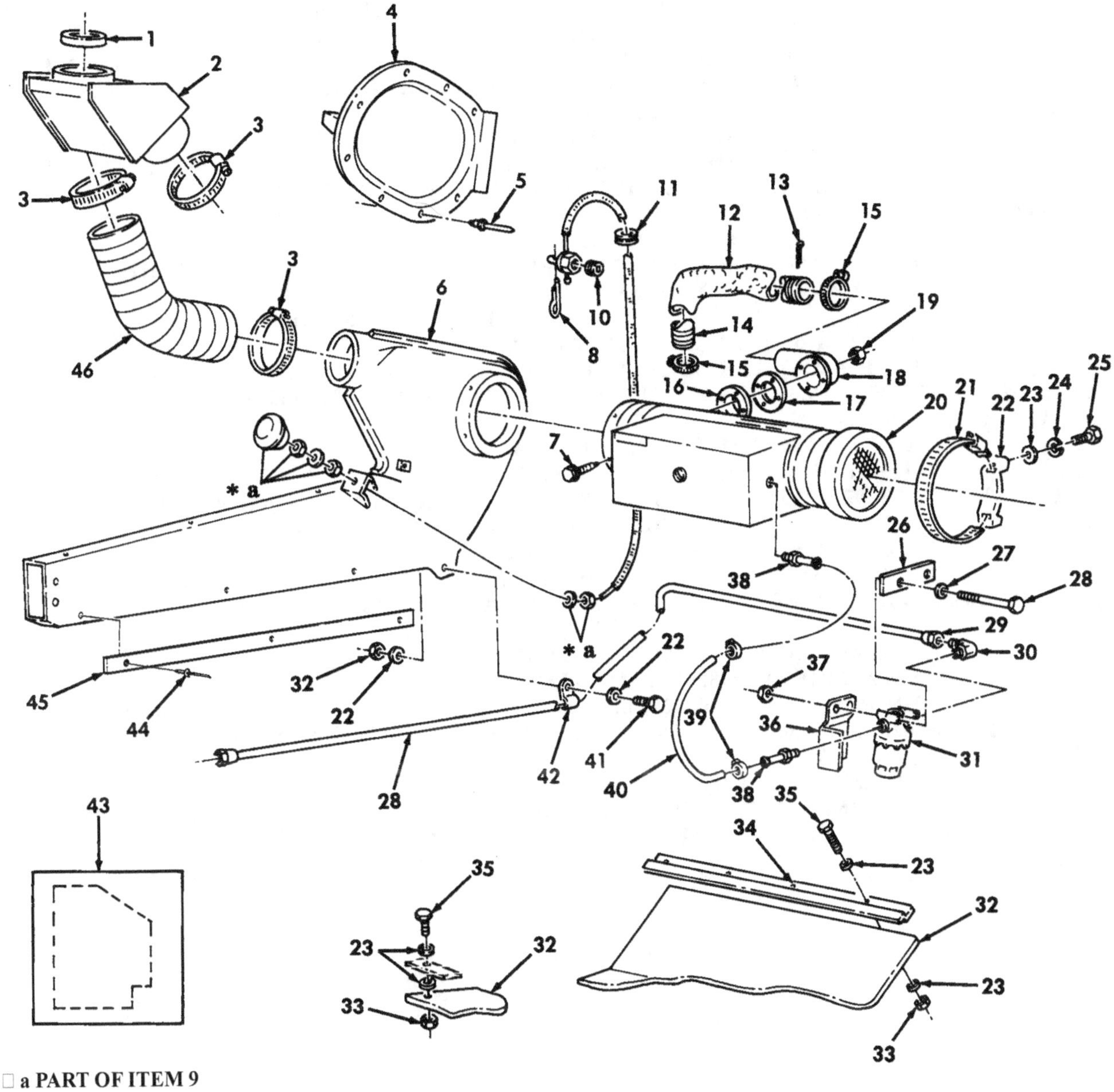

□ a PART OF ITEM 9

Figure 368. Arctic Kit, Heater Assembly, Plenum Assembly, and Related Components.

SECTION II   TM9-2320-280-24P

(1) NO	(2) SMR CODE	(3) NSN	(4) CAGEC	(5) PART NUMBER	(7) ITEM DESCRIPTION AND USABLE ON CODES(UOC)	QTY

GROUP 3303 WINTERIZATION KITS

FIG. 368 ARCTIC KIT, HEATER ASSEMBLY, PLENUM ASSEMBLY, AND RELATED COMPONENTS

1 PAOZZ 5330012039187 34623 5582936 SEAL,NONMETALLIC PART OF KIT P/N 5705698.................... 1
UOC:AVY,A11,A13,A14,A15,A20,A24,A25,A26,A27,B16,B17,B18,HVY,H11,H13,H14,H15,H16,H17,H18,H20,H21,H24,H25,H26,H27,H28,MMM

2 PFOZZ 2540012936926 19207 12340432 REGULATOR,AIR DUCT PART OF KIT P/N 5705698.................... 1
UOC:AVY,A11,A13,A14,A15,A20,A24,A25,A26,A27,B16,B17,B18,HVY,H11,H13,H14,H15,H16,H17,H18,H20,H21,H24,H25,H26,H27,H28,MMM

3 PAOZZ 4730009086292 96906 MS35842-14 CLAMP,HOSE PART OF KIT P/N 5705698.. 3
UOC:AVY,A11,A13,A14,A15,A20,A24,A25,A26,A27,B16,B17,B18,HVY,H11,H13,H14,H15,H16,H17,H18,H20,H21,H24,H25,H26,H27,H28,MMM

4 PFOZZ 2540011975450 19207 12340434 COVER ASSEMBLY,COWL PART OF KIT P/N 5705698........................ 1
UOC:AVY,A11,A13,A14,A15,A20,A24,A25,A26,A27,B16,B17,B18,HVY,H11,H13,H14,H15,H16,H17,H18,H20,H21,H24,H25,H26,H27,H28,MMM

5 PFOZZ 5320011357319 80205 NAS9301BNS-6-03 RIVET,BLIND PART OF KIT P/N 5705698. 8
UOC:AVY,A11,A13,A14,A15,A20,A24,A25,A26,A27,B16,B17,B18,HVY,H11,H13,H14,H15,H16,H17,H18,H20,H21,H24,H25,H26,H27,H28,MMM

6 PFOZZ 4140012592175 19207 12340477 HOUSING,CENTRIFUGAL PART OF KIT P/N 5705698........................ 1
UOC:AVY,A11,A13,A14,A15,A20,A24,A25,A26,A27,B16,B17,B18,HVY,H11,H13,H14,H15,H16,H17,H18,H20,H21,H24,H25,H26,H27,H28,MMM

7 PAOZZ 5305012596322 34623 12342499-1 SCREW,MACHINE #10-32 X 1/2 PART OF KIT P/N 5705698.................... 1
UOC:AVY,A11,A13,A14,A15,A20,A24,A25,A26,A27,B16,B17,B18,HVY,H11,H13,H14,H15,H16,H17,H18,H20,H21,H24,H25,H26,H27,H28,MMM

8 PAOZZ 5315008291480 80205 MS24665-208 PIN,COTTER PART OF KIT P/N 5705698.. 1
UOC:AVY,A11,A13,A14,A15,A20,A24,A25,A26,A27,B16,B17,B18,HVY,H11,H13,H14,H15,H16,H17,H18,H20,H21,H24,H25,H26,H27,H28,MMM

9 PAOZZ 5305007246783 96906 MS51965-29 SETSCREW #8-32 X 1/4 PART OF KIT P/N 5705698...................... 1

SECTION II   TM9-2320-280-24P

(1) ITEM NO	(2) SMR CODE	(3) NSN	(4) CAGEC	(5) PART NUMBER	(6) DESCRIPTION AND USABLE ON CODES(UOC)	(7) QTY
10	PAOZZ	5325001849846	81349	C3030	GROMMET,NONMETALLIC PART OF KIT P/N 05698............................ UOC:AVY,A11,A13,A14,A15,A20,A24,A25, A26,A27,B16,B17,B18,HVY,H11,H13,H14, H15,H16,H17,H18,H20,H21,H24,H25,H26, H27,H28,MMM	1
11	PAOZZ	2540013946167	19207	12342087-1	SHIELD,HEATER DUCT PART OF KIT P/N 05698............................ UOC:A11,A13,A14,A20,A24,A25,A26,A27, B17,B18,B20,B24,B25,C17,H11,H13,H14, H15,H16,H17,H18,H20,H21,H24,H25,H26, H27,H28,MMM	1
12	PAOZZ	5315002857161	80205	MS24665-377	PIN,COTTER PART OF KIT P/N 5705698.. UOC:AVY,A11,A13,A14,A15,A20,A24,A25, A26,A27,B16,B17,B18,HVY,H11,H13,H14, H15,H16,H17,H18,H20,H21,H24,H25,H26, H27,H28,MMM	3
13	PAOZZ	2990013229881	19207	12342091	CONNECTOR,EXHAUST P PART OF KIT P/N 05698............................ UOC:AVY,A11,A13,A14,A15,A20,A24,A25, A26,A27,B16,B17,B18,HVY,H11,H13,H14, H15,H16,H17,H18,H20,H21,H24,H25,H26, H27,H28,MMM	1
14	PAOZZ	4730009083193	66295	WWD48-58H	CLAMP,HOSE PART OF KIT P/N 5705698.. UOC:AVY,A11,A13,A14,A15,A20,A24,A25, A26,A27,B16,B17,B18,HVY,H11,H13,H14, H15,H16,H17,H18,H20,H21,H24,H25,H26, H27,H28,MMM	2
15	PAOZZ	4730013229871	19207	12342083	FLANGE,PIPE PART OF KIT P/N 5705698. UOC:AVY,A11,A13,A14,A15,A20,A24,A25, A26,A27,B16,B17,B18,HVY,H11,H13,H14, H15,H16,H17,H18,H20,H21,H24,H25,H26, H27,H28,MMM	1
16	PAOZZ	2540011231218	19207	12275161	THERMASEAL,HEATER PART OF KIT P/N 05698............................ UOC:AVY,A11,A13,A14,A15,A20,A24,A25, A26,A27,B16,B17,B18,HVY,H11,H13,H14, H15,H16,H17,H18,H20,H21,H24,H25,H26, H27,H28,MMM	1
17	PAOZZ	2990013229879	19207	12342093	CONNECTOR,EXHAUST P PART OF KIT P/N 05698............................ UOC:AVY,A11,A13,A14,A15,A20,A24,A25, A26,A27,B16,B17,B18,HVY,H11,H13,H14, H15,H16,H17,H18,H20,H21,H24,H25,H26, H27,H28,MMM	1
18	PAOZZ	5310009359022	96906	MS51943-32	NUT,SELF-LOCKING,HE 1/4-20 PART OF KIT P/N 5705698..................... UOC:AVY,A11,A13,A14,A15,A20,A24,A25, A26,A27,B16,B17,B18,HVY,H11,H13,H14,	4

SECTION II   TM9-2320-280-24P

(1) NO	(2) SMR CODE	(3) NSN	(4) CAGEC	(5) PART NUMBER	(6) DESCRIPTION AND USABLE ON CODES(UOC)	(7) QTY
19	PAOFF	2540011943323	81349	MIL-PRF-62550/3	HEATER,VEHICULAR,CO................. UOC:AVY,A11,A13,A14,A15,A20,A24,A25, A26,A27,B16,B17,B18,HVY,H11,H13,H14, H15,H16,H17,H18,H20,H21,H24,H25,H26, H27,H28,MMM	1
20	PAOZZ	4730009086294	17576	A-A-52506	CLAMP,HOSE PART OF KIT P/N 5705698 UOC:AVY,A11,A13,A14,A15,A20,A24,A25, A26,A27,B16,B17,B18,HVY,H11,H13,H14, H15,H16,H17,H18,H20,H21,H24,H25,H26, H27,H28,MMM	2
21	PAOZZ	5340012043904	19207	12340448	BRACKET,MOUNTING PART OF KIT P/N 5705698.......................... UOC:AVY,A11,A13,A14,A15,A20,A24,A25, A26,A27,B16,B17,B18,HVY,H11,H13,H14, H15,H16,H17,H18,H20,H21,H24,H25,H26, H27,H28,MMM	2
22	PAOZZ	5310008094058	96906	MS27183-10	WASHER,FLAT 1/4 PART OF KIT P/N 5705698.......................... UOC:AVY,A11,A13,A14,A15,A20,A24,A25, A26,A27,B16,B17,B18,HVY,H11,H13,H14, H15,H16,H17,H18,H20,H21,H24,H25,H26, H27,H28,MMM	22
23	PAOZZ	5310005825965	96906	MS35338-44	WASHER,LOCK 1/4 PART OF KIT P/N 5705698.......................... UOC:AVY,A11,A13,A14,A15,A20,A24,A25, A26,A27,B16,B17,B18,HVY,H11,H13,H14, H15,H16,H17,H18,H20,H21,H24,H25,H26, H27,H28,MMM	4
24	PAOZZ	5305000712506	80204	B1821BH025C050N	SCREW,CAP,HEXAGON H 1/4-20 X 1/2 PART OF KIT P/N 5705698............. UOC:AVY,A11,A13,A14,A15,A20,A24,A25, A26,A27,B16,B17,B18,HVY,H11,H13,H14, H15,H16,H17,H18,H20,H21,H24,H25,H26, H27,H28,MMM	4
25	PAOZZ	5365013261153	19207	12342092	SPACER,PLATE PART OF KIT P/N 5705698 UOC:AVY,A11,A13,A14,A15,A20,A24,A25, A26,A27,B16,B17,B18,HVY,H11,H13,H14, H15,H16,H17,H18,H20,H21,H24,H25,H26, H27,H28,MMM	1
26	PAOZZ	5310000145850	96906	MS27183-42	WASHER,FLAT 7/32 PART OF KIT P/N 5705698.......................... UOC:AVY,A11,A13,A14,A15,A20,A24,A25, A26,A27,B16,B17,B18,HVY,H11,H13,H14, H15,H16,H17,H18,H20,H21,H24,H25,H26, H27,H28,MMM	2
27	PAOZZ	5305009956311	96906	MS35207-271	SCREW,MACHINE #10-24 X 1.75 PART OF KIT P/N 5705698.................... UOC:AVY,A11,A13,A14,A15,A20,A24,A25, A26,A27,B16,B17,B18,HVY,H11,H13,H14, H15,H16,H17,H18,H20,H21,H24,H25,H26, H27,H28,MMM	2

SECTION II   TM9-2320-280-24P

(1) NO	(2) SMR CODE	(3) NSN	(4) CAGEC	(5) PART NUMBER	(6) DESCRIPTION AND USABLE ON CODES(UOC)	(7) QTY
28	PAOZZ	4710012031304	34623	5583352	TUBE ASSEMBLY,METAL PART OF KIT P/N 705698............................. UOC:AVY,A11,A13,A14,A15,A20,A24,A25, A26,A27,B16,B17,B18,HVY,H11,H13,H14, H15,H16,H17,H18,H20,H21,H24,H25,H26, H27,H28,MMM	1
29	PAOZZ	4730012045457	81343	5-2 010203CA	ELBOW,PIPE TO TUBE PART OF KIT P/N 705698............................. UOC:AVY,A11,A13,A14,A15,A20,A24,A25, A26,A27,B16,B17,B18,HVY,H11,H13,H14, H15,H16,H17,H18,H20,H21,H24,H25,H26, H27,H28,MMM	1
30	PAOZZ	2910014458097	58536	A52972-1	FILTER,FLUID PART OF KIT P/N 5705698 UOC:AVY,A11,A13,A14,A15,A20,A24,A25, A26,A27,B16,B17,B18,HVY,H11,H13,H14, H15,H16,H17,H18,H20,H21,H24,H25,H26, H27,H28,MMM	1
31	PAOZZ	2510013761092	34623	12339909	INSULATION PART OF KIT P/N 5705698.. UOC:AVY,A11,A13,A14,A15,A20,A24,A25, A26,A27,B16,B17,B18,HVY,H11,H13,H14, H15,H16,H17,H18,H20,H21,H24,H25,H26, H27,H28,MMM	1
32	PAOZZ	5310000614650	96906	MS51943-31	NUT,SELF-LOCKING,HE 1/4-20 PART OF KIT P/N 5705698..................... UOC:AVY,A11,A13,A14,A15,A20,A24,A25, A26,A27,B16,B17,B18,HVY,H11,H13,H14, H15,H16,H17,H18,H20,H21,H24,H25,H26, H27,H28,MMM	9
33	PAOZZ	2590012633254	19207	12338838-2	BEZEL,AUTOMOTIVE TR PART OF KIT P/N 705698............................. UOC:AVY,A11,A13,A14,A15,A20,A24,A25, A26,A27,B16,B17,B18,HVY,H11,H13,H14, H15,H16,H17,H18,H20,H21,H24,H25,H26, H27,H28,MMM	1
34	PAOZZ	5305002253843	80204	B1821BH025C100N	SCREW,CAP,HEXAGON H 1/4-20 X 1.00 PART OF KIT P/N 5705698.............. UOC:AVY,A11,A13,A14,A15,A20,A24,A25, A26,A27,B16,B17,B18,HVY,H11,H13,H14, H15,H16,H17,H18,H20,H21,H24,H25,H26, H27,H28,MMM	8
35	PAOZZ	2910013230123	19207	12342086	GUARD,FUEL FILTER PART OF KIT P/N 705698............................. UOC:AVY,A11,A13,A14,A15,A20,A24,A25, A26,A27,B16,B17,B18,HVY,H11,H13,H14, H15,H16,H17,H18,H20,H21,H24,H25,H26, H27,H28,MMM	1
36	PAOZZ	5310008775797	80205	MS21044N3	NUT,SELF-LOCKING,HE 3/16-32 PART OF KIT P/N 5705698..................... UOC:AVY,A11,A13,A14,A15,A20,A24,A25, A26,A27,B16,B17,B18,HVY,H11,H13,H14, H15,H16,H17,H18,H20,H21,H24,H25,H26, H27,H28,MMM	2
37	PAOZZ	4730013948345	34623	12340331	ADAPTER,STRAIGHT,PI PART OF KIT P/N	2

```
          SECTION II            TM9-2320-280-24P C01
    (1)     (2)     (3)             (4)         (5)
(7)       ITEM    SMR                           PART                       (6)
  NO      CODE    NSN             CAGEC        NUMBER    DESCRIPTION AND USABLE ON CODES(UOC)
QTY
```

```
                                                          5705698............................
UOC:AVY,A11,A13,A14,A15,A20,A24,A25,
                                                          A26,A27,B16,B17,B18,HVY,H11,H13,H14,
                                                          H15,H16,H17,H18,H20,H21,H24,H25,H26,
H27,H28,MMM
       38 PAOZZ 4730000243971 19207 12339884-1             CLAMP,HOSE PART OF KIT P/N 5705698..      2
UOC:AVY,A11,A13,A14,A15,A20,A24,A25,
                                                          A26,A27,B16,B17,B18,HVY,H11,H13,H14,
                                                          H15,H16,H17,H18,H20,H21,H24,H25,H26,
H27,H28,MMM
       39 PAOZZ 4720013162538 9C234 12338553-1             HOSE,NONMETALLIC PART OF KIT P/N          1
5705698............................                                                                 UOC
:AVY,A11,A13,A14,A15,A20,A24,A25,
                                                          A26,A27,B16,B17,B18,HVY,H11,H13,H14,
                                                          H15,H16,H17,H18,H20,H21,H24,H25,H26,
H27,H28,MMM
       40 PAOZZ 5305000680508 80204 B1821BH025C075N        SCREW,CAP,HEXAGON H  1/4-20 X .75         1
PART OF KIT P/N 5705698.............
                                                          UOC:AVY,A11,A13,A14,A15,A20,A24,A25,
                                                          A26,A27,B16,B17,B18,HVY,H11,H13,H14,
                                                          H15,H16,H17,H18,H20,H21,H24,H25,H26,
H27,H28,MMM
       41 PAOZZ 3040011975510 34623 12340829-1             CABLE,CONTROL PART OF KIT P/N             1
5705698............................                                                                 UOC
:AVY,A11,A13,A14,A15,A20,A24,A25,
                                                          A26,A27,B16,B17,B18,HVY,H11,H13,H14,
                                                          H15,H16,H17,H18,H20,H21,H24,H25,H26,
H27,H28,MMM
       42 PAOZZ 5340009936207 96906 MS21333-99             CLAMP,LOOP PART OF KIT P/N 5705698..      1
UOC:AVY,A11,A13,A14,A15,A20,A24,A25,
                                                          A26,A27,B16,B17,B18,HVY,H11,H13,H14,
                                                          H15,H16,H17,H18,H20,H21,H24,H25,H26,
H27,H28,MMM
       43 KFFZZ                 19207 12340706             TEMPLATE PART OF KIT P/N 5705698....      1
UOC:AVY,A11,A13,A14,A15,A20,A24,A25,
                                                          A26,A27,B16,B17,B18,HVY,H11,H13,H14,
                                                          H15,H16,H17,H18,H20,H21,H24,H25,H26,
H27,H28,MMM
*      44 PAOZZ 5320008503282 81349 M24243/1-A408          RIVET,BLIND.........................      6
UOC:AVY,A11,A13,A14,A15,A20,A24,A25,
                                                          A26,A27,B16,B17,B18,HVY,H11,H13,H14,
                                                          H15,H16,H17,H18,H20,H21,H24,H25,H26,
H27,H28,MMM
*      45 PAOZZ 5340014989619 19207 12340589-1             RETAINER  R.H.......................      1
UOC:AVY,A11,A13,A14,A15,A20,A24,A25,
                                                          A26,A27,B16,B17,B18,HVY,H11,H13,H14,
                                                          H15,H16,H17,H18,H20,H21,H24,H25,H26,
H27,H28,MMM
*      45 PAOZZ 5340014989625 19207 12340589-2             RETAINER  L.H.......................      1
UOC:AVY,A11,A13,A14,A15,A20,A24,A25,
                                                          A26,A27,B16,B17,B18,HVY,H11,H13,H14,
                                                          H15,H16,H17,H18,H20,H21,H24,H25,H26,
H27,H28,MMM
```

SECTION II   TM9-2320-280-24P C01
  (1)   (2)   (3)   (4)   (5)   (6)
(7) ITEM SMR   PART
    NO   CODE   NSN   CAGEC   NUMBER   DESCRIPTION AND USABLE ON CODES(UOC)
QTY

*   46 PAOZZ 4720013946170 19207 12339265-10   HOSE,AIR DUCT PART OF KIT P/N   1
705698............................   UOC
AVY,A11,A13,A14,A15,A20,A24,A25,
                                                   A26,A27,B16,B17,B18,HVY,H11,H13,H14,
                                                   H15,H16,H17,H18,H20,H21,H24,H25,H26,
27,H28,MMM

Section II. TM 9-2320-280-24P-2

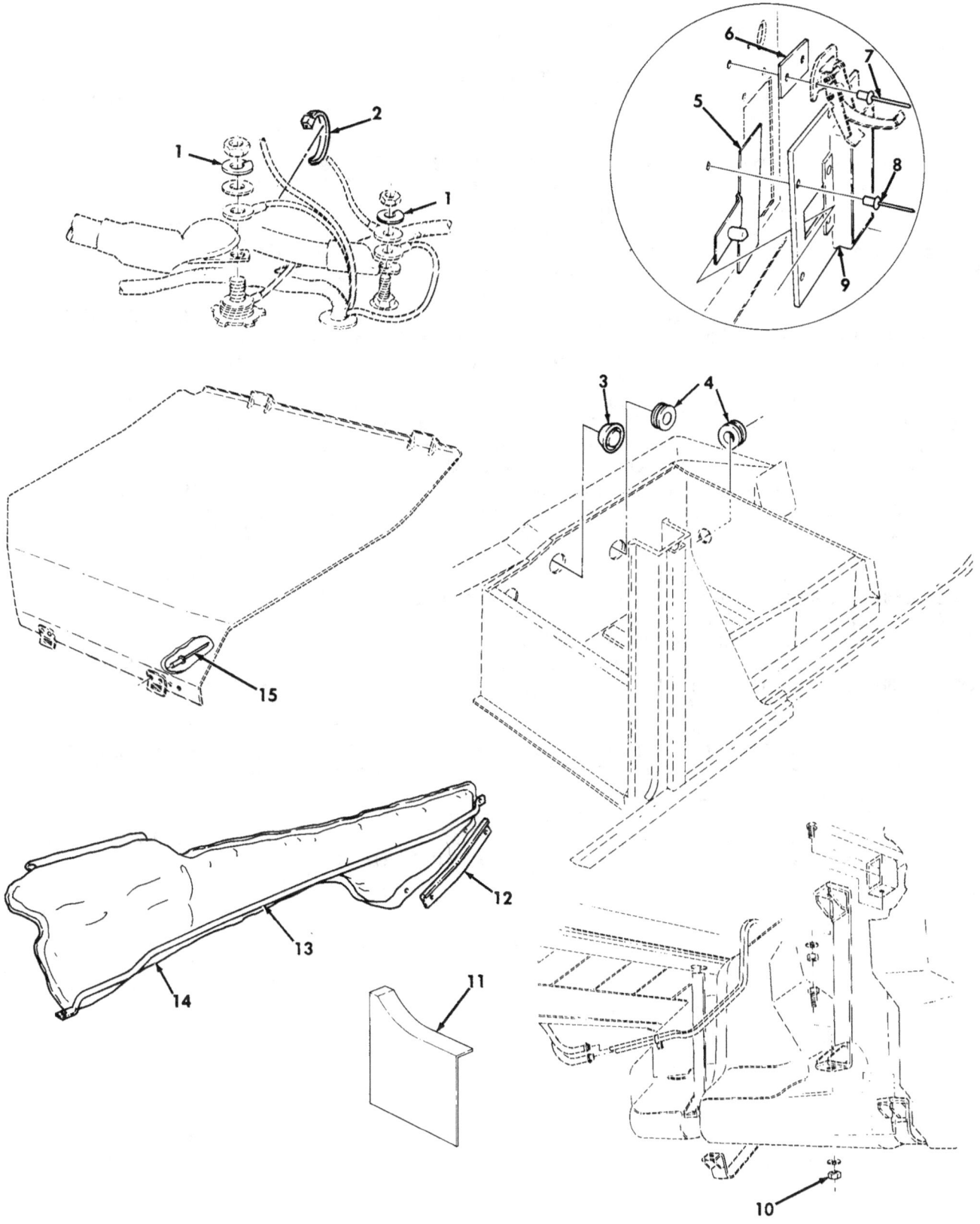

Figure 369. Arctic Kit, Battery Box Components, Insulation, and Control Box Connector.

SECTION II   TM9-2320-280-24P

(1) NO	(2) SMR CODE	(3) NSN	(4) CAGEC	(5) PART NUMBER	(7) ITEM DESCRIPTION AND USABLE ON CODES(UOC)	QTY

GROUP 3303 WINTERIZATION KITS

FIG. 369 ARCTIC KIT,BATTERY BOX COMPONENTS,INSULATION,AND CONTROL BOX CONNECTOR

(1)	(2)	(3)	(4)	(5)	(7)	QTY
1	PAOZZ	5310005845272	96906	MS35338-48	WASHER,LOCK 1/2,PART OF KIT P/N 5705698................. UOC:AVY,A11,A13,A14,A15,A20,A24,A25, A26,A27,B16,B17,B18,HVY,H11,H13,H14, H15,H16,H17,H18,H20,H21,H24,H25,H26, H27,H28,MMM	2
2	PAOZZ	5975010345871	96906	MS3367-7-0	STRAP,TIEDOWN,ELECT PART OF KIT P/N 5705698................. UOC:AVY,A11,A13,A14,A15,A20,A24,A25, A26,A27,B16,B17,B18,HVY,H11,H13,H14, H15,H16,H17,H18,H20,H21,H24,H25,H26, H27,H28,MMM	1
3	PAOZZ	5340013950812	19207	12340791	PLUG,EXPANSION PART OF KIT P/N 5705698................. UOC:AVY,A11,A13,A14,A15,A20,A24,A25, A26,A27,B16,B17,B18,HVY,H11,H13,H14, H15,H16,H17,H18,H20,H21,H24,H25,H26, H27,H28,MMM	3
4	PAOZZ	5325002766091	96906	MS35489-19	GROMMET,NONMETALLIC PART OF KIT P/N 5705698................. UOC:AVY,A11,A13,A14,A15,A20,A24,A25, A26,A27,B16,B17,B18,HVY,H11,H13,H14, H15,H16,H17,H18,H20,H21,H24,H25,H26, H27,H28,MMM	2
5	PAOZZ	2540014073296	34623	12342572	DOOR ASSEMBLY,ARCTI PART OF KIT P/N 5705698................. UOC:AVY,A11,A13,A14,A15,A20,A24,A25, A26,A27,B16,B17,B18,HVY,H11,H13,H14, H15,H16,H17,H18,H20,H21,H24,H25,H26, H27,H28,MMM	1
6	PAOZZ	5365013942394	19207	12342752	SPACER,PLATE PART OF KIT P/N 5705698 UOC:AVY,A11,A13,A14,A15,A20,A24,A25, A26,A27,B16,B17,B18,HVY,H11,H13,H14, H15,H16,H17,H18,H20,H21,H24,H25,H26, H27,H28,MMM	1
7	PAOZZ	5320012716357	19207	12339355-1	RIVET,BLIND PART OF KIT P/N 5705698. UOC:AVY,A11,A13,A14,A15,A20,A24,A25, A26,A27,B16,B17,B18,HVY,H11,H13,H14, H15,H16,H17,H18,H20,H21,H24,H25,H26, H27,H28,MMM	2
8	PAOZZ	5320011357319	80205	NAS9301BNS-6-03	RIVET,BLIND PART OF KIT P/N 5705698. UOC:AVY,A11,A13,A14,A15,A20,A24,A25, A26,A27,B16,B17,B18,HVY,H11,H13,H14, H15,H16,H17,H18,H20,H21,H24,H25,H26, H27,H28,MMM	4
9	PAOZZ	5340013947853	19207	12342571	PLATE,MOUNTING PART OF KIT P/N 5705698.................	1

SECTION II   TM9-2320-280-24P

(1) ITEM NO	(2) SMR CODE	(3) NSN	(4) CAGEC	(5) PART NUMBER	(6) DESCRIPTION AND USABLE ON CODES(UOC)	(7) QTY
10	PAOZZ	5310009359021	96906	MS51943-35	NUT,SELF-LOCKING,HE 3/8-16 PART OF KIT P/N 5705698.................... UOC:AVY,A11,A13,A14,A15,A20,A24,A25, A26,A27,B16,B17,B18,HVY,H11,H13,H14, H15,H16,H17,H18,H20,H21,H24,H25,H26, H27,H28,MMM	4
11	PAOZZ	5340013939371	19207	12342573	BRACKET,DOUBLE ANGL RIGHT BODY SIDE RAIL PART OF KIT P/N 5705698... UOC:AVY,A11,A13,A14,A15,A20,A24,A25, A26,A27,B16,B17,B18,HVY,H11,H13,H14, H15,H16,H17,H18,H20,H21,H24,H25,H26, H27,H28,MMM	1
12	PAOZZ	2590012616851	19207	12339014-2	BEZEL,AUTOMOTIVE TR................. UOC:AVY,A11,A13,A14,A15,A20,A24,A25, A26,A27,B16,B17,B18,HVY,H11,H13,H14, H15,H16,H17,H18,H20,H21,H24,H25,H26, H27,H28,MMM	1
13	PAOZZ	2590012578787	19207	12340591	RETAINER,INTERIOR PART OF KIT P/N 5705698............................. UOC:AVY,A11,A13,A14,A20,A24,A25,A26, A27,B16,B17,B18,HVY,H11,H13,H14,H16, H17,H18,H20,H21,H24,H25,H26,H27,H28, MMM	1
13	PAOZZ	2540014525034	19207	12340591-1	RETAINER,INTERIOR PART OF KIT P/N 5705698............................. UOC:A15,H15	1
14	PAOZA	2540014611134	19207	12340593-1	INSULATION,THERMAL, PART OF KIT P/N 5705698............................. UOC:AVY,A11,A13,A14,A20,A24,A25,A26, A27,B16,B17,B18,HVY,H11,H13,H14,H16, H17,H18,H20,H21,H24,H25,H26,H27,H28, MMM	1
14	PAOZA	2540014611134	34623	12340593-1	INSULATION,THERMAL, PART OF KIT P/N 5705698............................. UOC:A15,H15	1
15	PAOZZ	5320011435079	80205	NAS9301BNS-4-04	RIVET,BLIND PART OF KIT P/N 5705698. UOC:A11,A13,A14,A20,A24,A25,A26,A27, B17,B18,B20,B24,B25,C17,H11,H13,H14, H15,H16,H17,H18,H20,H21,H24,H25,H26, H27,H28,MMM	2

END OF FIGURE

Section II. TM 9-2320-280-24P-2

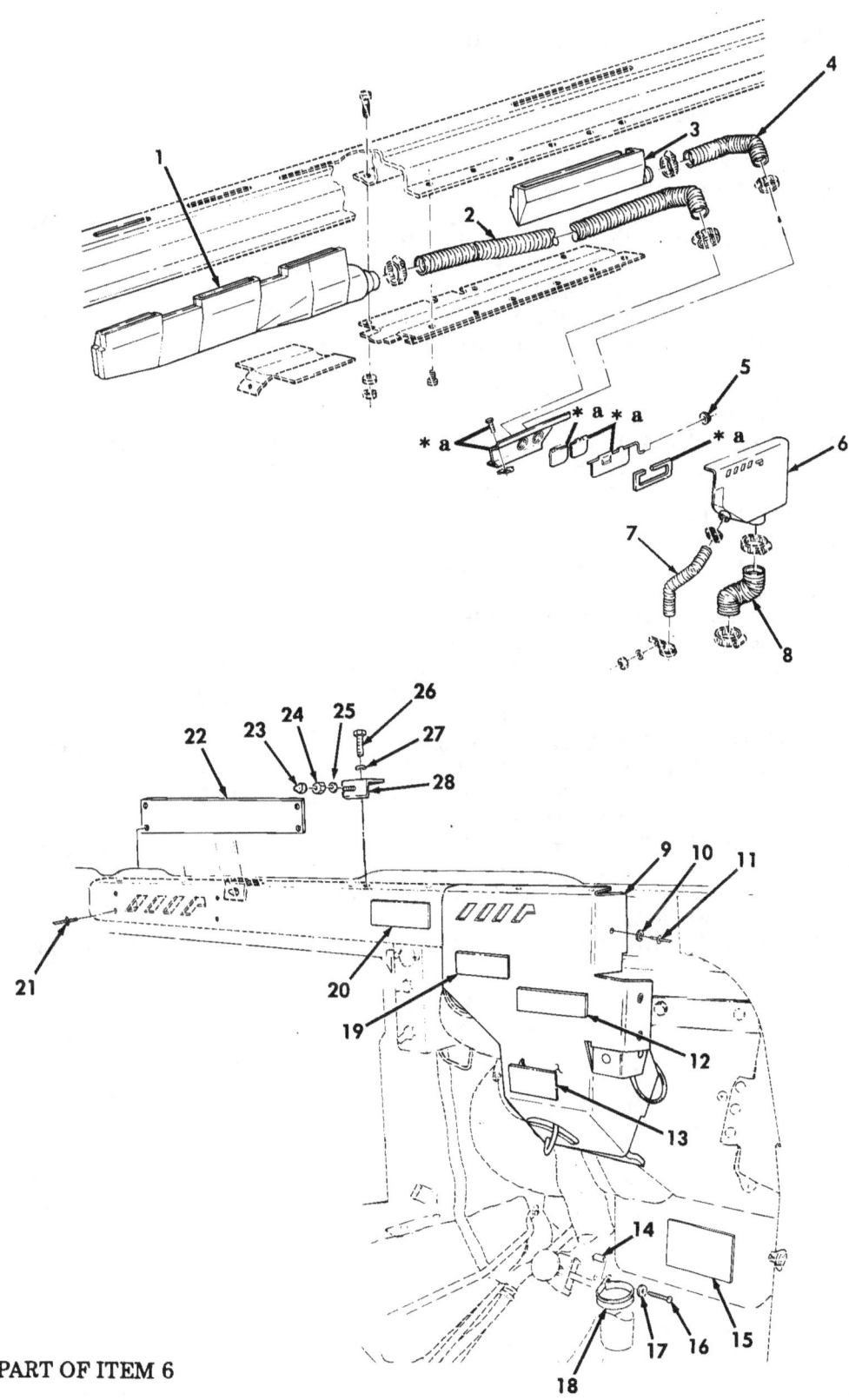

* a PART OF ITEM 6

Figure 370. Arctic Kit, Diverter Box, Defroster, Heater Ducting, and Radio Rack Bracket.

SECTION II TM9-2320-280-24P

(1) NO	(2) SMR CODE	(3) NSN	(4) CAGEC	(5) PART NUMBER	(6) DESCRIPTION AND USABLE ON CODES(UOC)	(7) QTY

GROUP 3303 WINTERIZATION KITS

FIG. 370 ARCTIC KIT, DIVERTER BOX, DEFROSTER, HEATER DUCTING, AND RADIO RACK BRACKET

1	PFOZZ	2540012573742	34623	12339289-2	VENTILATOR,AIR CIRC L.H.,PART OF KIT P/N 5705698..................... UOC:AVY,A11,A13,A14,A15,A20,A24,A25, A26,A27,B16,B17,B18,HVY,H11,H13,H14, H15,H16,H17,H18,H20,H21,H24,H25,H26, H27,H28,MMM	1
2	PAOZZ	4720013946166	19207	12339265-6	HOSE,AIR DUCT PART OF KIT P/N 5705698......................... UOC:AVY,A11,A13,A14,A15,A20,A24,A25, A26,A27,B16,B17,B18,HVY,H11,H13,H14, H15,H16,H17,H18,H20,H21,H24,H25,H26, H27,H28,MMM	57
3	PAOZZ	2540013905711	34623	5591935	VENTILATOR,AIR CIRC R.H.,PART OF KIT P/N 5705698..................... UOC:AVY,A11,A13,A14,A15,A20,A24,A25, A26,A27,B16,B17,B18,HVY,H11,H13,H14, H15,H16,H17,H18,H20,H21,H24,H25,H26, H27,H28,MMM	1
4	PAOZZ	4720013256985	19207	12339265-7	HOSE,AIR DUCT PART OF KIT P/N 5705698............................. UOC:AVY,A11,A13,A14,A15,A20,A24,A25, A26,A27,B16,B17,B18,HVY,H11,H13,H14, H15,H16,H17,H18,H20,H21,H24,H25,H26, H27,H28,MMM	57
5	PAOZZ	5340013327599	19207	12340258	BUTTON,PLUG PART OF KIT P/N 5705698 UOC:AVY,A11,A13,A14,A15,A20,A24,A25, A26,A27,B16,B17,B18,HVY,H11,H13,H14, H15,H16,H17,H18,H20,H21,H24,H25,H26, H27,H28,MMM	1
6	PFOZZ	2540013954202	19207	12342570	VENTILATOR,AIR CIRC PART OF KIT P/N 5705698............................. UOC:AVY,A11,A13,A14,A15,A20,A24,A25, A26,A27,B16,B17,B18,HVY,H11,H13,H14, H15,H16,H17,H18,H20,H21,H24,H25,H26, H27,H28,MMM	1
7	PAOZZ	4720014510894	19207	12339265-16	HOSE,AIR DUCT PART OF KIT P/N 5705698............................. UOC:AVY,A11,A13,A14,A15,A20,A24,A25, A26,A27,B16,B17,B18,HVY,H11,H13,H14, H15,H16,H17,H18,H20,H21,H24,H25,H26, H27,H28,MMM	1
8	PAOZZ	4720013946170	19207	12339265-10	HOSE,AIR DUCT PART OF KIT P/N 5705698............................. UOC:AVY,A11,A13,A14,A15,A20,A24,A25, A26,A27,B16,B17,B18,HVY,H11,H13,H14, H15,H16,H17,H18,H20,H21,H24,H25,H26, H27,H28,MMM	1

SECTION II   TM9-2320-280-24P

(1) NO	(2) CODE	(3) NSN	(4) CAGEC	(5) NUMBER	(6) DESCRIPTION AND USABLE ON CODES(UOC)	(7) ITEM SMR PART QTY

9 PAOZZ 2540012546511 19207 12340470   ARCTIC COVER,DIVERT PART OF KIT P/N   1
05698............................   UOC:AVY,A11,A13,A14,A15,A20,A24,A25,
                                        A26,A27,B16,B17,B18,HVY,H11,H13,H14,
                                        H15,H16,H17,H18,H20,H21,H24,H25,H26,   H27,H28,MMM
10 PAOZZ 2540012546511 24617 11500362   WASHER,FLAT PART OF KIT P/N 5705698   1
OC:AVY,A11,A13,A14,A15,A20,A24,A25,
                                        A26,A27,B16,B17,B18,HVY,H11,H13,H14,
                                        H15,H16,H17,H18,H20,H21,H24,H25,H26,   H27,H28,MMM
11 PAOZZ 5320010232529 81349 M24243/1-A404   RIVET,BLIND PART OF KIT P/N 5705698   4
OC:AVY,A11,A13,A14,A15,A20,A24,A25,
                                        A26,A27,B16,B17,B18,HVY,H11,H13,H14,
                                        H15,H16,H17,H18,H20,H21,H24,H25,H26,   H27,H28,MMM
12 PAOZZ 9905013973196 19207 12342648   PLATE,INSTRUCTION PART OF KIT P/N   1
05698............................   UOC:AVY,A11,A13,A14,A15,A20,A24,A25,
                                        A26,A27,B16,B17,B18,HVY,H11,H13,H14,
                                        H15,H16,H17,H18,H20,H21,H24,H25,H26,   H27,H28,MMM
13 PAOZZ 9905012487656 19207 12339108   PLATE,INSTRUCTION PART OF KIT P/N   1
05698............................   UOC:AVY,A11,A13,A14,A15,A20,A24,A25,
                                        A26,A27,B16,B17,B18,HVY,H11,H13,H14,
                                        H15,H16,H17,H18,H20,H21,H24,H25,H26,   H27,H28,MMM
14 PAOZZ 5365013940440 19207 12342751   SPACER,SLEEVE PART OF KIT P/N   1   57
698............................   UOC:AVY,A11,A13,A14,A15,A20,A24,A25,
                                        A26,A27,B16,B17,B18,HVY,H11,H13,H14,
                                        H15,H16,H17,H18,H20,H21,H24,H25,H26,   H27,H28,MMM   15
AOZZ 7690014078248 19207 12342806   MARKER,IDENTIFICATI PART OF KIT P/N   1   57
698............................   UOC:AVY,A11,A13,A14,A15,A20,A24,A25,
                                        A26,A27,B16,B17,B18,HVY,H11,H13,H14,
                                        H15,H16,H17,H18,H20,H21,H24,H25,H26,   H27,H28,MMM
16 PAOZZ 5305013943543 24617 9415778   SCREW,CAP,HEXAGON H PART OF KIT P/N   1
05698............................   UOC:AVY,A11,A13,A14,A15,A20,A24,A25,
                                        A26,A27,B16,B17,B18,HVY,H11,H13,H14,
                                        H15,H16,H17,H18,H20,H21,H24,H25,H26,   H27,H28,MMM
17 PAOZZ 5310000453296 30379 120217   WASHER,LOCK #10,PART OF KIT P/N   1
05698............................   UOC:A11,A13,A14,A20,A24,A25,A26,A27,
                                        B17,B18,B20,B24,B25,C17,H11,H13,H14,
                                        H15,H16,H17,H18,H20,H21,H24,H25,H26,   H27,H28,MMM
18 PAOZZ 5340001934111 96906 MS21333-86   CLAMP,LOOP PART OF KIT P/N 5705698.   1

SECTION II   TM9-2320-280-24P

(1) NO	(2) SMR CODE	(3) NSN	(4) CAGEC	(5) PART NUMBER	(6) DESCRIPTION AND USABLE ON CODES(UOC)	(7) QTY
19	PAOZZ	9905012039994	19207	12339100	PLATE,INSTRUCTION PART OF KIT P/N 5705698.............................. UOC:AVY,A11,A13,A14,A15,A20,A24,A25, A26,A27,B16,B17,B18,HVY,H11,H13,H14, H15,H16,H17,H18,H20,H21,H24,H25,H26, H27,H28,MMM	1
20	PAOZZ	9905011934065	19207	12339106	PLATE,INSTRUCTION PART OF KIT P/N 5705698.............................. UOC:AVY,A11,A13,A14,A15,A20,A24,A25, A26,A27,B16,B17,B18,HVY,H11,H13,H14, H15,H16,H17,H18,H20,H21,H24,H25,H26, H27,H28,MMM	1
21	PAOZZ	5320010195694	81349	M24243/1A402	RIVET,BLIND PART OF KIT P/N 5705698 UOC:AVY,A11,A13,A14,A15,A20,A24,A25, A26,A27,B16,B17,B18,HVY,H11,H13,H14, H15,H16,H17,H18,H20,H21,H24,H25,H26, H27,H28,MMM	4
22	PAOZZ	5340013947288	19207	12342754	COVER,ACCESS PART OF KIT P/N 5705698 UOC:AVY,A11,A13,A14,A15,A20,A24,A25, A26,A27,B16,B17,B18,HVY,H11,H13,H14, H15,H16,H17,H18,H20,H21,H24,H25,H26, H27,H28,MMM	1
23	PAOZZ	5340011744894	81349	M5501/11-F1	CAP,PROTECTIVE DUST PART OF KIT P/N 5705698.............................. UOC:AVY,A11,A13,A14,A15,A20,A24,A25, A26,A27,B16,B17,B18,HVY,H11,H13,H14, H15,H16,H17,H18,H20,H21,H24,H25,H26, H27,H28,MMM	2
24	PAOZZ	5310000120367	24617	120367	NUT,PLAIN,HEXAGON PART OF KIT P/N 5705698.............................. UOC:AVY,A11,A13,A14,A15,A20,A24,A25, A26,A27,B16,B17,B18,HVY,H11,H13,H14, H15,H16,H17,H18,H20,H21,H24,H25,H26, H27,H28,MMM	2
25	PAOZZ	5310000446188	24617	446188	WASHER,FLAT PART OF KIT P/N 5705698 UOC:A11,A13,A14,A20,A24,A25,A26,A27, B17,B18,B20,B24,B25,C17,H11,H13,H14, H15,H16,H17,H18,H20,H21,H24,H25,H26, H27,H28,MMM	2
26	PAOZZ	5305005434372	80204	B1821BH038C075N	SCREW,CAP,HEXAGON H PART OF KIT P/N 5705698.............................. UOC:AVY,A11,A13,A14,A15,A20,A24,A25, A26,A27,B16,B17,B18,HVY,H11,H13,H14, H15,H16,H17,H18,H20,H21,H24,H25,H26, H27,H28,MMM	2
27	PAOZZ	5310000446212	24617	446212	WASHER,FLAT PART OF KIT P/N 5705698 UOC:AVY,A11,A13,A14,A15,A20,A24,A25, A26,A27,B16,B17,B18,HVY,H11,H13,H14, H15,H16,H17,H18,H20,H21,H24,H25,H26,	2

SECTION II   TM9-2320-280-24P

(1) NO	(2) CODE	(3) NSN	(4) CAGEC	(5) NUMBER	(6) DESCRIPTION AND USABLE ON CODES(UOC)	(7) ITEM SMR PART QTY

28 PAOZZ 2590014398268 34623 12446763 BRACKET,ANGLE PART OF
5705698............................. UOC:AVY,A11,A13,A14,A1
IT P/N 2 A20,A24,A25,
A26,A27,B16,B17,B18,HVY,H11,H13,H14,
H15,H16,H17,H18,H20,H21,H24,H25,H26, H27,H28,MMM

END OF FIGURE

370-4

Section II.

TM 9-2320-280-24P-2

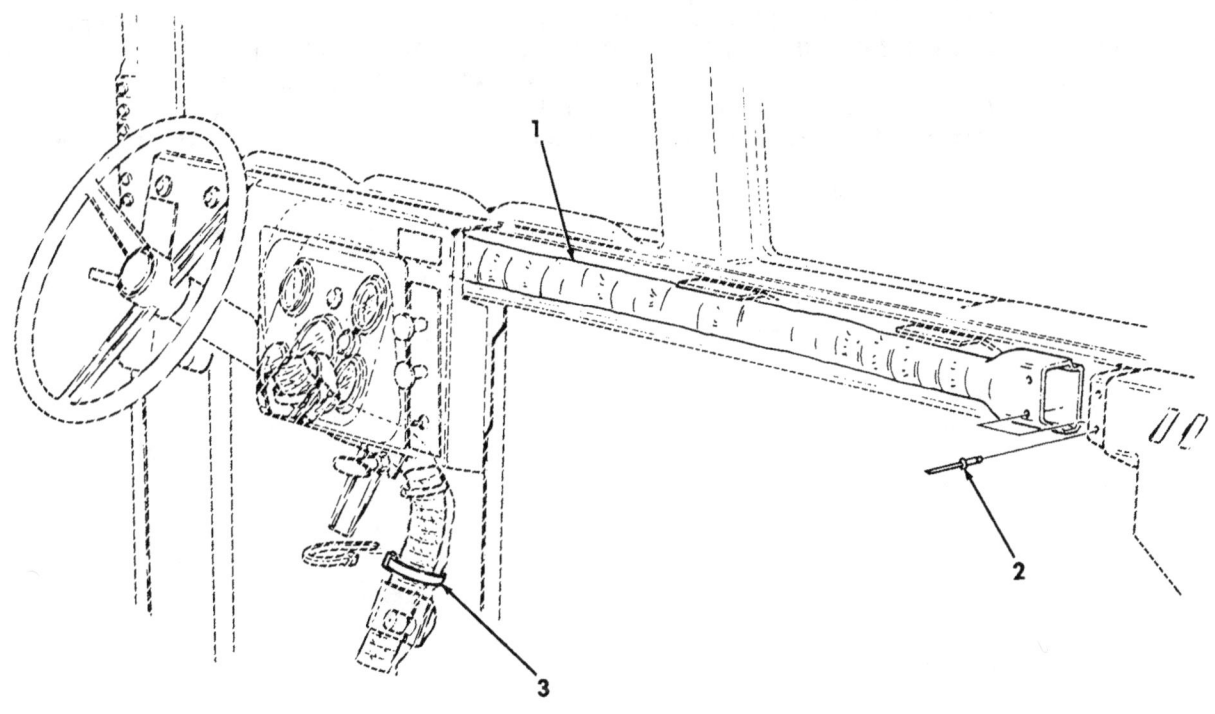

Figure 371. Arctic Kit, Heater Duct Assembly.

SECTION II TM9-2320-280-24P

(1) NO	(2) CODE	(3) NSN	(4) CAGEC	(5) NUMBER	(6) DESCRIPTION AND USABLE ON CODES(UOC)	(7) ITEM SMR QTY	PART

GROUP 3303 WINTERIZATION KITS

FIG. 371 ARCTIC KIT, HEATER DUCT ASSEMBLY

1 PAOZZ 19207 12447054 DUCT ASSEMBLY PART OF KIT P/N 1 5705698..............................
UOC:AVY,A11,A13,A14,A15,A20,A24,A25,
A26,A27,B16,B17,B18,HVY,H11,H13,H14,
H15,H16,H17,H18,H20,H21,H24,H25,H26, H27,H28,MMM

2 PAOZZ 5320010195694 81349 M24243/1A402 RIVET,BLIND PART OF KIT P/N 5705698 2
UOC:AVY,A11,A13,A14,A15,A20,A24,A25,
A26,A27,B16,B17,B18,HVY,H11,H13,H14,
H15,H16,H17,H18,H20,H21,H24,H25,H26, H27,H28,MMM

3 PAOZZ 5975010345871 96906 MS3367-7-0 STRAP,TIEDOWN,ELECT PART OF KIT P/N 3
5705698.............................
UOC:AVY,A11,A13,A14,A15,A20,A24,A25,
A26,A27,B16,B17,B18,HVY,H11,H13,H14,
H15,H16,H17,H18,H20,H21,H24,H25,H26, H27,H28,MMM

END OF FIGURE

Section II. TM 9-2320-280-24P-2

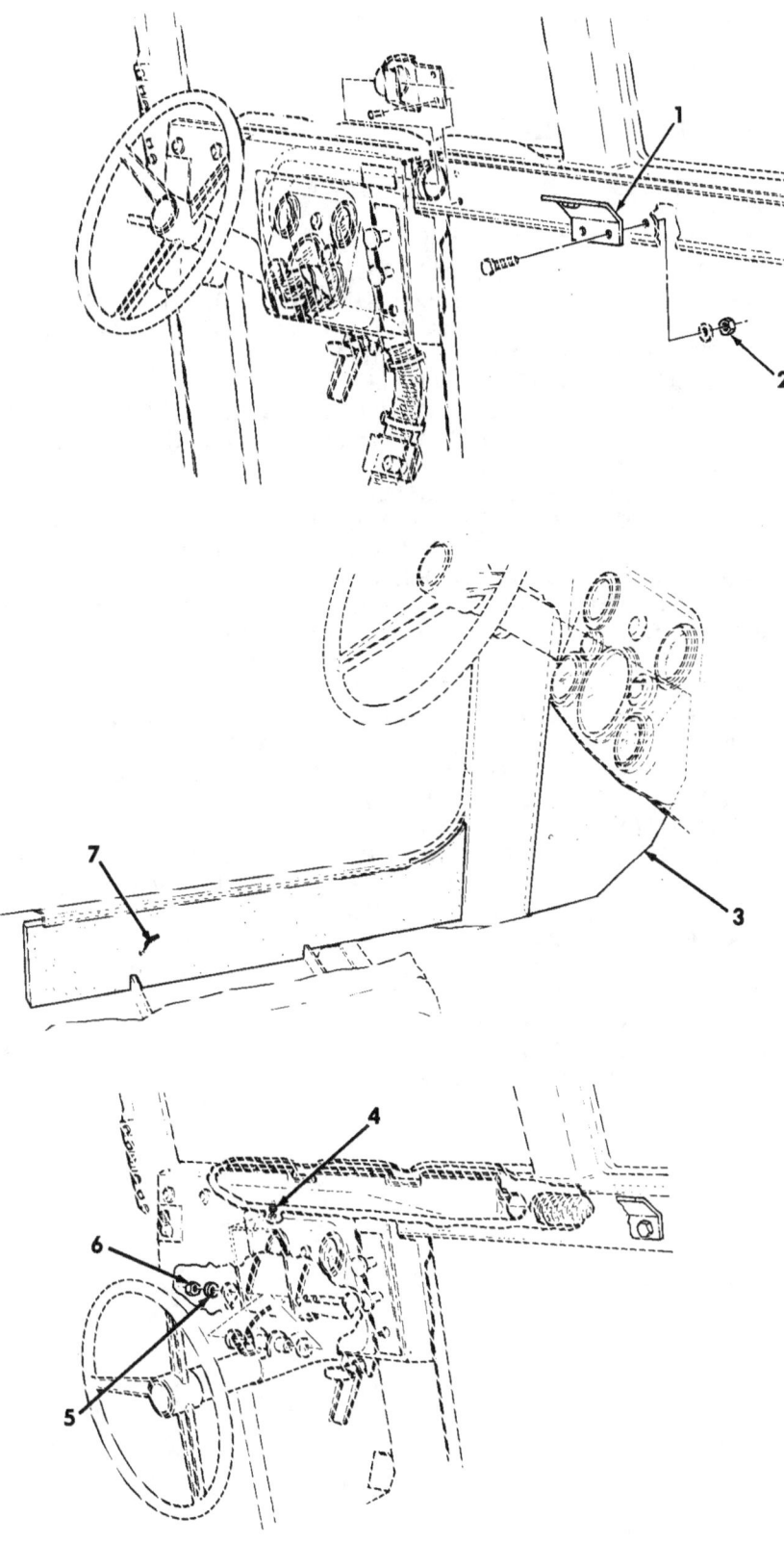

Figure 372. Arctic Kit, Diverter Box, Ducting, Insulation, and Radio Mounting Brackets.

SECTION II   TM9-2320-280-24P

(1) NO	(2) SMR CODE	(3) NSN	(4) CAGEC	(5) PART NUMBER	(7) ITEM DESCRIPTION AND USABLE ON CODES(UOC)	QTY

GROUP 3303 WINTERIZATION KITS

FIG. 372 ARCTIC KIT,DIVERTER BOX, DUCTING,INSULATION,AND RADIO MOUNTING BRACKETS

1 PFOZZ 5340014764374 19207 12446765 BRACKET,DOUBLE ANGL................. 1
UOC:AVY,A11,A13,A14,A15,A20,A24,A25,
A26,A27,B16,B17,B18,HVY,H11,H13,H14,
H15,H16,H17,H18,H20,H21,H24,H25,H26,
H27,H28,MMM

2 PAOZZ 5310014326727 24617 9419471 NUT,SELF-LOCKING,HE PART OF KIT P/N 4
5705698...........................
UOC:AVY,A11,A13,A14,A15,A20,A24,A25,
A26,A27,B16,B17,B18,HVY,H11,H13,H14,
H15,H16,H17,H18,H20,H21,H24,H25,H26,
H27,H28,MMM

3 PAOZZ 2540013953979 9C234 12342755-2 INSULATION,THERMAL PART OF KIT P/N 1
5705698...........................
UOC:AVY,A11,A13,A14,A15,A20,A24,A25,
A26,A27,B16,B17,B18,HVY,H11,H13,H14,
H15,H16,H17,H18,H20,H21,H24,H25,H26,
H27,H28,MMM

4 PAOZZ 5325012097843 19207 12339402 GROMMET,NONMETALLIC PART OF KIT P/N 1
5705698...........................
UOC:AVY,A11,A13,A14,A15,A20,A24,A25,
A26,A27,B16,B17,B18,HVY,H11,H13,H14,
H15,H16,H17,H18,H20,H21,H24,H25,H26,
H27,H28,MMM

5 PAOZZ 5310006379541 96906 MS35338-46 WASHER,LOCK 3/8 PART OF KIT P/N 1
5705698...........................
UOC:AVY,A11,A13,A14,A15,A20,A24,A25,
A26,A27,B16,B17,B18,HVY,H11,H13,H14,
H15,H16,H17,H18,H20,H21,H24,H25,H26,
H27,H28,MMM

6 PAOZZ 5310009359021 96906 MS51943-35 NUT,SELF-LOCKING,HE PART OF KIT P/N 1
5705698...........................
UOC:AVY,A11,A13,A14,A15,A20,A24,A25,
A26,A27,B16,B17,B18,HVY,H11,H13,H14,
H15,H16,H17,H18,H20,H21,H24,H25,H26,
H27,H28,MMM

7 PAOZZ 2540013952230 9C234 12342756 INSULATION,THERMAL PART OF KIT P/N 1
5705698...........................
UOC:AVY,A11,A13,A14,A15,A20,A24,A25,
A26,A27,B16,B17,B18,HVY,H11,H13,H14,
H15,H16,H17,H18,H20,H21,H24,H25,H26,
H27,H28,MMM

KIT PDFZZ 2540013170728 34623 5705698 PARTS KIT,VEHICULAR................. 1
UOC:AVY,A11,A13,A14,A15,A20,A24,A25,
A26,A27,B16,B17,B18,HVY,H11,H13,H14,
H15,H16,H17,H18,H20,H21,H24,H25,H26,
H27,H28,MMM
ADAPTER,CONNECTOR (1) 366-31

SECTION II TM9-2320-280-24P

(1) ITEM NO	(2) SMR CODE	(3) NSN	(4) CAGEC	(5) PART NUMBER	(6) DESCRIPTION AND USABLE ON CODES(UOC)	(7) QTY
					ADAPTER,CONNECTOR	(1) 366-32
					ADAPTER,STRAIGHT,PI	(1) 366-12
					ADAPTER,STRAIGHT,PI	(1) 366-14
					ADAPTER,STRAIGHT,PI	(1) 366-24
					ADAPTER,STRAIGHT,PI	(1) 368-37
					ANCHOR,STRAP	(1) 365-19
					ARCTIC COVER,DIVERT	(1) 370-9
					BEZEL,AUTOMOTIVE TR	(1) 368-33
					BRACKET,DOUBLE ANGL	(1) 366-27
					BRACKET,DOUBLE ANGL	(1) 369-11
					BRACKET,DOUBLE ANGL	(2) 370-28
					BRACKET,MOUNTING	(2) 368-21
					BUTTON,PLUG	(1) 370-5
					CABLE,CONTROL	(1) 368-41
					CAP,FILLER,OPENING	(1) 366-1
					CAP,PROTECTIVE	(2) 370-23
					CAP,TUBE	(1) 366-15
					CLAMP,HOSE	(2) 364-3
					CLAMP,HOSE	(2) 365-10
					CLAMP,HOSE	(2) 366-21
					CLAMP,HOSE	(3) 368-3
					CLAMP,HOSE	(2) 368-14
					CLAMP,HOSE	(2) 368-20
					CLAMP,HOSE	(2) 368-38
					CLAMP,LOOP	(1) 365-21
					CLAMP,LOOP	(1) 366-5
					CLAMP,LOOP	(1) 366-7
					CLAMP,LOOP	(1) 366-20
					CLAMP,LOOP	(1) 366-28
					CLAMP,LOOP	(1) 368-42
					CLAMP,LOOP	(1) 370-18
					CONNECTOR,EXHAUST P	(1) 365-11
					CONNECTOR,EXHAUST P	(1) 368-13
					CONNECTOR,EXHAUST P	(1) 368-17
					CONTROL BOX,ELECTRI	(1) 366-2
					COUPLING,PIPE	(1) 366-10
					COVER ACCESS	(1) 370-22
					COVER,ASSEMBLY,COWL	(1) 368-4
					DOOR ASSEMBLY,BATTE	(1) 369-5
					DIVERTER ASSEMBLY	(1) 365-6
					DUCT ASSEMBLY	(1) 371-1
					DUMMY CONNECTOR,PLU	(1) 366-34
					ELBOW,PIPE TO TUBE	(2) 366-9
					ELBOW,PIPE TO TUBE	(1) 368-29
					FILTER,FLUID	(1) 368-30
					FLANGE,PIPE	(1) 368-15
					GASKET,AIR HORN	(1) 364-2
					GROMMET,NONMETALLIC	(1) 368-10
					GROMMET,NONMETALLIC	(2) 369-4
					GROMMET,NONMETALLIC	(1) 372-4
					GUARD,FUEL FILTER	(1) 368-35
					HEATER,VEHICULAR,CO	(1) 368-19
					HOSE,AIR DUCT	(1) 368-44

```
             SECTION II         TM9-2320-280-24P C01
         (1)     (2)     (3)         (4)        (5)                         (6)
     (7)    ITEM    SMR                        PART
         NO     CODE    NSN        CAGEC      NUMBER      DESCRIPTION AND USABLE ON CODES(UOC)
     QTY
```

HOSE,AIR DUCT (1)	368-46
HOSE,AIR DUCT (1)	370-2
HOSE,AIR DUCT (1)	370-4
HOSE,AIR DUCT (1)	370-7
HOSE,AIR DUCT (1)	370-8
HOSE,NONMETALLIC (2)	364-4
HOSE,NONMETALLIC (1)	366-22
HOSE,NONMETALLIC (1)	368-39
HOSE,PREFORMED (1)	364-9
HOUSING,CENTRIFUGAL(1)	368-6
INSULATION (1)	372 3
INSULATION,BLANKET (1)	368-31
INSULATION,THERMAL (1)	372-7
INSULATION,TUNNEL (1)	369-14
INSULATION,TUNNEL (1)	369-14
LEAD,ELECTRICAL (1)	366-29
MARKER,IDENTIFICATI(1)	370-15
NUT,PLAIN,HEXAGON (2)	370-24
NUT,SELF-LOCKING,HE(2)	368-36
NUT,SELF-LOCKING,HE(1)	364-7
NUT,SELF-LOCKING,HE(10)	365-18
NUT,SELF-LOCKING,HE(1)	365-22
NUT,SELF-LOCKING,HE(2)	366-30
NUT,SELF-LOCKING,HE(4)	368-18
NUT,SELF-LOCKING,HE(9)	368-32
NUT,SELF-LOCKING,HE(2)	368-36
NUT,SELF-LOCKING,HE(4)	369-10
NUT,SELF-LOCKING,HE(4)	372-2
NUT,SELF-LOCKING,HE(1)	372-6
PIN,COTTER (4)	365-8
PIN,COTTER (3)	368-12
PIN,COTTER (1)	368-8
PIPE,EXHAUST (1)	365-7
PIPE,EXHAUST (1)	365-9
PIPE,EXHAUST (1)	365-12
PLATE,IDENTIFICATIO(1)	370-20
PLATE,INSTRUCTION (1)	370-12
PLATE,INSTRUCTION (1)	370-13
PLATE,INSTRUCTION (1)	370-19
PLATE,MOUNTING (1)	369-9
PLUG,EXPANSION (3)	369-3
PUMP,FUEL,ELECTRICA(1)	366-25
REGULATOR,AIR DUCT (1)	368-2
RETAINER,INTERIOR I(1)	369-13
RETAINER,INTERIOR I(1)	369-13
RIVET,BLIND (2)	365-1
RIVET,BLIND (4)	365-14
RIVET,BLIND (4)	365-26
RIVET,BLIND (8)	368-5
RIVET,BLIND (2)	369-7
RIVET,BLIND (4)	369-8
RIVET,BLIND (2)	369-15
RIVET,BLIND (4)	370-11

(1)	(2)	(3)	(4)	(5)	(6)
ITEM NO	SMR CODE	NSN	CAGEC	PART NUMBER	DESCRIPTION AND USABLE ON CODES (UOC) QTY

SECTION II   TM9-2320-280-24P C01

```
                                    RIVET,BLIND            (  4) 370-21
                                    RIVET,BLIND            (  2) 371-3
                                    SCREW,CAP,HEXAGON H(   2) 365-4
                                    SCREW,CAP,HEXAGON H(   1) 365-13
                                    SCREW,CAP,HEXAGON H(   2) 366-17
                                    SCREW,CAP,HEXAGON H(   4) 368-24
                                    SCREW,CAP,HEXAGON H(   8) 368-34
                                    SCREW,CAP,HEXAGON H(   1) 368-40
                                    SCREW,CAP,HEXAGON H(   1) 370-16
                                    SCREW,CAP,HEXAGON H(   2) 370-26
                                    SCREW,MACHINE          (  1) 365-20
                                    SCREW,MACHINE          (  1) 368-7
                                    SCREW,MACHINE          (  2) 368-27
                                    SCREW,TAPPING          (  2) 366-3
                                    SCREW,TAPPING          (  2) 366-6
                                    SEAL,NONMETALLIC       (  1) 368-1
                                    SETSCREW               (  1) 368-9
                                    SHELL,ELECTRICAL CO(   1) 366-33
                                    SHIELD,HEATER DUCT     (  1) 368-11
                                    SHROUD,ARTIC           (  1) 365-23
                                    SPACER,PLATE           (  1) 368-25
                                    SPACER,PLATE           (  1) 369-6
                                    SPACER,SLEEVE          (  1) 370-14
                                    STRAP,TIEDOWN,ELECT(   1) 369-2
                                    STRAP,TIEDOWN,ELECT(   3) 371-3
                                    STUD,SHOULDERED        (  7) 365-17
                                    TEE,PIPE               (  1) 366-13
                                    TEMPLATE               (  1) 365-2
                                    TEMPLATE               (  2) 365-3
                                    TEMPLATE               (  1) 365-15
                                    TEMPLATE               (  1) 366-8
                                    TEMPLATE               (  1) 366-35
                                    TEMPLATE               (  1) 366-36
                                    TEMPLATE               (  1) 368-43
                                    TEMPLATE,FUEL PUMP     (  1) 366-37
                                    THERMASEAL,HEATER      (  1) 368-16
                                    TUBE ASSEMBLY          (  1) 366-11
                                    TUBE ASSEMBLY,METAL(   1) 364-5
                                    TUBE ASSEMBLY,METAL(   1) 366-15
                                    TUBE ASSEMBLY,METAL(   1) 366-18
                                    TUBE ASSEMBLY,METAL(   1) 366-19
                                    TUBE ASSEMBLY,METAL(   1) 366-23
                                    TUBE ASSEMBLY,METAL(   1) 368-28
                                    TUBE,BENT,METALLIC     (  1) 364-8
                                    VENTILATOR,AIR CIRC(   1) 370-1
                                    VENTILATOR,AIR CIRC(   1) 370-3
                                    VENTILATOR,AIR CIRC(   1) 370-6
                                    WASHER,FLAT            (  2) 364-1
                                    WASHER,FLAT            ( 10) 365-16
                                    WASHER,FLAT            (  4) 366-16
                                    WASHER,FLAT            ( 22) 368-22
                                    WASHER,FLAT            (  2) 368-26
                                    WASHER,FLAT            (  1) 370-10
```

SECTION II   TM9-2320-280-24P C01
  (1)   (2)   (3)   (4)   (5)   (6)
(7) ITEM SMR   PART
  NO CODE   NSN   CAGEC   NUMBER   DESCRIPTION AND USABLE ON CODES(UOC)
QTY

                                         WASHER,FLAT   (2) 370-25
                                         WASHER,FLAT   (2) 370-27
                                         WASHER,LOCK   (1) 364-6
                                         WASHER,LOCK   (2) 365-5
                                         WASHER,LOCK   (2) 366-4
                                         WASHER,LOCK   (4) 368-23
                                         WASHER,LOCK   (2) 369-1
                                         WASHER,LOCK   (1) 370-17
                                         WASHER,LOCK   (1) 372-5

Section II.   TM 9-2320-280-24P-2

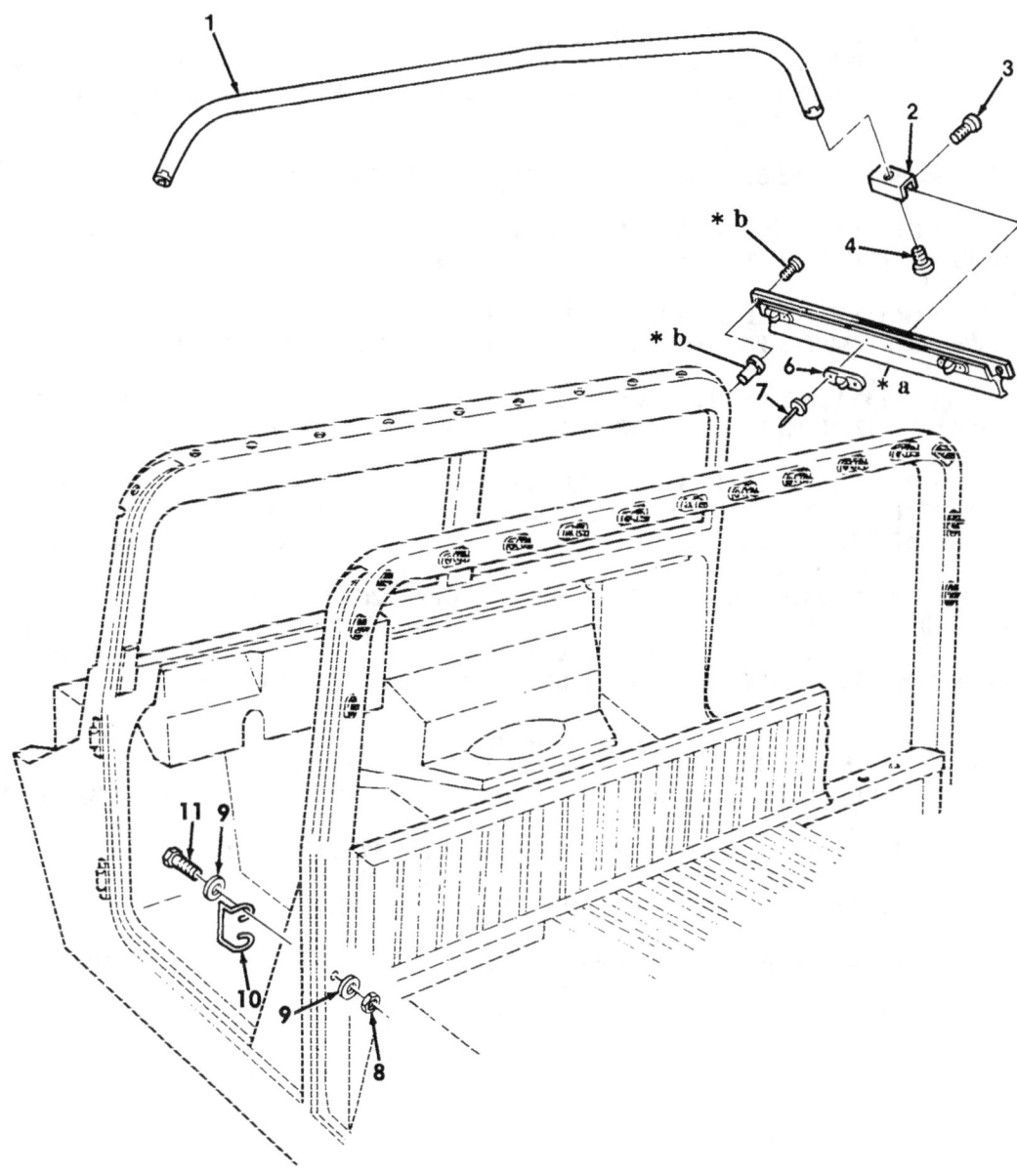

* a PART OF ITEM 5
* b SEE FIGURE 225

Figure 373. Arctic Kit, Two Man Crew, Bow, and Mounting Components.

SECTION II TM9-2320-280-24P

(1) NO	(2) SMR CODE	(3) NSN	(4) CAGEC	(5) PART NUMBER	(7) DESCRIPTION AND USABLE ON CODES(UOC)	ITEM QTY

GROUP 3303 WINTERIZATION KITS

FIG. 373 ARCTIC KIT, TWO MAN CREW, BOW, AND MOUNTING COMPONENTS

1	PAOZZ	2540012064115	19207	12340713	BOW,VEHICULAR TOP PART OF KIT P/N 57K0137.............................. UOC:AVY,A13,A14,BVY,HVY,H13,H14,H21,H28,NNN	1
2	PAOZZ	5340012043862	19207	12340239	STRAP,RETAINING PART OF KIT P/N 57K0137.............................. UOC:AVY,A13,A14,BVY,HVY,H13,H14,H21,H28,NNN	4
3	PAOZZ	5305000593659	96906	MS51958-63	SCREW,MACHINE #10-32 X 1/2 PART OF KIT P/N 57K0137..................... UOC:AVY,A13,A14,BVY,HVY,H13,H14,H21,H28,NNN	4
4	PAOZZ	5305000711313	96906	MS51957-77	SCREW,MACHINE 1/4-20 X 3/8 PART OF KIT P/N 57K0137..................... UOC:AVY,A13,A14,BVY,HVY,H13,H14,H21,H28,NNN	2
5	PAOZZ	2590011982895	19207	12340603-1	RAIL ASSEMBLY,HAND L.H. PART OF KIT P/N 57K0137..................... UOC:AVY,A13,A14,BVY,HVY,H13,H14,H21,H28,NNN	1
5	PAOZZ	2590011974898	19207	12340603-2	RAIL ASSEMBLY,HAND R.H. PART OF KIT P/N 57K0137..................... UOC:AVY,A13,A14,BVY,HVY,H13,H14,H21,H28,NNN	1
6	PAOZZ	5325008235999	13940	XB78323-05001	.STUD,TURNBUTTON FAS PART OF KIT P/N 57K0137.............................. UOC:AVY,A13,A14,BVY,HVY,H13,H14,H21,H28,NNN	3
7	PAOZZ	5320000835009	81349	M24243/1-A403	.RIVET,BLIND 1/8 DIA.,.126-.187 GRIP PART OF KIT P/N 57K0137........ UOC:AVY,A13,A14,BVY,HVY,H13,H14,H21,H28,NNN	6
8	PAOZZ	5310002416658	96906	MS51943-34	NUT,SELF-LOCKING,HE 5/16-24 PART OF KIT P/N 57K0137..................... UOC:AVY,A13,A14,BVY,HVY,H13,H14,H21,H28,NNN	4
9	PAOZZ	5310004079566	96906	MS35338-45	WASHER,LOCK 5/16 PART OF KIT P/N 57K0137.............................. UOC:AVY,A13,A14,BVY,HVY,H13,H14,H21,H28,NNN	4
10	PAOZZ	5340012070717	19207	12339187	BUCKLE PART OF KIT P/N 57K0137..... UOC:AVY,A13,A14,BVY,HVY,H13,H14,H21,H28,NNN	2
11	PAOZZ	5306001822027	19207	7346712	BOLT,MACHINE 5/16-24 PART OF KIT P/N 57K0137........................ UOC:AVY,A13,A14,BVY,HVY,H13,H14,H21,H28,NNN	4

END OF FIGURE 373-1

Section II. TM 9-2320-280-24P-2

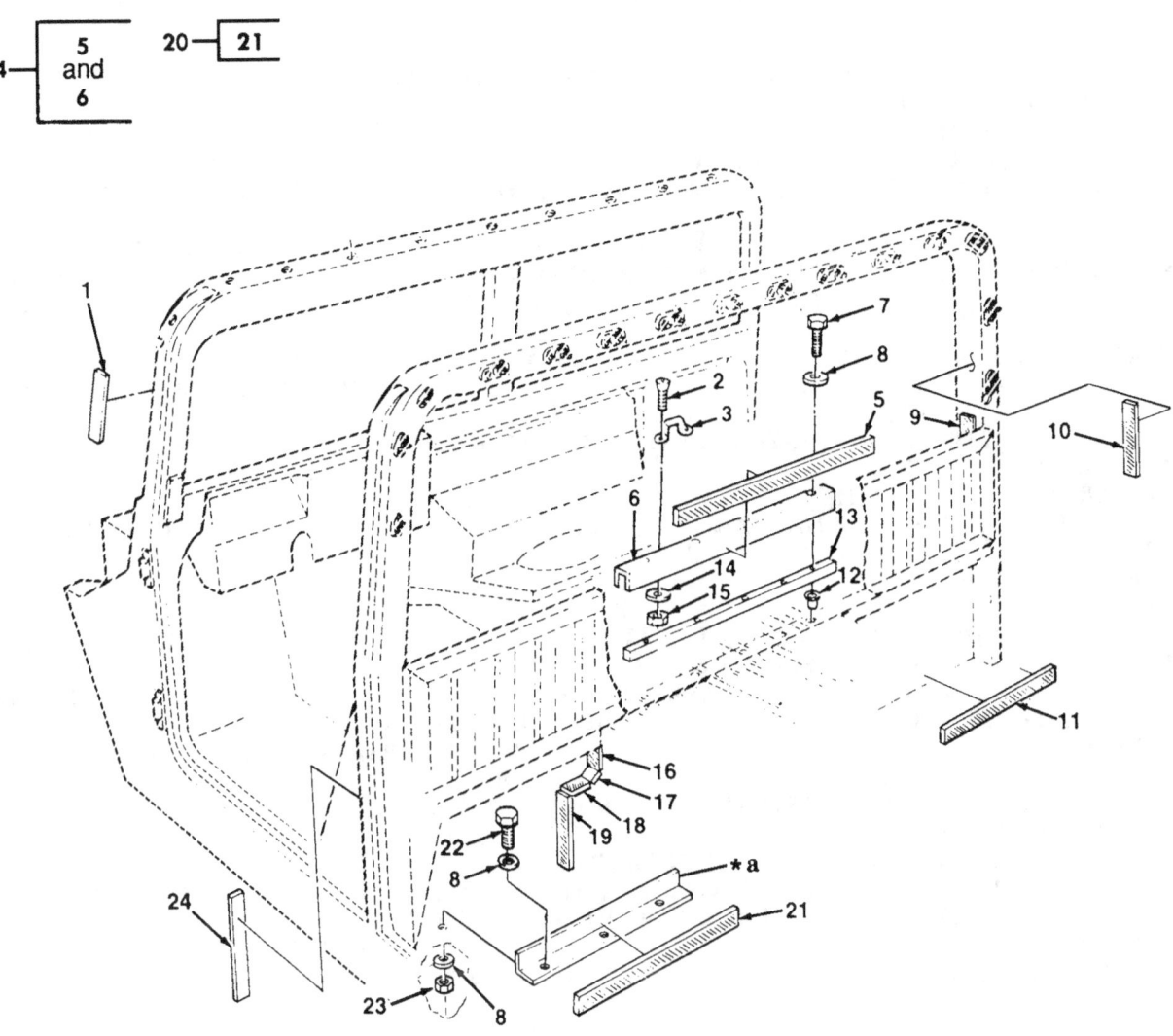

* a PART OF ITEM 20

Figure 374. Arctic Kit, Two Man Crew Curtain Fasteners.

SECTION II   TM9-2320-280-24P

(1) ITEM NO	(2) SMR CODE	(3) NSN	(4) CAGEC	(5) PART NUMBER	(6) DESCRIPTION AND USABLE ON CODES(UOC)	(7) QTY

GROUP 3303 WINTERIZATION KITS

FIG. 374 ARCTIC KIT, TWO MAN CREW CURTAIN FASTENERS

Item	SMR	NSN	CAGEC	Part Number	Description	Qty
1	MOOZZ		34623	5589275	INSULATION,FOAM MAKE FROM PLASTIC STRIP,P/N 5741221,12 INCHES LONG PART OF KIT P/N 57K0137............. UOC:AVY,A13,A14,BVY,HVY,H13,H14,H21,H28,NNN	2
2	PAOZZ	5305000888332	96906	MS35190-272	SCREW,MACHINE #10-24 X 5/8 PART OF KIT P/N 57K0137..................... UOC:AVY,A13,A14,BVY,HVY,H13,H14,H21,H28,NNN	4
3	PAOZZ	5340004895684	19220	6411-B	LOOP,STRAP FASTENER RART OF KIT P/N 57K0137....................... UOC:AVY,A13,A14,BVY,HVY,H13,H14,H21,H28,NNN	2
4	PFOZA	5340014665243	19207	12340708	BRACKET,DOUBLE ANGL PART OF KIT P/N 57K0137............................ UOC:AVY,A13,A14,BVY,HVY,H13,H14,H21,H28,NNN	1
5	MOOZZ		19207	12340700-12	.FASTENER,TAPE HOOK MAKE FROM FASTENER,TAPE,HOOK P/N 20197272, 31.75 INCHES....................... UOC:AVY,A13,A14,BVY,HVY,H13,H14,H21,H28,NNN	1
6	XAOZZ		19207	12340708-1	.CHANNEL............................ UOC:AVY,A13,A14,BVY,HVY,H13,H14,H21,H28,NNN	1
7	PAOZZ	5306012077487	7X677	9422042	BOLT,MACHINE 5/16-18 X 2.00 PART OF KIT P/N 57K0137............. UOC:AVY,A13,A14,BVY,HVY,H13,H14,H21,H28,NNN	4
8	PAOZZ	5310011191024	24617	2436162	WASHER,FLAT 5/16 PART OF KIT P/N 57K0137........... .... UOC:AVY,A13,A14,BVY,HVY,H13,H14,H21,H28,NNN	10
9	MOOZZ		19207	12340700-7	FASTENER,TAPE,HOOK MAKE FROM FASTENER,TAPE,HOOK P/N 20197272, 6.50 INCHES PART OF KIT P/N 57K0137. UOC:AVY,A13,A14,BVY,HVY,H13,H14,H21,H28,NNN	2
10	MOOZZ		19207	12340700-10	FASTENER,TAPE,HOOK MAKE FROM FASTENER,TAPE,HOOK P/N 20197272,24 INCHES PART OF KIT P/N 57K0137...... UOC:AVY,A13,A14,BVY,HVY,H13,H14,H21,H28,NNN	2
11	MOOZZ		19207	12340700-11	FASTENER,TAPE,HOOK MAKE FROM FASTENER,TAPE,HOOK P/N 20197272, 19.5 INCHES PART OF KIT P/N 57K0137. UOC:AVY,A13,A14,BVY,HVY,H13,H14,H21,H28,NNN	4

SECTION II   TM9-2320-280-24P

(1) ITEM NO	(2) SMR CODE	(3) NSN	(4) CAGEC	(5) PART NUMBER	(6) DESCRIPTION AND USABLE ON CODES(UOC)	(7) QTY
12	PAOZZ	5310009523567	03481	A31-125	NUT,PLAIN,BLIND RIV 5/16-18 PART OF KIT P/N 57K0137...................... UOC:AVY,A13,A14,BVY,HVY,H13,H14,H21,H28,NNN	4
13	PAOZZ	30011975458	19207	12340697	PLASTIC MATERIA,CE PART OF KIT P/N 57K0137........................ UOC:AVY,A13,A14,BVY,HVY,H13,H14,H21,H28,NNN	1
14	PAOZZ	5310000145850	96906	MS27183-42	WASHER,FLAT #10 PART OF KIT P/N 57K0137........................ UOC:AVY,A13,A14,BVY,HVY,H13,H14,H21,H28,NNN	
15	PAOZZ	5310011870678	7X677	116000	NUT,PLAIN,HEXAGON #10-24 PART OF KIT P/N 57K0137..................... UOC:AVY,A13,A14,A25,A26,A27,BVY,B15,B16,B17,B18,B20,NNN	4
16	MOOZZ		19207	12340700-6	FASTENER,TAPE,HOOK R.H.,MAKE FROM FASTENER,TAPE,HOOK,P/N 20197272, 6.38 IN PART OF KIT P/N 57K0137..... UOC:AVY,A13,A14,BVY,HVY,H13,H14,H21,H28,NNN	1
16	MOOZZ		19207	12340700-4	FASTENER,TAPE,HOOK L.H.,MAKE FROM FASTENER,TAPE,HOOK P/N 20197272, 5.31 IN PART OF KIT P/N 57K0137..... UOC:AVY,A13,A14,BVY,HVY,H13,H14,H21,H28,NNN	1
17	MOOZZ		19207	12340700-1	FASTENER,TAPE,HOOK MAKE FROM FASTENER,TAPE,HOOK P/N 20197272, 1.88 INCHES PART OF KIT P/N 57K0137. UOC:AVY,A13,A14,BVY,HVY,H13,H14,H21,H28,NNN	2
18	MOOZZ		19207	12340700-2	FASTENER,TAPE,HOOK R.H.,MAKE FROM FASTENER,TAPE,HOOK P/N 20197272,3.12 PART OF KIT P/N 57K0137............ UOC:AVY,A13,A14,BVY,HVY,H13,H14,H21,H28,NNN	1
18	MOOZZ		19207	12340700-3	FASTENER,TAPE,HOOK L.H.,MAKE FROM FASTENER,TAPE,HOOK P/N 20197272, 3.19 IN PART OF KIT P/N 57K0137..... UOC:AVY,A13,A14,BVY,HVY,H13,H14,H21,H28,NNN	1
19	MOOZZ		19207	12340700-9	FASTENER,TAPE,HOOK MAKE FROM FASTENER,TAPE,HOOK P/N 20197272,8.5 INCHES LONG PART OF KIT P/N 57K0137. UOC:AVY,A13,A14,BVY,HVY,H13,H14,H21,H28,NNN	2
20	PFOZZ	2540011975463	34623	5590630	BRACE,ARTIC CURTAI PART OF KIT P/N 57K0137............................. UOC:AVY,A13,A14,BVY,HVY,H13,H14,H21,H28,NNN	
21	MOOZZ		19207	12340700-8	.FASTENER,TAPE HOOK MAKE FROM FASTENER,TAPE,HOOK P/N 20197272,19 INCHES LONG........................	1

374-2

SECTION II   TM9-2320-280-24P

(1) NO	(2) SMR CODE	(3) NSN	(4) CAGEC	(5) PART NUMBER	(6) DESCRIPTION AND USABLE ON CODES(UOC)	(7) QTY
					UOC:AVY,A13,A14,BVY,HVY,H13,H14,H21, H28,NNN	
22	PAOZZ	5306002264828	80204	B1821BH031C113N	BOLT,MACHINE 5/16-18 X 1.125 PART OF KIT P/N 57K0137..................	3
					UOC:AVY,A13,A14,BVY,HVY,H13,H14,H21, H28,NNN	
23	PAOZZ	5310012085252	34623	5590556	NUT,SELF-LOCKING,HE 5/16-18 PART OF KIT P/N 57K0137..................	3
					UOC:AVY,A13,A14,BVY,HVY,H13,H14,H21, H28,NNN	
24	MOOZZ		19207	12340700-11	TAPE,FASTENER HOOK MAKE FROM FASTENER,TAPE,HOOK P/N 20197272 PART OF KIT P/N 57K0137................	2
					UOC:AVY,A13,A14,BVY,HVY,H13,H14,H21, H28,NNN	

END OF FIGURE

Section II.   TM 9-2320-280-24P-2

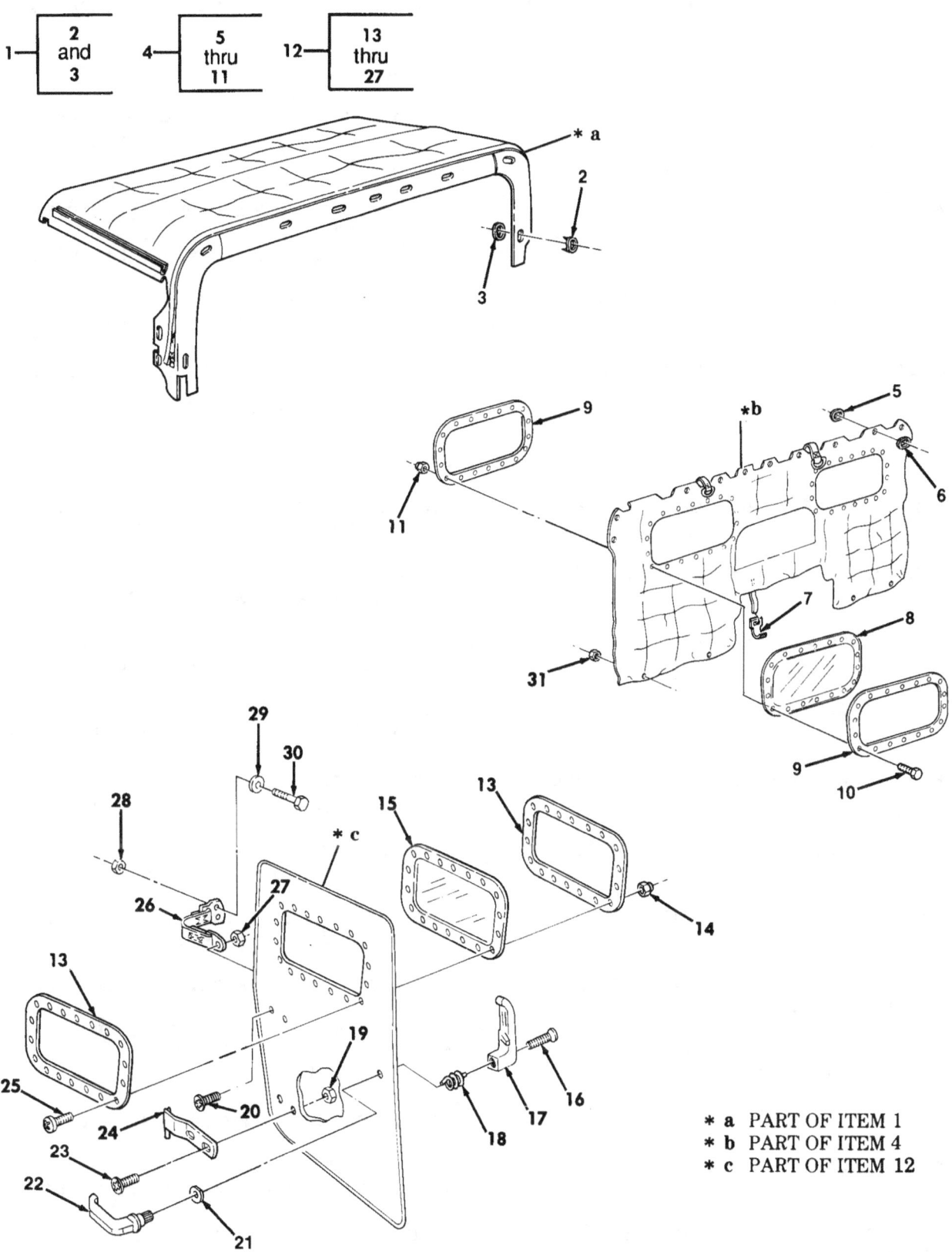

* a PART OF ITEM 1
* b PART OF ITEM 4
* c PART OF ITEM 12

Figure 375. Arctic Kit, Two Man Crew, Top Curtain, Rear Curtain, and Door Assemblies.

SECTION II TM9-2320-280-24P C01

(1) ITEM NO	(2) SMR CODE	(3) NSN	(4) CAGEC	(5) PART NUMBER	(6) DESCRIPTION AND USABLE ON CODES (UOC)	(7) QTY

GROUP 3303 WINTERIZATION KITS

FIG. 375 ARCTIC KIT, TWO MAN CREW TOP CURTAIN, REAR CURTAIN, AND DOOR ASSEMBLIES

1	PAOFF	2540013958785	34623	12342441	COVER,FITTED,VEHICU PART OF KIT P/N 57K0137............................. UOC:AVY,A13,A14,BVY,HVY,H13,H14,H21,H28,NNN	1
2	PAFZZ	5325006049662	21450	426686	.CLINCH PLATE,TURNBU................ UOC:AVY,A13,A14,BVY,HVY,H13,H14,H21,H28,NNN	8
3	PAFZZ	5325002818643	21450	426687	.SOCKET,TURNBUTTON F................ UOC:AVY,NNN	8
4	PAOFF	2540011975528	19207	12342442	CURTAIN,VEHICULAR PART OF KIT P/N 57K0137............................. UOC:AVY,A13,A14,BVY,HVY,H13,H14,H21,H28,NNN	1
5	PAFZZ	5325006049662	21450	426686	.CLINCH PLATE,TURNBU................ UOC:AVY,A13,A14,BVY,HVY,H13,H14,H21,H28,NNN	12
6	PAFZZ	5325002818643	21450	426687	.SOCKET,TURNBUTTON F................ UOC:AVY,A13,A14,BVY,HVY,H13,H14,H21,H28,NNN	12
7	PAFZZ	5340012036542	31272	44702-10	.BUCKLE............................. UOC:AVY,A13,A14,BVY,HVY,H13,H14,H21,H28,NNN	2
* 8	PAOZZ	2510013258741	19207	12340608	.WINDOW,VEHICULAR................... UOC:AVY,A13,A14,BVY,HVY,H13,H14,H21,H28,NNN	2
9	PFOZZ	5340011980686	19207	12340710	.PLATE,RETAINING,WIN................ UOC:AVY,A13,A14,BVY,HVY,H13,H14,H21,H28,NNN	4
10	PAOZZ	5305008132785	80205	NAS601-12P	.SCREW,MACHINE #8-32 X 3/4......... UOC:AVY,A13,A14,BVY,HVY,H13,H14,H21,H28,NNN	36
11	PAOZZ	5310010389579	24617	454748	.NUT,SELF-LOCKING,HE #8-32......... UOC:AVY,A13,A14,BVY,HVY,H13,H14,H21,H28,NNN	36
12	PAOOO	2510011975546	34623	SF5581723	DOOR,VEHICULAR FRONT,L.H. PART OF KIT P/N 57K0137..................... UOC:AVY,A13,A14,BVY,HVY,H13,H14,H21,H28,NNN	1
12	PAOOO	2510011975547	34623	SF5581724	DOOR,VEHICULAR FRONT,R.H. PART OF KIT P/N 57K0137..................... UOC:AVY,A13,A14,BVY,HVY,H13,H14,H21,H28,NNN	1
13	PFOZZ	5340011980686	19207	12340710	.PLATE,RETAINING,WIN................ UOC:AVY,A13,A14,BVY,HVY,H13,H14,H21,H28,NNN	2
14	PAOZZ	5310010389579	24617	454748	.NUT,SELF-LOCKING,HE #8-32......... UOC:AVY,A13,A14,BVY,HVY,H13,H14,H21,H28,NNN	18

SECTION II TM9-2320-280-24P

(1) ITEM NO	(2) SMR CODE	(3) NSN	(4) CAGEC	(5) PART NUMBER	(6) DESCRIPTION AND USABLE ON CODES(UOC)	(7) QTY
15	PFOZZ	2510013258741	19207	12340608	WINDOW,VEHICULAR UOC:AVY,A13,A14,BVY,HVY,H13,H14,H21,H28,NNN	1
16	PAOZZ	5310000711327	96906	MS51960-70	SCREW,MACHINE #10-32 X 1/4 UOC:AVY,A13,A14,BVY,HVY,H13,H14,H21,H28,NNN	1
17	PAOZZ	2540012037721	19207	12301363-2	HANDLE,DOOR,VEHICUL INSIDE,R.H. PART OF KIT P/N 5705619 UOC:AVY,A13,A14,BVY,HVY,H13,H14,H21,H28,NNN	1
17	PAOZZ	2540012001994	19207	12301363-1	HANDLE,DOOR,VEHICUL INSIDE,L.H. PART OF KIT P/N 5705618 UOC:AVY,A13,A14,BVY,HVY,H13,H14,H21,H28,NNN	1
18	PAOZZ	5360013157212	19207	12356764-1	SPRING,HELICAL,TORS L.H. PART OF KIT P/N 5705619 UOC:AVY,A13,A14,BVY,HVY,H13,H14,H21,H28,NNN	1
18	PAOZZ	5360013157211	19207	12356764-2	SPRING,HELICAL,TORS R.H. PART OF KIT P/N 5705618 PART OF KIT P/N 5705619 UOC:AVY,A13,A14,BVY,HVY,H13,H14,H21,H28,NNN	1
19	PAOZZ	5310009971888	96906	MS35649-2252	NUT,PLAIN,HEXAGON 1/4-20 UOC:AVY,A13,A14,BVY,HVY,H13,H14,H21,H28,NNN	3
20	PAOZZ	5305012067217	16941	PL25D040P8	SCREW,CAP,SOCKET HE 1/4-20 X 3/4 UOC:AVY,A13,A14,BVY,HVY,H13,H14,H21,H28,NNN	1
21	PAOZZ	5310012059056	34623	5581321	WASHER,FLAT PART OF KIT P/N 5705618 PART OF KIT P/N 5705619 UOC:AVY,A13,A14,BVY,HVY,H13,H14,H21,H28,NNN	1
22	PAOZZ	2540012001995	19207	12301365	HANDLE,DOOR,VEHICUL OUTSIDE,L.H., R.H. PART OF KIT P/N 5705618 PART OF KIT P/N 5705619 UOC:AVY,A13,A14,BVY,HVY,H13,H14,H21,H28,NNN	1
23	PAOZZ	5305000711315	96906	MS51957-79	SCREW,MACHINE 1/4-20 X 1/2 UOC:AVY,A13,A14,BVY,HVY,H13,H14,H21,H28,NNN	3
24	PAOZZ	5340012059021	19207	12340193-1	LEAF,BUTT HINGE L.H. UOC:AVY,A13,A14,BVY,HVY,H13,H14,H21,H28,NNN	2
24	PAOZZ	5340012059022	19207	12340193-2	LEAF,BUTT HINGE R.H. UOC:AVY,A13,A14,BVY,HVY,H13,H14,H21,H28,NNN	2
25	PAOZZ	5305008132785	80205	NAS601-12P	SCREW,MACHINE #8-32 X 3/4 UOC:AVY,A13,A14,BVY,HVY,H13,H14,H21,H28,NNN	18
26	PAOZZ	5340011993510	31272	44877	STRAP,WEBBING UOC:AVY,A13,A14,BVY,HVY,H13,H14,H21,	1

SECTION II   TM9-2320-280-24P

(1) NO	(2) SMR CODE	(3) NSN	(4) CAGEC	(5) PART NUMBER	(7) DESCRIPTION AND USABLE ON CODES(UOC)	ITEM QTY

27 PAOZZ 5310009971888 96906 MS35649-2252 .NUT,PLAIN,HEXAGON 1/4-20.......... 1
UOC:AVY,A13,A14,BVY,HVY,H13,H14,H21,H28,NNN

28 PAOZZ 5310000614650 96906 MS51943-31 NUT,SELF-LOCKING,HE 1/4-20 PART OF KIT P/N 57K0137.................... 2
UOC:AVY,A13,A14,BVY,HVY,H13,H14,H21,H28,NNN

29 PAOZZ 5310008094058 96906 MS27183-10 WASHER,FLAT 1/4................... 2
UOC:AVY,A13,A14,BVY,HVY,H13,H14,H21,H28,NNN

30 PAOZZ 5305000712506 80204 B1821BH025C050N SCREW,CAP,HEXAGON H 1/4-20 X 1/2... 2
UOC:AVY,A13,A14,BVY,HVY,H13,H14,H21,H28,NNN

31 PAOZZ 5310008140673 96906 MS51943-33 NUT,SELF-LOCKING,HE 5/16-18 PART OF KIT P/N 57K0137.................... 4
UOC:AVY,A13,A14,BVY,HVY,H13,H14,H21,H28,NNN

KIT PDFZZ 2540013944454 19207 57K0137 WINTERIZATION KIT,V................. 1
UOC:AVY,A13,A14,BVY,HVY,H13,H18,H21,H28,NNN

```
BOLT,MACHINE        ( 4) 373-11
BOLT,MACHINE        ( 4) 374-7
BOLT,MACHINE        ( 3) 374-22
BOW,VEHICULAR TOP   ( 1) 373-1
BRACE,ARCTIC CURTAI ( 1) 374-20
BRACKET,DOUBLE ANGL ( 1) 374-4
BUCKLE              ( 2) 373-10
COVER,FITTED,VEHICU ( 1) 375-1
CURTAIN,VEHICULAR   ( 1) 375-4
DOOR,VEHICULAR      ( 1) 375-12
DOOR,VEHICULAR      ( 1) 375-12
FASTENER,TAPE HOOK  ( 2) 374-9
FASTENER,TAPE HOOK  ( 2) 374-10
FASTENER,TAPE HOOK  ( 4) 374-11
FASTENER,TAPE HOOK  ( 1) 374-16
FASTENER,TAPE HOOK  ( 1) 374-16
FASTENER,TAPE HOOK  ( 2) 374-17
FASTENER,TAPE HOOK  ( 1) 374-18
FASTENER,TAPE HOOK  ( 1) 374-18
FASTENER,TAPE HOOK  ( 2) 374-19
FASTENER,TAPE HOOK  ( 2) 374-24
INSULATION,FOAM     ( 2) 374-1
LOOP,STRAP FASTENER ( 2) 374-3
NUT,PLAIN,BLIND RIV ( 4) 374-12
NUT,PLAIN,HEXAGON   ( 4) 374-15
NUT,SELF-LOCKING,HE ( 4) 373-8
NUT,SELF-LOCKING,HE ( 3) 374-23
NUT,SELF-LOCKING,HE ( 2) 375-28
NUT,SELF-LOCKING,HE ( 4) 375-31
PLASTIC,MATERIAL,CE ( 1) 374-13
RAIL ASSEMBLY,HAND  ( 1) 373-5
RAIL ASSEMBLY,HAND  ( 1) 373-5
```

SECTION II   TM9-2320-280-24P

(1) NO	(2) CODE	(3) NSN	(4) CAGEC	(5) PART NUMBER	(6) DESCRIPTION AND USABLE ON CODES(UOC)	(7) ITEM SMR QTY
					RIVET,BLIND	(6) 373-7
					SCREW,MACHINE	(4) 373-3
					SCREW,MACHINE	(2) 373-4
					SCREW,MACHINE	(4) 374-2
					STRAP,RETAINING	(4) 373-2
					STUD,TURNBUTTON FAS	(3) 373-6
					WASHER,FLAT	(10) 374-8
					WASHER,FLAT	(4) 374-14
					WASHER,LOCK	(8) 373-9

END OF FIGURE

Section II. TM 9-2320-280-24P-2

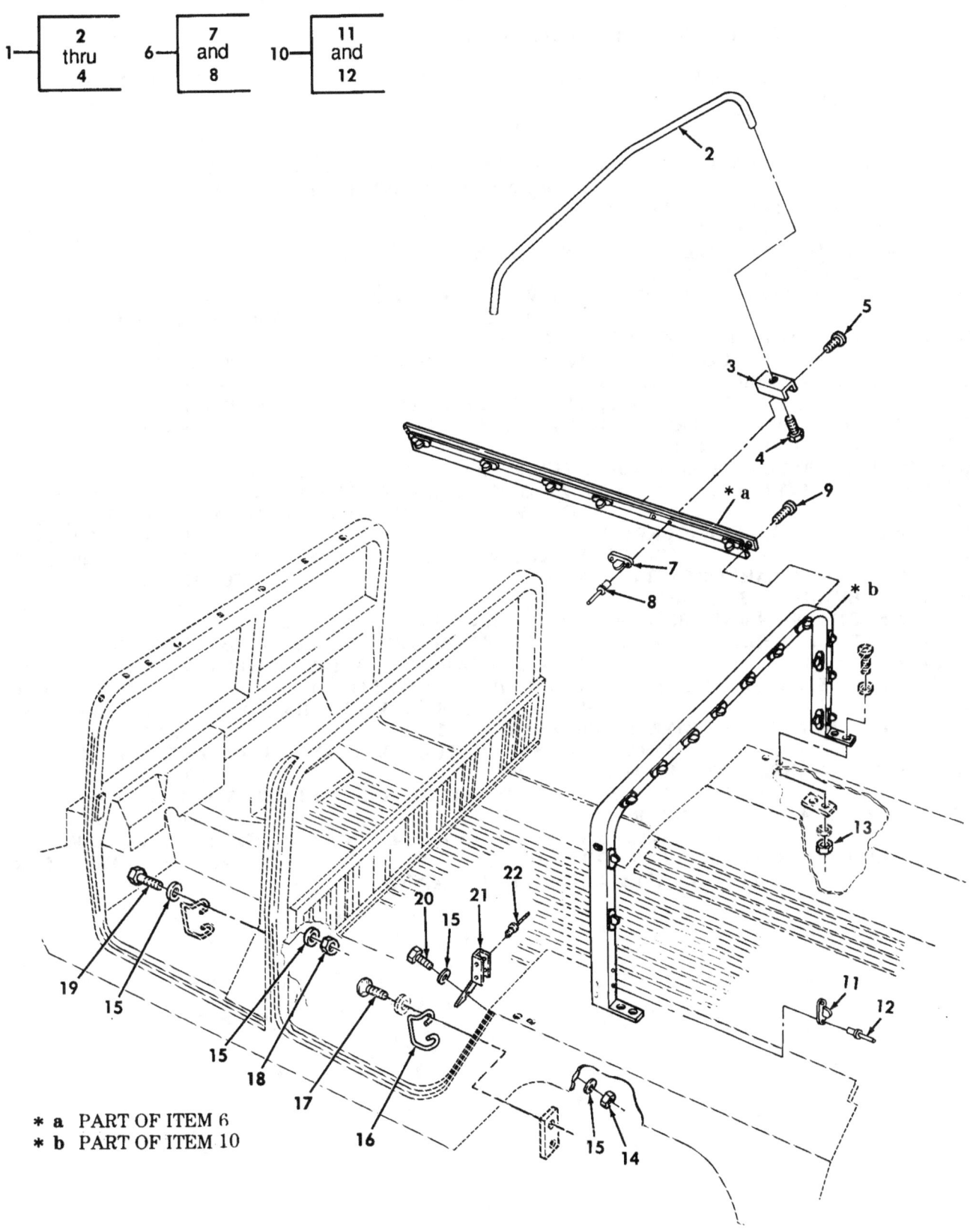

* a PART OF ITEM 6
* b PART OF ITEM 10

Figure 376. Arctic Kit, Four Man Crew, Top Enclosure, Bows, and Mounting Components.

SECTION II TM9-2320-280-24P

(1) NO	(2) SMR CODE	(3) NSN	(4) CAGEC	(5) PART NUMBER	(7) ITEM DESCRIPTION AND USABLE ON CODES(UOC)	QTY

GROUP 3303 WINTERIZATION KITS

FIG. 376 ARCTIC KIT,FOUR MAN CREW, TOP ENCLOSURE,BOWS,AND MOUNTING COMPONENTS

1	PFOZZ	2540011996759	19207	12340682	BOW,VEHICULAR TOP PART OF KIT P/N 5705692............................ UOC:AVY,A13,A14,BVY,HVY,H13,H14,NNN	2
2	XAOZZ		19207	12340713	.BOW,VEHICULAR TOP.................. UOC:AVY,A13,A14,BVY,HVY,H13,H14,NNN	1
3	PFOZZ	5340012043862	19207	12340239	.STRAP,RETAINING.................... UOC:AVY,A13,A14,BVY,HVY,H13,H14,NNN	2
4	PAOZZ	5305000711313	96906	MS51957-77	.SCREW,MACHINE 1/4-20 X 3/8........ UOC:AVY,A13,A14,BVY,HVY,H13,H14,NNN	2
5	PAOZZ	5305000593659	96906	MS51958-63	SCREW,MACHINE #10-32 X 1/2 PART OF KIT P/N 5705692.................... UOC:AVY,A13,A14,BVY,HVY,H13,H14,NNN	8
6	PFOZZ	2590011961290	19207	12340222-1	HANDRAIL,VEHICULAR L.H. PART OF KIT P/N 5705692............................ UOC:AVY,A13,A14,BVY,HVY,H13,H14,NNN	1
6	PFOZZ	2590011965314	19207	12340222-2	HANDRAIL,VEHICULAR R.H. PART OF KIT P/N 5705692........................ UOC:AVY,A13,A14,BVY,HVY,H13,H14,NNN	1
7	PAOZZ	5325008235999	13940	XB78323-05001	.STUD,TURNBUTTON FAS................ UOC:AVY,A13,A14,BVY,HVY,H13,H14,NNN	6
8	PAOZZ	5320000835009	81349	M24243/1-A403	.RIVET,BLIND 1/8 DIA,.126-.187 GRIP UOC:AVY,A13,A14,BVY,HVY,H13,H14,NNN	12
9	PAOZZ	5305011173396	80205	NAS-1635-3LE12	SCREW,CAP,SOCKET HE #10-32 X 1/2 PART OF KIT P/N 5705692............. UOC:AVY,A13,A14,BVY,HVY,H13,H14,NNN	2
10	PFOZZ	2540011961291	19207	12340679	BOW,VEHICULAR TOP PART OF KIT P/N 5705692............................ UOC:AVY,A13,A14,BVY,HVY,H13,H14,NNN	1
11	PAOZZ	5325008235999	13940	XB78323-05001	.STUD,TURNBUTTON FAS................ UOC:AVY,A13,A14,BVY,HVY,H13,H14,NNN	17
12	PAOZZ	5320000835009	81349	M24243/1-A403	.RIVET,BLIND 1/8 DIA,.126-.187 GRIP UOC:AVY,A13,A14,BVY,HVY,H13,H14,NNN	34
13	PAOZZ	5310000614650	96906	MS51943-31	NUT,SELF-LOCKING,HE 1/4-20 PART OF KIT P/N 5705692.................... UOC:AVY,A13,A14,BVY,HVY,H13,H14,NNN	4
14	PAOZZ	5310004838791	96906	MS17829-4F	NUT,SELF-LOCKING,HE #10-32 PART OF KIT P/N 5705692.................... UOC:AVY,A13,A14,BVY,HVY,H13,H14,NNN	2
15	PAOZZ	5310000814219	96906	MS27183-12	WASHER,FLAT 5/16 PART OF KIT P/N 5705692................... UOC:AVY,A13,A14,BVY,HVY,H13,H14,NNN	12
16	PAOZZ	5340012087670	19207	12340182-1	STRIKE,CATCH REAR DOOR,L.H. PART OF KIT P/N 5705692....... UOC:AVY,A13,A14,BVY,HVY,H13,H14,NNN	1
16	PAOZZ	5342012049610	34623	5581364	STRIKER,LATCH,DOOR REAR DOOR,R.H. PART OF KIT P/N 5705692........ UOC:AVY,A13,A14,BVY,HVY,H13,H14,NNN	1

SECTION II TM9-2320-280-24P

(1) ITEM NO	(2) SMR CODE	(3) NSN	(4) CAGEC	(5) PART NUMBER	(6) DESCRIPTION AND USABLE ON CODES(UOC)	(7) QTY
17	PAOZZ	5305009881727	96906	MS35206-283	SCREW,MACHINE 1/4-20 X 3/4 PART OF KIT P/N 5705692.................... UOC:AVY,A13,A14,BVY,HVY,H13,H14,NNN	4
18	PAOZZ	5310002416658	96906	MS51943-34	NUT,SELF-LOCKING,HE 5/16-24 PART OF KIT P/N 5705692.................... UOC:AVY,A13,A14,BVY,HVY,H13,H14,NNN	4
19	PAOZZ	5306001822027	88044	AN5H10A	BOLT,MACHINE PART OF KIT P/N 5705692 UOC:AVY,A13,A14,BVY,HVY,H13,H14,NNN	4
20	PAOZZ	5305009932457	96906	MS35207-284	SCREW,MACHINE 1/4-28 X 1-1/4 PART OF KIT P/N 5705692.................... UOC:AVY,A13,A14,BVY,HVY,H13,H14,NNN	2
21	PAOZA	5340014309239	19207	12447124-1	BRACKET,MOUNTING C-PILLAR,RH PART OF KIT P/N 5705692.................... UOC:AVY,A13,A14,BVY,HVY,H13,H14,NNN	1
21	PAOZA	5340014309240	19207	12447123-1	BRACKET,MOUNTING C-PILLAR,LH PART OF KIT P/N 5705692.................... UOC:AVY,A13,A14,BVY,HVY,H13,H14,NNN	1
22	PAOZZ	5320010341884	11815	TR3243-6-5	RIVET,BLIND PART OF KIT P/N 5705692. UOC:AVY,A13,A14,BVY,HVY,H13,H14,NNN	8

END OF FIGURE

Section II. TM 9-2320-280-24P-2

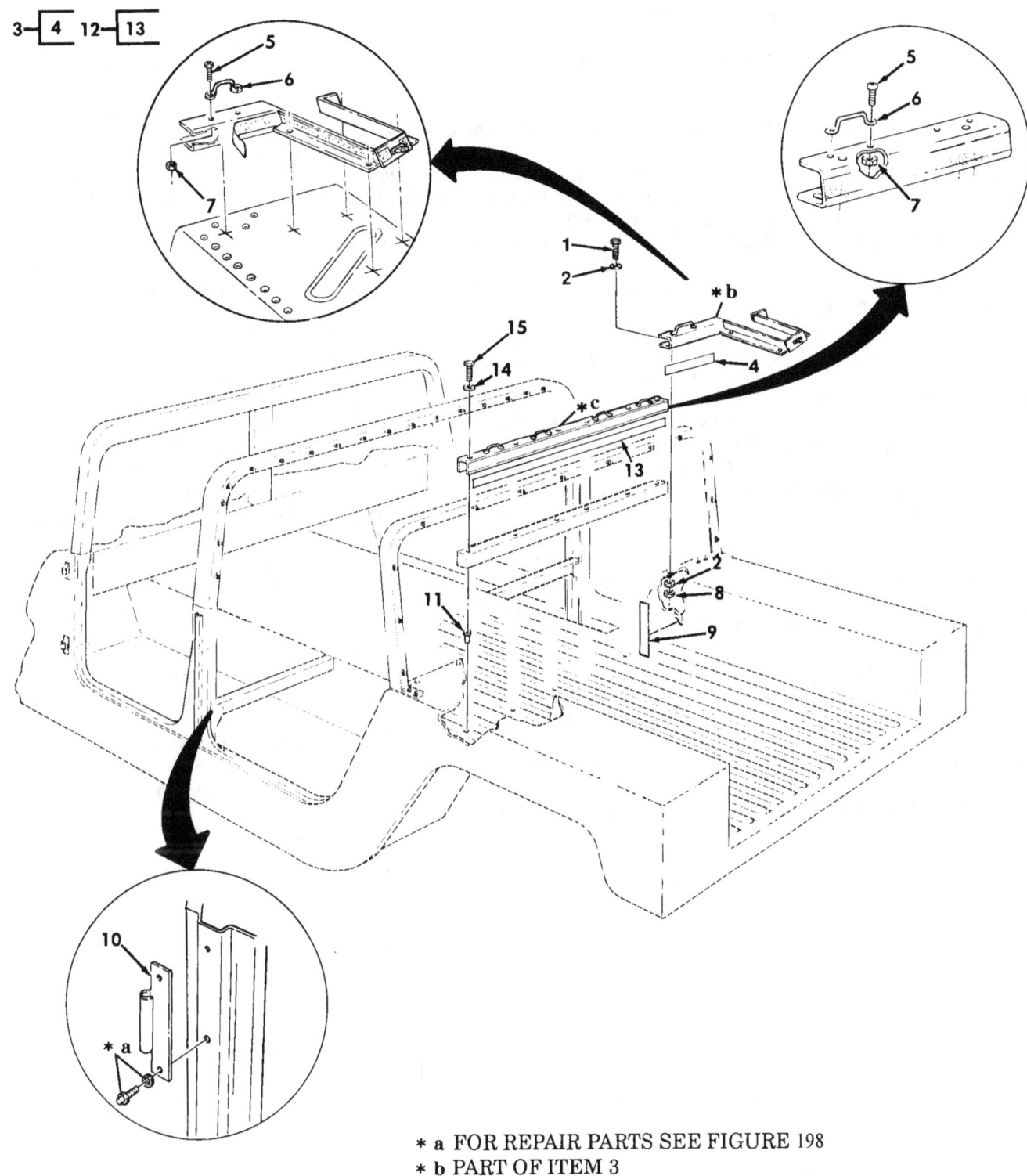

* a FOR REPAIR PARTS SEE FIGURE 198
* b PART OF ITEM 3
* c PART OF ITEM 11

Figure 377. Arctic Kit, Door Hinges, and Curtain Channels.

SECTION II   TM9-2320-280-24P

(1) NO	(2) SMR CODE	(3) NSN	(4) CAGEC	(5) PART NUMBER	(7) DESCRIPTION AND USABLE ON CODES(UOC)	ITEM QTY

GROUP 3303 WINTERIZATION KITS

FIG. 377 ARCTIC KIT, DOOR HINGES, AND CURTAIN CHANNELS

1	PAOZZ	5305000680502	96906	MS90725-6	SCREW,CAP,HEXAGON H 1/4-20 X 3/4 PART OF KIT P/N 5705692............. UOC:AVY,A13,A14,BVY,HVY,H13,H14,NNN	10
2	PAOZZ	5310011992293	7X677	9421394	WASHER,FLAT 1/4 PART OF KIT P/N 5705692.................................. UOC:AVY,A13,A14,BVY,HVY,H13,H14,NNN	20
3	PAOZZ	5340013968322	19207	12342496	BRACKET,MOUNTING R.H. PART OF KIT P/N 5705692......................... UOC:AVY,A13,A14,BVY,HVY,H13,H14,NNN	1
3	PAOZZ	5340013850057	19207	12342495	BRACKET,MOUNTING L.H. PART OF KIT P/N 5705692......................... UOC:AVY,A13,A14,BVY,HVY,H13,H14,NNN	1
4	MOOZZ		19207	12340700-14	.TAPE,FASTENER MAKE FROM FASTENER, TAPE,HOOK P/N 20197272,14 INCHES LONG................................ UOC:AVY,A13,A14,BVY,HVY,H13,H14,NNN	2
5	PAOZZ	5305009847341	96906	MS35191-273	SCREW,MACHINE #10-32 X 5/8 PART OF KIT P/N 5705692..................... UOC:AVY,A13,A14,BVY,HVY,H13,H14,NNN	12
6	PAOZZ	5340011978673	19207	12338839-4	HANDLE,BOW PART OF KIT P/N 5705692.. UOC:AVY,A13,A14,BVY,HVY,H13,H14,NNN	6
7	PAOZZ	5310001641790	72582	454749	NUT,SELF-LOCKING,HE #10-32 PART OF KIT P/N 5705692...................... UOC:AVY,A13,A14,BVY,HVY,H13,H14,NNN	12
8	PAOZZ	5310014326727	24617	9419471	NUT,SELF-LOCKING,HE 1/4-20 PART OF KIT P/N 5705692...................... UOC:AVY,A13,A14,BVY,HVY,H13,H14,NNN	10
9	MOOZZ		19207	12340700-16	TAPE,FASTENER HOOK MAKE FROM FASTENER,TAPE,HOOK P/N 20197272,16 INCHES LONG PART OF KIT P/N 5705692. UOC:AVY,A13,A14,BVY,HVY,H13,H14,NNN	2
10	PAOZZ	5342012032822	19207	12338650	MOUNT,RESILIENT PART OF KIT P/N 5705692..................................... UOC:AVY,A13,A14,BVY,HVY,H13,H14,NNN	4
11	PAOZZ	5310010851208	96906	MS27130-S71	NUT,PLAIN,BLIND RIV .030-.125 PART OF KIT P/N 5705692................. UOC:AVY,A13,A14,BVY,HVY,H13,H14,NNN	6
12	PAOZZ	2590013332999	80212	12342152	BEZEL,AUTOMOTIVE TR PART OF KIT P/N 5705692............................. UOC:AVY,A13,A14,BVY,HVY,H13,H14,NNN	1
13	MOOZZ		19207	12340700-15	.TAPE,FASTENER MAKE FROM FASTENER, TAPE,HOOK P/N 20197272,15 INCHES LONG................................ UOC:AVY,A13,A14,BVY,HVY,H13,H14,NNN	1
14	PAOZZ	5310000814219	96906	MS27183-12	WASHER,FLAT 11/32 PART OF KIT P/N 5705692..................................... UOC:AVY,A13,A14,BVY,HVY,H13,H14,NNN	6
15	PAOZZ	5306002264833	80204	B1821BH031C200N	BOLT,MACHINE 5/16-18 X 2.00 PART OF	6

(1) NO	(2) CODE	(3) NSN	(4) CAGEC	(5) NUMBER	(6) PART DESCRIPTION AND USABLE ON CODES(UOC)	(7) ITEM SMR QTY
					KIT P/N 5705692.................... UOC:AVY,A13,A14,BVY,H-Y,H13,H14,NNN	
					END OF FIGURE	

Section II. TM 9-2320-280-24P-2

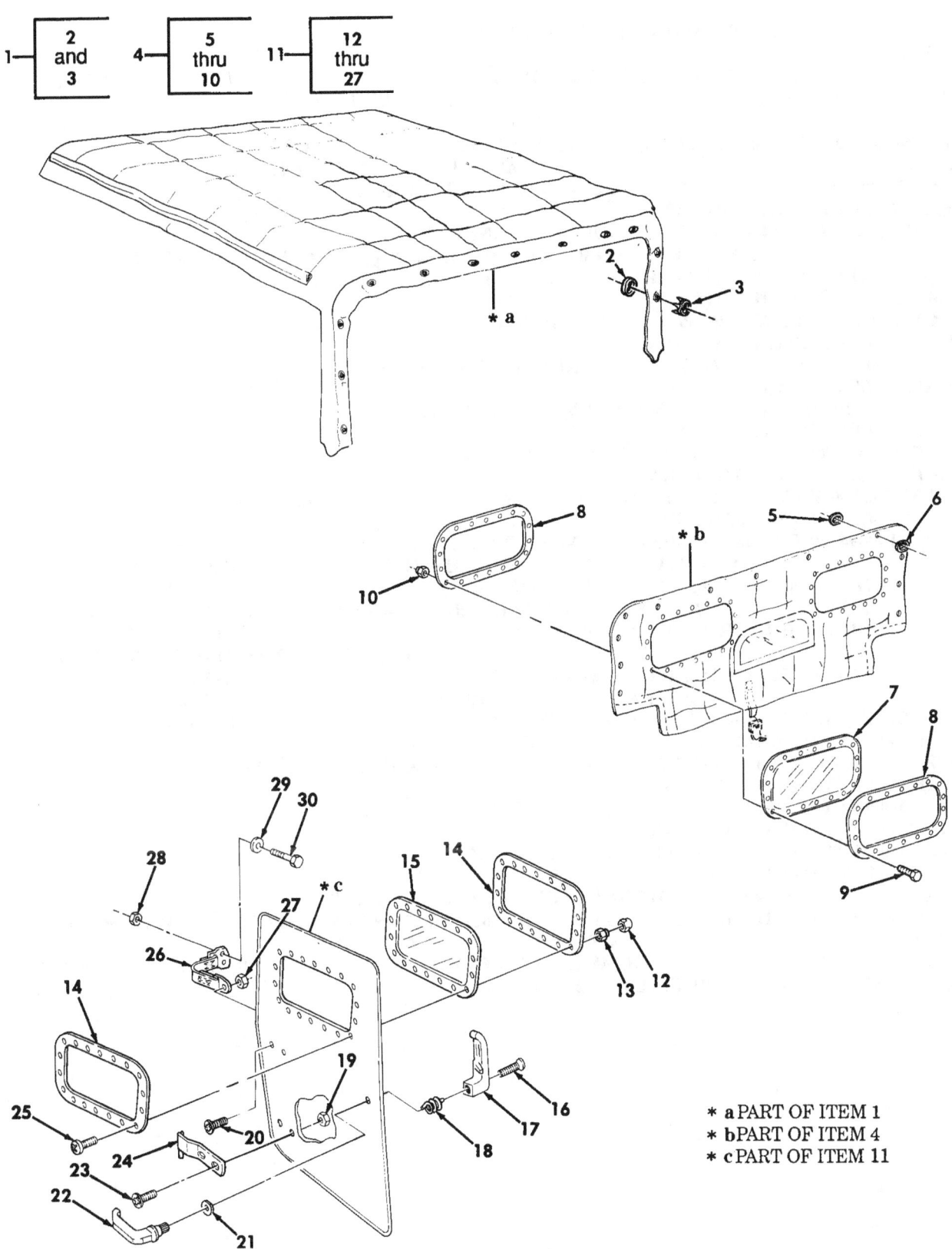

Figure 378. Arctic Kit, Four Man Crew, Top Curtain, Rear Curtain, and Door Assemblies.

SECTION II   TM9-2320-280-24P

(1) NO	(2) SMR CODE	(3) NSN	(4) CAGEC	(5) PART NUMBER	(7) DESCRIPTION AND USABLE ON CODES (UOC)	QTY

GROUP 3303 WINTERIZATION KITS

FIG. 378 ARCTIC KIT, FOUR MAN CREW, TOP CURTAIN, REAR CURTAIN, AND DOOR ASSEMBLIES

1	PAOFF	2540013298649	19207	12342153-1	COVER,FITTED,VEHICU PART OF KIT P/N 5705692.......................... UOC:AVY,A13,A14,BVY,HVY,H13,H14,NNN	1
2	PAFZZ	5325006049662	21450	426686	.CLINCH PLATE,TURNBU................ UOC:AVY,A13,A14,BVY,HVY,H13,H14,NNN	31
3	PAFZZ	5325002818643	21450	426687	.SOCKET,TURNBUTTON F................ UOC:AVY,A13,A14,BVY,HVY,H13,H14,NNN	31
4	PAOFF	2540013360791	19207	12342153-2	CURTAIN,VEHICULAR PART OF KIT P/N 5705692.......................... UOC:AVY,A13,A14,BVY,HVY,H13,H14,NNN	1
5	PAFZZ	5325006049662	21450	426686	.CLINCH PLATE,TURNBU................ UOC:AVY,A13,A14,BVY,HVY,H13,H14,NNN	13
6	PAFZZ	5325002818643	21450	426687	.SOCKET,TURNBUTTON F................ UOC:AVY,A13,A14,BVY,HVY,H13,H14,NNN	13
7	XBFZZ		19207	12340604	.WINDOW,VEHICULAR................ UOC:AVY,A13,A14,BVY,HVY,H13,H14,NNN	2
8	PFOZZ	5340011980686	19207	12340710	.PLATE,RETAINING,WIN................ UOC:AVY,A13,A14,BVY,HVY,H13,H14,NNN	4
9	PAOZZ	5305009846195	96906	MS35206-247	.SCREW,MACHINE #8-32 X 3/4......... UOC:AVY,A13,A14,BVY,HVY,H13,H14,NNN	36
10	PAOZZ	5310009416019	96906	MS21083N08	.NUT,SELF-LOCKING,HE #8-32......... UOC:AVY,A13,A14,BVY,HVY,H13,H14,NNN	36
11	PAOOO	2510011975547	34623	SF5581724	DOOR,VEHICULAR FRONT,R.H. PART OF KIT P/N 5705692......... UOC:AVY,A13,A14,BVY,HVY,H13,H14,NNN	1
11	PAOOO	2510011975546	34623	SF5581723	DOOR,VEHICULAR FRONT,L.H. PART OF KIT P/N 5705692......... UOC:AVY,A13,A14,BVY,HVY,H13,H14,NNN	1
11	PAOOO	2510013302249	19207	12342157	DOOR,VEHICULAR REAR,L.H. PART OF KIT P/N 5705692........... UOC:AVY,A13,A14,BVY,HVY,H13,H14,NNN	1
11	PAOOO	2510013302250	19207	12342155	DOOR,VEHICULAR REAR,R.H. PART OF KIT P/N 5705692................... UOC:AVY,A13,A14,BVY,HVY,H13,H14,NNN	1
12	PAOZZ	5340010263250	81349	M5501/7-F2	.CAP-PLUG,PROTECTIVE................ UOC:AVY,A13,A14,BVY,HVY,H13,H14,NNN	18
13	PAOZZ	5310009416019	96906	MS21083N08	.NUT,SELF-LOCKING,HE #8-32......... UOC:AVY,A13,A14,BVY,HVY,H13,H14,NNN	18
14	PFOZZ	5340011980686	19207	12340710	.PLATE,RETAINING,WIN................ UOC:AVY,A13,A14,BVY,HVY,H13,H14,NNN	2
15	XBOZZ		19207	12340605	.WINDOW,VEHICULAR................... UOC:AVY,A13,A14,BVY,HVY,H13,H14,NNN	1
16	PAOZZ	5305000711327	96906	MS51960-70	.SCREW,MACHINE #10-32 X 1/4 PART OF KIT P/N 5705618 PART OF KIT P/N 5705619.... .... UOC:AVY,A13,A14,BVY,HVY,H13,H14,NNN	1
17	PAOZZ	2540012037721	19207	12301363-2	.HANDLE,DOOR,VEHICUL L.H. PART OF	1

378-1

SECTION II   TM9-2320-280-24P

(1) NO	(2) SMR CODE	(3) NSN	(4) CAGEC	(5) PART NUMBER	(6) DESCRIPTION AND USABLE ON CODES(UOC)	(7) ITEM QTY
					KIT P/N 5705619..................... UOC:AVY,A13,A14,BVY,HVY,H13,H14,NNN	
17	PAOZZ	2540012001994	19207	12301363-1	.HANDLE,DOOR,VEHICUL R.H. PART OF KIT P/N 5705618..................... UOC:AVY,A13,A14,BVY,HVY,H13,H14,NNN	1
18	PAOZZ	5360013157212	19207	12356764-1	.SPRING,HELICAL,TORS L.H. PART OF KIT P/N 5705619..................... UOC:AVY,A13,A14,BVY,HVY,H13,H14,NNN	1
18	PAOZZ	5360013157211	19207	12356764-2	.SPRING,HELICAL,TORS R.H. PART OF KIT P/N 5705618................. UOC:AVY,A13,A14,BVY,HVY,H13,H14,NNN	1
19	PAOZZ	5310011520598	24617	271172	.NUT,SELF-LOCKING,AS................ UOC:AVY,A13,A14,BVY,HVY,H13,H14,NNN	13
20	PAOZZ	5305009881725	96906	MS35206-281	.SCREW,MACHINE 1/4-20 X 3/4........ UOC:AVY,A13,A14,BVY,HVY,H13,H14,NNN	
21	PAOZZ	5310012059056	34623	5581321	.WASHER,FLAT 5/8 PART OF KIT P/N 5705618 PART OF KIT P/N 5705619..... UOC:AVY,A13,A14,BVY,HVY,H13,H14,NNN	1
22	PAOZZ	2540012001995	19207	12301365	.HANDLE,DOOR,VEHICUL L.H.,R.H. PART OF KIT P/N 5705618 PART OF KIT P/N 5705619....................... UOC:AVY,A13,A14,BVY,HVY,H13,H14,NNN	1
23	PAOZZ	5305000711315	96906	MS51957-79	.SCREW,MACHINE 1/4-20 X 1/2........ UOC:AVY,A13,A14,BVY,HVY,H13,H14,NNN	
24	PAOZZ	5340012059022	19207	12340193-2	.LEAF,BUTT HINGE FRONT AND REAR, L.H............................ UOC:AVY,A13,A14,BVY,HVY,H13,H14,NNN	2
24	PAOZZ	5340012059021	19207	12340193-1	.LEAF,BUTT HINGE FRONT AND REAR, R.H............................ UOC:AVY,A13,A14,BVY,HVY,H13,H14,NNN	2
25	PAOZZ	5305009846195	96906	MS35206-247	.SCREW,MACHINE #8-32 X 3/4......... UOC:AVY,A13,A14,BVY,HVY,H13,H14,NNN	18
26	PAOZZ	5340012547189	19207	12340201	.LOOP,STRAP FASTENER................ UOC:AVY,A13,A14,BVY,HVY,H13,H14,NNN	1
27	PAOZZ	5310009971888	96906	MS35649-2252	.NUT,PLAIN,HEXAGON 1/4-20.......... UOC:AVY,A13,A14,BVY,HVY,H13,H14,NNN	
28	PAOZZ	5310002416658	96906	MS51943-34	NUT,SELF-LOCKING,HE 5/16-20 PART OF KIT P/N 5705692........ UOC:AVY,A13,A14,BVY,HVY,H13,H14,NNN	2
29	PAOZZ	5310000814219	96906	MS27183-12	WASHER,FLAT 11/32 PART OF KIT P/N 5705692...................... UOC:AVY,A13,A14,BVY,HVY,H13,H14,NNN	2
30	PAOZZ	5305000514076	80204	B1821BH-31F100N	SCREW,CAP,HEXAGON H 5/16-24 X 1.00 PART OF KIT P/N 5705692............. UOC:AVY,A13,A14,BVY,HVY,H13,H14,NNN	2
	KIT PDFZZ	2540011992390	19207	5705692	WINTERIZATION KIT,V................ UOC:A13,A14,HVY,H13,H14	1

```
BEZEL,AUTO          ( 1) 377-12
BOLT,MACHINE        ( 4) 376-19
BOLT,MACHINE        ( 6) 377-15
BOW,VEHICULAR TOP   ( 2) 376-1
BOW,VEHICULAR TOP   ( 1) 376-10
BRACKET,MOUNTING    ( 1) 376-21
```

(1) NO	(2) CODE	(3) NSN	(4) CAGEC	(5) PART NUMBER	(7) ITEM DESCRIPTION AND USABLE ON CODES(UOC)	(6) SMR QTY
					BRACKET,MOUNTING	(1) 376-21
					BRACKET,MOUNTING	(1) 377-3
					BRACKET,MOUNTING	(1) 377-3
					COVER,FITTED,VEHICU	(1) 378-1
					CURTAIN,VEHICULAR	(1) 378-4
					DOOR,VEHICULAR	(1) 378-11
					DOOR,VEHICULAR	(1) 378-11
					DOOR,VEHICULAR	(1) 378-11
					DOOR,VEHICULAR	(1) 378-11
					FASTENER,TAPE,HOOK	(2) 377-9
					HANDLE,BOW	(6) 377-6
					HANDRAIL,VEHICULAR	(1) 376-6
					HANDRAIL,VEHICULAR	(1) 376-6
					MOUNT	(4) 377-10
					NUT,PLAIN,BLIND RIV	(6) 377-11
					NUT,SELF-LOCKING,HE	(4) 376-13
					NUT,SELF-LOCKING,HE	(2) 376-14
					NUT,SELF-LOCKING,HE	(4) 376-18
					NUT,SELF-LOCKING,HE	(12) 377-7
					NUT,SELF-LOCKING,HE	(10) 377-8
					NUT,SELF-LOCKING.HE	(2) 378-28
					RIVET,BLIND	(8) 376-22
					SCREW,CAP,HEXAGON H	(10) 377-1
					SCREW,CAP,HEXAGON H	(2) 378-30
					SCREW,CAP,SOCKET HE	(2) 376-9
					SCREW,MACHINE	(8) 376-5
					SCREW,MACHINE	(4) 376-17
					SCREW,MACHINE	(2) 376-20
					SCREW,MACHINE	(12) 377-5
					STRIKE,CATCH	(1) 376-16
					STRIKE,CATCH	(1) 376-16
					WASHER,FLAT	(12) 376-15
					WASHER,FLAT	(20) 377-2
					WASHER,FLAT	(6) 377-14
					WASHER,FLAT	(2) 378-29

END OF FIGURE

Section II.

TM 9-2320-280-24P-2

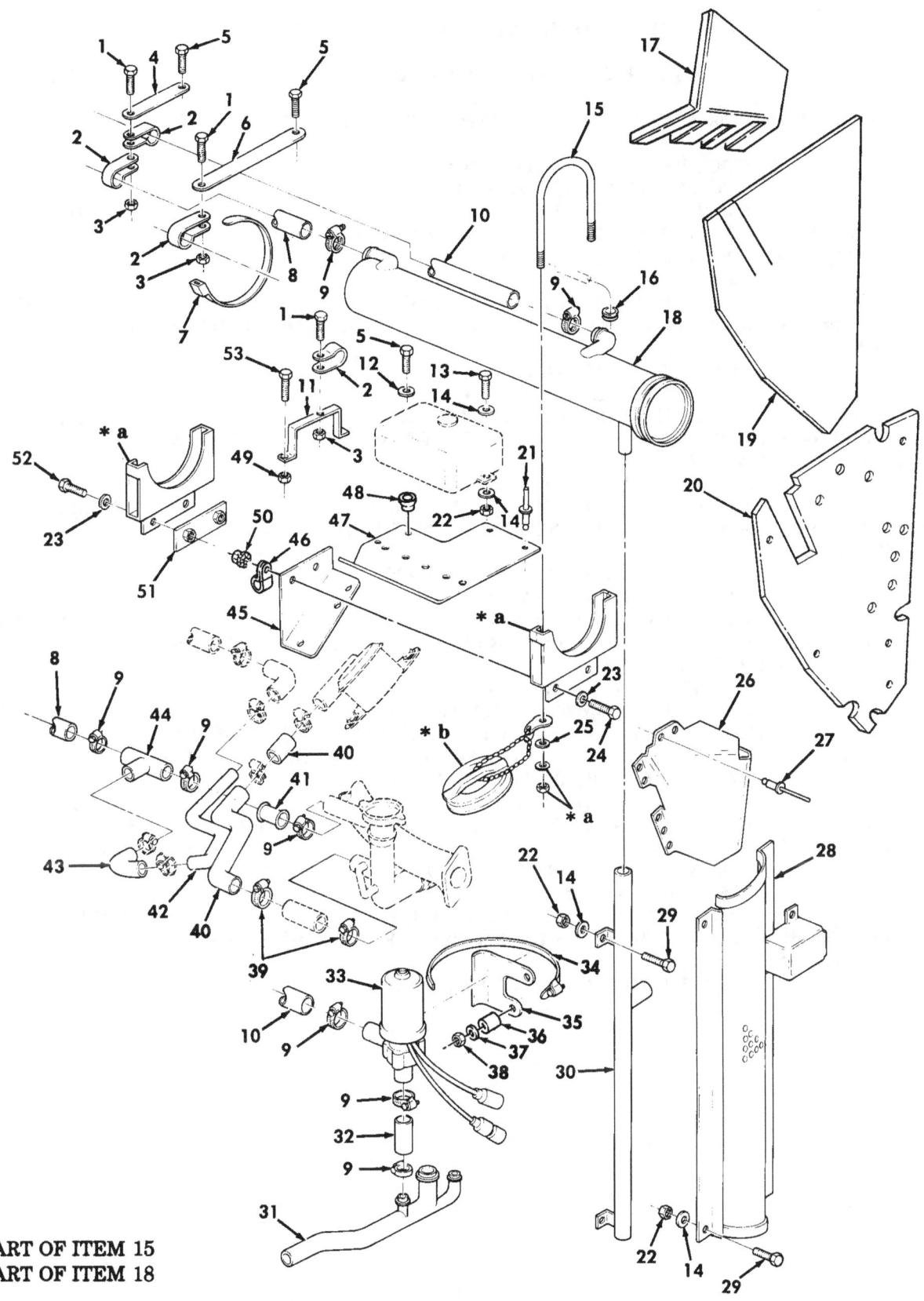

* a PART OF ITEM 15
* b PART OF ITEM 18

Figure 379. Swingfire Heater Kit and Related Parts.

SECTION II   TM9-2320-280-24P

(1) NO	(2) CODE	(3) NSN	(4) CAGEC	(5) NUMBER	(6) DESCRIPTION AND USABLE ON CODES(UOC)	(7) ITEM SMR PART QTY

GROUP 3303 WINTERIZATION KITS

FIG. 379 SWINGFIRE HEATER KIT AND RELATED PARTS

1 PAOZZ 5305009846194 96906 MS35206-246   SCREW,MACHINE #8-32 X 5/8 PART OF   3
KIT P/N 5705615.....................
UOC:AVY,A11,A13,A14,A15,A20,A24,A25,
A26,A27,B16,B17,B18,HVY,H11,H13,H14,
H15,H16,H17,H18,H20,H21,H24,H25,H26,   H27,H28,MMM

2 PAOZZ 5340009226300 96906 MS21333-77   CLAMP,LOOP PART OF KIT P/N 5705615..   4
UOC:AVY,A11,A13,A14,A15,A20,A24,A25,
A26,A27,B16,B17,B18,HVY,H11,H13,H14,
H15,H16,H17,H18,H20,H21,H24,H25,H26,   H27,H28,MMM

3 PAOZZ 5310008113494 96906 MS21044-N08   NUT,SELF-LOCKING,HE #8-32 PART OF   3
KIT P/N 5705615.....................
UOC:AVY,A11,A13,A14,A15,A20,A24,A25,
A26,A27,B16,B17,B18,HVY,H11,H13,H14,
H15,H16,H17,H18,H20,H21,H24,H25,H26,   H27,H28,MMM

4 PAOZZ 5365013145544 19207 12368397-2   SPACER,PLATE PART OF KIT P/N 5705615   1
UOC:AVY,A11,A13,A14,A15,A20,A24,A25,
A26,A27,B16,B17,B18,HVY,H11,H13,H14,
H15,H16,H17,H18,H20,H21,H24,H25,H26,   H27,H28,MMM

5 PAOZZ 5305007578287 96906 MS51851-43   SCREW,TAPPING #8-12 X 3/8 PART OF   3
KIT P/N 5705615.....................
UOC:AVY,A11,A13,A14,A15,A20,A24,A25,
A26,A27,B16,B17,B18,HVY,H11,H13,H14,
H15,H16,H17,H18,H20,H21,H24,H25,H26,   H27,H28,MMM

6 PAOZZ 5365013145544 19207 12368397-2   SPACER,PLATE PART OF KIT P/N 5705615   1
UOC:AVY,A11,A13,A14,A15,A20,A24,A25,
A26,A27,B16,B17,B18,HVY,H11,H13,H14,
H15,H16,H17,H18,H20,H21,H24,H25,H26,   H27,H28,MMM

7 PAOZZ 5975005709598 96906 MS3367-7-9   STRAP,TIEDOWN,ELECT PART OF KIT P/N   9
5705615............................   UOC:AVY,A11,A13,A14,A15,A20,A24,A25,
A26,A27,B16,B17,B18,HVY,H11,H13,H14,
H15,H16,H17,H18,H20,H21,H24,H25,H26,   H27,H28,M-
MM   8 MOOZZ   96906 MS521301A203R 74 HOSE,NONMETALLIC MAKE FROM HOSE,P/N   1
MS521301A203R,74 INCHES LONG PART OF   KIT P/N 5705615....................
UOC:AVY,A11,A13,A14,A15,A20,A24,A25,
A26,A27,B16,B17,B18,HVY,H11,H13,H14,
H15,H16,H17,H18,H20,H21,H24,H25,H26,   H27,H28,MMM

9 PAOZZ 4730009083194 96906 MS35842-11   CLAMP,HOSE PART OF KIT P/N 5705615..   8
UOC:AVY,A11,A13,A14,A15,A20,A24,A25,
A26,A27,B16,B17,B18,HVY,H11,H13,H14,

379-1

SECTION II TM9-2320-280-24P

(1) NO	(2) SMR CODE	(3) NSN	(4) CAGEC	(5) PART NUMBER	(6) DESCRIPTION AND USABLE ON CODES(UOC)	(7) QTY
10	MOOZZ		96906	MS521301A203R 58	HOSE,NONMETALLIC MAKE FROM HOSE,P/N MS521301A203R,58 INCHES LONG PART OF KIT P/N 5705615..................... UOC:AVY,A11,A13,A14,A15,A20,A24,A25, A26,A27,B16,B17,B18,HVY,H11,H13,H14, H15,H16,H17,H18,H20,H21,H24,H25,H26, H27,H28,MMM	1
11	PAOZZ	5340013145567	19207	12368392	BRACKET,MULTIPLE AN PART OF KIT P/N 5705615............................. UOC:AVY,A11,A13,A14,A15,A20,A24,A25, A26,A27,B16,B17,B18,HVY,H11,H13,H14, H15,H16,H17,H18,H20,H21,H24,H25,H26, H27,H28,M-MM	1
12	PAOZZ	5310005590070	96906	MS35333-38	WASHER,LOCK PART OF KIT P/N 5705615. UOC:AVY,A11,A13,A14,A15,A20,A24,A25, A26,A27,B16,B17,B18,HVY,H11,H13,H14, H15,H16,H17,H18,H20,H21,H24,H25,H26, H27,H28,MMM	1
13	PAOZZ	5306000680514	80204	B1821BH025F088N	BOLT,MACHINE 1/4-28 X 7/8 PART OF KIT P/N 5705615..................... UOC:AVY,A11,A13,A14,A15,A20,A24,A25, A26,A27,B16,B17,B18,HVY,H11,H13,H14, H15,H16,H17,H18,H20,H21,H24,H25,H26, H27,H28,MMM	2
14	PAOZZ	5310008094058	96906	MS27183-10	WASHER,FLAT 1/4 PART OF KIT P/N 5705615.......................... UOC:AVY,A11,A13,A14,A15,A20,A24,A25, A26,A27,B16,B17,B18,HVY,H11,H13,H14, H15,H16,H17,H18,H20,H21,H24,H25,H26, H27,H28,MMM	16
15	PAOZZ	5340013154955	19207	12368375	CLAMP,LOOP PART OF KIT P/N 5705615.. UOC:AVY,A11,A13,A14,A15,A20,A24,A25, A26,A27,B16,B17,B18,HVY,H11,H13,H14, H15,H16,H17,H18,H20,H21,H24,H25,H26, H27,H28,MMM	2
16	PAOZZ	5325013097164	96906	MS35489-149	GROMMET,NONMETALLIC .875 PART OF KIT P/N 5705615..................... UOC:AVY,A11,A13,A14,A15,A20,A24,A25, A26,A27,B16,B17,B18,HVY,H11,H13,H14, H15,H16,H17,H18,H20,H21,H24,H25,H26, H27,H28,MMM	1
17	PAOZZ	2540014456029	19207	12342880	INSULATION,THERMAL, L.H. PART OF KIT P/N 5705615..................... UOC:AVY,A11,A13,A14,A15,A20,A24,A25, A26,A27,B16,B17,B18,HVY,H11,H13,H14, H15,H16,H17,H18,H20,H21,H24,H25,H26, H27,H28,M-MM	1
18	PFOZZ	2540013509380	19207	12342038	WATER JACKET,VEHICU PART OF KIT P/N 5705615............................. UOC:AVY,A11,A13,A14,A15,A20,A24,A25, A26,A27,B16,B17,B18,HVY,H11,H13,H14, H15,H16,H17,H18,H20,H21,H24,H25,H26,	1

SECTION II TM9-2320-280-24P

(1) NO	(2) SMR CODE	(3) NSN	(4) CAGEC	(5) PART NUMBER	(6) DESCRIPTION AND USABLE ON CODES(UOC)	(7) QTY
19	PAOZZ	2835014657620	19207	12342881-1	INSULATION,COWL LEF L.H. PART OF KIT P/N 5705615.................... UOC:AVY,A11,A13,A14,A15,A20,A24,A25, A26,A27,B16,B17,B18,HVY,H11,H13,H14, H15,H16,H17,H18,H20,H21,H24,H25,H26, H27,H28,MMM	1
20	PAOZZ	2540014491718	34623	12342878-1	ARMOR,SUPPLEMENTAL, R.H. PART OF KIT P/N 5705615.................... UOC:AVY,A11,A13,A14,A15,A20,A24,A25, A26,A27,B16,B17,B18,HVY,H11,H13,H14, H15,H16,H17,H18,H20,H21,H24,H25,H26, H27,H28,MMM	1
20	PAOZZ	2540014519345	19207	12342878-2	ARMOR,SUPPLEMENTAL, L.H. PART OF KIT P/N 5705615.................... UOC:AVY,A11,A13,A14,A15,A20,A24,A25, A26,A27,B16,B17,B18,HVY,H11,H13,H14, H15,H16,H17,H18,H20,H21,H24,H25,H26, H27,H28,MMM	1
21	PAOZZ	5320011357319	80205	NAS9301BNS-6-03	RIVET,BLIND 3/16 X 3/16 PART OF KIT P/N 5705615....................... UOC:AVY,A11,A13,A14,A15,A20,A24,A25, A26,A27,B16,B17,B18,HVY,H11,H13,H14, H15,H16,H17,H18,H20,H21,H24,H25,H26, H27,H28,MMM	2
22	PAOZZ	5310009359022	96906	MS51943-32	NUT,SELF-LOCKING,HE 1/4-20 PART OF KIT P/N 5705615.................... UOC:AVY,A11,A13,A14,A15,A20,A24,A25, A26,A27,B16,B17,B18,HVY,H11,H13,H14, H15,H16,H17,H18,H20,H21,H24,H25,H26, H27,H28,MMM	14
23	PAOZZ	5310006379541	96906	MS35338-46	WASHER,LOCK 1/4 PART OF KIT P/N 5705615.................... UOC:AVY,A11,A13,A14,A15,A20,A24,A25, A26,A27,B16,B17,B18,HVY,H11,H13,H14, H15,H16,H17,H18,H20,H21,H24,H25,H26, H27,H28,MMM	4
24	PAOZZ	5306000680513	60285	6893-2	BOLT,MACHINE 1/4-20 X 3/4 PART OF KIT P/N 5705615.................... UOC:AVY,A11,A13,A14,A15,A20,A24,A25, A26,A27,B16,B17,B18,HVY,H11,H13,H14, H15,H16,H17,H18,H20,H21,H24,H25,H26, H27,H28,MMM	4
25	PAOZZ	5310012805796	96906	MS27183-57	WASHER,FLAT 3/8 PART OF KIT P/N 5705615.................... UOC:AVY,A11,A13,A14,A15,A20,A24,A25, A26,A27,B16,B17,B18,HVY,H11,H13,H14, H15,H16,H17,H18,H20,H21,H24,H25,H26, H27,H28,MMM	1
26	PAOZZ	2540013141130	19207	12368398	HOUSING,HEATER COMP PART OF KIT P/N 5705615.................... UOC:AVY,A11,A13,A14,A15,A20,A24,A25, A26,A27,B16,B17,B18,HVY,H11,H13,H14, H15,H16,H17,H18,H20,H21,H24,H25,H26,	1

SECTION II   TM9-2320-280-24P

(1) ITEM NO	(2) SMR CODE	(3) NSN	(4) CAGEC	(5) PART NUMBER	(6) DESCRIPTION AND USABLE ON CODES(UOC)	(7) QTY
27	PAOZZ	5320012751998	11815	BAPKTR-64	RIVET,BLIND 1/8 PART OF KIT P/N 5705615.................... UOC:AVY,A11,A13,A14,A15,A20,A24,A25, A26,A27,B16,B17,B18,HVY,H11,H13,H14, H15,H16,H17,H18,H20,H21,H24,H25,H26, H27,H28,MMM	11
28	PFOZZ	2540013173309	34623	12368400	GUARD ASSEMBLY,BRUS PART OF KIT P/N 5705615............................ UOC:AVY,A11,A13,A14,A15,A20,A24,A25, A26,A27,B16,B17,B18,HVY,H11,H13,H14, H15,H16,H17,H18,H20,H21,H24,H25,H26, H27,H28,MMM	1
29	PAOZZ	5306000680513	60285	6893-2	BOLT,MACHINE 1/4-20 X 3/4 PART OF KIT P/N 5705615..................... UOC:AVY,A11,A13,A14,A15,A20,A24,A25, A26,A27,B16,B17,B18,HVY,H11,H13,H14, H15,H16,H17,H18,H20,H21,H24,H25,H26, H27,H28,MMM	12
30	PAOZZ	2990013140151	19207	12368443	PIPE,EXHAUST PART OF KIT P/N 5705615 UOC:AVY,A11,A13,A14,A15,A20,A24,A25, A26,A27,B16,B17,B18,HVY,H11,H13,H14, H15,H16,H17,H18,H20,H21,H24,H25,H26, H27,H28,MMM	1
31	PFOZZ	4710013139340	19207	12342095	TUBE,BENT,METALLIC PART OF KIT P/N 5705615.................... UOC:AVY,A11,A13,A14,A15,A20,A24,A25, A26,A27,B16,B17,B18,HVY,H11,H13,H14, H15,H16,H17,H18,H20,H21,H24,H25,H26, H27,H28,MMM	1
32	MOOZZ		96906	MS521301A203R	HOSE,NONMETALLIC MAKE FROM HOSE,P/N MS521301A203R,7.50 INCHES LONG PART OF KIT P/N 5705615................. UOC:AVY,A11,A13,A14,A15,A20,A24,A25, A26,A27,B16,B17,B18,HVY,H11,H13,H14, H15,H16,H17,H18,H20,H21,H24,H25,H26, H27,H28,MMM	1
33	PAOZZ	2930013140145	19207	12342294	PUMP,COOLING SYSTEM PART OF KIT P/N 5705615............................ UOC:AVY,A11,A13,A14,A15,A20,A24,A25, A26,A27,B16,B17,B18,HVY,H11,H13,H14, H15,H16,H17,H18,H20,H21,H24,H25,H26, H27,H28,MMM	1
34	PAOZZ	4730009098627	01276	FF9311-36	CLAMP,HOSE PART OF KIT P/N 5705615.. UOC:AVY,A11,A13,A14,A15,A20,A24,A25, A26,A27,B16,B17,B18,HVY,H11,H13,H14, H15,H16,H17,H18,H20,H21,H24,H25,H26, H27,H28,MMM	1
35	PAOZZ	5340013142445	19207	12368374	BRACKET,MOUNTING PART OF KIT P/N 5705615............................ UOC:AVY,A11,A13,A14,A15,A20,A24,A25, A26,A27,B16,B17,B18,HVY,H11,H13,H14, H15,H16,H17,H18,H20,H21,H24,H25,H26, H27,H28,MMM	1

SECTION II   TM9-2320-280-24P

(1) NO	(2) CODE	(3) NSN	(4) CAGEC	(5) NUMBER	(6)	(7) ITEM SMR DESCRIPTION AND USABLE ON CODES(UOC)	PART QTY

36 PAOZZ 5365012550965 19207 12338521-14 SPACER,SLEEVE PART OF KIT P/N 5705615.......................... 2
UOC:AVY,A11,A13,A14,A15,A20,A24,A25,
A26,A27,B16,B17,B18,HVY,H11,H13,H14,
H15,H16,H17,H18,H20,H21,H24,H25,H26,
H27,H28,MMM

37 PAOZZ 5310012067306 7X677 11500207 WASHER,LOCK PART OF KIT P/N 5705615. 2
UOC:AVY,A11,A13,A14,A15,A20,A24,A25,
A26,A27,B16,B17,B18,HVY,H11,H13,H14,
H15,H16,H17,H18,H20,H21,H24,H25,H26,
H27,H28,MMM

38 PAOZZ 5310013151535 24617 11505922 NUT,PLAIN,HEXAGON M10-O.35 PART OF 2
KIT P/N 5705615.....................
UOC:AVY,A11,A13,A14,A15,A20,A24,A25,
A26,A27,B16,B17,B18,HVY,H11,H13,H14,
H15,H16,H17,H18,H20,H21,H24,H25,H26,
H27,H28,MMM

39 PAOZZ 4730009083193 7Z588 30024H CLAMP,HOSE PART OF KIT P/N 5705615.. 2
UOC:AVY,A11,A13,A14,A15,A20,A24,A25,
A26,A27,B16,B17,B18,HVY,H11,H13,H14,
H15,H16,H17,H18,H20,H21,H24,H25,H26,
H27,H28,MMM

40 PAOZZ 4720012111998 19207 12340492 HOSE,NONMETALLIC PART OF KIT P/N 2
5705615..........................
UOC:AVY,A11,A13,A14,A15,A20,A24,A25,
A26,A27,B16,B17,B18,HVY,H11,H13,H14,
H15,H16,H17,H18,H20,H21,H24,H25,H26,
H27,H28,MMM

41 MOOZZ 96906 521301A203R 5.50 HOSE,NONMETALLIC MAKE FROM HOSE,P/N 1
MS521301A203R,5.50 INCHES LONG PART OF KIT P/N 5705615.................
UOC:AVY,A11,A13,A14,A15,A20,A24,A25,
A26,A27,B16,B17,B18,HVY,H11,H13,H14,
H15,H16,H17,H18,H20,H21,H24,H25,H26,
H27,H28,MMM

42 PAOZZ 4710012572649 19207 12340498 TUBE ASSEMBLY,METAL PART OF KIT P/N 1
5705615..........................
UOC:AVY,A11,A13,A14,A15,A20,A24,A25,
A26,A27,B16,B17,B18,HVY,H11,H13,H14,
H15,H16,H17,H18,H20,H21,H24,H25,H26,
H27,H28,MMM

43 PAOZZ 4720012572655 19207 12340493 HOSE,PREFORMED PART OF KIT P/N 1
5705615..........................
UOC:AVY,A11,A13,A14,A15,A20,A24,A25,
A26,A27,B16,B17,B18,HVY,H11,H13,H14,
H15,H16,H17,H18,H20,H21,H24,H25,H26,
H27,H28,MMM

44 PAOZZ 4730001003918 96906 MS24521-7 TEE,HOSE PART OF KIT P/N 5705615.... 1
UOC:AVY,A11,A13,A14,A15,A20,A24,A25,
A26,A27,B16,B17,B18,HVY,H11,H13,H14,
H15,H16,H17,H18,H20,H21,H24,H25,H26,
H27,H28,MMM

45 PFOZZ 5340013201079 19207 12368387 BRACKET,ANGLE PART OF KIT P/N 1
5705615..........................

SECTION II TM9-2320-280-24P

(1) ITEM NO	(2) SMR CODE	(3) NSN	(4) CAGEC	(5) PART NUMBER	(6) DESCRIPTION AND USABLE ON CODES(UOC)	(7) QTY
46	PAOZZ	5340009936207	96906	MS21333-99	CLAMP,LOOP PART OF KIT P/N 5705615.. UOC:AVY,A11,A13,A14,A15,A20,A24,A25, A26,A27,B16,B17,B18,HVY,H11,H13,H14, H15,H16,H17,H18,H20,H21,H24,H25,H26, H27,H28,MMM	1
47	PAOZZ	5340013154120	19207	12368391	PLATE,MOUNTING PART OF KIT P/N 5705615............ UOC:AVY,A11,A13,A14,A15,A20,A24,A25, A26,A27,B16,B17,B18,HVY,H11,H13,H14, H15,H16,H17,H18,H20,H21,H24,H25,H26, H27,H28,MMM	1
48	PAOZZ	5975012113142	28520	SB-750-625-2840	BUSHING,ELECTRICAL PART OF KIT P/N 5705615............ UOC:AVY,A11,A13,A14,A15,A20,A24,A25, A26,A27,B16,B17,B18,HVY,H11,H13,H14, H15,H16,H17,H18,H20,H21,H24,H25,H26, H27,H28,MMM	1
49	PAOZZ	5310008775797	96906	MS21044-N3	NUT,SELF-LOCKING,HE #10-32 PART OF KIT P/N 5705615..................... UOC:AVY,A11,A13,A14,A15,A20,A24,A25, A26,A27,B16,B17,B18,HVY,H11,H13,H14, H15,H16,H17,H18,H20,H21,H24,H25,H26, H27,H28,MMM	2
50	PAOZZ	5310013335245	24617	271175	NUT,SELF-LOCKING,HE 1/4-20 PART OF KIT P/N 5705615..................... UOC:AVY,A11,A13,A14,A15,A20,A24,A25, A26,A27,B16,B17,B18,HVY,H11,H13,H14, H15,H16,H17,H18,H20,H21,H24,H25,H26, H27,H28,MMM	1
51	PAOZZ	5340013146838	19207	12368390	PLATE,MENDING PART OF KIT P/N 5705615............ UOC:AVY,A11,A13,A14,A15,A20,A24,A25, A26,A27,B16,B17,B18,HVY,H11,H13,H14, H15,H16,H17,H18,H20,H21,H24,H25,H26, H27,H28,MMM	1
52	PAOZZ	5305000680515	80204	B1821BH025F100N	SCREW,CAP,HEXAGON H 1/4-28 X 1.00 PART OF KIT P/N 5705615............. UOC:AVY,A11,A13,A14,A15,A20,A24,A25, A26,A27,B16,B17,B18,HVY,H11,H13,H14, H15,H16,H17,H18,H20,H21,H24,H25,H26, H27,H28,MMM	2
53	PAOZZ	5305009897435	96906	MS35207-264	SCREW,MACHINE #10-32 X 5/8 PART OF KIT P/N 5705615..................... UOC:AVY,A11,A13,A14,A15,A20,A24,A25, A26,A27,B16,B17,B18,HVY,H11,H13,H14, H15,H16,H17,H18,H20,H21,H24,H25,H26, H27,H28,MMM	2

END OF FIGURE

Section II. TM 9-2320-280-24P-2

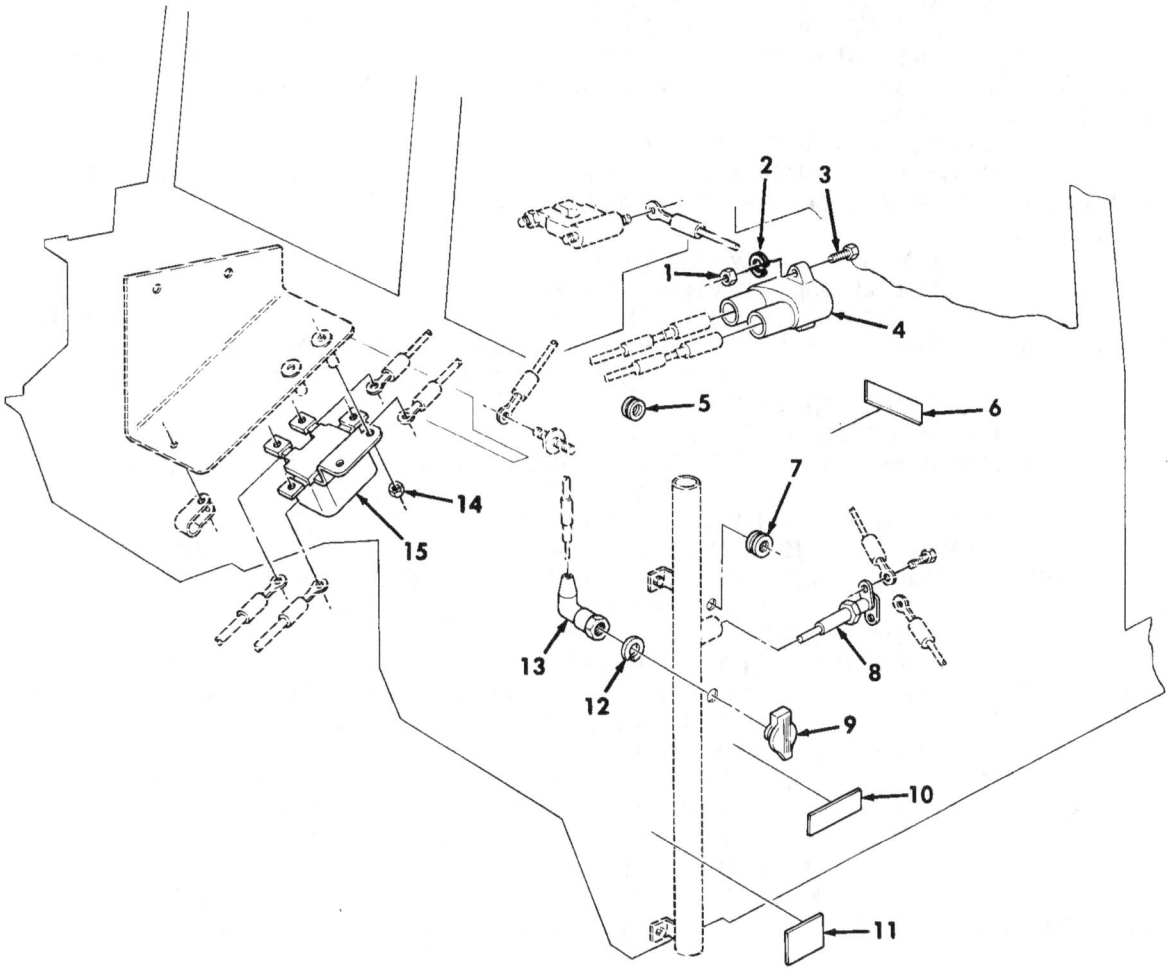

Figure 380. Swingfire Heater Kit, Circuit Breaker and Related Parts.

SECTION II   TM9-2320-280-24P

(1) NO	(2) SMR CODE	(3) NSN	(4) CAGEC	(5) PART NUMBER	(7) ITEM DESCRIPTION AND USABLE ON CODES(UOC)	QTY

GROUP 3303 WINTERIZATION KITS

FIG. 380 SWINGFIRE HEATER KIT, CIRCUIT BREAKER AND RELATED PARTS

1 PAOZZ 5310009349754 96906 MS35650-382 NUT,PLAIN,HEXAGON #8-32 PART OF KIT P/N 5705615......................
UOC:AVY,A11,A13,A14,A15,A20,A24,A25,
A26,A27,B16,B17,B18,HVY,H11,H13,H14,
H15,H16,H17,H18,H20,H21,H24,H25,H26,
H27,H28,MMM   2

2 PAOZZ 5310000453299 96906 MS35338-42 WASHER,LOCK PART OF KIT P/N 5705615.   2
UOC:AVY,A11,A13,A14,A15,A20,A24,A25,
A26,A27,B16,B17,B18,HVY,H11,H13,H14,
H15,H16,H17,H18,H20,H21,H24,H25,H26,
H27,H28,MMM

3 PAOZZ 5305009846195 96906 MS35206-247 SCREW,MACHINE #8-32 X 3/4 PART OF   2
KIT P/N 5705615.....................
UOC:AVY,A11,A13,A14,A15,A20,A24,A25,
A26,A27,B16,B17,B18,HVY,H11,H13,H14,
H15,H16,H17,H18,H20,H21,H24,H25,H26,
H27,H28,MMM

4 PAOZZ 5925014302318 58536 AA5557/01-001 CIRCUIT BREAKER PART OF KIT P/N   1
5705615..........................
UOC:AVY,A11,A13,A14,A15,A20,A24,A25,
A26,A27,B16,B17,B18,HVY,H11,H13,H14,
H15,H16,H17,H18,H20,H21,H24,H25,H26,
H27,H28,MMM

5 PAOZZ 5325002708891 96906 MS35489-21 GROMMET,NONMETALLIC PART OF KIT P/N   1
5705615...........................
UOC:AVY,A11,A13,A14,A15,A20,A24,A25,
A26,A27,B16,B17,B18,HVY,H11,H13,H14,
H15,H16,H17,H18,H20,H21,H24,H25,H26,
H27,H28,MMM

6 PAOZZ 9905013177987 19207 12342026 PLATE,INSTRUCTION PART OF KIT P/N   1
5705615...........................
UOC:AVY,A11,A13,A14,A15,A20,A24,A25,
A26,A27,B16,B17,B18,HVY,H11,H13,H14,
H15,H16,H17,H18,H20,H21,H24,H25,H26,
H27,H28,MMM

7 PAOZZ 5325002919366 96906 MS35489-11 GROMMET,NONMETALLIC PART OF KIT P/N   1
5705615...........................
UOC:AVY,A11,A13,A14,A15,A20,A24,A25,
A26,A27,B16,B17,B18,HVY,H11,H13,H14,
H15,H16,H17,H18,H20,H21,H24,H25,H26,
H27,H28,MMM

8 PAOZZ 5930013182809 19207 12356766 SWITCH,THERMOSTATIC PART OF KIT P/N   1
5705615...........................
UOC:AVY,A11,A13,A14,A15,A20,A24,A25,
A26,A27,B16,B17,B18,HVY,H11,H13,H14,
H15,H16,H17,H18,H20,H21,H24,H25,H26,
H27,H28,MMM

9 PAOZZ 5935013142084 19207 11669531 CONNECTOR SET,ELECT PART OF KIT P/N   1

SECTION II TM9-2320-280-24P

(1) ITEM NO	(2) SMR CODE	(3) NSN	(4) CAGEC	(5) PART NUMBER	(6) DESCRIPTION AND USABLE ON CODES(UOC)	(7) QTY
				5705615..........................	UOC:AVY,A11,A13,A14,A15,A20,A24,A25,A26,A27,B16,B17,B18,HVY,H11,H13,H14,H15,H16,H17,H18,H20,H21,H24,H25,H26,H27,H28,MMM	
10	PAOZZ	9905013177986	19207	12342027	PLATE,IDENTIFICATIO PART OF KIT P/N 5705615.......................... UOC:AVY,A11,A13,A14,A15,A20,A24,A25,A26,A27,B16,B17,B18,HVY,H11,H13,H14,H15,H16,H17,H18,H20,H21,H24,H25,H26,H27,H28,MMM	1
11	PAOZZ	7690013158540	19207	12342024	MARKER,IDENTIFICATI PART OF KIT P/N 5705615.......................... UOC:AVY,A11,A13,A14,A15,A20,A24,A25,A26,A27,B16,B17,B18,HVY,H11,H13,H14,H15,H16,H17,H18,H20,H21,H24,H25,H26,H27,H28,MMM	1
12	PAOZZ	5310005503714	96906	MS35333-47	WASHER,LOCK 3/4 PART OF KIT P/N 5705615.......................... UOC:AVY,A11,A13,A14,A15,A20,A24,A25,A26,A27,B16,B17,B18,HVY,H11,H13,H14,H15,H16,H17,H18,H20,H21,H24,H25,H26,H27,H28,MMM	1
13	PAOZZ	5975005536995	96906	MS25171-1S	CABLE NIPPLE,ELECTR PART OF KIT P/N 5705615.......................... UOC:AVY,A11,A13,A14,A15,A20,A24,A25,A26,A27,B16,B17,B18,HVY,H11,H13,H14,H15,H16,H17,H18,H20,H21,H24,H25,H26,H27,H28,MMM	1
14	PAOZZ	5310009359022	96906	MS51943-32	NUT,SELF-LOCKING,HE 1/4-28 PART OF KIT P/N 5705615................... UOC:AVY,A11,A13,A14,A15,A20,A24,A25,A26,A27,B16,B17,B18,HVY,H11,H13,H14,H15,H16,H17,H18,H20,H21,H24,H25,H26,H27,H28,MMM	2
15	PAOZZ	2920008483292	16764	1116968	CUTOUT RELAY,ENGINE 24 VOLTS PART OF KIT P/N 5705615.................. UOC:AVY,A11,A13,A14,A15,A20,A24,A25,A26,A27,B16,B17,B18,HVY,H11,H13,H14,H15,H16,H17,H18,H20,H21,H24,H25,H26,H27,H28,MMM	1

END OF FIGURE

Section II.

TM 9-2320-280-24P-2

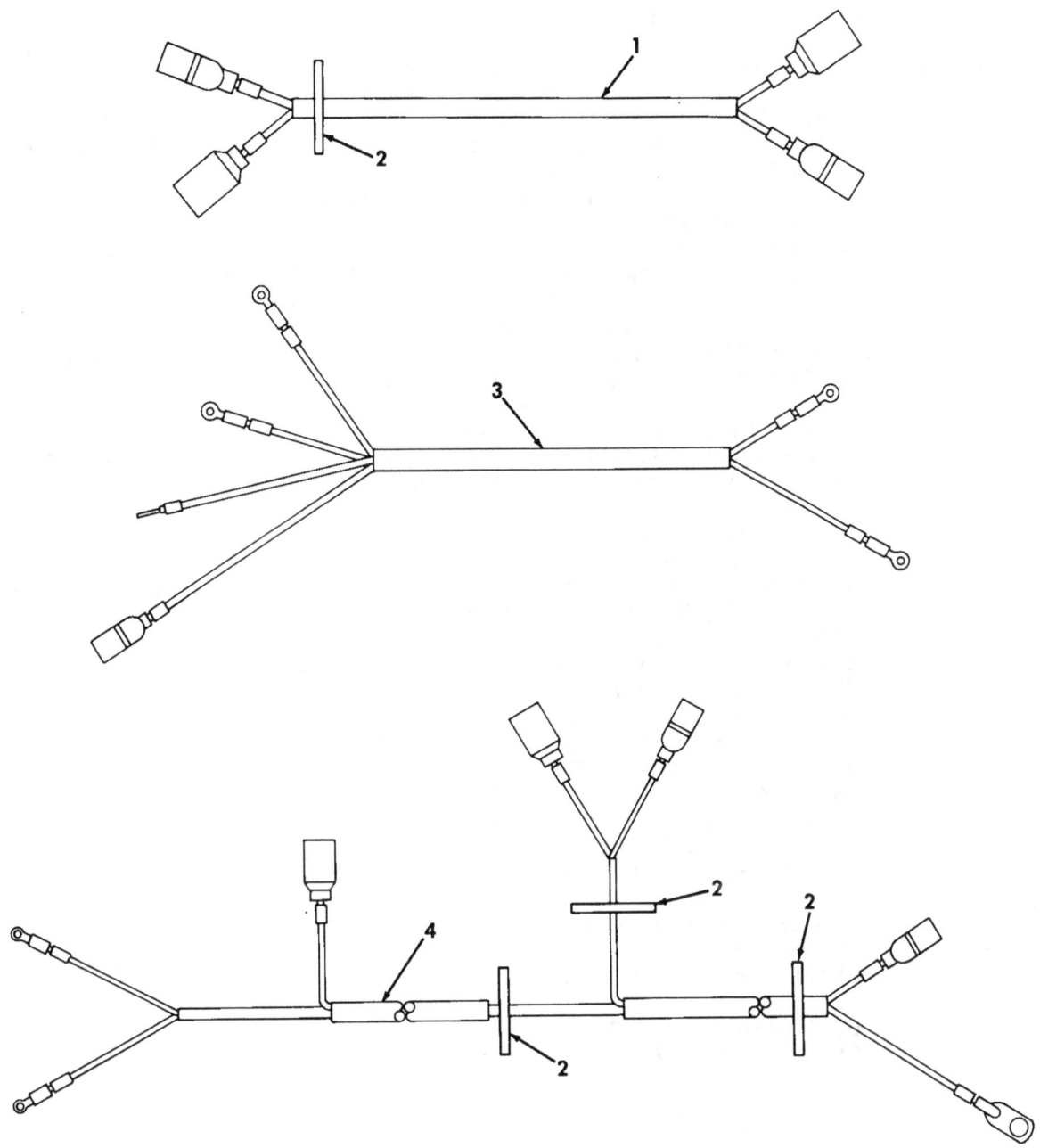

Figure 381. Swingfire Heater Kit, Harness Assemblies.

SECTION II   TM9-2320-280-24P

(1) NO	(2) SMR CODE	(3) NSN	(4) CAGEC	(5) PART NUMBER	(7) DESCRIPTION AND USABLE ON CODES(UOC)	QTY

(6) ITEM

GROUP 3303 WINTERIZATION KITS

FIG. 381 SWINGFIRE HEATER KIT, HARNESS ASSEMBLIES

```
1  PAOZZ 6150013145643 19207 12368396   CABLE ASSEMBLY,SPEC PART OF KIT P/N   1
5705615.............................               UOC:AVY,A11,A13,A14,A15,A20,A24,A25,
                                        A26,A27,B16,B17,B18,HVY,H11,H13,H14,
                                        H15,H16,H17,H18,H20,H21,H24,H25,H26,                H27,H28,MMM
2  PAOZZ 5975005709598 96906 MS3367-7-9  STRAP,TIEDOWN,ELECT PART OF KIT P/N   4
5705615.............................               UOC:AVY,A11,A13,A14,A15,A20,A24,A25,
                                        A26,A27,B16,B17,B18,HVY,H11,H13,H14,
                                        H15,H16,H17,H18,H20,H21,H24,H25,H26,                H27,H28,MMM
3  PAOZZ 6150013151148 19207 12368394   WIRING HARNESS,BRAN PART OF KIT P/N   1
5705615.............................               UOC:AVY,A11,A13,A14,A15,A20,A24,A25,
                                        A26,A27,B16,B17,B18,HVY,H11,H13,H14,
                                        H15,H16,H17,H18,H20,H21,H24,H25,H26,                H27,H28,MMM
4  PFOZZ 6150013145241 19207 12368395   CABLE ASSEMBLY,SPEC PART OF KIT P/N   1
5705615.............................               UOC:AVY,A11,A13,A14,A15,A20,A24,A25,
                                        A26,A27,B16,B17,B18,HVY,H11,H13,H14,
                                        H15,H16,H17,H18,H20,H21,H24,H25,H26,                H27,H28,MMM
KIT PDFZZ 2540013149320 19207 5705615    PARTS KIT,VEHICULAR  SWINGFIRE        1
                                         HEATER...............................
                                        UOC:AVY,A11,A13,A14,A15,A20,A24,A25,
                                        A26,A27,B16,B17,B18,HVY,H11,H13,H14,
                                        H15,H16,H17,H18,H20,H21,H24,H25,H26,                H27,H28,MMM
                                         ARMOR,SUPPLEMENTAL,( 1) 379-20
                                         ARMOR,SUPPLEMENTAL,( 1) 379-20
                                         BOLT,MACHINE     ( 12) 379-29
                                         BOLT,MACHINE     (  4) 379-24
                                         BOLT,MACHINE     (  2) 379-13
                                         BRACKET,MOUNTING ( 1) 379-35
                                         BRACKET,ANGLE    ( 1) 379-45
                                         BRACKET,MULTIPLE AN( 1) 379-11
                                         BUSHING,ELECTRICAL( 1) 379-48
                                         CABLE ASSEMBLY,SPEC( 1) 381-4
                                         CABLE ASSEMBLY,SPEC( 1) 381-1
                                         CABLE NIPPLE,ELECTR( 1) 380-13
                                         CIRCUIT BREAKER  ( 1) 380-4
                                         CLAMP,HOSE       ( 8) 379-9
                                         CLAMP,HOSE       ( 2) 379-39
                                         CLAMP,HOSE       ( 1) 379-34
                                         CLAMP,LOOP       ( 2) 379-15
                                         CLAMP,LOOP       ( 4) 379-2
```

SECTION II	TM9-2320-280-24P						
(1)	(2)	(3)	(4)	(5)	(6)	(7) ITEM SMR	PART
NO	CODE	NSN	CAGEC	NUMBER	DESCRIPTION AND USABLE ON CODES(UOC)		QTY

Description	Qty	Part
CLAMP,LOOP	(1)	379-46
CONNECTOR SET,ELECT	(1)	380-9
CUTOUT RELAY,ENGINE	(1)	380-15
GROMMET,NONMETALLIC	(1)	380-7
GROMMET,NONMETALLIC	(1)	380-5
GROMMET,NONMETALLIC	(1)	379-16
GUARD ASSEMBLY,BRUS	(1)	379-28
HOSE,NONMETALLIC	(1)	379-8
HOSE,NONMETALLIC	(1)	379-10
HOSE,NONMETALLIC	(1)	379-32
HOSE,NONMETALLIC	(2)	379-40
HOSE,NONMETALLIC	(1)	379-41
HOSE,PREFORMED	(1)	379-43
HOUSING,HEATER COMP	(1)	379-26
INSULATION,COWL LEF	(1)	379-19
INSULATION,THERMAL,	(1)	379-17
MARKER,IDENTIFICATI	(1)	380-11
NUT,PLAIN,HEXAGON	(2)	380-1
NUT,PLAIN,HEXAGON	(2)	379-38
NUT,SELF-LOCKING,HE	(2)	379-49
NUT,SELF-LOCKING,HE	(3)	379-3
NUT,SELF-LOCKING,HE	(14)	379-22
NUT,SELF-LOCKING,HE	(2)	380-14
NUT,SELF-LOCKING,HE	(1)	379-50
PIPE,EXHAUST	(1)	379-30
PLATE,IDENTIFICATIO	(1)	380-10
PLATE,INSTRUCTION	(1)	380-6
PLATE,MENDING	(1)	379-51
PLATE,MOUNTING	(1)	379-47
PUMP,COOLING SYSTEM	(1)	379-33
RIVET,BLIND	(11)	379-27
RIVET,BLIND	(2)	379-21
SCREW,CAP,HEXAGON H	(2)	379-52
SCREW,MACHINE	(3)	379-1
SCREW,MACHINE	(2)	380-3
SCREW,MACHINE	(2)	379-53
SCREW,TAPPING	(3)	379-5
SPACER,PLATE	(1)	379-6
SPACER,PLATE	(1)	379-4
SPACER,SLEEVE	(2)	379-36
STRAP,TIEDOWN,ELECT	(9)	379-7
STRAP,TIEDOWN,ELECT	(4)	381-2
SWITCH,THERMOSTATIC	(1)	380-8
TEE,HOSE	(1)	379-44
TUBE ASSEMBLY,METAL	(1)	379-42
TUBE,BENT,METALLIC	(1)	379-31
WASHER,FLAT	(16)	379-14
WASHER,LOCK	(1)	379-12
WASHER,LOCK	(1)	380-12
WASHER,LOCK	(2)	379-37
WASHER,LOCK	(4)	379-23
WASHER,LOCK	(1)	379-25
WASHER,LOCK	(2)	380-2

(1) NO	(2) CODE	(3) NSN	(4) CAGEC	(5) PART NUMBER	(6) DESCRIPTION AND USABLE ON CODES(UOC)	(7) QTY	ITEM SMR

SECTION II   TM9-2320-280-24P

WATER JACKET,VEHICU(1) 379-18
WIRING HARNESS,BRAN(1) 381-3

END OF FIGURE

Section II.

TM 9-2320-280-24P-2

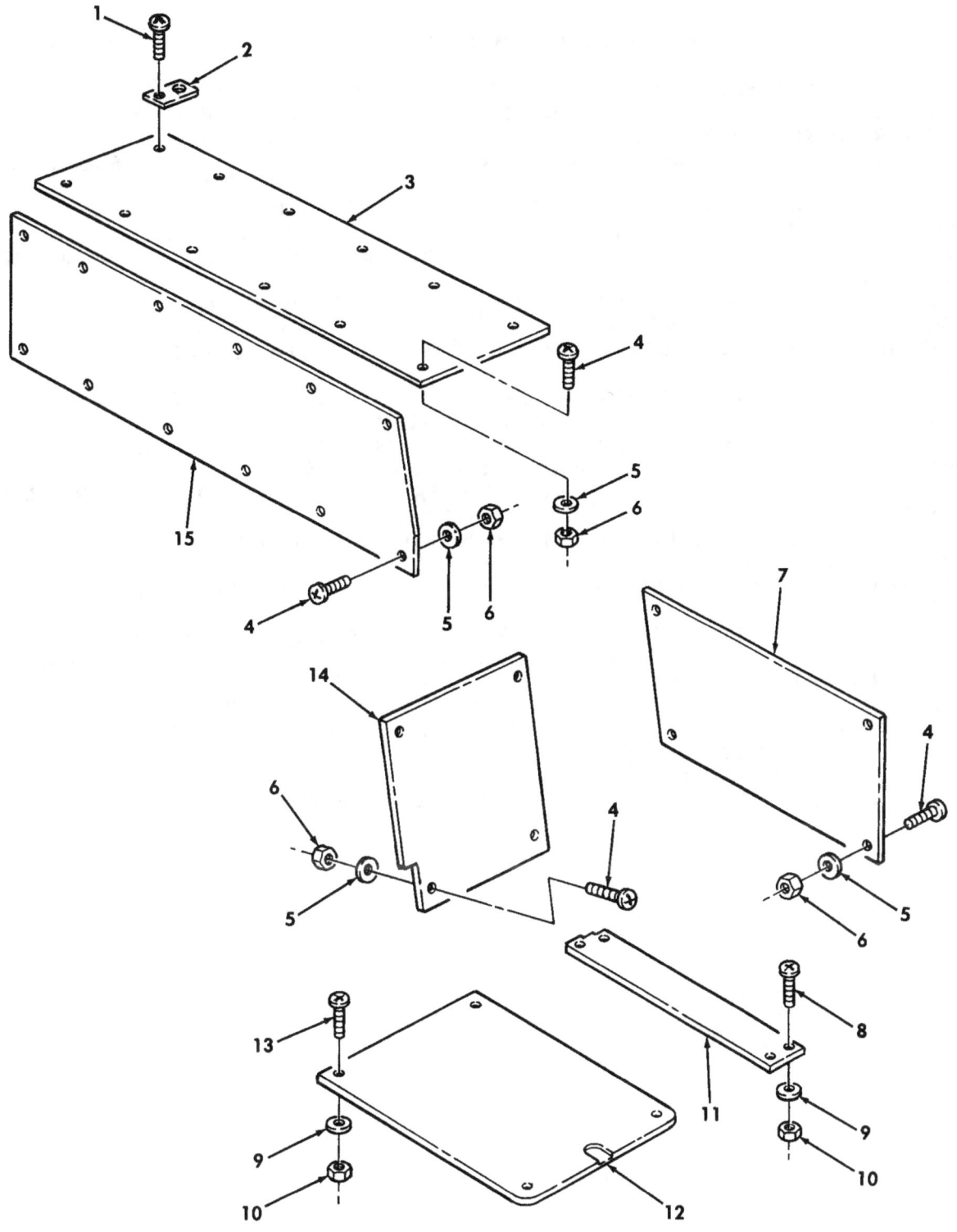

Figure 382. Two Door Seat and Floor Covers.

SECTION II   TM9-2320-280-24P

(1) NO	(2) SMR CODE	(3) NSN	(4) CAGEC	(5) PART NUMBER	(7) ITEM DESCRIPTION AND USABLE ON CODES (UOC)	QTY

GROUP 3303 WINTERIZATION KITS

FIG. 382 TWO DOOR SEAT AND FLOOR COVERS

1	PAOZZ	5305009002546	96906	MS35492-28	SCREW,WOOD #6 X .50,PART OF KIT P/N 5705611.................... UOC:A13,A14,H13,H14	12
2	PAOZZ	5365013168966	19207	12342210	SPACER,PLATE PART OF KIT P/N 5705611 UOC:A13,A14,H13,H14	12
3	KFFZZ		19207	12342181	TOP COVER ASSEMBLY L.H.,PART OF KIT P/N 5705611........................ UOC:A13,A14,H13,H14	1
3	KFFZZ		19207	12342182-2	TOP COVER ASSEMBLY R.H.,PART OF KIT P/N 5705611............................ UOC:A13,A14,H13,H14	1
4	PAOZZ	5305009932457	96906	MS35207-284	SCREW,MACHINE #10-32 X 1.50,PART OF KIT P/N 5705611..................... UOC:A13,A14,H13,H14	48
5	PAOZZ	5310008094058	96906	MS27183-10	WASHER,FLAT #10,PART OF KIT P/N 5705611........................ UOC:A13,A14,H13,H14	48
6	PAOZZ	5310004838791	96906	MS17829-4F	NUT,SELF-LOCKING,HE #10-32,PART OF KIT P/N 5705611..................... UOC:A13,A14,H13,H14	48
7	MFFZZ		19207	12342171-1	COVER R.H., MAKE FROM PLYWOOD,P/N NN-P-530 PART OF KIT P/N 5705611.... UOC:A13,A14,H13,H14	1
7	MFFZZ		19207	12342171-2	COVER L.H.,MAKE FROM PLYWOOD,P/N NN-P-530 PART OF KIT P/N 5705611.... UOC:A13,A14,H13,H14	1
8	PAOZZ	5305009592704	96906	MS35191-324	SCREW,MACHINE 3/8-24 X 1.75,PART OF KIT P/N 5705611..................... UOC:A13,A14,H13,H14	8
9	PAOZZ	5310010164871	88044	AN960-616	WASHER,FLAT 3/8,PART OF KIT P/N 5705611............................. UOC:A13,A14,H13,H14	16
10	PAOZZ	5310010583353	72582	192481	NUT,SELF-LOCKING,HE 3/8-24,PART OF KIT P/N 5705611..................... UOC:A13,A14,H13,H14	16
11	KFFZZ		19207	12342170	COVER,FLOOR L.H.,PART OF KIT P/N 5705611............................ UOC:A13,A14,H13,H14	1
11	KFFZZ		19207	12342265	COVER,FLOOR R.H.,PART OF KIT P/N 5705611.......................... UOC:A13,A14,H13,H14	1
12	KFFZZ		19207	12342184	SEAT BACK PART OF KIT P/N 5705611.. UOC:A13,A14,H13,H14	2
13	PAOZZ	5305008242004	96906	MS35191-326	SCREW,MACHINE 3/8-24 X 2.50,PART OF KIT P/N 5705611..................... UOC:A13,A14,H13,H14	8
14	KFFZZ		19207	12342172	COVER ASSEMBLY L.H.,PART OF KIT P/N 5705611............................	1

SECTION II   TM9-2320-280-24P

(1) NO	(2) CODE	(3) NSN	(4) CAGEC	(5) NUMBER	(6) DESCRIPTION AND USABLE ON CODES(UOC)	(7) ITEM QTY	SMR	PART
14	KFFZZ		19207	12342183	COVER ASSEMBLY R.H.,PART OF KIT P/N 5705611............ UOC:A13,A14,H13,H14	1		
15	KFFZZ		19207	12342180	COVER ASSEMBLY L.H.,SIDE,PART OF KIT P/N 5705611............ UOC:A13,A14,H13,H14	1		
15	KFFZZ		19207	12342257	COVER ASSEMBLY R.H.,SIDE,PART OF KIT P/N 5705611............ UOC:A13,A14,H13,H14	1		

END OF FIGURE

Section II. TM 9-2320-280-24P-2

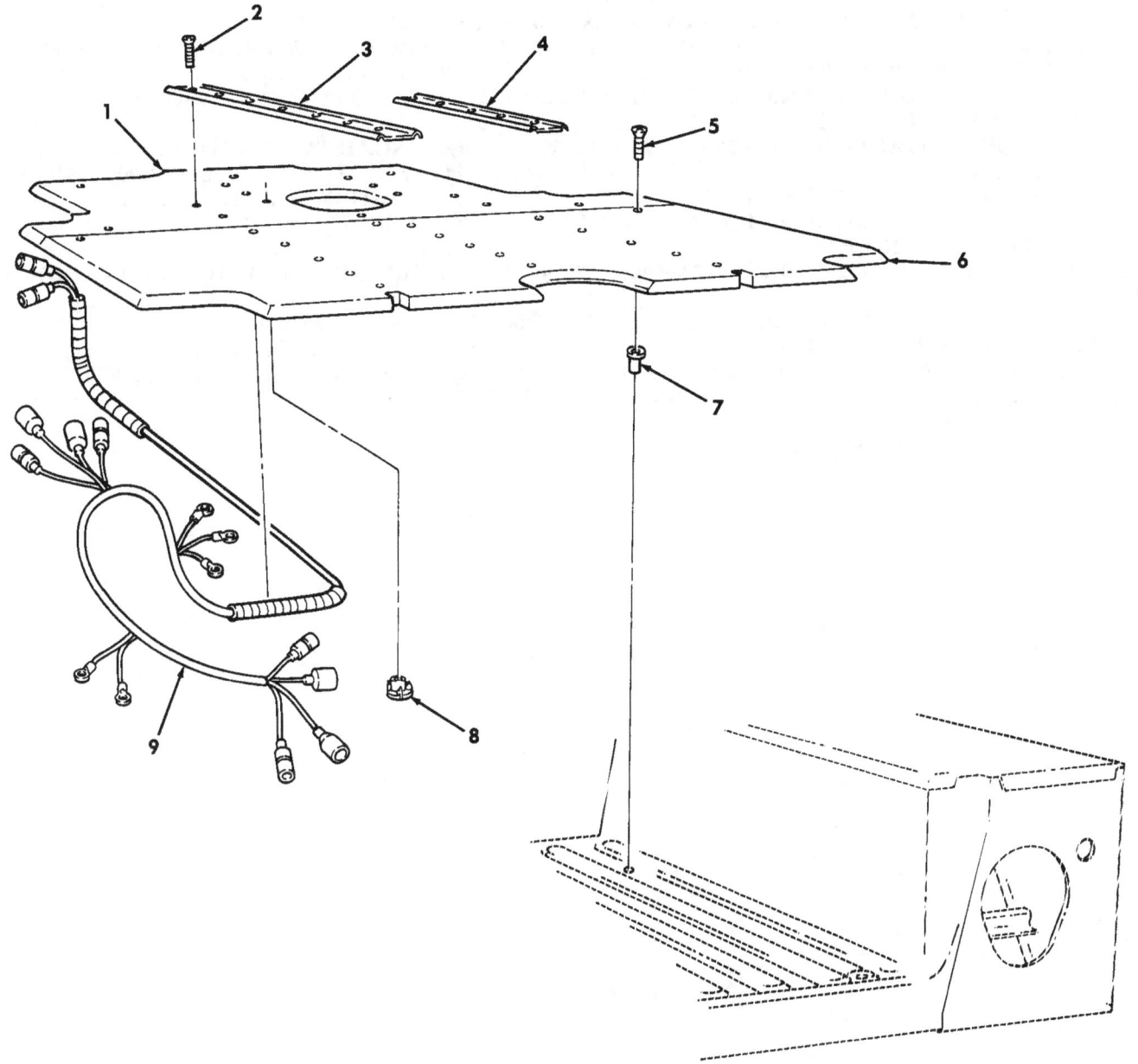

Figure 383. Cargo Floor Cover and Harness Assembly.

SECTION II   TM9-2320-280-24P

(1) NO	(2) SMR CODE	(3) NSN	(4) CAGEC	(5) PART NUMBER	(7) DESCRIPTION AND USABLE ON CODES(UOC)	ITEM QTY

GROUP 3303 WINTERIZATION KITS

FIG. 383 CARGO FLOOR COVER AND HARNESS ASSEMBLY

1	KFFZZ		19207	12342178	FLOOR COVER ASSEMBL PART OF KIT P/N 5705611.......................... UOC:A13,A14,H13,H14	57
2	PAOZZ	5305009026643	96906	MS35492-76	SCREW,WOOD #10 X .75 PART OF KIT P/N 5705611....................... UOC:A13,A14,H13,H14	19
3	PAOZZ	2590013174856	34623	12342179	MOLDING,METAL PART OF KIT P/N 5705611........................... UOC:A13,A14,H13,H14	2
4	PAOZZ	2590013173953	34623	12342203	MOLDING,METAL PART OF KIT P/N 5705611...................... UOC:A13,A14,H13,H14	1
5	PAOZZ	5305009585262	96906	MS35190-324	SCREW,MACHINE 3/8-16 X 2.25 PART OF KIT P/N 5705611.................... UOC:A13,A14,H13,H14	9
6	KFFZZ		19207	12342177	FLOOR COVER ASSEMBL PART OF KIT P/N 5705611......................... UOC:A13,A14,H13,H14	1
7	PAOZZ	5310012838482	96906	MS27130-CR31	NUT,PLAIN,BLIND RIV 3/8-16 PART OF KIT P/N 5705611.................... UOC:A13,A14,H13,H14	9
8	PAOZZ	5310010256444	96906	MS51941-10	NUT,PLAIN,PLATE PART OF KIT P/N 5705611........................... UOC:A13,A14,H13,H14	6
9	PFFZZ	6150013246356	19207	12342255	CABLE ASSEMBLY,SPEC PART OF KIT P/N 5705611...................... UOC:A13,A14,H13,H14	1

END OF FIGURE

Section II. TM 9-2320-280-24P-2

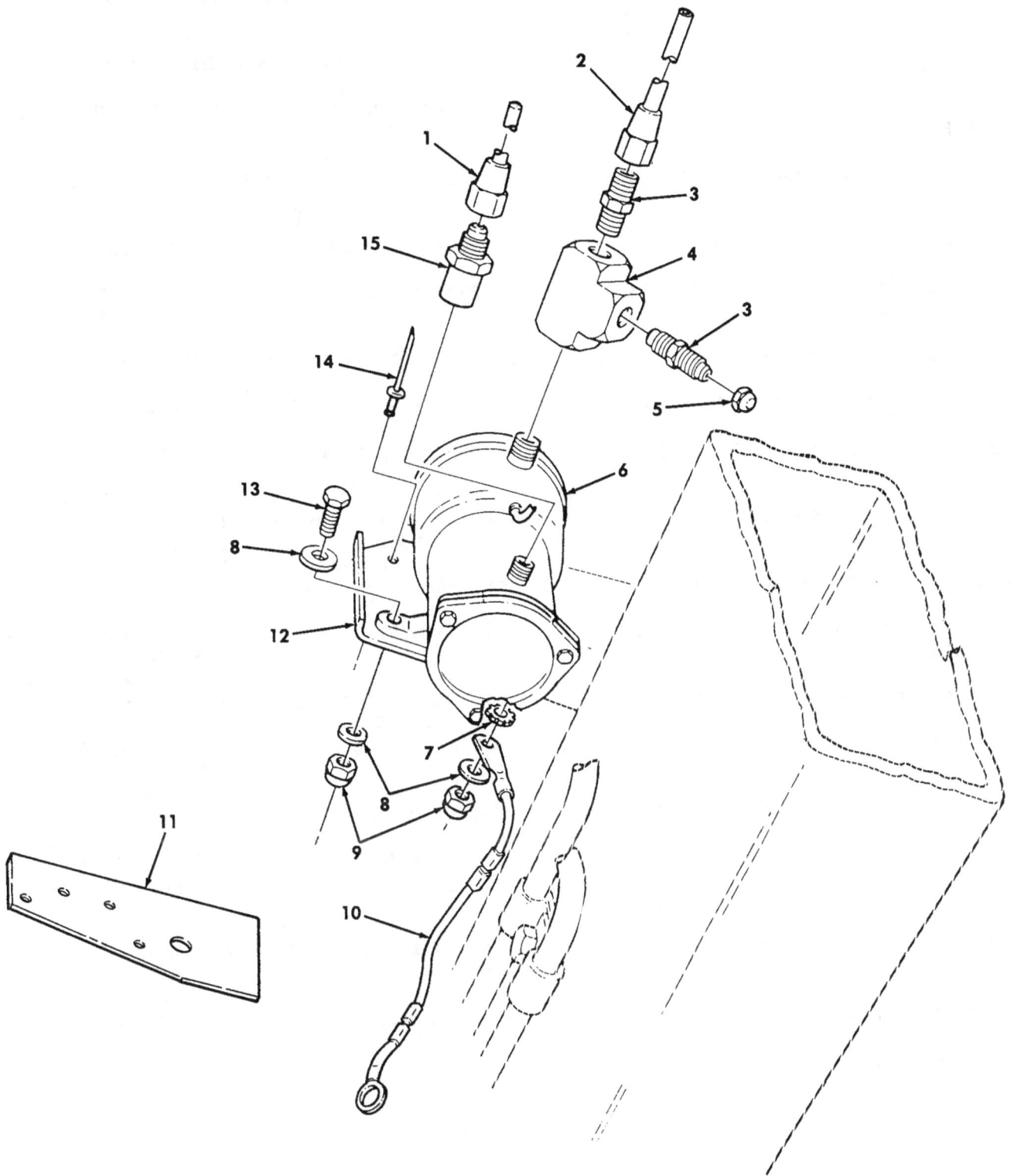

Figure 384. Winterization Kit, Fuel Pump, and Attaching Hardware.

SECTION II TM9-2320-280-24P C01

(1) ITEM NO	(2) SMR CODE	(3) NSN	(4) CAGEC	(5) PART NUMBER	(6) DESCRIPTION AND USABLE ON CODES (UOC)	(7) QTY
					GROUP 3303 WINTERIZATION KITS	
					FIG. 384 WINTERIZATION KIT,FUEL PUMP AND ATTACHING HARDWARE	
1	PFOZZ	4710012030608	34623	5588696	TUBE ASSEMBLY,METAL PART OF KIT P/N 5705611............................ UOC:A13,A14,H13,H14	1
2	PFOZZ	4710013171046	34623	12342277	TUBE ASSEMBLY,METAL PART OF KIT P/N 5705611............................ UOC:A13,A14,H13,H14	1
3	PAOZZ	4730005291237	13899	191410	ADAPTER,STRAIGHT,-PI PART OF KIT P/N 5705611................ UOC:A13,A14,H13,H14	2
4	PAOZZ	4730010489739	81343	4-4-4 140438C	TEE,PIPE PART OF KIT P/N 5705611... UOC:A13,A14,H13,H14	1
5	PAOZZ	4730002608285	81343	5010112B(N5)	CAP,TUBE PART OF KIT P/N 5705611... UOC:A13,A14,H13,H14	1
* 6	PAOZZ	2910009309367	53711	2590174	PUMP,FUEL,ELECTRICA PART OF KIT P/N 5705611............................ UOC:A13,A14,H13,H14	1
7	PAOZZ	5310005501130	96906	MS35338-40	WASHER,LOCK 1/4,PART OF KIT P/N 5705611............................ UOC:A13,A14,H13,H14	1
8	PAOZZ	5310008094058	96906	MS27183-10	WASHER,FLAT 1/4,PART OF KIT P/N 5705611............................ UOC:A13,A14,H13,H14	4
9	PAOZZ	5310000614650	96906	MS51943-31	NUT,SELF-LOCKING,HE PART OF KIT P/N 5705611............ UOC:A13,A14,H13,H14	2
10	PAOZZ	6150012608000	34623	5595746	LEAD,ELECTRICAL PART OF KIT P/N 5705611............................ UOC:A13,A14,H13,H14	1
11	PFFZZ	2910013206645	34623	12340696	TEMPLATE,FUEL PART OF KIT P/N 5705611............................ UOC:A13,A14,H13,H14	1
12	PFOZZ	5340012043903	19207	12338559	BRACKET,DOUBLE ANGL PART OF KIT P/N 5705611............................ UOC:A13,A14,H13,H14	1
13	PAOZZ	5305000680502	96906	MS90725-6	SCREW,CAP,HEXAGON H 1/4-20 X .75, PART OF KIT P/N 5705611............. UOC:A13,A14,H13,H14	2
* 14	PAOZZ	5320011357319	11815	CR-213-6-3	RIVET,BLIND .126-.187,PART OF KIT P/N 5705611...................... UOC:A13,A14,H13,H14	4
15	PAOZZ	4730002776347	34623	5583386	ADAPTER,STRAIGHT,PI PART OF KIT P/N 5705611........................ UOC:A13,A14,H13,H14	1

END OF FIGURE

384-1

Section II. TM 9-2320-280-24P-2

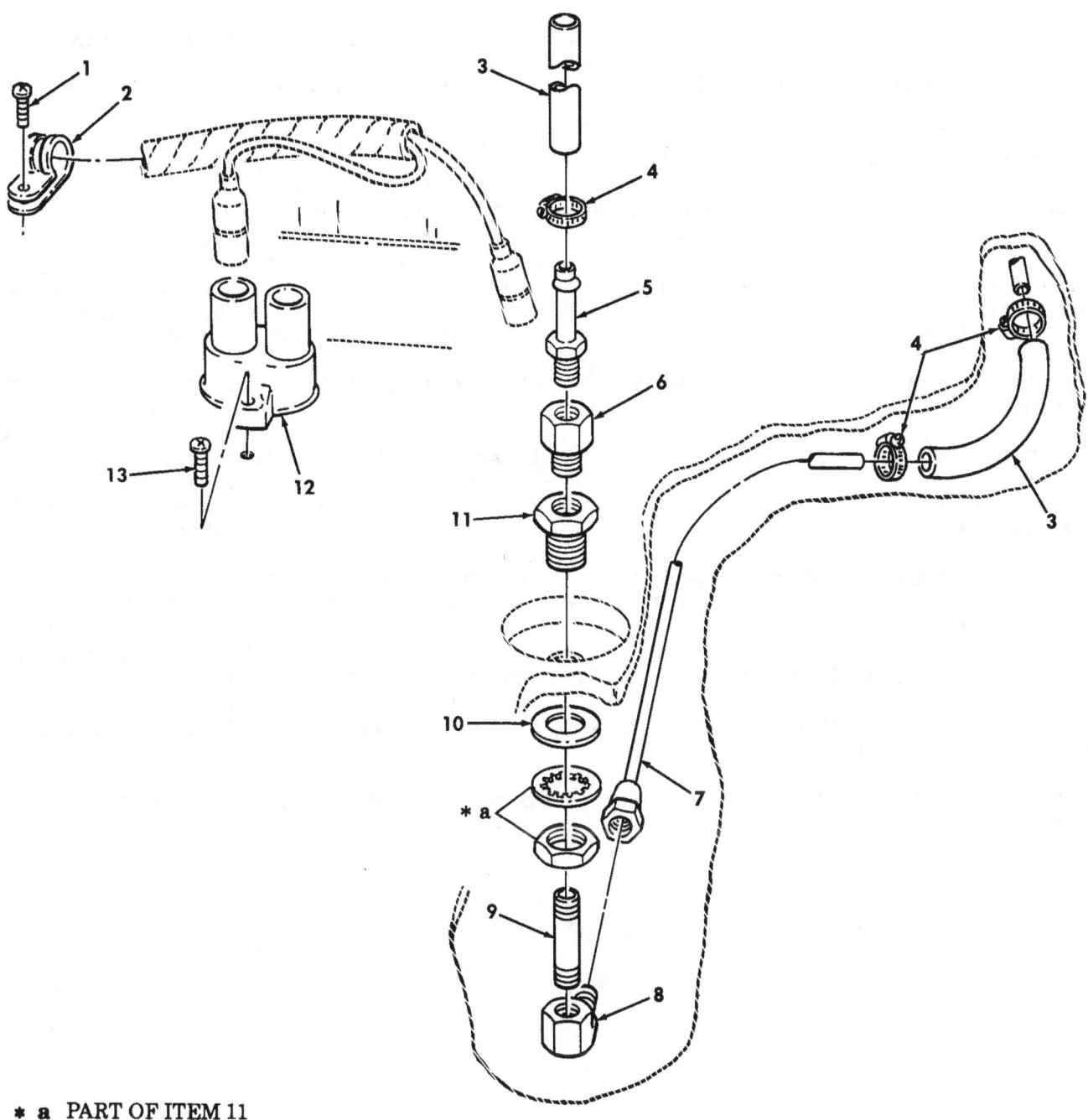

* a PART OF ITEM 11

Figure 385. Fuel Pump Tube Assembly.

SECTION II   TM9-2320-280-24P C01

(1) ITEM NO	(2) SMR CODE	(3) NSN	(4) CAGEC	(5) PART NUMBER	(6) DESCRIPTION AND USABLE ON CODES (UOC)	(7) QTY
					GROUP 3303 WINTERIZATION KITS	
					FIG. 385 FUEL PUMP TUBE ASSEMBLY	
1	PAOZZ	5305000567879	96906	MS35493-74	SCREW,WOOD #10 X .50,PART OF KIT P/N 5705611......................... UOC:A13,A14,H13,H14	1
2	PAOZZ	5340009899222	96906	MS21333-74	CLAMP,LOOP PART OF KIT P/N 5705611. UOC:A13,A14,H13,H14	1
* 3	MOOZZ		7X677	94383154-3	TUBE MAKE FROM HOSE,P/N G3336US-10, 7 INCHES LONG PART OF KIT P/N 5705611............................. UOC:A13,A14,H13,H14	2
4	PAOZZ	4730002782523	88907	B54-32780	CLAMP,LOOP PART OF KIT P/N 5705611. UOC:A13,A14,H13,H14	3
5	PAOZZ	4730013576523	4A439	BN252	ADAPTER,ATRAIGHT,PI PART OF KIT P/N 5705611............................. UOC:A13,A14,H13,H14	1
6	PAOZZ	4730013477342	81343	4-4 130139C	COUPLING,PIPE PART OF KIT P/N 5705611........... .... UOC:A13,A14,H13,H14	1
7	PAOZZ	4710013171045	34623	12342270	TUBE ASEMBLY,METAL PART OF KIT P/N 5705611............................. UOC:A13,A14,H13,H14	1
8	PFOZZ	4730012031025	81343	5010203CA	ELBOW,PIPE TO TUBE PART OF KIT P/N 5705611........... .... UOC:A13,A14,H13,H14	1
9	PAOZZ	4730013153242	96906	MS51873-32	NIPPLE,PIPE PART OF KIT P/N 5705611 UOC:A13,A14,H13,H14	1
10	PAOZZ	5310008098533	96906	MS27183-23	WASHER,FLAT 1-1/2,PART OF KIT P/N 5705611... .... UOC:A13,A14,H13,H14	1
11	PAOZZ	4730008975497	79470	W21204	COUPLING,PIPE PART OF KIT P/N 5705611............................. UOC:A13,A14,H13,H14	1
12	PAOZZ	5925000264767	81349	M13516/1-1	CIRCUIT BREAKER PART OF KIT P/N 5705611......... .... UOC:A13,A14,H13,H14	1
13	PAOZZ	5305009012134	96906	MS35493-55	SCREW,WOOD #8 X 1.00,PART OF KIT P/N 5705611....................... UOC:A13,A14,H13,H14	2

END OF FIGURE

385-1

Section II.

TM 9-2320-280-24P-2

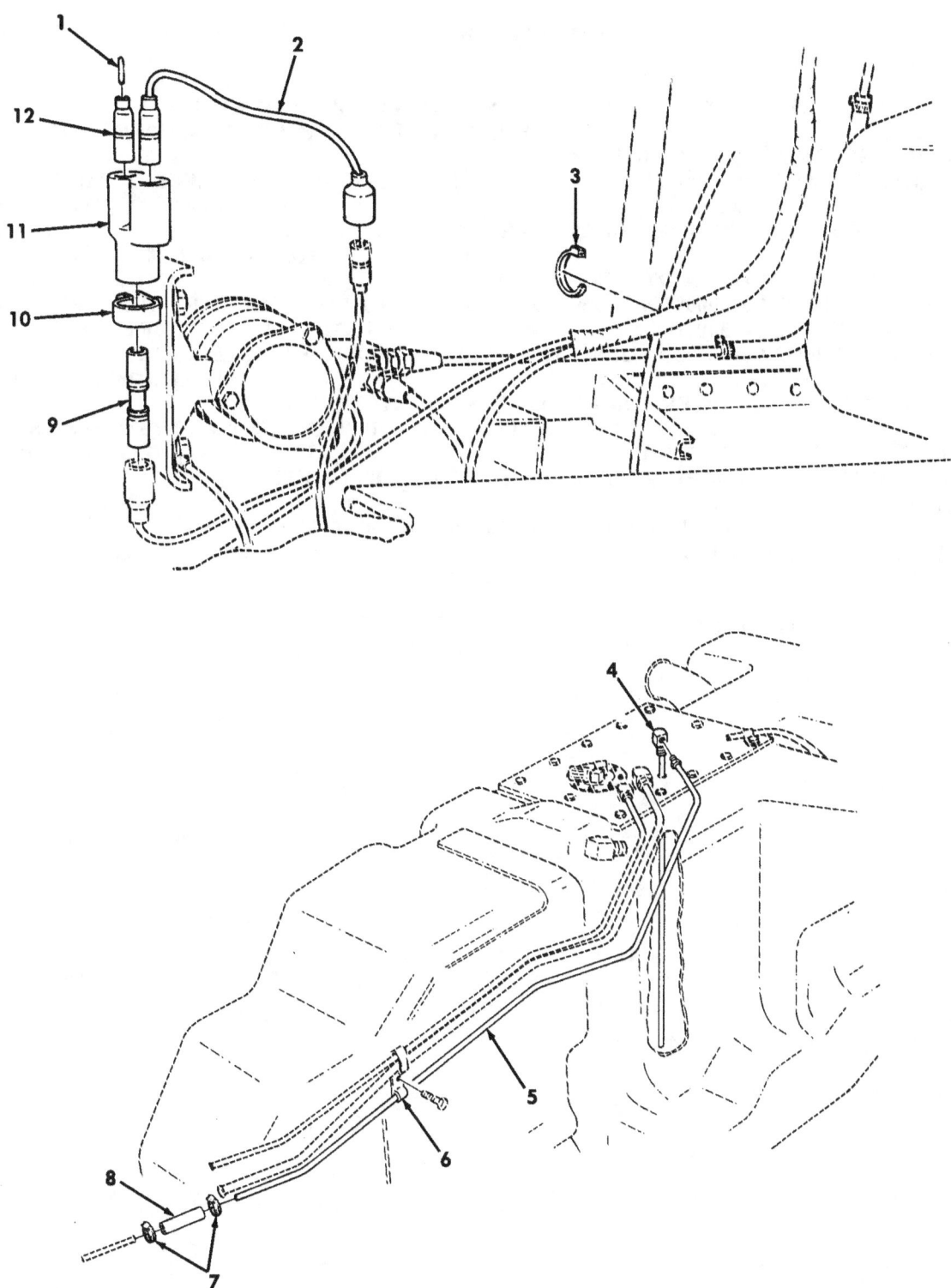

Figure 386. Winterization Kit, Fuel Line and Jumper Cable.

SECTION II TM9-2320-280-24P

(1) ITEM NO	(2) SMR CODE	(3) NSN	(4) CAGEC	(5) PART NUMBER	(6) DESCRIPTION AND USABLE ON CODES (UOC)	(7) QTY
					GROUP 3303 WINTERIZATION KITS	
					FIG. 386 WINTERIZATION KIT, FUEL LINE AND JUMPER CABLE	
1	PAOZZ	5935002140904	19207	7982907	DUMMY,CONNECTOR,PLU PART OF KIT P/N 5705611 UOC:A13,A14,H13,H14	1
2	PAOZZ	6150013960906	19207	12342902	LEAD,ELECTRICAL PART OF KIT P/N 5705611 UOC:A13,A14,H13,H14	1
3	PAOZZ	5975001338687	96906	MS3367-5-0	STRAP,TIEDOWN,ELECT PART OF KIT P/N 5705611 UOC:A13,A14,H13,H14	1
4	PAOZZ	4710012030615	34623	5578694	TUBE ASSEMBLY,METAL PART OF KIT P/N 5705611 UOC:A13,A14,H13,H14	1
5	PAOZZ	4710012030607	34623	5588694	TUBE ASSEMBLY,METAL PART OF KIT P/N 5705611 UOC:A13,A14,H13,H14	1
6	PAOZZ	5340009936207	96906	MS21333-99	CLAMP,LOOP PART OF KIT P/N 5705611. UOC:A13,A14,H13,H14	1
7	PAOZZ	4730011188278	76599	MMF4-SS	CLAMP,HOSE PART OF KIT P/N 57T5611. UOC:A13,A14,H13,H14	2
8	PAOZZ	4720013162538	9C234	12338553-1	HOSE,NONMETALLIC PART OF KIT P/N 5705611 UOC:A13,A14,H13,H14	1
9	PAOZZ	5935008074109	19207	8741492	ADAPTER,CONNECTOR PART OF KIT P/N 5705611 UOC:A13,A14,H13,H14	1
10	PAOZZ	5340002827537	96906	MS21333-41	CLAMP,LOOP PART OF KIT P/N 5705611. UOC:A13,A14,H13,H14	1
11	PAOZZ		5935009006281	96906	MS27147-1 ADAPTER,CONNECTOR 3-WAY,PART OF KIT P/N 5705611 UOC:A13,A14,H13,H14	1
12	PAOZZ	5935008338561	19207	8338561	SHELL,ELECTRICAL CO PART OF KIT P/N 5705611 UOC:A13,A14,H13,H14	1

END OF FIGURE

Section II. TM 9-2320-280-24P-2

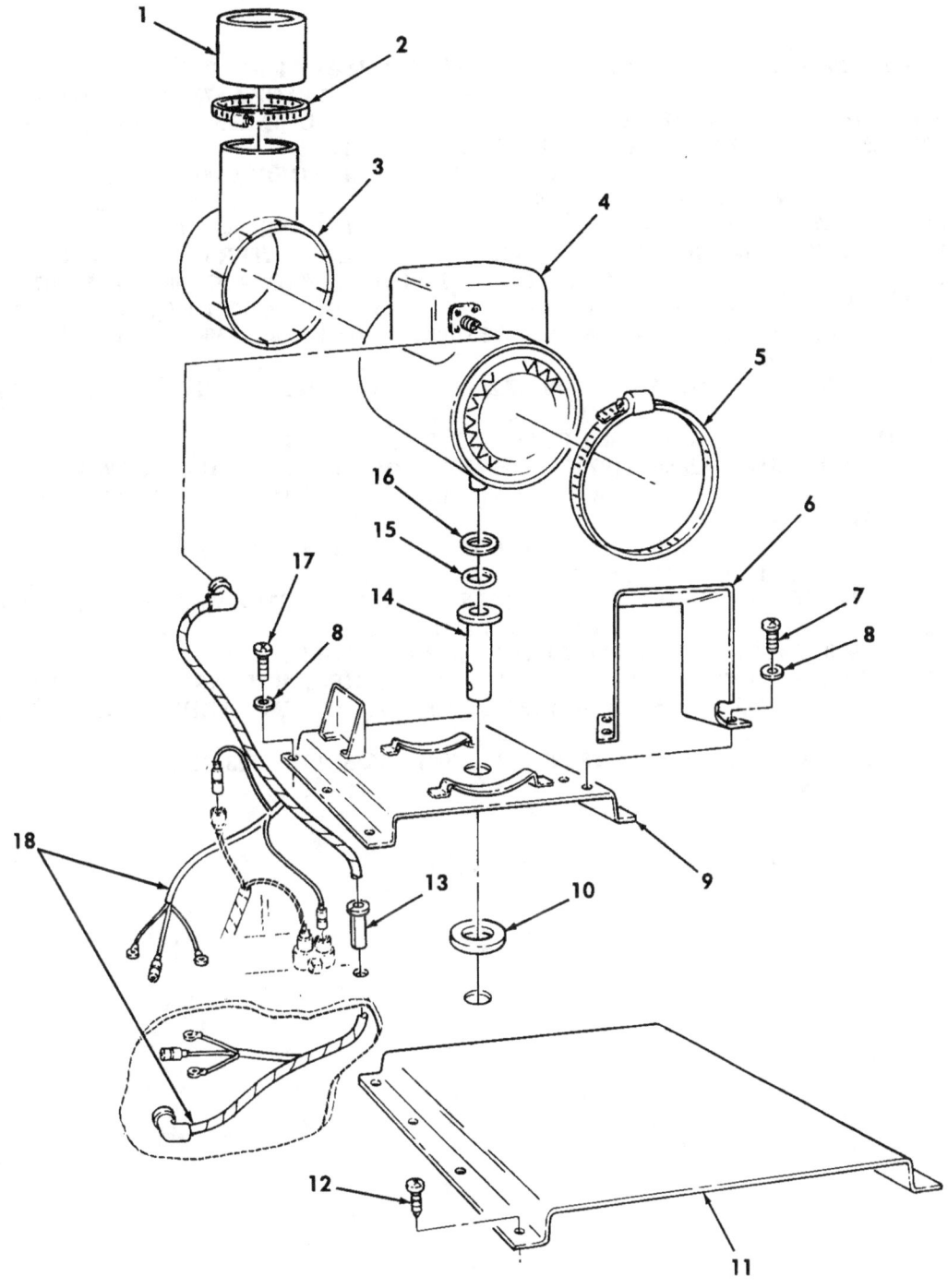

Figure 387. Winterization Kit, Heater and Exhaust Diverter Assembly.

SECTION II   TM9-2320-280-24P

(1) ITEM NO	(2) SMR CODE	(3) NSN	(4) CAGEC	(5) PART NUMBER	(6) DESCRIPTION AND USABLE ON CODES(UOC)	(7) QTY

GROUP 3303 WINTERIZATION KITS

FIG. 387 WINTERIZATION KIT, HEATER AND EXHAUST DIVERTER ASSEMBLY

(1)	(2)	(3)	(4)	(5)	(6)	(7)
1	PFOZZ	5340014507521	19207	12446982	SLEEVE,COUPLING,CLA PART OF KIT P/N 5705611.............. UOC:A13,A14,H13,H14	1
2	PAOZZ	4730009086293	96906	MS35842-15	CLAMP,HOSE PART OF KIT P/N 5705611.. UOC:A13,A14,H13,H14	1
3	PFOZZ	4520013170929	19207	12342256	ELBOW,AIR CONDITION PART OF KIT P/N 5705611............. UOC:A13,A14,H13,H14	1
4	PAOFF	2540011943323	19207	11669898	HEATER,VEHICULAR,CO PART OF KIT P/N 5705611............. UOC:A13,A14,H13,H14	1
5	PAOZZ	4730009086294	34623	5578947	CLAMP,HOSE PART OF KIT P/N 5705611.. UOC:A13,A14,H13,H14	2
6	PAOZZ	2540013162597	34623	12342213	SHIELD,HEATER DUCT PART OF KIT P/N 5705611............. UOC:A13,A14,H13,H14	1
7	PAOZZ	5305002678952	80204	B1821BH025F050N	SCREW,CAP,HEXAGON H 1/4-28 X .50, PART OF KIT P/N 5705611............. UOC:A13,A14,H13,H14	4
8	PAOZZ	5310002092946	12204	120380	WASHER,LOCK 1/4,PART OF KIT P/N 5705611............. UOC:A13,A14,H13,H14	10
9	PFOZZ	4520013162585	19207	12342191	BASE,PLATE PART OF KIT P/N 5705611.. UOC:A13,A14,H13,H14	1
10	PAOZZ	5330013197302	19207	12356789	GASKET PART OF KIT P/N 5705611...... UOC:A13,A14,H13,H14	1
11	PAOZZ	5340013208662	19207	11609942	BRACKET,MULTIPLE AN PART OF KIT P/N 5705611............. UOC:A13,A14,H13,H14	1
12	PAOZZ	5325000149926	96906	MS35493-76	SCREW,WOOD #10 X .75,PART OF KIT P/N 5705611............. UOC:A13,A14,H13,H14	8
13	PAOZZ	5325011757442	19207	12300715-1	GROMMET,NONMETALLIC PART OF KIT P/N 5705611............. UOC:A13,A14,H13,H14	1
14	PAOZZ	4710013171044	34623	12342276	TUBE ASSEMBLY,METAL PART OF KIT P/N 5705611............. UOC:A13,A14,H13,H14	1
15	PAOZZ	5331007700242	19207	7700242	O-RING PART OF KIT P/N 5705611...... UOC:A13,A14,H13,H14	1
16	PAOZZ	5310007700243	19207	7700243	WASHER,FLAT PART OF KIT P/N 5705611. UOC:A13,A14,H13,H14	1
17	PAOZZ	5305000680515	80204	B1821BH025F100N	SCREW,CAP,HEXAGON H 1/4-28 X 1.00, PART OF KIT P/N 5705611............. UOC:A13,A14,H13,H14	6
18	PFOZZ	6150013246355	19207	12342193	CABLE ASSEMBLY,SPEC PART OF KIT P/N 5705611............. UOC:A13,A14,H13,H14	1

END OF FIGURE

Section II.   TM 9-2320-280-24P-2

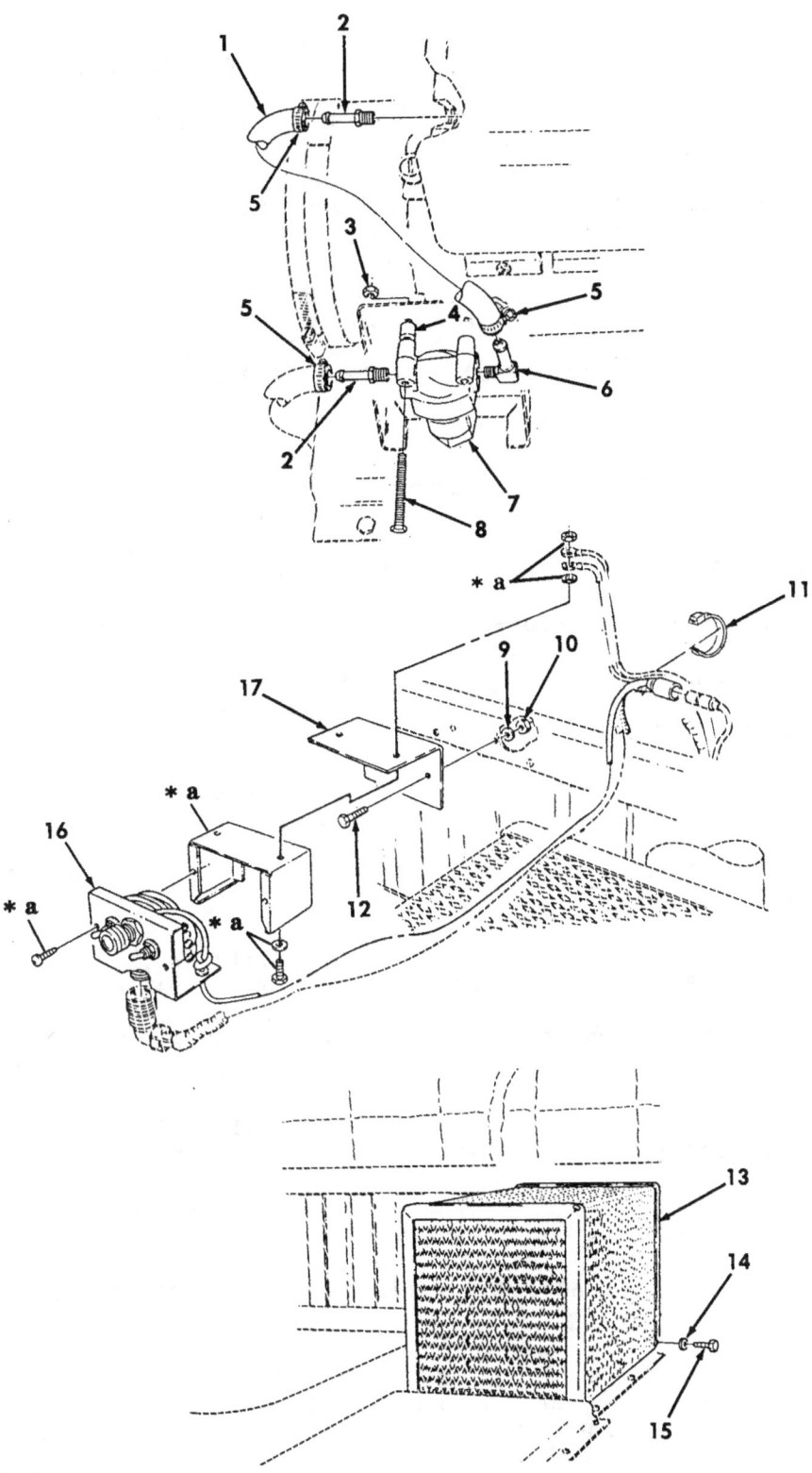

* a PART OF ITEM 16

Figure 388. Winterization Kit, Control Box, Filter, and Cover Assembly.

SECTION II TM9-2320-280-24P C01

GROUP 3303 WINTERIZATION KITS

FIG. 388 WINTERIZATION KIT, CONTROL BOX, FILTER, AND COVER ASSEMBLY

(1) ITEM NO	(2) SMR CODE	(3) NSN	(4) CAGEC	(5) PART NUMBER	(6) DESCRIPTION AND USABLE ON CODES (UOC)	(7) QTY
* 1	MOOZZ		19207	12338553-2	HOSE,NONMETALLIC MAKE FROM HOSE,P/N 9438315,14 INCHES LONG,PART OF KIT P/N 5705611....................... UOC:A13,A14,H13,H14	1
2	PAOZZ	4730013575651	81343	5-2 430160C	ADAPTER,STRAIGHT,PI PART OF KIT P/N 5705611................................. UOC:A13,A14,H13,H14	2
3	PAOZZ	5310008775797	96906	MS21044N3	NUT,SELF-LOCKING,HE #10-32,PART OF KIT P/N 5705611..................... UOC:A13,A14,H13,H14	2
4	PAOZZ	5365013182066	19207	8359764	SPACER,SLEEVE PART OF KIT P/N 5705611.......................... UOC:A13,A14,H13,H14	2
5	PAOZZ	4730002782523	88907	B54-32780	CLAMP,HOSE PART OF KIT P/N 5705611.. UOC:A13,A14,H13,H14	3
6	PAOZZ	4730013486231	81343	5-2 430260	ELBOW PART OF KIT P/N 5705611....... UOC:A13,A14,H13,H14	1
* 7	PAOZZ	2910014458097	58536	A52472-1	FILTER ELEMENT,FLUI PART OF KIT P/N 5705611 UOC:A13,A14,H13,H14	1
8	PAOZZ	5305011204363	54132	9415779	SCREW,CAP,HEXAGON H #10-32 X 2.00 PART OF KIT P/N 5705611............. UOC:A13,A14,H13,H14	2
9	PAOZZ	5310008094058	96906	MS27183-10	WASHER,FLAT 1/4,PART OF KIT P/N 5705611............................. UOC:A13,A14,H13,H14	2
10	PAOZZ	5310009359022	96906	MS51943-32	NUT,SELF-LOCKING,HE 1/4-28,PART OF KIT P/N 5705611..................... UOC:A13,A14,H13,H14	2
11	PAOZZ	5975001338687	96906	MS3367-5-0	STRAP,TIEDOWN,ELECT PART OF KIT P/N 5705611................................. UOC:A13,A14,H13,H14	1
12	PAOZZ	5305002678953	80204	B1821BH025F063N	SCREW,CAP,HEXAGON H 1/4-28 X .625 PART OF KIT P/N 5705611............. UOC:A13,A14,H13,H14	2
13	PAOZZ	2540013208918	19207	12342189	SHIELD,HEATER DUCT PART OF KIT P/N 5705611.......................... UOC:A13,A14,H13,H14	1
14	PAOZZ	5310005825965	96906	MS35338-44	WASHER,LOCK 1/4,PART OF KIT P/N 5705611............................. UOC:A13,A14,H13,H14	4
15	PAOZZ	5306000680513	80204	B1821BH025F075N	BOLT,MACHINE 1/4-28 X .75,PART OF KIT P/N 5705611..................... UOC:A13,A14,H13,H14	4
16	PAOZZ	2540011256154	19207	11669705	CONTROL BOX PART OF KIT P/N 5705611. UOC:A13,A14,H13,H14	1
17	PFOZZ	5340013172675	19207	12342260	BRACKET,ANGLE PART PF KIT P/N 5705611.................... UOC:A13,A14,H13,H14	1

END OF FIGURE

Section II. TM 9-2320-280-24P-2

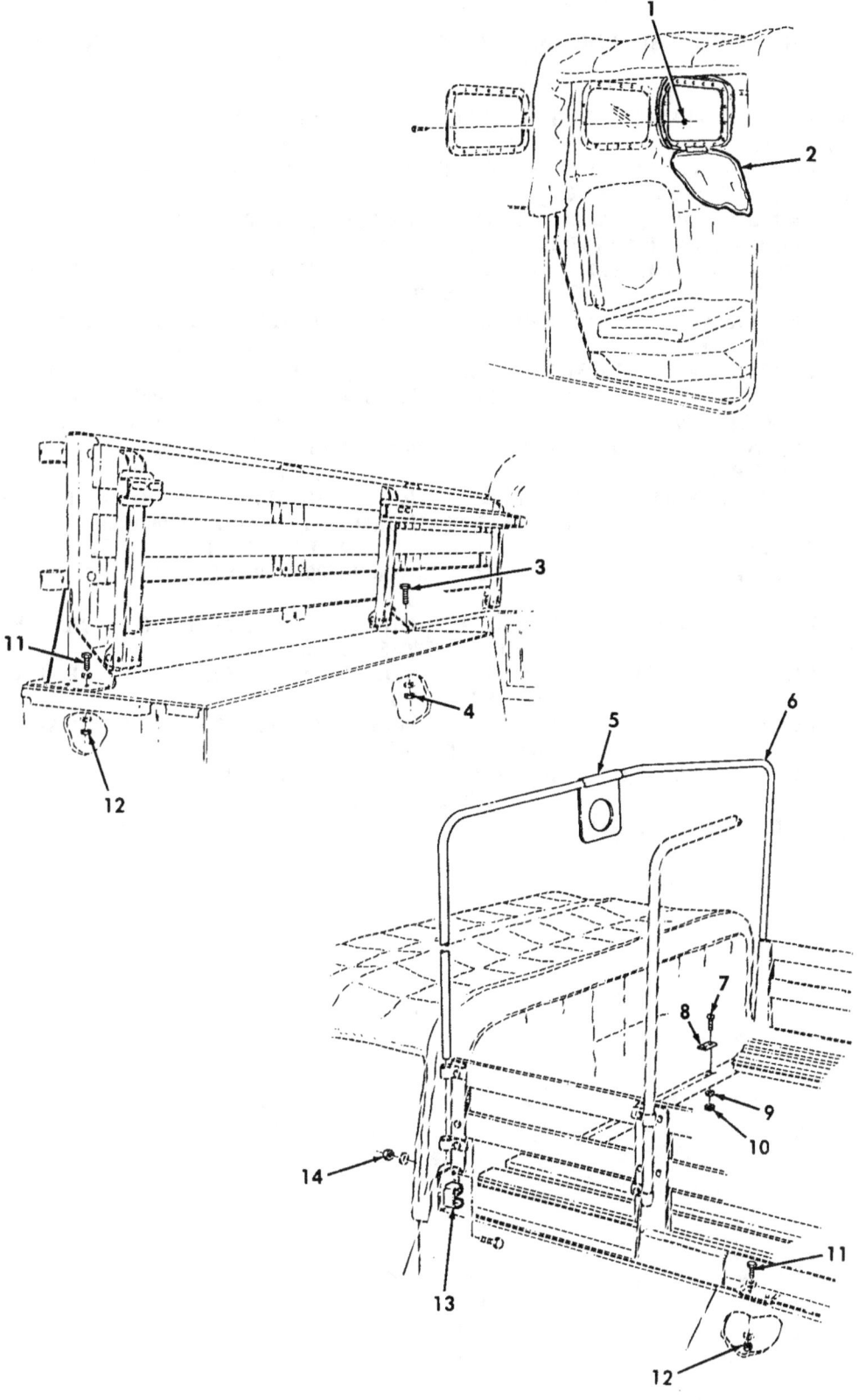

Figure 389. Seat Support Assemblies and Mounting Hardware.

SECTION II   TM9-2320-280-24P

(1) NO	(2) SMR CODE	(3) NSN	(4) CAGEC	(5) PART NUMBER	(7) ITEM DESCRIPTION AND USABLE ON CODES(UOC)	QTY

GROUP 3303 WINTERIZATION KITS

FIG. 389 SEAT SUPPORT ASSEMBLIES AND MOUNTING HARDWARE

1 PAOZZ 5310009416019 96906 MS21083N08 NUT,SELF-LOCKING,HE #10-24 PART OF KIT P/N 5705611..................... 36
UOC:A13,A14,H13,H14

2 PAOZZ 2540001042889 19207 11609693-1 CURTAIN,VEHICULAR,B PART OF KIT P/N 5705611........................... 2
UOC:A13,A14,H13,H14

3 PFOZZ 5305000712056 80204 B1821BH044C175N SCREW,CAP,HEXAGON H 7/16-14 X 1.75 PART OF KIT P/N 5705611............. 4
UOC:A13,A14,H13,H14

4 PAOZZ 5310013185237 24617 9419456 NUT,SELF-LOCKING,HE 7/16-14,PART OF KIT P/N 5705611.................... 4
UOC:A13,A14,H13,H14

5 PFOZZ 5340013177501 19207 12342266 BRACKET,MOUNTING PART OF KIT P/N 5705611............................ 1
UOC:A13,A14,H13,H14

6 PFOZZ 2540011996760 34623 5581331 BOW,VEHICULAR TOP PART OF KIT P/N 5705611............................ 3
UOC:A13,A14,H13,H14

7 PFOZZ 5305009932457 96906 MS35207-284 SCREW,MACHINE 1/4-28 X 1.25,PART OF KIT P/N 5705611.................... 4
UOC:A13,A14,H13,H14

8 PAOZZ 2510013160216 19207 12342166 BLOCK,PLYWOOD PART OF KIT P/N 5705611............................ 2
UOC:A13,A14,H13,H14

9 PFOZZ 5310008094058 96906 MS27183-10 WASHER,FLAT 1/4,PART OF KIT P/N 5705611............................ 4
UOC:A13,A14,H13,H14

10 PAOZZ 5310004838791 96906 MS17829-4F NUT,SELF-LOCKING,HE 1/4-28,PART OF KIT P/N 5705611.................... 4
UOC:A13,A14,H13,H14

11 PAOZZ 5305000712509 80204 B1821BH025C150N SCREW,CAP,HEXAGON H 1/4-20 X 1.50 PART OF KIT P/N 5705611............. 18
UOC:A13,A14,H13,H14

12 PAOZZ 5310010666759 72962 21NE-040 NUT,SELF-LOCKING,HE 1/4-20,PART OF KIT P/N 5705611.................... 18
UOC:A13,A14,H13,H14

13 PAOZZ 2590013170534 34623 12342175 BRACKET,VEHICULAR C R.H.,PART OF KIT P/N 5705611.................... 1
UOC:A13,A14,H13,H14

13 PAOZZ 2590013170535 34623 12342264 BRACKET,VEHICULAR C L.H.,PART OF KIT P/N 5705611.................... 1
UOC:A13,A14,H13,H14

14 PAOZZ 5310002416658 96906 MS51943-34 NUT,SELF-LOCKING,HE 5/16-24,PART OF KIT P/N 5705611.................... 6
UOC:A13,A14,H13,H14

END OF FIGURE

Section II. TM 9-2320-280-24P-2

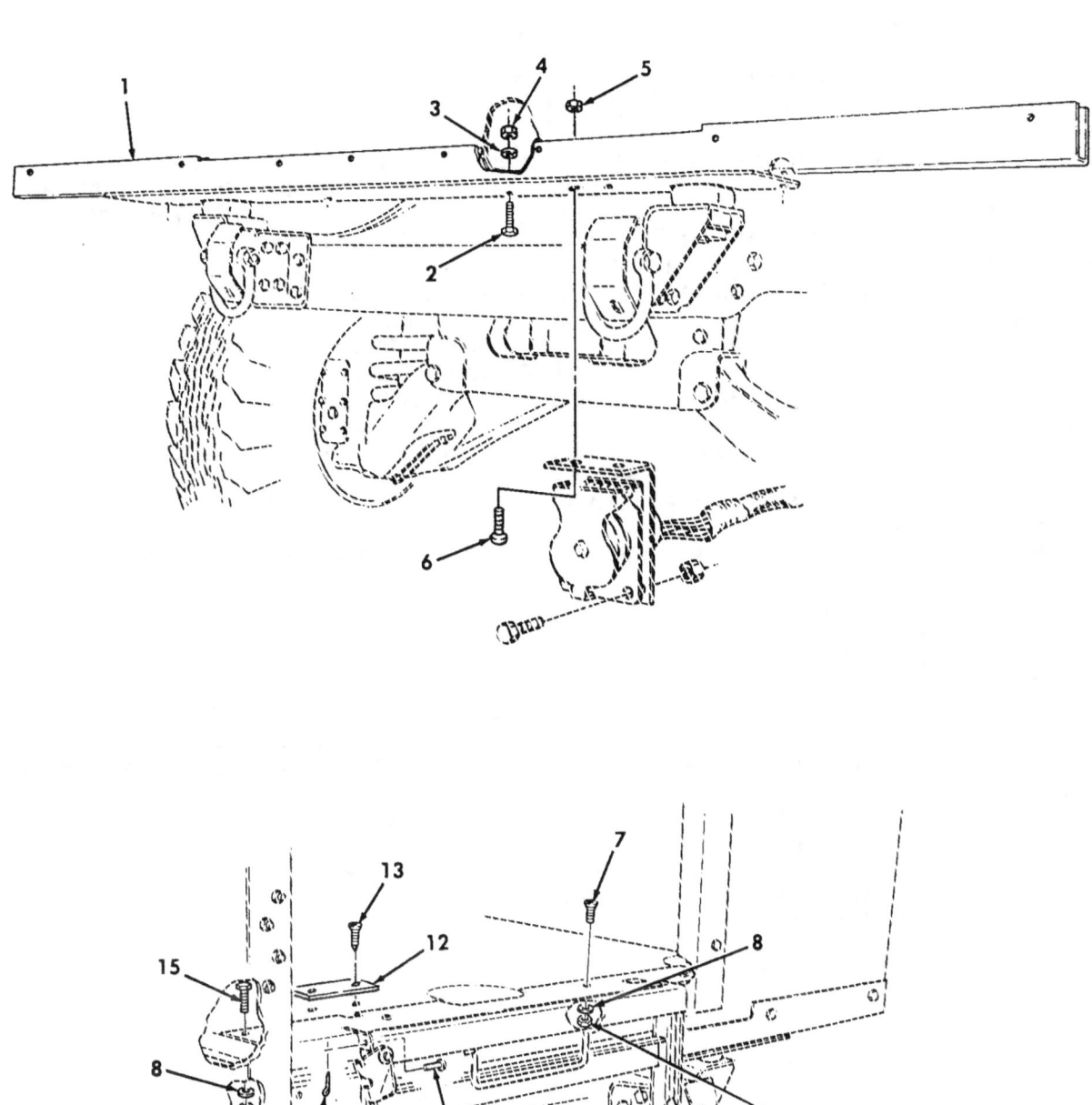

Figure 390. Step Assembly, Support Assembly, and Mounting Hardware.

SECTION II TM9-2320-280-24P

(1) ITEM NO	(2) SMR CODE	(3) NSN	(4) CAGEC	(5) PART NUMBER	(6) DESCRIPTION AND USABLE ON CODES(UOC)	(7) QTY
					GROUP 3303 WINTERIZATION KITS	
					FIG. 390 STEP ASSEMBLY, SUPPORT ASSEMBLY, AND MOUNTING HARDWARE	
1	PFOZZ	2540013170820	34623	12342332	SUPPORT ASSEMBLY,BO PART OF KIT P/N 5705611.................... UOC:A13,A14,H13,H14	1
2	PAOZZ	5305007195221	80204	B1821BH-050F150N	SCREW,CAP,HEXAGON H 1/2-20 X 1.50, PART OF KIT P/N 5705611............. UOC:A13,A14,H13,H14	4
3	PAOZZ	5310008095998	96906	MS27183-18	WASHER,FLAT 1/2 PART OF KIT P/N 5705611........................ UOC:A13,A14,H13,H14	4
4	PAOZZ	5310007749073	72582	190171	NUT,SELF-LOCKING,CA 1/2-20 PART OF KIT P/N 5705611.................... UOC:A13,A14,H13,H14	4
5	PAOZZ	5310008140673	96906	MS51943-33	NUT,SELF-LOCKING,HE 5/16-18 PART OF KIT P/N 5705611.................... UOC:A13,A14,H13,H14	2
6	PAOZZ	5306002264829	80204	B1821BH031C125N	BOLT,MACHINE 5/16-18 X 1.25 PART OF KIT P/N 5705611.................... UOC:A13,A14,H13,H14	2
7	PAOZZ	5305009017250	96906	MS24693-S350	SCREW,MACHINE 3/8-24 X 1.25,PART OF KIT P/N 5705611.................... UOC:A13,A14,H13,H14	2
8	PAOZZ	5310001670821	61465	M2025103	WASHER,FLAT 3/8,PART OF KIT P/N 5705611.................... UOC:A13,A14,H13,H14	4
9	PAOZZ	5310010583353	72582	192481	NUT,SELF-LOCKING,HE 3/8-24,PART OF KIT P/N 5705611.................... UOC:A13,A14,H13,H14	4
10	PAOZZ	2540013171024	34623	12342301	LADDER,VEHICLE BOAR PART OF KIT P/N 5705611............................ UOC:A13,A14,H13,H14	1
11	PAOZZ	5315008903461	96906	MS20392-6C17	PIN,STRAIGHT,HEADER 7/16 X .531, PART OF KIT P/N 5705611............. UOC:A13,A14,H13,H14	2
12	PAOZZ	5340013942409	19207	12342903	PLATE,MOUNTING PART OF KIT P/N 5705611..................... UOC:A13,A14,H13,H14	1
13	PAOZZ	5305009026643	96906	MS35492-76	SCREW,WOOD #10 X .75,PART OF KIT P/N 5705611........................ UOC:A13,A14,H13,H14	2
14	PAOZZ	5315008423044	96906	MS24665-283	PIN,COTTER 1/16,PART OF KIT P/N 5705611............................ UOC:A13,A14,H13,H14	2
15	PAOZZ	5305002693242	80204	B1821BH038F200N	SCREW,CAP,HEXAGON H 3/8-24 X 2.00, PART OF KIT P/N 5705611............. UOC:A13,A14,H13,H14	2

END OF FIGURE

Section II. TM 9-2320-280-24P-2

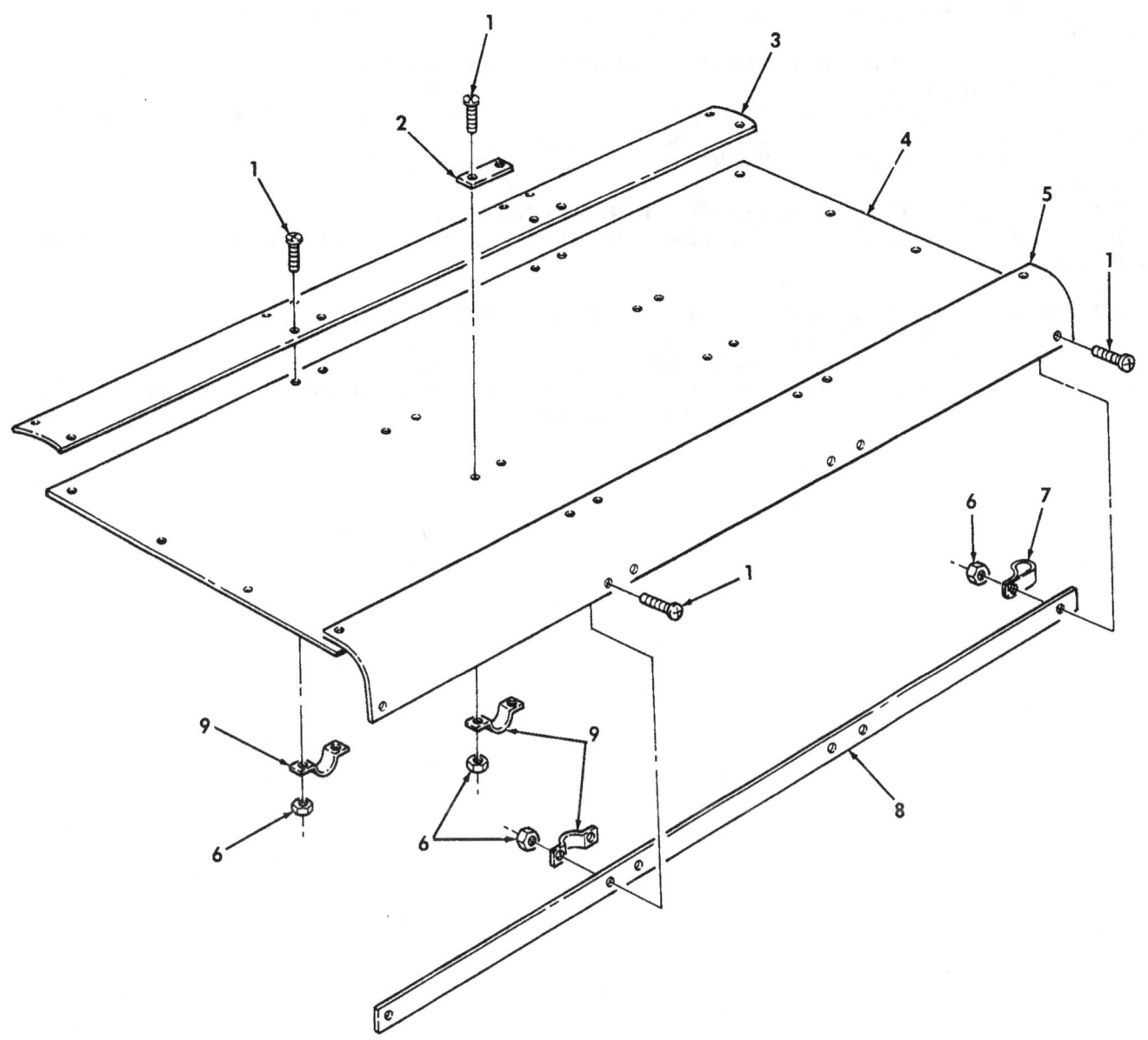

Figure 391. Top Plate Support and Attaching Hardware.

SECTION II   TM9-2320-280-24P

(1) ITEM NO	(2) SMR CODE	(3) NSN	(4) CAGEC	(5) PART NUMBER	(7) DESCRIPTION AND USABLE ON CODES(UOC)	QTY

GROUP 3303 WINTERIZATION KITS

FIG. 391 TOP PLATE SUPPORT AND ATTACHING HARDWARE

Item	SMR	NSN	CAGEC	Part Number	Description	QTY
1	PAOZZ	5305009932460	96906	MS35207-282	SCREW,MACHINE 1/4-28 X .875 PART OF KIT P/N 5705611............ UOC:A13,A14,H13,H14	50
2	PAOZZ	5340013182040	19207	12342218	PLATE,MENDING PART OF KIT P/N 5705611........................... UOC:A13,A14,H13,H14	8
3	KFFZZ		19207	12342226	PLATE,SEAM PART OF KIT P/N 5705611.. UOC:A13,A14,H13,H14	1
4	KFFZZ		19207	12342220-1	SUPPORT PART OF KIT P/N 5705611..... UOC:A13,A14,H13,H14	1
5	KFFZZ		19207	12342221	SUPPORT PART OF KIT P/N 5705611..... UOC:A13,A14,H13,H14	1
6	PAOZZ	5310013174022	24617	9413553	NUT,PLAIN,EXTENDED 1/4-28 PART OF KIT P/N 5705611.................... UOC:A13,A14,H13,H14	50
7	PAOZZ	5340013179086	19207	12342234	CLAMP,LOOP PART OF KIT P/N 5705611.. UOC:A13,A14,H13,H14	10
8	KFFZZ		19207	12342220-2	SUPPORT PART OF KIT P/N 5705611..... UOC:A13,A14,H13,H14	1
9	PAOZZ	5340013199426	19207	12342235	STRAP,RETAINING,PART OF KIT P/N 5705611........................... UOC:A13,A14,H13,H14	20

END OF FIGURE

Section II. TM 9-2320-280-24P-2

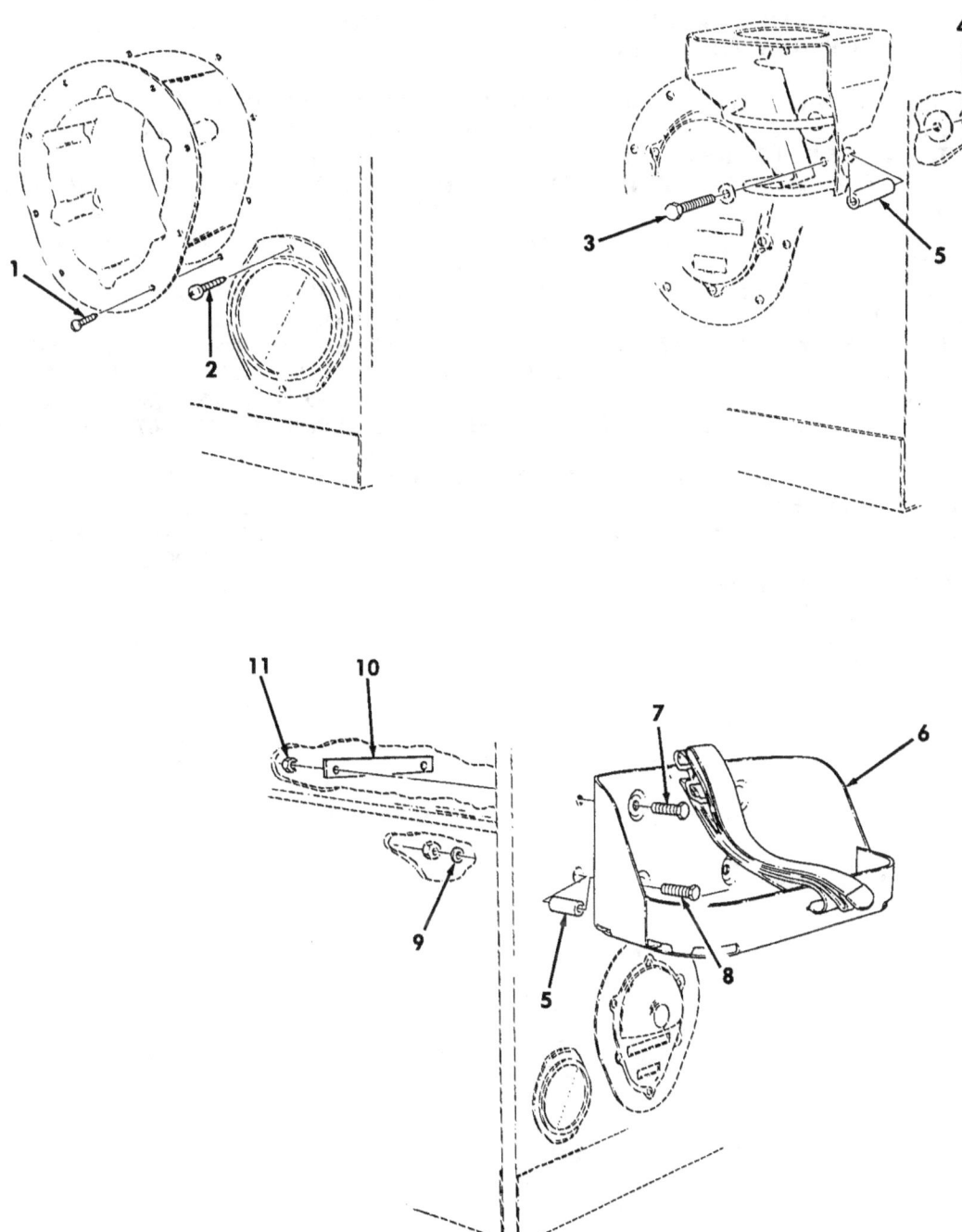

Figure 392. Fuel Can Bracket, Antenna Mounting Bracket, and Composite Light Housing Mounting Hardware.

SECTION II   TM9-2320-280-24P

(1) ITEM NO	(2) SMR CODE	(3) NSN	(4) CAGEC	(5) PART NUMBER	(7) DESCRIPTION AND USABLE ON CODES(UOC)	(6) QTY

GROUP 3303 WINTERIZATION KITS

FIG. 392 FUEL CAN BRACKET, ANTENNA MOUNTING BRACKET, AND COMPOSITE LIGHT HOUSING MOUNTING HARDWARE

1	PAOZZ	5305000149926	96906	MS35493-76	SCREW,WOOD #10 X .875,PART OF KIT P/N 5705611............... UOC:A13,A14,H13,H14	16
2	PAOZZ	5305013159512	96906	MS35495-124	SCREW,WOOD #14 X .75,PART OF KIT P/N 5705611............... UOC:A13,A14,H13,H14	4
3	PAOZZ	5306002259097	96906	MS90726-42	BOLT,MACHINE 5/16-24 X 2.50,PART OF KIT P/N 5705611............... UOC:A13,A14,H13,H14	3
4	PAOZZ	5310000880553	24617	190139	NUT,SELF-LOCKING,HE 5/16-24,PART OF KIT P/N 5705611............... UOC:A13,A14,H13,H14	3
5	PAOZZ	5365013168985	19207	12342237	SPACER,SLEEVE PART OF KIT P/N 5705611............... UOC:A13,A14,H13,H14	5
6	PAOZZ	2590004736331	07860	C21452	BRACKET ASSEMBLY,LI PART OF KIT P/N 5705611............... UOC:A13,A14,H13,H14	1
7	PAOZZ	5305009146131	96906	MS18153-63	SCREW,CAP,HEXAGON H 3/8-24 X 1.75, PART OF KIT P/N 5705611............. UOC:A13,A14,H13,H14	2
8	PAOZZ	5305002692811	96906	MS90726-67	SCREW,CAP,HEXAGON H 3/8-24 X 2.50, PART OF KIT P/N 5705611............. UOC:A13,A14,H13,H14	2
9	PAOZZ	5310000806004	96906	MS27183-14	WASHER,FLAT 3/8,PART OF KIT P/N 5705611............... UOC:A13,A14,H13,H14	2
10	PAOZZ	5340013182041	19207	12342219	PLATE,MENDING PART OF KIT P/N 5705611............... UOC:A13,A14,H13,H14	1
11	PAOZZ	5310010583353	72582	192481	NUT,SELF-LOCKING,HE 3/8-24,PART OF KIT P/N 5705611............... UOC:A13,A14,H13,H14	4

END OF FIGURE

Section II. TM 9-2320-280-24P-2

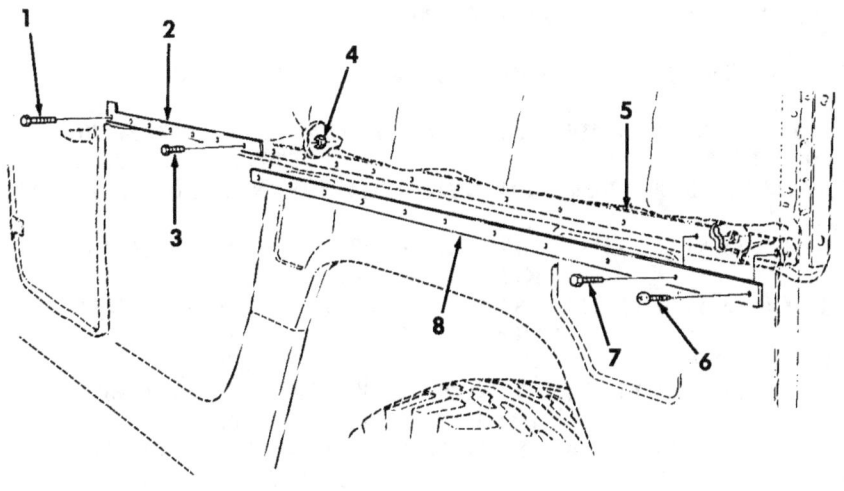

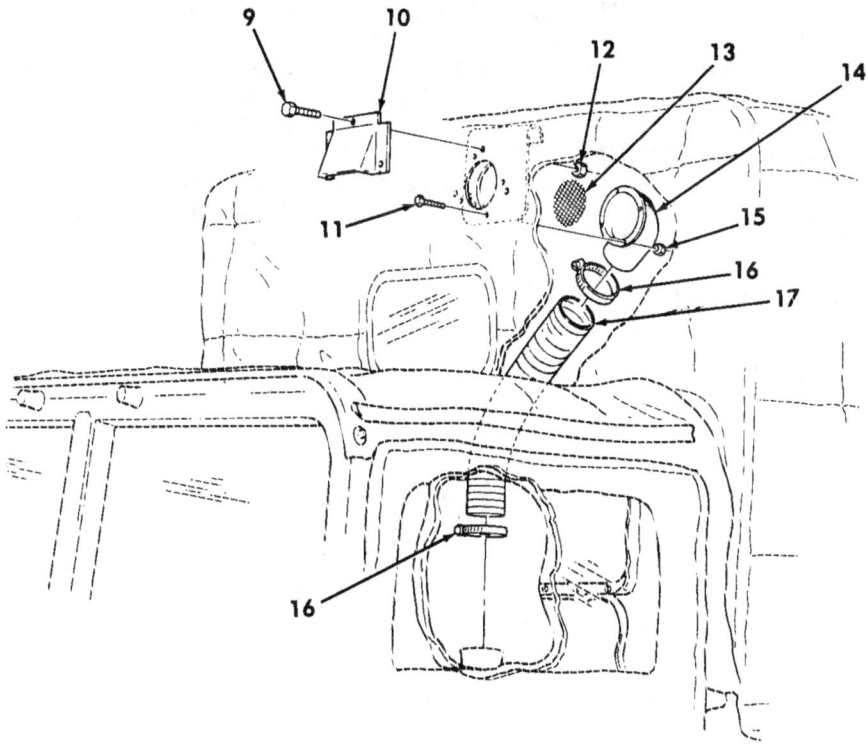

Figure 393. Mounting Plates and Ventilator Exhaust Ducting.

SECTION II   TM9-2320-280-24P

(1) NO	(2) SMR CODE	(3) NSN	(4) CAGEC	(5) PART NUMBER	(7) DESCRIPTION AND USABLE ON CODES(UOC)	(6) QTY

GROUP 3303 WINTERIZATION KITS

FIG. 393 MOUNTING PLATES AND VENTILATOR EXHAUST DUCTING

(1) NO	(2) SMR	(3) NSN	(4) CAGEC	(5) PART NUMBER	(7) DESCRIPTION AND UOC	(6) QTY
1	PAOZZ	5305000712237	96906	MS90725-14	SCREW,CAP,HEXAGON H 1/4-20 X 2.00, PART OF KIT P/N 5705611............. UOC:A13,A14,H13,H14	1
2	PAOZZ	5340013202060	19207	12342201	PLATE,MOUNTING PART OF KIT P/N 5705611............. UOC:A13,A14,H13,H14	1
3	PAOZZ	5305000712056	80204	B1821BH044C175N	SCREW,CAP,HEXAGON H 1/4-20 X 1.75, PART OF KIT P/N 5705611............. UOC:A13,A14,H13,H14	6
4	PAOZZ	5310010128962	24617	9416918	NUT,PLAIN,EXTENDED 1/4-20,PART OF KIT P/N 5705611.............. UOC:A13,A14,H13,H14	17
5	PAOZZ	5330013175392	19207	12342248	RUBBER,STRIP PART OF KIT P/N 5705611 UOC:A13,A14,H13,H14	1
6	PAOZZ	5305013170409	96906	MS35495-127	SCREW,WOOD #14 X 1.25,PART OF KIT P/N 5705611............ UOC:A13,A14,H13,H14	1
7	PAOZZ	5305000680502	96906	MS90725-6	SCREW,CAP,HEXAGON H 1/4-20 X .75, PART OF KIT P/N 5705611............. UOC:A13,A14,H13,H14	10
8	PAOZZ	2590013174854	19207	12342173	MOLDING,METAL PART OF KIT P/N 5705611................. UOC:A13,A14,H13,H14	1
9	PAOZZ	5305000680515	80204	B1821BH025F100N	SCREW,CAP,HEXAGON H 1/4-28 X 1.00, PART OF KIT P/N 5705611............. UOC:A13,A14,H13,H14	3
10	PFOZZ	2540013162588	34623	7390378	VENTILATOR,AIR,CIRC PART OF KIT P/N 5705611............. UOC:A13,A14,H13,H14	1
11	PAOZZ	5305009896265	96906	MS35207-262	SCREW,MACHINE #10-32 X .375,PART OF KIT P/N 5705611................. UOC:A13,A14,H13,H14	4
12	PAOZZ	5310009359022	96906	MS51943-32	NUT,SELF-LOCKING,HE 1/4-28,PART OF KIT P/N 5705611.................. UOC:A13,A14,H13,H14	3
13	PAOZZ	4730005671630	78358	484487	STRAINER ELEMENT,SE PART OF KIT P/N 5705611............................ UOC:A13,A14,H13,H14	1
14	PFOZZ	2540001778108	19207	7951057	ADAPTER,VEHICLE STO PART OF KIT P/N 5705611............................ UOC:A13,A14,H13,H14	1
15	PAOZZ	5310008775797	96906	MS21044N3	NUT,SELF-LOCKING,HE #10-32,PART OF KIT P/N 5705611.................... UOC:A13,A14,H13,H14	4
16	PAOZZ	4730009086293	66295	60HS	CLAMP.HOSE PART OF KIT P/N 5705611.. UOC:A13,A14,H13,H14	2
17	PAOZZ	4720014764656	19207	12342261-1	HOSE,AIR DUCT 44 INCHES LONG PART OF KIT P/N 5705611................. UOC:A13,A14,H13,H14	1

END OF FIGURE   393-1

Section II. TM 9-2320-280-24P-2

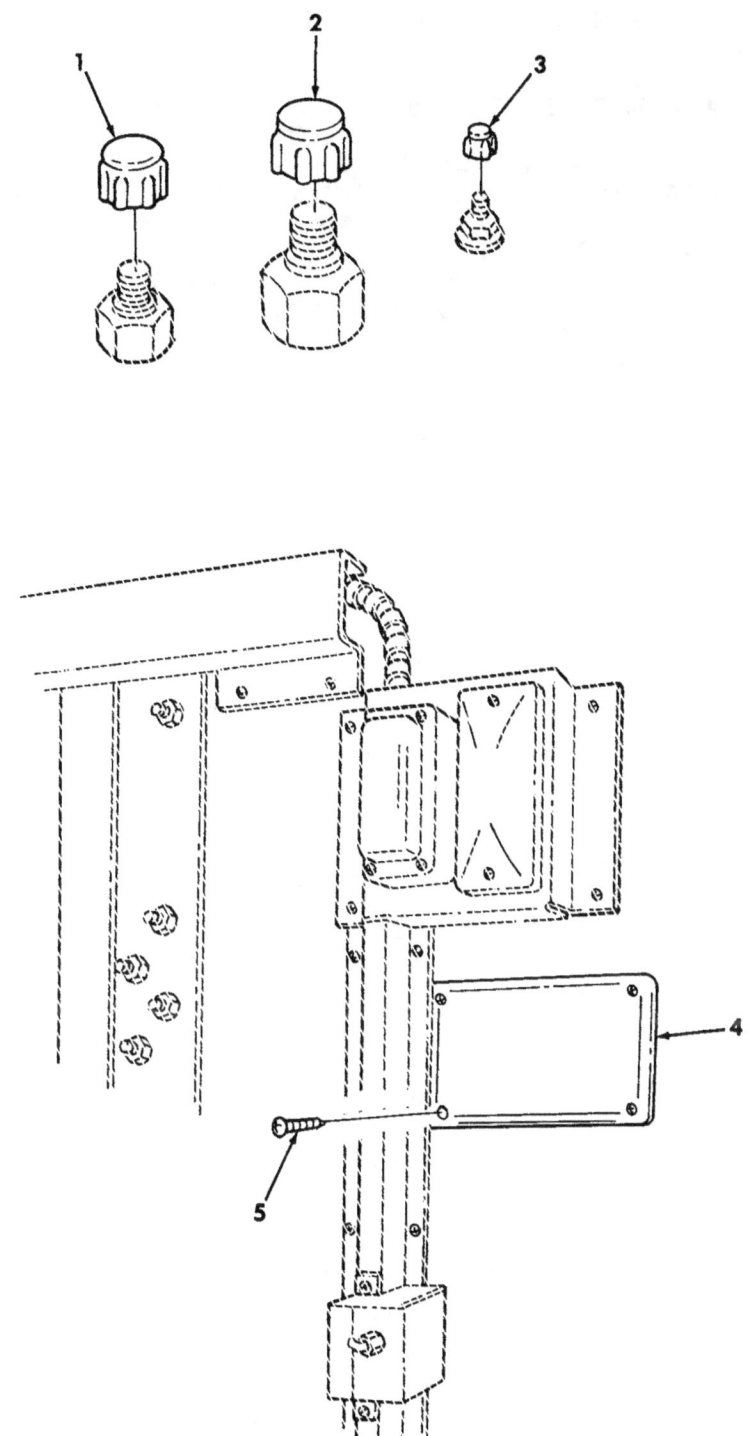

Figure 394. Screw Caps and Winterization Instruction Plate.

SECTION II   TM9-2320-280-24P

(1) NO	(2) SMR CODE	(3) NSN	(4) CAGEC	(5) PART NUMBER	(7) ITEM DESCRIPTION AND USABLE ON CODES(UOC)	QTY

GROUP 3303 WINTERIZATION KITS

FIG. 394 SCREW CAPS AND WINTERIZATION INSTRUCTION PLATE

1	PAOZZ	5340007274774	81349	M5501/11-F5	CAP,PROTECTIVE,DUST PART OF KIT P/N 5705611............................ UOC:A13,A14,H13,H14	23
2	PAOZZ	5340011679314	81349	M5501/11-F7	CAP,PROTECTIVE,DUST PART OF KIT P/N 5705611............................ UOC:A13,A14,H13,H14	2
3	PAOZZ	5340011744894	81349	M5501/11-F1	CAP.PROTECTIVE,DUST PART OF KIT P/N 5705611............................ UOC:A13,A14,H13,H14	52
4	PFOZZ	9905010327002	19207	10896651	PLATE,INSTRUCTION PART OF KIT P/N 5705611............................ UOC:A13,A14,H13,H14	1
5	PAOZZ	5305009012099	96906	MS35493-51	SCREW,WOOD #8 X .50,PART OF KIT P/N 5705611............................ UOC:A13,A14,H13,H14	4

END OF FIGURE

Section II. TM 9-2320-280-24P-2

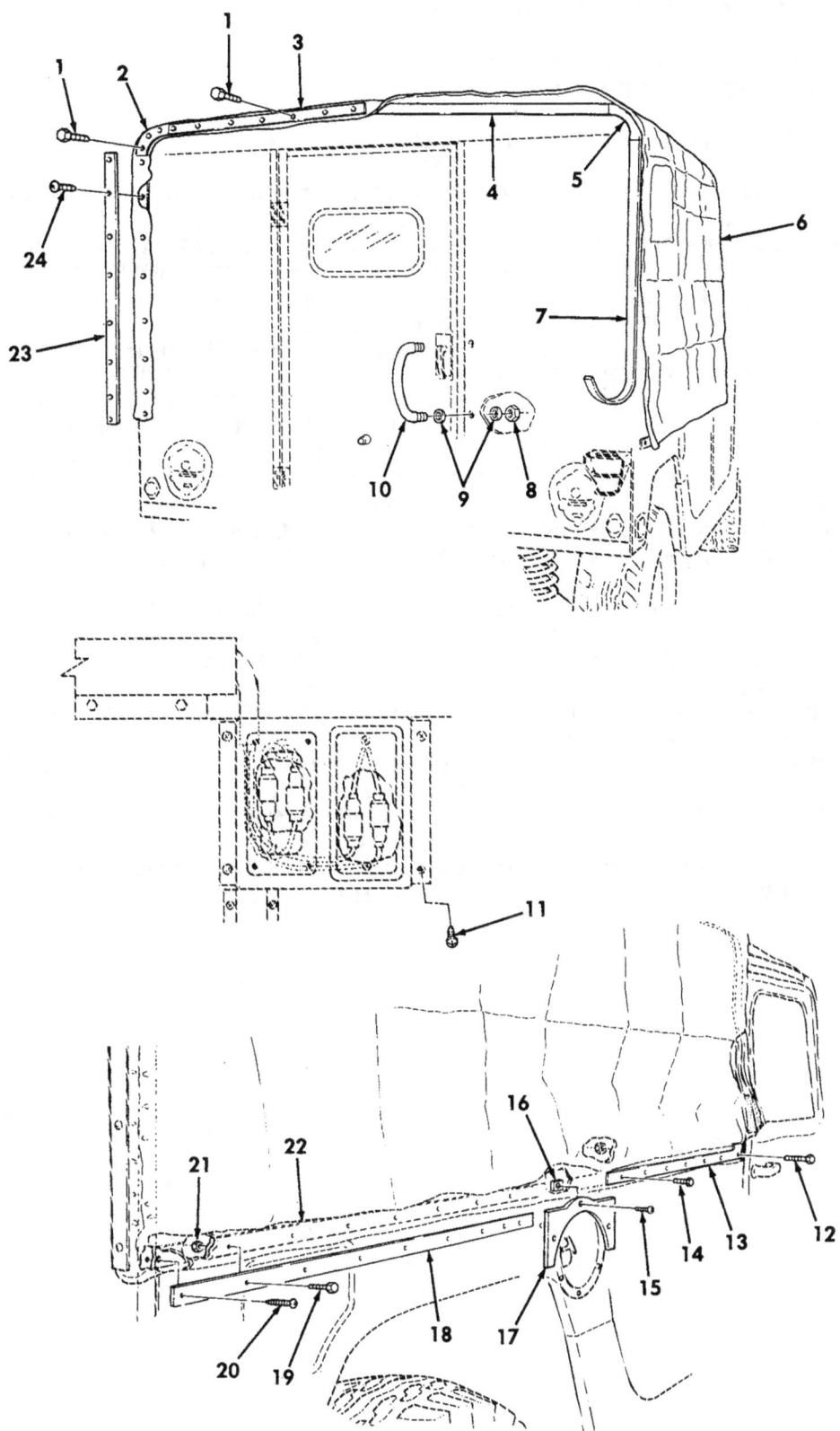

Figure 395. Top Cover Assembly and Mounting Hardware.

SECTION II   TM9-2320-280-24P

(1) NO	(2) SMR CODE	(3) NSN	(4) CAGEC	(5) PART NUMBER	(7) DESCRIPTION AND USABLE ON CODES(UOC)	ITEM QTY

GROUP 3303 WINTERIZATION KITS

FIG. 395 TOP COVER ASSEMBLY AND MOUNTING HARDWARE

1	PAOZZ	5305004324253	96906	MS51861-67	SCREW,TAPPING 1/4-14 X .75,PART OF KIT P/N 5705611..................... UOC:A13,A14,H13,H14	20
2	PAOZZ	2590013173952	19207	12342195	MOLDING,METAL PART OF KIT P/N 5705611............................ UOC:A13,A14,H13,H14	2
3	PAOZZ	2590013178292	19207	12342216	MOLDING,METAL PART OF KIT P/N 5705611......................... UOC:A13,A14,H13,H14	2
4	PAOZZ	5330013175393	19207	12342345	RUBBER STRIP PART OF KIT P/N 5705611 UOC:A13,A14,H13,H14	2
5	PAOZZ	5330013181998	19207	12342343	SEAL,NONMETALLIC ST PART OF KIT P/N 5705611..................... UOC:A13,A14,H13,H14	2
6	PAOFF	2540013153762	19207	12342187	COVER,FITTED,VEHICU PART OF KIT P/N 5705611............................ UOC:A13,A14,H13,H14	1
7	PAOZZ	5330013189780	19207	12342344	RUBBER,STRIP PART OF KIT P/N 5705611 UOC:A13,A14,H13,H14	2
8	PAOZZ	5310013178164	24617	272739	NUT,SELF-LOCKING,HE 5/8-18,PART OF KIT P/N 5705611.................... UOC:A13,A14,H13,H14	2
9	PAOZZ	5310004986675	10001	1788116	WASHER,FLAT 5/8,PART OF KIT P/N 5705611.............. UOC:A13,A14,H13,H14	4
10	PAOZZ	2540013174289	34623	11602903	HANDLE,DOOR PART OF KIT P/N 5705611 UOC:A13,A14,H13,H14	1
11	PAOZZ	5305000567879	96906	MS35493-74	SCREW,WOOD #10 X .50,PART OF KIT P/N 5705611..................... UOC:A13,A14,H13,H14	8
12	PAOZZ	5305000712237	96906	MS90725-14	SCREW,CAP,HEXAGON H 1/4-20 X 2.00, PART OF KIT P/N 5705611............ UOC:A13,A14,H13,H14	1
13	PAOZZ	5340013202060	19207	12342201	PLATE,MOUNTING PART OF KIT P/N 5705611............................ UOC:A13,A14,H13,H14	1
14	PAOZZ	5305000712056	80204	B1821BH044C175N	SCREW,CAP,HEXAGON H 7/16-14 X 1.75 PART OF KIT P/N 5705611............. UOC:A13,A14,H13,H14	6
15	PAOZZ	5305013159524	96906	MS51850-69	SCREW,TAPPING #10 X 1.25,PART OF KIT P/N 5705611.................... UOC:A13,A14,H13,H14	3
16	PAOZZ	5310011101145	96906	MS90724-39	NUT,SHEET SPRING #10-39,PART OF KIT P/N 5705611......................... UOC:A13,A14,H13,H14	3
17	PAOZZ	2510013199384	19207	12342258	FRAME SECTION,STRUC PART OF KIT P/N 5705611............................ UOC:A13,A14,H13,H14	1

395-1

SECTION II   TM9-2320-280-24P

(1) ITEM NO	(2) SMR CODE	(3) NSN	(4) CAGEC	(5) PART NUMBER	(6) DESCRIPTION AND USABLE ON CODES(UOC)	(7) QTY
18	PAOZZ	2590013174855	19207	12342185	MOLDING,METAL PART OF KIT P/N 5705611............................ UOC:A13,A14,H13,H14	1
19	PAOZZ	5305000680502	96906	MS90725-6	SCREW,CAP,HEXAGON H 1/4-20 X .75 PART OF KIT P/N 5705611............ UOC:A13,A14,H13,H14	8
20	PAOZZ	5305013170409	96906	MS35489-127	SCREW,WOOD 14 X 1.25,PART OF KIT P/N 5705611........................ UOC:A13,A14,H13,H14	1
21	PAOZZ	5310010128962	24617	9416918	NUT,PLAIN,EXTENDED 1/4-20,PART OF KIT P/N 5705611..................... UOC:A13,A14,H13,H14	15
22	PAOZZ	5330013175392	19207	12342248	RUBBER STRIP PART OF KIT P/N 5705611 UOC:A13,A14,H13,H14	1
23	PAOZZ	2590013177561	34623	12342215	MOLDING,METAL PART OF KIT P/N 5705611............................ UOC:A13,A14,H13,H14	2
24	PAOZZ	5305013159882	96906	MS35495-126	SCREW,WOOD #14 X 1.00,PART OF KIT P/N 5705611........................ UOC:A13,A14,H13,H14	24

END OF FIGURE

Section II. TM 9-2320-280-24P-2

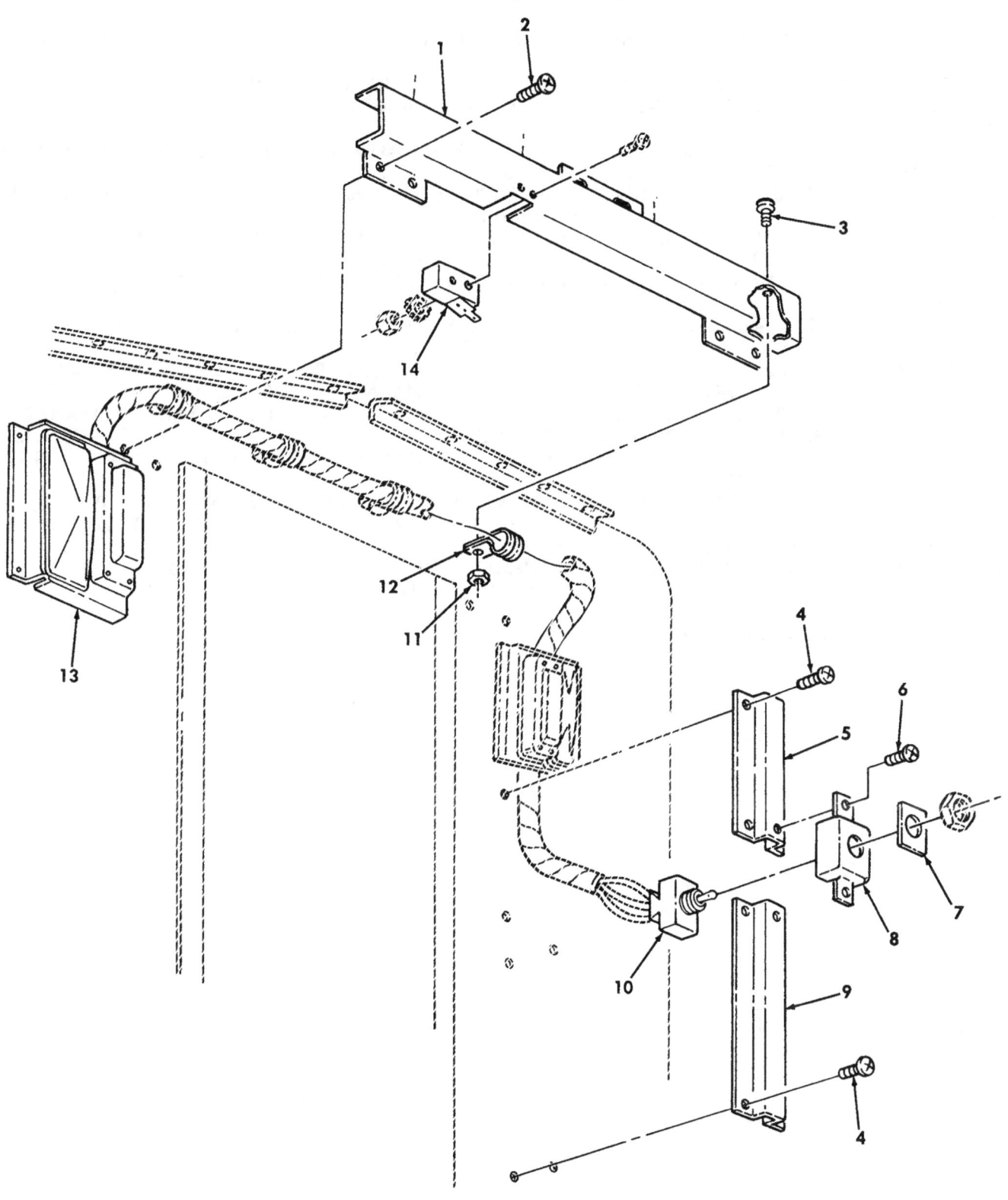

Figure 396. Wiring Harness Channel and Hardware.

SECTION II   TM9-2320-280-24P

(1) NO	(2) SMR CODE	(3) NSN	(4) CAGEC	(5) PART NUMBER	(7) ITEM DESCRIPTION AND USABLE ON CODES(UOC)	QTY

GROUP 3303 WINTERIZATION KITS

FIG. 396 WIRING HARNESS CHANNEL AND HARDWARE

1	PFOZZ	2510013178258	19207	12342214	CHANNEL,BODY PART OF KIT P/N 5705611 UOC:A13,A14,H13,H14	1
2	PAOZZ	5305004324253	96906	MS51861-67	SCREW,TAPPING 1/4-14 X .75,PART OF KIT P/N 5705611 UOC:A13,A14,H13,H14	6
3	PAOZZ	5305009897434	96906	MS35207-263	SCREW,MACHINE #10-32 X .50,PART OF KIT P/N 5705611 UOC:A13,A14,H13,H14	4
4	PAOZZ	5305000567879	96906	MS35493-74	SCREW,WOOD #10 X .50,PART OF KIT P/N 5705611 UOC:A13,A14,H13,H14	14
5	PFOZZ	2540013173350	19207	12342199	CHANNEL,LIFT,VEHICL PART OF KIT P/N 5705611 UOC:A13,A14,H13,H14	1
6	PAOZZ	5305009906444	96906	MS35207-261	SCREW,MACHINE #10-32 X .375,PART OF KIT P/N 5705611 UOC:A13,A14,H13,H14	2
7	PFOZZ	9905013240886	19207	12342194	PLATE,INSTRUCTION PART OF KIT P/N 5705611 UOC:A13,A14,H13,H14	1
8	PFOZZ	5975013226373	19207	12342198	JUNCTION BOX PART OF KIT P/N 5705611 UOC:A13,A14,H13,H14	1
9	PFOZZ	2540013162697	34623	12342200	CHANNEL,LIFT,VEHICL PART OF KIT P/N 5705611 UOC:A13,A14,H13,H14	1
10	PAOZZ	5930006814727	96906	MS24523-21	SWITCH,TOGGLE PART OF KIT P/N 5705611 UOC:A13,A14,H13,H14	1
11	PAOZZ	5310008775797	96906	MS21044N3	NUT,SELF-LOCKING,HE #10-32,PART OF KIT P/N 5705611 UOC:A13,A14,H13,H14	4
12	PAOZZ	5340007647051	96906	MS21333-69	CLAMP,LOOP PART OF KIT P/N 5705611. UOC:A13,A14,H13,H14	4
13	PAOZZ	6220013230431	19207	12342403	STOP LIGHT-TAILLIGH PART OF KIT P/N 5705611 UOC:A13,A14,H13,H14	2
14	PFOZZ	5930012663919	34623	5598011	SWITCH,SENSITIVE PART OF KIT P/N 5705611 UOC:A13,A14,H13,H14	1

END OF FIGURE

Section II. TM 9-2320-280-24P-2 C01

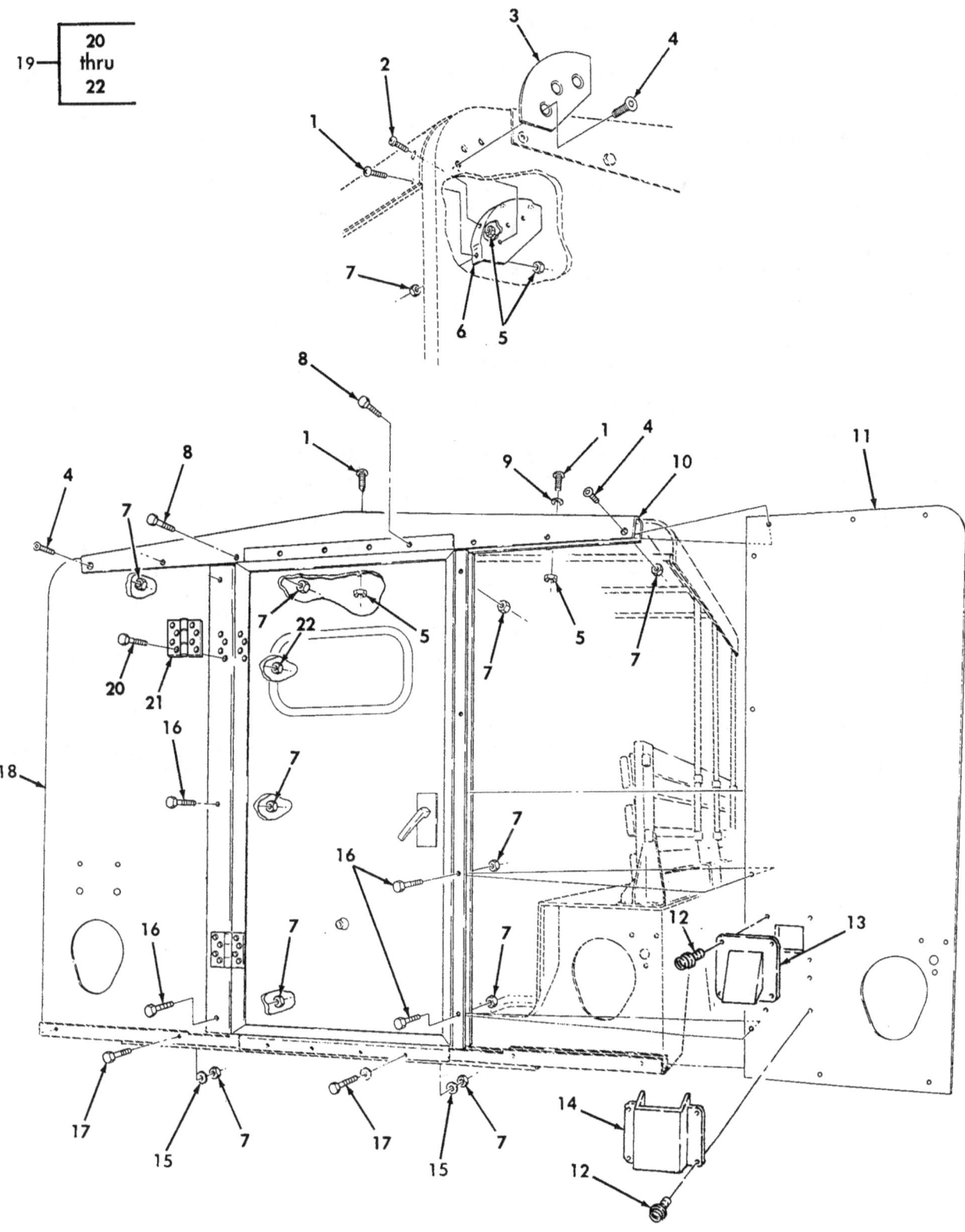

Figure 397. End Closures, Door Assembly, and Mounting Hardware.

SECTION II   TM9-2320-280-24P

(1) NO	(2) SMR CODE	(3) NSN	(4) CAGEC	(5) PART NUMBER	(7) DESCRIPTION AND USABLE ON CODES(UOC)	(6) QTY

GROUP 3303 WINTERIZATION KITS

FIG. 397 END CLOSURES, DOOR ASSEMBLY, AND MOUNTING HARDWARE

1	PAOZZ	5305009932460	96906	MS35207-282	SCREW,MACHINE 1/4-28 X .875,PART OF KIT P/N 5705611..................... UOC:A13,A14,H13,H14	14
2	PAOZZ	5305009932738	96906	MS35207-280	SCREW,MACHINE 1/4-28 X .75,PART OF KIT P/N 5705611..................... UOC:A13,A14,H13,H14	4
3	PAOZZ	5340013182180	19207	12342324	PLATE,MOUNTING PART OF KIT P/N 5705611........................... UOC:A13,A14,H13,H14	2
4	PAOZZ	5305013159683	96906	MS35191-348	SCREW,MACHINE 1/2-20 X 2.00 PART OF KIT P/N 5705611..................... UOC:A13,A14,H13,H14	8
5	PAOZZ	5310004838791	96906	MS17829-4F	NUT,SELF-LOCKING,HE 1/4-28 PART OF KIT P/N 5705611..................... UOC:A13,A14,H13,H14	8
6	PAOZZ	5340013184893	19207	12342322	BRACKET,MOUNTING PART OF KIT P/N 5705611........................... UOC:A13,A14,H13,H14	2
7	PAOZZ	5310007749073	72582	190171	NUT,SELF-LOCKING,CA 1/2-20 PART OF KIT P/N 5705611..................... UOC:A13,A14,H13,H14	28
8	PAOZZ	5305007195239	96906	MS90727-116	SCREW,CAP,HEXAGON H 1/2-20 X 2.25, PART OF KIT P/N 5705611............. UOC:A13,A14,H13,H14	8
9	PAOZZ	5310008094058	96906	MS27183-10	WASHER,FLAT 1/4 PART OF KIT P/N 5705611.......................... UOC:A13,A14,H13,H14	8
10	PAFZZ	2510013195952	19207	12342317	FRAME,SECTION,STRUC PART OF KIT P/N 5705611........................... UOC:A13,A14,H13,H14	1
11	KFFZZ		19207	12342223-2	CLOSURE END R.H. PART OF KIT P/N 5705611........................... UOC:A13,A14,H13,H14	1
12	PAOZZ	5305013159882	96906	MS35495-126	SCREW,WOOD #14 X 1.00 PART OF KIT P/N 5705611........................ UOC:A13,A14,H13,H14	8
13	PFOZZ	2540014767827	19207	12446983	VENTILATOR PART OF KIT P/N 5705611.. UOC:A13,A14,H13,H14	1
14	PFOZZ	2540014770242	19207	12446981	DIVERTER,LIGHT PART OF KIT P/N 5705611........................... UOC:A13,A14,H13,H14	1
15	PAOZZ	5310008095998	96906	MS27183-18	WASHER,FLAT 1/2 PART OF KIT P/N 5705611........................... UOC:A13,A14,H13,H14	12
16	PAOZZ	5305007195238	80204	B1821BH-050F200N	SCREW,CAP,HEXAGON H 1/2-20 X 2.00 PART OF KIT P/N 5705611............. UOC:A13,A14,H13,H14	4

SECTION II   TM9-2320-280-24P C01
 (1)   (2)   (3)   (4)   (5)   (6)
                                     PART
 ITEM SMR
 NO   CODE   NSN   CAGEC   NUMBER   DESCRIPTION AND USABLE ON CODES(UOC)

  17 PAOZZ 5305007195241 80204 B1821BH050F275N SCREW,CAP,HEXAGON H 1/2-20 X 2.75,   8
                                                PART OF KIT P/N 5705611.............
                                                UOC:A13,A14,H13,H14
  18 KFFZZ   19207 12342223-1   CLOSURE END   L.H. PART OF KIT P/N
                                                5705611.............................
                                                UOC:A13,A14,H13,H14
  19 PAOZZ 2510013144892 19207 12342330   DOOR,VEHICULAR PART OF KIT P/N
                                                5705611.............................
                                                UOC:A13,A14,H13,H14
  20 PAOZZ 5305007195241 80204 B1821BH050F275N .SCREW,CAP HEXAGON H...............   16
                                                UOC:A13,A14,H13,H14
  21 PAOZZ 2510014988000 19207 12342327   .HINGE,DOOR,VEHICLA.................   2
                                                UOC:A13,A14,H13,H14
  22 PAOZZ 5310007749073 72582 190171   .NUT,SELF-LOCKING,HE...............   16
                                                UOC:A13,A14,H13,H14
 KIT PDFZZ 2540013140142 19207 5705611   WINTERIZATION KIT...................   1
                                                UOC:A13,A14,H13,H14
                                                  ADAPTER,CONNECTOR   (1) 386-9
                                                  ADAPTER,CONNECTOR   (1) 386-11
                                                  ADAPTER,STRAIGHT,PI(2) 384-3
                                                  ADAPTER,STRAIGHT,PI(1) 384-15
                                                  ADAPTER,STRAIGHT,PI(1) 385-5
                                                  ADAPTER,STRAIGHT,PI(2) 388-2
                                                  ADAPTER,VEHICLE STO(1) 393-14
                                                  BASE PLATE,SPACE HE(1) 387-9
                                                  BLOCK,PLYWOOD   (2) 389-8
                                                  BOLT,MACHINE   (4) 388-15
                                                  BOLT,MACHINE   (2) 390-6
                                                  BOLT,MACHINE   (3) 392-3
                                                  BOW,VEHICULAR,TOP   (3) 389-6
                                                  BRACKET,ANGLE   (1) 388-17
                                                  BRACKET,DOUBLE ANGL(1) 384-12
                                                  BRACKET,MOUNTING   (1) 389-5
                                                  BRACKET,MOUNTING   (2) 397-6
                                                  BRACKET,MULTIPLE AN(1) 387-11
                                                  BRACKET,VEHICULAR C(1) 389-13
                                                  BRACKET,VEHICULAR C(1) 389-13
                                                  BRACKET,VEHICULAR C(1) 392-6
                                                  CABLE ASSEMBLY,SPEC(1) 383-9
                                                  CABLE ASSEMBLY,SPEC(1) 387-18
                                                  CAP,PROTECTIVE,DUST(23) 394-1
                                                  CAP,PROTECTIVE,DUST(2) 394-2
                                                  CAP,PROTECTIVE,DUST(52) 394-3
                                                  CAP,TUBE   (1) 384-5
                                                  CHANNEL,BODY   (1) 396-1
                                                  CHANNEL,LIFT,VEHICU(1) 396-5
                                                  CHANNEL,LIFT,VEHICU(1) 396-9
                                                  CIRCUIT,BREAKER   (1) 385-12
                                                  CLAMP,HOSE   (3) 385-4
                                                  CLAMP,HOSE   (2) 386-7
                                                  CLAMP,HOSE   (1) 387-2
                                                  CLAMP,HOSE   (2) 387-5
                                                  CLAMP,HOSE   (3) 388-5

```
SECTION II            TM9-2320-280-24P C01
    (1)    (2)    (3)      (4)      (5)                    (6)
(7)  ITEM   SMR                     PART
    NO    CODE   NSN      CAGEC    NUMBER    DESCRIPTION AND USABLE ON CODES(UOC)
QTY
```

CLAMP,HOSE	(2) 393-16
CLAMP,LOOP	(1) 385-2
CLAMP,LOOP	(1) 386-6
CLAMP,LOOP	(1) 386-10
CLAMP,LOOP	(10) 391-7
CLAMP,LOOP	(4) 396-12
CLOSURE END,R.H.	(1) 397-11
CLOSURE END,L.H.	(1) 397-18
CONTROL BOX,ELECTRI	(1) 388-16
COUPLING,PIPE	(1) 385-6
COUPLING,PIPE	(1) 385-11
COVER ASSEMBLY	(1) 382-14
COVER ASSEMBLY	(1) 382-14
COVER ASSEMBLY	(1) 382-15
COVER ASSEMBLY,SIDE	(1) 382-15
COVER,FITTED,VEHICU	(1) 395-6
COVER,FLOOR,L.H.	(1) 382-11
COVER,FLOOR,R.H.	(1) 382-11
COVER,L.H.	(1) 382-7
COVER,R.H.	(1) 382-7
CURTAIN,VEHICULAR,B	(2) 389-2
DIVERTER,LIGHT	(1) 397-14
DOOR,VEHICULAR	(1) 397-19
DUMMY CONNECTOR,PLU	(1) 386-1
ELBOW,AIR CONDITION	(1) 387-3
ELBOW,PIPE TO HOSE	(1) 388-6
ELBOW,PIPE TO TUBE	(1) 385-8
FILTER,FLUID	(1) 388-7
FLOOR COVER ASSEMBL	(1) 383-1
FLOOR COVER ASSEMBL	(1) 383-6
FRAME SECTION,STRUC	(1) 395-17
FRAME SECTION,STRUC	(1) 397-10
GASKET	(1) 387-10
GROMMET,NONMETALLIC	(1) 387-13
HANDLE,DOOR	(1) 395-10
HEATER,VEHICULAR,CO	(1) 387-4
HOSE,AIR DUCT	(1) 393-17
HOSE,NONMETALLIC	(1) 386-8
HOSE,NONMETALLIC	(1) 388-1
JUNCTION BOX	(1) 396-8
LADDER,VEHICLE BOA	(1) 390-10
LEAD,ELECTRICAL	(1) 384-10
LEAD,ELECTRICAL	(1) 386-2
MOLDING,METAL	(2) 383-3
MOLDING,METAL	(1) 383-4
MOLDING,METAL	(1) 393-8
MOLDING,METAL	(2) 395-2
MOLDING,METAL	(2) 395-3
MOLDING,METAL	(1) 395-18
MOLDING,METAL	(2) 395-23
NIPPLE,PIPE	(1) 385-9
NUT,PLAIN,BLIND RIV	(9) 383-7
NUT,PLAIN,EXTENDED	(50) 391-6

(1)	(2)	(3)	(4)	(5)	(6)	(7)
ITEM NO	SMR CODE	NSN	CAGEC	PART NUMBER	DESCRIPTION AND USABLE ON CODES (UOC)	QTY

SECTION II TM9-2320-280-24P C01

```
                                    NUT,PLAIN,EXTENDED   ( 17) 393-4
                                    NUT,PLAIN,EXTENDED   ( 15) 395-21
                                    NUT,PLAIN,PLATE      (  6) 393-8
                                    NUT,SELF-LOCKING,CA  (  4) 390-4
                                    NUT,SELF-LOCKING,CA  ( 28) 397-7
                                    NUT,SELF-LOCKING,HE  ( 48) 382-6
                                    NUT,SELF-LOCKING,HE  ( 16) 382-10
                                    NUT,SELF-LOCKING,HE  (  2) 384-9
                                    NUT,SELF-LOCKING,HE  (  2) 388-3
                                    NUT,SELF-LOCKING,HE  (  2) 388-10
                                    NUT,SELF-LOCKING,HE  ( 36) 389-1
                                    NUT,SELF-LOCKING,HE  (  4) 389-4
                                    NUT,SELF-LOCKING,HE  (  4) 389-10
                                    NUT,SELF-LOCKING,HE  ( 18) 389-12
                                    NUT,SELF-LOCKING,HE  (  6) 389-14
                                    NUT,SELF-LOCKING,HE  (  2) 390-5
                                    NUT,SELF-LOCKING,HE  (  4) 390-9
                                    NUT,SELF-LOCKING,HE  (  3) 393-12
                                    NUT,SELF-LOCKING,HE  (  3) 392-4
                                    NUT,SELF-LOCKING,HE  (  4) 392-11
                                    NUT,SELF-LOCKING,HE  (  4) 393-15
                                    NUT,SELF-LOCKING,HE  (  2) 395-8
                                    NUT,SELF-LOCKING,HE  (  4) 396-11
                                    NUT,SELF-LOCKING,HE  (  8) 397-5
                                    NUT,SHEET,SPRING     (  3) 395-16
                                    O-RING               (  1) 387-15
                                    PIN,COTTER           (  2) 390-14
                                    PIN,STRAIGHT,HEADED  (  2) 390-11
                                    PLATE,INSTRUCTION    (  1) 394-4
                                    PLATE,INSTRUCTION    (  1) 396-7
                                    PLATE,MENDING        (  8) 391-2
                                    PLATE,MENDING        (  1) 392-10
                                    PLATE,MENDING        (  1) 390-12
                                    PLATE,MOUNTING       (  1) 393-2
                                    PLATE,MOUNTING       (  1) 395-13
                                    PLATE,MOUNTING       (  2) 397-3
                                    PLATE,SEAM           (  1) 391-3
                                    PUMP,FUEL            (  1) 384-6
                                    RIVET,BLIND          (  4) 384-14
                                    RUBBER STRIP         (  1) 393-5
                                    RUBBER STRIP         (  2) 395-4
                                    RUBBER,STRIP         (  2) 395-7
                                    RUBBER STRIP         (  1) 395-22
                                    SCREW,CAP,HEXAGON H  (  2) 384-13
                                    SCREW,CAP,HEXAGON H  (  4) 387-7
                                    SCREW,CAP,HEXAGON H  (  6) 387-17
                                    SCREW,CAP,HEXAGON H  (  2) 388-8
                                    SCREW,CAP,HEXAGON H  (  2) 388-12
                                    SCREW,CAP,HEXAGON H  (  4) 389-3
                                    SCREW,CAP,HEXAGON H  ( 18) 389-11
                                    SCREW,CAP,HEXAGON H  (  6) 395-14
                                    SCREW,CAP,HEXAGON H  (  4) 390-2
                                    SCREW,CAP,HEXAGON H  (  2) 390-15
```

```
       SECTION II           TM9-2320-280-24P C01
      (1)     (2)     (3)         (4)         (5)                      (6)
(7)   ITEM    SMR                            PART
      NO      CODE    NSN         CAGEC      NUMBER     DESCRIPTION AND USABLE ON CODES(UOC)
QTY
```

```
                                                 SCREW,CAP,HEXAGON H(  2) 392-7
                                                 SCREW,CAP,HEXAGON H(  2) 392-8
                                                 SCREW,CAP,HEXAGON H(  1) 393-1
                                                 SCREW,CAP,HEXAGON H(  6) 393-3
                                                 SCREW,CAP,HEXAGON H( 10) 393-7
                                                 SCREW,CAP,HEXAGON H(  3) 393-9
                                                 SCREW,CAP,HEXAGON H(  1) 395-12
                                                 SCREW,CAP,HEXAGON H(  6) 395-14
                                                 SCREW,CAP,HEXAGON H(  8) 395-19
                                                 SCREW,CAP,HEXAGON H(  8) 397-8
                                                 SCREW,CAP,HEXAGON H(  4) 397-16
                                                 SCREW,CAP,HEXAGON H(  8) 397-17
                                                 SCREW,MACHINE       ( 48) 382-4
                                                 SCREW,MACHINE       (  8) 382-8
                                                 SCREW,MACHINE       (  8) 382-13
                                                 SCREW,MACHINE       (  9) 383-5
                                                 SCREW,MACHINE       (  4) 389-7
                                                 SCREW,MACHINE       (  2) 390-7
                                                 SCREW,MACHINE       ( 50) 391-1
                                                 SCREW,MACHINE       (  4) 393-11
                                                 SCREW,MACHINE       (  4) 396-3
                                                 SCREW,MACHINE       (  2) 396-6
                                                 SCREW,MACHINE       ( 14) 397-1
                                                 SCREW,MACHINE       (  4) 397-2
                                                 SCREW,MACHINE       (  8) 397-4
                                                 SCREW,TAPPING       ( 20) 395-1
                                                 SCREW,TAPPING       (  3) 395-15
                                                 SCREW,TAPPING       (  6) 396-2
                                                 SCREW,WOOD          ( 12) 382-1
                                                 SCREW,WOOD          ( 19) 383-2
                                                 SCREW,WOOD          (  1) 385-1
                                                 SCREW,WOOD          (  2) 385-13
                                                 SCREW,WOOD          (  8) 387-12
                                                 SCREW,WOOD          (  2) 390-13
                                                 SCREW,WOOD          ( 16) 392-1
                                                 SCREW,WOOD          (  4) 392-2
                                                 SCREW,WOOD          (  1) 393-6
                                                 SCREW,WOOD          (  4) 394-5
                                                 SCREW,WOOD          (  8) 395-11
                                                 SCREW,WOOD          (  1) 395-20
                                                 SCREW,WOOD          ( 24) 395-24
                                                 SCREW,WOOD          ( 14) 396-4
                                                 SCREW,WOOD          (  8) 397-12
                                                 SEAL,NONMETALLIC    (  2) 395-5
                                                 SEAT BACK           (  2) 382-12
                                                 SHELL,ELECTRICAL CO(  1) 386-12
                                                 SHIELD,HEATER DUCT  (  1) 387-6
                                                 SHIELD,HEATER DUCT  (  1) 388-13
                                                 SLEEVE,COUPLING     (  1) 387-1
                                                 SPACER,PLATE        ( 12) 382-2
                                                 SPACER,SLEEVE       (  2) 388-4
                                                 SPACER,SLEEVE       (  5) 392-5
                                                 STOP LIGHT-TAILLIGH(  2) 396-13
```

SECTION II   TM9-2320-280-24P C01
 (1)   (2)   (3)   (4)   (5)   (6)
                                   PART
 ITEM SMR
 NO   CODE NSN   CAGEC   NUMBER   DESCRIPTION AND USABLE ON CODES(UOC)

                                             STRAINER ELEMENT,SE(1) 393-13
                                             STRAP,RETAINING   (20) 391-9
                                             STRAP,TIEDOWN,ELECT(1) 386-3
                                             STRAP,TIEDOWN,ELECT(1) 388-11
                                             SUPPORT   (1) 391-4
                                             SUPPORT   (1) 391-5
                                             SUPPORT   (1) 391-8
                                             SUPPORT ASSEMBLY,BO(1) 390-1
                                             SWITCH,SENSITIVE   (1) 396-14
                                             SWITCH,TOGGLE   (1) 396-10
                                             TEE,PIPE   (1) 384-4
                                             TEMPLATE,FUEL PUMP (1) 384-11
                                             TOP COVER ASSEMBLY (1) 382-3
                                             TOP COVER ASSEMBLY (1) 382-3
                                             TUBE   (2) 385-3
                                             TUBE ASSEMBLY,METAL(1) 384-1
                                             TUBE ASSEMBLY,METAL(1) 384-2
                                             TUBE ASSEMBLY,METAL(1) 385-7
                                             TUBE ASSEMBLY,METAL(1) 386-4
                                             TUBE ASSEMBLY,METAL(1) 386-5
                                             TUBE ASSEMBLY,METAL(1) 387-14
                                             VENTILATOR   (1) 397-13
                                             VENTILATOR,AIR CIRC(1) 393-10
                                             WASHER,FLAT   (48) 382-5
                                             WASHER,FLAT   (16) 382-9
                                             WASHER,FLAT   (4) 384-8
                                             WASHER,FLAT   (1) 385-10
                                             WASHER,FLAT   (1) 387-16
                                             WASHER,FLAT   (2) 388-9
                                             WASHER,FLAT   (4) 389-9
                                             WASHER,FLAT   (4) 390-3
                                             WASHER,FLAT   (4) 390-8
                                             WASHER,FLAT   (2) 392-9
                                             WASHER,FLAT   (4) 395-9
                                             WASHER,FLAT   (8) 397-9
                                             WASHER,FLAT   (12) 397-15
                                             WASHER,LOCK   (1) 384-7
                                             WASHER,LOCK   (10) 387-8
                                             WASHER,LOCK   (4) 388-14

                                   END OF FIGURE

Section II. TM 9-2320-280-24P-2 C01

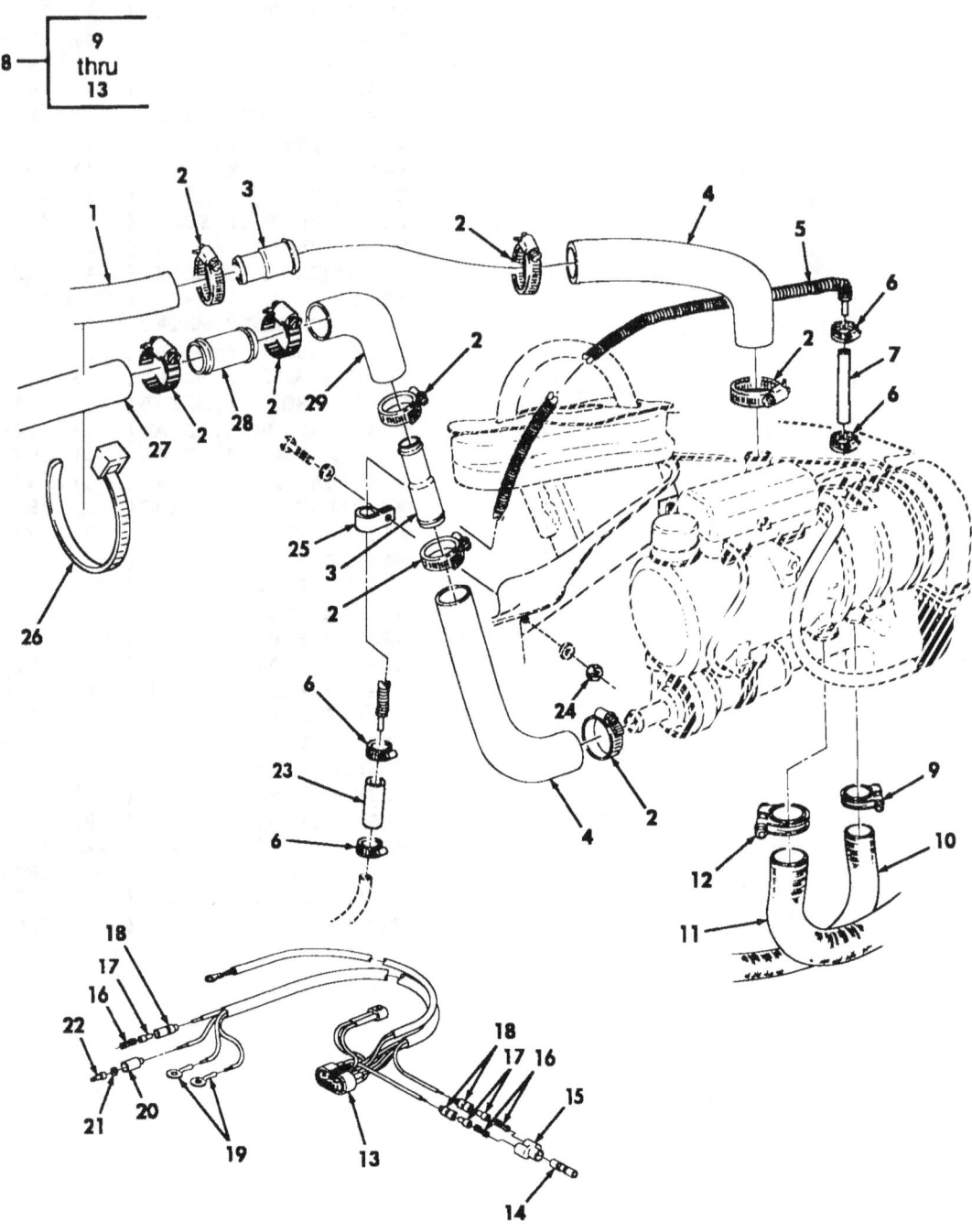

SEE SHEET 2 FOR HEATER DOWNPARTS

Figure 397A. Engine and Crew Compartment Heater and Related Parts (Sheet 1 of 2).

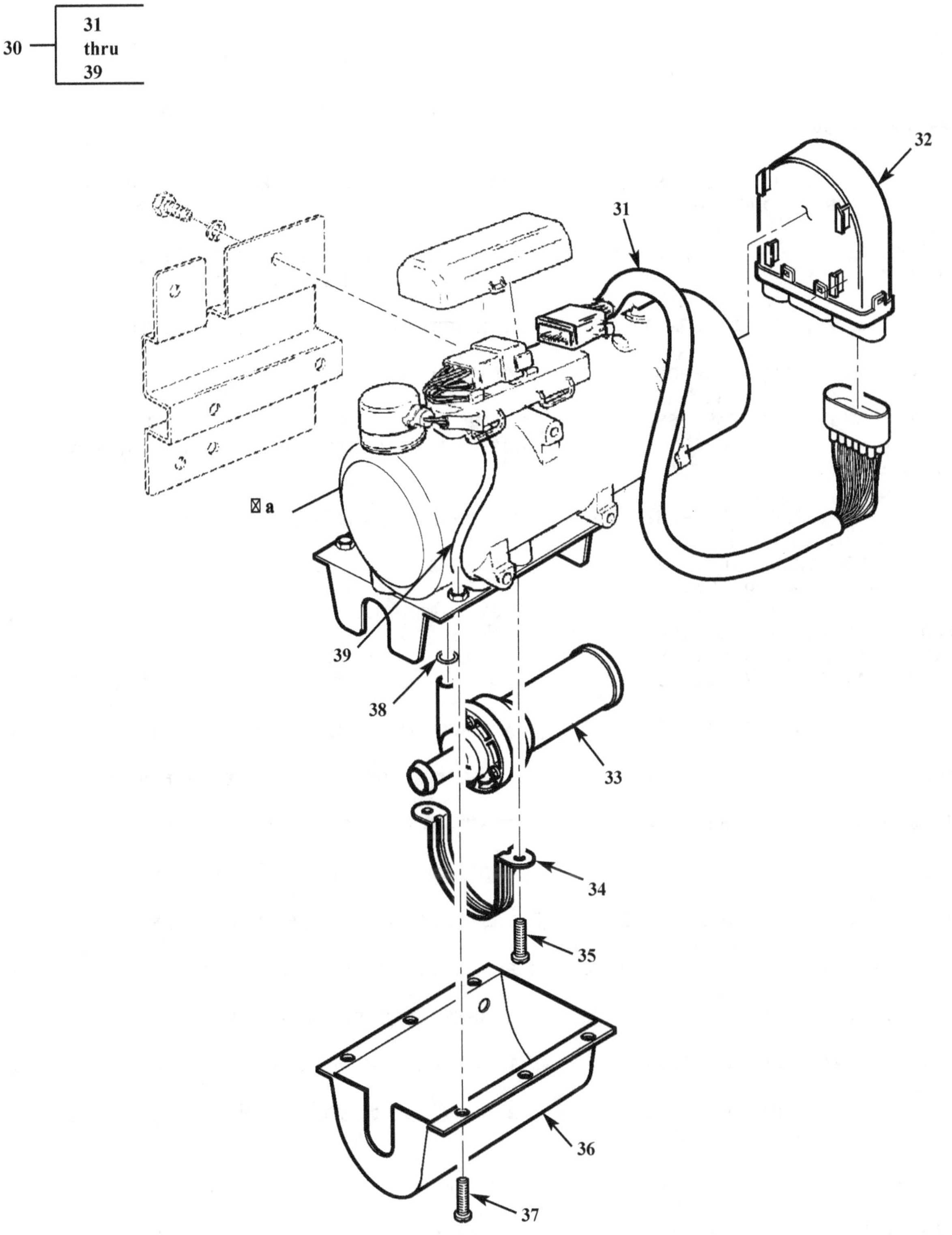

Section II. TM 9-2320-280-24P-2 C01

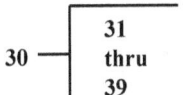

☒ a PART OF ITEM 30

Figure 397A. Engine and Crew Compartment Heater and Related Parts (Sheet 2 of 2).

SECTION II TM9-2320-280-24P C01

(1) ITEM NO	(2) SMR CODE	(3) NSN	(4) CAGEC	(5) PART NUMBER	(6) DESCRIPTION AND USABLE ON CODES (UOC)	(7) QTY

GROUP 3303 SPECIAL PURPOSE KITS

FIG. 397A ENGINE AND CREW COMPARTMENT

HEATER AND RELATED PARTS

Item	SMR	NSN	CAGEC	Part Number	Description	Qty
* 1	MOOZZ		19207	12339155-3-25	HOSE,NONMETALLIC MAKE FROM HOSE P/N MS521304A203R 25.0 INCHES LONG PART OF KIT P/N 57K3497..................	1
* 2	PAOZZ	4730009083194	96906	MS35842-11	CLAMP,HOSE PART OF KIT P/N 57K3497	8
* 3	PAOZZ	4730014751191	19207	12469041	REDUCER,TUBE PART OF KIT P/N 57K3497.............................	2
* 4	PAOZZ	4720014749105	19207	12339251-1	HOSE,PREFORMED PART OF KIT P/N 57K3497.............................	2
* 5	PAOZZ	4710014751227	19207	12469054	TUBE ASSEMBLY,METAL PART OF KIT P/N 57K3497.............................	1
* 6	PAOZZ	4730011188278	76599	MMF4-SS	CLAMP,HOSE PART OF KIT P/N 57K3497.	4
* 7	MOOZZ		24617	9436711-2.5	HOSE,NONMETALLIC MAKE FROM HOSE P/N 9438381 2.50 INCHES LONG PART OF KIT 57K3497.......................	1 P/N
* 8	AOOOO		19207	12469058	HEATER ASSEMBLY PART OF KIT P/N 57K3497................................	1
* 9	PAOZZ	5340012855558	62380	100.424	.CLAMP,LOOP......................	1
* 10	PAOZZ	4720014752616	62380	29842A	.PIPE,INTAKE.....................	1
* 11	PAOZZ	2540014766062	62380	442437	.PIPE,EXHAUST....................	1
* 12	PAOZZ	5340014753164	62380	367400	.CLAMP,LOOP......................	1
* 13	PAOZZ	6150014751843	62380	905.782	.CABLE ASSEMBLY,SPEC..............	1
* 14	PAOZZ	5935008074109	19207	8741492	ADAPTER,CONNECTOR PART OF KIT P/N 57K3497.............................	1
* 15	PAOZZ	5935009006281	96906	MS27147-1	ADAPTER,CONNECTOR PART OF KIT P/N 57K3497.............................	1
* 16	PAOZZ	5940003996676	19207	8338564	TERMINAL,QUICK SE PART OF KIT P/N 57K3497	3
* 17	PAOZZ	5970008338562	19207	8338562	INSULATOR,BUSHING PART OF KIT P/N 57K3497........................	3
* 18	PAOZZ	5975006605962	19207	8724494	CABLE,NIPPLE PART OF KIT P/N 57K3497.............................	3
* 19	PAOZZ	5940002835280	96906	MS25036-106	TERMINAL,LUG PART OF KIT P/N 57K3497.............................	2
* 20	PAOZZ	5935006915591	97403	13207E6498-2	SHELL,ELECTRICAL PART OF KIT P/N 57K3497.............................	1
* 21	PAOZZ	5310006560067	19207	8724497	WASHER,SLOTTED PART OF KIT P/N 57K3497.............................	1
* 22	PAOZZ	5999004263144	96906	MS27148-3	TERMINAL,LUG PART OF KIT P/N 57K3497.............................	1
* 23	MOOZZ		24617	9436711-2.0	HOSE,NONMETALLIC MAKE FROM HOSE P/N 9438381 2.00 INCHES LONG PART OF KIT 57K3497.......................	1 P/N
* 24	PAOZZ	5310000614650	96906	MS51943-31	NUT,SELF-LOCKING PART OF KIT P/N... 57K3497.............................	1
* 25	PAOZZ	5340009936207	96906	MS21333-99	CLAMP,LOOP PART OF KIT P/N 57K3497.	1
* 26	PAOZZ	5975010345871	96906	MS3367-7-0	STRAP,TIEDOWN ELEC PART OF KIT P/N 57K3497.............................	5

397A-1

SECTION II   TM9-2320-280-24P C01
 (1)   (2)   (3)   (4)   (5)   (6)
       ITEM   SMR   PART
        NO   CODE   NSN   CAGEC   NUMBER   DESCRIPTION AND USABLE ON CODES(UOC)   QTY

 27 MOOZZ   19207 12339155-3-15 HOSE,NONMETALLIC MAKE FROM HOSE P/N   1
 M521304B203R 15.0 INCHES LONG PART   OF
 KIT P/N 57K3497.................
 28 PAOZZ 4730013948129 96906 MS24520-7   MENDER,HOSE PART OF KIT P/N 57K3497   1
 29 PAOZZ 4720011892218 19207 12339251   HOSE,PREFORMED PART OF KIT P/N   1
 57K3497.........................
 30 PAOZZ 2540015012690 62380 67582B   HEATER,VEHICULAR,CO.................   1
 31 PAOZZ 5999014750215 62380 67134A   .HARNESS,ELECTRICAL.................   1
 32 PAOZZ 2540014753944 62380 89093B   .CONTROL BOX,ELECTRI................   1
 33 PAOZZ 2930014753976 62380 65759B   .PUMP,COOLING SYSTEM................   1
 34 PAOZZ 5340014753171 62380 38634A   .STRAP,RETAINING....................   1
 35 PAOZZ 5305014753402 62380 35283A   .SCREW,TAPPING......................   2
 36 PAOZZ   62380 82436A   .SHIELD,FUEL PUMP...................   1
 37 PAOZZ   62380 147818   .SCREW,MACHINE......................   6
 38 PAOZZ   62380 22405A   .O-RING.............................   1
 39 PAOZZ 5999014763674 62380 67133A   .HARNESS,ELECTRICAL E...............   1

Section II. TM 9-2320-280-24P-2 C01

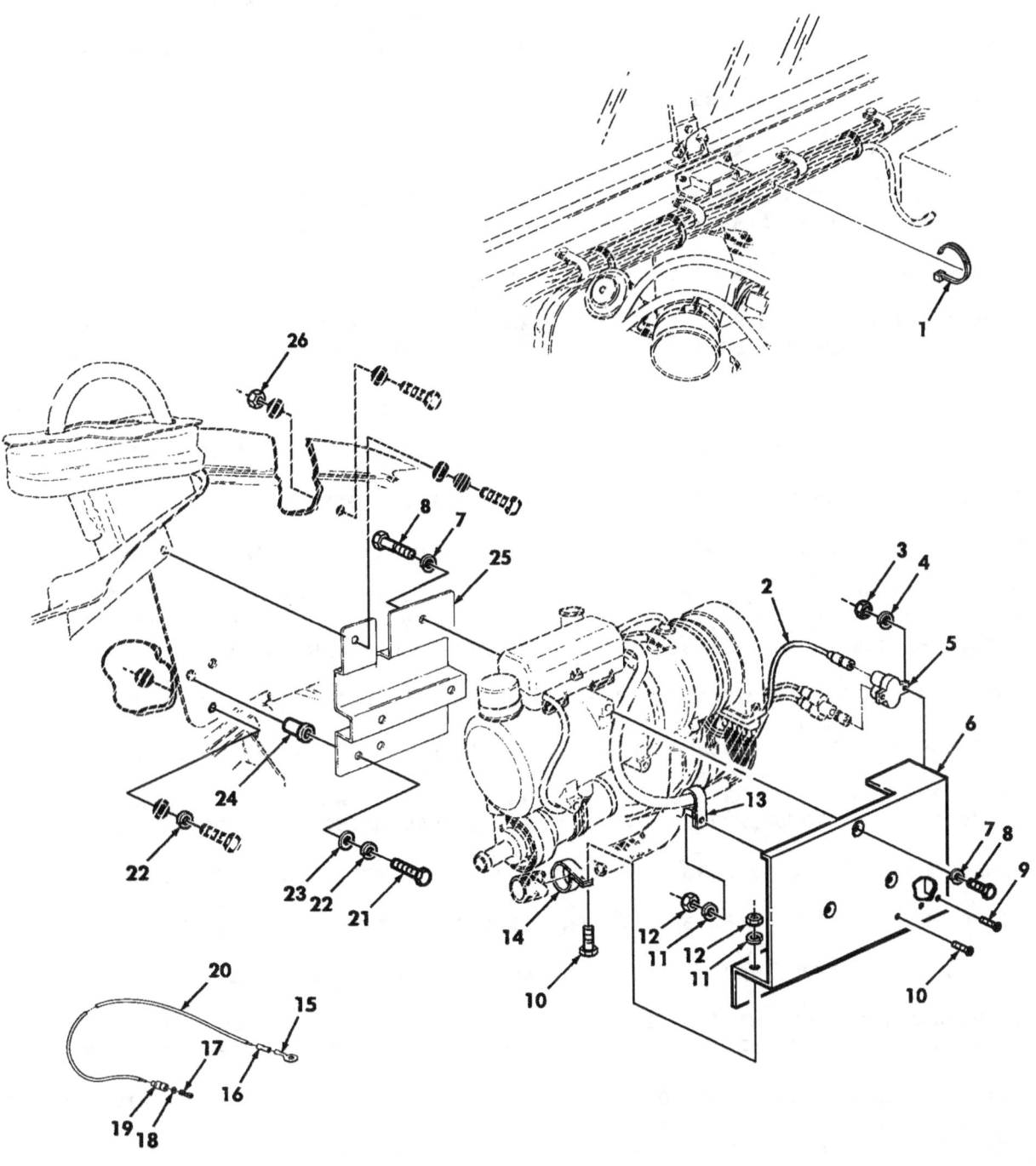

Figure 397B. Engine and Crew Compartment Heater Support Bracket, Heat Shield, and Mounting Hardware.

```
SECTION II              TM9-2320-280-24P C01
   (1)    (2)   (3)          (4)         (5)                    (6)
(7)  ITEM   SMR                         PART
   NO   CODE   NSN         CAGEC       NUMBER      DESCRIPTION AND USABLE ON CODES(UOC)
QTY
```

GROUP 3303 SPECIAL PURPOSE KITS

FIG. 397B ENGINE AND CREW COMPARTMENT HEATER SUPPORT BRACKET, HEAT SHIELD, AND MOUNTING HARDWARE

(1) ITEM NO	(2) SMR CODE	(3) NSN	(4) CAGEC	(5) PART NUMBER	(6) DESCRIPTION AND USABLE ON CODES (UOC)	(7) QTY
* 1	PAOZZ	5975009846582	96906	MS3367-1-0	STRAP,TIEDOWN ELEC PART OF KIT P/N 57K3497..........................	4
* 2	PAOZZ	6145007056678	81349	M13486-1-7	WIRE,ELECTRICAL PART OF KIT P/N 57K3497..........................	1
* 3	PAOZZ	5310014741368	34623	114653	NUT,PLAIN,HEXAGON PART OF KIT P/N 57K3497..........................	2
* 4	PAOZZ	5310000453299	96906	MS35338-42	WASHER,LOCK PART OF KIT P/N 57K3497	2
* 5	PAOZZ	5925014302318	58536	AA5571/01-001	BREAKER,CIRCUIT PART OF KIT P/N 57K3497..........................	1
* 6	PAOZZ	2990014739420	19207	12460427	GUARD,MUFFLER EXHAU PART OF KIT P/N 57K3497..........................	1
* 7	PAOZZ	5310014785620	80204	B18212HRCZ080	WASHER,LOCK PART OF KIT P/N 57K3497	6
* 8	PAOZZ	5305011491938	7X677	11504336	SCREW,CAP,HEXAGON H PART OF KIT P/N 57K3497..........................	6
* 9	PAOZZ	5305009846197	96906	MS35206-249	SCREW,MACHINE PART OF KIT P/N 57K3497..........................	2
* 10	PAOZZ	5305009846210	96906	MS35206-263	SCREW,MACHINE PART OF KIT P/N 57K3497..........................	2
* 11	PAOZZ	5310000453296	96906	MS35338-43	WASHER,LOCK PART OF KIT P/N 57K3497..........................	2
* 12	PAOZZ	5310014222147	24617	116003	NUT,PLAIN,HEXAGON PART OF KIT P/N 57K3497..........................	2
* 13	PAOZZ	5340007247038	96906	MS21333-76	CLAMP,LOOP PART OF KIT P/N 57K3497.	1
* 14	PAOZZ	5340001773711	80205	MS21333-16	CLAMP,LOOP PART OF KIT P/N 57K3497.	1
* 15	PAOZZ	5940006822445	96906	MS25036-158	TERMINAL,LUG PART OF KIT P/N 57K3497..........................	1
* 16	PAOZZ	5970007872331	81349	M23053/5-106-2	INSULATION SLEEVING PART OF KIT P/N 57K3497..........................	1
* 17	PAOZZ	5940003996676	19207	8338564	TERMINAL,SET QUICK PART OF KIT P/N 57K3497..........................	1
* 18	PAOZZ	5310002988903	19207	8338570	WASHER,FLAT PART OF KIT P/N 57K3497	1
* 19	PAOZZ	2590006959076	77060	5297084	SHELL,HEAD LIGHT PART OF KIT P/N 57K3497..........................	1
* 20	PAOZZ	6145007056678	81349	M13486-1-7	WIRE,ELECTRICAL PART OF KIT P/N 57K3497..........................	1
* 21	PAOZZ	5306002264829	80204	B1821BH031C125N	BOLT,MACHINE PART OF KIT P/N 57K3497..........................	2
* 22	PAOZZ	5310004079566	96906	MS35338-45	WASHER,LOCK PART OF KIT P/N 57K3497..........................	3
* 23	PAOZZ	5310000814219	96906	MS27183-12	WASHER,FLAT PART OF KIT P/N 57K3497	2
* 24	PAOZZ	5325015053122	19207	12446871-11	INSERT,SCREW THREAD PART OF KIT P/N 57K3497..........................	2
* 25	PAOZZ	5340014736897	19207	12460430	BRACKET,MULTIPLE AN PART OF KIT P/N 57K3497..........................	1
* 26	PAOZZ	5310008140673	96906	MS51943-33	NUT,SELF-LOCKING PART OF KIT P/N 57K3497..........................	2

END OF FIGURE

Section II. TM 9-2320-280-24P-2 C01

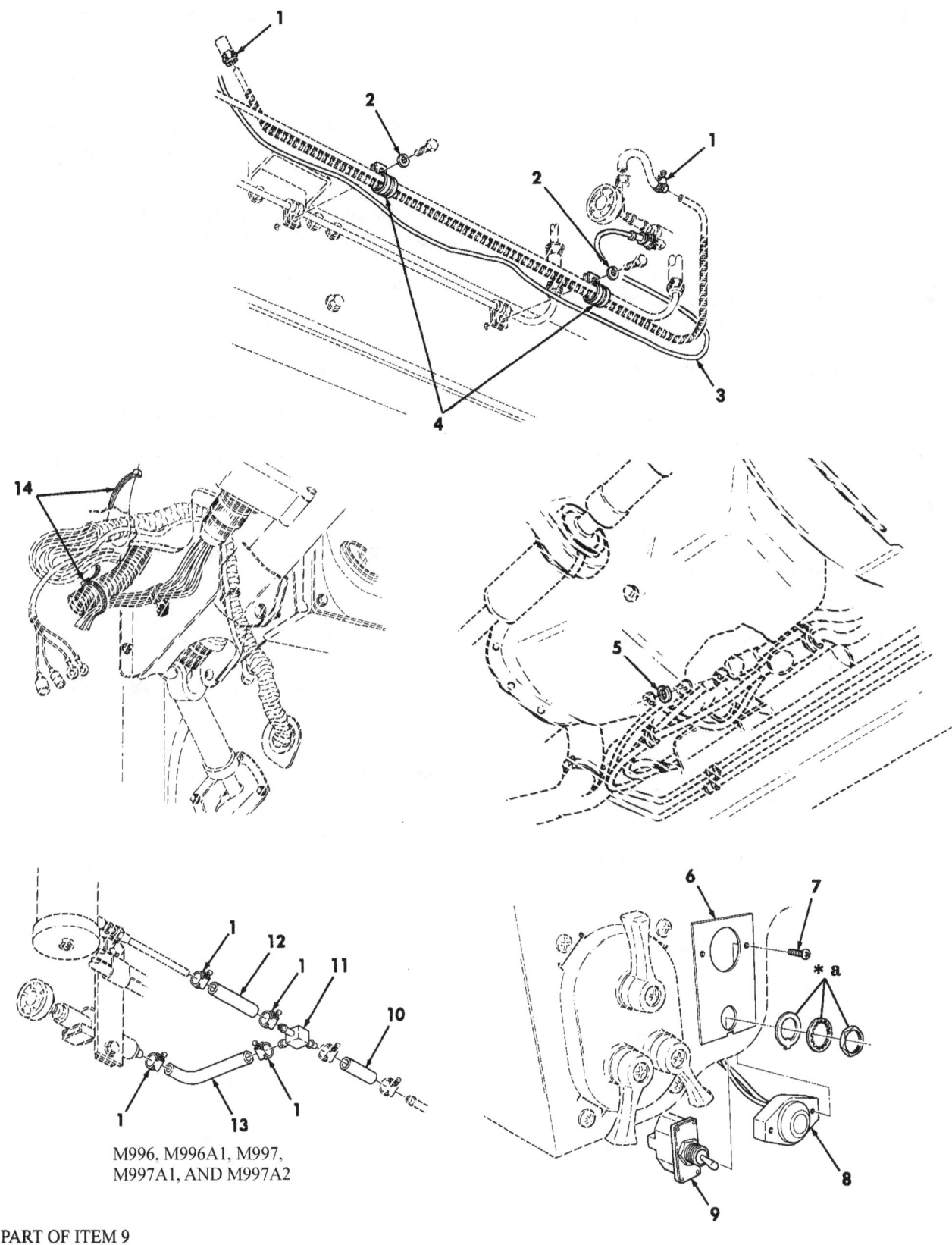

M996, M996A1, M997,
M997A1, AND M997A2

☒ a PART OF ITEM 9

Figure 397C. Engine and Crew Compartment Heater Switch, Fuel Lines, and Mounting Hardware.

```
    SECTION II           TM9-2320-280-24P C01
  (1)    (2)   (3)           (4)        (5)                          (6)
(7)   ITEM   SMR                       PART
   NO   CODE   NSN            CAGEC   NUMBER      DESCRIPTION AND USABLE ON CODES(UOC)
QTY
```

GROUP 3303 SPECIAL PURPOSE KITS

FIG. 397C ENGINE AND CREW COMPART-
MENT HEATER SWITCH,FUEL LINES,AND
MOUNTING HARDWARE

```
*  1 PAOZZ 4730011188278 76599 MMF4-SS       CLAMP,HOSE   PART OF KIT P/N 57K3497.    6
*  2 PAOZZ 5310006379541 81718 H2525M        WASHER,LOCK  PART OF KIT P/N             2
57K3497............................
*  3 PAOZZ 5999014763674 62380 905.794       HARNESS,ELECTRICAL..................    1
*  4 PAOZZ 5340007255280 96903 MS21333-125   CLAMP,LOOP   PART OF KIT P/N 57K3497     2
*  5 PAOZZ 5310000116121 29510 120384        WASHER,LOCK  PART OF KIT P/N             1
57K3497............................
*  6 PAOZZ 7690014751202 19207 12469040      MARKER,INDENTIFI  PART OF KIT P/N        1
57K3497............................
*  7 PAOZZ 5305011389540 96906 MS51863-33    SCREW,TAPPING  PART OF KIT P/N           2
57K3497............................
*  8 PAOZZ 5980014746679 19207 12356703-4    DIODE,LIGHT EMITTI  PART OF KIT P/N      1
57K3497............................
*  9 PAOZZ 5930006831628 34355 106450        SWITCH,TOGGLE   PART OF KIT P/N          1
57K3497............................
* 10 MOOZZ                24617 9436356-2    HOSE,NONMETALLIC   MAKE FROM HOSE P/N    1
9438315 2.00 INCHES LONG PART OF KIT                                                P/N
57K3497............................
                                                UOC:A15,B15,B16,H15,H16
* 11 PAOZZ 4730014754524 19207 12460439      TEE,HOSE   PART OF KIT P/N 57K3497...    1
UOC:A15,B15,B16,H15,H16
* 12 MOOZZ                24617 9436356-3    HOSE,NONMETALLIC   MAKE FROM HOSE P/N    1
9438315 3.00 INCHES LONG PART OF KIT                                                P/N
57K3497............................
                                                UOC:A15,B15,B16,H15,H16
* 13 MOOZZ                24617 9436711-5.5  HOSE,NONMETALLIC   MAKE FROM HOSE P/N    1
9438381 5.50 INCHES LONG PART OF KIT                                                P/N
57K3497............................
                                                UOC:A15,B15,B16,H15,H16
* 14 PAOZZ 5975010345871 96906 MS3367-7-2    STRAP,TIEDOWN ELEC  PART OF KIT P/N      2
57K3497............................
```

END OF FIGURE

Section II. TM 9-2320-280-24P-2 C01

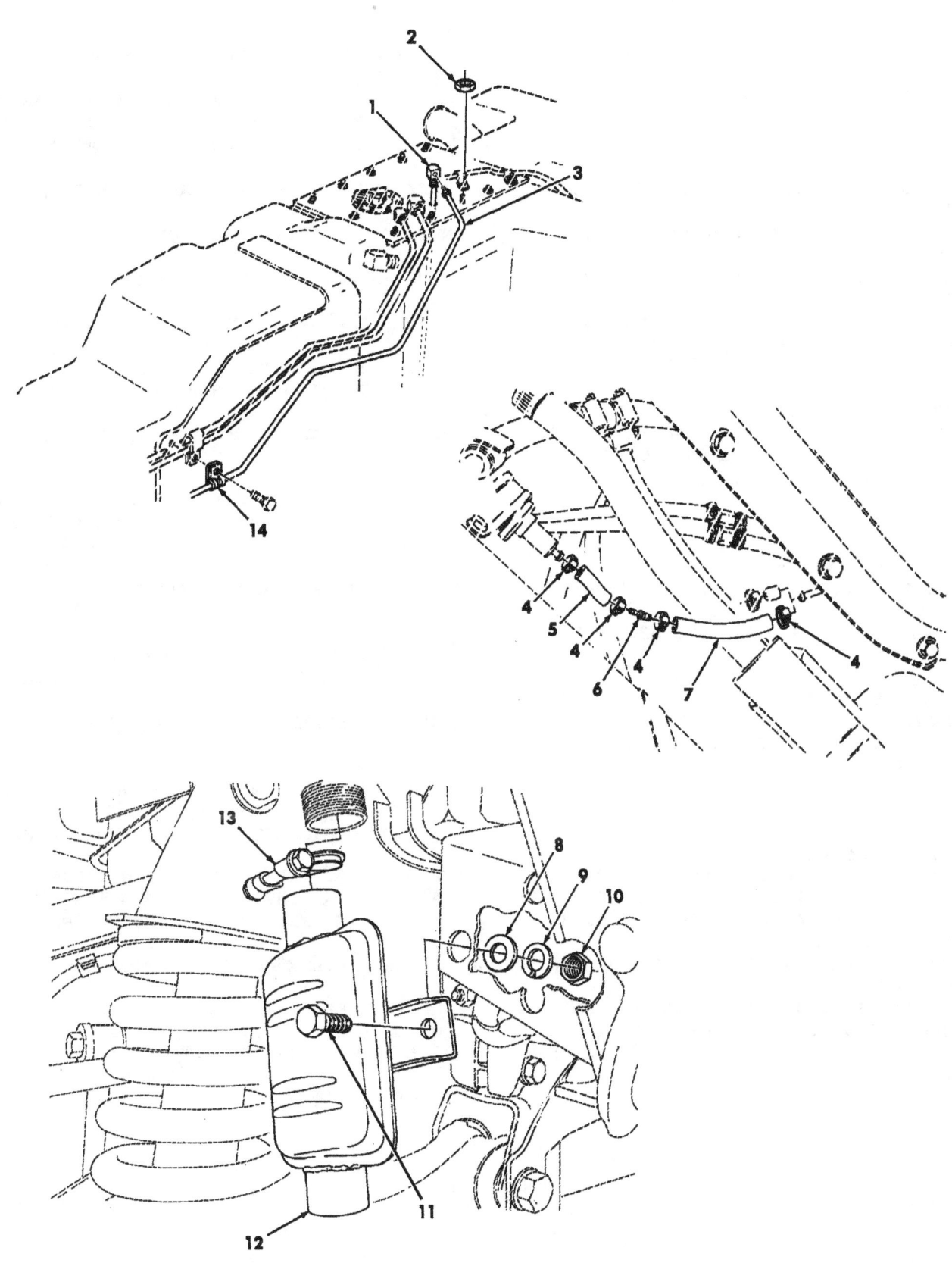

Figure 397D. Engine and Crew Compartment Heater Fuel Supply and Return Lines, Muffler, and Mounting Hardware.

```
      SECTION II             TM9-2320-280-24P C01
    (1)     (2)     (3)         (4)        (5)                        (6)
(7)  ITEM    SMR                                PART
     NO     CODE    NSN        CAGEC         NUMBER       DESCRIPTION AND USABLE ON CODES(UOC)
QTY
```

GROUP 3303 SPECIAL PURPOSE KITS

FIG. 397D ENGINE AND CREW COMPARTMENT HEATER FUEL SUPPLY AND RETURN LINES, MUFFLER, AND MOUNTING HARDWARE

```
* 1 PAOZZ 4710013592956 19207 12338548-2    TUBE ASSEMBLY,METAL   PART OF KIT P/N    1
57K3497........................
* 2 PAOZZ 5310000614650 96906 MS51943-31    NUT,SELF-LOCKING   PART OF KIT P/N       1
57K3497........................
* 3 PAOZZ 4710013586397 19207 12342747      TUBE ASSEMBLY,METAL   PART OF KIT P/N    1
57K3497........................
* 4 PAOZZ 4730011181278 76599 MMF4-SS       CLAMP,HOSE   PART OF KIT P/N 57K3497.    4
UOC:AVY,A11,A13,A14,A20,A24,A25,A26,
                                              A27,BVY,B17,B18,B20,B24,B25,C17,HVY,
                                              H11,H13,H14,H17,H18,H20,H21,H24,H25,
H26,H27,H28,MMM,NNN
* 5 MOOZZ                24617 9436711-2.0  HOSE,NONMETALLIC   MAKE FROM HOSE,P/N    1
9438381,2.00 INCHES LONG PART OF KIT                                                 P/N
57K3497........................

                                              UOC:AVY,A11,A13,A14,A20,A24,A25,A26,
                                              A27,BVY,B17,B18,B20,B24,B25,C17,HVY,
                                              H11,H13,H14,H17,H18,H20,H21,H24,H25,
H26,H27,H28,MMM,NNN
* 6 PAOZZ 4730014746940 19207 12460438      REDUCER,TUBE   PART OF KIT P/N           1
57K3497........................                                                      UOC
:AVY,A11,A13,A14,A20,A24,A25,A26,
                                              A27,BVY,B17,B18,B20,B24,B25,C17,HVY,
                                              H11,H13,H14,H17,H18,H20,H21,H24,H25,
H26,H27,H28,MMM,NNN
* 7 MOOZZ                24617 9436356-AR   HOSE,NONMETALLIC   MAKE FROM HOSE,P/N    1
9438315,CUT LENGTH TO FIT PART OF                                                    KIT
P/N 57K3497...................

                                              UOC:AVY,A11,A13,A14,A20,A24,A25,A26,
                                              A27,BVY,B17,B18,B20,B24,B25,C17,HVY,
                                              H11,H13,H14,H17,H18,H20,H21,H24,H25,
H26,H27,H28,MMM,NNN
* 8 PAOZZ 5310014086466 24617 9417098       WASHER,FLAT   PART OF KIT P/N 57K3497    1
* 9 PAOZZ 5310000116121 29510 120384        WASHER,LOCK   PART OF KIT P/N 57K3497    1
* 10 PAOZZ 5310014654525 24617 9418753      NUT,PLAIN,HEXAGON   PART OF KIT P/N      1
57K3497........................
* 11 PAOZZ 5305009474351 80204 B1821BH075C100N  SCREW,CAP,HEXAGON H   PART OF KIT    1
P/N 57K3497....................
* 12 PAOZZ 2990014753969 62380 19562D       MUFFLER,EXHAUST....................      1
* 13 PAOZZ 5340014753164 62380 367400       CLAMP,LOOP..........................     1
* 14 PAOZZ 5340009936207 96906 MS21333-99   CLAMP,LOOP PART OF KIT P/N 57K3497..     1
```

END OF FIGURE

Section II. TM 9-2320-280-24P-2 C01

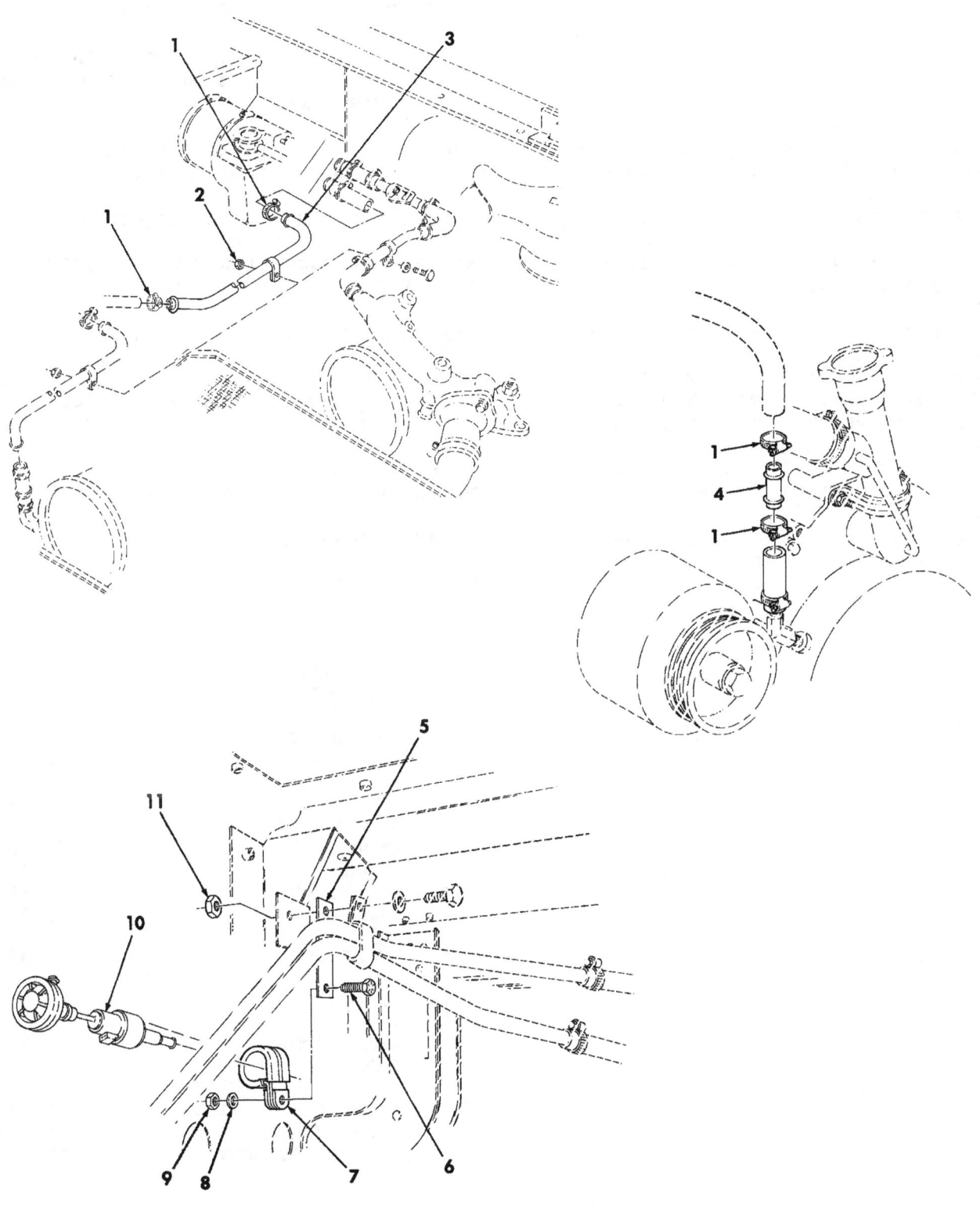

Figure 397E. Crew and Engine Compartment Heater Fuel Pump, Lines, and Mounting Hardware.

```
SECTION II            TM9-2320-280-24P C01
    (1)    (2)   (3)         (4)       (5)                        (6)
(7)      ITEM   SMR                             PART
   NO    CODE   NSN         CAGEC     NUMBER      DESCRIPTION AND USABLE ON CODES(UOC)
QTY
```

GROUP 3303 SPECIAL PURPOSE KITS

FIG. 397E ENGINE AND CREW COMPART-
     HEATER FUEL PUMP,LINES,AND MOUNT-
HARDWARE

MENT
ING

```
*  1 PAOZZ 4730009083194 96906 MS35842-11        CLAMP,HOSE   PART OF KIT P/N 57K3497.    4
*  2 PAOZZ 5310000614650 96906 MS51943-31        NUT,SELF-LOCKING   PART OF KIT P/N       1
57K3497..............................
*  3 PAOZZ 4710014746875 19207 12460437          TUBE,BENT METALLIC   PART OF KIT P/N     1
57K3497..............
*  4 PAOZZ 4730013948192 96906 MS24520-7         MENDER,HOSE   PART OF KIT P/N 57K3497    1
*  5 PAOZZ 5340014753147 19207 12469052          STRAP,RETAINING   PART OF KIT P/N        1
57K3497..............................
*  6 PAOZZ 5306002264825 80204 B1821BH031C075N   BOLT,MACHINE   PART OF KIT P/N           1
57K3497..............................
*  7 PAOZZ 5340014753159 62380 21499A            CLAMP,LOOP.........................      1
*  8 PAOZZ 5310004079566 96906 MS35338-45        WASHER,LOCK   PART OF KIT P/N 57K3497    1
*  9 PAOZZ 5310012097702 7X677 9429048           NUT,PLAIN,HEXAGON   PART OF KIT P/N      1
57K3497..............................
* 10 PAOZZ 2910014753937 62380 91473A            PUMP,FUEL...........................     1
* 11 PAOZZ 5310000614650 96906 MS51943-31        NUT,SELF-LOCKING   PART OF KIT P/N       1
57K3497..............................
* KIT PAOZZ 2540014754919 19207 57K3497             PARTS KIT,VEHICULAR................   1
ADAPTER,CONNECTOR           ( 1) 397A-14
                                                    ADAPTER,CONNECTOR      ( 1) 397A-
15
                                                    BOLT,MACHINE           ( 2) 397B-
21
                                                    BOLT,MACHINE           ( 1) 397E-6
                                                    BRACKET,MULTIPLE AN    ( 1) 397B-
25
                                                    BREAKER,CIRCUIT        ( 1) 397B-5
                                                    CABLE,NIPPLE           ( 3) 397A-
18
                                                    CLAMP,HOSE             ( 8) 397A-2
                                                    CLAMP,HOSE             ( 4) 397A-6
                                                    CLAMP,HOSE             ( 6) 397C-1
                                                    CLAMP,HOSE             ( 4) 397D-4
                                                    CLAMP,HOSE             ( 4) 397E-1
                                                    CLAMP,LOOP             ( 1) 397A-
25
                                                    CLAMP,LOOP             ( 1) 397B-
13
                                                    CLAMP,LOOP             ( 1) 397B-
14
                                                    CLAMP,LOOP             ( 2) 397C-4
                                                    CLAMP,LOOP             ( 1) 397D-
14
                                                    DIODE,LIGHT EMITTING   ( 1) 397C-8
                                                    GUARD,MUFFLER EXHAU    ( 1) 397B-6
                                                    HEATER ASSEMBLY        ( 1) 397A-8
```

SECTION II   TM9-2320-280-24P C01
(1)   (2)   (3)   (4)   (5)   (6)
(7) ITEM   SMR   PART
    NO   CODE   NSN   CAGEC   NUMBER   DESCRIPTION AND USABLE ON CODES(UOC)
TY

                                              HOSE,NONMETALLIC   (1) 397D-5
                                              HOSE,NONMETALLIC   (1) 397D-7
                                              HOSE,PREFORMED   (2) 397A-4
                                              HOSE,PREFORMED   (1) 397A-
9
                                              INSERT,SCREW THREAD   (2) 397B-
4
                                              INSULATOR,BUSHING   (3) 397A-
7
                                              INSULATION,SLEEVING   (1) 397B-
6
                                              MARKER,IDENTIFICATION (1) 397C-6
                                              MENDER,HOSE   (1) 397A-
8
                                              MENDER,HOSE   (1) 397E-4
                                              NUT,PLAIN,HEXAGON   (2) 397B-3
                                              NUT,PLAIN,HEXAGON   (2) 397B-
2
                                              NUT,PLAIN,HEXAGON   (1) 397D-
0
                                              NUT,PLAIN,HEXAGON   (1) 397E-9
                                              NUT,SELF-LOCKING   (1) 397A-
4
                                              NUT,SELF-LOCKING   (2) 397B-
6
                                              NUT,SELF-LOCKING   (1) 397D-2
                                              NUT,SELF-LOCKING   (1) 397E-2
                                              NUT,SELF-LOCKING   (1) 397E-
1
                                              REDUCER,TUBE   (2) 397A-3
                                              REDUCER,TUBE   (1) 397D-6
                                              SCREW,CAP,HEXAGON   (6) 397B-8
                                              SCREW,CAP HEXAGON H   (1) 397D-
1
                                              SCREW,MACHINE   (2) 397B-9
                                              SCREW,MACHINE   (2) 397B-
0
                                              SCREW,TAPPING   (2) 397C-7
                                              SHELL,ELECTRICAL   (1) 397A-
0
                                              SHELL,HEAD LIGHT   (1) 397B-
9
                                              STRAP,RETAINING   (1) 397E-5
                                              STRAP,TIEDOWN ELEC   (5) 397A-
6
                                              STRAP,TIEDOWN ELEC   (4) 397B-1
                                              STRAP,TIEDOWN ELEC   (2) 397C-
4
                                              SWITCH,TOGGLE   (1) 397C-9
                                              TEE,HOSE   (1) 397C-
1
                                              TERMINAL,LUG   (2) 397A-
9
                                              TERMINAL,LUG   (1) 397A-

397E-2

SECTION II   TM9-2320-280-24P C01
  (1)   (2)   (3)   (4)   (5)   (6)
(7)   ITEM SMR   PART
  NO   CODE NSN   CAGEC   NUMBER   DESCRIPTION AND USABLE ON CODES(UOC)
QTY

                                                WASHER,LOCK   (1) 397C-5
                                                WASHER,LOCK   (1) 397D-9
                                                WASHER,LOCK   (1) 397E-8
                                                WASHER,SLOTTED   (1) 397A-
21

                                                WIRE,ELECTRICAL   (1) 397B-2
                                                WIRE,ELECTRICAL   (1) 397B-

Section II. TM 9-2320-280-24P-2

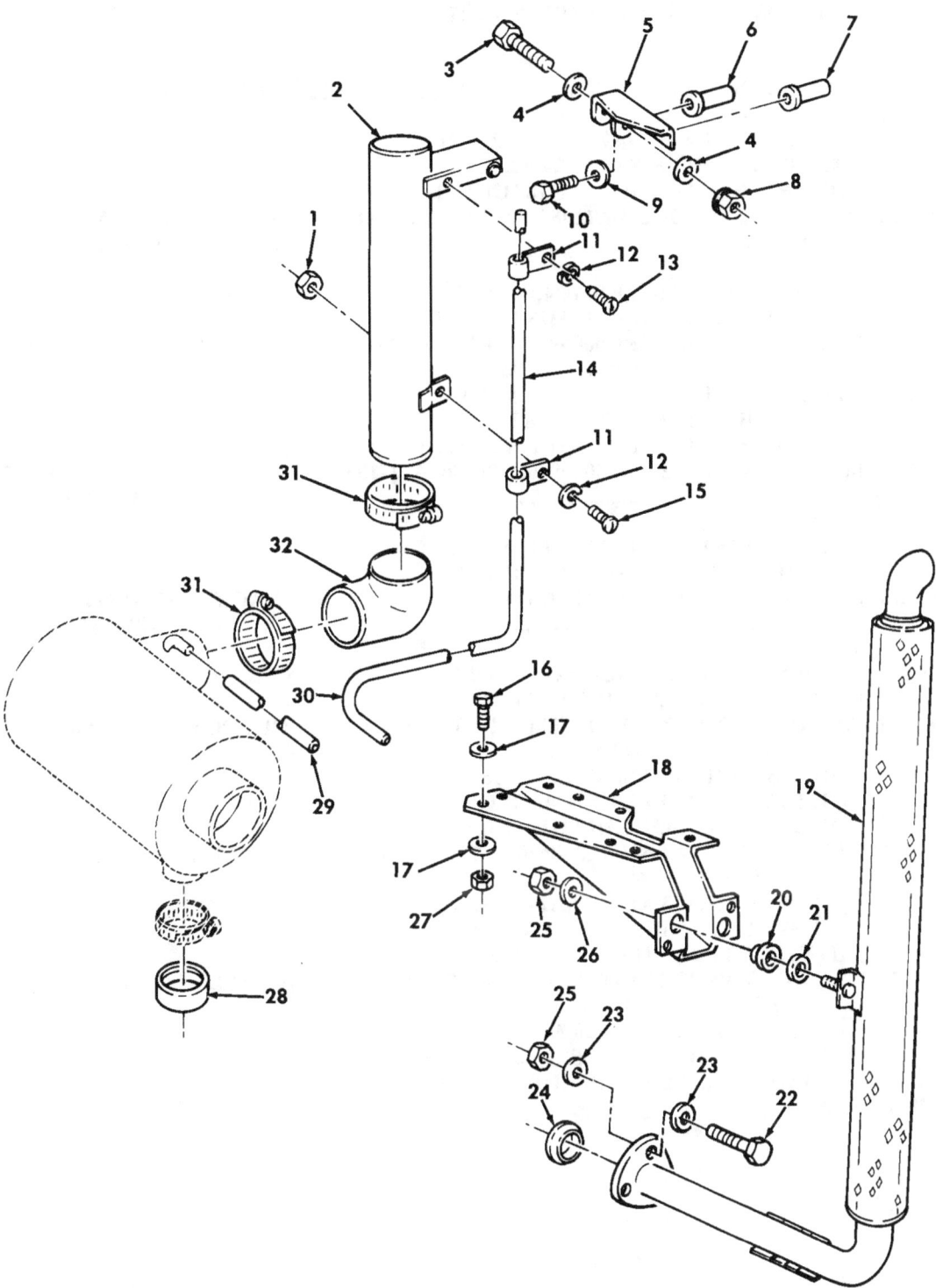

Figure 398. Deep Water Fording Kit, Intake and Exhaust Systems.

SECTION II TM9-2320-280-24P

(1) NO	(2) SMR CODE	(3) NSN	(4) CAGEC	(5) PART NUMBER	(7) ITEM DESCRIPTION AND USABLE ON CODES(UOC)	QTY

GROUP 3305 DEEP WATER FORDING KITS

FIG. 398 DEEP WATER FORDING KIT, INTAKE AND EXHAUST SYSTEMS

1 PAOZZ 5310012490899 96906 MS51470-03 NUT,PLAIN,HEXAGON #10-32 PART OF KIT P/N 5705693..................... 1
UOC:AVY,A11,A13,A14,A20,A24,A25,A26,
A27,BVY,B16,B17,B18,B20,B24,B25,C17,
HVY,H11,H13,H14,H16,H17,H18,H20,H21,
H24,H25,H26,H27,H28,MMM,NNN

2 PFOZZ 2510011980333 19207 12339843 STACK ASSEMBLY PART OF KIT P/N 5705693............................. 1
UOC:AVY,A11,A13,A14,A20,A24,A25,A26,
A27,BVY,B16,B17,B18,B20,B24,B25,C17,
HVY,H11,H13,H14,H16,H17,H18,H20,H21,
H24,H25,H26,H27,H28,MMM,NNN

3 PAOZZ 5305000695578 96906 MS90725-94 SCREW,CAP,HEXAGON H 7/16-14 X 2.75 PART OF KIT P/N 5705693............. 1
UOC:AVY,A11,A13,A14,A20,A24,A25,A26,
A27,BVY,B16,B17,B18,B20,B24,B25,C17,
HVY,H11,H13,H14,H16,H17,H18,H20,H21,
H24,H25,H26,H27,H28,MMM,NNN

4 PAOZZ 5310008094085 96906 MS27183-16 WASHER,FLAT 7/16 PART OF KIT P/N 5705693............................. 2
UOC:AVY,A11,A13,A14,A20,A24,A25,A26,
A27,BVY,B16,B17,B18,B20,B24,B25,C17,
HVY,H11,H13,H14,H16,H17,H18,H20,H21,
H24,H25,H26,H27,H28,MMM,NNN

5 PFOZZ 5340012389543 19207 12339832 BRACKET,MOUNTING PART OF KIT P/N 5705693............................. 1
UOC:AVY,A11,A13,A14,A20,A24,A25,A26,
A27,BVY,B16,B17,B18,B20,B24,B25,C17,
HVY,H11,H13,H14,H16,H17,H18,H20,H21,
H24,H25,H26,H27,H28,MMM,NNN

6 PFOZZ 5310012317455 19207 12339397-3 NUT,PLAIN,BLIND RIV 1/4-20 PART OF KIT P/N 5705693..................... 1
UOC:AVY,A11,A13,A14,A20,A24,A25,A26,
A27,BVY,B16,B17,B18,B20,B24,B25,C17,
HVY,H11,H13,H14,H16,H17,H18,H20,H21,
H24,H25,H26,H27,H28,MMM,NNN

7 PAOZZ 5310013886205 19207 12339397-6 NUT,PLAIN,BLIND RIV 1/4-20 PART OF KIT P/N 5705693..................... 1
UOC:AVY,A11,A13,A14,A20,A24,A25,A26,
A27,BVY,B16,B17,B18,B20,B24,B25,C17,
HVY,H11,H13,H14,H16,H17,H18,H20,H21,
H24,H25,H26,H27,H28,MMM,NNN

8 PAOZZ 5310011504003 24617 9422299 NUT,SELF-LOCKING,HE 7/16-14 PART OF KIT P/N 5705693..................... 1
UOC:AVY,A11,A13,A14,A20,A24,A25,A26,
A27,BVY,B16,B17,B18,B20,B24,B25,C17,
HVY,H11,H13,H14,H16,H17,H18,H20,H21,
H24,H25,H26,H27,H28,MMM,NNN

SECTION II   TM9-2320-280-24P

(1) NO	(2) SMR CODE	(3) NSN	(4) CAGEC	(5) PART NUMBER	(6) DESCRIPTION AND USABLE ON CODES(UOC)	(7) QTY
9	PAOZZ	5310008775975	19207	10918603	WASHER,FLAT 1/4 PART OF KIT P/N 5705693............................. UOC:AVY,A11,A13,A14,A20,A24,A25,A26, A27,BVY,B16,B17,B18,B20,B24,B25,C17, HVY,H11,H13,H14,H16,H17,H18,H20,H21, H24,H25,H26,H27,H28,MMM,NNN	2
10	PAOZZ	5305000680509	80204	B1821BH025C125N	SCREW,CAP,HEXAGON H 1/4-20 X 1.25 PART OF KIT P/N 5705693............. UOC:AVY,A11,A13,A14,A20,A24,A25,A26, A27,BVY,B16,B17,B18,B20,B24,B25,C17, HVY,H11,H13,H14,H16,H17,H18,H20,H21, H24,H25,H26,H27,H28,MMM,NNN	2
11	PAOZZ	5340000797837	96906	MS21333-67	CLAMP,LOOP PART OF KIT P/N 5705693.. UOC:AVY,A11,A13,A14,A20,A24,A25,A26, A27,BVY,B16,B17,B18,B20,B24,B25,C17, HVY,H11,H13,H14,H16,H17,H18,H20,H21, H24,H25,H26,H27,H28,MMM,NNN	2
12	PAOZZ	5310007217809	96906	MS35340-43	WASHER,LOCK #10 PART OF KIT P/N 5705693............................. UOC:AVY,A11,A13,A14,A20,A24,A25,A26, A27,BVY,B16,B17,B18,B20,B24,B25,C17, HVY,H11,H13,H14,H16,H17,H18,H20,H21, H24,H25,H26,H27,H28,MMM,NNN	2
13	PAOZZ	5305012542461	34623	5592899	SCREW,TAPPING,THREA #10-32 X 1/2 PART OF KIT P/N 5705693............. UOC:AVY,A11,A13,A14,A20,A24,A25,A26, A27,BVY,B16,B17,B18,B20,B24,B25,C17, HVY,H11,H13,H14,H16,H17,H18,H20,H21, H24,H25,H26,H27,H28,MMM,NNN	1
14	PFOZZ	4710012096746	34623	12339833	TUBE,BENT,METALLIC PART OF KIT P/N 5705693............................. UOC:AVY,A11,A13,A14,A20,A24,A25,A26, A27,BVY,B16,B17,B18,B20,B24,B25,C17, HVY,H11,H13,H14,H16,H17,H18,H20,H21, H24,H25,H26,H27,H28,MMM,NNN	1
15	PAOZZ	5305009897434	96906	MS35207-263	SCREW,MACHINE #10-32 X 1/2 PART OF KIT P/N 5705693.................... UOC:AVY,A11,A13,A14,A20,A24,A25,A26, A27,BVY,B16,B17,B18,B20,B24,B25,C17, HVY,H11,H13,H14,H16,H17,H18,H20,H21, H24,H25,H26,H27,H28,MMM,NNN	1
16	PAOZZ	5305000680502	96906	MS90725-6	SCREW,CAP,HEXAGON H 1/4-20 X 3/4 PART OF KIT P/N 5705693............. UOC:AVY,A11,A13,A14,A20,A24,A25,A26, A27,BVY,B16,B17,B18,B20,B24,B25,C17, HVY,H11,H13,H14,H16,H17,H18,H20,H21, H24,H25,H26,H27,H28,MMM,NNN	8
17	PAOZZ	5310011023270	24617	2436161	WASHER,FLAT 1/4 PART OF KIT P/N 5705693............................. UOC:AVY,A11,A13,A14,A20,A24,A25,A26, A27,BVY,B16,B17,B18,B20,B24,B25,C17, HVY,H11,H13,H14,H16,H17,H18,H20,H21, H24,H25,H26,H27,H28,MMM,NNN	6

57

SECTION II TM9-2320-280-24P

(1) NO	(2) SMR CODE	(3) NSN	(4) CAGEC	(5) PART NUMBER	(6) DESCRIPTION AND USABLE ON CODES (UOC)	(7) QTY
18	PFOZZ	5340011857959	19207	12339848	BRACKET,PIPE PART OF KIT P/N 5705693 UOC:AVY,A11,A13,A14,A20,A24,A25,A26,A27,BVY,B16,B17,B18,B20,B24,B25,C17,HVY,H11,H13,H14,H16,H17,H18,H20,H21,H24,H25,H26,H27,H28,MMM,NNN	1
19	PAOZZ	2990012102176	19207	12339849	PIPE,EXHAUST PART OF KIT P/N 5705693 UOC:AVY,A11,A13,A14,A20,A24,A25,A26,A27,BVY,B16,B17,B18,B20,B24,B25,C17,HVY,H11,H13,H14,H16,H17,H18,H20,H21,H24,H25,H26,H27,H28,MMM,NNN	1
20	PFOZZ	5365012135739	34623	5584575-B	BUSHING,NONMETALLIC PART OF KIT P/N 5705693............................ UOC:AVY,A11,A13,A14,A20,A24,A25,A26,A27,BVY,B16,B17,B18,B20,B24,B25,C17,HVY,H11,H13,H14,H16,H17,H18,H20,H21,H24,H25,H26,H27,H28,MMM,NNN	2
21	PAOZZ	5310012987770	3M915	5596167	WASHER,FLAT 3/8 PART OF KIT P/N 5705693............................ UOC:AVY,A11,A13,A14,A20,A24,A25,A26,A27,BVY,B16,B17,B18,B20,B24,B25,C17,HVY,H11,H13,H14,H16,H17,H18,H20,H21,H24,H25,H26,H27,H28,MMM,NNN	2
22	PAOZZ	5305007829489	80204	B1821BH038C200N	SCREW,CAP,HEXAGON H 3/8-16 X 2.00 PART OF KIT P/N 5705693............. UOC:AVY,A11,A13,A14,A20,A24,A25,A26,A27,BVY,B16,B17,B18,B20,B24,B25,C17,HVY,H11,H13,H14,H16,H17,H18,H20,H21,H24,H25,H26,H27,H28,MMM,NNN	3
23	PAOZZ	5310000877493	96906	MS27183-13	WASHER,FLAT 3/8 PART OF KIT P/N 5705693............................ UOC:AVY,A11,A13,A14,A20,A24,A25,A26,A27,BVY,B16,B17,B18,B20,B24,B25,C17,HVY,H11,H13,H14,H16,H17,H18,H20,H21,H24,H25,H26,H27,H28,MMM,NNN	6
24	PFOZZ	5330012000466	19207	12338339	GASKET PART OF KIT P/N 5705693...... UOC:AVY,A11,A13,A14,A20,A24,A25,A26,A27,BVY,B16,B17,B18,B20,B24,B25,C17,HVY,H11,H13,H14,H16,H17,H18,H20,H21,H24,H25,H26,H27,H28,MMM,NNN	1
25	PAOZZ	5310009359021	96906	MS51943-35	NUT,SELF-LOCKING,HE 3/8-16 PART OF KIT P/N 5705693..................... UOC:AVY,A11,A13,A14,A20,A24,A25,A26,A27,BVY,B16,B17,B18,B20,B24,B25,C17,HVY,H11,H13,H14,H16,H17,H18,H20,H21,H24,H25,H26,H27,H28,MMM,NNN	5
26	PAOZZ	5310012581536	0MAY0	5593221	WASHER,FLAT 3/8 PART OF KIT P/N 5705693............................ UOC:AVY,A11,A13,A14,A20,A24,A25,A26,A27,BVY,B16,B17,B18,B20,B24,B25,C17,HVY,H11,H13,H14,H16,H17,H18,H20,H21,H24,H25,H26,H27,H28,MMM,NNN	2
27	PAOZZ	5310000614650	96906	MS51943-31	NUT,SELF-LOCKING,HE 1/4-20 PART OF KIT P/N 5705693.....................	8

398-3

SECTION II   TM9-2320-280-24P

(1) NO	(2) SMR CODE	(3) NSN	(4) CAGEC	(5) PART NUMBER	(6) DESCRIPTION AND USABLE ON CODES(UOC)	(7) QTY

UOC:AVY,A11,A13,A14,A20,A24,A25,A26,
A27,BVY,B16,B17,B18,B20,B24,B25,C17,
HVY,H11,H13,H14,H16,H17,H18,H20,H21,
H24,H25,H26,H27,H28,MMM,NNN   28 PAOZZ 5340011972382 19207 12339996   CAP,PROTECTIVE,DUST PART OF KIT P/N 5705693.............................   1

UOC:AVY,A11,A13,A14,A20,A24,A25,A26,
A27,BVY,B16,B17,B18,B24,B25,C17,HVY,
H11,H13,H14,H15,H16,H17,H18,H20,H21,
H24,H25,H26,H27,H28,MMM,NNN   29 MOOZZ   34623 5591707   TUBE MAKE FROM TUBE,P/N CPR104420-1,70 INCHES LONG PART OF KIT P/N 5705693.............................

UOC:AVY,A11,A13,A14,A20,A24,A25,A26,
A27,BVY,B16,B17,B18,B20,B24,B25,C17,
HVY,H11,H13,H14,H16,H17,H18,H20,H21,
H24,H25,H26,H27,H28,MMM,NNN   30 MOOZZ   34623 5591210   TUBE,NONMETALLIC MAKE FROM TUBE,P/N CPR104420-2,11 INCHES LONG PART OF KIT P/N 5705693.....................   1

UOC:AVY,A11,A13,A14,A20,A24,A25,A26,
A27,BVY,B16,B17,B18,B20,B24,B25,C17,
HVY,H11,H13,H14,H16,H17,H18,H20,H21,
H24,H25,H26,H27,H28,MMM,NNN   31 PAOZZ 4730003599487 66295 C72P   CLAMP,HOSE PART OF KIT P/N 5705693..   2

UOC:AVY,A11,A13,A14,A20,A24,A25,A26,
A27,BVY,B16,B17,B18,B20,B24,B25,C17,
HVY,H11,H13,H14,H16,H17,H18,H20,H21,
H24,H25,H26,H27,H28,MMM,NNN   32 PAOZZ 4720011945338 34623 5578943   HOSE,PREFORMED PART OF KIT P/N 5705693.............................   1

UOC:AVY,A11,A13,A14,A20,A24,A25,A26,
A27,BVY,B16,B17,B18,B20,B24,B25,C17,
HVY,H11,H13,H14,H16,H17,H18,H20,H21,
H24,H25,H26,H27,H28,MMM,NNN

END OF FIGURE

Section II. TM 9-2320-280-24P-2

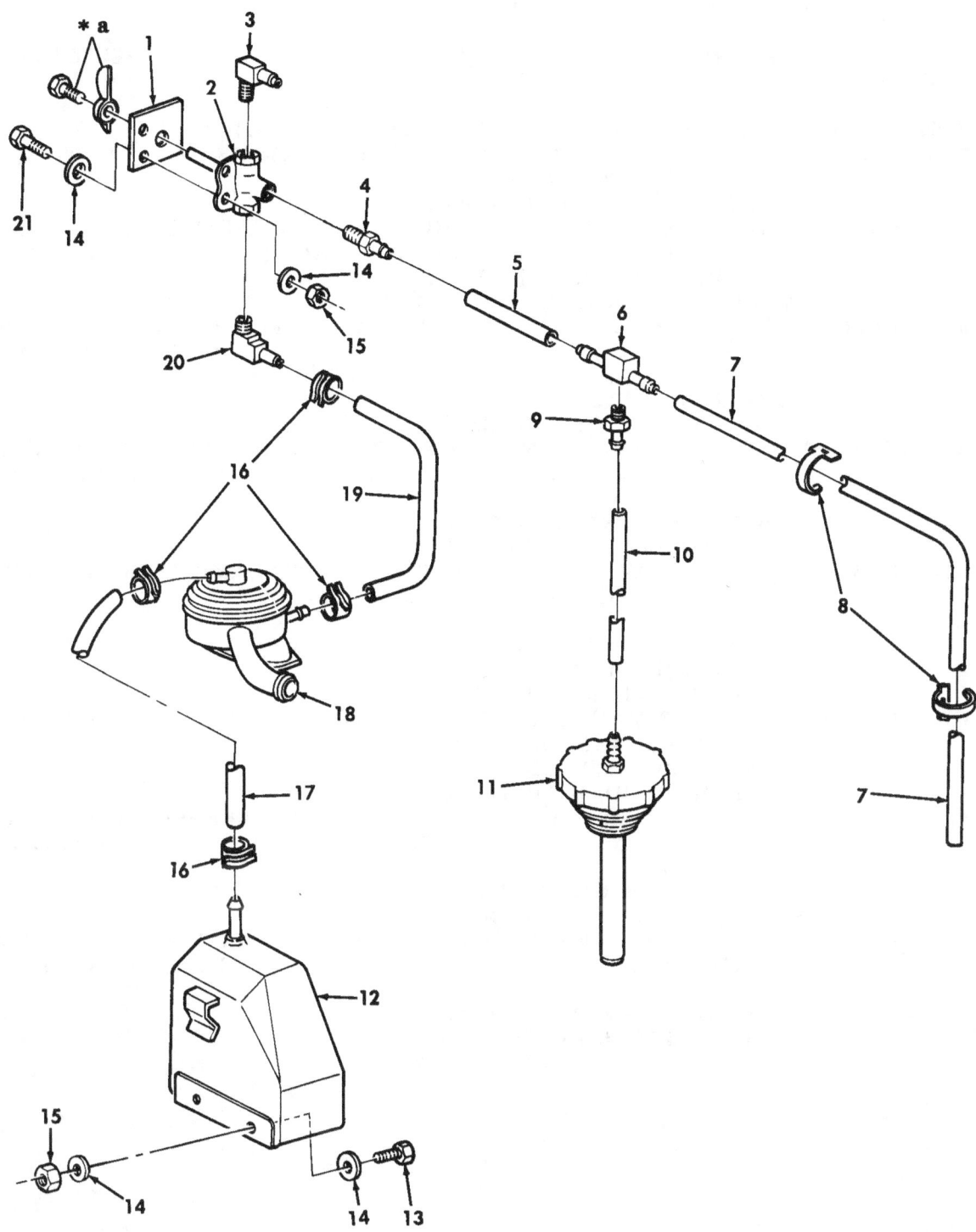

* a PART OF ITEM 2

Figure 399. Deep Water Fording Kit, Selector Valve, Sensor Cup, and Vent Lines.

SECTION II   TM9-2320-280-24P

(1) ITEM NO	(2) SMR CODE	(3) NSN	(4) CAGEC	(5) PART NUMBER	(6) DESCRIPTION AND USABLE ON CODES(UOC)	(7) QTY

GROUP 3305 DEEP WATER FORDING KITS

FIG. 399 DEEP WATER FORDING KIT, SELECTOR VALVE, SENSOR CUP, AND VENT LINES

(1)	(2)	(3)	(4)	(5)	(6)	(7)
1	PAOZZ	9905012058635	19207	12339109	PLATE,IDENTIFICATIO PART OF KIT P/N 5705693................... UOC:AVY,A11,A13,A14,A20,A24,A25,A26, A27,BVY,B16,B17,B18,B20,B24,B25,C17, HVY,H11,H13,H14,H16,H17,H18,H20,H21, H24,H25,H26,H27,H28,MMM,NNN	1
2	PAOZZ	4820011928030	70411	220C-J	VALVE,CROSS PART OF KIT P/N 5705693. UOC:AVY,A11,A13,A14,A20,A24,A25,A26, A27,BVY,B16,B17,B18,B20,B24,B25,C17, HVY,H11,H13,H14,H16,H17,H18,H20,H21, H24,H25,H26,H27,H28,MMM,NNN	1
3	PAOZZ	4730010707680	01989	1069X4	ELBOW,PIPE TO HOSE PART OF KIT P/N 5705693........................ UOC:AVY,A11,A13,A14,A20,A24,A25,A26, A27,BVY,B16,B17,B18,B20,B24,B25,C17, HVY,H11,H13,H14,H16,H17,H18,H20,H21, H24,H25,H26,H27,H28,MMM,NNN	1
4	PAOZZ	4730011855348	01989	1068X4	ADAPTER,STRAIGHT,PI PART OF KIT P/N 5705693........................ UOC:AVY,A11,A13,A14,A20,A24,A25,A26, A27,BVY,B16,B17,B18,B20,B24,B25,C17, HVY,H11,H13,H14,H16,H17,H18,H20,H21, H24,H25,H26,H27,H28,MMM,NNN	1
5	MOOZZ		34623	5584373	TUBE,NONMETALLIC MAKE FROM TUBE,P/N CPR104420-1, 26 INCHES LONG PART OF KIT P/N 5705693................ UOC:AVY,A11,A13,A14,A20,A24,A25,A26, A27,BVY,B16,B17,B18,B20,B24,B25,C17, HVY,H11,H13,H14,H16,H17,H18,H20,H21, H24,H25,H26,H27,H28,MMM,NNN	1
6	PAOZZ	4730010035105	79470	1064X4	TEE,HOSE PART OF KIT P/N 5705693.... UOC:AVY,A11,A13,A14,A20,A24,A25,A26, A27,BVY,B16,B17,B18,B20,B24,B25,C17, HVY,H11,H13,H14,H16,H17,H18,H20,H21, H24,H25,H26,H27,H28,MMM,NNN	1
7	MOOZZ		34623	5591253	TUBE MAKE FROM TUBE, P/N CPR104420-1, 41 INCHES LONG PART OF KIT P/N 5705693........................ UOC:AVY,A11,A13,A14,A20,A24,A25,A26, A27,BVY,B16,B17,B18,B20,B24,B25,C17, HVY,H11,H13,H14,H16,H17,H18,H20,H21, H24,H25,H26,H27,H28,MMM,NNN	1
8	PAOZZ	5340011494434	7X677	3816659	STRAP,LINE SUPPORTI PART OF KIT P/N 5705693........................ UOC:AVY,A11,A13,A14,A20,A24,A25,A26, A27,BVY,B16,B17,B18,B20,B24,B25,C17, HVY,H11,H13,H14,H16,H17,H18,H20,H21,	3

SECTION II   TM9-2320-280-24P

(1) ITEM NO	(2) SMR CODE	(3) NSN	(4) CAGEC	(5) PART NUMBER	(6) DESCRIPTION AND USABLE ON CODES(UOC)	(7) QTY
9	PAOZZ	4730005951078	24869	AM-2-B	ADAPTER,STRAIGHT,PI PART OF KIT P/N 5705693............. UOC:AVY,A11,A13,A14,A20,A24,A25,A26, A27,BVY,B16,B17,B18,B20,B24,B25,C17, HVY,H11,H13,H14,H16,H17,H18,H20,H21, H24,H25,H26,H27,H28,MMM,NNN	1
10	MOOZZ		34623	5582644	TUBE MAKE FROM TUBE, P/N CPR104420-1, 20 INCHES LONG PART OF KIT P/N 5705693............. UOC:AVY,A11,A13,A14,A20,A24,A25,A26, A27,BVY,B16,B17,B18,B20,B24,B25,C17, HVY,H11,H13,H14,H16,H17,H18,H20,H21, H24,H25,H26,H27,H28,MMM,NNN	1
11	PAOZZ	2590011924425	19207	12339997	CAP,FILLER OPENING PART OF KIT P/N 5705693............. UOC:AVY,A11,A13,A14,A20,A24,A25,A26, A27,BVY,B16,B17,B18,B20,B24,B25,C17, HVY,H11,H13,H14,H16,H17,H18,H20,H21, H24,H25,H26,H27,H28,MMM,NNN	1
12	PAOZZ	2540011924502	19207	12339845	CUP,SENSOR,DEEP WAT PART OF KIT P/N 5705693............. UOC:AVY,A11,A13,A14,A20,A24,A25,A26, A27,BVY,B16,B17,B18,B20,B24,B25,C17, HVY,H11,H13,H14,H16,H17,H18,H20,H21, H24,H25,H26,H27,H28,MMM,NNN	1
13	PAOZZ	5305000680502	96906	MS90725-6	SCREW,CAP,HEXAGON H 1/4-20 X 4 PART OF KIT P/N 5705693............. UOC:AVY,A11,A13,A14,A20,A24,A25,A26, A27,BVY,B16,B17,B18,B20,B24,B25,C17, HVY,H11,H13,H14,H16,H17,H18,H20,H21, H24,H25,H26,H27,H28,MMM,NNN	2
14	PAOZZ	5310011023270	24617	2436161	WASHER,FLAT PART OF KIT P/N 5705693. UOC:AVY,A11,A13,A14,A20,A24,A25,A26, A27,BVY,B16,B17,B18,B20,B24,B25,C17, HVY,H11,H13,H14,H16,H17,H18,H20,H21, H24,H25,H26,H27,H28,MMM,NNN	8
15	PAOZZ	5310000614650	96906	MS51943-31	NUT,SELF-LOCKING,HE 1/4-20 PART OF KIT P/N 5705693............. UOC:AVY,A11,A13,A14,A20,A24,A25,A26, A27,BVY,B16,B17,B18,B20,B24,B25,C17, HVY,H11,H13,H14,H16,H17,H18,H20,H21, H24,H25,H26,H27,H28,MMM,NNN	4
16	PAOZZ	4730001389201	81348	WW-C-4408	CLAMP,HOSE PART OF KIT P/N 5705693.. UOC:AVY,A11,A13,A14,A20,A24,A25,A26, A27,BVY,B16,B17,B18,B20,B24,B25,C17, HVY,H11,H13,H14,H16,H17,H18,H20,H21, H24,H25,H26,H27,H28,MMM,NNN	4
17	MOOZZ		34623	12339981-2	HOSE,NONMETALLIC MAKE FROM HOSE,P/N RB1450-1-4IDX1-20D, 25 INCHES LONG PART OF KIT P/N 5705693............. UOC:AVY,A11,A13,A14,A20,A24,A25,A26, A27,BVY,B16,B17,B18,B20,B24,B25,C17,	1

SECTION II   TM9-2320-280-24P

(1) NO	(2) SMR CODE	(3) NSN	(4) CAGEC	(5) PART NUMBER	(6) DESCRIPTION AND USABLE ON CODES(UOC)	(7) QTY
18	PAOZZ	4820011927678	70040	25043364	VALVE,VENT PART OF KIT P/N 5705693.. UOC:AVY,A11,A13,A14,A20,A24,A25,A26, A27,BVY,B16,B17,B18,B20,B24,B25,C17, HVY,H11,H13,H14,H16,H17,H18,H20,H21, H24,H25,H26,H27,H28,MMM,NNN	1
19	MOOZZ		34623	5583855	HOSE MAKE FROM HOSE, P/N RB1450-1-4IDX1-20D, 42 INCHES LONG PART OF KIT P/N 5705693..................... UOC:AVY,A11,A13,A14,A20,A24,A25,A26, A27,BVY,B16,B17,B18,B20,B24,B25,C17, HVY,H11,H13,H14,H16,H17,H18,H20,H21, H24,H25,H26,H27,H28,MMM,NNN	1
20	PAOZZ	4730012004277	79470	1582	ELBOW,PIPE TO HOSE PART OF KIT P/N 5705693............................ UOC:AVY,A11,A13,A14,A20,A24,A25,A26, A27,BVY,B16,B17,B18,B20,B24,B25,C17, HVY,H11,H13,H14,H16,H17,H18,H20,H21, H24,H25,H26,H27,H28,MMM,NNN	1
21	PAOZZ	5305000680508	80204	B1821BH025C075N	SCREW,CAP,HEXAGON H PART OF KIT P/N 5705693............................ UOC:AVY,A11,A13,A14,A20,A24,A25,A26, A27,BVY,B16,B17,B18,B20,B24,B25,C17, HVY,H11,H13,H14,H16,H17,H18,H20,H21, H24,H25,H26,H27,H28	2

END OF FIGURE

Section II.

TM 9-2320-280-24P-2

1—[2] 5—[6]

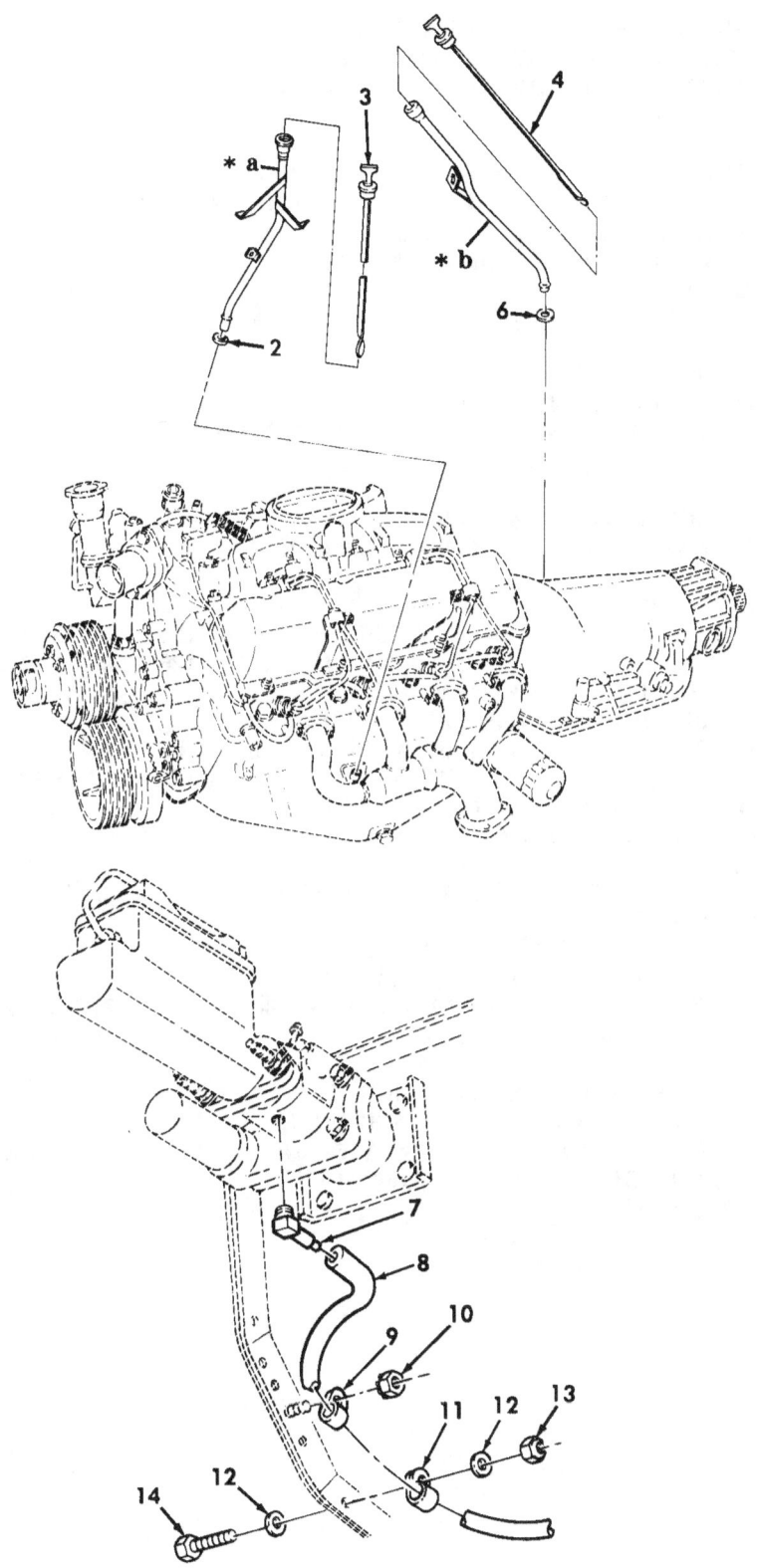

* **a** PART OF ITEM 1
* **b** PART OF ITEM 5

*Figure 400. Deep Water Fording Kit, Hydro-Boost Vent Line,
Engine and Transmission Dipstick, and Tube Assemblies*

SECTION II   TM9-2320-280-24P

(1) NO	(2) SMR CODE	(3) NSN	(4) CAGEC	(5) PART NUMBER	(6) DESCRIPTION AND USABLE ON CODES(UOC)	(7) QTY

GROUP 3305 DEEP WATER FORDING KITS

FIG. 400 DEEP WATER FORDING KIT, HYDRO-BOOST VENT LINE, ENGINE AND TRANSMISSION DIPSTICK, AND TUBE ASSEMBLIES

1 PFOZZ 4710012028245 19207 12339844   TUBE ASSEMBLY,METAL PART OF KIT P/N   1 5705693............................ UOC:AVY,A11,A13,A14,A20,A24,A25,A26, A27,BVY,B16,B17,B18,B20,B24,B25,C17, HVY,H11,H13,H14,H16,H17,H18,H20,H21, H24,H25,H26,H27,H28,MMM,NNN

2 PAOZZ 5331009359136 11862 274244   .O-RING............................ 1   UOC:AVY,A11,A13,A14,A20,A24,A25,A26, A27,BVY,B16,B17,B18,B20,B24,B25,C17, HVY,H11,H13,H14,H16,H17,H18,H20,H21, H24,H25,H26,H27,H28,MMM,NNN

3 PAOZZ 6680011853202 19207 12339846   GAGE ROD-CAP,LIQUID PART OF KIT P/N   1   5705693............................   UOC:AVY,A11,A13,A14,A20,A24,A25,A26, A27,BVY,B16,B17,B18,B20,B24,B25,C17, HVY,H11,H13,H14,H16,H17,H18,H20,H21, H24,H25,H26,H27,H28,MMM,NNN

4 PAOZZ 6680014667238 19207 12460314   GAGE ROD-CAP,LIQUID................ 1   UOC:BVY,B20,B24,B25,C17,NNN

4 PAOZZ 6680011794349 34623 5582843   GAGE ROD-CAP,LIQUID PART OF KIT P/N   1 5705693............................   UOC:AVY,A11,A13,A14,A20,A24,A25,A26, A27,B16,B17,B18,HVY,H11,H13,H14,H16, H17,H18,H20,H21,H24,H25,H26,H27,H28,   MMM

5 PAOZZ 4710011794346 19207 12339147-2   TUBE,BENT,METALLIC PART OF KIT P/N   1 5705693............................   UOC:AVY,A11,A13,A14,A20,A24,A25,A26, A27,BVY,B16,B17,B18,B20,B24,B25,C17, HVY,H11,H13,H14,H16,H17,H18,H20,H21, H24,H25,H26,H27,H28,MMM,NNN

6 PAOZZ 5330011846492 02697 2-113N497-70   .PACKING,PREFORMED................. 1   UOC: AVY,A11,A13,A14,A20,A24,A25,A26, A27,BVY,B16,B17,B18,B20,B24,B25,C17, HVY,H11,H13,H14,H16,H17,H18,H20,H21, H24,H25,H26,H27,H28,MMM,NNN

7 PAOZZ 4730011953803 01989 1069X4X1   ELBOW,PIPE TO HOSE PART OF KIT P/N   1   5705693............................   UOC:AVY,A11,A13,A14,A20,A24, A25,A26, A27,BVY,B16,B17,B18,B20,B24,B25,C17, HVY,H11,H13,H14,H16,H17,H18,H20,H21, H24,H25,H26,H27,H28,MMM,NNN

8 MOOZZ   34623 5582644   TUBE MAKE FROM TUBE, P/N CPR104420-1, 20 INCHES LONG PART OF KIT P/N   5705693............................ UOC:AVY,A11,A13,A14,A20,A24,A25,A26,

SECTION II   TM9-2320-280-24P

(1) NO	(2) SMR CODE	(3) NSN	(4) CAGEC	(5) PART NUMBER	(6) DESCRIPTION AND USABLE ON CODES(UOC)	(7) QTY
9	PAOZZ	5340000797837	96906	MS21333-67	CLAMP,LOOP PART OF KIT P/N 5705693.. UOC:AVY,A11,A13,A14,A20,A24,A25,A26, A27,BVY,B16,B17,B18,B20,B24,B25,C17, HVY,H11,H13,H14,H16,H17,H18,H20,H21, H24,H25,H26,H27,H28,MMM,NNN	1
10	PAOZZ	5310001249265	72582	271169	NUT,PLAIN,ASSEMBLED PART OF KIT P/N 5705693............................. UOC:AVY,A11,A13,A14,A20,A24,A25,A26, A27,BVY,B16,B17,B18,B20,B24,B25,C17, HVY,H11,H13,H14,H16,H17,H18,H20,H21, H24,H25,H26,H27,H28,MMM,NNN	1
11	PAOZZ	5340008091490	96906	MS21333-98	CLAMP,LOOP PART OF KIT P/N 5705693.. UOC:AVY,A11,A13,A14,A20,A24,A25,A26, A27,BVY,B16,B17,B18,B20,B24,B25,C17, HVY,H11,H13,H14,H16,H17,H18,H20,H21, H24,H25,H26,H27,H28,MMM,NNN	1
12	PAOZZ	5310011023270	24617	2436161	WASHER,FLAT 1/4 PART OF KIT P/N 5705693............................. UOC:AVY,A11,A13,A14,A20,A24,A25,A26, A27,BVY,B16,B17,B18,B20,B24,B25,C17, HVY,H11,H13,H14,H16,H17,H18,H20,H21, H24,H25,H26,H27,H28,MMM,NNN	1
13	PAOZZ	5310000614650	96906	MS51943-31	NUT,SELF-LOCKING,HE 1/4-20 PART OF KIT P/N 5705693...................... UOC:AVY,A11,A13,A14,A20,A24,A25,A26, A27,BVY,B16,B17,B18,B20,B24,B25,C17, HVY,H11,H13,H14,H16,H17,H18,H20,H21, H24,H25,H26,H27,H28,MMM,NNN	1
14	PAOZZ	5305000680502	96906	MS90725-6	SCREW,CAP,HEXAGON H 1/4-20 X 4 PART OF KIT P/N 5705693............. UOC:AVY,A11,A13,A14,A20,A24,A25,A26, A27,BVY,B16,B17,B18,B20,B24,B25,C17, HVY,H11,H13,H14,H16,H17,H18,H20,H21, H24,H25,H26,H27,H28,MMM,NNN	1
KIT	PDOZZ	2590013195435	34623	5705693	MODIFICATION KIT,VE................. UOC:AVY,A11,A13,A14,A20,A24,A25,A26, A27,BVY,B16,B17,B18,B20,B24,B25,C17, HVY,H11,H13,H14,H16,H17,H18,H20,H21, H24,H25,H26,H27,H28,MMM,NNN	1

```
ADAPTER,STRAIGHT,PI( 1) 399-4
ADAPTER,STRAIGHT,PI( 1) 399-9
BRACKET,MOUNTING   ( 1) 398-5
BRACKET,PIPE       ( 1) 398-18
BUSHING,NONMETALLIC( 2) 398-20
CAP,FILLER OPENING ( 1) 399-11
CAP,PROTECTIVE,DUST( 1) 398-28
CLAMP,HOSE         ( 2) 398-31
CLAMP,HOSE         ( 4) 399-16
CLAMP,LOOP         ( 1) 400-9
CLAMP,LOOP         ( 1) 400-11
```

SECTION II   TM9-2320-280-24P

(1) NO	(2) CODE	(3) NSN	(4) CAGEC	(5) NUMBER	(6)	(7) DESCRIPTION AND USABLE ON CODES(UOC)	ITEM SMR	PART QTY
						CLAMP,LOOP	(2)	398-11
						CUP,SENSOR,DEEP WAT	(1)	399-12
						ELBOW,PIPE TO HOSE	(1)	399-3
						ELBOW,PIPE TO HOSE	(1)	399-20
						ELBOW,PIPE TO HOSE	(1)	400-7
						GAGE ROD-CAP,LIQUID	(1)	400-3
						GAGE ROD-CAP,LIQUID	(1)	400-4
						GAGE ROD-CAP,LIQUID	(1)	400-4
						GASKET	(1)	398-24
						HOSE	(1)	399-19
						HOSE,NONMETALLIC	(1)	399-17
						HOSE,PREFORMED	(1)	398-32
						NUT,PLAIN,ASSEMBLED	(1)	400-10
						NUT,PLAIN,HEXAGON	(1)	398-1
						NUT,PLAIN,BLIND RIV	(1)	398-6
						NUT,PLAIN,BLIND RIV	(1)	398-7
						NUT,SELF-LOCKING,HE	(1)	400-13
						NUT,SELF-LOCKING,HE	(1)	398-8
						NUT,SELF-LOCKING,HE	(4)	399-15
						NUT,SELF-LOCKING,HE	(5)	398-25
						NUT,SELF-LOCKING,HE	(8)	398-27
						PIPE,EXHAUST	(1)	398-19
						PLATE,IDENTIFICATIO	(1)	399-1
						SCREW,CAP,HEXAGON H	(1)	398-3
						SCREW,CAP,HEXAGON H	(2)	398-10
						SCREW,CAP,HEXAGON H	(2)	399-13
						SCREW,CAP,HEXAGON H	(2)	399-21
						SCREW,CAP,HEXAGON H	(1)	400-14
						SCREW,CAP,HEXAGON H	(8)	398-16
						SCREW,CAP,HEXAGON H	(3)	398-22
						SCREW,MACHINE	(1)	398-15
						SCREW,TAPPING,THREA	(1)	398-13
						STACK ASSEMBLY	(1)	398-2
						STRAP,LINE SUPPORTI	(3)	399-8
						TEE,HOSE	(1)	399-6
						TUBE	(1)	398-29
						TUBE	(1)	400-8
						TUBE	(1)	399-10
						TUBE	(1)	399-7
						TUBE ASSEMBLY,METAL	(1)	400-1
						TUBE,BENT,METALLIC	(1)	400-5
						TUBE,BENT,METALLIC	(1)	398-14
						TUBE,NONMETALLIC	(1)	399-5
						TUBE,NONMETALLIC	(1)	398-30
						VALVE,CROSS	(1)	399-2
						VALVE,VENT	(1)	399-18
						WASHER,FLAT	(2)	400-12
						WASHER,FLAT	(2)	398-21
						WASHER,FLAT	(2)	398-9
						WASHER,FLAT	(2)	398-4
						WASHER,FLAT	(8)	399-14
						WASHER,FLAT	(6)	398-23
						WASHER,FLAT	(2)	398-26

(1) NO	(2) CODE	(3) NSN	(4) CAGEC	(5) NUMBER	(6) PART	(7) ITEM SMR DESCRIPTION AND USABLE ON CODES(UOC)	QTY
						WASHER,FLAT	(16) 398-17
						WASHER,LOCK	(2) 398-12

END OF FIGURE

SECTION II   TM9-2320-280-24P

Section II. TM 9-2320-280-24P-2

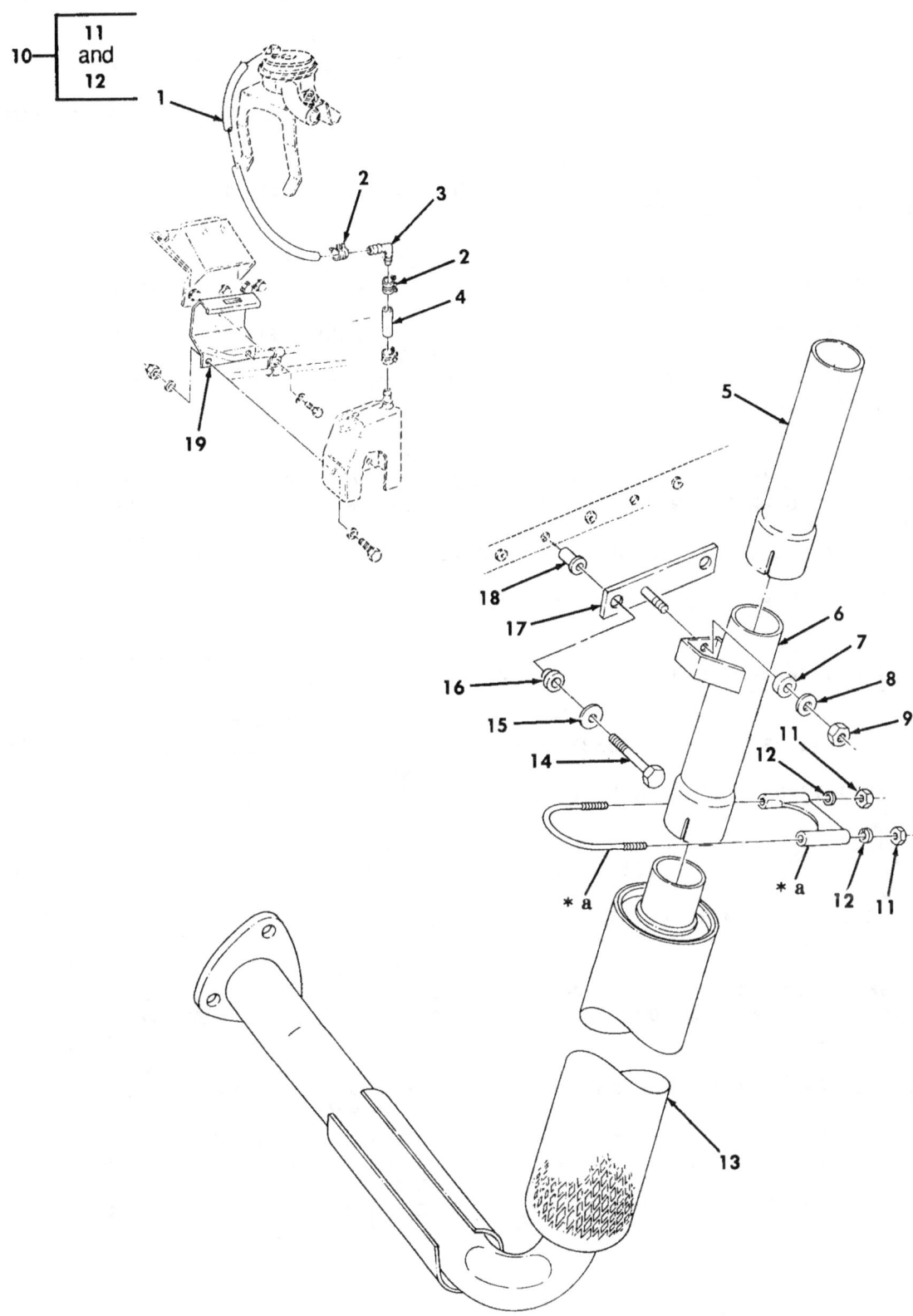

* a PART OF ITEM 10

Figure 401. Supplemental Deep Water Fording Kit.

SECTION II TM9-2320-280-24P C01

(1) ITEM NO	(2) SMR CODE	(3) NSN	(4) CAGEC	(5) PART NUMBER	(6) DESCRIPTION AND USABLE ON CODES (UOC)	(7) QTY
					GROUP 3305 DEEP WATER FORDING KITS	
					FIG. 401 SUPPLEMENTAL DEEP WATER FORDING KIT	
* 1	MOOZZ		19207	12339981-5	HOSE,NONMETALLIC MAKE FROM HOSE, P/N 4219-3681, 42 INCHES LONG PART OF KIT P/N 5705703.................... UOC:A15,B15,H15	1
2	PAOZA	4730002678945	81348	WW-C-440B TYPE E 1/2	CLAMP,HOSE PART OF KIT P/N 5705703.. UOC:A15,B15,H15	2
3	PAOZZ	4730012984363	61424	P4EUB-4	ELBOW,HOSE PART OF KIT P/N 5705703.. UOC:A15,B15,H15	1
* 4	MOOZZ		19207	12339981-6	HOSE,NONMETALLIC MAKE FROM HOSE,P/N 4219-3681, 2 INCHES LONG PART OF KIT P/N 5705703........................ UOC:A15,B15,H15	1
5	PAOZZ	2990013858988	19207	12342567-2	PIPE,EXHAUST PART OF KIT P/N 5705703 UOC:A15,B15,H15	1
6	PAOZZ	2990013828796	19207	12342566	PIPE,EXHAUST PART OF KIT P/N 5705703 UOC:A15,B15,H15	1
7	PAOZZ	5365013811783	19207	12342569	SPACER,SLEEVE PART OF KIT P/N 5705703............................ UOC:A15,B15,H15	1
8	PAOZZ	5310000877493	81495	330 2000	WASHER,FLAT 5/16 PART OF KIT P/N 5705703.............. UOC:A15,B15,H15	1
9	PAOZZ	5310008140673	96906	MS51943-33	NUT,SELF-LOCKING,HE 5/16-18 PART OF KIT P/N 5705703.................... UOC:A15,B15,H15	1
10	PAOZZ	5340013832397	19207	12338343-3	CLAMP,LOOP PART OF KIT P/N 5705703.. UOC:A15,B15,H15	2
11	PAOZZ	5310007320558	96906	MS51967-8	.NUT,PLAIN,HEXAGON 3/8-16.......... UOC:A15,B15,H15	2
12	PAOZZ	5310006379541	81718	H2525M	.WASHER,LOCK 3/8.................. UOC:A15,B15,H15	2
13	PAOZZ	2990012998820	34623	12342565	PIPE,EXHAUST PART OF KIT P/N 5705703 UOC:A15,B15,H15	1
14	PAOZZ	5305000712236	96906	MS90725-15	SCREW,CAP,HEXAGON H 1/4-20 X 2.25 PART OF KIT P/N 5705703............. UOC:A15,B15,H15	2
15	PAOZZ	5310001670893	88044	AN8013-2	WASHER,FLAT PART OF KIT P/N 5705703. UOC:A15,B15,H15	2
16	PAOZZ	2990013835689	34623	12341027	HANGER,ENGINE EXHAU PART OF KIT P/N 5705703......... .... UOC:A15,B15,H15	2
17	PAOZZ	5340013812045	19207	12342795	BRACKET,MOUNTING PART OF KIT P/N 5705703............................ UOC:A15,B15,H15	1
18	PAOZZ	4931012028692	19207	12339998-13	PLUSNUT PART OF KIT P/N 5705703..... UOC:A15,B15,H15	2
19	PAOZZ	5340013811266	19207	12341641	STRAP,RETAINING PART OF KIT P/N 5705703............................	1

SECTION II   TM9-2320-280-24P

(1) NO	(2) SMR CODE	(3) NSN	(4) CAGEC	(5) PART NUMBER	(6) DESCRIPTION AND USABLE ON CODES(UOC)	(7) QTY
					UOC:A15,B15,H15	
KIT	PDOZZ	2540013208731	19207	5705703	MODIFICATION KIT,VE.................	1
					UOC:A15,B15,H15	
					BRACKET,MOUNTING (1) 401-17	
					CLAMP,HOSE (2) 401-2	
					CLAMP,LOOP (2) 401-10	
					ELBOW,HOSE (1) 401-3	
					HANGER,ENGINE EXHAU(2) 401-16	
					HOSE,NONMETALLIC (1) 401-1	
					HOSE,NONMETALLIC (1) 401-4	
					NUT,SELF-LOCKING,HE(1) 401-9	
					PIPE,EXHAUST (1) 401-5	
					PIPE,EXHAUST (1) 401-6	
					PIPE,EXHAUST (1) 401-13	
					PLUSNUT (2) 401-18	
					SCREW,CAP,HEXAGON H(2) 401-14	
					SPACER,SLEEVE (1) 401-7	
					STRAP,RETAINING (1) 401-19	
					WASHER,FLAT (1) 401-8	
					WASHER,FLAT (2) 401-15	

END OF FIGURE

Section II.   TM 9-2320-280-24P-2

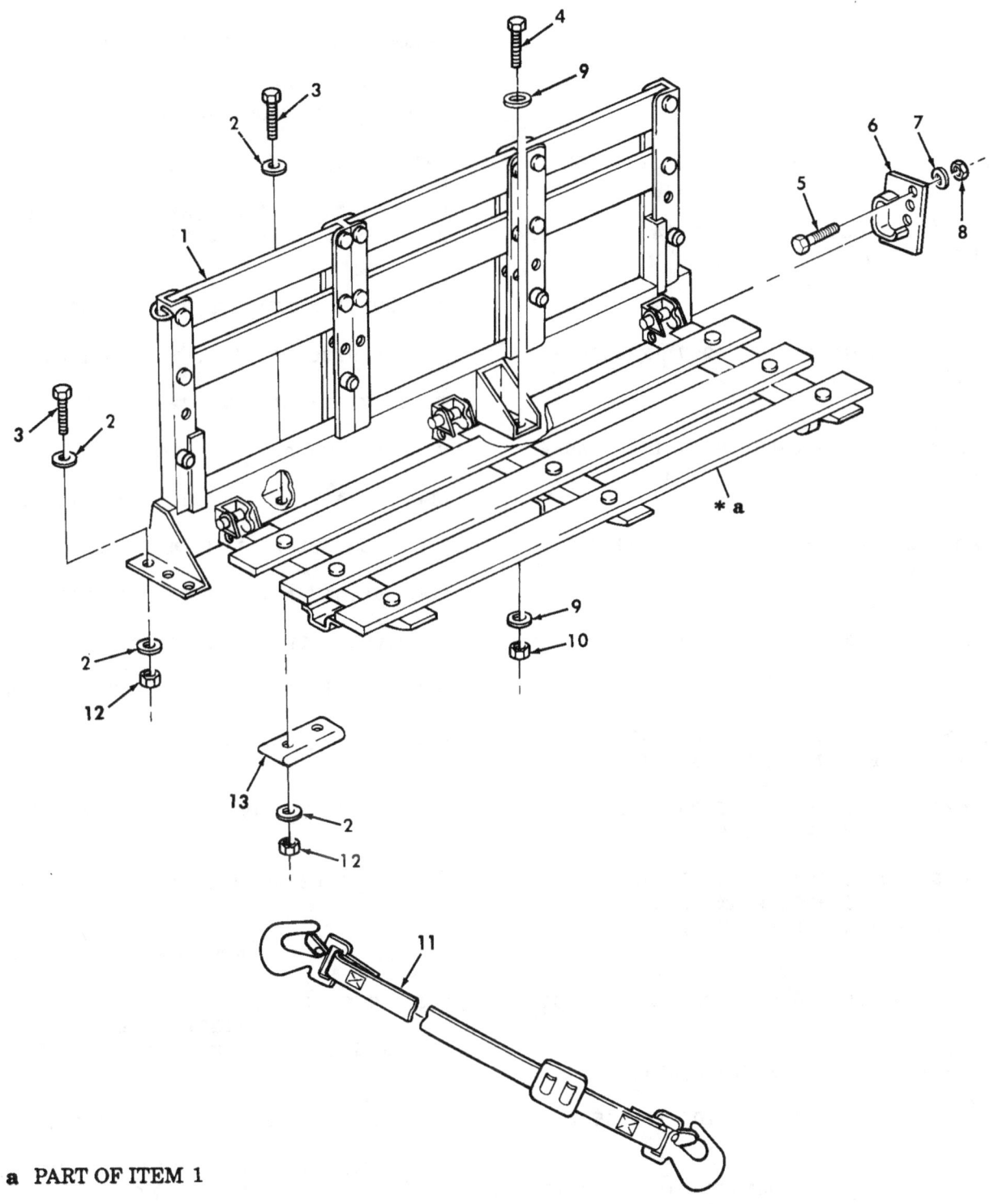

* a PART OF ITEM 1

Figure 402. Troop Seat Kit, Safety Strap, and Mounting Hardware.

SECTION II TM9-2320-280-24P C01

(1) ITEM NO	(2) SMR CODE	(3) NSN	(4) CAGEC	(5) PART NUMBER	(6) DESCRIPTION AND USABLE ON CODES (UOC)	(7) QTY
1	PAOOO	2540011858016	19207	12340560-1	SEAT,VEHICULAR L.H.,GREEN,PART OF KIT P/N 12338105.................... UOC:AVY,A13,A14,BVY,HPM,HVY,H13,H14,KTV,NNN	1
1	PAOOO	2540011853214	19207	12340560-2	SEAT,VEHICULAR R.H.,GREEN,PART OF KIT P/N 12338105.................... UOC:AVY,A13,A14,BVY,HPM,HVY,H13,H14,KTV,NNN	1
1	PAOOO	2540013967759	19207	12340560-3	SEAT,VEHICUALR L.H.,TAN,PART OF KIT P/N 57K0216....................... UOC:AVY,A13,A14,BVY,HPM,HVY,H13,H14,KTV,NNN	1
1	PAOOO	2540013990785	19207	12340560-4	SEAT,VEHICULAR R.H.,TAN,PART OF KIT P/N 57K0216....................... UOC:AVY,A13,A14,BVY,HPM,HVY,H13,H14,KTV,NNN	1
1	PAOOO	2540013967756	19207	12340560-5	SEAT,VEHICULAR L.H.,WHITE,PART OF KIT P/N 57K0220..................... UOC:AVY,A13,A14,BVY,HPM,HVY,H13,H14,KTV,NNN	1
1	PAOOO	2540013967757	19207	12340560-6	SEAT,VEHICULAR R.H.,WHITE,PART OF KIT P/N 57K0220..................... UOC:AVY,A13,A14,BVY,HPM,HVY,H13,H14,KTV,NNN	1
2	PAOZZ	5310008094058	96906	MS27183-10	WASHER,FLAT 1/4................... UOC:AVY,A13,A14,BVY,HPM,HVY,H13,H14,KTV,NNN	36
3	PAOZZ	5305000180024	96906	MS35297-10	SCREW,CAP,HEXAGON H 1/4-20 X 1.25.. UOC:AVY,A13,A14,BVY,HPM,HVY,H13,H14,KTV,NNN	18
* 4	PAOZZ	5305013157066	24617	9425339	SCREW,CAP,HEXAGON H 7/16-20 X 3/4, P/N PART OF KIT P/N 12338105,PART OF KIT 57K0216,PART OF KIT P/N 57K0220. UOC:AVY,A13,A14,BVY,HPM,HVY,H13,H14,KTV,NNN	4
5	PAOZZ	5306002258499	96906	MS90725-34	BOLT,MACHINE 5/16-18 X 1.00,PART OF KIT P/N 12338105,PART OF KIT P/N 57K0216,PART OF KIT P/N 57K0220..... UOC:AVY,A13,A14,BVY,HPM,HVY,H13,H14,KTV,NNN	6
6	PAOZZ	2540013949682	19207	12340542-1	SUPPORT,SEAT,VEHICU L.H.,GREEN,PART OF KIT P/N 12338105................ UOC:AVY,A13,A14,BVY,HPM,HVY,H13,H14,KTV,NNN	1
6	PAOZZ	2540013949683	19207	12340542-2	SUPPORT,SEAT,VEHI- CU R.H.,GREEN,PART OF KIT P/N 12338105................ UOC:AVY,A13,A14,BVY,HPM,HVY,H13,H14,KTV,NNN	1

GROUP 3307 SPECIAL PURPOSE KITS

FIG. 402 TROOP SEAT KIT,SAFETY STRAP, AND MOUNTING HARDWARE

SECTION II   TM9-2320-280-24P

(1) ITEM NO	(2) SMR CODE	(3) NSN	(4) CAGEC	(5) PART NUMBER	(6) DESCRIPTION AND USABLE ON CODES(UOC)	(7) QTY
6	PAOZZ	2540013944788	19207	12340542-3	SUPPORT,SEAT,VEHICU L.H.,TAN,PART OF KIT P/N 57K0216................... UOC:AVY,A13,A14,BVY,HPM,HVY,H13,H14,KTV,NNN	1
6	PAOZZ	2540013944787	19207	12340542-4	SUPPORT,SEAT,VEHICU R.H.,TAN,PART OF KIT P/N 57K0216................... UOC:AVY,A13,A14,BVY,HPM,HVY,H13,H14,KTV,NNN	1
6	PAOZZ	2540013944792	19207	12340542-5	SUPPORT,SEAT,VEHICU L.H.,WHITE,PART OF KIT P/N 57K0220................... UOC:AVY,A13,A14,BVY,HPM,HVY,H13,H14,KTV,NNN	1
6	PAOZZ	2540013949679	19207	12340542-6	SUPPORT,SEAT,VEHICU R.H.,WHITE,PART OF KIT P/N 57K0220................... UOC:AVY,A13,A14,BVY,HPM,HVY,H13,H14,KTV,NNN	1
7	PAOZZ	5310000814219	96906	MS27183-12	WASHER,FLAT 5/16.................. UOC:AVY,A13,A14,BVY,HPM,HVY,H13,H14,KTV,NNN	6
8	PAOZZ	5310002416658	96906	MS51943-34	NUT,SELF-LOCKING,HE 5/16-18,PART OF KIT P/N 12338105,PART OF KIT P/N 57K0216,PART OF KIT P/N 57K0220..... UOC:AVY,A13,A14,BVY,HPM,HVY,H13,H14,KTV,NNN	6
9	PAOZZ	5310008094085	96906	MS27183-16	WASHER,FLAT 7/16,PART OF KIT P/N 12338105,PART OF KIT P/N 57K0216, PART OF KIT P/N 57K0220............. UOC:AVY,A13,A14,BVY,HPM,HVY,H13,H14,KTV,NNN	8
10	PAOZZ	5310013153403	24617	9422300	NUT,PLAIN,HEXAGON 7/16-20,PART OF KIT P/N 12338105,PART OF KIT P/N 57K0216,PART OF KIT P/N 57K0220..... UOC:AVY,A13,A14,BVY,HPM,HVY,H13,H14,KTV,NNN	4
11	PAOZZ	5340011147712	19207	11682088-1	STRAP,WEBBING PART OF KIT P/N 12338105,PART OF KIT P/N 57K0216, PART OF KIT P/N 57K0220............. UOC:AVY,A13,A14,BVY,HPM,HVY,H13,H14,KTV,NNN	1
12	PAOZZ	5310000614650	96906	MS51943-31	NUT,SELF-LOCKING,HE 1/4-20,PART OF KIT P/N 12338105,PART OF KIT P/N 57K0216,PART OF KIT P/N 57K0220..... UOC:AVY,A13,A14,BVY,HPM,HVY,H13,H14,KTV,NNN	8
13	PFOZZ	5340012559510	19207	12339009	PLATE,MENDING....................... UOC:AVY,A13,A14,BVY,HPM,HVY,H13,H14,KTV,NNN	6
KIT	PDOZZ	2540011853216	19207	12338105	PARTS KIT,SEAT GREEN............... UOC:AVY,A13,A14,BVY,HPM,HVY,H13,H14,KTV,NNN	1

```
BOLT,MACHINE       ( 6) 402-5
NUT,PLAIN,HEXAGON  ( 4) 402-10
NUT,SELF-LOCKING,HE( 6) 402-8
```

SECTION II   TM9-2320-280-24P

(1) NO	(2) CODE	(3) NSN	(4) CAGEC	(5) NUMBER	(6) PART	(7) ITEM SMR DESCRIPTION AND USABLE ON CODES(UOC)	QTY

```
                    NUT,SELF-LOCKING,HE( 18) 402-12
                    SCREW,CAP,HEXAGON H(  4) 402-4
                    SEAT,VEHICULAR      ( 1) 402-1
                    SEAT,VEHICULAR      ( 1) 402-1
                    STRAP,WEBBING       ( 1) 402-11
                    SUPPORT,SEAT,VEHICU( 1) 402-6
                    SUPPORT,SEAT,VEHICU( 1) 402-6
                    WASHER,FLAT         ( 8) 402-9
KIT PDOZZ 2540013388081 19207 57K0216    PARTS KIT,SEAT  TAN................. 1
                    UOC:AVY,A13,A14,BVY,HPM,HVY,H13,H14,                        KTV,NNN
                    BOLT,MACHINE        ( 6) 402-5
                    NUT,PLAIN,HEXAGON   ( 4) 402-10
                    NUT,SELF-LOCKING,HE( 6) 402-8
                    NUT,SELF-LOCKING,HE( 18) 402-12
                    SCREW,CAP,HEXAGON H(  4) 402-4
                    SEAT,VEHICULAR      ( 1) 402-1
                    SEAT,VEHICULAR      ( 1) 402-1
                    STRAP,WEBBING       ( 1) 402-11
                    SUPPORT,SEAT,VEHICU( 1) 402-6
                    SUPPORT,SEAT,VEHICU( 1) 402-6
                    WASHER,FLAT         ( 8) 402-9
KIT PDOZZ 2540013958771 19207 57K0220    PARTS KIT,SEAT  WHITE............... 1
                    UOC:AVY,A13,A14,BVY,HPM,HVY,H13,H14,                        KTV,NNN
                    BOLT,MACHINE        ( 6) 402-5
                    NUT,PLAIN,HEXAGON   ( 4) 402-10
                    NUT,SELF-LOCKING,HE( 6) 402-8
                    NUT,SELF-LOCKING,HE( 18) 402-12
                    SCREW,CAP,HEXAGON H(  4) 402-4
                    SEAT,VEHICULAR      ( 1) 402-1
                    SEAT,VEHICULAR      ( 1) 402-1
                    STRAP,WEBBING       ( 1) 402-11
                    SUPPORT,SEAT,VEHICU( 1) 402-6
                    SUPPORT,SEAT,VEHICU( 1) 402-6
                    WASHER,FLAT         ( 8) 402-9
```

END OF FIGURE

Section II. TM 9-2320-280-24P-2

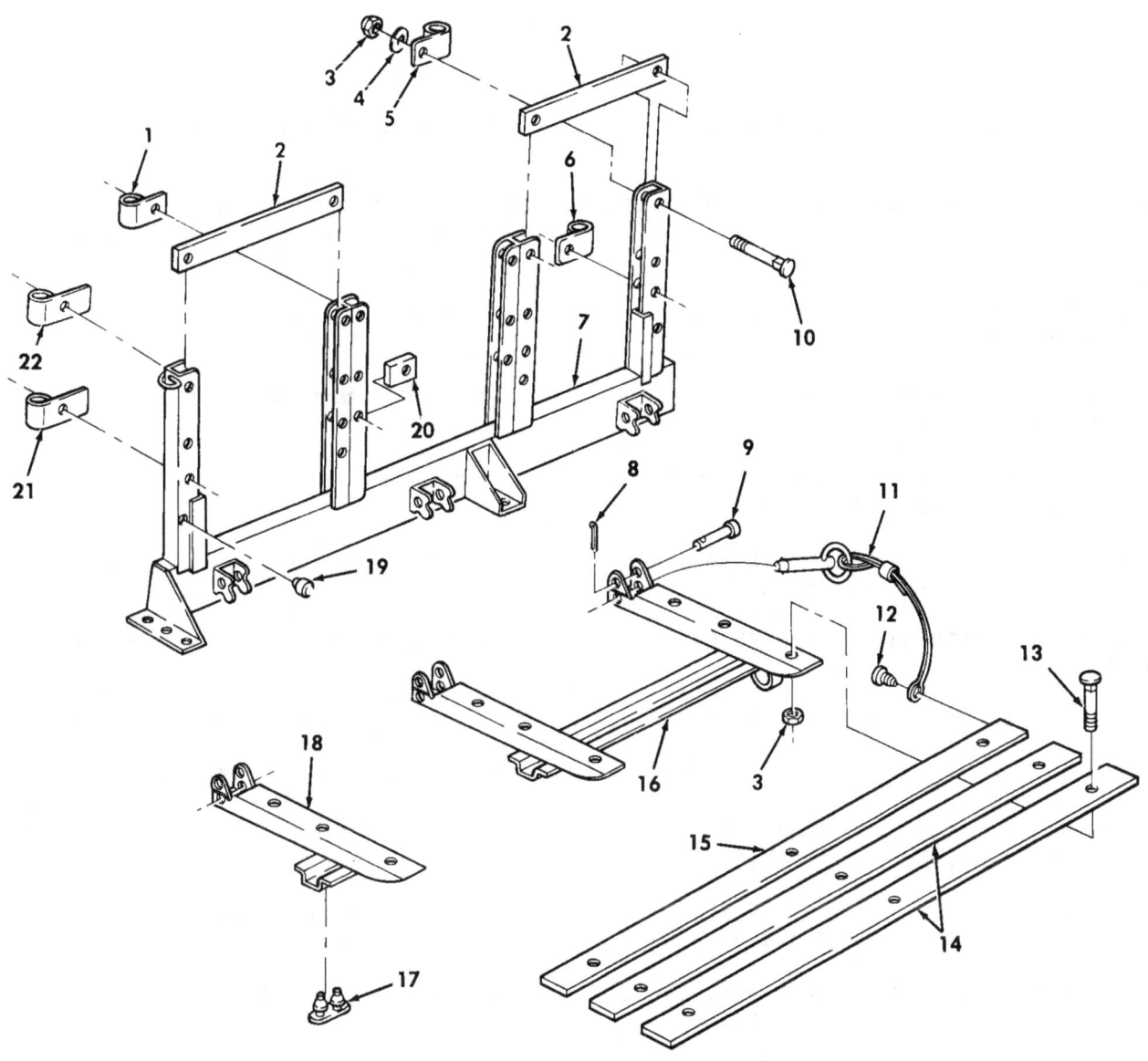

Figure 403. Troop Seat Assembly, Left and Right.

SECTION II   TM9-2320-280-24P C01

(1) ITEM NO	(2) SMR CODE	(3) NSN	(4) CAGEC	(5) PART NUMBER	(6) DESCRIPTION AND USABLE ON CODES (UOC)	(7) QTY
					GROUP 3307 SPECIAL PURPOSE KITS	
					FIG. 403 TROOP SEAT ASSEMBLY, LEFT AND RIGHT	
1	PAOZZ	2540011929712	34623	5578466	RETAINER ASSEMBLY,B USE ON L.H. SIDE RACK ASSY..................... UOC:AVY,A13,A14,BVY,HPM,HVY,H13,H14, KTV,NNN	2
1	PAOZZ	2540011929713	19207	12340545-2	RETAINER ASSEMBLY,B USE ON R.H. SIDE RACK ASSY..................... UOC:AVY,A13,A14,BVY,HPM,HVY,H13,H14, KTV,NNN	2
2	PAOZZ	2540011857951	19207	12340544	BOARD SEAT .... UOC:AVY,A13,A14,BVY,HPM,HVY,H13,H14, KT-V,NNN	12
3	PAOZZ	5310009334310	24617	271184	NUT,PLAIN ASSEMBLED 5/16-18........ UOC:AVY,A13,A14,BVY,HPM,HVY,H13,H14, KT-V,NNN	50
* 4	PAOZZ	5310011191024	24617	2436162	WASHER,FLAT 5/16................... UOC:AVY,A13,A14,BVY,HPM,HVY,H13,H14, KT-V,NNN	32
5	PFOZZ	2540011929711	19207	12340546-1	RETAINER,BOW,VEHICU................ UOC:AVY,A13,A14,BVY,HPM,HVY,H13,H14, KT-V,NNN	3
6	PAOZZ	2540011929712	34623	5578466	RETAINER ASSEMBLY,B USE ON R.H. SIDE RACK ASSY..................... UOC:AVY,A13,A14,BVY,HPM,HVY,H13,H14, KTV,NNN	1
6	PAOZZ	2540011929713	19207	12340545-2	RETAINER ASSEMBLY,B USE ON L.H. SIDE RACK ASSY..................... UOC:AVY,A13,A14,BVY,HPM,HVY,H13,H14, KTV,NNN	1
7	XAOZZ		34623	5591333	CHANNEL ASSEMBLY R.H.............. UOC:AVY,A13,A14,BVY,HPM,HVY,H13,H14, KT-V,NNN	1
7	XAOZZ		34623	5591334	CHANNEL ASSEMBLY L.H.............. UOC:AVY,A13,A14,BVY,HPM,HVY,H13,H14, KT-V,NNN	1
8	PAOZZ	5315012677570	96906	MS24665-319	PIN,COTTER 3/32 X 1.00............ UOC:AVY,A13,A14,BVY,HPM,HVY,H13,H14, KT-V,NNN	6
9	PAOZZ	5315007370134	19207	7370134	PIN,STRAIGHT,HEADED 3/8 X 2.00..... UOC:AVY,A13,A14,BVY,HPM,HVY,H13,H14, KT-V,NNN	6
10	PAOZZ	5306000120231	96906	MS35751-44	BOLT,SQUARE NECK 5/16-18 X 1.75.... UOC:AVY,A13,A14,BVY,HPM,HVY,H13,H14, KT-V,NNN	32
11	PAOZZ	5315010949025	19207	12255608	PIN,QUICK RELEASE.................. UOC:AVY,A13,A14,BVY,HPM,HVY,H13,H14, KT-V,NNN	2
12	PAOZZ	5305004324170	96906	MS51861-35	SCREW,TAPPING #8-18 X 1/2.......... UOC:AVY,A13,A14,BVY,HPM,HVY,H13,H14,	2

SECTION II   TM9-2320-280-24P

(1) NO	(2) SMR CODE	(3) NSN	(4) CAGEC	(5) PART NUMBER	(7) DESCRIPTION AND USABLE ON CODES(UOC)	ITEM QTY
13	PAOZZ	5306007536996	96906	MS35751-43	BOLT,SQUARE NECK 5/16-18 X 50..... UOC:AVY,A13,A14,BVY,HPM,HVY,H13,H14,KTV,NNN	18
14	PAOZZ	2540011858015	19207	12340553	BOARD SEAT UOC:AVY,A13,A14,BVY,HPM,HVY,H13,H14,KTV,NNN	4
15	PFOZZ	2540011924501	19207	12340549	SEAT,VEHICULAR..................... UOC:AVY,A13,A14,BVY,HPM,HVY,H13,H14,KTV,NNN	2
16	XAOZZ		19207	12340562-1	CHANNEL ASSEMBLY L.H................ UOC:AVY,A13,A14,BVY,HPM,HVY,H13,H14,KTV,NNN	1
16	XAOZZ		19207	12340562-2	CHANNEL ASSEMBLY R.H............... UOC:AVY,A13,A14,BVY,HPM,HVY,H13,H14,KTV,NNN	1
17	PAOZZ	5340011968028	19207	12339961-1	BUMPER,NONMETALLIC.................. UOC:AVY,A13,A14,BVY,HPM,HVY,H13,H14,KTV,NNN	6
18	PFOZZ	2540013344333	34623	5588656	SUPPORT,SEAT,VEHICU REAR........... UOC:AVY,A13,A14,BVY,HPM,HVY,H13,H14,KTV,NNN	2
19	PAOZZ	5340011962573	34623	5584977	BUMPER,NONMETALLIC.................. UOC:AVY,A13,A14,BVY,HPM,HVY,H13,H14,KTV,NNN	4
20	PAOZZ	5365011934479	34623	5584999	SPACER............................ UOC:AVY,A13,A14,BVY,HPM,HVY,H13,H14,KTV,NNN	4
21	PFOZZ	2540011929715	19207	12340535-1	RETAINER ASSEMBLY,B L.H............ UOC:AVY,A13,A14,BVY,HPM,HVY,H13,H14,KTV,NNN	1
21	PFOZZ	2540011929714	19207	12340535-2	RETAINER ASSEMBLY,B R.H............ UOC:AVY,A13,A14,BVY,HPM,HVY,H13,H14,KTV,NNN	1
22	PAOZZ	2540011932711	19207	12340536-1	RETAINER,BOW,VEHICU................. UOC:AVY,A13,A14,BVY,HPM,HVY,H13,H14,KTV,NNN	1

END OF FIGURE

Section II. TM 9-2320-280-24P-2 C01

DELETED

Figure 404. Alternator, 100 AMP Kit (PRESTOLITE).

```
      SECTION II          TM9-2320-280-24P C01
    (1)    (2)    (3)         (4)       (5)                      (6)
(7)  ITEM   SMR                         PART
    NO    CODE   NSN         CAGEC     NUMBER    DESCRIPTION AND USABLE ON CODES(UOC)
QTY

                                                 GROUP 3307 SPECIAL PURPOSE KITS

                                                 FIG. 404 ALTERNATOR,100 AMP KIT
```

DELETED

Section II. TM 9-2320-280-24P-2 C01

DELETED

Figure 405. Alternator, 100 AMP Kit (NIEHOFF).

(1) ITEM NO	(2) SMR CODE	(3) NSN	(4) CAGEC	(5) PART NUMBER	(6) DESCRIPTION AND USABLE ON CODES (UOC)	(7) QTY

SECTION II   TM9-2320-280-24P C01

GROUP 3307 SPECIAL PURPOSE KITS

FIG. 405 ALTERNATOR, 100 AMP KIT

DELETED

Section II. TM 9-2320-280-24P-2 C01

DELETED

Figure 406. Alternator, 100 AMP (PRESTOLITE).

```
     SECTION II         TM9-2320-280-24P C01
   (1)    (2)    (3)         (4)       (5)                        (6)
(7)    ITEM   SMR                             PART
   NO   CODE   NSN         CAGEC      NUMBER       DESCRIPTION AND USABLE ON CODES(UOC)
QTY
```

GROUP 3307 SPECIAL PURPOSE KITS

FIG. 406 ALTERNATOR,100 AMP KIT

DELETED

Section II. TM 9-2320-280-24P-2 C01

DELETED

Figure 407. Alternator, 100 AMP (NIEHOFF).

```
    SECTION II        TM9-2320-280-24P C01
  (1)    (2)    (3)       (4)      (5)                    (6)
(7)    ITEM   SMR                        PART
  NO   CODE   NSN        CAGEC    NUMBER    DESCRIPTION AND USABLE ON CODES(UOC)
QTY
```

GROUP 3307 SPECIAL PURPOSE KITS

FIG. 407 ALTERNATOR,100 AMP

DELETED

Section II. TM 9-2320-280-24P-2 C01

DELETED

Figure 408. Alternator, 200 AMP Kit.

(1) ITEM NO	(2) SMR CODE	(3) NSN	(4) CAGEC	(5) PART NUMBER	(6) DESCRIPTION AND USABLE ON CODES (UOC)	(7) QTY

SECTION II   TM9-2320-280-24P C01

GROUP 3307 SPECIAL PURPOSE KITS

DELETED

Section II. TM 9-2320-280-24P-2

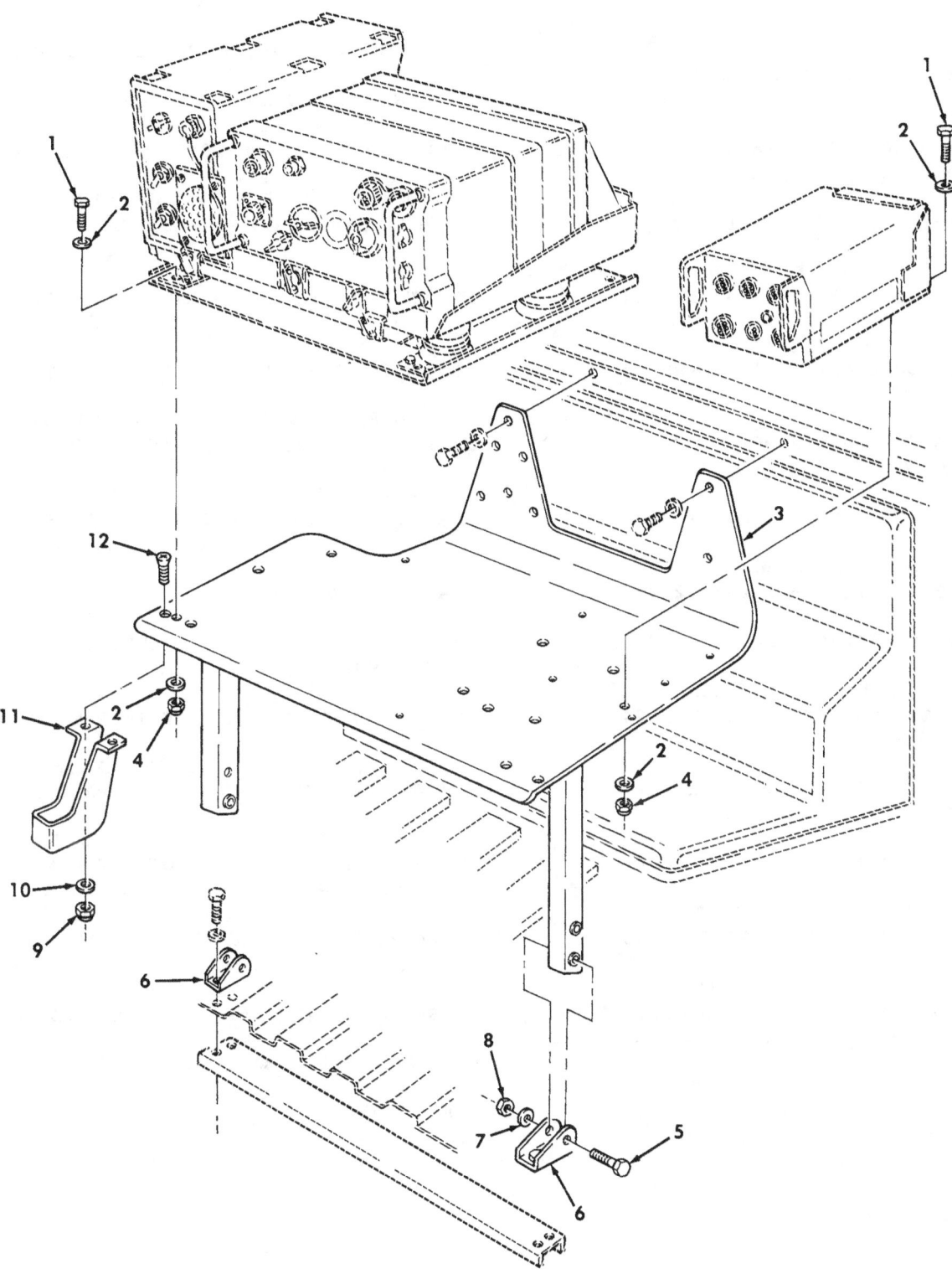

Figure 409. Communications Kit, Radio Rack Assembly, and Mounting Hardware, AN/GRC-160 and TSEC/KY-57.

SECTION II   TM9-2320-280-24P C01

(1) ITEM NO	(2) SMR CODE	(3) NSN	(4) CAGEC	(5) PART NUMBER	(6) DESCRIPTION AND USABLE ON CODES (UOC)	(7) QTY
					GROUP 3307 SPECIAL PURPOSE KITS	
					FIG. 409 COMMUNICATIONS KIT, RADIO AN/GRC-160 AND TSEC/KY-57	
					RACK ASSEMBLY, AND MOUNTING HARDWARE,	
1	PAOZZ	5306002258499	96906	MS90725-34	BOLT,MACHINE 5/16-18 X 1.00,PART OF KIT P/N 5590244............ UOC:AVY,A13,A14,BVY,HVY,H13,H14	9
2	PAOZZ	5310011191024	24617	2436162	WASHER,FLAT 5/16 PART OF KIT P/N 5590244............ UOC:AVY,A13,A14,BVY,HVY,H13,H14	18
3	PFOZZ	5975011975505	19207	12340167	MOUNTING BASE,ELECT PART OF KIT P/N 5590244............ UOC:AVY,A13,A14,BVY,HVY,H13,H14	1
4	PAOZZ	5310008140673	96906	MS51943-33	NUT,SELF-LOCKING,HE 5/16-18 PART OF KIT P/N 5590244............ UOC:AVY,A13,A14,BVY,HVY,H13,H14	9
5	PAOZZ	5306010684592	96906	MS35764-1303	BOLT,SELF-LOCKING 3/8-16 X 2.25 PART OF KIT P/N 5590244............ UOC:AVY,A13,A14,BVY,HVY,H13,H14	2
6	PFOZZ	2590012028546	19207	12340134	BRACKET,VEHICULAR C PART OF KIT P/N 5590244............ UOC:AVY,A13,A14,HVY,H13,H14	2
6	PAOZZ	5340014120866	19207	12446766	BRACKET,DOUBLE ANGL.......... UOC:BVY	2
7	PAOZZ	5310014124013	24617	2436163	WASHER,FLAT 3/8 PART OF KIT P/N 5590244............ UOC:AVY,A13,A14,BVY,HVY,H13,H14	2
8	PAOZZ	5310009359021	96906	MS51943-35	NUT,SELF-LOCKING,HE 3/8-16,PART OF KIT P/N 5590244............ UOC:AVY,A13,A14,BVY,HVY,H13,H14	2
9	PAOZZ	5310000614650	96906	MS51943-31	NUT,SELF-LOCKING,HE 1/4-20 PART OF KIT P/N 5590244............ UOC:AVY,A13,A14,BVY,HVY,H13,H14	2
10	PAOZZ	5310011023270	24617	2436161	WASHER,FLAT 1/4 PART OF KIT P/N 5590244............ UOC:AVY,A13,A14,BVY,HVY,H13,H14	2
11	PFOZZ	2540012028545	19207	12340131	BRACKET,HEAD PHONE PART OF KIT P/N 5590244............ UOC:AVY,A13,A14,BVY,HVY,H13,H14	1
12	PAOZZ	5305007195021	96906	MS51959-81	SCREW,MACHINE 1/4-20 X .75 PART OF KIT P/N 5590244............ UOC:AVY,A13,A14,BVY,HVY,H13,H14	2

END OF FIGURE

Section II. TM 9-2320-280-24P-2 C01

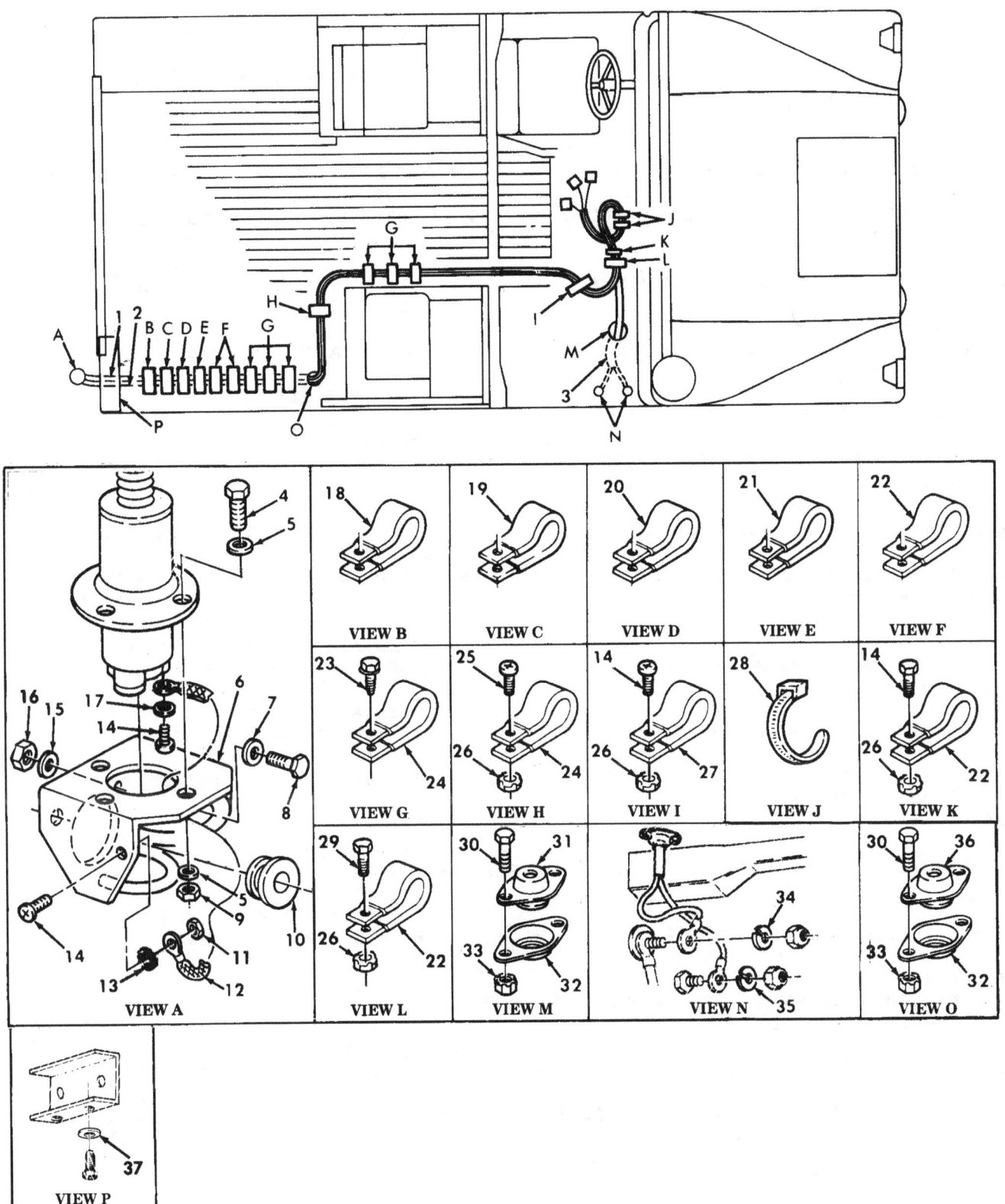

Figure 410. Communications Kit, Radio Cables, AN/GRC-160, AN/VRC-46, and TSEC/KY-57.

SECTION II   TM9-2320-280-24P C01

(1) ITEM NO	(2) SMR CODE	(3) NSN	(4) CAGEC	(5) PART NUMBER	(6) DESCRIPTION AND USABLE ON CODES (UOC)	(7) QTY
					GROUP 3307 SPECIAL PURPOSE KITS	
					FIG. 410 COMMUNICATIONS KIT, RADIO CABLES, AN/GRC-160, AN/VRC-46, AND TSEC/KY-57	
1	PAOZZ	5995012017495	19207	12338086-3	CABLE ASSEMBLY,RADI PART OF KIT P/N 5590244............................ UOC:A13,A14,H13,H14	1
2	PAOZZ	5995012014129	19207	12338089-3	CABLE ASSEMBLY,SPEC PART OF KIT P/N 5590244............. UOC:A13,A14,H13,H14	1
* 3	PAOZZ	5995012014128	34623	5577765-B	CABLE ASSEMBLY,SPEC PART OF KIT P/N 5590244............................ UOC:A13,A14,H13,H14	1
* 4	PAOZZ	5305002693216	80205	MS90725-66	SCREW,CAP,HEXAGON H 3/8-16 X 2.00, PART OF KIT P/N 5590244............. UOC:A13,A14,H13,H14	4
5	PAOZZ	5310014124013	24617	2436163	WASHER,FLAT 3/8,PART OF KIT P/N 5590244............................ UOC:A13,A14,H13,H14	8
6	PFOZZ	5985012159404	80063	A3046166	SUPPORT,ANTENNA PART OF KIT P/N 5590244......... .... UOC:A13,A14,H13,H14	1
* 7	PAOZZ	5310011191024	24617	2436162	WASHER,FLAT 5/16,PART OF KIT P/N 5590244............................ UOC:A13,A14,H13,H14	3
8	PAOZZ	5306002258499	96906	MS90725-34	BOLT,MACHINE 5/16-18 X 3/4,PART OF KIT P/N 5590244................... UOC:A13,A14,H13,H14	3
9	PAOZZ	5310009359021	96906	MS51943-35	NUT,SELF-LOCKING,HE 3/8-16,PART OF KIT P/N 5590244................... UOC:A13,A14,H13,H14	4
10	PAOZZ	5325012085424	19207	12338098	GROMMET,NONMETALLIC PART OF KIT P/N 5590244............................ UOC:A13,A14,H13,H14	1
11	PAOZZ	5310012510726	34623	5593053	NUT,SELF-LOCKING,HE #10-32,PART OF KIT P/N 5590244................... UOC:A13,A14,H13,H14	1
12	PAOZZ	5999012720018	34623	5598121	STRIP,ELECTRICAL GR PART OF KIT P/N 5590244......................... UOC:A13,A14,H13,H14	1
13	PAOZZ	5310000453296	96906	MS35338-43	WASHER,LOCK #10,PART OF KIT P/N 5590244............................ UOC:A13,A14,H13,H14	1
* 14	PAOZZ	5305000593659	96906	MS51958-63	SCREW,MACHINE #10-32 X 1/2,PART OF KIT P/N 5590244................... UOC:A13,A14,H13,H14	4
15	PFOZZ	5310012087576	19207	12338095	WASHER,FLAT 11/32,PART OF KIT P/N 5590244............................ UOC:A13,A14,H13,H14	3
16	PAOZZ	5310012531615	34623	5592958	NUT,SELF-LOCKING,HE 5/16-18,PART OF KIT P/N 5590244...................	3

SECTION II TM9-2320-280-24P C01

(1) ITEM NO	(2) SMR CODE	(3) NSN	(4) CAGEC	(5) PART NUMBER	(6) DESCRIPTION AND USABLE ON CODES (UOC)	(7) QTY
17	PAOZZ	5310000453296	96906	MS35338-43	WASHER,LOCK #10,PART OF KIT P/N 5590244. UOC:A13,A14,H13,H14	
18	PAOZZ	5340008338476	96906	MS21333-122	CLAMP,LOOP PART OF KIT P/N 5590244. UOC:A13,A14,H13,H14	1
19	PAOZZ	5340000572906	96906	MS21333-73	CLAMP,LOOP PART OF KIT P/N 5590244. UOC:A13,A14,H13,H14	1
20	PAOZZ	5340009226300	96906	MS21333-77	CLAMP,LOOP PART OF KIT P/N 5590244. UOC:A13,A14,H13,H14	1
21	PAOZZ	5340007247038	96906	MS21333-76	CLAMP,LOOP PART OF KIT P/N 5590244. UOC:A13,A14,H13,H14	1
22	PAOZZ	5340000502740	96906	MS21333-75	CLAMP,LOOP PART OF KIT P/N 5590244. UOC:A13,A14,H13,H14	4
23	PAOZZ	5305012854234	24617	9414241	SCREW,MACHINE #10-32 X .75,PART OF KIT P/N 5590244. UOC:A13,A14,H13,H14	6
24	PAOZZ	5340000572904	96906	MS21333-71	CLAMP,LOOP PART OF KIT P/N 5590244. UOC:A13,A14,H13,H14	7
25	PAOZZ	5305000593661	96906	MS51958-65	SCREW,MACHINE #10-32 X 3/4,PART OF KIT P/N 5590244. UOC:A13,A14,H13,H14	1
26	PAOZZ	5310001249265	72582	271169	NUT,PLAIN,ASSEMBLED #10-32,PART OF KIT P/N 5590244. UOC:A13,A14,H13,H14	4
27	PAOZZ	5340000886655	96906	MS21333-101	CLAMP,LOOP PART OF KIT P/N 5590244. UOC:A13,A14,H13,H14	1
28	PAOZZ	5975010345871	96906	MS3367-7-0	STRAP,TIEDOWN,ELECT PART OF KIT P/N 5590244. UOC:A13,A14,H13,H14	
29	PAOZZ	5305000711322	96906	MS51960-65	SCREW,MACHINE #10-32 X 1/2,PART OF KIT P/N 5590244. UOC:A13,A14,H13,H14	1
30	PAOZZ	5305000680502	96906	MS90725-6	SCREW,CAP,HEXAGON H 1/4-20 X 3/4, PART OF KIT P/N 5590244. UOC:A13,A14,H13,H14	4
31	PAOZZ	5325012055378	19207	12338084	GROMMET PART OF KIT P/N 5590244. UOC:A13,A14,H13,H14	1
32	PAOZZ	2590004543620	34623	809223	RETAINER ASSEMBLY PART OF KIT P/N 5590244. UOC:A13,A14,H13,H14	2
33	PAOZZ	5310011520598	24617	271172	NUT,SELF-LOCKING 1/4-20,PART OF KIT P/N 5590244. UOC:A13,A14,H13,H14	4
34	PAOZZ	5310012538440	34623	5596566	WASHER,LOCK 1/2,PART OF KIT P/N 5590244. UOC:A13,A14,H13,H14	
35	PAOZZ	5310005845272	96906	MS35338-48	WASHER,LOCK 1/2,PART OF KIT P/N 5590244. UOC:A13,A14,H13,H14	
36	PAOZZ	5325012053203	19207	12338085	GROMMET PART OF KIT P/N 5590244. UOC:A13,A14,H13,H14	1
37	PAOZZ	5310005825965	96906	MS35338-44	WASHER,LOCK 1/4,PART OF KIT P/N 5590244.	1

SECTION II   TM9-2320-280-24P C01
   (1)   (2)   (3)   (4)   (5)   (6)
(7)   ITEM   SMR   PART
   NO   CODE   NSN   CAGEC   NUMBER   DESCRIPTION AND USABLE ON CODES(UOC)
QTY

                                                5590244.........................
  .   KIT PDOZZ 5999011975465 34623 5590244   PARTS KIT,ELECTRONI RAIDO,AN/GRC-
1   160,AN/VRC-46,AND TSEC KY-57,CARGO..
UOC:A13,A14,H13,H14

                                                   BOLT,MACHINE   (9) 409-1
                                                   BOLT,MACHINE   (3) 410-8
                                                   BOLT,SELF-LOCKING   (2) 409-5
                                                   BRACKET,HEAD PHONE   (1) 409-11
                                                   BRACKET,RADIO   (2) 409-6
                                                   CABLE ASSEMBLY,RADI(1) 410-1
                                                   CABLE ASSEMBLY,SPEC(1) 410-2
                                                   CABLE ASSEMBLY,SPEC(1) 410-3
                                                   CLAMP,LOOP   (7) 410-24
                                                   CLAMP,LOOP   (1) 410-20
                                                   CLAMP,LOOP   (1) 410-19
                                                   CLAMP,LOOP   (1) 410-27
                                                   CLAMP,LOOP   (1) 410-21
                                                   CLAMP,LOOP   (4) 410-22
                                                   CLAMP,LOOP   (1) 410-18
                                                   GROMMET   (1) 410-36
                                                   GROMMET   (1) 410-31
                                                   GROMMET,NONMETALLIC(1) 410-10
                                                   MOUNTING BASE,ELECT(1) 409-3
                                                   NUT,PLAIN,ASSEMBLED(4) 410-26
                                                   NUT,SELF-LOCKING   (2) 409-9
                                                   NUT,SELF-LOCKING   (4) 410-33
                                                   NUT,SELF-LOCKING   (9) 409-4
                                                   NUT,SELF-LOCKING   (2) 409-8
                                                   NUT,SELF-LOCKING   (3) 410-16
                                                   NUT,SELF-LOCKING   (4) 410-9
                                                   NUT,SELF-LOCKING   (1) 410-11
                                                   RETAINER ASSEMBLY   (2) 410-32
                                                   SCREW,CAP,HEXAGON H(4) 410-30
                                                   SCREW,CAP,HEXAGON H(4) 410-4
                                                   SCREW,MACHINE   (4) 410-14
                                                   SCREW,MACHINE   (2) 409-12
                                                   SCREW,MACHINE   (6) 410-23
                                                   SCREW,MACHINE   (1) 410-25
                                                   SCREW,MACHINE   (1) 410-29
                                                   STRAP   (2) 410-28
                                                   STRIP,ELECTRIC,GR   (1) 410-12
                                                   SUPPORT,ANTENNA   (1) 410-6
                                                   WASHER,FLAT   (3) 410-15
                                                   WASHER,FLAT   (3) 410-7
                                                   WASHER,FLAT   (18) 409-2
                                                   WASHER,FLAT   (2) 409-7
                                                   WASHER,FLAT   (2) 409-10
                                                   WASHER,FLAT   (8) 410-5
                                                   WASHER,LOCK   (1) 410-17
                                                   WASHER,LOCK   (2) 410-34
                                                   WASHER,LOCK   (1) 410-13
                                                   WASHER,LOCK   (2) 410-35

(1)	(2)	(3)	(4)	(5)	(6)	(7)
ITEM NO	SMR CODE	NSN	CAGEC	PART NUMBER	DESCRIPTION AND USABLE ON CODES (UOC)	QTY
					WASHER,LOCK (1) 410-37	

SECTION II TM9-2320-280-24P C01

Section II. TM 9-2320-280-24P-2

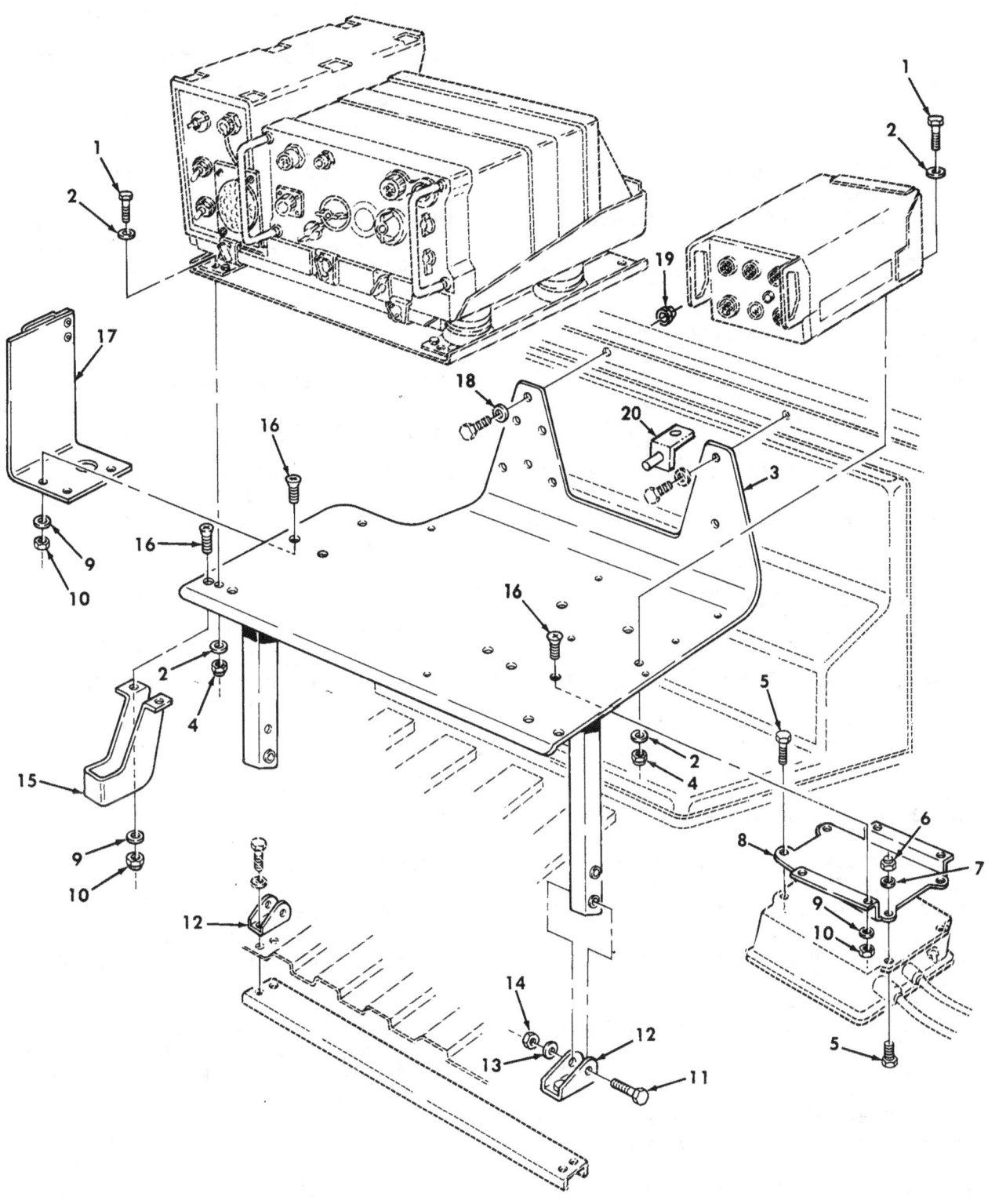

Figure 411. Communications Kit, Radio Rack Assembly, and Mounting Hardware, AN/GRC-160 and TSEC/KY-57, TOW Carrier.

```
SECTION II            TM9-2320-280-24P C01
  (1)   (2)   (3)           (4)         (5)                    (6)
(7) ITEM  SMR                          PART
    NO   CODE   NSN         CAGEC     NUMBER      DESCRIPTION AND USABLE ON CODES(UOC)
QTY
```

GROUP 3307 SPECIAL PURPOSE KITS

FIG. 411 COMMUNICATIONS KIT,RADIO RACK ASSEMBLY,AND MOUNTING HARDWARE, AN/GRC-160 AND TSEC/KY-57,TOW CARRIER

```
        1 PAOZZ 5306002258499 96906 MS90725-34      BOLT,MACHINE  5/16-18 X 1.00.......   8
UOC:A11,A24,A27,B24,H11,H24,H27,MMM
        2 PAOZZ 5310011023270 24617 2436161         WASHER,FLAT  1/4...................
16                                                    UOC:A11,A24,A27,B24,H11,H24,H27,M-
MM      3 PFOZZ 5975011975505 19207 12340167          MOUNTING BASE,ELECT  PART OF KIT P/N
1                                                     5705689............................
UOC:A11,A24,A27,B24,H11,H24,H27,MMM
        4 PAOZZ 5310008140673 96906 MS51943-33     NUT,SELF-LOCKING,HE  5/16-18........   8
UOC:A11,A24,A27,B24,H11,H24,H27,MMM
        5 PAOZZ 5305002400194 96906 MS51849-76     SCREW,MACHINE  #10-24 X 3/4.........   4
UOC:A11,A24,A27,B24,H11,H24,H27,MMM
        6 PAOZZ 5310012699245 24617 190254         NUT,SELF-LOCKING,HE  #10-24.........   4
UOC:A11,A24,A27,B24,H11,H24,H27,MMM
        7 PAOZZ 5310000145850 96906 MS27183-42     WASHER,FLAT  #10...................
4                                                     UOC:A11,A24,A27,B24,H11,H24,H27,M-
MM      8 PAOZZ 2540012028548 19207 12340153         BRACE,RADIO MOUNTIN  PART OF KIT P/N
1                                                     5705689............................
UOC:A11,A24,A27,B24,H11,H24,H27,MMM
        9 PAOZZ 5310011023270 24617 2436161        WASHER,FLAT  1/4...................   9
UOC:A11,A24,A27,B24,H11,H24,H27,MMM
       10 PAOZZ 5310000131245 21450 131245         NUT,SELF-LOCKING,HE  1/4-20.........
9                                                     UOC:A11,A24,A27,B24,H11,H24,H27,M-
MM     11 PAOZZ 5306010684592 96906 MS35764-1303     BOLT,SELF-LOCKING  3/8-16 X 2.25,
2                                                     PART OF KIT P/N 5705689............
UOC:A11,A24,A27,B24,H11,H24,H27,MMM
*      12 PFOZZ 2590012028546 19207 12340134       BRACKET,RADIO  PART OF KIT P/N
2                                                     5705689............................
UOC:A11,A24,A27,H11,H24,H27,MMM
*      12 PAOZZ 5340014120866 19207 12446766       BRACKET,DOUBLE ANGL................   2
UOC:B24
       13 PAOZZ 5310014124013 24617 2436163        WASHER,FLAT  3/8...................   2
UOC:A11,A24,A27,B24,H11,H24,H27,MMM
       14 PAOZZ 5310009359021 96906 MS51943-35     NUT,SELF-LOCKING,HE  3/8-16.........
2                                                     UOC:A11,A24,A27,B24,H11,H24,H27,M-
MM     15 PFOZZ 2540012028545 19207 12340131         BRACKET,HEAD PHONE  PART OF KIT P/N
1                                                     5705689............................
UOC:A11,A24,A27,B24,H11,H24,H27,MMM
       16 PAOZZ 5305007195021 96906 MS51959-81     SCREW,MACHINE  1/4-20 X .75........
9                                                     UOC:A11,A24,A27,B24,H11,H24,H27,M-
MM     17 PFOZZ 5340012097386 19207 12340151         BRACKET,ANGLE  PART OF KIT P/N
1                                                     5705689............................
UOC:A11,A24,A27,B24,H11,H24,H27,MMM
       18 PAOZZ 5310011708705 24617 9419265        WASHER,FLAT........................   2
UOC:B24
*      19 PAOZZ 5310009359022 81349 M45913/3-4FG8C NUT,SELF-LOCKING,HE................   2
UOC:B24
```

411-1

(1)	(2)	(3)	(4)	(5)	(6)	
ITEM NO	SMR CODE	NSN	CAGEC	PART NUMBER	DESCRIPTION AND USABLE ON CODES (UOC)	QTY
20	PFOZZ	2590014398268	19207	12446763	BRACKET,ANGLE.................... UOC:B24	1

Section II.

TM 9-2320-280-24P-2

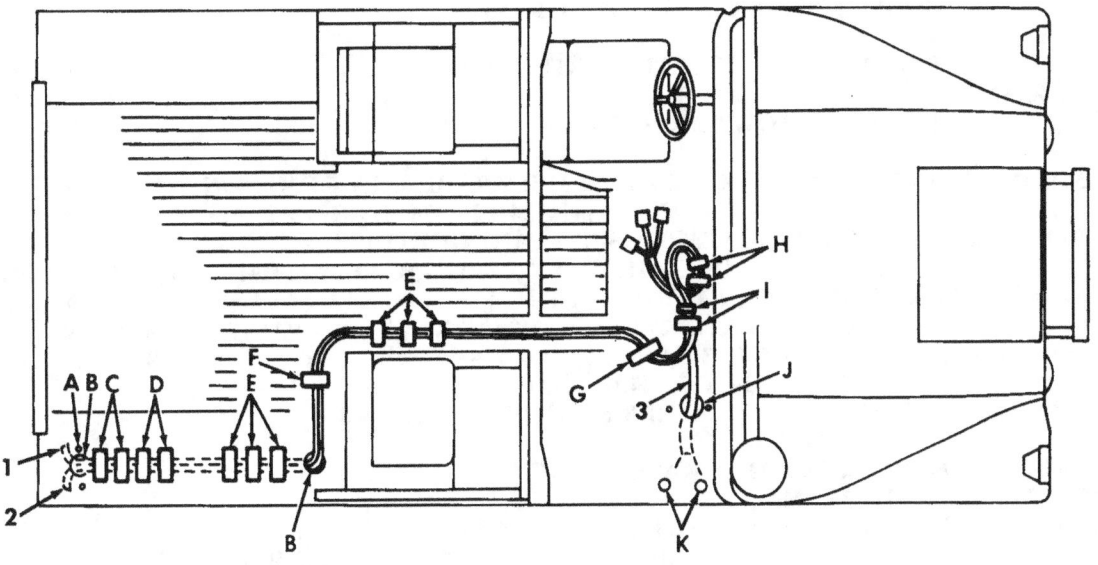

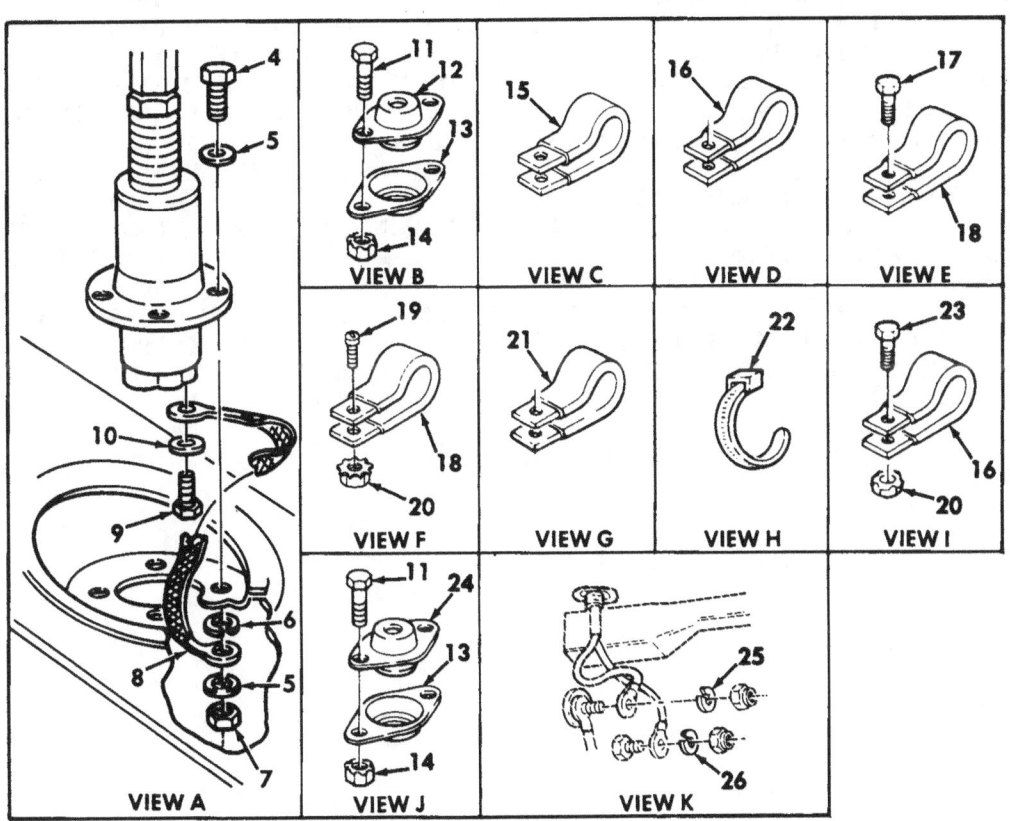

Figure 412. Communications Kit, Radio Cables, AN/GRC-160, AN/VRC-46, and TSEC/KY-57, TOW Carrier.

SECTION II   TM9-2320-280-24P

(1) NO	(2) SMR CODE	(3) NSN	(4) CAGEC	(5) PART NUMBER	(6) DESCRIPTION AND USABLE ON CODES(UOC)	(7) QTY

GROUP 3307 SPECIAL PURPOSE KITS

FIG. 412 COMMUNICATIONS KIT,RADIO CABLES,AN/GRC-160,AN/VRC-46,AND TSEC/KY-57,TOW CARRIER

1	PAOZZ	5995012017495	34623	5589167	CABLE ASSEMBLY,RADI PART OF KIT P/N 5705689 UOC:A11,A24,A27,B24,H11,H24,H27,MMM	1
2	PAOZZ	5995012014129	19207	12338089-3	CABLE ASSEMBLY,SPEC PART OF KIT P/N 5705689 UOC:A11,A24,A27,B24,H11,H24,H27,MMM	1
3	PAOZZ	5995012014128	19207	12338088	CABLE ASSEMBLY,SPEC PART OF KIT P/N 5705689 UOC:A11,A24,A27,B24,H11,H24,H27,MMM	1
4	PAOZZ	5305007829489	80204	B1821BH038C200N	SCREW,CAP,HEXAGON H 3/8-16 X 2.00, PART OF KIT P/N 5705689 UOC:A11,A24,A27,B24,H11,H24,H27,MMM	4
5	PAOZZ	5310014124013	24617	2436163	WASHER,FLAT 3/8 UOC:A11,A24,A27,B24,H11,H24,H27,MMM	8
6	PAOZZ	5310000453299	96906	MS35333-42	WASHER,LOCK 3/8 UOC:A11,A24,A27,B24,H11,H24,H27,MMM	1
7	PAOZZ	5310009359021	96906	MS51943-35	NUT,SELF-LOCKING,HE 3/8-16 UOC:A11,A24,A27,B24,H11,H24,H27,MMM	4
8	PAOZZ	5940012048830	34623	5591200	TERMINAL STRIP,GROU PART OF KIT P/N 5705689 UOC:A11,A24,A27,B24,H11,H24,H27,MMM	1
9	PAOZZ	5305000593659	96906	MS51958-63	SCREW,MACHINE #10-32 X 1/2,PART OF KIT P/N 5705689 UOC:A11,A24,A27,B24,H11,H24,H27,MMM	2
10	PAOZZ	5310000453296	96906	MS35338-43	WASHER,LOCK 3/8,PART OF KIT P/N 5705689 UOC:A11,A24,A27,B24,H11,H24,H27,MMM	1
11	PAOZZ	5305000680502	96906	MS90725-6	SCREW,CAP,HEXAGON H 1/4-20 X 3/4, PART OF KIT P/N 5705689 UOC:A11,A24,A27,B24,H11,H24,H27,MMM	6
12	PAOZZ	5325012053203	19207	12338085	GROMMET PART OF KIT P/N 5705689 UOC:A11,A24,A27,B24,H11,H24,H27,MMM	2
13	PAOZZ	2590004543620	19207	809223	RETAINER ASSEMBLY PART OF KIT P/N 5705689 UOC:A11,A24,A27,B24,H11,H24,H27,MMM	3
14	PAOZZ	5310011520598	24617	271172	NUT,SELF-LOCKING,HE 1/4-20,PART OF KIT P/N 5705689 UOC:A11,A24,A27,B24,H11,H24,H27,MMM	6
15	PAOZZ	5340008338476	96906	MS21333-122	CLAMP,LOOP PART OF KIT P/N 5705689. UOC:A11,A24,A27,B24,H11,H24,H27,MMM	2
16	PAOZZ	5340000502740	96906	MS21333-75	CLAMP,LOOP PART OF KIT P/N 5705689. UOC:A11,A24,A27,B24,H11,H24,H27,MMM	4
17	PAOZZ	5305012596322	34623	5593913	SCREW,MACHINE #10-32 X 3/4,PART OF KIT P/N 5705689 UOC:A11,A24,A27,B24,H11,H24,H27,MMM	6
18	PAOZZ	5340000572904	96906	MS21333-71	CLAMP,LOOP PART OF KIT P/N 5705689. UOC:A11,A24,A27,B24,H11,H24,H27,MMM	7

SECTION II   TM9-2320-280-24P C01
 (1)   (2)   (3)   (4)   (5)   (6)
7)   ITEM SMR   PART
      NO   CODE   NSN   CAGEC   NUMBER   DESCRIPTION AND USABLE ON CODES(UOC)
TY

       19 PAOZZ 5305000593661   96906   MS51958-65   SCREW,MACHINE #10-32 X 3/4,PART OF
                                                          KIT P/N 5705689.....................
OC:A11,A24,A27,B24,H11,H24,H27,MMM
       20 PAOZZ 5310012510726   34623   5593053   NUT,SELF-LOCKING,HE #10-32.........   3
OC:A11,A24,A27,B24,H11,H24,H27,MMM
       21 PAOZZ 5340008546729   96906   MS21333-103   CLAMP,LOOP..........................   1
OC:A11,A24,A27,B24,H11,H24,H27,MMM
       22 PAOZZ 5975010345871   96906   MS3367-7-0   STRAP PART OF KIT P/N 5705689......
                                                          UOC:A11,A24,A27,B24,H11,H24,H27,M-
M   23 PAOZZ 5305009847363   96906   MS35191-272   SCREW,MACHINE #10-32 X 1/2,PART
                                                          KIT P/N 5705689.................
  2   UOC:A11,A24,A27,B24,H11,H24,H27,M-
...   GROMMET PART OF KIT P/N 5705689....
M   24 PAOZZ 5325012055378   19207   12338084   UOC:A11,A24,A27,B24,H11,H24,H27,MMM

       25 PFOZZ 5310012538440   34623   5596566   WASHER,LOCK 1/2...................   1
OC:A11,A24,A27,B24,H11,H24,H27,MMM
       26 PAOZZ 5310000034094   96906   MS35338-48   WASHER,LOCK 1/2...................   1
OC:A11,A24,A27,B24,H11,H24,H27,MMM
    KIT   PDOZZ 5895011975467   34623   05705689   PARTS KIT,ELECTRONI RADIO,AN/GRC-
                                                        160,AN/VRC-46,AND TSEC/KY-57,TOW
ARRIER...........................
OC:A11,A24,A27,B24,H11,H24,H27,MMM
OLT,SELF-LOCKING   (2) 411-11

                                                         BRACE,RADIO MOUNTIN(1) 411-8
                                                         BRACKET,HEAD PHONE (1) 411-15
                                                         BRACKET,ANGLE   (1) 411-17
                                                         BRACKET,RADIO   (2) 411-12
                                                         CABLE ASSEMBLY,RADI(1) 412-1
                                                         CABLE ASSEMBLY,SPEC(1) 412-3
                                                         CABLE ASSEMBLY,SPEC(1) 412-2
                                                         CLAMP,LOOP   (2) 412-15
                                                         CLAMP,LOOP   (4) 412-16
                                                         CLAMP,LOOP   (7) 412-18
                                                         GROMMET   (1) 412-24
                                                         GROMMET   (2) 412-12
                                                         MOUNTING BASE,ELECT(1) 411-3
                                                         RETAINER ASSEMBLY (3) 411-13
                                                         NUT,SELF-LOCKING   (6) 412-14
                                                         SCREW,CAP,HEXAGON H(6) 412-11
                                                         SCREW,CAP,HEXAGON H(4) 412-4
                                                         SCREW,MACHINE   (2) 412-23
                                                         SCREW,MACHINE   (2) 412-9
                                                         SCREW,MACHINE   (6) 412-17
                                                         SCREW,MACHINE   (1) 412-19
                                                         STRAP   (2) 412-22
                                                         TERMINAL STRIP   (1) 412-8
                                                         WASHER,LOCK   (1) 412-10

                              END OF FIGURE

Section II. TM 9-2320-280-24P-2

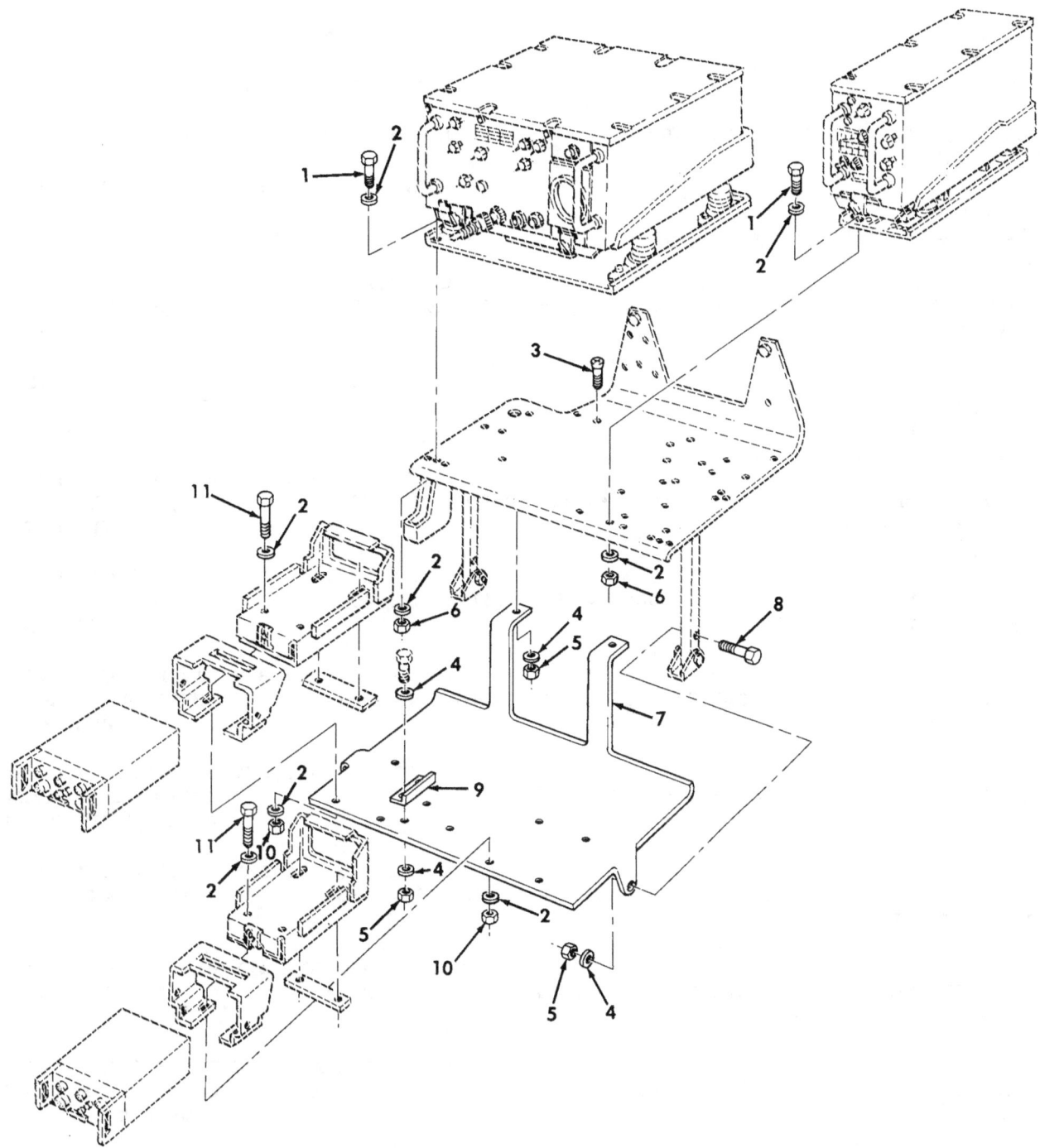

Figure 413. Communications Kit, Radio Rack Assembly, AN/VRC-47 and (2) TSEC/KY-57 Assemblies.

SECTION II TM9-2320-280-24P C01

(1) ITEM NO	(2) SMR CODE	(3) NSN	(4) CAGEC	(5) PART NUMBER	(6) DESCRIPTION AND USABLE ON CODES (UOC)	(7) QTY
					GROUP 3307 SPECIAL PURPOSE KITS	
					FIG. 413 COMMUNICATIONS KIT, RADIO TSEC/KY-57 ASSEMBLIES	
					RACK ASSEMBLY, AN/VRC-47 AND	(2)
1	PAOZZ	5306002258499	96906	MS90725-34	BOLT, MACHINE 5/16-18 X 1.00, PART OF KIT P/N PPL-10323................. UOC:AVY,A13,A14,BVY,B17,B18,C17,HVY, H13,H14,H17,H18,NNN	8
2	PAOZZ	5310011191024	24617	2436162	WASHER, FLAT 5/16, PART OF IT P/N PPL-10323....................... UOC:AVY,A13,A14,BVY,B17,B18,C17,HVY, H13,H14,H17,H18,NNN	32
3	PAOZZ	5305007195021	96906	MS51959-81	SCREW, MACHINE 1/4-20 X .75, PART OF KIT P/N PPL-10323................ UOC:AVY,A13,A14,BVY,B17,B18,C17,HVY, H13,H14,H17,H18,NNN	2
4	PAOZZ	5310011023270	24617	2436161	WASHER, FLAT 1/4, PART OF KIT P/N PPL-10323..................... UOC:AVY,A13,A14,BVY,B17,B18,C17,HVY, H13,H14,H17,H18,NNN	8
5	PAOZZ	5310000614650	96906	MS51943-31	NUT, SELF-LOCKING, HE 1/4-20, PART OF KIT P/N PPL-10323................. UOC:AVY,A13,A14,BVY,B17,B18,C17,HVY, H13,H14,H17,H18,NNN	6
6	PAOZZ	5310008140673	96906	MS51943-33	NUT, SELF-LOCKING, HE 5/16-18, PART OF KIT P/N PPL-10323............... UOC:AVY,A13,A14,BVY,B17,B18,C17,HVY, H13,H14,H17,H18,NNN	8
7	PAOZZ	2540012491589	19207	12340641	SHELF, AUXILIARY PART OF KIT P/N PPL-10323..................... UOC:AVY,A13,A14,BVY,B17,B18,C17,HVY, H13,H14,H17,H18,NNN	1
8	PAOZZ	5305000712237	96906	MS90725-14	SCREW, CAP, HEXAGON H 1/4-20 X 2.00, PART OF KIT P/N PPL-10323.......... UOC:AVY,A13,A14,BVY,B17,B18,C17,HVY, H13,H14,H17,H18,NNN	2
9	PAOZZ	5340012551109	19207	12340635	BRACKET, ANGLE PART OF KIT P/N PPL-10323..................... UOC:AVY,A13,A14,BVY,B17,B18,C17,HVY, H13,H14,H17,H18,NNN	1
10	PAOZZ	5310012559452	34623	5593027	NUT, PLAIN, HEXAGON 5/16-24, PART OF KIT P/N PPL-10323................. UOC:AVY,A13,A14,BVY,B17,B18,C17,HVY, H13,H14,H17,H18,NNN	8
11	PAOZZ	5306000514077	80204	B1821BH031F113N	BOLT, MACHINE 5/16-24 X 1.125, PART OF KIT P/N 5705689................. UOC:AVY,A13,A14,BVY,B17,B18,C17,HVY, H13,H14,H17,H18,NNN	8

END OF FIGURE

413-1

Section II. TM 9-2320-280-24P-2

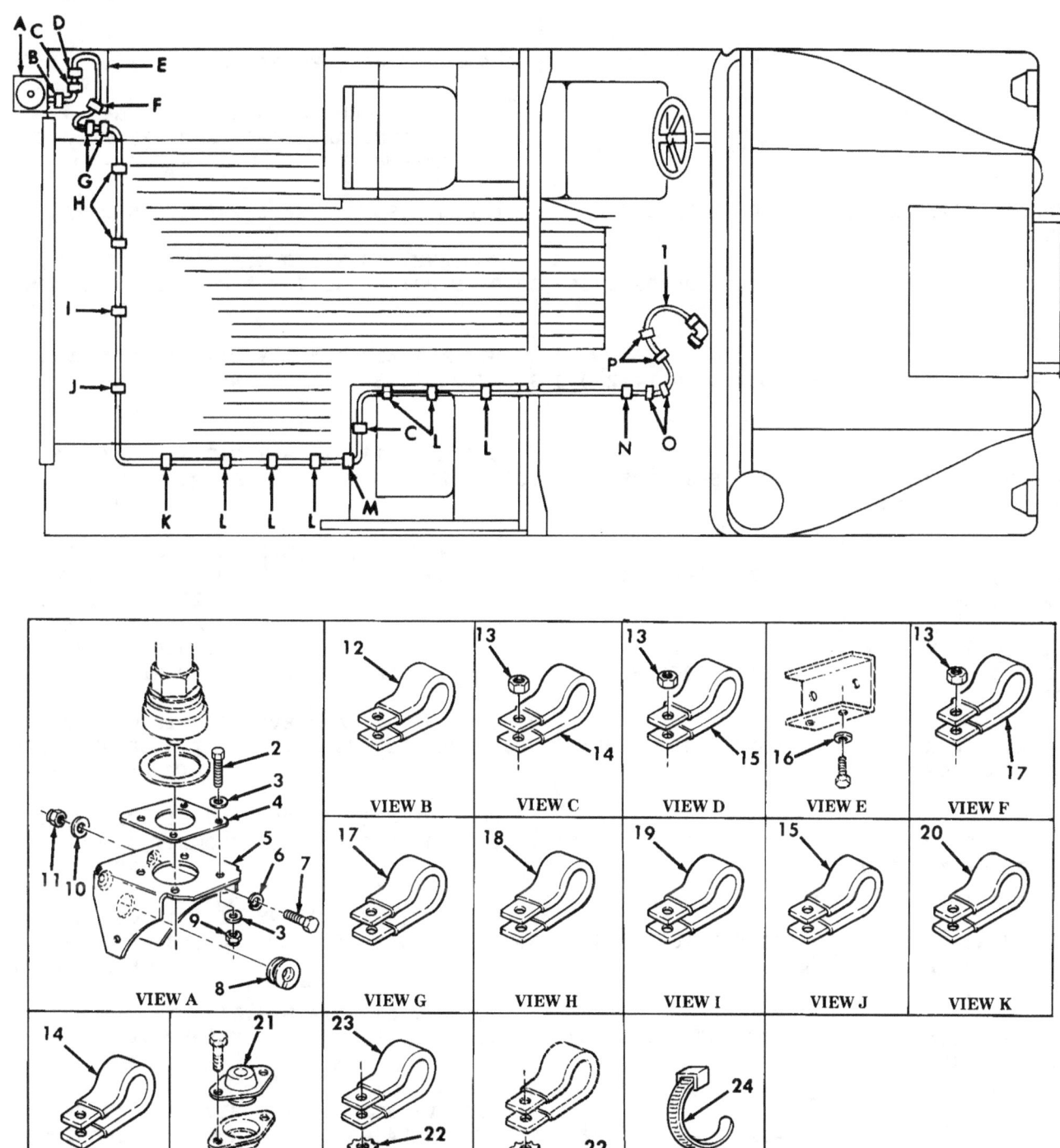

Figure 414. Communications Kit, Radio Cables, AN/VRC-47, and (2) TSEC/KY-57 Assemblies.

SECTION II   TM9-2320-280-24P C01

GROUP 3307 SPECIAL PURPOSE KITS

FIG. 414 COMMUNICATIONS KIT,RADIO CABLES,AN/VRC-47 AND (2) TSEC/KY-57 ASSEMBLIES

(1) ITEM NO	(2) SMR CODE	(3) NSN	(4) CAGEC	(5) PART NUMBER	(6) DESCRIPTION AND USABLE ON CODES (UOC)	(7) QTY
1	PAOZZ	5995012519316	19207	12338086-4	CABLE ASSEMBLY,RADI PART OF KIT P/N PPL-10323............ UOC:AVY,A13,A14,BVY,B17,B18,C17,HVY,H13,H14,H17,H18,NNN	1
2	PAOZZ	5305008213869	80204	B1821BH038C175N	SCREW,CAP,HEXAGON H 3/8-16 X 1.75 PART OF KIT P/N PPL-10323.......... UOC:AVY,A13,A14,BVY,B17,B18,C17,HVY,H13,H14,H17,H18,NNN	4
3	PAOZZ	5310014124013	24617	2436163	WASHER,FLAT 3/8 PART OF KIT P/N PPL-10323............ UOC:AVY,A13,A14,BVY,B17,B18,C17,HVY,H13,H14,H17,H18,NNN	8
4	PAOZZ	5985012533514	19207	12340637	ADAPTER,ANTENNA TO PART OF KIT P/N PPL-10323............ UOC:AVY,A13,A14,BVY,B17,B18,C17,HVY,H13,H14,H17,H18,NNN	1
5	PAOZZ	5985012159404	80063	A3046166	SUPPORT,ANTENNA PART OF KIT P/N PPL-10323............ UOC:AVY,A13,A14,BVY,B17,B18,C17,HVY,H13,H14,H17,H18,NNN	1
* 6	PAOZZ	5310011191024	24617	2436162	WASHER,FLAT 5/16 PART OF KIT P/N PPL-10323............ UOC:AVY,A13,A14,BVY,B17,B18,C17,HVY,H13,H14,H17,H18,NNN	3
7	PAOZZ	5306002258499	96906	MS90725-34	BOLT,MACHINE 5/16-18 X 1.00 PART OF KIT P/N PPL-10323................ UOC:AVY,A13,A14,BVY,B17,B18,C17,HVY,H13,H14,H17,H18,NNN	3
8	PAOZZ	5325012085424	19207	12338098	GROMMET,NONMETALLIC PART OF KIT P/N PPL-10323............ UOC:AVY,A13,A14,BVY,B17,B18,C17,HVY,H13,H14,H17,H18,NNN	1
9	PAOZZ	5310009359021	96906	MS51943-35	NUT,SELF-LOCKING,HE 3/8-16 PART OF KIT P/N PPL-10323................ UOC:AVY,A13,A14,BVY,B17,B18,C17,HVY,H13,H14,H17,H18,NNN	4
10	PAOZZ	5310012087576	19207	12338095	WASHER,FLAT 5/16 PART OF KIT P/N PPL-10323............ UOC:AVY,A13,A14,BVY,B17,B18,C17,HVY,H13,H14,H17,H18,NNN	3
11	PAOZZ	5310008140673	96906	MS51943-33	NUT,SELF-LOCKING,HE 5/16-18 PART OF KIT P/N PPL-10323................ UOC:AVY,A13,A14,BVY,B17,B18,C17,HVY,H13,H14,H17,H18,NNN	3
12	PAOZZ	5340000509077	96906	MS21333-119	CLAMP,LOOP PART OF KIT P/N PPL-10323 UOC:AVY,A13,A14,BVY,B17,B18,C17,HVY,H13,H14,H17,H18,NNN	1

414-1

SECTION II   TM9-2320-280-24P C01

(1) ITEM NO	(2) SMR CODE	(3) NSN	(4) CAGEC	(5) PART NUMBER	(6) DESCRIPTION AND USABLE ON CODES (UOC)	(7) QTY
13	PAOZZ	5310001249265	72582	271169	NUT,PLAIN,ASSEMBLED #10-32 PART OF KIT P/N PPL-10323................... UOC:AVY,A13,A14,BVY,B17,B18,C17,HVY,H13,H14,H17,H18,NNN	6
14	PAOZZ	5340000572904	96906	MS21333-71	CLAMP,LOOP PART OF KIT P/N PPL-10323 UOC:AVY,A13,A14,BVY,B17,B18,C17,HVY,H13,H14,H17,H18,NNN	8
15	PAOZZ	5340009899222	96906	MS21333-74	CLAMP,LOOP PART OF KIT P/N PPL-10323 UOC:AVY,A13,A14,BVY,B17,B18,C17,HVY,H13,H14,H17,H18,NNN	2
16	PAOZZ	5310005825965	96906	MS35338-44	WASHER,LOCK 1/4 PART OF KIT P/N PPL-10323............................ UOC:AVY,A13,A14,BVY,B17,B18,C17,HVY,H13,H14,H17,H18,NNN	2
17	PAOZZ	5340000572906	96906	MS21333-73	CLAMP,LOOP PART OF KIT P/N PPL-10323 UOC:AVY,A13,A14,BVY,B17,B18,C17,HVY,H13,H14,H17,H18,NNN	2
18	PAOZZ	5340007247038	96906	MS21333-76	CLAMP,LOOP PART OF KIT P/N PPL-10323 UOC:AVY,A13,A14,BVY,B17,B18,C17,HVY,H13,H14,H17,H18,NNN	1
19	PAOZZ	5340009226300	96906	MS21333-77	CLAMP,LOOP PART OF KIT P/N PPL-10323 UOC:AVY,A13,A14,BVY,B17,B18,C17,HVY,H13,H14,H17,H18,NNN	1
20	PAOZZ	5340000502740	96906	MS21333-75	CLAMP,LOOP PART OF KIT P/N PPL-10323 UOC:AVY,A13,A14,BVY,B17,B18,C17,HVY,H13,H14,H17,H18,NNN	1
21	PAOZZ	5325012555062	34623	5590029	GROMMET,NONMETALLIC PART OF KIT P/N PPL-10323............................ UOC:AVY,A13,A14,BVY,B17,B18,C17,HVY,H13,H14,H17,H18,NNN	1
22	PAOZZ	5310011520598	24617	271172	NUT,SELF-LOCKING,AS 1/4-20 PART OF KIT P/N PPL-10323................... UOC:AVY,A13,A14,BVY,B17,B18,C17,HVY,H13,H14,H17,H18,NNN	4
23	PAOZZ	5340009848540	96906	MS21333-102	CLAMP,LOOP PART OF KIT P/N PPL-10323 UOC:AVY,A13,A14,BVY,B17,B18,C17,HVY,H13,H14,H17,H18,NNN	1
24	PAOZZ	5975010345871	96906	MS3367-7-0	STRAP,TIEDOWN,ELECT PART OF KIT P/N PPL-10323............................ UOC:AVY,A13,A14,BVY,B17,B18,C17,HVY,H13,H14,H17,H18,NNN	2
KIT	PDOZZ	5820012221107	80063	PPL-10323	TRANSITION KIT,ELEC RADIO RACK ASSEMBLY,AN/VRC-47 AND TSEC/KY-57... UOC:AVY,A13,A14,BVY,B17,B18,C17,HVY,H13,H14,H17,H18,NNN	1

```
                          ADAPTER,ANTENNA TO  ( 1) 414-4
                          BOLT,MACHINE        ( 8) 413-1
                          BOLT,MACHINE        ( 8) 413-11
                          BOLT,MACHINE        ( 3) 414-7
                          BRACKET,ANGLE       ( 1) 413-9
                          CABLE ASSEMBLY,RADI ( 1) 414-1
                          CLAMP,LOOP          ( 1) 414-12
                          CLAMP,LOOP          ( 8) 414-14
```

(1)	(2)	(3)	(4)	(5)	(6)	(7) ITEM	
NO	SMR CODE	NSN	CAGEC	PART NUMBER	DESCRIPTION AND USABLE ON CODES(UOC)		QTY

SECTION II TM9-2320-280-24P

```
CLAMP,LOOP          ( 2) 414-15
CLAMP,LOOP          ( 2) 414-17
CLAMP,LOOP          ( 2) 414-18
CLAMP,LOOP          ( 1) 414-19
CLAMP,LOOP          ( 1) 414-20
CLAMP,LOOP          ( 1) 414-23
GROMMET,NONMETALLIC ( 1) 414-8
GROMMET,NONMETALLIC ( 1) 414-21
NUT,PLAIN,HEXAGON   ( 8) 413-10
NUT,PLAIN,ASSEMBLED ( 6) 414-13
NUT,SELF-LOCKING,HE ( 6) 413-5
NUT,SELF-LOCKING,HE ( 8) 413-6
NUT,SELF-LOCKING,HE ( 4) 414-9
NUT,SELF-LOCKING,HE ( 3) 414-11
NUT,SELF-LOCKING,HE ( 4) 414-22
SCREW,CAP,HEXAGON H ( 2) 413-8
SCREW,CAP,HEXAGON H ( 4) 414-2
SCREW,MACHINE       ( 2) 413-3
SHELF,AUXILIARY     ( 1) 413-7
STRAP,TIEDOWN,ELECT ( 2) 414-24
SUPPORT,ANTENNA     ( 1) 414-5
WASHER,FLAT         (32) 413-2
WASHER,FLAT         ( 8) 413-4
WASHER,FLAT         ( 8) 414-3
WASHER,FLAT         ( 3) 414-6
WASHER,FLAT         ( 3) 414-10
WASHER,LOCK         ( 2) 414-16
```

END OF FIGURE

Section II.

TM 9-2320-280-24P-2

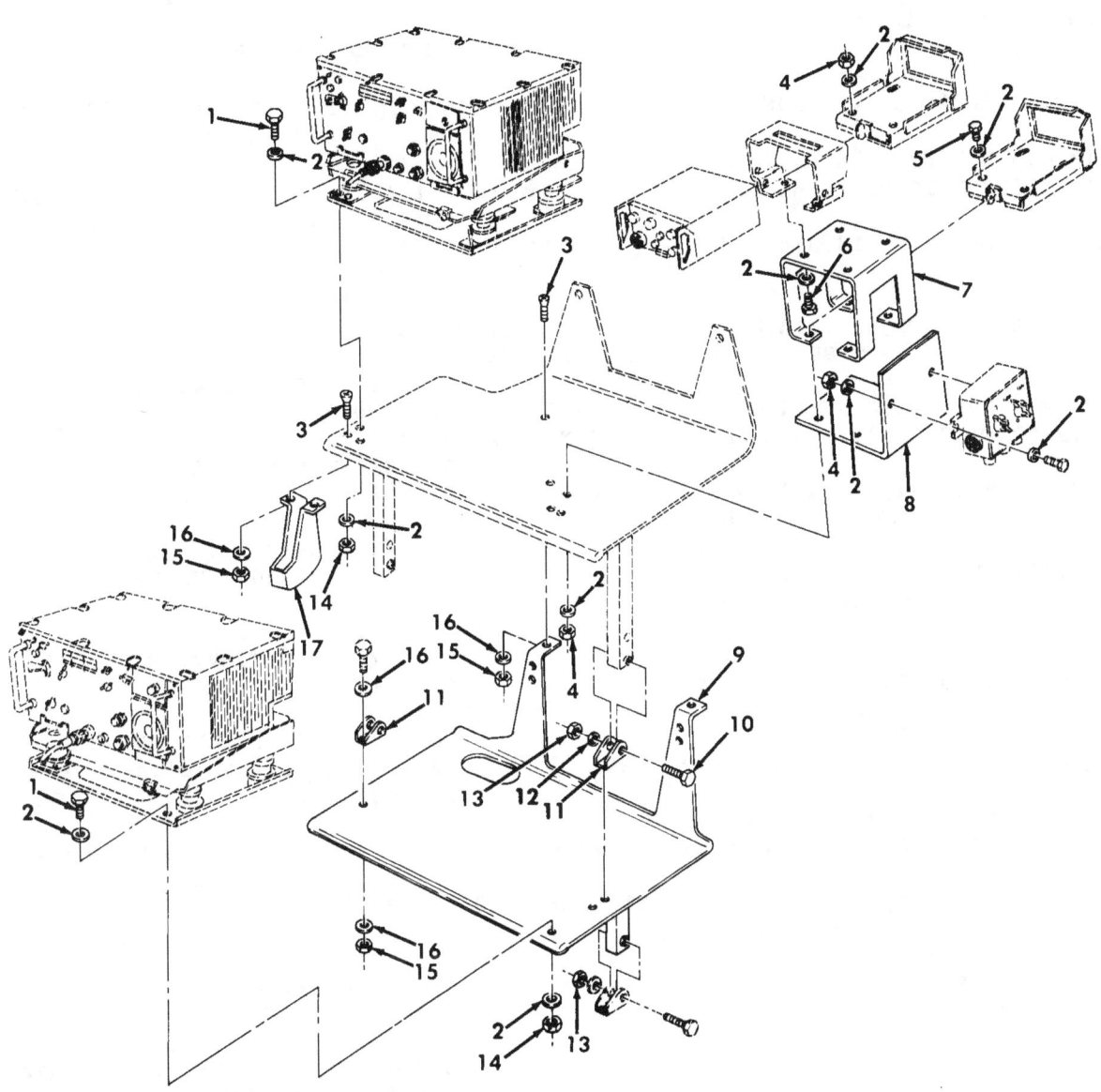

Figure 415. Communications Kit, Radio Rack Assembly, AN/VRC-49, and (2) TSEC/KY-57 Assemblies.

SECTION II   TM9-2320-280-24P C01

GROUP 3307 SPECIAL PURPOSE KITS

FIG. 415 COMMUNICATIONS KIT, RADIO TSEC/KY-57 ASSEMBLIES

(1) ITEM NO	(2) SMR CODE	(3) NSN	(4) CAGEC	(5) PART NUMBER	(6) DESCRIPTION AND USABLE ON CODES (UOC)	(7) QTY
1	PAOZZ	5306002558499	96906	MS90725-34	BOLT,MACHINE 5/16-18 X 1.00 PART OF KIT P/N PPL-10324................... UOC:AVY,A13,A14,BVY,HVY,H13,H14,NNN	10
2	PAOZZ	5310011191024	24617	2436162	WASHER,FLAT 5/16 PART OF KIT P/N PPL-10324.......................... UOC:AVY,A13,A14,BVY,HVY,H13,H14,NNN	40
3	PAOZZ	5305007195021	96906	MS51959-81	SCREW,MACHINE 1/4-20 X .75 PART OF KIT P/N PPL-10324................... UOC:AVY,A13,A14,BVY,HVY,H13,H14,NNN	6
4	PAOZZ	5310002416658	96906	MS51943-34	NUT,SELF-LOCKING,HE 5/16-24 PART OF KIT P/N PPL-10324................... UOC:AVY,A13,A14,BVY,HVY,H13,H14,NNN	10
5	PAOZZ	5306000514077	80204	B1821BH031F113N	BOLT,MACHINE 5/16-24 X 1.125 PART OF KIT P/N PPL-10324................. UOC:AVY,A13,A14,BVY,HVY,H13,H14,NNN	4
6	PAOZZ	5306012552662	34623	5593359	BOLT,MACHINE 5/16-24 X .88 PART OF KIT P/N PPL-10324.................... UOC:AVY,A13,A14,BVY,HVY,H13,H14,NNN	4
7	PAOZZ	5340012491590	19207	12340639	BRACKET,MOUNTING PART OF KIT P/N PPL-10324............................ UOC:AVY,A13,A14,BVY,HVY,H13,H14,NNN	1
8	PFOZZ	5340012559514	19207	12340636	BRACKET,ANGLE PART OF KIT P/N PPL-10324............................ UOC:AVY,A13,A14,BVY,HVY,H13,H14,NNN	1
9	PAOZZ	2540012507590	19207	12340581	RACK ASSEMBLY,RADIO PART OF KIT P/N PPL-10324.......................... UOC:AVY,A13,A14,BVY,HVY,H13,H14,NNN	1
10	PAOZZ	5306010684592	96906	MS35764-1303	BOLT,SELF-LOCKING 3/8-16 X 2.25 PART OF KIT P/N PPL-10324........... UOC:AVY,A13,A14,BVY,HVY,H13,H14,NNN	2
11	PFOZZ	2590012028546	19207	12340134	BRACKET,VEHICULAR C PART OF KIT P/N PPL-10324......................... UOC:AVY,A13,A14,BVY,HVY,H13,H14,NNN	2
12	PAOZZ	5310014124013	24617	2436163	WASHER,FLAT 3/8 PART OF KIT P/N PPL-10324............................ UOC:AVY,A13,A14,BVY,HVY,H13,H14,NNN	2
13	PAOZZ	5310009359021	96906	MS51943-35	NUT,SELF-LOCKING,HE 3/8-16 PART OF KIT P/N PPL-10324................... UOC:AVY,A13,A14,BVY,HVY,H13,H14,NNN	4
14	PAOZZ	5310008140673	96906	MS51943-33	NUT,SELF-LOCKING,HE 5/16-18 PART OF KIT P/N PPL-10324................... UOC:AVY,A13,A14,BVY,HVY,H13,H14,NNN	10
15	PAOZZ	5310000614650	96906	MS51943-31	NUT,SELF-LOCKING,HE 1/4-20 PART OF KIT P/N PPL-10324................... UOC:AVY,A13,A14,BVY,HVY,H13,H14,NNN	10
16	PAOZZ	5310011023270	24617	2436161	WASHER,FLAT 1/4 PART OF KIT P/N PPL-10324............................	14

RACK ASSEMBLY,AN/VRC-49 AND (2)

415-1

SECTION II   TM9-2320-280-24P

(1) NO	(2) CODE	(3) NSN	(4) CAGEC	(5) NUMBER	(6)	(7) ITEM DESCRIPTION AND USABLE ON CODES(UOC)	SMR	PART QTY
17	PFOZZ	2540012028545	19207	12340131 PPL-10324..........................		BRACKET,HEAD PHONE PART OF KIT P/N UOC:AVY,A13,A14,BVY,HVY,H13,H14,NNN UOC:AVY,A13,A14,BVY,HVY,H13,H14		1

END OF FIGURE

Section II.

TM 9-2320-280-24P-2

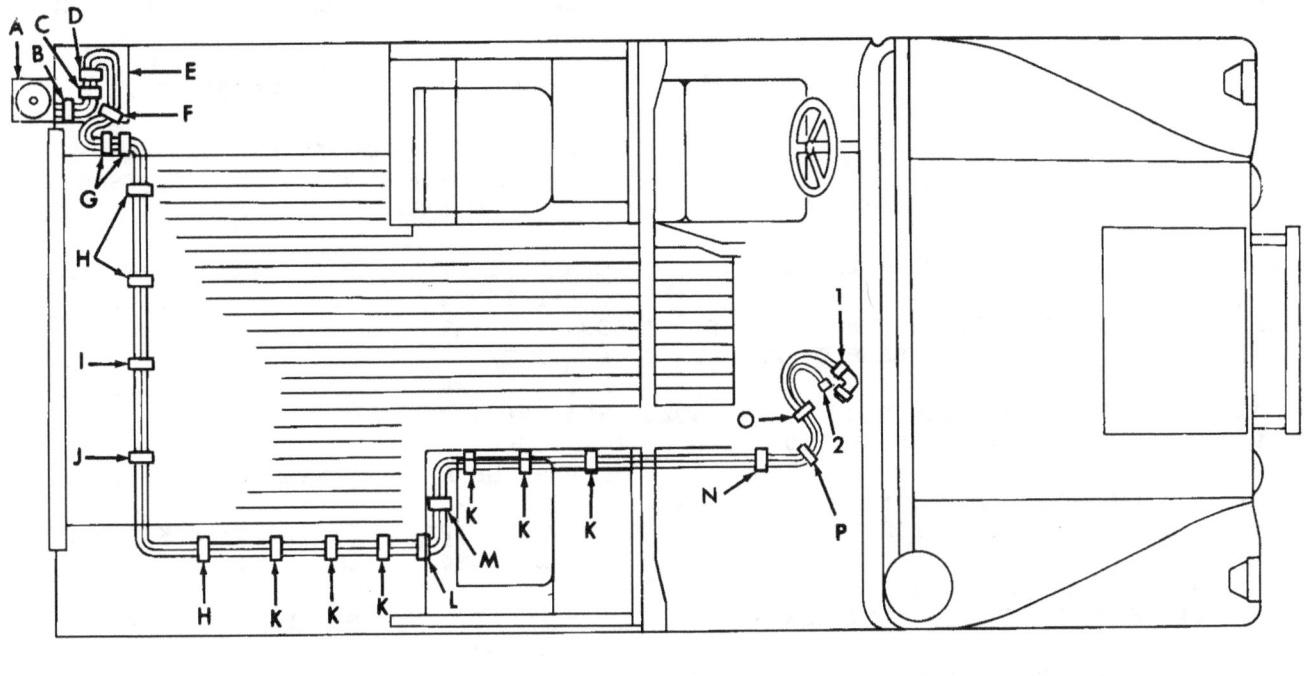

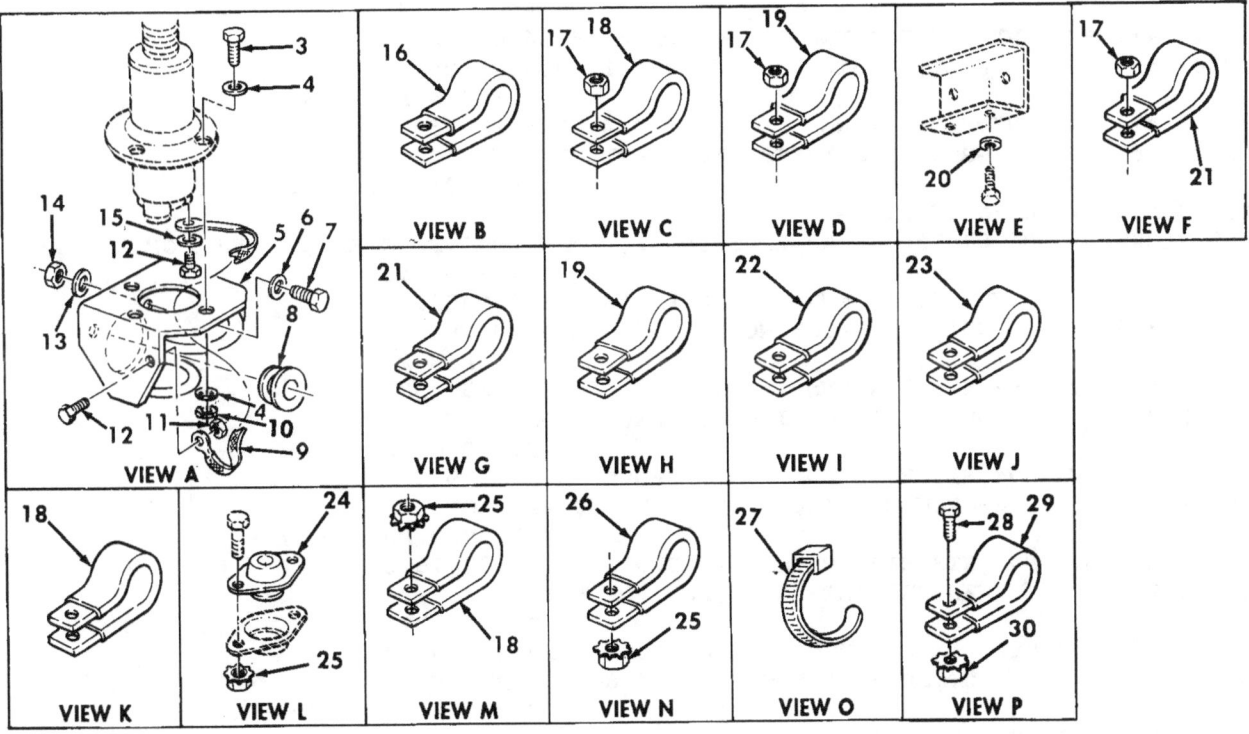

Figure 416. Communications Kit, Radio Cables, AN/VRC-49 and (2) TSEC/KY-57 Assemblies.

SECTION II   TM9-2320-280-24P C01
  (1)   (2)   (3)   (4)   (5)   (6)
(7) ITEM SMR   PART
 NO   CODE NSN   CAGEC   NUMBER   DESCRIPTION AND USABLE ON CODES(UOC)
QTY

GROUP 3307 SPECIAL PURPOSE KITS

FIG. 416 COMMUNICATIONS KIT,RA-
DIO   CABLES,AN/VRC-49 AND (2) TSEC/KY-
57   ASSEMBLIES

    1 PAOZZ 5995012516785 19207 12338089-4   CABLE ASSEMBLY,SPEC   PART OF KIT P/N
1   PPL-10324............
UOC:AVY,A13,A14,BVY,HVY,H13,H14,NNN   2 PAOZZ 5995012519316
19207 12338086-4   CABLE ASSEMBLT,RADI   PART OF KIT P/N
1   PPL-10324............
UOC:AVY,A13,A14,BVY,HVY,H13,H14,NNN   3 PAOZZ 5305000680502
96906 MS90725-6   SCREW,CAP,HEXAGON H 3/8-16 X 2.00,
4   PART OF KIT P/N PPL-10324...........
UOC:AVY,A13,A14,BVY,HVY,H13,H14,N-
NN   4 PAOZZ 5310014124013 19207 2436163   WASHER,FLAT 3/8,PART OF KIT P/N
8   PPL-10324............
UOC:AVY,A13,A14,BVY,HVY,H13,H14,N-
NN   5 PFOZZ 5340011975470 19207 12340364   BRACKET,ANTENNA PART OF KIT P/N
1   PPL-10324............
UOC:AVY,A13,A14,BVY,HVY,H13,H14,NNN
  *   6 PAOZZ 5310011191024 24617 2436162   WASHER,FLAT 5/16,PART OF KIT P/N
3   PPL-10324............
UOC:AVY,A13,A14,BVY,HVY,H13,H14,NNN   7 PAOZZ 5306002258499
96906 MS90725-34   BOLT,MACHINE 5/16-18 X 1.00,PART OF
3   KIT P/N PPL-10324................
UOC:AVY,A13,A14,BVY,HVY,H13,H14,NNN   8 PAOZZ 5325012085424
19207 12338098   GROMMET,NONMETALLIC PART OF KIT P/N
1   PPL-10324............
UOC:AVY,A13,A14,BVY,HVY,H13,H14,N-
NN   9 PAOZZ 6150012078120 19207 12338099-1 LEAD,ELECTRICAL PART OF KIT P/N
1   PPL-10324............
UOC:AVY,A13,A14,BVY,HVY,H13,H14,NNN   10 PAOZZ 5310009359021
96906 MS51943-35   NUT,SELF-LOCKING,HE 3/8-16,PART OF
4   KIT P/N PPL-10324................
UOC:AVY,A13,A14,BVY,HVY,H13,H14,NNN   11 PAOZZ 5310002514503
24617 9411893   NUT,SELF-LOCKING,HE #10-32,PART OF
1   KIT P/N PPL-10324................
UOC:AVY,A13,A14,BVY,HVY,H13,H14,NNN   12 PAOZZ 5305000593659
96906 MS51958-63   SCREW,MACHINE #10-32 X .50,PART OF
2   KIT P/N PPL-10324................
UOC:AVY,A13,A14,BVY,HVY,H13,H14,NNN   13 PAOZZ 5310012087576
19207 12338095   WASHER,FLAT 5/16,PART OF KIT P/M
3   PPL-10324............
UOC:AVY,A13,A14,BVY,HVY,H13,H14,NNN   14 PAOZZ 5310008140673
96906 MS51943-33   NUT,SELF-LOCKING,HE 5/16-18,PART OF
3   KIT P/N PPL-10323................
UOC:AVY,A13,A14,BVY,HVY,H13,H14,N-
NN   15 PAOZZ 5310000453296 96906 MS35338-43   WASHER,LOCK #10,PART OF KIT P/N
1   PPL-10324............
UOC:AVY,A13,A14,BVY,HVY,H13,H14,NNN   16 PAOZZ 5340008546730
96906 MS21333-124   CLAMP,LOOP PART OF KIT P/N PPL-10324   1
UOC:AVY,A13,A14,BVY,HVY,H13,H14,NNN

416-1

SECTION II   TM9-2320-280-24P C01
  (1)   (2)   (3)   (4)   (5)   (6)
(7) ITEM   SMR   PART
     NO   CODE   NSN   CAGEC   NUMBER   DESCRIPTION AND USABLE ON CODES (UOC)
QTY

     17 PFOZZ 5310001249265 72582 271169   NUT,SELF-LOCKING,HE PART OF KIT P/N
                                                    PPL-10324........................
UOC:AVY,A13,A14,BVY,HVY,H13,H14,NNN   18 PAOZZ 5340000572906
96906 MS21333-73   CLAMP,LOOP PART OF KIT P/N PPL-10324
                                                    UOC:AVY,A13,A14,BVY,HVY,H13,H14,N-
     19 PAOZZ 5340009226300 96906 MS21333-77   CLAMP,LOOP PART OF KIT P/N PPL-10324
                                                    UOC:AVY,A13,A14,BVY,HVY,H13,H14,N-
     20 PAOZZ 5310005825965 96906 MS35338-44   WASHER,LOCK 1/4,PART OF KIT P/N
                                                    PPL-10324........................
UOC:AVY,A13,A14,BVY,HVY,H13,H14,NNN   21 PAOZZ 5340007247038
96906 MS21333-76   CLAMP,LOOP PART OF KIT P/N PPL-10324
                                                    UOC:AVY,A13,A14,BVY,HVY,H13,H14,N-
     22 PFOZZ 5340009588457 96906 MS21333-78   CLAMP,LOOP PART OF KIT P/N PPL-10324
                                                    UOC:AVY,A13,A14,BVY,HVY,H13,H14,N-
     23 PAOZZ 5340000502740 96906 MS21333-75   CLAMP,LOOP PART OF KIT P/N PPL10324
                                                    UOC:AVY,A13,A14,BVY,HVY,H13,H14,N-
     24 PAOZZ 5325012555063 19207 12340631   GROMMET,NONMETALLIC PART OF KIT P/N
                                                    PPL-10324........................
UOC:AVY,A13,A14,BVY,HVY,H13,H14,NNN   25 PAOZZ 5310011520598
24623 5592818   NUT,SELF-LOCKING,HE 1/4-20,PART OF
                                                    KIT P/N PPL-10324................
UOC:AVY,A13,A14,BVY,HVY,H13,H14,NNN   26 PAOZZ 5340000881254
96906 MS21333-104   CLAMP,LOOP PART OF KIT P/N PPL-10324   1
UOC:AVY,A13,A14,BVY,HVY,H13,H14,NNN
   * 27 PAOZZ 5975010345871 96906 MS3367-7-0   STRAP,TIEDOWN,ELECT PART OF KIT P/N
                                                    PPL-10324........................
UOC:AVY,A13,A14,BVY,HVY,H13,H14,NNN
     28 PAOZZ 5305009897434 96906 MS35207-263   SCREW,MACHINE....................   1
UOC:AVY,A13,A14,BVY,HVY,H13,H14,NNN
     29 PAOZZ 5340000502740 96906 MS21333-75   CLAMP,LOOP.......................   1
UOC:AVY,A13,A14,BVY,HVY,H13,H14,NNN
     30 PAOZZ 5310001249265 24617 271169   NUT,PLAIN,ASSEMBLED..............
                                                    UOC:AVY,A13,A14,BVY,HVY,H13,H14,NNN
KIT PDOZZ 5820012221106 80036 PPL-10324   TRANSITION KIT,ELEC RADIO RACK   1
                                                    ASSEMBLY,AN/VRC-49 RADIO AND TSCE/..
                                                    KY-57 ASSEMBLIES.................
                                                    UOC:AVY,A13,A14,BVY,HVY,H13,H14,NNN
                                                      BOLT,MACHINE   (10) 415-1
                                                      BOLT,MACHINE   (4) 415-5
                                                      BOLT,MACHINE   (4) 415-6
                                                      BOLT,MACHINE   (3) 416-7
                                                      BOLT,SELF-LOCKING   (2) 415-10
                                                      BRACKET,ANGLE   (1) 415-8
                                                      BRACKET,ANTENNA   (1) 416-5
                                                      BRACKET,HEAD PHONE (1) 415-17
                                                      BRACKET,MOUNTING   (1) 415-7
                                                      BRACKET,VEHICULAR C (2) 415-11
                                                      CABLE ASSEMBLY,RADI (1) 416-2
                                                      CABLE ASSEMBLY,SPEC (1) 416-1
                                                      CLAMP,LOOP   (1) 416-16
                                                      CLAMP,LOOP   (8) 416-18
                                                      CLAMP,LOOP   (4) 416-19

SECTION II   TM9-2320-280-24P
(1) (2) (3)   (4)   (5)   (6)   (7) ITEM SMR   PART
NO CODE NSN CAGEC NUMBER   DESCRIPTION AND USABLE ON CODES(UOC) QTY

                                  CLAMP,LOOP   (3) 416-21
                                  CLAMP,LOOP   (1) 416-22
                                  CLAMP,LOOP   (1) 416-23
                                  CLAMP,LOOP   (1) 416-26
                                  GROMMET,NONMETALLIC(1) 416-8
                                  GROMMET,NONMETALLIC(1) 416-24
                                  LEAD,ELECTRICAL   (1) 416-9
                                  NUT,SELF-LOCKING,HE(4) 416-25
                                  NUT,SELF-LOCKING,HE(10) 415-4
                                  NUT,SELF-LOCKING,HE(4) 415-13
                                  NUT,SELF-LOCKING,HE(10) 415-14
                                  NUT,SELF-LOCKING,HE(10) 415-15
                                  NUT,SELF-LOCKING,HE(4) 416-10
                                  NUT,SELF0LOCKING,HE(1) 416-11
                                  NUT,SELF-LOCKING,HE(3) 416-14
                                  NUT,SELF-LOCKING,HE(4) 416-17
                                  RACK ASSEMBLY,RADIO(1) 415-9
                                  SCREW,CAP,HEXAGON H(6) 415-3
                                  SCREW,CAP,HEXAGON H(4) 416-3
                                  SCREW,MACHINE   (2) 416-12
                                  STRAP,TIEDOWN,ELECT(1) 416-27
                                  WASHER,FLAT   (40) 415-2
                                  WASHER,FLAT   (2) 415-12
                                  WASHER,FLAT   (14) 415-16
                                  WASHER,FLAT   (8) 416-4
                                  WASHER,FLAT   (3) 416-6
                                  WASHER,FLAT   (3) 416-13
                                  WASHER,LOCK   (1) 416-15
                                  WASHER,LOCK   (2) 416-20

END OF FIGURE

Section II. TM 9-2320-280-24P-2

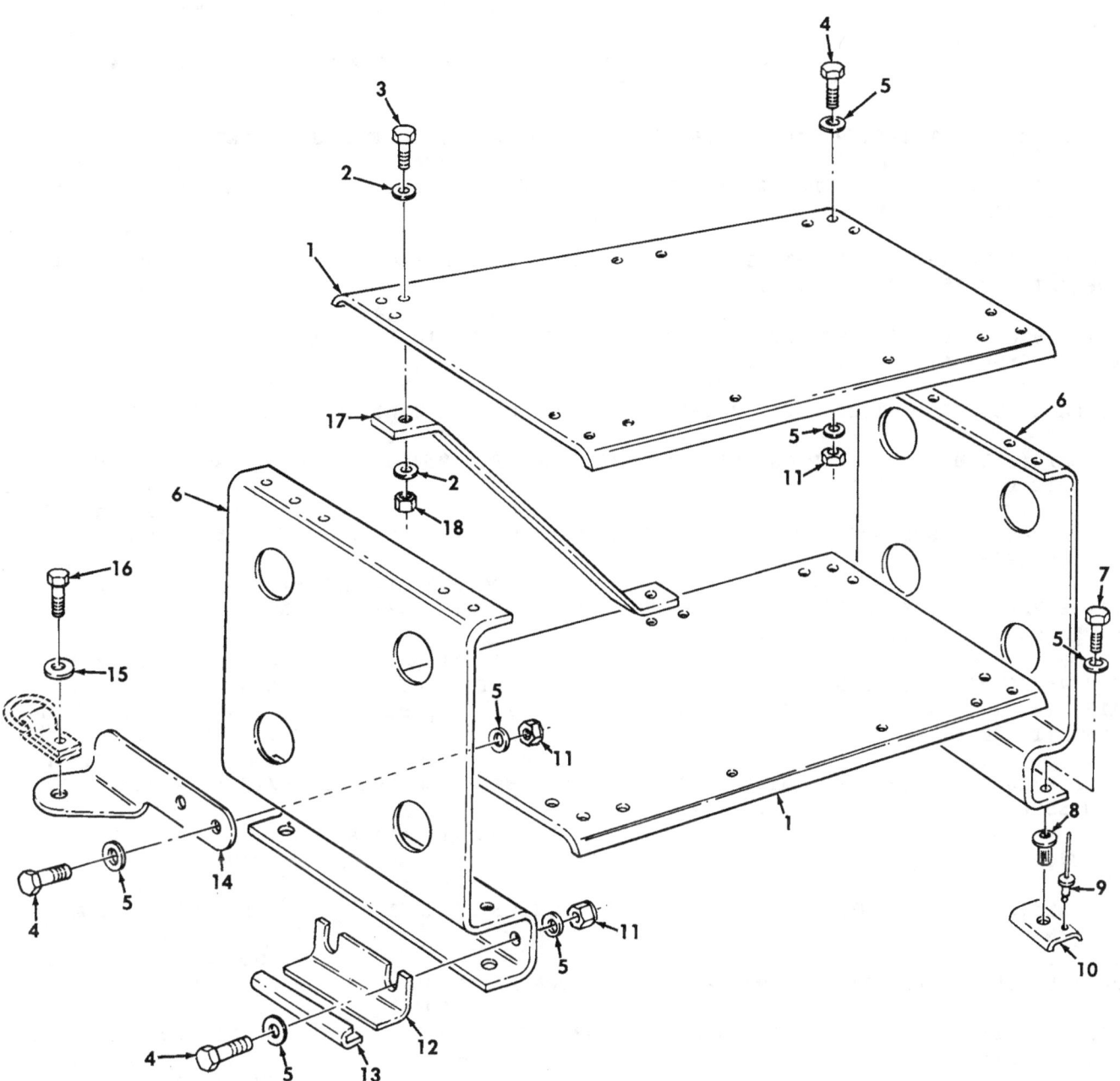

Figure 417. Communications Kit, Radio Rack Assembly.

SECTION II   TM9-2320-280-24P C01
   (1)   (2)   (3)   (4)   (5)   (6)
(7) ITEM   SMR   PART
  NO   CODE NSN   CAGEC   NUMBER   DESCRIPTION AND USABLE ON CODES(UOC)
QTY

GROUP 3307 SPECIAL PURPOSE KITS

FIG. 417 COMMUNICATIONS KIT,RADIO RACK ASSEMBLY

1 PFOZZ 5975013125469 34623 5596795   SHELF,ELECTRICAL EQ PART OF KIT P/N
2   5596946...........................
UOC:A13,A14,H13,H14   2 PAOZZ 5310012510748 34623 5593691   WASHER,FLAT 1/4
PART OF KIT P/N   8   5596946.........
....   UOC:A13,A14,H13,H14
3 PAOZZ 5305002253843 80204 B1821BH025C100N SCREW,CAP,HEXAGON H 1/4-20 X 1.00   4
PART OF KIT P/N 5596946............
                                          UOC:A13,A14,H13,H14   4 PAOZZ
5306002264828 80204 B1821BH031C113N BOLT,MACHINE 5/16-18 X 1.125 PART   14
OF KIT P/N 5596946..................
                                          UOC:A13,A14,H13,H14
  * 5 PAOZZ 5310011191024 24617 2436162   WASHER,FLAT 5/16 PART OF KIT P/N
32   5596946...........................
UOC:A13,A14,H13,H14   6 PAOZZ 5340012902279 34623 5596794   BRACKET,MOUNTING
PART OF KIT P/N   2   5596946.........
....   UOC:A13,A14,H13,H14
7 PAOZZ 5306002264829 80204 B1821BH031C125N BOLT,MACHINE 5/16-18 X 1.25 PART OF   4
KIT P/N 5596946....................
                                          UOC:A13,A14,H13,H14   8 PAOZZ
5310011981724 03481 S31P280   NUT,PLAIN,BLIND RIV 5/16-18 PART OF   4
KIT P/N 5596946....................
                                          UOC:A13,A14,H13,H14   9 PAOZZ
5320012544251 34623 5593050   RIVET,BLIND PART OF KIT P/N 5596946.   8
UOC:A13,A14,H13,H14   10 PAOZZ 5340012915711 34623 5588193   BRACKET,DOUBLE ANGL
PART OF KIT P/N   4   5596946..........
....   UOC:A13,A14,H13,H14
11 PAOZZ 5310008140673 96906 MS51943-33   NUT,SELF-LOCKING,HE 5/16-18 PART OF   14
KIT P/N 5596946..................
                                          UOC:A13,A14,H13,H14   12
PAOZZ 5340012914537 34623 5597356   BRACKET,ANGLE PART OF KIT P/N
1   5596946...........................
UOC:A13,A14,H13,H14   13 PAOZZ 2590012911033 34623 5595551   PROTECTOR,EDGE PART
OF KIT P/N   1   5596946..........
....   UOC:A13,A14,H13,H14
14 PAOZZ 5340012989691 34623 5597372   BRACKET,ANGLE R.H. PART OF KIT P/N
1   5596946...........................
UOC:A13,A14,H13,H14   14 PAOZZ 5340012906342 34623 5597350   BRACKET,ANGLE L.H.
PART OF KIT P/N   1   5596946..........
....   UOC:A13,A14,H13,H14
15 PAOZZ 5310012538948 34623 5593320   WASHER,FLAT .50 PART OF KIT P/N
2   5596946...........................
UOC:A13,A14,H13,H14   16 PAOZZ 5305000712069 52304 113324   SCREW,CAP,HEXAGON H
1/2-13 X 1.50   2

417-1

SECTION II   TM9-2320-280-24P

(1) ITEM NO	(2) SMR CODE	(3) NSN	(4) CAGEC	(5) PART NUMBER	(7) DESCRIPTION AND USABLE ON CODES(UOC)	QTY
					PART OF KIT P/N 5596946............. UOC:A13,A14,H13,H14	
17	PAOZZ	5340012908390	34623	5597086	BRACKET,MOUNTING PART OF KIT P/N 5596946............................ UOC:A13,A14,H13,H14	2
18	PAOZZ	5310000614650	96906	MS51943-31	NUT,SELF-LOCKING,HE 1/4-20 PART OF KIT P/N 5596946..................... UOC:A13,A14,H13,H14	4

END OF FIGURE

Section II.

TM 9-2320-280-24P-2

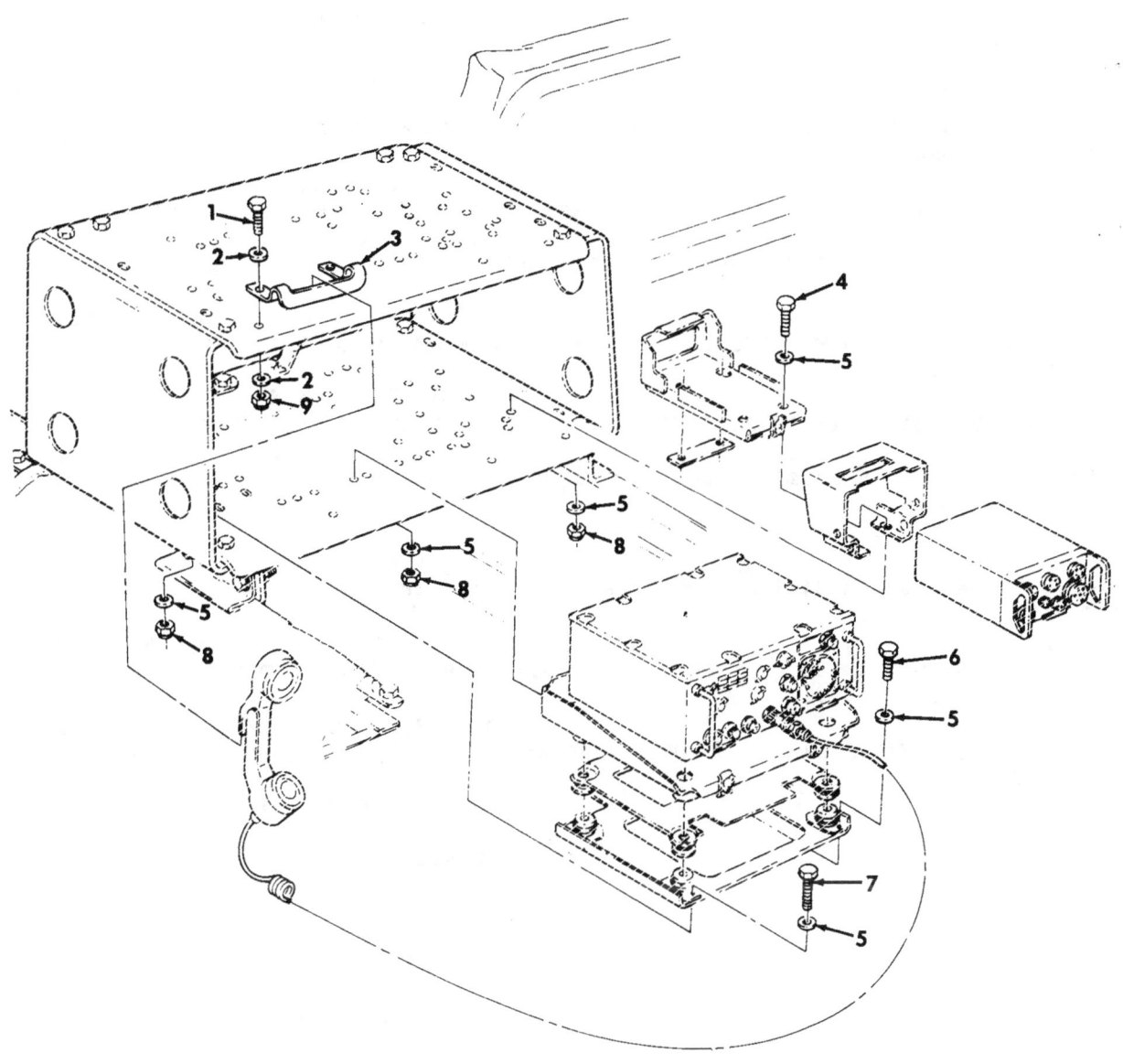

Figure 418. Communications Kit, Radio Rack Assembly Mounting Hardware, AN/VRC-46.

SECTION II   TM9-2320-280-24P C01

(1) ITEM NO	(2) SMR CODE	(3) NSN	(4) CAGEC	(5) PART NUMBER	(6) DESCRIPTION AND USABLE ON CODES (UOC)	(7) QTY
					GROUP 3307 SPECIAL PURPOSE KITS	
					FIG. 418 COMMUNICATIONS KIT, RADIO AN/VRC-46 RACK ASSEMBLY MOUNTING HARDWARE,	
1	PAOZZ	5305000712505	80204	MS90728-7	SCREW,CAP,HEXAGON H 1/4-20 X 7/8 PART OF KIT P/N 5596946............. UOC:A13,A14,H13,H14	2
2	PAOZZ	5310012510748	34623	5593691	WASHER,FLAT 1/4 PART OF KIT P/N 5596946............................ UOC:A13,A14,H13,H14	4
3	PAOZZ	5340012918915	34623	5597357	BRACKET,MOUNTING PART OF KIT P/N 5596946................................. UOC:A13,A14,H13,H14	1
4	PAOZZ	5306002258498	96906	MS90725-33	BOLT,MACHINE 5/16-18 X 7/8 PART OF KIT P/N 5596946..................... UOC:A13,A14,H13,H14	4
* 5	PAOZZ	5310011191024	24617	2436162	WASHER,FLAT 5/16 PART OF KIT P/N 5596946............................. UOC:A13,A14,H13,H14	18
6	PAOZZ	5306002258499	96906	MS90725-34	BOLT,MACHINE 5/16-18 X 1.00 PART OF KIT P/N 5596946.................... UOC:A13,A14,H13,H14	3
7	PAOZZ	5306002264828	80204	B1821BH031C113N	BOLT,MACHINE 5/16-18 X 1.125 PART OF KIT P/N 5596946................... UOC:A13,A14,H13,H14	2
8	PAOZZ	5310008140673	96906	MS51943-33	NUT,SELF-LOCKING,HE 5/16-18 PART OF KIT P/N 5596946.................... UOC:A13,A14,H13,H14	9
9	PAOZZ	5310000614650	96906	MS51943-31	NUT,SELF-LOCKING,HE 1/4-20 PART OF KIT P/N 5596946.................... UOC:A13,A14,H13,H14	2

END OF FIGURE

Section II. TM 9-2320-280-24P-2

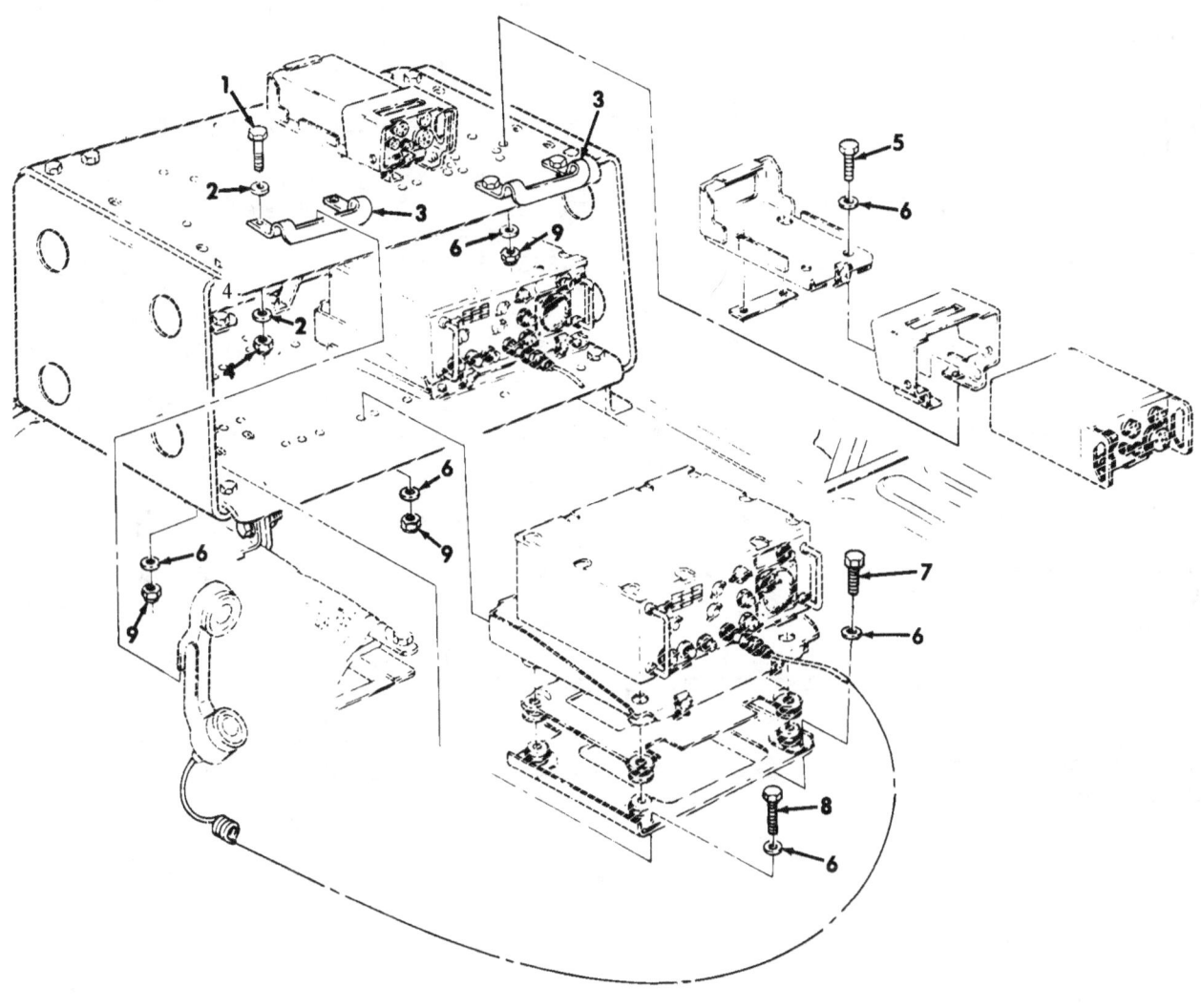

Figure 419. Communications Kit, Radio Rack Assembly Mounting Hardware, (2) AN/VRC-46.

SECTION II   TM9-2320-280-24P C01

(1) ITEM NO	(2) SMR CODE	(3) NSN	(4) CAGEC	(5) PART NUMBER	(6) DESCRIPTION AND USABLE ON CODES (UOC)	(7) QTY
					GROUP 3307 SPECIAL PURPOSE KITS	
					FIG. 419 COMMUNICATIONS KIT, RADIO AN/VRC-46	
					RACK ASSEMBLY MOUNTING HARDWARE,	
1	PAOZZ	5305000712505	80204	B1821BH025CO88N	SCREW,CAP,HEXAGON H 1/4-20 X 7/8 PART OF KIT P/N 5596946............. UOC:A13,A14,H13,H14	4
2	PAOZZ	5310012510748	34623	5593691	WASHER,FLAT 1/4 PART OF KIT P/N 5596946................ UOC:A13,A14,H13,H14	8
3	PAOZZ	5340012918915	34623	5597357	BRACKET,MOUNTING PART OF KIT P/N 5596946......... UOC:A13,A14,H13,H14	2
4	PAOZZ	5310000614650	96906	MS51943-31	NUT,SELF-LOCKING,HE 1/4-20 PART OF KIT P/N 5596946................. UOC:A13,A14,H13,H14	4
5	PAOZZ	5306002258498	96906	MS90725-33	BOLT,MACHINE 5/16-18 X 7/8 PART OF KIT P/N 5596946................. UOC:A13,A14,H13,H14	8
* 6	PAOZZ	5310011191024	24617	2436162	WASHER,FLAT 5/16 PART OF KIT P/N 5596946............. UOC:A13,A14,H13,H14	36
7	PAOZZ	5306002258499	96906	MS90725-34	BOLT,MACHINE 5/16-18 X 1.00 PART OF KIT P/N 5596946................. UOC:A13,A14,H13,H14	6
8	PAOZZ	5306002264828	80204	B1821BH031C113N	BOLT,MACHINE 5/16-18 X 1.125 PART OF KIT P/N 5596946................. UOC:A13,A14,H13,H14	4
9	PAOZZ	5310008140673	96906	MS51943-33	NUT,SELF-LOCKING,HE 5/16-18 PART OF KIT P/N 5596946................. UOC:A13,A14,H13,H14	18

END OF FIGURE

419-1

Section II.   TM 9-2320-280-24P-2

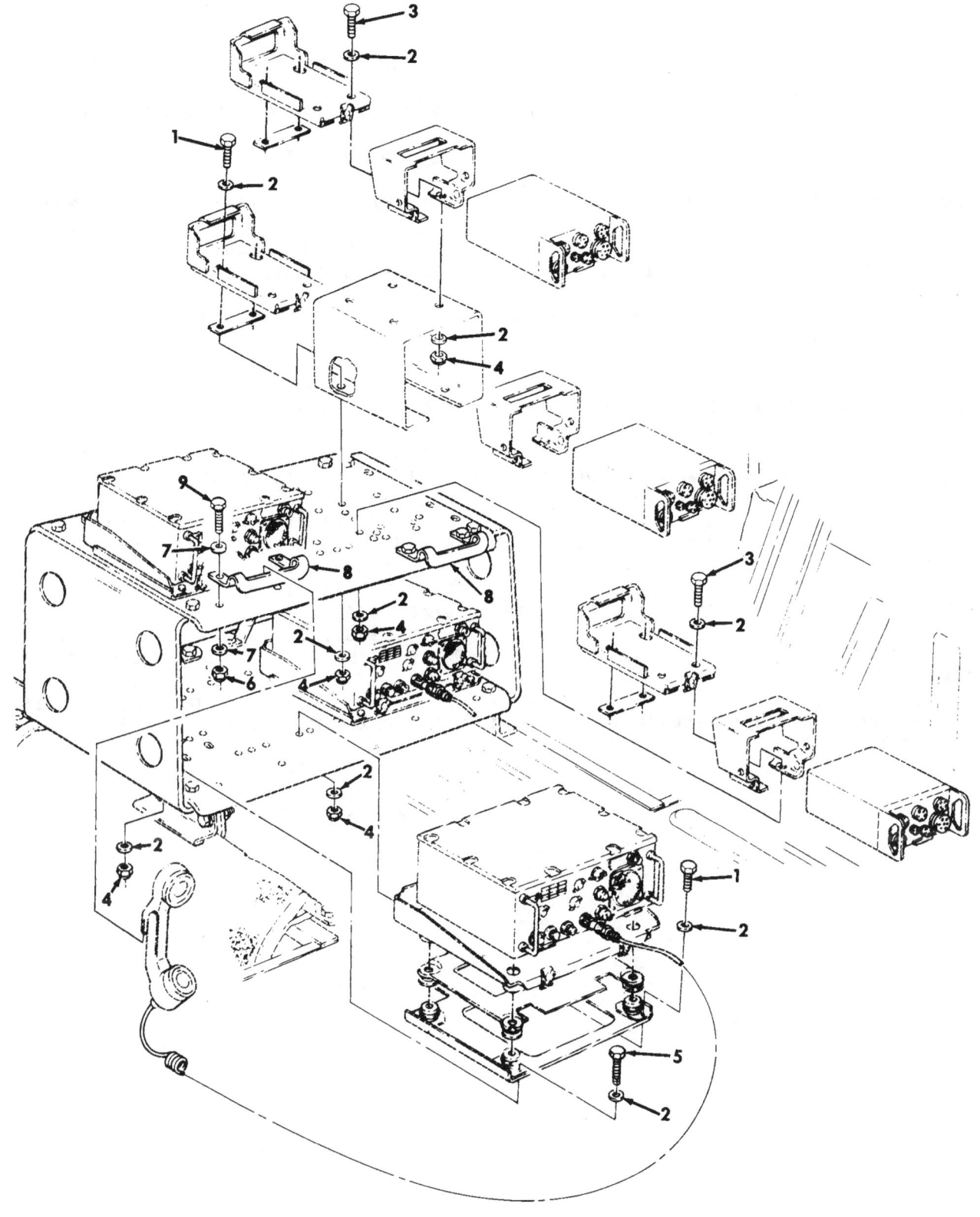

Figure 420. Communications Kit, Radio Rack Assembly Mounting Hardware, (3) AN/VRC-46.

SECTION II TM9-2320-280-24P C01

(1) ITEM NO	(2) SMR CODE	(3) NSN	(4) CAGEC	(5) PART NUMBER	(6) DESCRIPTION AND USABLE ON CODES (UOC)	(7) QTY

GROUP 3307 SPECIAL PURPOSE KITS

FIG. 420 COMMUNICATIONS KIT, RADIO (3) AN/VRC-46

(1)	(2)	(3)	(4)	(5)	(6)	(7)
1	PAOZZ	5306002258499	96906	MS90725-34	BOLT,MACHINE 5/16-18 X 1.00 PART OF KIT P/N 5596946................... UOC:A13,A14,H13,H14	13
* 2	PAOZZ	5310011191024	24617	2436162	WASHER,FLAT 5/16 PART OF KIT P/N 5596946............................ UOC:A13,A14,H13,H14	54
3	PAOZZ	5306002258498	96906	MS90725-33	BOLT,MACHINE 5/16-18 X .88 PART OF KIT P/N 5596946................... UOC:A13,A14,H13,H14	8
4	PAOZZ	5310008140673	96906	MS51943-33	NUT,SELF-LOCKING,HE 5/16-18 PART OF KIT P/N 5596946................... UOC:A13,A14,H13,H14	27
5	PAOZZ	5306002264828	80204	B1821BH031C113N	BOLT,MACHINE 5/16-18 X 1.125 PART OF KIT P/N 5596946................. UOC:A13,A14,H13,H14	6
6	PAOZZ	5310000614650	96906	MS51943-31	NUT,SELF-LOCKING,HE 1/4-20 PART OF KIT P/N 5596946................... UOC:A13,A14,H13,H14	4
7	PAOZZ	5310012510748	34623	5593691	WASHER,FLAT 1/4 PART OF KIT P/N 5596946............................ UOC:A13,A14,H13,H14	8
8	PAOZZ	5340012918915	34623	5597357	BRACKET,MOUNTING PART OF KIT P/N 5596946......... .... UOC:A13,A14,H13,H14	2
9	PAOZZ	5305000712505	80204	MS90728-7	SCREW,CAP,HEXAGON H 1/4-20 X .88 PART OF KIT P/N 5596946............. UOC:A13,A14,H13,H14	4

RACK ASSEMBLY MOUNTING HARDWARE,

END OF FIGURE

Section II. TM 9-2320-280-24P-2

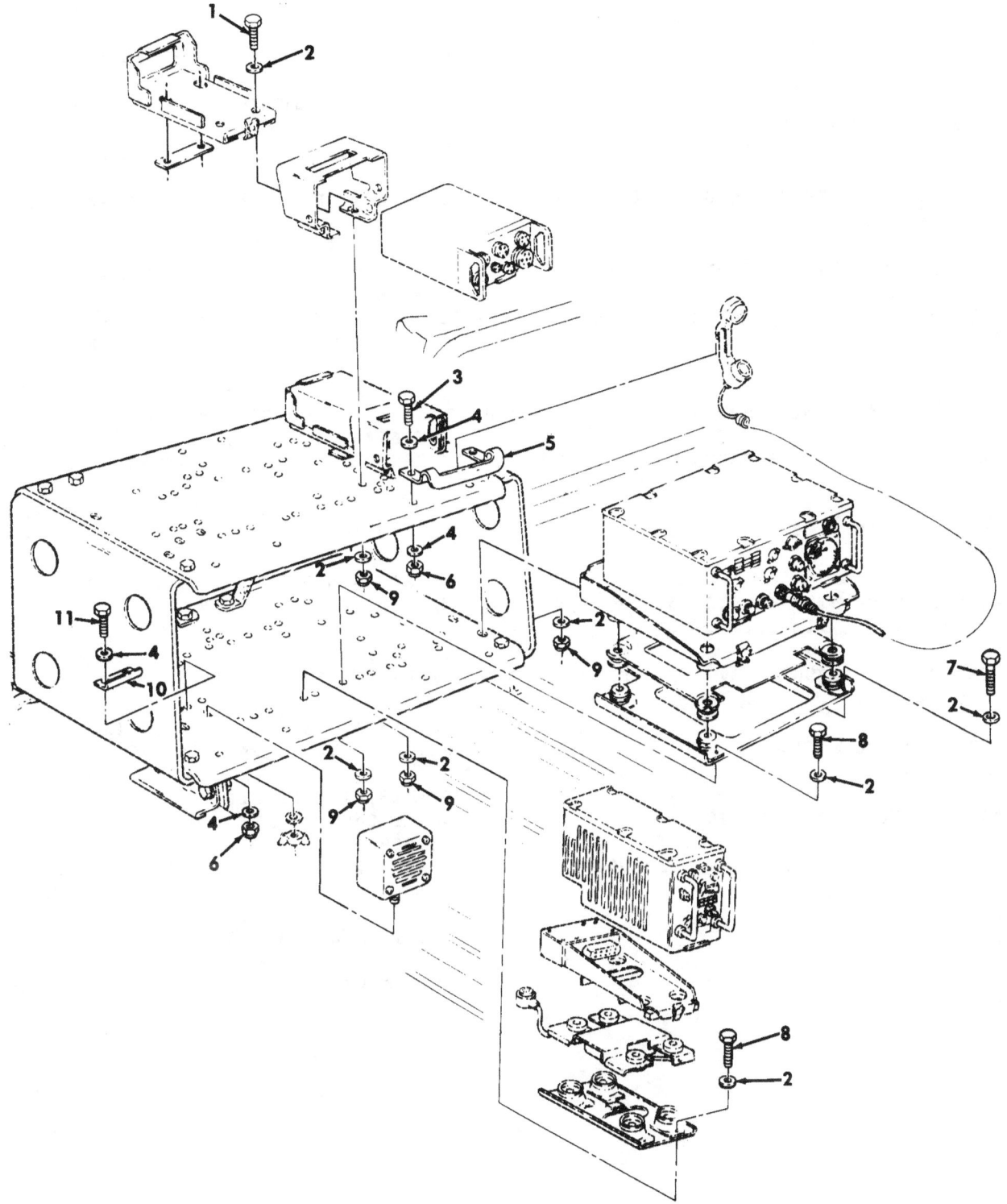

Figure 421. Communications Kit, Radio Rack Assembly Mounting Hardware, AN/VRC-47.

SECTION II   TM9-2320-280-24P C01
  (1)   (2)   (3)   (4)   (5)   (6)
(7)   ITEM   SMR   PART
  NO   CODE   NSN   CAGEC   NUMBER   DESCRIPTION AND USABLE ON CODES(UOC)
QTY

GROUP 3307 SPECIAL PURPOSE KITS

FIG. 421 COMMUNICATIONS KIT,RADIO

RACK ASSEMBLY MOUNTING HARDWARE,

AN/VRC-47

   1 PAOZZ 5306002258498 96906 MS90725-33   BOLT,MACHINE 5/16-18 X .88 PART OF   8
KIT P/N 5596946...................
                                              UOC:A13,A14,H13,H14
   * 2 PAOZZ 5310011191024 24617 2436162   WASHER,FLAT 5/16 PART OF KIT P/N
34   5596946...........................
UOC:A13,A14,H13,H14   3 PAOZZ 5305000712505 80204 MS90728-7   SCREW,CAP,HEXAGON
H 1/4-20 X .88   2   PART OF KIT P/N
5596946............
                                              UOC:A13,A14,H13,H14   4
PAOZZ 5310012510748 34623 5593691   WASHER,FLAT 1/4 PART OF KIT P/N
8   5596946...........................
UOC:A13,A14,H13,H14   5 PAOZZ 5340012918915 34623 5597357   BRACKET,MOUNTING
PART OF KIT P/N   1   5596946.........
....   UOC:A13,A14,H13,H14
 6 PAOZZ 5310000614650 96906 MS51943-31   NUT,SELF-LOCKING,HE 1/4-20 PART OF   4
KIT P/N 5596946...................
                                              UOC:A13,A14,H13,H14   7 PAOZZ
5306002264828 80204 B1821BH031C113N BOLT,MACHINE 5/16-18 X 1.125 PART   2
OF KIT P/N 5596946.................
                                              UOC:A13,A14,H13,H14   8 PAOZZ
5306002258499 96906 MS90725-34   BOLT,MACHINE 5/16-18 X 1.00 PART OF   7
KIT P/N 5596946...................
                                              UOC:A13,A14,H13,H14   9 PAOZZ
5310004722963 96906 MS51943-3   NUT,SELF-LOCKING,HE 5/16-18 PART OF   17
KIT P/N 5596946...................
                                              UOC:A13,A14,H13,H14   10
PAOZZ 5340012551109 19207 12340635   BRACKET,ANGLE PART OF KIT P/N
1   5596946...........................
UOC:A13,A14,H13,H14   11 PAOZZ 5305000680502 96906 MS90725-6   SCREW,CAP,HEXAGON
H 1/4-20 X .75   2   PART OF KIT P/N
5596946............
                                              UOC:A13,A14,H13,H14

                                   END OF FIGURE

421-1

Section II. TM 9-2320-280-24P-2

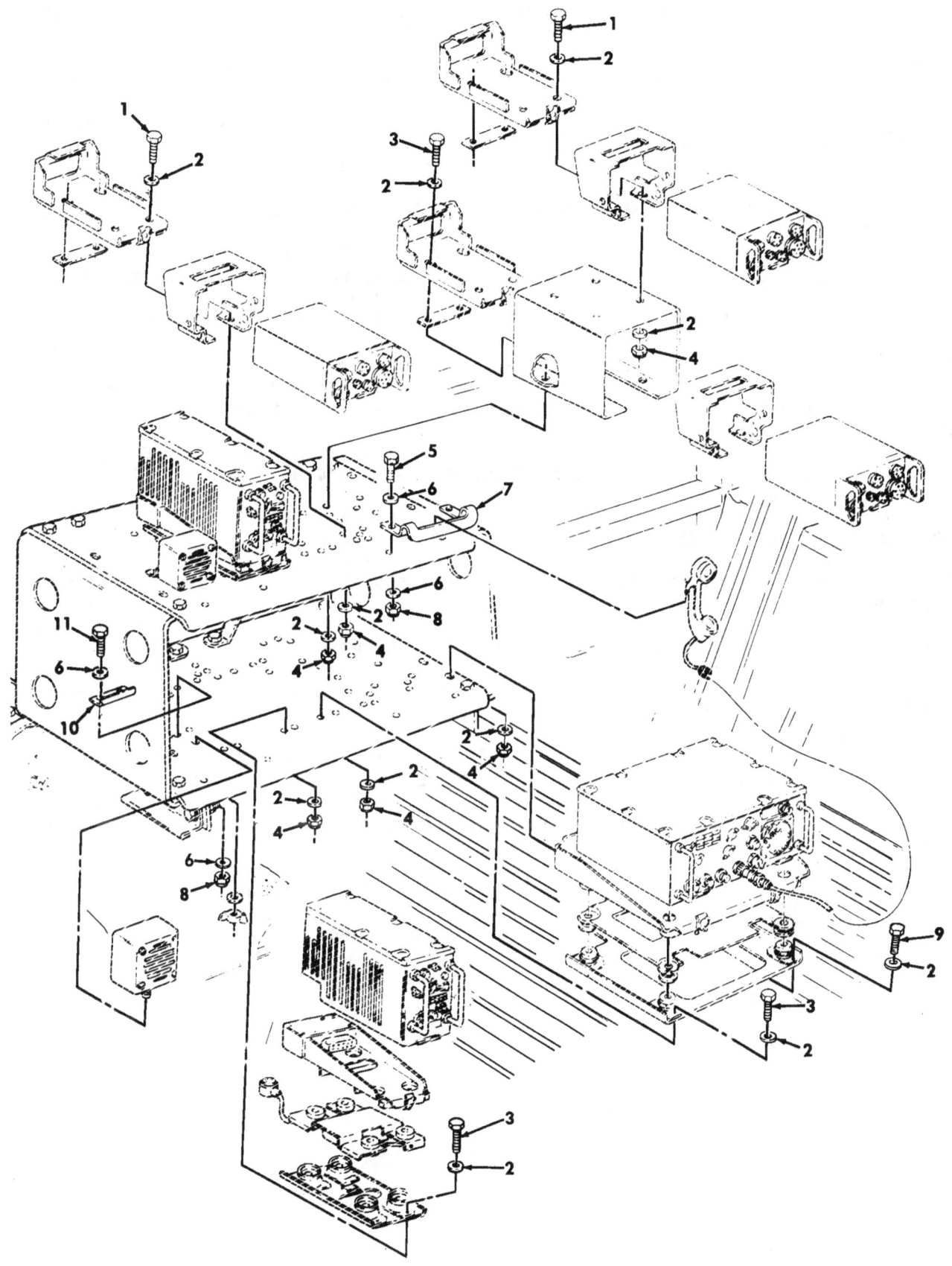

Figure 422. Communications Kit, Radio Rack Assembly Mounting Hardware, AN/VRC-48.

SECTION II   TM9-2320-280-24P C01

(1) ITEM NO	(2) SMR CODE	(3) NSN	(4) CAGEC	(5) PART NUMBER	(6) DESCRIPTION AND USABLE ON CODES (UOC)	(7) QTY

GROUP 3307 SPECIAL PURPOSE KITS

FIG. 422 COMMUNICATIONS KIT, RADIO AN/VRC-48

RACK ASSEMBLY MOUNTING HARDWARE,

1	PAOZZ	5306002258498	96906	MS90725-33	BOLT,MACHINE 5/16-18 X .88 PART OF KIT P/N 5596946................... UOC:A13,A14,H13,H14	8
* 2	PAOZZ	5310011191024	24617	2436162	WASHER,FLAT 5/16 PART OF KIT P/N 5596946............................ UOC:A13,A14,H13,H14	50
3	PAOZZ	5306002258499	96906	MS90725-34	BOLT,MACHINE 5/16-18 X 1.00 PART OF KIT P/N 5596946................... UOC:A13,A14,H13,H14	15
4	PAOZZ	5310004722963	96906	MS51943-3	NUT,SELF-LOCKING,HE 5/16-18 PART OF KIT P/N 5596946................... UOC:A13,A14,H13,H14	25
5	PAOZZ	5305000712505	80204	MS90728-7	SCREW,CAP,HEXAGON H 1/4-20 X .88 PART OF KIT P/N 5596946............. UOC:A13,A14,H13,H14	2
6	PAOZZ	5310012510748	34623	5593691	WASHER,FLAT 1/4 PART OF KIT P/N 5596946............................ UOC:A13,A14,H13,H14	12
7	PAOZZ	5340012918915	34623	5597357	BRACKET,MOUNTING PART OF KIT P/N 5596946......... UOC:A13,A14,H13,H14	1
8	PAOZZ	5310000614650	96906	MS51943-31	NUT,SELF-LOCKING,HE 1/4-20 PART OF KIT P/N 5596946................... UOC:A13,A14,H13,H14	6
9	PAOZZ	5306002264828	80204	B1821BH031C113N	BOLT,MACHINE 5/16-18 X 1.125 PART OF KIT P/N 5596946................ UOC:A13,A14,H13,H14	2
10	PAOZZ	5340012551109	19207	12340635	BRACKET,ANGLE PART OF KIT P/N 5596946............................ UOC:A13,A14,H13,H14	2
11	PAOZZ	5305000680502	96906	MS90725-6	SCREW,CAP,HEXAGON H 1/4-20 X .75 PART OF KIT P/N 5596946............. UOC:A13,A14,H13,H14	4

END OF FIGURE

Section II. TM 9-2320-280-24P-2

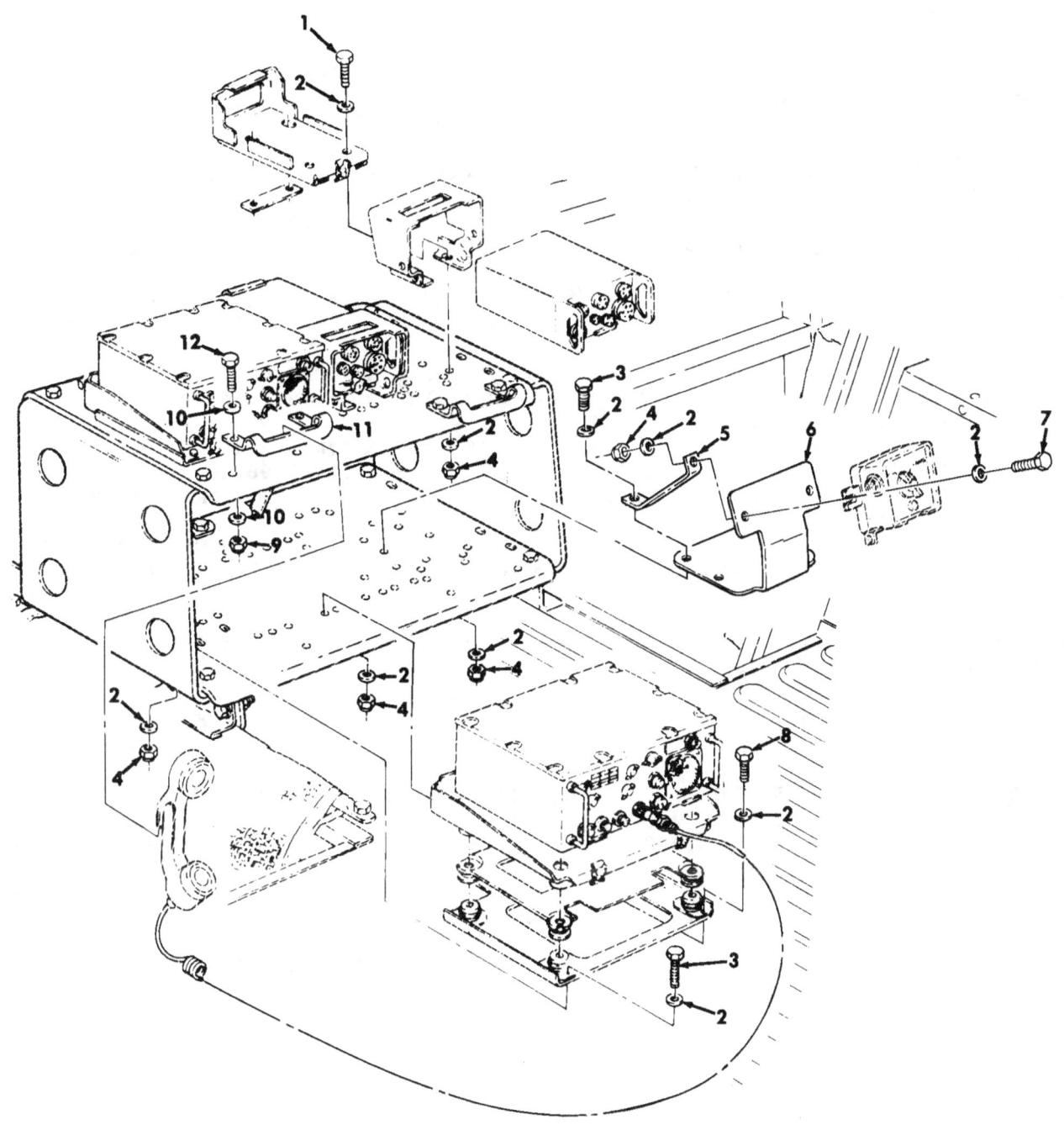

Figure 423. Communications Kit, Radio Rack Assembly Mounting Hardware, AN/VRC-49.

SECTION II TM9-2320-280-24P C01

(1) ITEM NO	(2) SMR CODE	(3) NSN	(4) CAGEC	(5) PART NUMBER	(6) DESCRIPTION AND USABLE ON CODES (UOC)	(7) QTY

GROUP 3307 SPECIAL PURPOSE KITS

FIG. 423 COMMUNICATIONS KIT, RADIO AN/VRC-49

RACK ASSEMBLY MOUNTING HARDWARE,

1	PAOZZ	5306002258498	96906	MS90725-33	BOLT,MACHINE 5/16-18 X .88 PART OF KIT P/N 5596946.................... UOC:A13,A14,H13,H14	8
*2	PAOZZ	5310011191024	24617	2436162	WASHER,FLAT 5/16 PART OF KIT P/N 5596946........................... UOC:A13,A14,H13,H14	48
3	PAOZZ	5306002264828	80204	B1821BH031C113N	BOLT,MACHINE 5/16-18 X 1.125 PART OF KIT P/N 5596946................. UOC:A13,A14,H13,H14	8
4	PAOZZ	5310008140673	96906	MS51943-33	NUT,SELF-LOCKING,HE 5/16-18 PART OF KIT P/N 5596946................... UOC:A13,A14,H13,H14	24
5	PAOZZ	2510012905675	34623	5597595	BRACE,RACK PART OF KIT P/N 5596946.. UOC:A13,A14,H13,H14	2
6	PAOZZ	5340012908473	34623	5596797	BRACKET,DOUBLE ANGL PART OF KIT P/N 5596946.......... .... UOC:A13,A14,H13,H14	1
7	PAOZZ	5306002264829	80204	B1821BH031C125N	BOLT,MACHINE 5/16-18 X 1.25 PART OF KIT P/N 5596946.................... UOC:A13,A14,H13,H14	2
8	PAOZZ	5306002258499	96906	MS90725-34	BOLT,MACHINE 5/16-18 X 1.00 PART OF KIT P/N 5596946.................... UOC:A13,A14,H13,H14	6
9	PAOZZ	5310000614650	96906	MS51943-31	NUT,SELF-LOCKING,HE 1/4-20 PART OF KIT P/N 5596946.................... UOC:A13,A14,H13,H14	4
10	PAOZZ	5310012510748	34623	5593691	WASHER,FLAT 1/4 PART OF KIT P/N 5596946........................... UOC:A13,A14,H13,H14	8
11	PAOZZ	5340012918915	34623	5597357	BRACKET,MOUNTING PART OF KIT P/N 5596946......... .... UOC:A13,A14,H13,H14	2
12	PAOZZ	5305000712505	80204	MS90728-7	SCREW,CAP,HEXAGON H 1/4-20 X .88 PART OF KIT P/N 5596946............. UOC:A13,A14,H13,H14	4

END OF FIGURE

Section II.

TM 9-2320-280-24P-2

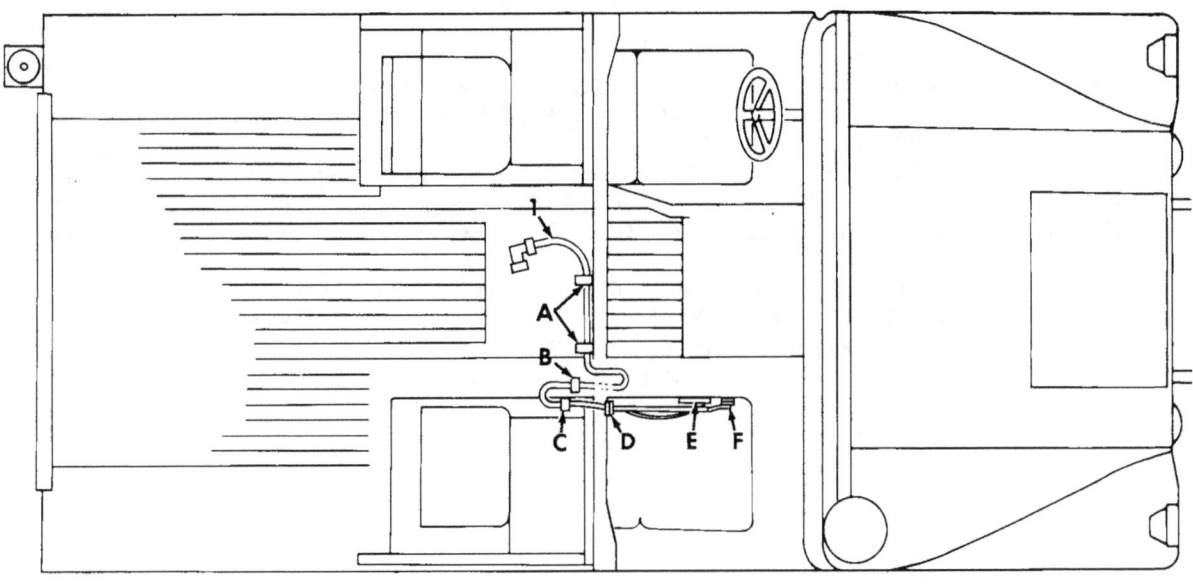

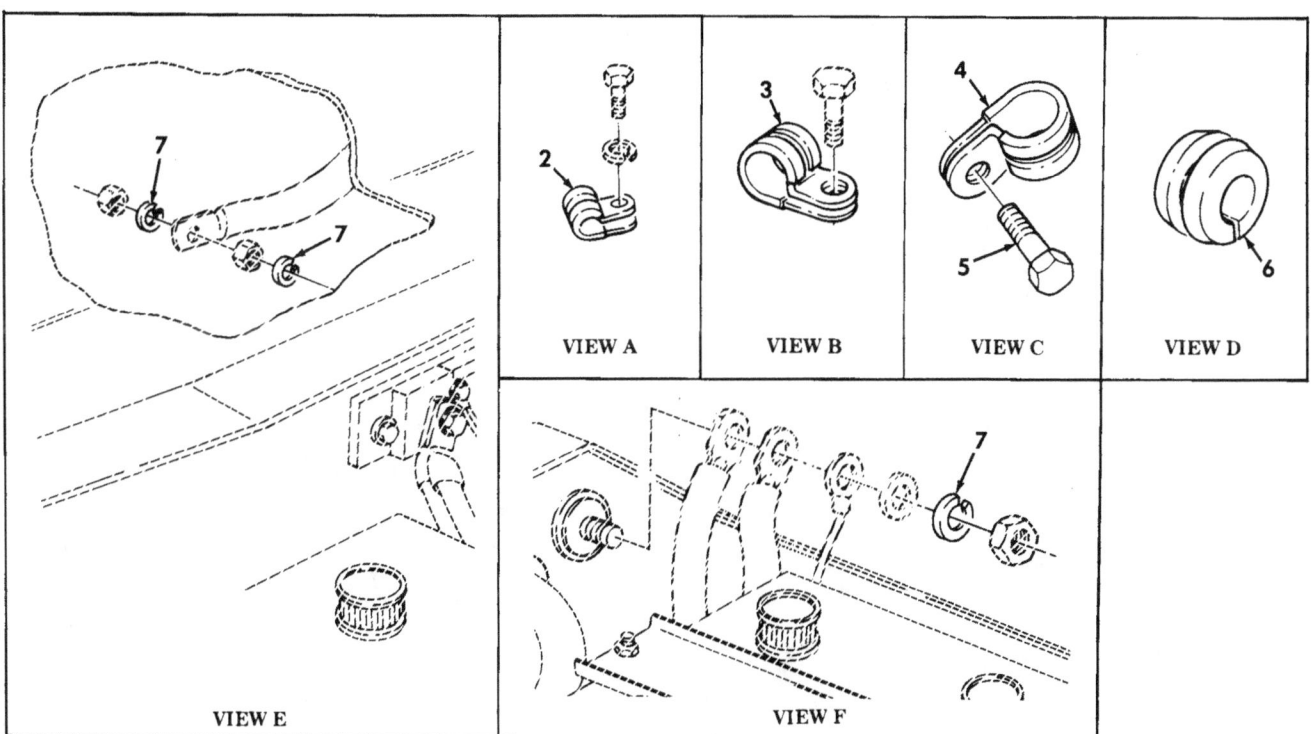

Figure 424. Communications Kit, Radio Cable, Main Power.

SECTION II   TM9-2320-280-24P

(1) NO	(2) SMR CODE	(3) NSN	(4) CAGEC	(5) PART NUMBER	(6) DESCRIPTION AND USABLE ON CODES(UOC)	(7) QTY

GROUP 3307 SPECIAL PURPOSE KITS

FIG. 424 COMMUNICATIONS KIT,RADIO CABLE,MAIN POWER

1 PAOZZ 6150012894761 34623 5597136-C CABLE ASSEMBLY,POWE PART OF KIT P/N 5596946............................ 1
 UOC:A13,A14,H13,H14

2 PAOZZ 5340000573043 96906 MS21333-112 CLAMP,LOOP PART OF KIT P/N 5596946.. 2
 UOC:A13,A14,H13,H14

3 PAOZZ 5340000502740 96906 MS21333-75 CLAMP,LOOP PART OF KIT P/N 5596946.. 1
 UOC:A13,A14,H13,H14

4 PAOZZ 5340000913790 96906 MS21333-72 CLAMP,LOOP PART OF KIT P/N 5596946.. 1
 UOC:A13,A14,H13,H14

5 PAOZZ 5305012596322 34623 12342499-1 SCREW,MACHINE #10-32 X 3/4 PART OF KIT P/N 5596946..................... 1
 UOC:A13,A14,H13,H14

6 PAOZZ 5325002858363 96906 MS35489-45 GROMMET,NONMETALLIC PART OF KIT P/N 5596946............................ 1
 UOC:A13,A14,H13,H14

7 PAOZZ 5310005845272 96906 MS35338-48 WASHER,LOCK 1/2 PART OF KIT P/N 5596946............................ 3
 UOC:A13,A14,H13,H14

END OF FIGURE

Section II. TM 9-2320-280-24P-2

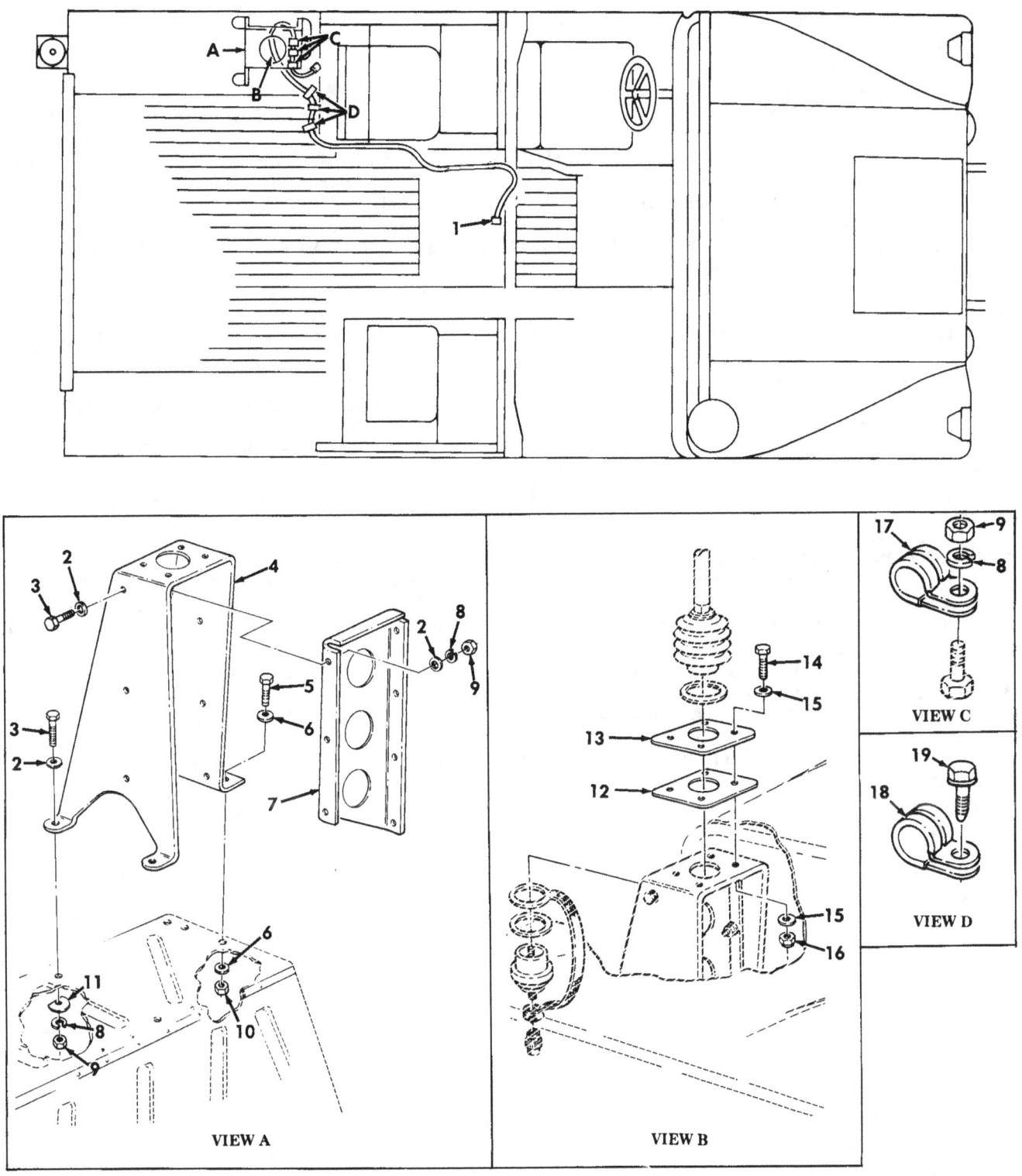

Figure 425. Communications Kit, Radio Cables, AN/VRC-47 and AN/VRC-48.

SECTION II   TM9-2320-280-24P

(1) NO	(2) SMR CODE	(3) NSN	(4) CAGEC	(5) PART NUMBER	(6) DESCRIPTION AND USABLE ON CODES(UOC)	(7) QTY

GROUP 3307 SPECIAL PURPOSE KITS

FIG. 425 COMMUNICATIONS KIT,RADIO CABLES,AN/VRC-47 AND AN/VRC-48

1	PAOZZ	5995012916377	34623	5597138	CABLE ASSEMBLY,SPEC PART OF KIT P/N 5596946.......................... UOC:A13,A14,H13,H14	1
2	PAOZZ	5310012510748	34623	5593691	WASHER,FLAT 1/4 PART OF KIT P/N 5596946.............................. UOC:A13,A14,H13,H14	16
3	PAOZZ	5305000680509	80204	B1821BH025C125N	SCREW,CAP,HEXAGON H 1/4-20 X 1.25 PART OF KIT P/N 5596946............. UOC:A13,A14,H13,H14	9
4	PAOZZ	5340012904887	34623	5596843	BRACKET,MOUNTING PART OF KIT P/N 5596946.......................... UOC:A13,A14,H13,H14	1
5	PAOZZ	5305000711788	80204	B1821BH044C125N	SCREW,CAP,HEXAGON H 7/16-14 X 1.25 PART OF KIT P/N 5596946............. UOC:A13,A14,H13,H14	2
6	PAOZZ	5310013034701	96906	MS51412-1	WASHER,FLAT 15/32 PART OF KIT P/N 5596946.............................. UOC:A13,A14,H13,H14	4
7	PAOZZ	5340012915186	34623	5597150	BRACKET,MOUNTING PART OF KIT P/N 5596946.......................... UOC:A13,A14,H13,H14	1
8	PAOZZ	5310005825965	96906	MS35338-44	WASHER,LOCK 1/4 PART OF KIT P/N 5596946.......................... UOC:A13,A14,H13,H14	12
9	PAOZZ	5310009050762	96906	MS51967-3	NUT,PLAIN,HEXAGON 1/4-20 PART OF KIT P/N 5596946.................. UOC:A13,A14,H13,H14	12
10	PAOZZ	5310008960903	96906	MS51967-12	NUT,PLAIN,HEXAGON 7/16-14 PART OF KIT P/N 5596946.................. UOC:A13,A14,H13,H14	2
11	PAOZZ	5310012087576	19207	12338095	WASHER,FLAT 1/4 PART OF KIT P/N 5596946.............................. UOC:A13,A14,H13,H14	2
12	PAOZZ	5330012899231	34623	5597371	GASKET PART OF KIT P/N 5596946...... UOC:A13,A14,H13,H14	1
13	PAOZZ	5985012533514	19207	12340637	ADAPTER,ANTENNA TO PART OF KIT P/N 5596946.......................... UOC:A13,A14,H13,H14	1
14	PAOZZ	5305005432419	80204	B1821BH038C113N	SCREW,CAP,HEXAGON H 3/8-16 X 1.125 PART OF KIT P/N 5596946............. UOC:A13,A14,H13,H14	4
15	PAOZZ	5310014124013	24617	2436163	WASHER,FLAT 3/8 PART OF KIT P/N 5596946.............................. UOC:A13,A14,H13,H14	8
16	PAOZZ	5310009359021	96906	MS51943-35	NUT,SELF-LOCKING,HE 3/8-16 PART OF KIT P/N 5596946..................... UOC:A13,A14,H13,H14	4
17	PAOZZ	5340008546729	96906	MS21333-103	CLAMP,LOOP PART OF KIT P/N 5596946..	3

425-1

SECTION II   TM9-2320-280-24P

(1) ITEM NO	(2) SMR CODE	(3) NSN	(4) CAGEC	(5) PART NUMBER	(6) DESCRIPTION AND USABLE ON CODES(UOC)	(7) QTY
18	PAOZZ	5340000797837	96906	MS21333-67	CLAMP,LOOP PART OF KIT P/N 5596946.. UOC:A13,A14,H13,H14	3
19	PAOZZ	5305012596322	34623 93913	SCREW,MACHINE #10-32 X 3/4 PART OF KIT P/N 5596946..................... UOC:A13,A14,H13,H14	3

END OF FIGURE

Section II. TM 9-2320-280-24P-2

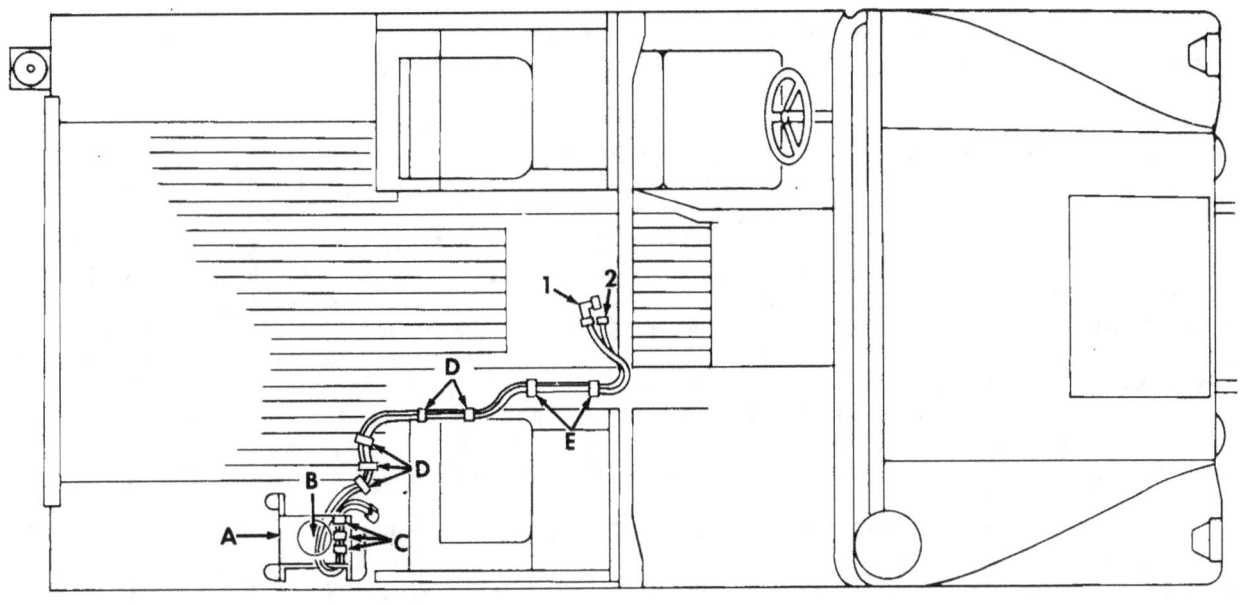

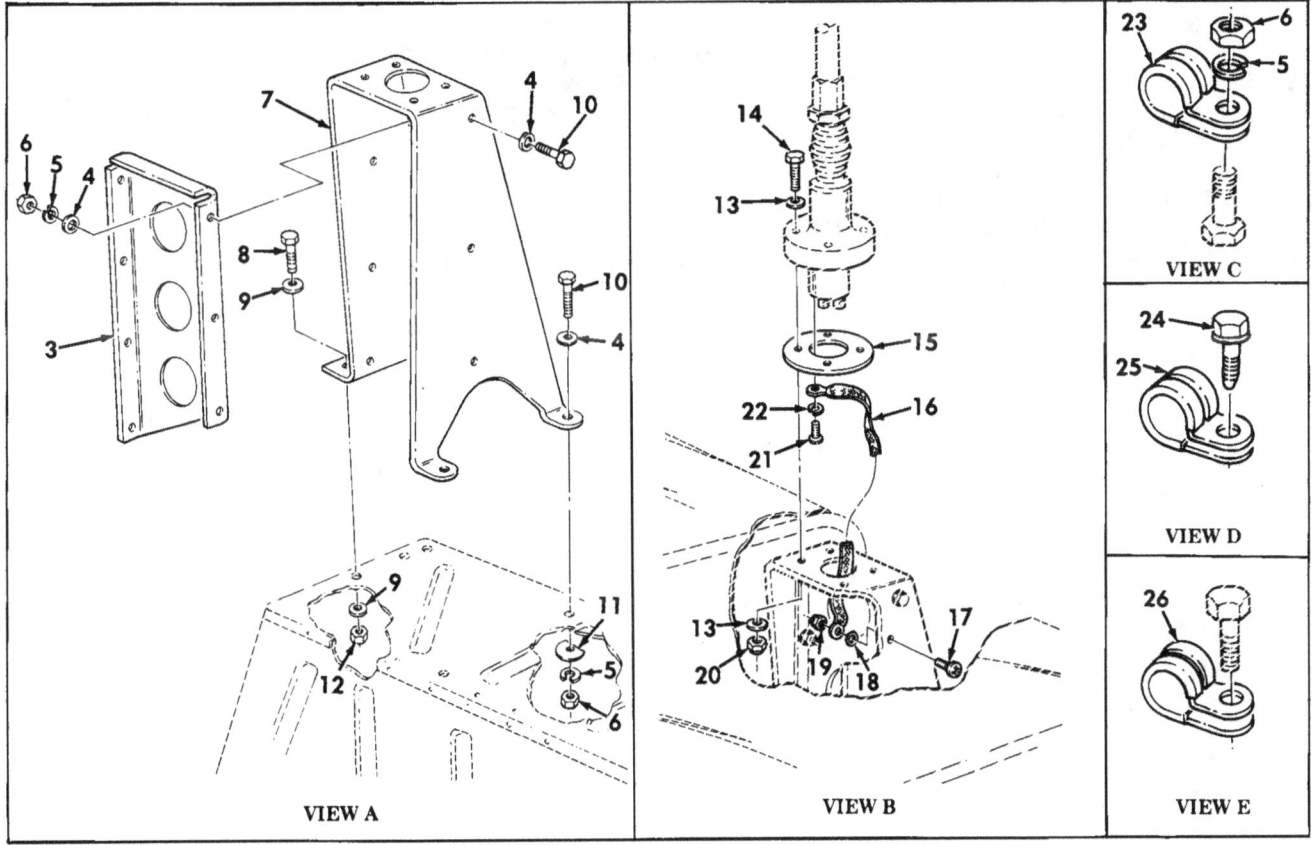

Figure 426. Communications Kit, Radio Cables, AN/VRC-46, (2) AN/VRC-46, (3) AN/VRC-46, AN/VRC-47, AN/VRC-48, AND AN/VRC-49.

SECTION II   TM9-2320-280-24P

(1) NO	(2) SMR CODE	(3) NSN	(4) CAGEC	(5) PART NUMBER	(7) DESCRIPTION AND USABLE ON CODES(UOC)	(6) QTY

GROUP 3307 SPECIAL PURPOSE KITS

FIG. 426 COMMUNICATIONS KIT, RADIO CABLES, AN/VRC-46, (2) AN/VRC-46, (3) AN/VRC-46, AN/VRC-47, AN/VRC-48, AND AN/VRC-49

(1)	(2)	(3)	(4)	(5)	(7)	(6)
1	PAOZZ	5995012901719	34623	5597137	CABLE ASSEMBLY,RADI PART OF KIT P/N 5596946............... UOC:A13,A14,H13,H14	1
2	PAOZZ	5995012916377	34623	5597138	CABLE ASSEMBLY,SPEC PART OF KIT P/N 5596946............... UOC:A13,A14,H13,H14	1
3	PAOZZ	5340012906343	34623	5597151	BRACKET,MOUNTING PART OF KIT P/N 5596946............... UOC:A13,A14,H13,H14	1
4	PAOZZ	5310012510748	34623	5593691	WASHER,FLAT 1/4 PART OF KIT P/N 5596946............... UOC:A13,A14,H13,H14	16
5	PAOZZ	5310005825965	96906	MS35338-44	WASHER,LOCK 1/4 PART OF KIT P/N 5596946............... UOC:A13,A14,H13,H14	12
6	PAOZZ	5310009050762	96906	MS51967-3	NUT,PLAIN,HEXAGON 1/4-20 PART OF KIT P/N 5596946............... UOC:A13,A14,H13,H14	12
7	PAOZZ	5340012904888	34623	5597155	BRACKET,MOUNTING PART OF KIT P/N 5596946............... UOC:A13,A14,H13,H14	1
8	PAOZZ	5305000711788	80204	B1821BH044C125N	SCREW,CAP,HEXAGON H 7/16-14 X 1.25 PART OF KIT P/N 5596946............ UOC:A13,A14,H13,H14	2
9	PAOZZ	5310013034701	96906	MS51412-1	WASHER,FLAT 15/32 PART OF KIT P/N 5596946............ UOC:A13,A14,H13,H14	4
10	PAOZZ	5305000680509	80204	B1821BH025C125N	SCREW,CAP,HEXAGON H 1/4-20 X 1.25 PART OF KIT P/N 5596946............ UOC:A13,A14,H13,H14	9
11	PAOZZ	5310012087576	19207	12338095	WASHER,FLAT 1/4 PART OF KIT P/N 5596946............... UOC:A13,A14,H13,H14	2
12	PAOZZ	5310008960903	96906	MS51967-12	NUT,PLAIN,HEXAGON 7/16-14 PART OF KIT P/N 5596946............... UOC:A13,A14,H13,H14	2
13	PAOZZ	5310014124013	24617	2436163	WASHER,FLAT 3/8 PART OF KIT P/N 5596946............... UOC:A13,A14,H13,H14	8
14	PAOZZ	5305012608881	34623	5593315	SCREW,CAP,HEXAGON H 3/8-16 X 2.25 PART OF KIT P/N 5596946............ UOC:A13,A14,H13,H14	4
15	PAOZZ	5330012708315	19207	12341876	GASKET PART OF KIT P/N 5596946...... UOC:A13,A14,H13,H14	1
16	PAOZZ	6150012078120	34623	12338099-1	LEAD,ELECTRICAL PART OF KIT P/N 5596946...............	1

SECTION II   TM9-2320-280-24P

(1) ITEM NO	(2) SMR CODE	(3) NSN	(4) CAGEC	(5) PART NUMBER	(6) DESCRIPTION AND USABLE ON CODES(UOC)	(7) QTY
17	PAOZZ	5305000593660	96906	MS51958-64	SCREW,MACHINE 10-32 X .625 PART OF KIT P/N 5596946.................... UOC:A13,A14,H13,H14	1
18	PAOZZ	5310000809786	96906	MS45904-5596946..........................	WASHER,LOCK #10 PART OF KIT P/N 5596946 UOC:A13,A14,H13,H14	1
19	PAOZZ	5310012510726	34623	5593053	NUT,SELF-LOCKING,HE #10-32 PART OF KIT P/N 5596946..................... UOC:A13,A14,H13,H14	1
20	PAOZZ	5310009359021	96906	MS51943-35	NUT,SELF-LOCKING,HE 3/8-16 PART OF KIT P/N 5596946..................... UOC:A13,A14,H13,H14	4
21	PAOZZ	5305000593659	96906	MS51958-63	SCREW,MACHINE 10-32 X .50 PART OF KIT P/N 5596946..................... UOC:A13,A14,H13,H14	1
22	PAOZZ	5310000453296	96906	MS35338-5596946..........................	WASHER,LOCK #10 PART OF KIT P/N 5596946 UOC:A13,A14,H13,H14	1
23	PAOZZ	5340008546729	96906	MS21333-103	CLAMP,LOOP PART OF KIT P/N 5596946.. UOC:A13,A14,H13,H14	1
24	PAOZZ	5305012596322	34623	12342499-1	SCREW,MACHINE 10-32 X 3/4 PART OF KIT P/N 5596946..................... UOC:A13,A14,H13,H14	5
25	PAOZZ	5340000913790	96906	MS21333-72	CLAMP,LOOP PART OF KIT P/N 5596946.. UOC:A13,A14,H13,H14	5
26	PAOZZ	5340000573043	96906	MS21333-112	CLAMP,LOOP PART OF KIT P/N 5596946.. UOC:A13,A14,H13,H14	2

END OF FIGURE

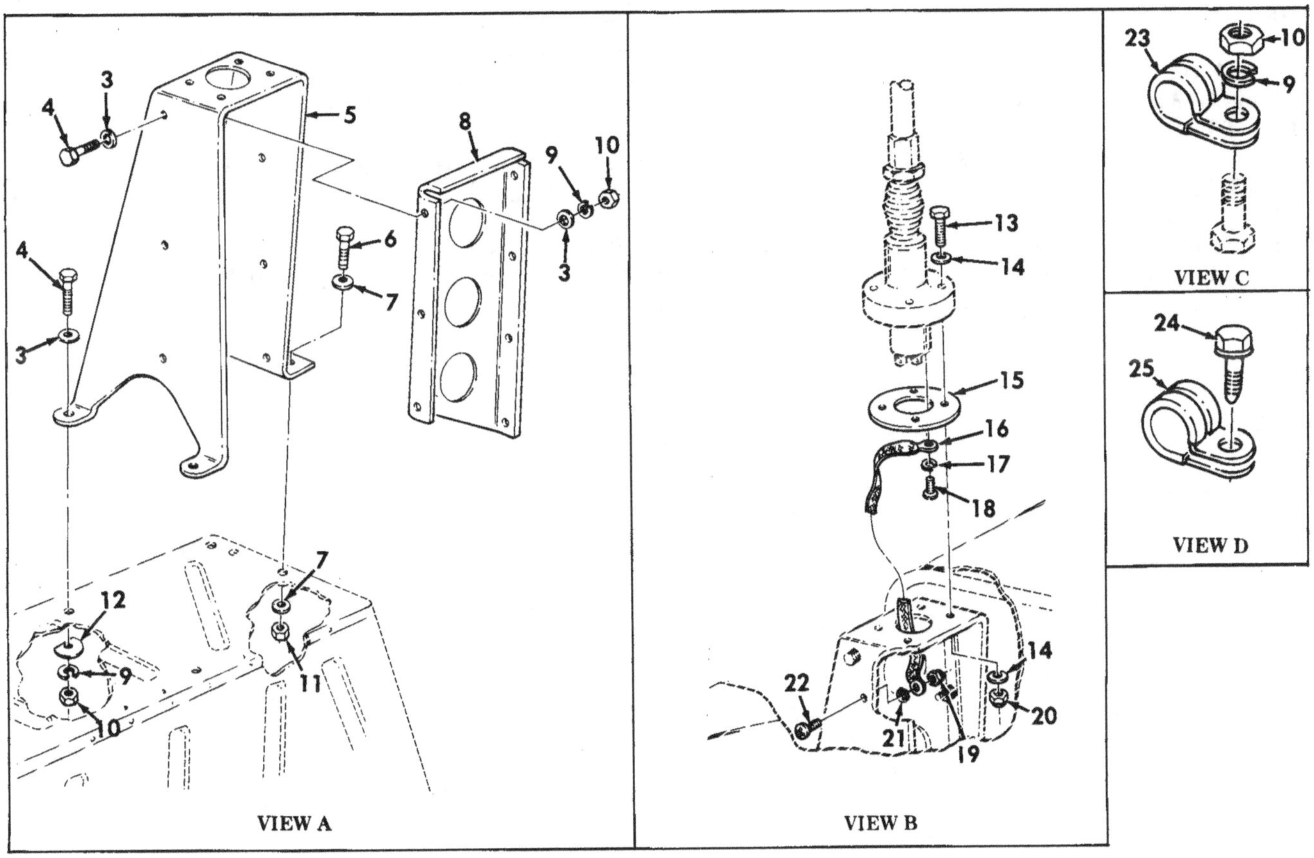

Section II. TM 9-2320-280-24P-2

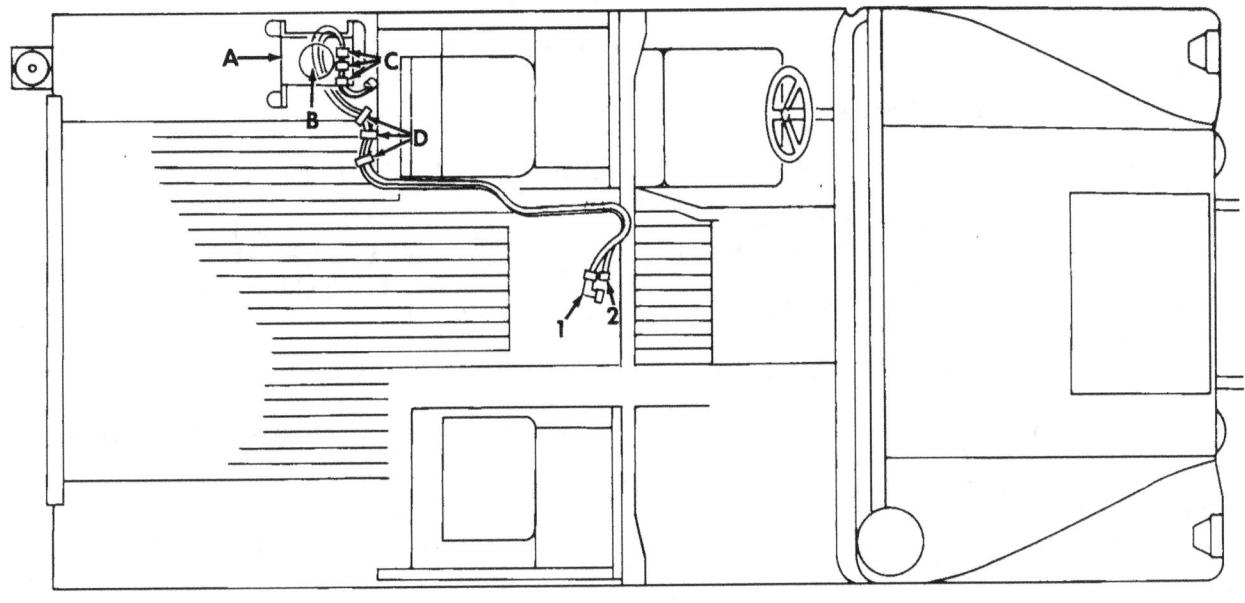

Figure 427. Communications Kit, Radio Cables, (2) AN/VRC-46, (3) AN/VRC-47, and AN/VRC-49.

SECTION II TM9-2320-280-24P

(1) ITEM NO	(2) SMR CODE	(3) NSN	(4) CAGEC	(5) PART NUMBER	(6) DESCRIPTION AND USABLE ON CODES(UOC)	(7) QTY

GROUP 3307 SPECIAL PURPOSE KITS

FIG. 427 COMMUNICATIONS KIT,RADIO CABLES,(2) AN/VRC-46,(3) AN/VRC-47, AND AN/VRC-49

Item	SMR	NSN	CAGEC	Part Number	Description	Qty
1	PAOZZ	5995012901719	34623	5597137	CABLE ASSEMBLY,RADI PART OF KIT P/N 5596946............... UOC:A13,A14,H13,H14	1
2	PAOZZ	5995012916377	34623	5597138	CABLE ASSEMBLY,SPEC PART OF KIT P/N 5596946............... UOC:A13,A14,H13,H14	1
3	PAOZZ	5310012510748	34623	5593691	WASHER,FLAT 1/4 PART OF KIT P/N 5596946............... UOC:A13,A14,H13,H14	16
4	PAOZZ	5305000680509	80204	B1821BH025C125N	SCREW,CAP,HEXAGON H 1/4-20 X 1.25 PART OF KIT P/N 5596946............. UOC:A13,A14,H13,H14	9
5	PAOZZ	5340012904887	34623	5596843	BRACKET,MOUNTING PART OF KIT P/N 5596946............... UOC:A13,A14,H13,H14	1
6	PAOZZ	5305000711788	80204	B1821BH044C125N	SCREW,CAP,HEXAGON H 7/16-14 X 1.25 PART OF KIT P/N 5596946............. UOC:A13,A14,H13,H14	2
7	PAOZZ	5310013034701	96906	MS51412-1	WASHER,FLAT 15/32 PART OF KIT P/N 5596946............... UOC:A13,A14,H13,H14	4
8	PAOZZ	5340012915186	34623	5597150	BRACKET,MOUNTING PART OF KIT P/N 5596946............... UOC:A13,A14,H13,H14	1
9	PAOZZ	5310005825965	96906	MS35338-44	WASHER,LOCK 1/4 PART OF KIT P/N 5596946............... UOC:A13,A14,H13,H14	12
10	PAOZZ	5310009050762	96906	MS51967-3	NUT,PLAIN,HEXAGON 1/4-20 PART OF KIT P/N 5596946............... UOC:A13,A14,H13,H14	12
11	PAOZZ	5310008960903	96906	MS51967-12	NUT,PLAIN,HEXAGON 7/16-14 PART OF KIT P/N 5596946............... UOC:A13,A14,H13,H14	2
12	PAOZZ	5310012087576	19207	12338095	WASHER,FLAT 1/4 PART OF KIT P/N 5596946............... UOC:A13,A14,H13,H14	2
13	PAOZZ	5305012608881	34623	5593315	SCREW,CAP,HEXAGON H 3/8-16 X 2.25 PART OF KIT P/N 5596946............. UOC:A13,A14,H13,H14	4
14	PAOZZ	5310014124013	24617	2436163	WASHER,FLAT 3/8 PART OF KIT P/N 5596946............... UOC:A13,A14,H13,H14	8
15	PAOZZ	5330012708315	19207	12341876	GASKET PART OF KIT P/N 5596946...... UOC:A13,A14,H13,H14	1
16	PAOZZ	6150012078120	34623	12338099-1	LEAD,ELECTRICAL PART OF KIT P/N 5596946............... UOC:A13,A14,H13,H14	1

SECTION II TM9-2320-280-24P

(1) ITEM NO	(2) SMR CODE	(3) NSN	(4) CAGEC	(5) PART NUMBER	(6) DESCRIPTION AND USABLE ON CODES(UOC)	(7) QTY
17	PAOZZ	5310000453296	96906	MS35338-43	WASHER,LOCK #10 PART OF KIT P/N 5596946............ UOC:A13,A14,H13,H14	1
18	PAOZZ	5305000593659	96906	MS51958-63	SCREW,MACHINE #10-32 X .50 PART OF KIT P/N 5596946.............. UOC:A13,A14,H13,H14	1
19	PAOZZ	5310012510726	34623	5593053	NUT,SELF-LOCKING,HE #10-32 PART OF KIT P/N 5596946.............. UOC:A13,A14,H13,H14	4
20	PAOZZ	5310009359021	96906	MS51943-35	NUT,SELF-LOCKING,HE 3/8-16 PART OF KIT P/N 5596946.............. UOC:A13,A14,H13,H14	1
21	PAOZZ	5310000809786	96906	MS45904-	WASHER,LOCK #10 PART OF KIT P/N 5596946............ UOC:A13,A14,H13,H14	1
22	PAOZZ	5305000593660	96906	MS51958-64	SCREW,MACHINE #10-32 X .625 PART OF KIT P/N 5596946.............. UOC:A13,A14,H13,H14	1
23	PAOZZ	5340008546729	96906	MS21333-103	CLAMP,LOOP PART OF KIT P/N 5596946.. UOC:A13,A14,H13,H14	3
24	PAOZZ	5305012596322	34623	342499-1	SCREW,MACHINE #10-32 X 3/4 PART OF KIT P/N 5596946............... UOC:A13,A14,H13,H14	3
25	PAOZZ	5340000913790	96906	MS21333-72	CLAMP,LOOP PART OF KIT P/N 5596946.. UOC:A13,A14,H13,H14	3

END OF FIGURE

Section II.

TM 9-2320-280-24P-2

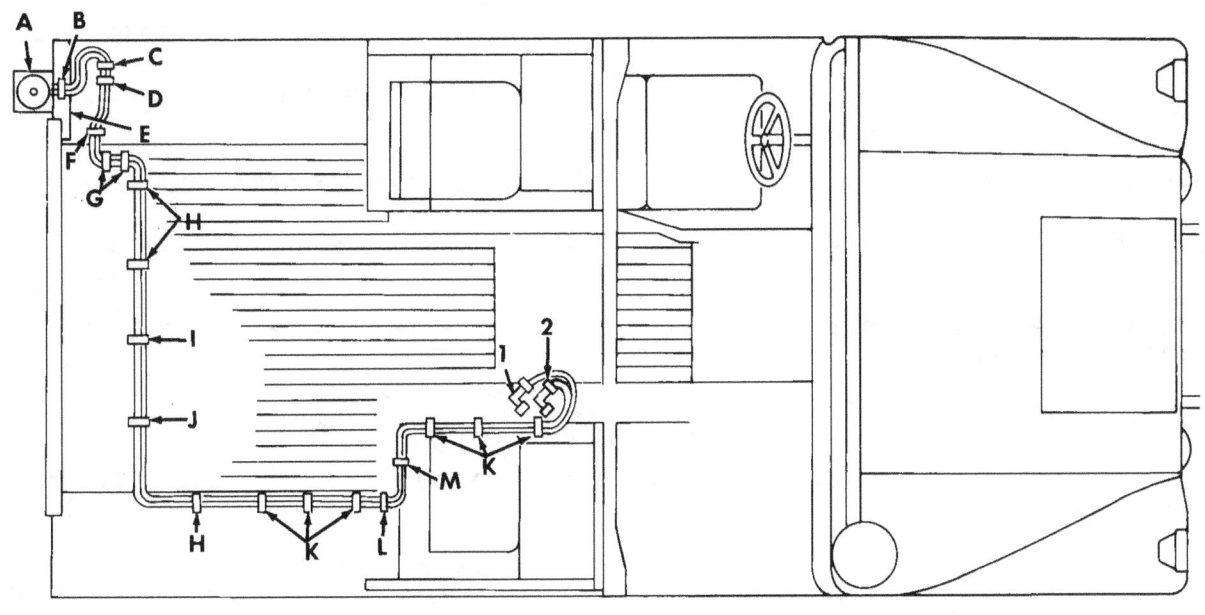

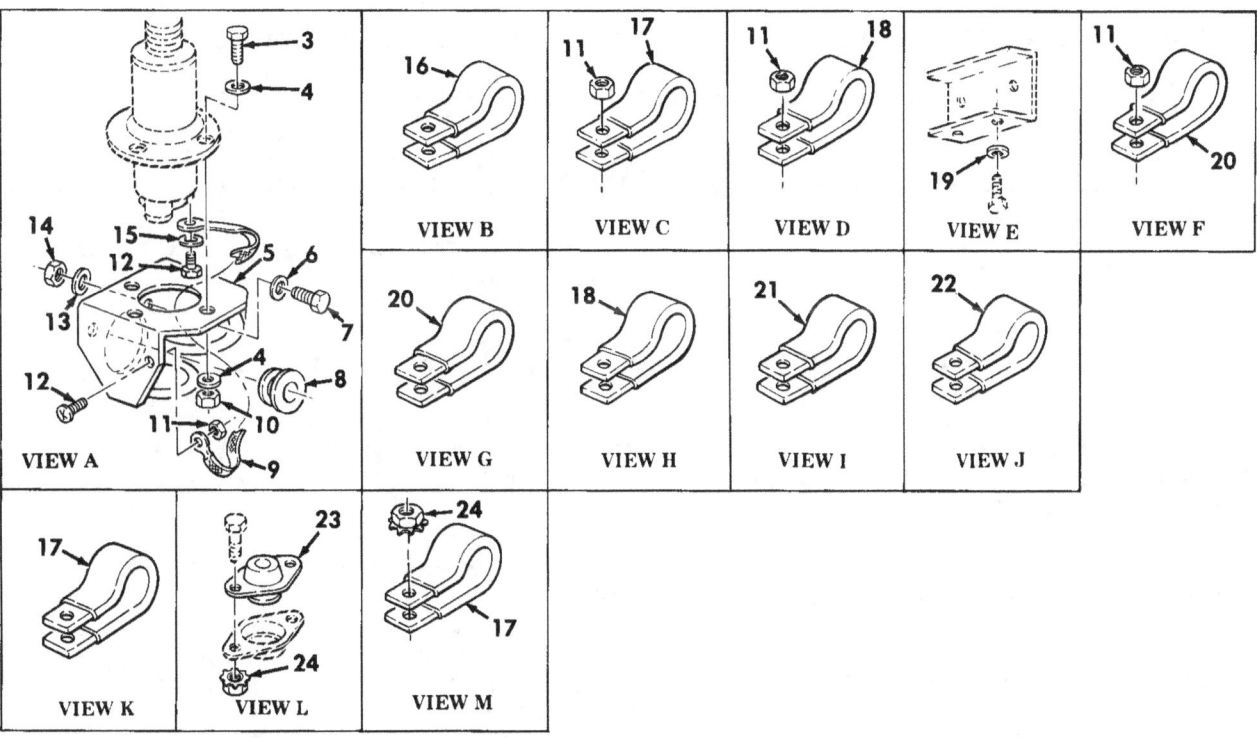

Figure 428. Communications Kit, Radio Cables, (3) AN/VRC-46.

SECTION II TM9-2320-280-24P C01

GROUP 3307 SPECIAL PURPOSE KITS

FIG. 428 COMMUNICATIONS KIT, RADIO CABLES, (3) AN/VRC-46

(1) ITEM NO	(2) SMR CODE	(3) NSN	(4) CAGEC	(5) PART NUMBER	(6) DESCRIPTION AND USABLE ON CODES (UOC)	(7) QTY
1	PAOZZ	5995012916384	34623	5597496	CABLE ASSEMBLY,RADI PART OF KIT P/N 5596946............................... UOC:A13,A14,H13,H14	1
2	PAOZZ	5995012516785	19201	12338089-4	CABLE ASSEMBLY,SPEC PART OF KIT P/N 5596946........... .... UOC:A13,A14,H13,H14	1
3	PAOZZ	5305007829489	80204	B1821BH038C200N	SCREW,CAP,HEXAGON H 3/8-16 X 2.00 PART OF KIT P/N 5596946............. UOC:A13,A14,H13,H14	4
4	PAOZZ	5310014124013	24617	2436163	WASHER,FLAT 3/8 PART OF KIT P/N 5596946............................... UOC:A13,A14,H13,H14	8
5	PAOZZ	5985012159404	80063	A3046166	SUPPORT,ANTENNA LEFT REAR PART OF KIT P/N 5596946..................... UOC:A13,A14,H13,H14	1
* 6	PAOZZ	5310011191024	24617	2436162	WASHER,FLAT 5/16 PART OF KIT P/N 5596946............................... UOC:A13,A14,H13,H14	3
7	PAOZZ	5306002258499	96906	MS90725-34	BOLT,MACHINE 5/16-18 X 1.00 PART OF KIT P/N 5596946..................... UOC:A13,A14,H13,H14	3
8	PAOZZ	5325007541155	96906	MS35489-50	GROMMET,NONMETALLIC PART OF KIT P/N 5596946............................... UOC:A13,A14,H13,H14	1
9	PAOZZ	6150012078120	34623	12338099-1	LEAD,ELECTRICAL PART OF KIT P/N 5596946........... .... UOC:A13,A14,H13,H14	1
10	PAOZZ	5310009359021	96906	MS51943-35	NUT,SELF-LOCKING,HE 3/8-16 PART OF KIT P/N 5596946..................... UOC:A13,A14,H13,H14	4
11	PAOZZ	5310012510726	34623	5593053	NUT,SELF-LOCKING,HE #10-32 PART OF KIT P/N 5596946..................... UOC:A13,A14,H13,H14	4
12	PAOZZ	5305000593659	96906	MS51958-63	SCREW,MACHINE #10-32 X .50 PART OF KIT P/N 5596946..................... UOC:A13,A14,H13,H14	2
13	PAOZZ	5310012087576	19207	12338095	WASHER,FLAT PART OF KIT P/N 5596946. UOC:A13,A14,H13,H14	3
14	PAOZZ	5310008140673	96906	MS51943-33	NUT,SELF-LOCKING,HE 5/16-18 PART OF KIT P/N 5596946..................... UOC:A13,A14,H13,H14	3
15	PAOZZ	5310000453296	96906	MS35338-43	WASHER,LOCK #10 PART OF KIT P/N 5596946........................ ...... UOC:A13,A14,H13,H14	1
16	PAOZZ	5340008546730	96906	MS21333-124	CLAMP,LOOP PART OF KIT P/N 5596946.. UOC:A13,A14,H13,H14	1
17	PAOZZ	5340000572906	96906	MS21333-73	CLAMP,LOOP PART OF KIT P/N 5596946.. UOC:A13,A14,H13,H14	8

SECTION II   TM9-2320-280-24P C01

(1) ITEM NO	(2) SMR CODE	(3) NSN	(4) CAGEC	(5) PART NUMBER	(6) DESCRIPTION AND USABLE ON CODES (UOC)	(7) QTY
18	PAOZZ	5340009226300	96906	MS21333-77	CLAMP,LOOP PART OF KIT P/N 5596946.. UOC:A13,A14,H13,H14	4
19	PAOZZ	5310005825965	96906	MS35338-44	WASHER,LOCK 1/4 PART OF KIT P/N 5596946......... UOC:A13,A14,H13,H14	2
20	PAOZZ	5340007247038	96906	MS21333-76	CLAMP,LOOP PART OF KIT P/N 5596946.. UOC:A13,A14,H13,H14	1
21	PAOZZ	5340009588457	96906	MS21333-78	CLAMP,LOOP PART OF KIT P/N 5596946.. UOC:A13,A14,H13,H14	1
22	PAOZZ	5340000502740	96906	MS21333-75	CLAMP,LOOP PART OF KIT P/N 5596946.. UOC:A13,A14,H13,H14	1
23	PAOZZ	5325012555063	34623	5590155	GROMMET,NONMETALLIC PART OF KIT P/N 5596946............ UOC:A13,A14,H13,H14	1
* 24	PAOZZ	5310011520598	24617	271172	NUT,SEL-LOCKING,AS 1/4-20 PART OF KIT P/N 5596946.................... UOC:A13,A14,H13,H14	2
KIT	PDOZZ	5820012879612	34623	5596946	INSTALLATION KIT,EL................ UOC:A13,A14,H13,H14	1

```
ADAPTER,ANTENNA TO  (  1) 425-13
BOLT,MACHINE        ( 14) 417-4
BOLT,MACHINE        (  4) 417-7
BOLT,MACHINE        (  4) 418-4
BOLT,MACHINE        (  2) 418-7
BOLT,MACHINE        (  3) 418-6
BOLT,MACHINE        (  8) 419-5
BOLT,MACHINE        (  6) 419-7
BOLT,MACHINE        (  4) 419-8
BOLT,MACHINE        ( 13) 420-1
BOLT,MACHINE        (  8) 420-3
BOLT,MACHINE        (  6) 420-5
BOLT,MACHINE        (  8) 421-1
BOLT,MACHINE        (  2) 421-7
BOLT,MACHINE        (  7) 421-8
BOLT,MACHINE        (  8) 422-1
BOLT,MACHINE        ( 15) 422-3
BOLT,MACHINE        (  2) 422-9
BOLT,MACHINE        (  8) 423-1
BOLT,MACHINE        (  8) 423-3
BOLT,MACHINE        (  2) 423-7
BOLT,MACHINE        (  6) 423-8
BOLT,MACHINE        (  3) 428-7
BRACE,RACK          (  2) 423-5
BRACKET,ANGLE       (  1) 417-12
BRACKET,ANGLE       (  1) 417-14
BRACKET,ANGLE       (  1) 417-14
BRACKET,ANGLE       (  1) 421-10
BRACKET,ANGLE       (  2) 422-10
BRACKET,DOUBLE ANGL (  4) 417-10
BRACKET,DOUBLE ANGL (  1) 423-6
BRACKET,MOUNTING    (  2) 417-6
BRACKET,MOUNTING    (  2) 417-17
BRACKET,MOUNTING    (  1) 418-3
BRACKET,MOUNTING    (  2) 419-3
```

SECTION II   TM9-2320-280-24P

(1) NO	(2) CODE	(3) NSN	(4) CAGEC	(5) NUMBER	(6) DESCRIPTION AND USABLE ON CODES(UOC)	(7) ITEM SMR PART QTY
					BRACKET,MOUNTING	(2) 420-8
					BRACKET,MOUNTING	(1) 421-5
					BRACKET,MOUNTING	(1) 422-7
					BRACKET,MOUNTING	(2) 423-11
					BRACKET,MOUNTING	(1) 425-4
					BRACKET,MOUNTING	(1) 425-7
					BRACKET,MOUNTING	(1) 426-3
					BRACKET,MOUNTING	(1) 426-7
					BRACKET,MOUNTING	(1) 427-5
					BRACKET,MOUNTING	(1) 427-8
					CABLE ASSEMBLY,POWE	(1) 424-1
					CABLE ASSEMBLY,RADI	(1) 426-1
					CABLE ASSEMBLY,RADI	(1) 427-1
					CABLE ASSEMBLY,RADI	(1) 428-1
					CABLE ASSEMBLY,SPEC	(1) 425-1
					CABLE ASSEMBLY,SPEC	(1) 426-2
					CABLE ASSEMBLY,SPEC	(1) 427-2
					CABLE ASSEMBLY,SPEC	(1) 428-2
					CLAMP,LOOP	(2) 424-2
					CLAMP,LOOP	(1) 424-3
					CLAMP,LOOP	(1) 424-4
					CLAMP,LOOP	(3) 425-17
					CLAMP,LOOP	(3) 425-18
					CLAMP,LOOP	(3) 426-23
					CLAMP,LOOP	(5) 426-25
					CLAMP,LOOP	(2) 426-26
					CLAMP,LOOP	(3) 427-23
					CLAMP,LOOP	(3) 427-25
					CLAMP,LOOP	(1) 428-16
					CLAMP,LOOP	(8) 428-17
					CLAMP,LOOP	(4) 428-18
					CLAMP,LOOP	(3) 428-20
					CLAMP,LOOP	(1) 428-21
					CLAMP,LOOP	(1) 428-22
					GASKET	(1) 425-12
					GASKET	(1) 427-15
					GASKET	(1) 426-15
					GROMMET,NONMETALLIC	(1) 424-6
					GROMMET,NONMETALLIC	(1) 428-8
					GROMMET,NONMETALLIC	(1) 428-23
					LEAD,ELECTRICAL	(1) 426-16
					LEAD,ELECTRICAL	(1) 427-16
					LEAD,ELECTRICAL	(1) 428-9
					NUT	(2) 428-24
					NUT,PLAIN,BLIND RIV	(4) 417-8
					NUT,PLAIN,HEXAGON	(12) 425-9
					NUT,PLAIN,HEXAGON	(2) 425-10
					NUT,PLAIN,HEXAGON	(12) 426-6
					NUT,PLAIN,HEXAGON	(2) 426-12
					NUT,PLAIN,HEXAGON	(12) 427-10
					NUT,PLAIN,HEXAGON	(2) 427-11
					NUT,SELF-LOCKING,HE	(14) 417-11
					NUT,SELF-LOCKING,HE	(4) 417-18

(1) NO	(2) SMR CODE	(3) NSN	(4) CAGEC	(5) PART NUMBER	(6) DESCRIPTION AND USABLE ON CODES(UOC)	(7) QTY	ITEM
					NUT,SELF-LOCKING,HE	(9)	418-8
					NUT,SELF-LOCKING,HE	(2)	418-9
					NUT,SELF-LOCKING,HE	(4)	419-4
					NUT,SELF-LOCKING,HE	(18)	419-9
					NUT,SELF-LOCKING,HE	(27)	420-4
					NUT,SELF-LOCKING,HE	(4)	420-6
					NUT,SELF-LOCKING,HE	(4)	421-6
					NUT,SELF-LOCKING,HE	(17)	421-9
					NUT,SELF-LOCKING,HE	(25)	422-4
					NUT,SELF-LOCKING,HE	(6)	422-8
					NUT,SELF-LOCKING,HE	(24)	423-4
					NUT,SELF-LOCKING,HE	(4)	423-9
					NUT,SELF-LOCKING,HE	(4)	425-16
					NUT,SELF-LOCKING,HE	(1)	426-19
					NUT,SELF-LOCKING,HE	(4)	426-20
					NUT,SELF-LOCKING,HE	(4)	427-19
					NUT,SELF-LOCKING,HE	(1)	427-20
					NUT,SELF-LOCKING,HE	(4)	428-10
					NUT,SELF-LOCKING,HE	(4)	428-11
					NUT,SELF-LOCKING,HE	(3)	428-14
					PROTECTOR,EDGE	(1)	417-13
					RIVET,BLIND	(8)	417-9
					SCREW,CAP,HEXAGON H	(4)	417-3
					SCREW,CAP,HEXAGON H	(2)	417-16
					SCREW,CAP,HEXAGON H	(2)	418-1
					SCREW,CAP,HEXAGON H	(4)	419-1
					SCREW,CAP,HEXAGON H	(4)	420-9
					SCREW,CAP,HEXAGON H	(2)	421-3
					SCREW,CAP,HEXAGON H	(2)	421-11
					SCREW,CAP,HEXAGON H	(2)	422-5
					SCREW,CAP,HEXAGON H	(4)	422-11
					SCREW,CAP,HEXAGON H	(4)	423-12
					SCREW,CAP,HEXAGON H	(9)	425-3
					SCREW,CAP,HEXAGON H	(2)	425-5
					SCREW,CAP,HEXAGON H	(4)	425-14
					SCREW,CAP,HEXAGON H	(9)	426-10
					SCREW,CAP,HEXAGON H	(4)	426-14
					SCREW,CAP,HEXAGON H	(9)	427-4
					SCREW,CAP,HEXAGON H	(2)	427-6
					SCREW,CAP,HEXAGON H	(4)	427-13
					SCREW,CAP,HEXAGON H	(4)	428-3
					SCREW,CAP,HEXAGON H	(2)	426-8
					SCREW,MACHINE	(1)	424-5
					SCREW,MACHINE	(3)	425-19
					SCREW,MACHINE	(1)	426-17
					SCREW,MACHINE	(1)	426-21
					SCREW,MACHINE	(5)	426-24
					SCREW,MACHINE	(1)	427-18
					SCREW,MACHINE	(1)	427-22
					SCREW,MACHINE	(3)	427-24
					SCREW,MACHINE	(2)	428-12
					SHELF,ELECTRICAL EQ	(2)	417-1
					SUPPORT,ANTENNA	(1)	428-5

SECTION II   TM9-2320-280-24P

(1) NO	(2) CODE	(3) NSN	(4) CAGEC	(5) PART NUMBER	(6) DESCRIPTION AND USABLE ON CODES(UOC)	(7) ITEM SMR QTY
					WASHER,FLAT	(8) 417-2
					WASHER,FLAT	(32) 417-5
					WASHER,FLAT	(2) 417-15
					WASHER,FLAT	(4) 418-2
					WASHER,FLAT	(18) 418-5
					WASHER,FLAT	(8) 419-2
					WASHER,FLAT	(36) 419-6
					WASHER,FLAT	(54) 420-2
					WASHER,FLAT	(8) 420-7
					WASHER,FLAT	(34) 421-2
					WASHER,FLAT	(8) 421-4
					WASHER,FLAT	(50) 422-2
					WASHER,FLAT	(12) 422-6
					WASHER,FLAT	(48) 423-2
					WASHER,FLAT	(8) 423-10
					WASHER,FLAT	(16) 425-2
					WASHER,FLAT	(4) 425-6
					WASHER,FLAT	(2) 425-11
					WASHER,FLAT	(8) 425-15
					WASHER,FLAT	(16) 426-4
					WASHER,FLAT	(4) 426-9
					WASHER,FLAT	(2) 426-11
					WASHER,FLAT	(8) 426-13
					WASHER,FLAT	(16) 427-3
					WASHER,FLAT	(4) 427-7
					WASHER,FLAT	(2) 427-12
					WASHER,FLAT	(8) 427-14
					WASHER,FLAT	(8) 428-4
					WASHER,FLAT	(3) 428-6
					WASHER,FLAT	(3) 428-13
					WASHER,LOCK	(3) 424-7
					WASHER,LOCK	(12) 425-8
					WASHER,LOCK	(12) 426-5
					WASHER,LOCK	(1) 426-18
					WASHER,LOCK	(1) 426-22
					WASHER,LOCK	(12) 427-9
					WASHER,LOCK	(1) 427-17
					WASHER,LOCK	(1) 427-21
					WASHER,LOCK	(1) 428-15
					WASHER,LOCK	(2) 428-19

END OF FIGURE

Section II. TM 9-2320-280-24P-2

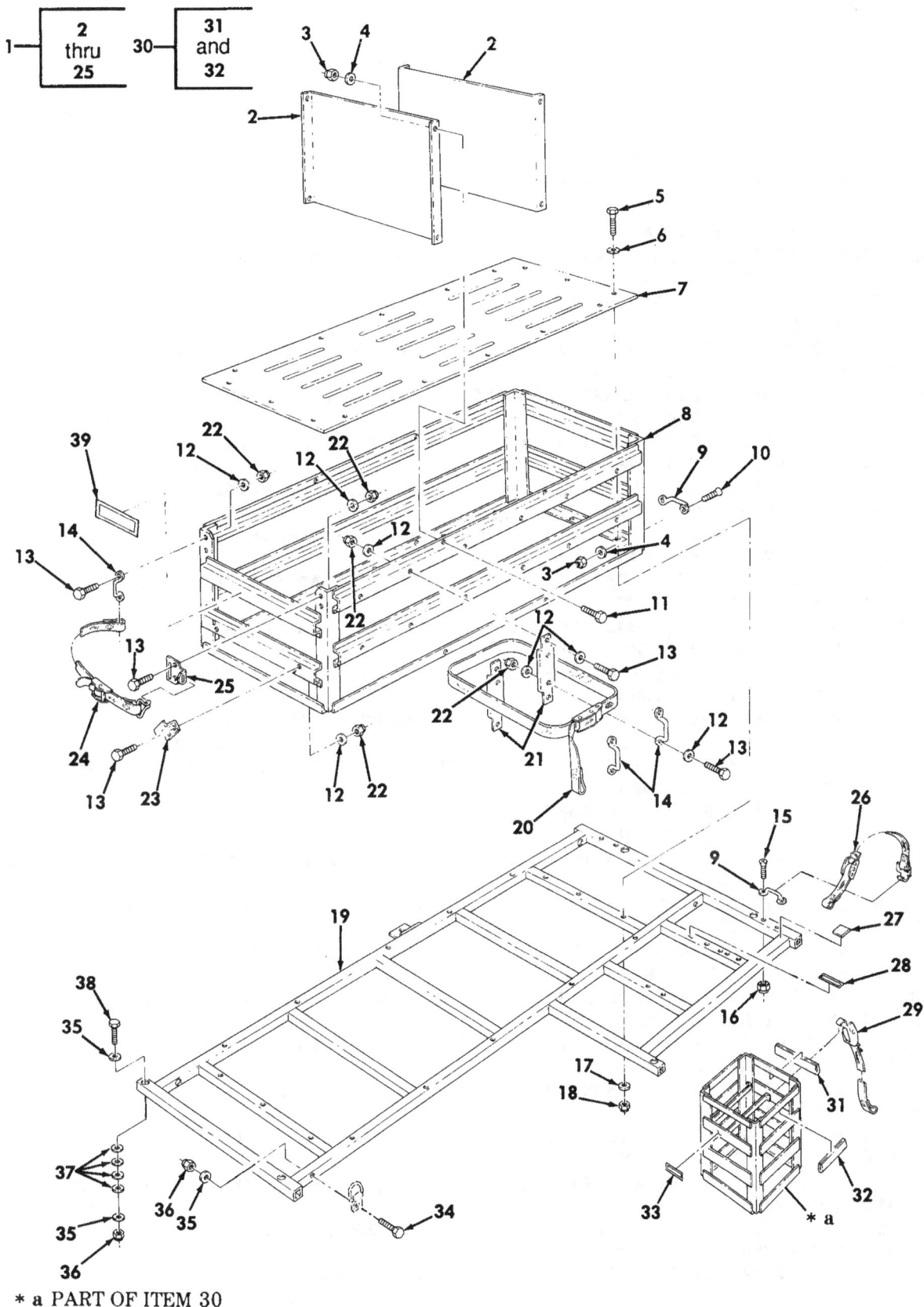

Figure 429. 81 MM Mortar Kit, Container Assembly, and Stowage Rack.

```
      SECTION II              TM9-2320-280-24P C01
   (1)    (2)     (3)           (4)         (5)                     (6)
(7) ITEM   SMR                                PART
    NO    CODE    NSN          CAGEC        NUMBER     DESCRIPTION AND USABLE ON CODES(UOC)
QTY

                                                       GROUP 3307 SPECIAL PURPOSE KITS

                                                       FIG. 429 81 MM MORTAR KIT,CONTAINER
ASSEMBLY,AND STOWAGE RACK

         1 PFOOO 2540012507589 19207 12340422          BOX,AMMUNITION STOW PART OF KIT P/N
1                                                          12338103.........................
UOC:AVY,A13,A14,BVY,HVY,H13,H14,NNN
         2 PFOZZ 5340012893243 34623 5594249           .BRACKET,MOUNTING..................         2
UOC:AVY,A13,A14,BVY,HVY,H13,H14,NNN
         3 PAOZZ 5310012033217 34623 5584710           .NUT,SELF-LOCKING,HE   #10-24........        22
UOC:AVY,A13,A14,BVY,HVY,H13,H14,NNN
         4 PAOZZ 5310000145850 96906 MS27183-42        .WASHER,FLAT   #10..................         22
UOC:AVY,A13,A14,BVY,HVY,H13,H14,NNN
         5 PAOZZ 5306002264835 80204 B1821BH031C250N   .BOLT,MACHINE   5/16-18 X 2.50.......         16
UOC:AVY,A13,A14,BVY,HVY,H13,H14,NNN
         6 PAOZZ 5310000814219 96906 MS27183-12        .WASHER,FLAT   5/16.................         16
UOC:AVY,A13,A14,BVY,HVY,H13,H14,NNN
         7 PFOZZ 9515012491591 34623 5594252           .PLATE,FLOOR,METAL..................         1
UOC:AVY,A13,A14,BVY,HVY,H13,H14,NNN
         8 PFOZZ 2540012491592 19207 12340413          .FRAME ASSEMBLY MORT................         1
UOC:AVY,A13,A14,BVY,HVY,H13,H14,NNN
  *      9 PAOZZ 5340013145957 19207 12338839-1        .LOOP,STRAP FASTENER................         6
UOC:AVY,A13,A14,BVY,HVY,H13,H14,NNN
        10 PAOZZ 5305000888332 80205 MS35190-272       .SCREW,MACHINE   #10-24 X .62.......         2
UOC:AVY,A13,A14,BVY,HVY,H13,H14,NNN
        11 PAOZZ 5305010062053 96906 MS51849-75        .SCREW,MACHINE   #10-24 X .62.......         20
UOC:AVY,A13,A14,BVY,HVY,H13,H14,NNN
        12 PAOZZ 5310008094058 96906 MS27183-10        .WASHER,FLAT   1/4..................
58                                                         UOC:AVY,A13,A14,BVY,HVY,H13,H14,NNN
13 PAOZZ 5305012566870 34623 5592817                   .SCREW,CAP,HEXAGON H   1/4-20 X 3/4.. 32
UOC:AVY,A13,A14,BVY,HVY,H13,H14,NNN
        14 PAOZZ 5340011978674 34623 5583568           .HANDLE,BOW........................         7
UOC:AVY,A13,A14,BVY,HVY,H13,H14,NNN
        15 PAOZZ 5305009847343 96906 MS35191-276       .SCREW,MACHINE   #10-32 X 1.00.......         10
UOC:AVY,A13,A14,BVY,HVY,H13,H14,NNN
        16 PAOZZ 5310011981723 19207 12339998-14       .NUT,PLAIN,BLIND RIV   #10-32........         4
UOC:AVY,A13,A14,BVY,HVY,H13,H14,NNN
        17 PAOZZ 5310000877493 96906 MS27183-13        .WASHER,FLAT   5/16.................         16
UOC:AVY,A13,A14,BVY,HVY,H13,H14,NNN
        18 PAOZZ 5310008140673 96906 MS51943-33        .NUT,SELF-LOCKING,HE   5/16-18.......         16
UOC:AVY,A13,A14,BVY,HVY,H13,H14,NNN
        19 PFOZZ 2540012491593 19207 12340419          .BASE ASSEMBLY,MORT.................         1
UOC:AVY,A13,A14,BVY,HVY,H13,H14,NNN
        20 PAOZZ 5340012570908 31272 45503-10          .STRAP,WEBBING......................         3
UOC:AVY,A13,A14,BVY,HVY,H13,H14,NNN
        21 PFOZZ 5340012491594 19207 12340382          .BRACKET,MOUNTING..................         6
UOC:AVY,A13,A14,BVY,HVY,H13,H14,NNN
        22 PAOZZ 5310000614650 96906 MS51943-31        .NUT,SELF-LOCKING,HE   1/4-20........         32
UOC:AVY,A13,A14,BVY,HVY,H13,H14,NNN
        23 PFOZZ 5340012552729 19207 12340366          .BRACKET,DOUBLE ANGL................         2
UOC:AVY,A13,A14,BVY,HVY,H13,H14,NNN
        24 PAOZZ 5340012609026 31272 45502-10          .STRAP,WEBBING......................         1
```

SECTION II TM9-2320-280-24P

(1) ITEM NO	(2) SMR CODE	(3) NSN	(4) CAGEC	(5) PART NUMBER	(6) DESCRIPTION AND USABLE ON CODES(UOC)	(7) QTY
25	PFOZZ	2590012563628	31272	44931-11	.BRACKET,AMMO CONTAI................ UOC:AVY,A13,A14,BVY,HVY,H13,H14,NNN	
26	PAOZZ	2590011854390	31272	44672-2	STRAP ASSEMBLY,TOOL PART OF KIT P/N 12338103.......................... UOC:AVY,A13,A14,BVY,HVY,H13,H14,NNN	4
27	PAOZZ	5330011983521	19207	12340721	RUBBER STRIP PART OF KIT P/N 12338103.......................... UOC:AVY,A13,A14,BVY,HVY,H13,H14,NNN	5
28	PAOZZ	7690012487679	19207	12340377	DECAL PART OF KIT P/N 12338103...... UOC:AVY,A13,A14,BVY,HVY,H13,H14,NNN	1
29	PAOZZ	5340012547191	31272	45410-10	STRAP,WEBBING PART OF KIT P/N 12338103.......................... UOC:AVY,A13,A14,BVY,HVY,H13,H14,NNN	2
30	PFOZZ	2540012491595	19207	12340401	RACK ASSEMBLY,AMMO PART OF KIT P/N 12338103.......................... UOC:AVY,A13,A14,BVY,HVY,H13,H14,NNN	1
31	MOOZZ		34623	5595995	.U-CHANNEL MAKE FROM TUBING,P/N M6855/4-16H029,8.3 INCHES LONG...... UOC:AVY,A13,A14,BVY,HVY,H13,H14,NNN	2
32	MOOZZ		34623	594278	.U-CHANNEL MAKE FROM TUBING,P/N M6855/4-16H029,6.3 INCHES LONG...... UOC:AVY,A13,A14,BVY,HVY,H13,H14,NNN	2
33	PAOZZ	7690012487680	19207	12340372	DECAL PART OF KIT P/N 12338103...... UOC:AVY,A13,A14,BVY,HVY,H13,H14,NNN	
34	PAOZZ	5305000712067	80204	B1821BH050C125N	SCREW,CAP,HEXAGON H 1/2-13 X 1.25 PART OF KIT P/N 12338103............ UOC:AVY,A13,A14,BVY,HVY,H13,H14,NNN	4
35	PAOZZ	5310001670823	88044	AN960-816	WASHER,FLAT 1/2 PART OF KIT P/N 12338103................ UOC:AVY,A13,A14,BVY,HVY,H13,H14,NNN	12
36	PAOZZ	5310004883889	96906	MS51943-39	NUT,SELF-LOCKING,HE 1/2-13 PART OF KIT P/N 12338103.................... UOC:AVY,A13,A14,BVY,HVY,H13,H14,NNN	6
37	PAOZZ	5310008095997	96906	MS27183-17	WASHER,FLAT 1/2 PART OF KIT P/N 12338103.......................... UOC:AVY,A13,A14,BVY,HVY,H13,H14,NNN	24
38	PAOZZ	5305000712070	80204	B1821BH050C175N	SCREW,CAP,HEXAGON H 1/2-13 X 1.75 PART OF KIT P/N 12338103............ UOC:AVY,A13,A14,BVY,HVY,H13,H14,NNN	
39	PAOZZ	7690012487681	19207	12340379	DECAL PART OF KIT P/N 12338103...... UOC:AVY,A13,A14,BVY,HVY,H13,H14,NNN	1

END OF FIGURE

Section II.

TM 9-2320-280-24P-2

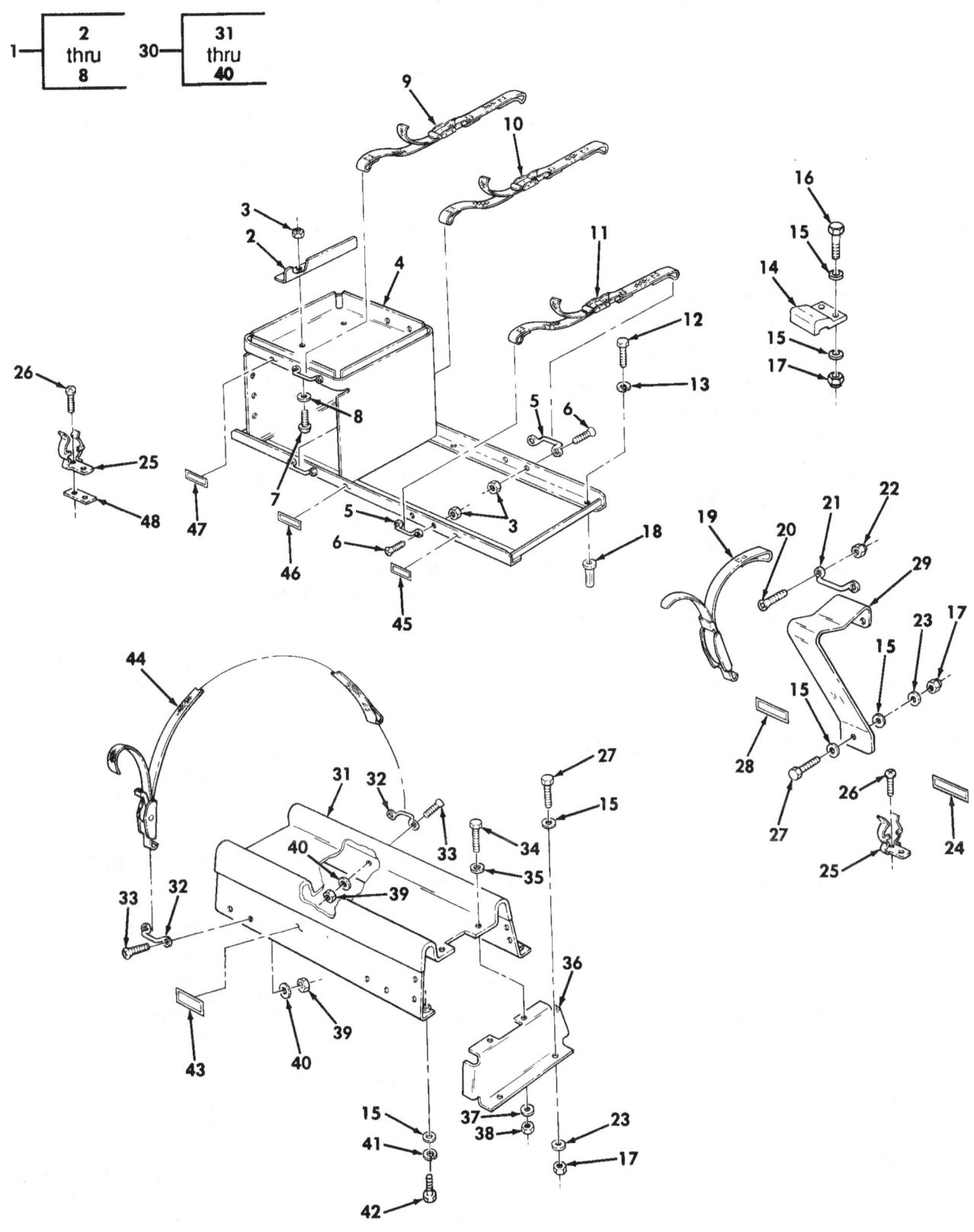

Figure 430. 81 MM Mortar Kit, Equipment Rack Assembly, Barrel Stowage Bracket, and Bipod Stowage Tray Assembly.

SECTION II   TM9-2320-280-24P C01

(1) ITEM NO	(2) SMR CODE	(3) NSN	(4) CAGEC	(5) PART NUMBER	(6) DESCRIPTION AND USABLE ON CODES (UOC)	(7) QTY

GROUP 3307 SPECIAL PURPOSE KITS

FIG. 430 81 MM MORTAR KIT, EQUIPMENT BRACKET RACK ASSEMBLY, BARREL STOWAGE, AND BIPOD STOWAGE TRAY ASSEMBLY

ITEM NO	SMR CODE	NSN	CAGEC	PART NUMBER	DESCRIPTION AND USABLE ON CODES (UOC)	QTY
1	PFOOO	2590012491596	19207	12340414	RACK,AMMUNITION STO PART OF KIT P/N 12338103................ UOC:AVY,A13,A14,BVY,HVY,H13,H14,NNN	1
2	PFOZZ	5340012559515	19207	12340390	.BRACKET,ANGLE.................... UOC:AVY,A13,A14,BVY,HVY,H13,H14,NNN	1
3	PAOZZ	5310012033217	34623	5584710	.NUT,SELF-LOCKING,HE #10-24....... UOC:AVY,A13,A14,BVY,HVY,H13,H14,NNN	14
4	PFOZZ	2540012491597	19207	12340421	.RACK ASSEMBLY,81MM............... UOC:AVY,A13,A14,BVY,HVY,H13,H14,NNN	1
* 5	PAOZZ	5340013145957	19207	12338839-1	.LOOP,STRAP FASTENER.............. UOC:AVY,A13,A14,BVY,HVY,H13,H14,NNN	6
6	PAOZZ	5305000888332	80205	MS35190-272	.SCREW,MACHINE #10-24 X .62...... UOC:AVY,A13,A14,BVY,HVY,H13,H14,NNN	12
7	PAOZZ	5305009846210	08484	WA-1429-10	.SCREW,MACHINE #10-24 X 1/2...... UOC:AVY,A13,A14,BVY,HVY,H13,H14,NNN	2
8	PAOZZ	5310000145850	96906	MS27183-42	.WASHER,FLAT #10................. UOC:AVY,A13,A14,BVY,HVY,H13,H14,NNN	2
9	PAOZZ	5340012552812	31272	44648-26	.STRAP,WEBBING PART OF KIT P/N 12338103............ UOC:AVY,A13,A14,BVY,HVY,H13,H14,NNN	1
10	PAOZZ	5340012572644	31272	45096-13	.STRAP,WEBBING PART OF KIT P/N 12338103............ UOC:AVY,A13,A14,BVY,HVY,H13,H14,NNN	1
11	PAOZZ	5340012560942	31272	44648-25	.STRAP,WEBBING PART OF KIT P/N 12338103............ UOC:AVY,A13,A14,BVY,HVY,H13,H14,NNN	1
12	PAOZZ	5306010758519	96906	MS90725-36	.BOLT,MACHINE 5/16-18 X 1.25 PART OF KIT P/N 12338103.... UOC:AVY,A13,A14,BVY,HVY,H13,H14,NNN	6
13	PAOZZ	5310004079566	96906	MS35338-45	.WASHER,LOCK 5/16 PART OF KIT P/N 12338103........... UOC:AVY,A13,A14,BVY,HVY,H13,H14,NNN	6
14	PAOZZ	5340012646108	34623	5597186	.STRAP,RETAINING PART OF KIT P/N 12338103............. UOC:AVY,A13,A14,BVY,HVY,H13,H14,NNN	2
15	PAOZZ	5310008094058	96906	MS27183-10	.WASHER,FLAT 1/4 PART OF KIT P/N 12338103......... UOC:AVY,A13,A14,BVY,HVY,H13,H14,NNN	24
16	PAOZZ	5305002253842	80204	B1821BH025C113N	.SCREW,CAP,HEXAGON H 1/4-20 X 1.125 PART OF KIT P/N 12338103............ UOC:AVY,A13,A14,BVY,HVY,H13,H14,NNN	4
17	PAOZZ	5310000614650	96906	MS51943-31	.NUT,SELF-LOCKING,HE 1/4-20 PART OF KIT P/N 12338103.. UOC:AVY,A13,A14,BVY,HVY,H13,H14,NNN	15
18	PAOZZ	5310011981724	03481	S31P280	.NUT,PLAIN,BLIND RIV 5/16-18 PART OF KIT P/N 12338103... UOC:AVY,A13,A14,BVY,HVY,H13,H14,NNN	6

430-1

SECTION II   TM9-2320-280-24P C01

(1) ITEM NO	(2) SMR CODE	(3) NSN	(4) CAGEC	(5) PART NUMBER	(6) DESCRIPTION AND USABLE ON CODES (UOC)	(7) QTY
19	PAOZZ	5340012547191	31272	45410-10	STRAP,WEBBING PART OF KIT P/N 12338103............................ UOC:AVY,A13,A14,BVY,HVY,H13,H14,NNN	
20	PAOZZ	5305000888332	80205	MS35190-272	SCREW,MACHINE #10-24 X .62 PART OF KIT P/N 12338103............... UOC:AVY,A13,A14,BVY,HVY,H13,H14,NNN	
21	PAOZZ	5340013145957	19207	12338839-1	LOOP,STRAP FASTENER PART OF KIT P/N 12338103............................ UOC:AVY,A13,A14,BVY,HVY,H13,H14,NNN	
22	PAOZZ	5310012033217	34623	5584710	NUT,SELF-LOCKING,HE #10-24 PART KIT P/N 12338103.............. UOC:AVY,A13,A14,BVY,HVY,H13,H14,NNN	
23	PAOZZ	5310011023270	24617	2436161	WASHER,FLAT 1/4 PART OF KIT P/N 12338103............................ UOC:AVY,A13,A14,BVY,HVY,H13,H14,NNN	
24	PAOZZ	7690012487682	19207	12340374	DECAL PART OF KIT P/N 12338103...... UOC:AVY,A13,A14,BVY,HVY,H13,H14,NNN	
25	PAOZZ	5340012573778	19207	12340387	CLIP,SPRING TENSION PART OF KIT P/N 12338103............................ UOC:AVY,A13,A14,BVY,HVY,H13,H14,NNN	
26	PAOZZ	5305012560406	34623	5593006	SCREW,ASSEMBLED WAS #10-32 X 3/4 PART OF KIT P/N 12338103.......... UOC:AVY,A13,A14,BVY,HVY,H13,H14,NNN	
27	PAOZZ	5305012566870	34623	5592817	SCREW,CAP,HEXAGON H 1/4-20 X 3/4 PART OF KIT P/N 12338103.......... UOC:AVY,A13,A14,BVY,HVY,H13,H14,NNN	
28	PAOZZ	7690012484907	19207	12340423	MARKER,IDENTIFICATI PART OF KIT P/N 12338103............................ UOC:AVY,A13,A14,BVY,HVY,H13,H14,NNN	
29	PFOZZ	5342012607853	34623	5593990	BRACKET,SPECIAL PART OF KIT P/N 12338103............................ UOC:AVY,A13,A14,BVY,HVY,H13,H14,NNN	
30	PFOZZ	2540012491598	19207	12340411	TRAY ASSEMBLY,BIPOD PART OF KIT P/N 12338103............................ UOC:AVY,A13,A14,BVY,HVY,H13,H14,NNN	
31	PFOZZ	2540012507592	19207	12340420	.TRAY,STOWAGE,BIPOD................ UOC:AVY,A13,A14,BVY,HVY,H13,H14,NNN	1
32	PAOZZ	5340013145957	19207	12338839-1	.LOOP,STRAP FASTENER................ UOC:AVY,A13,A14,BVY,HVY,H13,H14,NNN	4
33	PAOZZ	5305000888332	80205	MS35190-272	.SCREW,MACHINE #10-24 X .64........ UOC:AVY,A13,A14,BVY,HVY,H13,H14,NNN	
34	PAOZZ	5305012566870	34623	5592817	.SCREW,CAP,HEXAGON H 1/4-20 X 3/4.. UOC:AVY,A13,A14,BVY,HVY,H13,H14,NNN	6
35	PAOZZ	5310008093078	96906	MS27183-11	.WASHER,FLAT 1/4................... UOC:AVY,A13,A14,BVY,HVY,H13,H14,NNN	6
36	PFOZZ	5340012487649	19207	12340409	.BRACKET,DOUBLE ANGL............... UOC:AVY,A13,A14,BVY,HVY,H13,H14,NNN	1
37	PAOZZ	5310011023270	24617	2436161	.WASHER,FLAT 1/4................... UOC:AVY,A13,A14,BVY,HVY,H13,H14,NNN	6
38	PAOZZ	5310000614650	96906	MS51943-31	.NUT,SELF-LOCKING,HE 1/4-20........ UOC:AVY,A13,A14,BVY,HVY,H13,H14,NNN	6
39	PAOZZ	5310012033217	34623	5584710	.NUT,SELF-LOCKING,HE #10-24........ UOC:AVY,A13,A14,BVY,HVY,H13,H14,NNN	8
40	PAOZZ	5310000145850	96906	MS27183-42	.WASHER,FLAT #10...................	8

SECTION II TM9-2320-280-24P

(1) ITEM NO	(2) SMR CODE	(3) NSN	(4) CAGEC	(5) PART NUMBER	(6) DESCRIPTION AND USABLE ON CODES(UOC)	(7) QTY
41	PAOZZ	5310005825965	96906	MS35338-44	WASHER,LOCK 1/4 PART OF KIT P/N 12338103 UOC:AVY,A13,A14,BVY,HVY,H13,H14,NNN	3
42	PAOZZ	5305002253843	80204	B1821BH025C100N	SCREW,CAP,HEXAGON H 1/4-20 X 1.00 PART OF KIT P/N 12338103 UOC:AVY,A13,A14,BVY,HVY,H13,H14,NNN	3
43	PAOZZ	7690012484908	19207	12340373	DECAL PART OF KIT P/N 12338103 UOC:AVY,A13,A14,BVY,HVY,H13,H14,NNN	1
44	PAOZZ	5340012547190	31272	45410-11	STRAP,WEBBING PART OF KIT P/N 12338103 UOC:AVY,A13,A14,BVY,HVY,H13,H14,NNN	2
45	PAOZZ	7690012484909	19207	12340375	DECAL PART OF KIT P/N 12338103 UOC:AVY,A13,A14,BVY,HVY,H13,H14,NNN	1
46	PAOZZ	7690012487683	19207	12340378	DECAL PART OF KIT P/N 12338103 UOC:AVY,A13,A14,BVY,HVY,H13,H14,NNN	1
47	PAOZZ	7690012501101	19207	12340376	DECAL PART OF KIT P/N 12338103 UOC:AVY,A13,A14,BVY,HVY,H13,H14,NNN	1
48	PFOZZ	5975011975505	19207	12340167	MOUNTING BASE,ELECT PART OF KIT P/N 12338103 UOC:AVY,A13,A14,BVY,HVY,H13,H14,NNN	1
KIT	PDOZZ	2540012507591	34623	12338103	INSTALLATION AND EQ	1

```
BOLT,MACHINE        ( 6) 430-12
BOX,AMMUNITION STOW ( 1) 429-1
BRACKET,SPECIAL     ( 2) 430-29
CLIP,SPRING TENSION ( 2) 430-25
DECAL               ( 1) 429-28
DECAL               ( 1) 429-33
DECAL               ( 1) 429-39
DECAL               ( 1) 430-24
DECAL               ( 1) 430-43
DECAL               ( 1) 430-45
DECAL               ( 1) 430-46
DECAL               ( 1) 430-47
HANDLE,BOW          ( 2) 430-21
MARKER,IDENTIFICATI ( 1) 430-28
MOUNTING BASE,ELECT ( 1) 430-48
NUT,PLAIN,BLIND RIV ( 6) 430-18
NUT,SELF-LOCKING,HE ( 6) 429-36
NUT,SELF-LOCKING,HE (15) 430-17
NUT,SELF-LOCKING,HE ( 4) 430-22
RACK,AMMUNITION STO ( 1) 430-1
RACK ASSEMBLY,AMMO  ( 1) 429-30
RUBBER STRIP        ( 5) 429-27
SCREW,ASSEMBLED WAS ( 4) 430-26
SCREW,CAP,HEXAGON H ( 4) 429-34
SCREW,CAP,HEXAGON H ( 6) 429-38
SCREW,CAP,HEXAGON H ( 4) 430-16
SCREW,CAP,HEXAGON H (11) 430-27
SCREW,CAP,HEXAGON H ( 3) 430-42
SCREW,MACHINE       ( 4) 430-20
STRAP ASSEMBLY,TOOL ( 2) 429-26
```

SECTION II   TM9-2320-280-24P

(1) NO	(2) CODE	(3) NSN	(4) CAGEC	(5) PART NUMBER	(6) DESCRIPTION AND USABLE ON CODES(UOC)	(7) QTY	ITEM SMR
					STRAP,RETAINING	(2)	430-14
					STRAP,WEBBING	(2)	429-29
					STRAP,WEBBING	(1)	430-9
					STRAP,WEBBING	(1)	430-10
					STRAP,WEBBING	(1)	430-11
					STRAP,WEBBING	(2)	430-19
					STRAP,WEBBING	(2)	430-44
					TRAY ASSEMBLY,BIPOD	(1)	430-30
					WASHER,FLAT	(12)	429-35
					WASHER,FLAT	(24)	429-37
					WASHER,FLAT	(24)	430-15
					WASHER,FLAT	(11)	430-23
					WASHER,LOCK	(6)	430-13
					WASHER,LOCK	(3)	430-41

END OF FIGURE

Section II.

TM 9-2320-280-24P-2

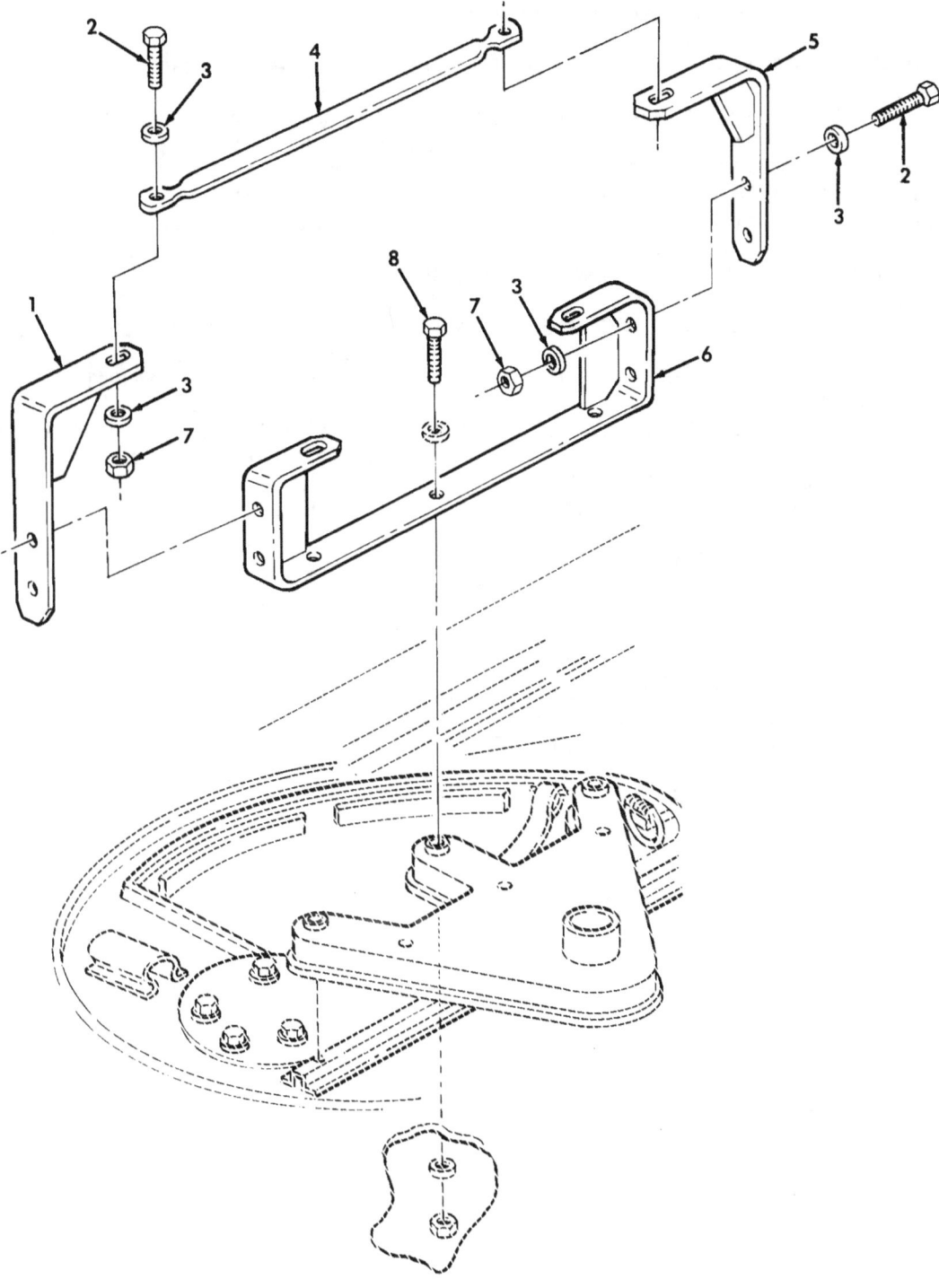

Figure 431. Traversing Bar Kit.

SECTION II   TM9-2320-280-24P

(1) NO	(2) SMR CODE	(3) NSN	(4) CAGEC	(5) PART NUMBER	(7) ITEM DESCRIPTION AND USABLE ON CODES(UOC)	QTY

GROUP 3307 SPECIAL PURPOSE KITS

FIG. 431 TRAVERSING BAR KIT

1	PAOZZ	5340012958169	19207	12342044-1	BRACKET,ANGLE UPPER,L.H. PART OF KIT P/N 12342040................... UOC:A25,A26,B25,H25,H26	1
2	PAOZZ	5305000680511	80204	B1821BH038C125N	SCREW,-CAP,HEXAGON H 3/8-16 X 1.25 PART OF KIT P/N 12342040............ UOC:A25,A26,B25,H25,H26	6
3	PAOZZ	5310000806004	96906	MS27183-14	WASHER,FLAT 3/8 PART OF KIT P/N 12342040......................... UOC:A25,A26,B25,H25,H26	12
4	PAOZZ	3040013271426	19207	12342045	CONNECTING LINK,RIG PART OF KIT P/N 12342040............................ UOC:A25,A26,B25,H25,H26	1
5	PAOZZ	5340012951088	19207	12342044-2	BRACKET,ANGLE UPPER R.H. PART OF KIT P/N 12342040.................. UOC:A25,A26,B25,H25,H26	1
6	PAOZZ	5340013241076	19207	12342043	BRACKET,DOUBLE ANGL LOWER PART OF KIT P/N 12342040................... UOC:A25,A26,B25,H25,H26	1
7	PAOZZ	5310009359021	96906	MS51943-35	NUT,-SELF-LOCKING,HE 3/8-16 PART OF KIT P/N 12342040................... UOC:A25,A26,B25,H25,H26	6
8	PAOZZ	5305013869052	80204	B1821BH038C600N	SCREW,-CAP,HEXAGON H LOWER PART OF KIT P/N 12342040................... UOC:A25,A26,B25,H25,H26	3
KIT	PDOZZ	1005013216774	9C234	12342040	TRAVERSING MECHANIS................ UOC:A25,A26,B25,H25,H26	1

```
BRACKET,ANGLE         ( 1) 431-1
BRACKET,ANGLE         ( 1) 431-5
BRACKET,DOUBLE ANGL   ( 1) 431-6
CONNECTING LINK,RIG   ( 1) 431-4
NUT,SELF-LOCKING,HE   ( 6) 431-7
SCREW,CAP,HEXAGON H   ( 3) 431-8
SCREW,CAP,HEXAGON H   ( 6) 431-2
WASHER,FLAT           (12) 431-3
```

END OF FIGURE

Section II. TM 9-2320-280-24P-2

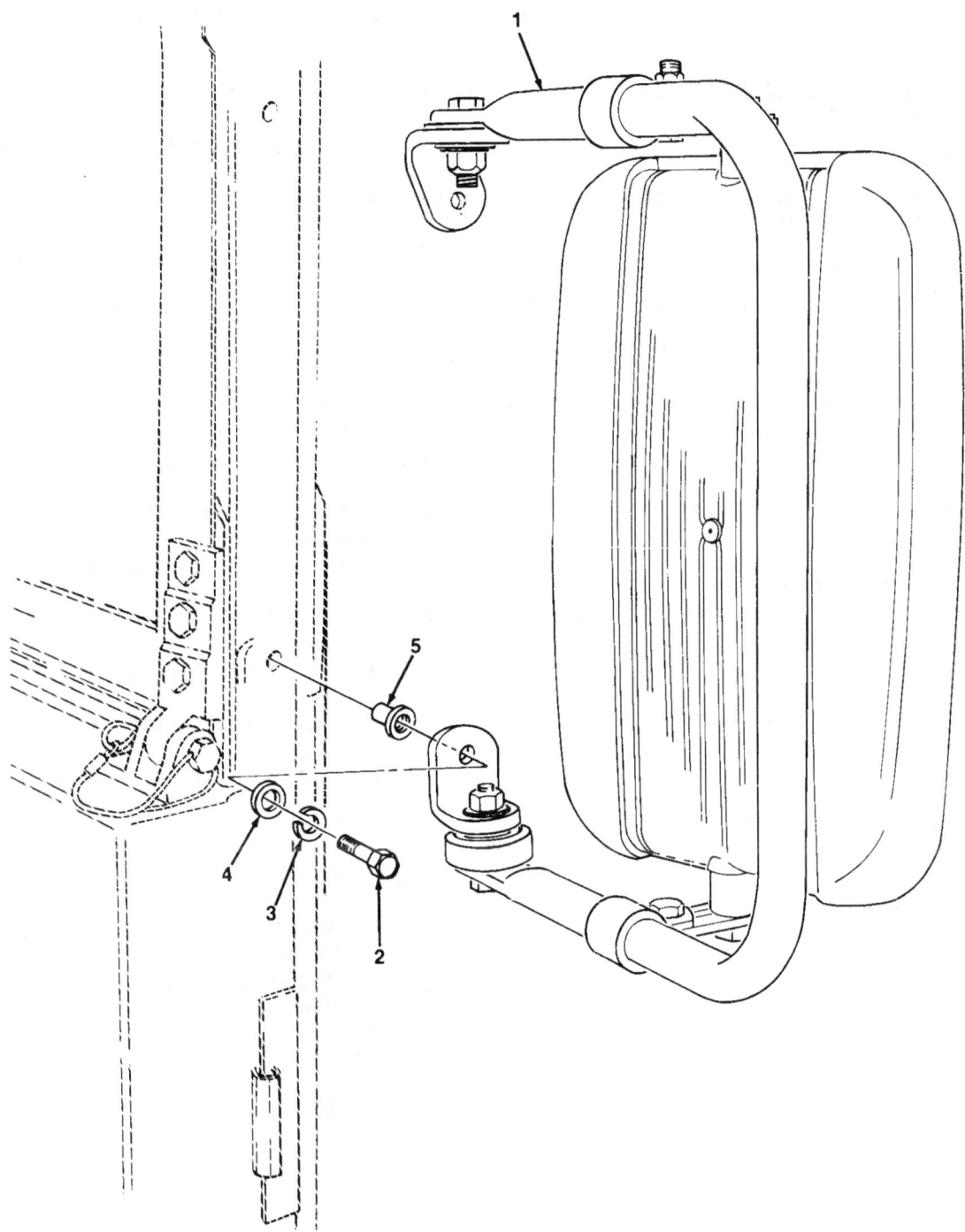

Figure 432. Rearview Mirror Mounting Kit.

```
SECTION II                TM9-2320-280-24P C01
   (1)     (2)     (3)         (4)        (5)                              (6)
(7)      ITEM    SMR                               PART
 NO      CODE    NSN           CAGEC      NUMBER        DESCRIPTION AND USABLE ON CODES(UOC)
QTY
```

GROUP 3307 SPECIAL PURPOSE KITS

FIG. 432 REARVIEW MIRROR MOUNTING KIT

```
* 1  PAOZZ 2540013149379  34623  12342130      MIRROR ASSEMBLY,REA   REARVIEW,R.H.         1
PART OF KIT P/N 57K3214............
                                                   UOC:H11,H13,H14,H15,H16,H17,H18,H20,
H21,H24,H25,H26,H27,H28,MMM
* 1  PAOZZ 2540014112130  34623  12343061      MIRROR ASSEMBLY,REA   REARVIEW,L.H.,        1
GREEN PART OF KIT P/N 57K3214.......
                                                   UOC:H11,H13,H14,H15,H16,H17,H18,H20,
H21,H24,H25,H26,H27,H28,MMM
* 1  PAOZZ 2540014873673  70082  1411          MIRROR ASSEMBLY,REA   REARVIEW,L.H.,        1
TAN.............................
* 1  PAOZZ 2540014379932  70082  1302          MIRROR ASSEMBLY,REA   REARVIEW,R.H.,        1
TAN.............................
* 2  PAOZZ 5306002264829  80204  B1821BH031C125N  BOLT,MACHINE  5/16-18 X 1.25 PART OF     4
KIT P/N 57K3214..................
                                                   UOC:H11,H13,H14,H15,H16,H17,H18,H20,
H21,H24,H25,H26,H27,H28,MMM
* 3  PAOZZ 5310004079566  96906  MS35338-45    WASHER,LOCK  5/16 PART OF KIT P/N
4                                                     57K3214............................
.                                                  UOC:H11,H13,H14,H15,H16,H17,H18,H20,
H21,H24,H25,H26,H27,H28,MMM
* 4  PAOZZ 5310000814219  96906  MS27183-12    WASHER,FLAT  5/16 PART OF KIT P/N
4                                                     57K3214............................
.                                                  UOC:H11,H13,H14,H15,H16,H17,H18,H20,
H21,H24,H25,H26,H27,H28,MMM
* 5  PAOZZ 5310014133276  78276  ALS4-518-150  NUT,PLAIN,BLIND RIV  PART OF KIT P/N
4                                                     57K3214............................
                                                   UOC:H11,H13,H14,H15,H16,H17,H18,H20,
H21,H24,H25,H26,H27,H28,MMM
* KIT PAOZZ 2540014361626  19207  57K3214        PARTS KIT,MIRROR AS   USBL EFF 1          1
THRU 99,999......................
                                                   UOC:H11,H13,H14,H15,H16,H17,H18,H20,
H21,H24,H25,H26,H27,H28,MMM
                                                      BOLT,MACHINE        (  4) 432-2
                                                      MIRROR ASSEMBLY,REA(  1) 432-1
                                                      MIRROR ASSEMBLY,REA(  1) 432-1
                                                      NUT,PLAIN,BLIND RIV(  4) 432-5
                                                      WASHER,FLAT         (  4) 432-4
                                                      WASHER,LOCK         (  4) 432-3
```

END OF FIGURE

Section II.

TM 9-2320-280-24P-2

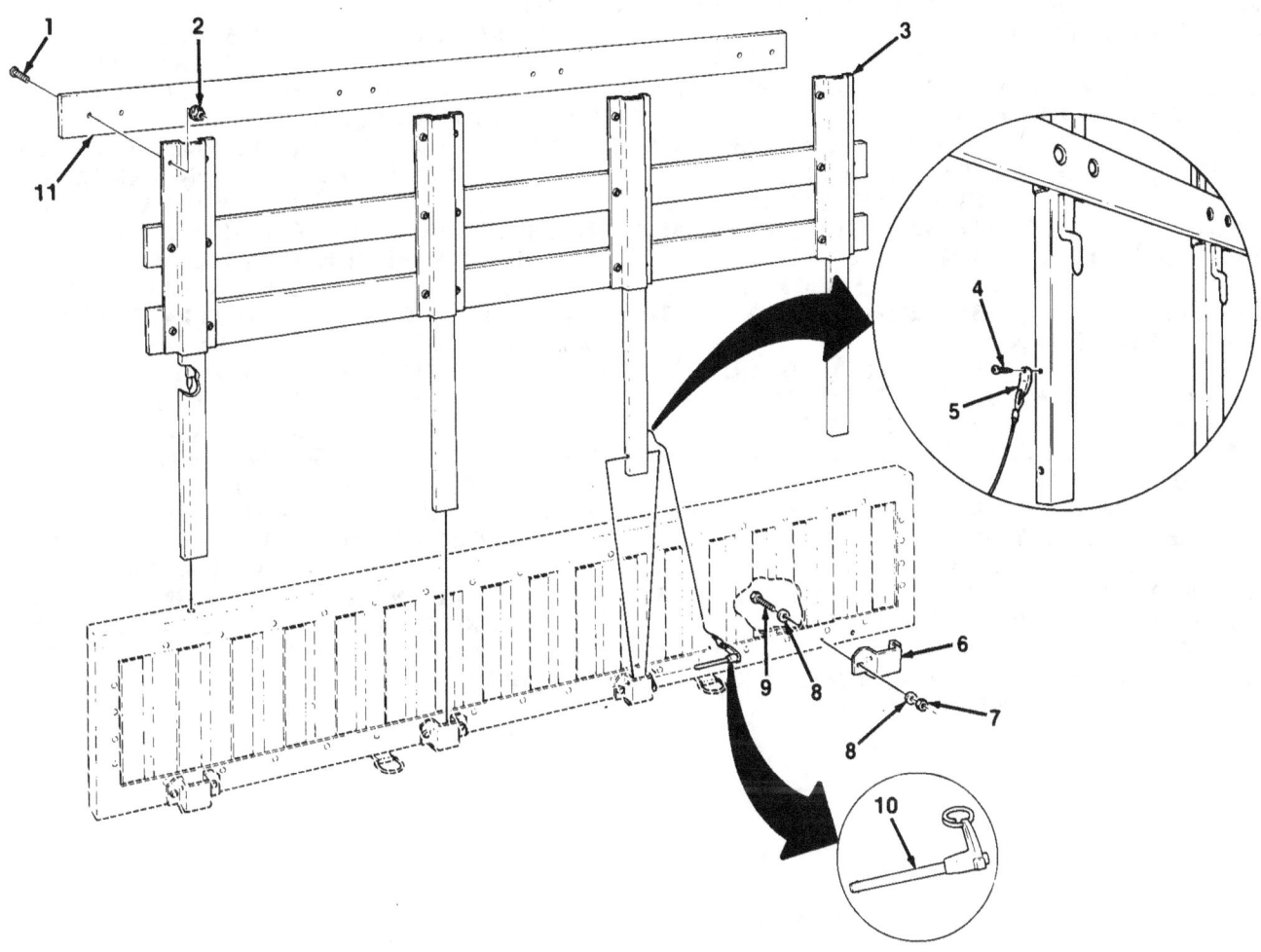

Figure 433. Cargo Barrier Extension Kit.

SECTION II   TM9-2320-280-24P

(1) NO	(2) SMR CODE	(3) NSN	(4) CAGEC	(5) PART NUMBER	(7) DESCRIPTION AND USABLE ON CODES(UOC)	ITEM QTY

GROUP 3307 SPECIAL PURPOSE KITS

FIG. 433 CARGO BARRIER EXTENSION KIT

1	PAOZZ	5306009115005	80205	MS35751-42	BOLT,SQUARE NECK 5/16-18 X 1.25 PART OF KIT P/N 57K0218 PART OF KIT P/N 57K0222 PART OF KIT P/N 5705613. UOC:AVY,A13,A14,BVY,HVY,H13,H14,NNN	24
2	PAOZZ	5310009334310	24617	271184	NUT,PLAIN,ASSEMBLED 5/16-18 PART OF KIT P/N 57K0218 PART OF KIT P/N 57K0222 PART OF KIT P/N 5705613..... UOC:AVY,A13,A14,BVY,HVY,H13,H14,NNN	24
3	PAOZZ	5340013183366	19207	12342284	BRACKET,MOUNTING GREEN(PAINT TAN OR WHITE IF REQUIRED)PART OF KIT P/N 57K0218 PART OF KIT P/N 57K0222 PART OF KIT P/N 5705613............. UOC:AVY,A13,A14,BVY,HVY,H13,H14,NNN	4
4	PAOZZ	5305004324170	96906	MS51861-35	SCREW,TAPPING #8-18 X .50 PART OF KIT P/N 57K0218 PART OF KIT P/N 57K0222 PART OF KIT P/N 5705613..... UOC:AVY,A13,A14,BVY,HVY,H13,H14,NNN	1
5	PAOZZ	4010010721167	84256	LT1504-C6-08	WIRE ROPE ASSEMBLY, PART OF KIT P/N 57K0218 PART OF KIT P/N 57K0222 PART OF KIT P/N 5705613.................. UOC:AVY,A13,A14,BVY,HVY,H13,H14,NNN	1
6	PAOZZ	5340013252891	19207	12342281	STRAP,RETAINING GREEN(PAINT TAN OR WHITE IF REQUIRED)PART OF KIT P/N 57K0218 PART OF KIT P/N 57K0222 PART OF KIT P/N 5705613.................. UOC:AVY,A13,A14,BVY,HVY,H13,H14,NNN	4
7	PAOZZ	5310002416658	96906	MS51943-34	NUT,SELF-LOCKING,HE 5/16-24 PART OF KIT P/N 57K0218 PART OF KIT P/N 57K0222 PART OF KIT P/N 5705613..... UOC:AVY,A13,A14,BVY,HVY,H13,H14,NNN	8
8	PAOZZ	5310000814219	96906	MS27183-12	WASHER,FLAT 5/16 PART OF KIT P/N 57K0218 PART OF KIT P/N 57K0222 PART OF KIT P/N 5705613.................. UOC:AVY,A13,A14,BVY,HVY,H13,H14,NNN	16
9	PAOZZ	5306000501238	80204	B1821BH031F075N	BOLT,MACHINE 5/16-24 X 3/4 PART OF KIT P/N 57K0218 PART OF KIT P/N 57K0222 PART OF KIT P/N 5705613..... UOC:AVY,A13,A14,BVY,HVY,H13,H14,NNN	8
10	PAOZZ	5315003769034	96906	MS17986-C423	PIN,QUICK RELEASE PART OF KIT P/N 57K0218 PART OF KIT P/N 57K0222 PART OF KIT P/N 5705613.................. UOC:AVY,A13,A14,BVY,HVY,H13,H14,NNN	1
11	PAOZZ	5340013180212	19207	12342285	BRACKET,MOUNTING GREEN(PAINT TAN OR WHITE IF REQUIRED)PART OF KIT P/N 57K0218 PART OF KIT P/N 57K0222 PART OF KIT P/N 5705613................... UOC:AVY,A13,A14,BVY,HVY,H13,H14,NNN	3
KIT	PAOZZ	2590013374071	19207	5705613	MODIFICATION KIT,VE CARGO BARRIER EXTENSION,GREEN.....................	1

SECTION II   TM9-2320-280-24P

(1) (2) (3)   (4)   (5)   (6)   (7) ITEM SMR   PART
NO CODE NSN   CAGEC   NUMBER   DESCRIPTION AND USABLE ON CODES(UOC) QTY

                    UOC:AVY,A13,A14,BVY,HVY,H13,H14,NNN   BOLT,MACHINE
(8) 433-9
                       BOLT,SQUARE NECK   (24) 433-1
                       BRACKET,MOUNTING   (4) 433-3
                       BRACKET,MOUNTING   (3) 433-11
                       NUT,PLAIN ASSEMBLED(24) 433-2
                       NUT,SELF-LOCKING,HE(8) 433-7
                       PIN,QUICK RELEASE (1) 433-10
                       SCREW,TAPPING   (1) 433-4
                       STRAP,RETAINING   (4) 433-6
                       WASHER,FLAT   (16) 433-8
                       WIRE ROPE ASSEMBLY (1) 433-5
KIT PDOZZ 2590013424918 19207 57K0218   MODIFICATION KIT,VE CARGO BARRIER   1
                       EXTENSION,TAN......................
                    UOC:AVY,A13,A14,BVY,HVY,H13,H14,NNN
                       BOLT,MACHINE   (8) 433-9
                       BOLT,SQUARE NECK   (24) 433-1
                       BRACKET,MOUNTING   (4) 433-3
                       BRACKET,MOUNTING   (3) 433-11
                       NUT,PLAIN ASSEMBLED(24) 433-2
                       NUT,SELF-LOCKING,HE(8) 433-7
                       PIN,QUICK RELEASE (1) 433-10
                       SCREW,TAPPING   (1) 433-4
                       STRAP,RETAINING   (4) 433-6
                       WASHER,FLAT   (16) 433-8
                       WIRE ROPE ASSEMBLY (1) 433-5
KIT PDOZZ   19207 57K0222   MODIFICATION KIT,VE CARGO BARRIER   1
                       EXTENSION,WHITE....................
                    UOC:AVY,A13,A14,BVY,HVY,H13,H14,NNN
                       BOLT,MACHINE   (8) 433-9
                       BOLT,SQUARE NECK   (24) 433-1
                       BRACKET,MOUNTING   (4) 433-3
                       BRACKET,MOUNTING   (3) 433-11
                       NUT,PLAIN ASSEMBLED(24) 433-2
                       NUT,SELF-LOCKING,HE(8) 433-7
                       PIN,QUICK RELEASE (1) 433-10
                       SCREW,TAPPING   (1) 433-4
                       STRAP,RETAINING   (4) 433-6
                       WASHER,FLAT   (16) 433-8
                       WIRE ROPE ASSEMBLY (1) 433-5

                       END OF FIGURE

Section II. TM 9-2320-280-24P-2

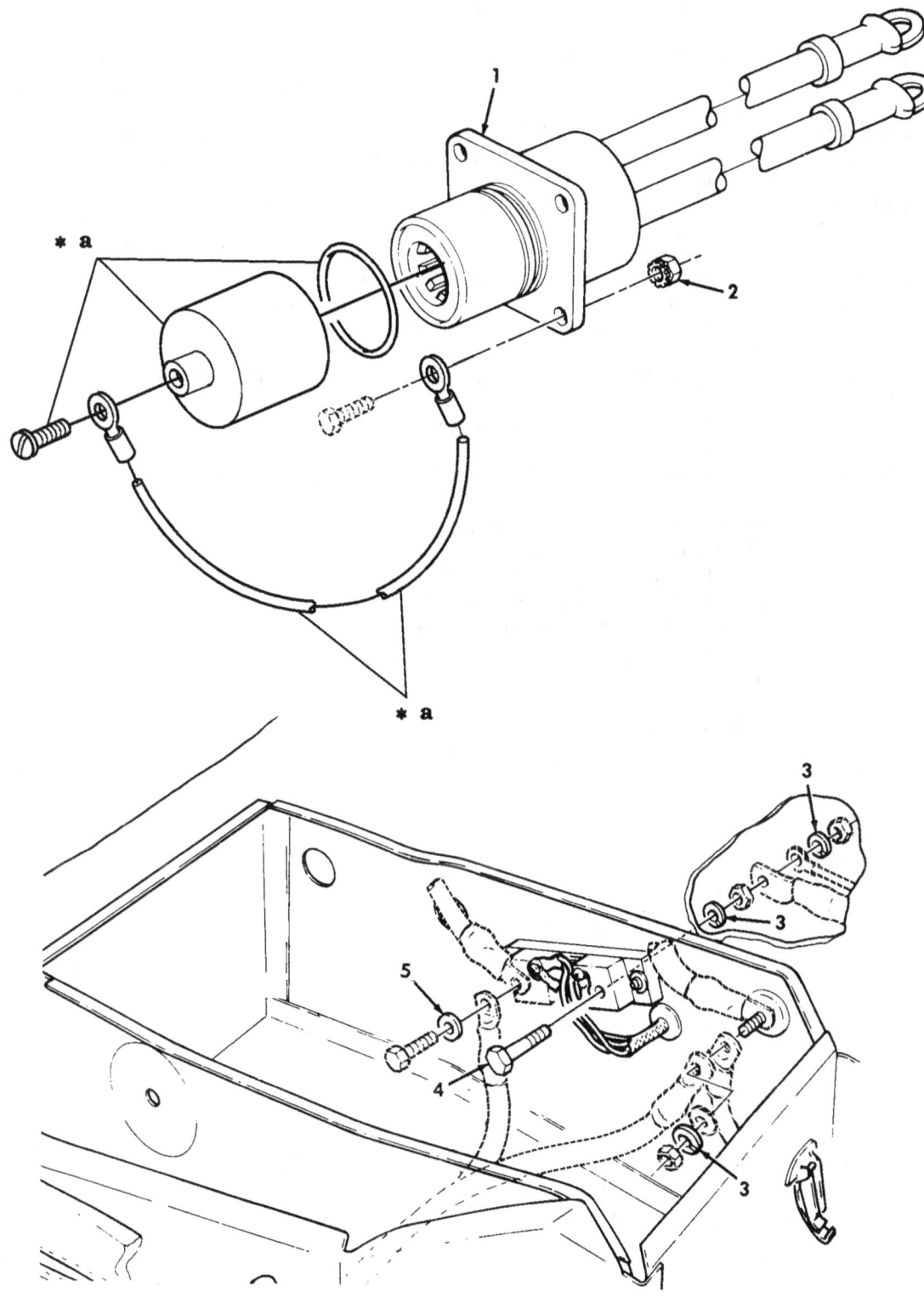

* a PART OF FIGURE 72

Figure 434. Slave Receptacle Kit.

SECTION II   TM9-2320-280-24P

(1) NO	(2) SMR CODE	(3) NSN	(4) CAGEC	(5) PART NUMBER	(7) ITEM DESCRIPTION AND USABLE ON CODES(UOC)	QTY

GROUP 3307 SPECIAL PURPOSE KITS

FIG. 434 SLAVE RECEPTACLE KIT

1 PAOZZ 6150013175853 19207 12342303 CABLE ASSEMBLY,POWE USBL EFF 1 THRU 99,999 PART OF KIT P/N 5705623...... UOC:H13,H14,H15,H16,H17,H18,H20,H21,H25,H26,H28 1

2 PAOZZ 5310012041039 7X677 272474 NUT,PLAIN,ASSEMBLED PART OF KIT P/N 5705623............................. UOC:H13,H14,H15,H16,H17,H18,H20,H21,H25,H26,H28 4

3 PAOZZ 5310005845272 31007 10332B WASHER,LOCK PART OF KIT P/N 5705623. UOC:H13,H14,H15,H16,H17,H18,H20,H21,H25,H26,H28 3

4 PAOZZ 5305013158649 19207 12342338 SCREW,CAP,HEXAGON H 1/2-13 X 3.00, USBL EFF 1 THRU 99,999 PART OF KIT P/N 5705623........................ UOC:H13,H14,H15,H16,H17,H18,H20,H21,H25,H26,H28 1

5 PAOZZ 5310001848971 96906 MS35338-103 WASHER,LOCK USBL EFF 1 THRU 99,999 PART OF KIT P/N 5705623............. UOC:H13,H14,H15,H16,H17,H18,H20,H21,H25,H26,H28 1

KIT PDOZZ 2920013788035 19207 5705623 PARTS KIT,TERMINAL, USBL EFF 1 THRU 99,999........................ UOC:H13,H14,H15,H16,H17,H18,H20,H21,H25,H26,H28 1

    CABLE ASSEMBLY,POWE(1) 434-1
    NUT,PLAIN,ASSEMBLED(4) 434-2
    SCREW,CAP,HEXAGON H(1) 434-4
    WASHER,LOCK   (3) 434-3
    WASHER,LOCK   (1) 434-5

END OF FIGURE

Section II. TM 9-2320-280-24P-2

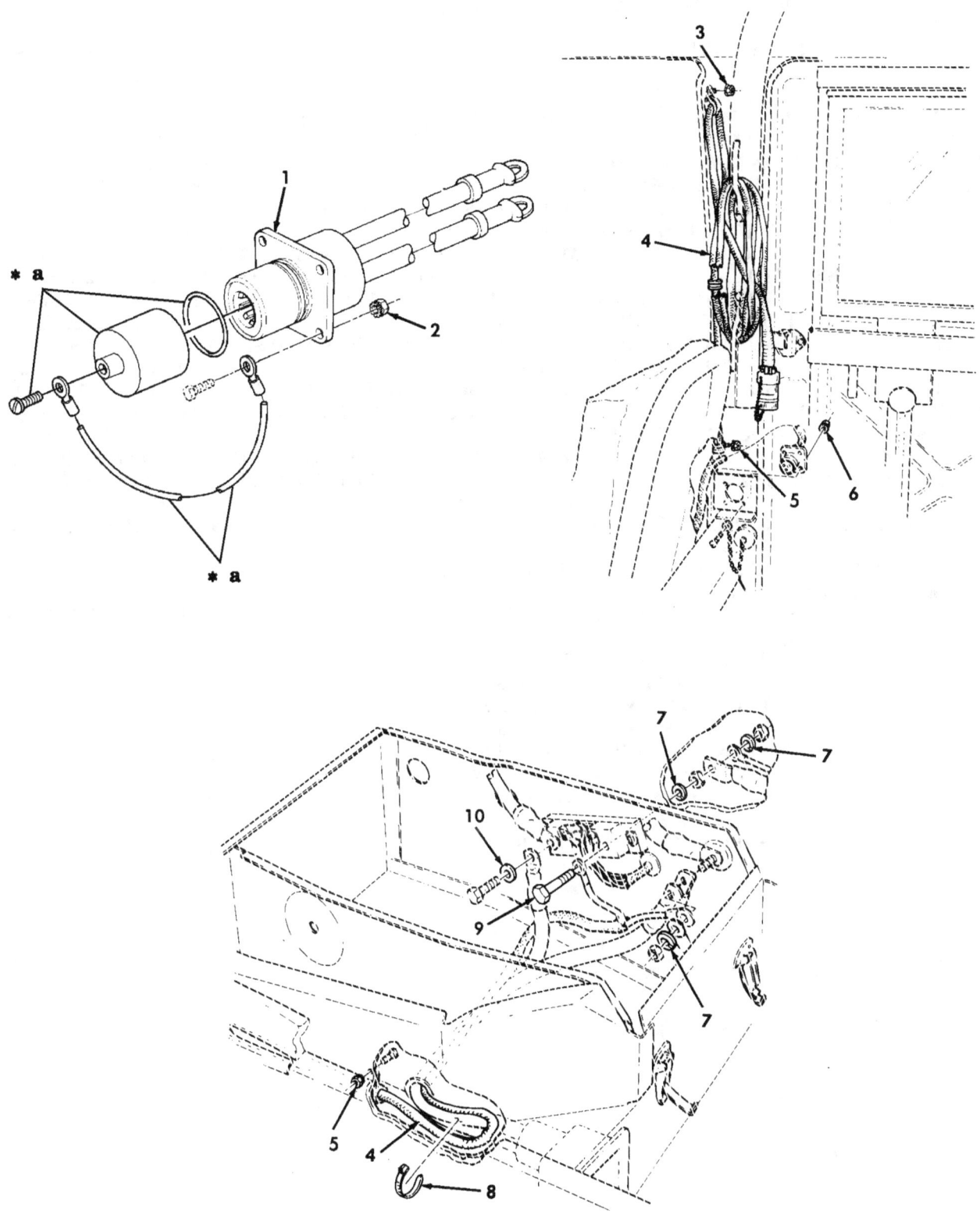

* a PART OF FIGURE 72

Figure 435. Slave Receptacle with Power Cable Assembly Kit.

SECTION II   TM9-2320-280-24P

(1) NO	(2) SMR CODE	(3) NSN	(4) CAGEC	(5) PART NUMBER	(7) DESCRIPTION AND USABLE ON CODES(UOC)	ITEM QTY

GROUP 3307 SPECIAL PURPOSE KITS

FIG. 435 SLAVE RECEPTACLE WITH POWER CABLE ASSEMBLY KIT

(1)	(2)	(3)	(4)	(5)	(6)	(7)
1	PFOZZ	6150013175853	19207	12342303	CABLE ASSEMBLY,POWE USBL EFF 1 THRU 99,999 PART OF KIT P/N 5705624. UOC:H11,H24,H27,MMM	1
2	PAOZZ	5310012041039	7X677	272474	NUT,PLAIN,ASSEMBLED 7/16-24,USBL EFF 1 THRU 99,999 PART OF KIT P/N 5705624............... UOC:H11,H24,H27,MMM	4
3	PAOZZ	5310009334310	24617	271184	NUT,PLAIN,ASSEMBLED 5/16-18,USBL EFF 1 THRU 99,999 PART OF KIT P/N 5705624............ UOC:H11,H24,H27,MMM	4
4	PFOZZ	5995013390190	19207	12340655-1	WIRING HARNESS,BRAN USBL EFF 1 THRU 99,999 PART OF KIT P/N 5705624. UOC:H11,H24,H27,MMM	1
5	PAOZZ	5310010695243	72582	271163	NUT,PLAIN,ASSEMBLED 8-32,USBL EFF 1 THRU 99,999 PART OF KIT P/N 5705624........................ UOC:H11,H24,H27,MMM	4
6	PAOZZ	5310001249265	72582	271169	NUT,PLAIN ,ASSEMBLED #10-32,USBL EFF 1 THRU 99,999,PART OF KIT P/N 5705624......................... UOC:H11,H24,H27,MMM	4
7	PAOZZ	5310005845272	31007	10332B	WASHER,LOCK 1/2,USBL EFF 1 THRU 99, 999 PART OF KIT P/N 5705624......... UOC:H11,H24,H27,MMM	3
8	PAOZZ	5975004515001	96906	MS3367-3-9	STRAP,TIEDOWN,ELECT USBL EFF 1 THRU 99,999 PART OF KIT P/N 5705624. UOC:H11,H24,H27,MMM	1
9	PAOZZ	5305013158649	19207	12342338	SCREW,CAP,HEXAGON H 1/2-13 X 3.00, USBL EFF 1 THRU 99,999 PART OF KIT P/N 5705624...................... UOC:H11,H24,H27,MMM	1
10	PAOZZ	5310001848971	96906	MS35338-103	WASHER,LOCK 3/8,USBL EFF 1 THRU 99, 999 PART OF KIT P/N 5705624. .... UOC:H11,H24,H27,MMM	1
KIT	PAOZZ	5975014173267	19207	5705624	ELECTRICAL KIT,SHOP SLAVE RECEPTACLE WITH POWER CABLE ASSEMBLY,USBL EFF 1 THRU 99,999..... UOC:H11,H24,H27,MMM	1

```
              CABLE ASSEMBLY,POWE( 1) 435-1
              NUT,PLAIN,ASSEMBLED( 4) 435-2
              NUT,PLAIN,ASSEMBLED( 4) 435-3
              NUT,PLAIN,ASSEMBLED( 4) 435-5
              NUT,PLAIN,ASSEMBLED( 4) 435-6
              SCREW,CAP,HEXAGON H( 1) 435-9
              STRAP,TIEDOWN,ELECT( 1) 435-8
              WASHER,LOCK        ( 3) 435-7
              WASHER,LOCK        ( 1) 435-10
```

SECTION II   TM9-2320-280-24P

(1) NO	(2) CODE	(3) NSN	(4) CAGEC	(5) NUMBER	(6) DESCRIPTION AND USABLE ON CODES(UOC)	(7) ITEM QTY	SMR	PART

WIRING HARNESS,BRAN(1) 435-4

END OF FIGURE

Section II. TM 9-2320-280-24P-2

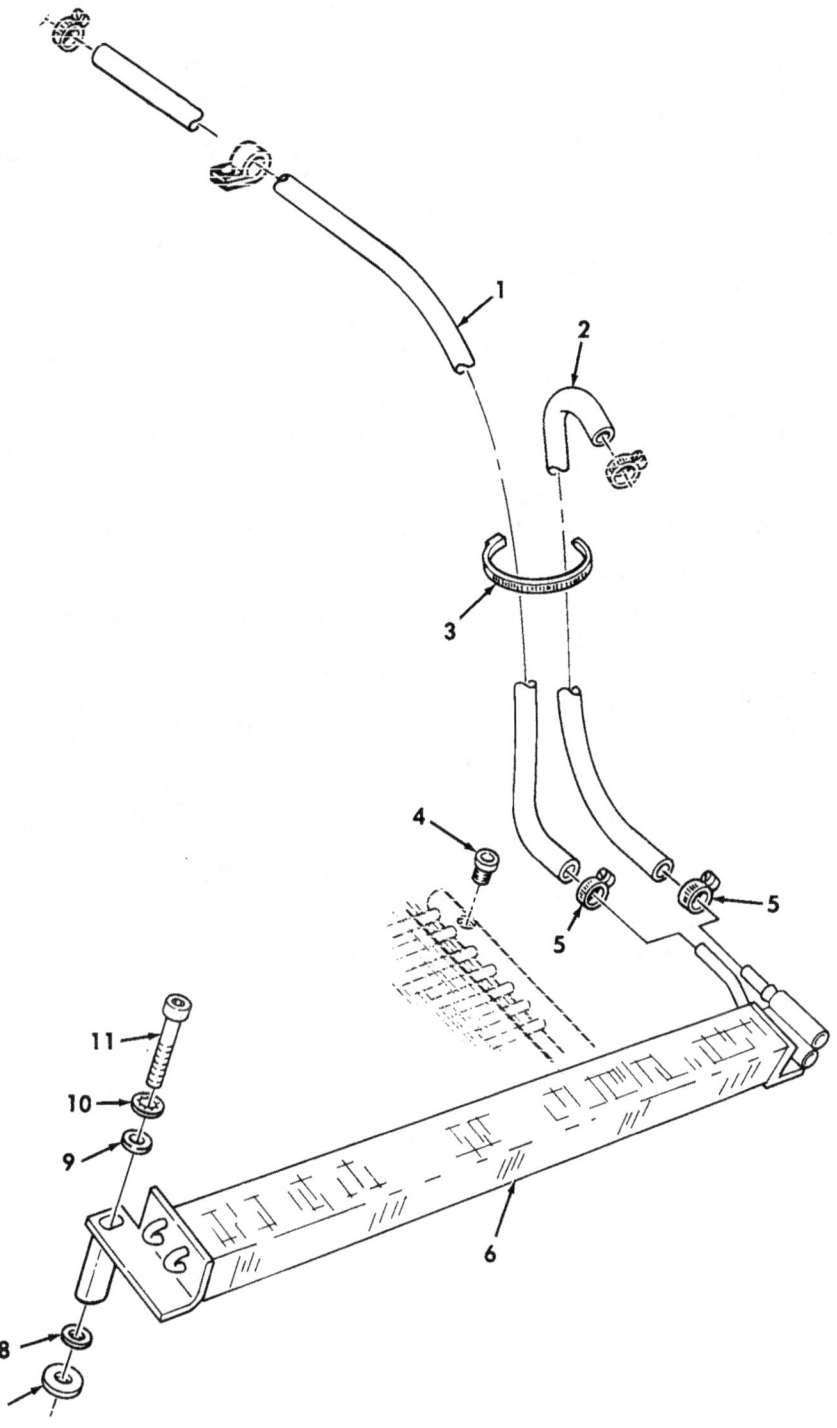

Figure 436. Power Steering Oil Cooler Kit.

SECTION II   TM9-2320-280-24P

(1) NO	(2) SMR CODE	(3) NSN	(4) CAGEC	(5) PART NUMBER	(7) DESCRIPTION AND USABLE ON CODES (UOC)	QTY

GROUP 3307 SPECIAL PURPOSE KITS

FIG. 436 POWER STEERING OIL COOLER KIT

1	MOOZZ		19207	12338330-6	HOSE,NONMETALLIC MAKE FROM HOSE,P/N 6490610019,29 INCHES LONG,USBL EFF 1 THRU 99,999 PART OF KIT P/N 5705612. UOC:H11,H13,H14,H15,H16,H17,H18,H20,H21,H24,H25,H26,H27,H28,MMM	1
2	MOOZZ		19207	12338330-7	HOSE,NONMETALLIC MAKE FROM HOSE,P/N 6490610019,45 INCHES LONG,USBL EFF 1 THRU 99,999 PART OF KIT P/N 5705612............... UOC:H11,H13,H14,H15,H16,H17,H18,H20,H21,H24,H25,H26,H27,H28,MMM	1
3	PAOZZ	5975009856630	96906	MS3367-3-0	STRAP,TIEDOWN,ELECT USBL EFF 1 THRU 99,999 PART OF KIT P/N 5705612. UOC:H11,H13,H14,H15,H16,H17,H18,H20,H21,H24,H25,H26,H27,H28,MMM	1
4	PAOZZ	4730009541281	81348	WW-P-471ACABCB	PLUG,PIPE USBL EFF 1 THRU 99,999 PART OF KIT P/N 5705612............. UOC:H11,H13,H14,H15,H16,H17,H18,H20,H21,H24,H25,H26,H27,H28,MMM	1
5	PAOZZ	4730008776298	46717	L-3604-13	CLAMP,HOSE USBL EFF 1 THRU 99,999 PART OF KIT P/N 5705612............. UOC:H11,H13,H14,H15,H16,H17,H18,H20,H21,H24,H25,H26,H27,H28,MMM	2
6	PAOZZ	2520013162630	34623	12342142	COOLER,FLUID,TRANSM USBL EFF 1 THRU 99,999 PART OF KIT P/N 5705612. UOC:H11,H13,H14,H15,H16,H17,H18,H20,H21,H24,H25,H26,H27,H28,MMM	1
7	PAOZZ	5310011857214	19207	12339052	WASHER,FLAT 5/16,USBL EFF 1 THRU 99,999 PART OF KIT P/N 5705612...... UOC:H11,H13,H14,H15,H16,H17,H18,H20,H21,H24,H25,H26,H27,H28,MMM	2
8	PAOZZ	5310000877493	96906	MS27183-13	WASHER,FLAT 5/16,USBL EFF 1 THRU 99,999 PART OF KIT P/N 5705612...... UOC:H11,H13,H14,H15,H16,H17,H18,H20,H21,H24,H25,H26,H27,H28,MMM	2
9	PAOZZ	5310000814219	96906	MS27183-12	WASHER,FLAT 5/16,USBL EFF 1 THRU 99,999 PART OF KIT P/N 5705612...... UOC:H11,H13,H14,H15,H16,H17,H18,H20,H21,H24,H25,H26,H27,H28,MMM	2
10	PAOZZ	5310001670721	96906	MS35333-41	WASHER,LOCK 5/16,USBL EFF 1 THRU 99,999 PART OF KIT P/N 5705612...... UOC:H11,H13,H14,H15,H16,H17,H18,H20,H21,H24,H25,H26,H27,H28,MMM	2
11	PAOZZ	5305009226143	80205	MS16995-72	SCREW,CAP,SOCKET HE 5/16-18 X 2-1/2,USBL EFF 1 THRU 99,999 PART OF KIT P/N 5705612...................... UOC:H11,H13,H14,H15,H16,H17,H18,H20,H21,H24,H25,H26,H27,H28,MMM	2

SECTION II   TM9-2320-280-24P

(1) NO	(2) SMR CODE	(3) NSN	(4) CAGEC	(5) PART NUMBER	(6) DESCRIPTION AND USABLE ON CODES(UOC)	(7) ITEM QTY
KIT	PDOZZ	2530013214497	19207	5705612	PARTS KIT,POWER STE ,POWER STEERING OIL COOLER USBL EFF 1 THRU 99,999.............................. UOC:H11,H13,H14,H15,H16,H17,H18,H20, H21,H24,H25,H26,H27,H28,MMM	1
					CLAMP,HOSE (2) 436-5	
					COOLER,FLUID,TRANSM(1) 436-6	
					HOSE,NONMETALLIC (1) 436-1	
					HOSE,NONMETALLIC (1) 436-2	
					PLUG,PIPE (1) 436-4	
					SCREW,CAP,SOCKET HE(2) 436-11	
					STRAP,TIEDOWN,ELECT(1) 436-3	
					WASHER,FLAT (2) 436-7	
					WASHER,FLAT (2) 436-8	
					WASHER,FLAT (2) 436-9	
					WASHER,LOCK (2) 436-10	

END OF FIGURE

Section II. TM 9-2320-280-24P-2

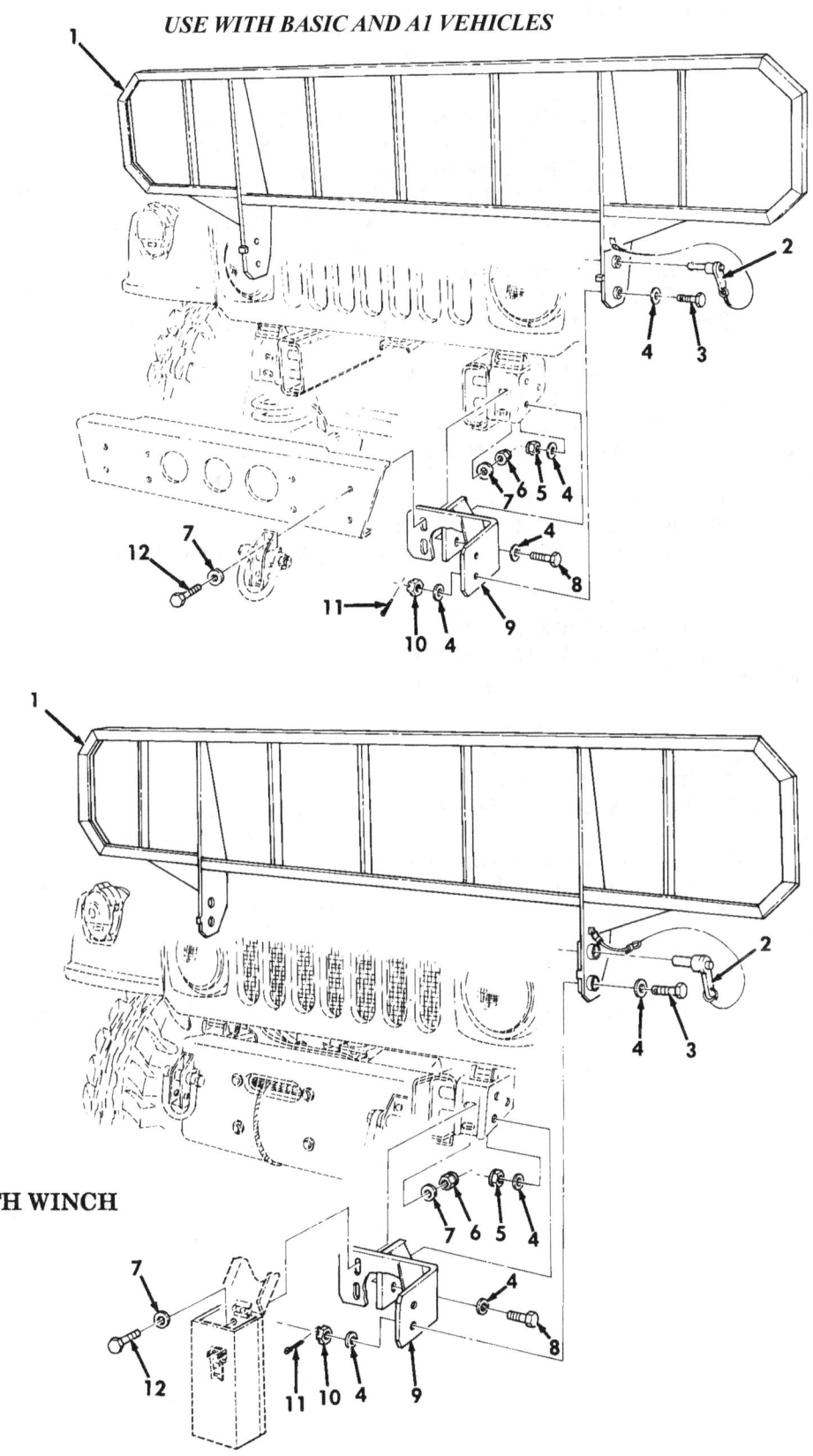

Figure 437. Brushguard Kit (Basic and A1 Series).

SECTION II   TM9-2320-280-24P

(1) ITEM NO	(2) SMR CODE	(3) NSN	(4) CAGEC	(5) PART NUMBER	(6) DESCRIPTION AND USABLE ON CODES(UOC)	(7) QTY

GROUP 3307 SPECIAL PURPOSE KITS

FIG. 437 BRUSHGUARD KIT (BASIC AND A1 SERIES)

1 KFOZZ 19207 12342412 BRUSHGUARD ASSEMBLY PART OF KIT P/N 57K0106............................ 1
UOC:AVY,A11,A13,A14,A15,A20,A24,A25,
A26,A27,B16,B17,B18,HVY,H11,H13,H14,
H15,H16,H17,H18,H20,H21,H24,H25,H26,
H27,H28,MMM

2 PAOZZ 5315012490555 96906 MS17986-C1013 PIN,QUICK RELEASE PART OF KIT P/N 57K0106............................ 2
UOC:AVY,A11,A13,A14,A15,A20,A24,A25,
A26,A27,B16,B17,B18,HVY,H11,H13,H14,
H15,H16,H17,H18,H20,H21,H24,H25,H26,
H27,H28,MMM

3 PAOZZ 5305013316284 96906 MS51108-139 SCREW,CAP,HEXAGON H 5/8-18 X 2.00 PART OF KIT P/N 57K0106............. 2
UOC:AVY,A11,A13,A14,A15,A20,A24,A25,
A26,A27,B16,B17,B18,HVY,H11,H13,H14,
H15,H16,H17,H18,H20,H21,H24,H25,H26,
H27,H28,MMM

4 PAOZZ 5310008238803 96906 MS27183-21 WASHER,FLAT 5/8 PART OF KIT P/N 57K0106............................ 12
UOC:AVY,A11,A13,A14,A15,A20,A24,A25,
A26,A27,B16,B17,B18,HVY,H11,H13,H14,
H15,H16,H17,H18,H20,H21,H24,H25,H26,
H27,H28,MMM

5 PAOZZ 5310001986691 7X677 9422306 NUT,SELF-LOCKING,HE 5/8-18 PART OF KIT P/N 57K0106.................... 4
UOC:AVY,A11,A13,A14,A15,A20,A24,A25,
A26,A27,B16,B17,B18,HVY,H11,H13,H14,
H15,H16,H17,H18,H20,H21,H24,H25,H26,
H27,H28,MMM

6 PAOZZ 5310000103028 96906 MS35690-824 NUT,PLAIN,HEXAGON 1/2-20 PART OF KIT P/N 57K0106.................... 4
UOC:AVY,A11,A13,A14,A15,A20,A24,A25,
A26,A27,B16,B17,B18,HVY,H11,H13,H14,
H15,H16,H17,H18,H20,H21,H24,H25,H26,
H27,H28,MMM

7 PAOZZ 5310011211703 06032 2310-0143-001 WASHER,FLAT 1/2 PART OF KIT P/N 57K0106............................ 8
UOC:AVY,A11,A13,A14,A15,A20,A24,A25,
A26,A27,B16,B17,B18,HVY,H11,H13,H14,
H15,H16,H17,H18,H20,H21,H24,H25,H26,
H27,H28,MMM

8 PAOZZ 5305007272283 96906 MS90726-162 SCREW,CAP,HEXAGON H 5/8-18 X 1 3/4 PART OF KIT P/N 57K0106............. 4
UOC:AVY,A11,A13,A14,A15,A20,A24,A25,
A26,A27,B16,B17,B18,HVY,H11,H13,H14,
H15,H16,H17,H18,H20,H21,H24,H25,H26,
H27,H28,MMM

SECTION II   TM9-2320-280-24P

(1) NO	(2) SMR CODE	(3) NSN	(4) CAGEC	(5) PART NUMBER	(6) DESCRIPTION AND USABLE ON CODES(UOC)	(7) QTY
9	PFOZZ	5340013327516	19207	12342418	BRACKET,MOUNTING PART OF KIT P/N 57K0106..................... UOC:AVY,A11,A13,A14,A15,A20,A24,A25, A26,A27,B16,B17,B18,HVY,H11,H13,H14, H15,H16,H17,H18,H20,H21,H24,H25,H26, H27,H28,MMM	1
10	PAOZZ	5310008507004	96906	MS35692-54	NUT,PLAIN,SLOTTED,H 5/8-18 PART OF KIT P/N 57K0106..................... UOC:AVY,A11,A13,A14,A15,A20,A24,A25, A26,A27,B16,B17,B18,HVY,H11,H13,H14, H15,H16,H17,H18,H20,H21,H24,H25,H26, H27,H28,MMM	2
11	PAOZZ	5315002981481	80205	MS24665-357	PIN,COTTER 5/64 X 2.00 PART OF KIT P/N 57K0106......................... UOC:AVY,A11,A13,A14,A15,A20,A24,A25, A26,A27,B16,B17,B18,HVY,H11,H13,H14, H15,H16,H17,H18,H20,H21,H24,H25,H26, H27,H28,MMM	2
12	PAOZZ	5305007195235	24617	9423995	SCREW,CAP,HEXAGON H 1/2-20 X 1 3/4, WITH WINCH PART OF KIT P/N 57K0106.. UOC:AVY,A11,A13,A14,A15,A20,A24,A25, A26,A27,B16,B17,B18,HVY,H11,H13,H14, H15,H16,H17,H18,H20,H21,H24,H25,H26, H27,H28,MMM	4
12	PAOZZ	5305007195239	96906	MS90727-116	SCREW,CAP,HEXAGON H 1/2-20 X 2 1/4, WITHOUT WINCH PART OF KIT P/N 57K0106........................... UOC:AVY,A11,A13,A14,A15,A20,A24,A25, A26,A27,B16,B17,B18,HVY,H11,H13,H14, H15,H16,H17,H18,H20,H21,H24,H25,H26, H27,H28,MMM	
KIT	PDOZZ	2590013282904	19207	57K0106	MODIFICATION KIT,VE BRUSHGUARD..... UOC:AVY,A11,A13,A14,A15,A20,A24,A25, A26,A27,B16,B17,B18,HVY,H11,H13,H14, H15,H16,H17,H18,H20,H21,H24,H25,H26, H27,H28,MMM	1

BRACKET,MOUNTING (1) 437-9
BRUSHGUARD ASSEMBLY(1) 437-1
NUT,PLAIN,HEXAGON (4) 437-6
NUT,PLAIN,SLOTTED (2) 437-10
NUT,SELF-LOCKING,HE(4) 437-5
PIN,COTTER   (2) 437-11
PIN,QUICK RELEASE (2) 437-2
SCREW,CAP,HEXAGON H(2) 437-3
SCREW,CAP,HEXAGON H(4) 437-12
SCREW,CAP,HEXAGON H(4) 437-12
SCREW,CAP,HEXAGON H(4) 437-8
WASHER,FLAT   (12) 437-4
WASHER,FLAT   (8) 437-7

END OF FIGURE

Section II. TM 9-2320-280-24P-2 C01

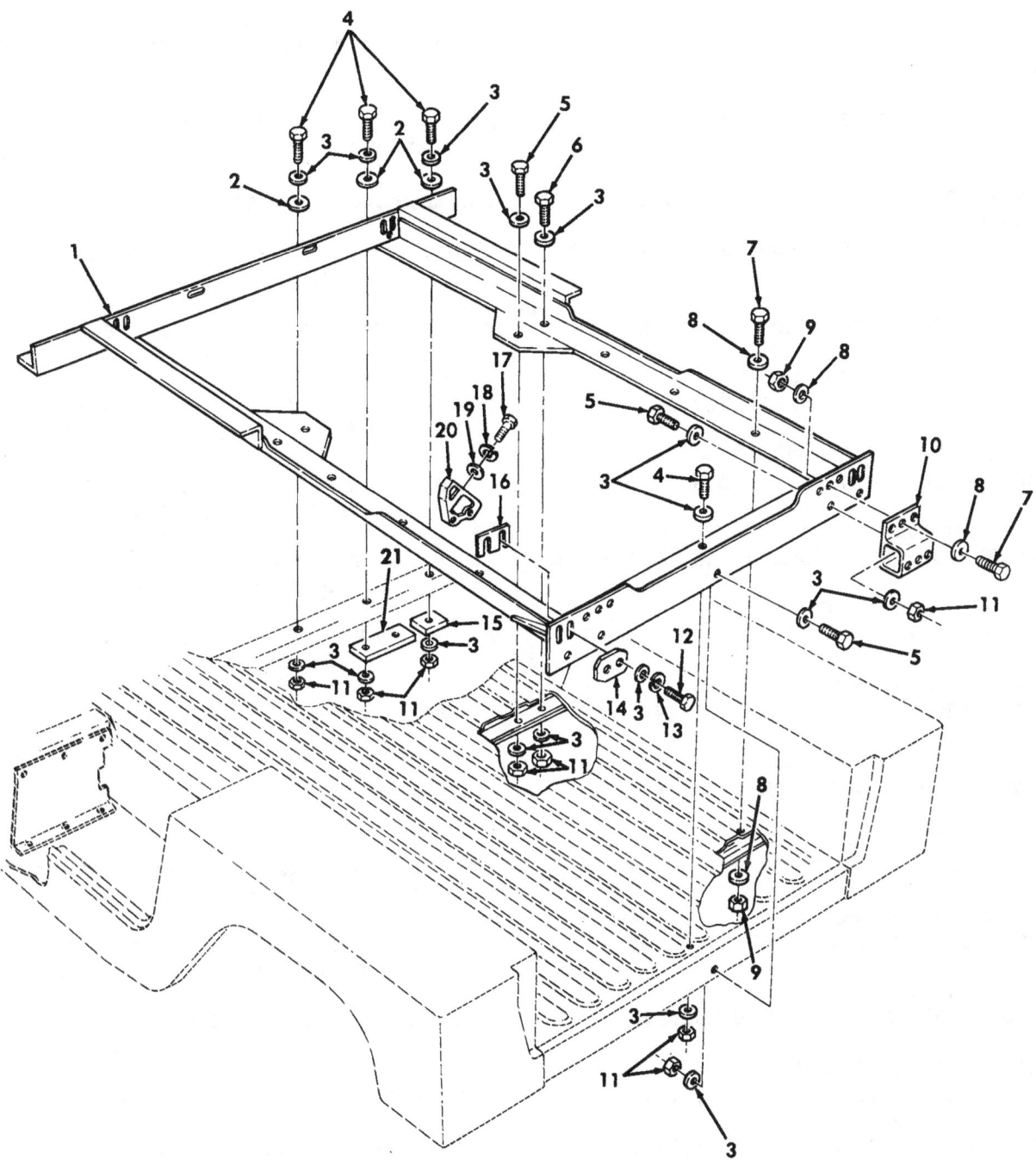

Figure 438. Kit, S250 Shelter Mounting Components.

SECTION II TM9-2320-280-24P C01

(1) ITEM NO	(2) SMR CODE	(3) NSN	(4) CAGEC	(5) PART NUMBER	(6) DESCRIPTION AND USABLE ON CODES (UOC)	(7) QTY

GROUP 3307 SPECIAL PURPOSE KITS

FIG. 438 KIT, S250 SHELTER MOUNTING

COMPONENTS

* 1 PAOZZ 2540012138057 19207 12339875 SUPPORT ASSEMBLY,MO PART OF KIT P/N 57K3542.............. 1
UOC:AVY,BVY,HVY,H21,H28,NNN

* 2 PAOZZ 5310012110698 19207 12339862-2 WASHER,FLAT PART OF KIT P/N 57K3542 8
UOC:AVY,BVY,HVY,H21,H28,NNN

* 3 PAOZZ 5310011211703 06032 2310-0143-001 WASHER,FLAT 1/2,PART OF KIT P/N 57K3542.............. 44
UOC:AVY,BVY,HVY,H21,H28,NNN

* 4 PAOZZ 5306013600926 3M915 17C836R BOLT,MACHINE 1/2-13 X 1.50,PART OF KIT P/N 57K3542.............. 9
UOC:AVY,BVY,HVY,H21,H28,NNN

* 5 PAOZZ 5306008440036 7X677 427566 BOLT,MACHINE 1/2-13 X 2.25,PART OF KIT P/N 57K3542.............. 9
UOC:AVY,BVY,HVY,H21,H28,NNN

* 6 PAOZZ 5305000712070 80204 B1821BH050C175N SCREW,CAP,HEXAGON 1/2-13 X 1.75 PART OF KIT P/N 57K3542.............. 5
UOC:AVY,BVY,HVY,H21,H28,NNN

* 7 PAOZZ 5305006882111 80204 B1821BH038C138N SCREW,CAP,HEXAGON H 3/8-16 X 1.375,PART OF KIT P/N 57K3542.............. 10
UOC:AVY,BVY,HVY,H21,H28,NNN

* 8 PAOZZ 5310014124013 24617 2436163 WASHER,FLAT 3/8,PART OF KIT P/N 57K3542.............. 20
UOC:AVY,BVY,HVY,H21,H28,NNN

* 9 PAOZZ 5310013611152 24617 9413266 NUT,SELF-LOCKING,HE 3/8-16,PART OF KIT P/N 57K3542.............. 16
UOC:AVY,BVY,HVY,H21,H28,NNN

* 10 PAOZZ 5340012144675 19207 12338972 BRACKET,MOUNTING PART OF KIT P/N 57K3542.............. 2
UOC:AVY,BVY,HVY,H21,H28,NNN

* 11 PAOZZ 5310000200358 72962 21NE083 NUT,SELF-LOCKING,HE 1/2-13,PART OF KIT P/N 57K3542.............. 17
UOC:AVY,BVY,HVY,H21,H28,NNN

* 12 PAOZZ 5306012761621 34623 5597349 BOLT,SELF-LOCKING 1/2-13 X 2.50, PART OF KIT P/N 57K3542.............. 8
UOC:AVY,BVY,HVY,H21,H28,NNN

* 13 PAOZZ 5310008347606 96906 MS35340-48 WASHER,LOCK 1/2,PART OF KIT P/N 57K3542.............. 8
UOC:AVY,BVY,HVY,H21,H28,NNN

* 14 PAOZZ 5365012772387 19207 12341968 SPACER,PLATE PART OF KIT P/N 57K3542 4
UOC:AVY,BVY,HVY,H21,H28,NNN

* 15 PAOZZ 2590012131629 19207 12339857 PLATE,REINFORCEMENT PART OF KIT P/N 57K3542.............. 2
UOC:AVY,BVY,HVY,H21,H28,NNN

* 16 PAOZZ 5365012658937 19207 12341967-1 SPACER,PLATE 0.09,PART OF KIT P/N 57K3542.............. 2

SECTION II TM9-2320-280-24P C01

(1)	(2)	(3)	(4)	(5)	(6)
ITEM NO	SMR CODE	NSN	CAGEC	PART NUMBER	DESCRIPTION AND USABLE ON CODES (UOC)
(7) QTY					

```
 16  PAOZZ  5365012658938  19207  12341967-2   SPACER,PLATE  0.12,PART OF KIT P/N
                                                57K3542..............................
UOC:AVY,BVY,HVY,H21,H28,NNN

 16  PAOZZ  5365012658939  19207  12341967-3   SPACER,PLATE  0.03,PART OF KIT P/N
                                                57K3542..............................
UOC:AVY,BVY,HVY,H21,H28,NNN

 17  PAOZZ  5306013601123  24617  9423557      BOLT,MACHINE  7/16-14 X 2.50,PART OF
                                                KIT P/N 57K3542.....................
UOC:AVY,BVY,HVY,H21,H28,NNN

 18  PAOZZ  5310006559370  96906  MS35340-47   WASHER,LOCK  29/64,PART OF KIT P/N
                                                57K3542..............................
UOC:AVY,BVY,HVY,H21,H28,NNN

 19  PAOZZ  5310008094085  96906  MS27183-16   WASHER,FLAT  7/16,PART OF KIT P/N
                                                57K3542..............................
UOC:AVY,BVY,HVY,H21,H28,NNN

 20  PAOZZ  5340012658905  19207  12341969     BRACKET,MOUNTING  PART OF KIT P/N
                                                57K3542..............................
UOC:AVY,BVY,HVY,H21,H28,NNN

 21  PAOZZ  2590012131628  19207  12339856     PLATE,REINFORCEMENT  B-BEAM,PART OF
                                                KIT P/N 57K3542.....................
UOC:AVY,BVY,HVY,H21,H28,NNN
```

Section II. TM 9-2320-280-24P-2

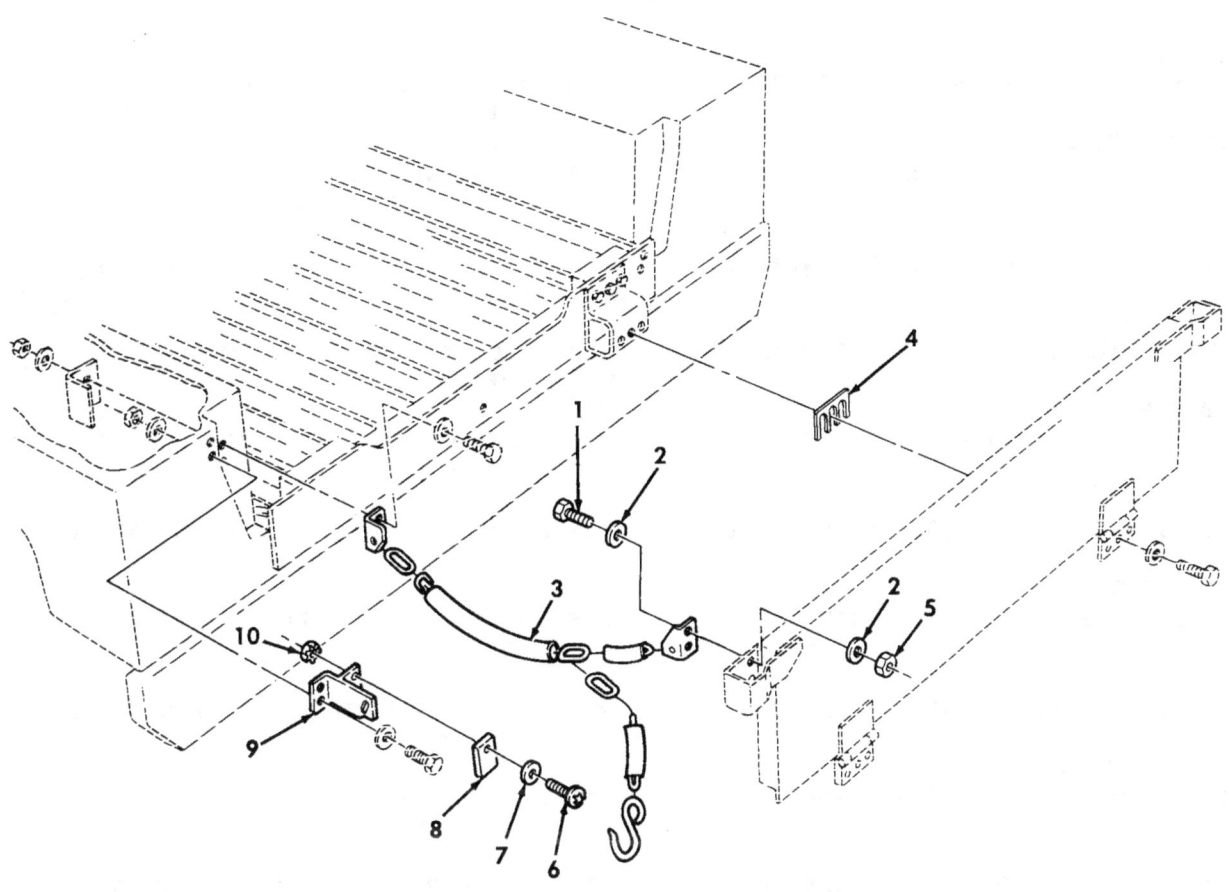

Figure 439. Kit, S250 Shelter Tailgate Mounting Components.

```
       SECTION II           TM9-2320-280-24P C01
      (1)   (2)    (3)           (4)         (5)                        (6)
(7)   ITEM  SMR                                PART
      NO    CODE   NSN           CAGEC        NUMBER      DESCRIPTION AND USABLE ON CODES(UOC)
QTY
```

GROUP 3307 SPECIAL PURPOSE KITS

FIG. 439 KIT,S250 SHELTER TAILGATE

MOUNTING COMPONENTS

```
* 1 PAOZZ 5306002264828 80204 B1821BH031C113N  BOLT,MACHINE  5/16-18 X 1.13,PART OF
4                                                KIT P/N 57K3542....................
UOC:AVY,BVY,HVY,NNN
* 2 PAOZZ 5310011191024 24617 2436162          WASHER,FLAT  5/16,PART OF KIT P/N
8                                                57K3542............................
UOC:AVY,BVY,HVY,NNN
* 3 PAOZZ 4010012177086 34623 12339229-1       CHAIN ASSEMBLY,SING  L.H.,PART OF
1                                                KIT P/N 57K3542....................
UOC:AVY,BVY,HVY,NNN
* 3 PAOZZ 4010012157466 19207 12339229-2       CHAIN ASSEMBLY,SING  R.H.,PART OF
1                                                KIT P/N 57K3542....................
UOC:AVY,BVY,HVY,NNN
* 4 PAOZZ 5365012712491 19207 12340045         SPACER,PLATE  PART OF KIT P/N
2                                                57K3542............................
UOC:AVY,BVY,HVY,NNN
* 5 PAOZZ 5310011193668 24617 9422295          NUT,SELF-LOCKING,HE  5/16-18,PART OF
4                                                KIT P/N 57K3542....................
UOC:AVY,BVY,HVY,NNN
* 6 PAOZZ 5305009846212 96906 MS35206-265      SCREW,MACHINE  #10-24 X 0.75,PART OF
4                                                KIT P/N 57K3542....................
UOC:AVY,BVY,HVY,NNN
* 7 PAOZZ 5310013611163 24617 9417373          WASHER,FLAT  #10,PART OF KIT P/N
4                                                57K3542............................
UOC:AVY,BVY,HVY,NNN
* 8 PAOZZ 5342012147809 19207 12338976         MOUNTING,RESILIENT  PART OF KIT P/N
2                                                57K3542............................
UOC:AVY,BVY,HVY,NNN
* 9 PAOZZ 2540013141121 34623 5590669          LATCH,DOOR,VEHICULA  L.H.,PART OF
1                                                KIT P/N 57K3542....................
UOC:AVY,BVY,HVY,NNN
* 9 PAOZZ 2540012096875 19207 12338978-2       LATCH,DOOR,VEHICULA  R.H.,PART OF
1                                                KIT P/N 57K3542....................
UOC:AVY,BVY,HVY,NNN
* 10 PAOZZ 5310012110691 24617 9422771         NUT,PLAIN,ASSEMBLED  #10-24,PART OF
4                                                KIT P/N 57K3542....................
UOC:AVY,BVY,HVY,NNN
* KIT PAOZZ 5340014558700 19207 57K3542        HARDWARE KIT,MECHAN  S250 SHELTER
1                                                MOUNTING...........................
UOC:AVY,BVY,HVY,NNN

                                                 BOLT,MACHINE         (  9) 438-4
                                                 BOLT,MACHINE         (  9) 438-5
                                                 BOLT,MACHINE         (  4) 439-1
                                                 BOLT,MACHINE         (  4) 438-17
                                                 BOLT,SELF-LOCKING    (  8) 438-12
                                                 BRACKET,MOUNTING     (  2) 438-10
                                                 BRACKET,MOUNTING     (  4) 438-20
                                                 CHAIN ASSEMBLY,SING(   1) 439-3
                                                 CHAIN ASSEMBLY,SING(   1) 439-3
```

```
     SECTION II           TM9-2320-280-24P C01
   (1)    (2)    (3)          (4)         (5)                    (6)
7)   ITEM   SMR                          PART
     NO    CODE    NSN        CAGEC     NUMBER      DESCRIPTION AND USABLE ON CODES(UOC)
TY
                                           LATCH,DOOR,VEHICULA(   1) 439-9
                                           LATCH,DOOR,VEHICULA(   1) 439-9
                                           MOUNT,RESILIENT    (   2) 439-8
                                           NUT,PLAIN,ASSEMBLED(   4) 439-10
                                           NUT,SELF-LOCKING   (   4) 439-5
                                           NUT,SELF-LOCKING,HE(  16) 438-9
                                           NUT,SELF-LOCKING,HE(  17) 438-11
                                           PLATE,REINFORCEMENT(   2) 438-15
                                           PLATE,REINFORCEMENT(   2) 438-21
                                           SCREW,CAP,HEXAGON  (  10) 438-7
                                           SCREW,MACHINE      (   5) 438-6
                                           SCREW,MACHINE      (   4) 439-6
                                           SPACER,PLATE       (   4) 438-14
                                           SPACER,PLATE       (   2) 438-16
                                           SPACER,PLATE       (   2) 438-16
                                           SPACER,PLATE       (   2) 438-16
                                           SPACER,PLATE       (   2) 439-4
                                           SUPPORT ASSEMBLY,MO(   1) 438-1
                                           WASHER,FLAT        (   8) 438-2
                                           WASHER,FLAT        (  44) 438-3
                                           WASHER,FLAT        (  20) 438-8
                                           WASHER,FLAT        (   4) 438-19
                                           WASHER,FLAT        (   8) 439-2
                                           WASHER,FLAT        (   4) 439-7
                                           WASHER,LOCK        (   8) 438-13
                                           WASHER,LOCK        (   4) 438-18

                              END OF FIGURE
```

Section II.   TM 9-2320-280-24P-2

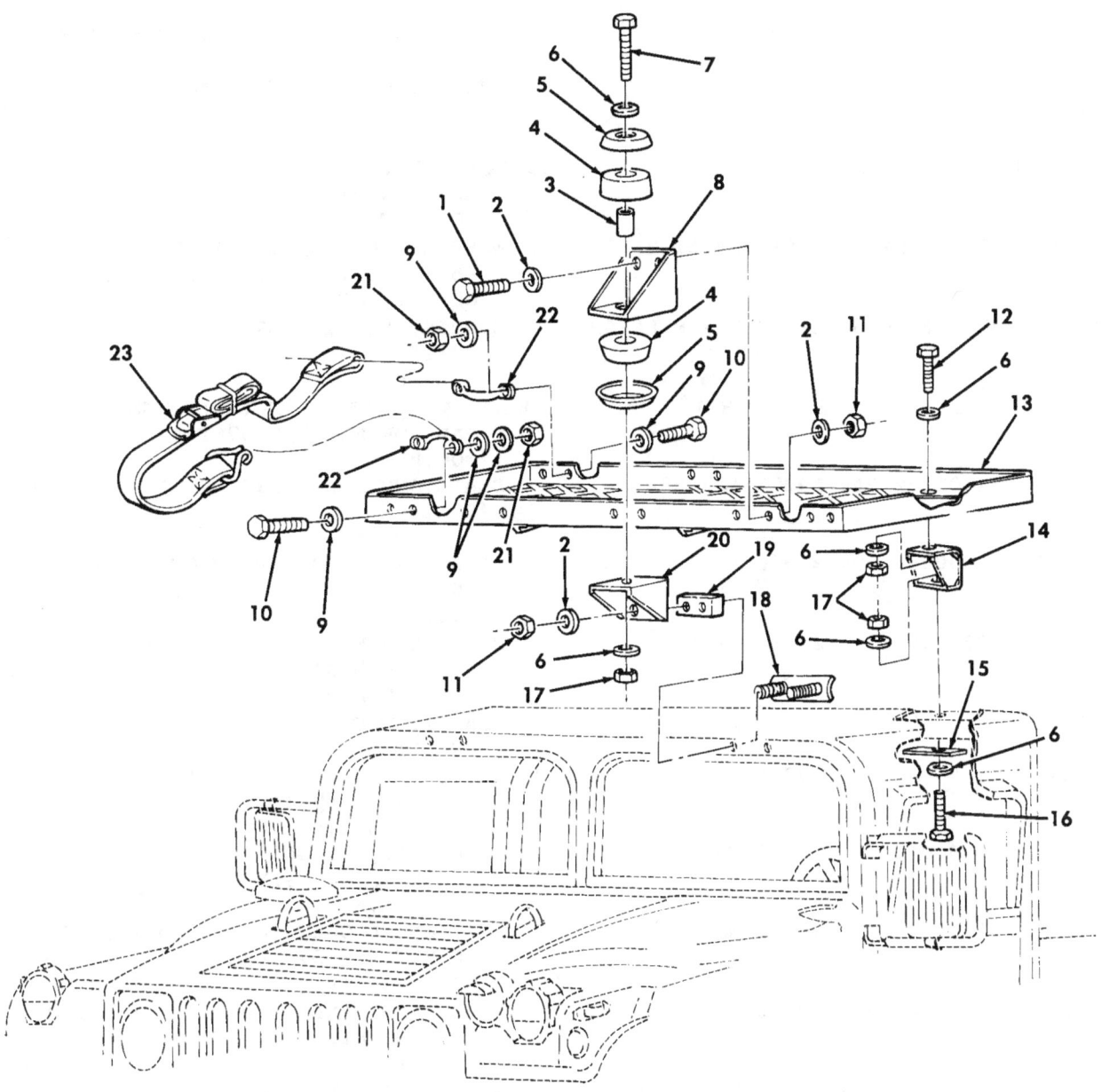

Figure 440. Kit, Camouflage Rack and Stowage Rack Assembly.

```
           SECTION II         TM9-2320-280-24P C01
       (1)    (2)    (3)          (4)         (5)                    (6)
(7)   ITEM   SMR                                 PART
      NO    CODE    NSN         CAGEC         NUMBER    DESCRIPTION AND USABLE ON CODES(UOC)
QTY
```

GROUP 3307 SPECIAL PURPOSE KITS

FIG. 440 KIT,CAMOUFLAGE RACK AND STOWAGE RACK ASSEMBLY

```
*  1 PAOZZ 5306002264825 29510 25228R1       BOLT,MACHINE   5/16-18 X .75,PART OF
4                                               KIT P/N 57K1641....................
                                                UOC:HPM,KTV
   2 PAOZZ 5310011191024 24617 2436162       WASHER,FLAT   5/16,PART OF KIT P/N
12                                              57K1641............................
                                                UOC:HPM,KTV
   3 PAOZZ 5365012160364 19207 12338186-52   SPACER,SLEEVE   PART OF KIT P/N
2                                               57K1641............................
                                                UOC:HPM,KTV
*  4 PAOZZ 5340012110838 34623 5584891       MOUNT,RESILIENT   UPPER AND LOWER,
4                                               PART OF KIT P/N 57K1641............
                                                UOC:HPM,KTV
   5 PAOZZ 2540011925949 19207 12339561      PLATE,SNUBBER,VEHIC   PART OF KIT P/N
4                                               57K1641............................
                                                UOC:HPM,KTV
   6 PAOZZ 5310000806004 96906 MS27183-14    WASHER,FLAT   3/8,PART OF KIT P/N
12                                              57K1641............................
                                                UOC:HPM,KTV
   7 PAOZZ 5306013601129 24617 427583        BOLT,MACHINE   3/8-16 X 3.25,PART OF
2                                               KIT P/N 57K1641....................
                                                UOC:HPM,KTV
   8 PAOZZ 5340013599474 19207 12342559      BRACKET,MOUNTING   UPPER FRONT,PART
2                                               OF KIT P/N 57K1641.................
                                                UOC:HPM,KTV
   9 PAOZZ 5310011023270 24617 2436161       WASHER,FLAT   1/4,PART OF KIT
30                                              P/N 57K1641........................
                                                UOC:HPM,KTV
  10 PAOZZ 5305000680502 96906 MS90725-6     SCREW,CAP,HEXAGON H   1/4-20 X .75,
12                                              PART OF KIT P/N 57K1641............
                                                UOC:HPM,KTV
  11 PAOZZ 5310008140673 96906 MS51943-33    NUT,SELF-LOCKING,HE   5/16-18,PART
8                                               OF KIT P/N 57K1641.................
                                                UOC:HPM,KTV
  12 PAOZZ 5305009890830 72582 186678        SCREW,CAP,HEXAGON H   3/8-16 X 1.00,
2                                               PART OF KIT P/N 57K1641............
                                                UOC:HPM,KTV
  13 PAOZZ 2540013875578 19207 12342558      RACK ASSEMBLY   PART OF KIT P/N
1                                               57K1641............................
                                                UOC:HPM,KTV
  14 PAOZZ 5340013602053 19207 12342561      BRACKET,MOUNTING   REAR,PART
2                                               OF KIT P/N 57K1641.................
                                                UOC:HPM,KTV
  15 PAOZZ 5340013602021 19207 12342666      PLATE,MOUNTING   PART OF KIT P/N
2                                               57K1641............................
                                                UOC:HPM,KTV
  16 PAOZZ 5306013601130 24617 456958        BOLT,MACHINE   3/8-16 X 4.00,PART OF
2                                               KIT P/N 57K1641....................
                                                UOC:HPM,KTV
```

440-1

SECTION II   TM9-2320-280-24P
 (1)   (2)   (3)   (4)   (5)   (6)
(7) ITEM SMR   PART
      NO   CODE   NSN   CAGEC   NUMBER   DESCRIPTION AND USABLE ON CODES (UOC)
QTY

      17 PAOZZ 5310009359021 96906 MS51943-35   NUT,SELF-LOCKING,HE 3/8-16,PART OF
                                                       KIT P/N 57K1641..................
                                                    UOC:HPM,KTV
      18 PAOZZ 5340013611212 19207 12342623   BRACKET,DOUBLE ANGL PART OF KIT P/N
                                                       57K1641..........................
                                                    UOC:HPM,KTV
      19 PAOZZ 5365013599406 19207 12342557   SPACER,STRAIGHT PART OF KIT P/N
                                                       57K1641..........................
UOC:HPM,KTV   20 PAOZZ 5340013598933 19207 12342560   BRACKET,ANGLE LOWER
FRONT,PART OF   2   KIT P/N 57K1641.....
....   UOC:HPM,KTV
      21 PAOZZ 5310000614650 96906 MS51943-31   NUT,SELF-LOCKING,HE 1/4-20,PART OF
2   KIT P/N 57K1641..................
                                                    UOC:HPM,KTV
      22 PAOZZ 5340011978674 19207 12338839-6   HANDLE,BOW PART OF KIT P/N 57K1641.   6
UOC:HPM,KTV
      23 PAOZZ 5340013601886 19207 12342562   STRAP,RETAINING PART OF KIT P/N
                                                       57K1641..........................
                                                    UOC:HPM,KTV

     KIT KFOZZ   19207 57K1641   KIT,CAMOUFLAGE RACK ASSEMBLY PART   1
OF KIT P/N 57K1628 PART OF KIT P/N   57K0
35..............................   UOC:HP-
M,KTV
                                                    BOLT,MACHINE   (4) 440-1
                                                    BOLT,MACHINE   (2) 440-7
                                                    BOLT,MACHINE   (2) 440-16
                                                    BRACKET,ANGLE   (2) 440-20
                                                    BRACKET,DOUBLE ANGL (2) 440-18
                                                    BRACKET,MOUNTING   (2) 440-8
                                                    BRACKET,MOUNTING   (2) 440-14
                                                    HANDLE,BOW   (6) 440-22
                                                    MOUNT,RESILIENT   (4) 440-4
                                                    NUT,SELF-LOCKING,HE (8) 440-11
                                                    NUT,SELF-LOCKING,HE (6) 440-17
                                                    NUT,SELF-LOCKING,HE (12) 440-21
                                                    PLATE,MOUNTING   (2) 440-15
                                                    PLATE,SNUBBER,VEHIC (4) 440-5
                                                    RACK ASSEMBLY   (1) 440-13
                                                    SCREW,CAP,HEXAGON H (12) 440-10
                                                    SCREW,CAP,HEXAGON H (2) 440-12
                                                    SPACER,SLEEVE   (2) 440-3
                                                    SPACER,STRAIGHT   (2) 440-19
                                                    STRAP,RETAINING   (3) 440-23
                                                    WASHER,FLAT   (12) 440-2
                                                    WASHER,FLAT   (12) 440-6
                                                    WASHER,FLAT   (30) 440-9

                                      END OF FIGURE

Section II.

TM 9-2320-280-24P-2

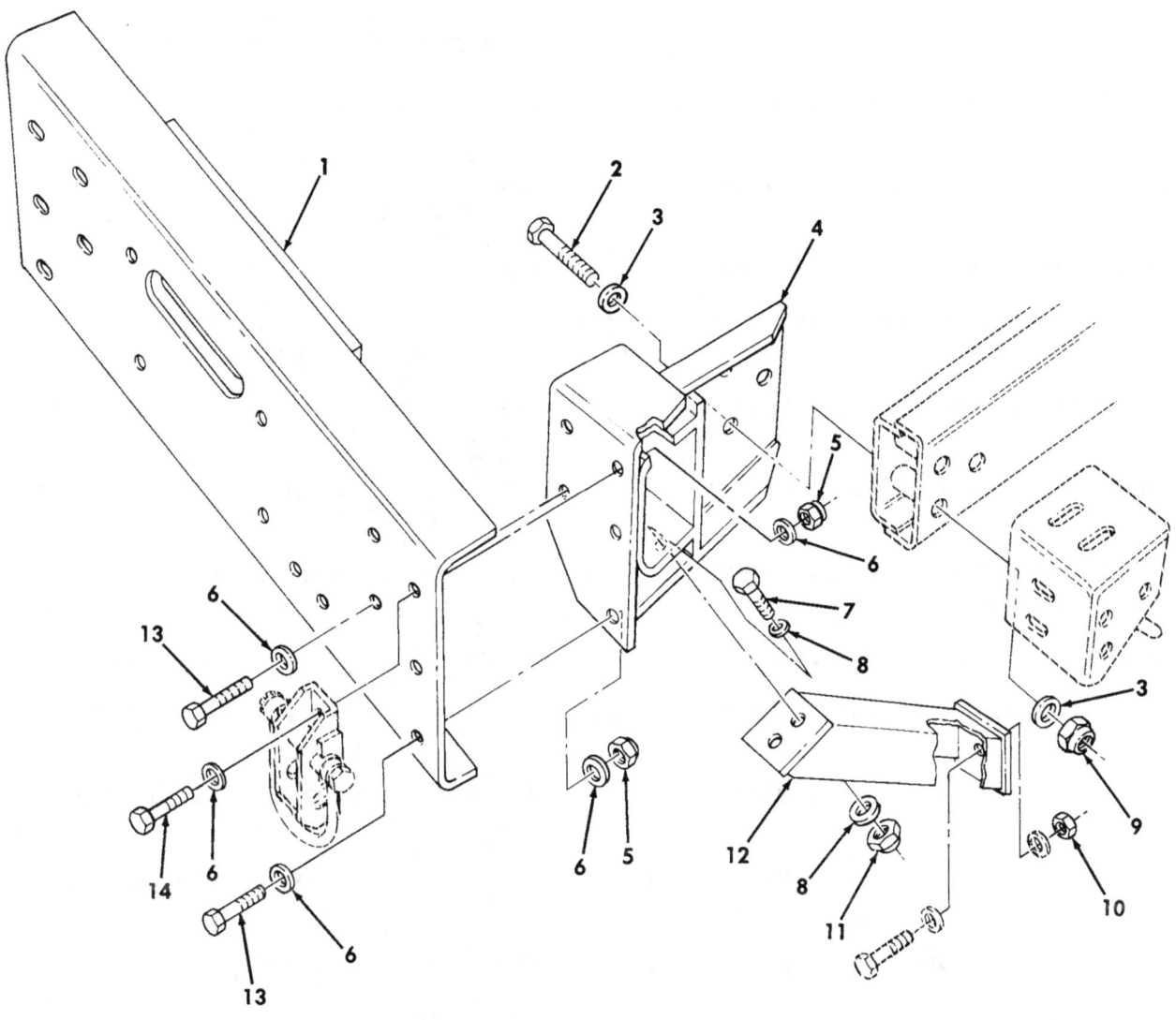

Figure 441. Kit, Prime Mover, Lt. Howitzer, 105 MM L119, Front Bumper.

SECTION II TM9-2320-280-24P C01

(1) ITEM NO	(2) SMR CODE	(3) NSN	(4) CAGEC	(5) PART NUMBER	(6) DESCRIPTION AND USABLE ON CODES (UOC)	(7) QTY
					GROUP 3307 SPECIAL PURPOSE KITS	
					FIG. 441 KIT,PRIME MOVER,LT. HOWITZER,105 MM L119,FRONT BUMPER	
1	PAOZZ	2540013618206	19207	12342395	BUMPER,VEHICULAR PART OF KIT P/N 57K1628.............. UOC:HPM	1
2	PAOZZ	5305011112774	24617	9428039	SCREW,CAP,HEXAGON H 5/8-11 X 3.75, PART OF KIT P/N 57K1628........... UOC:HPM	6
3	PAOZZ	5310011517347	24617	2436167	WASHER,FLAT 5/8,PART OF KIT P/N 57K1628.............. UOC:HPM	12
4	PFOZZ	2510013581178	19207	12342387	FRAME SECTION,STRU L.H.,PART OF KIT P/N 57K1628.............. UOC:HPM	1
4	PFOZZ	2510013578789	19207	12342388	FRAME SECTION,STRU R.H.,PART OF KIT P/N 57K1628.............. UOC:HPM	1
5	PAOZZ	5310004883889	96906	MS51943-39	NUT,SELF-LOCKING,HE 1/2-13,PART OF KIT P/N 57K1628.............. UOC:HPM	10
* 6	PAOZZ	5310011211703	06032	2310-0143-001	WASHER,FLAT 1/2,PART OF KIT P/N 57K1628.............. UOC:HPM	16
* 7	PAOZZ	5305011044846	24617	9427321	SCREW,CAP,HEXAGON H 3/8-16 X 1/8, PART OF KIT P/N 57K1628............. UOC:HPM	4
8	PAOZZ	5310000806004	96906	MS27183-14	WASHER,FLAT 3/8,PART OF KIT P/N 57K1628.............. UOC:HPM	8
9	PAOZZ	5310000614651	96906	MS51943-43	NUT,SELF-LOCKING,HE 5/8-11,PART OF KIT P/N 57K1628.............. UOC:HPM	6
10	PAOZZ	5310004093333	96906	MS51943-45	NUT,PLAIN,HEXAGON 3/4-10,PART OF KIT P/N 57K1628.............. UOC:HPM	2
11	PAOZZ	5310009359021	96906	MS51943-35	NUT,SELF-LOCKING,HE 3/8-16,PART OF KIT P/N 57K1628.............. UOC:HPM	4
12	PFOZZ	2510013581176	19207	12342389	FRAME SECTION,STRU L.H.,PART OF KIT P/N 57K1628.............. UOC:HPM	1
12	PFOZZ	2510013581177	19207	12342390	FRAME SECTION,STRU R.H.,PART OF KIT P/N 57K1628.............. UOC:HPM	1
13	PAOZZ	5306013601125	24617	455000	BOLT,MACHINE 1/2-13 X 1.25,PART OF KIT P/N 57K1628.............. UOC:HPM	6
14	PAOZZ	5305000712070	80204	B1821BH050C175N	SCREW,CAP,HEXAGON H 1/2-13 X 1.75, PART OF KIT P/N 57K1628............. UOC:HPM	4

END OF FIGURE

Section II. TM 9-2320-280-24P-2

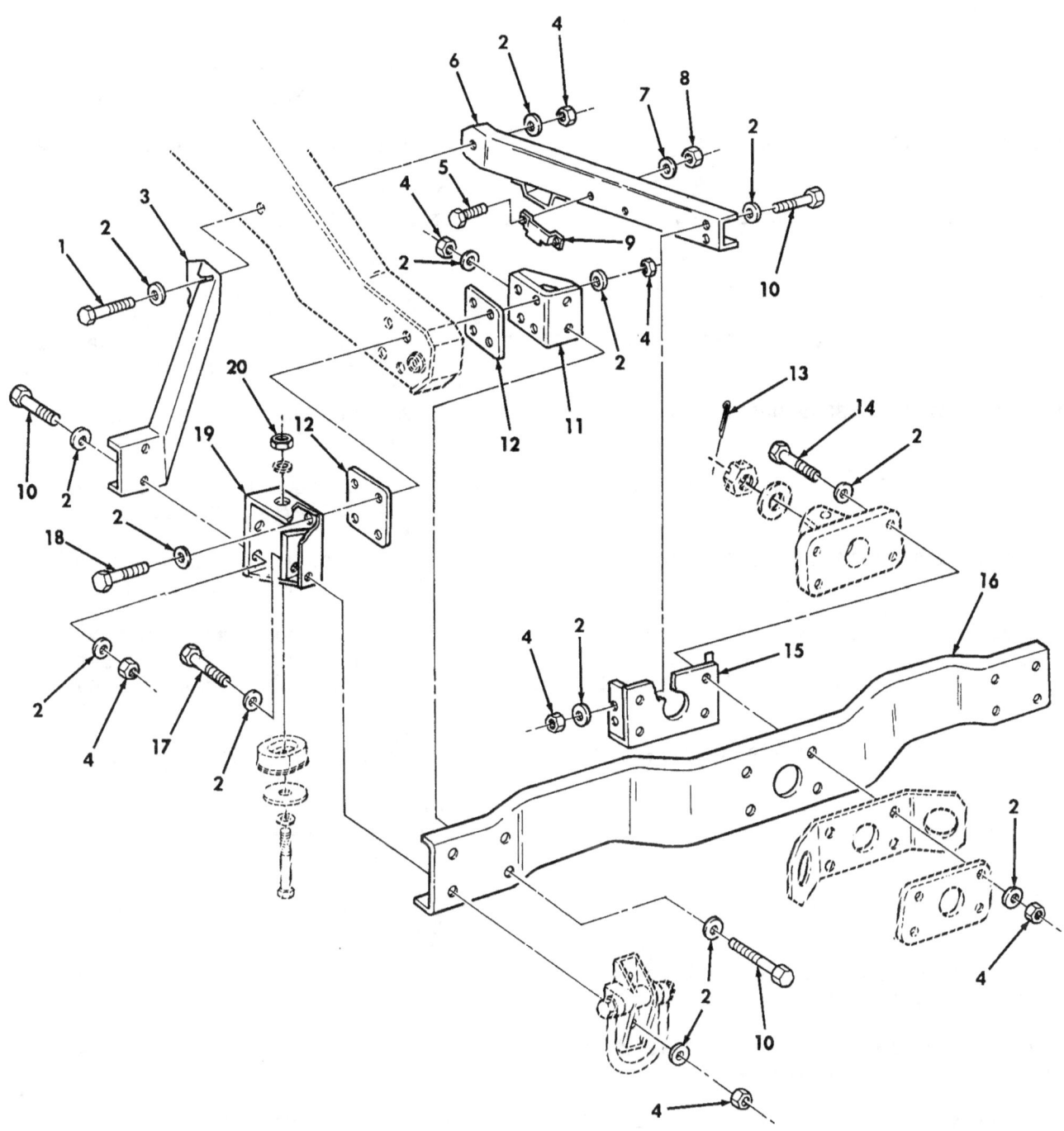

Figure 442. Kit, Prime Mover, Lt. Howitzer, 105 MM L119, Rear Bumper.

SECTION II TM9-2320-280-24P C01

(1) ITEM NO	(2) SMR CODE	(3) NSN	(4) CAGEC	(5) PART NUMBER	(6) DESCRIPTION AND USABLE ON CODES (UOC)	(7) QTY

GROUP 3307 SPECIAL PURPOSE KITS

FIG. 442 KIT,PRIME MOVER,LT. HOWITZER,105 MM L119,REAR BUMPER

1	PAOZZ	5306013601128	24617	455006	BOLT,MACHINE 1/2-13 X 3.75,PART OF KIT P/N 57K1628.................... UOC:HPM	2
*2	PAOZZ	5310011211703	06032	2310-0143-001	WASHER,FLAT 1/2,PART OF KIT P/N 57K1628............................. UOC:HPM	60
*3	PAOZZ	2590013618087	34623	12342522	BRACKET,VEHICULAR REAR BUMPER, OUTER,L.H.,PART OF KIT P/N 57K1628.. UOC:HPM	1
3	PAOZZ	2590013618088	19207	12342540	BRACKET,VEHICULAR REAR BUMPER, OUTER,R.H.,PART OF KIT P/N 57K1628. UOC:HPM	1
4	PAOZZ	5310004883889	96906	MS51943-39	NUT,SELF-LOCKING,HE 1/2-13,PART OF KIT P/N 57K1628.................... UOC:HPM	30
5	PAOZZ	5305011858674	24617	9421294	SCREW,CAP,HEXAGON H #10-24 X 1/2, PART OF KIT P/N 57K1628............ UOC:HPM	4
6	PAOZZ	2590013605414	19207	12342516	BRACKET,VEHICULAR C REAR BUMPER, INNER,L.H.,PART OF KIT P/N 57K1628. UOC:HPM	1
6	PAOZZ	2590013605413	19207	12342517	BRACKET,VEHICULAR C REAR BUMPER, INNER,R.H.,PART OF KIT P/N 57K1628.. UOC:HPM	1
7	PAOZZ	5310000453296	96906	MS35338-43	WASHER,LOCK #10,PART OF KIT P/N 57K1628............................. UOC:HPM	4
8	PAOZZ	5310007349758	96906	MS35649-202	NUT,PLAIN,HEXAGON #10-24,PART OF KIT P/N 57K1628.................... UOC:HPM	4
9	PAOZZ	2590013609537	19207	12342539	BRACKET,VEHICULAR C PART OF KIT P/N 57K1628............................. UOC:HPM	2
10	PAOZZ	5306013601125	24617	455000	BOLT,MACHINE 1/2-13 X 1.25,PART OF KIT P/N 57K1628.................... UOC:HPM	12
11	PAOZZ	2590013605412	19207	12342520	BRACKET,VEHICULAR C REAR BUMPER MOUNT,INNER,L.H.,PART OF KIT P/N 57K1628............................. UOC:HPM	1
11	PAOZZ	2590013609536	19207	12342521	BRACKET,VEHICULAR C REAR BUMPER MOUNT,INNER,R.H.,PART OF KIT P/N 57K1628............................. UOC:HPM	1
12	PAOZZ	5340013602020	19207	12342750	PLATE,MOUNTING PART OF KIT P/N 57K1628............................. UOC:HPM	4
13	PAOZZ	5315008460126	96906	MS24665-628	PIN,COTTER 1/4 X 3.00,PART OF KIT	1

442-1

SECTION II   TM9-2320-280-24P

(1) NO	(2) SMR CODE	(3) NSN	(4) CAGEC	(5) PART NUMBER	(7) ITEM DESCRIPTION AND USABLE ON CODES(UOC)	QTY
14	PAOZZ	5306013601126	24617	455002	BOLT,MACHINE 1/2-13 X 2.50,PART OF KIT P/N 57K1628.................... UOC:HPM	4
15	PAOZZ	2590013641534	19207	12342523	BRACKET,VEHICULAR PART OF KIT P/N 57K1628.................... UOC:HPM	1
16	PFOZZ	2540013602482	19207	12342515	BUMPER,VEHICULAR REAR,PART OF KIT P/N 57K1628.................... UOC:HPM	1
17	PAOZZ	5306013601127	24617	455004	BOLT,MACHINE 1/2-13 X 3.00,PART OF KIT P/N 57K1628.................... UOC:HPM	
18	PAOZZ	5306011865369	7X677	9415560	BOLT,MACHINE 1/2-13 X 3.00,PART OF KIT P/N 57K1628.................... UOC:HPM	8
19	PFOZZ	2590013605411	19207	12342518	BRACKET,VEHICULAR C REAR BUMPER MOUNT,OUTER,L.H.,PART OF KIT P/N 57K1628.................... UOC:HPM	
19	PFOZZ	2590013632084	19207	12342519	BRACKET,VEHICULAR C REAR BUMPER MOUNT,OUTER,R.H.,PART OF KIT P/N 57K1628.................... UOC:HPM	
20	PAOZZ	5310000614651	96906	MS51943-43	NUT,SELF-LOCKING 5/8-11,PART OF KIT P/N 57K1628.................... UOC:HPM	2

END OF FIGURE

Section II. TM 9-2320-280-24P-2

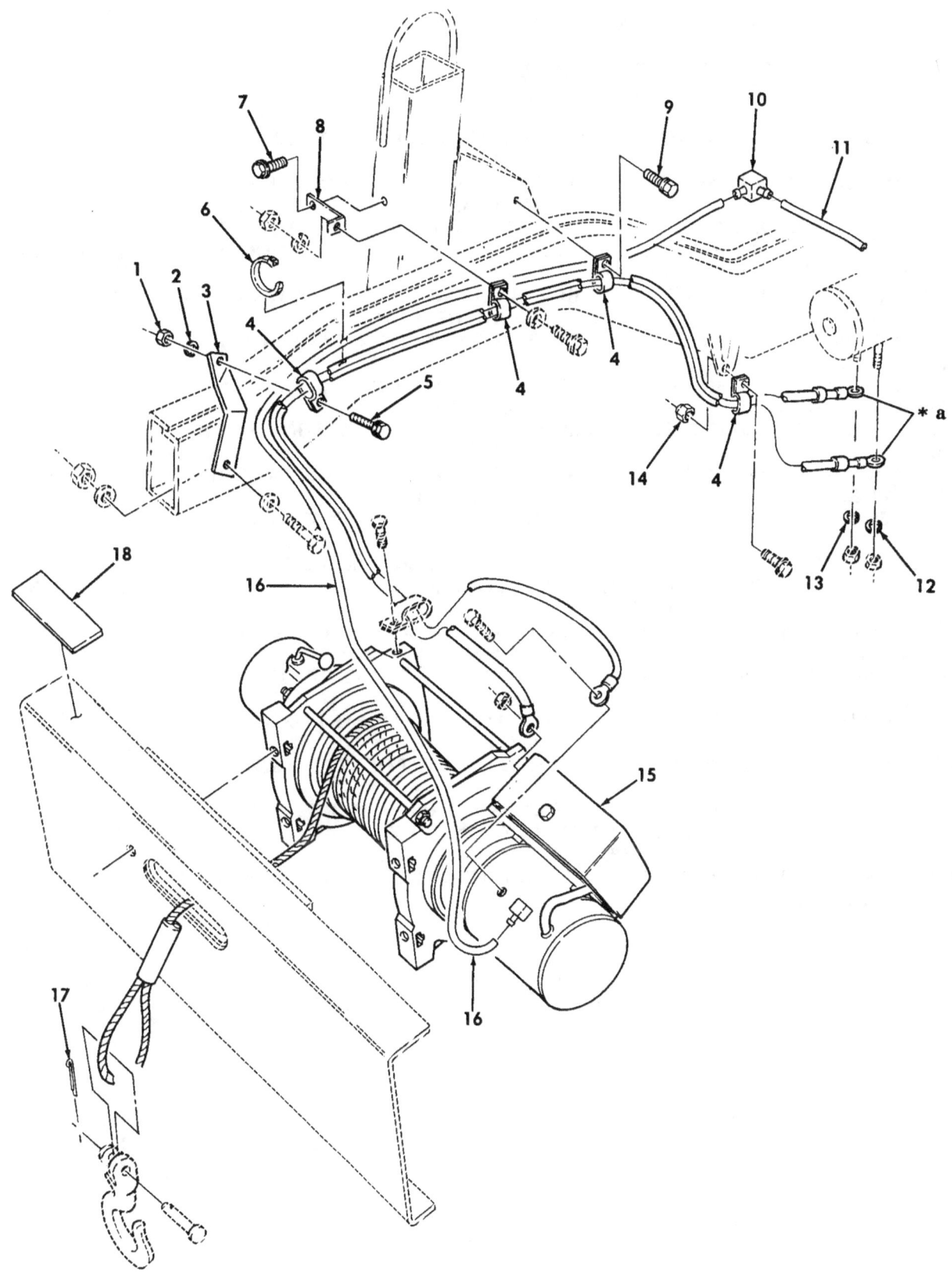

* a PART OF ITEM 15

Figure 443. Kit, Prime Mover, Lt. Howitzer, 105 MM L119, Winch.

SECTION II TM9-2320-280-24P

(1) NO	(2) SMR CODE	(3) NSN	(4) CAGEC	(5) PART NUMBER	(7) ITEM DESCRIPTION AND USABLE ON CODES(UOC)	QTY

GROUP 3307 SPECIAL PURPOSE KITS

FIG. 443 KIT,PRIME MOVER,LT. HOWITZER,105 MM L119,WINCH

1	PAOZZ	5310009050762	96906	MS51967-3	NUT,PLAIN,HEXAGON 1/4-20,PART OF KIT P/N 57K1628................. UOC:HPM	2
2	PAOZZ	5310005825965	96906	MS35338-44	WASHER,LOCK 1/4,PART OF KIT P/N 57K1628............. UOC:HPM	2
3	PFOZZ	5340012669173	19207	12338481	BRACKET,DOUBLE ANGL PART OF KIT P/N 57K1628............. UOC:HPM	1
4	PAOZZ	5340004381833	19207	7721654	CLAMP,LOOP PART OF KIT P/N 57K1628. UOC:HPM	4
5	PAOZZ	5305004890743	24617	9415477	SCREW,MACHINE 1/4-20 X 0.75,PART OF KIT P/N 57K1628................. UOC:HPM	2
6	PAOZZ	5975012055379	96906	MS3367-7	STRAP,TIEDOWN,ELECT PART OF KIT P/N 57K1628............. UOC:HPM	4
7	PAOZZ	5305011013312	96906	MS51851-126	SCREW,TAPPING 3/8-16 X 0.75,PART OF KIT P/N 57K1628................. UOC:HPM	1
8	PAOZZ	5340011922256	19207	12338663	BRACKET PART OF KIT P/N 57K1628.... UOC:HPM	1
9	PAOZZ	5305001913640	96906	MS51851-85	SCREW,TAPPING 1/4-20 X 0.62,PART OF KIT P/N 57K1628................. UOC:HPM	1
10	PAOZZ	4730010035105	79470	1064X4	TEE,HOSE PART OF KIT P/N 57K1628... UOC:HPM	1
11	MOOZZ		19207	CPR104420-1-2	TUBING,NONMETALLIC MAKE FROM TUBE, P/N CPR104420-1,2 INCHES LONG,PART OF KIT P/N 57K1628... UOC:HPM	1
12	PAOZZ	5310006379541	96906	MS35338-46	WASHER,LOCK 3/8,PART OF KIT P/N 57K1628............. UOC:HPM	1
13	PAOZZ	5310012167390	96906	MS51415-9	WASHER,LOCK 1/2,PART OF KIT P/N 57K1628............. UOC:HPM	1
14	PAOZZ	5310011520598	24617	271172	NUT,SELF-LOCKING,AS PART OF KIT P/N 57K1628............. UOC:HPM	1
15	PAOFF	2590011797602	19207	12339723	WINCH,DRUM PART OF KIT P/N 57K1628. UOC:HPM	1
16	MOOZZ		19207	CPR104420-1-70	TUBING,NONMETALLIC MAKE FROM TUBE, P/N CPR104420-1,70 INCHES LONG,PART OF KIT P/N 57K1628. UOC:HPM	1
17	PAOZZ	5315008395822	96906	MS24665-353	PIN,COTTER 0.120 X 1.13,PART OF KIT P/N 57K1628..................	1

SECTION II   TM9-2320-280-24P

(1) NO	(2) CODE	(3) NSN	(4) CAGEC	(5) NUMBER	(6)	(7) ITEM DESCRIPTION AND USABLE ON CODES(UOC)	SMR	PART QTY

UOC:HPM

18 PAOZZ 7690012646536 19207 12340923   DECAL PART OF KIT P/N 57K1628...... 1
UOC:HPM

END OF FIGURE

443-2

Section II.

TM 9-2320-280-24P-2

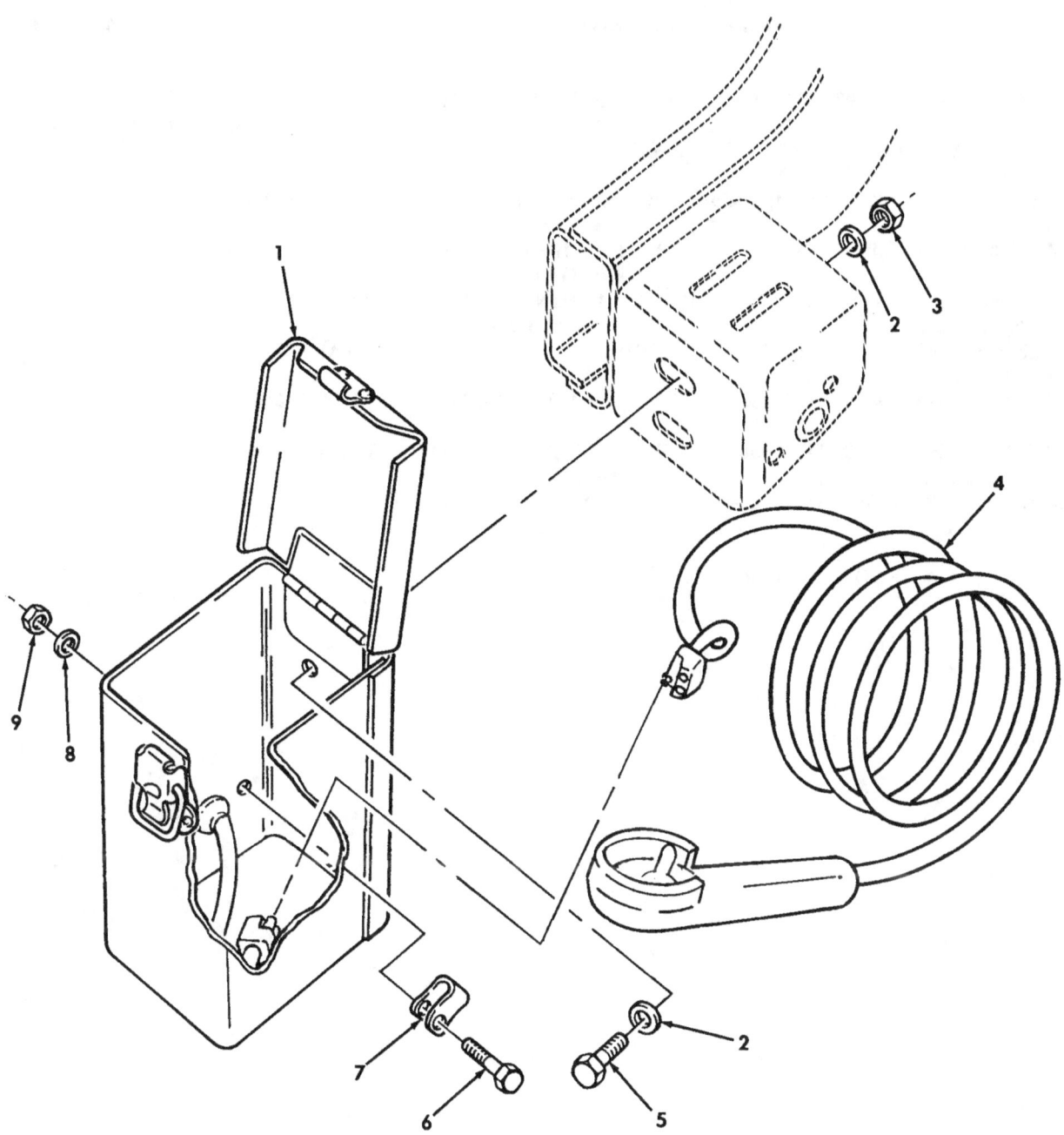

Figure 444. Kit, Prime Mover, Lt. Howitzer, 105 MM L119, Winch Control and Storage Box.

SECTION II TM9-2320-280-24P

(1) NO	(2) SMR CODE	(3) NSN	(4) CAGEC	(5) PART NUMBER	(7) ITEM DESCRIPTION AND USABLE ON CODES (UOC)	QTY

GROUP 3307 SPECIAL PURPOSE KITS

FIG. 444 KIT,PRIME MOVER,LT. HOWITZER,105 MM L119,WINCH CONTROL AND STORAGE BOX

1	PAOZZ	2540012096838	19207	12338489-1	BOX ACCESSORIES STO PART OF KIT P/N 57K1628............... UOC:HPM	1
2	PAOZZ	5310011211703	24617	2436165	WASHER,FLAT 1/2,PART OF KIT P/N 57K1628............. UOC:HPM	4
3	PAOZZ	5310004883889	96906	MS51943-39	NUT,SELF-LOCKING 1/2-13,PART OF KIT P/N 57K1628............ UOC:HPM	2
4	PAOZZ	2590012141566	19207	12338495	CONTROL HANDLE PART OF KIT P/N 57K1628............ UOC:HPM	1
5	PAOZZ	5306008440036	7X677	427566	BOLT,MACHINE 1/2-13 X 1.50,PART OF KIT P/N 57K1628............ UOC:HPM	2
6	PAOZZ	5305004890743	24617	9415477	SCREW,MACHINE 1/4-20 X .750,PART OF KIT P/N 57K1628............ UOC:HPM	1
7	PAOZZ	5340008091492	96906	MS21333-100	CLAMP,LOOP PART OF KIT P/N 57K1628. UOC:HPM	1
8	PAOZZ	5310005825965	96906	MS35338-44	WASHER,LOCK 1/4,PART OF KIT P/N 57K1628............ UOC:HPM	1
9	PAOZZ	5310009050762	96906	MS51967-3	NUT,PLAIN,HEXAGON 1/4-20,PART OF KIT P/N 57K1628............ UOC:HPM	1

END OF FIGURE

Section II. TM 9-2320-280-24P-2

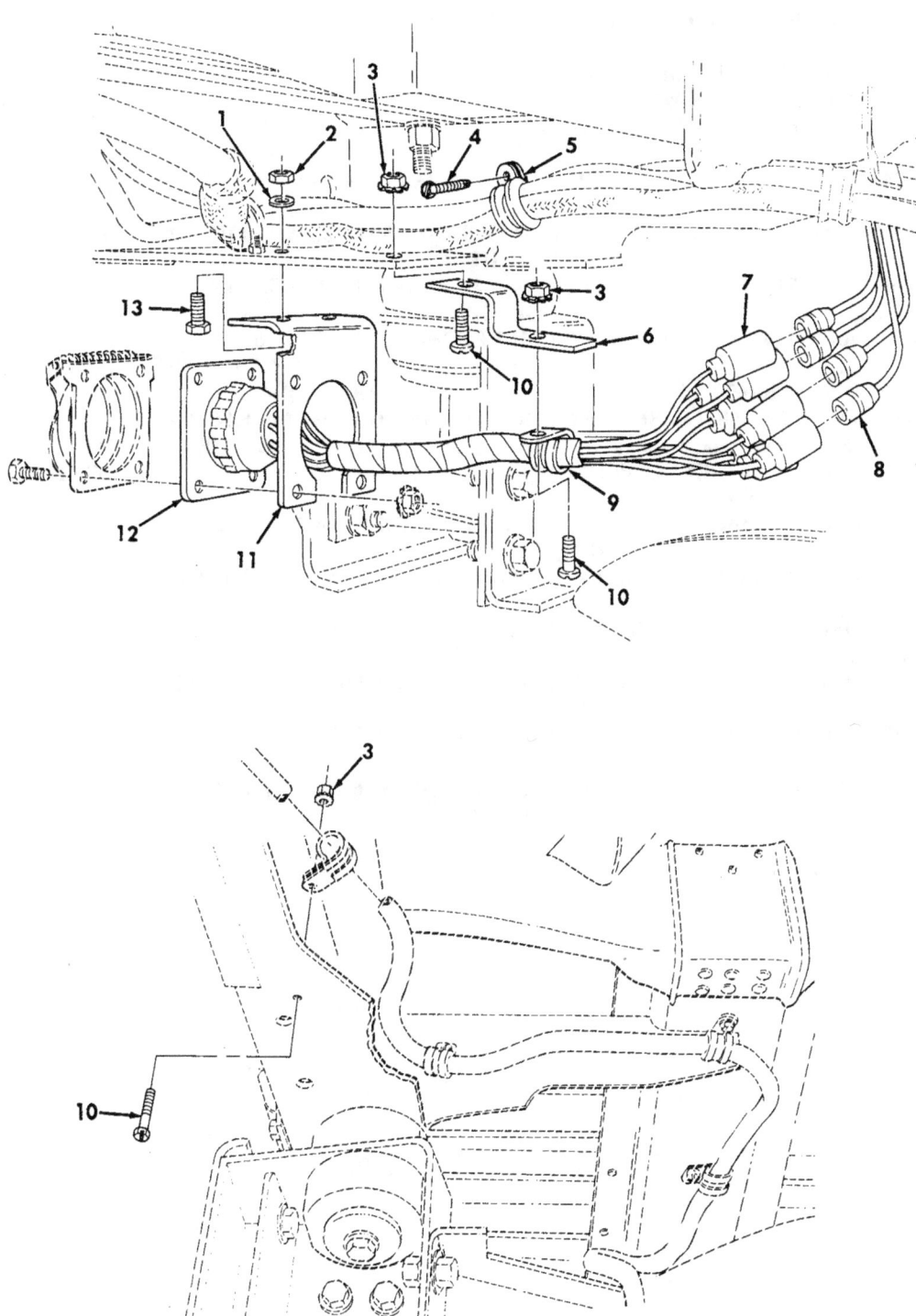

Figure 445. Kit, Prime Mover, Lt. Howitzer, 105 MM L119, Trailer Receptacle and Body Wiring Harness.

SECTION II TM9-2320-280-24P

(1) ITEM NO	(2) SMR CODE	(3) NSN	(4) CAGEC	(5) PART NUMBER	(6) DESCRIPTION AND USABLE ON CODES(UOC)	(7) QTY
					GROUP 3307 SPECIAL PURPOSE KITS	
					FIG. 445 KIT,PRIME MOVER,LT. HOWITZER,105 MM L119,TRAILER RECEPTACLE AND BODY WIRING HARNESS	
1	PAOZZ	5310011191024	24617	2436162	WASHER,FLAT 5/16,PART OF KIT P/N 57K1628............... UOC:HPM	2
2	PAOZZ	5310008140673	96906	MS51943-33	NUT,SELF-LOCKING 5/16-18,PART OF KIT P/N 57K1628.................... UOC:HPM	2
3	PAOZZ	5310001249265	24617	271169	NUT,PLAIN ASSEMBLED #10-32,PART OF KIT P/N 57K1628.................... UOC:HPM	3
4	PAOZZ	5305012854234	24617	9414241	BOLT,ASSEMBLED WASH #10-32 X .62, PART OF KIT P/N 57K1628............. UOC:HPM	1
5	PAOZZ	5340009899222	96906	MS21333-74	CLAMP,LOOP PART OF KIT P/N 57K1628. UOC:HPM	1
6	PAOZZ	5340013601189	19207	12342668	BRACKET,DOUBLE ANGL PART OF KIT P/N 57K1628........................... UOC:HPM	1
7	PAOZZ	5935004626603	96906	MS27142-2	CONNECTOR,PLUG,ELEC PART OF KIT P/N 57K1628......................... UOC:HPM	2
8	PAOZZ	5935001677775	96906	MS27144-1	CONNECTOR,PLUG,ELEC PART OF KIT P/N 57K1628......................... UOC:HPM	7
9	PAOZZ	5340000913790	96906	MS21333-72	CLAMP,LOOP PART OF KIT P/N 57K1628. UOC:HPM	1
10	PAOZZ	5305000187838	24617	187838	SCREW,CAP,HEXAGON H #10-32 X 5/8, PART OF KIT P/N 57K1628............. UOC:HPM	3
11	PAOZZ	5340013144951	19207	12338696-1	BRACKET,ANGLE PART OF KIT P/N 57K1628............................ UOC:HPM	1
12	PAOZZ	6150013662916	19207	12342667	WIRING,HARNESS PART OF IT P/N 57K1628............................ UOC:HPM	1
13	PAOZZ	5306002264827	80204	B1821BH031C100N	BOLT,MACHINE 5/16-18 X 1 PART OF KIT P/N 57K1628.................... UOC:HPM	2

END OF FIGURE

Section II. TM 9-2320-280-24P-2

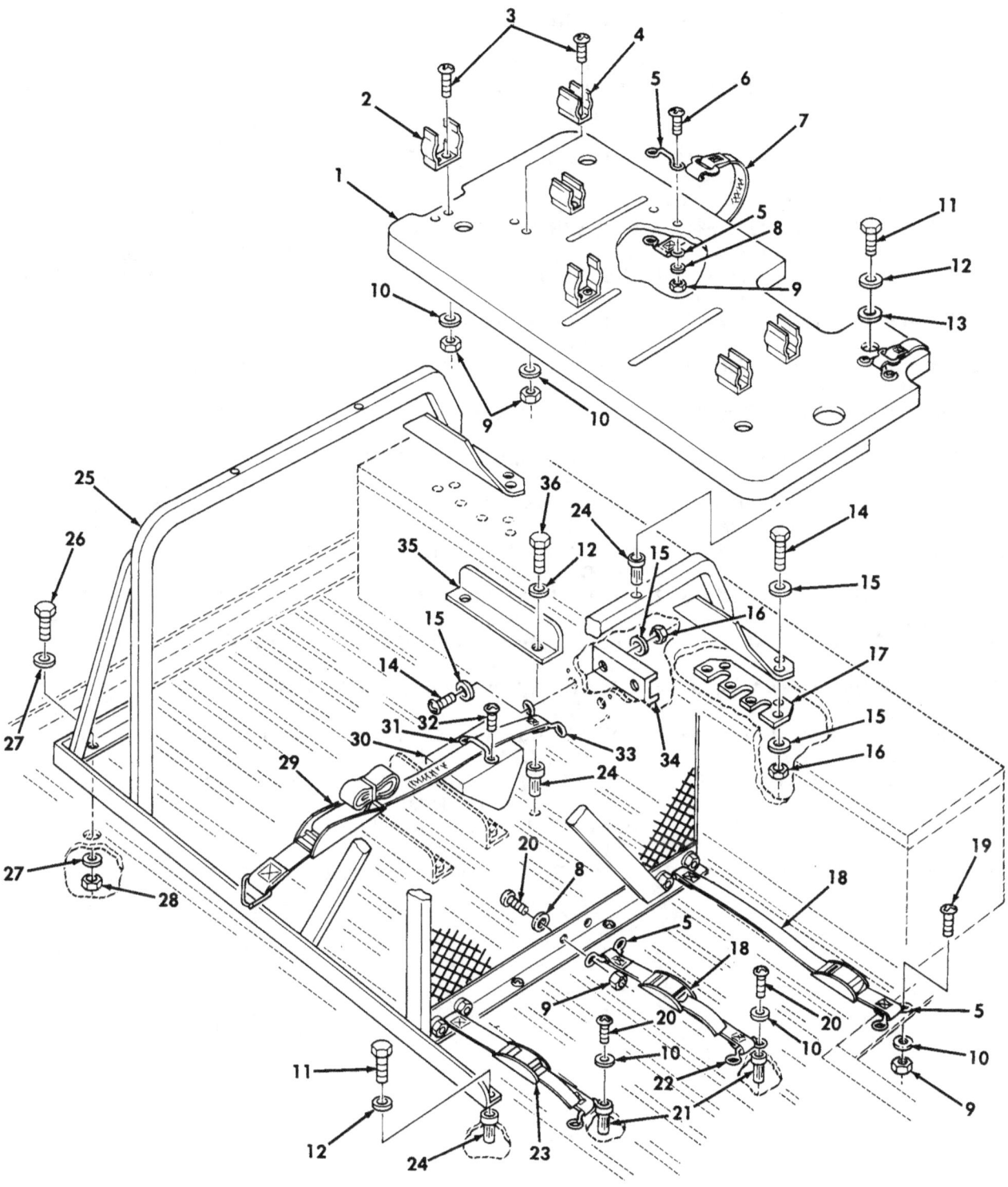

Figure 446. Kit, Prime Mover, Lt. Howitzer, 105 MM L119, Ammo Rack and Sight Boxes Stowage.

SECTION II   TM9-2320-280-24P

(1) ITEM NO	(2) SMR CODE	(3) NSN	(4) CAGEC	(5) PART NUMBER	(6) DESCRIPTION AND USABLE ON CODES(UOC)	(7) QTY

GROUP 3307 SPECIAL PURPOSE KITS

FIG. 446 KIT,PRIME MOVER,LT. HOWITZER,105 MM L119,AMMO RACK AND SIGHT BOXES STOWAGE

Item	SMR	NSN	CAGEC	Part Number	Description	Qty
1	MOOZZ		19207	12342595	TRAY,AMMO GUARD TOP MAKE FROM PLYWOOD,P/N MIL-P-18066,.750 THICK, PART OF KIT P/N 57K1628............. UOC:HPM	1
2	PAOZZ	5340013601727	19207	12342584	CLIP,SPRING TENSION PART OF KIT P/N 57K1628............ UOC:HPM	2
3	PAOZZ	5305009953442	96906	MS35207-268	SCREW,MACHINE #10-32 X 1.250,PART OF KIT P/N 57K1628......... UOC:HPM	12
4	PAOZZ	5340013598819	19207	12342585	CLIP,SPRING TENSION PART OF KIT P/N 57K1628............ UOC:HPM	4
5	PAOZZ	5340013145957	19207	12338839-1	LOOP,STRAP FASTENER PART OF KIT P/N 57K1628............ UOC:HPM	8
6	PAOZZ	5305009847352	96906	MS35191-278	SCREW,MACHINE #10-32 X 1.50,PART OF KIT P/N 57K1628............ UOC:HPM	4
7	PAOZZ	5340013604909	19207	12339816-3	STRAP,WEBBING PART OF KIT P/N 57K1628............ UOC:HPM	2
8	PAOZZ	5310008098546	96906	MS27183-8	WASHER,FLAT #10,PART OF KIT P/N 57K1628............ UOC:HPM	16
9	PAOZZ	5310002514503	24617	9411893	NUT,LOCK #10-32,PART OF KIT P/N 57K1628............ UOC:HPM	24
10	PAOZZ	5310000145850	96906	MS27183-42	WASHER,FLAT #10,PART OF KIT P/N 57K1628............ UOC:HPM	18
11	PAOZZ	5340013235509	24617	454838	BOLT,MACHINE PART OF KIT P/N 57K1628 UOC:HPM	10
12	PAOZZ	5310005825965	96906	MS35338-44	WASHER,LOCK 1/4,PART OF KIT P/N 57K1628............ UOC:HPM	16
13	PAOZZ	5310008093078	96906	MS27183-11	WASHER,FLAT 1/4,PART OF KIT P/N 57K1628............ UOC:HPM	4
14	PAOZZ	5305002253843	OAT62	35A2C5	SCREW,CAP,HEXAGON H 1/4-20 X 1.00, PART OF KIT P/N 57K1628............ UOC:HPM	14
15	PAOZZ	5310011023270	24617	2436161	WASHER,FLAT 1/4,PART OF KIT P/N 57K1628............ UOC:HPM	28
16	PAOZZ	5310000614650	96906	MS51943-31	NUT,SELF-LOCKING,HE 1/4-20,PART OF	14

SECTION II   TM9-2320-280-24P

(1) NO	(2) SMR CODE	(3) NSN	(4) CAGEC	(5) PART NUMBER	(6) DESCRIPTION AND USABLE ON CODES(UOC)	(7) QTY
					KIT P/N 57K1628 UOC:HPM	
17	PFOZZ	5340013602979	19207	12342586	PLATE,MOUNTING PART OF KIT P/N 57K1628 UOC:HPM	2
18	PAOZZ	5340012552812	19207	12340529-16	STRAP,WEBBING PART OF KIT P/N 57K1628 UOC:HPM	2
19	PAOZZ	5305009847342	96906	MS35191-274	SCREW,MACHINE #10-32 X .750,PART OF KIT P/N 57K1628 UOC:HPM	2
20	PAOZZ	5305009931848	96906	MS35207-265	SCREW,MACHINE #10-32 X .750,PART OF KIT P/N 57K1628 UOC:HPM	2
21	PAOZZ	5310011981723	03481	S10P175	NUT,PLAIN,BLIND RIV #10-32,PART OF KIT P/N 57K1628 UOC:HPM	4
22	PAOZZ	2510011995748	19207	12338839-8	LOOP,FOOTMAN,FLAT PART OF KIT P/N 57K1628 UOC:HPM	2
23	PAOZZ	5340012032708	19207	12340529-7	STRAP,WEBBING PART OF KIT P/N 57K1628 UOC:HPM	1
24	PAOZZ	5310011981722	19207	12339998-13	PLUSNUT 1/4-20,PART OF KIT P/N 57K1628 UOC:HPM	16
25	PAOZZ	2590013602417	19207	12342574	RACK,AMMUNITION STO PART OF KIT P/N 57K1628 UOC:HPM	1
26	PAOZZ	5305000712070	81204	B1821BH050C175N	BOLT,HEXAGON HEAD 1/2-13 X 1.75, PART OF KIT P/N 57K1628 UOC:HPM	1
27	PAOZZ	5310001670823	88044	AN960-816	WASHER,FLAT 1/2,PART OF KIT P/N 57K1628 UOC:HPM	2
28	PAOZZ	5310004883889	96906	MS51943-39	NUT,SELF-LOCKING 1/2-13,PART OF KIT P/N 57K1628 UOC:HPM	1
29	PAOZZ	5340013601887	19207	12342594	STRAP,WEBBING PART OF KIT P/N 57K1628 UOC:HPM	1
30	MOOZZ		19207	12342596	WEDGE MAKE FROM WOOD,P/N 13219E0079, 1.00 THICK,PART OF KIT P/N 57K1628 UOC:HPM	1
31	PAOZZ	5340011978674	19207	12338839-6	HANDLE,BOW PART OF KIT P/N 57K1628. UOC:HPM	7
32	PAOZZ	5305004324254	96906	MS51861-69	SCREW,TAPPING 1/4-13 X 1.00,PART OF KIT P/N 57K1628 UOC:HPM	2
33	PAOZZ	5340013609070	19207	12342587	LAMP,BRIDGE PART OF KIT P/N 57K1628 UOC:HPM	1
34	PAOZZ	5340013601188	19207	12342583	BRACKET,DOUBLE ANGL PART OF KIT P/N 57K1628 UOC:HPM	1

446-2

SECTION II   TM9-2320-280-24P

(1) ITEM NO	(2) SMR CODE	(3) NSN	(4) CAGEC	(5) PART NUMBER	(6) DESCRIPTION AND USABLE ON CODES(UOC)	(7) QTY
35	PAOZZ	5340013601784	19207	12342580	BRACKET,ANGLE PART OF KIT P/N 57K1628............... UOC:HPM	3
36	PAOZZ	5305000680508	07482	1C817P63	BOLT,HEXAGON 1/4-20 X .750,PART OF KIT P/N 57K1628................. UOC:HPM	6

END OF FIGURE

Section II.

TM 9-2320-280-24P-2

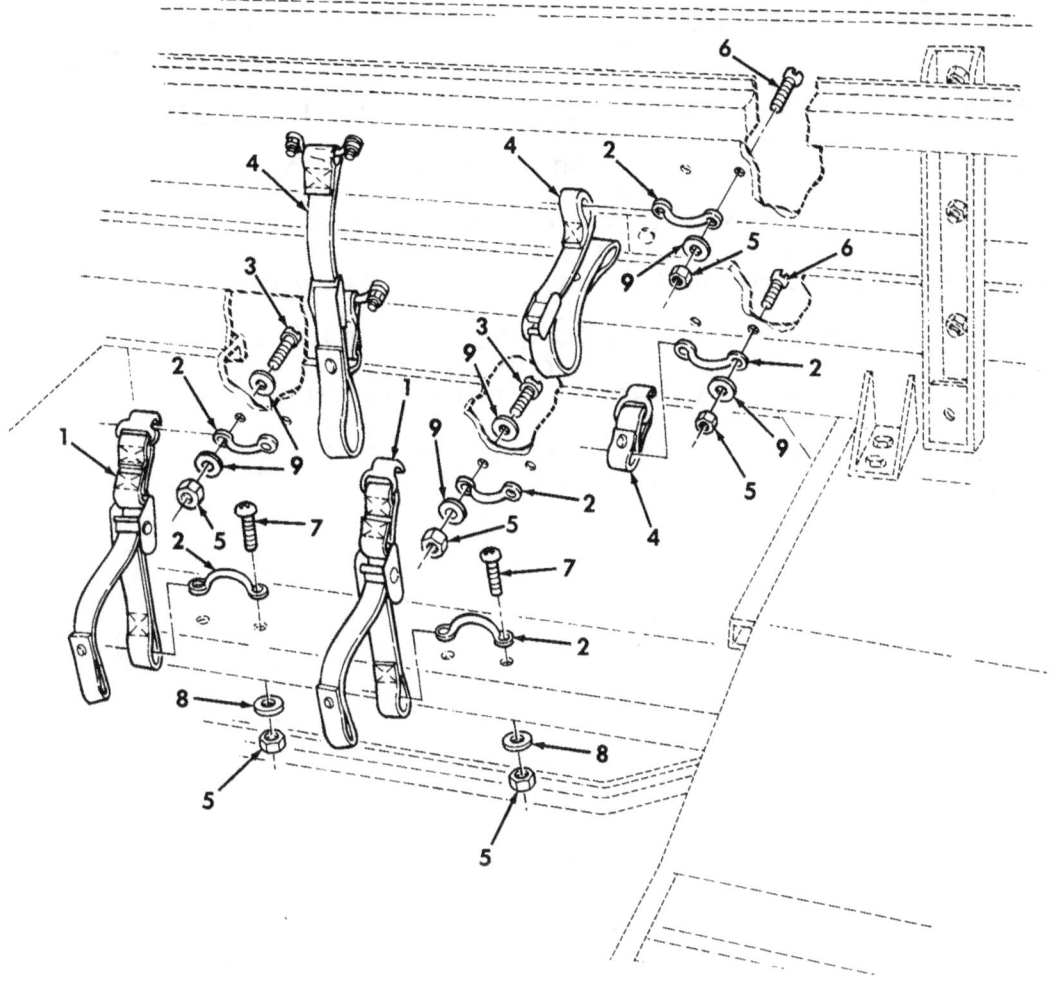

Figure 447. Kit, Prime Mover, Lt. Howitzer, 105 MM L119, Telephone, Remote, and Tripod Stowage.

SECTION II TM9-2320-280-24P

(1) NO	(2) SMR CODE	(3) NSN	(4) CAGEC	(5) PART NUMBER	(7) ITEM DESCRIPTION AND USABLE ON CODES(UOC)	(6) QTY

GROUP 3307 SPECIAL PURPOSE KITS

FIG. 447 KIT,PRIME MOVER,LT. HOWITZER,105 MM L119,TELEPHONE, REMOTE,AND TRIPOD STOWAGE

1	PAOZZ	5340013604909	19207	12339816-3	STRAP ASSEMBLY PART OF KIT P/N 57K1628............................. UOC:HPM	2
2	PAOZZ	5340013145957	19207	12338839-1	LOOP,STRAP,FASTENER PART OF KIT P/N 57K1628............................ UOC:HPM	8
3	PAOZZ	5305009897435	96906	MS35207-264	SCREW,MACHINE PART OF KIT P/N 57K1628............................ UOC:HPM	4
4	PAOZZ	5340012658923	19207	12339817-4	STRAP,WEBBING PART OF KIT P/N 57K1628............................ UOC:HPM	2
5	PAOZZ	5310002514503	24617	9411893	NUT,LOCK #10-32,PART OF KIT P/N 57K1628............................ UOC:HPM	16
6	PAOZZ	5305009572635	96906	MS35191-277	SCREW,MACHINE #10-32 X 1.25,PART OF KIT P/N 57K1628.................... UOC:HPM	8
7	PAOZZ	5305009847341	96906	MS35191-273	SCREW,MACHINE #10-32 X 0.625,PART OF KIT P/N 57K1628.................. UOC:HPM	4
8	PAOZZ	5310000145850	96906	MS27183-42	WASHER,FLAT #10,PART OF KIT P/N 57K1628............................ UOC:HPM	4
9	PAOZZ	5310008098546	96906	MS27183-8	WASHER,FLAT #10,PART OF KIT P/N 57K1628............................ UOC:HPM	16

END OF FIGURE

Section II. TM 9-2320-280-24P-2

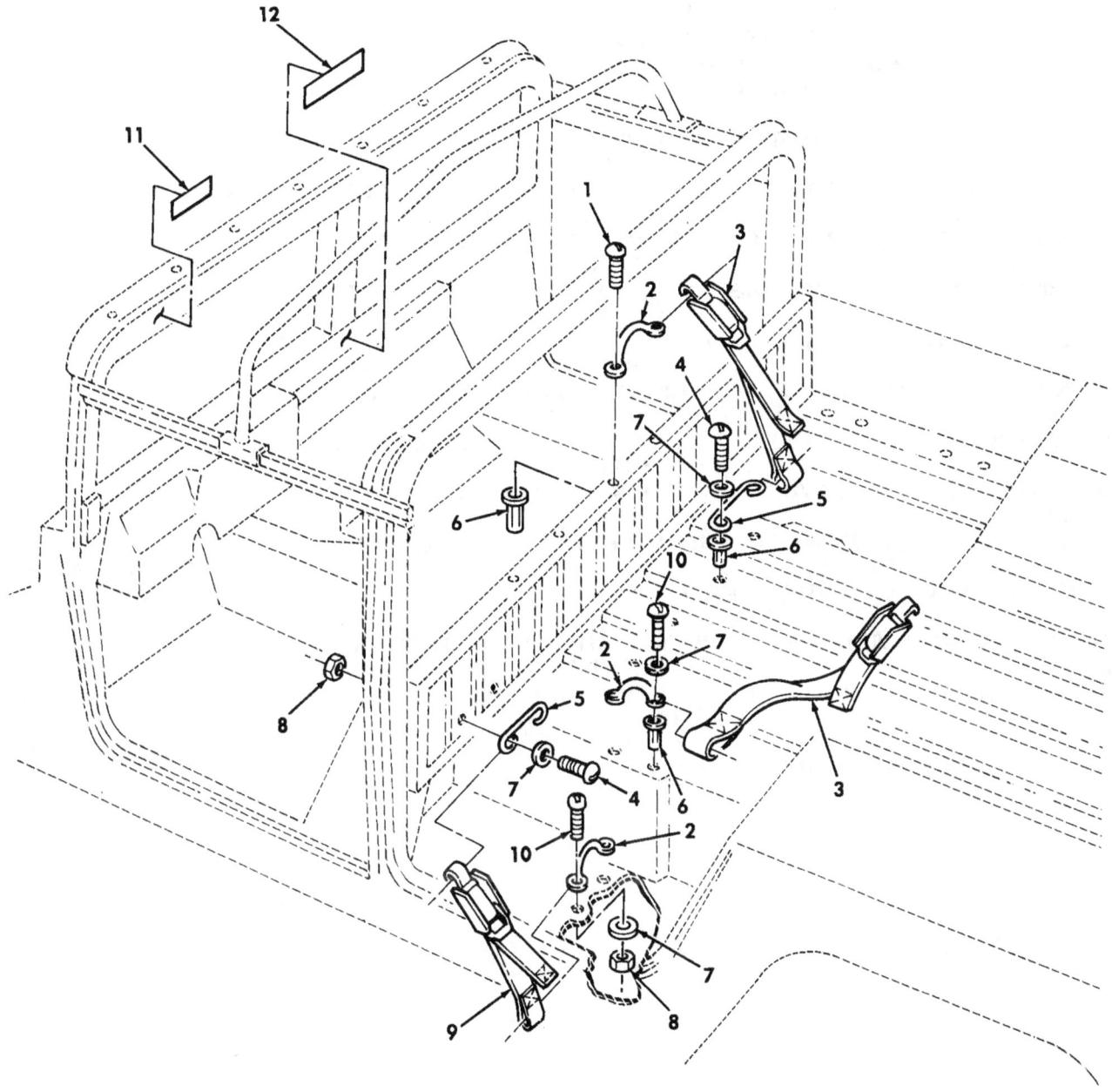

Figure 448. Kit, Prime Mover, Lt. Howitzer, 105 MM L119, Fuel Can, Water Can, Section Chest, and Spade Stowage.

SECTION II   TM9-2320-280-24P

(1) NO	(2) SMR CODE	(3) NSN	(4) CAGEC	(5) PART NUMBER	(7) ITEM DESCRIPTION AND USABLE ON CODES(UOC)	QTY

GROUP 3307 SPECIAL PURPOSE KITS

FIG. 448 KIT,PRIME MOVER,LT. HOWITZER,105 MM L119,FUEL CAN, WATER CAN,SECTION CHEST,AND SPADE STOWAGE

1	PAOZZ	5305009847342	96906	MS35191-274	SCREW,MACHINE #10-32 X 0.75,PART OF KIT P/N 57K1628..................... UOC:HPM	4
2	PAOZZ	5340013145957	19207	12338839-1	LOOP,STRAP,FASTENER PART OF KIT P/N 57K1628............................. UOC:HPM	5
3	PAOZZ	5340012584529	19207	12340507-4	STRAP,WEBBING PART OF KIT P/N 57K1628............................. UOC:HPM	4
4	PAOZZ	5305009931848	96906	MS35207-265	SCREW,MACHINE PART OF KIT P/N 57K1628............................. UOC:HPM	8
5	PAOZZ	2510011995748	19207	12338839-8	LOOP,FOOTMAN,FLAT PART OF KIT P/N 57K1628............................. UOC:HPM	4
6	PAOZZ	5310011981723	03481	S10P175	NUT,PLAIN,BLIND RIV #10-32,PART OF KIT P/N 57K1628..................... UOC:HPM	12
7	PAOZZ	5310000145850	96906	MS27183-42	WASHER,FLAT #10,PART OF KIT P/N 57K1628............................. UOC:HPM	14
8	PAOZZ	5310002514503	24617	9411893	NUT,LOCK #10-32,PART OF KIT P/N 57K1628............................. UOC:HPM	6
9	PAOZZ	5340013617879	19207	12340507-6	STRAP,RETAINING PART OF KIT P/N 57K1628............................. UOC:HPM	1
10	PAOZZ	5305009847341	96906	MS35191-273	SCREW,MACHINE #10-32 X 5/8,PART OF KIT P/N 57K1628.............. UOC:HPM	6
11	PAOZZ	9905013658849	19207	12342670	PLATE,INSTRUCTION PART OF KIT P/N 57K1628............................. UOC:HPM	1
12	PAOZZ		19207	12342669	DECAL CAUTION PART OF KIT P/N 57K1628............................. UOC:HPM	1

END OF FIGURE

Section II.

TM 9-2320-280-24P-2

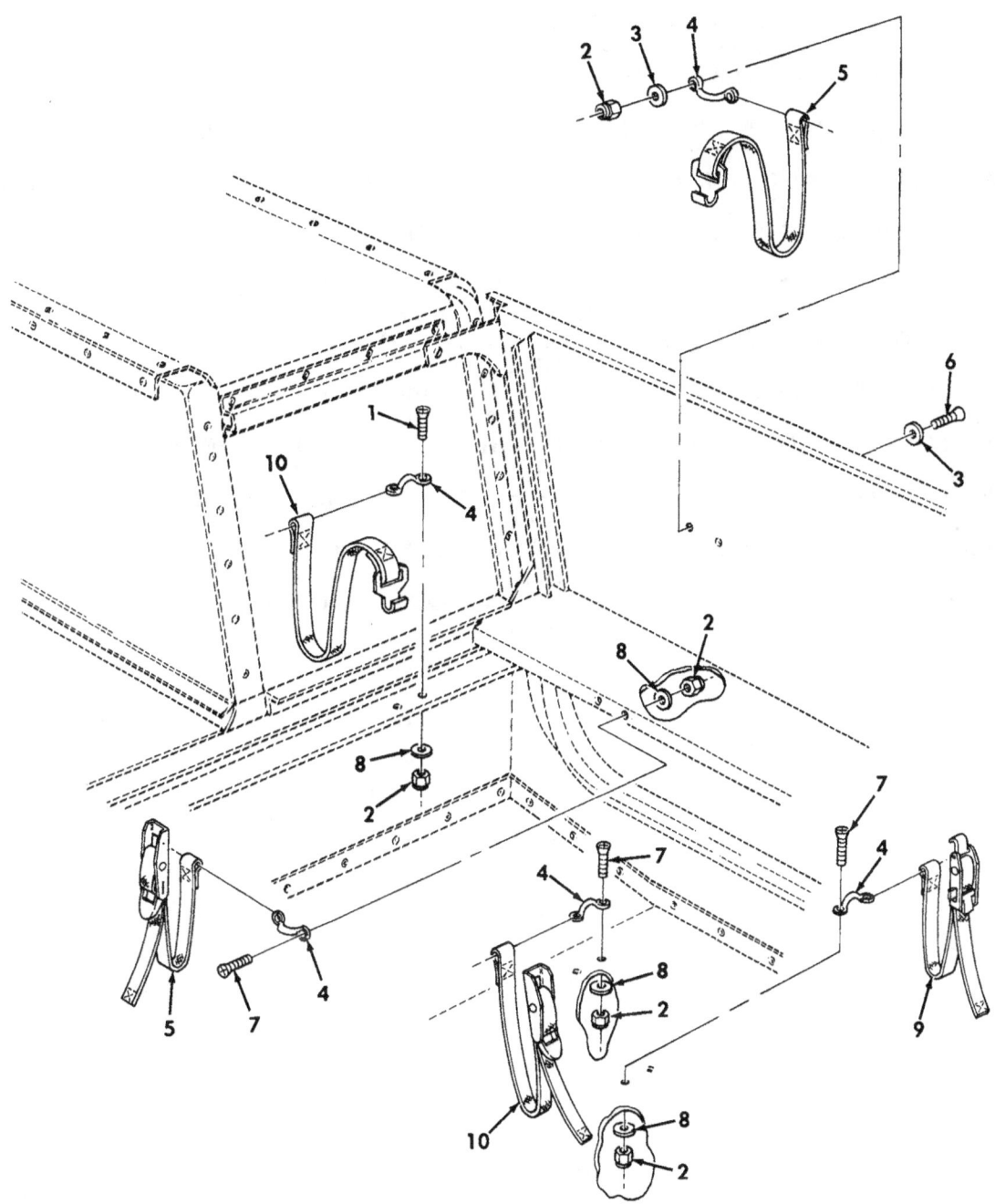

Figure 449. Kit, Prime Mover, Lt. Howitzer, 105 MM L119, G.D.U. and G.D.U. Battery Stowage.

SECTION II TM9-2320-280-24P

(1) NO	(2) SMR CODE	(3) NSN	(4) CAGEC	(5) PART NUMBER	(6) DESCRIPTION AND USABLE ON CODES(UOC)	(7) QTY

GROUP 3307 SPECIAL PURPOSE KITS

FIG. 449 KIT,PRIME MOVER,LT. HOWITZER,105 MM L119,G.D.U. AND G.D.U. BATTERY STOWAGE

(1)	(2)	(3)	(4)	(5)	(6)	(7)
1	PAOZZ	5305009847343	96906	MS35191-276	SCREW,MACHINE #10-32 X 1.00,PART OF KIT P/N 57K1628.............. UOC:HPM	2
2	PAOZZ	5310002514503	24617	9411893	NUT,LOCK #10-32,PART OF KIT P/N 57K1628............. UOC:HPM	10
3	PAOZZ	5310008098546	96906	MS27183-8	WASHER,FLAT #10,PART OF KIT P/N 57K1628......................... UOC:HPM	4
4	PAOZZ	5340013145957	19207	12338839-1	LOOP,STRAP,FASTENER PART OF KIT P/N 57K1628......................... UOC:HPM	5
5	PAOZZ	5340012032708	19207	12340529-7	STRAP,WEBBING PART OF KIT P/N 57K1628......................... UOC:HPM	2
6	PAOZZ	5305009897435	96906	MS35207-264	SCREW,MACHINE #10-32 X 5/8,PART OF KIT P/N 57K1628.................. UOC:HPM	2
7	PAOZZ	5305009847341	96906	MS35191-273	SCREW,MACHINE #10-32 X 5/8,PART OF KIT P/N 57K1628.................. UOC:HPM	6
8	PAOZZ	5310000145850	96906	MS27183-42	WASHER,FLAT #10,PART OF KIT P/N 57K1628......................... UOC:HPM	8
9	PAOZZ	5340013617879	19207	12340507-6	STRAP,WEBBING PART OF KIT P/N 57K1628......................... UOC:HPM	1
10	PAOZZ	5340012552812	19207	12340529-16	STRAP,WEBBING PART OF KIT P/N 57K1628......................... UOC:HPM	2

END OF FIGURE

Section II. TM 9-2320-280-24P-2

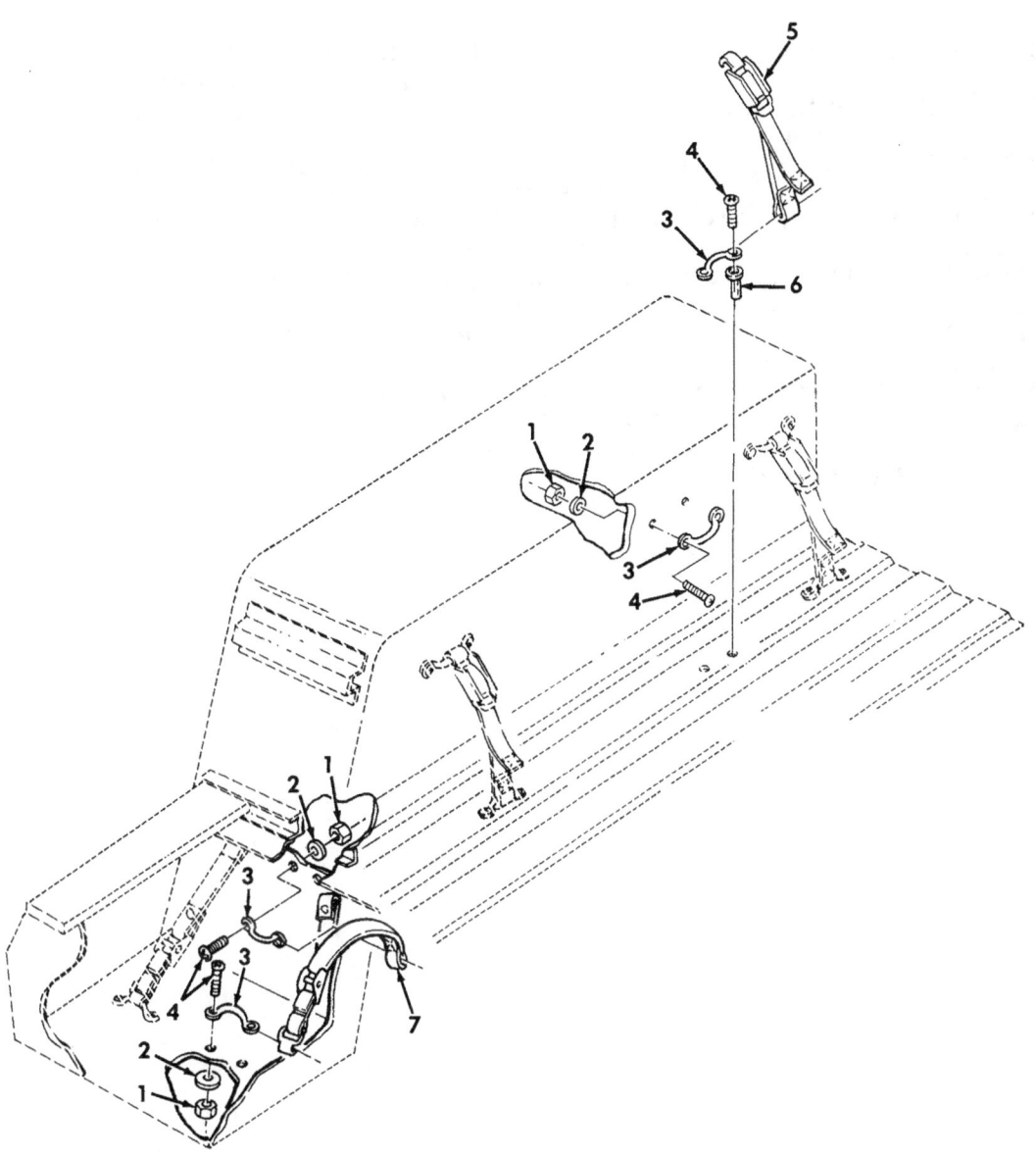

Figure 450. Kit, Prime Mover, Lt. Howitzer, 105 MM L119, Aiming Post and Cable Reel Stowage.

SECTION II   TM9-2320-280-24P

(1) NO	(2) SMR CODE	(3) NSN	(4) CAGEC	(5) PART NUMBER	(7) ITEM DESCRIPTION AND USABLE ON CODES(UOC)	QTY

GROUP 3307 SPECIAL PURPOSE KITS

FIG. 450 KIT,PRIME MOVER,LT. HOWITZER,105 MM L119,AIMIMG POST AND CABLE REEL STOWAGE

1	PAOZZ	5310002514503	24617	9411893	NUT,LOCK #10-32,PART OF KIT P/N 57K1628............................. UOC:HPM	19
2	PAOZZ	5310000145850	96906	MS27183-42	WASHER,FLAT #10,PART OF KIT P/N 57K1628............................. UOC:HPM	19
3	PAOZZ	5340013145957	19207	12338839-1	LOOP,STRAP,FASTENER PART OF KIT P/N 57K1628............................. UOC:HPM	10
4	PAOZZ	5305009847341	96906	MS35191-273	SCREW,MACHINE #10-32 X 5/8,PART OF KIT P/N 57K1628..................... UOC:HPM	20
5	PAOZZ	5340012584529	19207	12340507-4	STRAP,WEBBING PART OF KIT P/N 57K1628............................. UOC:HPM	3
6	PAOZZ	5310011981723	03481	S10P175	NUT,PLAIN,BLIND RIV #10-32,PART OF KIT P/N 57K1628..................... UOC:HPM	1
7	PAOZZ	5340011985165	31272	45096-10	STRAP,WEBBING PART OF KIT P/N 57K1628............................. UOC:HPM	2

END OF FIGURE

Section II. TM 9-2320-280-24P-2

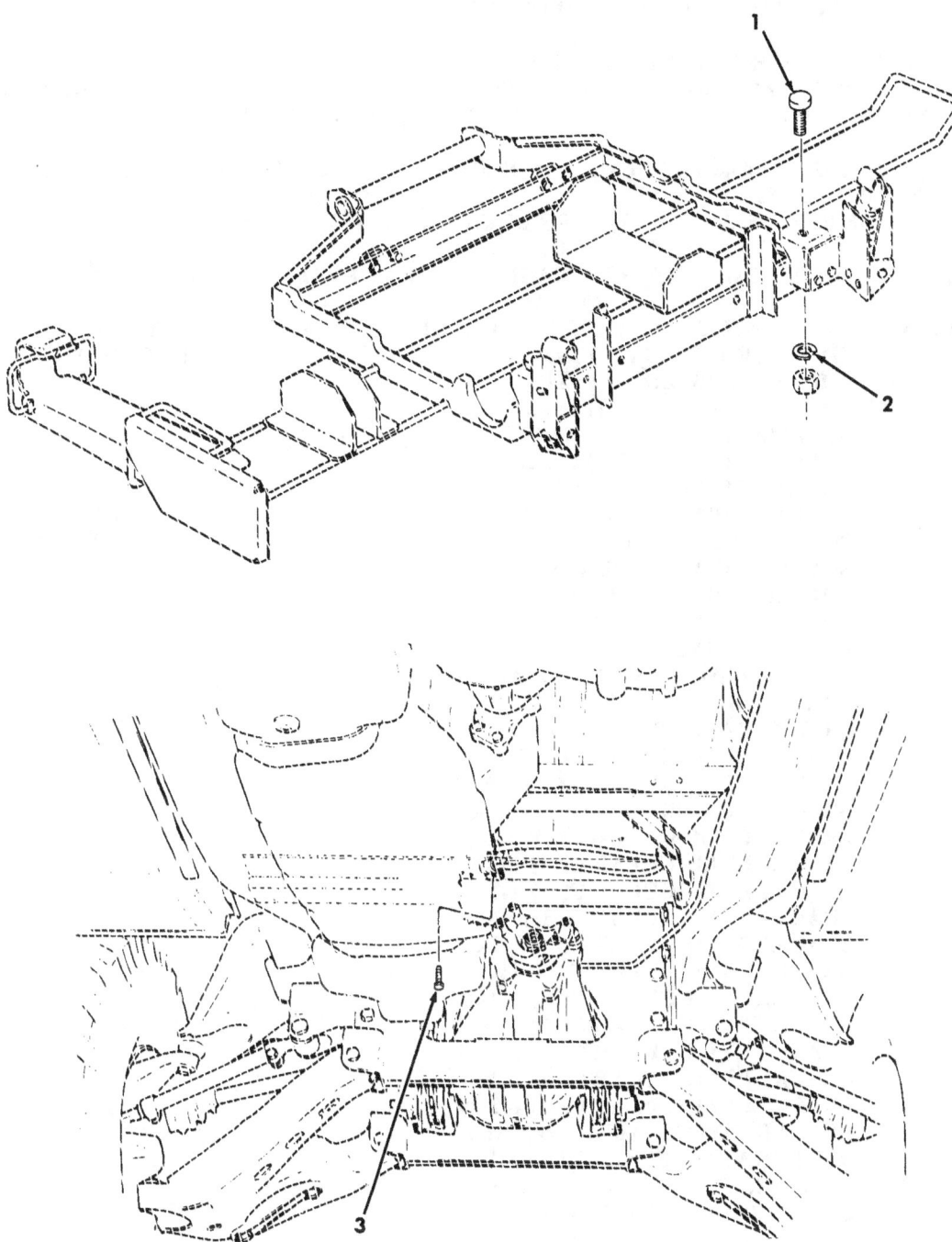

Figure 451. Kit, Prime Mover, Lt. Howitzer, 105 MM L119, Pioneer Tool Stowage Tray Hardware and Parking Brake Hardware.

SECTION II   TM9-2320-280-24P

(1) NO	(2) SMR CODE	(3) NSN	(4) CAGEC	(5) PART NUMBER	(7) ITEM DESCRIPTION AND USABLE ON CODES(UOC)	QTY

GROUP 3307 SPECIAL PURPOSE KITS

FIG. 451 KIT,PRIME MOVER,LT. HOWITZER,105 MM L119,PIONEER TOOL STOWAGE TRAY HARDWARE AND PARKING BRAKE HARDWARE

1	PAOZZ	5340011858820	19207	12340846-1	BUMPER,NONMETALLIC PIONEER TOOL TRAY,STOP,PART OF KIT P/N 57K1628... UOC:HPM	2
2	PAOZZ	5310005825965	96906	MS35338-44	WASHER,LOCK 1/4,PART OF KIT P/N 57K1628.......................... UOC:HPM	2
3	PAOZZ	5305012854234	24617	9414241	SCREW,MACHINE PART OF KIT P/N 57K1628.......................... UOC:HPM	1
KIT	PDOZZ	2590013489748	19207	57K1628	MODIFICATION KIT,VE PRIME MOVER,LT. HOWITZER 105 MM L119................ UOC:HPM	1

```
                    BOLT,ASSEMBLED    ( 1) 445-4
                    BOLT,HEXAGON      ( 1) 446-26
                    BOLT,MACHINE      ( 2) 442-1
                    BOLT,MACHINE      (12) 442-10
                    BOLT,MACHINE      ( 4) 442-14
                    BOLT,MACHINE      ( 4) 442-17
                    BOLT,MACHINE      ( 8) 442-18
                    BOLT,MACHINE      ( 2) 444-5
                    BOLT,MACHINE      (10) 446-11
                    BOLT,MACHINE      ( 6) 441-13
                    BOLT,MACHINE      ( 6) 446-36
                    BOX,ACCESSORIES STO( 1) 444-1
                    BRACKET           ( 1) 443-8
                    BRACKET,ANGLE     ( 1) 445-11
                    BRACKET,ANGLE     ( 3) 446-35
                    BRACKET,DOUBLE ANGL( 1) 446-34
                    BRACKET,DOUBLE ANGL( 1) 443-3
                    BRACKET,DOUBLE ANGL( 1) 445-6
                    BRACKET,VEHICULAR C( 1) 442-3
                    BRACKET,VEHICULAR C( 1) 442-6
                    BRACKET,VEHICULAT C( 3) 442-6
                    BRACKET,VEHICULAR C( 1) 442-3
                    BRACKET,VEHICULAR C( 2) 442-9
                    BRACKET,VEHICULAR C( 1) 442-11
                    BRACKET,VEHICULAR C( 1) 442-15
                    BRACKET,VEHICULAR C( 1) 442-19
                    BRACKET,VEHICULAR C( 1) 442-11
                    BRACKET,VEHICULAR C( 1) 442-19
                    BUMPER,NONMETALLIC ( 2) 451-1
                    BUMPER,VEHICULAR  ( 1) 441-1
                    BUMPER,VEHICULAR  ( 1) 442-16
                    CLAMP,BRIDGE      ( 1) 446-33
                    CLAMP,LOOP        ( 4) 443-4
                    CLAMP,LOOP        ( 1) 444-7
```

SECTION II   TM9-2320-280-24P

(1) NO	(2) CODE	(3) NSN	(4) CAGEC	(5) PART NUMBER	(6) DESCRIPTION AND USABLE ON CODES(UOC)	(7) ITEM SMR QTY
					CLAMP,LOOP	(1) 445-5
					CLAMP,LOOP	(1) 445-9
					CLIP,SPRING	(4) 446-4
					CLIP,SPRING	(2) 446-2
					CONNECTOR PLUG ELEC	(2) 445-7
					CONNECTOR PLUG ELEC	(7) 445-8
					CONTROL HANDLE,WINC	(1) 444-4
					DECAL	(1) 443-18
					FRAME SECTION,STRUC	(1) 441-4
					FRAME SECTION,STRUC	(1) 441-4
					FRAME SECTION,STRUC	(1) 441-12
					FRAME SECTION,STRUC	(1) 441-12
					HANDLE,BOW	(7) 446-31
					LOOP,FOOTMAN	(2) 446-22
					LOOP,STRAP,FASTENER	(4) 448-5
					LOOP,STRAP,FASTENER	(8) 446-5
					LOOP,STRAP,FASTENER	(8) 447-2
					LOOP,STRAP,FASTENER	(5) 448-2
					LOOP,STRAP,FASTENER	(5) 449-4
					LOOP,STRAP,FASTENER	(10) 450-3
					NUT,LOCK	(24) 446-9
					NUT,LOCK	(16) 447-5
					NUT,LOCK	(6) 448-8
					NUT,LOCK	(10) 449-2
					NUT,LOCK	(19) 450-1
					NUT,PLAIN,ASSEMBLED	(3) 445-3
					NUT,PLAIN,BLIND RIV	(4) 446-21
					NUT,PLAIN,BLIND RIV	(12) 448-6
					NUT,PLAIN,BLIND RIV	(1) 450-6
					NUT,PLAIN,HEXAGON	(2) 441-10
					NUT,PLAIN,HEXAGON	(4) 442-8
					NUT,PLAIN,HEXAGON	(2) 443-1
					NUT,SELF-LOCKING	(10) 441-5
					NUT,SELF-LOCKING	(30) 442-4
					NUT,SELF-LOCKING	(2) 442-20
					NUT,SELF-LOCKING	(1) 443-14
					NUT,SELF-LOCKING	(2) 444-3
					NUT,SELF-LOCKING	(2) 445-2
					NUT,SELF-LOCKING	(14) 446-16
					NUT,SELF-LOCKING	(1) 444-9
					NUT,SELF-LOCKING	(1) 446-28
					NUT,SELF-LOCKING,HE	(6) 441-9
					NUT,SELF-LOCKING,HE	(4) 441-11
					PIN,COTTER	(1) 442-13
					PIN,COTTER	(1) 443-17
					PLATE,INSTRUCTION	(1) 448-11
					PLATE,INSTRUCTION	(1) 448-12
					PLATE,MOUNTING	(4) 442-12
					PLATE,MOUNTING	(2) 446-17
					PLUSNUT	(16) 446-24
					RACK,AMMUNITION STO	(1) 446-25
					SCREW,CAP,HEXAGON H	(6) 441-2
					SCREW,CAP,HEXAGON H	(4) 441-7

SECTION II   TM9-2320-280-24P

(1) NO	(2) CODE	(3) NSN	(4) CAGEC	(5) NUMBER	(6)	(7) ITEM DESCRIPTION AND USABLE ON CODES(UOC)	SMR	PART QTY
						SCREW,CAP,HEXAGON H	(4)	441-14
						SCREW,CAP,HEXAGON H	(4)	442-5
						SCREW,CAP,HEXAGON H	(3)	445-10
						SCREW,CAP,HEXAGON H	(2)	445-13
						SCREW,CAP,HEXAGON H	(14)	446-14
						SCREW,MACHINE	(6)	448-10
						SCREW,MACHINE	(2)	443-5
						SCREW,MACHINE	(12)	446-3
						SCREW,MACHINE	(4)	446-6
						SCREW,MACHINE	(2)	446-19
						SCREW,MACHINE	(10)	446-20
						SCREW,MACHINE	(4)	447-3
						SCREW,MACHINE	(8)	447-6
						SCREW,MACHINE	(4)	447-7
						SCREW,MACHINE	(4)	448-1
						SCREW,MACHINE	(8)	448-4
						SCREW,MACHINE	(2)	449-1
						SCREW,MACHINE	(2)	449-6
						SCREW,MACHINE	(6)	449-7
						SCREW,MACHINE	(20)	450-4
						SCREW,MACHINE	(1)	451-3
						SCREW,MACHINE	(1)	444-6
						SCREW,TAPPING	(1)	443-7
						SCREW,TAPPING	(1)	443-9
						SCREW,TAPPING	(2)	446-32
						STRAP ASSEMBLY	(2)	447-1
						STRAP,RETAINING	(1)	446-29
						STRAP,RETAINING	(1)	448-9
						STRAP,RETAINING	(2)	446-7
						STRAP,TIEDOWN,ELECT	(4)	443-6
						STRAP,WEBBING	(2)	446-18
						STRAP,WEBBING	(1)	446-23
						STRAP,WEBBING	(2)	447-4
						STRAP,WEBBING	(4)	448-3
						STRAP,WEBBING	(2)	449-5
						STRAP,WEBBING	(2)	450-7
						STRAP,WEBBING	(1)	449-9
						STRAP,WEBBING	(2)	449-10
						STRAP,WEBBING	(3)	450-5
						TEE,HOSE	(1)	443-10
						TRAY,AMMO	(1)	446-1
						TUBING,NONMETALLIC	(1)	443-11
						TUBING,NONMETALLIC	(1)	443-16
						WASHER,FLAT	(12)	441-3
						WASHER,FLAT	(18)	441-6
						WASHER,FLAT	(8)	441-8
						WASHER,FLAT	(4)	444-2
						WASHER,FLAT	(2)	445-1
						WASHER,FLAT	(16)	446-8
						WASHER,FLAT	(18)	446-10
						WASHER,FLAT	(4)	446-13
						WASHER,FLAT	(28)	446-15
						WASHER,FLAT	(2)	446-27

SECTION II   TM9-2320-280-24P

(1) NO	(2) CODE	(3) NSN	(4) CAGEC	(5) PART NUMBER	(7) DESCRIPTION AND USABLE ON CODES(UOC)	ITEM SMR QTY
					WASHER,FLAT	(4) 447-8
					WASHER,FLAT	(16) 447-9
					WASHER,FLAT	(14) 448-7
					WASHER,FLAT	(4) 449-3
					WASHER,FLAT	(8) 449-8
					WASHER,FLAT	(19) 450-2
					WASHER,FLAT	(60) 442-2
					WASHER,LOCK	(4) 442-7
					WASHER,LOCK	(2) 443-2
					WASHER,LOCK	(1) 443-13
					WASHER,LOCK	(1) 443-12
					WASHER,LOCK	(1) 444-8
					WASHER,LOCK	(16) 446-12
					WASHER,LOCK	(2) 451-2
					WEDGE	(1) 446-30
					WINCH,DRUM,VEHICLE	(1) 443-15
					WIRING ASSEMBLY	(1) 445-12

END OF FIGURE

Section II. TM 9-2320-280-24P-2

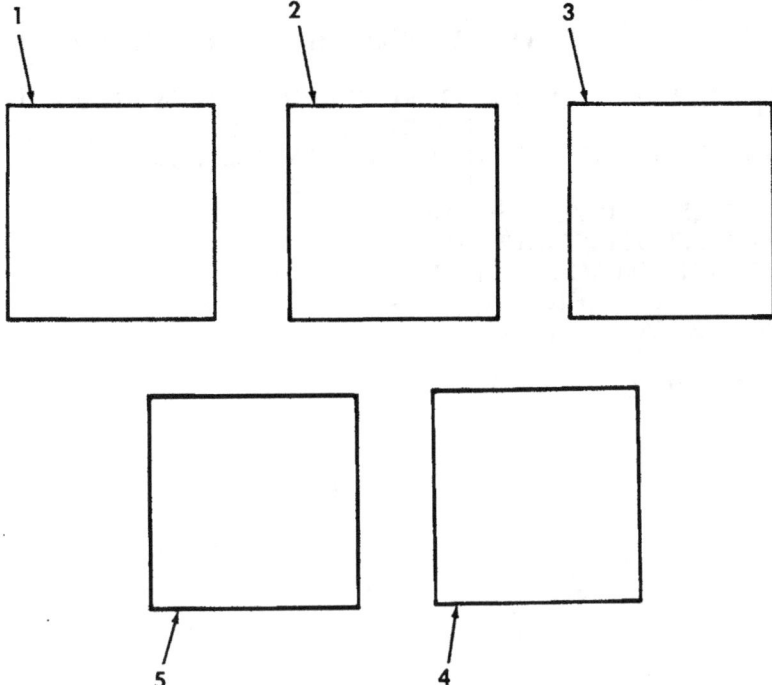

Figure 452. Kit, Towed Vulcan System.

SECTION II   TM9-2320-280-24P

(1) NO	(2) SMR CODE	(3) NSN	(4) CAGEC	(5) PART NUMBER	(7) ITEM DESCRIPTION AND USABLE ON CODES(UOC)	QTY

GROUP 3307 SPECIAL PURPOSE KITS

FIG. 452 KIT, TOWED VULCAN SYSTEM

1	KFOZZ		19207	57K1641	KIT,CAMOUFLAGE RACK PART OF KIT P/N 57K0235............ UOC:KTV	1
2	PAOZZ	2540011853216	19207	12338105	PARTS KIT,SEAT PART OF KIT P/N 57K0235............ UOC:KTV	1
3	PAOZZ	2510012468287	19207	12338119	PANEL,BODY,VEHICULA PART OF KIT P/N 57K0235............ UOC:KTV	1
4	PAOZZ	2540013160892	19207	57K0112	COVER,FITTED,VEHICU PART OF KIT P/N 57K0235............ UOC:KTV	1
5	PAOZZ	2540011910973	19207	12340744	COVER,FITTED,VEHICU PART OF KIT P/N 57K0235............ UOC:KTV	1
KIT	PDOZZ	2590013573127	19207	57K0235	KIT,TOWED VULCAN SY................. UOC:KTV	1

COVER,FITTED,VEHICU(1) 452-4
COVER,FITTED,VEHICU(1) 452-5
KIT,CAMOUFLAGE   (1) 452-1
PANEL,BODY,VEHICULA(1) 452-3
PARTS KIT,SEAT   (1) 452-2

END OF FIGURE

Section II.

TM 9-2320-280-24P-2

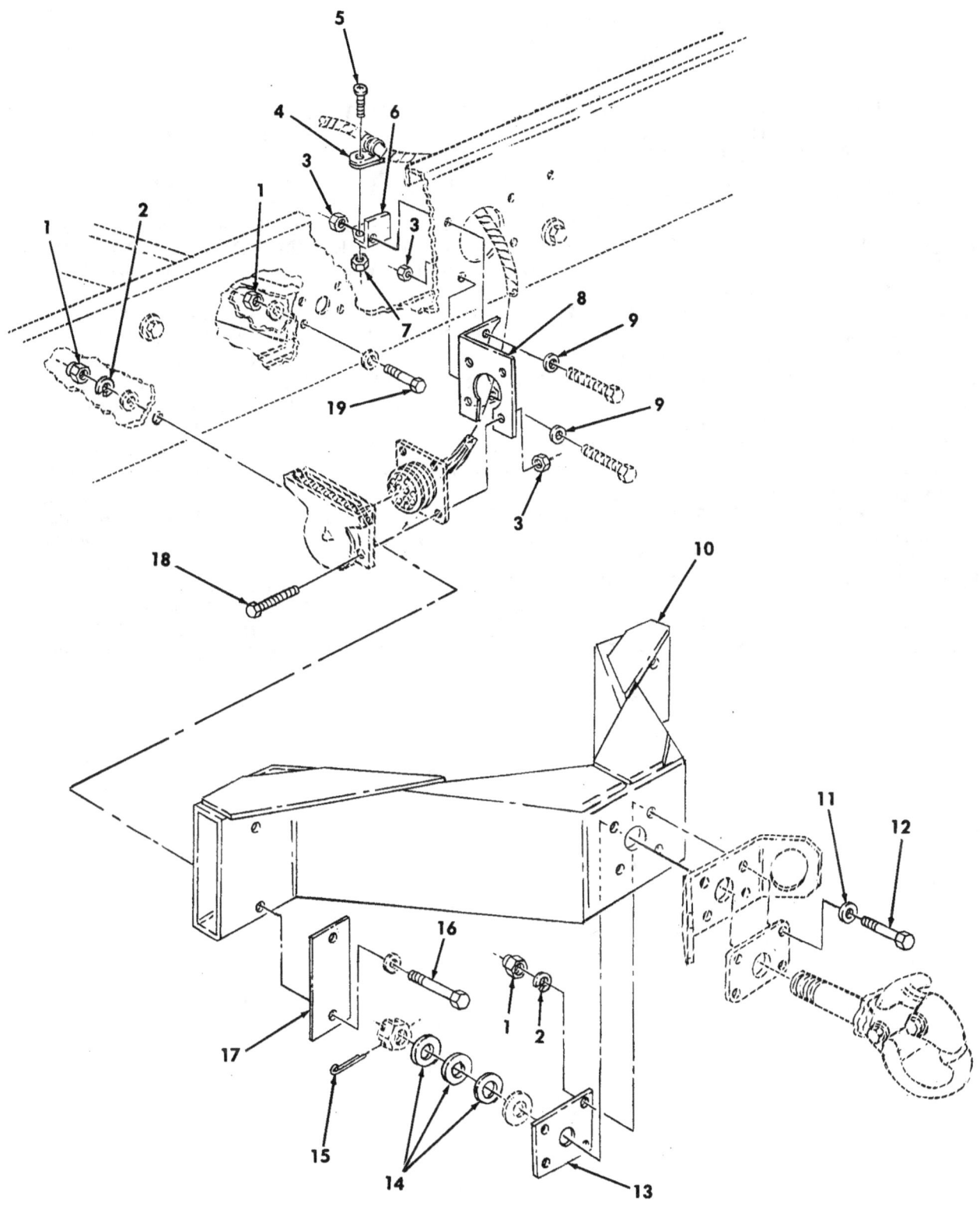

Figure 453. Kit, Pintle Extension.

SECTION II TM9-2320-280-24P

(1) ITEM NO	(2) SMR CODE	(3) NSN	(4) CAGEC	(5) PART NUMBER	(6) DESCRIPTION AND USABLE ON CODES(UOC)	(7) QTY

GROUP 3307 SPECIAL PURPOSE KITS

FIG. 453 KIT, PINTLE EXTENSION

(1)	(2)	(3)	(4)	(5)	(6)	(7)
1	PAOZZ	5310004883889	96906	MS51943-39	NUT,SELF-LOCKING,HE 1/2-13 PART OF KIT P/N 57K1640.................... UOC:AVY,BVY,HVY,NNN	12
2	PAOZZ	5310005845272	02978	ERNA245	WASHER,LOCK 1/2 PART OF KIT P/N 57K1640............................ UOC:AVY,BVY,HVY,NNN	8
3	PAOZZ	5310011520598	24617	271172	NUT,SELF-LOCKING,AS 1/4-20 PART OF KIT P/N 57K1640.................... UOC:AVY,BVY,HVY,NNN	6
4	PAOZZ	5340000572906	96906	MS21333-73	CLAMP,LOOP PART OF KIT P/N 57K1640.. UOC:AVY,BVY,HVY,NNN	1
5	PAOZZ	5305009846193	96906	MS35206-245	SCREW,MACHINE .164-32 X 1/2 PART OF KIT P/N 57K1640................... UOC:AVY,BVY,HVY,NNN	1
6	PAOZZ	5340014067418	19207	12342664	BRACKET,ANGLE PART OF KIT P/N 57K1640............................ UOC:AVY,BVY,HVY,NNN	1
7	PAOZZ	5310010695243	72582	271163	NUT,PLAIN,ASSEMBLED #8-32 PART OF KIT P/N 57K1640................... UOC:AVY,BVY,HVY,NNN	1
8	PAOZZ	5340014066965	19207	12342663	BRACKET,ANGLE PART OF KIT P/N 57K1640.......................... UOC:AVY,BVY,HVY,NNN	1
9	PAOZZ	5310011023270	24617	2436161	WASHER,FLAT 1/4 PART OF KIT P/N 57K1640............................ UOC:AVY,BVY,HVY,NNN	2
10	PAOZZ	2540014117519	19207	12342661	PINTLE EXTENSION PART OF KIT P/N 57K1640........................... UOC:AVY,BVY,HVY,NNN	1
11	PAOZZ	5310011211703	06032	2310-0143-001	WASHER,FLAT 1/2 PART OF KIT P/N 57K1640.......................... UOC:AVY,BVY,HVY,NNN	4
12	PAOZZ	5306013601128	24617	455006	BOLT,MACHINE 1/2-13 X 3.75 PART OF KIT P/N 57K1640.................... UOC:AVY,BVY,HVY,NNN	4
13	PAOZZ	5340014066964	19207	12342662	PLATE,MOUNTING PART OF KIT P/N 57K1640.......................... UOC:AVY,BVY,HVY,NNN	1
14	PAOZZ	5310005158776	96906	MS20002-20	WASHER,FLAT 1-1/4 X 2.125 PART OF KIT P/N 57K1640..................... UOC:AVY,BVY,HVY,NNN	3
15	PAOZZ	5315008460126	80205	MS24665-628	PIN,COTTER 1/4 X 3.00 PART OF KIT P/N 57K1640...................... UOC:AVY,BVY,HVY,NNN	1
16	PAOZZ	5305014114455	24617	9430516	SCREW,CAP,HEXAGON H 1/2-13 X 5-3/4 PART OF KIT P/N 57K1640............ UOC:AVY,BVY,HVY,NNN	4
17	PAOZZ	5340014077192	19207	12342665	PLATE,MOUNTING PART OF KIT P/N 57K1640............................	2

SECTION II   TM9-2320-280-24P

(1) (2) (3) (4) (5) (6) (7) ITEM SMR PART
NO CODE NSN CAGEC NUMBER DESCRIPTION AND USABLE ON CODES(UOC) QTY

UOC:AVY,BVY,HVY,NNN 18 PAOZZ 5305010164344 96906 MS51849- SCREW,MACHINE 1/4-20 X 5/8 PART OF 4 KIT P/N 57K1640....................
UOC:AVY,BVY,HVY,NNN 19 PAOZZ 5306013601125 24617 455000 BOLT,MACHINE 1/2-13 X 1.25 PART OF 4 KIT P/N 57K1640....................
UOC:AVY,BVY,HVY,NNN
KIT PAOZZ 5340014107036 19207 57K1640 HARDWARE KIT,MECHAN................. 1
UOC:AVY,BVY,HVY

BOLT,MACHINE	(4)	453-19
BOLT,MACHINE	(4)	453-12
BRACKET,ANGLE	(1)	453-8
BRACKET,ANGLE	(1)	453-6
CLAMP,LOOP	(1)	453-4
NUT,PLAIN,ASSEMBLED	(1)	453-7
NUT,SELF-LOCKING,HE	(12)	453-1
NUT,SELF-LOCKING,AS	(6)	453-3
PIN,COTTER	(1)	453-15
PINTLE EXTENSION	(1)	453-10
PLATE,MOUNTING	(1)	453-13
PLATE,MOUNTING	(2)	453-17
SCREW,CAP,HEXAGON H	(4)	453-16
SCREW,MACHINE	(4)	453-18
SCREW,MACHINE	(1)	453-5
WASHER,FLAT	(3)	453-14
WASHER,FLAT	(2)	453-9
WASHER,FLAT	(4)	453-11
WASHER,LOCK	(8)	453-2

END OF FIGURE

Section II.

TM 9-2320-280-24P-2

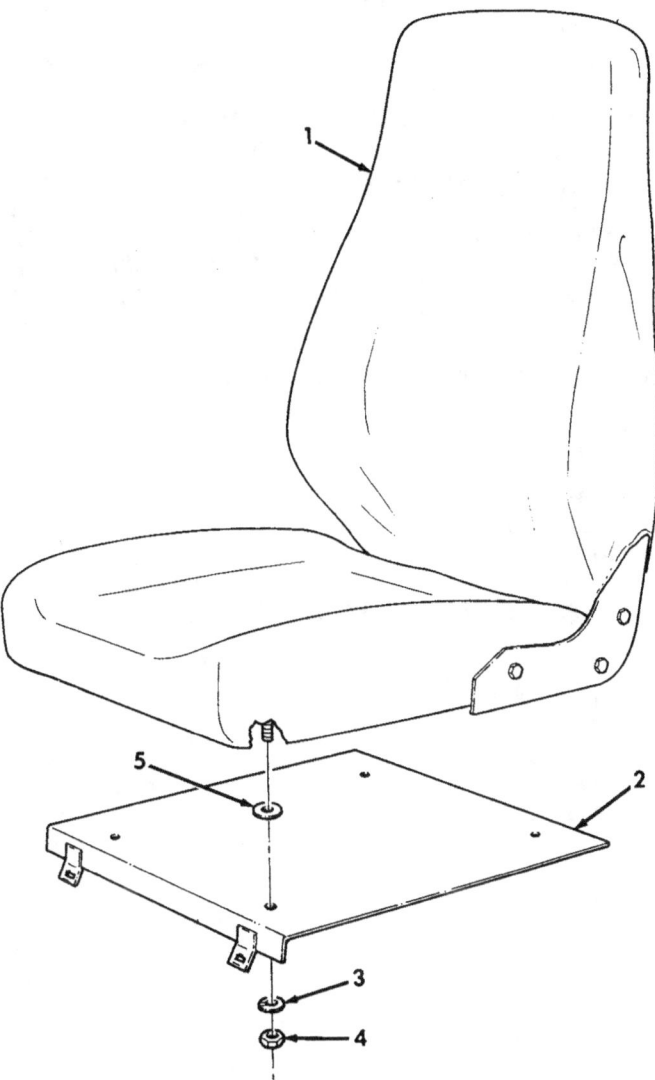

Figure 454. Kit, Rear Seat Assembly.

SECTION II   TM9-2320-280-24P

(1) NO	(2) SMR CODE	(3) NSN	(4) CAGEC	(5) PART NUMBER	(6) DESCRIPTION AND USABLE ON CODES(UOC)	(7) QTY

GROUP 3307 SPECIAL PURPOSE KITS

FIG. 454 KIT,REAR SEAT ASSEMBLY

1 PAOZZ 2540014364175 62226 8.6272-01   SEAT,VEHICULAR REAR PART OF KIT P/N 57K3196............................   UOC:A13,A14,BVY,H13,H14,NNN   2

1 PAOZZ 2540014769353 62226 8.6272-02   SEAT,VEHICULAR REAR PART OF KIT P/N 57K3197............................   UOC:A13,A14,BVY,H13,H14,NNN   2

2 PBOZZ 2540014770162 19207 12446966   PLATE,REAR SEAT,MNT LEFT,PART OF KIT P/N 57K3196,PART OF KIT P/N 57K3197............................   UOC:A13,A14,BVY,H13,H14,NNN   1

2 PBOZZ 2540014770184 19207 12446967   PLATE,REAR SEAT,MNT RIGHT,PART OF KIT P/N 57K3196,PART OF KIT P/N 57K3197............................   UOC:A13,A14,BVY,H13,H14,NNN   1

3 PAOZZ 5310004079566 96906 MS35338-45   WASHER,LOCK PART OF KIT P/N 57K3196, PART OF KIT P/N 57K3197............   UOC:A13,A14,BVY,H13,H14,NNN   8

4 PAOZZ 5310013815328 24617 9413583   NUT,PLAIN,HEXAGON PART OF KIT P/N 57K3196 PART OF KIT P/N 57K3197.....   UOC:A13,A14,BVY,H13,H14,NNN   8

5 PAOZZ 5365013820986 19207 12343055   SPACER,RING PART OF KIT P/N 57K3196   PART OF KIT P/N 57K3197.............   UOC:A13,A14,BVY,H13,H14,NNN   8

KIT PDOZZ 2540014107035 19207 57K3196   PARTS KIT,SEAT,VEHI GREEN..........   UOC:A13,A14,BVY,H13,H14,NNN   1

        PLATE,REAR   (1) 454-2
        PLATE,REAR   (1) 454-2
        NUT   (8) 454-4
        SEAT,REAR   (2) 454-1
        SPACER   (8) 454-5
        WASHER   (8) 454-3

KIT PDOZZ 2540014107034 19207 57K3197   PARTS KIT,SEAT,VEHI TAN............   UOC:A13,A14,BVY,H13,H14,NNN   1

        PLATE,REAR   (1) 454-2
        PLATE,REAR   (1) 454-2
        NUT   (8) 454-4
        SEAT,REAR   (2) 454-1
        SPACER   (8) 454-5
        WASHER   (8) 454-3

END OF FIGURE

Section II. TM 9-2320-280-24P-2

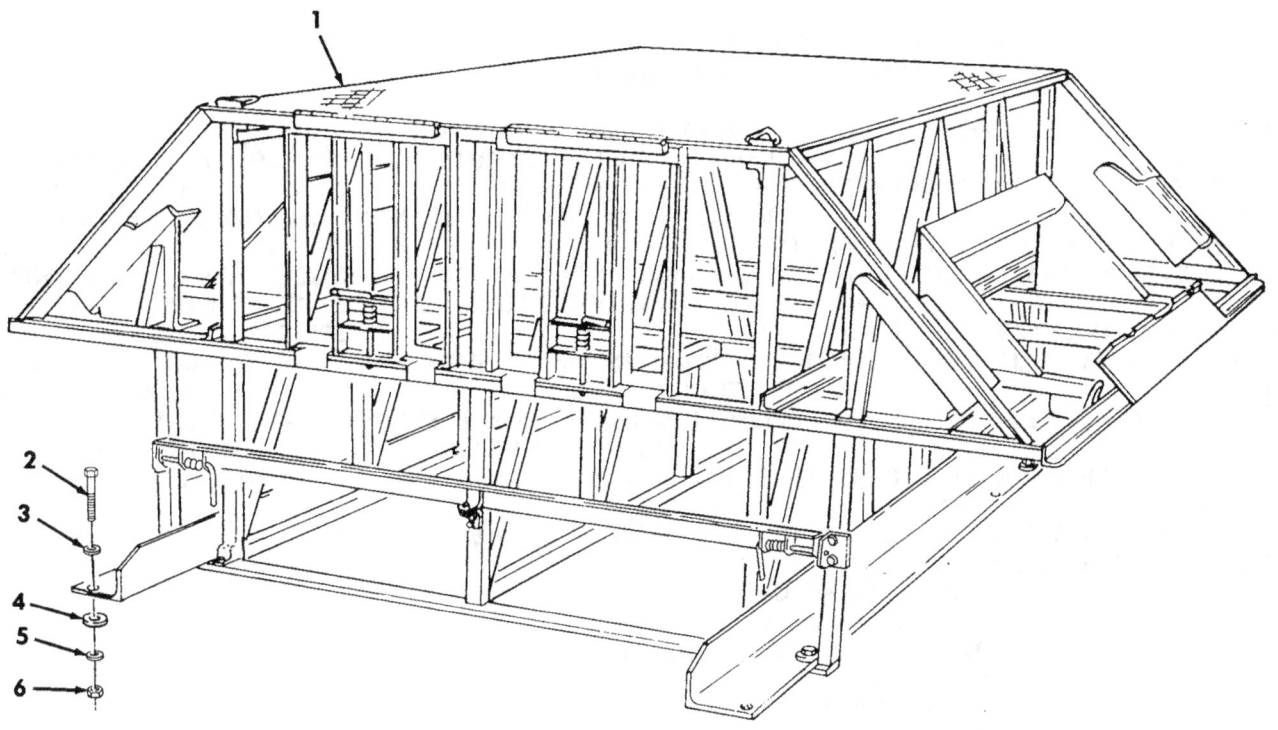

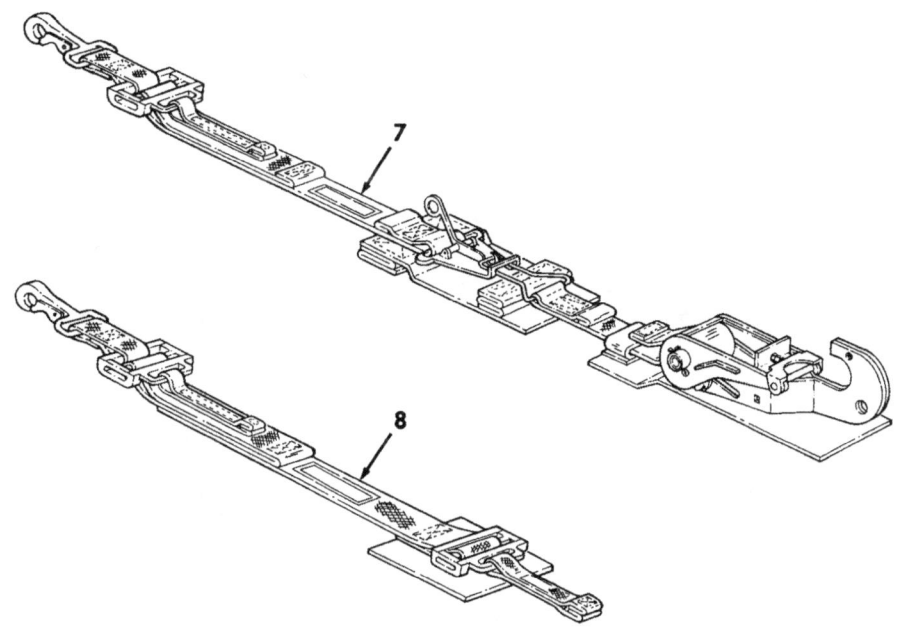

Figure 455. Kit, Man Portable Air-Defense System (MANPADS).

SECTION II TM9-2320-280-24P

(1) ITEM NO	(2) SMR CODE	(3) NSN	(4) CAGEC	(5) PART NUMBER	(7) DESCRIPTION AND USABLE ON CODES (UOC)	QTY

GROUP 3307 SPECIAL PURPOSE KITS

FIG. 455 KIT, MAN PORTABLE AIR-DEFENSE SYSTEM (MANPADS)

Item	SMR	NSN	CAGEC	P/N	Description	Qty
1	KFOFF		19207	12342855	RACK ASSEMBLY, MANPA PART OF KIT P/N 57K0227............... UOC:A13,A14,BVY,HVY,H13,H14,NNN	1
2	PAOZZ	5305000712071	80204	B1821BH050C200N	SCREW,CAP,HEXAGON H 1/2-13 X 2.00, PART OF KIT P/N 57K0227............. UOC:A13,A14,BVY,HVY,H13,H14,NNN	4
3	PAOZZ	5310008095998	96906	MS27183-18	WASHER,FLAT 1/2,PART OF KIT P/N 57K0227............................ UOC:A13,A14,BVY,HVY,H13,H14,NNN	4
4	PAOZZ	5365014066336	19207	12342859	SPACER,RING PART OF KIT P/N 57K0227 UOC:A13,A14,BVY,HVY,H13,H14,NNN	4
5	PAOZZ	5310008093079	96906	MS27183-19	WASHER,FLAT 1/2,PART OF KIT P/N 57K0227............................ UOC:A13,A14,BVY,HVY,H13,H14,NNN	4
6	PAOZZ	5310004883889	96906	MS51943-39	NUT,SELF-LOCKING,HE 1/2-13,PART OF KIT P/N 57K0227..................... UOC:A13,A14,BVY,HVY,H13,H14,NNN	4
7	PAOZZ	5340014088526	19207	12342884	STRAP,WEBBING PART OF KIT P/N 57K0227............................ UOC:A13,A14,BVY,HVY,H13,H14,NNN	4
8	PAOZZ	5340014067413	19207	12342883	STRAP,WEBBING PART OF KIT P/N 57K0227............................ UOC:A13,A14,BVY,HVY,H13,H14,NNN	2
KIT	PDOFF	2510012744234	19207	57K0227	RACK ASSEMBLY,MANPA................. UOC:A13,A14,BVY,HVY,H13,H14,NNN	1

```
NUT,SELF-LOCKING,HE  ( 4) 455-6
RACK ASSEMBLY,MANPA  ( 1) 455-1
SCREW,CAP,HEXAGON H  ( 4) 455-2
SPACER,RING          ( 4) 455-4
STRAP,WEBBING        ( 2) 455-7
STRAP,WEBBING        ( 4) 455-8
WASHER,FLAT          ( 4) 455-3
WASHER,FLAT          ( 4) 455-5
```

END OF FIGURE

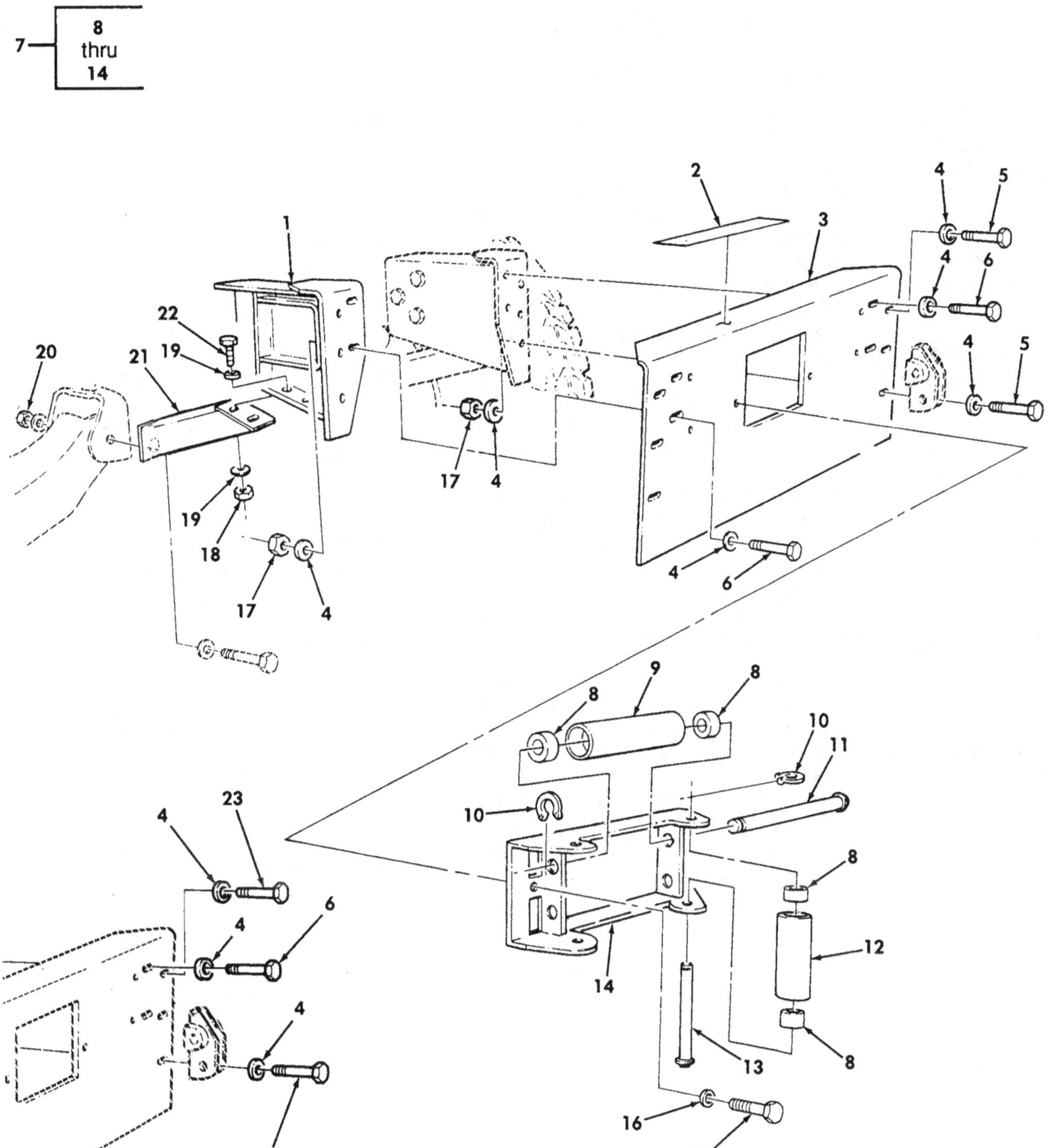

Section II.

TM 9-2320-280-24P-2

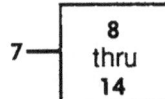

Figure 456. Kit, 9,000 Lb Winch.

SECTION II   TM9-2320-280-24P

(1) NO	(2) SMR CODE	(3) NSN	(4) CAGEC	(5) PART NUMBER	(6) DESCRIPTION AND USABLE ON CODES(UOC)	(7) QTY

GROUP 3307 SPECIAL PURPOSE KITS

FIG. 456 KIT, 9,000 LB WINCH

1 PAOZZ 2510013581178 19207 12342387   FRAME SECTION,STRUC L.H. PART OF   1
KIT P/N 57K0285 PART OF KIT P/N   57K3217............................   UO
C:AVY,A11,A13,A14,A24,A25,A26,A27,
           BVY,B17,B18,B24,B25,C17,HVY,H11,H13,
           H14,H17,H18,H21,H24,H25,H26,H27,H28,   MMM,NNN

1 PAOZZ 2510013578789 19207 12342388   FRAME SECTION,STRUC R.H. PART OF   1
KIT P/N 57K0285 PART OF KIT P/N   57K3217............................   UO
C:AVY,A11,A13,A14,A24,A25,A26,A27,
           BVY,B17,B18,B24,B25,C17,HVY,H11,H13,
           H14,H17,H18,H21,H24,H25,H26,H27,H28,   MMM,NNN

2 PAOZZ 7690014450444 19207 12447102   MARKER,IDENTIFICATI PART OF KIT P/N 1
57K0285 PART OF KIT P/N 57K3217.....   UOC:AVY,A11,A13,A14,A24,A25,A26,A27,
           BVY,B17,B18,B24,B25,C17,HVY,H11,H13,
           H14,H17,H18,H21,H24,H25,H26,H27,H28,   MMM,NNN

3 PAOZZ 2540014128610 19207 12447091   BUMPER,VEHICULAR PART OF KIT P/N   1
57K3217............................   UOC:AVY,A11,A13,A14,A24,A25,A26,A27,
           BVY,B17,B18,B24,B25,C17,HVY,H11,H13,
           H14,H17,H18,H21,H24,H25,H26,H27,H28,   MMM,NNN

4 PAOZZ 5310011211703 06032 2310-0143-001   WASHER,FLAT 1/2 PART OF KIT P/N   18
57K0285 PART OF KIT P/N 57K3217.....   UOC:AVY,A11,A13,A14,A24,A25,A26,A27,
           BVY,B17,B18,B24,B25,C17,HVY,H11,H13,
           H14,H17,H18,H21,H24,H25,H26,H27,H28,   MMM,NNN

5 PAOZZ 5305000712071 80204 B1821BH050C200N   SCREW,CAP,HEXAGON H 1/2-13 X 2.00   6
PART OF KIT P/N 57K3217,PART OF KIT   P/N57K0285..........................
           UOC:AVY,A11,A13,A14,A24,A25,A26,A27,
           BVY,B17,B18,B24,B25,C17,HVY,H11,H13,
           H14,H17,H18,H21,H24,H25,H26,H27,H28,   MMM,NNN

6 PAOZZ 5305000712067 80204 B1821BH050C125N   SCREW,CAP,HEXAGON H 1/2-13 X 1.25   4
PART OF KIT P/N 57K0285 PART OF KIT   P/N 57K3217.........................
           UOC:AVY,A11,A13,A14,A24,A25,A26,A27,
           BVY,B17,B18,B24,B25,C17,HVY,H11,H13,
           H14,H17,H18,H21,H24,H25,H26,H27,H28,   MMM,NNN

7 PAOZZ 2590014486351 19207 12447104   BRACKET,VEHICULAR C PART OF KIT P/N 2
57K0285,PART OF KIT P/N 57K3217.....   UOC:AVY,A11,A13,A14,A24,A25,A26,A27,

SECTION II   TM9-2320-280-24P

(1) NO	(2) SMR CODE	(3) NSN	(4) CAGEC	(5) PART NUMBER	(6) DESCRIPTION AND USABLE ON CODES(UOC)	(7) QTY
8	PAOZZ	3120014400043	19207	12447058	.BUSHING,SLEEVE UOC:AVY,A11,A13,A14,A24,A25,A26,A27,BVY,B17,B18,B24,B25,C17,HVY,H11,H13,H14,H17,H18,H21,H24,H25,H26,H27,H28,MMM,NNN	8
9	PAOZZ	3120014486269	19207	12447057-1	.ROLLER,LINEAR-ROTAR UOC:AVY,A11,A13,A14,A24,A25,A26,A27,BVY,B17,B18,B24,B25,C17,HVY,H11,H13,H14,H17,H18,H21,H24,H25,H26,H27,H28,MMM,NNN	2
10	PAOZZ	5325013882363	96906	MS3217-2075	.RING,RETAINING UOC:AVY,A11,A13,A14,A24,A25,A26,A27,BVY,B17,B18,B24,B25,C17,HVY,H11,H13,H14,H17,H18,H21,H24,H25,H26,H27,H28,MMM,NNN	8
11	PAOZZ	3040014779645	19207	12447060-1	.SHAFT,STRAIGHT UOC:AVY,A11,A13,A14,A24,A25,A26,A27,BVY,B17,B18,B24,B25,C17,HVY,H11,H13,H14,H17,H18,H21,H24,H25,H26,H27,H28,MMM,NNN	2
12	PAOZZ	3120014486268	19207	12447057-2	.ROLLER,LINEAR-ROTAR UOC:AVY,A11,A13,A14,A24,A25,A26,A27,BVY,B17,B18,B24,B25,C17,HVY,H11,H13,H14,H17,H18,H21,H24,H25,H26,H27,H28,MMM,NNN	2
13	PAOZZ	3040014443350	34623	12447060-2	.SHAFT,STRAIGHT UOC:AVY,A11,A13,A14,A24,A25,A26,A27,BVY,B17,B18,B24,B25,C17,HVY,H11,H13,H14,H17,H18,H21,H24,H25,H26,H27,H28,MMM,NNN	2
14	PAOZZ	2590014443399	34623	12447088	.BRACKET,VEHICULAR C UOC:AVY,A11,A13,A14,A24,A25,A26,A27,BVY,B17,B18,B24,B25,C17,HVY,H11,H13,H14,H17,H18,H21,H24,H25,H26,H27,H28,MMM,NNN	1
15	PAOZZ	5305006602832	96906	MS35307-389	SCREW,CAP,HEXAGON H 7/16-14 X 1.50 PART OF KIT P/N 57K0285 PART OF KIT P/N 57K3217 UOC:AVY,A11,A13,A14,A24,A25,A26,A27,BVY,B17,B18,B24,B25,C17,HVY,H11,H13,H14,H17,H18,H21,H24,H25,H26,H27,H28,NNN,MMM	4
16	PAOZZ	5310008094085	24617	9420022	WASHER,FLAT 7/16 PART OF KIT P/N 57K0285 PART OF KIT P/N 57K3217 UOC:AVY,A11,A13,A14,A24,A25,A26,A27,BVY,B17,B18,B24,B25,C17,HVY,H11,H13,H14,H17,H18,H21,H24,H25,H26,H27,H28,MMM,NNN	4
17	PAOZZ	5310004883889	96906	MS51943-39	NUT,SELF-LOCKING,HE 1/2-13 PART OF KIT P/N 57K0285 PART OF KIT P/N	8

456-2

SECTION II   TM9-2320-280-24P

(1) NO	(2) SMR CODE	(3) NSN	(4) CAGEC	(5) PART NUMBER	(6) DESCRIPTION AND USABLE ON CODES (UOC)	(7) QTY
				57K3217...............	UOC:AVY,A11,A13,A14,A24,A25,A26,A27,BVY,B17,B18,B24,B25,C17,HVY,H11,H13,H14,H17,H18,H21,H24,H25,H26,H27,H28,MMM,NNN	
18	PAOZZ	5310009359021	96906	MS51943-35	NUT,SELF-LOCKING,HE 3/8-16 PART OF KIT P/N 57K0285 PART OF KIT P/N 57K3217............... UOC:AVY,A11,A13,A14,A24,A25,A26,A27,BVY,B17,B18,B24,B25,C17,HVY,H11,H13,H14,H17,H18,H21,H24,H25,H26,H27,H28,MMM,NNN	4
19	PAOZZ	5310014124013	24617	2436163	WASHER,FLAT 3/8 PART OF KIT P/N 57K0285 PART OF KIT P/N 57K3217..... UOC:AVY,A11,A13,A14,A24,A25,A26,A27,BVY,B17,B18,B24,B25,C17,HVY,H11,H13,H14,H17,H18,H21,H24,H25,H26,H27,H28,MMM,NNN	8
20	PAOZZ	5310004093333	96906	MS51943-45	NUT,SELF-LOCKING,HE 3/4-10 PART OF KIT P/N 57K0285 PART OF KIT P/N 57K3217............... UOC:AVY,A11,A13,A14,A24,A25,A26,A27,BVY,B17,B18,B24,B25,C17,HVY,H11,H13,H14,H17,H18,H21,H24,H25,H26,H27,H28,MMM,NNN	2
21	PAOZZ	2510013581176	19207	12342389	FRAME SECTION,STRUC L.H. PART OF KIT P/N 57K0285 PART OF KIT P/N 57K3217............... UOC:AVY,A11,A13,A14,A24,A25,A26,A27,BVY,B17,B18,B24,B25,C17,HVY,H11,H13,H14,H17,H18,H21,H24,H25,H26,H27,H28,MMM,NNN	1
21	PAOZZ	2510013581177	19207	12342390	FRAME SECTION,STRUC R.H. PART OF KIT P/N 57K0285 PART OF KIT P/N 57K3217............... UOC:AVY,A11,A13,A14,A24,A25,A26,A27,BVY,B17,B18,B24,B25,C17,HVY,H11,H13,H14,H17,H18,H21,H24,H25,H26,H27,H28,MMM,NNN	1
22	PAOZZ	5305009422196	80204	B1821BH038C100D	SCREW,CAP,HEXAGON H 3/8-16 X 1.00 PART OF KIT P/N 57K0285 PART OF KIT P/N 57K3217............... UOC:AVY,A11,A13,A14,A24,A25,A26,A27,BVY,B17,B18,B24,B25,C17,HVY,H11,H13,H14,H17,H18,H21,H24,H25,H26,H27,H28,MMM,NNN	4
23	PAOZZ	5305000712069	80204	B1821BH050C150N	SCREW,CAP,HEXAGON H 1/2-13 X 1.50 PART OF KIT P/N 57K0285 PART OF KIT P/N 57K3217............... UOC:AVY,A11,A13,A14,A24,A25,A26,A27,BVY,B17,B18,B24,B25,C17,HVY,H11,H13,H14,H17,H18,H21,H24,H25,H26,H27,H28,MMM,NNN	2

SECTION II   TM9-2320-280-24P

(1) NO	(2) CODE	(3) NSN	(4) CAGEC	(5) NUMBER	(6) DESCRIPTION AND USABLE ON CODES(UOC)	(7) ITEM SMR QTY	PART

24 PAOZZ 5305000712070 80204 B1821BH050C175N SCREW,CAP,HEXAGON H 1/2-13 X 1.75   4
ART OF KIT P/N 57K0285 PART OF KIT   P/N 57K3217........................
        UOC:AVY,A11,A13,A14,A24,A25,A26,A27,
        BVY,B17,B18,B24,B25,C17,HVY,H11,H13,
        H14,H17,H18,H21,H24,H25,H26,H27,H28,   MMM,NNN

END OF FIGURE

Section II.

TM 9-2320-280-24P-2

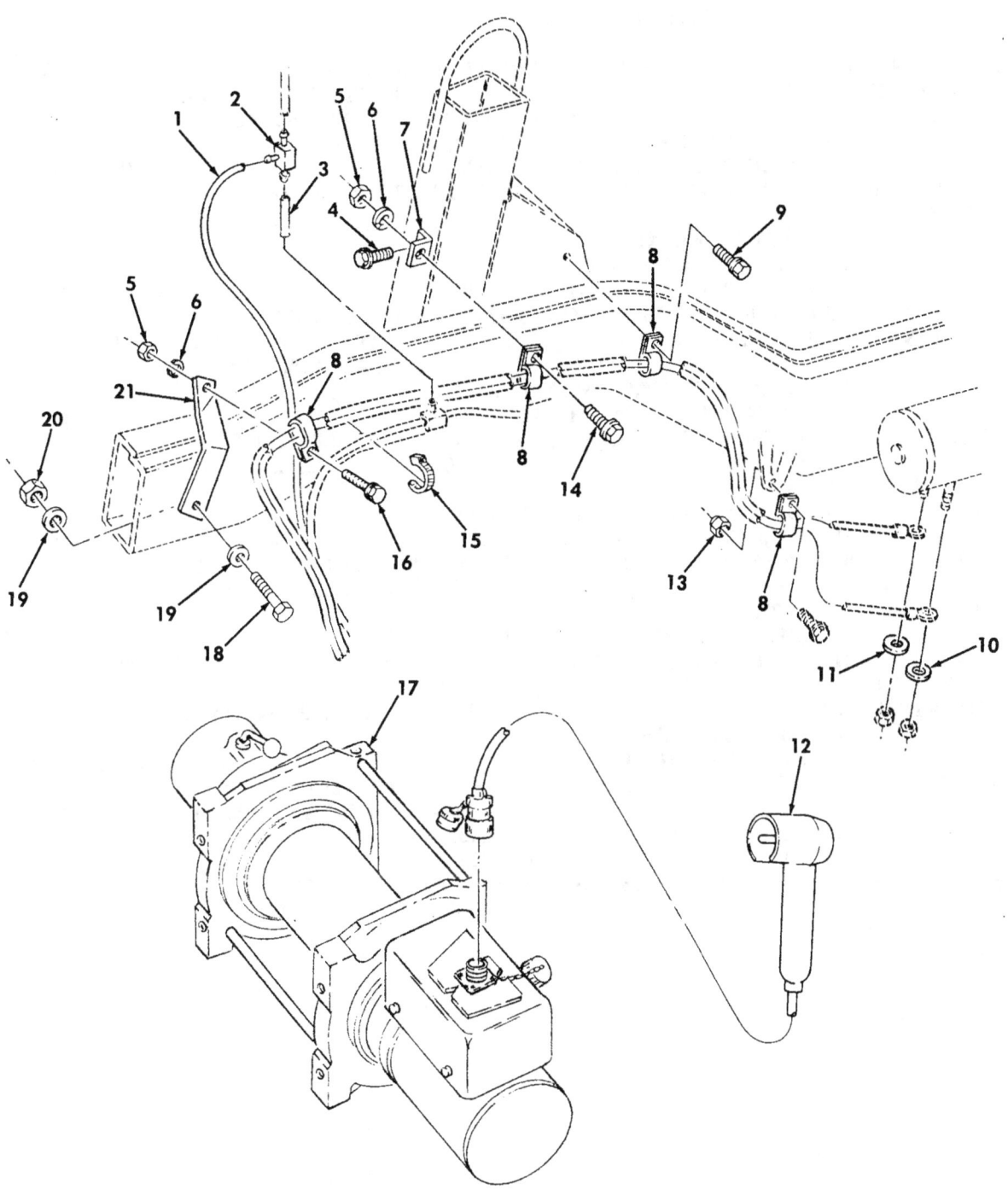

Figure 457. Kit, 9,000 Lb. Winch.

SECTION II TM9-2320-280-24P

(1) NO	(2) SMR CODE	(3) NSN	(4) CAGEC	(5) PART NUMBER	(7) DESCRIPTION AND USABLE ON CODES(UOC)	ITEM QTY

GROUP 3307 SPECIAL PURPOSE KITS

FIG. 457 KIT,9,000 LB WINCH

1 MOOZZ 19207 CPR104420-1-70 TUBING,NONMETALLIC MAKE FROM TUBING ,P/N CPR104420-1,70 INCHES LONG,PART OF KIT P/N 57K3217,PART OF KIT P/N 57K0285............................. UOC:AVY,A11,A13,A14,A24,A25,A26,A27, BVY,B17,B18,B24,B25,C17,HVY,H11,H13, H14,H17,H18,H21,H24,H25,H26,H27,H28, MMM,NNN   1

2 PAOZZ 4730010035105 19207 12339973-1 TEE,HOSE PART OF KIT P/N 57K3217, PART OF KIT P/N 57K0285............. UOC:AVY,A11,A13,A14,A24,A25,A26,A27, BVY,B17,B18,B24,B25,C17,HVY,H11,H13, H14,H17,H18,H21,H24,H25,H26,H27,H28, MMM,NNN   1

3 MOOZZ 19207 CPR104420-1-4 TUBING,NONMETALLIC MAKE FROM TUBING ,P/N CPR104420-1,4 INCHES LONG,PART OF KIT P/N 57K3217,PART OF KIT P/N 57K0285............................ UOC:AVY,A11,A13,A14,A24,A25,A26,A27, BVY,B17,B18,B24,B25,C17,HVY,H11,H13, H14,H17,H18,H21,H24,H25,H26,H27,H28, MMM,NNN   1

4 PAOZZ 5305011013312 96906 MS51851-126 SCREW,TAPPING 3/8-16 X 3/4,PART OF KIT P/N 57K3217,PART OF KIT P/N 57K0285............ UOC:AVY,A11,A13,A14,A24,A25,A26,A27, BVY,B17,B18,B24,B25,C17,HVY,H11,H13, H14,H17,H18,H21,H24,H25,H26,H27,H28, MMM,NNN   1

5 PAOZZ 5310009050762 96906 MS51967-3 NUT,PLAIN,HEXAGON 1/4-20,PART OF KIT P/N 57K3217,PART OF KIT P/N 57K0285............ UOC:AVY,A11,A13,A14,A24,A25,A26,A27, BVY,B17,B18,B24,B25,C17,HVY,H11,H13, H14,H17,H18,H21,H24,H25,H26,H27,H28, MMM,NNN   2

6 PAOZZ 5310005432705 96906 MS35338-27 WASHER,LOCK 1/4,PART OF KIT P/N 57K3217,PART OF KIT 57K0285......... UOC:AVY,A11,A13,A14,A24,A25,A26,A27, BVY,B17,B18,B24,B25,C17,HVY,H11,H13, H14,H17,H18,H21,H24,H25,H26,H27,H28, MMM,NNN   2

7 PAOZZ 5340011922256 19207 12338663 BRACKET,ANGLE PART OF KIT P/N 57K3217,PART OF KIT P/N 57K0285..... UOC:AVY,A11,A13,A14,A24,A25,A26,A27, BVY,B17,B18,B24,B25,C17,HVY,H11,H13, H14,H17,H18,H21,H24,H25,H26,H27,H28, MMM,NNN   1

8 PAOZZ 5340007385182 19207 7385182 CLAMP,LOOP PART OF KIT P/N 57K3217,   4

SECTION II   TM9-2320-280-24P

(1) NO	(2) CODE	(3) NSN	(4) CAGEC	(5) NUMBER	(6) DESCRIPTION AND USABLE ON CODES(UOC)	(7) ITEM SMR PART QTY

PART OF KIT P/N 57K0285.............
UOC:AVY,A11,A13,A14,A24,A25,A26,A27,
BVY,B17,B18,B24,B25,C17,HVY,H11,H13,
H14,H17,H18,H21,H24,H25,H26,H27,H28, MMM,NNN

9 PAOZZ 5305001913640 96906 MS51851-85   SCREW,TAPPING 1/4-20 X 5/8,PART OF   1
KIT P/N 57K3217,PART OF KIT P/N   57K0285.............   UO
C:AVY,A11,A13,A14,A24,A25,A26,A27,
BVY,B17,B18,B24,B25,C17,HVY,H11,H13,
H14,H17,H18,H21,H24,H25,H26,H27,H28, MMM,NNN

10 PAOZZ 5310012187137 96906 MS51415-7   WASHER,LOCK 3/8,PART OF KIT P/N   1
57K3217,PART OF KIT P/N 57K0285.....   UOC:AVY,A11,A13,A14,A24,A25,A26,A27,
BVY,B17,B18,B24,B25,C17,HVY,H11,H13,
H14,H17,H18,H21,H24,H25,H26,H27,H28, MMM,NNN

11 PAOZZ 5310012167390 96906 MS51415-9   WASHER,LOCK 1/2,PART OF KIT P/N   1
57K3217,PART OF KIT P/N 57K0285.....   UOC:AVY,A11,A13,A14,A24,A25,A26,A27,
BVY,B17,B18,B24,B25,C17,HVY,H11,H13,
H14,H17,H18,H21,H24,H25,H26,H27,H28, MMM,N-

N 12 PAOZZ 2590014481105 34623 12447171   HANDLE ASSEMBLY,CON PART OF KIT P/N   1
57K3217,PART OF KIT P/N 57K0285.....   UOC:AVY,A11,A13,A14,A24,A25,A26,A27,
BVY,B17,B18,B24,B25,C17,HVY,H11,H13,
H14,H17,H18,H21,H24,H25,H26,H27,H28, MMM,NNN

13 PAOZZ 5310011520598 24617 271172   NUT,SELF-LOCKING,AS 1/4-20,PART OF   1
KIT P/N 57K3217,PART OF KIT P/N   57K0285.............   UO
C:AVY,A11,A13,A14,A24,A25,A26,A27,
BVY,B17,B18,B24,B25,C17,HVY,H11,H13,
H14,H17,H18,H21,H24,H25,H26,H27,H28, MMM,NNN

14 PAOZZ 5305000680508 80204 B1821BH025C075N SCREW,CAP,HEXAGON H 1/4-20 X 3/4,   1
PART OF KIT P/N 57K3217,PART OF KIT   P/N 57K0285........
UOC:AVY,A11,A13,A14,A24,A25,A26,A27,
BVY,B17,B18,B24,B25,C17,HVY,H11,H13,
H14,H17,H18,H21,H24,H25,H26,H27,H28, MMM,NNN

15 PAOZZ 5975009856630 96906 MS3367-3-0   STRAP,TIEDOWN,ELECT PART OF KIT P/N   4
57K3217,PART OF KIT P/N 57K0285.....   UOC:AVY,A11,A13,A14,A24,A25,A26,A27,
BVY,B17,B18,B24,B25,C17,HVY,H11,H13,
H14,H17,H18,H21,H24,H25,H26,H27,H28, MMM,NNN

16 PAOZZ 5305004890743 24617 9415477   SCREW,CAP,HEXAGON H 1/4-20 X 3/4,   1
PART OF KIT P/N 57K3217,PART OF KIT   P/N 57K0285........

```
         SECTION II            TM9-2320-280-24P C01
      (1)      (2)      (3)          (4)        (5)                              (6)
  (7)   ITEM    SMR                             PART
        NO     CODE     NSN          CAGEC     NUMBER        DESCRIPTION AND USABLE ON CODES(UOC)
  QTY
```

                                                            UOC:AVY,A11,A13,A14,A24,A25,A26,A27,
                                                            BVY,B17,B18,B24,B25,C17,HVY,H11,H13,
                                                            H14,H17,H18,H21,H24,H25,H26,H27,H28,

MMM,NNN
  * 17 PAOFF 2590014322691 27647 27855   WINCH,DRUM,VEHICLE 9000 LB,PART OF   1
KIT P/N 57K3217,PART OF KIT P/N   57K0285.
....   UOC:AVY,A11
,A13,A14,A24,A25,A26,A27,

                                                            BVY,B17,B18,B24,B25,C17,HVY,H11,H13,
                                                            H14,H17,H18,H21,H24,H25,H26,H27,H28,

MMM,NNN
    18 PAOZZ 5305007247248 80204 B1821BH063C375N   SCREW,CAP,HEXAGON H 5/8-11 X 3.75,   6
PART OF KIT P/N 57K3217,PART OF KIT   P/N
57K0285.........................

                                                            UOC:AVY,A11,A13,A14,A24,A25,A26,A27,
                                                            BVY,B17,B18,B24,B25,C17,HVY,H11,H13,
                                                            H14,H17,H18,H21,H24,H25,H26,H27,H28,

MMM,NNN
    19 PAOZZ 5310011517347 24617 2436167   WASHER,FLAT   5/8,PART OF KIT P/N   12
57K3217,PART OF KIT P/N 57K0285.....   UOC
:AVY,A11,A13,A14,A24,A25,A26,A27,

                                                            BVY,B17,B18,B24,B25,C17,HVY,H11,H13,
                                                            H14,H17,H18,H21,H24,H25,H26,H27,H28,

MMM,NNN
    20 PAOZZ 5310000614651 96906 MS51943-43   NUT,SELF-LOCKING,HE 5/8-11,PART OF   6
KIT P/N 57K3217,PART OF KIT P/N   57K0285.
....   UOC:AVY,A11
,A13,A14,A24,A25,A26,A27,

                                                            BVY,B17,B18,B24,B25,C17,HVY,H11,H13,
                                                            H14,H17,H18,H21,H24,H25,H26,H27,H28,

MMM,NNN
    21 PAOZZ 5340012669173 19207 12338481   BRACKET,DOUBLE ANGL PART OF KIT P/N   1
57K3217,PART OF KIT P/N 57K0285.....   UOC
:AVY,A11,A13,A14,A24,A25,A26,A27,

                                                            BVY,B17,B18,B24,B25,C17,HVY,H11,H13,
                                                            H14,H17,H18,H21,H24,H25,H26,H27,H28,

MMM,NNN
       KIT PDOZZ 2590014180310 19207 57K0285   ACCESSORY KIT,WINCH 9000 LB........   1
UOC:BVY,B24,B25,C17,NNN
                                                              BRACKET   (1) 457-7
                                                              BRACKET,DOUBLE ANGL(1) 457-21
                                                              BRACKET,WIRE ROPE G(1) 456-7
                                                              CLAMP,LOOP   (4) 457-8
                                                              DECAL,WARNING   (1) 456-2
                                                              FRAME SECTION,STRUC(1) 456-1
                                                              FRAME SECTION,STRUC(1) 456-1
                                                              FRAME SECTION,STRUC(1) 456-21
                                                              FRAME SECTION,STRUC(1) 456-21
                                                              HANDLE ASSEMBLY,CON(1) 457-12
                                                              NUT,PLAIN,HEXAGON   (2) 457-5
                                                              NUT,SELF-LOCKING,AS(1) 457-13
                                                              NUT,SELF-LOCKING,HE(4) 456-18
                                                              NUT,SELF-LOCKING,HE(8) 456-17

SECTION II   TM9-2320-280-24P C01
  (1)   (2)   (3)   (4)   (5)   (6)
7)   ITEM   SMR   PART
     NO   CODE   NSN   CAGEC   NUMBER   DESCRIPTION AND USABLE ON CODES(UOC)
TY

                                               NUT,SELF-LOCKING,HE (2) 456-20
                                               SCREW,CAP,HEXAGON H (1) 457-16
                                               SCREW,CAP,HEXAGON H (6) 456-5
                                               SCREW,CAP,HEXAGON H (4) 456-15
                                               SCREW,CAP,HEXAGON H (4) 456-22
                                               SCREW,CAP,HEXAGON H (2) 456-23
                                               SCREW,CAP,HEXAGON H (4) 456-24
                                               SCREW,CAP,HEXAGON H (1) 457-14
                                               SCREW,CAP,HEXAGON H (4) 456-6
                                               SCREW,CAP,HEXAGON H (6) 457-18
                                               SCREW,TAPPING   (1) 457-9
                                               SCREW,TAPPING   (1) 457-4
                                               STRAP,TIEDOWN,ELECT (4) 457-15
                                               TEE,HOSE   (1) 457-2
                                               TUBING,NONMETALLIC (1) 457-1
                                               TUBING,NONMETALLIC (1) 457-3
                                               WASHER,FLAT   (8) 456-19
                                               WASHER,FLAT   (4) 456-16
                                               WASHER,FLAT   (18) 456-4
                                               WASHER,FLAT   (12) 457-19
                                               WASHER,LOCK   (1) 457-11
                                               WASHER,LOCK   (1) 457-10
                                               WASHER,LOCK   (2) 457-6
                                               WINCH   (1) 457-17

*   AOOZZ 2590014182135 19207 57K3217   ACCESSORY KIT,WINCH 9000 LB........   1
OC:AVY,A11,A13,A14,A24,A25,A26,A27,
                                               B17,B18,HVY,H11,H13,H14,H17,H18,H21,
24,H25,H26,H27,H28,MMM
                                               BRACKET   (1) 457-7
                                               BRACKET,DOUBLE ANGL (1) 457-21
                                               BRACKET,WIRE ROPE G (1) 456-7
                                               BUMPER,FRONT   (1) 456-3
                                               CLAMP,LOOP   (4) 457-8
                                               DECAL,WARNING   (1) 456-2
                                               FRAME SECTION,STRUC (1) 456-1
                                               FRAME SECTION,STRUC (1) 456-1
                                               FRAME SECTION,STRUC (1) 456-21
                                               FRAME SECTION,STRUC (1) 456-21
                                               HANDLE ASSEMBLY,CON (1) 457-12
                                               NUT,PLAIN,HEXAGON   (2) 457-5
                                               NUT,SELF-LOCKING,AS (1) 457-13
                                               NUT,SELF-LOCKING,HE (4) 456-18
                                               NUT,SELF-LOCKING,HE (8) 456-17
                                               NUT,SELF-LOCKING,HE (6) 457-20
                                               NUT,SELF-LOCKING,HE (2) 456-20
                                               SCREW,CAP,HEXAGON H (6) 456-5
                                               SCREW,CAP,HEXAGON H (1) 457-16
                                               SCREW,CAP,HEXAGON H (4) 456-15
                                               SCREW,CAP,HEXAGON H (4) 456-22
                                               SCREW,CAP,HEXAGON H (2) 456-23
                                               SCREW,CAP,HEXAGON H (4) 456-24
                                               SCREW,CAP,HEXAGON H (1) 457-14
                                               SCREW,CAP,HEXAGON H (4) 456-6

```
       SECTION II          TM9-2320-280-24P
      (1)    (2)    (3)       (4)         (5)                   (6)
(7)   ITEM   SMR                          PART
 NO   CODE   NSN            CAGEC        NUMBER     DESCRIPTION AND USABLE ON CODES(UOC)
QTY
```

                                                    SCREW,CAP,HEXAGON H(1) 457-14
                                                    SCREW,CAP,HEXAGON H(4) 456-6
                                                    SCREW,CAP,HEXAGON H(6) 457-18
                                                    SCREW,TAPPING   (1) 457-9
                                                    SCREW,TAPPING   (1) 457-4
                                                    STRAP,TIEDOWN,ELECT(4) 457-15
                                                    TEE,HOSE   (1) 457-2
                                                    TUBING,NONMETALLIC (1) 457-1
                                                    TUBING,NONMETALLIC (1) 457-3
                                                    WASHER,FLAT   (8) 456-19
                                                    WASHER,FLAT   (4) 456-16
                                                    WASHER,FLAT   (18) 456-4
                                                    WASHER,FLAT   (12) 457-19
                                                    WASHER,LOCK   (1) 457-11
                                                    WASHER,LOCK   (1) 457-10
                                                    WASHER,LOCK   (2) 457-6
                                                    WINCH   (1) 457-17

                                          END OF FIGURE

Section II. TM 9-2320-280-24P-2 C01

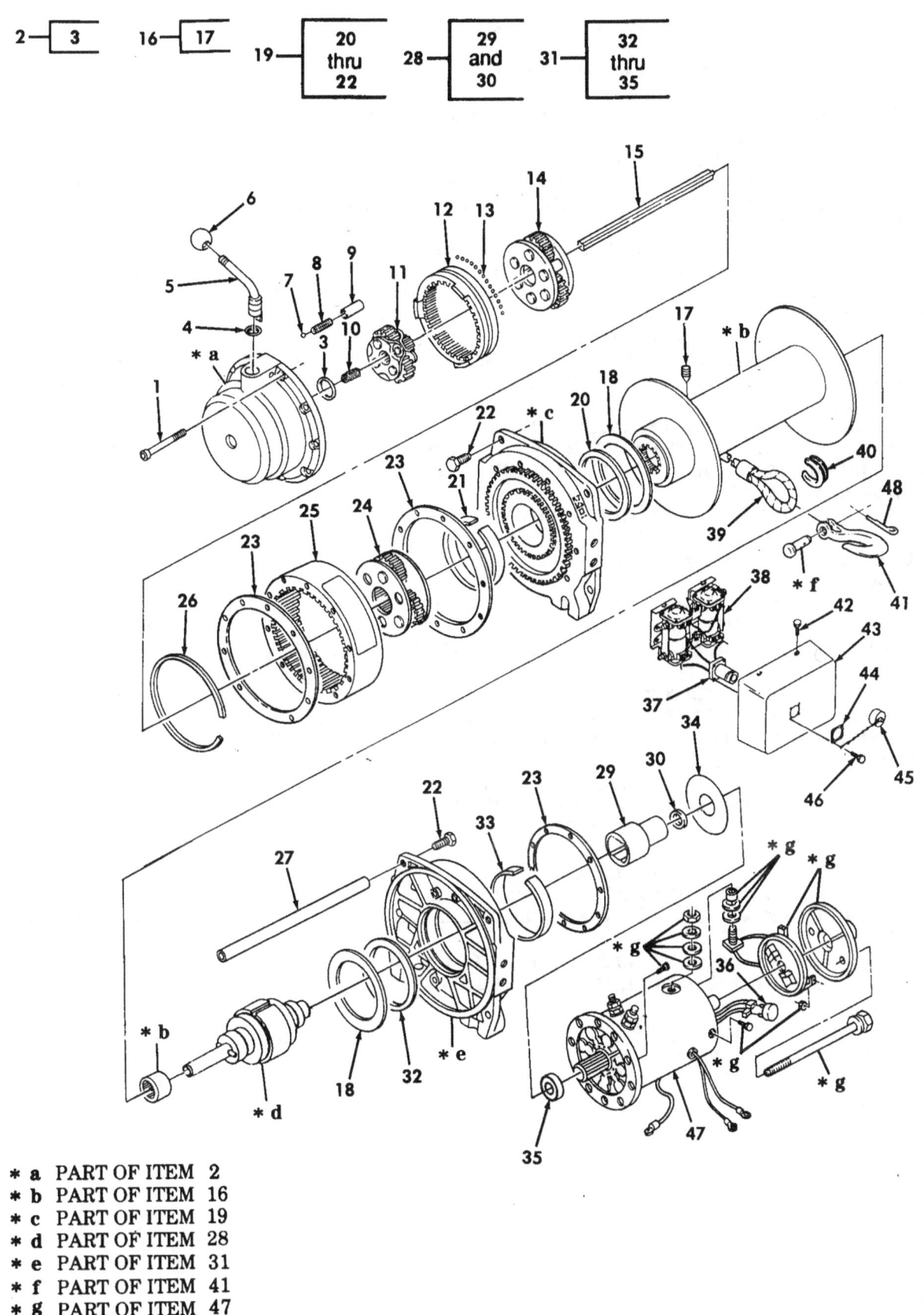

* a PART OF ITEM 2
* b PART OF ITEM 16
* c PART OF ITEM 19
* d PART OF ITEM 28
* e PART OF ITEM 31
* f PART OF ITEM 41
* g PART OF ITEM 47

Figure 458. 9,000 Lb Winch Assembly.

SECTION II TM9-2320-280-24P

(1) NO	(2) SMR CODE	(3) NSN	(4) CAGEC	(5) PART NUMBER	(7) ITEM DESCRIPTION AND USABLE ON CODES(UOC)	QTY

GROUP 3307 SPECIAL PURPOSE KITS

FIG. 458 9,000 LB WINCH ASSEMBLY

1 PAFZZ 5305013091521 27647 15603 SCREW,CAP,SOCKET HE 1/4-20 X 3.00.. 10
UOC:AVY,A11,A13,A14,A24,A25,A26,A27,
BVY,B17,B18,B24,B25,C17,HVY,H11,H13,
H14,H17,H18,H21,H24,H25,H26,H27,H28,
MMM,NNN

2 PAFZZ 3040014441756 27647 34384 HOUSING,MECHANICAL.................. 1
UOC:AVY,A11,A13,A14,A24,A25,A26,A27,
BVY,B17,B18,B24,B25,C17,HVY,H11,H13,
H14,H17,H18,H21,H24,H25,H26,H27,H28,
MMM,NNN

3 PAFZZ 3040013076157 27647 21009 .THRUST PLATE,CLUTCH................ 1
UOC:AVY,A11,A13,A14,A24,A25,A26,A27,
BVY,B17,B18,B24,B25,C17,HVY,H11,H13,
H14,H17,H18,H21,H24,H25,H26,H27,H28,
MMM,NNN

4 PAFZZ 5330014455285 27647 15605 O-RING............................... 1
UOC:AVY,A11,A13,A14,A24,A25,A26,A27,
BVY,B17,B18,B24,B25,C17,HVY,H11,H13,
H14,H17,H18,H21,H24,H25,H26,H27,H28,
MMM,NNN

5 PAFZZ 3040014463689 27647 31262 LEVER,MANUAL CONTRO................. 1
UOC:AVY,A11,A13,A14,A24,A25,A26,A27,
BVY,B17,B18,B24,B25,C17,HVY,H11,H13,
H14,H17,H18,H21,H24,H25,H26,H27,H28,
MMM,NNN

6 PAFZZ 5355012116991 27647 15604 KNOB................................ 1
UOC:AVY,A11,A13,A14,A24,A25,A26,A27,
BVY,B17,B18,B24,B25,C17,HVY,H11,H13,
H14,H17,H18,H21,H24,H25,H26,H27,H28,
MMM,NNN

7 PAFZZ 3110014444020 27647 1834 BALL,BEARING 1/4 DIA............... 1
UOC:AVY,A11,A13,A14,A24,A25,A26,A27,
BVY,B17,B18,B24,B25,C17,HVY,H11,H13,
H14,H17,H18,H21,H24,H25,H26,H27,H28,
MMM,NNN

8 PAFZZ 5360014200480 27646 1833 SPRING,HELICAL,COMP................. 1
UOC:AVY,A11,A13,A14,A24,A25,A26,A27,
BVY,B17,B18,B24,B25,C17,HVY,H11,H13,
H14,H17,H18,H21,H24,H25,H26,H27,H28,
MMM,NNN

9 PAFZZ 5365013069955 27647 15686 SPACER,SLEEVE...................... 1
UOC:AVY,A11,A13,A14,A24,A25,A26,A27,
BVY,B17,B18,B24,B25,C17,HVY,H11,H13,
H14,H17,H18,H21,H24,H25,H26,H27,H28,
MMM,NNN

10 PAFZZ 3020014441359 27647 24832 GEAR,SPUR.......................... 1
UOC:AVY,A11,A13,A14,A24,A25,A26,A27,
BVY,B17,B18,B24,B25,C17,HVY,H11,H13,
H14,H17,H18,H21,H24,H25,H26,H27,H28,

SECTION II TM9-2320-280-24P

(1) NO	(2) SMR CODE	(3) NSN	(4) CAGEC	(5) PART NUMBER	(6) DESCRIPTION AND USABLE ON CODES(UOC)	(7) ITEM QTY
11	PAFZZ	3010014444018	27647	15643	CARRIER,GEAR ASSEMB................. UOC:AY,A11,A13,A14,A24,A25,A26,A27,BVY,B17,B18,B24,B25,C17,HVY,H11,H13,H14,H17,H18,H21,H24,H25,H26,H27,H28,MMM,NNN	1
12	PAFZZ	3020014464387	27647	14916	GEAR,INTERNAL...................... UOC:AVY,A11,A13,A14,A24,A25,A26,A27,BVY,B17,B18,B24,B25,C17,HVY,H11,H13,H14,H17,H18,H21,H24,H25,H26,H27,H28,MMM,NNN	1
13	PAFZZ	3110014444100	27647	22350	BALL,BEARING 3/16 DIA (85 BALLS)... UOC:AVY,A11,A13,A14,A24,A25,A26,A27,BVY,B17,B18,B24,B25,C17,HVY,H11,H13,H14,H17,H18,H21,H24,H25,H26,H27,H28,MMM,NNN	1
14	PAFZZ	3010014444019	27647	24563	CARRIER,GEAR ASSEMB................. UOC:AY,A11,A13,A14,A24,A25,A26,A27,BVY,B17,B18,B24,B25,C17,HVY,H11,H13,H14,H17,H18,H21,H24,H25,H26,H27,H28,MMM,NNN	1
15	PAFZZ	3040014463684	27647	32209	SHAFT,STRAIGHT OUTPUT............. UOC:AY,A11,A13,A14,A24,A25,A26,A27,BVY,B17,B18,B24,B25,C17,HVY,H11,H13,H14,H17,H18,H21,H24,H25,H26,H27,H28,MMM,NNN	1
16	PAFZZ	2590014465962	27647	32231	REEL,CABLE........................ UOC:AVY,A11,13,A14,A24,A25,A26,A27,BVY,B17,B18,B24,B25,C17,HVY,H11,H13,H14,H17,H18,H21,H24,H25,H26,H27,H28,MMM,NNN	1
17	PAFZZ	5305014485074	27647	36974	.SETSCREW M10..................... UOC:AVY,A1,A13,A14,A24,A25,A26,A27,BVY,B17,B18,B24,B25,C17,HVY,H11,H13,H14,H17,H18,H21,H24,H25,H26,H27,H28,MMM,NNN	1
18	PAFZZ	3120014478663	27647	30277	BEARING,WASHER,THRU................. UOC:AY,A11,A13,A14,A24,A25,A26,A27,BVY,B17,B18,B24,B25,C17,HVY,H11,H13,H14,H17,H18,H21,H24,H25,H26,H27,H28,MMM,NNN	2
19	PAFZZ	2590014465957	27647	32237	HOLDER,CABLE REEL................... UOC:AV,A11,A13,A14,A24,A25,A26,A27,BVY,B17,B18,B24,B25,C17,HVY,H11,H13,H14,H17,H18,H21,H24,H25,H26,H27,H28,MMM,NNN	1
20	PAFZA	5330014464696	27647	30275	.PACKING ASSEMBLY................. UOC:AVY,11,A13,A14,A24,A25,A26,A27,BVY,B17,B18,B24,B25,C17,HVY,H11,H13,H14,H17,H18,H21,H24,H25,H26,H27,H28,MMM,NNN	2
21	PAFZZ	3120014478658	27647	30274	.BUSHING,SLEEVE.................... UOC:AVY,A,A13,A14,A24,A25,A26,A27,	2

SECTION II   TM9-2320-280-24P

(1) NO	(2) SMR CODE	(3) NSN	(4) CAGEC	(5) PART NUMBER	(6) DESCRIPTION AND USABLE ON CODES(UOC)	(7) QTY

					BVY,B17,B18,B24,B25,C17,HVY,H11,H13, H14,H17,H18,H21,H24,H25,H26,H27,H28, MMM,NNN	
22	PAFZZ	5305014485080	27647	30328	.SCREW,MACHINE 3/8-16 X .75........ UOC:AVY,A11,A13,A14,A24,A25,A26,A27,	6
					BVY,B17,B18,B24,B25,C17,HVY,H11,H13, H14,H17,H18,H21,H24,H25,H26,H27,H28, MMM,NNN	
23	PAFZZ	5330013067887	27647	14964	GASKET.............. UOC:AVY,A11,A13,A14,A24,A25,A26,A27,	3
					BVY,B17,B18,B24,B25,C17,HVY,H11,H13, H14,H17,H18,H21,H24,H25,H26,H27,H28, MMM,NNN	
24	PAFZZ	3010013064113	27647	15647	GEAR ASSEMBLY,SPEED................ UOC:AVY,A11,A13,A14,A24,A25,A26,A27,	1
					BVY,B17,B18,B24,B25,C17,HVY,H11,H13, H14,H17,H18,H21,H24,H25,H26,H27,H28, MMM,NNN	
25	PAFZZ	3020014463695	27647	32239	GEAR,INTERNAL...................... UOC:AVY,A11,A13,A14,A24,A25,A26,A27,	1
					BVY,B17,B18,B24,B25,C17,HVY,H11,H13, H14,H17,H18,H21,H24,H25,H26,H27,H28, MMM,NNN	
26	PAFZZ	5365014266643	27647	30326	SPACER,RING........................ UOC:AVY,A11,A13,A14,A24,A25,A26,A27,	1
					BVY,B17,B18,B24,B25,C17,HVY,H11,H13, H14,H17,H18,H21,H24,H25,H26,H27,H28, MMM,NNN	
27	PAFZZ	5365014340186	27647	32208	SPACER,SLEEVE...................... UOC:AVY,A11,A13,A14,A24,A25,A26,A27,	3
					BVY,B17,B18,B24,B25,C17,HVY,H11,H13, H14,H17,H18,H21,H24,H25,H26,H27,H28, MMM,NNN	
28	PAFZZ	3950014351632	34623	5714499	BRAKE AND CLUTCH AS ELECTRIC....... UOC:AVY,A11,A13,A14,A24,A25,A26,A27,	1
					BVY,B17,B18,B24,B25,C17,HVY,H11,H13, H14,H17,H18,H21,H24,H25,H26,H27,H28, MMM,NNN	
29	PAFZZ	3010014295333	27647	30260	.COUPLING,SHAFT,RIGI................ UOC:AVY,A11,A13,A14,A24,A25,A26,A27,	1
					BVY,B17,B18,B24,B25,C17,HVY,H11,H13, H14,H17,H18,H21,H24,H25,H26,H27,H28, MMM,NNN	
30	PAFZZ	3120014352560	27647	15271	.BEARING,WASHER,THRU................ UOC:AVY,A11,A13,A14,A24,A25,A26,A27,	1
					BVY,B17,B18,B24,B25,C17,HVY,H11,H13, H14,H17,H18,H21,H24,H25,H26,H27,H28, MMM,NNN	
31	PAFZZ	3950014351014	27647	32236	HOUSING PART,MECHAN................ UOC:AVY,A11,A13,A14,A24,A25,A26,A27,	1
					BVY,B17,B18,B24,B25,C17,HVY,H11,H13, H14,H17,H18,H21,H24,H25,H26,H27,H28, MMM,NNN	

SECTION II   TM9-2320-280-24P

(1) NO	(2) CODE	(3) NSN	(4) CAGEC	(5) PART NUMBER	(6) DESCRIPTION AND USABLE ON CODES(UOC)	(7) ITEM SMR QTY

32 PAFZA 5330014464696 27647 30275 .PACKING ASSEMBLY................... 2   UOC:AVY,A11,A13,A14,A24,A25,A26,A27,
BVY,B17,B18,B24,B25,C17,HVY,H11,H13,
H14,H17,H18,H21,H24,H25,H26,H27,H28, MMM,NNN

33 PAFZZ 3120014478658 27647 30274 .BUSHING,SLEEVE..................... 2   UOC:AVY,A11,A13,A14,A24,A25,A26,A27,
BVY,B17,B18,B24,B25,C17,HVY,H11,H13,
H14,H17,H18,H21,H24,H25,H26,H27,H28, MMM,NNN

34 PAFZZ 3120014352495 27647 32062 .BEARING,WASHER,THRU................ 1   UOC:AVY,A11,A13,A14,A24,A25,A26,A27,
BVY,B17,B18,B24,B25,C17,HVY,H11,H13,
H14,H17,H18,H21,H24,H25,H26,H27,H28, MMM,NNN

35 PAFZZ 3110001089247 19207 900529 .BEARING,BALL,ANNULA................ 1   UOC:AVY,A11,A13,A14,A24,A25,A26,A27,
BVY,B17,B18,B24,B25,C17,HVY,H11,H13,
H14,H17,H18,H21,H24,H25,H26,H27,H28, MMM,NNN

36 PAFZZ 6110014345562 27647 24800 PROTECTOR,THERMAL-O................ 1   UOC:AVY,A11,A13,A14,A24,A25,A26,A27,
BVY,B17,B18,B24,B25,C17,HVY,H11,H13,
H14,H17,H18,H21,H24,H25,H26,H27,H28, MMM,NNN

37 PAFZZ 5935014315656 27647 31597 CONNECTOR,RECEPTACL................ 1   UOC:AVY,A11,A13,A14,A24,A25,A26,A27,
BVY,B17,B18,B24,B25,C17,HVY,H11,H13,
H14,H17,H18,H21,H24,H25,H26,H27,H28, MMM,NNN

38 PAFZZ 5945014315195 27647 31604 SOLENOID ASSEMBLY 24 VDC........... 1   UOC:AVY,A11,A13,A14,A24,A25,A26,A27,
BVY,B17,B18,B24,B25,C17,HVY,H11,H13,
H14,H17,H18,H21,H24,H25,H26,H27,H28, MMM,NNN

39 PAFZZ 4010014264536 27647 27569 WIRE ROPE ASSEMBLY, 3/8" EIPS X   1   UOC:AVY,A11,A13,A14,A24,A25,A26,A27,
BVY,B17,B18,B24,B25,C17,HVY,H11,H13,
H14,H17,H18,H21,H24,H25,H26,H27,H28, MMM,NNN

40 PAFZZ 4030012560471 75535 G-408-3/8 THIMBLE,ROPE....................... 1   UOC:AVY,A11,A13,A14,A24,A25,A26,A27,
BVY,B17,B18,B24,B25,C17,HVY,H11,H13,
H14,H17,H18,H21,H24,H25,H26,H27,H28, MMM,NNN

41 PAFZZ 4030014264537 27647 29519 HOOK,HOIST W/CLEVIS AND COTTER PINS 1   UOC:AVY,A11,A13,A14,A24,A25,A26,A27,
BVY,B17,B18,B24,B25,C17,HVY,H11,H13,
H14,H17,H18,H21,H24,H25,H26,H27,H28, MMM,NNN

42 PAFZZ 5305012102309 27647 8548 SCREW,TAPPING 10-24 X 3/8.......... 3   UOC:AVY,A11,A13,A14,A24,A25,A26,A27,
BVY,B17,B18,B24,B25,C17,HVY,H11,H13,

SECTION II   TM9-2320-280-24P C01

(1)	(2)	(3)	(4)	(5)	(6)	(7)
ITEM NO	SMR CODE	NSN	CAGEC	PART NUMBER	DESCRIPTION AND USABLE ON CODES (UOC)	QTY

```
    43  PAFZA  5340014300037  27647  31585        COVER,PROTECTIVE,DU................  1
        UOC:AVY,A11,A13,A14,A24,A25,A26,A27,
            BVY,B17,B18,B24,B25,C17,HVY,H11,H13,
            H14,H17,H18,H21,H24,H25,H26,H27,H28,
            MMM,NNN

  * 44  PAOZZ  5340014861005  27647  31293        CLIP,RETAINING....................  1
        UOC:AVY,A11,A13,A14,A24,A25,A26,A27,
            BVY,B17,B18,B24,B25,C17,HVY,H11,H13,
            H14,H17,H18,H21,H24,H25,H26,H27,H28,
            MMM,NNN

    45  PAOZZ  5935014605508  71468  CA121003-3   COVER,ELECTRICAL CO...............  1
        UOC:AVY,A11,A13,A14,A24,A25,A26,A27,
            BVY,B17,B18,B24,B25,C17,HVY,H11,H13,
            H14,H17,H18,H21,H24,H25,H26,H27,H28,
            MMM,NNN

    46  PAFZZ  5305014286635  27647  31625        SCREW,MACHINE  M4 X 10............  4
        UOC:AVY,A11,A13,A14,A24,A25,A26,A27,
            BVY,B17,B18,B24,B25,C17,HVY,H11,H13,
            H14,H17,H18,H21,H24,H25,H26,H27,H28,
            MMM,NNN

    47  PAFZZ  6105014341729  27647  32235        MOTOR,DIRECT CURREN  24 VDC.......  1
        UOC:AVY,A11,A13,A14,A24,A25,A26,A27,
            BVY,B17,B18,B24,B25,C17,HVY,H11,H13,
            H14,H17,H18,H21,H24,H25,H26,H27,H28,
            MMM,NNN

  * 48  PAFZZ  5315001879384  18631  PDG6120-1/8-1-1/  PIN,COTTER......................  4
                                              UOC:AVY,A11,A13,A14,A24,A25,A26,A27,
            BVY,B17,B18,B24,B25,C17,HVY,H11,H13,
            H14,H17,H18,H21,H24,H25,H26,H27,H28,
            MMM,NNN

                            END OF FIGURE
```

458-5

Section II. TM 9-2320-280-24P-2

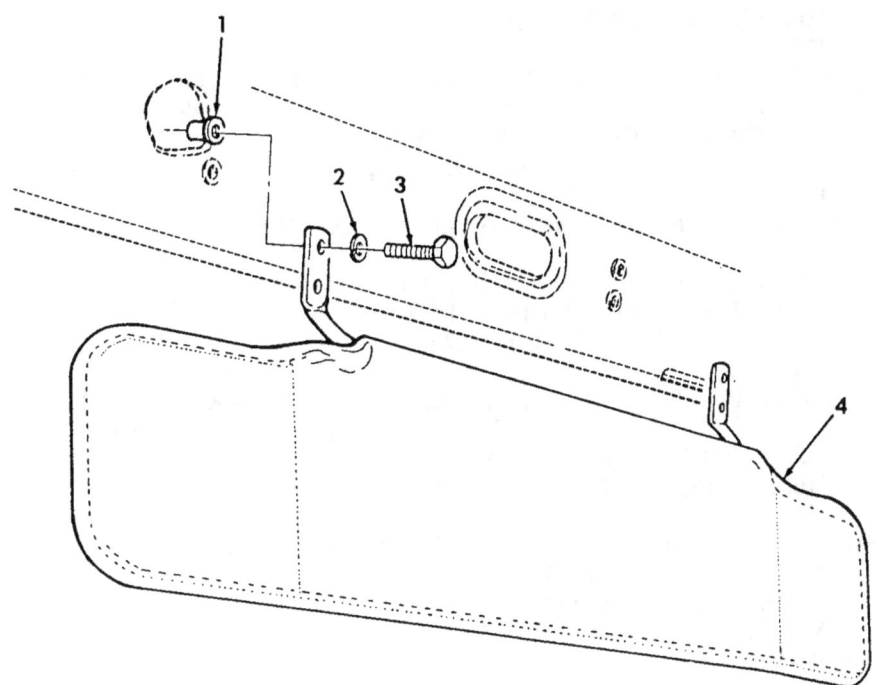

Figure 459. Modification Kit, Sun Visor.

SECTION II TM9-2320-280-24P

| (1) NO | (2) CODE | (3) NSN | (4) CAGEC | (5) NUMBER | (6) | (7) ITEM SMR DESCRIPTION AND USABLE ON CODES(UOC) | PART QTY |

GROUP 3307 SPECIAL PURPOSE KITS

FIG. 459 MODIFICATION KIT, SUN VISOR

1 PAOZZ 5310014113422 78276 ALS4-420-165 NUT,PLAIN,BLIND RIV 1/4-20 PART OF 8
KIT P/N 57K3209.....................
UOC:AVY,A11,A13,A14,A15,A20,A24,A25,
A26,A27,B16,B17,B18,HVY,H13,H14,H15,
H16,H17,H18,H20,H21,H24,H25,H26,H27,
H28

2 PAOZZ 5310006379541 96906 MS35338-46 WASHER,LOCK 1/4 PART OF KIT P/N 8
57K3209........................... UOC:AVY,A11,A13,A14,A15,A20,A24,A25,
A26,A27,B16,B17,B18,HVY,H13,H14,H15,
H16,H17,H18,H20,H21,H24,H25,H26,H27,
H28

3 PAOZZ 5305002253843 80204 B1821BH025C100N SCREW,CAP,HEXAGON H 1/4-20 PART OF 8
KIT P/N 57K3209.....................
UOC:AVY,A11,A13,A14,A15,A20,A24,A25,
A26,A27,B16,B17,B18,HVY,H13,H14,H15,
H16,H17,H18,H20,H21,H24,H25,H26,H27,
H28

4 PAOZZ 2540014108793 34623 12446819 VISOR,SUN,VEHICLE PART OF KIT P/N 2
57K3209............................ UOC:AVY,A11,A13,A14,A15,A20,A24,A25,
A26,A27,B16,B17,B18,HVY,H13,H14,H15,
H16,H17,H18,H20,H21,H24,H25,H26,H27,
H28

  KIT PDOZZ 2540014319182 19207 57K3209 VISOR,SUN,VEHICLE................... 1
           NUT,PLAIN,BLIND RIV(8) 459-1
           SCREW,CAP,HEXAGON H(8) 459-3
           VISOR,SUN,VEHICLE (2) 459-4
           WASHER,LOCK   (8) 459-2

END OF FIGURE

459-1

Section II. TM 9-2320-280-24P-2

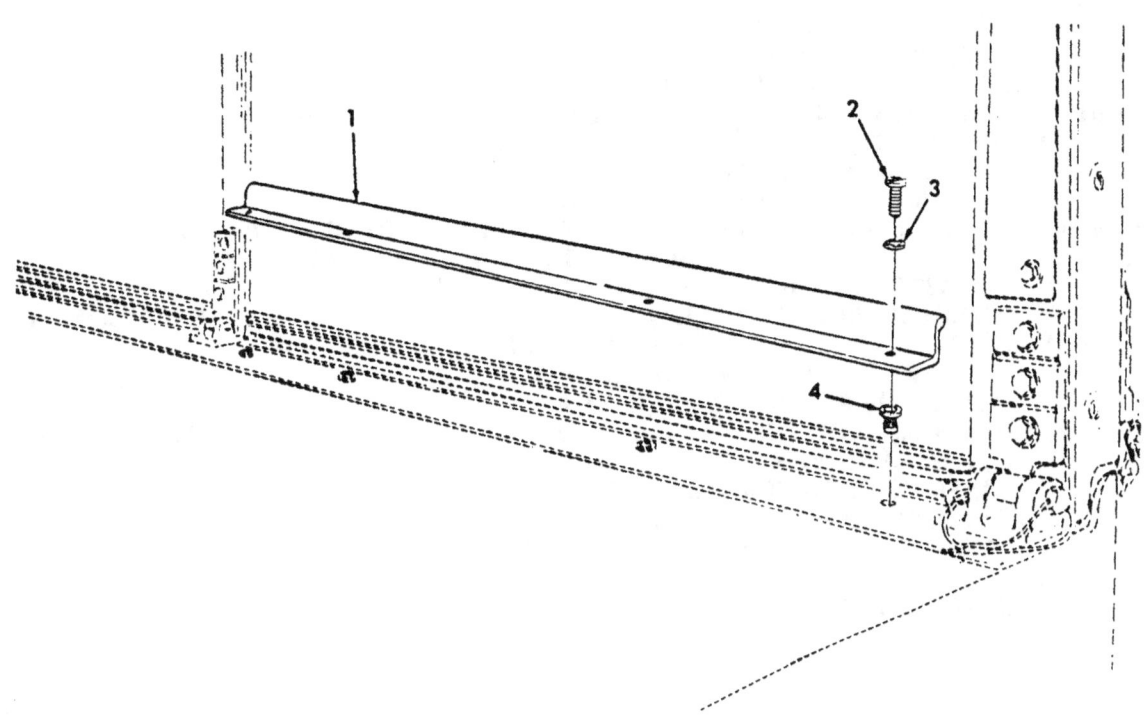

Figure 460. Modification Kit, Windshield Retainer.

SECTION II TM9-2320-280-24P

(1) NO	(2) SMR CODE	(3) NSN	(4) CAGEC	(5) PART NUMBER	(7) ITEM DESCRIPTION AND USABLE ON CODES(UOC)	QTY

GROUP 3307 SPECIAL PURPOSE KITS

FIG. 460 MODIFICATION KIT, WINDSHIELD RETAINER

1 PAOZZ 2510014110652 34623 12447029 FRAME,WINDOW,VEHICU PART OF KIT P/N 57K3206............................ 2
UOC:AVY,A11,A13,A14,A15,A20,A24,A25,
A26,A27,B16,B17,B18,HVY,H11,H13,H14,
H15,H16,H17,H18,H20,H21,H24,H25,H26,
H27,H28,MMM

2 PAOZZ 5305002400194 96906 MS51849-76 SCREW,MACHINE #10-24 X 3/4 PART OF KIT P/N 57K3206..................... 8
UOC:AVY,A11,A13,A14,A15,A20,A24,A25,
A26,A27,B16,B17,B18,HVY,H11,H13,H14,
H15,H16,H17,H18,H20,H21,H24,H25,H26,
H27,H28,MMM

3 PAOZZ 5310000453296 96906 MS35338-43 WASHER,LOCK PART OF KIT P/N 57K3206. 8
UOC:AVY,A11,A13,A14,A15,A20,A24,A25,
A26,A27,B16,B17,B18,HVY,H11,H13,H14,
H15,H16,H17,H18,H20,H21,H24,H25,H26,
H27,H28,MMM

4 PAOZZ 5325014110066 78276 ALS4-1024-130 INSERT,SCREW THREAD PART OF KIT P/N 57K3206............................ 8
UOC:AVY,A11,A13,A14,A15,A20,A24,A25,
A26,A27,B16,B17,B18,HVY,H11,H13,H14,
H15,H16,H17,H18,H20,H21,H24,H25,H26,
H27,H28,MMM

KIT PDOZZ 2510014311339 19207 57K3206 PARTS KIT,WINDSHIEL................. 1
UOC:AVY,A11,A13,A14,A15,A20,A24,A25,
A26,A27,B16,B17,B18,HVY,H11,H13,H14,
H15,H16,H17,H18,H20,H21,H24,H25,H26,
H27,H28,MMM
  FRAME,WINDOW,VEHICU(2) 460-1
  INSERT,SCREW THREAD(8) 460-4
  SCREW,MACHINE   (8) 460-2
  WASHER,LOCK   (8) 460-3

END OF FIGURE

Section II.

TM 9-2320-280-24P-2

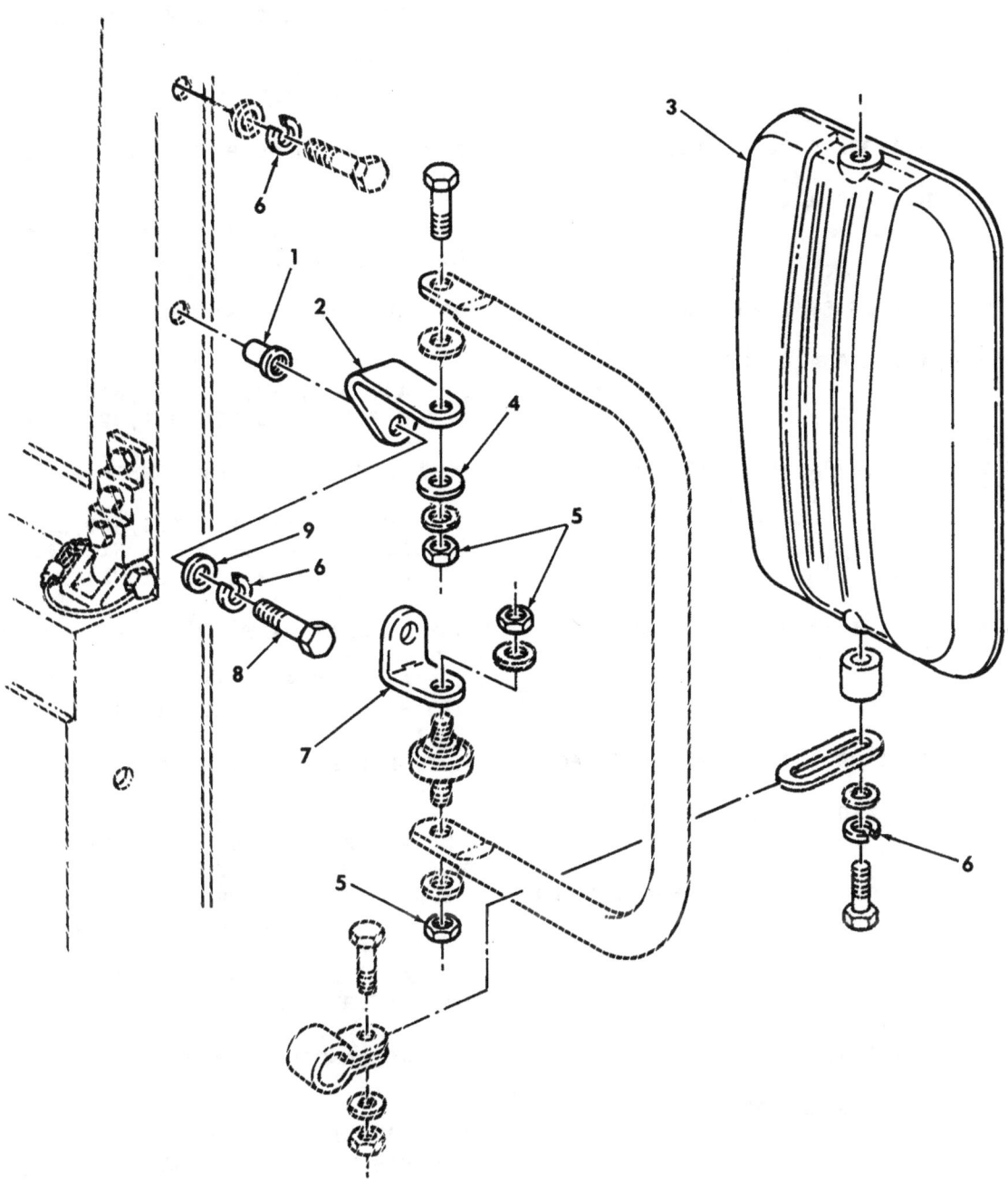

Figure 461. Left Rearview Mirror Relocation Kit.

```
      SECTION II           TM9-2320-280-24P C01
  (1)    (2)     (3)           (4)         (5)                        (6)
(7)    ITEM    SMR                         PART
     NO      CODE   NSN       CAGEC       NUMBER    DESCRIPTION AND USABLE ON CODES(UOC)
QTY
```

GROUP 3307 SPECIAL PURPOSE KITS

FIG. 461 LEFT REARVIEW MIRROR RELOCATION KIT

```
     1 PAOZZ 5310014133276 78276 ALS4-518-150   NUT,PLAIN,BLIND RIV PART OF KIT P/N   2
57K3213............................                                                UOC
:AVY,A11,A13,A14,A15,A20,A24,A25,

                                                A26,A27,B16,B17,B18,HVY,H11,H13,H14,
                                                H15,H16,H17,H18,H20,H21,H24,H25,H26,
H27,H28,MMM
     2 PAOZZ 5340014086456 19207 12343062       BRACKET,ANGLE PART OF KIT P/N        1
57K3213............................                                                UOC
:AVY,A11,A13,A14,A15,A20,A24,A25,

                                                A26,A27,B16,B17,B18,HVY,H11,H13,H14,
                                                H15,H16,H17,H18,H20,H21,H24,H25,H26,
H27,H28,MMM
   * 3 PAOZZ 2540014108794 70082 1351           MIRROR HEAD,VEHICUL  GREEN PART OF   1
KIT P/N 57K3213....................
                                                UOC:AVY,A11,A13,A14,A15,A20,A24,A25,
                                                A26,A27,B16,B17,B18,HVY,H11,H13,H14,
                                                H15,H16,H17,H18,H20,H21,H24,H25,H26,
H27,H28,MMM
     4 PAOZZ 5310013133562 96906 MS51859-20     WASHER,FLAT PART OF KIT P/N 57K3213. 2
UOC:AVY,A11,A13,A14,A15,A20,A24,A25,

                                                A26,A27,B16,B17,B18,HVY,H11,H13,H14,
                                                H15,H16,H17,H18,H20,H21,H24,H25,H26,
H27,H28,MMM
     5 PAOZZ 5310012490904 81349 M45913/3-6CG8P NUT,SELF-LOCKING,HE PART OF KIT P/N  3
57K3213............................                                                UOC
:AVY,A11,A13,A14,A15,A20,A24,A25,

                                                A26,A27,B16,B17,B18,HVY,H11,H13,H14,
                                                H15,H16,H17,H18,H20,H21,H24,H25,H26,
H27,H28,MMM    6 PAOZZ 5310004079566 96906 MS35338-45    WASHER,LOCK PART OF KIT P/N
57K3213.   6                                             UOC:AVY,A11,A13,A14,A15,A20
,A24,A25,

                                                A26,A27,B16,B17,B18,HVY,H11,H13,H14,
                                                H15,H16,H17,H18,H20,H21,H24,H25,H26,
H27,H28,MMM
     7 PAOZZ 5340014086460 19207 12343063       BRACKET,ANGLE PART OF KIT P/N        1
57K3213............................                                                UOC
:AVY,A11,A13,A14,A15,A20,A24,A25,

                                                A26,A27,B16,B17,B18,HVY,H11,H13,H14,
                                                H15,H16,H17,H18,H20,H21,H24,H25,H26,
H27,H28,MMM
     8 PAOZZ 5306002264829 80204 B1821BH031C125N BOLT,MACHINE PART OF KIT P/N 57K3213 2
UOC:AVY,A11,A13,A14,A15,A20,A24,A25,

                                                A26,A27,B16,B17,B18,HVY,H11,H13,H14,
                                                H15,H16,H17,H18,H20,H21,H24,H25,H26,
H27,H28,MMM
     9 PAOZZ 5310000806004 96906 MS27183-14     WASHER,FLAT PART OF KIT P/N 57K3213. 2
UOC:AVY,A11,A13,A14,A15,A20,A24,A25,

                                                A26,A27,B16,B17,B18,HVY,H11,H13,H14,
```

SECTION II   TM9-2320-280-24P

(1) NO	(2) SMR CODE	(3) NSN	(4) CAGEC	(5) PART NUMBER	(6) DESCRIPTION AND USABLE ON CODES(UOC)	(7) QTY
KIT	PDOZZ	2540014247363	19207	57K3213	PARTS KIT,MIRROR AS................ UOC:AVY,A11,A13,A14,A15,A20,A24,A25, A26,A27,B16,B17,B18,HVY,H11,H13,H14, H15,H16,H17,H18,H20,H21,H24,H25,H26, H27,H28,MMM	1
				461-8	BOLT,MACHINE	2
				461-2	BRACKET,ANGLE	1
				461-7	BRACKET,ANGLE	1
				461-3	MIRROR HEAD,VEHICUL	1
				461-1	NUT,PLAIN,BLIND RIV	2
				461-5	NUT,SELF-LOCKING,HE	3
				461-4	WASHER,FLAT	2
				461-9	WASHER,FLAT	2
				461-6	WASHER,LOCK	6

END OF FIGURE

Section II. TM 9-2320-280-24P-2

USE WITH A2 VEHICLES

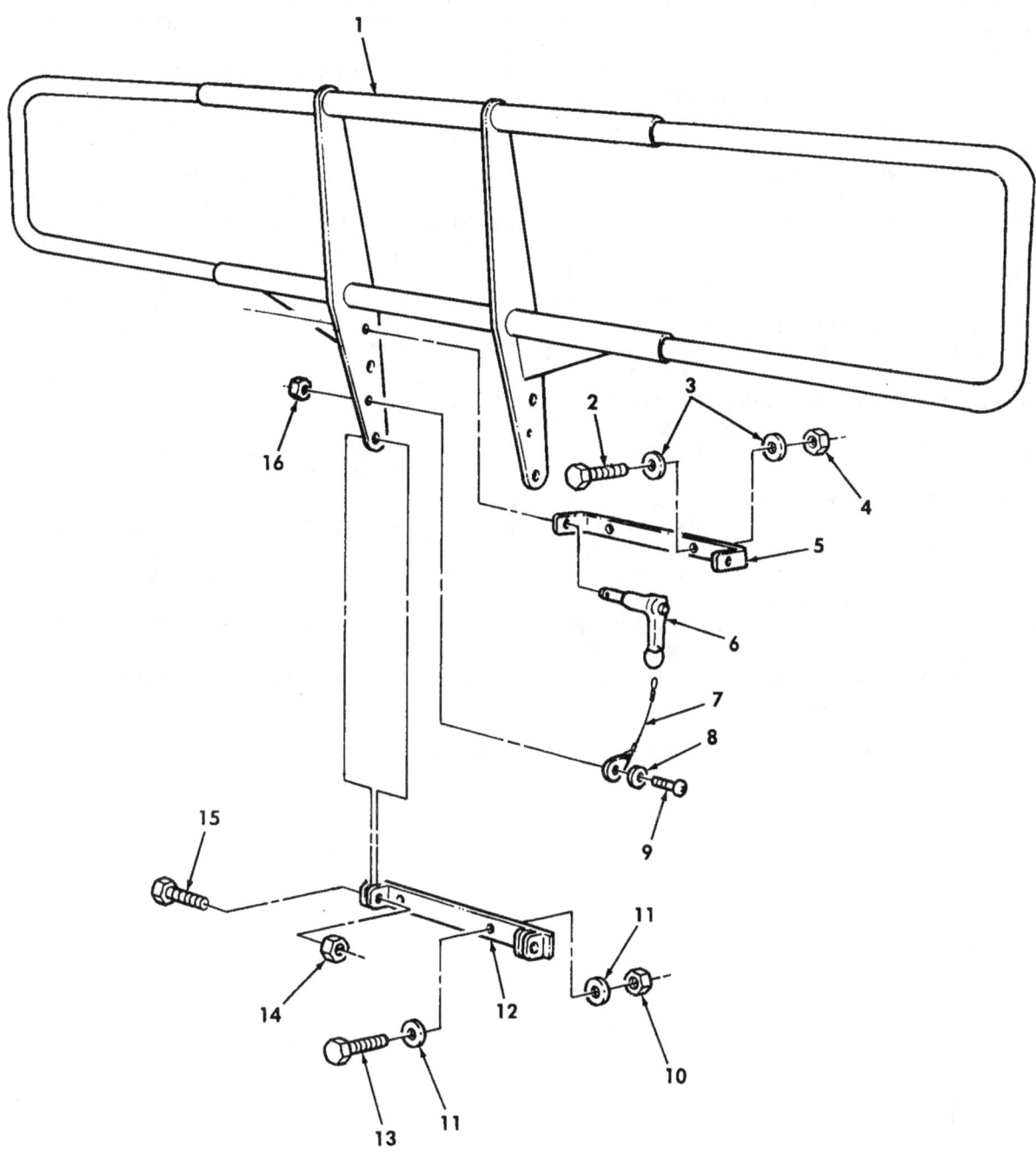

Figure 462. Brushguard Kit.

SECTION II TM9-2320-280-24P

(1) NO	(2) SMR CODE	(3) NSN	(4) CAGEC	(5) PART NUMBER	(6) DESCRIPTION AND USABLE ON CODES(UOC)	(7) QTY

GROUP 3307 SPECIAL PURPOSE KITS

FIG. 462 BRUSHGUARD KIT

1	PAOZZ	2540014666114	34623	12460550	GUARD,BRUSH,VEHICUL PART OF KIT P/N 57K3490.............................. UOC:BVY,B15,B20,B24,B25,C17,NNN	1
2	PAOZZ	5305000680511	80204	B1821BH038C125N	SCREW,CAP,HEXAGON H 7/16-14 X 2.00 PART OF KIT P/N 57K3490............. UOC:BVY,B15,B20,B24,B25,C17,NNN	2
3	PAOZZ	5310014124013	24617	2436163	WASHER,FLAT 3/8 PART OF KIT P/N 57K3490............................ UOC:BVY,B15,B20,B24,B25,C17,NNN	4
4	PAOZZ	5310004582382	24617	9418969	NUT,SELF-LOCKING,HE 3/8-16 PART OF KIT P/N 57K3490..................... UOC:BVY,B15,B20,B24,B25,C17,NNN	2
5	PAOZA	5340014661978	19207	12460280	BRACKET,MOUNTING PART OF KIT P/N 57K3490.............................. UOC:BVY,B15,B20,B24,B25,C17,NNN	1
6	PAOZZ	5315014731590	80205	MS17986C1009	PIN,QUICK RELEASE PART OF KIT P/N 57K3490.............................. UOC:BVY,B15,B20,B24,B25,C17,NNN	2
7	PAOZZ	4010014660849	19207	12342287	WIRE ROPE ASSEMBLY, PART OF KIT P/N 57K3490.......................... UOC:BVY,B15,B20,B24,B25,C17,NNN	2
8	PAOZZ	5310013611163	24617	9417373	WASHER,FLAT #10 PART OF KIT P/N 57K3490............................ UOC:BVY,B15,B20,B24,B25,C17,NNN	2
9	PAOZZ	5305011858647	7X677	160057	SCREW,MACHINE #10-32 X 3/4 PART OF KIT P/N 57K3490............ UOC:BVY,B15,B20,B24,B25,C17,NNN	2
10	PAOZZ	5310014664852	24617	9419477	NUT,SELF-LOCKING,HE 7/16-14 W/O WINCH PART OF KIT P/N 57K3490....... UOC:BVY,B15,B20,B24,B25,C17,NNN	2
11	PAOZZ	5310008094085	24617	9420022	WASHER,FLAT 7/16 PART OF KIT P/N 57K3490............................ UOC:BVY,B15,B20,B24,B25,C17,NNN	4
12	PAOZZ	2590014665095	19207	12460279	BRACKET,VEHICULAR C PART OF KIT P/N 57K3490......................... UOC:BVY,B15,B20,B24,B25,C17,NNN	1
13	PAOZZ	5305000712057	80204	B1821BH044C200N	SCREW,CAP,HEXAGON H PART OF KIT P/N 57K3490......................... UOC:BVY,B15,B20,B24,B25,C17,NNN	2
14	PAOZZ	5310014090897	24617	9419482	NUT,SELF-LOCKING,HE PART OF KIT P/N 57K3490......................... UOC:BVY,B15,B20,B24,B25,C17,NNN	2
15	PAOZZ	5305007247222	80204	B1821BH063C200N	SCREW,CAP,HEXAGON H 5/8-11 X 2.00 PART OF KIT P/N 57K3490............. UOC:BVY,B15,B20,B24,B25,C17,NNN	2
16	PAOZZ	5310001249265	72582	271169	NUT,PLAIN,ASSEMBLED #10-32 PART OF KIT P/N 57K3490..................... UOC:BVY,B15,B20,B24,B25,C17,NNN	2
KIT	PDOZZ	2510014617075	19207	57K3490	KIT,BRUSHGUARD.....................	1

SECTION II   TM9-2320-280-24P

(1) NO	(2) CODE	(3) NSN	(4) CAGEC	(5) PART NUMBER	(6) DESCRIPTION AND USABLE ON CODES(UOC)	(7) QTY

UOC:BVY,B15,B20,B24,B25,C17,NNN
BRACKET,VEHICULAR C(1) 462-12
BRACKET,MOUNTING (1) 462-5
GUARD,BRUSH,VEHICUL(1) 462-1
NUT,PLAIN,ASSEMBLED(2) 462-16
NUT,SELF-LOCKING,HE(2) 462-4
NUT,SELF-LOCKING,HE(2) 462-10
NUT,SELF-LOCKING,HE(2) 462-14
PIN,QUICK RELEASE (2) 462-6
SCREW,CAP,HEXAGON H(2) 462-2
SCREW,CAP,HEXAGON H(2) 462-13
SCREW,CAP,HEXAGON H(2) 462-15
SCREW,MACHINE   (2) 462-9
WASHER,FLAT   (4) 462-3
WASHER,FLAT   (4) 462-11
WASHER,FLAT   (2) 462-8
WIRE ROPE ASSEMBLY,(2) 462-7

END OF FIGURE

Section II. TM 9-2320-280-24P-2

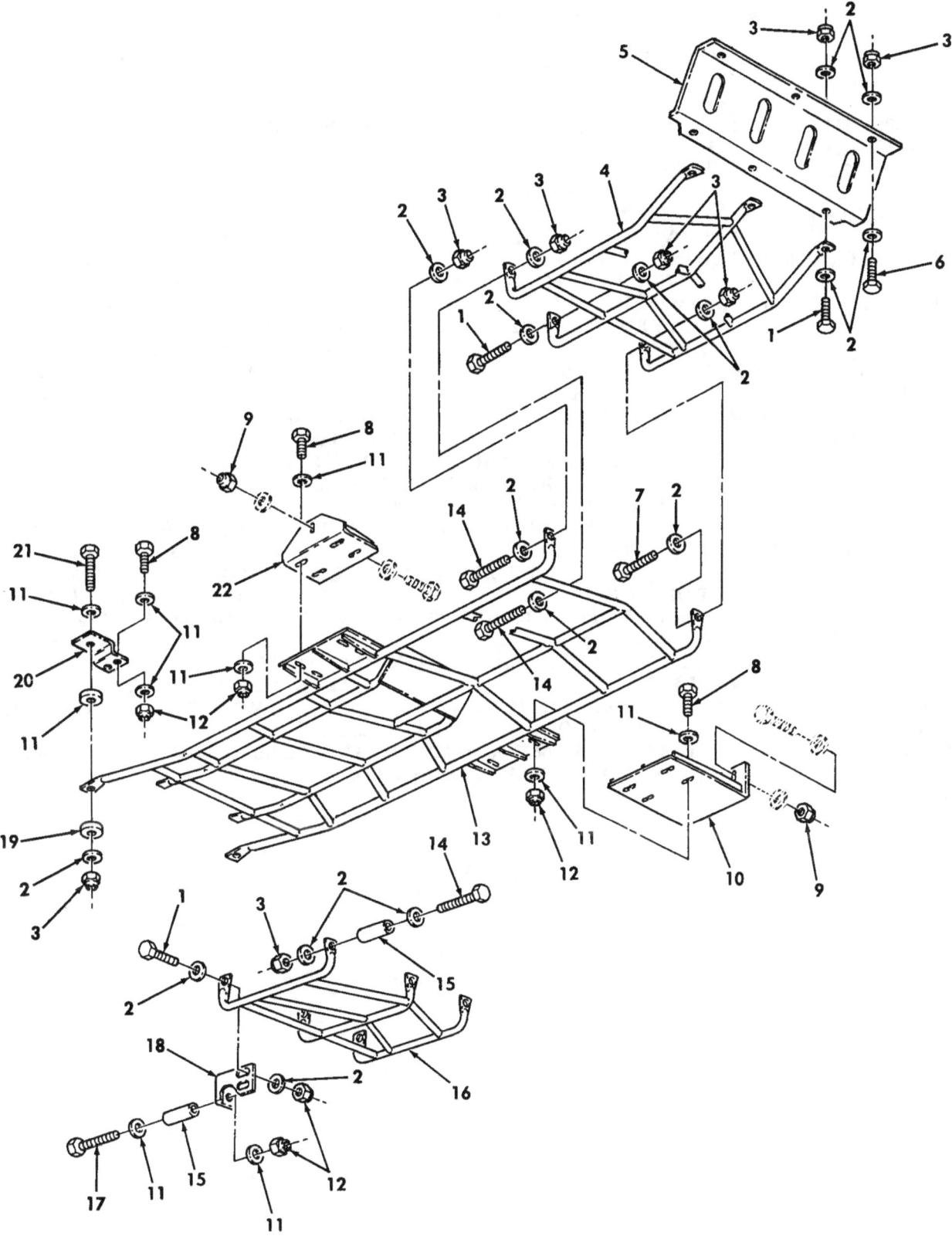

Figure 463. Underbody Protection Kit With and Wtihout Winch.

SECTION II   TM9-2320-280-24P

(1) ITEM NO	(2) SMR CODE	(3) NSN	(4) CAGEC	(5) PART NUMBER	(6) DESCRIPTION AND USABLE ON CODES(UOC)	(7) QTY
					GROUP 3307 SPECIAL PURPOSE KITS	
					FIG. 463 UNDERBODY PROTECTION KIT WITH AND WITHOUT WINCH	
1	PAOZZ	5305005432419	80204	B1821BH038C113N	SCREW,CAP,HEXAGON H PART OF KIT P/N 57K3492 PART OF KIT P/N 57K3495..... UOC:BVY,B15,B20,B24,B25,C17,NNN	7
2	PAOZZ	5310014124013	24617	2436163	WASHER,FLAT W/WINCH PART OF KIT P/N 57K3492......................... UOC:BVY,B24,B25,C17,NNN	34
2	PAOZZ	5310014124013	24617	2436163	WASHER,FLAT W/O WINCH PART OF KIT P/N 57K3495................... UOC:BVY,B15,B20,B24,B25,C17,NNN	36
3	PAOZZ	5310009359021	96906	MS51943-35	NUT,SELF-LOCKING,HE W/WINCH PART OF KIT PART OF KIT P/N 57K3492......... UOC:BVY,B24,B25,C17,NNN	17
3	PAOZZ	5310009359021	96906	MS51943-35	NUT,SELF-LOCKING,HE W/O WINCH PART OF KIT P/N 57K3495............ UOC:BVY,B15,B20,B24,B25,C17,NNN	18
4	PAOZZ	2540014663849	19207	12460415	GUARD,BRUSH,VEHICUL PART OF KIT P/N 57K3492 PART OF KIT P/N 57K3495..... UOC:BVY,B15,B20,B24,B25,C17,NNN	1
5	PAOZZ	2590014657756	19207	12447163-2	PLATE ASSEMBLY SKID W/WINCH PART OF KIT P/N 57K3492..................... UOC:BVY,B24,B25,C17,NNN	1
5	PAOZZ	5340014779397	19207	12460414	BRACKET,MULTIPLE AN W/O WINCH PART OF KIT P/N 57K3495................... UOC:BVY,B15,B20,B24,B25,C17,NNN	1
6	PAOZZ	5305000680510	80204	B1821BH038C100N	SCREW,CAP,HEXAGON H PART OF KIT P/N 57K3495..................... UOC:BVY,B15,B20,B24,B25,C17,NNN	3
6	PAOZZ	5305000680510	80204	B1821BH038C100N	SCREW,CAP,HEXAGON H PART OF KIT P/N 57K3492........................ UOC:BVY,B15,B20,B24,B25,C17,NNN	2
7	PAOZZ	5305011983440	7X677	9430724	SCREW,CAP,HEXAGON H PART OF KIT P/N 57K3492 PART OF KIT P/N 57K3495..... UOC:BVY,B15,B20,B24,B25,C17,NNN	1
8	PAOZZ	5306011300457	24617	9430677	BOLT,MACHINE PART OF KIT P/N 57K3492 PART OF KIT P/N 57K3495..... UOC:BVY,B15,B20,B24,B25,C17,NNN	12
9	PAOZZ	5310004883889	96906	MS51943-39	NUT,SELF-LOCKING,HE PART OF KIT P/N 57K3492 PART OF KIT P/N 57K3495..... UOC:BVY,B15,B20,B24,B25,C17,NNN	4
10	PAOZZ	5340014661982	19207	12447146	BRACKET,MOUNTING L.H. PART OF KIT P/N 57K3492 PART OF KIT P/N 57K3495..... UOC:BVY,B15,B20,B24,B25,C17,NNN	1
11	PAOZZ	5310011191024	24617	2436162	WASHER,FLAT PART OF KIT P/N 57K3492 PART OF KIT P/N 57K3495............. UOC:BVY,B15,B20,B24,B25,C17,NNN	36
12	PAOZZ	5310008140673	96906	MS51943-33	NUT,SELF-LOCKING,HE PART OF KIT P/N 57K3492 PART OF KIT P/N 57K3495..... UOC:BVY,B15,B20,B24,B25,C17,NNN	18

Section II. TM 9-2320-280-24P-2

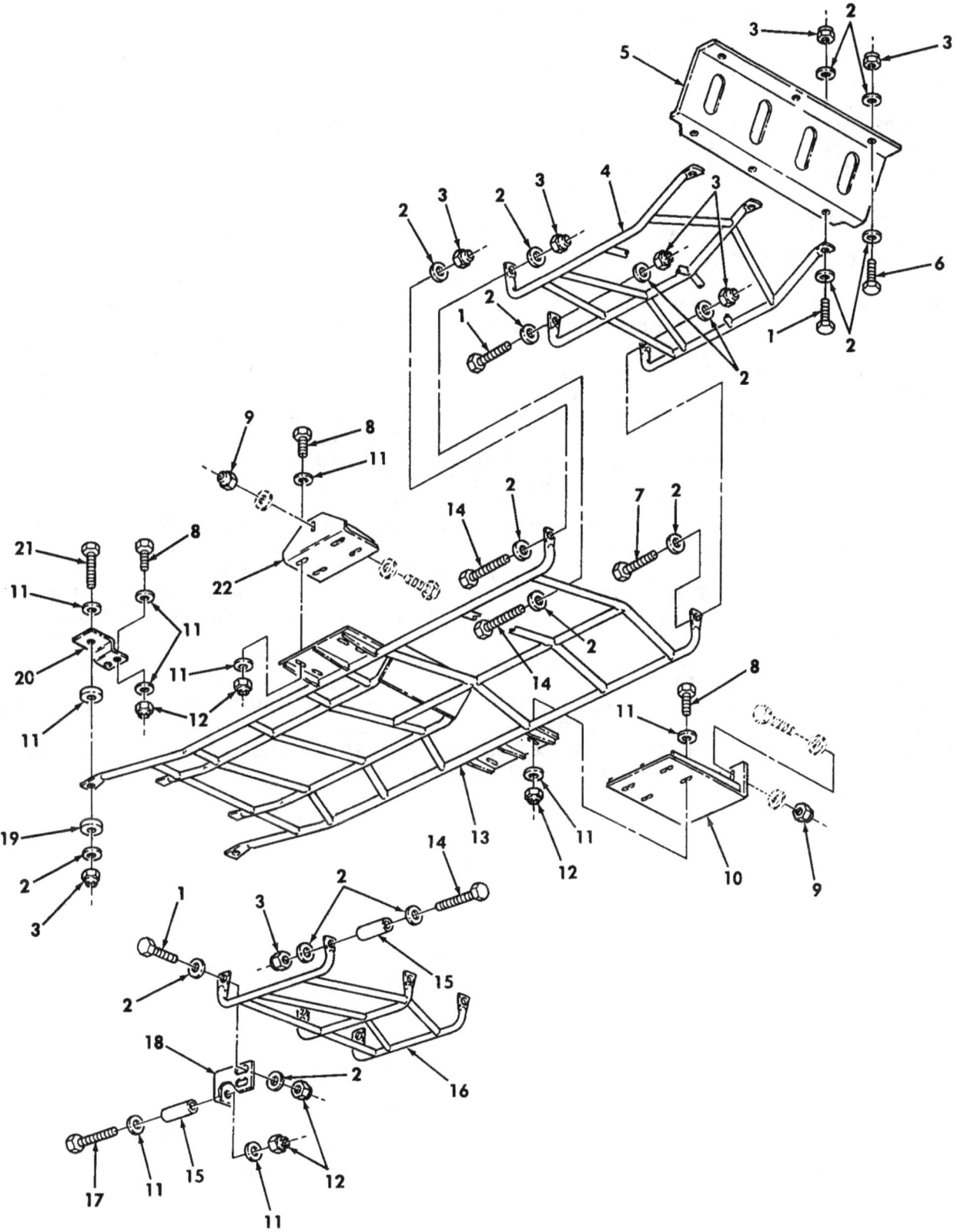

Figure 463. Underbody Protection Kit With and Wtihout Winch.

SECTION II TM9-2320-280-24P

(1) NO	(2) CODE	(3) NSN	(4) CAGEC	(5) NUMBER	(6) DESCRIPTION AND USABLE ON CODES(UOC)	(7) ITEM SMR PART QTY

GROUP 3307 SPECIAL PURPOSE KITS

FIG. 463 UNDERBODY PROTECTION KIT WITH AND WITHOUT WINCH

Item	SMR	NSN	CAGEC	Part Number	Description	QTY
1	PAOZZ	5305005432419	80204	B1821BH038C113N	SCREW,CAP,HEXAGON H PART OF KIT P/N 57K3492 PART OF KIT P/N 57K3495..... UOC:BVY,B15,B20,B24,B25,C17,NNN	7
2	PAOZZ	5310014124013	24617	2436163	WASHER,FLAT W/WINCH PART OF KIT P/N 57K3492......... UOC:BVY,B24,B25,C17,NNN	34
2	PAOZZ	5310014124013	24617	2436163	WASHER,FLAT W/O WINCH PART OF KIT P/N 57K3495............ UOC:BVY,B15,B20,B24,B25,C17,NNN	36
3	PAOZZ	5310009359021	96906	MS51943-35	NUT,SELF-LOCKING,HE W/WINCH PART OF KIT PART OF KIT P/N 57K3492......... UOC:BVY,B24,B25,C17,NNN	17
3	PAOZZ	5310009359021	96906	MS51943-35	NUT,SELF-LOCKING,HE W/O WINCH PART OF KIT P/N 57K3495.............. UOC:BVY,B15,B20,B24,B25,C17,NNN	18
4	PAOZZ	2540014663849	19207	12460415	GUARD,BRUSH,VEHICUL PART OF KIT P/N 57K3492 PART OF KIT P/N 57K3495..... UOC:BVY,B15,B20,B24,B25,C17,NNN	1
5	PAOZZ	2590014657756	19207	12447163-2	PLATE ASSEMBLY SKID W/WINCH PART OF KIT P/N 57K3492............ UOC:BVY,B24,B25,C17,NNN	1
5	PAOZZ	5340014779397	19207	12460414	BRACKET,MULTIPLE AN W/O WINCH PART OF KIT P/N 57K3495.................. UOC:BVY,B15,B20,B24,B25,C17,NNN	1
6	PAOZZ	5305000680510	80204	B1821BH038C100N	SCREW,CAP,HEXAGON H PART OF KIT P/N 57K3495............ UOC:BVY,B15,B20,B24,B25,C17,NNN	3
6	PAOZZ	5305000680510	80204	B1821BH038C100N	SCREW,CAP,HEXAGON H PART OF KIT P/N 57K3492........................ UOC:BVY,B15,B20,B24,B25,C17,NNN	2
7	PAOZZ	5305011983440	7X677	9430724	SCREW,CAP,HEXAGON H PART OF KIT P/N 57K3492 PART OF KIT P/N 57K3495..... UOC:BVY,B15,B20,B24,B25,C17,NNN	1
8	PAOZZ	5306011300457	24617	9430677	BOLT,MACHINE PART OF KIT P/N 57K3492 PART OF KIT P/N 57K3495..... UOC:BVY,B15,B20,B24,B25,C17,NNN	12
9	PAOZZ	5310004883889	96906	MS51943-39	NUT,SELF-LOCKING,HE PART OF KIT P/N 57K3492 PART OF KIT P/N 57K3495..... UOC:BVY,B15,B20,B24,B25,C17,NNN	4
10	PAOZZ	5340014661982	19207	12447146	BRACKET,MOUNTING L.H.PART OF KIT P/N 57K3492 PART OF KIT P/N 57K3495..... UOC:BVY,B15,B20,B24,B25,C17,NNN	1
11	PAOZZ	5310011191024	24617	2436162	WASHER,FLAT PART OF KIT P/N 57K3492 PART OF KIT P/N 57K3495............. UOC:BVY,B15,B20,B24,B25,C17,NNN	36
12	PAOZZ	5310008140673	96906	MS51943-33	NUT,SELF-LOCKING,HE PART OF KIT P/N 57K3492 PART OF KIT P/N 57K3495..... UOC:BVY,B15,B20,B24,B25,C17,NNN	18

Section II. TM 9-2320-280-24P-2

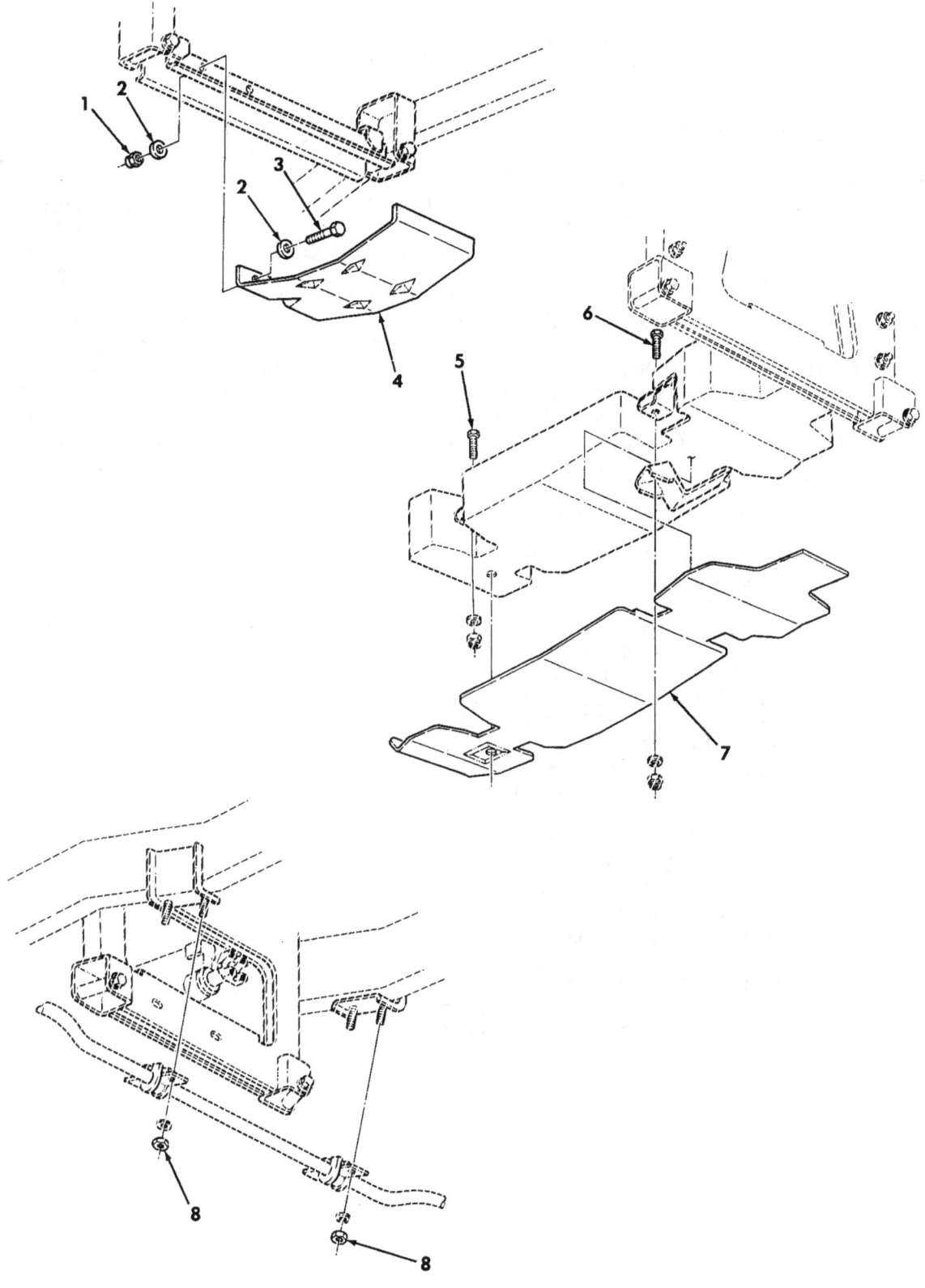

Figure 464. Underbody Protection Kit, Transfer Case, Fuel Tank Shields, and Sway Bar Link Nuts.

SECTION II   TM9-2320-280-24P

(1) ITEM NO	(2) SMR CODE	(3) NSN	(4) CAGEC	(5) PART NUMBER	(7) DESCRIPTION AND USABLE ON CODES (UOC)	QTY

GROUP 3307 SPECIAL PURPOSE KITS

FIG. 464 UNDERBODY PROTECTION KIT, TRANSFER CASE, FUEL TANK SHIELDS, AND SWAY BAR LINK NUTS

Item	SMR	NSN	CAGEC	Part Number	Description	Qty	UOC
1	PAOZZ	5310009359021	96906	MS51943-35	NUT,SELF-LOCKING,HE PART OF KIT P/N 57K3492,PART OF KIT P/N 57K3495.....	2	BVY,B15,B20,B24,B25,C17,NNN
2	PAOZZ	5310014124013	24617	2436163	WASHER,FLAT PART OF KIT P/N 57K3492, PART OF KIT P/N 57K3495.............	4	BVY,B15,B20,B24,B25,C17,NNN
3	PAOZZ	5305000680510	80204	B1821BH038C100N	SCREW,CAP,HEXAGON H PART OF KIT P/N 57K3492,PART OF KIT P/N 57K3495.....	2	BVY,B15,B20,B24,B25,C17,NNN
4	PAOZZ	2590014663824	19207	12447151	SHIELD,TRANSFER CAS PART OF KIT P/N 57K3492,PART OF KIT P/N 57K3495.....	1	BVY,B15,B20,B24,B25,C17,NNN
5	PAOZZ	5305014125995	19207	12460145-2	SCREW,CAP,HEXAGON H PART OF KIT P/N 57K3492,PART OF KIT P/N 57K3495.....	1	BVY,B15,B20,B24,B25,C17,NNN
6	PAOZZ	5306014112338	19207	12460145-1	BOLT,MACHINE PART OF KIT P/N 57K3492 PART OF KIT P/N 57K3495.............	1	BVY,B15,B20,B24,B25,C17,NNN
7	PAOZZ	2590014663866	19207	12447150	SHIELD,FUEL TANK PART OF KIT P/N 57K3492,PART OF KIT P/N 57K3495.....	1	BVY,B15,B20,B24,B25,C17,NNN
8	PAOZZ	5310011504003	24617	9422299	NUT,SELF-LOCKING,HE PART OF KIT P/N 57K3492,PART OF KIT P/N 57K3495.....	4	BVY,B15,B20,B24,B25,C17,NNN
KIT	PDOZZ	2540014678313	19207	57K3492	KIT WITH WINCH......................	1	BVY,B15,B20,B24,B25,C17,NNN

```
BOLT,MACHINE      ( 12) 463-8
BOLT,MACHINE      (  1) 464-6
BRACKET,ANGLE     (  3) 463-18
BRACKET,MOUNTING  (  1) 463-10
BRACKET,MOUNTING  (  1) 463-22
BRACKET,MULTIPLE AN(  2) 463-20
GUARD,BRUSH,VEHICUL(  1) 463-4
GUARD,BRUSH,VEHICUL(  1) 463-16
NUT,SELF-LOCKING,HE( 17) 463-3
NUT,SELF-LOCKING,HE(  4) 463-9
NUT,SELF-LOCKING,HE( 18) 463-12
NUT,SELF-LOCKING,HE(  2) 464-1
NUT,SELF-LOCKING,HE(  4) 464-8
PLATE ASSEMBLY SKID(  1) 463-5
SCREW,CAP,HEXAGON H(  7) 463-1
SCREW,CAP,HEXAGON H(  2) 463-6
SCREW,CAP,HEXAGON H(  1) 463-7
SCREW,CAP,HEXAGON H(  5) 463-14
SCREW,CAP,HEXAGON H(  6) 463-17
SCREW,CAP,HEXAGON H(  2) 463-21
SCREW,CAP,HEXAGON H(  2) 464-3
```

SECTION II   TM9-2320-280-24P

(1) NO	(2) CODE	(3) NSN	(4) CAGEC	(5) PART NUMBER	(6) DESCRIPTION AND USABLE ON CODES(UOC)	(7) ITEM SMR QTY

SCREW,CAP,HEXAGON H(1) 464-5
SHIELD ASSEMBLY,DRI(1) 463-13
SHIELD,FUEL TANK   (1) 464-7
SHIELD,TRANSFER CAS(1) 464-4
SPACER,SLEEVE   (9) 463-15
WASHER,FLAT   (34) 463-2
WASHER,FLAT   (36) 463-11
WASHER,FLAT   (4) 463-19
WASHER,FLAT   (4) 464-2

PDOZZ 2540014678341 19207 57K3495   KIT WITH OUT WINCH.................. 1   UOC:B-Y,B15,B20,B24,B25,C17 ,NNN

BOLT,MACHINE   (12) 463-8
BOLT,MACHINE   (1) 464-6
BRACKET,ANGLE   (1) 463-18
BRACKET,MOUNTING (1) 463-10
BRACKET,MOUNTING (1) 463-22
BRACKET,MULTIPLE AN(1) 463-5
BRACKET,MULTIPLE AN(2) 463-20
GUARD,BRUSH,VEHICUL(1) 463-4
GUARD,BRUSH,VEHICUL(1) 463-16
NUT,SELF-LOCKING,HE(18) 463-3
NUT,SELF-LOCKING,HE(4) 463-9
NUT,SELF-LOCKING,HE(18) 463-12
NUT,SELF-LOCKING,HE(2) 464-1
NUT,SELF-LOCKING,HE(4) 464-8
SCREW,CAP,HEXAGON H(7) 463-1
SCREW,CAP,HEXAGON H(3) 463-6
SCREW,CAP,HEXAGON H(1) 463-7
SCREW,CAP,HEXAGON H(5) 463-14
SCREW,CAP,HEXAGON H(6) 463-17
SCREW,CAP,HEXAGON H(2) 463-21
SCREW,CAP,HEXAGON H(2) 464-3
SCREW,CAP,HEXAGON H(1) 464-5
SHIELD ASSEMBLY,DRI(1) 463-13
SHIELD,FUEL TANK   (1) 464-7
SHIELD,TRANSFER CAS(1) 464-4
SPACER,SLEEVE   (9) 463-15
WASHER,FLAT   (36) 463-2
WASHER,FLAT   (36) 463-11
WASHER,FLAT   (4) 463-19
WASHER,FLAT   (4) 464-2

END OF FIGURE

Section II. TM 9-2320-280-24P-2

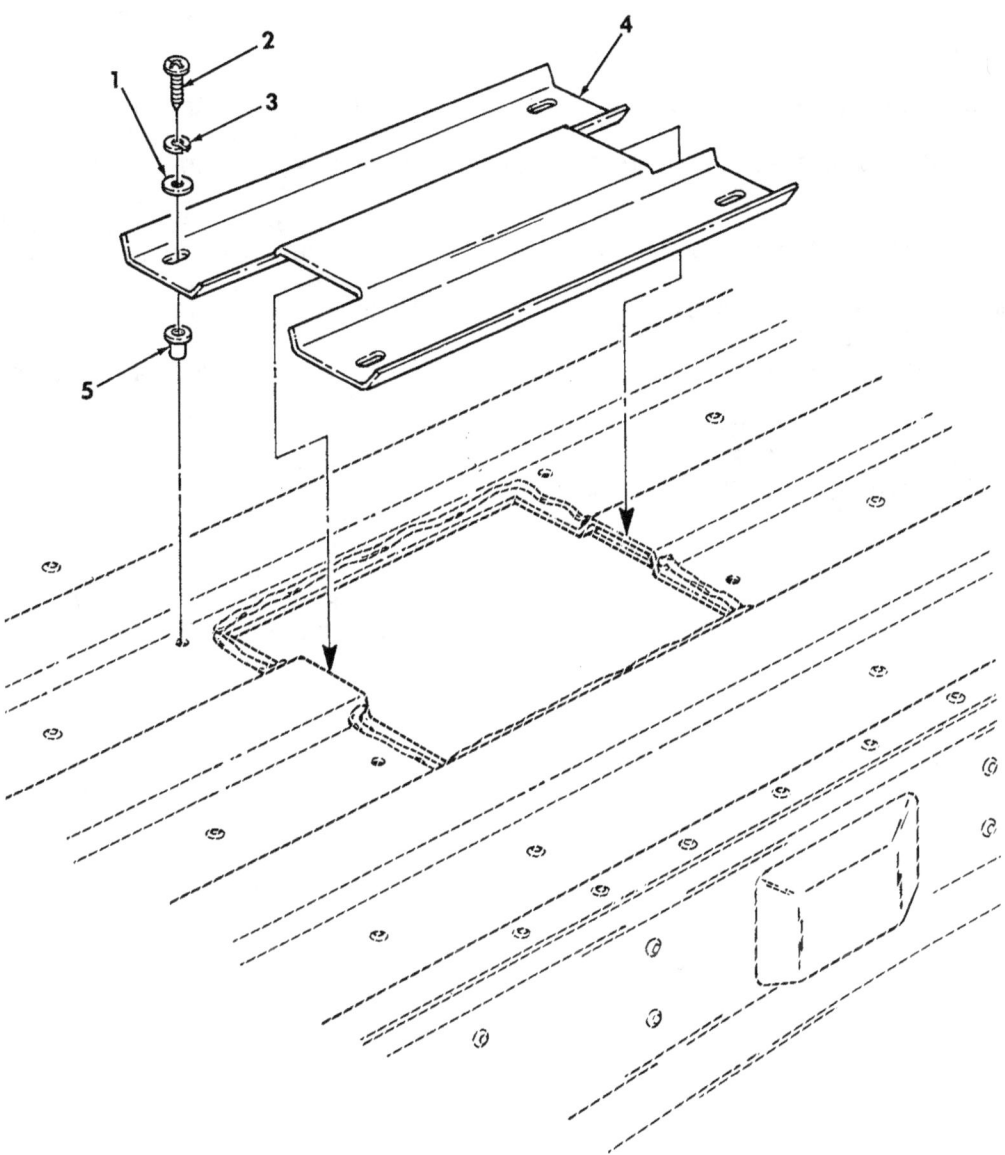

Figure 465. Cover, Access Cargo Floor Modification Kit.

SECTION II TM9-2320-280-24P C01

(1) ITEM NO	(2) SMR CODE	(3) NSN	(4) CAGEC	(5) PART NUMBER	(6) DESCRIPTION AND USABLE ON CODES (UOC)	(7) QTY

GROUP 3307 SPECIAL PURPOSE KITS

FIG. 465 COVER,ACCESS CARGO FLOOR MODIFICATION KIT

(1)	(2)	(3)	(4)	(5)	(6)	(7)
1	PAOZZ	5310013611163	24617	9417373	WASHER,FLAT PART OF KIT P/N 57K3529. UOC:BVY,B15,B20,B24,B25,C17,NNN	4
2	PAOZZ	5305010831591	96906	MS51957-64B	SCREW,MACHINE PART OF KIT P/N 57K3529............ UOC:BVY,B15,B20,B24,B25,C17,NNN	4
3	PAOZZ	5310009222017	30379	120217	WASHER,LOCK PART OF KIT P/N 57K3529. UOC:BVY,B15,B20,B24,B25,C17,NNN	4
4	PAOZZ	5340014570804	19207	12460506	COVER,ACCESS PART OF KIT P/N 57K3529 UOC:BVY,B15,B20,B24,B25,C17,NNN	1
* 5	PAOZZ	5325014608350	19207	12446871-10	INSERT,SCREW THREAD PART OF KIT P/N 57K3529............ UOC:BVY,B15,B20,B24,B25,C17,NNN	4
KIT	PDOZZ	2510014547077	19207	57K3529	PARTS KIT,FLOOR VEH........ UOC:BVY,B15,B20,B24,B25,C17,NNN	1

```
COVER,ACCESS         ( 1) 465-4
INSERT,SCREW,THREAD  ( 4) 465-5
SCREW,MACHINE        ( 4) 465-2
WASHER,FLAT          ( 4) 465-1
WASHER,LOCK          ( 4) 465-3
```

Section II. TM 9-2320-280-24P-2 C01

DELETED

Figure 466. FR12 Refrigerant Parts Kit.

SECTION II			TM9-2320-280-24P C01			
(1)	(2)	(3)	(4)	(5)	(6)	(7)
ITEM NO	SMR CODE	NSN	CAGEC	PART NUMBER	DESCRIPTION AND USABLE ON CODES (UOC)	QTY

GROUP 3307 SPECIAL PURPOSE KITS

DELETED

Section II. TM 9-2320-280-24P-2

DELETED

Figure 467. R-12 to FR12 Refrigerant Conversion Retrofit Kit.

```
SECTION II            TM9-2320-280-24P C01
   (1)    (2)   (3)         (4)        (5)                              (6)
(7)   ITEM   SMR                      PART
  NO   CODE  NSN           CAGEC     NUMBER      DESCRIPTION AND USABLE ON CODES(UOC)
QTY
```

GROUP 3307 SPECIAL PURPOSE KITS

FIG. 467 R-12 TO FR12 REFRIGERANT

DELETED

Section II. TM 9-2320-280-24P-2

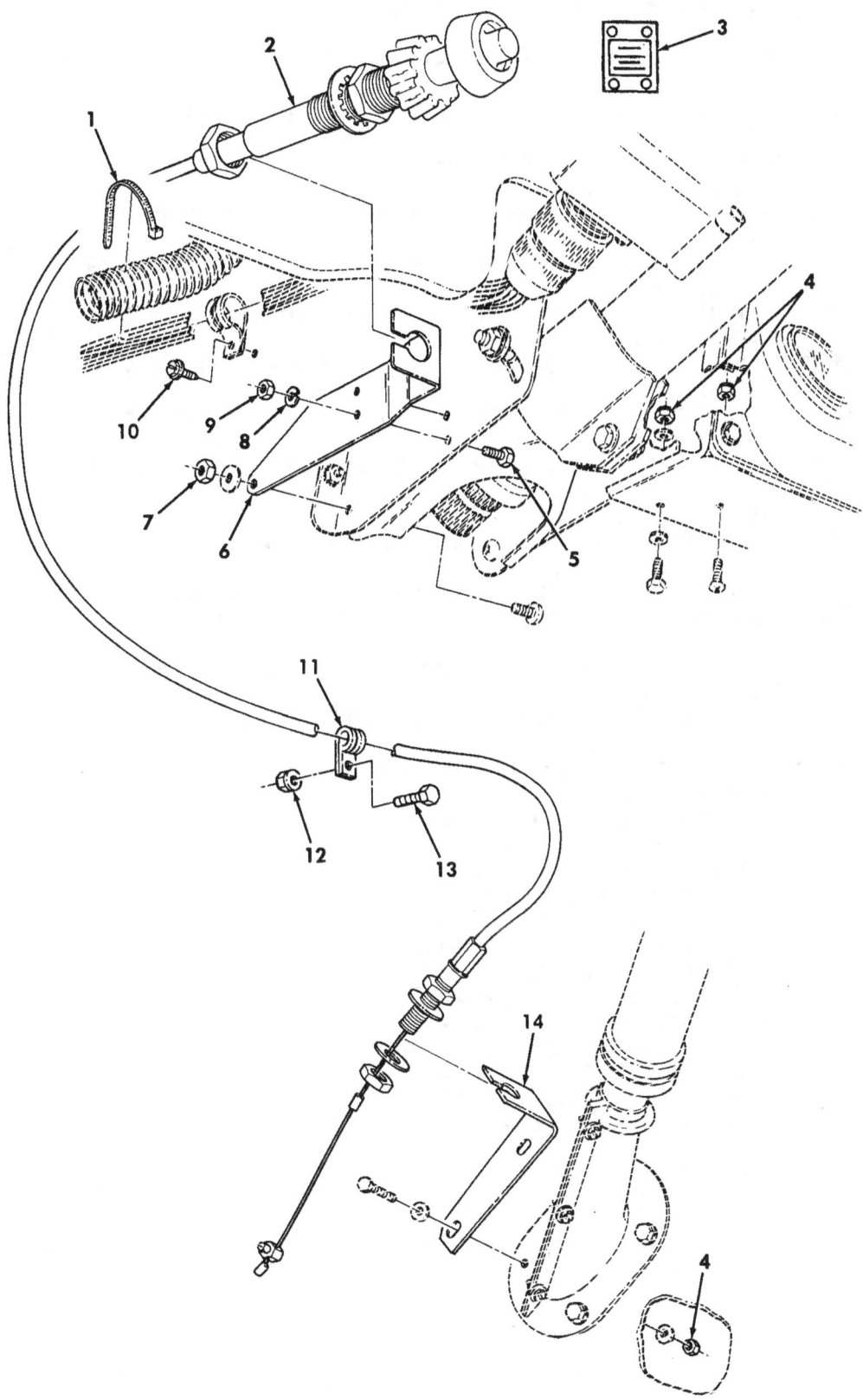

Figure 468. Hand Throttle Control Kit.

SECTION II   TM9-2320-280-24P

(1) NO	(2) SMR CODE	(3) NSN	(4) CAGEC	(5) PART NUMBER	(7) ITEM DESCRIPTION AND USABLE ON CODES(UOC)	QTY

GROUP 3307 SPECIAL PURPOSE KITS

FIG. 468 HAND THROTTLE CONTROL KIT

1	PAOZZ	5975010345871	96906	MS3367-7-0	STRAP,TIEDOWN,ELECT PART OF KIT P/N 57K3531............................	1
2	PAOZZ	2590014664397	60602	36162-23	CONTROL ASSEMBLY,PU PART OF KIT P/N 57K3531............................	1
3	PAOZZ	7690014665217	19207	12460548	MARKER,IDENTIFICATI PART OF KIT P/N 57K3531............................	1
4	PAOZZ	5310000614650	96906	MS51943-31	NUT,SELF-LOCKING,HE PART OF KIT P/N 57K3531............................	4
5	PAOZZ	5305000680508	80204	B1821BH025C075N	SCREW,CAP,HEXAGON H PART OF KIT P/N 57K3531............................	2
6	PAOZZ	2590014665250	34623	12460547	BRACKET,VEHICULAR C PART OF KIT P/N 57K3531............................	1
7	PAOZZ	5310008140673	96906	MS51943-33	NUT,SELF-LOCKING,HE PART OF KIT P/N 57K3531............................	1
8	PAOZZ	5310006379541	96906	MS35338-46	WASHER,LOCK PART OF KIT P/N 57K3531.	2
9	PAOZZ	5310005845005	12204	120375	NUT,PLAIN,HEXAGON PART OF KIT P/N 57K3531............................	2
10	PAOZZ	5305012596322	34623	12342499-1	SCREW,MACHINE PART OF KIT P/N 57K3531............................	1
11	PAOZZ	5340007647051	96906	MS21333-69	CLAMP,LOOP PART OF KIT P/N 57K3531..	1
12	PAOZZ	5310001249265	72582	271169	NUT,PLAIN,ASSEMBLED PART OF KIT P/N 57K3531............................	1
13	PAOZZ	5305005829501	82386	410-63	SCREW,CAP,HEXAGON H PART OF KIT P/N 57K3531............................	1
14	PAOZZ	2590013079303	57958	C5136359	BRACKET,VEHICULAR C PART OF KIT P/N 57K3531............................	1
KIT	PAOZZ	2910014666264	34623	57K3531	PARTS KIT,THROTTLE..................	1

```
BRACKET,VEHICULAR C( 1) 468-6
BRACKET,VEHICULAR C( 1) 468-14
CLAMP,LOOP         ( 1) 468-11
CONTROL ASSEMBLY,PU( 1) 468-2
MARKER,IDENTIFICATI( 1) 468-3
NUT,PLAIN,HEXAGON  ( 2) 468-9
NUT,PLAIN,ASSEMBLED( 1) 468-12
NUT,SELF-LOCKING,HE( 4) 468-4
NUT,SELF-LOCKING,HE( 1) 468-7
SCREW,CAP,HEXAGON H( 2) 468-5
SCREW,CAP,HEXAGON H( 1) 468-13
SCREW,MACHINE      ( 1) 468-10
STRAP,TIEDOWN,ELECT( 1) 468-1
WASHER,LOCK        ( 2) 468-8
```

END OF FIGURE

468-1

Section II.

TM 9-2320-280-24P-2

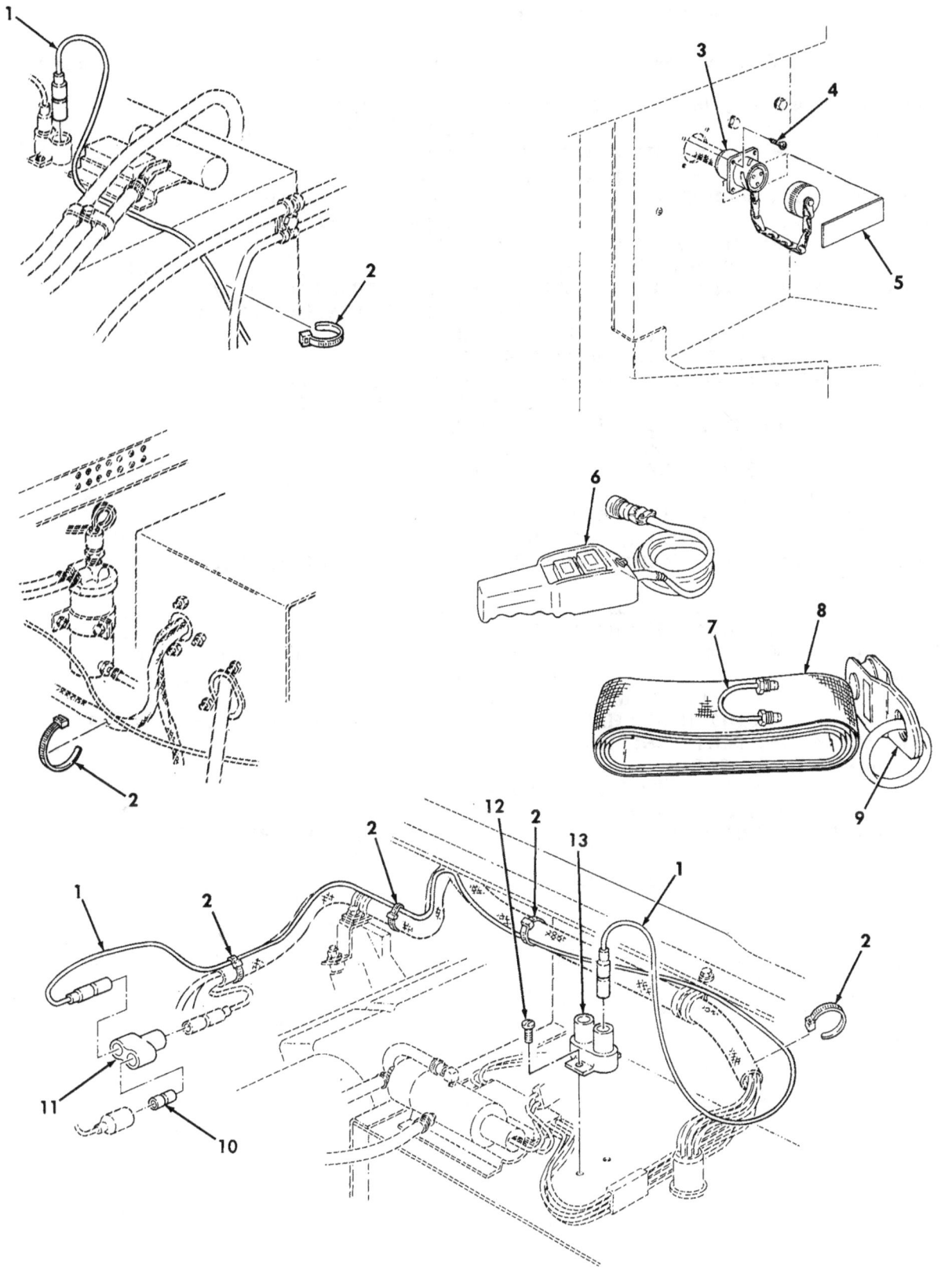

Figure 469. 10,500 Lb Hydraulic Winch Accessory Kit.

SECTION II   TM9-2320-280-24P

(1) NO	(2) SMR CODE	(3) NSN	(4) CAGEC	(5) PART NUMBER	(7) DESCRIPTION AND USABLE ON CODES(UOC)	(6) QTY

GROUP 3307 SPECIAL PURPOSE KITS

FIG. 469 10,500 LB HYDRAULIC WINCH ACCESSORY KIT

1	PAOZZ	6150014731169	19207	12469065-1	LEAD,ELECTRICAL PART OF KIT P/N 57K3541............ UOC:AVY,A13,A14,A24,A25,A26,A27,BVY, B16,B17,B18,B24,B25,C17,HVY,H13,H14, H16,H17,H18,H21,H24,H25,H26,H27,H28, NNN	1
2	PAOZZ	5975009846582	96906	MS3367-1-0	STRAP,TIEDOWN,ELECT PART OF KIT P/N 57K3541............ UOC:AVY,A13,A14,A24,A25,A26,A27,BVY, B16,B17,B18,B24,B25,C17,HVY,H13,H14, H16,H17,H18,H21,H24,H25,H26,H27,H28, NNN	6
3	PAOZZ	6150014731178	19207	12469064	CABLE ASSEMBLY,SPEC PART OF KIT P/N 57K3541............ UOC:AVY,A13,A14,A24,A25,A26,A27,BVY, B16,B17,B18,B24,B25,C17,HVY,H13,H14, H16,H17,H18,H21,H24,H25,H26,H27,H28, NNN	1
4	PAOZZ	5305014740988	24617	9417323	SCREW,TAPPING PART OF KIT P/N 57K3541............ UOC:AVY,A13,A14,A24,A25,A26,A27,BVY, B16,B17,B18,B24,B25,C17,HVY,H13,H14, H16,H17,H18,H21,H24,H25,H26,H27,H28, NNN	4
5	PAOZZ		19207	12469066	DECAL PART OF KIT P/N 57K3541....... UOC:AVY,A13,A14,A24,A25,A26,A27,BVY, B16,B17,B18,B24,B25,C17,HVY,H13,H14, H16,H17,H18,H21,H24,H25,H26,H27,H28, NNN	1
6	PAOZZ	6150014758835	0GZB7	983-74-1001	CONTROL,REEL PART OF KIT P/N 57K3541 UOC:AVY,A13,A14,A24,A25,A26,A27,BVY, B16,B17,B18,B24,B25,C17,HVY,H13,H14, H16,H17,H18,H21,H24,H25,H26,H27,H28, NNN	1
7	PAOZZ	4710014753753	0GZB7	983-90-50201	TUBE ASSEMBLY,METAL PART OF KIT P/N 57K3541............ UOC:AVY,A13,A14,A24,A25,A26,A27,BVY, B16,B17,B18,B24,B25,C17,HVY,H13,H14, H16,H17,H18,H21,H24,H25,H26,H27,H28	1
8	PAOZA	5340014753650	0GZB7	983-60-50089	STRAP,RETAINING PART OF KIT P/N 57K3541............ UOC:AVY,A13,A14,A24,A25,A26,A27,BVY, B16,B17,B18,B24,B25,C17,HVY,H13,H14, H16,H17,H18,H21,H24,H25,H26,H27,H28, NNN	1
9	PAOZZ	3940014754983	0GZB7	983-60-50085-A	BLOCK,TACKLE PART OF KIT P/N 57K3541 UOC:AVY,A13,A14,A24,A25,A26,A27,BVY, B16,B17,B18,B24,B25,C17,HVY,H13,H14,	1

SECTION II TM9-2320-280-24P

(1) NO	(2) SMR CODE	(3) NSN	(4) CAGEC	(5) PART NUMBER	(6) DESCRIPTION AND USABLE ON CODES(UOC)	(7) QTY
10	PAOZZ	5935008074109	19207	8741492	ADAPTER,CONNECTOR PART OF KIT P/N 7K3541............. UOC:AVY,A13,A14,A24,A25,A26,A27,BVY, B16,B17,B18,B24,B25,C17,HVY,H13,H14, H16,H17,H18,H21,H24,H25,H26,H27,H28,NNN	1
11	PAOZZ	5935009006281	96906	MS27147-1	ADAPTER,CONNECTOR PART OF KIT P/N 7K3541............. UOC:AVY,A13,A14,A24,A25,A26,A27,BVY, B16,B17,B18,B24,B25,C17,HVY,H13,H14, H16,H17,H18,H21,H24,H25,H26,H27,H28,NNN	1
12	PAOZZ	5305014783387	24617	449617	SCREW,TAPPING PART OF KIT P/N K3541............. UOC:AVY,A13,A14,A24,A25,A26,A27,BVY, B16,B17,B18,B24,B25,C17,HVY,H13,H14, H16,H17,H18,H21,H24,H25,H26,H27,H28,NNN	2
13	PAOZZ	5925014302318	58536	AA55571/01-001	CIRCUIT BREAKER PART OF KIT P/N 7K3541............. UOC:AVY,A13,A14,A24,A25,A26,A27,BVY, B16,B17,B18,B24,B25,C17,HVY,H13,H14, H16,H17,H18,H21,H24,H25,H26,H27,H28,NNN	1

END OF FIGURE

Section II.

TM 9-2320-280-24P-2

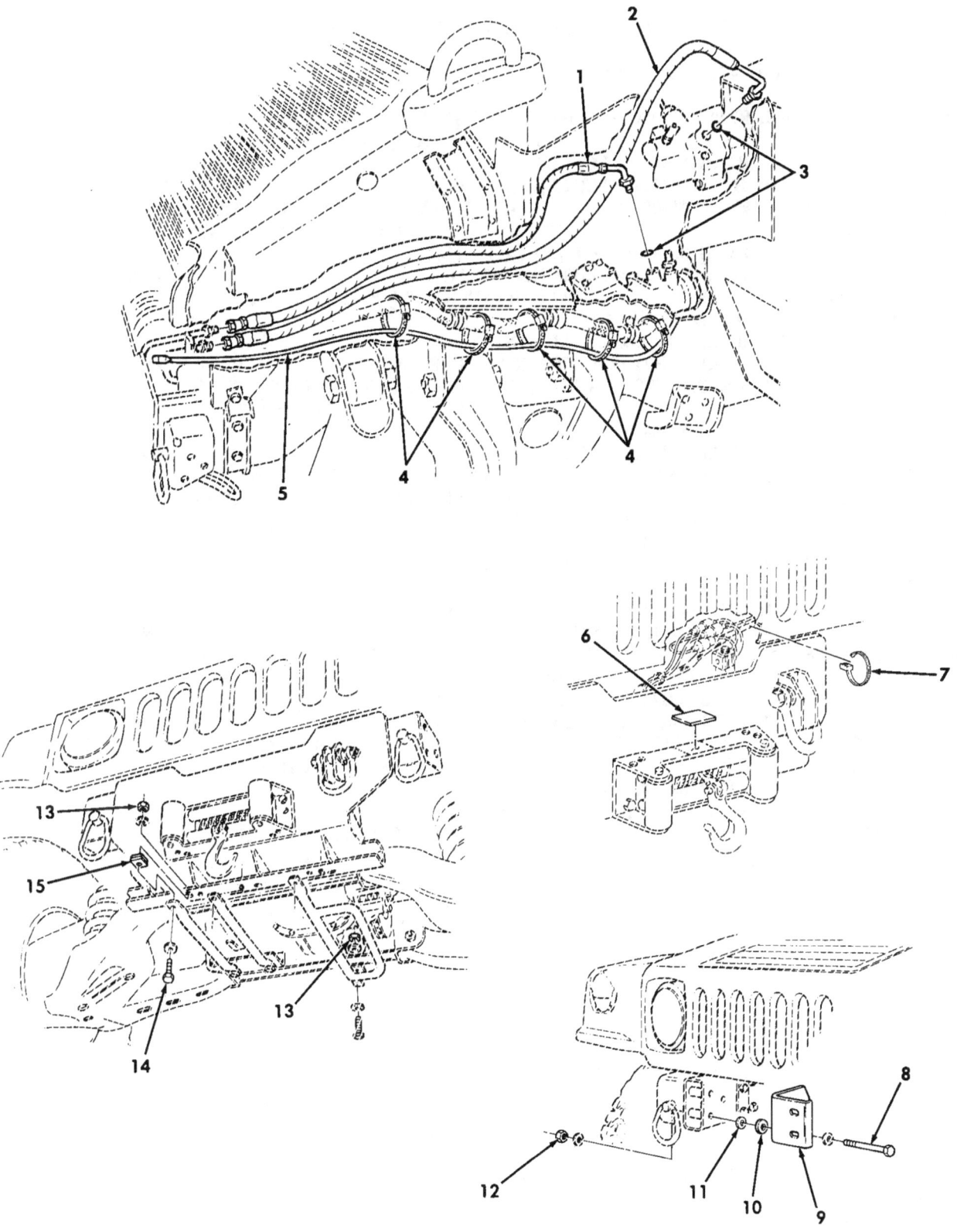

Figure 470. 10,500 Lb Hydraulic Winch Accessory Kit.

SECTION II   TM9-2320-280-24P

(1) ITEM NO	(2) SMR CODE	(3) NSN	(4) CAGEC	(5) PART NUMBER	(6) DESCRIPTION AND USABLE ON CODES(UOC)	(7) QTY
					GROUP 3307 SPECIAL PURPOSE KITS	
					FIG. 470 10,500 LB HYDRAULIC WINCH ACCESSORY KIT	
1	PAOZZ	4720014753757	0GZB7	983-88-00C36	HOSE ASSEMBLY,NONME PART OF KIT P/N 57K3541............ UOC:AVY,A13,A14,A24,A25,A26,A27,BVY, B16,B17,B18,B24,B25,C17,HVY,H13,H14, H16,H17,H18,H21,H24,H25,H26,H27,H28, NNN	1
2	PAOZZ	4720014754641	0GZB7	983-88-00C47	HOSE ASSEMBLY,NONME PART OF KIT P/N 57K3541............ UOC:AVY,A13,A14,A24,A25,A26,A27,BVY, B16,B17,B18,B24,B25,C17,HVY,H13,H14, H16,H17,H18,H21,H24,H25,H26,H27,H28, NNN	1
3	PAOZZ	5331005806586	41387	2938-2	O-RING PART OF KIT P/N 57K3541...... UOC:AVY,A13,A14,A24,A25,A26,A27,BVY, B16,B17,B18,B24,B25,C17,HVY,H13,H14, H16,H17,H18,H21,H24,H25,H26,H27,H28, NNN	2
4	PAOZZ	5975010345871	96906	MS3367-7-0	STRAP,TIEDOWN,ELECT PART OF KIT P/N 57K3541............ UOC:AVY,A13,A14,A24,A25,A26,A27,BVY, B16,B17,B18,B24,B25,C17,HVY,H13,H14, H16,H17,H18,H21,H24,H25,H26,H27,H28, NNN	5
5	PAOZZ	6150014731174	19207	12469065-2	LEAD,ELECTRICAL PART OF KIT P/N 57K3541............ UOC:AVY,A13,A14,A24,A25,A26,A27,BVY, B16,B17,B18,B24,B25,C17,HVY,H13,H14, H16,H17,H18,H21,H24,H25,H26,H27,H28, NNN	1
6	PAOZZ		19207	12469067	DECAL PART OF KIT P/N 57K3541....... UOC:AVY,A13,A14,A24,A25,A26,A27,BVY, B16,B17,B18,B24,B25,C17,HVY,H13,H14, H16,H17,H18,H21,H24,H25,H26,H27,H28, NNN	1
7	PAOZZ	5975009846582	56501	TY25M0	STRAP,TIEDOWN,ELECT PART OF KIT P/N 57K3541............ UOC:AVY,A13,A14,A24,A25,A26,A27,BVY, B16,B17,B18,B24,B25,C17,HVY,H13,H14, H16,H17,H18,H21,H24,H25,H26,H27,H28, NNN	2
8	PAOZZ	5305007247254	80204	B1821BH063C400N	SCREW,CAP,HEXAGON H PART OF KIT P/N 57K3541............ UOC:AVY,A13,A14,A24,A25,A26,A27,BVY, B16,B17,B18,B24,B25,C17,HVY,H13,H14, H16,H17,H18,H21,H24,H25,H26,H27,H28, NNN	6
9	PAOZZ	2540014739040	34623	6001553	BRACKET,R.H. TOW HOOK PART OF KIT P/N 57K3541..................	1

SECTION II   TM9-2320-280-24P

(1) NO	(2) SMR CODE	(3) NSN	(4) CAGEC	(5) PART NUMBER	(7) DESCRIPTION AND USABLE ON CODES(UOC)	ITEM QTY
9	PAOZZ	2540014739045	34623	6001552	BRACKET,L.H. TOW HOOK PART OF KIT P/N 57K3541.......... UOC:AVY,A13,A14,A24,A25,A26,A27,BVY, B16,B17,B18,B24,B25,C17,HVY,H13,H14, H16,H17,H18,H21,H24,H25,H26,H27,H28,	NNN 1
10	PAOZZ	5310011474052	24617	2436168	WASHER,FLAT PART OF KIT P/N 57K3541. UOC:AVY,A13,A14,A24,A25,A26,A27,BVY, B16,B17,B18,B24,B25,C17,HVY,H13,H14, H16,H17,H18,H21,H24,H25,H26,H27,H28,	NNN 2
11	PAOZZ	5310011517347	24617	2436167	WASHER,FLAT PART OF KIT P/N 57K3541. UOC:AVY,A13,A14,A24,A25,A26,A27,BVY, B16,B17,B18,B24,B25,C17,HVY,H13,H14, H16,H17,H18,H21,H24,H25,H26,H27,H28,	NNN 2
12	PAOZZ	5310000614651	19207	12387349-43	NUT,SELF-LOCKING,HE PART OF KIT P/N 57K3541.......... UOC:AVY,A13,A14,A24,A25,A26,A27,BVY, B16,B17,B18,B24,B25,C17,HVY,H13,H14, H16,H17,H18,H21,H24,H25,H26,H27,H28,	NNN 6
13	PAOZZ	5310009359021	96906	MS51943-35	NUT,SELF-LOCKING,HE PART OF KIT P/N 57K3541.......... UOC:AVY,A13,A14,A24,A25,A26,A27,BVY, B16,B17,B18,B24,B25,C17,HVY,H13,H14, H16,H17,H18,H21,H24,H25,H26,H27,H28,	NNN 4
14	PAOZZ	5305008213869	80204	B1821BH038C175N	SCREW,CAP,HEXAGON H PART OF KIT P/N 57K3541.......... UOC:AVY,A13,A14,A24,A25,A26,A27,BVY, B16,B17,B18,B24,B25,C17,HVY,H13,H14, H16,H17,H18,H21,H24,H25,H26,H27,H28,	NNN 3
15	PAOZZ		0GZB7	983-92-52000	SPACER PART OF KIT P/N 57K3541...... UOC:AVY,A13,A14,A24,A25,A26,A27,BVY, B16,B17,B18,B24,B25,C17,HVY,H13,H14, H16,H17,H18,H21,H24,H25,H26,H27,H28,	NNN 3

END OF FIGURE

470-2

Section II.

TM 9-2320-280-24P-2

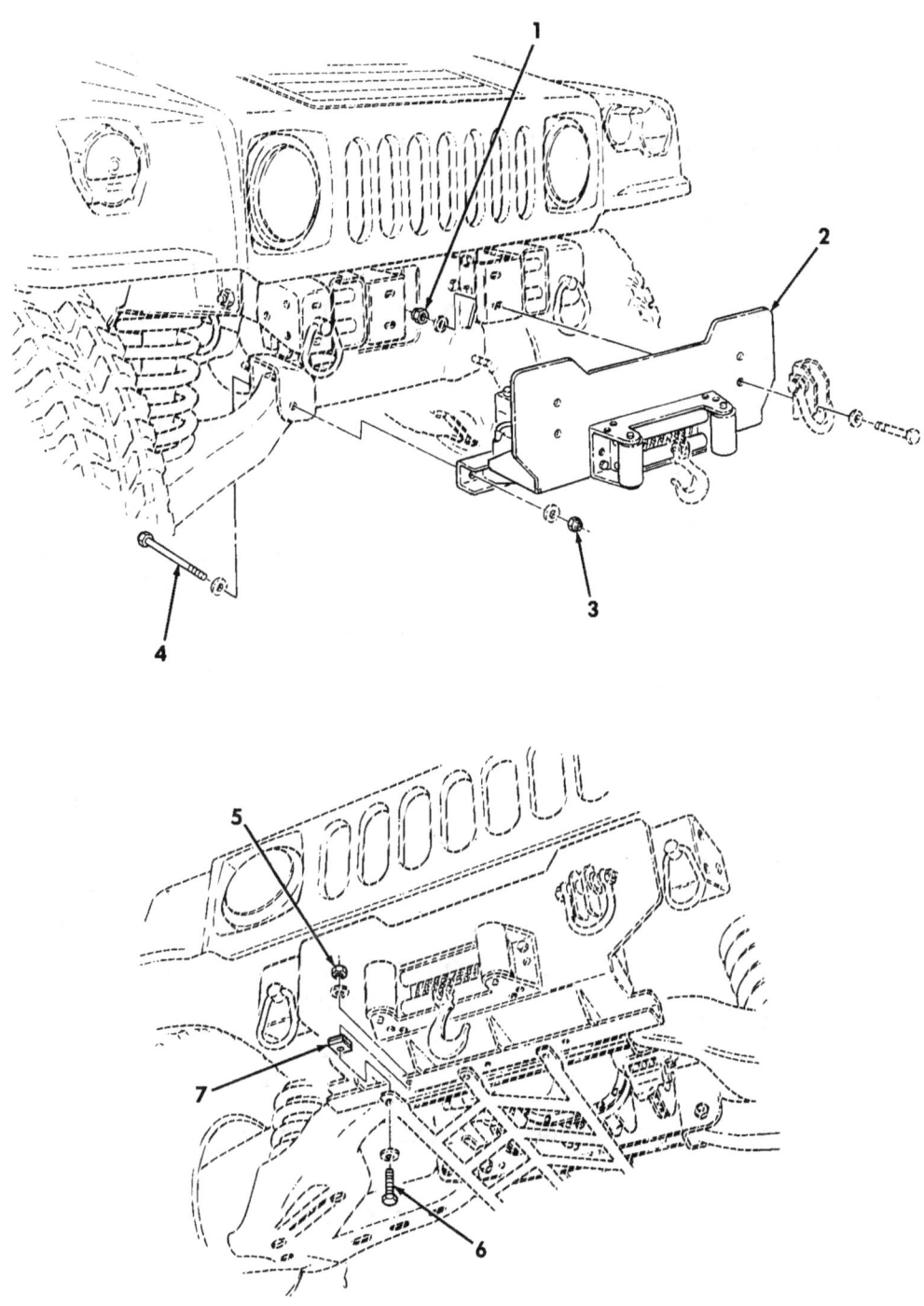

Figure 471. 10,500 Lb Hydraulic Winch Accessory Kit.

SECTION II TM9-2320-280-24P C01

(1) ITEM NO	(2) SMR CODE	(3) NSN	(4) CAGEC	(5) PART NUMBER	(6) DESCRIPTION AND USABLE ON CODES (UOC)	(7) QTY
					GROUP 3307 SPECIAL PURPOSE KITS	
					FIG. 471 10,500 LB HYDRAULIC WINCH ACCESSORY KIT	
1	PAOZZ	5310004883889	96906	MS51943-39	NUT,SELF-LOCKING,HE PART OF KIT P/N 57K3541................................ UOC:AVY,A13,A14,A24,A25,A26,A27,BVY, B16,B17,B18,B24,B25,C17,HVY,H13,H14, H16,H17,H18,H21,H24,H25,H26,H27,H28, NNN	4
* 2	PAOFF	2590014743326	0GZB7	983-75-50050CV	WINCH,DRUM,VEHICULA PART OF KIT P/N 57K3541................................ UOC:AVY,A13,A14,A24,A25,A26,A27,BVY, B16,B17,B18,B24,B25,C17,HVY,H13,H14, H16,H17,H18,H21,H24,H25,H26,H27,H28, NNN	1
3	PAOZZ	5310004093333	96906	MS51943-45	NUT,SELF-LOCKING,HE PART OF KIT P/N 57K3541................................ UOC:AVY,A13,A14,A24,A25,A26,A27,BVY, B16,B17,B18,B24,B25,C17,HVY,H13,H14, H16,H17,H18,H21,H24,H25,H26,H27,H28, NNN	2
4	PAOZZ	5305009474363	80204	B1821BH075C550N	SCREW,CAP,HEXAGON H PART OF KIT P/N 57K3541................................ UOC:AVY,A13,A14,A24,A25,A26,A27,BVY, B16,B17,B18,B24,B25,C17,HVY,H13,H14, H16,H17,H18,H21,H24,H25,H26,H27,H28, NNN	2
5	PAOZZ	5310009359021	96906	MS51943-35	NUT,SELF-LOCKING,HE PART OF KIT P/N 57K3541................................ UOC:AVY,A13,A14,A24,A25,A26,A27,BVY, B16,B17,B18,B24,B25,C17,HVY,H13,H14, H16,H17,H18,H21,H24,H25,H26,H27,H28, NNN	4
6	PAOZZ	5305008213869	80204	B1821BH038C175N	SCREW,CAP,HEXAGON H PART OF KIT P/N 57K3541................................ UOC:AVY,A13,A14,A24,A25,A26,A27,BVY, B16,B17,B18,B24,B25,C17,HVY,H13,H14, H16,H17,H18,H21,H24,H25,H26,H27,H28, NNN	3
* 7	PAOZZ	5365014741057	0GZB7	983-92-52000	SPACER,PLATE PART OF KIT P/N 57K3541................................ UOC:AVY,A13,A14,A24,A25,A26,A27,BVY, B16,B17,B18,B24,B25,C17,HVY,H13,H14, H16,H17,H18,H21,H24,H25,H26,H27,H28, NNN	3
KIT	PAOZZ	2590014567879	19207	57K3541	PARTS KIT,WINCH.................... UOC:AVY,A13,A14,A24,A25,A26,A27,BVY, B16,B17,B18,B24,B25,C17,HVY,H13,H14, H16,H17,H18,H21,H24,H25,H26,H27,H28, NNN	1
					ADAPTER,CONNECTOR (1) 469-10	

SECTION II   TM9-2320-280-24P

(1) NO	(2) CODE	(3) NSN	(4) CAGEC	(5) PART NUMBER	(7) ITEM SMR DESCRIPTION AND USABLE ON CODES(UOC)	QTY
					ADAPTER,CONNECTOR	(1) 469-11
					BLOCK,TACKLE	(1) 469-9
					BRACKET,R.H.TOW HOO	(1) 470-9
					BRACKET,L.H.TOW HOO	(1) 470-9
					CABLE ASSEMBLY,SPEC	(1) 469-3
					CIRCUIT BREAKER	(1) 469-13
					CONTROL,REEL	(1) 469-6
					DECAL	(1) 469-5
					DECAL	(1) 470-6
					HOSE ASSEMBLY,NONME	(1) 470-1
					HOSE ASSEMBLY,NONME	(1) 470-2
					LEAD,ELECTRICAL	(1) 469-1
					LEAD,ELECTRICAL	(1) 470-5
					NUT,SELF-LOCKING,HE	(4) 471-1
					NUT,SELF-LOCKING,HE	(6) 470-12
					NUT,SELF-LOCKING,HE	(2) 471-3
					NUT,SELF-LOCKING,HE	(4) 470-13
					NUT,SELF-LOCKING,HE	(4) 471-5
					O-RING	(2) 470-3
					SCREW,CAP,HEXAGON H	(3) 470-14
					SCREW,CAP,HEXAGON H	(6) 470-8
					SCREW,CAP,HEXAGON H	(2) 471-4
					SCREW,CAP,HEXAGON H	(3) 471-6
					SCREW,TAPPING	(2) 469-12
					SCREW,TAPPING	(4) 469-4
					SPACER	(3) 470-15
					SPACER,PLATE	(3) 471-7
					STRAP,RETAINING	(1) 469-8
					STRAP,TIEDOWN,ELECT	(6) 469-2
					STRAP,TIEDOWN,ELECT	(2) 470-7
					STRAP,TIEDOWN,ELECT	(5) 470-4
					TUBE ASSEMBLY,METAL	(1) 469-7
					WASHER,FLAT	(2) 470-10
					WASHER,FLAT	(2) 470-11
					WINCH,DRUM,VEHICULA	(1) 471-2

END OF FIGURE

Section II.

TM 9-2320-280-24P-2

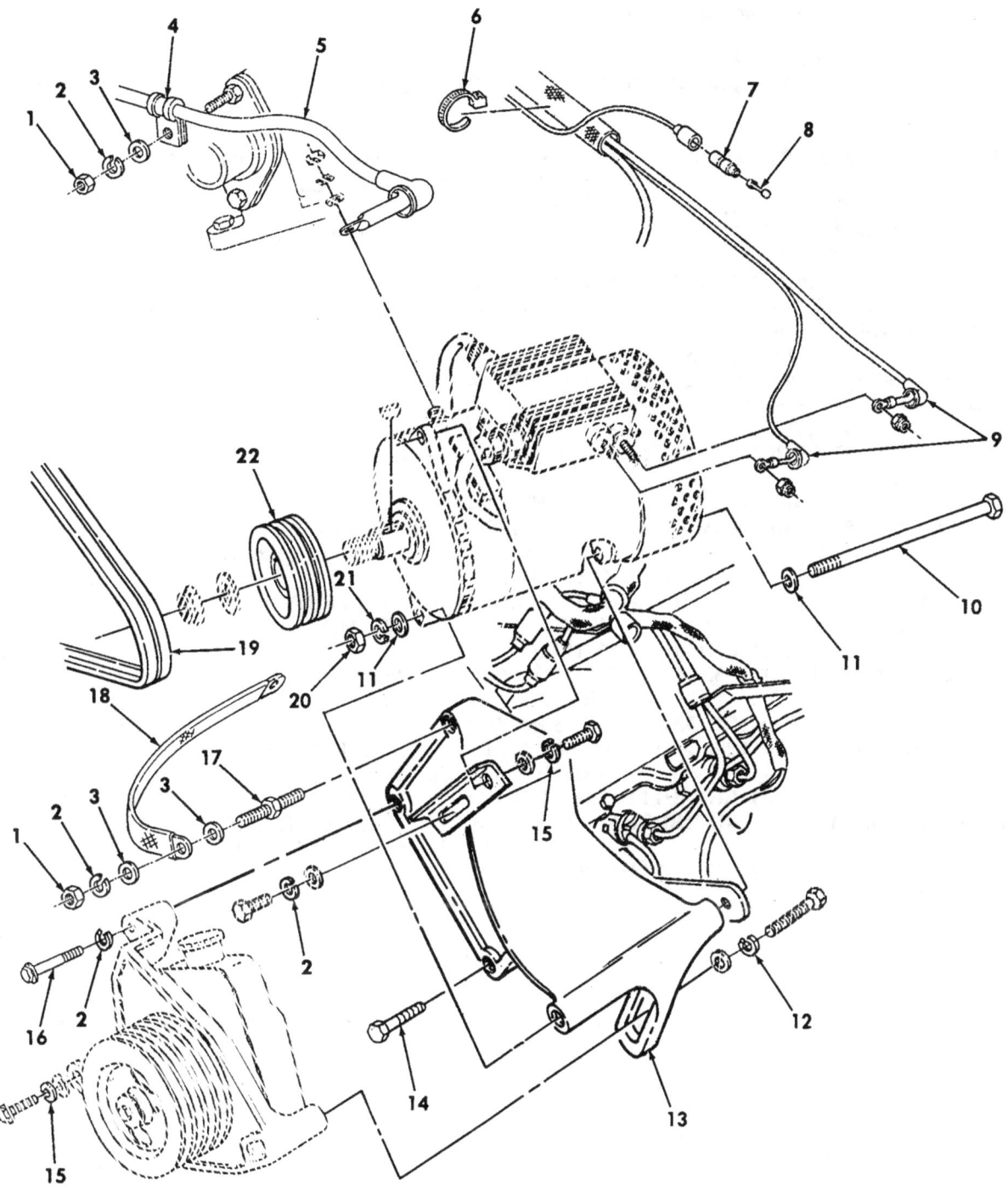

Figure 472. 100 AMP Dual Voltage Alternator and Regulator Mounting Hardware Kit.

SECTION II   TM9-2320-280-24P

(1) NO	(2) SMR CODE	(3) NSN	(4) CAGEC	(5) PART NUMBER	(7) ITEM DESCRIPTION AND USABLE ON CODES(UOC)	QTY

GROUP 3307 SPECIAL PURPOSE KITS

FIG. 472 100 AMP DUAL VOLTAGE ALTERNATOR AND REGULATOR MOUNTING HARDWARE KIT

1	PAOZZ	5310013958747	80204	B18241B100F	NUT,PLAIN,HEXAGON PART OF KIT P/N 57K3519............ UOC:A11,A13,A14,A20,A24,A25,A26,A27, B17,B18,H11,H13,H14,H17,H18,H20,H21, H24,H25,H26,H27,H28,MMM	2
2	PAOZZ	5310012067306	7X677	11500207	WASHER,LOCK PART OF KIT P/N 57K3519. UOC:A11,A13,A14,A20,A24,A25,A26,A27, B17,B18,H11,H13,H14,H17,H18,H20,H21, H24,H25,H26,H27,H28,MMM	4
3	PAOZZ	5310012101587	7X677	11500256	WASHER,FLAT PART OF KIT P/N 57K3519. UOC:A11,A13,A14,A20,A24,A25,A26,A27, B17,B18,H11,H13,H14,H17,H18,H20,H21, H24,H25,H26,H27,H28,MMM	4
4	PAOZZ	5340000538994	96906	MS21333-126	CLAMP,LOOP PART OF KIT P/N 57K3519.. UOC:A11,A13,A14,A20,A24,A25,A26,A27, B17,B18,H11,H13,H14,H17,H18,H20,H21, H24,H25,H26,H27,H28,MMM	3
5	PAOZZ	6150014676716	19207	12446821-2	CABLE ASSEMBLY,SPEC PART OF KIT P/N 57K3519............ UOC:A11,A13,A14,A20,A24,A25,A26,A27, B17,B18,H11,H13,H14,H17,H18,H20,H21, H24,H25,H26,H27,H28,MMM	1
6	PAOZZ	5975010345871	96906	MS3367-7-0	STRAP,TIEDOWN,ELECT PART OF KIT P/N 57K3519............ UOC:A11,A13,A14,A20,A24,A25,A26,A27, B17,B18,H11,H13,H14,H17,H18,H20,H21, H24,H25,H26,H27,H28,MMM	2
7	PAOZZ	5935008338561	19207	8338561	SHELL,ELECTRICAL CO PART OF KIT P/N 57K3519............ UOC:A11,A13,A14,A20,A24,A25,A26,A27, B17,B18,H11,H13,H14,H17,H18,H20,H21, H24,H25,H26,H27,H28,MMM	1
8	PAOZZ	5935002140904	19207	7982907	DUMMY CONNECTOR,PLU PART OF KIT P/N 57K3519............ UOC:A11,A13,A14,A20,A24,A25,A26,A27, B17,B18,H11,H13,H14,H17,H18,H20,H21, H24,H25,H26,H27,H28,MMM	1
9	PAOZZ	5975014695558	77060	0298 4033	CABLE NIPPLE,ELECTR PART OF KIT P/N 57K3519............ UOC:A11,A13,A14,A20,A24,A25,A26,A27, B17,B18,H11,H13,H14,H17,H18,H20,H21, H24,H25,H26,H27,H28,MMM	2
10	PAOZZ	5305014299149	80204	B1821BH044F975N	SCREW,CAP,HEXAGON H PART OF KIT P/N 57K3519............ UOC:A11,A13,A14,A20,A24,A25,A26,A27, B17,B18,H11,H13,H14,H17,H18,H20,H21, H24,H25,H26,H27,H28,MMM	1

SECTION II   TM9-2320-280-24P

(1) NO	(2) SMR CODE	(3) NSN	(4) CAGEC	(5) PART NUMBER	(6) DESCRIPTION AND USABLE ON CODES(UOC)	(7) QTY
11	PAOZZ	5310008094085	24617	9420022	WASHER,FLAT PART OF KIT P/N 57K3519. UOC:A11,A13,A14,A20,A24,A25,A26,A27, B17,B18,H11,H13,H14,H17,H18,H20,H21, H24,H25,H26,H27,H28,M-MM	2
12	PAOZZ	5310000116121	29510	120384	WASHER,LOCK PART OF KIT P/N 57K3519. UOC:A11,A13,A14,A20,A24,A25,A26,A27, B17,B18,H11,H13,H14,H17,H18,H20,H21, H24,H25,H26,H27,H28,M-MM	1
13	PAOZZ	5340014484245	19207	12342036	BRACKET,MOUNTING PART OF KIT P/N 57K3519............................ UOC:A11,A13,A14,A20,A24,A25,A26,A27, B17,B18,H11,H13,H14,H17,H18,H20,H21, H24,H25,H26,H27,H28,M-MM	1
14	PAOZZ	5306012638889	19207	12340845-2	BOLT,MACHINE PART OF KIT P/N 57K3519 UOC:A11,A13,A14,A20,A24,A25,A26,A27, B17,B18,H11,H13,H14,H17,H18,H20,H21, H24,H25,H26,H27,H28,M-MM	1
15	PAOZZ	5310006379541	96906	MS35338-46	WASHER,LOCK PART OF KIT P/N 57K3519. UOC:A11,A13,A14,A20,A24,A25,A26,A27, B17,B18,H11,H13,H14,H17,H18,H20,H21, H24,H25,H26,H27,H28,M-MM	1
16	PAOZZ	5306012643531	19207	12340845-4	BOLT,MACHINE PART OF KIT P/N 57K3519 UOC:A11,A13,A14,A20,A24,A25,A26,A27, B17,B18,H11,H13,H14,H17,H18,H20,H21, H24,H25,H26,H27,H28,M-MM	1
17	PAOZZ	5307014655796	19207	12339406-4	STUD,SHOULDERED PART OF KIT P/N 57K3519............................ UOC:A11,A13,A14,A20,A24,A25,A26,A27, B17,B18,H11,H13,H14,H17,H18,H20,H21, H24,H25,H26,H27,H28,M-MM	1
18	PAOZZ	6150014667528	19207	12446821-6	CABLE ASSEMBLY,SPEC PART OF KIT P/N 57K3519............................ UOC:A11,A13,A14,A20,A24,A25,A26,A27, B17,B18,H11,H13,H14,H17,H18,H20,H21, H24,H25,H26,H27,H28,M-MM	1
19	PAOZZ	3030012826968	34623	12339359-12	BELTS,V,MATCHED SET PART OF KIT P/N 57K3519............................ UOC:A11,A13,A14,A20,A24,A25,A26,A27, B17,B18,H11,H13,H14,H17,H18,H20,H21, H24,H25,H26,H27,H28,M-MM	1
20	PAOZZ	5310014573171	24617	9417028	NUT,PLAIN,HEXAGON PART OF KIT P/N 57K3519............................ UOC:A11,A13,A14,A20,A24,A25,A26,A27, B17,B18,H11,H13,H14,H17,H18,H20,H21, H24,H25,H26,H27,H28,M-MM	1
21	PAOZZ	5310002090965	96906	MS35338-47	WASHER,LOCK PART OF KIT P/N 57K3519. UOC:A11,A13,A14,A20,A24,A25,A26,A27, B17,B18,H11,H13,H14,H17,H18,H20,H21, H24,H25,H26,H27,H28,M-MM	1
22	PAOZZ	3020012048132	19207	12339412	PULLEY,GROOVE PART OF KIT P/N 57K3519............................ UOC:A11,A13,A14,A20,A24,A25,A26,A27, B17,B18,H11,H13,H14,H17,H18,H20,H21,	1

SECTION II   TM9-2320-280-24P

| (1) NO | (2) CODE | (3) NSN | (4) CAGEC | (5) NUMBER | (6) DESCRIPTION AND USABLE ON CODES(UOC) | (7) ITEM QTY | SMR | PART |

H24,H25,H26,H27,H28,MMM

END OF FIGURE

Section II.

TM 9-2320-280-24P-2

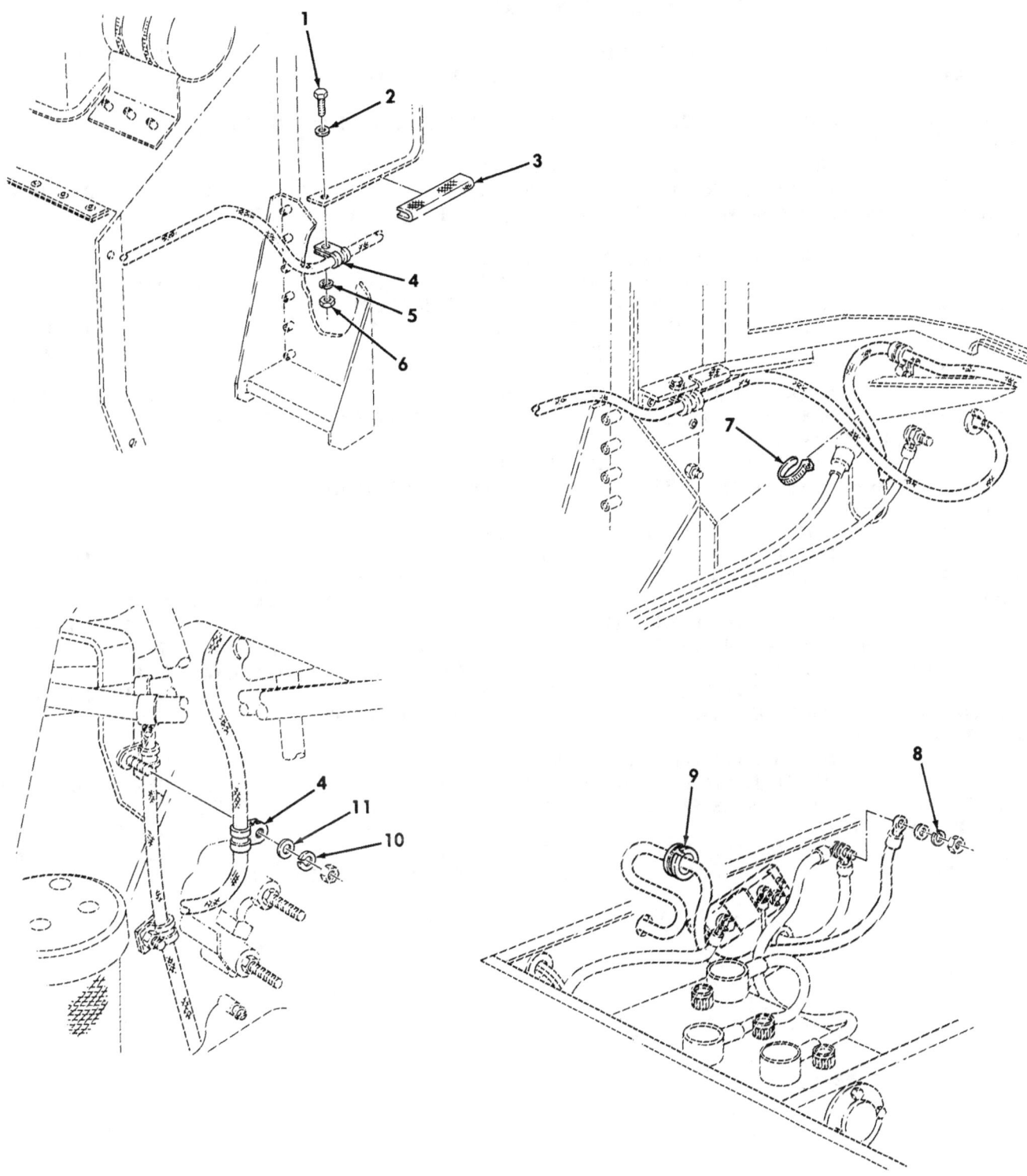

Figure 473. 100/200 AMP Dual Voltage Alternator and Regulator Mounting Hardware Kit.

SECTION II   TM9-2320-280-24P

(1) NO	(2) SMR CODE	(3) NSN	(4) CAGEC	(5) PART NUMBER	(7) DESCRIPTION AND USABLE ON CODES(UOC)	QTY

GROUP 3307 SPECIAL PURPOSE KITS

FIG. 473 100/200 AMP DUAL VOLTAGE ALTERNATOR AND REGULATOR MOUNTING HARDWARE KIT

1 PAOZZ 5306002264827 80204 B1821BH031C100N BOLT,MACHINE PART OF KIT P/N 57K3520 PART OF KIT P/N 57K3519............. 1
  UOC:A11,A13,A14,A20,A24,A25,A26,A27,
  B17,B18,H11,H13,H14,H17,H18,H20,H21,H24,H25,H26,H27,H28,M-MM

2 PAOZZ 5310000814219 96906 MS27183-12 WASHER,FLAT PART OF KIT P/N 57K3520 PART OF KIT P/N 57K3519............. 1
  UOC:A11,A13,A14,A20,A24,A25,A26,A27,
  B17,B18,H11,H13,H14,H17,H18,H20,H21,H24,H25,H26,H27,H28,M-MM

3 MOOZZ 19207 12340103-4 BEZEL,AUTOMOTIVE TR MAKE FROM BEZEL, P/N 12340103,4 INCHES LONG PART OF KIT P/N 57K3519 PART OF KIT P/N 57K3520............. 1
  UOC:A11,A13,A14,A20,A24,A25,A26,A27,
  B17,B18,H11,H13,H14,H17,H18,H20,H21,H24,H25,H26,H27,H28,M-MM

4 PAOZZ 5340007022848 96906 MS21333-128 CLAMP,LOOP PART OF KIT P/N 57K3520.. 2
  UOC:A11,A13,A14,A20,A24,A25,A26,A27,
  B17,B18,H11,H13,H14,H17,H18,H20,H21,H24,H25,H26,H27,H28,M-MM

5 PAOZZ 5310004079566 96906 MS-35338-45 WASHER,LOCK PART OF KIT P/N 57K3520 PART OF KIT P/N 57K3519............. 1
  UOC:A11,A13,A14,A20,A24,A25,A26,A27,
  B17,B18,H11,H13,H14,H17,H18,H20,H21,H24,H25,H26,H27,H28,M-MM

6 PAOZZ 5310009318167 96906 MS51967-6 NUT,PLAIN,HEXAGON PART OF KIT P/N 57K3520 PART OF KIT P/N 57K3519..... 1
  UOC:A11,A13,A14,A20,A24,A25,A26,A27,
  B17,B18,H11,H13,H14,H17,H18,H20,H21,H24,H25,H26,H27,H28,M-MM

7 PAOZZ 5975010345871 96906 MS3367-7-0 STRAP,TIEDOWN,ELECT PART OF KIT P/N 57K3520 PART OF KIT P/N 57K3519..... 1
  UOC:A11,A13,A14,A20,A24,A25,A26,A27,
  B17,B18,H11,H13,H14,H17,H18,H20,H21,H24,H25,H26,H27,H28,M-MM

8 PAOZZ 5310011867066 13445 85031 WASHER,LOCK PART OF KIT P/N 57K3520 PART OF KIT P/N 57K3519............. 1
  UOC:A11,A13,A14,A20,A24,A25,A26,A27,
  B17,B18,H11,H13,H14,H17,H18,H20,H21,H24,H25,H26,H27,H28,M-MM

9 PAOZZ 5325002766343 96906 MS35489-23 GROMMET,NONMETALLIC PART OF KIT P/N 57K3520 PART OF KIT P/N 57K3519..... 1
  UOC:A11,A13,A14,A20,A24,A25,A26,A27,
  B17,B18,H11,H13,H14,H17,H18,H20,H21,H24,H25,H26,H27,H28,M-MM

10 PAOZZ 5310012067306 7X677 11500207 WASHER,LOCK PART OF KIT P/N 57K3520 1

SECTION II TM9-2320-280-24P

(1) NO	(2) CODE	(3) NSN	(4) CAGEC	(5) NUMBER	(6) PART	(7) ITEM DESCRIPTION AND USABLE ON CODES(UOC)	SMR	PART QTY
						PART OF KIT P/N 57K3519............. UOC:A11,A13,A14,A20,A24,A25,A26,A27, B17,B18,H11,H13,H14,H17,H18,H20,H21, H24,H25,H26,H27,H28,MMM		
11	PAOZZ	5310012101587	7X677	11500256		WASHER,FLAT PART OF KIT P/N 57K3520 PART OF KIT P/N 57K3519............. UOC:A11,A13,A14,A20,A24,A25,A26,A27, B17,B18,H11,H13,H14,H17,H18,H20,H21, H24,H25,H26,H27,H28,MMM		1

END OF FIGURE

473-2

Section II. TM 9-2320-280-24P-2

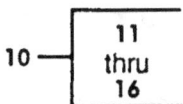

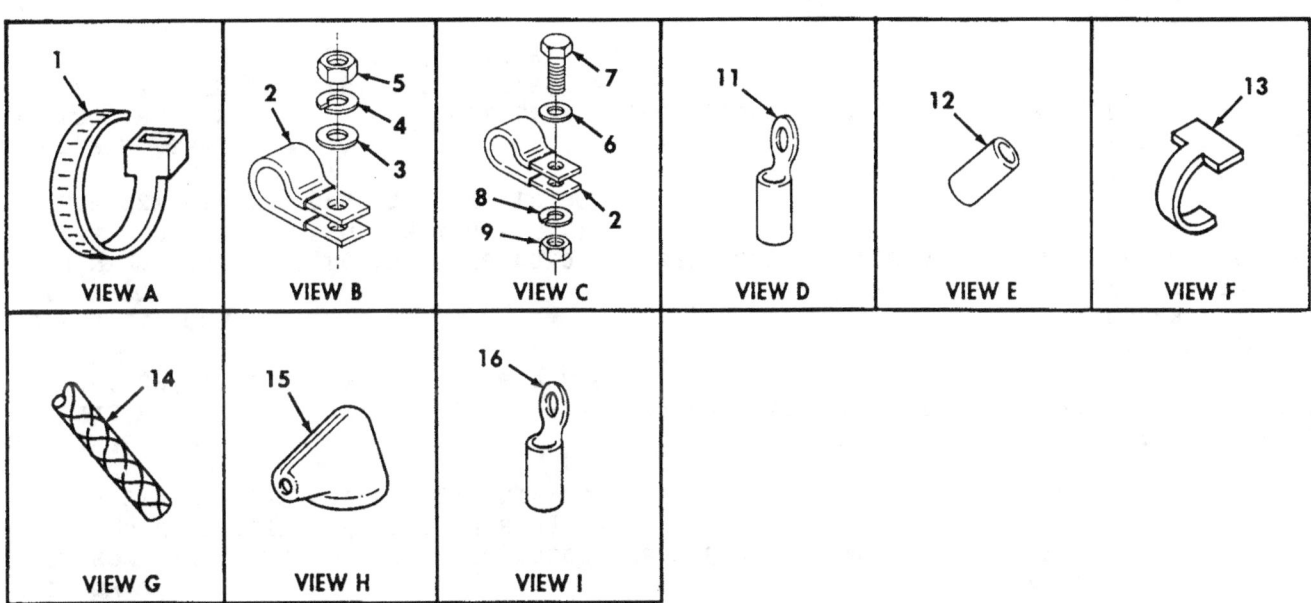

Figure 474. 100/200 AMP Dual Voltage Alternator Wiring Harness.

SECTION II   TM9-2320-280-24P C01

(1) ITEM NO	(2) SMR CODE	(3) NSN	(4) CAGEC	(5) PART NUMBER	(6) DESCRIPTION AND USABLE ON CODES (UOC)	(7) QTY

GROUP 3307 SPECIAL PURPOSE KITS

FIG. 474 100/200 AMP DUAL VOLTAGE ALTERNATOR WIRING HARNESS

1	PAOZZ	5975010345871	96906	MS3367-7-0	STRAP,TIEDOWN,ELECT PART OF KIT P/N 57K3519,PART OF KIT P/N 57K3520..... UOC:A11,A13,A14,A20,A24,A25,A26,A27, B17,B18,H11,H13,H14,H17,H18,H20,H21, H24,H25,H26,H27,H28,MMM	4
2	PAOZZ	5340000538994	96906	MS21333-126	CLAMP,LOOP 100AMP PART OF KIT P/N 57K3519... UOC:A11,A13,A14,A20,A24,A25,A26,A27, B17,B18,H11,H13,H14,H17,H18,H20,H21, H24,H25,H26,H27,H28,MMM	3
2	PAOZZ	5340007022848	96906	MS21333-128	CLAMP,LOOP 200AMP PART OF KIT P/N 57K3520... UOC:A11,A13,A14,A20,A24,A25,A26,A27, B17,B18,H11,H13,H14,H17,H18,H20,H21, H24,H25,H26,H27,H28,MMM	3
3	PAOZZ	5310012101587	7X677	11500256	WASHER,FLAT PART OF KIT P/N 57K3519, PART OF KIT P/N 57K3520............ UOC:A11,A13,A14,A20,A24,A25,A26,A27, B17,B18,H11,H13,H14,H17,H18,H20,H21, H24,H25,H26,H27,H28,MMM	2
4	PAOZZ	5310012067306	7X677	11500207	WASHER,LOCK PART OF KIT P/N 57K3519, PART OF KIT P/N 57K3520............ UOC:A11,A13,A14,A20,A24,A25,A26,A27, B17,B18,H11,H13,H14,H17,H18,H20,H21, H24,H25,H26,H27,H28,MMM	2
5	PAOZZ	5310013958747	80204	B18241B100F	NUT,PLAIN,HEXA- GON PART OF KIT P/N 57K3519,PART OF KIT P/N 57K3520..... UOC:A11,A13,A14,A20,A24,A25,A26,A27, B17,B18,H11,H13,H14,H17,H18,H20,H21, H24,H25,H26,H27,H28,MMM	2
6	PAOZZ	5310000814219	96906	MS27183-12	WASHER,FLAT PART OF KIT P/N 57K3519, PART OF KIT P/N 57K3520............ UOC:A11,A13,A14,A20,A24,A25,A26,A27, B17,B18,H11,H13,H14,H17,H18,H20,H21, H24,H25,H26,H27,H28,MMM	1
7	PAOZZ	5306002264827	80204	B1821BH031C100N	BOLT,MACHINE ,PART OF KIT PART OF KIT P/N 57K3519 P/N 57K3520............ UOC:A11,A13,A14,A20,A24,A25,A26,A27, B17,B18,H11,H13,H14,H17,H18,H20,H21, H24,H25,H26,H27,H28,MMM	1
8	PAOZZ	5310004079566	96906	MS35338-45	WASHER,LOCK PART OF KIT P/N 57K3519, PART OF KIT P/N 57K3520............ UOC:A11,A13,A14,A20,A24,A25,A26,A27, B17,B18,H11,H13,H14,H17,H18,H20,H21, H24,H25,H26,H27,H28,MMM	1
9	PAOZZ	5310009318167	96906	MS51967-6	NUT,PLAIN,HEXA- GON PART OF KIT P/N 57K3519,PART OF KIT P/N 57K3520..... UOC:A11,A13,A14,A20,A24,A25,A26,A27,	1

474-1

SECTION II   TM9-2320-280-24P C01

(1) ITEM NO	(2) SMR CODE	(3) NSN	(4) CAGEC	(5) PART NUMBER	(6) DESCRIPTION AND USABLE ON CODES (UOC)	(7) QTY
10	PAOZZ	6150014173411	19207	12446821-2	CABLE ASSEMBLY,SPEC 100 AMP PART OF KIT P/N 57K3519..................... UOC:A11,A13,A14,A20,A24,A25,A26,A27, B17,B18,H11,H13,H14,H17,H18,H20,H21, H24,H25,H26,H27,H28,MMM	1
10	PAOZZ	6150014167899	19207	12446825	LEAD,ELECTRICAL 200AMP PART OF KIT P/N 57K3520..................... UOC:A11,A13,A14,A20,A24,A25,A26,A27, B17,B18,H11,H13,H14,H17,H18,H20,H21, H24,H25,H26,H27,H28,MMM	1
11	PAOZZ	5940011704956	19207	8689221	.TERMINAL LUG 100AMP PART OF KIT P/N 57K3519..................... UOC:A11,A13,A14,A20,A24,A25,A26,A27, B17,B18,H11,H13,H14,H17,H18,H20,H21, H24,H25,H26,H27,H28,MMM	1
11	PAOZZ	5940007355520	56501	TG33	.TERMINAL,LUG 200AMP PART OF KIT P/N 57K3520..................... UOC:A11,A13,A14,A20,A24,A25,A26,A27, B17,B18,H11,H13,H14,H17,H18,H20,H21, H24,H25,H26,H27,H28,MMM	1
12	MOOZZ		81349	M23053/4-304-2-2	.INSULATION SLEEVING MAKE FROM INSULATION,P/N M23053/4-304-2,2 INCHES LONG,PART OF KIT P/N 57K3519, PART OF KIT P/N 57K3520............. UOC:A11,A13,A14,A20,A24,A25,A26,A27, B17,B18,H11,H13,H14,H17,H18,H20,H21, H24,H25,H26,H27,H28,MMM	2
13	PAOZZ	9905008933570	81349	M43436/1-3	.BAND,MARKER PART OF KIT P/N 57K3519,PART OF KIT P/N 57K3520..... UOC:A11,A13,A14, A20,A24,A25,A26,A27, B17,B18,H11,H13,H14,H17,H18,H20,H21, H24,H25,H26,H27,H28,MMM	2
14	MOOZZ		9D117	SAEJ562	.LOOM NON-METALLIC MAKE FROM TUBING, AE208-10,LENGTH TO SUIT,PART OF P/N KIT P/N 57K3519,PART OF KIT P/N 57K3520......... .... UOC:A11,A13,A14,A20 ,A24,A25,A26,A27, B17,B18,H11,H13,H14,H17,H18,H20,H21, H24,H25,H26,H27,H28,MMM	1
15	PAOZZ	5340014496854	20722	9609A383-002	.BOOT,DUST AND MOIST 100 AMP PART OF KIT P/N 57K3519.................... UOC:A11,A13,A14,A20,A24,A25,A26,A27, B17,B18,H11,H13,H14,H17,H18,H20,H21, H24,H25,H26,H27,H28,MMM	1
15	PAOZZ	5975012089618	19207	12339710	.CABLE NIPPLE,ELECTR 200AMP PART OF KIT P/N 57K3520.................... UOC:A11,A13,A14,A20,A24,A25,A26,A27, B17,B18,H11,H13,H14,H17,H18,H20,H21, H24,H25,H26,H27,H28,MMM	1
16	PAOZZ	5940001152678	96906	MS20659-111	.TERMINAL,LUG 100AMP PART OF KIT P/N 57K3519.....................	1

```
           SECTION II             TM9-2320-280-24P C01
      (1)     (2)      (3)           (4)         (5)                      (6)
(7)   ITEM    SMR                                PART
      NO     CODE      NSN          CAGEC       NUMBER     DESCRIPTION AND USABLE ON CODES(UOC)
QTY
                                                           UOC:A11,A13,A14,A20,A24,A25,A26,A27,
                                                           B17,B18,H11,H13,H14,H17,H18,H20,H21,
H24,H25,H26,H27,H28,MMM        16 PAOZZ 5940001155006 96906 MS25036-133     .TERMINAL,LUG
200AMP PART OF KIT P/N 1                                                57K3520.......
......................                                     UOC:A11,A13,A14,A
20,A24,A25,A26,A27,

H24,H25,H26,H27,H28,MMM                                    B17,B18,H11,H13,H14,H17,H18,H20,H21,
         KIT PAOZZ 2920014551626 19207 57K3519             PARTS KIT ENGINE.GE................   1
                                                           UOC:A11,A13,A14,A20,A24,A25,A26,A27,
                                                           B17,B18,H11,H13,H14,H17,H18,H20,H21,
H24,H25,H26,H27,H28,MMM
                                                              BAND,MARKER         (  2) 474-13
                                                              BELTS,V,MATCHED SET(  1) 472-19
                                                              BEZEL,AUTOMOTIVE TR(  1) 473-3
                                                              BOLT,MACHINE        (  1) 472-14
                                                              BOLT,MACHINE        (  1) 472-16
                                                              BOLT,MACHINE        (  1) 473-1
                                                              BOLT,MACHINE        (  1) 474-7
                                                              BOOT,DUST AND MOIST(  1) 474-15
                                                              BRACKET,MOUNTING    (  1) 472-13
                                                              CABLE ASSEMBLY,SPEC(  1) 472-5
                                                              CABLE ASSEMBLY,SPEC(  1) 472-18
                                                              CABLE ASSEMBLY,SPEC(  1) 474-10
                                                              CABLE NIPPLE,ELECTR(  2) 472-9
                                                              CLAMP,LOOP          (  3) 472-4
                                                              CLAMP,LOOP          (  3) 474-2
                                                              DUMMY CONNECTOR,PLU(  1) 472-8
                                                              GROMMET,NONMETALLIC(  1) 473-9
                                                              INSULATION SLEEVING(  2) 474-12
                                                              LOOM NONMETALLIC    (  1) 474-14
                                                              NUT,PLAIN,HEXAGON   (  2) 472-1
                                                              NUT,PLAIN,HEXAGON   (  1) 472-20
                                                              NUT,PLAIN,HEXAGON   (  1) 473-6
                                                              NUT,PLAIN,HEXAGON   (  2) 474-5
                                                              NUT,PLAIN,HEXAGON   (  1) 474-9
                                                              PULLEY,GROOVE       (  1) 472-22
                                                              SCREW,CAP,HEXAGON H(  1) 472-10
                                                              SHELL,ELECTRICAL CO(  1) 472-7
                                                              STRAP,TIEDOWN,ELECT(  2) 472-6
                                                              STRAP,TIEDOWN,ELECT(  1) 473-7
                                                              STRAP,TIEDOWN,ELECT(  4) 474-1
                                                              STUD,SHOULDERED     (  1) 472-17
                                                              TERMINAL,LUG        (  1) 474-11
                                                              TERMINAL,LUG        (  1) 474-16
                                                              WASHER,FLAT         (  4) 472-3
                                                              WASHER,FLAT         (  2) 472-11
                                                              WASHER,FLAT         (  1) 473-2
                                                              WASHER,FLAT         (  1) 473-11
                                                              WASHER,FLAT         (  2) 474-3
                                                              WASHER,FLAT         (  1) 474-6
                                                              WASHER,LOCK         (  4) 472-2
                                                              WASHER,LOCK         (  1) 472-12
```

SECTION II   TM9-2320-280-24P

(1) NO	(2) CODE	(3) NSN	(4) CAGEC	(5) PART NUMBER	(6) DESCRIPTION AND USABLE ON CODES(UOC)	(7) ITEM SMR QTY
					WASHER,LOCK	(1) 472-15
					WASHER,LOCK	(1) 472-21
					WASHER,LOCK	(1) 473-5
					WASHER,LOCK	(1) 473-8
					WASHER,LOCK	(1) 473-10
					WASHER,LOCK	(2) 474-4
					WASHER,LOCK	(1) 474-8

END OF FIGURE

Section II. TM 9-2320-280-24P-2

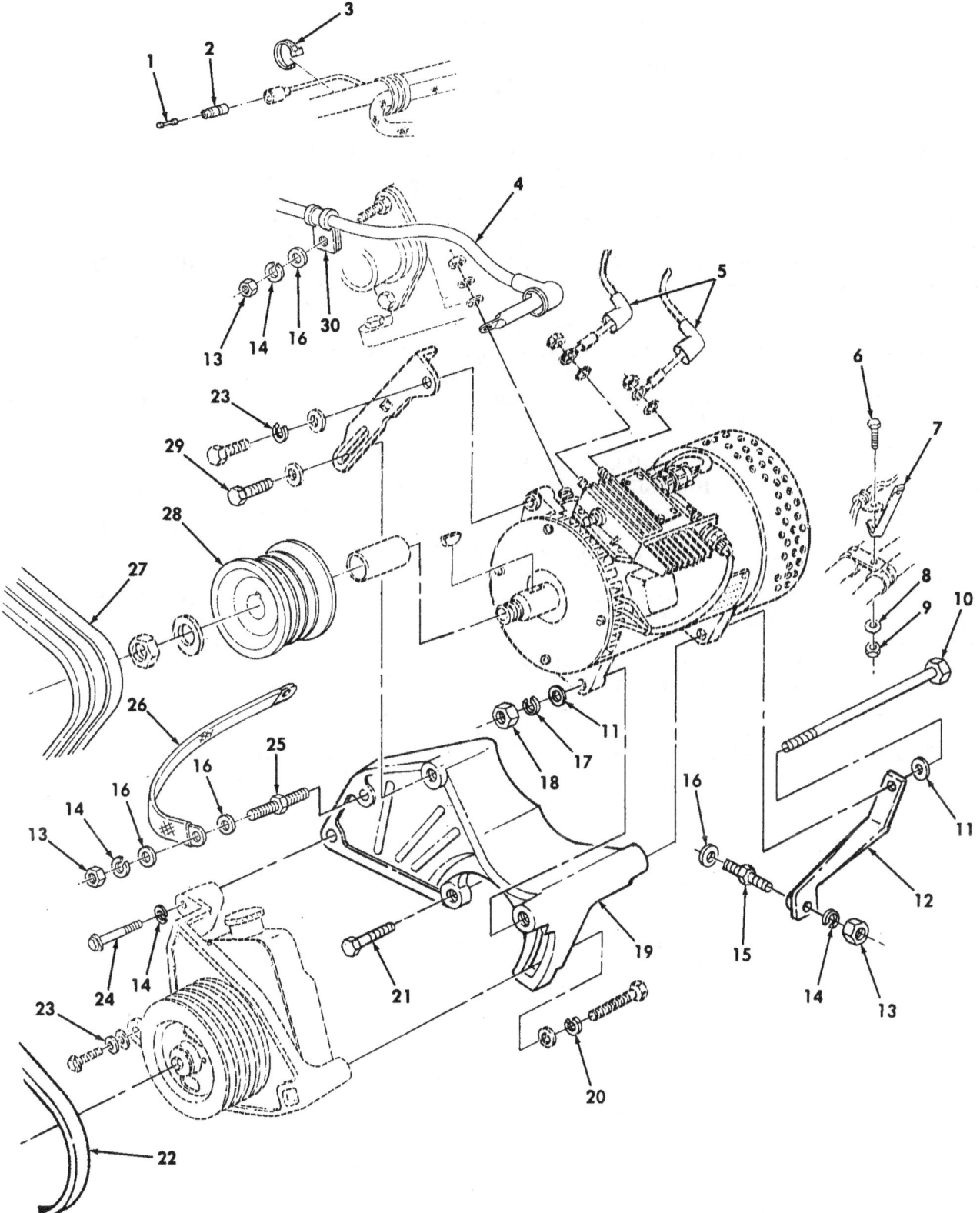

Figure 475. 200 AMP Dual Voltage Alternator and Regulator Mounting Hardware Kit.

SECTION II   TM9-2320-280-24P

(1) NO	(2) SMR CODE	(3) NSN	(4) CAGEC	(5) PART NUMBER	(6) DESCRIPTION AND USABLE ON CODES(UOC)	(7) QTY

GROUP 3307 SPECIAL PURPOSE KITS

FIG. 475 200 AMP DUAL VOLTAGE ALTERNATOR AND REGULATOR MOUNTING HARDWARE KIT

1	PAOZZ	5935002140904	19207	7982907	DUMMY CONNECTOR,PLU PART OF KIT P/N 57K3520............................ UOC:A11,A13,A14,A20,A24,A25,A26,A27, B17,B18,H11,H13,H14,H17,H18,H20,H21, H24,H25,H26,H27,H28,MMM	1
2	PAOZZ	5935008338561	19207	8338561	SHELL,ELECTRICAL CO PART OF KIT P/N 57K3520............................ UOC:A11,A13,A14,A20,A24,A25,A26,A27, B17,B18,H11,H13,H14,H17,H18,H20,H21, H24,H25,H26,H27,H28,MMM	1
3	PAOZZ	5975010345871	96906	MS3367-7-0	STRAP,TIEDOWN,ELECT PART OF KIT P/N 57K3520............................ UOC:A11,A13,A14,A20,A24,A25,A26,A27, B17,B18,H11,H13,H14,H17,H18,H20,H21, H24,H25,H26,H27,H28,MMM	1
4	PAOZZ	6150014167899	19207	12446825	LEAD,ELECTRICAL PART OF KIT P/N 57K3520............................ UOC:A11,A13,A14,A20,A24,A25,A26,A27, B17,B18,H11,H13,H14,H17,H18,H20,H21, H24,H25,H26,H27,H28,MMM	1
5	PAOZZ	5975014695558	77060	0298 4033	CABLE NIPPLE,ELECTR PART OF KIT P/N 57K3520............................ UOC:A11,A13,A14,A20,A24,A25,A26,A27, B17,B18,H11,H13,H14,H17,H18,H20,H21, H24,H25,H26,H27,H28,MMM	2
6	PAOZZ	5305002253843	80204	B1821BH025C100N	SCREW,CAP,HEXAGON H PART OF KIT P/N 57K3520............................ UOC:A11,A13,A14,A20,A24,A25,A26,A27, B17,B18,H11,H13,H14,H17,H18,H20,H21, H24,H25,H26,H27,H28,MMM	1
7	PAOZZ	5340012925402	19207	12342076	BRACKET,DOUBLE ANGL PART OF KIT P/N 57K3520............................ UOC:A11,A13,A14,A20,A24,A25,A26,A27, B17,B18,H11,H13,H14,H17,H18,H20,H21, H24,H25,H26,H27,H28,MMM	1
8	PAOZZ	5310008094058	96906	MS27183-10	WASHER,FLAT PART OF KIT P/N 57K3520. UOC:A11,A13,A14,A20,A24,A25,A26,A27, B17,B18,H11,H13,H14,H17,H18,H20,H21, H24,H25,H26,H27,H28,MMM	1
9	PAOZZ	5310000614650	96906	MS51943-31	NUT,SELF-LOCKING,HE PART OF KIT P/N 57K3520............................ UOC:A11,A13,A14,A20,A24,A25,A26,A27, B17,B18,H11,H13,H14,H17,H18,H20,H21, H24,H25,H26,H27,H28,MMM	1
10	PAOZZ	5305014299149	80204	B1821BH044F975N	SCREW,CAP,HEXAGON H PART OF KIT P/N 57K3520............................ UOC:A11,A13,A14,A20,A24,A25,A26,A27,	1

SECTION II   TM9-2320-280-24P

(1) ITEM NO	(2) SMR CODE	(3) NSN	(4) CAGEC	(5) PART NUMBER	(6) DESCRIPTION AND USABLE ON CODES(UOC)	(7) QTY
11	PAOZZ	5310008094085	96906	MS27183-16	WASHER,FLAT PART OF KIT P/N 57K3520. UOC:A11,A13,A14,A20,A24,A25,A26,A27,B17,B18,H11,H13,H14,H17,H18,H20,H21,H24,H25,H26,H27,H28,M-	2
12	PAOZZ	5340012928404	34623	5935092	BRACKET,MOUNTING PART OF KIT P/N 57K3520. UOC:A11,A13,A14,A20,A24,A25,A26,A27,B17,B18,H11,H13,H14,H17,H18,H20,H21,H24,H25,H26,H27,H28,M-	1
13	PAOZZ	5310013958747	80204	B18241B100F	NUT,PLAIN,HEXAGON PART OF KIT P/N 57K3520. UOC:A11,A13,A14,A20,A24,A25,A26,A27,B17,B18,H11,H13,H14,H17,H18,H20,H21,H24,H25,H26,H27,H28,M-	3
14	PAOZZ	5310012067306	7X677	11500207	WASHER,LOCK PART OF KIT P/N 57K3520. UOC:A11,A13,A14,A20,A24,A25,A26,A27,B17,B18,H11,H13,H14,H17,H18,H20,H21,H24,H25,H26,H27,H28,M-	4
15	PAOZZ	5307013153597	19207	12339406-3	STUD,SHOULDERED PART OF KIT P/N 57K3520. UOC:A11,A13,A14,A20,A24,A25,A26,A27,B17,B18,H11,H13,H14,H17,H18,H20,H21,H24,H25,H26,H27,H28,M-	1
16	PAOZZ	5310012101587	7X677	11500256	WASHER,FLAT PART OF KIT P/N 57K3520. UOC:A11,A13,A14,A20,A24,A25,A26,A27,B17,B18,H11,H13,H14,H17,H18,H20,H21,H24,H25,H26,H27,H28,M-	4
17	PAOZZ	5310002090965	96906	MS35338-47	WASHER,LOCK PART OF KIT P/N 57K3520. UOC:A11,A13,A14,A20,A24,A25,A26,A27,B17,B18,H11,H13,H14,H17,H18,H20,H21,H24,H25,H26,H27,H28,M-	1
18	PAOZZ	5310014573171	24617	9417028	NUT,PLAIN,HEXAGON PART OF KIT P/N 57K3520. UOC:A11,A13,A14,A20,A24,A25,A26,A27,B17,B18,H11,H13,H14,H17,H18,H20,H21,H24,H25,H26,H27,H28,M-	1
19	PAOZZ	2590014444365	34623	12338786-1	BRACKET,VEHICULAR C PART OF KIT P/N 57K3520. UOC:A11,A13,A14,A20,A24,A25,A26,A27,B17,B18,H11,H13,H14,H17,H18,H20,H21,H24,H25,H26,H27,H28,M-	1
20	PAOZZ	5310000116121	29510	120384	WASHER,LOCK PART OF KIT P/N 57K3520. UOC:A11,A13,A14,A20,A24,A25,A26,A27,B17,B18,H11,H13,H14,H17,H18,H20,H21,H24,H25,H26,H27,H28,M-	1
21	PAOZZ	5306012638889	19207	12340845-2	BOLT,MACHINE PART OF KIT P/N 57K3520 UOC:A11,A13,A14,A20,A24,A25,A26,A27,B17,B18,H11,H13,H14,H17,H18,H20,H21,H24,H25,H26,H27,H28,M-	1
22	PAOZZ	3030014526591	19207	12339359-14	BELTING,V PART OF KIT P/N 57K3520... UOC:A11,A13,A14,A20,A24,A25,A26,A27,B17,B18,H11,H13,H14,H17,H18,H20,H21,H24,H25,H26,H27,H28,M-	1

SECTION II TM9-2320-280-24P

(1) NO	(2) SMR CODE	(3) NSN	(4) CAGEC	(5) PART NUMBER	(7) ITEM DESCRIPTION AND USABLE ON CODES(UOC)	QTY
MMM 23	PAOZZ	5310006379541	96906	MS35338-46	WASHER,LOCK PART OF KIT P/N 57K3520. UOC:A11,A13,A14,A20,A24,A25,A26,A27, B17,B18,H11,H13,H14,H17,H18,H20,H21, H24,H25,H26,H27,H28,MMM	2
MMM 24	PAOZZ	5306012643531	19207	12340845-4	BOLT,MACHINE PART OF KIT P/N 57K3520 UOC:A11,A13,A14,A20,A24,A25,A26,A27, B17,B18,H11,H13,H14,H17,H18,H20,H21, H24,H25,H26,H27,H28,MMM	1
MMM 25	PAOZZ	5307014084029	19207	12342545	STUD,SHOULDERED PART OF KIT P/N 57K3520............. UOC:A11,A13,A14,A20,A24,A25,A26,A27, B17,B18,H11,H13,H14,H17,H18,H20,H21, H24,H25,H26,H27,H28,MMM	1
MMM 26	PAOZZ	5999013725601	19207	12341151	STRIP,ELECTRICAL GR PART OF KIT P/N 57K3520............. UOC:A11,A13,A14,A20,A24,A25,A26,A27, B17,B18,H11,H13,H14,H17,H18,H20,H21, H24,H25,H26,H27,H28,MMM	1
MMM 27	PAOZZ	3030013214482	19207	12339359-18	BELTS,V,MATCHED SET PART OF KIT P/N 57K3520............. UOC:A11,A13,A14,A20,A24,A25,A26,A27, B17,B18,H11,H13,H14,H17,H18,H20,H21, H24,H25,H26,H27,H28,MMM	1
MMM 28	PAOZZ	3020012329629	19207	12339392	PULLEY,GROOVE PART OF KIT P/N 57K3520............. UOC:A11,A13,A14,A20,A24,A25,A26,A27, B17,B18,H11,H13,H14,H17,H18,H20,H21, H24,H25,H26,H27,H28,MMM	1
29	PAOZZ	5306011140963	80204	B1821BH038C100L	BOLT,MACHINE....................... UOC:A11,A13,A14,A20,A24,A25,A26,A27, B17,B18,H11,H13,H14,H17,H18,H20,H21, H24,H25,H26,H27,H28,MMM	1
MMM 30	PAOZZ	5340007022848	96906	MS21333-128	CLAMP,LOOP PART OF KIT P/N 57K3520.. UOC:A11,A13,A14,A20,A24,A25,A26,A27, B17,B18,H11,H13,H14,H17,H18,H20,H21, H24,H25,H26,H27,H28,MMM	2
KIT	PAOZZ	2920014551630	19207	57K3520	PARTS KIT,ENGINE GE 200-AMP DUAL VOLTAGE ALTERNATOR................. UOC:A11,A13,A14,A20,A24,A25,A26,A27, B17,B18,H11,H13,H14,H17,H18,H20,H21, H24,H25,H26,H27,H28,MMM	1

```
BAND,MARKER          ( 2) 474-13
BELTING,V            ( 1) 475-22
BELTS,V,MATCHED SET  ( 1) 475-27
BEZEL,AUTOMOTIVE TR  ( 1) 473-3
BOLT,MACHINE         ( 1) 473-1
BOLT,MACHINE         ( 1) 474-7
BOLT,MACHINE         ( 1) 475-21
BOLT,MACHINE         ( 1) 475-24
BRACKET,DOUBLE ANGL  ( 1) 475-7
BRACKET,MOUNTING     ( 1) 475-12
```

SECTION II   TM9-2320-280-24P C01
 (1)   (2)   (3)   (4)   (5)   (6)
(7)   ITEM   SMR   PART
      NO   CODE   NSN   CAGEC   NUMBER   DESCRIPTION AND USABLE ON CODES(UOC)
QTY

                                          BRACKET,VEHICULAR C(1) 475-19
                                          CABLE NIPPLE,ELECTR(1) 474-15
                                          CABLE NIPPLE,ELECTR(2) 475-5
                                          CLAMP,LOOP   (2) 473-4
                                          CLAMP,LOOP   (3) 474-2
                                          CLAMP,LOOP   (2) 475-30
                                          DUMMY CONNECTOR,PLU(1) 475-1
                                          GROMMET,NONMETALLIC(1) 473-9
                                          INSULATION SLEEVING(2) 474-12
                                          LEAD,ELECTRICAL   (1) 474-10
                                          LEAD,ELECTRICAL   (1) 475-4
                                          LOOM NONMETALLIC   (1) 474-14
                                          NUT,PLAIN,HEXAGON (1) 473-6
                                          NUT,PLAIN,HEXAGON (2) 474-5
                                          NUT,PLAIN,HEXAGON (1) 474-9
                                          NUT,PLAIN,HEXAGON (1) 475-18
                                          NUT,PLAIN,HEXAGON (3) 475-13
                                          NUT,SELF-LOCKING,HE(1) 475-9
                                          PULLEY,GROOVE   (1) 475-28
                                          SCREW,CAP,HEXAGON H(1) 475-6
                                          SCREW,CAP,HEXAGON H(1) 475-10
                                          SHELL,ELECTRICAL CO(1) 475-2
                                          STRAP,TIEDOWN,ELECT(1) 473-7
                                          STRAP,TIEDOWN,ELECT(4) 474-1
                                          STRAP,TIEDOWN,ELECT(1) 475-3
                                          STRIP,ELECTRICAL GR(1) 475-26
                                          STUD,SHOULDERED   (1) 475-15
                                          STUD,SHOULDERED   (1) 475-25
                                          TERMINAL,LUG   (1) 474-11
                                          TERMINAL,LUG   (1) 474-16
                                          WASHER,FLAT   (1) 473-2
                                          WASHER,FLAT   (1) 473-11
                                          WASHER,FLAT   (2) 474-3
                                          WASHER,FLAT   (1) 474-6
                                          WASHER,FLAT   (1) 475-8
                                          WASHER,FLAT   (2) 475-11
                                          WASHER,FLAT   (4) 475-16
                                          WASHER,LOCK   (1) 473-5
                                          WASHER,LOCK   (1) 473-8
                                          WASHER,LOCK   (1) 473-10
                                          WASHER,LOCK   (2) 474-4
                                          WASHER,LOCK   (1) 474-8
                                          WASHER,LOCK   (1) 475-17
                                          WASHER,LOCK   (4) 475-14
                                          WASHER,LOCK   (1) 475-20
                                          WASHER,LOCK   (2) 475-23

                                 END OF FIGURE

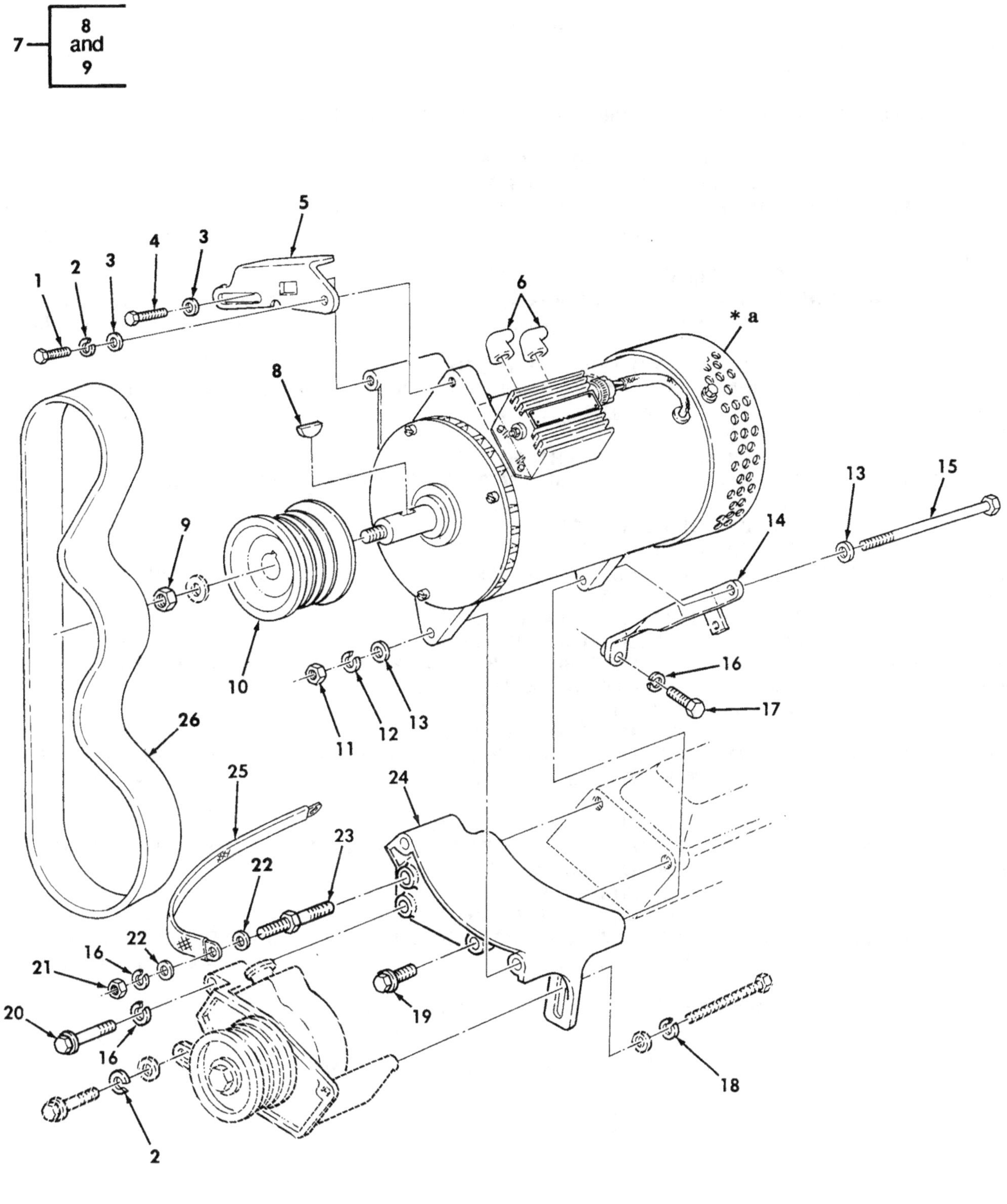

Section II.	TM 9-2320-280-24P-2

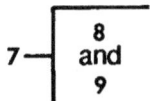

* a PART OF ITEM 7

Figure 476. 400 AMP Alternator Kit.

SECTION II TM9-2320-280-24P

(1) ITEM NO	(2) SMR CODE	(3) NSN	(4) CAGEC	(5) PART NUMBER	(6) DESCRIPTION AND USABLE ON CODES (UOC)	(7) QTY

GROUP 3307 SPECIAL PURPOSE KITS

FIG. 476 400 AMP ALTERNATOR KIT

(1)	(2)	(3)	(4)	(5)	(6)	(7)
1	PAOZZ	5305000680510	80204	B1821BH038C100N	SCREW,CAP,HEXAGON H 3/8-16X1.00 PART OF KIT P/N 57K0273............. UOC:AVY,A11,A13,A14,A20,A24,A25,A26,A27,B17,B18,HVY,H11,H13,H14,H17,H18,H20,H21,H24,H25,H26,H27,H28,MMM	1
2	PAOZZ	5310006379541	96906	MS35338-46	WASHER,LOCK 3/8 PART OF KIT P/N 57K0273............. UOC:AVY,A11,A13,A14,A20,A24,A25,A26,A27,B17,B18,HVY,H11,H13,H14,H17,H18,H20,H21,H24,H25,H26,H27,H28,MMM	2
3	PAOZZ	5310014124013	24617	2436163	WASHER,FLAT 3/8 PART OF KIT P/N 57K0273............. UOC:AVY,A11,A13,A14,A20,A24,A25,A26,A27,B17,B18,HVY,H11,H13,H14,H17,H18,H20,H21,H24,H25,H26,H27,H28,MMM	2
4	PAOZZ	5306011140963	80204	B1821BH038C100L	BOLT,MACHINE 3/8 PART OF KIT P/N 57K0273............. UOC:AVY,A11,A13,A14,A20,A24,A25,A26,A27,B17,B18,HVY,H11,H13,H14,H17,H18,H20,H21,H24,H25,H26,H27,H28,MMM	1
5	PAOZA	5342014666276	19207	12446758	BRACKET,ENGINE ACCE 3/8 PART OF KIT P/N 57K0273............. UOC:AVY,A11,A13,A14,A20,A24,A25,A26,A27,B17,B18,HVY,H11,H13,H14,H17,H18,H20,H21,H24,H25,H26,H27,H28,MMM	1
6	PAOZZ	5975014695558	77060	0298 4033	CABLE NIPPLE,ELECTR 3/8 PART OF KIT P/N 57K0273............. UOC:AVY,A11,A13,A14,A20,A24,A25,A26,A27,B17,B18,HVY,H11,H13,H14,H17,H18,H20,H21,H24,H25,H26,H27,H28,MMM	2
7	PAOZZ	2920014661855	76761	N1602-1	GENERATOR,ENGINE AC 400 AMP PART OF KIT P/N 57K0273............. UOC:AVY,A11,A13,A14,A20,A24,A25,A26,A27,B17,B18,HVY,H11,H13,H14,H17,H18,H20,H21,H24,H25,H26,H27,H28,MMM	1
8	PAOZZ	5315006165501	80205	MS35756-20	.KEY,WOODRUFF 3/8 PART OF KIT P/N 57K0273............. UOC:AVY,A11,A13,A14,A20,A24,A25,A26,A27,B17,B18,HVY,H11,H13,H14,H17,H18,H20,H21,H24,H25,H26,H27,H28,MMM	1
9	PAOZZ	5310004190876	96906	MS21245-L12	.NUT,SELF-LOCKING,HE 3/4-16X1.00 PART OF KIT P/N 57K0273............. UOC:AVY,A11,A13,A14,A20,A24,A25,A26,A27,B17,B18,HVY,H11,H13,H14,H17,H18,H20,H21,H24,H25,H26,H27,H28,MMM	1
10	PAOZZ	3020014663603	34623	12446753	PULLEY,GROOVE PART OF KIT P/N 57K0273............. UOC:AVY,A11,A13,A14,A20,A24,A25,A26,A27,B17,B18,HVY,H11,H13,H14,H17,H18,	1

SECTION II   TM9-2320-280-24P

(1) NO	(2) SMR CODE	(3) NSN	(4) CAGEC	(5) PART NUMBER	(6) DESCRIPTION AND USABLE ON CODES(UOC)	(7) QTY
11	PAOZZ	5310007638906	96906	MS51968-8	NUT,PLAIN,HEXAGON PART OF KIT P/N 57K0273............................ UOC:AVY,A11,A13,A14,A20,A24,A25,A26, A27,B17,B18,HVY,H11,H13,H14,H17,H18, H20,H21,H24,H25,H26,H27,H28,MMM	1
12	PAOZZ	5310001670680	96906	MS35338-49	WASHER,LOCK PART OF KIT P/N 57K0273. UOC:AVY,A11,A13,A14,A20,A24,A25,A26, A27,B17,B18,HVY,H11,H13,H14,H17,H18, H20,H21,H24,H25,H26,H27,H28,MMM	1
13	PAOZZ	5310006560114	96906	MS15795-819	WASHER,FLAT PART OF KIT P/N 57K0273. UOC:AVY,A11,A13,A14,A20,A24,A25,A26, A27,B17,B18,HVY,H11,H13,H14,H17,H18, H20,H21,H24,H25,H26,H27,H28,MMM	2
14	PAOZA	5342014666280	19207	12460502	BRACKET,ENGINE ACCE PART OF KIT P/N 57K0273............................ UOC:AVY,A11,A13,A14,A20,A24,A25,A26, A27,B17,B18,HVY,H11,H13,H14,H17,H18, H20,H21,H24,H25,H26,H27,H28,MMM	1
15	PAOZZ	5305014753014	80204	B1821BH056F950N	BOLT,SHOULDER PART OF KIT P/N 57K0273............................ UOC:AVY,A11,A13,A14,A20,A24,A25,A26, A27,B17,B18,HVY,H11,H13,H14,H17,H18, H20,H21,H24,H25,H26,H27,H28,MMM	1
16	PAOZZ	5310012067306	7X677	11500207	WASHER,LOCK M10 PART OF KIT P/N 57K0273............................ UOC:AVY,A11,A13,A14,A20,A24,A25,A26, A27,B17,B18,HVY,H11,H13,H14,H17,H18, H20,H21,H24,H25,H26,H27,H28,MMM	3
17	PAOZZ	5305012433759	7X677	14060130	SCREW,CAP,HEXAGON H M10 X 1.5 X 40 PART OF KIT P/N 57K0273............. UOC:AVY,A11,A13,A14,A20,A24,A25,A26, A27,B17,B18,HVY,H11,H13,H14,H17,H18, H20,H21,H24,H25,H26,H27,H28,MMM	1
18	PAOZZ	5310000116121	29510	120384	WASHER,LOCK 1/2 PART OF KIT P/N 57K0273............................ UOC:AVY,A11,A13,A14,A20,A24,A25,A26, A27,B17,B18,HVY,H11,H13,H14,H17,H18, H20,H21,H24,H25,H26,H27,H28,MMM	1
19	PAOZZ	5306012638889	19207	12340845-2	BOLT,MACHINE M10-1.50 X 30 PART OF KIT P/N 57K0273.................... UOC:AVY,A11,A13,A14,A20,A24,A25,A26, A27,B17,B18,HVY,H11,H13,H14,H17,H18, H20,H21,H24,H25,H26,H27,H28,MMM	1
20	PAOZZ	5306012643531	19207	12340845-4	BOLT,MACHINE M10-1.5 X 100 PART OF KIT P/N 57K0273.................... UOC:AVY,A11,A13,A14,A20,A24,A25,A26, A27,B17,B18,HVY,H11,H13,H14,H17,H18, H20,H21,H24,H25,H26,H27,H28,MMM	1
21	PAOZZ	5310013958747	80204	B18241B100F	NUT,PLAIN,HEXAGON PART OF KIT P/N 57K0273............................ UOC:AVY,A11,A13,A14,A20,A24,A25,A26, A27,B17,B18,HVY,H11,H13,H14,H17,H18, H20,H21,H24,H25,H26,H27,H28,MMM	1

SECTION II   TM9-2320-280-24P

(1) NO	(2) CODE	(3) NSN	(4) CAGEC	(5) NUMBER	(6) PART	(7) ITEM SMR DESCRIPTION AND USABLE ON CODES(UOC)	QTY

22 PAOZZ 5310012101587 7X677 11500256 WASHER,FLAT 10MM PART OF KIT P/N 57K0273............................ 2
UOC:AVY,A11,A13,A14,A20,A24,A25,A26,A27,B17,B18,HVY,H11,H13,H14,H17,H18,H20,H21,H24,H25,H26,H27,H28,MMM

23 PAOZZ 5307014084029 19207 12342545 STUD,SHOULDERED PART OF KIT P/N 57K0273............................ 1
UOC:AVY,A11,A13,A14,A20,A24,A25,A26,A27,B17,B18,HVY,H11,H13,H14,H17,H18,H20,H21,H24,H25,H26,H27,H28,MMM

24 PAOZA 5342014666283 19207 12446755 BRACKET,ENGINE ACCE PART OF KIT P/N 57K0273............................ 1
UOC:AVY,A11,A13,A14,A20,A24,A25,A26,A27,B17,B18,HVY,H11,H13,H14,H17,H18,H20,H21,H24,H25,H26,H27,H28,MMM

25 PAOZZ 5999013725601 19207 12341151 STRIP,ELECTRICAL GR PART OF KIT P/N 57K0273............................ 1
UOC:AVY,A11,A13,A14,A20,A24,A25,A26,A27,B17,B18,HVY,H11,H13,H14,H17,H18,H20,H21,H24,H25,H26,H27,H28,MMM

26 PAOZA 3030014665387 19207 12339359-28 V-BELT SET 58 INCHES LONG PART OF KIT P/N 57K0273.................... 1
UOC:AVY,A11,A13,A14,A20,A24,A25,A26,A27,B17,B18,HVY,H11,H13,H14,H17,H18,H20,H21,H24,H25,H26,H27,H28,MMM

END OF FIGURE

Section II. TM 9-2320-280-24P-2

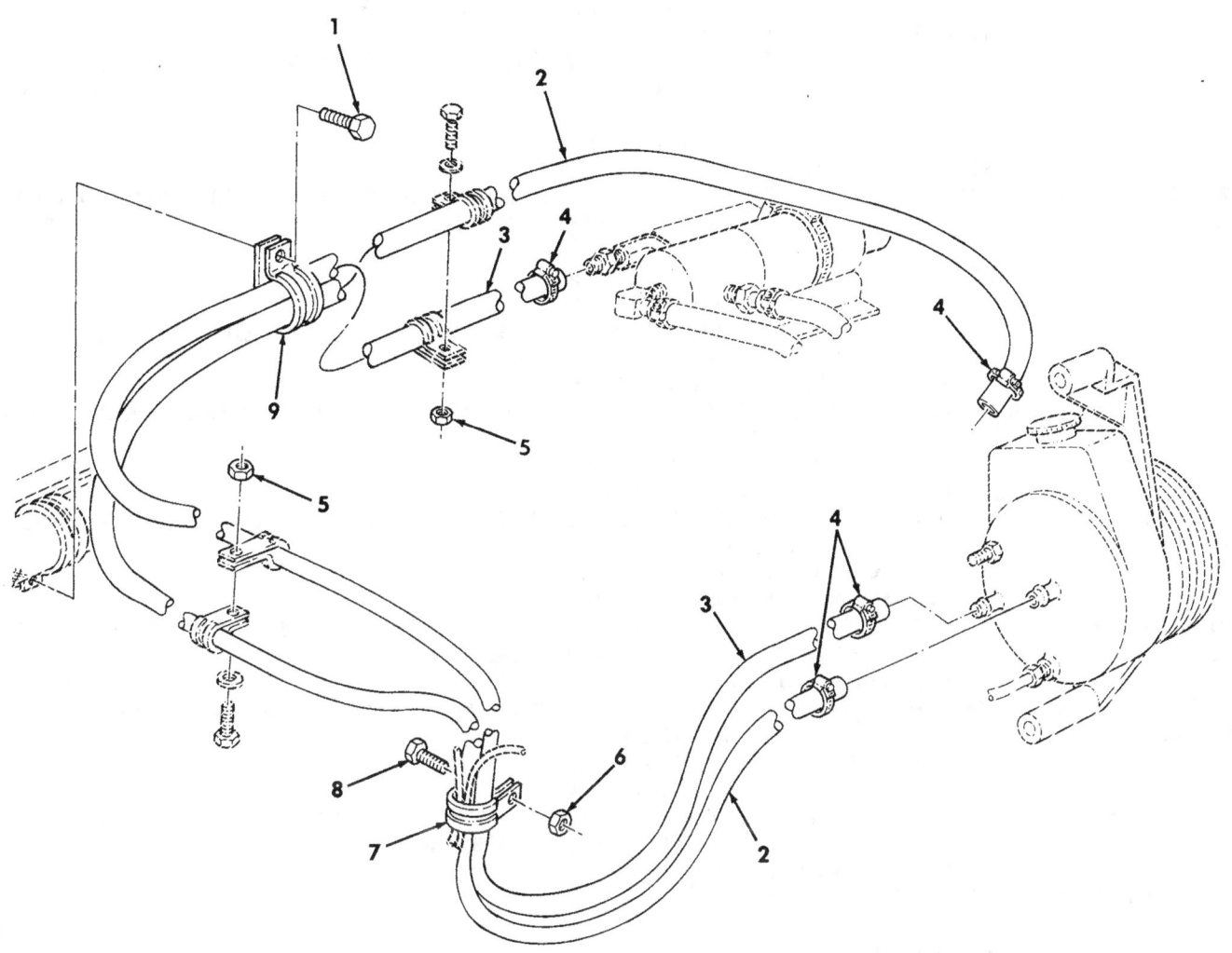

Figure 477. Power Steering Pump Hoses-400 AMP Alternator Kit.

SECTION II   TM9-2320-280-24P

(1) NO	(2) SMR CODE	(3) NSN	(4) CAGEC	(5) PART NUMBER	(7) DESCRIPTION AND USABLE ON CODES(UOC)	ITEM QTY

GROUP 3307 SPECIAL PURPOSE KITS

FIG. 477 POWER STEERING PUMP
HOSES-400 AMP ALTERNATOR KIT

1 PAOZZ 5305012068401 7X677 448655 SCREW,TAPPING PART OF KIT P/N 57K0273.............................. 1
UOC:AVY,A11,A13,A14,A20,A24,A25,A26,A27,B17,B18,HVY,H11,H13,H14,H17,H18,H20,H21,H24,H25,H26,H27,H28,MMM

2 MOOZZ   19207 12338330-9 HOSE,NONMETALLIC MAKE FROM HOSE,P/N 6490610019,51 INCHES LONG PART OF KIT P/N 57K0273..................... 1
UOC:AVY,A11,A13,A14,A20,A24,A25,A26,A27,B17,B18,HVY,H11,H13,H14,H17,H18,H20,H21,H24,H25,H26,H27,H28,MMM

3 MOOZZ   19207 12338330-8 HOSE,NONMETALLIC MAKE FROM HOSE,P/N 6490610019,37 INCHES LONG PART OF KIT P/N 57K0273..................... 1
UOC:AVY,A11,A13,A14,A20,A24,A25,A26,A27,B17,B18,HVY,H11,H13,H14,H17,H18,H20,H21,H24,H25,H26,H27,H28,MMM

4 PAOZZ 4730008776298 18876 8486066 CLAMP,HOSE PART OF KIT P/N 57K0273.. 4
UOC:AVY,A11,A13,A14,A20,A24,A25,A26,A27,B17,B18,HVY,H11,H13,H14,H17,H18,H20,H21,H24,H25,H26,H27,H28,MMM

5 PAOZZ 5310000614650 96906 MS51943-31 NUT,SELF-LOCKING,HE 1/4-20 PART OF KIT P/N 57K0273..................... 2
UOC:AVY,A11,A13,A14,A20,A24,A25,A26,A27,B17,B18,HVY,H11,H13,H14,H17,H18,H20,H21,H24,H25,H26,H27,H28,MMM

6 PAOZZ 5310001249265 72582 271169 NUT,PLAIN,ASSEMBLED #10-32 PART OF KIT P/N 57K0273..................... 1
UOC:AVY,A11,A13,A14,A20,A24,A25,A26,A27,B17,B18,HVY,H11,H13,H14,H17,H18,H20,H21,H24,H25,H26,H27,H28,MMM

7 PAOZZ 5340009291794 96906 MS21334-31 CLAMP,LOOP PART OF KIT P/N 57K0273.. 1
UOC:AVY,A11,A13,A14,A20,A24,A25,A26,A27,B17,B18,HVY,H11,H13,H14,H17,H18,H20,H21,H24,H25,H26,H27,H28,MMM

8 PAOZZ 5305011432328 72582 454065 SCREW,CAP,HEXAGON H #10-32X3/4 PART OF KIT P/N 57K0273................. 1
UOC:AVY,A11,A13,A14,A20,A24,A25,A26,A27,B17,B18,HVY,H11,H13,H14,H17,H18,H20,H21,H24,H25,H26,H27,H28,MMM

9 PAOZZ 5340004381836 19207 7397780 CLAMP,LOOP PART OF KIT P/N 57K0273.. 1
UOC:AVY,A11,A13,A14,A20,A24,A25,A26,A27,B17,B18,HVY,H11,H13,H14,H17,H18,H20,H21,H24,H25,H26,H27,H28,MMM

END OF FIGURE

Section II. TM 9-2320-280-24P-2

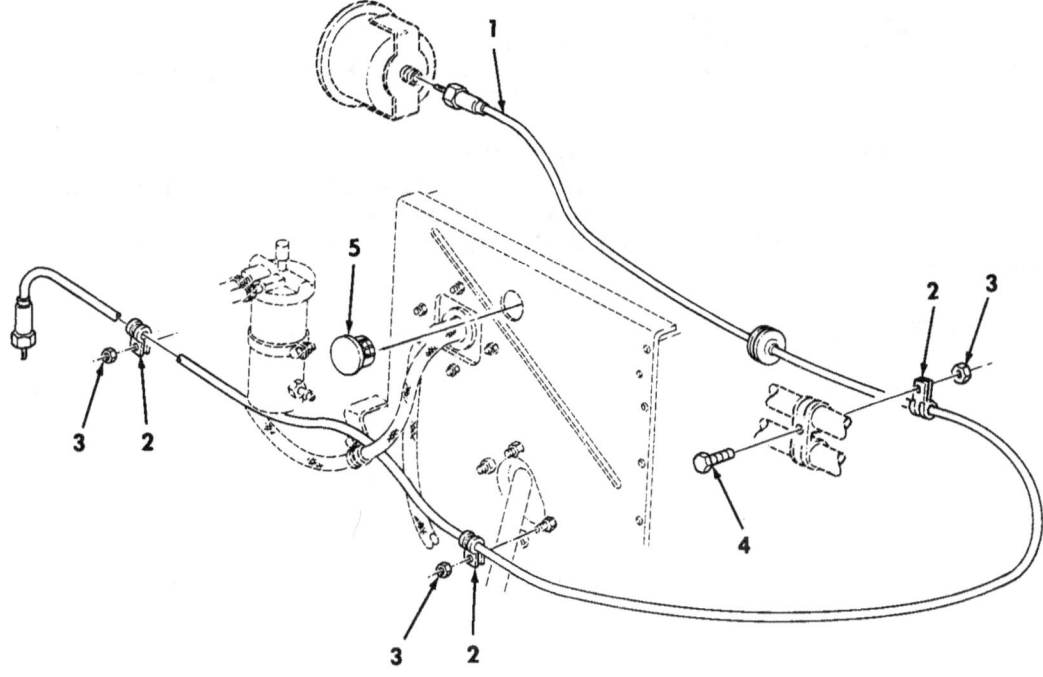

Figure 478. Speedometer Cable Assembly-400 AMP Alternator Kit.

SECTION II TM9-2320-280-24P

(1) NO	(2) SMR CODE	(3) NSN	(4) CAGEC	(5) PART NUMBER	(6) DESCRIPTION AND USABLE ON CODES(UOC)	(7) QTY

GROUP 3307 SPECIAL PURPOSE KITS

FIG. 478 SPEEDOMETER CABLE ASSEMBLY- 400 AMP ALTERNATOR KIT

1 PAOZZ 6680014429413 19200 12338428-2 SHAFT ASSEMBLY,FLEX PART OF KIT P/N 57K0273............................ 1
UOC:AVY,A11,A13,A14,A20,A24,A25,A26,A27,B17,B18,HVY,H11,H13,H14,H17,H18,H20,H21,H24,H25,H26,H27,H28,MMM

2 PAOZZ 5340008091492 81348 CMDX2-3PT573036 CLAMP,LOOP PART OF KIT P/N 57K0273.. 3
UOC:AVY,A11,A13,A14,A20,A24,A25,A26,A27,B17,B18,HVY,H11,H13,H14,H17,H18,H20,H21,H24,H25,H26,H27,H28,MMM

3 PAOZZ 5310000614650 96906 MS51943-31 NUT,SELF-LOCKING,HE 1/4-20 PART OF KIT P/N 57K0273..................... 3
UOC:AVY,A11,A13,A14,A20,A24,A25,A26,A27,B17,B18,HVY,H11,H13,H14,H17,H18,H20,H21,H24,H25,H26,H27,H28,MMM

4 PAOZZ 5305000680508 80204 B1821BH025C075N SCREW,CAP,HEXAGON H 1/4-20 X 3/4 PART OF KIT P/N 57K0273............. 1
UOC:AVY,A11,A13,A14,A20,A24,A25,A26,A27,B17,B18,HVY,H11,H13,H14,H17,H18,H20,H21,H24,H25,H26,H27,H28,MMM

5 PAOZZ 5340011104036 19207 12298661 PLUG,VENT PART OF KIT P/N 57K0273... 1
UOC:AVY,A11,A13,A14,A20,A24,A25,A26,A27,B17,B18,HVY,H11,H13,H14,H17,H18,H20,H21,H24,H25,H26,H27,H28,MMM

END OF FIGURE

Section II.

TM 9-2320-280-24P-2

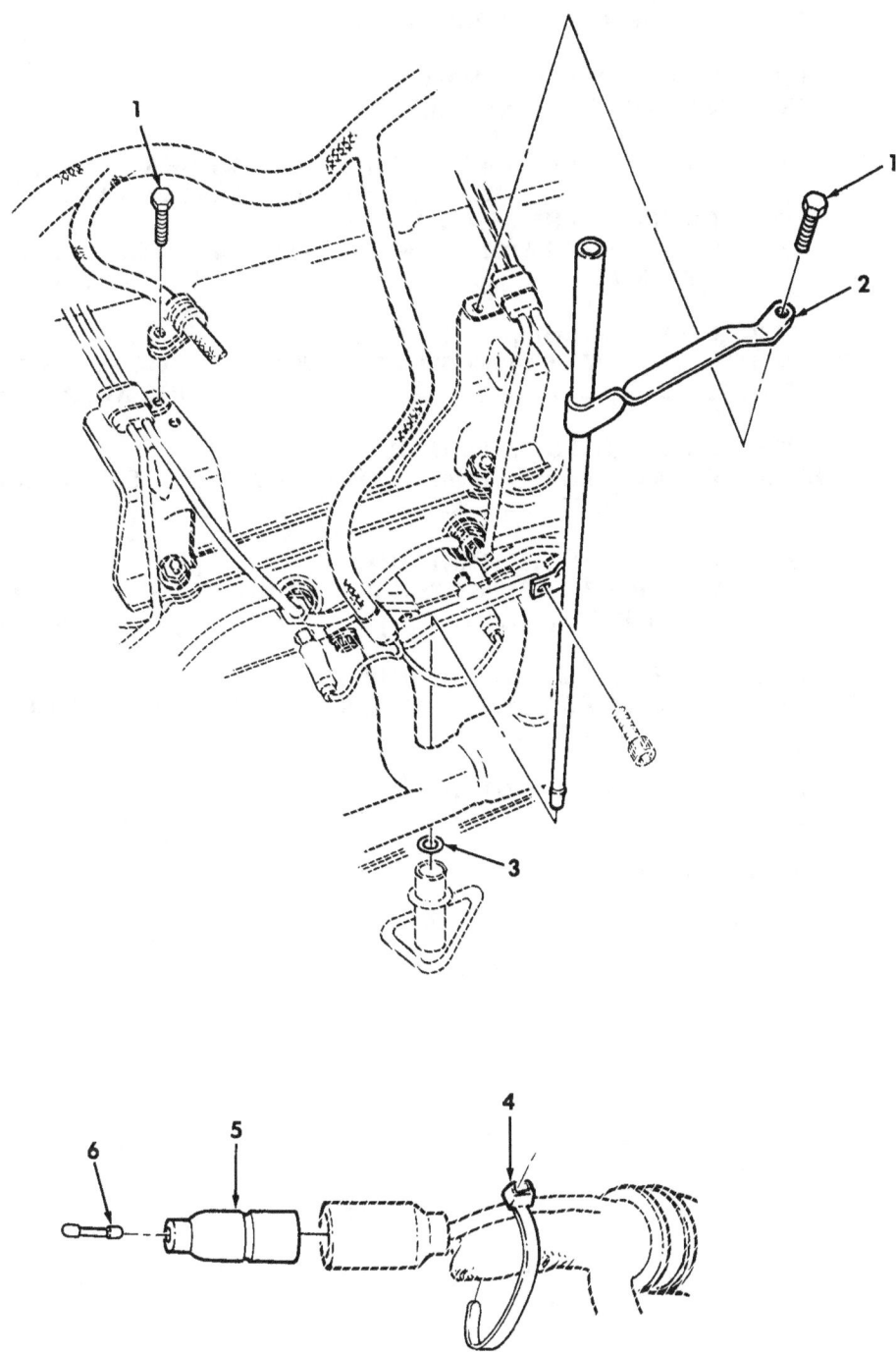

Figure 479. Dipstick and Dummy Connector Plug-400 AMP Alternator Kit.

SECTION II TM9-2320-280-24P

(1) NO	(2) SMR CODE	(3) NSN	(4) CAGEC	(5) PART NUMBER	(6) DESCRIPTION AND USABLE ON CODES(UOC)	(7) QTY

GROUP 3307 SPECIAL PURPOSE KITS

FIG. 479 DIPSTICK AND DUMMY CONNECTOR PLUG-400 AMP ALTERNATOR KIT

No	SMR	NSN	CAGEC	PART NUMBER	DESCRIPTION	QTY
1	PAOZZ	5305012155174	19207	12340515	SCREW,ASSEMBLED WAS PART OF KIT P/N 57K0273............................. UOC:AVY,A11,A13,A14,A20,A24,A25,A26,A27,B17,B18,HVY,H11,H13,H14,H17,H18,H20,H21,H24,H25,H26,H27,H28,MMM	2
2	PAOZZ	2590014665447	34623	12446754	TUBE,DIPSTICK PART OF KIT P/N 57K0273............................. UOC:AVY,A11,A13,A14,A20,A24,A25,A26,A27,B17,B18,HVY,H11,H13,H14,H17,H18,H20,H21,H24,H25,H26,H27,H28,MMM	1
3	PAOZZ	5331009359136	24617	274244	O-RING .355"I.D PART OF KIT P/N 57K0273............................. UOC:AVY,A11,A13,A14,A20,A24,A25,A26,A27,B17,B18,HVY,H11,H13,H14,H17,H18,H20,H21,H24,H25,H26,H27,H28,MMM	1
4	PAOZZ	5975010345871	96906	MS3367-7-0	STRAP,TIEDOWN,ELECT PART OF KIT P/N 57K0273............................. UOC:AVY,A11,A13,A14,A20,A24,A25,A26,A27,B17,B18,HVY,H11,H13,H14,H17,H18,H20,H21,H24,H25,H26,H27,H28,MMM	1
5	PAOZZ	5935008338561	19207	8338561	SHELL,ELECTRICAL CO PART OF KIT P/N 57K0273............................. UOC:AVY,A11,A13,A14,A20,A24,A25,A26,A27,B17,B18,HVY,H11,H13,H14,H17,H18,H20,H21,H24,H25,H26,H27,H28,MMM	1
6	PAOZZ	5935002140904	19207	7982907	DUMMY CONNECTOR,PLU PART OF KIT P/N 57K0273............................. UOC:AVY,A11,A13,A14,A20,A24,A25,A26,A27,B17,B18,HVY,H11,H13,H14,H17,H18,H20,H21,H24,H25,H26,H27,H28,MMM	1

END OF FIGURE

Section II. TM 9-2320-280-24P-2

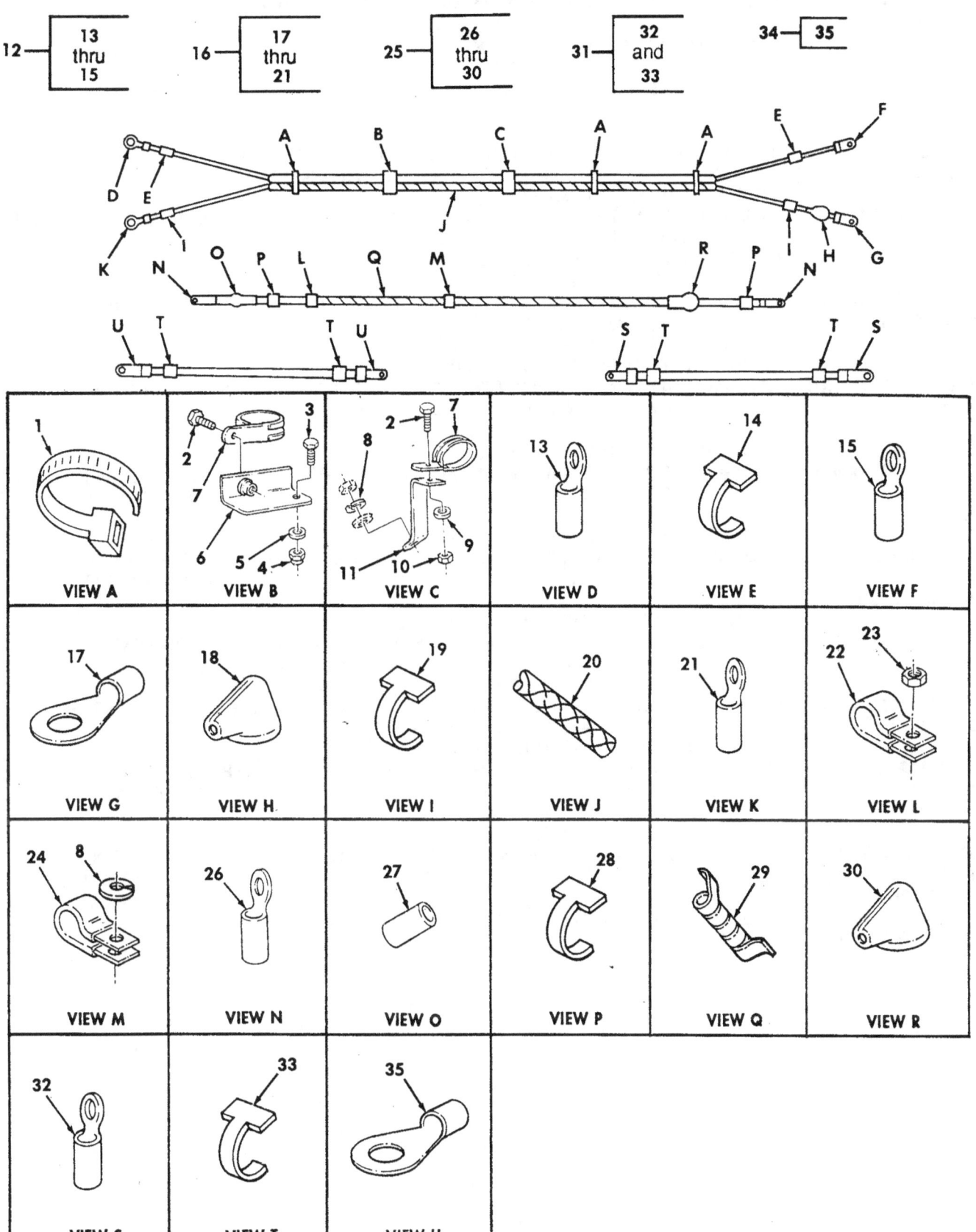

Figure 480. 400 AMP Alternator Wiring Harness Cable Assemblies.

SECTION II TM9-2320-280-24P

(1) NO	(2) SMR CODE	(3) NSN	(4) CAGEC	(5) PART NUMBER	(7) DESCRIPTION AND USABLE ON CODES (UOC)	(6) QTY

GROUP 3307 SPECIAL PURPOSE KITS

FIG. 480 400 AMP ALTERNATOR WIRING HARNESS CABLE ASSEMBLIES

1 PAOZZ 5975010345871 96906 MS3367-7-0 STRAP,TIEDOWN,ELECT PART OF KIT P/N 57K0273............................ 3
UOC:AVY,A11,A13,A14,A20,A24,A25,A26,A27,B17,B18,HVY,H11,H13,H14,H17,H18,H20,H21,H24,H25,H26,H27,H28,MMM

2 PAOZZ 5305011432328 72582 454065 SCREW,CAP,HEXAGON H #10-32 X 3/4 PART OF KIT P/N 57K0273............. 2
UOC:AVY,A11,A13,A14,A20,A24,A25,A26,A27,B17,B18,HVY,H11,H13,H14,H17,H18,H20,H21,H24,H25,H26,H27,H28,MMM

3 PAOZZ 5306002264825 80204 B1821BH031C075N BOLT,MACHINE 1/4-20 X 3/4 PART OF KIT P/N 57K0273..................... 1
UOC:AVY,A11,A13,A14,A20,A24,A25,A26,A27,B17,B18,HVY,H11,H13,H14,H17,H18,H20,H21,H24,H25,H26,H27,H28,MMM

4 PAOZZ 5310008140673 96906 MS51943-33 NUT,SELF-LOCKING,HE 5/16-18 PART OF KIT P/N 57K0273.................... 1
UOC:AVY,A11,A13,A14,A20,A24,A25,A26,A27,B17,B18,HVY,H11,H13,H14,H17,H18,H20,H21,H24,H25,H26,H27,H28,MMM

5 PAOZZ 5310000814219 96906 MS27183-12 WASHER,FLAT 5/16 PART OF KIT P/N 57K0273............................ 1
UOC:AVY,A11,A13,A14,A20,A24,A25,A26,A27,B17,B18,HVY,H11,H13,H14,H17,H18,H20,H21,H24,H25,H26,H27,H28,MMM

6 PAOZA 5340014662468 19207 12460496 BRACKET,ANGLE PART OF KIT P/N 57K0273............................ 1
UOC:AVY,A11,A13,A14,A20,A24,A25,A26,A27,B17,B18,HVY,H11,H13,H14,H17,H18,H20,H21,H24,H25,H26,H27,H28,MMM

7 PAOZZ 5340007319611 96906 MS21334-34 CLAMP,LOOP PART OF KIT P/N 57K0273.. 2
UOC:AVY,A11,A13,A14,A20,A24,A25,A26,A27,B17,B18,HVY,H11,H13,H14,H17,H18,H20,H21,H24,H25,H26,H27,H28,MMM

8 PAOZZ 5310012067306 7X677 11500207 WASHER,LOCK PART OF KIT P/N 57K0273. 2
UOC:AVY,A11,A13,A14,A20,A24,A25,A26,A27,B17,B18,HVY,H11,H13,H14,H17,H18,H20,H21,H24,H25,H26,H27,H28,MMM

9 PAOZZ 5310000145850 96906 MS27183-42 WASHER,FLAT #10 PART OF KIT P/N 57K0273............................ 1
UOC:AVY,A11,A13,A14,A20,A24,A25,A26,A27,B17,B18,HVY,H11,H13,H14,H17,H18,H20,H21,H24,H25,H26,H27,H28,MMM

10 PAOZZ 5310002089255 96906 MS21044-C3 NUT,SELF-LOCKING,HE PART OF KIT P/N 57K0273............................ 1
UOC:AVY,A11,A13,A14,A20,A24,A25,A26,A27,B17,B18,HVY,H11,H13,H14,H17,H18,H20,H21,H24,H25,H26,H27,H28,MMM

SECTION II   TM9-2320-280-24P

(1) ITEM NO	(2) SMR CODE	(3) NSN	(4) CAGEC	(5) PART NUMBER	(6) DESCRIPTION AND USABLE ON CODES (UOC)	(7) QTY
11	PAOZA	5340014662620	19207	12460495	BRACKET,DOUBLE ANGL PART OF KIT P/N 57K0273............... UOC:AVY,A11,A13,A14,A20,A24,A25,A26, A27,B17,B18,HVY,H11,H13,H14,H17,H18, H20,H21,H24,H25,H26,H27,H28,MMM	1
12	PAOZZ	6150014707014	19207	12460498-4	CABLE ASSEMBLY,SPEC PART OF KIT P/N 57K0273............... UOC:AVY,A11,A13,A14,A20,A24,A25,A26, A27,B17,B18,HVY,H11,H13,H14,H17,H18, H20,H21,H24,H25,H26,H27,H28,MMM	1
13	PAOZZ	5940001155023	96906	MS20659-124	.TERMINAL,LUG PART OF KIT P/N 57K0273............... UOC:AVY,A11,A13,A14,A20,A24,A25,A26, A27,B17,B18,HVY,H11,H13,H14,H17,H18, H20,H21,H24,H25,H26,H27,H28,MMM	1
14	PAOZZ	9905009457637	81349	M43436/1-4	.BAND,MARKER PART OF KIT P/N 57K0273 UOC:AVY,A11,A13,A14,A20,A24,A25,A26, A27,B17,B18,HVY,H11,H13,H14,H17,H18, H20,H21,H24,H25,H26,H27,H28,MMM	2
15	PAOZA	5940001155012	96906	MS20659-157	.TERMINAL,LUG PART OF KIT P/N 57K0273............... UOC:AVY,A11,A13,A14,A20,A24,A25,A26, A27,B17,B18,HVY,H11,H13,H14,H17,H18, H20,H21,H24,H25,H26,H27,H28,MMM	1
16	MOOZZ		19207	12460498-3	CABLE ASSEMBLY PART OF KIT P/N 57K0273 MAKE FROM WIRE,P/N M13486/1-4,78,INCHES LONG........... A27,B17,B18,HVY,H11,H13,H14,H17,H18, A27,B17,B18,HVY,H11,H13,H14,H17,H18, H20,H21,H24,H25,H26,H27,H28,MMM	
17	PAOZZ	5940001155006	96906	MS25036-133	.TERMINAL,LUG PART OF KIT P/N 57K0273............... UOC:AVY,A11,A13,A14,A20,A24,A25,A26, A27,B17,B18,HVY,H11,H13,H14,H17,H18, H20,H21,H24,H25,H26,H27,H28,MMM	1
18	PAOZZ	5975012089618	9Z043	12339710	.CABLE NIPPLE,ELECTR PART OF KIT P/N 57K0273............... UOC:AVY,A11,A13,A14,A20,A24,A25,A26, A27,B17,B18,HVY,H11,H13,H14,H17,H18, H20,H21,H24,H25,H26,H27,H28,MMM	1
19	PAOZZ	9905008933570	81349	M43436/1-3	.BAND,MARKER PART OF KIT P/N 57K0273 UOC:AVY,A11,A13,A14,A20,A24,A25,A26, A27,B17,B18,HVY,H11,H13,H14,H17,H18, H20,H21,H24,H25,H26,H27,H28,MMM	2
20	MOOZZ		9D117	SAEJ562	.LOOM,NONMETALLIC 3/4 PART OF KIT P/N 57K0273 MAKE FROM TUBING,P/N AE208-,LENGTH AS REQUIRED......... UOC:AVY,A11,A13,A14,A20,A24,A25,A26, A27,B17,B18,HVY,H11,H13,H14,H17,H18, H20,H21,H24,H25,H26,H27,H28,MMM	1
21	PAOZZ	5940001155001	96906	MS20659-135	.TERMINAL,LUG PART OF KIT P/N 57K0273............... UOC:AVY,A11,A13,A14,A20,A24,A25,A26,	1

SECTION II TM9-2320-280-24P C01

(1) ITEM NO	(2) SMR CODE	(3) NSN	(4) CAGEC	(5) PART NUMBER	(6) DESCRIPTION AND USABLE ON CODES (UOC)	(7) QTY
22	PAOZZ	5340008091494	96906	MS21333-105	CLAMP,LOOP PART OF KIT P/N 57K0273.. UOC:AVY,A11,A13,A14,A20,A24,A25,A26,A27,B17,B18,HVY,H11,H13,H14,H17,H18,H20,H21,H24,H25,H26,H27,H28,MMM	1
23	PAOZZ	5310011520598	24617	271172	NUT,SELF-LOCKING,AS 1/4-20 PART OF KIT P/N 57K0273.................... UOC:AVY,A11,A13,A14,A20,A24,A25,A26,A27,B17,B18,HVY,H11,H13,H14,H17,H18,H20,H21,H24,H25,H26,H27,H28,MMM	1
24	PAOZZ	5340000538994	96906	MS21333-126	CLAMP,LOOP PART OF KIT P/N 57K0273.. UOC:AVY,A11,A13,A14,A20,A24,A25,A26,A27,B17,B18,HVY,H11,H13,H14,H17,H18,H20,H21,H24,H25,H26,H27,H28,MMM	1
25	MOOZZ		19207	12339317	LEAD,ELECTRICAL PART OF KIT P/N 57K0273 MAKE FROM WIRE,P/N M13486/1-14, 58 INCHES LONG........ UOC:AVY,A11,A13,A14,A20,A24,A25,A26,A27,B17,B18,HVY,H11,H13,H14,H17,H18,H20,H21,H24,H25,H26,H27,H28,MMM	1
26	PAOZZ	5940007355520	56501	TG33	.TERMINAL,LUG PART OF KIT P/N 57K0273............... UOC:AVY,A11,A13,A14,A20,A24,A25,A26,A27,B17,B18,HVY,H11,H13,H14,H17,H18,H20,H21,H24,H25,H26,H27,H28,MMM	2
* 27	PAOZZ	5970011749449	19207	7056640	.INSULATOR PART OF KIT P/N 57K0273.. UOC:AVY,A11,A13,A14,A20,A24,A25,A26,A27,B17,B18,HVY,H11,H13,H14,H17,H18,H20,H21,H24,H25,H26,H27,H28,MMM	1
28	PAOZZ	9905008933570	81349	M43436/1-3	.BAND,MARKER PART OF KIT P/N 57K0273 UOC:AVY,A11,A13,A14,A20,A24,A25,A26,A27,B17,B18,HVY,H11,H13,H14,H17,H18,H20,H21,H24,H25,H26,H27,H28,MMM	2
29	MOOZZ		19207	12339317-1-AR	.WRAP,NYLON MAKE FROM WRAP,P/N 12339317-1,LENGTH TO SUIT PART OF KIT P/N 57K0273.................... UOC:AVY,A11,A13,A14,A20,A24,A25,A26,A27,B17,B18,HVY,H11,H13,H14,H17,H18,H20,H21,H24,H25,H26,H27,H28,MMM	1
30	PAOZZ	5975012089618	9Z043	12339710	.CABLE NIPPLE,ELECTR PART OF KIT P/N 57K0273......................... UOC:AVY,A11,A13,A14,A20,A24,A25,A26,A27,B17,B18,HVY,H11,H13,H14,H17,H18,H20,H21,H24,H25,H26,H27,H28,MMM	1
31	MOOZZ		19207	12460498-2	CABLE ASSEMBLY,SPEC PART OF KIT P/N 57K0273 MAKE FROM WIRE,P/N M13486/1-14,11 INCHES LONG......... UOC:AVY,A11,A13,A14,A20,A24,A25,A26,A27,B17,B18,HVY,H11,H13,H14,H17,H18,H20,H21,H24,H25,H26,H27,H28,MMM	1
32	PAOZZ	5940001152683	96906	MS20659-151	.TERMINAL,LUG PART OF KIT P/N 57K0273...............	2

480-3

SECTION II   TM9-2320-280-24P

(1) NO	(2) CODE	(3) NSN	(4) CAGEC	(5) NUMBER	(6)	(7) ITEM DESCRIPTION AND USABLE ON CODES(UOC)	SMR	PART QTY
						UOC:AVY,A11,A13,A14,A20,A24,A25,A26, A27,B17,B18,HVY,H11,H13,H14,H17,H18, H20,H21,H24,H25,H26,H27,H28,MMM		
33	PAOZZ	9905008933570	81349	M43436/1-3		.BAND,MARKER PART OF KIT P/N 57K0273		4
						UOC:AVY,A11,A13,A14,A20,A24,A25,A26, A27,B17,B18,HVY,H11,H13,H14,H17,H18, H20,H21,H24,H25,H26,H27,H28,MMM		
34	MOOZZ		19207	12460498-1		CABLE ASSEMBLY,SPEC PART OF KIT P/N 57K0273 MAKE FROM WIRE,P/N M13486/1-14 11 INCHES LONG..........		1
						UOC:AVY,A11,A13,A14,A20,A24,A25,A26, A27,B17,B18,HVY,H11,H13,H14,H17,H18, H20,H21,H24,H25,H26,H27,H28,MMM		
35	PAOZZ	5940001152684	96906	MS20659-118		.TERMINAL,LUG PART OF KIT P/N 57K0273............... UOC:AVY,A11,A13,A14,A20,A24,A25,A26, A27,B17,B18,HVY,H11,H13,H14,H17,H18, H20,H21,H24,H25,H26,H27,H28,MMM		2

END OF FIGURE

Section II.

TM 9-2320-280-24P-2

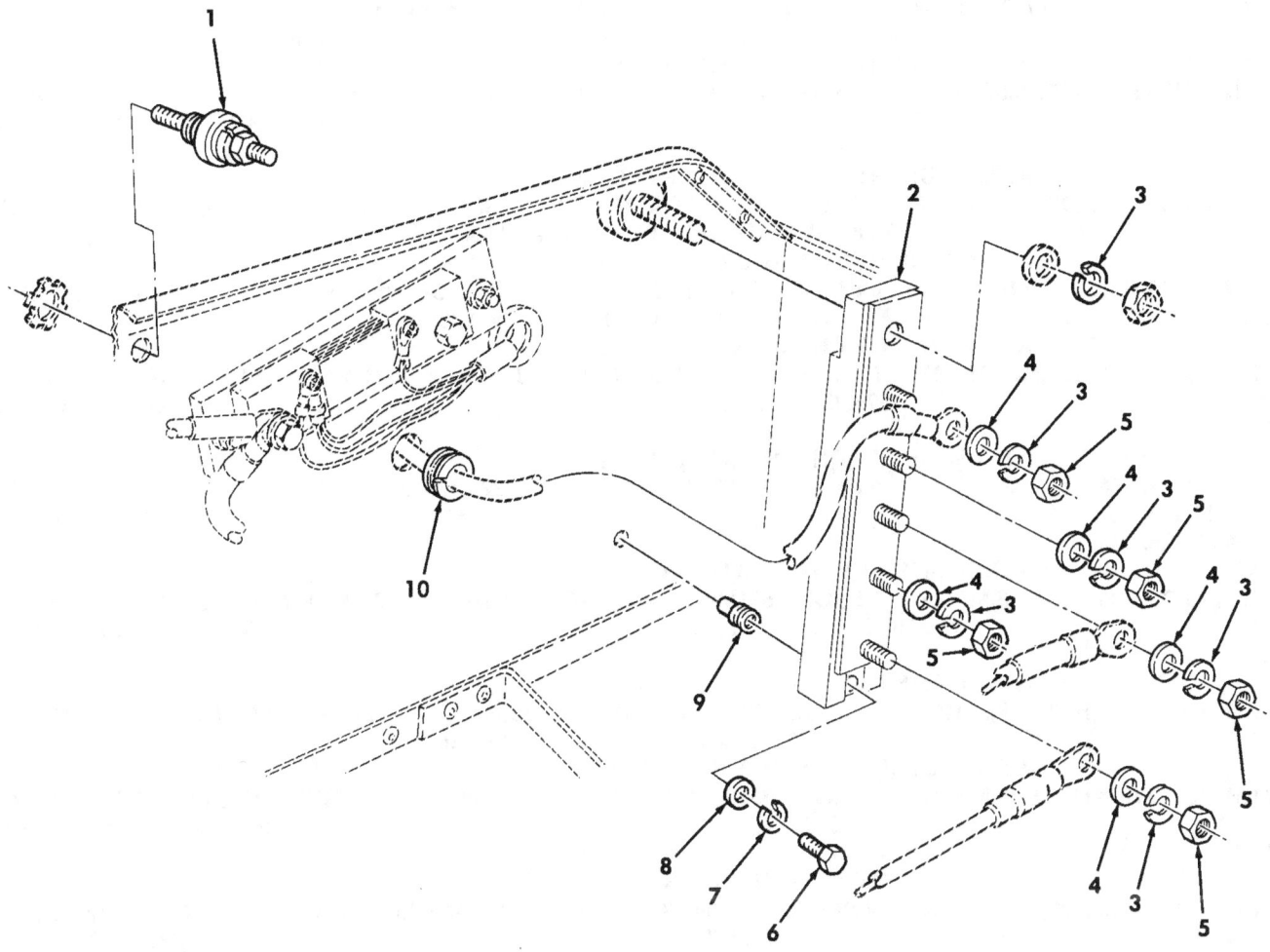

Figure 481. Buss Assembly-400 AMP Alternator Kit.

SECTION II   TM9-2320-280-24P

(1) NO	(2) SMR CODE	(3) NSN	(4) CAGEC	(5) PART NUMBER	(7) ITEM DESCRIPTION AND USABLE ON CODES (UOC)	QTY

GROUP 3307 SPECIAL PURPOSE KITS

FIG. 481 BUSS ASSEMBLY-400 AMP ALTERNATOR KIT

1 PAOZZ 5940011803655 19207 12340594   TERMINAL ASSEMBLY PART OF KIT P/N 57K0273............................   1
UOC:AVY,A11,A13,A14,A20,A24,A25,A26,A27,B17,B18,HVY,H11,H13,H14,H17,H18,H20,H21,H24,H25,H26,H27,H28,MMM

2 PAOZZ 6150014137946 19207 12460070   BUSS,CONDUCTOR PART OF KIT P/N 57K0273............................   1
UOC:AVY,A11,A13,A14,A20,A24,A25,A26,A27,B17,B18,HVY,H11,H13,H14,H17,H18,H20,H21,H24,H25,H26,H27,H28,MMM

3 PAOZZ 5310000116121 29510 120384   WASHER,LOCK PART OF KIT P/N 57K0273   6
UOC:AVY,A11,A13,A14,A20,A24,A25,A26,A27,B17,B18,HVY,H11,H13,H14,H17,H18,H20,H21,H24,H25,H26,H27,H28,MMM

4 PAOZZ 5310000455218 96906 MS15795-918   WASHER,FLAT PART OF KIT P/N 57K0273.   5
UOC:AVY,A11,A13,A14,A20,A24,A25,A26,A27,B17,B18,HVY,H11,H13,H14,H17,H18,H20,H21,H24,H25,H26,H27,H28,MMM

5 PAOZZ 5310000115774 24617 115774   NUT,PLAIN,HEXAGON PART OF KIT P/N 57K0273............................   5
UOC:AVY,A11,A13,A14,A20,A24,A25,A26,A27,B17,B18,HVY,H11,H13,H14,H17,H18,H20,H21,H24,H25,H26,H27,H28,MMM

6 PAOZZ 5305000680508 80204 B1821BH025C075N   SCREW,CAP,HEXAGON H PART OF KIT P/N 57K0273............................   1
UOC:AVY,A11,A13,A14,A20,A24,A25,A26,A27,B17,B18,HVY,H11,H13,H14,H17,H18,H20,H21,H24,H25,H26,H27,H28,MMM

7 PAOZZ 5310002092946 24617 120380   WASHER,LOCK 1/4 PART OF KIT P/N 57K0273............................   1
UOC:AVY,A11,A13,A14,A20,A24,A25,A26,A27,B17,B18,HVY,H11,H13,H14,H17,H18,H20,H21,H24,H25,H26,H27,H28,MMM

8 PAOZZ 5310001870146 24617 120392   WASHER,FLAT PART OF KIT P/N 57K0273   1
UOC:AVY,A11,A13,A14,A20,A24,A25,A26,A27,B17,B18,HVY,H11,H13,H14,H17,H18,H20,H21,H24,H25,H26,H27,H28,MMM

9 PAOZZ 5310014113422 19207 12446871-2   NUT,PLAIN,BLIND RIV PART OF KIT P/N 57K0273............................   1
UOC:AVY,A11,A13,A14,A20,A24,A25,A26,A27,B17,B18,HVY,H11,H13,H14,H17,H18,H20,H21,H24,H25,H26,H27,H28,MMM

10 PAOZZ 5325002766343 96906 MS35489-23   GROMMET,NONMETALLIC PART OF KIT P/N 57K0273............................   1
UOC:AVY,A11,A13,A14,A20,A24,A25,A26,A27,B17,B18,HVY,H11,H13,H14,H17,H18,H20,H21,H24,H25,H26,H27,H28,MMM

KIT PAOZZ 2920014651096 19207 57K0273   400 AMP ALTERNATOR..................   1

SECTION II   TM9-2320-280-24P

(1) NO	(2) CODE	(3) NSN	(4) CAGEC	(5) PART NUMBER	(6) DESCRIPTION AND USABLE ON CODES(UOC)	(7) ITEM SMR QTY

UOC:AVY,A11,A13,A14,A20,A24,A25,A26,
A27,B17,B18,HVY,H11,H13,H14,H17,H18,
20,H21,H24,H25,H26,H27,H28,MMM

```
BAND,MARKER         ( 2) 480-14
BAND,MARKER         ( 2) 480-19
BAND,MARKER         ( 2) 480-28
BAND,MARKER         ( 4) 480-33
BOLT,MACHINE        ( 1) 476-4
BOLT,MACHINE        ( 1) 476-19
BOLT,MACHINE        ( 1) 476-20
BOLT,MACHINE        ( 1) 480-3
BOLT,SHOULDER       ( 1) 476-15
BRACKET ADJUSTMENT  ( 1) 476-5
BRACKET ASSEMBLY    ( 1) 480-6
BRACKET,CLAMP       ( 1) 480-11
BRACKET,ENGINE,ACCE ( 1) 476-14
BRACKET,LOWER MOUNT ( 1) 476-24
BUSS,CONDUCTOR      ( 1) 481-2
CABLE ASSEMBLY      ( 1) 480-12
CABLE ASSEMBLY      ( 1) 480-16
CABLE ASSEMBLY      ( 1) 480-34
CABLE,NIPPLE,ELECTR ( 2) 476-6
CABLE NIPPLE,ELECTR ( 1) 480-18
CABLE NIPPLE,ELECTR ( 1) 480-30
CABLE,ASSEMBLY      ( 1) 480-31
CLAMP,HOSE          ( 4) 477-4
CLAMP,LOOP          ( 1) 477-7
CLAMP,LOOP          ( 1) 477-9
CLAMP,LOOP          ( 3) 478-2
CLAMP,LOOP          ( 2) 480-7
CLAMP,LOOP          ( 1) 480-22
CLAMP,LOOP          ( 1) 480-24
DUMMY CONNECTOR,PLU ( 1) 479-6
GENERATOR,ENGINE    ( 1) 476-7
GROMMET,NONMETALLIC ( 1) 481-10
HOSE,NONMETALLIC    ( 1) 477-2
HOSE,NONMETALLIC    ( 1) 477-3
INSULATOR           ( 1) 480-27
KEY,WOODRUFF        ( 1) 476-8
LEAD,ELECTRICAL     ( 1) 480-25
LOOM,NONMETALLIC    ( 1) 480-20
NUT,BRASS           ( 5) 481-5
NUT,PLAIN,HEXAGON   ( 1) 476-11
NUT,PLAIN,HEXAGON   ( 1) 476-21
NUT,PLAIN,ASSEMBLED ( 1) 477-6
NUT,PLAIN,BLIND RIV ( 1) 481-9
NUT,SELF-LOCKING,HE ( 1) 476-9
NUT,SELF-LOCKING,HE ( 2) 477-5
NUT,SELF-LOCKING,HE ( 3) 478-3
NUT,SELF-LOCKING,HE ( 1) 480-4
NUT,SELF-LOCKING,HE ( 1) 480-10
NUT,SELF-LOCKING,AS ( 1) 480-23
O-RING              ( 1) 479-3
```

SECTION II   TM9-2320-280-24P

(1) NO	(2) CODE	(3) NSN	(4) CAGEC	(5) PART NUMBER	(6) DESCRIPTION AND USABLE ON CODES(UOC)	(7) ITEM SMR QTY
					PLUG,VENT	(1) 478-5
					PULLEY	(1) 476-10
					SCREW,ASSEMBLED WAS	(2) 479-1
					SCREW,CAP,HEXAGON H	(1) 476-1
					SCREW,CAP,HEXAGON H	(1) 476-17
					SCREW,CAP,HEXAGON H	(1) 477-8
					SCREW,CAP,HEXAGON H	(1) 478-4
					SCREW,CAP,HEXAGON H	(1) 481-6
					SCREW,CAP,HEXAGON H	(2) 480-2
					SCREW,TAPPING	(1) 477-1
					SHAFT ASSEMBLY,FLEX	(1) 478-1
					SHELL,ELECTRICAL CO	(1) 479-5
					STRAP,TIEDOWN,ELECT	(1) 479-4
					STRAP,TIEDOWN,ELECT	(3) 480-1
					STRIP,ELECTRICAL GR	(1) 476-25
					STUD,SHOULDERED	(1) 476-23
					TERMINAL ASSEMBLY	(1) 481-1
					TERMINAL,LUG	(1) 480-13
					TERMINAL LUG	(1) 480-15
					TERMINAL,LUG	(1) 480-17
					TERMINAL,LUG	(1) 480-21
					TERMINAL,LUG	(2) 480-26
					TERMINAL,LUG	(2) 480-32
					TERMINAL,LUG	(2) 480-35
					TUBE,DIPSTICK	(1) 479-2
					V-BELT SET	(1) 476-26
					WASHER,FLAT	(2) 476-3
					WASHER,FLAT	(2) 476-13
					WASHER,FLAT	(2) 476-22
					WASHER,FLAT	(5) 481-4
					WASHER,FLAT	(1) 481-8
					WASHER,FLAT	(1) 480-5
					WASHER,FLAT	(1) 480-9
					WASHER,LOCK	(2) 476-2
					WASHER,LOCK	(1) 476-12
					WASHER,LOCK	(3) 476-16
					WASHER,LOCK	(1) 476-18
					WASHER,LOCK	(6) 481-3
					WASHER,LOCK	(1) 481-7
					WASHER,LOCK	(2) 480-8
					WRAP,NYLON	(1) 480-29

END OF FIGURE

Section II. TM 9-2320-280-24P-2 C01

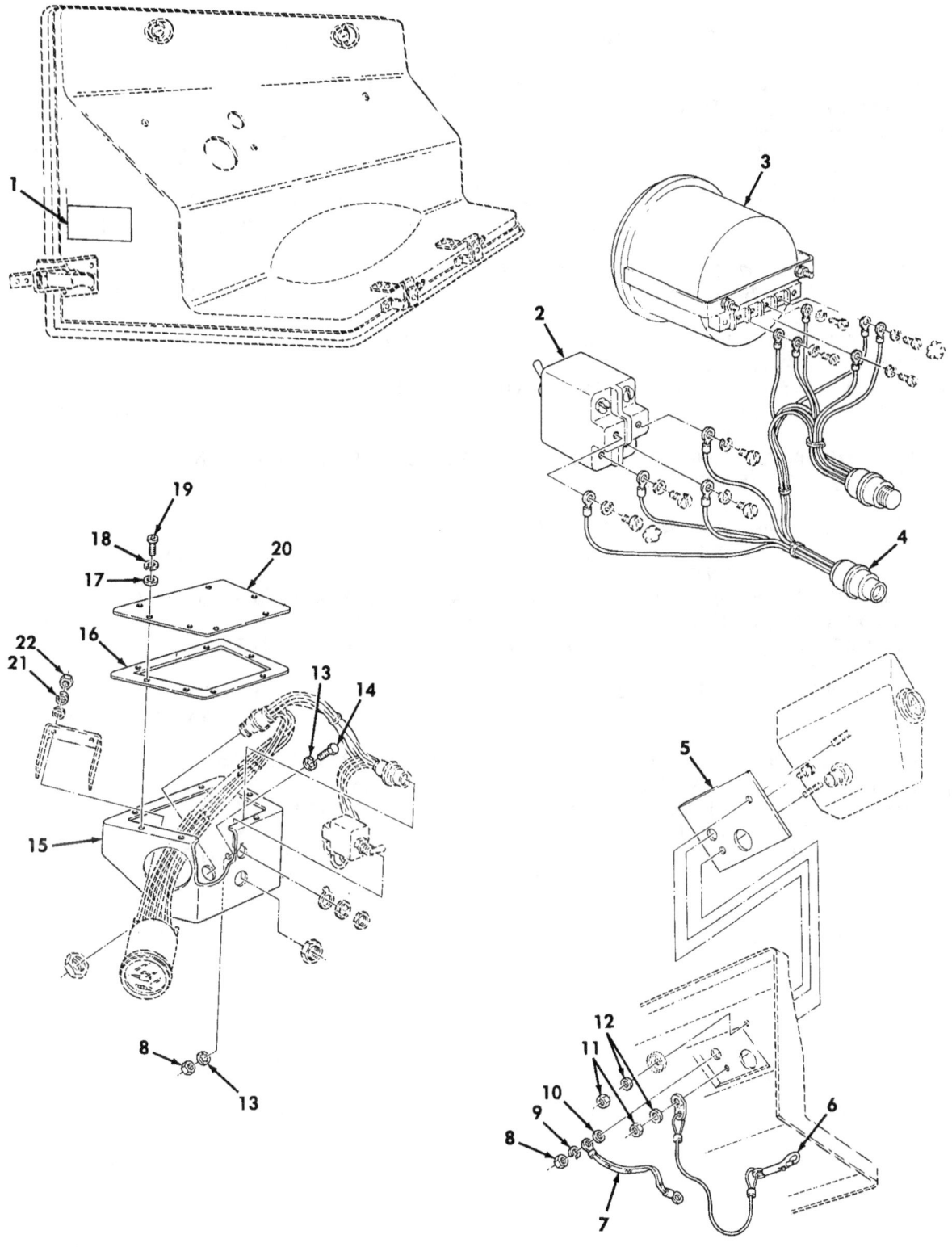

Figure 482. Tachometer Parts Kit.

SECTION II   TM9-2320-280-24P

(1) ITEM NO	(2) SMR CODE	(3) NSN	(4) CAGEC	(5) PART NUMBER	(7) DESCRIPTION AND USABLE ON CODES (UOC)	QTY

GROUP 3307 SPECIAL PURPOSE KITS

FIG. 482 TACHOMETER PARTS KIT

ITEM NO	SMR CODE	NSN	CAGEC	PART NUMBER	DESCRIPTION	QTY
1	PAOZA	7690014751197	19207	12469032	DECAL PART OF KIT P/N 57K3532 UOC:AVY,A13,A14,BVY,HVY,H13,H14,H21,NNN	1
2	PAOZZ	5930008561654	96906	MS24659-23G	SWITCH,TOGGLE PART OF KIT P/N 57K3532 UOC:AVY,A13,A14,BVY,HVY,H13,H14,H21,NNN	1
3	PAOZZ	6680014740384	06046	535K1-49383	TACHOMETER,ELECTRIC PART OF KIT P/N 57K3532 UOC:AVY,A13,A14,BVY,HVY,H13,H14,H21,NNN	1
4	PAOZZ	6150014749676	19207	12469043	CABLE,ASSEMBLY,SPEC PART OF KIT P/N 57K3532 UOC:AVY,A13,A14,BVY,HVY,H13,H14,H21,NNN	1
5	PAOZZ	5330014751671	19207	12469035	GASKET PART OF KIT P/N 57K3532 UOC:AVY,A13,A14,BVY,HVY,H13,H14,H21,NNN	1
6	PAOZZ	4010013089168	57958	C5136341	WIRE ROPE ASSEMBLY, PART OF KIT P/N 57K3532 UOC:AVY,A13,A14,BVY,HVY,H13,H14,H21,NNN	1
7	PAOZZ	5999014710710	19207	12469047	STRIP,ELECTRICAL GR PART OF KIT P/N 57K3532 UOC:AVY,A13,A14,BVY,HVY,H13,H14,H21,NNN	1
8	PAOZZ	5310004776768	96906	MS35649-2384	NUT,PLAIN,HEXAGON PART OF KIT P/N 57K3532 UOC:AVY,A13,A14,BVY,HVY,H13,H14,H21,NNN	2
9	PAOZZ	5310009847042	96906	MS35338-141	WASHER,LOCK PART OF KIT P/N 57K3532 UOC:AVY,A13,A14,BVY,HVY,H13,H14,H21,NNN	1
10	PAOZZ	5310008024701	96906	MS15795-813	WASHER,FLAT PART OF KIT P/N 57K3532 UOC:AVY,A13,A14,BVY,HVY,H13,H14,H21,NNN	1
11	PAOZZ	5310002089255	96906	MS21044-C3	NUT,SELF-LOCKING,HE PART OF KIT P/N 57K3532 UOC:AVY,A13,A14,BVY,HVY,H13,H14,H21,NNN	2
12	PAOZZ	5310006191148	96906	MS15795-808	WASHER,FLAT PART OF KIT P/N 57K3532 UOC:AVY,A13,A14,BVY,HVY,H13,H14,H21,NNN	2
13	PAOZZ	5310005957237	61208	M2046S	WASHER,LOCK PART OF KIT P/N 57K3532 UOC:AVY,A13,A14,BVY,HVY,H13,H14,H21,NNN	2
14	PAOZZ	5305007215665	96906	MS35307-361	SCREW,CAP,HEXAGON H PART OF KIT P/N 57K3532 UOC:AVY,A13,A14,BVY,HVY,H13,H14,H21,	1

```
    SECTION II          TM9-2320-280-24P C01
   (1)   (2)    (3)           (4)       (5)                        (6)
(7)    ITEM   SMR                              PART
 NO    CODE   NSN            CAGEC     NUMBER      DESCRIPTION AND USABLE ON CODES(UOC)
QTY
                                                   NNN
       15 PAOZZ 2540014753129 19207 12469034       HOUSING,TACHOMETER PART OF KIT P/N
                                                     57K3532.............................
UOC:AVY,A13,A14,BVY,HVY,H13,H14,H21,
                                                                                       NNN
       16 PAOZZ 5330013094340 57958 C5136338       GASKET PART OF KIT P/N 57K3532......  1
UOC:AVY,A13,A14,BVY,HVY,H13,H14,H21,
                                                                                       NNN
       17 PAOZZ 5310002255328 96906 MS15795-841    WASHER,FLAT PART OF KIT P/N 57K3532.  7
UOC:AVY,A13,A14,BVY,HVY,H13,H14,H21,
                                                                                       NNN
       18 PAOZZ 5310009338119 81100 110401-8SS     WASHER,LOCK PART OF KIT P/N 57K3532.  7
UOC:AVY,A13,A14,BVY,HVY,H13,H14,H21,
                                                                                       NNN
       19 PAOZZ 5305000546670 96906 MS51957-45     SCREW,MACHINE PART OF KIT P/N
                                                     57K3532.............................
UOC:AVY,A13,A14,BVY,HVY,H13,H14,H21,
                                                                                       NNN
       20 PAOZZ 5340013082168 57958 C5136336       COVER,ACCESS PART OF KIT P/N 57K3532  1
UOC:AVY,A13,A14,BVY,HVY,H13,H14,H21,
                                                                                       NNN
     * 21 PAOZZ 5310000453296 96906 MS35338-43     WASHER,LOCK.........................  1
UOC:AVY,A13,A14,BVY,HVY,H13,H14,H21,
                                                                                       NNN
     * 22 PAOZZ 5310009349751 96906 MS35650-302    NUT,PLAIN,HEXAGON...................  1
UOC:AVY,A13,A14,BVY,HVY,H13,H14,H21,
                                                                                       NNN

                                                   END OF FIGURE
```

Section II. TM 9-2320-280-24P-2

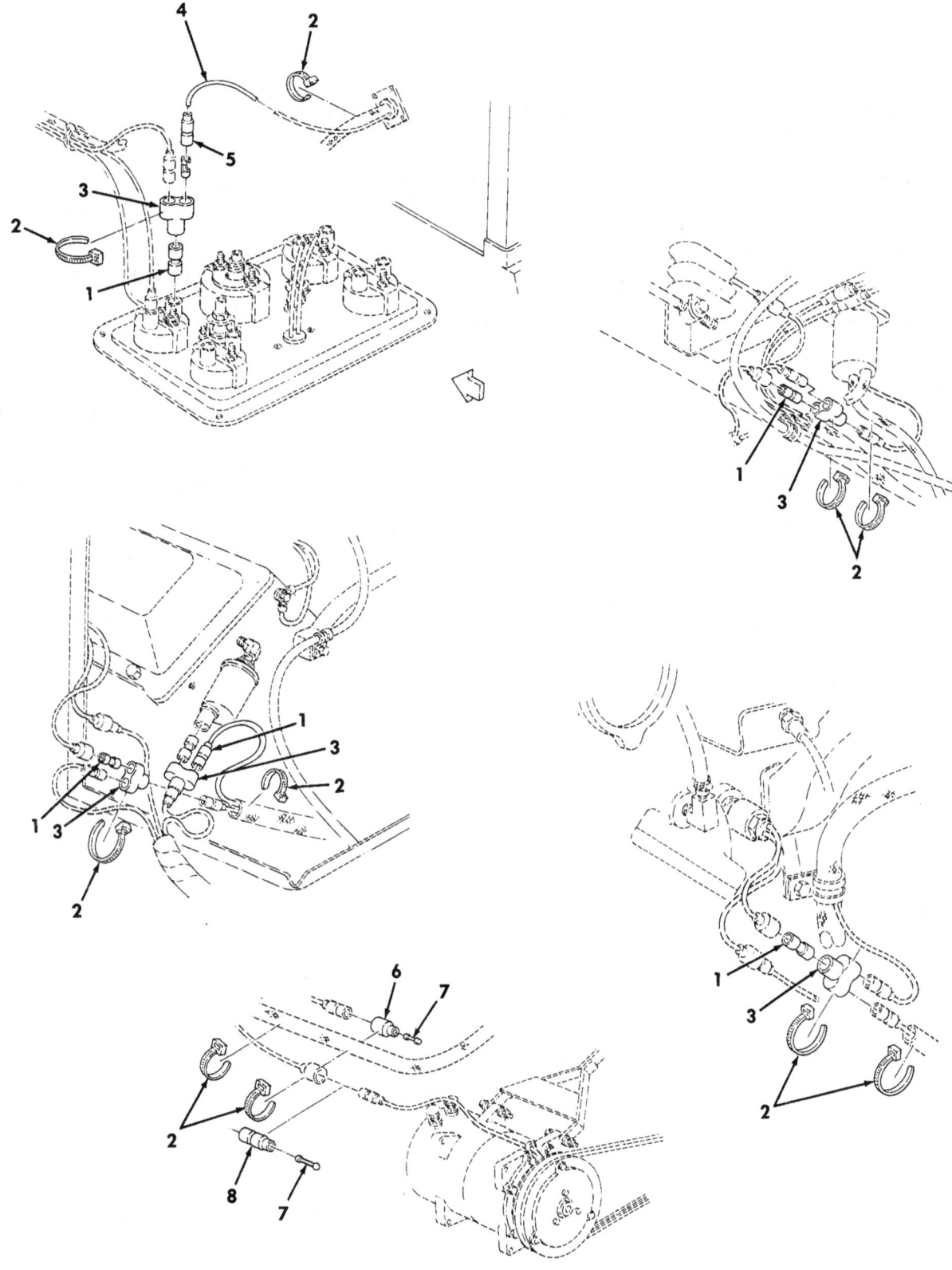

Figure 483. Tachometer Parts Kit.

SECTION II   TM9-2320-280-24P

| (1) NO | (2) CODE | (3) NSN | (4) CAGEC | (5) NUMBER | (7) ITEM DESCRIPTION AND USABLE ON CODES(UOC) | PART SMR QTY |

GROUP 3307 SPECIAL PURPOSE KITS

FIG. 483 TACHOMETER PARTS KIT

Item	SMR	NSN	CAGEC	Part Number	Description	Qty
1	PAOZZ	5935008074109	19207	8741492	ADAPTER,CONNECTOR PART OF KIT P/N 57K3532.............. UOC:AVY,A13,A14,BVY,HVY,H13,H14,H21,NNN	5
2	PAOZZ	5975009856630	96906	MS3367-3-0	STRAP,TIEDOWN,ELECT PART OF KIT P/N 57K3532.............. UOC:AVY,A13,A14,BVY,HVY,H13,H14,H21,NNN	12
3	PAOZZ	5935009006281	96906	MS27147-1	ADAPTER,CONNECTOR PART OF KIT P/N 57K3532.............. UOC:AVY,A13,A14,BVY,HVY,H13,H14,H21,NNN	5
4	PAOZZ	6150014738663	19207	12469029	WIRING HARNESS,BRAN PART OF KIT P/N 57K3532.............. UOC:AVY,A13,A14,BVY,HVY,H13,H14,H21,NNN	1
5	PAOZA	5935001152307	96906	MS27144-2	CONNECTOR,PLUG,ELEC PART OF KIT P/N 57K3532.............. UOC:AVY,A13,A14,BVY,HVY,H13,H14,H21,NNN	1
6	PAOZZ	5935005729180	19207	8338566	SHELL,ELECTRICAL CO PART OF KIT P/N 57K3532.............. UOC:AVY,A13,A14,BVY,HVY,H13,H14,H21,NNN	1
7	PAOZZ	5935002140904	19207	7982907	DUMMY CONNECTOR,PLU PART OF KIT P/N 57K3532.............. UOC:AVY,A13,A14,BVY,HVY,H13,H14,H21,NNN	2
8	PAOZZ	5935008338561	19207	8338561	SHELL,ELECTRICAL CO PART OF KIT P/N 57K3532.............. UOC:AVY,A13,A14,BVY,HVY,H13,H14,H21,NNN	1

END OF FIGURE

Section II.

TM 9-2320-280-24P-2

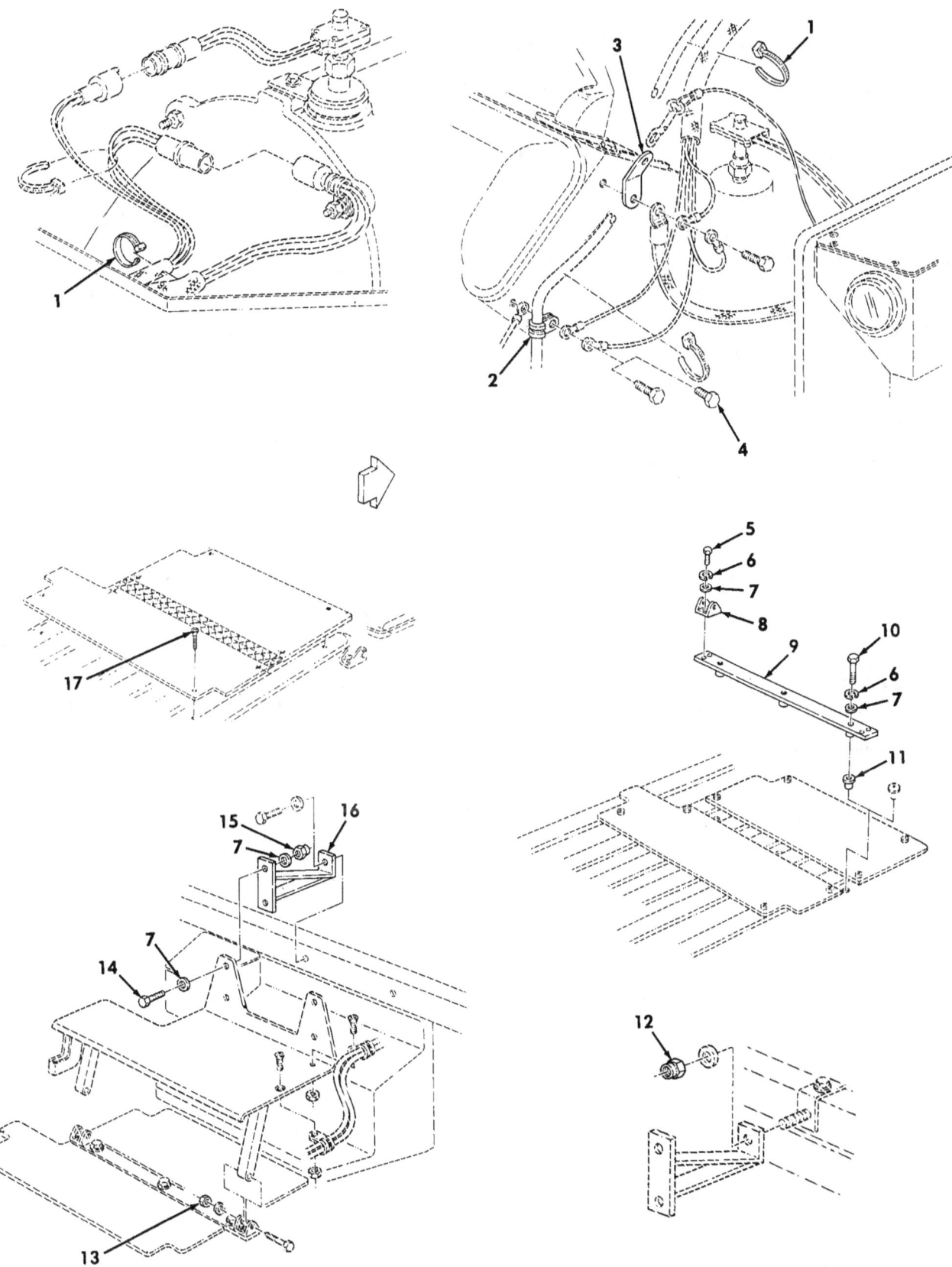

Figure 484. Tachometer Parts Kit.

SECTION II   TM9-2320-280-24P

(1) (2) (3)   (4)   (5)   (6)   (7) ITEM SMR   PART
NO CODE NSN   CAGEC NUMBER DESCRIPTION AND USABLE ON CODES(UOC) QTY

GROUP 3307 SPECIAL PURPOSE KITS

FIG. 484 TACHOMETER PARTS KIT

1 PAOZZ 5975009856630 96906 MS3367-3-0 STRAP,TIEDOWN,ELECT PART OF KIT P/N 57K3532............................ 4 UOC:AVY,A13,A14,BVY,HVY,H13,H14,H21, NNN

2 PAOZZ 5340009891771 96906 MS21333-123 CLAMP,LOOP PART OF KIT P/N 57K3532.. 1 UOC:AVY,A13,A14,BVY,HVY,H13,H14,H21, NNN

3 PAOZZ 5340013083851 57958 C5136342 BRACKET,ANGLE PART OF KIT P/N 57K3532............................ 1 UOC:AVY,A13,A14,BVY,HVY,H13,H14,H21, NNN

4 PAOZZ 5305013900547 80204 B18231B10020NF SCREW,CAP,HEXAGON H PART OF KIT P/N 57K3532............................ 1 UOC:AVY,A13,A14,BVY,HVY,H13,H14,H21, NNN

5 PAOZZ 5305000680502 96906 MS90725-6 SCREW,CAP,HEXAGON H PART OF KIT P/N 57K3532............................ 4 UOC:AVY,A13,A14,BVY,HVY,H13,H14,H21, NNN

6 PAOZZ 5310005501130 96906 MS35333-40 WASHER,LOCK PART OF KIT P/N 57K3532. 7 UOC:AVY,A13,A14,BVY,HVY,H13,H14,H21, NNN

7 PAOZZ 5310011023270 24617 2436161 WASHER,FLAT PART OF KIT P/N 57K3532. 15 UOC:AVY,A13,A14,BVY,HVY,H13,H14,H21, NNN

8 PAOZZ 5340014743279 19207 12460543 BRACKET,DOUBLE ANGL PART OF KIT P/N 57K3532............................ 2 UOC:AVY,A13,A14,BVY,HVY,H13,H14,H21, NNN

9 PAOZZ 5340014740954 19207 12460544 PLATE,MOUNTING PART OF KIT P/N 57K3532............................ 1 UOC:AVY,A13,A14,BVY,HVY,H13,H14,H21, NNN

10 PAOZZ 5305000712510 80204 B1821BH025C175N SCREW,CAP,HEXAGON H PART OF KIT P/N 57K3532............................ 3 UOC:AVY,A13,A14,BVY,HVY,H13,H14,H21, NNN

11 PAOZZ 5310014113422 78276 ALS4-420-165 NUT,PLAIN,BLIND RIV PART OF KIT P/N 57K3532............................ 3 UOC:AVY,A13,A14,BVY,HVY,H13,H14,H21, NNN

12 PAOZZ 5310009359022 96906 MS51943-32 NUT,SELF-LOCKING,HE PART OF KIT P/N 57K3532............................ 2 UOC:AVY,A13,A14,BVY,HVY,H13,H14,H21, NNN

13 PAOZZ 5310009359021 96906 MS51943-35 NUT,SELF-LOCKING,HE PART OF KIT P/N 57K3532............................ 2 UOC:AVY,A13,A14,BVY,HVY,H13,H14,H21, NNN

SECTION II   TM9-2320-280-24P

(1) NO	(2) CODE	(3) NSN	(4) CAGEC	(5) PART NUMBER	(7) DESCRIPTION AND USABLE ON CODES(UOC)	ITEM SMR	QTY
14	PAOZZ	5305002253843	80204	B1821BH025C100N	SCREW,CAP,HEXAGON H PART OF KIT P/N 57K3532.................... UOC:AVY,A13,A14,BVY,HVY,H13,H14,H21,		NNN
15	PAOZZ	5310000614650	96906	MS51943-31	NUT,SELF-LOCKING,HE PART OF KIT P/N 57K3532.................... UOC:AVY,A13,A14,BVY,HVY,H13,H14,H21,		NNN
16	PAOZZ	5340014330936	19207	12460542-2	BRACKET,MOUNTING PART OF KIT P/N 57K3532.................... UOC:AVY,A13,A14,BVY,HVY,H13,H14,H21,		NNN
17	PAOZZ	5305012707225	19207	12341822	SCREW,TAPPING THREA PART OF KIT P/N 57K3532.................... UOC:AVY,A13,A14,BVY,HVY,H13,H14,H21,		NNN
KIT	PAOZZ	5999014743213	19207	57K3532	PARTS KIT,TACHOMETE................. UOC:AVY,A13,A14,BVY,HVY,H13,H14,H21,		1 NNN

```
                ADAPTER,CONNECTOR  ( 5) 483-1
                ADAPTER,CONNECTOR  ( 5) 483-3
                BRACKET,MOUNTING   ( 2) 484-16
                BRACKET,DOUBLE ANGL( 2) 484-8
                BRACKET,ANGLE      ( 1) 484-3
                CABLE,ASSEMBLY,SPEC( 1) 482-4
                CLAMP,LOOP         ( 1) 484-2
                CONNECTOR,PLUG,ELEC( 1) 483-5
                COVER,ACCESS       ( 1) 482-20
                DECAL              ( 1) 482-1
                DUMMY CONNECTOR,PLU( 2) 483-7
                GASKET             ( 1) 482-5
                GASKET             ( 1) 482-16
                HOUSING,TACHOMETER ( 1) 482-15
                NUT,PLAIN,BLIND RIV( 3) 484-11
                NUT,PLAIN,HEXAGON  ( 2) 482-8
                NUT,SELF-LOCKING,HE( 2) 482-11
                NUT,SELF-LOCKING,HE( 4) 484-15
                NUT,SELF-LOCKING,HE( 2) 484-12
                NUT,SELF-LOCKING,HE( 2) 484-13
                PLATE,MOUNTING     ( 1) 484-9
                SCREW,CAP,HEXAGON H( 1) 482-14
                SCREW,CAP,HEXAGON H( 4) 484-5
                SCREW,CAP,HEXAGON H( 4) 484-14
                SCREW,CAP,HEXAGON H( 3) 484-10
                SCREW,CAP,HEXAGON H( 1) 484-4
                SCREW,MACHINE      ( 7) 482-19
                SCREW,TAPPING THREA( 8) 484-17
                SHELL,ELECTRICAL CO( 1) 483-8
                SHELL,ELECTRICAL CO( 1) 483-6
                STRAP,TIEDOWN,ELECT(12) 483-2
                STRAP,TIEDOWN,ELECT( 4) 484-1
                STRIP,ELECTRICAL GR( 1) 482-7
                SWITCH,TOGGLE      ( 1) 482-2
```

SECTION II	TM9-2320-280-24P						
(1) NO	(2) CODE	(3) NSN	(4) CAGEC	(5) NUMBER	(6)	(7) ITEM DESCRIPTION AND USABLE ON CODES(UOC)	SMR PART QTY

```
                    TACHOMETER,ELECTRIC( 1) 482-3
                    WASHER,FLAT      ( 2) 482-12
                    WASHER,FLAT      ( 1) 482-10
                    WASHER,FLAT      ( 7) 482-17
                    WASHER,FLAT      ( 15) 484-7
                    WASHER,LOCK      ( 2) 482-13
                    WASHER,LOCK      ( 7) 482-18
                    WASHER,LOCK      ( 1) 482-9
                    WASHER,LOCK      ( 7) 484-6
                    WIRE ROPE ASSEMBLY,( 1) 482-6
                    WIRING HARNESS,BRAN( 1) 483-4

                    END OF FIGURE
```

Section II. TM 9-2320-280-24P-2

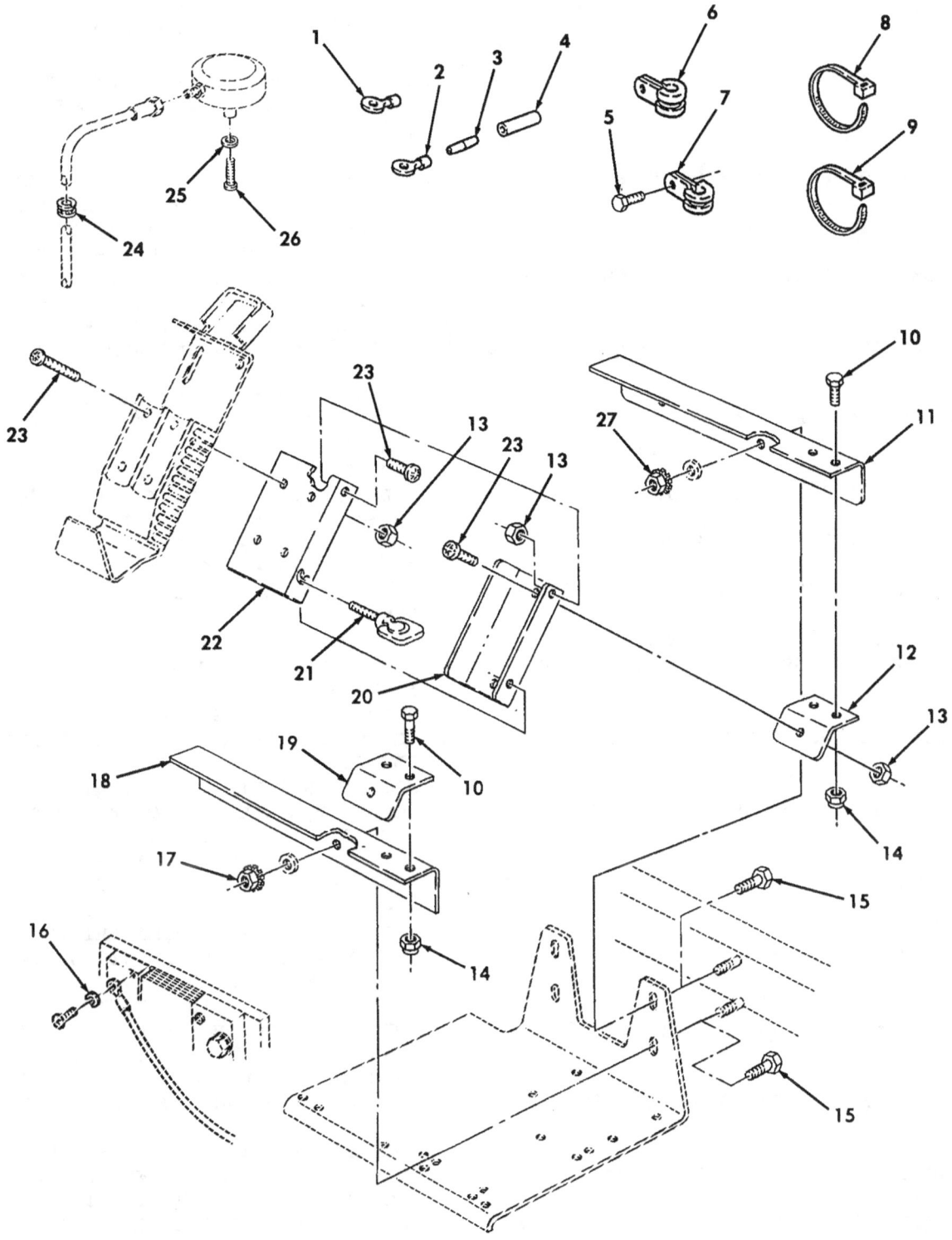

Figure 485. Precision Lightweight Global Positioning System Receiver (PLGR).

```
    SECTION II            TM9-2320-280-24P C01
  (1)    (2)    (3)         (4)       (5)                          (6)
(7)    ITEM   SMR                               PART
  NO    CODE   NSN          CAGEC    NUMBER     DESCRIPTION AND USABLE ON CODES(UOC)
QTY

                                                GROUP 3307 SPECIAL PURPOSE KITS

                                                FIG. 485 PRECISION LIGHTWEIGHT
                                                GLOBAL POSTIONING SYSTEM RECEIVER
                                                (PLGR)


       1 PAOZZ 5940009834067 16528  7728777     TERMINAL,LUG PART OF KIT P/N 57K3233
1
                                                UOC:A11,B17,B18,C17,H11,H17,H18,M-
MM     2 PAOZZ 5940010354212 19207  7728780     TERMINAL,LUG PART OF KIT P/N 57K3233
1
                                                UOC:A11,B17,B18,C17,H11,H17,H18,M-
MM     3 PAOZZ 5940001139420 80205  NAS1387-4   SPLICE,CONDUCTOR PART OF KIT P/N
2                                               57K3233..........................
..
                                                UOC:A11,B17,B18,C17,H11,H17,H18,M-
MM     4 PAOZZ 5970010489378 81349  M23053/4-103-0 INSULATION SLEEVING PART OF KIT P/N
1                                               57K3233..........................
.
                                                UOC:A11,B17,B18,C17,H11,H17,H18,M-
MM     5 PAOZZ 5305004324201 96906  MS51861-45  SCREW,TAPPING PART OF KIT P/N
1                                               57K3233..........................
.
                                                UOC:A11,B17,B18,C17,H11,H17,H18,M-
MM     6 PAOZZ 5340009050790 96906  MS21333-65  CLAMP,LOOP PART OF KIT P/N 57K3233..
1                                               UOC:A11,B17,B18,C17,H11,H17,H18,M-
MM     7 PAOZZ 5340000881255 96906  MS21333-96  CLAMP,LOOP PART OF KIT P/N 57K3233..
1                                               UOC:A11,B17,B18,C17,H11,H17,H18,M-
MM     8 PAOZZ 5975000742072 96906  MS3367-1-9  STRAP,TIEDOWN,ELECT PART OF KIT P/N
2                                               57K3233..........................
UOC:A11,B17,B18,C17,H11,H17,H18,MMM
     * 9 PAOZZ 5975010345871 96906  MS3367-7-0  STRAP,TIEDOWN,ELECT PART OF KIT P/N
1                                               57K3233..........................
..                                              UOC:A11,B17,B18,C17,H11,H17,H18,M-
MM    10 PAOZZ 5305000680508 80204  B1821BH025C075N SCREW,CAP,HEXAGON H PART OF KIT P/N
4                                               57K3233..........................
.                                               UOC:A11,B17,B18,C17,H11,H17,H18,M-
MM    11 PAOZA 5340014662451 19207  12460221    BRACKET,ANGLE PART OF KIT P/N
1                                               57K3233..........................
.                                               UOC:A11,B17,B18,C17,H11,H17,H18,M-
MM    12 PAOZA 5340014711040 19207  12460220-1  BRACKET,ANGLE PART OF KIT P/N
1                                               57K3233..........................
.                                               UOC:A11,B17,B18,C17,H11,H17,H18,M-
MM    13 PAOZZ 5310001249265 72582  271169      NUT,PLAIN,ASSEMBLED  #10-32 PART
OF     8                                        KIT P/N 57K3233.................
....                                            UOC:A11,B17,B18,C17,H11,H17,H18,M-
MM    14 PAOZZ 5310000614650 96906  MS51943-31  NUT,SELF-LOCKING,HE PART OF KIT P/N
4                                               57K3233..........................
..                                              UOC:A11,B17,B18,C17,H11,H17,H18,M-
MM    15 PAOZZ 5305002678953 80204  B1821BH025F063N SCREW,CAP,HEXAGON H PART OF KIT P/N
2                                               57K3233..........................
.                                               UOC:A11,B17,B18,C17,H11,H17,H18,M-
MM    16 PAOZZ 5310000453296 96906  MS35338-43  WASHER,LOCK PART OF KIT P/N 57K3233.
1                                               UOC:A11,B17,B18,C17,H11,H17,H18,M-
MM    17 PAOZZ 5310009359022 96906  MS51943-32  NUT,SELF-LOCKING,HE PART OF KIT P/N
2                                               57K3233..........................
.                                               UOC:A11,B17,B18,C17,H11,H17,H18,MMM
    18 PAOZA 5340014662438 19207  12460222      BRACKET,ANGLE PART OF KIT P/N         1
```

485-1

SECTION II   TM9-2320-280-24P

(1) ITEM NO	(2) SMR CODE	(3) NSN	(4) CAGEC	(5) PART NUMBER	(6) DESCRIPTION AND USABLE ON CODES(UOC)	(7) QTY
19	PAOZA	5340014661985	19207	12460220-2	BRACKET,ANGLE PART OF KIT P/N 57K3233.................... UOC:A11,B17,B18,C17,H11,H17,H18,MMM	1
20	PAOZA	5340014662606	19207	12460219	BRACKET,DOUBLE ANGL PART OF KIT P/N 57K3233.................... UOC:A11,B17,B18,C17,H11,H17,H18,MMM	1
21	PAOZZ	5305014717623	19207	12460223	THUMBSCREW PART OF KIT P/N 57K3233.................... UOC:A11,B17,B18,C17,H11,H17,H18,MMM	1
22	PAOZA	5340014662611	19207	12460218	BRACKET,DOUBLE ANGL PART OF KIT P/N 57K3233.................... UOC:A11,B17,B18,C17,H11,H17,H18,MMM	1
23	PAOZZ	5305014077713	24617	60052	SCREW,MACHINE PART OF KIT P/N 57K3233.................... UOC:A11,B17,B18,C17,H11,H17,H18,MMM	8
24	PAOZZ	5325012927558	96906	MS35489-43	GROMMET,NONMETALLIC PART OF KIT P/N 57K3233.................... UOC:A11,B17,B18,C17,H11,H17,H18,MMM	1
25	PAOZZ	5310013611163	24617	9417373	WASHER,FLAT PART OF KIT P/N 57K3233. UOC:A11,B17,B18,C17,H11,H17,H18,MMM	1
26	PAOZZ	5305009906444	96906	MS35207-261	SCREW,MACHINE PART OF KIT P/N 57K3233.................... UOC:A11,B17,B18,C17,H11,H17,H18,MMM	1
27	PAOZZ	5310013335245	24617	71175	NUT,SELF-LOCKING,HE PART OF KIT P/N 57K3233.................... UOC:A11,B17,B18,C17,H11,H17,H18,MMM	2
KIT	PAOZZ	2590014204179	19207	57K3233	PARTS KIT,BRACKET,V................ UOC:A11,B17,B18,C17,H11,H17,H18,MMM	1

```
BRACKET,ANGLE       ( 1) 485-11
BRACKET,ANGLE       ( 1) 485-12
BRACKET,ANGLE       ( 1) 485-18
BRACKET,ANGLE       ( 1) 485-19
BRACKET,DOUBLE ANGL ( 1) 485-20
BRACKET,DOUBLE ANGL ( 1) 485-22
CLAMP,LOOP          ( 1) 485-6
CLAMP,LOOP          ( 1) 485-7
GROMMET,NONMETALLIC ( 1) 485-24
INSULATION SLEEVING ( 1) 485-4
NUT,PLAIN,ASSEMBLED ( 8) 485-13
NUT,SELF-LOCKING,HE ( 4) 485-14
NUT,SELF-LOCKING,HE ( 2) 485-17
NUT,SELF-LOCKING,HE ( 2) 485-27
SCREW,CAP,HEXAGON H ( 4) 485-10
SCREW,CAP,HEXAGON H ( 2) 485-15
SCREW,MACHINE       ( 8) 485-23
SCREW,MACHINE       ( 1) 485-26
SCREW,TAPPING       ( 1) 485-5
SPLICE,CONDUCTOR    ( 2) 485-3
STRAP,TIEDOWN,ELECT ( 2) 485-8
STRAP,TIEDOWN,ELECT ( 1) 485-9
TERMINAL,LUG        ( 1) 485-1
TERMINAL,LUG        ( 1) 485-2
```

SECTION II   TM9-2320-280-24P

(1) NO	(2) CODE	(3) NSN	(4) CAGEC	(5) NUMBER	(6) PART	(7) ITEM SMR DESCRIPTION AND USABLE ON CODES(UOC) QTY
					THUMBSCREW	(1) 485-21
					WASHER,FLAT	(1) 485-25
					WASHER,LOCK	(1) 485-16

END OF FIGURE

Section II. TM 9-2320-280-24P-2

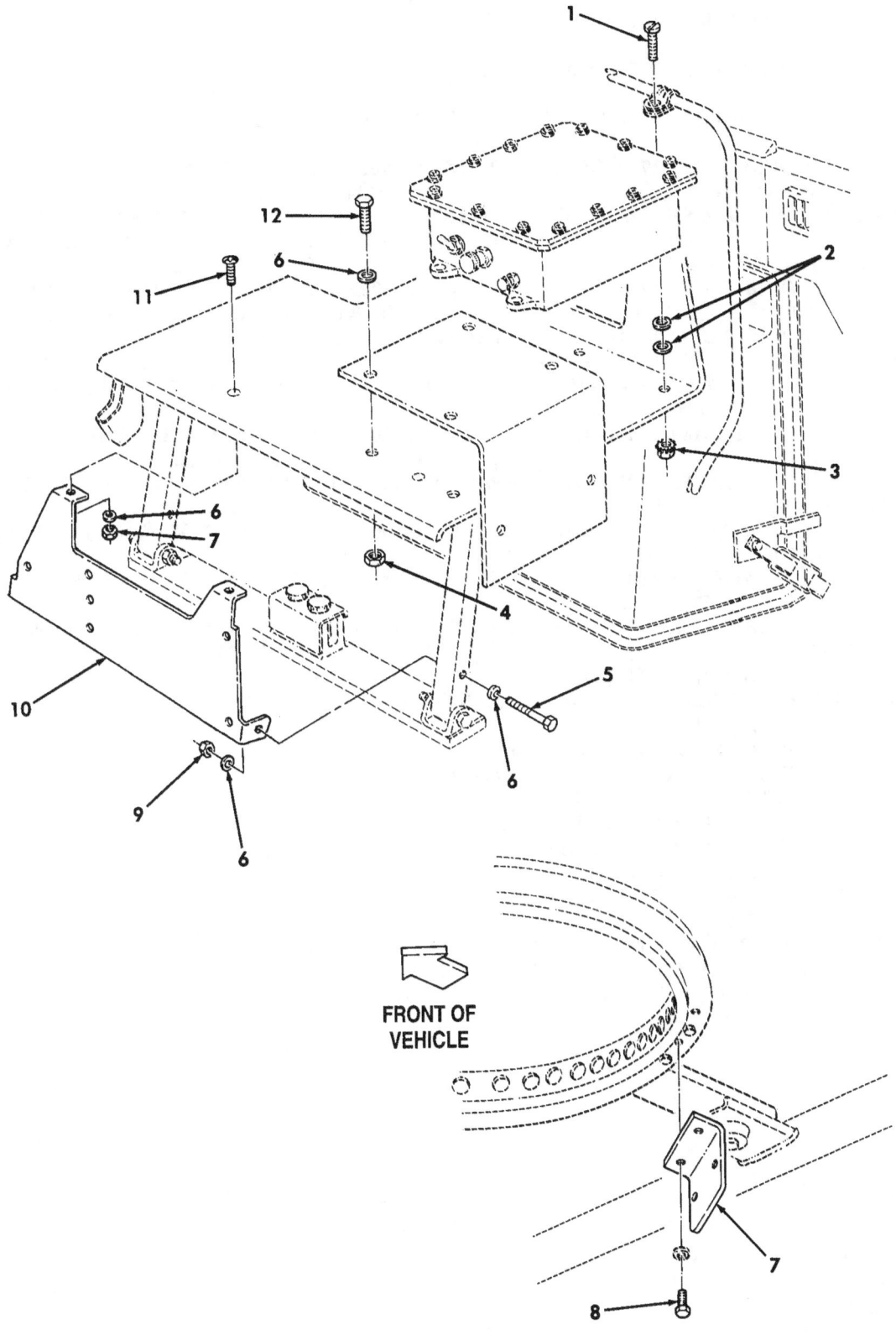

Figure 486. Kit, Intercom Bracket.

SECTION II   TM9-2320-280-24P

(1) NO	(2) SMR CODE	(3) NSN	(4) CAGEC	(5) PART NUMBER	(7) ITEM DESCRIPTION AND USABLE ON CODES(UOC)	QTY

GROUP 3307 SPECIAL PURPOSE KITS

FIG. 486 KIT, INTERCOM BRACKET

1	PAOZZ	5305014672711	24617	160069	SCREW,MACHINE #10-32X1 PART OF KIT P/N 57K3488.................... UOC:B17,B18,C17,H17,H18	4
2	PAOZZ	5310012175205	24617	9417714	WASHER,FLAT #10FLAT PART OF KIT P/N 57K3488.................... UOC:B17,B18,C17,H17,H18	8
3	PAOZZ	5310002514503	24617	9411893	NUT,LOCK #10-32 PART OF KIT P/N 57K3488.................... UOC:B17,B18,C17,H17,H18	4
4	PAOZZ	5310014326727	24617	9419471	NUT,SELF-LOCKING,HE 1/4-20 PART OF KIT P/N 57K3488.................... UOC:B17,B18,C17,H17,H18	2
5	PAOZZ	5305000712237	96906	MS90725-14	SCREW,CAP,HEXAGON H 1/4-20X2.00 PART OF KIT P/N 57K3488............. UOC:B17,B18,C17,H17,H18	2
6	PAOZZ	5310011023270	24617	2436161	WASHER,FLAT 1/4 PART OF KIT P/N 57K3488.................... UOC:B17,B18,C17,H17,H18	8
7	PAOZZ	5340014674153	80063	A3157855	BRACKET,ANGLE #10-32X1 PART OF KIT P/N 57K3488.................... UOC:B17,B18,C17,H17,H18	1
8	PAOZZ	5305008213869	80204	B1821BH038C175N	SCREW,CAP,HEXAGON H 3/8-16 X 1.75 PART OF KIT P/N 57K3488............. UOC:B17,B18,C17,H17,H18	2
9	PAOZZ	5310000614650	96906	MS51943-31	NUT,SELF-LOCKING,HE 1/4-20 PART OF KIT P/N 57K3488............. UOC:B17,B18,C17,H17,H18	4
10	PAOZZ	5340012708489	19207	12340879	BRACKET,MOUNTING PART OF KIT P/N 57K3488.................... UOC:B17,B18,C17,H17,H18	1
11	PAOZZ	5305007195021	96906	MS51959-81	SCREW,MACHINE 1/4-20X.75 PART OF KIT P/N 57K3488.................... UOC:B17,B18,C17,H17,H18	2
12	PAOZZ	5305002253843	80204	B1821BH025C100N	SCREW,CAP,HEXAGON H 1/4-20X1.00 PART OF KIT P/N 57K3488............. UOC:B17,B18,C17,H17,H18	2
KIT	PAOZZ	2590014293008	19207	57K3488	PARTS KIT,BRACKET,V................. UOC:B17,B18,C17,H17,H18	1

    BRACKET,ANGLE   (1) 486-7
    BRACKET,MOUNTING   (1) 486-10
    NUT,LOCK   (4) 486-3
    NUT,SELF-LOCKING,HE   (2) 486-4
    NUT,SELF-LOCKING,HE   (4) 486-9
    SCREW,CAP,HEXAGON H   (2) 486-5
    SCREW,CAP,HEXAGON H   (2) 486-8
    SCREW,CAP,HEXAGON H   (2) 486-12
    SCREW,MACHINE   (4) 486-1
    SCREW,MACHINE   (2) 486-11
    WASHER,FLAT   (8) 486-2

SECTION II   TM9-2320-280-24P

(1) NO	(2) CODE	(3) NSN	(4) CAGEC	(5) NUMBER	(6) DESCRIPTION AND USABLE ON CODES(UOC)	(7) ITEM SMR PART QTY
					WASHER,FLAT	(8) 486-6

END OF FIGURE

Section II.

TM 9-2320-280-24P-2

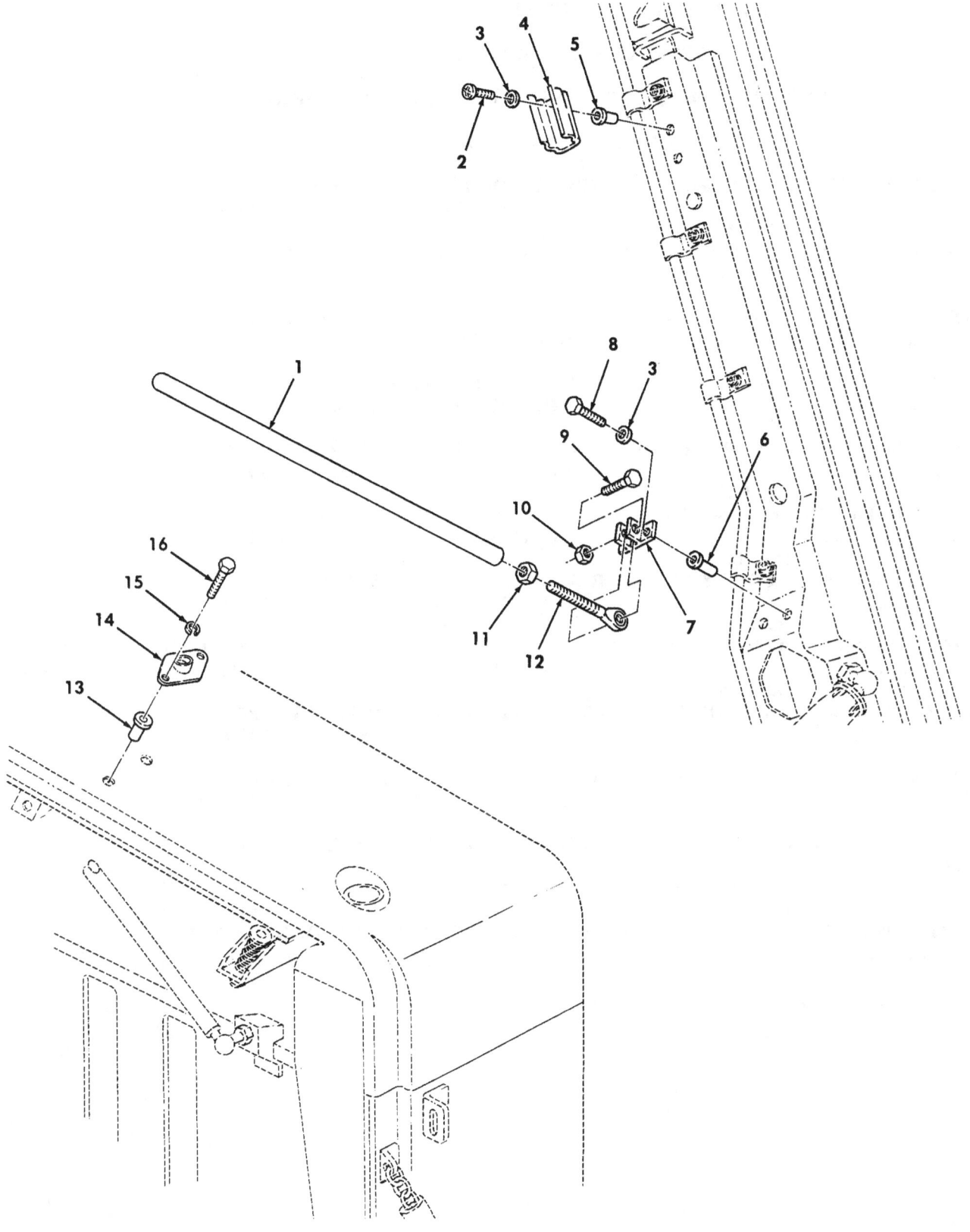

Figure 487. Kit, Rear Hatch Support

SECTION II   TM9-2320-280-24P

(1) NO	(2) SMR CODE	(3) NSN	(4) CAGEC	(5) PART NUMBER	(7) ITEM DESCRIPTION AND USABLE ON CODES(UOC)	QTY

GROUP 3307 SPECIAL PURPOSE KITS

FIG. 487 KIT, REAR HATCH SUPPORT

1 PAOZZ 5342014594992 19207 12460330 EXTENSION,CONTROL R PART OF KIT P/N 57K3243.......................... 1
UOC:A25,A26,B17,B18,B25,C17,H17,H18,H25,H26

2 PAOZZ 5305014077713 24617 160052 SCREW,MACHINE PART OF KIT P/N 57K3243............................ 2
UOC:A25,A26,B17,B18,B25,C17,H17,H18,H25,H26

3 PAOZZ 5310009222017 30379 120217 WASHER,LOCK PART OF KIT P/N 57K3243. 4
UOC:A25,A26,B17,B18,B25,C17,H17,H18,H25,H26

4 PAOZZ 5340014508693 81343 NAS1464-050-10T CLIP,SPRING TENSION PART OF KIT P/N 57K3243............................. 1
UOC:A25,A26,B17,B18,B25,C17,H17,H18,H25,H26

5 PAOZZ 5310009901361 96906 MS27130-S25 NUT,PLAIN,BLIND RIV PART OF KIT P/N 57K3243............................. 4
UOC:A25,A26,B17,B18,B25,C17,H17,H18,H25,H26

6 PAOZZ 5310010465371 96906 MS27130-CR27 NUT,PLAIN,BLIND RIV PART OF KIT P/N 57K3243............................. 2
UOC:A25,A26,B17,B18,B25,C17,H17,H18,H25,H26

7 PAOZZ 5340014695774 19207 12460331 BRACKET,MULTIPLE AN PART OF KIT P/N 57K3243............................. 1
UOC:A25,A26,B17,B18,B25,C17,H17,H18,H25,H26

8 PAOZZ 5305000187838 24617 187838 SCREW,CAP,HEXAGON H PART OF KIT P/N 57K3243............................. 2
UOC:A25,A26,B17,B18,B25,C17,H17,H18,H25,H26

9 PAOZZ 5305000680509 80204 B1821BH025C125N SCREW,CAP,HEXAGON H PART OF KIT P/N 57K3243............................. 1
UOC:A25,A26,B17,B18,B25,C17,H17,H18,H25,H26

10 PAOZZ 5310014326727 24617 9419471 NUT,SELF-LOCKING,HE PART OF KIT P/N 57K3243............................. 1
UOC:A25,A26,B17,B18,B25,C17,H17,H18,H25,H26

11 PAOZZ 5310007680319 96906 MS51968-2 NUT,PLAIN,HEXAGON PART OF KIT P/N 57K3243............................. 1
UOC:A25,A26,B17,B18,B25,C17,H17,H18,H25,H26

12 PAOZZ 81343 SAEJ1120 ROD END,THREADED SIZE4,TYPEB3 PART OF KIT P/N 57K3243.................. 1
UOC:A25,A26,B17,B18,B25,C17,H17,H18,H25,H26

13 PAOZZ 5310000802411 96906 MS27130-S43 NUT,PLAIN,BLIND RIV PART OF KIT P/N 57K3243............................. 2

57

SECTION II   TM9-2320-280-24P

(1)	(2)	(3)	(4)	(5)	(6)	(7) ITEM		
NO	CODE	NSN	CAGEC	NUMBER	DESCRIPTION AND USABLE ON CODES(UOC)		SMR	PART QTY

```
                        UOC:B17,B18,C17,H17,H18
13  PAOZZ 5310010805746 96906 MS27130-S44   NUT,PLAIN,BLIND RIV PART OF KIT P/N   2
                                             57K3243............................
                        UOC:A25,A26,B25,H25,H26
14  PAOZZ 5340014515903 19207 12460329      PLATE,MOUNTING PART OF KIT P/N        1
                                             57K3243............................
                        UOC:A25,A26,B17,B18,B25,C17,H17,H18,
                            H25,H26
15  PAOZZ 5310004079566 96906 MS-35338-45   WASHER,LOCK PART OF KIT P/N 57K3243.  2
                        UOC:A25,A26,B17,B18,B25,C17,H17,H18,
                            H25,H26
16  PAOZZ 5306002264825 80204 B1821BH031C075N BOLT,MACHINE PART OF KIT P/N 57K3243 2
                        UOC:A25,A26,B17,B18,B25,C17,H17,H18,
                            H25,H26
KIT PAOZZ 2510014338551 19207 57K3243        PARTS KIT,CARGO HAT.................  1
                        UOC:A25,A26,B17,B18,B25,C17,H17,H18,
                            H25,H26
                          BOLT,MACHINE       ( 2) 487-16
                          BRACKET,MULTIPLE AN( 1) 487-7
                          CLIP,SPRING TENSION( 1) 487-4
                          EXTENSION,CONTROL R( 1) 487-1
                          NUT,PLAIN,BLIND RIV( 4) 487-5
                          NUT,PLAIN,BLIND RIV( 2) 487-6
                          NUT,PLAIN,BLIND RIV( 2) 487-13
                          NUT,PLAIN.BLIND RIV( 2) 487-13
                          NUT,PLAIN,HEXAGON  ( 1) 487-11
                          NUT,SELF-LOCKING,HE( 1) 487-10
                          PLATE,MOUNTING     ( 1) 487-14
                          ROD,END            ( 1) 487-12
                          SCREW,CAP,HEXAGON H( 2) 487-8
                          SCREW,CAP,HEXAGON H( 1) 487-9
                          SCREW,MACHINE      ( 2) 487-2
                          WASHER,LOCK        ( 4) 487-3
                          WASHER,LOCK        ( 2) 487-15
```

END OF FIGURE

Section II. TM 9-2320-280-24P-2 C01

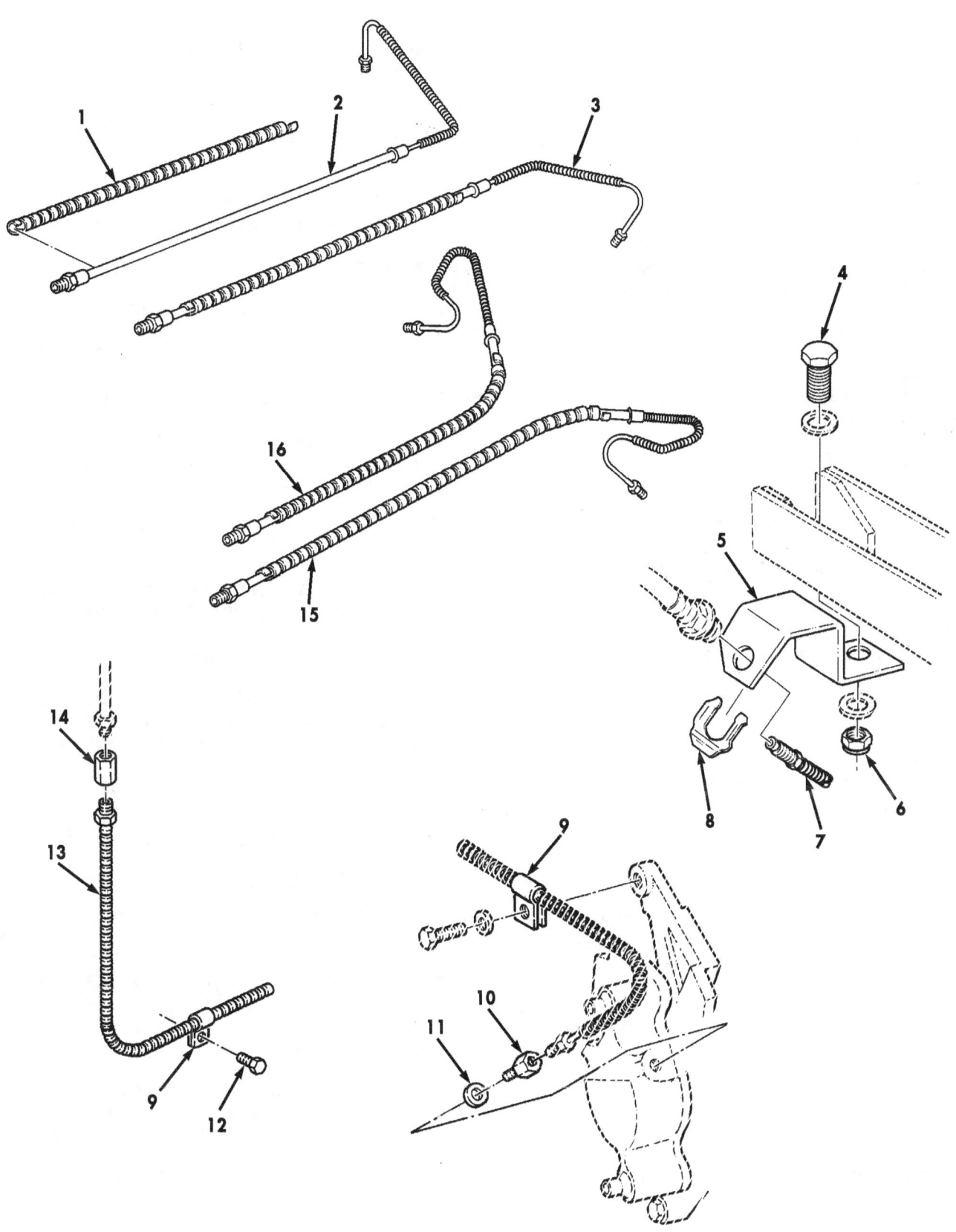

Figure 487A. Flexible Brake Line Kit.

```
     SECTION II           TM9-2320-280-24P C01
    (1)    (2)    (3)          (4)         (5)                    (6)
(7)  ITEM   SMR                                PART
     NO     CODE   NSN         CAGEC       NUMBER   DESCRIPTION AND USABLE ON CODES(UOC)
QTY
```

GROUP 3307 SPECIAL PURPOSE KITS

FIG. 487A FLEXIBLE BRAKE LINE KIT

```
* 1 PAOZZ 9330012521239 81349 B47207-II-1/2  TUBING,PLASTIC,SPRI  PART OF KIT P/N    4
57K3503.............................                                            UOC
:AVY,A11,A13,A14,A15,A20,A24,A25,

                                             A26,A27,B16,B17,B18,HVY,H11,H13,H14,
                                             H15,H16,H17,H18,H20,H21,H24,H25,H26,
H27,H28,MMM
* 2 PAOZZ 4720014433033 34623 RCSK18434-3    HOSE ASSEMBLY,NONME  L.H.,PART OF       1
KIT P/N 57K3503....................
                                             UOC:AVY,A11,A13,A14,A15,A20,A24,A25,
                                             A26,A27,B16,B17,B18,HVY,H11,H13,H14,
                                             H15,H16,H17,H18,H20,H21,H24,H25,H26,
H27,H28,MMM
* 3 PAOZZ 4720014429875 34623 RCSK18434-2    HOSE ASSEMBLY,NONME  R.H.,PART OF       1
KIT P/N 57K3503....................
                                             UOC:AVY,A11,A13,A14,A15,A20,A24,A25,
                                             A26,A27,B16,B17,B18,HVY,H11,H13,H14,
                                             H15,H16,H17,H18,H20,H21,H24,H25,H26,
H27,H28,MMM
* 4 PAOZZ 5305000712067 80204 B1821BH050C125N SCREW,CAP,HEXAGON H  PART OF KIT P/N   4
57K3503.............................                                            UOC
:AVY,A11,A13,A14,A15,A20,A24,A25,

                                             A26,A27,B16,B17,B18,HVY,H11,H13,H14,
                                             H15,H16,H17,H18,H20,H21,H24,H25,H26,
H27,H28,MMM
* 5 PAOZZ 5340014348109 34623 EX4228-2       BRACKET,MULTIPLE AN  L.H.,PART OF       2
KIT P/N 57K3503....................
                                             UOC:AVY,A11,A13,A14,A15,A20,A24,A25,
                                             A26,A27,B16,B17,B18,HVY,H11,H13,H14,
                                             H15,H16,H17,H18,H20,H21,H24,H25,H26,
H27,H28,MMM
* 5 PAOZZ 5340014328681 34623 EX4228-1       BRACKET,MULTIPLE AN  R.H.,PART OF       2
KIT P/N 57K3503....................
                                             UOC:AVY,A11,A13,A14,A15,A20,A24,A25,
                                             A26,A27,B16,B17,B18,HVY,H11,H13,H14,
                                             H15,H16,H17,H18,H20,H21,H24,H25,H26,
H27,H28,MMM
* 6 PAOZZ 5310004883889 96906 MS51943-39     NUT,SELF-LOCKING,HE  PART OF KIT P/N    4
57K3503.............................                                            UOC
:AVY,A11,A13,A14,A15,A20,A24,A25,

                                             A26,A27,B16,B17,B18,HVY,H11,H13,H14,
                                             H15,H16,H17,H18,H20,H21,H24,H25,H26,
H27,H28,MMM
* 7 PAOZZ 4710014343151 34623 EX4231-3       TUBE ASSEMBLY,METAL  PART OF KIT        2
P/N 57K3503........................
                                             UOC:AVY,A11,A13,A14,A15,A20,A24,A25,
                                             A26,A27,B16,B17,B18,HVY,H11,H13,H14,
                                             H15,H16,H17,H18,H20,H21,H24,H25,H26,
H27,H28,MMM
* 8 PAOZZ 5342006781753 19207 7331407        CLIP,HAND BRAKE      PART OF KIT P/N    4
```

487A-1

SECTION II   TM9-2320-280-24P C01

(1)	(2)	(3)	(4)	(5)	(6)	(7)
ITEM NO	SMR CODE	NSN	CAGEC	PART NUMBER	DESCRIPTION AND USABLE ON CODES (UOC)	QTY
					57K3503............................ UOC:AVY,A11,A13,A14,A15,A20,A24,A25, A26,A27,B16,B17,B18,HVY,H11,H13,H14, H15,H16,H17,H18,H20,H21,H24,H25,H26, H27,H28,MMM	
9	PAOZZ	5340011897640	7X677	179148	CLAMP,LOOP PART OF KIT P/N 57K3503. UOC:AVY,A11,A13,A14,A15,A20,A24,A25, A26,A27,B16,B17,B18,HVY,H11,H13,H14, H15,H16,H17,H18,H20,H21,H24,H25,H26, H27,H28,MMM	2
10	PAOZZ	4730011846971	34623	12339235	COUPLING,TUBE PART OF KIT P/N 57K3503............................ UOC: AVY,A11,A13,A14,A15,A20,A24,A25, A26,A27,B16,B17,B18,HVY,H11,H13,H14, H15,H16,H17,H18,H20,H21,H24,H25,H26, H27,H28,MMM	2
11	PAOZZ	5310011898476	34623	5582366	WASHER,FLAT PART OF KIT P/N 57K3503 UOC:AVY,A11,A13,A14,A15,A20,A24,A25, A26,A27,B16,B17,B18,HVY,H11,H13,H14, H15,H16,H17,H18,H20,H21,H24,H25,H26, H27,H28,MMM	2
12	PAOZZ	5305013936311	24617	271153	SCREW,TAPPING PART OF KIT P/N 57K3503............................ UOC: AVY,A11,A13,A14,A15,A20,A24,A25, A26,A27,B16,B17,B18,HVY,H11,H13,H14, H15,H16,H17,H18,H20,H21,H24,H25,H26, H27,H28,MMM	2
13	PAOZZ	4710014343037	34623	EX4233	TUBE ASSEMBLY,METAL PART OF KIT P/N 57K3503............................ UOC: AVY,A11,A13,A14,A15,A20,A24,A25, A26,A27,B16,B17,B18,HVY,H11,H13,H14, H15,H16,H17,H18,H20,H21,H24,H25,H26, H27,H28,MMM	2
14	PAOZZ	4730014570727	72582	442393	COUPLING,HOSE PART OF KIT P/N 57K3503............................ UOC: AVY,A11,A13,A14,A15,A20,A24,A25, A26,A27,B16,B17,B18,HVY,H11,H13,H14, H15,H16,H17,H18,H20,H21,H24,H25,H26, H27,H28,MMM	1
15	PAOZZ	4720014433481	34623	RCSK18435-3	TUBE ASSEMBLY,METAL L.H.,PART OF KIT P/N 57K3503..................... UOC:AVY,A11,A13,A14,A15,A20,A24,A25, A26,A27,B16,B17,B18,HVY,H11,H13,H14, H15,H16,H17,H18,H20,H21,H24,H25,H26, H27,H28,MMM	1
16	PAOZZ	4720014438487	34623	RCSK18435-2	HOSE ASSEMBLY,NONME R.H.,PART OF KIT P/N 57K3503..................... UOC:AVY,A11,A13,A14,A15,A20,A24,A25, A26,A27,B16,B17,B18,HVY,H11,H13,H14, H15,H16,H17,H18,H20,H21,H24,H25,H26, H27,H28,MMM	1

END OF FIGURE

Section II. TM 9-2320-280-24P-2 C01

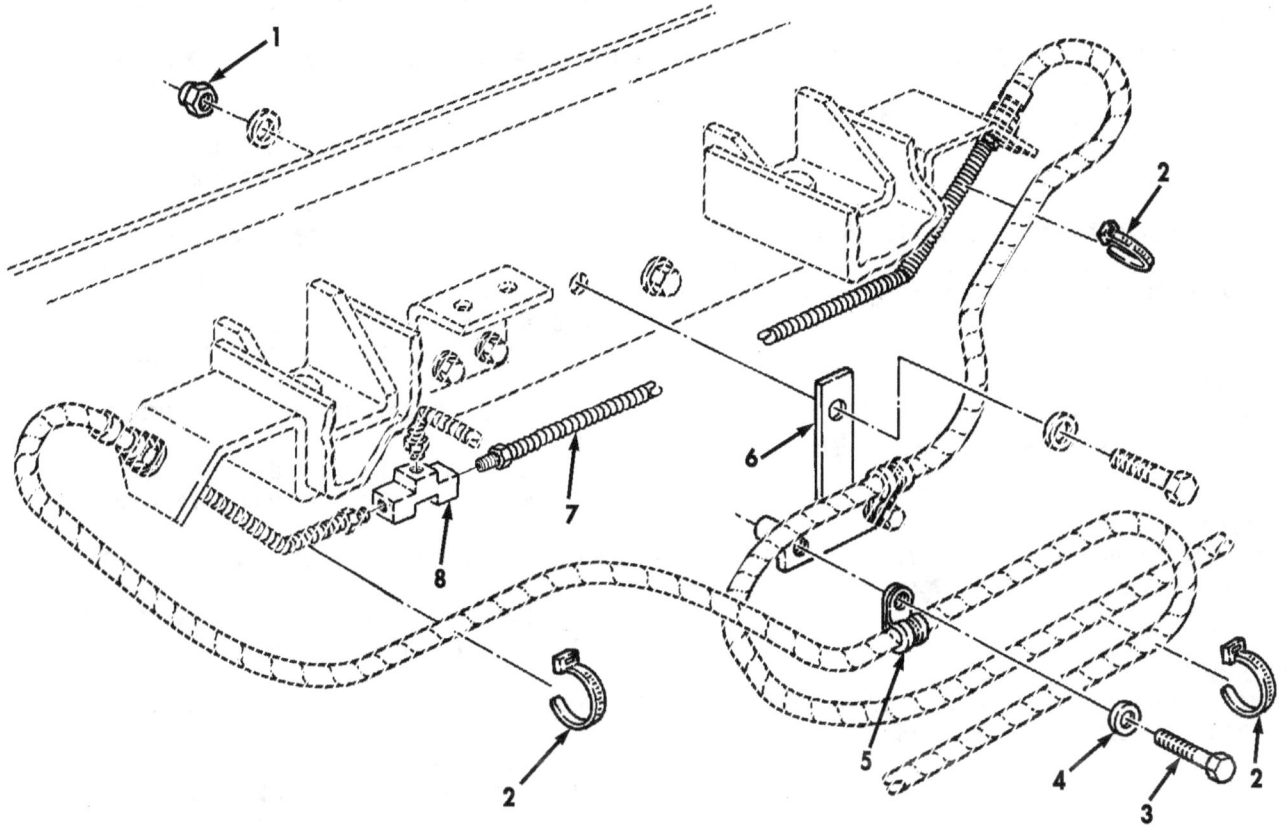

Figure 487B. Flexible Brake Line Kit.

```
SECTION II              TM9-2320-280-24P C01
  (1)    (2)    (3)           (4)       (5)                    (6)
(7)    ITEM   SMR                              PART
       NO     CODE   NSN           CAGEC       NUMBER      DESCRIPTION AND USABLE ON CODES(UOC)
QTY
```

GROUP 3307 SPECIAL PURPOSE KITS

FIG. 487B FLEXIBLE BRAKE LINE KIT

```
* 1 PAOZZ 5310004883889 96906 MS51943-39    NUT,SELF-LOCKING,HE  PART OF KIT P/N   1
57K3503..............................                                             UOC
:AVY,A11,A13,A14,A15,A20,A24,A25,
                                            A26,A27,B16,B17,B18,HVY,H11,H13,H14,
                                            H15,H16,H17,H18,H20,H21,H24,H25,H26,
H27,H28,MMM
* 2 PAOZZ 5975009846582 96906 MS3367-1-0    STRAP,TIEDOWN,ELECT  PART OF KIT P/N   3
57K3503..............................                                             UOC
:AVY,A11,A13,A14,A15,A20,A24,A25,
                                            A26,A27,B16,B17,B18,HVY,H11,H13,H14,
                                            H15,H16,H17,H18,H20,H21,H24,H25,H26,
H27,H28,MMM
* 3 PAOZZ 5306002264825 80204 B1821BH031C075N SCREW,CAP,HEXAGON H  PART OF KIT P/N   2
57K3503..............................                                             UOC
:AVY,A11,A13,A14,A15,A20,A24,A25,
                                            A26,A27,B16,B17,B18,HVY,H11,H13,H14,
                                            H15,H16,H17,H18,H20,H21,H24,H25,H26,
H27,H28,MMM
* 4 PAOZZ 5310000814219 96906 MS27183-12    WASHER,FLAT  PART OF KIT P/N 57K3503    2
UOC:AVY,A11,A13,A14,A15,A20,A24,A25,
                                            A26,A27,B16,B17,B18,HVY,H11,H13,H14,
                                            H15,H16,H17,H18,H20,H21,H24,H25,H26,
H27,H28,MMM
* 5 PAOZZ 5340009891771 96906 MS21333-123   CLAMP,LOOP  PART OF KIT P/N 57K3503.    2
UOC:AVY,A11,A13,A14,A15,A20,A24,A25,
                                            A26,A27,B16,B17,B18,HVY,H11,H13,H14,
                                            H15,H16,H17,H18,H20,H21,H24,H25,H26,
H27,H28,MMM
* 6 PAOZZ 5340014588236 34623 EXCE4718      BRACKET,T  PART OF KIT P/N 57K3503..    1
UOC:AVY,A11,A13,A14,A15,A20,A24,A25,
                                            A26,A27,B16,B17,B18,HVY,H11,H13,H14,
                                            H15,H16,H17,H18,H20,H21,H24,H25,H26,
H27,H28,MMM
* 7 PAOZZ 4710014343157 34623 EX4231-4      TUBE ASSEMBLY,METAL  R.H.,PART OF       2
KIT P/N 57K3503....................
                                               UOC:AVY,A11,A13,A14,A15,A20,A24,A25,
                                               A26,A27,B16,B17,B18,HVY,H11,H13,H14,
                                               H15,H16,H17,H18,H20,H21,H24,H25,H26,
H27,H28,MMM
* 8 PAOZZ 4730002871706 61208 178654        TEE,TUBE  PART OF KIT P/N 57K3503...    1
UOC:AVY,A11,A13,A14,A15,A20,A24,A25,
                                            A26,A27,B16,B17,B18,HVY,H11,H13,H14,
                                            H15,H16,H17,H18,H20,H21,H24,H25,H26,
H27,H28,MMM
```

END OF FIGURE

Section II. TM 9-2320-280-24P-2 C01

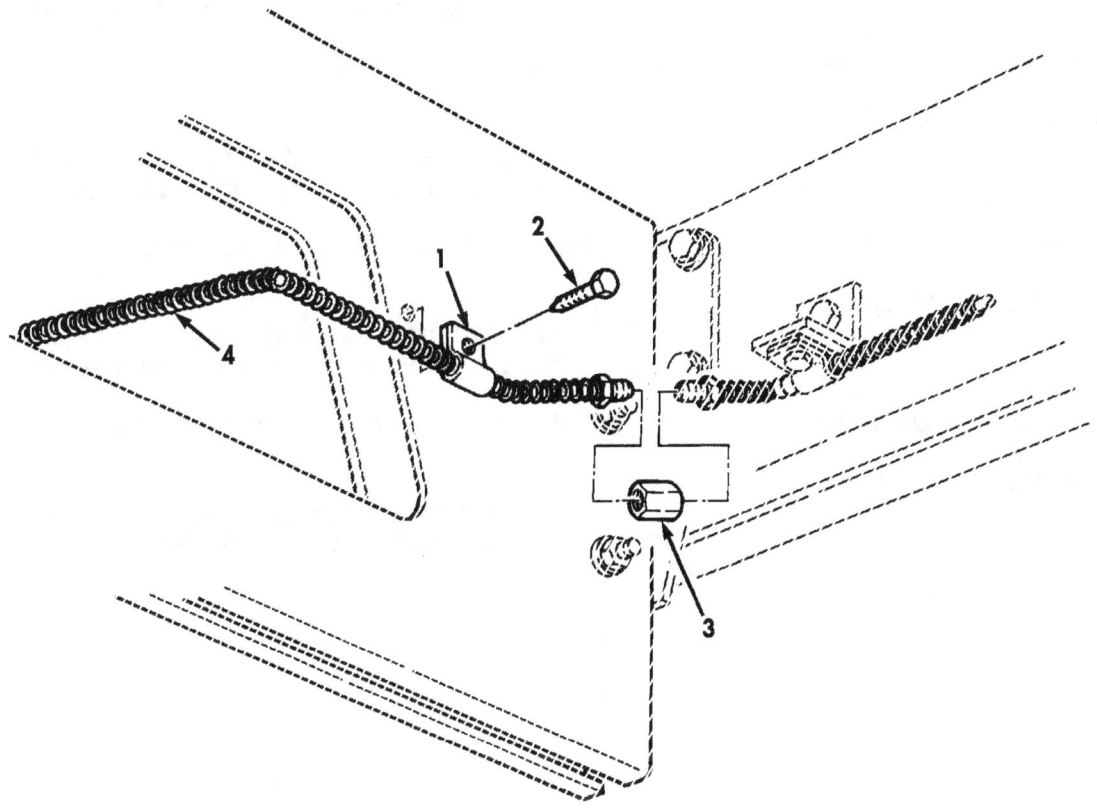

Figure 487C. Flexible Brake Line Kit.

```
    SECTION II         TM9-2320-280-24P C01
    (1)   (2)   (3)         (4)       (5)                   (6)
(7)       ITEM   SMR                            PART
    NO    CODE   NSN        CAGEC     NUMBER    DESCRIPTION AND USABLE ON CODES(UOC)
QTY
```

GROUP 3307 SPECIAL PURPOSE KITS

FIG. 487C FLEXIBLE BRAKE LINE KIT

```
* 1 PAOZZ 5340011897640 7X677 179148     CLAMP,LOOP   PART OF KIT P/N 57K3503.    2
UOC:AVY,A11,A13,A14,A15,A20,A24,A25,
                                             A26,A27,B16,B17,B18,HVY,H11,H13,H14,
                                             H15,H16,H17,H18,H20,H21,H24,H25,H26,
H27,H28,MMM
* 2 PAOZZ 5305013936311 24617 271153     SCREW,TAPPING   PART OF KIT P/N          2
57K3503.............................                                            UOC
:AVY,A11,A13,A14,A15,A20,A24,A25,
                                             A26,A27,B16,B17,B18,HVY,H11,H13,H14,
                                             H15,H16,H17,H18,H20,H21,H24,H25,H26,
H27,H28,MMM
* 3 PAOZZ 4730014570727 72582 442393     COUPLING,HOSE   PART OF KIT P/N          2
57K3503.............................                                            UOC
:AVY,A11,A13,A14,A15,A20,A24,A25,
                                             A26,A27,B16,B17,B18,HVY,H11,H13,H14,
                                             H15,H16,H17,H18,H20,H21,H24,H25,H26,
H27,H28,MMM
* 4 PAOZZ 4710014343159 34623 EX4234     TUBE ASSEMBLY,METAL  PART OF KIT P/N     1
57K3503.............................                                            UOC
:AVY,A11,A13,A14,A15,A20,A24,A25,
                                             A26,A27,B16,B17,B18,HVY,H11,H13,H14,
                                             H15,H16,H17,H18,H20,H21,H24,H25,H26,
H27,H28,MMM
```

Section II. TM 9-2320-280-24P-2 C01

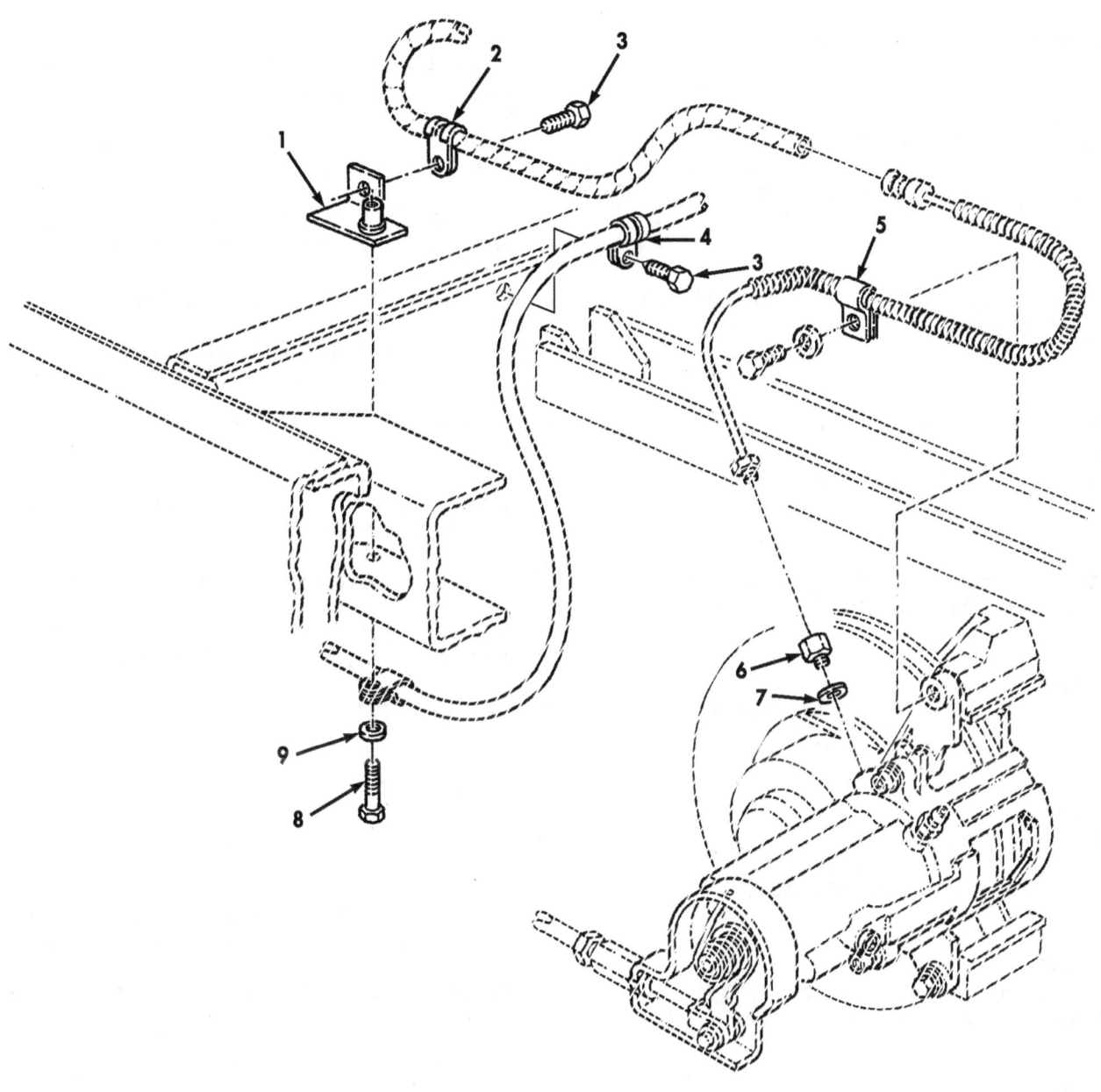

Figure 487D. Flexible Brake Line Kit.

```
            SECTION II         TM9-2320-280-24P C01
       (1)    (2)     (3)            (4)         (5)                        (6)
(7)   ITEM   SMR                                PART
       NO    CODE    NSN           CAGEC       NUMBER       DESCRIPTION AND USABLE ON CODES(UOC)
QTY
```

GROUP 3307 SPECIAL PURPOSE KITS

FIG. 487D FLEXIBLE BRAKE LINE KIT

```
* 1 PAOZZ 5340014587156 34623 EX4717-1        BRACKET,ANGLE   L.H.,PART OF KIT P/N      1
57K3503............................                                                   UOC
:AVY,A11,A13,A14,A15,A20,A24,A25,

                                              A26,A27,B16,B17,B18,HVY,H11,H13,H14,
                                              H15,H16,H17,H18,H20,H21,H24,H25,H26,

H27,H28,MMM
* 1 PAOZZ 5340014587147 34623 EX4717-2        BRACKET,ANGLE   R.H.,PART OF KIT P/N      1
57K3503............................                                                   UOC
:AVY,A11,A13,A14,A15,A20,A24,A25,

                                              A26,A27,B16,B17,B18,HVY,H11,H13,H14,
                                              H15,H16,H17,H18,H20,H21,H24,H25,H26,

H27,H28,MMM
* 2 PAOZZ 5340009891771 96906 MS21333-123     CLAMP,LOOP   PART OF KIT P/N 57K3503.     2
UOC:AVY,A11,A13,A14,A15,A20,A24,A25,

                                              A26,A27,B16,B17,B18,HVY,H11,H13,H14,
                                              H15,H16,H17,H18,H20,H21,H24,H25,H26,

H27,H28,MMM
* 3 PAOZZ 5305013936311 24617 271153          SCREW,TAPPING   PART OF KIT P/N           2
57K3503............................                                                   UOC
:AVY,A11,A13,A14,A15,A20,A24,A25,

                                              A26,A27,B16,B17,B18,HVY,H11,H13,H14,
                                              H15,H16,H17,H18,H20,H21,H24,H25,H26,

H27,H28,MMM
* 4 PAOZZ 5340009645267 96906 MS21333-120     CLAMP,LOOP   PART OF KIT P/N 57K3503.     2
UOC:AVY,A11,A13,A14,A15,A20,A24,A25,

                                              A26,A27,B16,B17,B18,HVY,H11,H13,H14,
                                              H15,H16,H17,H18,H20,H21,H24,H25,H26,

H27,H28,MMM
* 5 PAOZZ 5340011897640 7X677 179148          CLAMP,LOOP   PART OF KIT P/N 57K3503.     2
UOC:AVY,A11,A13,A14,A15,A20,A24,A25,

                                              A26,A27,B16,B17,B18,HVY,H11,H13,H14,
                                              H15,H16,H17,H18,H20,H21,H24,H25,H26,

H27,H28,MMM
* 6 PAOZZ 4730011846971 34623 12339235        COUPLING,TUBE   PART OF KIT P/N           2
57K3503............................                                                   UOC
:AVY,A11,A13,A14,A15,A20,A24,A25,

                                              A26,A27,B16,B17,B18,HVY,H11,H13,H14,
                                              H15,H16,H17,H18,H20,H21,H24,H25,H26,

H27,H28,MMM
* 7 PAOZZ 5310011898476 34623 5582366         WASHER,FLAT   PART OF KIT P/N 57K3503     2
UOC:AVY,A11,A13,A14,A15,A20,A24,A25,

                                              A26,A27,B16,B17,B18,HVY,H11,H13,H14,
                                              H15,H16,H17,H18,H20,H21,H24,H25,H26,

H27,H28,MMM
* 8 PAOZZ 5306002264828 80204 B1821BH031C113N BOLT,MACHINE   PART OF KIT P/N            2
57K3503............................                                                   UOC
:AVY,A11,A13,A14,A15,A20,A24,A25,

                                              A26,A27,B16,B17,B18,HVY,H11,H13,H14,
                                              H15,H16,H17,H18,H20,H21,H24,H25,H26,
```

```
SECTION II          TM9-2320-280-24P C01
 (1)   (2)   (3)         (4)       (5)                    (6)
(7)   ITEM   SMR                   PART
      NO     CODE  NSN       CAGEC NUMBER    DESCRIPTION AND USABLE ON CODES(UOC)
QTY
```

```
                                             H27,H28,MMM
  9 PAOZZ 5310000814219 96906 MS27183-12     WASHER,FLAT   PART OF KIT P/N 57K3503   2
UOC:AVY,A11,A13,A14,A15,A20,A24,A25,
                                             A26,A27,B16,B17,B18,HVY,H11,H13,H14,
                                             H15,H16,H17,H18,H20,H21,H24,H25,H26,
H27,H28,MMM
    KIT PAOZZ 4720014715063 19207 57K3503    PARTS KIT,HOSE ASSE  FLEXIBLE BRAKE    1
LINES............................
                                             UOC:AVY,A11,A13,A14,A15,A20,A24,A25,
                                             A26,A27,B16,B17,B18,HVY,H11,H13,H14,
                                             H15,H16,H17,H18,H20,H21,H24,H25,H26,
H27,H28,MMM
                                                  BOLT,MACHINE       ( 2) 487D-8
                                                  BRACKET,ANGLE      ( 1) 487D-1
                                                  BRACKET,ANGLE      ( 1) 487D-1
                                                  BRACKET,MULTIPLE AN( 2) 487A-5
                                                  BRACKET,MULTIPLE AN( 2) 487A-5
                                                  BRACKET,T          ( 1) 487B-6
                                                  CLAMP,LOOP         ( 2) 487A-9
                                                  CLAMP,LOOP         ( 2) 487B-5
                                                  CLAMP,LOOP         ( 2) 487C-1
                                                  CLAMP,LOOP         ( 2) 487D-2
                                                  CLAMP,LOOP         ( 2) 487D-4
                                                  CLAMP,LOOP         ( 2) 487D-5
                                                  CLIP,HAND BRAKE    ( 4) 487A-8
                                                  COUPLING,HOSE      ( 1) 487A-14
                                                  COUPLING,HOSE      ( 2) 487C-3
                                                  COUPLING,TUBE      ( 2) 487A-10
                                                  COUPLING,TUBE      ( 2) 487D-6
                                                  HOSE ASSEMBLY,NONME( 1) 487A-2
                                                  HOSE ASSEMBLY,NONME( 1) 487A-3
                                                  HOSE ASSEMBLY,NONME( 1) 487A-16
                                                  NUT,SELF-LOCKING,HE( 4) 487A-6
                                                  NUT,SELF-LOCKING,HE( 1) 487B-1
                                                  SCREW,CAP,HEXAGON H( 4) 487A-4
                                                  SCREW,CAP,HEXAGON H( 2) 487B-3
                                                  SCREW,TAPPING      ( 2) 487A-12
                                                  SCREW,TAPPING      ( 2) 487C-2
                                                  SCREW,TAPPING      ( 2) 487D-3
                                                  STRAP,TIEDOWN ELECT( 3) 487B-2
                                                  TEE,TUBE           ( 1) 487B-8
                                                  TUBE ASSEMBLY,METAL( 2) 487A-7
                                                  TUBE ASSEMBLY,METAL( 2) 487A-13
                                                  TUBE ASSEMBLY,METAL( 1) 487A-15
                                                  TUBE ASSEMBLY,METAL( 1) 487B-7
                                                  TUBE ASSEMBLY,METAL( 1) 487C-4
                                                  TUBING,PLASTIC,SPRI( 4) 487A-1
                                                  WASHER,FLAT        ( 2) 487A-11
                                                  WASHER,FLAT        ( 2) 487B-4
                                                  WASHER,FLAT        ( 2) 487D-7
                                                  WASHER,FLAT        ( 2) 487D-9

                                                  END OF FIGURE
```

Section II. TM 9-2320-280-24P-2 C01

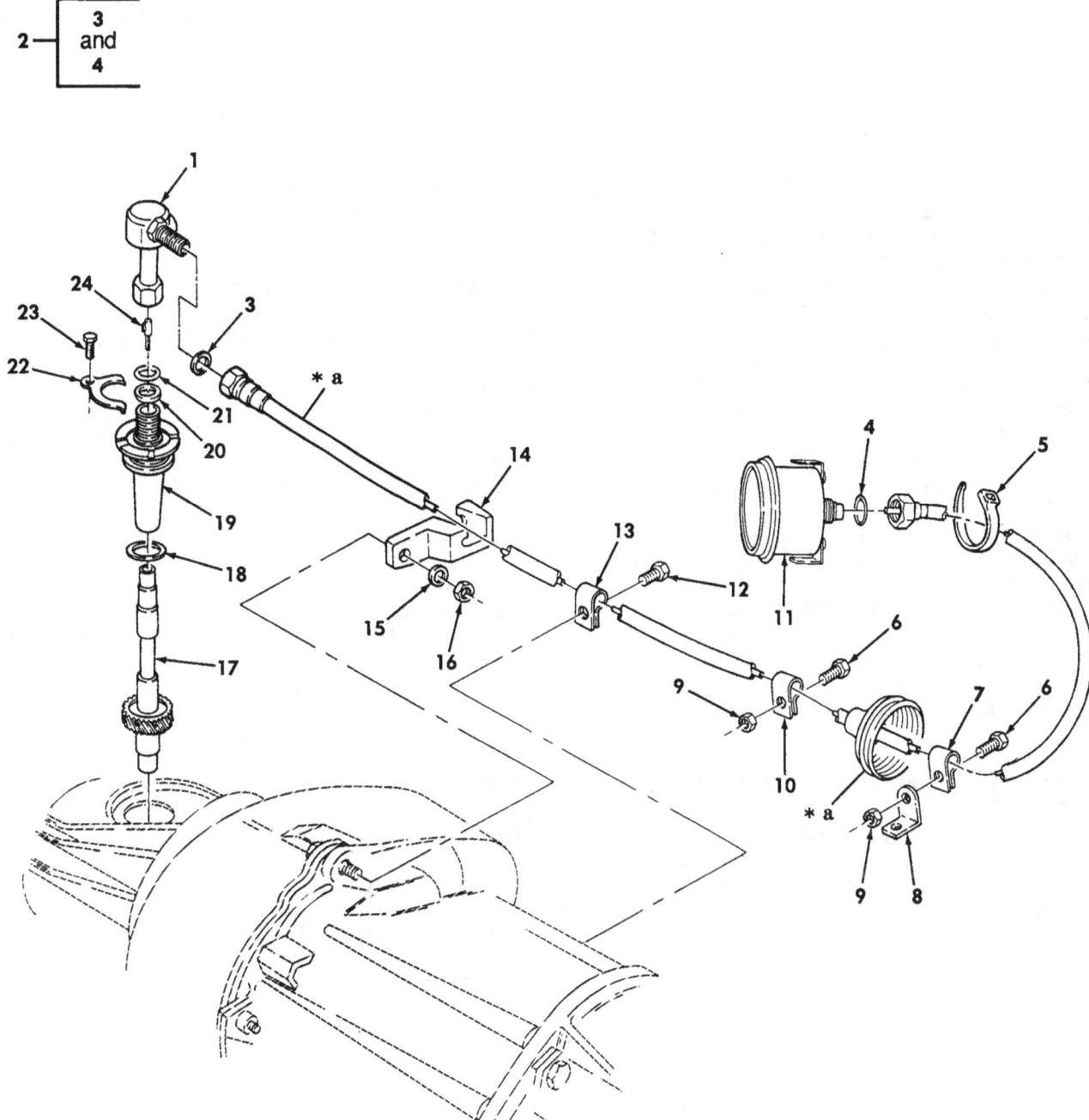

* a PART OF ITEM 2

Figure 488. Speedometer, Speedometer Drive, and Related Parts.

```
        SECTION II          TM9-2320-280-24P C01
    (1)    (2)    (3)         (4)        (5)                         (6)
(7)   ITEM   SMR                                     PART
       NO    CODE   NSN             CAGEC          NUMBER      DESCRIPTION AND USABLE ON CODES(UOC)
QTY

                                                              GROUP 47 GAGES,WEIGHTING AND MEASURE
                                                              -ING DEVICES
4701 INSTRUMENTS

                                                              FIG. 488 SPEEDOMETER,SPEEDOMETER
DRIVE,AND RELATED PARTS

       1 PAOZZ 2590012125868 7Z588 SP-311              ADAPTER,SPEEDOMETER..................  1
*      2 PAOZZ 6680014429413 19200 12338428-2          SHAFT ASSEMBLY,FLEX.................  1
       3 PAOZZ 5331012167392 19207 12338998              .O-RING...........................  1
       4 PAOZZ 5331012167393 19207 12338997              .O-RING...........................  1
       5 PAOZZ 5975001563253 96906 MS3367-2-9           STRAP,TIEDOWN,ELECT................  1
UOC:BVY,B15,B20,B24,B25,C17,NNN
*      6 PAOZZ 5305005829501 82386 410-63              SCREW,CAP,HEXAGON H  #10-32 X 1/2...  2
*      7 PAOZZ 5340006899514 96906 MS21333-70          CLAMP,LOOP.........................  2
*      8 PAOZZ 5340014481213 19207 12460199            BRACKET,ANGLE......................  2
*      9 PAOZZ 5310001249265 72582 271169              NUT,PLAIN,ASSEMBLED  #10-32........  2
      10 PAOZZ 5340001501658 17773 11176106-5          CLAMP,LOOP.........................  1
*     11 PAOZZ 6680011952146 19207 12338463            SPEEDOMETER........................  1
*     12 PAOZZ 5305007578122 96906 MS51851-64          SCREW,TAPPING  #10-24 X 3/8........  1
*     13 PAOZZ 5340000913790 96906 MS21333-72          CLAMP,LOOP.........................  1
*     14 PAOZZ 5342011937088 34623 5583576             BRACKET............................  1
*     15 PAOZZ 5310005956057 96906 MS15795-815         WASHER,FLAT  10MM..................  1
*     16 PAOZZ 5310012666392 34623 5593047             NUT,PLAIN,HEXAGON  10M-1.5.........  1
*     17 PAOZZ 3040012443646 92871 3212735             GEARSHAFT,HELICAL..................
1                                                          UOC:HVY,H11,H13,H14,H15,H16,H17,H18,
H20,H21,H24,H25,H26,H27,H28,MMM
*     17 PAOZZ 3040013587034 34623 12342602            GEARSHAFT,HELICAL..................  1
UOC:AVY,A11,A13,A14,A15,A20,A24,A25,
                                                           A26,A27,BVY,B15,B16,B17,B18,B20,B24,
B25,C17,HVY,NNN
*     18 PAOZZ 5331011958889 34623 12339002            O-RING.............................  1
*     19 PAOZZ 3040011923673 92871 J3212893            ADAPTER,SPEEDOMETER................  1
*     20 PAOZZ 5330012554986 34623 5595643             RETAINER,PACKING...................  1
*     21 PAOZZ 5330004095291 12204 4883434AA           SEAL...............................  1
*     22 PAOZZ 6680002690335 86403 2892399             STRAP,RETAINING....................  1
*     23 PAOZZ 5306002264824 80204 B1821BH031C063N     BOLT,MACHINE.......................  1
*     24 PAOZZ 5315014091662 7Z588 30448-D             PIN,SHOULDER,HEADLE................  1

                                                              END OF FIGURE
```

Section II. TM 9-2320-280-24P-2

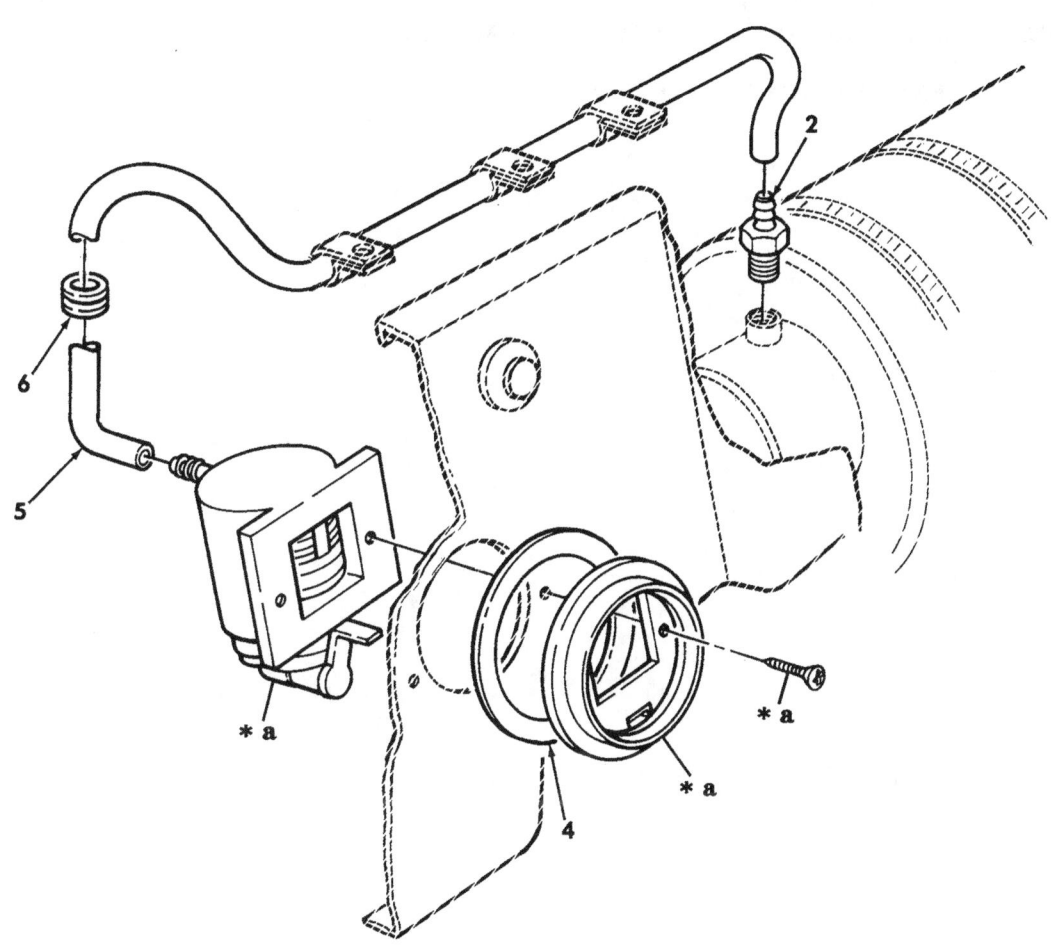

* a PART OF ITEM 3

Figure 489. Air Restriction Indicator Assembly.

SECTION II   TM9-2320-280-24P

(1) NO	(2) SMR CODE	(3) NSN	(4) CAGEC	(5) PART NUMBER	(6) DESCRIPTION AND USABLE ON CODES(UOC)	(7) QTY

GROUP 4702 GAGES, MOUNTING LINES AND FITTINGS

FIG. 489 AIR RESTRICTION INDICATOR ASSEMBLY

1	PAOOO	2940011761427	59150	4755-725	INDICATOR,FILTER WA.................	1
2	PAOZZ	4730011894600	59150	1675-182	.ADAPTER,STRAIGHT,PI................	1
3	XAOZZ		59150	1755-724	.INDICATOR..........................	1
4	PAOZZ	5330014769295	19207	12340078-3	..GASKET............................	1
5	MOOZZ		19207	12340078-5	.HOSE MAKE FROM HOSE,P/N 1775-826, 7.00 FEET LONG......................	1
6	PAOZZ	5325002766228	87585	ANA9-5	GROMMET,NONMETALLIC.................	1

END OF FIGURE

489-1

Section II. TM 9-2320-280-24P-2

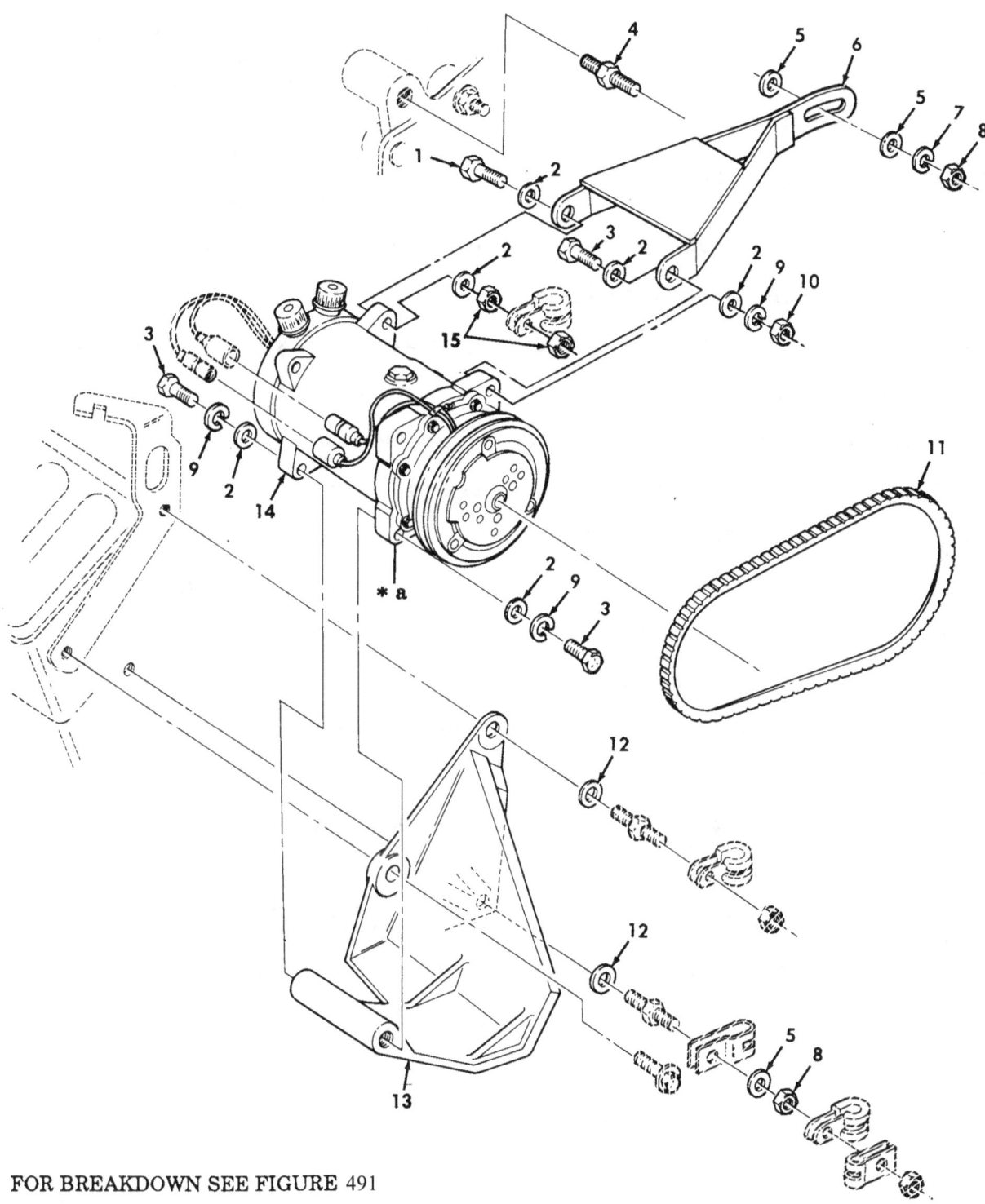

* a FOR BREAKDOWN SEE FIGURE 491

Figure 490. Air Conditioner Compressor and Mounting Hardware, M997 and M997A1 Ambulance.

```
SECTION II              TM9-2320-280-24P C01
   (1)    (2)   (3)           (4)         (5)                   (6)
(7)    ITEM   SMR                                PART
       NO    CODE   NSN       CAGEC       NUMBER      DESCRIPTION AND USABLE ON CODES(UOC)
QTY
```

GROUP 52 REFRIGERATION,AIR
CONDITIONER,HEATER AND AIR
CONDITIONING COMPONENTS

5203 AIR CONDITIONING COMPONENTS

AND MOUNTING HARDWARE,M997 AND

FIG. 490 AIR CONDITIONER COMPRESSOR

M997A1 AMBULANCE

```
     1 PAOZZ 5305007252317 80204 B1821BH038C150N   SCREW,CAP,HEXAGON H  3/8-16 X 1.50..    1
UOC:A15,H15
     2 PAOZZ 5310000877493 96906 MS27183-13        WASHER,FLAT  3/8....................    7
UOC:A15,H15
     3 PAOZZ 5305007829489 80204 B1821BH038C200N   SCREW,CAP,HEXAGON H  3/8-16 X 2.00..    3
UOC:A15,H15
     4 PAFZZ 5307012919018 19207 12339406-2        STUD,SHOULDERED  #10-1.5 X 30.......    2
UOC:A15,H15
     5 PAFZZ 5310011478743 7X677 3790768           WASHER,FLAT  #10....................    3
UOC:A15,H15
*    6 PAFZZ 5340012732368 34623 5598302           BRACKET,MOUNTING....................    1
UOC:A15,H15
     7 PAOZZ 5310012067306 7X677 11500207          WASHER,LOCK  10MM...................    1
UOC:A15,H15
     8 PAOZZ 5310013958747 80204 B18241B100F       NUT,PLAIN,HEXAGON  #10-1.5..........    2
UOC:A15,H15
     9 PAOZZ 5310006379541 96906 MS35338-46        WASHER,LOCK  3/8....................    3
UOC:A15,H15
    10 PAOZZ 5310009359021 96906 MS51943-35        NUT,SELF-LOCKING,HE  3/8-16.........    1
UOC:A15,H15
*   11 PAOZZ 3030008994888 7X677 9433741           BELT,V  3/8 X 43.75.................    1
UOC:A15,H15
*   11 PAOZZ 3030013328526 9C234 12339359-27       BELT,V  3/8 X 43.75,COLD WEATHER....    1
UOC:A15,H15
    12 PAFZZ 5310012046745 24617 11500324          WASHER,FLAT  #10....................    2
UOC:A15,H15
    13 PAFZZ 5340012726634 19207 12339906          BRACKET,MOUNTING....................    1
UOC:A15,H15
    14 PAFFF 2530011975502 19207 12340661          COMPRESSOR,RECIPROC.................    1
UOC:A15,H15
    15 PAOZZ 5310007320558 96906 MS51967-8         NUT,PLAIN,HEXAGON  3/8-16...........    2
UOC:A15,H15
```

END OF FIGURE

Section II. TM 9-2320-280-24P-2 C01

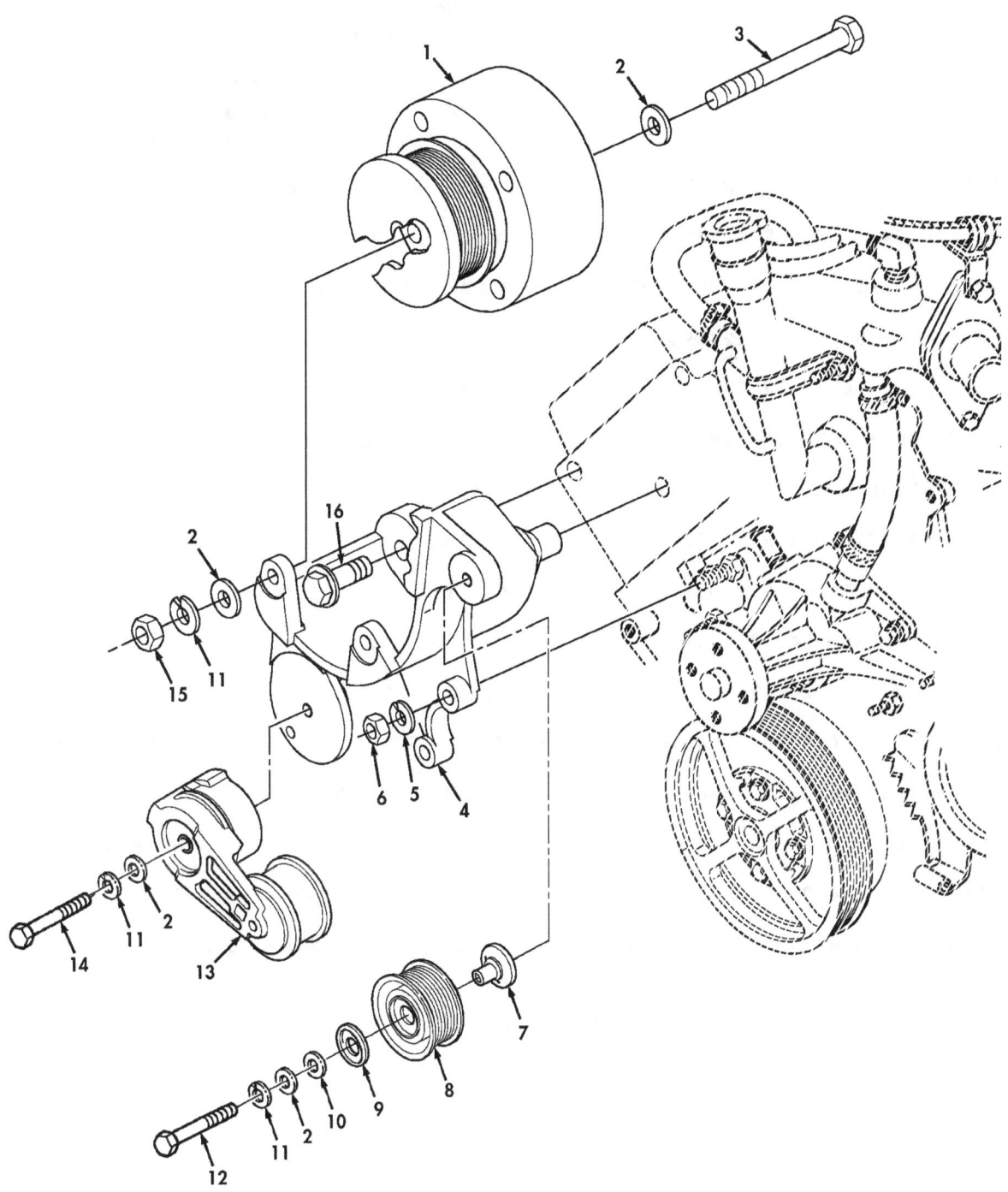

Figure 490A. A/C Compressor, Idler Pulley, Tensioner, and Related Parts (Serial Number 196901 and Above).

```
SECTION II            TM9-2320-280-24P C01
    (1)    (2)   (3)        (4)        (5)                    (6)
(7) ITEM   SMR                         PART
    NO    CODE   NSN        CAGEC      NUMBER     DESCRIPTION AND USABLE ON CODES(UOC)
QTY
```

GROUP 5203 AIR CONDITIONING COMPONENTS

FIG. 490A A/C COMRESSOR,IDLER PULLEY,TENSIONER,AND RELATED PARTS (SERIAL NUMBER 196901 AND ABOVE)

```
* 1  PAOZZ 4310014208306 34623 RCSK17567        COMPRESSOR,AIR COND................    1
UOC:B15
* 2  PAOZZ 5310014124013 24617 2436163          WASHER,FLAT........................    8
UOC:B15
* 3  PAOZZ 5305007813929 80204 B1821BH038C425N  SCREW,CAP HEXAGON H................    3
UOC:B15
* 4  PAOZZ 5340014886181 34623 12469498         BRACKET............................    1
UOC:B15
* 5  PAOZZ 5310012067306 7X677 11500207         WASHER,LOCK........................    2
UOC:B15
* 6  PAOZZ 5310013663539 80204 B18241B100       NUT,PLAIN,HEXAGON   M10-1.5........    2
UOC:B15
* 7  PAOZZ 5365014885640 34623 12469473         SPACER,STEPPED.....................    1
UOC:B15
* 8  PAOZZ 3020014910776 001Y9 600635           PULLEY,GROOVE......................    1
UOC:B15
* 9  PAOZZ 5340014912205 001Y9 600638           BOOT,DUST AND MOIST................    1
UOC:B15
* 10 PAOZZ 5310000120388 24617 120388           WASHER,FLAT........................    1
UOC:B15
* 11 PAOZZ 5310007747293 24617 120382           WASHER,LOCK........................    2
UOC:B15
* 12 PAOZZ 5305005432866 80204 B1821BH038C250N  SCREW,CAP HEXAGON H  .375-16 X 2.50.   1
UOC:B15
* 13 PAOZZ 2920014912011 001Y9 600637           ARM,ADJUSTING......................    1
UOC:B15
* 14 PAOZZ 5305007813927 80204 B1821BH038C350N  SCREW,CAP HEXAGON H  .375-16 X 3.50.   1
UOC:B15
* 15 PAOZZ 5310007610654 96906 MS51967-9        NUT,PLAIN,HEXAGON   .375-16........    3
UOC:B15
* 16 PAOZZ 5306012705448 19207 12340845-3       BOLT,MACHINE   M10-1.5 X 45........    2
UOC:B15
```

END OF FIGURE

Section II. TM 9-2320-280-24P-2 C01

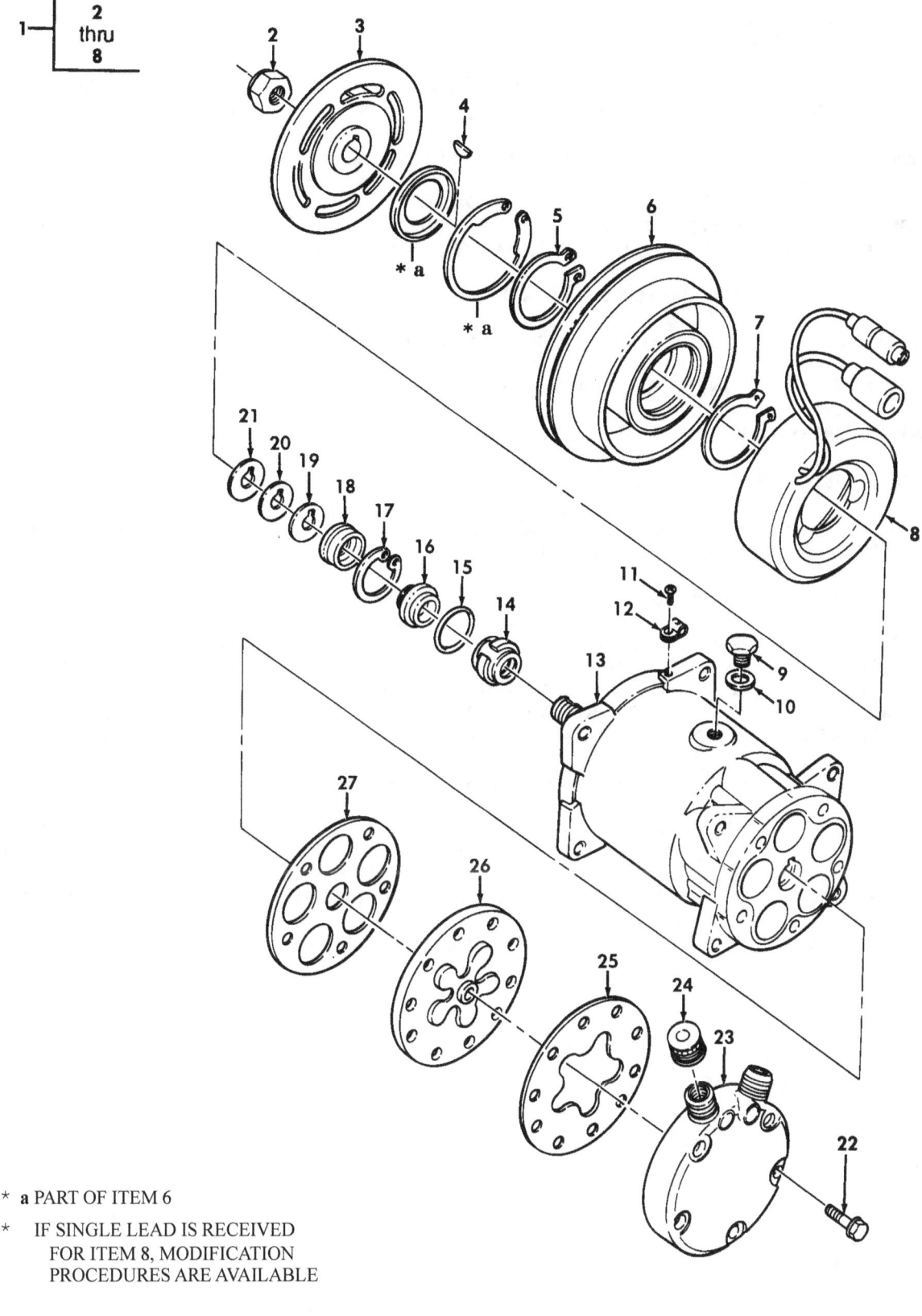

* **a** PART OF ITEM 6
* IF SINGLE LEAD IS RECEIVED FOR ITEM 8, MODIFICATION PROCEDURES ARE AVAILABLE

Figure 491. Air Conditioner Compressor Assembly, M997 and M997A1 Ambulance.

SECTION II TM9-2320-280-24P

(1) NO	(2) SMR CODE	(3) NSN	(4) CAGEC	(5) PART NUMBER	(7) DESCRIPTION AND USABLE ON CODES(UOC)	ITEM QTY

GROUP 5203 AIR CONDITIONING COMPONENTS

FIG. 491 AIR CONDITIONER COMPRESSOR ASSEMBLY-,M997 AND M997A1 AMBULANCE

1	PAFZZ	2520012899663	63829	9383-6010	CLUTCH ASSEMBLY,FRI................ UOC:A15,H15	1
2	KFFZZ		63829	XX1	.NUT PART OF KIT P/N 9103-9800...... UOC:A15,H15	1
3	KFFZZ		63829	X10	.PLATE PART OF KIT P/N 9140-9810.... UOC:A15,H15	1
4	KFFZZ		63829	9162-0121	.KEY,MACHINE PART OF KIT P/N 9103-9800...... UOC:A15,H15	1
5	KFFZZ		63829	XX6	.RING,SNAP PART OF KIT P/N 9103-9800 UOC:A15,H15	1
6	KFFZZ		63829	X11	.PULLEY PART OF KIT P/N 9140-9810... UOC:A15,H15	1
7	KFFZZ		63829	XX7	.RING,SNAP PART OF KIT P/N 9103-9800 UOC:A15,H15	1
8	PAFZZ	5950012896007	63829	8527-6041	.COIL,ELECTRICAL.................... UOC:A15,H15	1
9	PAFZZ	5365012878645	63829	8385-6410	PLUG,MACHINE THREAD................. UOC:A15,H15	1
10	PAFZZ	5331012881062	63829	8385-0520	O-RING.............................. UOC:A15,H15	1
11	KFFZZ		63829	XX9	SCREW PART OF KIT P/N 9103-9800..... UOC:A15,H15	1
12	KFFZZ		63829	XX8	CLAMP PART OF KIT P/N 9103-9800..... UOC:A15,H15	1
13	XAFZZ		63829	X18	BODY................................ UOC:A15,H15	1
14	KFFZZ		63829	X14	SEAL PART OF KIT P/N 5741908........ UOC:A15,H15	1
15	KFFZZ		63829	X15	SEAL PART OF KIT P/N 5741908........ UOC:A15,H15	1
16	KFFZZ		63829	X16	SEAL PART OF KIT P/N 5741908........ UOC:A15,H15	1
17	KFFZZ		63829	8385-0880	RING,RETAINING PART OF KIT P/N 5741908. UOC:A15,H15	1
18	KFFZZ		63829	X17	RING,FELT PART OF KIT P/N 5741908... UOC:A15,H15	1
19	KFFZZ		63829	XX5	SHIM PART OF KIT P/N 9103-9800...... UOC:A15,H15	1
20	KFFZZ		63829	XX4	SHIM PART OF KIT P/N 9103-9800...... UOC:A15,H15	1
21	KFFZZ		63829	XX3	SHIM PART OF KIT P/N 9103-9800...... UOC:A15,H15	1
22	PAFZZ	5306012968600	63829	4605-0470	BOLT,ASSEMBLED WASH................. UOC:A15,H15	5
23	PAFZZ	2815012903705	63829	8476-9631	CYLINDER HEAD,DIESE.................	1

SECTION II   TM9-2320-280-24P

(1) NO	(2) CODE	(3) NSN	(4) CAGEC	(5) PART NUMBER	(7) ITEM DESCRIPTION AND USABLE ON CODES(UOC)	SMR QTY

UOC:A15,H15

24 PAFZZ 5340012881242 63829 8363-6330   CAP,PROTECTIVE,DUST................. 2
UOC:A15,H15
25 KFFZZ   63829 X12   GASKET PART OF KIT P/N 8385-9611.... 1
UOC:A15,H15
26 PAFZZ 4320012900748 63829 6320-9620   VALVE PLATE,HYDRAUL................. 1
UOC:A15,H15
27 KFFZZ   63829 X13   GASKET PART OF KIT P/N 8385-9611.... 1
UOC:A15,H15

END OF FIGURE

Section II.

TM 9-2320-280-24P-2

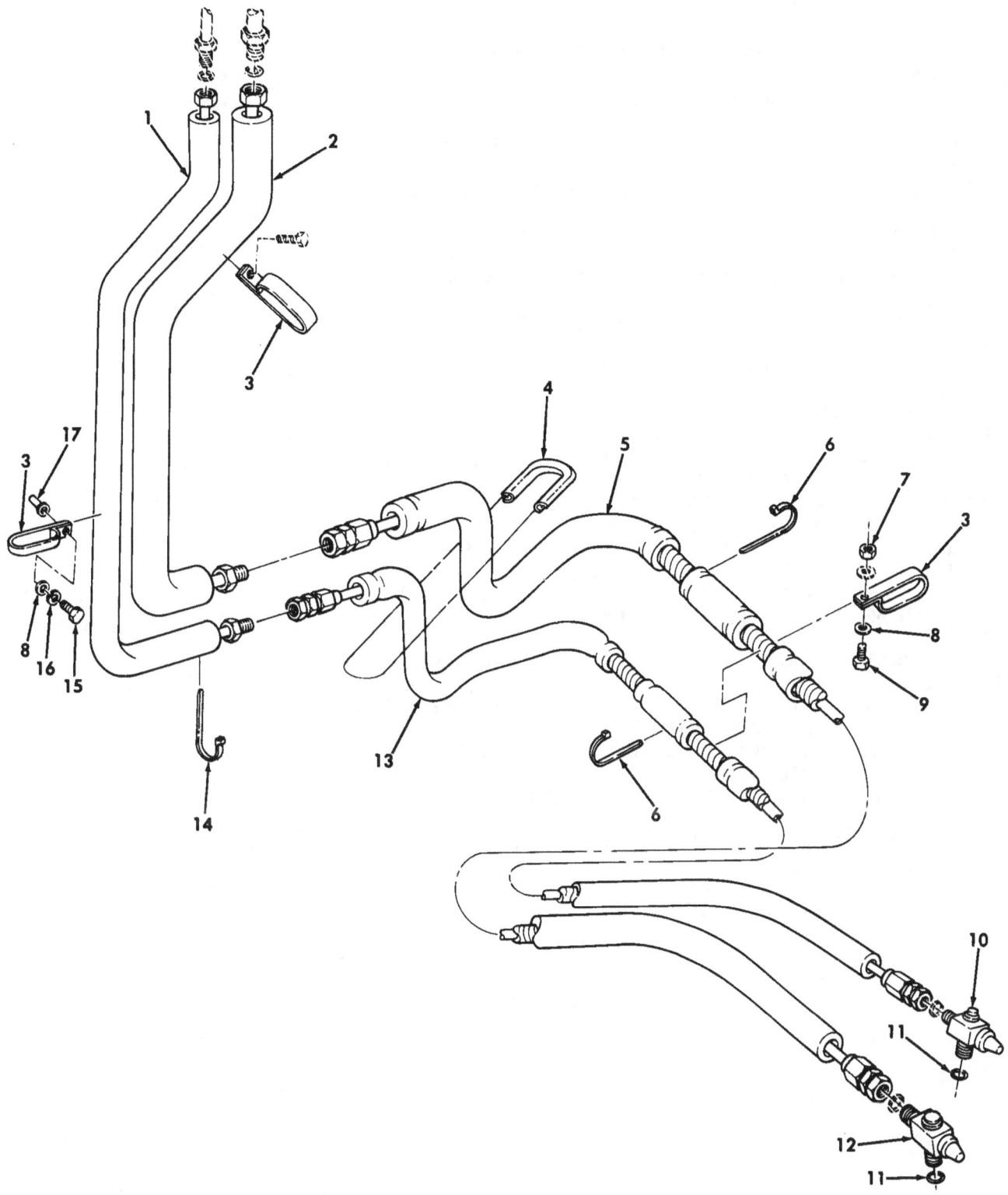

Figure 492. Air Conditioner Lines and Fittings, M997, M997A1, and M997A2 Ambulance.

SECTION II   TM9-2320-280-24P

(1) NO	(2) SMR CODE	(3) NSN	(4) CAGEC	(5) PART NUMBER	(7) ITEM DESCRIPTION AND USABLE ON CODES(UOC)	QTY

GROUP 5217 AIR CONDITIONING COMPONENTS

FIG. 492 AIR CONDITIONER LINES AND FITTINGS, M997, M997A1, AND M997A2 AMBULANCE

No	SMR	NSN	CAGEC	Part Number	Description	QTY
1	PFFZZ	4710012653233	34623	5597935	TUBE ASSEMBLY,METAL UOC:A15,B15,H15	1
2	PFFZZ	4710012653232	9C234	5597929	TUBE ASSEMBLY,METAL UOC:A15,B15,H15	1
3	PAFZZ	5340012731721	19207	12341730	CLAMP,LOOP UOC:A15,B15,H15	3
13	PAOZZ	5342012728391	34623	5597872	PROTECTOR,HANDLE ED UOC:A15,B15,H15	1
5	PFFZZ	4720012653224	34623	12341976	HOSE ASSEMBLY,NONME UOC:A15,B15,H15	1
6	PAFZZ	5975001113208	96906	MS3367-5-9	STRAP,TIEDOWN,ELECT UOC:A15,B15,H15	2
7	PAFZZ	5310000614650	96906	MS51943-31	NUT,SELF-LOCKING,HE 1/4-20 UOC:A15,B15,H15	1
8	PAFZZ	5310008094058	96906	MS27183-10	WASHER,FLAT 1/4 UOC:A15,B15,H15	2
9	PAFZZ	5305000680502	96906	MS90725-6	SCREW,CAP,HEXAGON H 1/4-20 X 3/4 UOC:A15,B15,H15	1
10	PAFZZ	4820012653156	24234	404187	VALVE ASSEMBLY,DIS UOC:A15,B15,H15	1
11	PAFZZ	5331012658809	34623	5597916	O-RING UOC:A15,B15,H15	2
12	PAFZZ	4820012653159	24234	404188	VALVE ASSEMBLY,SUC UOC:A15,B15,H15	1
5	PFFZZ	4720012653223	9C234	5597876	HOSE ASSEMBLY,NONME UOC:A15,B15,H15	1
14	PAOZZ	5975010345871	96906	MS3367-7-0	STRAP,TIEDOWN,ELECT UOC:A15,B15,H15	3
15	PAFZZ	5305002253843	80204	B1821BH025C100N	SCREW,CAP,HEXAGON H 1/4-20 X 1.00 UOC:A15,B15,H15	2
16	PAFZZ	5310005825965	96906	MS35338-44	WASHER,LOCK 1/4 UOC:A15,B15,H15	2
17	PAFZZ	5310010933493	96906	MS27130-31	NUT,PLAIN,BLIND RIV 1/4-20 UOC:A15,B15,H15	2

END OF FIGURE

Section II. TM 9-2320-280-24P-2

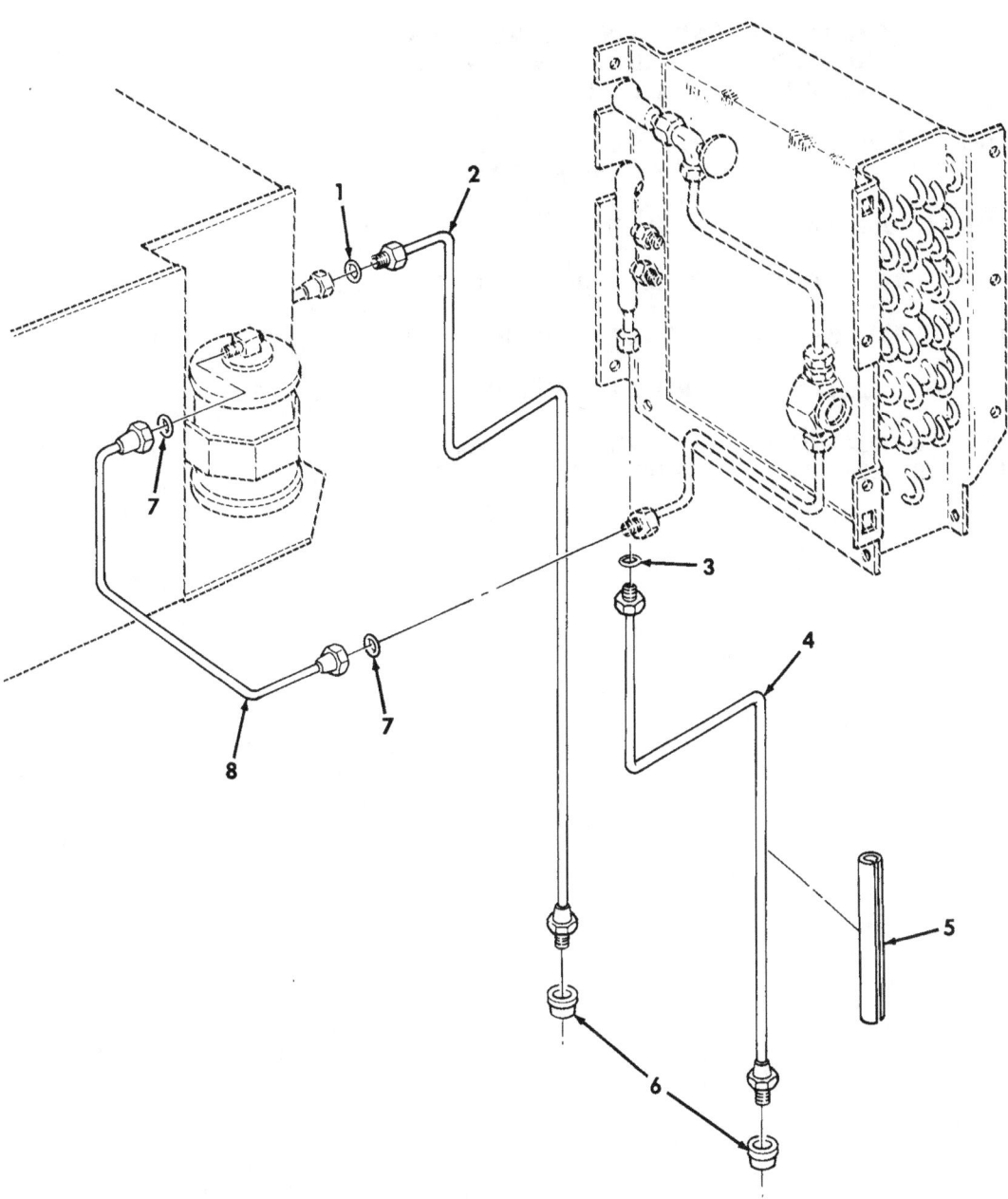

Figure 493. Air Conditioner Condenser and Evaporator Lines, M997, M997A1, and M997A2 Ambulance.

SECTION II   TM9-2320-280-24P

(1) NO	(2) CODE	(3) NSN	(4) CAGEC	(5) PART NUMBER	(7) ITEM DESCRIPTION AND USABLE ON CODES(UOC)	SMR QTY	

GROUP 5217 AIR CONDITIONING
COMPONENTS

FIG. 493 AIR CONDITIONER CONDENSER
AND EVAPORATOR LINES,M997,M997A1,
AND M997A2 AMBULANCE

```
  1 PAFZZ 5331011609857 24234 300442        O-RING............................  1         UOC:A15,B15,H15
  2 PAFZZ 4710012653277 24234 405684        TUBE AND FITTINGS,M.................  1
UOC:A15,B15,H15
  3 PAFZZ 5331011661712 02697 2-015C557-70  O-RING............................  1
UOC:A15,B15,H15
  4 PAFZZ 4710012653276 24234 405683        TUBE AND FITTINGS,M.................  1
UOC:A15,B15,H15
  5 PAFZZ 5640012885547 24234 300506-72.0   INSULATION SLEEVING.................  1
UOC:A15,B15,H15
  6 PAFZZ 5325012897859 28520 2400          GROMMET,NONMETALLIC.................  2
UOC:A15,B15,H15
  7 PAFZZ 5331010613000 24234 300448        O-RING............................  2         UOC:A15,B15,H15
  8 PAFZZ 4710012653230 24234 405682        TUBE ASSEMBLY,METAL.................  1
UOC:A15,B15,H15
```

END OF FIGURE

493-1

Section II.

TM 9-2320-280-24P-2

Figure 494. Heater and Air Conditioning Condenser, M997, M997A1, and M997A2 Ambulance.

SECTION II   TM9-2320-280-24P

(1) NO	(2) SMR CODE	(3) NSN	(4) CAGEC	(5) PART NUMBER	(7) DESCRIPTION AND USABLE ON CODES (UOC)	ITEM QTY

GROUP 5230 AIR CONDITIONING COMPONENTS

FIG. 494 HEATER AND AIR CONDITIONING CONDENSER, M997, M997A1, AND M997A2 AMBULANCE

1 PAOZZ 5305012967762 24234 201686 SCREW,CAP,HEXAGON H 1/4-20 X .75... 1
UOC:A15,B15,H15

2 PAOZZ 5310012702661 24234 202755 WASHER,LOCK 1/4.................... 7
UOC:A15,B15,H15

3 PFFZZ 4130013006350 24234 410909 CONDENSER,REFRIGERA................ 1
UOC:A15,B15,H15

4 PAOZZ 5310008238804 96906 MS27183-9 WASHER,FLAT 1/4.................... 10
UOC:A15,B15,H15

5 PAOZZ 5305000680502 96906 MS90725-6 SCREW,CAP,HEXAGON H 1/4-20 X 3/4... 5
UOC:A15,B15,H15

6 PAOZZ 5340012961736 24234 290308 CLIP,SPRING TENSION 1/4-20......... 15
UOC:A15,B15,H15

7 PAFZZ 4820012653149 24234 405475 VALVE............................... 1   UOC:A15,B15,H15

8 PAFZZ 5930012661809 24234 404179 SWITCH,PRESSURE.................... 1
UOC:A15,H15

9 PAOZZ 5310012723311 34623 5741427 NUT,PLAIN,CAP 1/4-20............... 8
UOC:A15,B15,H15

10 PAFZZ 5310007616882 96906 MS51967-2 NUT,PLAIN,HEXAGON 1/4-20........... 2
UOC:A15,B15,H15

11 PAOZZ 5305002253843 80204 B1821BH025C100N SCREW,CAP,HEXAGON H................ 2
UOC:A15,B15,H15

12 PAFZZ 5331010613000 24234 300448 O-RING............................. 1   UOC:A15,B15,H15

13 PAFZZ 4130011715997 99688 585810 RECEIVER-DEHYDRATOR................ 1
UOC:A15,H15

14 PAFZZ 5340012716465 24234 513395 BRACKET,MOUNTING................... 1
UOC:A15,B15,H15

15 PAOZZ 5340012716466 24234 513394 BRACKET,MOUNTING................... 2
UOC:A15,B15,H15

16 PAOZZ 5340012897793 24234 290309 CLAMP,LOOP......................... 2
UOC:A15,B15,H15

17 PAOZZ 5305012787118 24234 201685 SCREW,CAP,HEXAGON H 1/4-20 X 1-1/4. 8
UOC:A15,B15,H15

18 PAOZZ 4140012697101 24234 265330 FAN,VENTILATING.................... 2
UOC:A15,B15,H15

19 PAOZZ 6105012658634 24234 274278 MOTOR,DIRECT CURREN................ 2
UOC:A15,B15,H15

20 PAOZZ 5340012716455 24234 513396 STRAP,WEBBING...................... 2
UOC:A15,B15,H15

21 PAOZZ 9320010852889 99688 58701 RUBBER ROUND SECTIO................ 2
UOC:A15,B15,H15

END OF FIGURE

Section II.   TM 9-2320-280-24P-2

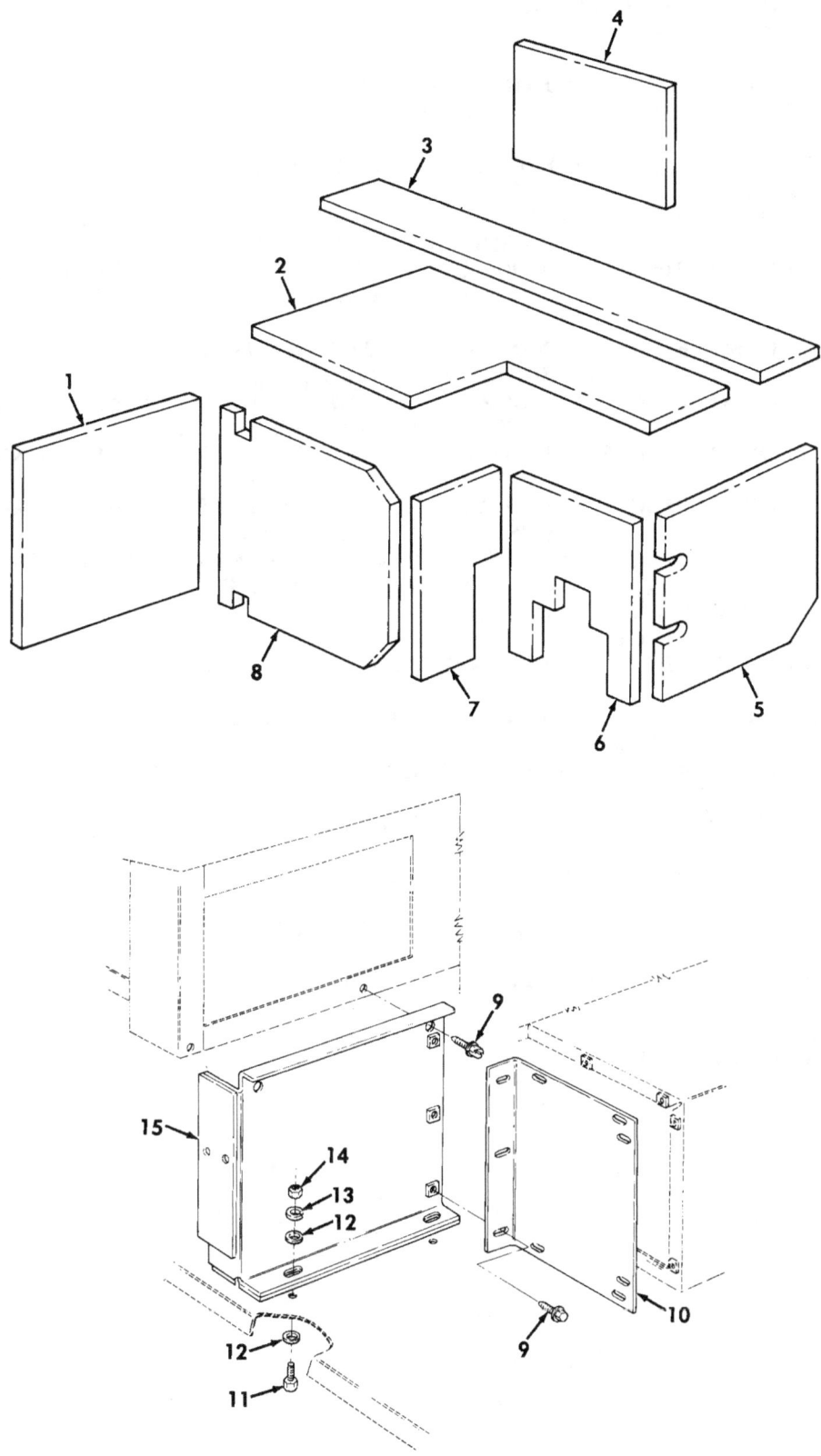

Figure 495. Heater and Air Conditioning Condenser, M997, M997A1, and M997A2 Ambulance.

SECTION II   TM9-2320-280-24P

(1) ITEM NO	(2) SMR CODE	(3) NSN	(4) CAGEC	(5) PART NUMBER	(7) DESCRIPTION AND USABLE ON CODES(UOC)	QTY

GROUP 5230 AIR CONDITIONING COMPONENTS

FIG. 495 HEATER AND AIR CONDITIONING CONDENSER, M997, M997A1, AND M997A2 AMBULANCE

(1)	(2)	(3)	(4)	(5)	(7)	QTY
1	MOOZZ		34623	SF5598816	PANEL FRONT CLOSE OUT, MAKE FROM PAD, CUSHIONING P/N CF40BH52......... UOC:A15,B15,H15	1
2	MOOZZ		34623	SF5598242	PANEL CONDENSER,TOP,-MAKE FROM PAD, CUSHIONING P/N CF40BH52............. UOC:A15,B15,H15	1
3	MOOZZ		19207	12341680	INSULATION BLANKET, CONDENSOR,TOP OUTER,MAKE FROM PAD CUSHIONING P/N CF40BH52............................ UOC:A15,B15,H15	1
4	MOOZZ		34623	SF5598821	PANEL HEATER SIDE,MAKE FROM PAD, CUSHIONING P/N CF40BH52............. UOC:A15,B15,H15	1
5	MOOZZ		34623	SF5598239	PANEL CONDENSER,REAR,-MAKE FROM PAD, CUSHIONING P/N CF40BH52............. UOC:A15,B15,H15	1
6	MOOZZ		34623	SF5598240	PANEL CONDENSER,SIDE,-MAKE FROM PAD, CUSHIONING P/N CF40BH52............. UOC:A15,B15,H15	1
7	MOOZZ		34623	SF5598815	PANEL CONDENSER REAR,-MAKE FROM PAD, CUSHIONING P/N CF40BH52............. UOC:A15,B15,H15	1
8	MOOZZ		34623	SF5598243	PANEL CONDENSER CLOSE OUT PANEL, MAKE FROM PAD,CUSHIONING P/N CF40BH52............................ UOC:A15,B15,H15	1
9	PAOZZ	5305012645874	19207	12340792	SCREW,TAPPING #10-12 X .75......... UOC:A15,B15,H15	11
10	PAOZZ	4130012789837	34623	5934219	COVER,AIR CONDITION................. UOC:A15,B15,H15	1
11	PAOZZ	5305000680502	96906	MS90725-6	SCREW,CAP,HEXAGON H 1/4-20 X 3/4... UOC:A15,B15,H15	5
12	PAOZZ	5310008238804	96906	MS27183-9	WASHER,FLAT 1/4.................... UOC:A15,B15,H15	10
13	PAOZZ	5310012702661	24234	202755	WASHER,LOCK........................ UOC:A15,B15,H15	7
14	PAOZZ	5310007616882	96906	MS51967-2	NUT,PLAIN,HEXAGON 1/4-20........... UOC:A15,B15,H15	2
15	PAOZZ	4420012765050	24234	410925	BAFFLE,FLUID COOLER FRONT.......... UOC:A15,B15,H15	1

END OF FIGURE

Section II. TM 9-2320-280-24P-2

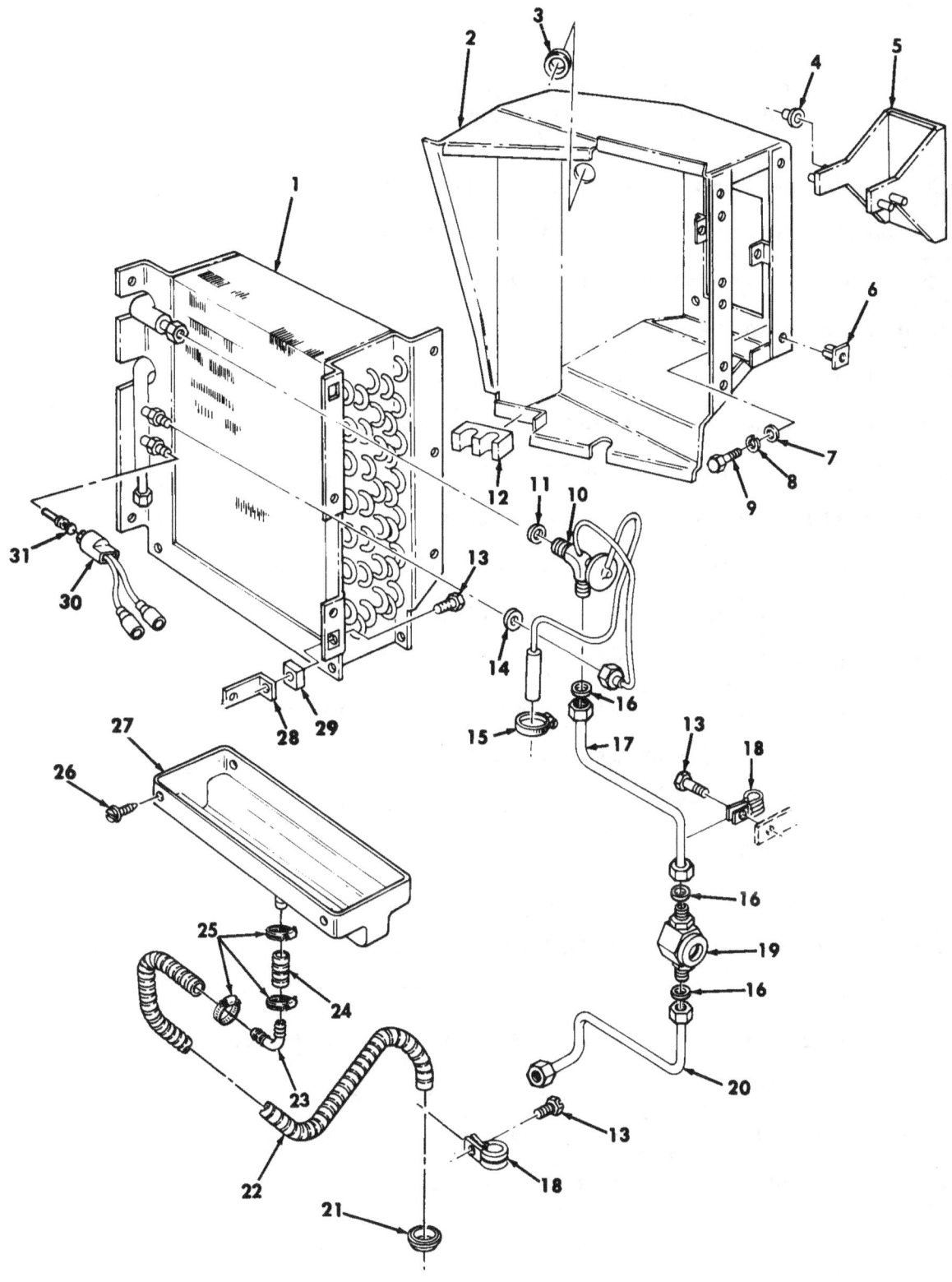

Figure 496. Air Conditioning Evaporator and Mounting Hardware, M997, M997A1, and M997A2 Ambulance

SECTION II TM9-2320-280-24P

(1) NO	(2) SMR CODE	(3) NSN	(4) CAGEC	(5) PART NUMBER	(7) ITEM DESCRIPTION AND USABLE ON CODES(UOC)	QTY

GROUP 5241 AIR CONDITIONING COMPONENTS

FIG. 496 AIR CONDITIONING EVAPORATOR AND MOUNTING HARDWARE, M997, M997A1, AND M997A2 AMBULANCE

1 PAFZZ 4130012658726 24234 400084 EVAPORATOR COIL,REF................ 1
UOC:A15,H15
2 PFFZZ 4130012821895 24234 410917 EVAPORATOR,AIR COND................ 1
UOC:A15,B15,H15
3 PAFZZ 5325002636648 96906 MS35489-135 GROMMET,NONMETALLIC................ 1
UOC:A15,B15,H15
4 PAFZZ 3120012702646 24234 204434 BUSHING,SLEEVE.................... 2
UOC:A15,B15,H15
5 PFFZZ 5340012817795 24234 410918 COVER,ACCESS...................... 1
UOC:A15,B15,H15
6 PAFZZ 5310012723311 34623 5741427 NUT,PLAIN,CAP 1/4-20.............. 5
UOC:A15,B15,H15
7 PAFZZ 5310008238804 96906 MS27183-9 WASHER,FLAT 1/4................... 6
UOC:A15,B15,H15
8 PAFZZ 5310012702661 24234 202755 WASHER,LOCK 1/4................... 6
UOC:A15,B15,H15
9 PAFZZ 5305000680502 96906 MS90725-6 SCREW,CAP,HEXAGON H 1/4-20 X 3/4... 6
UOC:A15,B15,H15
10 PFFZZ 4820011616435 99688 58583 VALVE,EXPANSION................... 1
UOC:A15,H15
11 PAFZZ 5331011609857 24234 300442 O-RING............................ 1
UOC:A15,B15,H15
12 PFFZZ 5330012822214 24234 354944 SEAL,NONMETALLIC ST LOWER RIGHT.... 1
UOC:A15,B15,H15
13 PAOZZ 5305000712506 80204 B1821BH025C050N SCREW,CAP,HEXAGON H 1/4-20 X 1/2... 5
UOC:A15,B15,H15
14 PAFZZ 5331001850075 28527 B2099690G1 O-RING............................ 1
UOC:A15,B15,H15
15 PAFZZ 4730009083193 7Z588 30024H CLAMP,HOSE........................ 1
UOC:A15,B15,H15
16 PAFZZ 5331010613000 24234 300448 O-RING............................ 3   UOC:A15,B15,H15
17 PAFZZ 4710012653152 24234 405680 TUBE UPPER........................ 1   UOC:A15,H15
18 PAFZZ 5340012766738 24234 290249 CLAMP,LOOP........................ 3
UOC:A15,B15,H15
19 PAFZZ 6680012658698 24234 405565 INDICATOR,SIGHT,LIQ............... 1
UOC:A15,H15
20 PFFZZ 4710012855123 24234 405681 TUBE ASSEMBLY,METAL LOWER......... 1
UOC:A15,H15
21 PFFZZ 5325002766090 96906 MS35489-18 GROMMET,NONMETALLIC............... 1
UOC:A15,B15,H15
22 PFOZZ 4720012820143 24234 300515 HOSE,NONMETALLIC .82 INCHES LONG... 1
UOC:A15,B15,H15
23 PFOZZ 4730014392814 34623 12342655 ELBOW,HOSE........................ 1
UOC:A15,B15,H15

SECTION II   TM9-2320-280-24P

(1) NO	(2) SMR CODE	(3) NSN	(4) CAGEC	(5) PART NUMBER	(7) ITEM DESCRIPTION AND USABLE ON CODES(UOC)	QTY
24	PFOZZ	4720012820160	24234	405695	HOSE ASSEMBLY,DRAIN.................. UOC:A15,B15,H15	1
25	PAOZZ	4730002302959	24234	290227	CLAMP,HOSE.......................... UOC:A15,B15,H15	3
26	PAOZZ	5305012705419	24234	200335	SCREW,TAPPING,THREA................. UOC:A15,B15,H15	4
27	PAFZZ	4130012694921	24234	410920	PAN,TRUCK........................... UOC:A15,B15,H15	1
28	PFOZZ	5340012716467	24234	513438	BRACKET,MOUNTING.................... UOC:A15,B15,H15	2
29	XBOZZ		24234	204531	SPACER,SLEEVE....................... UOC:A15,B15,H15	2
30	PAFZZ	5930012663917	24234	404180	SWITCH,PRESSURE..................... UOC:A15,B15,H15	1
31	PAFZZ	4820012653149	24234	405475	VALVE............................... UOC:A15,B15,H15	1

END OF FIGURE

Section II.

TM 9-2320-280-24P-2

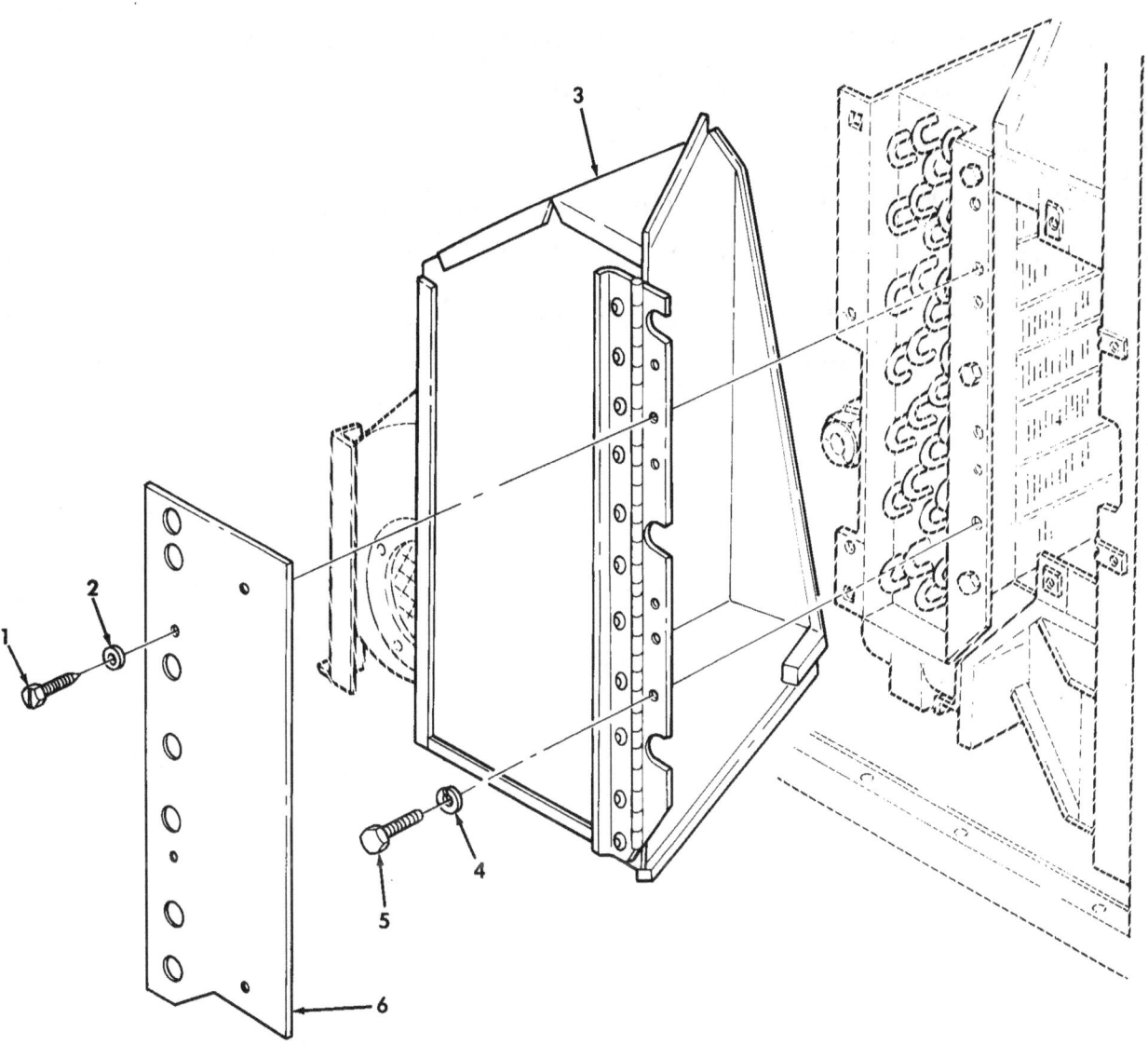

Figure 497. Evaporator Blower Motor Assembly Covers, M997, M997A1, and M997A2 Ambulance.

SECTION II   TM9-2320-280-24P

(1) NO	(2) SMR CODE	(3) NSN	(4) CAGEC	(5) PART NUMBER	(6) DESCRIPTION AND USABLE ON CODES(UOC)	(7) QTY

GROUP 5243 AIR CONDITIONING

FIG. 497 EVAPORATOR BLOWER MOTOR ASSEMBLY COVERS,M997,M997A1,AND M997A2 AMBULANCE

1 PAOZZ 5305012542460 34623 5592840 SCREW,TAPPING,THREA #10-16 X 1/2... 4
UOC:A15,B15,H15
2 PAOZZ 5310002748710 24234 202751 WASHER,LOCK #10.................... 4
UOC:A15,B15,H15
3 PAOZZ 5340014395636 34623 5598293 COVER,ACCESS........................ 1
UOC:A15,B15,H15
4 PAOZZ 5310008094058 96906 MS27183-10 WASHER,FLAT 1/4.................... 6
  UOC:A15,B15,H15
5 PAOZZ 5305000680502 96906 MS90725-6 SCREW,CAP,HEXAGON H 1/4-20 X 3/4... 6
  UOC:A15,B15,H15
6 XBOZZ   34623 5598307 COVER PLATE........................ 1   UOC:A15,B15,H15

END OF FIGURE

497-1

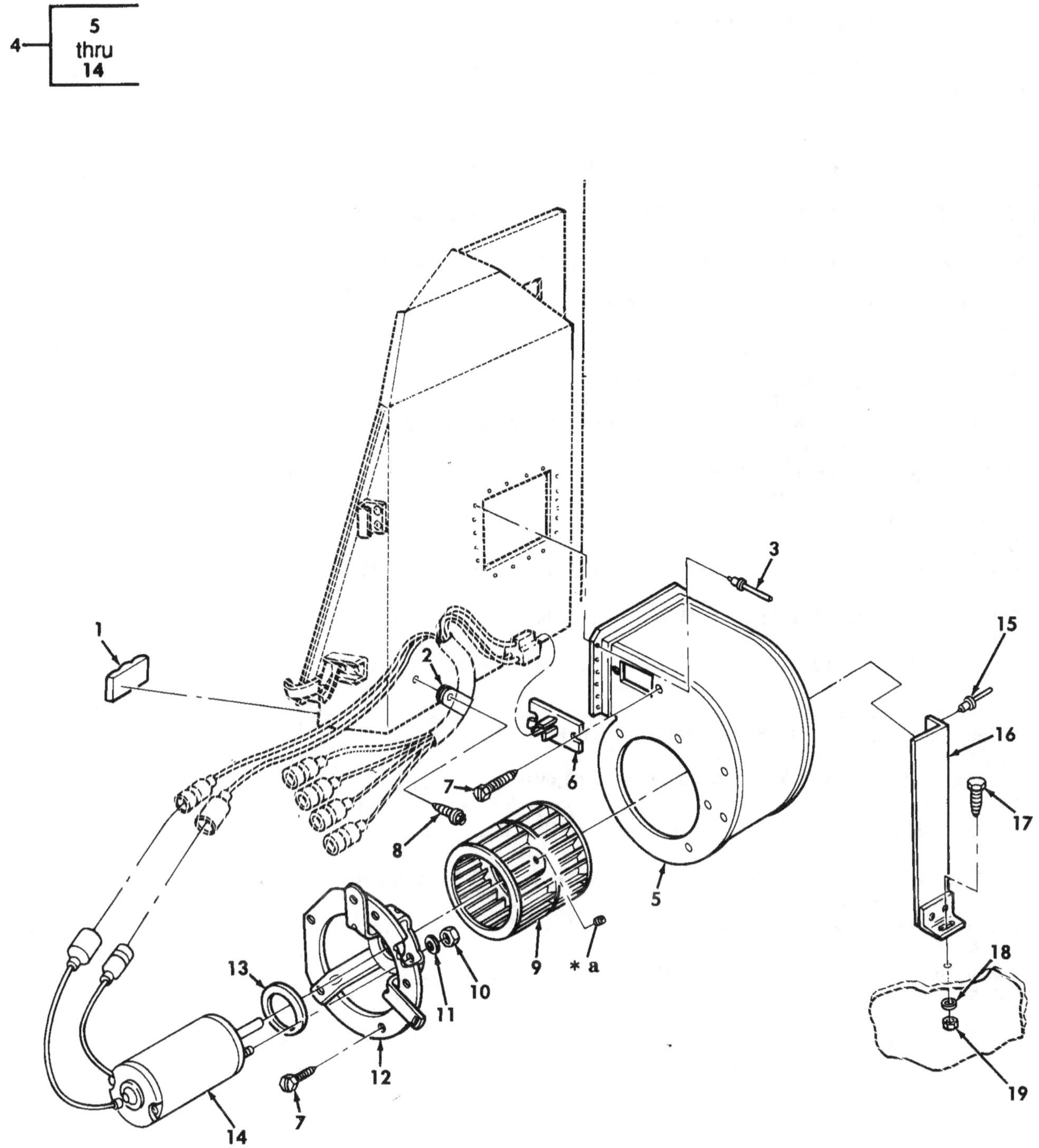

Section II. TM 9-2320-280-24P-2

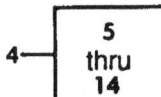

* a PART OF ITEM 9

Figure 498. Evaporator Blower Motor Assembly, M997, M997A1, and M997A2 Ambulance.

SECTION II TM9-2320-280-24P

(1) NO	(2) SMR CODE	(3) NSN	(4) CAGEC	(5) PART NUMBER	(7) ITEM DESCRIPTION AND USABLE ON CODES(UOC)	QTY

GROUP 5243 AIR CONDITIONING

FIG. 498 EVAPORATOR BLOWER MOTOR ASSEMBLY- ,M997,M997A1,AND M997A2 AMBULANCE

```
    1 PFOZZ 5330012822214 24234 354944         SEAL,NONMETALLIC ST................  1
UOC:A15,B15,H15
    2 PAOZZ 5340012820469 24234 290101         CLIP,SPRING TENSION................  1
UOC:A15,B15,H15
    3 PAOZZ 5320001196823 80205 MS20470B4-4    RIVET,SOLID........................  18
UOC:A15,B15,H15
    4 PFOOO 2540012856066 24234 410921         BLOWER ASSEMBLY,AIR................  1
UOC:A15,B15,H15
    5 XAOZZ                24234 513450        .HOUSING BLOWER....................  1         UOC:A15,B15,H15
    6 PAOZZ 5905011542354 99688 65158          .RESISTOR ASSEMBLY.................  1
              UOC:A15,B15,H15
    7 PAOZZ 5305012705419 24234 200335         .SCREW,TAPPING,THREA #10-16 X 1/2.. 7
              UOC:A15,B15,H15
    8 PAOZZ 5305000680508 80204 B1821BH025C075N .SCREW,CAP,HEXAGON H 1/4-20 X 3/4.. 1
              UOC:A15,B15,H15
    9 PFOZZ 2930012822524 24234 260790         .HUB,FAN CLUTCH....................  1
UOC:A15,B15,H15
   10 PAOZZ 5310008528593 96906 MS35649-103    .NUT,PLAIN,HEXAGON #10-32..........  2
UOC:A15,B15,H15
   11 PAOZZ 5310002748710 24234 202751         .WASHER,LOCK #10...................  2
UOC:A15,B15,H15
   12 PFOZZ 5340012817798 24234 513431         .BRACKET,MOUNTING..................  1
UOC:A15,B15,H15
   13 PFOZZ 5330012822208 24234 354674         .GASKET............................  1         UOC:A15,B15,H15
   14 PAOZZ 6105012658634 24234 274278         .MOTOR,DIRECT CURREN...............  1
UOC:A15,B15,H15
   15 PAOZZ 5320008658994 81349 M24243/1-D404  RIVET,BLIND.......................  2
UOC:A15,B15,H15
   16 PAOZZ 5340012893228 24234 514277         BRACKET,ANGLE.....................  1
              UOC:A15,B15,H15
   17 PAOZZ 5305000680502 96906 MS90725-6      SCREW,CAP,HEXAGON H 1/4-20 X .75... 1
              UOC:A15,B15,H15
   18 PAOZZ 5310005825965 96906 MS35338-44     WASHER,LOCK 1/4...................  1
UOC:A15,B15,H15
   19 PAOZZ 4931012028692 19207 12339998-13    PLUSNUT 1/4-20....................  1
UOC:A15,B15,H15
```

END OF FIGURE

SECTION II   TM9-2320-280-24P C01

(1) ITEM NO	(2) SMR CODE	(3) NSN	(4) CAGEC	(5) PART NUMBER	(6) DESCRIPTION AND USABLE ON CODES (UOC)	(7) QTY

GROUP 94 KITS
  9401 REPAIR PARTS KITS

FIG. KITS

| 1 | PAFZZ | 2920012058605 | 68505 | AMA-1046AS | PARTS KIT,ENGINE GE POSITIVE HEAT.. | 1 |

UOC:A11,A13,A14,A20,A24,A25,A26,A27,
B17,B18,H11,H13,H14,H17,H18,H20,H21,
H24,H25,H26,H27,H28,MMM

					INSULATOR,PLATE (1) 43-28	
					RECTIFIER,HEAT (1) 43-29	
2	PAFZZ	2920000095213	19728	AMA-19S	PARTS KIT,BRUSH SPR ALTERNATOR.....	1

UOC:A11,A13,A14,A20,A24,A25,A26,A27,
B17,B18,H11,H13,H14,H17,H18,H20,H21,
H24,H25,H26,H27,H28,MMM

| | | | | | SPRING   (2) 43-43 | |
| 3 | PAOZZ | 2910012184489 | 53964 | A630003 | PARTS KIT,FILTER.................. | 1 |

BOLT   (3) 19-3

| | | | | | WASHER   (3) 19-4 | |
| 4 | PAOZZ | 2530012063874 | 33116 | A98-09439 | PARTS KIT,HAND BRAK............... | 1 |

UOC:H11,H13,H14,H17,H18,H24,H25,H26,
H27,MMM

					DISC,BRAKE (2) 156-8	
					SPRING,HELICAL,EXTE(2) 156-4	
5	PAOZZ	4330011903579	53964	A910044	FILTER ELEMENT,FLUI...............	1

FILTER ELEMENT,FLUI(1) 19-7

| | | | | | O-RING   (1) 19-8 | |
| 6 | PAFZZ | 4520002013474 | 78385 | G704424 | BURNER ASSEMBLY SPACE HEATER...... | 1 |

UOC:A15,B15,H15

					BOLT,HOOK (2) 348-22	
					BURNER ASSEMBLY (1) 348-35	
					CLAMP,RIM CLENCHING(4) 348-23	
					GASKET (1) 348-20	
					O-RING (1) 348-21	
					NUT,PLAIN,HEXAGON (4) 348-3	
7	PAFZZ	4520012847099	78385	G706055	BURNER ASSEMBLY SPACE HEATER......	1

UOC:A15,B15,H15

					BOLT,HOOK (2) 348-22	
					BURNER ASSEMBLY (1) 348-35	
					CLAMP,RIM CLENCHING(4) 348-23	
					GASKET (1) 348-20	
					O-RING (1) 348-21	
8	PAFZZ	5977011894988	24975	MFJ-2012BS	BRUSH SET,ELECTRICA STARTER MOTOR..	1
					BRUSH STARTER MOTOR(4) 45-46	
9	PAFZZ	5360011907234	68505	MFY-19S	SPRING SET GENERATOR.............	1

SPRING BRUSH   (4) 45-48

| * 10 | PAFZZ | 2530014208025 | 34623 | RCSK17215 | BRAKE SHOE SET.................... | 1 |

UOC:BVY,B15,B20,B24,B25,C17,NNN

					DISK,BRAKE SHOE (4) 157-16	
					DISK,BRAKE SHOE (4) 158-11	
11	PAFZZ	2920011924474	68505	SAT-1004FS	PARTS KIT,ENGINE GE GENERATOR......	1

COVER AND CONTACT   (1) 45-31

| | | | | | GASKET   (1) 45-32 | |

KITS-1

SECTION II   TM9-2320-280-24P C01
  (1)   (2)   (3)   (4)   (5)   (6)
(7) ITEM SMR   PART
    NO   CODE   NSN   CAGEC   NUMBER   DESCRIPTION AND USABLE ON CODES(UOC)
 TY

   12 PAFZZ 5977000083338 76468 01-05479   BRUSH SET,ELECTRICA...............   1
UOC:A11,A13,A14,A20,A24,A25,A26,A27,
B17,B18,H11,H13,H14,H17,H18,H20,H21,
H24,H25,H26,H27,H28,MMM
                                              BRUSH   (2) 43-44
   13 PAOZZ 2520007227074 81221 1-0153   PARTS KIT,UNIVERSAL...............
                                              UOC:H11,H13,H14,H15,H16,H17,H18,H20,
H21,H24,H25,H26,H27,H28,MMM
                                              BEARING RACE   (4) 136-7
                                              BEARING RACE   (4) 136-20
                                              CROSS ASSEMBLY   (1) 136-8
                                              CROSS ASSEMBLY   (1) 136-21
                                              RING,RETAINING   (4) 136-10
                                              RING,RETAINING   (4) 136-23

   14 PAFZZ 2530011797511 34281 104870-01   PARTS KIT,BRAKE CAL SERVICE BRAKE..   1
BOOT,DUST,MOIST   (1) 158-10
                                              GASKET   (1) 158-8
   15 KFOZZ 4330011216350 11862 12337210   PARTS KIT,FLUID PRE OIL FILTER.....   1
UOC:AVY,A11,A13,A14,A15,A20,A24,A25,
A26,A27,B16,B17,B18,HVY,H11,H13,H14,
H15,H16,H17,H18,H20,H21,H24,H25,H26,
H27,H28,MMM
                                              GASKET   (1) 104-7
                                              PARTS KIT,FLUID PRE(1) 104-6
                                              SEAL   (1) 104-15
   16 PAOZZ 2530014599492 62828 2390077   PARTS KIT,AXLE,VEHI PLUNGE JOINT...   1
UOC:BVY,B15,B20,B24,B25,C17,NNN
                                              INSERT   (1) 139-12
                                              INSERT   (1) 146-12
                                              JOINT,PLUNGE   (1) 139-13
                                              JOINT,PLUNGE   (1) 146-13
                                              RING,RETAINING   (2) 139-6
                                              RING,RETAINING   (2) 146-6
                                              RING,SPACER   (2) 139-5
                                              RING,SPACER   (2) 146-5
                                              SPIDER ASSEMBLY (1) 139-11
                                              SPIDER ASSEMBLY (1) 146-11
   18 PAHZZ 4820014748911 7X677 24200418   PARTS KIT,PRESSURE................   1
UOC:BVY,B15,B20,B24,B25,C17,NNN
                                              PIN   (1) 117-1
                                              PISTON   (1) 117-4
                                              SEAL RING,METAL (1) 117-3
                                              WASHER,SLOTTED   (1) 117-2
   19 PAHZZ 2520014595454 7X677 24200789   PARTS KIT,TRANSMISS 4TH CLUTCH
                                              HOUSING..........................
UOC:BVY,B15,B20,B24,B25,C17,NNN
                                              BOLT,MACHINE   (1) 105-12
                                              HOUSING,FRICTION CL(1) 115-24
   20 PAHZZ 5330014107229 7X677 24202111   PARTS KIT,SEAL REPL 1995,96........   1
UOC:BVY,B15,B20,B24,B25,C17,NNN
                                              BOLT,MACHINE   (7) 122-1
                                              FILTER,FORCE FEED(1) 119-1
                                              GASKET   (1) 108-7

KITS-2

```
          SECTION II       TM9-2320-280-24P C01
         (1)   (2)   (3)        (4)        (5)                  (6)
  (7)   ITEM  SMR                         PART
        NO    CODE  NSN       CAGEC      NUMBER    DESCRIPTION AND USABLE ON CODES(UOC)
  QTY
```

	GASKET	(1)	116-10
	GASKET	(1)	117-15
	GASKET	(1)	117-18
	GASKET	(1)	120-8
	GASKET	(1)	120-10
	GASKET	(1)	120-25
	GASKET	(1)	122-30
	PARTS KIT,HYDRAULIC	(1)	105-15
	SEAL,PLAIN	(1)	122-4
	SEAL,PLAIN,ENCASED	(1)	105-14

```
* 21 PAHZZ 2520014854280 7X677 24204278    PAN KIT,HYDRUALIC  1997,98,99,00,01.    1
UOC:BVY,B15,B20,B24,B25,C17,NNN
```

	GASKET	(1)	105-20
	MAGNET,CLIP,COLLECT	(1)	105-16
	PAN,TRANSMISSION	(1)	105-18
	PLUG,DRAIN	(1)	105-21

```
* 21A PAHZZ 5330010865457 7X677 24204440   PARTS KIT,SEAL REPL  TRANSMISSION        1
OVERHUAL........................
                                           UOC:AVY,A11,A13,A14,A15,A20,A24,A25,
                                           A26,A27,B16,B17,B18,HVY,H11,H13,H14,
                                           H15,H16,H17,H18,H20,H21,H24,H25,H26,
H27,H28,MMM
```

	BOLT,MACHINE	(6)	121-21
	DISK,CLUTCH	(6)	109-3
	DISK,CLUTCH	(5)	111-16
	DISK,CLUTCH,VEHICUL	(3)	113-4
	FILTER,FLUID	(1)	104-6
	GASKET	(1)	100-10
	GASKET	(1)	104-7
	GASKET	(1)	116-10
	GASKET	(1)	116-16
	GASKET	(1)	116-20
	GASKET	(1)	118-1
	GASKET	(1)	118-1
	GASKET	(1)	118-13
	GASKET	(1)	118-13
	GASKET	(1)	121-19
	RETAINER,PACKING	(1)	100-2
	RETAINER,PACKING	(1)	100-5
	RETAINER,PACKING	(4)	113-12
	RETAINER,PACKING	(2)	121-18
	SEAL,CENTER	(1)	109-18
	SEAL,CENTER	(1)	111-3
	SEAL,INNER	(1)	109-8
	SEAL,INNER	(1)	111-5
	SEAL,INNER	(1)	113-14
	SEAL,NONMETALLIC RO	(1)	104-15
	SEAL,OUTER	(1)	109-19
	SEAL,OUTER	(1)	111-4
	SEAL,OUTER	(1)	113-13
	SEAL,PLAIN ENCASED	(1)	121-2
	SEAL,PLAIN ENCASED	(1)	121-22
	SEAL RING,METAL	(1)	116-1

```
    SECTION II           TM9-2320-280-24P C01
  (1)    (2)     (3)           (4)         (5)                        (6)
(7)   ITEM   SMR                                   PART
       NO   CODE    NSN              CAGEC       NUMBER      DESCRIPTION AND USABLE ON CODES(UOC)
QTY
```

* 22	PAHZZ	5330013983724	7X677	24204441	KIT,TRANSMISSION 1995,OVERHAUL......	1

UOC:BVY,B15,B20,B24,B25,C17,NNN

```
                                     BOLT,MACHINE       (  1) 105-12
                                     BOLT,MACHINE       (  1) 105-13
                                     BOLT,MACHINE       (  7) 122-1
                                     DISC,CLUTCH        (  5) 110-3
                                     DISK,CLUTCH        (  5) 112-16
                                     DISK,CLUTCH        (  4) 114-8
                                     DISK,CLUTCH        (  3) 115-18
                                     DISK,CLUTCH        (  4) 115-20
                                     FILTER,FORCE FEED  (  1) 119-1
                                     GASKET             (  1) 108-7
                                     GASKET             (  1) 117-18
                                     GASKET             (  1) 117-15
                                     GASKET             (  1) 120-8
                                     GASKET             (  1) 120-10
                                     GASKET             (  1) 120-25
                                     O-RING             (  1) 115-22
                                     O-RING             (  1) 115-26
                                     O-RING             (  2) 122-32
                                     PACKING,PREFORMED  (  1) 114-13
                                     PACKING,PREFORMED  (  1) 114-14
                                     PARTS KIT,HYDRAULIC(  1) 105-15
                                     RETAINER,PACKING   (  4) 114-15
                                     RING,SEAL,METAL    (  1) 117-3
                                     SEAL,DIRECT CLUTCH (  1) 110-5
                                     SEAL,INNER         (  1) 112-4
                                     SEAL,INNER         (  1) 112-5
                                     SEAL,OUTER         (  1) 112-6
                                     SEAL,PISTIN,INNER  (  1) 110-7
                                     SEAL,PISTON,OUTER  (  1) 110-6
                                     SEAL,PLAIN         (  1) 105-7
                                     SEAL,PLAIN         (  2) 115-4
                                     SEAL,PLAIN         (  1) 122-4
                                     SEAL,PLAIN,ENCASED (  1) 105-14
                                     SEAL,PLAIN,ENCASED (  2) 115-1
```

23	PAHZZ		7X677	24204449	PARTS KIT,PISTON...................	1

UOC:BVY,B15,B20,B24,B25,C17,NNN

```
                                     PACKING,PREFORMED  (  1) 114-13
                                     PACKING,PREFORMED  (  1) 114-14
```

24	PAFZZ	1005000499324	7X677	24204472	VERNIER ASSEMBLY...................	1

UOC:BVY,B15,B20,B24,B25,C17,NNN

```
                                     LEVER,REMOTE       (  1) 108-5
                                     SHAFT              (  1) 108-6
```

25	PAHZZ	3040014621286	7X677	24205250	PISTON,LINEAR ACTUA 1996,97,98.....	1

UOC:BVY,B15,B20,B24,B25,C17,NNN

```
                                     PISTON             (  1) 117-12
                                     RING,OIL SEAL      (  1) 117-11
                                     RING,OIL SEAL      (  1) 117-14
```

* 26	PAHZZ	5330014422874	7X677	24205251	PARTS KIT,SEAL REPL 1995,96 OVERHAUL...........................	1

UOC:BVY,B15,B20,B24,B25,C17,NNN

```
                                     BOLT,MACHINE       (  1) 105-12
```

KITS-4

```
     SECTION II            TM9-2320-280-24P C01
    (1)    (2)    (3)         (4)        (5)                   (6)
(7)  ITEM   SMR                          PART
 NO         CODE   NSN        CAGEC     NUMBER    DESCRIPTION AND USABLE ON CODES(UOC)
QTY
```

			BOLT,MACHINE	(1) 105-13
			BOLT,MACHINE	(7) 122-1
			DISK,CLUTCH	(6) 109-3
			DISK,CLUTCH	(5) 110-3
			DISK,CLUTCH	(5) 112-16
			DISK,CLUTCH	(1) 114-8
			DISK,CLUTCH	(3) 115-18
			DISK,CLUTCH	(4) 115-20
			FILTER,ELEMENT	(1) 119-1
			GASKET	(1) 108-7
			GASKET	(1) 116-10
			GASKET	(1) 117-15
			GASKET	(1) 117-18
			GASKET	(1) 120-8
			GASKET	(1) 120-25
			GASKET	(1) 122-30
			O-RING	(1) 115-22
			O-RING	(1) 115-26
			O-RING	(2) 122-32
			PACKING,PREFORMED	(1) 114-13
			PACKING,PREFORMED	(1) 114-14
			PARTS KIT,HYDRAULIC	(1) 105-15
			RETAINER,PACKING	(4) 114-15
			SEAL,DIRECT CLUTCH	(1) 110-5
			SEAL,INNER	(1) 112-4
			SEAL,INNER	(1) 112-5
			SEAL,OUTER	(1) 112-6
			SEAL,PISTON,INNER	(1) 110-7
			SEAL,PISTON,OUTER	(1) 110-6
			SEAL,PLAIN	(1) 105-7
			SEAL,PLAIN	(2) 115-4
			SEAL,PLAIN,ENCASED	(1) 105-14
			SEAL,PLAIN,ENCASED	(2) 115-1
			SEAL,RING,METAL	(1) 116-1
			SEAL,RING,METAL	(1) 117-3

```
     27 PAHZZ 2520014608301  7X677   24205348   PARTS KIT,MECHANICAL  1996..........   1
UOC:BVY,B15,B20,B24,B25,C17,NNN

                                                BEARING,CLUTCH       ( 1) 114-12
                                                SUPPORT              ( 1) 114-17
     28 PAHZZ 2520011497861  7X677   24205551   CARRIER ASSEMBLY,TR   1995,96........   1
UOC:BVY,B15,B20,B24,B25,C17,NNN

                                                CARRIER ASSEMBLY     ( 1) 107-9
                                                ROLLER ASSEMBLY      ( 1) 107-2
*    29 PAHZZ 2520014747051  7X677   24205552   TRANSMISSION,HYDRAU   1997,98,99,00,
1                                                  01......................................
UOC:BVY,B15,B20,B24,B25,C17,NNN

                                                CARRIER ASSEMBLY     ( 1) 107-9
                                                ROLLER ASSEMBLY      ( 1) 107-2
*    30 PAHZZ 5340014088525  7X677   24206749   SEAT,HELICAL COMPRE   ACCUMULATOR
1                                                  SEAL AND PISTON KIT.................
UOC:BVY,B15,B20,B24,B25,C17,NNN

                                                PACKING,PREFORMED    ( 1) 120-4
                                                PACKING,PREFORMED    ( 2) 120-6
```

```
   SECTION II          TM9-2320-280-24P C01
  (1)  (2)   (3)            (4)         (5)                    (6)
7)   ITEM  SMR                                PART
    NO   CODE   NSN        CAGEC        NUMBER    DESCRIPTION AND USABLE ON CODES(UOC)
TY
```

	PIN,STRAIGHT,HEADLE(	1) 120-24
	PISTON (	1) 120-23
	RING,RETAINING (	2) 120-2

 31 PAHZZ 3110014784266 7X677 24208848   BALL,BEARING 1997,98,99,00,01......   1
OC:BVY,B15,B20,B24,B25,C17,NNN

	BEARING,NEEDLE (	1) 107-11
	RACE,BEARING (	1) 107-10
	RACE,BEARING (	1) 107-12

 32 PAHZZ 2520014818460 7X677 24208849   PARTS KIT,CLUTCH DI 1997,98........   1
OC:BVY,B15,B20,B24,B25,C17,NNN

	BOLT,MACHINE (	1) 105-12
	SEAL (	1) 114-16
	SUPPORT (	1) 114-17

 33 PAHZZ 3010014807597 7X677 24210954   PARTS KIT,MECHANICA 1997,98,99,00,
                                            01..............................
OC:BVY,B15,B20,B24,B25,C17,NNN

	BOLT,MACHINE (	7) 122-1
	DISK,CLUTCH (	5) 110-3
	DISK,CLUTCH (	5) 112-16
	DISK,CLUTCH (	1) 114-8
	DISK,CLUTCH (	3) 115-18
	DISK,CLUTCH (	4) 115-20
	FILTER,ELEMENT (	1) 105-15
	FILTER,ELEMENT (	1) 119-1
	GASKET (	1) 117-15
	GASKET (	1) 117-18
	GASKET (	1) 120-8
	GASKET (	1) 120-10
	GASKET (	1) 120-25
	O-RING (	1) 115-22
	O-RING (	1) 115-26
	O-RING (	2) 122-32
	O-RING (	2) 122-32
	PACKING,PREFORMED (	1) 114-13
	RETAINER,PACKING (	4) 114-15
	RETAINER,PACKING (	1) 114-15
	RING,SEAL METAL (	1) 117-3
	SEAL,DIRECT CLUTCH (	1) 110-5
	SEAL,INNER (	1) 112-4
	SEAL,PLAIN (	1) 105-7
	SEAL,PLAIN (	2) 115-4
	SEAL,PLAIN,ENCASED (	1) 105-14
	SEAL,PLAIN,ENCASED (	2) 115-1

 34 PAHZZ 5330014800752 7X677 24210955   GASKET SET 1997,98,99,00,01........   1
OC:BVY,B15,B20,B24,B25,C17,NNN

	BOLT,MACHINE (	7) 122-1
	FILTER,ELEMENT (	1) 105-15
	FILTER,ELEMENT (	1) 119-1
	GASKET (	1) 108-7
	GASKET (	1) 117-15
	GASKET (	1) 117-18
	GASKET (	1) 120-8
	GASKET (	1) 120-10

SECTION II   TM9-2320-280-24P C01
   (1)   (2)   (3)   (4)   (5)   (6)
(7) ITEM   SMR   PART
    NO   CODE   NSN   CAGEC   NUMBER   DESCRIPTION AND USABLE ON CODES(UOC)
QTY

                                                   GASKET   (1) 120-25
                                                   SEAL,PLAIN,ENCASED (1) 105-14
* 35 PAHZZ 4330014965720 7X677 24210956   FILTER KIT,A/TRANS 1997,98,99,00,01   1
UOC:BVY,B15,B20,B24,B25,C17,NNN

                                                   FILTER,ELEMENT   (1) 105-15
                                                   GASKET   (1) 108-7
                                                   SEAL,PLAIN,ENCASED (1) 105-14
* 35A PAHZZ 2520014813230 7X677 24217454   GEAR,SUPPORT   1998,99,00,01.........   1
UOC:BVY,B15,B20,B24,B25,C17,NNN

                                                   BOLT,MACHINE   (1) 105-12
                                                   SEAL   (1) 114-16
                                                   SUPPORT   (1) 114-17
* 36 PAOZZ 2530011918741 62826 287-0006   PARTS KIT,DUST BOOT................. 1
                                                   UOC:H11,H13,H14 H15,H16,H17,H18,H20,
H21,H24,H25,H26 H27,H28,MMM

                                                   BOOT,FIXED   (1) 138-3
                                                   BOOT,FIXED   (2) 145-3
                                                   CLAMP,RACE   (2) 138-2
                                                   CLAMP,RACE   (2) 145-2
                                                   CLAMP,RACE   (2) 145-4
                                                   CLAMP,RIM   (2) 138-4
     37 PAOZZ 2530013946168 62826 2880019   PARTS KIT,DUST BOOT.................   1
UOC:AVY,A11,A13,A14,A15,A20,A24,A25,

                                                   A26,A27,B16,B17,B18,HVY,H11,H13,H14,
                                                   H15,H16,H17,H18,H20,H21,H24,H25,H26,

H27,H28,MMM
                                                   BOOT,FIXED   (1) 138-3
                                                   BOOT,FIXED   (1) 145-3
                                                   BOOT,PLUNGE   (1) 145-10
                                                   CLAMP,LOOP   (2) 138-2
                                                   CLAMP,LOOP   (1) 138-4
                                                   CLAMP,LOOP   (1) 138-15
                                                   CLAMP,LOOP   (1) 145-15
                                                   CLAMP,RACE   (1) 145-2
                                                   CLAMP,RACE   (1) 145-4
                                                   RING,RETAINER   (2) 138-13
                                                   RING,RETAINER   (2) 145-13
                                                   RING,STOP   (2) 138-13
                                                   SPACER,RING   (2) 138-14
                                                   SPACER,RING   (2) 145-14
     38 PAOZZ 2530014599493 62826 2990054   PARTS KIT,AXLE,VEHI................. 1
UOC:BVY,B15,B20,B24,B25,C17,NNN

                                                   BOOT   (2) 139-3
                                                   CLAMP,BOOT   (2) 139-2
                                                   CLAMP,BOOT   (2) 139-4
                                                   DEFLECTOR,DIRT   (1) 139-8
                                                   JOINT ASSEMBLY   (1) 139-7
                                                   JOINT ASSEMBLY   (1) 146-7
                                                   PARTS KIT,DUST BOOT (2) 146-4
                                                   RING,RETAINING   (2) 139-6
                                                   RING,RETAINING   (2) 146-6
                                                   RING,SPACER   (2) 139-5
                                                   SHAFT,SHOULDERED   (1) 139-9

SECTION II   TM9-2320-280-24P C01
  (1)   (2)   (3)   (4)   (5)   (6)
(7)   ITEM   SMR   PART
       NO   CODE   NSN   CAGEC   NUMBER   DESCRIPTION AND USABLE ON CODES(UOC)
QTY

       39   PAOZZ 2530014599494   62826   2990055   PARTS KIT,AXLE,VEHI................   1
UOC:BVY,B15,B20,B24,B25,C17,NNN

                                                       BOOT   (2) 139-3
                                                       CLAMP,BOOT   (2) 139-2
                                                       CLAMP,BOOT   (2) 139-4
                                                       DEFLECTOR,DIRT   (1) 139-8
                                                       JOINT ASSEMBLY   (1) 139-7
                                                       JOINT ASSEMBLY   (1) 146-7
                                                       RING,RETAINING   (2) 139-6
                                                       RING,SPACER   (2) 139-5
                                                       SHAFT,SHOULDERED   (1) 139-9

       40   PAOZZ 2530014599495   62826   2990056   PARTS KIT,AXLE,VEHI................   1
UOC:BVY,B15,B20,B24,B25,C17,NNN

                                                       BOOT   (2) 146-3
                                                       CLAMP,BOOT   (2) 146-2
                                                       DEFLECTOR,DIRT   (1) 146-8
                                                       JOINT ASSEMBLY   (1) 146-7
                                                       RING,SPACER   (2) 146-5
                                                       SHAFT,SHOULDERED   (1) 146-9

       41   PAFZZ 5330011907510   24975   36-630   GASKET SET.........................   1
GASKET,INTERMEDIATE(1) 45-6
       42   PAFZZ 5330011907509   68505   37-123A   GASKET SET.........................   1
GASKET   (1) 45 32
*   43   PAOZZ 2520011892135   95019   5-213X   PARTS KIT,UNIVERSAL................   1
UOC:AVY,A11,A13,A14,A15,A20,A24,A25,

                                                       A26,A27,BVY,B15,B16,B17,B18,B20,B24,
B25,C17,HVY,NNN

                                                       BEARING RACE   (4) 136-7
                                                       BEARING RACE   (4) 136-20
                                                       BEARING RACE   (8) 137-6
                                                       BEARING RACE   (8) 137-20
                                                       CROSS ASSEMBLY   (1) 136-8
                                                       CROSS ASSEMBLY   (2) 136-21
                                                       CROSS ASSEMBLY   (2) 137-7
                                                       CROSS ASSEMBLY   (2) 137-21
                                                       RING,RETAINING   (4) 136-10
                                                       RING,RETAINING   (8) 136-23
                                                       RING,RETAINING   (8) 137-5
                                                       RING,RETAINING   (8) 137-22
       44   PAHZZ 5365011753593   34623   5579445   SHIM ASSORTMENT PINION AXLE BEARING   1
                                                       SPACER,RING   (1) 148-17
                                                       SPACER,RING   (1) 148-17
                                                       SPACER,RING   (1) 148-17
                                                       SPACER,RING   (1) 148-17
                                                       SPACER,RING   (1) 148-17
                                                       SPACER,RING   (1) 148-17
                                                       SPACER,RING   (1) 148-17
                                                       SPACER,RING   (1) 148-17
                                                       SPACER,RING   (1) 148-17
                                                       SPACER,RING   (1) 148-17
                                                       SPACER,RING   (1) 148-17
                                                       SPACER,RING   (1) 148-17
                                                       SPACER,RING   (1) 148-17

```
              SECTION II           TM9-2320-280-24P C01
       (1)    (2)     (3)             (4)         (5)                       (6)
(7)   ITEM    SMR                                 PART
 NO   CODE   NSN                    CAGEC        NUMBER    DESCRIPTION AND USABLE ON CODES(UOC)
QTY
```

SPACER,RING	(1)	148-17
SPACER,RING	(1)	148-17
SPACER,RING	(1)	148-17
SPACER,RING	(1)	148-17
SPACER,RING	(1)	148-17
SPACER,RING	(1)	148-17
SPACER,RING	(1)	148-17
SPACER,RING	(1)	148-17
SPACER,RING	(1)	148-17
SPACER,RING	(1)	148-17
SPACER,RING	(1)	148-17
SPACER,RING	(1)	148-17
SPACER,RING	(1)	148-17

```
 45 PAHZZ 5365011748174 34623 5579453      SHIM ASSORTMENT   DIFFERENTIAL.......  1
SPACER,RING         (  1) 148-20
```

SPACER,RING	(1)	148-20
SPACER,RING	(1)	148-20
SPACER,RING	(1)	148-20
SPACER,RING	(1)	148-20
SPACER,RING	(1)	148-20
SPACER,RING	(1)	148-20
SPACER,RING	(1)	148-20
SPACER,RING	(1)	148-20
SPACER,RING	(1)	148-20
SPACER,RING	(1)	148-20
SPACER,RING	(1)	148-20
SPACER,RING	(1)	148-20
SPACER,RING	(1)	148-20
SPACER,RING	(1)	148-20
SPACER,RING	(1)	148-20
SPACER,RING	(1)	148-20
SPACER,RING	(1)	148-20
SPACER,RING	(1)	148-20
SPACER,RING	(1)	148-20
SPACER,RING	(1)	148-20
SPACER,RING	(1)	148-20
SPACER,RING	(1)	148-20
SPACER,RING	(1)	148-20
SPACER,RING	(1)	148-20
SPACER,RING	(1)	148-20
SPACER,RING	(1)	148-20
SPACER,RING	(1)	148-20
SPACER,RING	(1)	148-20
SPACER,RING	(1)	148-20
SPACER,RING	(1)	148-20
SPACER,RING	(1)	148-20
SPACER,RING	(1)	148-20
SPACER,RING	(1)	148-20

SECTION II   TM9-2320-280-24P C01
 (1) (2)   (3)   (4)   (5)   (6)
(7) ITEM SMR   PART
     NO   CODE   NSN   CAGEC   NUMBER   DESCRIPTION AND USABLE ON CODES(UOC)
QTY

                                                SPACER,RING   (1) 148-20
  * 46 PAOZZ 2510013643120 19207 57K0107   PARTS KIT,HINGE TAI..............   1
UOC:AVY,A11,A13,A14,A20,A24,A25,A26,

                               A27,BVY,B17,B18,B20,B24,B25,C17,HVY,
                               H11,H13,H14,H17,H18,H20,H24,H25,H26,
H27,MMM,NNN

                                                HINGE,BUTT   (2) 266-14
                                                NUT,SELF-LOCKING,HE (6) 266-16
                                                PIN,SPRINGHINGE TAI (2) 266-20
                                                SCREW,CAP,HEXAGON H (3) 266-21
                                                SCREW,CAP,HEXAGON H (3) 266-23
                                                SPACER,SLEEVE   (3) 266-22
                                                WASHER,FLAT   (12) 266-15
    47 PAOZZ 2540012885240 19207 57K0109   KIT,REPLACEMENT 60 OR 100AMP   1
ALTERNATOR MOUNTING BRACKET.........
                               UOC:H11,H13,H14,H17,H18,H20,H21,H24,
H25,H26,H27,H28,MMM

                                                BRACKET,DOUBLE ANGL (1) 42-9
                                                BRACKET,MOUNTING   (1) 42-1
                                                NUT,PLAIN,HEXAGON   (1) 42-11
                                                NUT,SELF-LOCKING,HE (1) 42-8
                                                SCREW,CAP,HEXAGON H (1) 42-5
                                                SCREW,CAP,HEXAGON H (1) 42-6
                                                STUD,PLAIN   (1) 42-2
                                                WASHER,FLAT   (1) 42-3
                                                WASHER,FLAT   (1) 42-7
                                                WASHER,FLAT   (1) 42-12
                                                WASHER,LOCK   (1) 42-4
                                                WASHER,LOCK   (1) 42-10

     48 PAFZZ 5340014120141 19207 57K0223   MODIFICATION KIT,BLAST SHIELD.......
                                                UOC:A11,A24,A27,B24,H11,H24,H27,MMM
CLIP,SPRING,TENSION(1) 278-2

                                                CUTTER,SPOTWELD   (1) 278-3
                                                RIVIT,BLIND   (28) 278-1
     49 PAOZZ 2640013718332 19207 57K0237   RUNFLAT KIT,INSERT................   1
UOC:AVY,A11,A13,A14,A15,A20,A24,A25,

                               A26,A27,B16,B17,B18,HVY,H11,H13,H14,
                               H15,H16,H17,H18,H20,H21,H24,H25,H26,
H27,H28,MMM

                                                BEDLOCK,TIRE RIM   (1) 167-18
                                                RUNFLAT,INSERT   (4) 167-17
    50 PAFZZ 5330013618013 19207 57K0239   PARTS KIT,SEAL REPL SHAFT AND   1
HOUSING..............................
                                                SEAL KIT   (1) 172-3
                                                SEAL,PITMAN SHAFT   (1) 172-19
    51 PAFZZ 5330013618014 19207 57K0240   PARTS KIT,SEAL REPL PISTON AND END 1
PLUG..................................
                                                PARTS KIT,STEERING (1) 172-9
                                                PARTS KIT,STEERING (1) 172-21
    52 PAFZZ 5330013618015 19207 57K0241   PARTS KIT,SEAL REPL DUST AND STUB   1
SHAFT.................................
                                                PARTS KIT,SEAL REPL (1) 172-12
                                                PARTS KIT,STEERING (1) 172-13

                                       KITS-10

SECTION II   TM9-2320-280-24P C01

(1)	(2)	(3)	(4)	(5)	(6)	(7)
ITEM NO	SMR CODE	NSN	CAGEC	PART NUMBER	DESCRIPTION AND USABLE ON CODES (UOC)	QTY
53	PAFZZ	2530013618012	19207	57K0242	PARTS KIT,POWER STE GEAR...........	1
					BEARING ASSEMBLY (1) 172-7	
					PARTS KIT,LINEAR AC(1) 172-8	
					PARTS KIT,SAFETY RE(1) 172-6	
					PARTS KIT,STEERING (1) 172-11	
* 54	PAOZZ	2530014590367	34623	57K0262	PARTS KIT,BRAKE SHO................	1
					UOC:H11,H13,H14,H15,H16,H17,H18,H20, H21,H24,H25,H26,H27,H28,MMM	
					DISC BRAKE SHOE,LH (4) 158-11	
					DISC BRAKE SHOE (4) 157-16	
* 55	PAOZZ	2530014073977	34623	57K0264	BRAKE SHOE SET.....................	1
					UOC:AVY,A13,A14,A15,A25,A26,A27,B16, B17,B18,HVY	
					BRAKE SHOE ASSEMBLY(4) 158-11	
					DISC BRAKE SHOE (4) 157-16	
56	PAOZZ	2530013943748	19207	57K0274	KIT,BOOT PLUNGE....................	1
					UOC:AVY,A11,A13,A14,A15,A20,A24,A25, A26,A27,B16,B17,B18,HVY	
					BOOT,PLUNGE (1) 138-10	
					BOOT,PLUNGE (1) 145-10	
					CLAMP,BOOT (1) 145-15	
					CLAMP,LOOP (2) 138-2	
					CLAMP,LOOP (1) 138-15	
					CLAMP,RACE (1) 145-2	
					RING,RETAINER (2) 138-13	
					RING,RETAINER (2) 145-13	
					RING,SPACER (2) 145-14	
					RING,STOP (2) 138-13	
					SPACER,RING (2) 138-14	
57	PAOZZ	2640014196205	19207	57K0297	VALVE STEM ASSEMBLY................	1
					UOC:AVY,A11,A13,A14,A15,A20,A24,A25, A26,A27,B16,B17,B18,HVY,H11,H13,H14, H15,H16,H17,H18,H20,H21,H24,H25,H26, H27,H28,MMM	
					ADAPTER,STRAIGHT,PI(1) 167-24	
					NUT,SELF-LOCKING,HE(1) 167-26	
					O-RING (1) 167-25	
					VALVE CORE (1) 167-22	
					VALVE,PNEUMATIC (1) 167-23	
58	PAOZZ	2530014138886	34623	57K3205	KIT,STEERING COLUMN................	1
					HOUSING GEAR (1) 169-8	
					NUT,SELF-LOCKING,HE(1) 169-1	
					NUT,SELF-LOCKING,HE(6) 169-27	
					NUT,SELF-LOCKING,HE(2) 169-33	
					PIN,STRAIGHT,THREAD(1) 169-5	
					SCREW,CAP,HEXAGON H(4) 169-30	
					SCREW,CAP,SOCKET HE(2) 169-31	
					WASHER,FLAT (4) 169-32	
					WASHER,KEY (1) 169-3	
* 59	PAOZZ	2540014311338	19207	57K3222	PARTS KIT,TURN SIGN................	1
					CONTROL,DIRECTIONAL(1) 49-1	
					CONTROL,DIRECTIONAL(1) 49-10	
					SCREW,TAPPING (4) 49-3	

SECTION II TM9-2320-280-24P C01

(1) ITEM NO	(2) SMR CODE	(3) NSN	(4) CAGEC	(5) PART NUMBER	(6) DESCRIPTION AND USABLE ON CODES (UOC)	(7) QTY
					SCREW,TAPPING (3) 49-11	
60	PAOZZ	5980014388939	19207	57K3238	REPLACEMENT KIT,LENSE AND LAMP......	1
					UOC:BVY,B15,B20,B24,B25,C17,NNN	
					SEMICONDUCTOR (2) 47-2	
					LENSE (2) 47-3	
61	PAOZZ	5330014596477	19207	57K3489	PARTS KIT,SEAL,PITMAN SHAFT AND HOUSING..............................	1
					UOC:BVY,B15,B20,B24,B25,C17,NNN	
					SEAL,KIT (1) 172-3	
					SEAL,PITMAN (1) 172-19	
62	PAFZZ	2520014305294	19207	57K3502	PARTS KIT,HYDRAULIC 1996..........	1
					UOC:BVY,B15,B20,B24,B25,C17,NNN	
					MICROCIRCUIT,LINEAR(1) 50-4	
					TRANSMISSION,HYDRAU(1) 105-1	
63	PAOZZ	2530014559330	19207	57K3512	PARTS KIT,CALIPER,D................	1
					BOOT,VEHICULAR COMP(1) 157-14	
					CAP,PISTON (1) 157-15	
					GASKET (1) 157-12	
					O-RING,ACTUATOR SHA(1) 157-11	
					O-RING,ANTI-ROTATIO(1) 157-7	
					O-RING,PIN RETAININ(1) 157-8	
					O-RING,THRUST SCREW(1) 157-13	
					SEAL,ACTUATOR SHAFT(1) 157-6	
64	PAOZZ	2530014571337	19207	57K3515	PARTS KIT,DUST BOOT................	1
					UOC:BVY,B15,B20,B24,B25,C17,NNN	
					BOOT (2) 139-3	
					BOOT (2) 146-3	
					CLAMP,BOOT (2) 139-2	
					CLAMP,BOOT (2) 139-4	
					CLAMP,BOOT (2) 146-2	
					PARTS KIT,AXLE (1) 139-10	
					PARTS KIT,AXLE (1) 146-10	
					PARTS KIT,DUST BOOT(2) 146-4	
					RING,RETAINING (2) 139-6	
					RING,RETAINING (2) 146-6	
					RING,SPACER (2) 139-5	
					RING,SPACER (2) 146-5	
65	PAFZZ	2520014396830	19207	57K3523	TRANSMISSION,HYDRAU 1997..........	1
					UOC:BVY,B15,B20,B24,B25,C17,NNN	
					COVER,ACCESS (2) 125-21	
					MICROCIRCUIT,LINEAR(1) 50-4	
					TRANSMISSION,HYDRAU(1) 105-1	
					TUBE,BENT,METALLIC (1) 125-3	
					TUBE,BENT,METALLIC (1) 125-4	
66	PAOZZ	2530014433405	19207	57K3527	PARTS KIT,VEHICULAR WHEEL..........	1
					ADAPTER,STRAIGHT,PI(1) 167-24	
					CAP,PNEUMATIC VALVE(1) 167-21	
					NUT,SELF-LOCKING (12) 167-20	
					NUT,SELF-LOCKING,HE(1) 167-26	
					O-RING (1) 167-16	
					O-RING (1) 167-25	
					RIM,WHEEL,PNEUMATIC(1) 167-14	
					RIM,WHEEL,PNEUMATIC(1) 167-19	

KITS-12

```
       SECTION II          TM9-2320-280-24P C01
    (1)    (2)    (3)         (4)       (5)               (6)
(7)    ITEM   SMR                              PART
   NO   CODE   NSN           CAGEC    NUMBER   DESCRIPTION AND USABLE ON CODES(UOC)
QTY
```

```
                                              VALVE CORE          (  1) 167-22
                                              VALVE,PNEUMATIC TIR(  1) 167-23
*  67 PAFZZ 2520014617072 19207 57K3539       TRANSMISSION,HYDRAU  1998..........   1
UOC:BVY,B15,B20,B24,B25,C17,NNN
                                              COVER,ACCESS        (  2) 125-21
                                              MICROCIRCUIT,LINEAR(  1) 50-4
                                              TRANSMISSION,HYDRAU(  1) 105-1
                                              TUBE,BENT,METALLIC  (  1) 125-3
                                              TUBE,BENT,METALLIC  (  1) 125-4
* 67A PAFZZ 2520014617074 19207 57K3549       TRANSMISSION,HYDRAU  1999..........   1
UOC:BVY,B15,B20,B24,B25,C17,NNN
                                              COVER,ACCESS        (  2) 125-21
                                              MICROCIRCUIT,LINEAR(  1) 50-4
                                              TRANSMISSION,HYDRAU(  1) 105-1
                                              TUBE,BENT,METALLIC  (  1) 125-3
                                              TUBE,BENT,METALLIC  (  1) 125-4
* 67B PAFZZ 2520014737410 19207 57K3558       TRANSMISSION,HYDRAU  2000..........   1
UOC:BVY,B15,B20,B24,B25,C17,NNN
                                              COVER,ACCESS        (  2) 125-21
                                              MICROCIRCUIT,LINEAR(  1) 50-4
                                              TRANSMISSION,HYDRAU(  1) 105-1
                                              TUBE,BENT,METALLIC  (  1) 125-3
                                              TUBE,BENT,METALLIC  (  1) 125-4
* 67C PAFZZ 2520014890849 19207 57K3569       TRANSMISSION,HYDRAU  2001..........   1
UOC:BVY,B15,B20,B24,B25,C17,NNN
                                              COVER,ACCESS        (  2) 125-21
                                              MICROCIRCUIT,LINEAR(  1) 50-4
                                              TRANSMISSION,HYDRAU(  1) 105-1
                                              TUBE,BENT,METALLIC  (  1) 125-3
                                              TUBE,BENT,METALLIC  (  1) 125-4
                                              TRANSMISSION,HYDRAU(  1) 105-1
   68 PAFZZ 2920004721723 19207 5703776       PARTS KIT,ENGINE GE................   1
UOC:A11,A13,A14,A20,A24,A25,A26,A27,

B17,B18,H11,H13,H14,H17,H18,H20,H21,
H24,H25,H26,H27,H28
                                              RECTIFIER AND LEAD (  1) 43A-26
                                              RECTIFIER AND LEAD (  1) 43A-28
   69 PAFZZ 2920012222183 19207 5703776-1     PARTS KIT,ENGINE GE................   1
UOC:A11,A13,A14,A20,A24,A25,A26,A27,

H24,H25,H26,H27,H28,MMM
                                              BRUSH,ELECTRICAL    (  2) 43A-18
                                              HOLDER AND STUD ASS(  2) 43A-16
                                              PACKING,PREFORMED   (  1) 43A-5
                                              PACKING,PREFORMED   (  2) 43A-9
                                              PACKING,PREFORMED   (  1) 43A-23
                                              SEAL,SINGLE LIP     (  2) 43A-4
                                              SEAL,SINGLE LIP     (  1) 43A-13
                                              SEAL,SINGLE LIP     (  1) 43A-21
   70 PFFZZ 2540002004249 19207 5704052       PARTS KIT,VEHICULAR  BURNER ASSY....   1
                                              UOC:A15,B15,B16,H15,H16
                                              BOLT,HOOK           (  4) 345-23
                                              BOLT,HOOK           (  2) 348-22
```

KITS-13

```
    SECTION II              TM9-2320-280-24P C01
  (1)   (2)    (3)           (4)       (5)                    (6)
7)   ITEM   SMR                                PART
     NO    CODE    NSN        CAGEC   NUMBER    DESCRIPTION AND USABLE ON CODES(UOC)
TY
```

				CLAMP,RIM CLENCHING(	4) 345-32
				CLAMP,RIM CLENCHING(	4) 348-23
				GASKET (	1) 345-20
				GASKET (	1) 348-20
				HEADER,PLATE ASSEMB(	1) 350-2
				HEADER,PLATE ASSEMB(	1) 351-1
				NUT,HEXAGON (	3) 350-1
				NUT,PLAIN HEXAGON (	4) 345-24
				NUT,PLAIN HEXAGON (	4) 348-3
				NUT,SELF-LOCKING (	4) 345-27
				O-RING (	1) 345-21
				O-RING (	1) 348-21
				SCREW (	3) 350-12
				SCREW,MACHINE (	1) 350-10
				SCREW,MACHINE (	1) 351-10
				SHIELD,FUEL VAPOR (	1) 350-9
				SHIELD,FUEL VAPOR (	1) 351-9
				VAPORIZER (	1) 350-6
				VAPORIZER (	1) 351-6
				WASHER (	1) 351-8
				WASHER (	1) 350-5
				WASHER (	2) 350-4
				WASHER (	1) 350-8
				WASHER,FIBER (	1) 351-5
				WASHER,FLAT (	1) 350-7
				WASHER,FLAT (	1) 351-7
				WICK (	1) 351-4
				WICK (	1) 350-3

```
 71 PAOZZ 2530011918740 34623 5705606    PARTS KIT,DUST BOOT..................
                                         UOC:H11,H13,H14,H15,H16,H17,H18,H20,
21,H24,H25,H26,H27,H28,MMM
```

				BOOT,PLUNGE (	1) 138-10
				BOOT,PLUNGE (	1) 145-10
				CLAMP,RACE (	2) 138-2
				CLAMP,RIM (	2) 138-4
				CLAMP,RACE (	2) 145-2
				CLAMP,RIM (	2) 145-4

```
* 72 PAOZZ 2540013008745 19207 5705618   PARTS KIT,DOOR LATCH RIGHT HAND.....
                                         UOC:AVY,A13,A14,BVY,HVY,H13,H14,H21,
28,NNN
```

				HANDLE,DOOR (	1) 316-9
				HANDLE,DOOR (	1) 316-14
				HANDLE,DOOR (	1) 320-9
				HANDLE,DOOR (	1) 320-12
				HANDLE,DOOR (	1) 320-21
				HANDLE,DOOR (	1) 320-24
				HANDLE,DOOR (	1) 320-28
				HANDLE,DOOR (	1) 324-12
				HANDLE,DOOR (	1) 324-19
				HANDLE,DOOR (	1) 324-26
				HANDLE,DOOR (	1) 324-31
				HANDLE,DOOR (	1) 375-17
				HANDLE,DOOR (	1) 375-22

SECTION II TM9-2320-280-24P C01

(1) ITEM NO	(2) SMR CODE	(3) NSN	(4) CAGEC	(5) PART NUMBER	(6) DESCRIPTION AND USABLE ON CODES(UOC)	(7) QTY
					HANDLE,DOOR (1) 378-17	
					HANDLE,DOOR (1) 378-22	
					LOOP,STRAP,FASTENER(1) 320-30	
					SCREW,MACHINE (1) 316-10	
					SCREW,MACHINE (1) 320-8	
					SCREW,MACHINE (1) 320-20	
					SCREW,MACHINE (1) 324-11	
					SCREW,MACHINE (1) 324-25	
					SCREW,MACHINE (1) 378-16	
					SPRING,HELICAL,TORS(1) 316-11	
					SPRING,HELICAL,TORS(1) 320-10	
					SPRING,HELICAL,TORS(1) 320-22	
					SPRING,HELICAL,TORS(1) 324-13	
					SPRING,HELICAL,TORS(1) 324-27	
					SPRING,HELICAL,TORS(1) 375-18	
					SPRING,HELICAL,TORS(1) 378-18	
					WASHER,FLAT (1) 316-13	
					WASHER,FLAT (1) 320-11	
					WASHER,FLAT (1) 320-23	
					WASHER,FLAT (1) 324-18	
					WASHER,FLAT (1) 375-21	
					WASHER,FLAT (1) 378-21	
73	PAOZZ	2540013008744	19207	5705619	HANDLE,DOOR,INSIDE LEFT HAND........ UOC:AVY,A13,A14,BVY,HVY,H13,H14,H21, H28,N-NN	1
					HANDLE,DOOR,VEHICUL(1) 316-9	
					HANDLE,DOOR,VEHICUL(1) 316-14	
					HANDLE,DOOR,VEHICUL(1) 320-9	
					HANDLE,DOOR,VEHICUL(1) 320-12	
					HANDLE,DOOR,VEHICUL(1) 320-21	
					HANDLE,DOOR,VEHICUL(1) 320-24	
					HANDLE,DOOR,VEHICUL(1) 324-12	
					HANDLE,DOOR,VEHICUL(1) 324-19	
					HANDLE,DOOR,VEHICUL(1) 324-26	
					HANDLE,DOOR,VEHICUL(1) 324-31	
					HANDLE,DOOR,VEHICUL(1) 375-17	
					HANDLE,DOOR,VEHICUL(1) 375-22	
					HANDLE,DOOR,VEHICUL(1) 378-17	
					HANDLE,DOOR,VEHICUL(1) 378-22	
					SCREW,MACHINE (1) 316-10	
					SCREW,MACHINE (1) 320-20	
					SCREW,MACHINE (1) 320-8	
					SCREW,MACHINE (1) 324-11	
					SCREW,MACHINE (1) 324-25	
					SCREW,MACHINE (1) 378-16	
					SPRING,HELICAL,TORS(1) 316-11	
					SPRING,HELICAL,TORS(1) 320-10	
					SPRING,HELICAL,TORS(1) 320-22	
					SPRING,HELICAL,TORS(1) 324-13	
					SPRING,HELICAL,TORS(1) 324-27	
					SPRING,HELICAL,TORS(1) 375-18	
					SPRING,HELICAL,TORS(1) 378-18	
					WASHER,FLAT (1) 316-13	

SECTION II TM9-2320-280-24P C01

(1) ITEM NO	(2) SMR CODE	(3) NSN	(4) CAGEC	(5) PART NUMBER	(6) DESCRIPTION AND USABLE ON CODES(UOC)	(7) QTY
					WASHER,FLAT (1) 320-11	
					WASHER,FLAT (1) 320-23	
					WASHER,FLAT (1) 324-18	
					WASHER,FLAT (1) 324-30	
					WASHER,FLAT (1) 375-21	
					WASHER,FLAT (1) 378-21	
					WASHER,FLAT (1) 324-30	
74	PAHZZ	5365011748184	34623	5740228	SHIM SET............................ UOC:H11,H13,H14,H15,H16,H17,H18,H20, H21,H24,H25,H26,H27,H28,MMM	1
					SHIM (1) 128-6	
					SHIM (1) 128-6	
					SHIM (1) 128-6	
					SHIM (1) 128-6	
					SHIM (1) 128-6	
					SHIM (1) 128-6	
75	PAHZZ	3110011753538	34623	5740229	BEARING,ROLLER,NEED................. UOC:H11,H13,H14,H15,H16,H17,H18,H20, H21,H24,H25,H26,H27,H28,MMM	1
					ROLLER,BEARING (82) 128-27	
*75A	PAFZZ		34623	5716414	KIT,POWER STEERING PUMP............. UOC:B-VY,B15,B20,B25,C17,NNN	1
					O-RING (1) 175A-4	
					SPRING (1) 175A-6	
					VALVE ASSEMBLY (1) 175A-5	
79	PAHZZ	2540012898330	34623	5741908	SEAL,KIT............................ UOC:A15,B15,H15	1
					RING,FELT (1) 491-18	
					RING,RETAINING (1) 491-17	
					SEAL (1) 491-14	
					SEAL (1) 491-15	
					SEAL (1) 491-16	
80	PAOZZ	2540012888567	34623	5742041	KIT,REPLACEMENT 200AMP ALTERNATOR MOUNTING BRACKET.................... UOC:HVY,H15,H16	1
					BRACKET,DOUBLR ANGL(1) 42-9	
					BRACKET,MOUNTING (1) 42-1	
					NUT,PLAIN,HEXAGON H(1) 42-11	
					NUT,SELF-LOCKING (1) 42-8	
					SCREW,CAP,HEXAGON H(1) 42-5	
					SCREW,CAP,HEXAGON H(1) 42-6	
					STUD,PLAIN (1) 42-2	
					WASHER,FLAT (1) 42-3	
					WASHER,FLAT (1) 42-7	
					WASHER,FLAT (1) 42-12	
					WASHER,LOCK (1) 42-4	
					WASHER,LOCK (1) 42-10	
81	PAOZZ	5340013889512	34623	5935425	KIT,BATTERY........................	1
					BOOT,DUST (3) 69-16	
					CAP,PROTECTIVE (3) 69-17	
82	PAFZZ	2530011238787	7X677	7846626	VALVE,STEERING SECT.................	1

```
             SECTION II             TM9-2320-280-24P C01
       (1)    (2)    (3)              (4)         (5)                (6)
 (7)         ITEM   SMR                                     PART
        NO   CODE   NSN             CAGEC        NUMBER          DESCRIPTION AND USABLE ON CODES(UOC)
 QTY
```

					BODY VALVE (1) 172-14	
					SHAFT, STEERING (1) 172-16	
					VALVE,LINEAR,DIRECT(1) 172-15	83
	PAFZZ	5330010440703	52788	7848522	PARTS KIT,SEAL REPL POWER STEERING	1

PUMP................................

					GASKET (1) 176-11	
					MAGNET,ERMANENT (1) 176-8	
					O-RING (2) 176-14	
					RING,SEAL (2) 176-4	
					SEAL (1) 176-3	
					SEAL (1) 176-7	
					SEAL,O-RING (1) 176-5	
84	PAFZZ	5330004933876	11083	8S4680	PACKING,PREFORMED...................	1

SEAL,O-RING   (3)   45-52

| 85 | PAHZZ | 5330012881307 | 63829 | 8385-9611 | GASKET SET......................... | 1 |

UOC:A15,B15,H15

					GASKET (1) 491-25	
					GASKET (1) 491-27	
86	PAHZZ	5330011385190	11862	8623917	SEAL,KIT...........................	1

UOC:AVY,A11,A13,A14,A15,A20,A24,A25,

A26,A27,B16,B17,B18,HVY,H11,H13,H14,
H15,H16,H17,H18,H20,H21,H24,H25,H26,

H27,H28,MMM

					SEAL,CENTER (1) 111-4	
					SEAL,INNER (1) 111-5	
					SEAL,OUTTER (1) 111-3	
87	PAHZZ	3120011674172	7X677	8623920	PARTS KIT,BEARING R................	1

UOC:AVY,A11,A13,A14,A15,A20,A24,A25,

A26,A27,B16,B17,B18,HVY,H11,H13,H14,
H15,H16,H17,H18,H20,H21,H24,H25,H26,

H27,H28,MMM

					BEARING,NEEDLE (1) 106-5	
					BEARING,NEEDLE (1) 107-5	
					RACE,BEARING (1) 106-4	
					RACE,BEARING (1) 106-6	
					RACE,BEARING (1) 107-4	
					RACE,BEARING (1) 107-6	
88	PAHZZ	3110011672443	7X677	8623921	BEARING,ROLLER,THRU................	1

BEARING,NEEDLE   (1) 106-25

					BEARING,NEEDLE (1) 107-11	
					RACE,BEARING (1) 106-24	
					RACE,BEARING (1) 106-26	
					RACE,BEARING (1) 107-10	
					RACE,BEARING (1) 107-12	
					RING,RETAINER (1) 106-19	
89	PAHZZ	3110011690734	7X677	8623922	BEARING,ROLLER,THRU TRANSMISSION...	1
					BEARING,NEEDLE (1) 106-21	
					BEARING,NEEDLE (1) 107-19	
					RACE,BEARING (1) 106-20	
					RACE,BEARING (1) 106-22	
					RACE,BEARING (1) 107-18	
					RACE,BEARING (1) 107-20	
90	PAHZZ	2520011513570	7X677	8625990	PARTS KIT,RACE ASSE................	1

SECTION II   TM9-2320-280-24P C01
 (1)   (2)   (3)   (4)   (5)   (6)
(7)   ITEM   SMR   PART
      NO   CODE   NSN   CAGEC   NUMBER   DESCRIPTION AND USABLE ON CODES(UOC)
TY

```
                                                UOC:AVY,A11,A13,A14,A15,A20,A24,A25,
                                                A26,A27,B16,B17,B18,HVY,H11,H12,H14,
                                                H15,H16,H17,H18,H20,H21,H24,H25,H26,
                                                H27,H28
                                                   GASKET              (  1) 116-20
      91   PAHZZ  2520011637213  7X677  8627989 PARTS KIT,TRANSMISS.................    1
OC:AVY,A11,A13,A14,A15,A20,A24,A25,
                                                A26,A27,B16,B17,B18,HVY,H11,H13,H14,
                                                H15,H16,H17,H18,H20,H21,H24,H25,H26,
                                                H27,H28
                                                   DISK,CLUTCH         (  1) 111-10
                                                   PARTS KIT,TRANSMISS (  1) 111-6
      93   PAFZZ  2520014108067  7X677  8680911 PARTS KIT,MECHANICA.................    1
OC:BVY,B15,B20,B24,B25,C17,NNN
                                                   RING,RETAINING      (  1) 115-12
      94   PAHZZ  3120011748153  7X677  8680914 KIT,TRANSMISSION...................    1
OLT,INTERNALLY REL(  1) 104-4
                                                   BOLT,MACHINE        (  1) 105-12
                                                   CLUTCH,SPRING       (  1) 113-16
                                                   PISTON              (  1) 114-12
                                                   PLUG,VENT           (  1) 113-11
                                                   PLUG,VENT           (  1) 114-16
                                                   SEAL ASSEMBLY       (  1) 113-15
                                                   SUPPORT             (  1) 113-10
                                                   SUPPORT             (  1) 114-17
      95   PAHZZ  2520014108077  7X677  8680915 KIT,TRANSMISSION...................    1
EARING,ROLLER,NEED(  1) 109-16
                                                   DISK,CLUTCH         (  1) 109-14
                                                   DISK,CLUTCH         (  1) 114-3
                                                   RACE                (  1) 114-2
                                                   RING                (  1) 110-14
                                                   RING,BEARING,OUTER  (  1) 109-15
                                                   SPRAG ASSEMBLY      (  1) 114-
      97   PAHZZ  2520011647157  7X677  8680929 PISTON AND RING ASSY REAR
                                                ACCUMULATOR.........................
OC:BVY,B15,B20,B24,B25,C17,NNN
                                                   PISTON              (  1) 117-12
                                                   RING,OIL SEAL       (  1) 117-11
                                                   RING,OIL SEAL,OUTER (  1) 117-14
      98   PAFZZ  2520014818344  7X677  8685921 PARTS KIT,OIL PAN..................    1
OC:BVY,B15,B20,B24,B25,C17,NNN
                                                   MAGNET,CLIP,COLLECT (  1) 105-16
                                                   PAN,TRANSMISSION    (  1) 105-18
                                                   PLUG,PROTECTIVE,DUS (  1) 105-21
                                                   WASHER,FLAT         (  1) 105-20
      99   PAFZZ  2920011922959  68505  90-816  PARTS KIT,SOLENOID.................    1
OOT                (  1)  45-27
                                                   RETAINER CORE       (  1)  45-24
                                                   RETAINER CORE       (  1)  45-26
                                                   SNAP RING CORE END  (  1)  45-23
                                                   SPRING,CORE         (  1)  45-25
                                                   WASHER,SOLENOID     (  1)  45-28
      100  PAFZZ  2920011684128  68505  90-2187 PARTS KIT,ENGINE GE................    1
```

KITS-18

```
         SECTION II            TM9-2320-280-24P C01
        (1)    (2)    (3)         (4)       (5)                       (6)
(7)    ITEM   SMR                                 PART
QTY     NO    CODE    NSN         CAGEC     NUMBER       DESCRIPTION AND USABLE ON CODES(UOC)
```

                                                         UOC:A11,A13,A14,A20,A24,A25,A26,A27,
                                                         UOC:A11,A13,A14,A20,A24,A25,A26,A27,
                                                         B17,B18,H11,H13,H14,H17,H18,H20,H21,
H24,H25,H26,H27,H28,MMM

                                                            BOLT,THRU,TORX   (6) 43-59
                                                            HOLDER,ELECTRICAL   (1) 43-40
                                                            PLUG,YOKE ADJUSTING (1) 43-56
                                                            SCREW,FLAT HEAD   (6) 43-60
                                                            SCREW,SELF-TAPPING (6) 43-26
                                                            SCREW,TORX   (6) 43-58
                                                            WASHER,LOCK   (6) 43-25
                                                            WASHER,LOCK   (1) 43-35
                                                            WASHER,LOCK   (12) 43-57

       101 PAFZZ 5310010279392   68505   90-2188   NUT,SELF-LOCKING,CA................   1
UOC:A11,A13,A14,A20,A24,A25,A26,A27,

H24,H25,H26,H27,H28,MMM   B17,B18,H11,H13,H14,H17,H18,H20,H21,

                                                            KEY,WOODRUFF   (1) 43-4
                                                            NUT   (1) 43-2
                                                            WASHER,FLAT   (1) 43-3

       102 PAFZZ 2920003026342   68505   90-2206   PARTS KIT,YOKE ADJU................   1
GASKET PLUG   (1) 45-15

                                                            NUT   (1) 45-22
                                                            PIN,STRAIGHT   (2) 45-13
                                                            PLUG,YOKE,ADJUSTING (1) 45-16
                                                            PLUG,YOKE,ADJUSTING (1) 45-14

       103 PAFZZ 2920010687182   68505   90-2225   PARTS KIT,ENGINE GE................   1
CONTACK ASSEMBLY   (1) 45-35

                                                            NUT,HEXAGON ELASTIC (1) 45-33
                                                            SPRING   (1) 45-36
                                                            WASHER,PLAIN   (1) 45-34
                                                            WASHER,STEEL   (1) 45-37

       104 PAFZZ 2920011684129   68505   90-2531   PARTS KIT,ENGINE GE................   1
UOC:A11,A13,A14,A20,A24,A25,A26,A27,

H24,H25,H26,H27,H28,MMM   B17,B18,H11,H13,H14,H17,H18,H20,H21,

                                                            BEARING,BALL   (1) 43-12
                                                            END HEAD   (1) 43-15
                                                            O-RING   (1) 43-10
                                                            RETAINER ASSEMBLY   (1) 43-6
                                                            SEAL,KIT   (1) 43-7
                                                            SEAL,KIT   (1) 43-13
                                                            SPACER   (1) 43-9
                                                            SPRING,SEAL BEARING (1) 43-8
                                                            WASHER   (1) 43-11

       105 PAFZZ 2920011684130   68505   90-2532   PARTS KIT,ENGINE GE................   1
UOC:A11,A13,A14,A20,A24,A25,A26,A27,

H24,H25,H26,H27,H28,MMM   B17,B18,H11,H13,H14,H17,H18,H20,H21,

                                                            BEARING,OUTER RACE (1) 43-17
                                                            RETAINER,BEARING   (1) 43-21
                                                            RETAINER,FELT   (1) 43-18
                                                            RETAINER,FELT   (1) 43-20

SECTION II   TM9-2320-280-24P C01
 (1)   (2)   (3)   (4)   (5)   (6)
(7)   ITEM   SMR   PART
       NO   CODE   NSN   CAGEC   NUMBER   DESCRIPTION AND USABLE ON CODES(UOC)
QTY

			RING,SEAL BEARING (1) 43-22
			RING,SEAL BEARING (1) 43-24
			SEAL (1) 43-16
			WASHER,FELT (1) 43-19

       106 PAFZZ 5305011889707 68505 90-2836   SCREW ASSORTMENT................... 1
SCREW,ASSORTMENT   (1)   45-65

                                                   SCREW,HEX HEAD   (7) 45-19
                                                   SCREW,HEX HEAD   (4) 45-76
       107 PAFZZ 2920011922956 68505 90-2837   PARTS KIT,ENGINE GE............... 1
BUSHING,INSULATING (1)   45-53

                                                   BUSHING,INSULATING (1) 45-72
                                                   NUT,HEXAGON   (1) 45-59
                                                   NUT,HEXAGON   (4) 45-61
                                                   NUT,HEXAGON   (1) 45-78
                                                   NUT,HEXAGON   (3) 45-81
                                                   STUD,TERMINAL   (1) 45-64
                                                   STUD,TERMINAL   (1) 45-66
                                                   STUD,TERMINAL   (1) 45-70
                                                   WASHER,CUP,STARTER (1) 45-56
                                                   WASHER,CUP,STARTER (1) 45-75
                                                   WASHER,INSULATING (1) 45-55
                                                   WASHER,INSULATING (1) 45-62
                                                   WASHER INSULATING (1) 45-71
                                                   WASHER,INSULATING (1) 45-74
                                                   WASHER,NEOPRENE   (1) 45-54
                                                   WASHER,NEOPRENE   (2) 45-73
                                                   WASHER,LOCK   (4) 45-58
                                                   WASHER,LOCK   (2) 45-80
                                                   WASHER,PLAIN   (1) 45-57
                                                   WASHER,PLAIN   (1) 45-63
                                                   WASHER,PLAIN   (1) 45-77
       108 PAFZZ 2920011924473 68505 90-2838   PARTS KIT,ENGINE GE............... 1
BEARING,ABS BRONZE (1)   45-69

                                                   FELT OIL RETAINER (1) 45-68
                                                   SCREW,FILL HEAD   (3) 45-5
       109 PAFZZ 3120011914637 19728 90-2840   PARTS KIT,BEARING R............... 1
SPACER   (1)   45-7

                                                   THRUST WASHER   (1) 45-39
                                                   THRUST WASHER   (8) 45-40
                                                   WASHER,FLAT   (1) 45-1
       110 PAFZZ 2920011916534 68505 90-2841   PARTS KIT,ENGINE GE............... 1
PINION STOP STARTER(1)   45-10

                                                   SNAP,RING   (1) 45-11
       111 PAFZZ 2520012893617 63829 9103-9800 PARTS KIT,CLUTCH DI............... 1
UOC:A15,B15,H15

                                                   CLAMP   (1) 491-12
                                                   KEY,MACHINE   (1) 491-4
                                                   NUT   (1) 491-2
                                                   RING,SNAP   (1) 491-5
                                                   RING,SNAP   (1) 491-7
                                                   SCREW   (1) 491-11
                                                   SHIM   (1) 491-19
                                                   SHIM   (1) 491-20

KITS-20

```
SECTION II          TM9-2320-280-24P C01
    (1)     (2)     (3)              (4)          (5)                         (6)
(7)     ITEM    SMR                                PART
   NO    CODE   NSN              CAGEC        NUMBER      DESCRIPTION AND USABLE ON CODES(UOC)
QTY

                                                           SHIM              (  1) 491-21
    112 PAFZZ 2520012903921 63829 9140-9810       PARTS KIT,CLUTCH DI................    1
UOC:AVY,BVY,HVY
                                                           PLATE             (  1) 491-3
                                                           PULLLEY           (  1) 491-6
```

KITS-21

SECTION II   TM9-2320-280-24P C01

(1) ITEM NO	(2) SMR CODE	(3) NSN	(4) CAGEC	(5) PART NUMBER	(6) DESCRIPTION AND USABLE ON CODES (UOC)	(7) QTY

GROUP 95 GENERAL USE STANDARDIZED
PARTS
9501 BULK MATERIAL

FIG. BULK

Item	SMR	NSN	CAGEC	Part Number	Description	
1	PAOZZ	2590014579922	19207	12340103	BEZEL,AUTOMOTIVE.................	V
2	PAOZZ	4010005852108	16004	C43974	CHAIN,WELDLESS TYPE"2",CLASS 2 STEEL,10 LINK...................	V
3	PAOZZ	4020002460688	81755	P5006-4G	CORD,FIBROUS...................	.
4	PAOZZ		7X677	20197272	FASTENER,TAPE,HOOK...............	V
5	PAOZZ	5330010419721	26051	R-451-N	GASKET 3/16 THICK, 42 INCHES WIDE, 60 FT. LONG.................	V
6	PAOZZ	5325000743301	96906	MS21266-2N	GROMMET,NONMETALLIC FLEXIBLE, .162 INCHES X .200 INCHES X 25.0 FEET LONG,BLACK.....................	V
*7	PAOZZ	4720011995717	7X677	9440949	HOSE,NONMETALLIC 3/4 O.D.,1/2 I.D...	
8	PAOZZ	4720001115096	21868	AD778-08B79	HOSE,AIR DUCT REINFORCED 2.00 I.D., 12 FT LONG,WIRE...............	V
9	PAOZZ	4720001393968	24161	4377	HOSE,NONMETALLIC 1-3/4 I.D.,1.00 O.D., 50 FT LONG................	V
*10	PAOZZ	4720000069541	24161	G3336US-10	HOSE,NONMETALLIC 1.00 O.D.,5/8 I.D..	V
*11	PAOZZ	4720011637833	24161	4219-3681	HOSE,NONMETALLIC.1/2 O.D.,1/4 I.D...	V
*12	PAOZZ	4720011560549	7X677	9439046	HOSE,NONMETALLIC.1/4 I.D...........	V
*13	PAOZZ	4720011560550	7X677	9438381	HOSE,NONMETALLIC 5/32 I.D..........	V
*14	PAOZZ	4720011482768	7X677	9438315	HOSE,NONMETALLIC 5/16 I.D..........	V
*17	PAOZZ	4720000658682	29510	1125218C1	HOSE,NONMETALLIC 3/16 I.D..........	V
18	PAOZZ	4720006844033	83065	RB1450-1-4IDX1-2	HOSE,NONMETALLIC RUBBER, INNER TUBE, OUTER COVER, FIBER BRAID,1/2 OD O.D.,1/4 I.D.,50 FT LONG...........	V
19	PAOZZ	4720011699891	61424	PFT-6B	HOSE,NONMETALLIC NYLON,3/8 O.D.,5/16 I.D.,50 FT LONG...............	V
*20	PAOZZ	4720010036706	61424	PFT-88-BLK-100	HOSE,NONMETALLIC.................	V
21	PAOZZ	4720011883190	59150	1775-826	HOSE,NONMETALLIC SYNTHETIC RUBBER INNER O.D., 1/4 I.D., 1/8, 50 FT LONG....................	V
22	PAOZZ	4720011856673	34623	ET029	HOSE,NONMETALLIC SPIRAL REINFORCED, .380 O.D.,290 I.D.,15 FT LONG.......	V
23	PAOZZ	4720011862358	81300	6490610019	HOSE,NONMETALLIC FUEL AND OIL RESISTANT,3/8 I.D.,5/8 O.D.,25 FT LONG..................................	V
24	PAOZZ	4720002414435	30327	C110-10	HOSE,NONMETALLIC.................	V
25	PAOZZ	4720011595796	7X677	9438373	HOSE,NONMETALLIC FUEL AND OIL RESISTANT,5/8 O.D.,3/8 I.D.,25 FT LONG..................................	V
26	PAOZZ	4720011557784	7X677	9438257	HOSE,NONMETALLIC 1/2 O.D. X 1/4 I.D.,50 FT LONG.....................	V
27	PAOZZ	4720004910102	96906	MS521301A203R	HOSE,NONMETALLIC STRAIGHT,ENGINE	V
*28	PAOZZ	4720005558015	96906	MS521302B206R	HOSE,NONMETALLIC COOLANT...........	V
29	PAOZZ		81349	M23053/4-202-2	INSULATION SLEEVING................	V
30	PAOZZ		81349	M23053/4-203-2	INSULATION SLEEVING................	V
*31	PAOZZ	5970011616796	81349	M23053/4-302-0	INSULATION SLEEVING................	V
32	PAOZZ		81349	M23053/4-302-2	INSULATION SLEEVING................	V

BULK-1

SECTION II   TM9-2320-280-24P C01
   (1)   (2)   (3)   (4)   (5)   (6)
(7)   ITEM   SMR   PART
        NO   CODE   NSN   CAGEC   NUMBER   DESCRIPTION AND USABLE ON CODES (UOC)
QTY

        33 PAOZZ 5970011422282 81349 M23053/4-303-0   INSULATION SLEEVING ELECTRICAL,   V
BLACK (12 OR 14-14 AWG)
        34 PAOZZ   81349 M23053/4-303-2   INSULATION SLEEVING.................. V
        35 PAOZZ 5970011631103 81349 M23053/4-304-0   INSULATION SLEEVING.................. V
        36 PAOZZ   81349 M23053/4-304-2   INSULATION SLEEVING.................. V
        37 PAOZZ 5970007241910 81349 M23053/6-103-6   INSULATION SLEEVING.................. V
        38 PAOZZ 5970010499948 81343 M23053/6-104-2   INSULATION SLEEVING.................. V
        39 PAOZZ 5970010310736 81349 M23053/6-107-0   INSULATION SLEEVING.................. V
        40 PAOZZ 5970007241907 81349 M23053/6-107-1   INSULATION SLEEVING.................. V
        41 PAOZZ 5970012767753 81343 M23053/15-112-0   INSULATION SLEEVING..................
        42 PAOZZ   81349 M23053/15-112-2   INSULATION SLEEVING..................
        43 PAOZZ 2590011967281 19207 12339902-1   PAD,CUSHIONING 3/8 THICK, 9/32
                                                           WIDE, 800 FT LONG....................
        44 PAOZZ 2590013127355 28818 CF40BH52   PAD,CUSHIONING 50 YARDS X 54
                                                           INCHES WIDE X 1 INCH THICK...........
        45 PAOZZ 4710012282177 81346 D1785   PIPE,PLASTIC.........................
        46 PAOZZ 5330012914615 34623 5741221   PLASTIC STRIP 1/4 THICK, 1/2 WIDE,
                                                           20 FT LONG...........................
        47 PAOZZ 5530006416069 81349 MIL-P-18066   PLYWOOD,CONSTRUCTIO 3/4 THICK.......
        48 PAOZZ 5530002628195 81348 NN-P-530   PLYWOOD,CONSTRUCTIO 1/2 THICK, 48   V
INCHES WIDE, 8 FT LONG...............
        49 PAOZZ 4010012930586 81349 MIL-W-83420/2-00 ROPE,WIRE .062-INCH DIAMETER....... V
                                                           2
        50 PAOZZ 4010009290041 39428 8930T31   ROPE,WIRE............................
        51 PAOZZ 5330011983521 19207 12340721   RUBBER STRIP ETHYLENE PROPYLENE   V
TERPOLYMER, ASTM 0-1056-68.38 X 40

                                                           X 10 INCHES..........................
        52 PAOZZ 5330012816523 82942 68-412121-1   SEAL,NONMETALLIC SP .250 THICK,1/2
                                                           WIDE, 50 FT LONG.....................
        53 PAOZZ 5330013218622 41024 21022103   SEAL,NONMETALLIC SP SPECIAL, EDPM
                                                           WIRE CARRIER, 396 FT LONG............
        54 PAOZZ 5330012131312 06968 1403   SEAL,NONMETALLIC SP SPECIAL, EDPM
                                                           RUBBER, WIRE CARRIER, 18 FT LONG.....
        55 PAOZZ 5330013117953 76385 IC006197   SEAL,NONMETALLIC SP..................
        56 PAOZZ   81343 R-495-T   SEAL,NONMETALLIC SP..................
        57 PAOZZ 5330014091981 19207 12460108-3   SEAL,NONMETALLIC SP..................
                                                           ASSEMBLY,FABRICATE FROM P/N Y-9485
                                                           13 INCHES LONG.......................
        58 PAOZZ 9320011552369 45152 4965   TAPE,ADHESIVE,RUBBE RUBBER DOUBLE   V
COATED, NEOPRENE, 3/4 X .45 X 108
                                                           FT LONG..............................
        59 PAOZZ   19207 12339376-1   TAPE,PRESSURE SENSI..................
        60 PAOZZ   81343 15618240   TUBE................................. V
        61 PAOZZ 4720010587213 19207 CPR104420-1   TUBING,NONMETALLIC PLASTIC, .170   V
        62 PAOZZ 4720010144915 19207 CPR104420-2   TUBING,NONMETALLIC................... V
INCH I.D., .250 INCH O.D............
        63 PAOZZ 4720000060048 81349 M6855/4-16H029   TUBING,NONMETALLIC...................
        64 PAOZZ 4720014219718 19207 12420924-004   TUBING,NONMETALLIC...................
        65 PAOZZ 9330000112000 00624 AE208-10   TUBING,PLASTIC,SPIR..................
        66 PAOZZ 9330012521239 81349 B47287-II-1/2   TUBING,PLASTIC,SPIR OIL COOLANT
                                                           LINES,.50 I.D.,.850 O.D.,100 FT LONG
        67 PAOZZ 6145013419591 81349 M13486/1-4   WIRE,ELECTRICAL...................... V

                                    BULK-2

SECTION II TM9-2320-280-24P C01

(1) ITEM NO	(2) SMR CODE	(3) NSN	(4) CAGEC	(5) PART NUMBER	(6) DESCRIPTION AND USABLE ON CODES(UOC)	(7) QTY
68	PAOZZ	6145005388222	81349	M13486/1-9	WIRE,ELECTRICAL.....................	V
69	PAOZZ	6145005388219	81349	M13486/1-11	WIRE,ELECTRICAL 4 AWG, SINGLE CONDUCTOR..........................	V
70	PAOZZ	6145007056674	81349	M13486/1-14	WIRE,ELECTRICAL 0 AWG, SINGLE CONDUCTOR..........................	V
71	PAOZZ	5510002706031	97403	13219E0079	WOOD LAMINATE,DECKI DECKING........	V
72	PAOZZ		19207	12339317-1	WRAP................................	V

END OF FIGURE

BULK-3

Section II.

TM 9-2320-280-24P-2

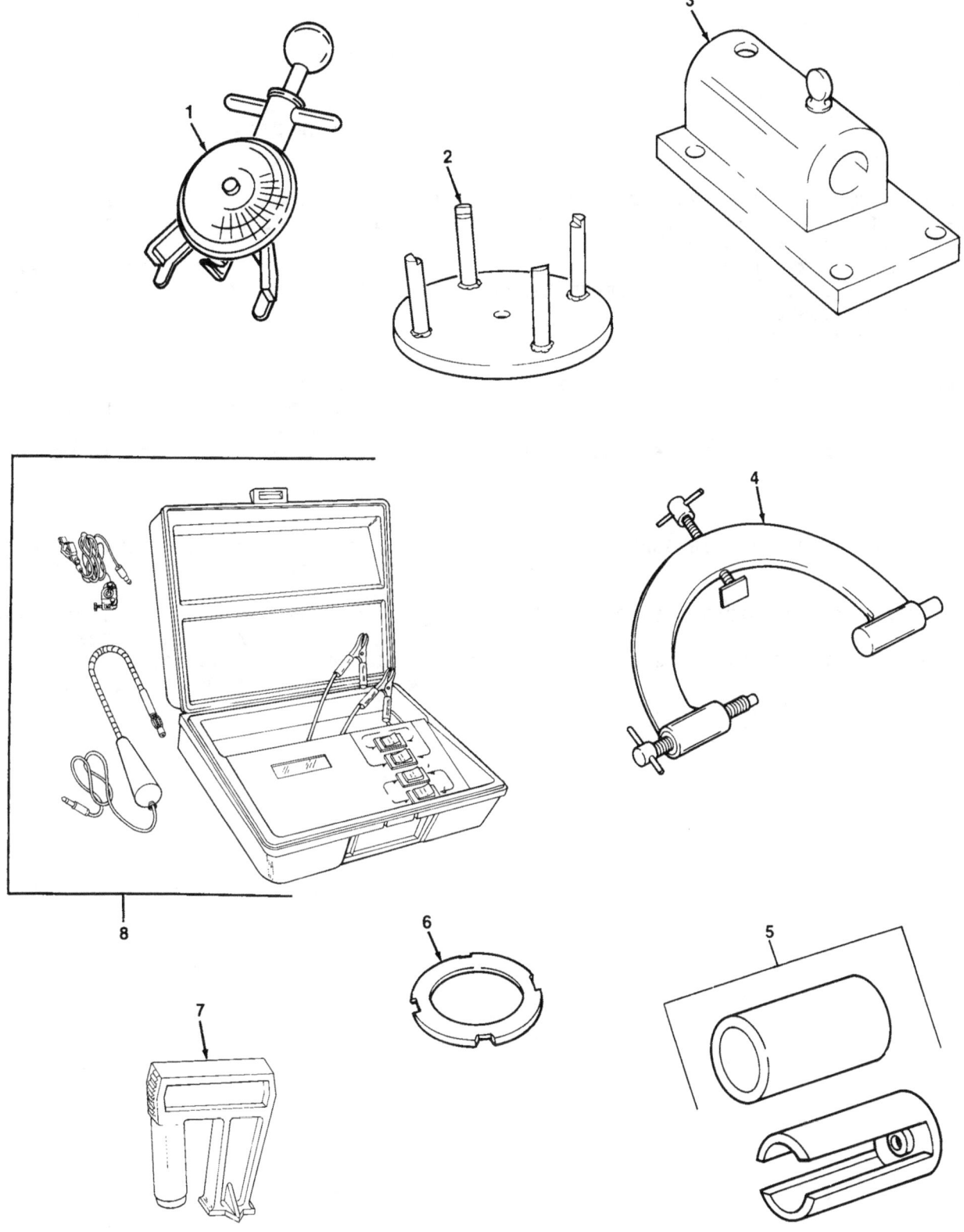

Figure 499. Special Tools, Drive Belts and Transmission.

SECTION III   TM9-2320-280-24P C01

(1) ITEM NO	(2) SMR CODE	(3) NSN	(4) CAGEC	(5) PART NUMBER	(6) DESCRIPTION AND USABLE ON CODES (UOC)	(7) QTY

GROUP 26 TOOL AND TEST EQUIPMENT

2604 SPECIAL TOOLS

FIG. 499 SPECIAL TOOLS, DRIVE BELTS AND TRANSMISSION

1	PEOZZ	6635010933710	33287	BT-33-73F	TENSIOMETER,DIAL IN BOI: SEE PARA. 5.I, PART OF KIT P/N 57K0267.......
2	PEHZZ	5120012108793	33287	J4670-01	COMPRESSOR,CLUTCH,S SPRING,BOI: SEE PARA. 5.I, PART OF KIT P/N 57K0266..................................
3	PEHZZ	5120011444484	33287	J3289-20	BASE,HOLDING FIXTUR TRANSMISSION, BOI: SEE PARA. 5.I, PART OF KIT P/N 57K0266.....................
* 4	PEHZZ	5120011987583	33287	J8763-0B	HOLDING FIXTURE,TRA TRANSMISSION, BOI: SEE PARA. 5.I, PART OF KIT P/N 57K0266.....................
5	PEHZZ	4910011788865	25341	J21795-02	HOLDING UNIT,GEAR TRANSMISSION,BOI: SEE PARA. 5.I, PART OF KIT P/N 57K0266.........................
6	PEHZZ	4910012101318	33287	J21664	ADAPTER,COMPRESSOR CLUTCH SPRING, BOI: SEE PARA.,5.I, PART OF KIT P/N 57K0266........................
7	PEFZZ	4910012313671	55719	MT95	GAGE,TIMING,DIESEL BOI: SEE PARA. 5.I, PART OF KIT P/N 57K0268.......
8	PEFZZ	5180011863114	33287	J33127	TOOL KIT,ENGINE TIM DYNAMIC,BOI: SEE PARA. 5.I, PART OF KIT P/N 57K0268.............................

499-1

Section II.

TM 9-2320-280-24P-2

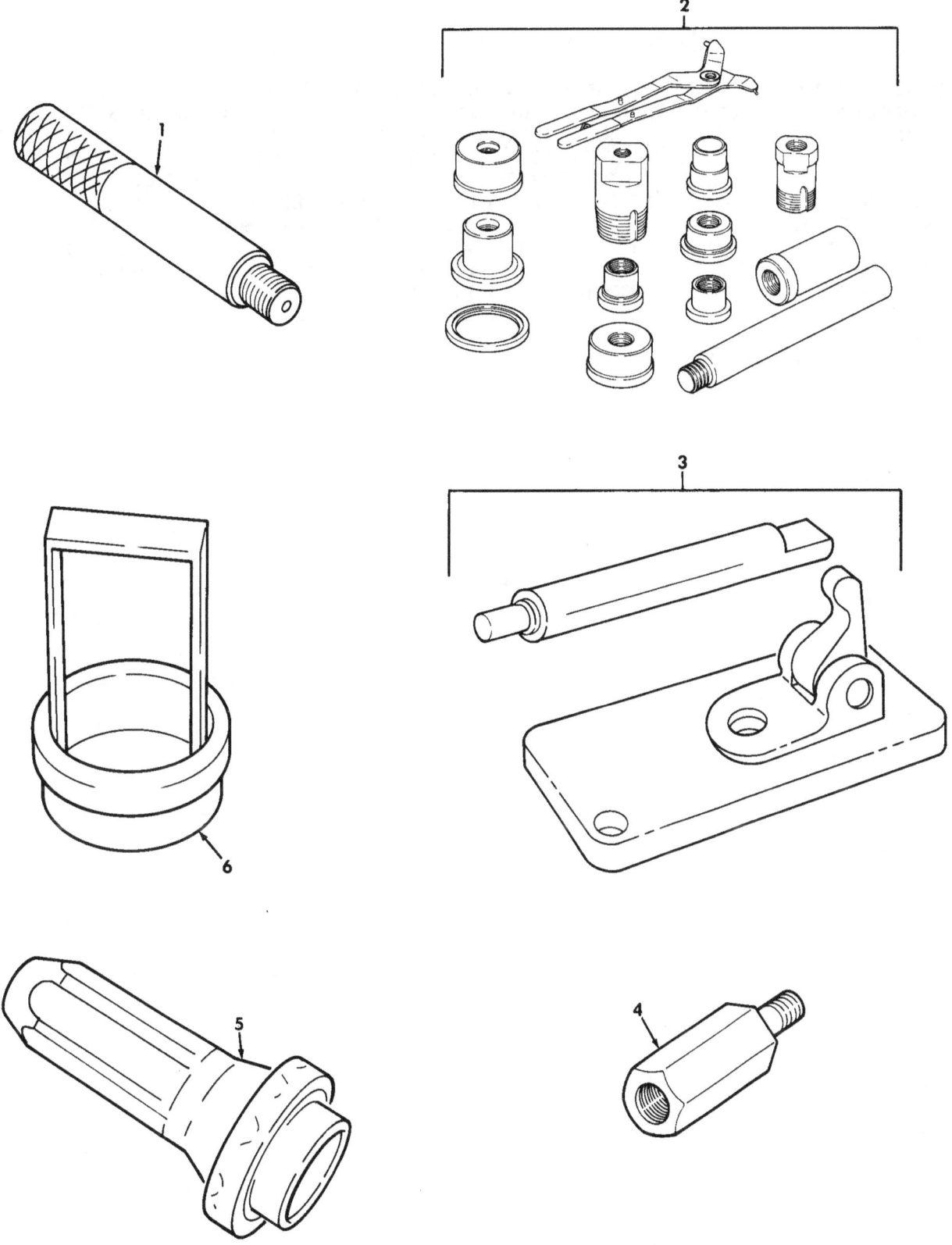

Figure 500. Special Tools, Transmission.

SECTION III TM9-2320-280-24P

(1) NO	(2) SMR CODE	(3) NSN	(4) CAGEC	(5) PART NUMBER	(7) DESCRIPTION AND USABLE ON CODES(UOC)	ITEM QTY

GROUP 2604 SPECIAL TOOLS

FIG. 500 SPECIAL TOOLS,TRANSMISSION

1	PEOZZ	5120006772259	33287	J8092	HANDLE,DRIVE TRANSMISSION,BOI: SEE PARA. 5.I, PART OF KIT P/N 57K0267.	
2	PEHZZ	5180011959777	33287	J21465-01	TOOL SET,BUSHING,SE TRANSMISSION, BOI: SEE PARA. 5.I, PART OF KIT P/N 57K0266	
3	PEHZZ	4910011780722	25341	J-21370	GAGE,PIN SELECTOR BAND APPLY,BOI: SEE PARA. 5.I, PART OF KIT P/N 57K0266	
4	PEHZZ	5120011308865	33287	J6471-2	HANDLE,SLIDE HAMMER SLIDE HAMMER, 5/8-16, 3/8-16,BOI: SEE PARA. 5.I, PART OF KIT P/N 57K0266	
5	PEFZZ	5120011761845	25341	J21359-A	INSTALLER,OIL SEAL TRANSMISSION OIL PUMP,BOI: SEE PARA. 5.I, PART OF KIT P/N 57K0268	
6	PEHZZ	4910012090729	33287	J24396	TOOL,ALIGNMENT,CLUT INTERMEDIATE, BOI: SEE PARA. 5.I, PART OF KIT P/N 57K0266	

END OF FIGURE

Section II. TM 9-2320-280-24P-2

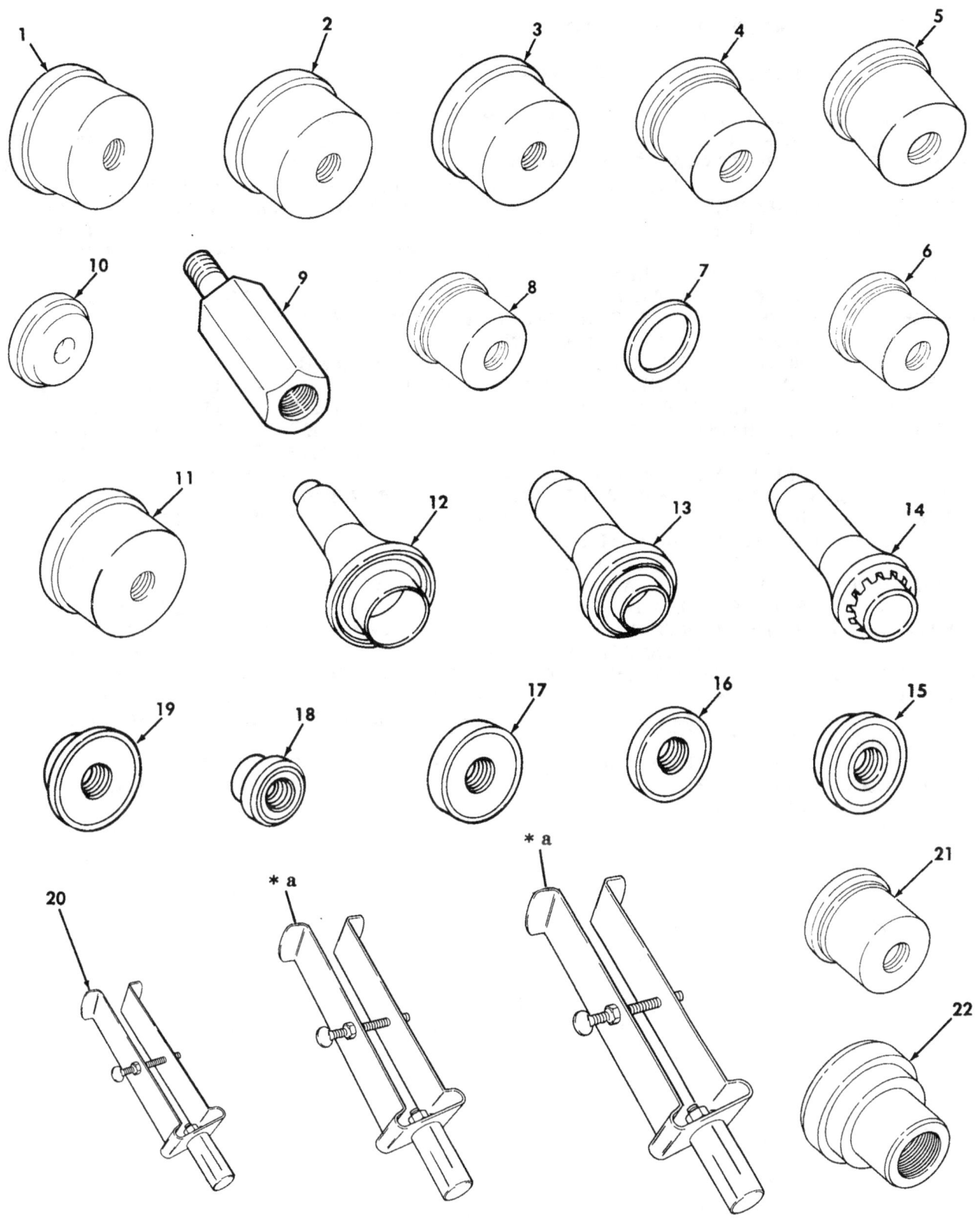

* a PART OF ITEM 20

Figure 501. Special Tools, Transfer Case.

SECTION III   TM9-2320-280-24P

(1) NO	(2) CODE	(3) NSN	(4) CAGEC	(5) NUMBER	(6)	(7) ITEM DESCRIPTION AND USABLE ON CODES(UOC)	PART SMR QTY

GROUP 2604 SPECIAL TOOLS

FIG. 501 SPECIAL TOOLS, TRANSFER CASE

1 PEHZZ 5120011960084 33287 J29166 INSERTER,BEARING AN REAR OUTPUT, BOI: SEE PARA. 5.I, PART OF KIT P/N 57K0266..............................

2 PEHZZ 5120011694876 25341 J-29168 INSERTER,BEARING AN FRONT OUTPUT, BOI: SEE PARA. 5.I, PART OF KIT P/N 57K0266..............................

3 PEHZZ 5120011703278 25341 J-29167 INSERTER,BEARING AN BOI: SEE PARA. 5.I,PART OF KIT P/N 57K0266.........

4 PEHZZ 5120011952721 33287 J29163 INSERTER,BEARING AN FRONT OUTPUT SHAFT REAR BEARING,BOI: SEE PARA. 5.I, PART OF KIT P/N 57K0266.......

5 PEHZZ 5120011857955 33287 J29169 INSERTER,BEARING AN INPUT GEAR,BOI: SEE PARA. 5.I, PART OF KIT P/N 57K0266..............................

6 PEHZZ 5120012654872 33287 J-35307 INSERTER,BEARING AN MAINSHAFT,BOI: SEE PARA. 5.I, PART OF KIT P/N 57K0266.........

7 PEHZZ 5120012476629 33287 J29185-2 INSTALLER,ANNULUS B ANNULUS GEAR BUSHING,BOI: SEE PARA. 5.I, ...

8 PEHZZ 5120011857956 33287 J29185 INSERTER,BEARING AN BOI: SEE PARA. 5.I PART OF KIT P/N 57K0266...

9 PEHZZ 5120013915131 33287 J6471-8 ADAPTER MECHANIAL SLIDE HAMMER,5/8-18 1/2-13,BOI: SEE PARA. 5.I,PART OF KIT P/N 57K0266.....................

10 PEHZZ 5120011858024 33287 J7818 INSERTER,BEARING AN RETAINER,BOI: SEE PARA. 5.I, PART OF KIT P/N 57K0266..............................

11 PEHZZ 5120011954551 33287 J29170 REMOVER,BEARING INPUT GEAR,BOI: SEE PARA. 5.I, PART OF KIT P/N 57K0266...........................

12 PEHZZ 5120013573632 33287 J-33831 INSERTER,SEAL INPUT GEAR SEAL,BOI: SEE PARA. 5.I PART OF KIT P/N 57K0266...........................

13 PEHZZ 5120012271680 33287 J-22661 INSTALLER,SEAL OUTPUT SHAFT,BOI: SEE PARA. 5.I,PART OF KIT P/N 57K0236,PART OF KIT P/N 57K0268.....

14 PEHZZ 5120013613101 33287 J-33843 INSERTER,SEAL EXTENSION HOUSING SEAL,BOI: SEE PARA. 5.I. PART OF KIT P/N 57K0236 PART OF KIT P/N 57K0266

15 PEHZZ 5120013573633 33287 J-33826 INSERTER AND REMOVE BEARING INSTALLER,BOI: SEE PARA. 5.I. PART OF KIT P/N 57K0236 PART OF KIT P/N 57K0266...........................

16 PEHZZ 5120013573629 33287 J-33829 INSERTER,BEARING AN MAINSHAFT PILOT BEARING,BOI: SEE PARA. 5.I. PART OF KIT P/N 57K0236.............

17 PEHZZ 5120013573630 33287 J-33833 INSERTER,BEARING AN OUTPUT SHAFT FRONT BEARING,BOI: SEE PARA. 5.I. PART OF KIT P/N 57K0236 PART OF KIT

SECTION III   TM9-2320-280-24P

(1) NO	(2) SMR CODE	(3) NSN	(4) CAGEC	(5) PART NUMBER	(6) DESCRIPTION AND USABLE ON CODES (UOC)	(7) QTY
18	PEHZZ	5120013573631	33287	J-33839	REMOVER,BEARING AND EXTENSION HOUSING BUSHING,BOI: SEE PARA. 5.I. PART OF KIT P/N 57K0236 PART OF KIT P/N 57K0266	
19	PEHZZ	5120013579123	33287	J-9276-3	INSERTER,BEARING AN BOI: SEE PARA. 5.I. PART OF KIT P/N 57K0236 PART OF KIT P/N 57K0266	
20	PEHZZ	5120012017857	33287	J29369	PULLER,MECHANICAL UNIVERSAL BUSHING AND BEARING,BOI: SEE PARA. 5.I, PART OF KIT P/N 57K0266	
21	PEHZZ	5120011857960	33287	J29174	INSERTER,BEARING AN BOI: SEE PARA. 5.1 PART OF KIT P/N 57K0266	
22	PEHZZ	5120013899992	33287	J39636	INSERTER,BEARING AN BOI: SEE PARA. 5.1, PART OF KIT P/N 57K0236 PART OF KIT P/N 57K0266	

END OF FIGURE

Section II.

TM 9-2320-280-24P-2

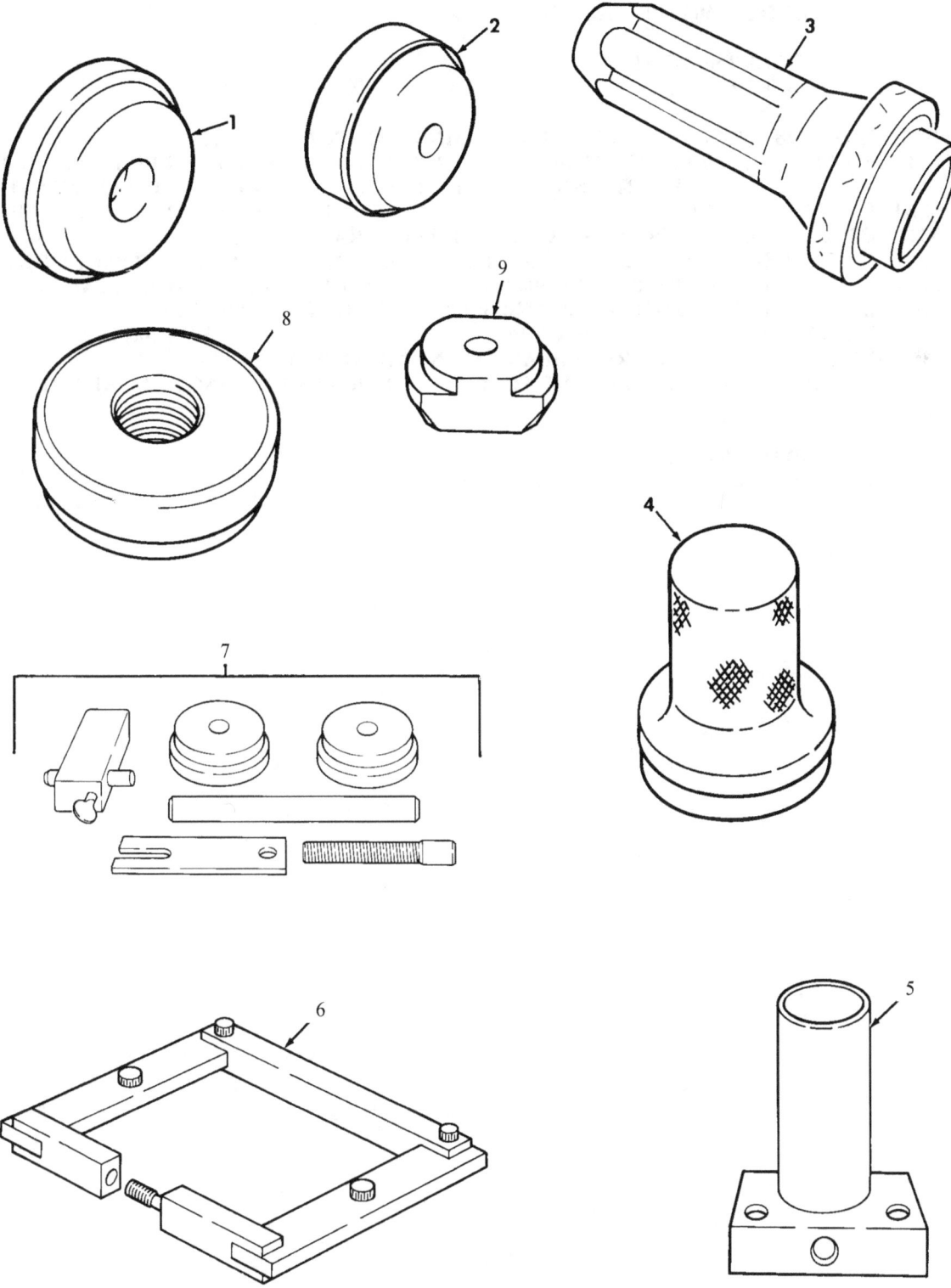

Figure 502. Special Tools, Axle Differential.

SECTION III   TM9-2320-280-24P

(1) NO	(2) SMR CODE	(3) NSN	(4) CAGEC	(5) PART NUMBER	(7) ITEM DESCRIPTION AND USABLE ON CODES(UOC)	QTY

GROUP 2604 SPECIAL TOOLS

FIG. 502 SPECIAL TOOLS, AXLE DIFFERENTIAL

1	PEHZZ	5120011873660	33287	J21787	REMOVER,BEARING AND BOI: SEE PARA. 5.I PART OF KIT P/N 57K0266	
2	PEHZZ	5120011857964	33287	J8611-01	INSERTER,BEARING AN BOI: SEE PARA. 5.I PART OF KIT P/N 57K0266	
3	PEFZZ	4910011795530	25341	J-29162	DRIVE TOOL,SEAL BOI: SEE PARA 5.I PART OF KIT P/N 57K0268	
4	PEFZZ	5120011873659	33287	J33142	INSERTER,BEARING AN BOI: SEE PARA. 5.I PART OF KIT P/N 57K0268	
5	PEHZZ	5120012188235	33287	J-33149-A	ADAPTER,AXLE FIXTUR PART OF KIT P/N 57K0266	
6	PEHZZ	4910001052823	80604	W-129-B	SPREADER,AXLE HOUSI PART OF KIT P/N 57K0266	
7	PEHZZ	5180012168643	33287	J-35199	TOOL KIT,PINION SETTING GAUGE,BOI: SEE PARA. 5.I,PART OF KIT P/N 57K0266	8
8	PEHZZ	5120011857962	33287	J8608	INSERTER,BEARING AN PART OF KIT P/N 57K0266	
9	PEHZZ	5120011857957	33287	J21786	REMOVER,BEARING AND PART OF KIT P/N 57K0266	

END OF FIGURE

502-1

Section II. TM 9-2320-280-24P-2

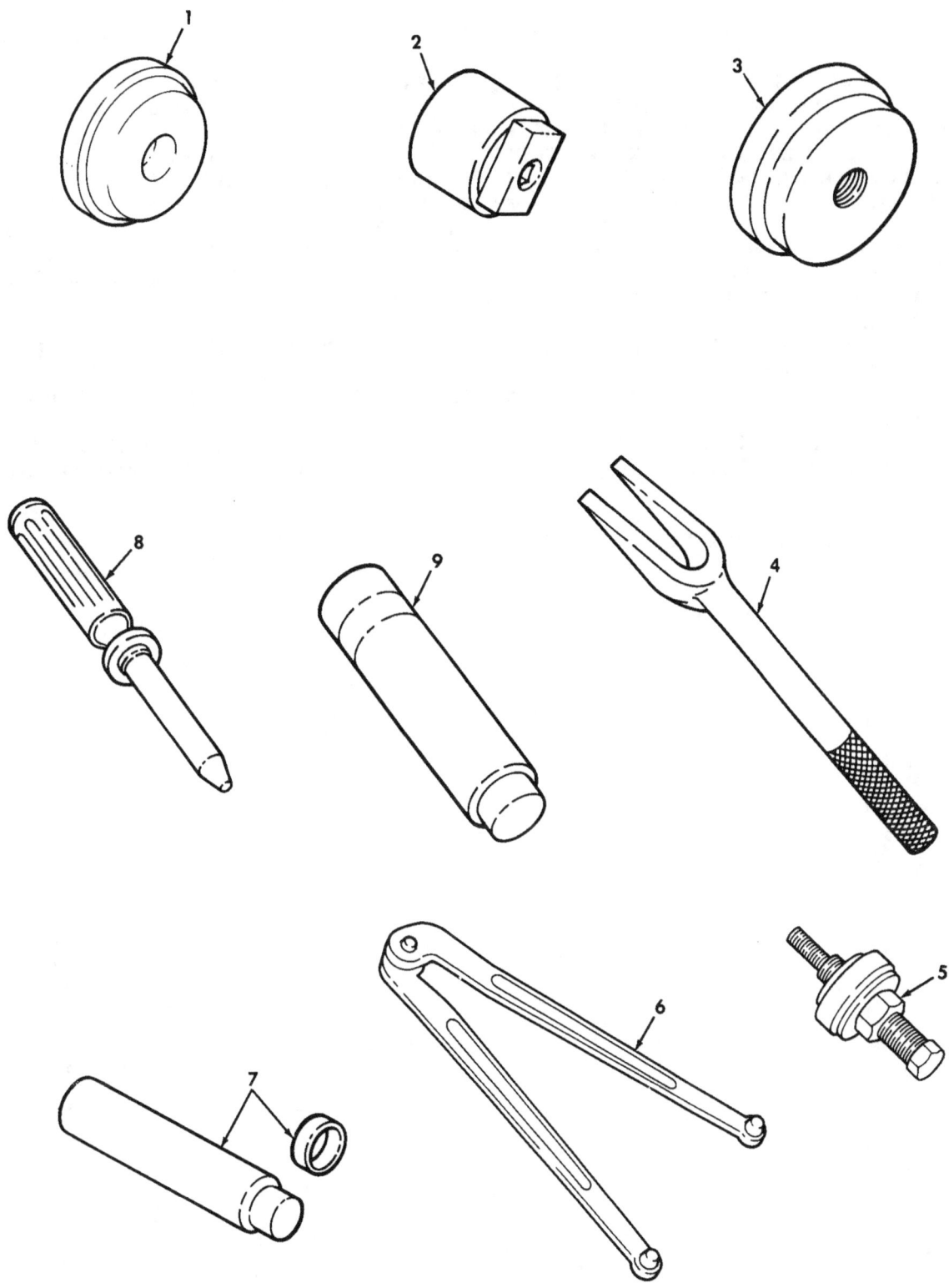

Figure 503. Special Tools, Geared Hub and Steering.

SECTION III   TM9-2320-280-24P

(1) NO	(2) SMR CODE	(3) NSN	(4) CAGEC	(5) PART NUMBER	(6) DESCRIPTION AND USABLE ON CODES(UOC)	(7) QTY

GROUP 2604 SPECIAL TOOLS

FIG. 503 SPECIAL TOOLS, GEARED HUB AND STEERING

1	PEOZZ	5120011873607	33287	J33143	INSERTER,SEAL GEARED HUB INPUT,BOI: SEE PARA. 5.I, PART OF KIT P/N 57K0267	2
2	PEOZZ	5120011857963	33287	J33144	SOCKET,SOCKET WRENC GEARED HUB,BOI: SEE PARA. 5.I, PART OF KIT P/N 57K0267	
3	PEOZZ	5120012290842	33287	J35184	INSERTER,SEAL GEARED HUB,BOI: SEE PARA. 5.I, PART OF KIT P/N 57K0267	
4	PEOZZ	5120008804268	19207	11595179	REMOVAL TOOL,TIE RO BALL JOINTS, BOI: SEE PARA. 5.I, PART OF KIT P/N 57K0267	
5	PEOZZ	4910011792517	33287	J25033-B	INSTALLER,POWER STE POWER STEERING PUMP PULLEY,BOI: SEE PARA. 5.I PART OF KIT P/N 57K0267	
6	PEFZZ	5120010826436	25341	J7624	WRENCH,SPANNER PRELOAD BEARING STEERING GEAR,BOI: SEE PARA. 5.I, PART OF KIT P/N 57K0268	
7	PEFZZ	5120010826447	25341	J-6278	INSTALLING AND REMO PITMAN SHAFT BEARING,BOI: SEE PARA. 5.I, PART OF KIT P/N 57K0268	
8	PEFZZ	5120011791032	33287	J-21552	PUNCH,HAND TOOL RACK PISTON STEERING GEAR,BOI: SEE PARA. 5.I, PART OF KIT P/N 57K0268	
9	PEFZZ	5120011857965	33287	J-6221	REMOVER,BEARING BEARING ADJUSTER PLUG,BOI: SEE PARA. 5.I,PART OF KIT P/N 57K0268	

END OF FIGURE

Section II.

TM 9-2320-280-24P-2

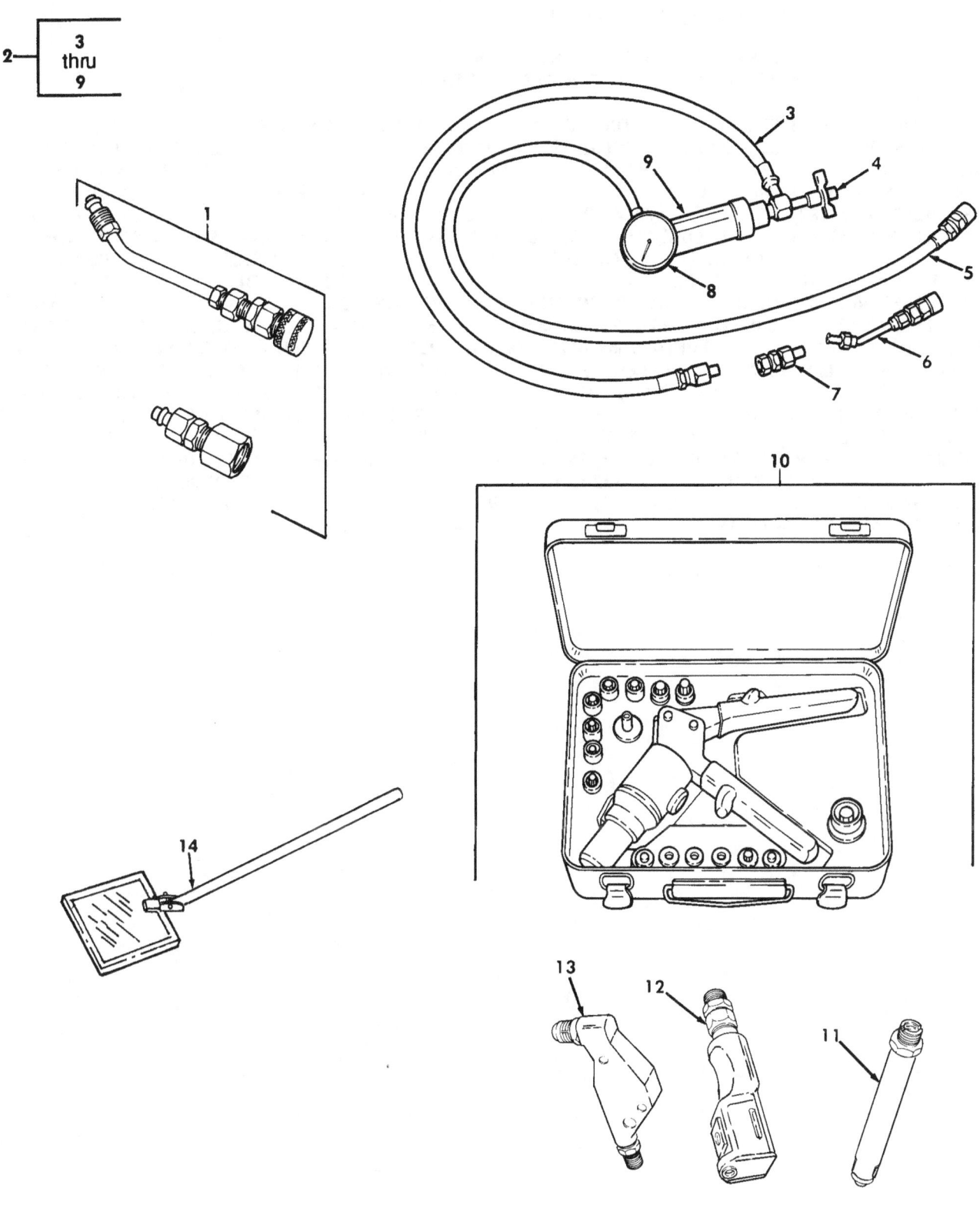

Figure 504. Special Tools, Steering Pump Body, and Speedometer Drive.

SECTION III   TM9-2320-280-24P

(1) NO	(2) SMR CODE	(3) NSN	(4) CAGEC	(5) PART NUMBER	(7) DESCRIPTION AND USABLE ON CODES(UOC)	ITEM QTY

GROUP 2604 SPECIAL TOOLS

FIG. 504 SPECIAL TOOLS,STEERING PUMP,BODY,AND SPEEDOMETER DRIVE

1 PEOZZ 5120012311709 33287 J33141 ADAPTER,ANALYZER BOI: SEE PARA. 5.I PART OF KIT P/N 5705568............

2 PEOOO 4910011857966 19207 12342943 ANALYZER,POWER STEE BOI:- SEE PARA. 5.I PART OF KIT P/N 57K0267.........

3 PEOZZ 4720012532809 33287 J25323-17 .HOSE ASSEMBLY,NONME BOI: SEE PARA. 5.I PART OF KIT P/N 12342943..

4 PEOZZ 4820012526169 33287 J25323-16 .VALVE,REGULATING,FL BOI: SEE PARA. 5.I PART OF KIT P/N 12342943..

5 PEOZZ 4720012532810 33287 J25323-18 .HOSE ASSEMBLY,NONME BOI: SEE PARA. 5.I PART OF KIT P/N 12342943..

6 PEOZZ 4730012546023 33287 J25323-2 .COUPLING HALF,QUICK BOI: SEE PARA. 5.1 PART OF KIT P/N 12342943..

7 PEOZZ 3020012552906 33287 J25323-1 .ADAPTER,GEAR BOI: SEE PARA. 5.I PART OF KIT P/N 12342943............

8 PEOZZ 6685012563406 33287 J25323-5 .GAGE,PRESSURE,DIAL BOI: SEE PARA. 5.I PART OF KIT P/N 12342943........

9 PEOZZ 6680012536782 33287 J25323-7 .INDICATOR,RATE OF F BOI: SEE PARA. 5.I PART OF KIT P/N 12342943..

10 PEOZZ 5180012014978 64878 D100-MIL-1 TOOL KIT,BLIND RIVE BOI: SEE PARA. 5.I, PART OF KIT P/N 57K0267.......

11 PEFZZ 5130011045370 11815 H749A-456 PULLING HEAD,RIVET BOI: SEE PARA. 5.I,PART OF KIT P/N 57K0268.........

12 PEFZZ 5130010447196 53551 RU818 PULLING HEAD,RIVET BOI: SEE PARA. 5.I,PART OF KIT P/N 57K0268........

13 PEFZZ 5130010453507 53551 RU817 PULLING HEAD,RIVET RIGHT ANGLE,BOI: SEE PARA. 5.I,PART OF KIT P/N 57K0268.......

14 PEOZZ 5120006186902 77335 71-510 MIRROR,INSPECTION SPEEDOMETER DRIVE,BOI: SEE PARA.5.I, PART OF KIT P/N 57K0267....................

END OF FIGURE

Section II.   TM 9-2320-280-24P-2

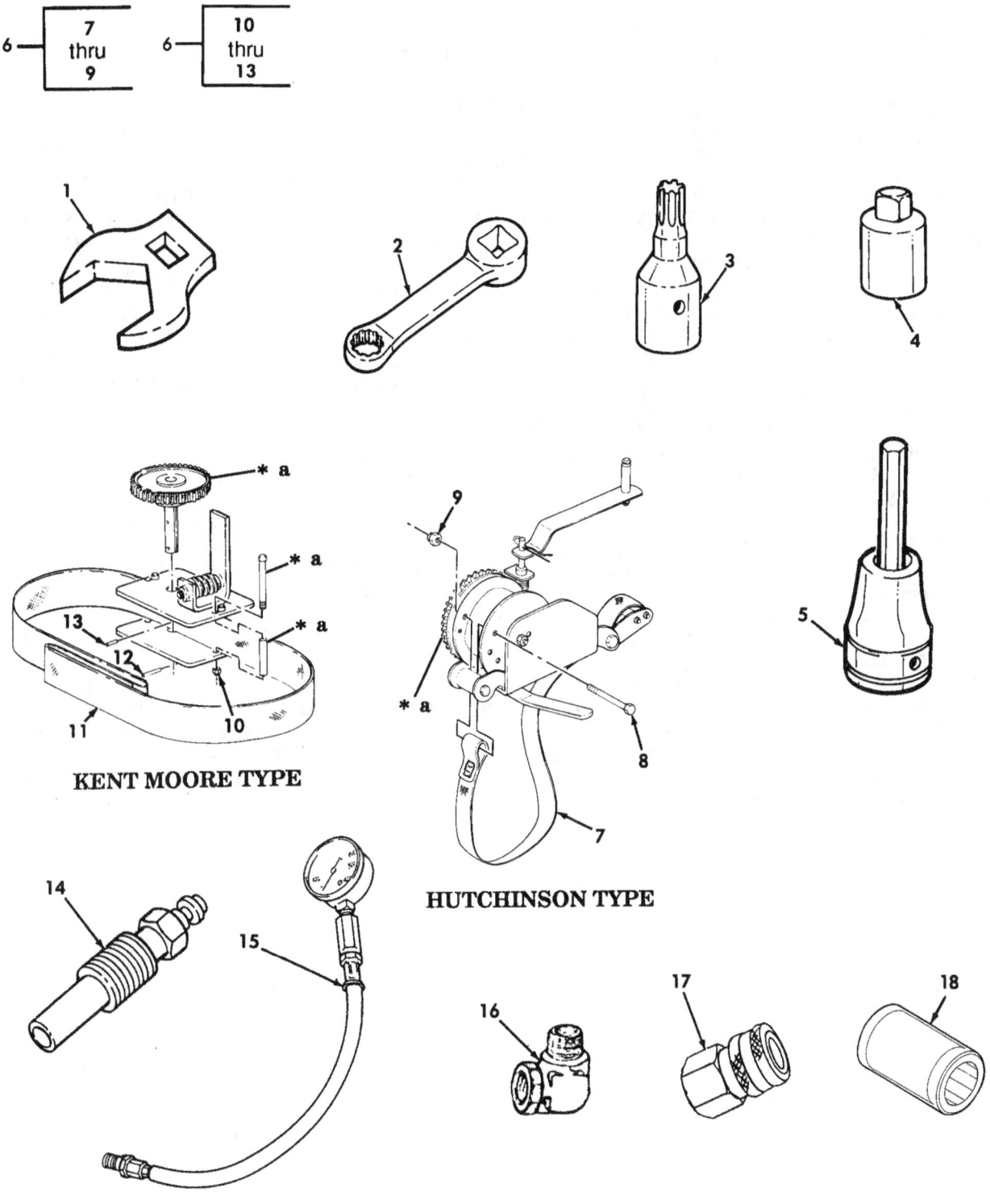

* a PART OF ITEM 6

Figure 505. Special Tools, Multipurpose.

SECTION III TM9-2320-280-24P

(1) NO	(2) SMR CODE	(3) NSN	(4) CAGEC	(5) PART NUMBER	(7) ITEM DESCRIPTION AND USABLE ON CODES(UOC)	(6) QTY

GROUP 2604 SPECIAL TOOLS

FIG. 505 SPECIAL TOOLS, MULTIPURPOSE

No	SMR	NSN	CAGEC	Part Number	Description	Qty
1	PEOZZ	5120013351094	55719	FC18B	CROWFOOT ATTACHMENT 9/16, 3/8 DRIVE, BOI: SEE PARA. 5.I PART OF KIT P/N 57K0267	1
1	PEOZZ	5120013351096	55719	FC22A	CROWFOOT ATTACHMENT 11/16, 3/8 DRIVE, BOI: SEE PARA. 5.I, PART OF KIT P/N 57K0267	1
1	PEOZZ	5120013351099	55719	FC28B	CROWFOOT ATTACHMENT 7/8, 3/8 DRIVE, BOI: SEE PARA.5.I, PART OF KIT P/N 57K0267	1
1	PEOZZ	5120010798023	33287	J-35158	CROWFOOT ATTACHMENT 1/2, 3/8 DRIVE, BOI: SEE PARA. 5.I, PART OF KIT P/N 57K0267	1
1	PEOZZ	5120013351100	55719	FC30B	CROWFOOT ATTACHMENT 15/16, 3/8 DRIVE, BOI: SEE PARA.5.I, PART OF KIT P/N 57K0267	1
1	PEOZZ	5120001848398	55719	AN8506-4	CROWFOOT ATTACHMENT 5/8, 3/8 DRIVE, BOI: SEE PARA.5.I, PART OF KIT P/N 57K0267	1
2	PEOZZ	5120013673582	55719	SRES18	ADAPTER, TORQUE WREN 9/16, BOI: SEE PARA.5.I, PART OF KIT P/N 57K0267	2
2	PEOZZ	5120013673585	55719	SRES24	ADAPTER, TORQUE WREN 3/4, BOI: SEE PARA.5.I, PART OF KIT P/N 57K0267	2
3	PEFZZ	5120012273159	33234	TLE60	SOCKET, SOCKET WRENC 1/4 DRIVE, BOI: SEE PARA.5.I, PART OF KIT P/N 57K0268	4
4	PEOZZ	5120002408702	19207	11655788-2	ADAPTER, SOCKET WREN 3/8 DRIVE TO 1/ DRIVE, BOI: SEE PARA.5.I, PART OF KIT P/N 57K0267	2
5	PEOZZ	5120006838597	96508	TW6B	DRIVER, HEXAGON HEAD 3/16, 3/8 DRIVE, BOI: SEE PARA. 5.I, PART OF KIT P/N 57K0267	5
5	PEOZZ	5120014373658	55719	FAM7E	DRIVER, HEXAGON HEAD 7 MM, 3/8 DRIVE, BOI: SEE PARA. 5.I, PART OF KIT P/N 57K0267	5
6	PEOZZ	5120013355847	19207	12342688	RUNFLAT TOOL COMPRESSION, BOI: SEE PARA. 5.I, PART OF KIT P/N 57K0267	6
7	KFOZZ		62161	528236-05	.STRAP ASSEMBLY RUNFLAT TOOL BOI: SEE PARA. 5.I, PART OF KIT P/N 528240	7
8	KFOZZ		62161	1118BAC1048	.BOLT BOI: SEE PARA. 5.I, PART OF KIT P/N 528240	8
9	KFOZZ		62161	1351BAC10SL	.NUT, SELF-LOCKING RUNFLAT TOOL BOI: SEE PARA. 5.I, PART OF KIT P/N 528240	9
10	KFOZZ		33287	RS-3400-135	.HEXAGON LOCKNUT BOI: SEE PARA. 5.I PART OF KIT P/N J-39295	10
11	KFOZZ		33287	J-39250-1	.STRAP ASSEMBLY BOI: SEE PARA. 5.I, PART OF KIT P/N J-39295	11
12	KFOZZ		33287	RS-5701-795	.PIN, SMALL BOI: SEE PARA. 5.I, PART OF KIT P/N J-39295	12
13	KFOZZ		33287	RS-700-125	.PIN, SHAFT BOI: SEE PARA. 5.I, PART	13

SECTION III TM9-2320-280-24P C01

(1) ITEM NO	(2) SMR CODE	(3) NSN	(4) CAGEC	(5) PART NUMBER	(6) DESCRIPTION AND USABLE ON CODES (UOC)	(7) QTY
14	PEFZZ	4910012382551	33287	J26999-30	ADAPTER,CYLINDER CO BOI: SEE PARA.5.1 PART OF KIT P/N 57K0268.... UOC:HVY,H11,H13,H14,H15,H16,H17,H18,H20,H21,H24,H25,H26,H27,H28	15
15	PEFZZ	4910007856437	33287	J6692	TESTER,CYLINDER COM CYLINDER COMPRESSION,BOI: SEE PARA.5.1 PART OF KIT P/N 57K0268..........	
16	PEFZZ	4730009854804	96906	MS51815-4P	ELBOW,PIPE TO TUBE BOI: SEE PARA.5.1 PART OF KIT P/N 57K0268....	
17	PEFZZ	4730010425266	33287	J-35209	COUPLING HALF,QUICK BOI: SEE PARA. 5.I,PART OF KIT P/N 57K0268.........	
18	PEOZZ	5120002771463	80244	5120-00-277-1463	SOCKET,SOCKET WRENCH 3/8,3/8 DRIVE DEEPWELL,BOI: SEE PARA. 5.I,PART OF KIT P/N 57K0267....................	
	PAOZZ	2530013383056	33287	J-39295	REPAIR KIT,BELT RUNFLAT............	
	PAOZZ	4310013455723	62161	528240	REPAIR KIT,BELT RUNFLAT............	

Section II. TM 9-2320-280-24P-2

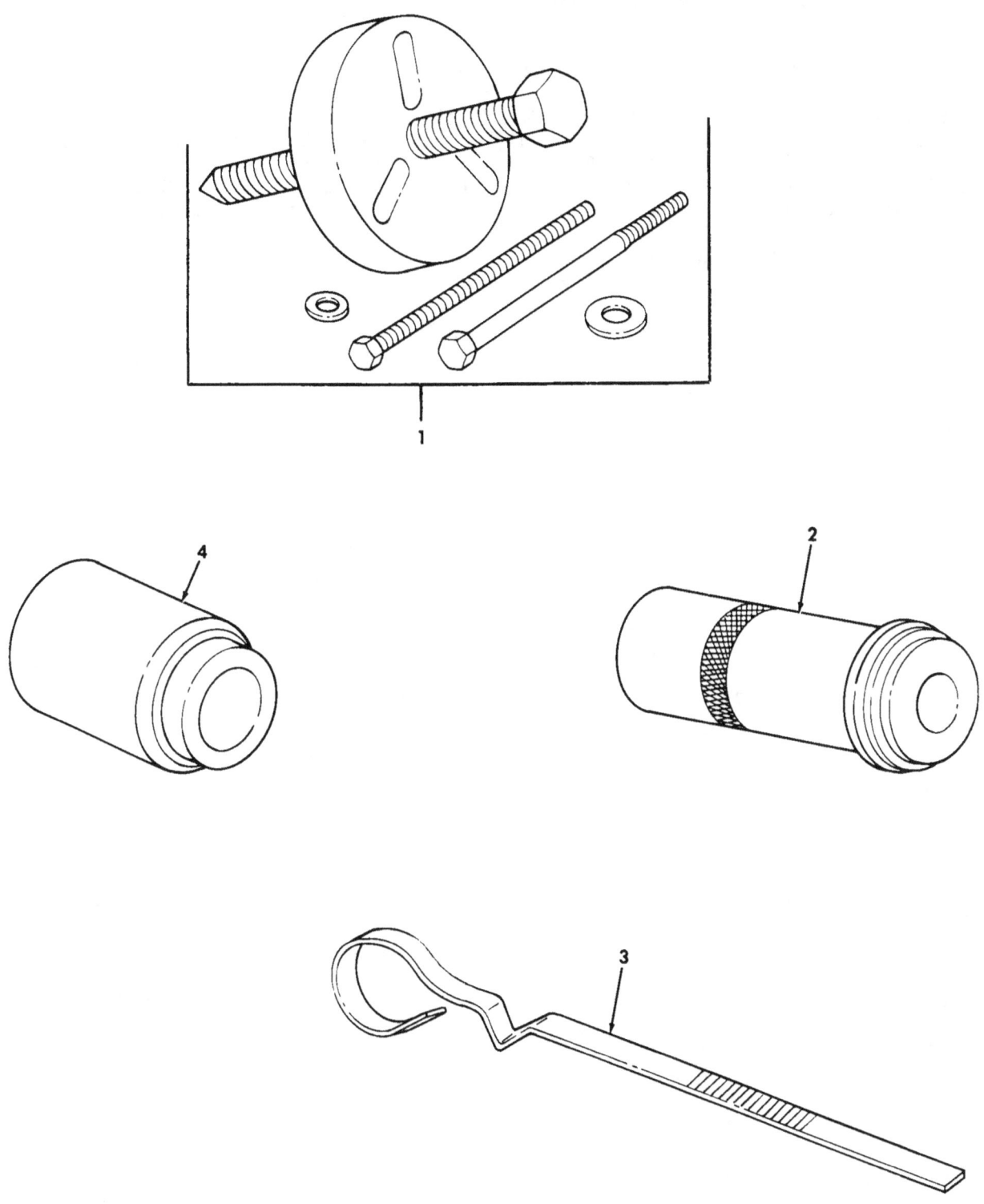

Figure 506. Special Tools, Air Conditioner Compressor.

SECTION III   TM9-2320-280-24P

(1) NO	(2) SMR CODE	(3) NSN	(4) CAGEC	(5) PART NUMBER	(6) DESCRIPTION AND USABLE ON CODES(UOC)	(7) ITEM QTY

GROUP 2604 SPECIAL TOOLS

FIG. 506 SPECIAL TOOLS, AIR CONDITIONER COMPRESSOR

1 PEFZZ 5120013734982 25341 43305 PULLER,MECHANICAL PART OF KIT P/N J-29642-C..........................

2 PEFZZ 5120013734989 25341 42109 INSERTER AND REMOVE PULLEY BEARING, PART OF IT P/N J-29642-C............

3 PEFZZ 6680013725748 25341 41102 GAGE ROD,LIQUID LEV PART OF KIT P/N J-29642-C..........................

4 PEFZZ 5120013735985 25341 43204 PILOT,PULLER PART OF KIT P/N J-29642-C..........................

END OF FIGURE

506-1

Section II. TM 9-2320-280-24P-2

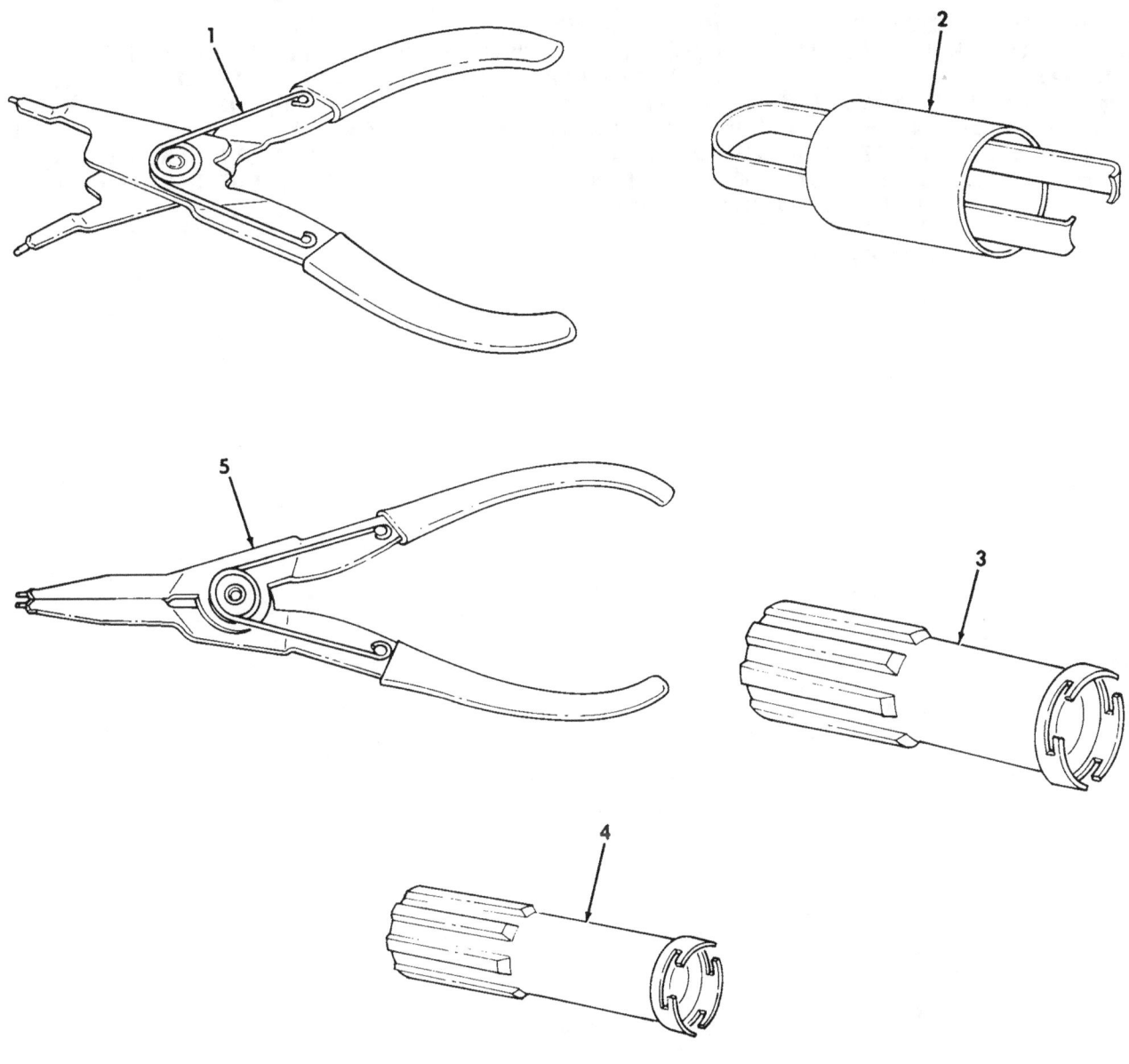

Figure 507. Special Tools, Air Conditioner Compressor.

SECTION III   TM9-2320-280-24P

(1) (2) (3)   (4)   (5)   (6)   (7) ITEM   SMR   PART
NO CODE NSN   CAGEC NUMBER   DESCRIPTION AND USABLE ON CODES(UOC) QTY

GROUP 2604 SPECIAL TOOLS

FIG. 507 SPECIAL TOOLS, AIR CONDITIONER COMPRESSOR

1 PEFZZ 5120013734979 25341 40210   PLIERS,RETAINING RI INTERNAL SNAP-RING,SMALL,PART OF KIT P/N J-29642-C

1 PEFZZ 5120013734980 85688 3R   PLIERS,RETAINING RI INTERNAL SNAP-RING,LARGE,PART OF KIT P/N J-29642-C

2 PEFZZ 5120013738836 25341 42106   REMOVER,SEAL SEAT,PART OF KIT P/N J-29642-C.......................... 3

PEFZZ 5120013734987 25341 43205   INSERTER AND REMOVE SEAL,PART OF KIT P/N J-29642-C....................

4 PEFZZ 5120013734988 25341 43501   INSERTER AND REMOVE SEAL,PART OF KIT P/N J-29642-C...................

5 PEFZZ 5120013734981 53800 9-46948   PLIERS,RETAINING RI EXTERNAL SNAP-RING,PART OF KIT P/N J-29642-C......

END OF FIGURE

Section II. TM 9-2320-280-24P-2

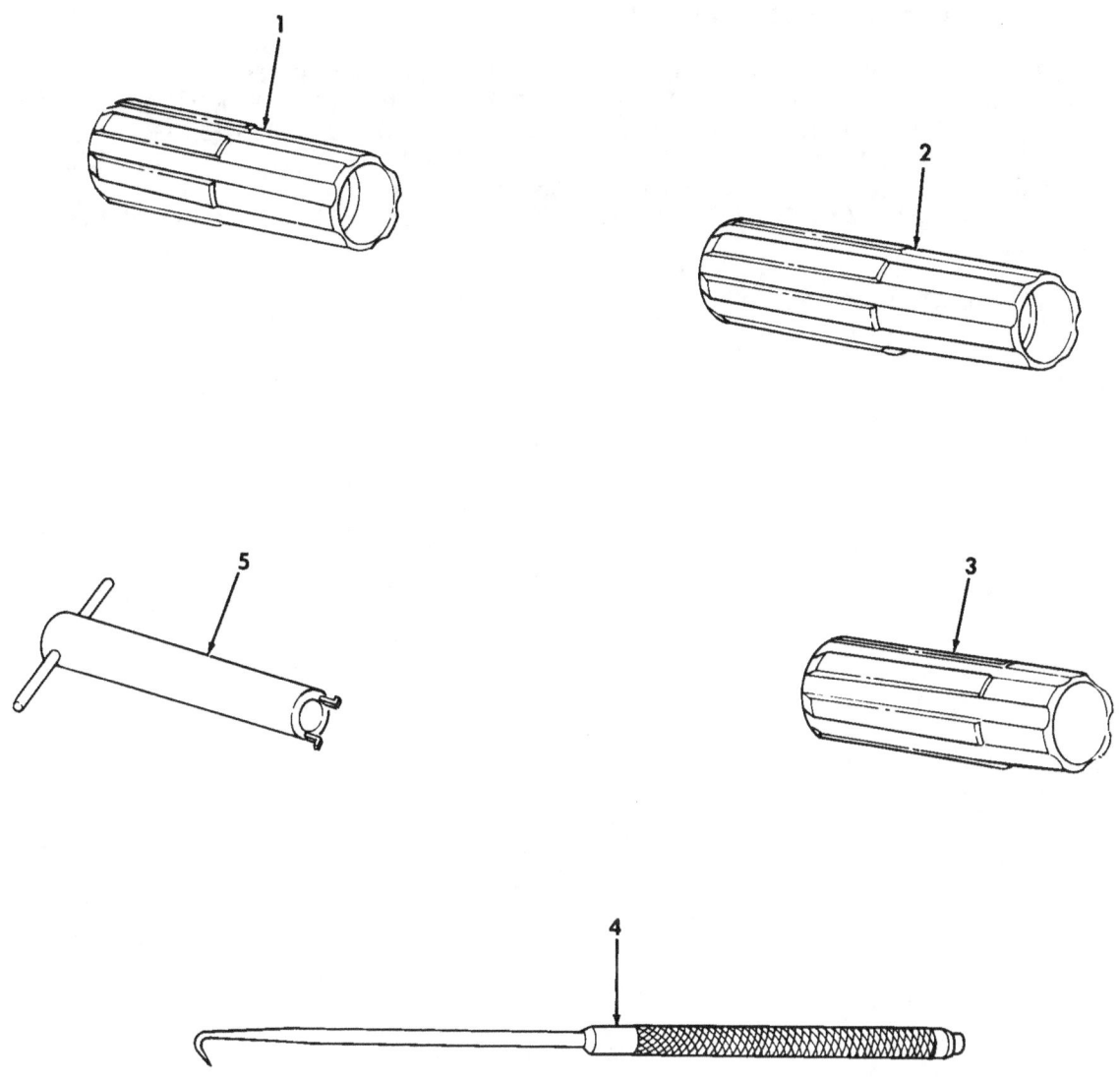

Figure 508. Special Tools, Air Conditioner Compressor.

SECTION III TM9-2320-280-24P

(1) NO	(2) CODE	(3) NSN	(4) CAGEC	(5) NUMBER	(6) DESCRIPTION AND USABLE ON CODES(UOC)	(7) ITEM SMR PART QTY

GROUP 2604 SPECIAL TOOLS

FIG. 508 SPECIAL TOOLS, AIR CONDITIONER COMPRESSOR

1 PEFZZ 5120013738840 25341 43210   INSTALLATION TOOL,R SEAL/SEAT,PART OF KIT P/N J-29642-C................. 2 PEFZZ 5120013734985 25341 42124   INSERTER,PREFORMED SNAP-RING,PART OF KIT P/N J-29642-C................. 3 PEFZZ 5120013735983 25341 42107   INSTALLER,SEAL SEAT,PART OF KIT P/N J-29642-C.......................... 4 PEFZZ 5120013735984 25341 42113   PICK,MINIATURE O-RING AND SEAL,PART OF KIT P/N J-29642-C................ 5 PEFZZ 5120013734984 25341 43208   REMOVER,SEAL PART OF KIT P/N J-29642-C..........................

END OF FIGURE

Section II. TM 9-2320-280-24P-2

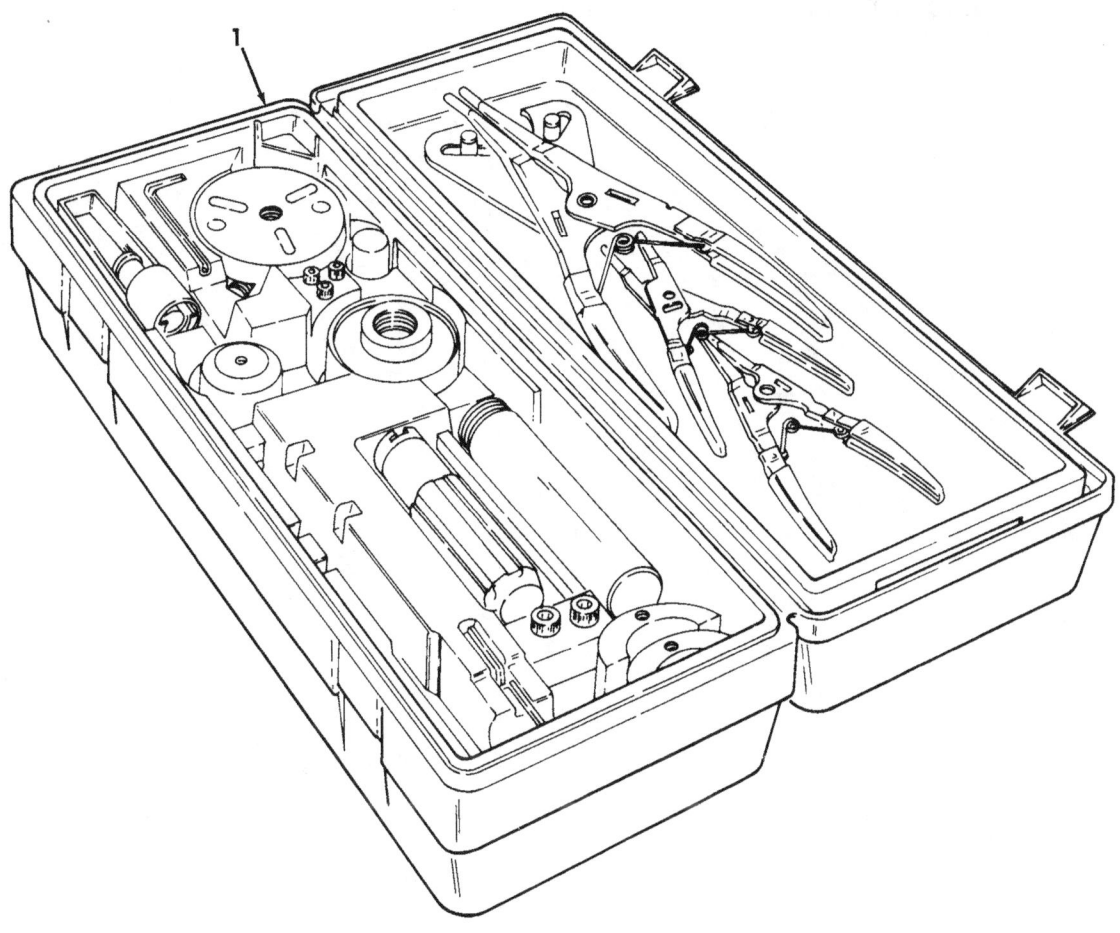

Figure 509. Special Tools, Air Compressor Tool Kit.

SECTION III   TM9-2320-280-24P

(1) NO	(2) CODE	(3) NSN	(4) CAGEC	(5) PART NUMBER	(7) ITEM DESCRIPTION AND USABLE ON CODES(UOC)	SMR	QTY

GROUP 2604 SPECIAL TOOLS

FIG. 509 SPECIAL TOOLS, AIR CONDITIONER COMPRESSOR TOOL KIT

1 PEFFF 5180012672907 25341 J-29642-C TOOL KIT,COMPRESSOR AIR CONDITIONER COMPRESSOR,BOI:SEE PARA.5.I............................

END OF FIGURE

Section II. TM 9-2320-280-24P-2 C01

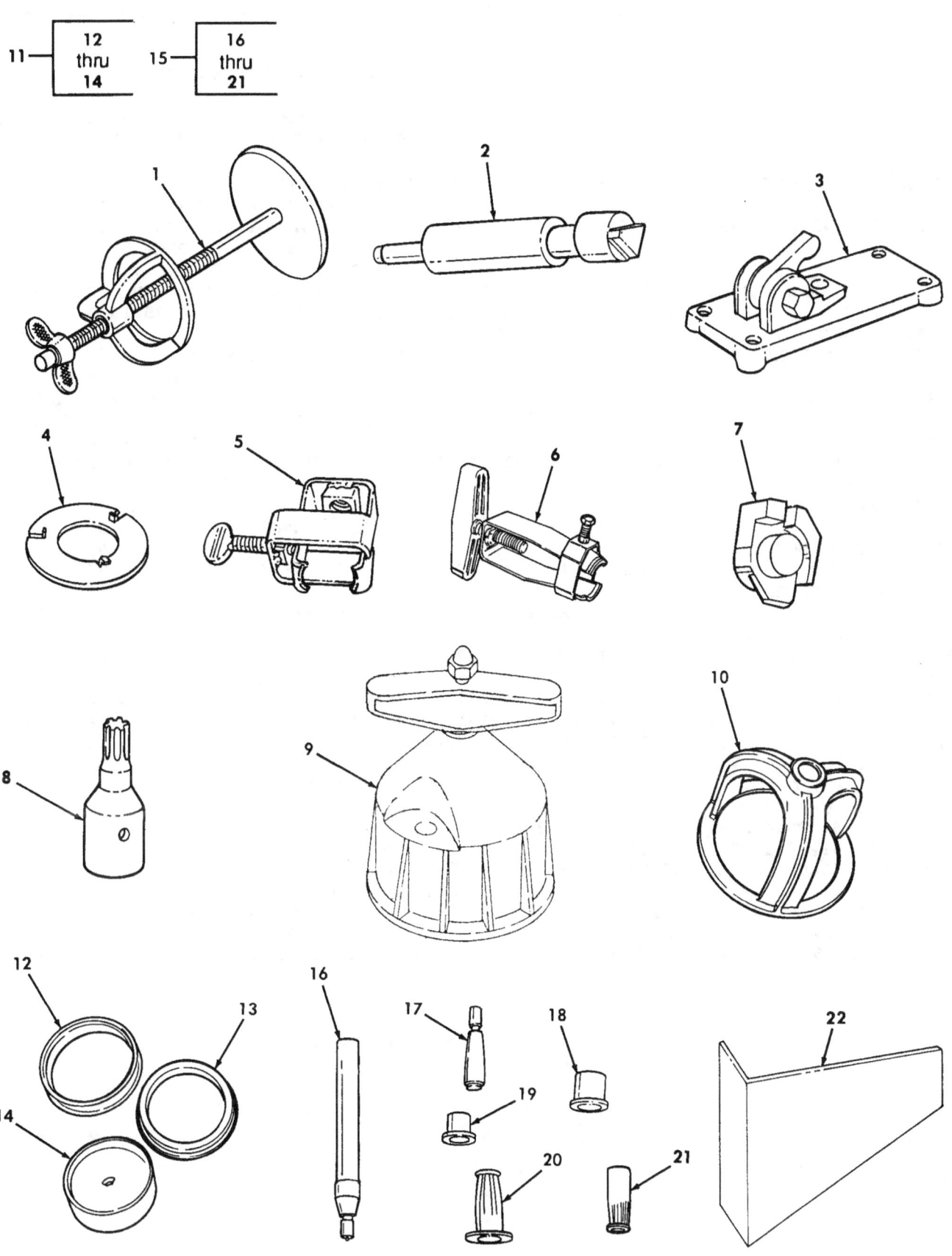

Figure 510. Special Tools, Transmission, Transfer Case, and Engine A2.

```
         SECTION III         TM9-2320-280-24P C01
    (1)    (2)    (3)          (4)         (5)                       (6)
(7) ITEM   SMR                              PART
    NO     CODE   NSN          CAGEC        NUMBER    DESCRIPTION AND USABLE ON CODES(UOC)
QTY
                                                   GROUP 2604 SPECIAL TOOLS

                                                   FIG. 510 SPECIAL TOOLS,TRANSMISSION,
                                                   TRANSFER CASE,AND ENGINE A2
```

* 1 PEHZZ 4910011780724 25341 J-23327   COMPRESSOR,CLUTCH S BOI: SEE PARA
                                                    5.I,PART OF KIT P/N 57K3218........
. 2 PEHZZ 5120014230032 33287 J-21370-10   GAUGE BAND TO APPLY PIN(USED WITH
                                                    J-38737) BOI: SEE PARA.5.I,PART
OF   KIT P/N 57K3218...............
..... 3 PEHZZ 5120014221313 33287 J-38737   TOOL,TRANS BAND PIN BOI:SEE
PARA.5.I PART OF KIT P/N 57K3218.... 4 PEHZZ 5120014108216 33287 J-25018-A
COMPRESSOR,CLUTCH S BOI:SEE   PARA.5.I
PART OF KIT P/N 57K3218.... 5 PEHZZ 5120014221300 33287 J-38868-A   TOOL,TRANS
SHAFT GEAR UNIT   ASSEMBLY BOI:-
SEE PARA.5.I PART OF   KIT P/N 57K32
18.................... 6 PEHZZ 5120014221308 33287 J-37789-A   TOOL,TRANS
PUMP ASS BOI:SEE   PARA.5.I PART OF
KIT P/N 57K3218....   7 PEHZZ 5120014221326 33287 J-38655   ADAPTER,TRANS
FIXTU TRANSMISSION   HOLDING FIX-
TURE BOI: SEE PARA.5.I   PART OF KIT
P/N 57K3218.............   8 PEHZZ 5120013673536 55719 FTX40E   SCREW-
DRIVER ATTACHM BOI:SEE   PARA.5.I PART
OF KIT P/N 57K3218....   9 PEHZZ 5120014220334 33287 J-39084   INSERTER,-
SEAL   REAR CRANKSHAFT SEAL   PART OF
KIT P/N 57K3218............. 10 PEHZZ 5120014221329 33287 J-38734   ADAPT-
ER,TRANS CLUTC PISTON   COMPRESSOR
PART OF KIT P/N 57K3218.. 11 ADHHH 5180014220138 33287 J-38731   TOOL SET,-
TRANSMISSI BOI:SEE   PARA.5.I PART
OF KIT P/N 57K3218.... 12 PEHZZ 5120014220173 33287 J-38731-1   .PROTECTOR,-
SEAL,INNE............... 13 PEHZZ 5120014220219 33287 J-38731-2   .PROTECTOR,-
SEAL,OUTE............... 14 PEHZZ 5120014220139 33287 J-38731-3   .BASE,TRANS
TOOL SET...............
* 15 PEHHH 5120014087051 33287 J-38736   INSTALLATION TOOL,S BOI:SEE
PARA.5.I PART OF KIT P/N 57K3218....
     16 PEHZZ 5120014221310 33287 J-38736-1   .LG SEAL RING PROTEC...............
     17 PEHZZ 5120014225406 33287 J-38736-2   .PROTECTOR,S RING,SM...............
     18 PEHZZ 5120014225411 33287 J-38736-3   .SIZER.............................
     19 PEHZZ 5120014221318 33287 J-38736-4   .SM SIZER,TRANS SEAL...............
     20 PEHZZ 5120014221324 33287 J-38736-5   .LRG PUSHER,TRANSMIS...............
* 21 PEHZZ 5120014087051 33287 J-38736-6   .INSTALLER,SEAL....................
* 22 PEHZZ 5210012490369 78514 23716   GAGE,THROTTLE BOI:SEE PARA. 5.I.....

                                                   END OF FIGURE

                                  510-1

Section II. TM 9-2320-280-24P-2

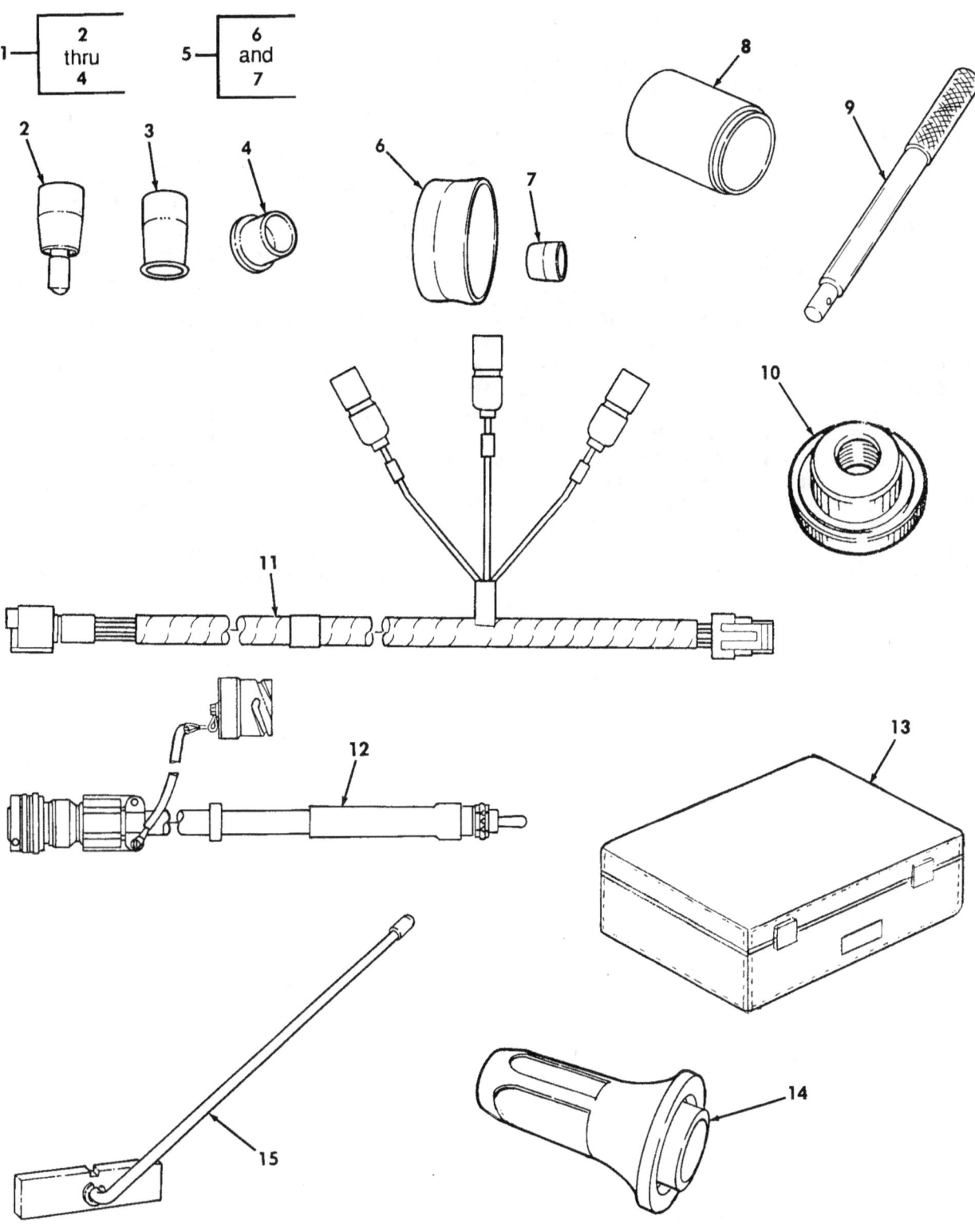

Figure 511. Special Tools, Transmission, Transfer Case, and Engine A2.

SECTION III   TM9-2320-280-24P

(1) NO	(2) SMR CODE	(3) NSN	(4) CAGEC	(5) PART NUMBER	(7) DESCRIPTION AND USABLE ON CODES(UOC)	ITEM QTY

GROUP 2604 SPECIAL TOOLS

FIG. 511 SPECIAL TOOLS, TRANSMISSION, TRANSFER CASE, AND ENGINE A2

1	ADHHH	5180014221294	33287	J-38739	TOOL SET, TRANS PUMP BOI: SEE PARA.5.I PART OF KIT P/N 57K3218	
2	PEHZZ	5120014221297	33287	J-38739-1	.PROTECTOR, TRANS PUM	
3	PEHZZ	5120014221302	33287	J-38739-2	.SIZER, TRANS PUMP AS	
4	PEHZZ	5120014221306	33287	J-38739-3	.PUSHER, TRANS PUMP A	
5	PEHZZ	5120014221301	33287	J-38732	TRANS SEAL PROTECTO BOI: SEE PARA.5.I PART OF KIT P/N 57K3218	
6	PEHZZ	5120014221293	33287	J-38732-1	.PROTECTOR, TRANS SEA	
7	PEHZZ	4910011786551	25341	J-21362	.PROTECTOR, INNER SEA PART OF KIT P/N J-38732	
8	PEOZZ	5120014370480	80604	6888	INSERTER, SEAL BOI:SEE PARA.5.I PART OF KIT P/N 57K3219	
9	PEOZZ	5120010261666	80604	C-4171	HANDLE DRIVER BOI:SEE PARA.5.I PART OF KIT P/N 57K3219	
10	PEOZZ		33287	J-33832	REMOVER AND INSTALLER PART OF KIT P/N 57K3219	
11	PFOZZ	6150014127774	19207	12460120	CABLE ASSEMBLY,SPEC	
12	PEOZZ	6150014108215	19207	12460137	CABLE ASSEMBLY-SWIT DIAGNOSTIC SWITCH,BOI: SEE PARA. 5.1	
13	PEOZZ	5120014088173	19207	12460136	CASE,DIAGNOSTIC CAB DIAGOSTIC CABLE ASSEMBLY,BOI: SEE PARA. 5.1	
14	PEOZZ	5120014141849	33287	J-38869	INSTALLER REAR EXTENSION SEAL,BOI: SEE PARA. 5.I, PART OF KIT P/N 57K3219	
15	PEOZZ	4820011794869	33287	J-33043	BLOCK, VALVE GAUGE BOI: SEE PARA. 5.I	
	ADHHH	5180014087050	19207	57K3218	TOOL KIT,VEHICULAR, REAR EXETNSION SEAL,BOI: SEE PARA. 5.I,PART OF KIT P/N 57K3219	
	ADOOO	5180014108467	19207	57K3219	TOOL KIT,VEHICULAR, REAR EXETNSION SEAL,BOI: SEE PARA. 5.I,PART OF KIT P/N 57K3218	

END OF FIGURE

511-1

Section II.

TM 9-2320-280-24P-2

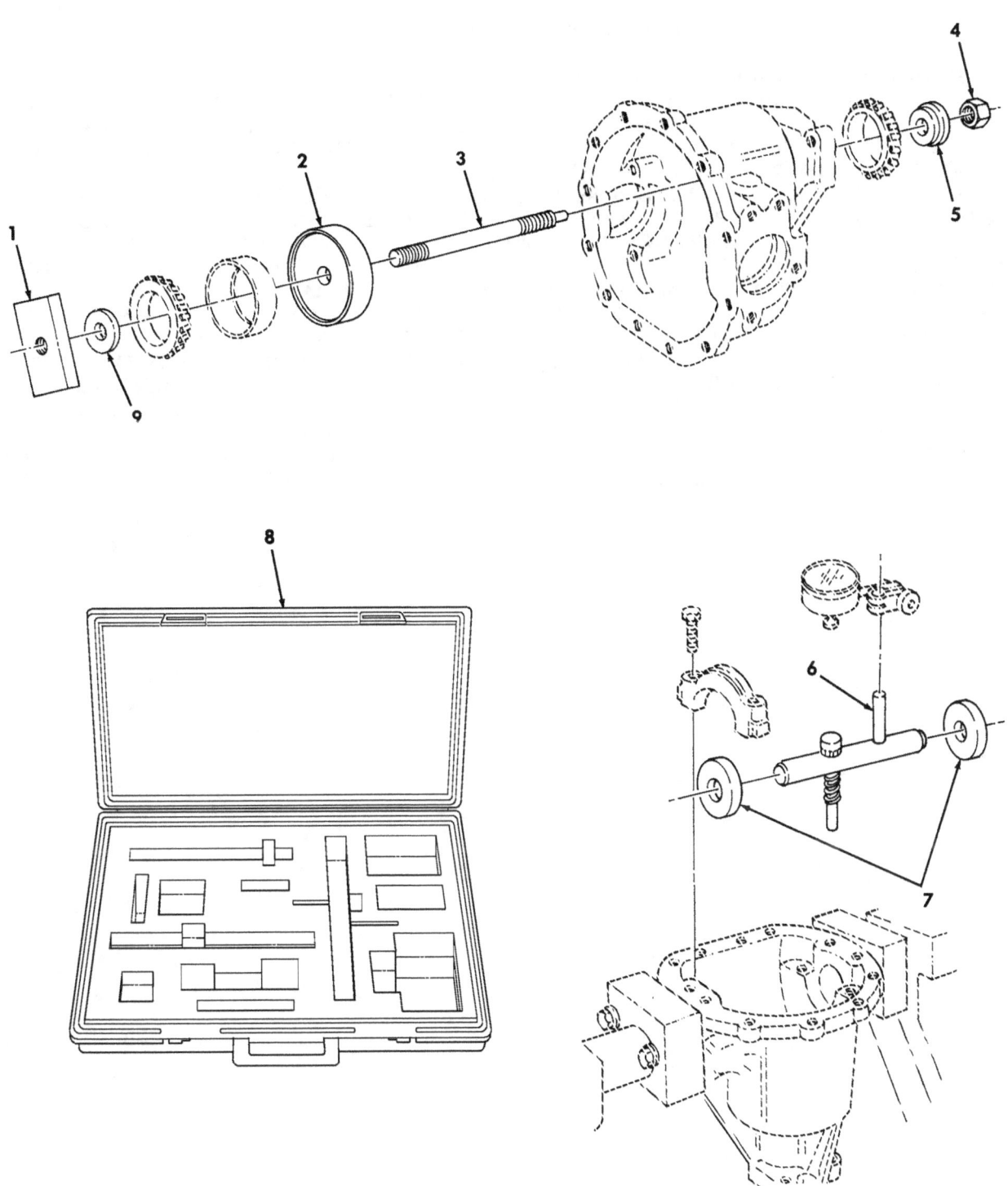

Figure 512. Pinion Setting Kit.

SECTION III TM9-2320-280-24P

(1) NO	(2) SMR CODE	(3) NSN	(4) CAGEC	(5) PART NUMBER	(7) DESCRIPTION AND USABLE ON CODES(UOC)	ITEM QTY

GROUP 2604 SPECIAL TOOLS

FIG. 512 PINION SETTING KIT

1	PEHZZ	5120013660893	33287	J-39524-2	BLOCK,RECTANGULAR GAUGE,BOI: SEE PARA. 5.I,PART OF KIT P/N 57K3229...	
2	PEHZZ	5120013662520	33287	J-39524-1	BLOCK,AXLE PINION BOI: SEE PARA. 5.I. PART OF KIT P/N J-39524........	
3	PEHZZ	5120013669468	33287	J-21777-43	STUD ASSEMBLY BOI: SEE PARA.5.I. PART OF KIT P/N J-39524.............	
4	PEHZZ	5310014654525	24617	9418753	NUT,PLAIN,HEXAGON 1/2-13,BOI: SEE PARA. 5.I,PART OF KIT P/N 57K3229...	
5	PEHZZ	5120013660895	33287	J-35506-3	PILOT WASHER BOI: SEE PARA.5.I. PART OF KIT P/N J-39524.............	
6	PEHZZ	5120013664713	33287	J-39524-5	AXLE PINION ARBOR BOI:SEE PARA.5.I. PART OF KIT P/N J-39524.............	
7	PEHZZ	5120013666260	33287	J-35506-2	SIDE BEARING DISC,A BOI SEE PARA.5.I. PART OF KIT P/N J-39524...	
8	PEHZZ	5140013660642	33287	J-39524-4	CASE,TOOL AND ACCES BOI: SEE PARA 5.I, PART OF KIT P/N J-39524.......	
9	PEHZZ	5120013660894	33287	J-34175-4	PILOT WASHER BOI: SEE PARA. 5.I PART OF KIT P/N J-39524.............	
	ADHHH	5180013638079	33287	J-39524	TOOL KIT,VEHICULAR, PART OF KIT P/N 57K0236 PART OF KIT P/N 57K0266.....	
	ADOOO	5180013875455	19207	57K0267	TOOL KIT,VEHICULAR..BOI: SEE PARA 5.I...............	
	ADFFF	5180013897560	19207	57K0268	TOOL KIT,VEHICULAR BOI: SEE PARA 5.I...............	
	PEHZZ	5180013579692	19207	57K0236	TOOL KIT,TRANSMISSS SUPPLEMENTAL KIT,-GENERAL SUPPORT................	
	ADHHH	5180013897561	19207	57K0266	TOOL KIT,VEHICULAR BOI: SEE PARA 5.I...............	

END OF FIGURE

SECTION IV TM9-2320-280-24P C01

CROSS REFERENCE INDEXES

NATIONAL STOCK NUMBER INDEX

STOCK NUMBER	FIGURE NO	ITEM NO	STOCK NUMBER	FIGURE NO	ITEM NO
			5315-00-012-0123	152	13
			5315-00-012-0123	170	1
			5315-00-012-0123	183	4
			5315-00-012-0123	185	15
			5315-00-012-0123	187	19
			5315-00-012-0123	188	13
			5315-00-012-0123	195	12
			5306-00-012-0231	403	10
			5310-00-012-0367	370	24
			5310-00-012-0388	33A	14
			5310-00-012-0388	33B	4
6220-00-001-1514	64	8	5310-00-012-1743	269	29
5330-00-001-1984	116	16	5315-00-012-4553	36	5
2840-00-001-4903	111	5	5315-00-012-4553	38	5
5330-00-001-4904	111	4	5315-00-012-4553	39	8
5310-00-003-6779	260	13	5315-00-012-4553	43B	26
3110-00-005-0873	109	6	4730-00-012-7951	143	18
3110-00-005-0873	111	8	4730-00-012-7951	151	18
4720-00-005-5008	385	3	5310-00-013-1482	215	10
4720-00-006-0048	BULK	63	5315-00-013-7228	156	14
5310-00-006-8291	225	44	4730-00-014-2432	10	12
4720-00-006-9541	BULK	10	5310-00-014-5850	63	6
5310-00-007-1607	283	12	5310-00-014-5850	93	25
5325-00-007-3052	109	1	5310-00-014-5850	95	39
5325-00-007-3052	110	1	5310-00-014-5850	96	29
5325-00-007-3052	111	11	5310-00-014-5850	169	20
5325-00-007-3052	112	10	5310-00-014-5850	199	33
4240-00-007-9453	311	10	5310-00-014-5850	199	57
4240-00-007-9453	312	6	5310-00-014-5850	202	42
5977-00-008-3338	KITS	12	5310-00-014-5850	203	31
2520-00-008-9987	109	2	5310-00-014-5850	205	39
2920-00-009-5213	KITS	2	5310-00-014-5850	206	15
5310-00-010-3028	437	6	5310-00-014-5850	213	15
9330-00-011-2000	BULK	65	5310-00-014-5850	214	11
4730-00-011-4627	348	16	5310-00-014-5850	222	11
5310-00-011-5093	138	17	5310-00-014-5850	227	49
5310-00-011-5093	139	14	5310-00-014-5850	232	2
5310-00-011-5093	145	17	5310-00-014-5850	241	8
5310-00-011-5093	146	14	5310-00-014-5850	243	2
5310-00-011-5093	164	3	5310-00-014-5850	244	18
5310-00-011-6121	70	7	5310-00-014-5850	246	8
5310-00-011-6121	127	5	5310-00-014-5850	246	13
5310-00-011-6121	472	12	5310-00-014-5850	247	12
5310-00-011-6121	475	20	5310-00-014-5850	248	4
5310-00-011-6121	476	18	5310-00-014-5850	249	41
5310-00-011-6121	481	3	5310-00-014-5850	268	25
5315-00-012-0123	144	13	5310-00-014-5850	272	15

SECTION IV TM9-2320-280-24P C01

CROSS REFERENCE INDEXES

NATIONAL STOCK NUMBER INDEX

STOCK NUMBER	FIGURE NO	ITEM NO	STOCK NUMBER	FIGURE NO	ITEM NO
5310-00-014-5850	293	12	5925-00-026-4767	385	12
5310-00-014-5850	299	2	5315-00-038-3059	121	6
5310-00-014-5850	303	10	2540-00-039-8258	56	13
5310-00-014-5850	331	22	2540-00-040-2129	271	14
5310-00-014-5850	333	15	4730-00-041-2526	30	19
5310-00-014-5850	337	17	5306-00-042-5570	84	5
5310-00-014-5850	339	27	5306-00-042-5570	88	46
5310-00-014-5850	347	4	5305-00-043-2682	279	23
5310-00-014-5850	368	26	4730-00-043-3750	345	34
5310-00-014-5850	374	14	4730-00-043-3750	348	34
5310-00-014-5850	411	7	5310-00-044-3340	40	7
5310-00-014-5850	429	4	5310-00-044-3342	127	13
5310-00-014-5850	430	8	5310-00-044-3342	166	5
5310-00-014-5850	430	40	5310-00-044-3342	180	17
5310-00-014-5850	446	10	4730-00-044-4577	15	7
5310-00-014-5850	447	8	5310-00-044-6188	370	25
5310-00-014-5850	448	7	5310-00-044-6212	370	27
5310-00-014-5850	449	8	6240-00-044-6914	60	7
5310-00-014-5850	450	2	6240-00-044-6914	61	7
5310-00-014-5850	480	9	6240-00-044-6914	62	3
5305-00-014-9926	387	12	6240-00-044-6914	64	11
5305-00-014-9926	392	1	5310-00-045-3296	39	33
5305-00-018-7838	88	44	5310-00-045-3296	54	12
5305-00-018-7838	254	7	5310-00-045-3296	55	21
5305-00-018-7838	254	10	5310-00-045-3296	59	17
5305-00-018-7838	445	10	5310-00-045-3296	240	19
5305-00-018-7838	487	8	5310-00-045-3296	260	10
5305-00-018-8370	345	37	5310-00-045-3296	283	29
5305-00-018-8370	348	27	5310-00-045-3296	327	35
6240-00-019-0877	63	4	5310-00-045-3296	335	13
6240-00-019-3093	60	6	5310-00-045-3296	336	10
6240-00-019-3093	61	6	5310-00-045-3296	337	18
5940-00-020-0072	79	12	5310-00-045-3296	338	11
5310-00-020-0358	438	11	5310-00-045-3296	410	13
4730-00-024-3971	13	10	5310-00-045-3296	410	17
4730-00-024-3971	14	16	5310-00-045-3296	412	10
4730-00-024-3971	19	21	5310-00-045-3296	416	15
4730-00-024-3971	27	6	5310-00-045-3296	426	22
4730-00-024-3971	125	9	5310-00-045-3296	427	17
4730-00-024-3971	366	21	5310-00-045-3296	428	15
4730-00-024-3971	368	38	5310-00-045-3296	43B	23
4730-00-024-3971	176A	15	5310-00-045-3296	442	7
5306-00-024-6580	104	8	5310-00-045-3296	460	3
2910-00-025-3493	335	9	5310-00-045-3296	482	21
2910-00-025-3493	336	6	5310-00-045-3296	485	16
2910-00-025-3493	388	7	5310-00-045-3299	39	77
5925-00-026-4767	380	4	5310-00-045-3299	46	13

I-2

SECTION IV TM9-2320-280-24P C01

CROSS REFERENCE INDEXES

NATIONAL STOCK NUMBER INDEX

STOCK NUMBER	FIGURE NO	ITEM NO	STOCK NUMBER	FIGURE NO	ITEM NO
5310-00-045-3299	48	2	4730-00-050-4203	152	9
5310-00-045-3299	52	22	4730-00-050-4203	152	19
5310-00-045-3299	64	18	4730-00-050-4203	170	5
5310-00-045-3299	380	2	4730-00-050-4203	170	8
5310-00-045-4007	39	30	4730-00-050-4205	170	4
5310-00-045-4007	56	22	4730-00-050-4205	170	12
5310-00-045-4007	84	25	4730-00-050-4205	170	18
5310-00-045-4007	213	18	4730-00-050-4205	195	7
5310-00-045-4007	214	14	4730-00-050-4207	191	25
5310-00-045-4007	333	29	5940-00-050-7106	79	11
5310-00-045-5207	69	32	5340-00-050-9077	123	3
5310-00-045-5218	70	6	5340-00-050-9077	135	6
5310-00-045-5218	481	4	5340-00-050-9077	141	24
5305-00-045-7603	224	8	5340-00-050-9077	149	8
5305-00-045-7603	268	26	5340-00-050-9077	414	12
4730-00-045-9769	142	30	5305-00-050-9229	347	2
4730-00-045-9769	143	39	5305-00-050-9233	280	16
4730-00-045-9769	150	30	5305-00-051-4076	324	7
4730-00-045-9769	151	39	5305-00-051-4076	378	30
1005-00-049-9324	KITS	24	5306-00-051-4077	266	4
2640-00-050-1229	167	22	5306-00-051-4077	321	29
5306-00-050-1238	433	9	5306-00-051-4077	321	34
5340-00-050-2740	59	9	5306-00-051-4077	325	29
5340-00-050-2740	83	21	5306-00-051-4077	413	11
5340-00-050-2740	84	12	5306-00-051-4077	415	5
5340-00-050-2740	88	18	5305-00-051-8605	141	11
5340-00-050-2740	91	11	5305-00-051-8605	149	9
5340-00-050-2740	94	9	5305-00-051-8605	159	11
5340-00-050-2740	96	27	5305-00-051-8605	161	4
5340-00-050-2740	282	5	5305-00-051-8605	162	4
5340-00-050-2740	338	16	5305-00-052-2232	199	18
5340-00-050-2740	339	12	5305-00-052-2232	251	19
5340-00-050-2740	410	22	5305-00-052-2232	270	11
5340-00-050-2740	412	16	5305-00-052-2232	296	1
5340-00-050-2740	414	20	5305-00-052-6456	34	17
5340-00-050-2740	416	23	5305-00-052-6921	28	2
5340-00-050-2740	416	29	5305-00-052-8881	469	4
5340-00-050-2740	424	3	5340-00-053-8994	75	39
5340-00-050-2740	428	22	5340-00-053-8994	472	4
4730-00-050-4203	136	9	5340-00-053-8994	474	2
4730-00-050-4203	136	18	5340-00-053-8994	480	24
4730-00-050-4203	136	22	5305-00-054-5642	238	21
4730-00-050-4203	137	8	5305-00-054-5650	282	14
4730-00-050-4203	137	11	5305-00-054-6652	214	15
4730-00-050-4203	137	23	5305-00-054-6670	482	19
4730-00-050-4203	144	9	5305-00-054-6672	64	17
4730-00-050-4203	144	19	5305-00-054-6672	252	7

I-3

SECTION IV TM9-2320-280-24P C01

CROSS REFERENCE INDEXES

NATIONAL STOCK NUMBER INDEX

STOCK NUMBER	FIGURE NO	ITEM NO	STOCK NUMBER	FIGURE NO	ITEM NO
3110-00-055-2100	128	45	5305-00-059-3659	426	21
5305-00-056-7879	385	1	5305-00-059-3659	427	18
5305-00-056-7879	395	11	5305-00-059-3659	428	12
5305-00-056-7879	396	4	5305-00-059-3659	43B	22
3110-00-056-9376	128	7	5305-00-059-3660	59	23
5340-00-057-2904	75	25	5305-00-059-3660	65	19
5340-00-057-2904	77	6	5305-00-059-3660	92	14
5340-00-057-2904	79	22	5305-00-059-3660	95	43
5340-00-057-2904	83	22	5305-00-059-3660	218	5
5340-00-057-2904	85	29	5305-00-059-3660	236	1
5340-00-057-2904	88	23	5305-00-059-3660	253	7
5340-00-057-2904	92	15	5305-00-059-3660	253	10
5340-00-057-2904	93	19	5305-00-059-3660	253	18
5340-00-057-2904	95	26	5305-00-059-3660	253	30
5340-00-057-2904	96	24	5305-00-059-3660	255	9
5340-00-057-2904	282	3	5305-00-059-3660	255	17
5340-00-057-2904	338	8	5305-00-059-3660	255	23
5340-00-057-2904	339	2	5305-00-059-3660	259	5
5340-00-057-2904	366	7	5305-00-059-3660	260	2
5340-00-057-2904	410	24	5305-00-059-3660	262	13
5340-00-057-2904	412	18	5305-00-059-3660	325	22
5340-00-057-2904	414	14	5305-00-059-3660	426	17
5340-00-057-2906	88	32	5305-00-059-3660	427	22
5340-00-057-2906	92	12	5305-00-059-3661	255	9
5340-00-057-2906	95	35	5305-00-059-3661	255	17
5340-00-057-2906	410	19	5305-00-059-3661	255	23
5340-00-057-2906	414	17	5305-00-059-3661	255	28
5340-00-057-2906	416	18	5305-00-059-3661	410	25
5340-00-057-2906	428	17	5305-00-059-3661	412	19
5340-00-057-2906	453	4	6240-00-060-4707	65	10
5340-00-057-3043	160	15	6240-00-060-4707	65	25
5340-00-057-3043	162	11	5310-00-061-0004	345	25
5340-00-057-3043	424	2	5310-00-061-0004	348	36
5340-00-057-3043	426	26	5310-00-061-4650	5	7
5340-00-057-3052	69	41	5310-00-061-4650	13	18
5340-00-057-3052	282	6	5310-00-061-4650	14	24
5970-00-057-9987	44	25	5310-00-061-4650	15	3
5305-00-058-1082	12	34	5310-00-061-4650	16	3
5305-00-058-1082	274	28	5310-00-061-4650	17	9
5305-00-059-3659	227	14	5310-00-061-4650	18	4
5305-00-059-3659	321	6	5310-00-061-4650	18	10
5305-00-059-3659	339	4	5310-00-061-4650	21	1
5305-00-059-3659	373	3	5310-00-061-4650	25	7
5305-00-059-3659	376	5	5310-00-061-4650	27	14
5305-00-059-3659	410	14	5310-00-061-4650	29	3
5305-00-059-3659	412	9	5310-00-061-4650	42	8
5305-00-059-3659	416	12	5310-00-061-4650	46	21

I-4

SECTION IV TM9-2320-280-24P C01

CROSS REFERENCE INDEXES

NATIONAL STOCK NUMBER INDEX

STOCK NUMBER	FIGURE NO	ITEM NO	STOCK NUMBER	FIGURE NO	ITEM NO
5310-00-061-4650	50	8	5310-00-061-4650	275	20
5310-00-061-4650	71	10	5310-00-061-4650	276	20
5310-00-061-4650	84	22	5310-00-061-4650	285	13
5310-00-061-4650	97	4	5310-00-061-4650	286	16
5310-00-061-4650	98	13	5310-00-061-4650	288	22
5310-00-061-4650	99	11	5310-00-061-4650	288	30
5310-00-061-4650	103	6	5310-00-061-4650	289	5
5310-00-061-4650	124	13	5310-00-061-4650	293	31
5310-00-061-4650	125	13	5310-00-061-4650	316	21
5310-00-061-4650	141	17	5310-00-061-4650	323	13
5310-00-061-4650	153	17	5310-00-061-4650	331	9
5310-00-061-4650	169	27	5310-00-061-4650	332	42
5310-00-061-4650	169	33	5310-00-061-4650	333	39
5310-00-061-4650	177	13	5310-00-061-4650	334	5
5310-00-061-4650	178	24	5310-00-061-4650	338	22
5310-00-061-4650	202	26	5310-00-061-4650	339	17
5310-00-061-4650	203	9	5310-00-061-4650	343	17
5310-00-061-4650	203	23	5310-00-061-4650	366	30
5310-00-061-4650	204	14	5310-00-061-4650	368	32
5310-00-061-4650	206	24	5310-00-061-4650	375	28
5310-00-061-4650	210	4	5310-00-061-4650	376	13
5310-00-061-4650	211	26	5310-00-061-4650	384	9
5310-00-061-4650	212	27	5310-00-061-4650	398	27
5310-00-061-4650	241	5	5310-00-061-4650	399	15
5310-00-061-4650	243	8	5310-00-061-4650	400	13
5310-00-061-4650	244	7	5310-00-061-4650	402	12
5310-00-061-4650	245	16	5310-00-061-4650	409	9
5310-00-061-4650	245	32	5310-00-061-4650	411	10
5310-00-061-4650	246	18	5310-00-061-4650	413	5
5310-00-061-4650	247	5	5310-00-061-4650	415	15
5310-00-061-4650	247	18	5310-00-061-4650	417	18
5310-00-061-4650	248	7	5310-00-061-4650	418	9
5310-00-061-4650	249	27	5310-00-061-4650	419	4
5310-00-061-4650	250	19	5310-00-061-4650	420	6
5310-00-061-4650	251	1	5310-00-061-4650	421	6
5310-00-061-4650	252	22	5310-00-061-4650	422	8
5310-00-061-4650	253	21	5310-00-061-4650	423	9
5310-00-061-4650	254	22	5310-00-061-4650	429	22
5310-00-061-4650	255	20	5310-00-061-4650	430	17
5310-00-061-4650	267	3	5310-00-061-4650	430	38
5310-00-061-4650	267	13	5310-00-061-4650	440	21
5310-00-061-4650	269	13	5310-00-061-4650	446	16
5310-00-061-4650	270	1	5310-00-061-4650	468	4
5310-00-061-4650	270	26	5310-00-061-4650	475	9
5310-00-061-4650	271	9	5310-00-061-4650	477	5
5310-00-061-4650	272	2	5310-00-061-4650	478	3
5310-00-061-4650	274	18	5310-00-061-4650	484	15

SECTION IV TM9-2320-280-24P C01

CROSS REFERENCE INDEXES

NATIONAL STOCK NUMBER INDEX

STOCK NUMBER	FIGURE NO	ITEM NO	STOCK NUMBER	FIGURE NO	ITEM NO
5310-00-061-4650	485	14	5305-00-068-0502	254	1
5310-00-061-4650	486	9	5305-00-068-0502	255	1
5310-00-061-4650	492	7	5305-00-068-0502	267	9
5310-00-061-4650	176A	21	5305-00-068-0502	269	11
5310-00-061-4651	185	8	5305-00-068-0502	270	5
5310-00-061-4651	186	11	5305-00-068-0502	294	4
5310-00-061-4651	187	6	5305-00-068-0502	311	2
5310-00-061-4651	194	12	5305-00-068-0502	313	9
5310-00-061-4651	441	9	5305-00-068-0502	331	25
5310-00-061-4651	442	20	5305-00-068-0502	333	9
5310-00-061-4651	457	20	5305-00-068-0502	338	30
5310-00-061-4651	470	12	5305-00-068-0502	339	28
5310-00-061-7326	365	22	5305-00-068-0502	342	4
5310-00-063-7360	345	42	5305-00-068-0502	343	14
5310-00-063-7360	348	39	5305-00-068-0502	377	1
4720-00-065-8682	BULK	17	5305-00-068-0502	384	13
9330-00-065-8682	BULK	1	5305-00-068-0502	393	7
5340-00-067-3868	135	2	5305-00-068-0502	395	19
5340-00-067-3868	141	22	5305-00-068-0502	399	13
5305-00-068-0501	18	13	5305-00-068-0502	410	30
5305-00-068-0501	118	12	5305-00-068-0502	412	11
5305-00-068-0501	286	19	5305-00-068-0502	416	3
5305-00-068-0501	307	9	5305-00-068-0502	421	11
5305-00-068-0502	19	16	5305-00-068-0502	422	11
5305-00-068-0502	21	4	5305-00-068-0502	440	10
5305-00-068-0502	50	11	5305-00-068-0502	484	5
5305-00-068-0502	53	3	5305-00-068-0502	492	9
5305-00-068-0502	58	9	5305-00-068-0502	494	5
5305-00-068-0502	68	1	5305-00-068-0502	495	11
5305-00-068-0502	70	9	5305-00-068-0502	496	9
5305-00-068-0502	84	24	5305-00-068-0502	497	5
5305-00-068-0502	93	16	5305-00-068-0502	498	17
5305-00-068-0502	124	10	5305-00-068-0508	15	20
5305-00-068-0502	125	10	5305-00-068-0508	17	12
5305-00-068-0502	141	14	5305-00-068-0508	29	6
5305-00-068-0502	155	2	5305-00-068-0508	88	50
5305-00-068-0502	180	19	5305-00-068-0508	208	2
5305-00-068-0502	203	11	5305-00-068-0508	233	6
5305-00-068-0502	206	4	5305-00-068-0508	241	12
5305-00-068-0502	208	14	5305-00-068-0508	243	13
5305-00-068-0502	241	3	5305-00-068-0508	252	3
5305-00-068-0502	244	16	5305-00-068-0508	269	33
5305-00-068-0502	245	20	5305-00-068-0508	270	28
5305-00-068-0502	250	16	5305-00-068-0508	275	5
5305-00-068-0502	251	21	5305-00-068-0508	283	14
5305-00-068-0502	252	1	5305-00-068-0508	287	12
5305-00-068-0502	253	1	5305-00-068-0508	327	45

SECTION IV TM9-2320-280-24P C01

CROSS REFERENCE INDEXES

NATIONAL STOCK NUMBER INDEX

STOCK NUMBER	FIGURE NO	ITEM NO	STOCK NUMBER	FIGURE NO	ITEM NO
5305-00-068-0508	331	35	5305-00-068-0510	275	1
5305-00-068-0508	332	5	5305-00-068-0510	276	24
5305-00-068-0508	365	4	5305-00-068-0510	277	47
5305-00-068-0508	368	40	5305-00-068-0510	286	29
5305-00-068-0508	398	16	5305-00-068-0510	38A	5
5305-00-068-0508	399	21	5305-00-068-0510	445	13
5305-00-068-0508	400	14	5305-00-068-0510	463	6
5305-00-068-0508	43B	41	5305-00-068-0510	464	3
5305-00-068-0508	446	36	5305-00-068-0510	471	11
5305-00-068-0508	457	14	5305-00-068-0510	476	1
5305-00-068-0508	468	5	5305-00-068-0511	24	8
5305-00-068-0508	478	4	5305-00-068-0511	138	16
5305-00-068-0508	481	6	5305-00-068-0511	139	15
5305-00-068-0508	485	10	5305-00-068-0511	145	16
5305-00-068-0508	492	9	5305-00-068-0511	146	15
5305-00-068-0508	498	8	5305-00-068-0511	189	9
5305-00-068-0509	49	6	5305-00-068-0511	190	7
5305-00-068-0509	198	7	5305-00-068-0511	266	18
5305-00-068-0509	251	28	5305-00-068-0511	272	33
5305-00-068-0509	252	11	5305-00-068-0511	283	6
5305-00-068-0509	270	8	5305-00-068-0511	431	2
5305-00-068-0509	285	3	5305-00-068-0511	462	2
5305-00-068-0509	288	24	5306-00-068-0513	68	6
5305-00-068-0509	288	32	5306-00-068-0513	379	24
5305-00-068-0509	321	11	5306-00-068-0513	379	29
5305-00-068-0509	398	10	5306-00-068-0513	388	15
5305-00-068-0509	402	3	5306-00-068-0513	66A	3
5305-00-068-0509	425	3	5306-00-068-0514	379	13
5305-00-068-0509	426	10	5305-00-068-0515	64	22
5305-00-068-0509	427	4	5305-00-068-0515	379	52
5305-00-068-0509	487	9	5305-00-068-0515	387	17
5305-00-068-0510	2	6	5305-00-068-0515	393	9
5305-00-068-0510	15	31	5310-00-068-5285	127	12
5305-00-068-0510	16	20	5310-00-068-5285	165	5
5305-00-068-0510	23	18	5310-00-068-5285	166	4
5305-00-068-0510	24	26	5310-00-068-5285	179	9
5305-00-068-0510	26	7	5310-00-068-5285	180	10
5305-00-068-0510	35	2	5305-00-068-7837	28	8
5305-00-068-0510	37	2	5305-00-068-7837	46	25
5305-00-068-0510	142	15	5305-00-068-7837	155	11
5305-00-068-0510	143	15	5305-00-068-7837	177	10
5305-00-068-0510	150	15	5305-00-068-7837	178	5
5305-00-068-0510	151	15	5305-00-068-7837	197	30
5305-00-068-0510	174	13	5305-00-068-7837	199	42
5305-00-068-0510	175	6	5305-00-068-7837	210	3
5305-00-068-0510	245	21	5305-00-068-7837	274	8
5305-00-068-0510	274	27	5305-00-068-7837	276	1

SECTION IV TM9-2320-280-24P C01

CROSS REFERENCE INDEXES

NATIONAL STOCK NUMBER INDEX

STOCK NUMBER	FIGURE NO	ITEM NO	STOCK NUMBER	FIGURE NO	ITEM NO
5305-00-068-7837	176A	17	5305-00-071-2067	147	5
5305-00-069-5578	398	3	5305-00-071-2067	150	16
5305-00-071-1313	373	4	5305-00-071-2067	151	16
5305-00-071-1313	376	4	5305-00-071-2067	179	17
5305-00-071-1315	316	16	5305-00-071-2067	180	27
5305-00-071-1315	320	15	5305-00-071-2067	182	20
5305-00-071-1315	320	27	5305-00-071-2067	183	19
5305-00-071-1315	375	23	5305-00-071-2067	184	7
5305-00-071-1315	378	23	5305-00-071-2067	237	9
5305-00-071-1318	214	4	5305-00-071-2067	429	34
5305-00-071-1318	246	24	5305-00-071-2067	456	6
5305-00-071-1318	251	7	5305-00-071-2067	487A	4
5305-00-071-1318	252	26	5305-00-071-2069	142	34
5305-00-071-1318	267	18	5305-00-071-2069	143	40
5305-00-071-1318	296	15	5305-00-071-2069	174	8
5305-00-071-1322	246	31	5305-00-071-2069	182	13
5305-00-071-1322	410	29	5305-00-071-2069	183	9
5305-00-071-1327	316	10	5305-00-071-2069	183	24
5305-00-071-1327	320	8	5305-00-071-2069	187	14
5305-00-071-1327	320	20	5305-00-071-2069	209	15
5305-00-071-1327	378	16	5305-00-071-2069	283	5
5305-00-071-1783	188	22	5305-00-071-2069	315	8
5305-00-071-1786	136	5	5305-00-071-2069	363	27
5305-00-071-1786	151	42	5305-00-071-2069	417	16
5305-00-071-1788	425	5	5305-00-071-2069	456	23
5305-00-071-1788	426	8	5305-00-071-2070	127	4
5305-00-071-1788	427	6	5305-00-071-2070	181	8
5305-00-071-2055	35	31	5305-00-071-2070	186	13
5305-00-071-2055	152	16	5305-00-071-2070	242	6
5305-00-071-2056	389	3	5305-00-071-2070	245	24
5305-00-071-2056	393	3	5305-00-071-2070	429	38
5305-00-071-2056	395	14	5305-00-071-2070	438	6
5305-00-071-2057	37	21	5305-00-071-2070	441	14
5305-00-071-2057	42	5	5305-00-071-2070	446	26
5305-00-071-2057	462	13	5305-00-071-2070	456	24
5305-00-071-2059	37	9	5305-00-071-2071	26	12
5305-00-071-2059	42	5	5305-00-071-2071	185	10
5305-00-071-2066	140	5	5305-00-071-2071	252	13
5305-00-071-2066	140	11	5305-00-071-2071	285	6
5305-00-071-2066	147	9	5305-00-071-2071	295	8
5305-00-071-2066	182	13	5305-00-071-2071	455	2
5305-00-071-2066	182	21	5305-00-071-2071	456	5
5305-00-071-2066	183	19	5305-00-071-2072	283	4
5305-00-071-2066	296	4	5305-00-071-2073	188	24
5305-00-071-2067	140	11	5305-00-071-2073	191	7
5305-00-071-2067	142	16	5305-00-071-2074	33	1
5305-00-071-2067	143	16	5305-00-071-2074	182	18

SECTION IV TM9-2320-280-24P C01

CROSS REFERENCE INDEXES

NATIONAL STOCK NUMBER INDEX

STOCK NUMBER	FIGURE NO	ITEM NO	STOCK NUMBER	FIGURE NO	ITEM NO
5305-00-071-2074	183	18	5305-00-071-2505	420	9
5305-00-071-2075	182	14	5305-00-071-2505	421	3
5305-00-071-2075	183	23	5305-00-071-2505	422	5
5305-00-071-2075	188	6	5305-00-071-2505	423	12
5305-00-071-2075	191	7	5305-00-071-2506	25	1
5305-00-071-2075	33A	1	5305-00-071-2506	28	8
5305-00-071-2076	147	9	5305-00-071-2506	55	12
5305-00-071-2076	182	8	5305-00-071-2506	103	4
5305-00-071-2076	182	16	5305-00-071-2506	156	16
5305-00-071-2076	188	9	5305-00-071-2506	365	13
5305-00-071-2077	191	5	5305-00-071-2506	365	26
5305-00-071-2078	181	7	5305-00-071-2506	368	24
5305-00-071-2078	184	4	5305-00-071-2506	375	30
5305-00-071-2078	188	1	5305-00-071-2506	496	13
5305-00-071-2079	181	3	5305-00-071-2509	83	29
5305-00-071-2079	182	8	5305-00-071-2509	88	46
5305-00-071-2079	188	24	5305-00-071-2509	225	41
5305-00-071-2079	191	7	5305-00-071-2509	246	35
5305-00-071-2081	2	11	5305-00-071-2509	248	27
5305-00-071-2081	179	13	5305-00-071-2509	249	20
5305-00-071-2081	180	18	5305-00-071-2509	252	10
5305-00-071-2081	184	17	5305-00-071-2509	253	13
5305-00-071-2082	179	6	5305-00-071-2509	253	20
5305-00-071-2082	180	7	5305-00-071-2509	255	19
5305-00-071-2083	363	16	5305-00-071-2509	289	9
5305-00-071-2236	266	23	5305-00-071-2509	389	11
5305-00-071-2236	401	14	5305-00-071-2510	153	11
5305-00-071-2237	338	21	5305-00-071-2510	211	11
5305-00-071-2237	339	16	5305-00-071-2510	212	7
5305-00-071-2237	393	1	5305-00-071-2510	246	29
5305-00-071-2237	395	12	5305-00-071-2510	246	38
5305-00-071-2237	413	8	5305-00-071-2510	249	9
5305-00-071-2237	486	5	5305-00-071-2510	484	10
5305-00-071-2505	80	13	5305-00-071-2511	248	20
5305-00-071-2505	98	15	5305-00-071-2512	293	26
5305-00-071-2505	99	6	5305-00-071-2513	248	25
5305-00-071-2505	202	16	5305-00-071-2513	249	36
5305-00-071-2505	209	6	5305-00-071-2514	251	10
5305-00-071-2505	245	37	2910-00-073-0165	334	1
5305-00-071-2505	247	15	5975-00-074-2072	12	9
5305-00-071-2505	280	54	5975-00-074-2072	98	8
5305-00-071-2505	281	54	5975-00-074-2072	345	7
5305-00-071-2505	311	4	5975-00-074-2072	485	8
5305-00-071-2505	325	13	5310-00-074-2351	292	1
5305-00-071-2505	326	13	5325-00-074-3301	BULK	6
5305-00-071-2505	418	1	5331-00-078-0265	43	22
5305-00-071-2505	419	1	2540-00-078-6633	191	9

SECTION IV TM9-2320-280-24P C01

CROSS REFERENCE INDEXES

NATIONAL STOCK NUMBER INDEX

STOCK NUMBER	FIGURE NO	ITEM NO	STOCK NUMBER	FIGURE NO	ITEM NO
2540-00-078-6633	191	17	5310-00-080-6004	329	13
5340-00-079-7837	21	16	5310-00-080-6004	332	34
5340-00-079-7837	398	11	5310-00-080-6004	392	9
5340-00-079-7837	400	9	5310-00-080-6004	431	3
5340-00-079-7837	425	18	5310-00-080-6004	438	8
5310-00-080-2411	432	5	5310-00-080-6004	440	6
5310-00-080-2411	487	13	5310-00-080-6004	441	8
5310-00-080-6004	2	3	5310-00-080-6004	461	9
5310-00-080-6004	8	13	5310-00-080-6004	471	10
5310-00-080-6004	12	21	5310-00-080-9786	88	27
5310-00-080-6004	23	2	5310-00-080-9786	215	9
5310-00-080-6004	24	7	5310-00-080-9786	222	10
5310-00-080-6004	26	5	5310-00-080-9786	339	8
5310-00-080-6004	30	7	5310-00-080-9786	346	6
5310-00-080-6004	36	17	5310-00-080-9786	347	3
5310-00-080-6004	38	17	5310-00-080-9786	426	18
5310-00-080-6004	41	13	5310-00-080-9786	427	21
5310-00-080-6004	66	6	5310-00-080-9786	467	1
5310-00-080-6004	79	20	5310-00-081-4219	5	4
5310-00-080-6004	97	7	5310-00-081-4219	74	2
5310-00-080-6004	128	54	5310-00-081-4219	142	28
5310-00-080-6004	129	23	5310-00-081-4219	143	37
5310-00-080-6004	133	5	5310-00-081-4219	150	28
5310-00-080-6004	134	2	5310-00-081-4219	151	37
5310-00-080-6004	142	14	5310-00-081-4219	154	7
5310-00-080-6004	143	14	5310-00-081-4219	160	14
5310-00-080-6004	150	14	5310-00-081-4219	162	13
5310-00-080-6004	151	14	5310-00-081-4219	178	14
5310-00-080-6004	155	18	5310-00-081-4219	180	11
5310-00-080-6004	163	6	5310-00-081-4219	189	4
5310-00-080-6004	164	2	5310-00-081-4219	190	2
5310-00-080-6004	175	9	5310-00-081-4219	198	20
5310-00-080-6004	187	9	5310-00-081-4219	199	2
5310-00-080-6004	189	1	5310-00-081-4219	204	7
5310-00-080-6004	190	4	5310-00-081-4219	209	3
5310-00-080-6004	245	22	5310-00-081-4219	211	14
5310-00-080-6004	251	5	5310-00-081-4219	212	12
5310-00-080-6004	266	25	5310-00-081-4219	225	30
5310-00-080-6004	270	32	5310-00-081-4219	226	15
5310-00-080-6004	272	25	5310-00-081-4219	230	4
5310-00-080-6004	273	2	5310-00-081-4219	235	7
5310-00-080-6004	274	11	5310-00-081-4219	236	7
5310-00-080-6004	275	2	5310-00-081-4219	243	24
5310-00-080-6004	277	2	5310-00-081-4219	244	10
5310-00-080-6004	280	25	5310-00-081-4219	273	8
5310-00-080-6004	281	25	5310-00-081-4219	273	17
5310-00-080-6004	328	18	5310-00-081-4219	277	33

SECTION IV TM9-2320-280-24P C01

CROSS REFERENCE INDEXES

NATIONAL STOCK NUMBER INDEX

STOCK NUMBER	FIGURE NO	ITEM NO	STOCK NUMBER	FIGURE NO	ITEM NO
5310-00-081-4219	280	46	5310-00-087-7493	160	10
5310-00-081-4219	281	46	5310-00-087-7493	205	2
5310-00-081-4219	288	19	5310-00-087-7493	205	36
5310-00-081-4219	321	19	5310-00-087-7493	206	1
5310-00-081-4219	321	31	5310-00-087-7493	206	9
5310-00-081-4219	324	8	5310-00-087-7493	206	21
5310-00-081-4219	325	27	5310-00-087-7493	206	29
5310-00-081-4219	329	6	5310-00-087-7493	238	22
5310-00-081-4219	333	23	5310-00-087-7493	240	7
5310-00-081-4219	363	15	5310-00-087-7493	268	18
5310-00-081-4219	376	15	5310-00-087-7493	270	14
5310-00-081-4219	377	14	5310-00-087-7493	285	5
5310-00-081-4219	378	29	5310-00-087-7493	286	10
5310-00-081-4219	402	7	5310-00-087-7493	288	7
5310-00-081-4219	429	6	5310-00-087-7493	292	17
5310-00-081-4219	432	4	5310-00-087-7493	293	5
5310-00-081-4219	433	8	5310-00-087-7493	398	23
5310-00-081-4219	436	9	5310-00-087-7493	401	8
5310-00-081-4219	473	2	5310-00-087-7493	429	17
5310-00-081-4219	474	6	5310-00-087-7493	436	8
5310-00-081-4219	480	5	5310-00-087-7493	490	2
5310-00-081-4219	487B	4	5310-00-088-0553	392	4
5310-00-081-4219	487D	9	5310-00-088-1251	50	13
5320-00-083-5009	52	9	5310-00-088-1251	169	12
5320-00-083-5009	197	1	5340-00-088-1254	75	33
5320-00-083-5009	208	6	5340-00-088-1254	84	17
5320-00-083-5009	211	4	5340-00-088-1254	88	15
5320-00-083-5009	270	17	5340-00-088-1254	155	1
5320-00-083-5009	296	12	5340-00-088-1254	416	26
5320-00-083-5009	311	5	5340-00-088-1254	176A	18
5320-00-083-5009	311	17	5340-00-088-1255	485	7
5320-00-083-5009	317	8	5325-00-088-6147	62	23
5320-00-083-5009	321	9	5340-00-088-6655	88	40
5320-00-083-5009	325	18	5340-00-088-6655	93	27
5320-00-083-5009	332	15	5340-00-088-6655	410	27
5320-00-083-5009	354	13	5305-00-088-8332	374	2
5320-00-083-5009	355	3	5305-00-088-8332	429	10
5320-00-083-5009	373	7	5305-00-088-8332	430	6
5320-00-083-5009	376	8	5305-00-088-8332	430	20
5320-00-083-5009	376	12	5305-00-088-8332	430	33
5310-00-087-7493	37	4	5330-00-089-0978	345	20
5310-00-087-7493	52	25	5330-00-089-0978	348	20
5310-00-087-7493	60	2	5331-00-089-0998	345	21
5310-00-087-7493	61	2	5331-00-089-0998	348	21
5310-00-087-7493	103	16	5365-00-090-5426	83	36
5310-00-087-7493	153	16	5365-00-090-5426	85	18
5310-00-087-7493	159	3	5365-00-090-5426	86	11

I-11

SECTION IV TM9-2320-280-24P C01

CROSS REFERENCE INDEXES

NATIONAL STOCK NUMBER INDEX

STOCK NUMBER	FIGURE NO	ITEM NO	STOCK NUMBER	FIGURE NO	ITEM NO
5340-00-091-3790	55	16	5940-00-113-9825	75	4
5340-00-091-3790	75	46	5940-00-113-9825	95	20
5340-00-091-3790	79	17	5940-00-113-9825	96	15
5340-00-091-3790	83	16	5940-00-113-9826	86	34
5340-00-091-3790	85	26	5940-00-114-1300	95	9
5340-00-091-3790	88	6	5940-00-114-1300	96	10
5340-00-091-3790	93	26	5940-00-114-1314	41	5
5340-00-091-3790	96	25	5940-00-114-1314	95	16
5340-00-091-3790	366	5	5940-00-114-1314	96	16
5340-00-091-3790	424	4	5940-00-114-1315	75	6
5340-00-091-3790	426	25	5940-00-114-1315	76	12
5340-00-091-3790	427	25	5940-00-114-1317	91	7
5340-00-091-3790	445	9	5935-00-115-2306	65	17
5340-00-091-3790	488	13	5935-00-115-2306	81	13
4730-00-100-3918	379	44	5935-00-115-2307	65	13
3110-00-100-5920	142	22	5935-00-115-2307	81	9
3110-00-100-5920	143	29	5935-00-115-2307	483	5
3110-00-100-5920	150	22	5940-00-115-2674	41	2
3110-00-100-5920	151	29	5940-00-115-2674	95	12
3110-00-100-5937	142	3	5940-00-115-2674	96	13
3110-00-100-5937	143	3	5940-00-115-2676	91	5
3110-00-100-5937	150	3	5940-00-115-2676	94	5
3110-00-100-5937	151	3	5940-00-115-2678	41	7
3110-00-100-6162	314	38	5940-00-115-2678	474	16
2540-00-104-2889	389	2	5940-00-115-2683	480	32
4910-00-105-2823	502	6	5940-00-115-2684	480	35
5940-00-107-1481	75	5	5940-00-115-2685	69	39
5940-00-107-1481	76	6	5940-00-115-4992	91	2
5940-00-107-1481	92	10	5940-00-115-4992	94	6
5940-00-107-1481	96	6	5940-00-115-5001	480	21
5330-00-107-3925	104	11	5940-00-115-5006	44	16
3110-00-108-9247	458	35	5940-00-115-5006	69	14
5975-00-111-3208	492	6	5940-00-115-5006	69	20
4720-00-111-5096	BULK	8	5940-00-115-5006	69	25
5310-00-112-2233	39	12	5940-00-115-5006	72	4
5940-00-113-0954	93	13	5940-00-115-5006	72	19
5940-00-113-3143	90	5	5940-00-115-5006	73	18
5940-00-113-3143	93	12	5940-00-115-5006	73	22
5940-00-113-3143	95	5	5940-00-115-5006	474	16
5940-00-113-3143	96	12	5940-00-115-5006	480	17
5940-00-113-3145	95	17	5940-00-115-5012	480	15
5940-00-113-3145	96	9	5940-00-115-5023	480	13
2540-00-113-4180	348	1	5305-00-115-9406	349	8
5940-00-113-9420	485	3	5305-00-115-9526	72	1
5940-00-113-9819	95	10	5305-00-115-9526	251	4
5940-00-113-9819	96	7	5305-00-115-9934	205	14
5940-00-113-9821	367	7	5320-00-117-6853	197	15

I-12

SECTION IV TM9-2320-280-24P C01

CROSS REFERENCE INDEXES

NATIONAL STOCK NUMBER INDEX

STOCK NUMBER	FIGURE NO	ITEM NO	STOCK NUMBER	FIGURE NO	ITEM NO
5320-00-117-6853	197	36	5310-00-124-9265	416	17
5320-00-117-6853	199	52	5310-00-124-9265	416	30
5320-00-117-6853	289	13	5310-00-124-9265	426	19
5320-00-117-6856	197	15	5310-00-124-9265	427	19
5320-00-117-6910	288	17	5310-00-124-9265	428	11
5320-00-117-7287	224	16	5310-00-124-9265	435	6
5320-00-117-7288	289	10	5310-00-124-9265	445	3
5320-00-117-7289	288	4	5310-00-124-9265	462	15
5320-00-117-7291	213	4	5310-00-124-9265	462	16
5320-00-119-6823	498	3	5310-00-124-9265	468	12
5310-00-124-9265	19	14	5310-00-124-9265	477	6
5310-00-124-9265	52	12	5310-00-124-9265	485	13
5310-00-124-9265	56	15	5310-00-124-9265	488	9
5310-00-124-9265	58	12	5310-00-126-3842	169	7
5310-00-124-9265	59	8	5310-00-126-5168	228	11
5310-00-124-9265	75	26	3040-00-127-5322	191	23
5310-00-124-9265	77	7	5930-00-130-5349	46	17
5310-00-124-9265	79	23	1640-00-132-9181	275	9
5310-00-124-9265	83	8	1640-00-132-9181	276	7
5310-00-124-9265	84	8	5975-00-133-8687	51	14
5310-00-124-9265	85	23	5975-00-133-8687	386	3
5310-00-124-9265	88	9	5975-00-133-8687	388	11
5310-00-124-9265	95	32	5930-00-134-5036	46	11
5310-00-124-9265	207	6	5305-00-135-3032	348	7
5310-00-124-9265	227	16	5330-00-135-6383	105	21
5310-00-124-9265	231	22	5331-00-137-3450	45	9
5310-00-124-9265	254	24	5330-00-138-0251	45	41
5310-00-124-9265	258	2	7690-00-138-5788	362	1
5310-00-124-9265	259	7	4720-00-139-3968	BULK	9
5310-00-124-9265	260	22	5315-00-140-1938	293	16
5310-00-124-9265	262	12	5310-00-141-3062	256	8
5310-00-124-9265	263	8	5310-00-141-3062	257	8
5310-00-124-9265	282	4	5310-00-141-3062	259	11
5310-00-124-9265	289	14	5310-00-141-3062	262	7
5310-00-124-9265	318	7	4730-00-142-2010	125	20
5310-00-124-9265	321	21	5940-00-143-4774	367	15
5310-00-124-9265	323	16	5940-00-143-4775	367	14
5310-00-124-9265	325	20	5940-00-143-4794	72	10
5310-00-124-9265	326	7	5940-00-143-4794	72	20
5310-00-124-9265	327	41	5940-00-143-4794	73	6
5310-00-124-9265	338	7	5340-00-150-1658	83	23
5310-00-124-9265	338	31	5340-00-150-1658	488	10
5310-00-124-9265	400	10	5315-00-150-2793	276	3
5310-00-124-9265	410	11	5305-00-150-3408	56	21
5310-00-124-9265	410	26	4730-00-150-6118	11	8
5310-00-124-9265	412	20	5340-00-151-9651	12	18
5310-00-124-9265	414	13	5340-00-151-9651	30	6

I-13

SECTION IV TM9-2320-280-24P C01

CROSS REFERENCE INDEXES

NATIONAL STOCK NUMBER INDEX

STOCK NUMBER	FIGURE NO	ITEM NO	STOCK NUMBER	FIGURE NO	ITEM NO
5340-00-151-9651	41	14	5935-00-167-7775	90	2
5120-00-152-2284	505	18	5935-00-167-7775	92	6
6145-00-152-6499	81	17	5935-00-167-7775	93	8
3110-00-155-6152	128	14	5935-00-167-7775	95	2
3110-00-155-6152	148	13	5935-00-167-7775	96	5
6240-00-155-7790	65	16	5935-00-167-7775	367	8
5975-00-156-3253	488	5	5935-00-167-7775	445	8
3110-00-162-3766	286	12	5310-00-171-1734	39	25
5310-00-164-1790	377	7	5331-00-171-6649	163	11
5305-00-164-7082	65	8	4730-00-172-0010	169	16
5305-00-164-7082	65	15	2520-00-172-1947	109	4
5310-00-166-8567	172	17	2520-00-172-1947	110	4
4730-00-166-9178	30	13	2520-00-172-1947	111	17
5310-00-167-0680	476	12	5325-00-174-9038	70	1
5310-00-167-0721	436	10	5325-00-174-9332	80	12
5310-00-167-0821	101	2	5325-00-174-9332	315	13
5310-00-167-0821	382	9	5325-00-174-9829	136	10
5310-00-167-0821	390	8	5325-00-174-9829	136	23
5310-00-167-0822	194	6	5325-00-174-9829	137	5
5310-00-167-0823	199	21	5310-00-177-7529	256	4
5310-00-167-0823	245	25	5310-00-177-7529	257	1
5310-00-167-0823	251	46	5310-00-177-7529	260	4
5310-00-167-0823	252	14	2540-00-177-8108	393	14
5310-00-167-0823	429	35	6220-00-179-4325	60	9
5310-00-167-0823	446	27	5305-00-179-8946	89	13
5310-00-167-0834	279	27	5305-00-179-8946	91	10
5310-00-167-0835	279	3	5305-00-179-8946	93	21
5310-00-167-0836	279	17	5305-00-179-8946	94	8
5310-00-167-0837	279	11	5305-00-179-8946	95	36
5310-00-167-0893	16	25	5306-00-180-2748	321	32
5310-00-167-0893	307	3	5305-00-180-4966	39	6
5310-00-167-0893	308	14	5305-00-180-4966	259	12
5310-00-167-0893	401	15	5305-00-180-4966	263	7
5935-00-167-7775	65	5	5305-00-180-4966	289	4
5935-00-167-7775	65	26	5305-00-180-4966	292	3
5935-00-167-7775	75	10	5305-00-180-4966	365	20
5935-00-167-7775	76	4	5306-00-182-2025	199	45
5935-00-167-7775	80	5	5306-00-182-2027	198	25
5935-00-167-7775	81	3	5306-00-182-2027	199	44
5935-00-167-7775	81	15	5306-00-182-2027	317	12
5935-00-167-7775	83	3	5306-00-182-2027	373	11
5935-00-167-7775	84	31	5306-00-182-2027	376	19
5935-00-167-7775	85	1	5305-00-182-9304	255	35
5935-00-167-7775	86	1	5935-00-184-6707	75	9
5935-00-167-7775	87	1	5935-00-184-6707	76	8
5935-00-167-7775	87	5	5935-00-184-6707	81	5
5935-00-167-7775	89	2	5935-00-184-6707	85	20

SECTION IV TM9-2320-280-24P C01

CROSS REFERENCE INDEXES

NATIONAL STOCK NUMBER INDEX

STOCK NUMBER	FIGURE NO	ITEM NO	STOCK NUMBER	FIGURE NO	ITEM NO
5935-00-184-6707	86	13	5310-00-208-1918	249	42
5120-00-184-8398	505	1	5310-00-208-1918	272	4
5310-00-184-8971	69	29	5310-00-208-1918	279	28
5310-00-184-8971	434	5	5310-00-208-1918	286	25
5310-00-184-8971	435	10	5306-00-208-3639	279	12
5325-00-184-9846	89	16	5310-00-208-9255	480	10
5325-00-184-9846	368	10	5310-00-208-9255	482	11
5325-00-185-0004	96	18	5310-00-209-0786	55	13
5331-00-185-0075	467	11	5310-00-209-0965	35	22
5331-00-185-0075	496	14	5310-00-209-0965	36	10
5325-00-185-0961	207	11	5310-00-209-0965	37	10
5315-00-187-9384	458	48	5310-00-209-0965	38	10
5305-00-191-3640	313	10	5310-00-209-0965	42	4
5305-00-191-3640	443	9	5310-00-209-0965	150	35
5305-00-191-3640	457	9	5310-00-209-0965	151	41
5330-00-193-0850	45	32	5310-00-209-0965	171	3
5340-00-193-4111	370	18	5310-00-209-0965	472	21
5310-00-194-1483	46	15	5310-00-209-0965	475	17
5310-00-194-1483	388	14	5935-00-214-0904	40	26
3110-00-198-0492	128	27	5935-00-214-0904	86	24
9905-00-198-2728	354	2	5935-00-214-0904	366	34
5310-00-198-6691	437	5	5935-00-214-0904	386	1
2540-00-200-4249	KITS	70	5935-00-214-0904	472	8
5342-00-200-7564	307	4	5935-00-214-0904	475	1
5342-00-200-7564	308	16	5935-00-214-0904	479	6
9905-00-202-3639	330	2	5935-00-214-0904	483	7
5325-00-202-4005	80	17	2540-00-216-5722	348	18
4730-00-204-3491	29	1	4520-00-217-5782	345	30
4730-00-204-3491	177	6	4520-00-217-5782	348	29
5305-00-204-4850	327	15	4730-00-221-2140	363	12
5940-00-204-8990	367	13	5970-00-221-5301	44	20
9905-00-205-2795	211	17	5970-00-221-5301	69	12
9905-00-205-2795	212	10	5970-00-221-5301	69	21
9905-00-205-2795	296	17	5970-00-221-5301	69	26
5305-00-206-2508	69	30	5970-00-221-5301	72	5
5306-00-207-2221	64	4	5970-00-221-5301	72	17
5306-00-207-2221	83	34	5970-00-221-5301	313	17
5930-00-207-9422	51	3	4010-00-224-482	BULK	49
5310-00-208-1918	88	4	5305-00-225-3842	5	16
5310-00-208-1918	202	41	5305-00-225-3842	202	10
5310-00-208-1918	203	29	5305-00-225-3842	203	5
5310-00-208-1918	205	38	5305-00-225-3842	204	5
5310-00-208-1918	206	16	5305-00-225-3842	308	9
5310-00-208-1918	243	28	5305-00-225-3842	430	16
5310-00-208-1918	244	17	5305-00-225-3843	21	17
5310-00-208-1918	248	5	5305-00-225-3843	42	6
5310-00-208-1918	249	18	5305-00-225-3843	95	42

SECTION IV TM9-2320-280-24P C01

CROSS REFERENCE INDEXES

NATIONAL STOCK NUMBER INDEX

STOCK NUMBER	FIGURE NO	ITEM NO	STOCK NUMBER	FIGURE NO	ITEM NO
5305-00-225-3843	153	5	5306-00-225-8499	153	27
5305-00-225-3843	169	10	5306-00-225-8499	154	10
5305-00-225-3843	169	30	5306-00-225-8499	203	20
5305-00-225-3843	203	1	5306-00-225-8499	205	31
5305-00-225-3843	203	26	5306-00-225-8499	206	2
5305-00-225-3843	204	1	5306-00-225-8499	206	18
5305-00-225-3843	227	28	5306-00-225-8499	268	4
5305-00-225-3843	247	1	5306-00-225-8499	329	11
5305-00-225-3843	249	2	5306-00-225-8499	333	24
5305-00-225-3843	250	5	5306-00-225-8499	339	20
5305-00-225-3843	252	12	5306-00-225-8499	402	5
5305-00-225-3843	254	13	5306-00-225-8499	409	1
5305-00-225-3843	266	21	5306-00-225-8499	410	8
5305-00-225-3843	267	5	5306-00-225-8499	411	1
5305-00-225-3843	277	23	5306-00-225-8499	413	1
5305-00-225-3843	285	14	5306-00-225-8499	414	7
5305-00-225-3843	286	36	5306-00-225-8499	415	1
5305-00-225-3843	292	11	5306-00-225-8499	416	7
5305-00-225-3843	312	1	5306-00-225-8499	418	6
5305-00-225-3843	334	6	5306-00-225-8499	419	7
5305-00-225-3843	343	18	5306-00-225-8499	420	1
5305-00-225-3843	352	3	5306-00-225-8499	421	8
5305-00-225-3843	366	17	5306-00-225-8499	422	3
5305-00-225-3843	368	34	5306-00-225-8499	423	8
5305-00-225-3843	417	3	5306-00-225-8499	428	7
5305-00-225-3843	430	42	5306-00-225-8503	329	8
5305-00-225-3843	446	14	5306-00-225-8511	12	13
5305-00-225-3843	459	3	5306-00-225-9089	280	45
5305-00-225-3843	475	6	5306-00-225-9089	281	45
5305-00-225-3843	484	14	5306-00-225-9097	392	3
5305-00-225-3843	486	12	5306-00-226-4822	35	16
5305-00-225-3843	492	15	5306-00-226-4823	143	36
5305-00-225-3843	494	11	5306-00-226-4823	151	36
5310-00-225-5328	79	14	5306-00-226-4824	273	19
5310-00-225-5328	482	17	5306-00-226-4824	43B	14
5310-00-225-6993	188	32	5306-00-226-4824	488	23
5305-00-225-7211	5	3	5306-00-226-4825	19	20
5306-00-225-8496	180	13	5306-00-226-4825	100	17
5306-00-225-8498	213	9	5306-00-226-4825	116	18
5306-00-225-8498	250	20	5306-00-226-4825	230	26
5306-00-225-8498	418	4	5306-00-226-4825	234	6
5306-00-225-8498	419	5	5306-00-226-4825	480	3
5306-00-225-8498	420	3	5306-00-226-4825	487	16
5306-00-225-8498	421	1	5306-00-226-4825	487B	3
5306-00-225-8498	422	1	5306-00-226-4826	230	19
5306-00-225-8498	423	1	5306-00-226-4826	231	12
5306-00-225-8499	74	1	5306-00-226-4826	273	7

SECTION IV TM9-2320-280-24P C01

CROSS REFERENCE INDEXES

NATIONAL STOCK NUMBER INDEX

STOCK NUMBER	FIGURE NO	ITEM NO	STOCK NUMBER	FIGURE NO	ITEM NO
5306-00-226-4826	277	7	5305-00-226-4831	288	23
5306-00-226-4826	328	8	5306-00-226-4832	121	8
5306-00-226-4827	83	30	5306-00-226-4833	211	15
5306-00-226-4827	204	6	5306-00-226-4833	212	11
5306-00-226-4827	225	29	5306-00-226-4833	242	4
5306-00-226-4827	226	14	5306-00-226-4833	377	15
5306-00-226-4827	242	10	5306-00-226-4834	211	16
5306-00-226-4827	287	20	5306-00-226-4834	212	13
5306-00-226-4827	329	18	5306-00-226-4835	209	16
5306-00-226-4827	473	1	5306-00-226-4835	338	24
5306-00-226-4827	474	7	5306-00-226-4835	429	5
5306-00-226-4828	160	13	5306-00-226-4838	189	3
5306-00-226-4828	162	14	5930-00-226-6429	46	6
5306-00-226-4828	225	2	5305-00-226-7237	12	37
5306-00-226-4828	226	13	5305-00-227-1543	333	31
5306-00-226-4828	284	1	4710-00-228-2177	BULK	45
5306-00-226-4828	288	26	4730-00-230-2959	496	25
5306-00-226-4828	292	13	5315-00-234-1861	268	9
5306-00-226-4828	340	1	4030-00-237-8741	314	18
5306-00-226-4828	346	1	5315-00-239-8032	186	8
5306-00-226-4828	363	23	5315-00-239-8032	188	19
5306-00-226-4828	374	22	5315-00-239-8032	188	35
5306-00-226-4828	417	4	5305-00-240-0194	67	2
5306-00-226-4828	418	7	5305-00-240-0194	88	10
5306-00-226-4828	419	8	5305-00-240-0194	169	22
5306-00-226-4828	420	5	5305-00-240-0194	197	33
5306-00-226-4828	421	7	5305-00-240-0194	199	59
5306-00-226-4828	422	9	5305-00-240-0194	205	4
5306-00-226-4828	423	3	5305-00-240-0194	206	7
5306-00-226-4828	439	1	5305-00-240-0194	225	50
5306-00-226-4828	487D	8	5305-00-240-0194	226	30
5306-00-226-4829	144	12	5305-00-240-0194	248	28
5306-00-226-4829	152	12	5305-00-240-0194	286	31
5306-00-226-4829	211	31	5305-00-240-0194	327	26
5306-00-226-4829	212	29	5305-00-240-0194	411	5
5306-00-226-4829	225	26	5305-00-240-0194	460	2
5306-00-226-4829	226	8	5315-00-240-0251	243	32
5306-00-226-4829	328	2	5305-00-240-6541	227	50
5306-00-226-4829	390	6	5305-00-240-6668	85	3
5306-00-226-4829	417	7	5305-00-240-6668	88	1
5306-00-226-4829	423	7	5305-00-240-6668	327	48
5306-00-226-4829	432	2	5120-00-240-8702	505	4
5306-00-226-4829	461	8	4720-00-241-4435	BULK	24
5305-00-226-4831	121	9	5310-00-241-6658	154	5
5305-00-226-4831	242	2	5310-00-241-6658	198	19
5305-00-226-4831	269	20	5310-00-241-6658	199	1
5305-00-226-4831	279	35	5310-00-241-6658	273	16

I-17

SECTION IV TM9-2320-280-24P C01

CROSS REFERENCE INDEXES

NATIONAL STOCK NUMBER INDEX

STOCK NUMBER	FIGURE NO	ITEM NO	STOCK NUMBER	FIGURE NO	ITEM NO
5310-00-241-6658	277	35	5310-00-251-4503	416	11
5310-00-241-6658	280	47	5310-00-251-4503	446	9
5310-00-241-6658	281	47	5310-00-251-4503	447	5
5310-00-241-6658	317	11	5310-00-251-4503	448	8
5310-00-241-6658	321	30	5310-00-251-4503	449	2
5310-00-241-6658	373	8	5310-00-251-4503	450	1
5310-00-241-6658	376	18	5310-00-251-4503	486	3
5310-00-241-6658	378	28	5905-00-251-7145	348	4
5310-00-241-6658	389	14	5305-00-253-5609	133	3
5310-00-241-6658	402	8	5305-00-253-5614	197	35
5310-00-241-6658	415	4	5305-00-253-5626	191	18
5310-00-241-6658	433	7	5305-00-253-5627	357	8
5310-00-241-6659	170	14	4730-00-254-6211	334	9
5310-00-241-6921	62	20	4730-00-254-6227	334	7
6150-00-242-4885	313	24	5315-00-255-5580	268	6
5315-00-243-1169	191	21	3120-00-255-5697	111	13
4020-00-246-0688	BULK	3	3120-00-255-5697	112	12
4810-00-248-1635	348	40	5905-00-258-6788	59	12
5325-00-249-6345	333	33	4730-00-260-8285	366	15
5310-00-251-4503	63	5	4730-00-260-8285	384	5
5310-00-251-4503	234	17	5310-00-261-7340	341	8
5310-00-251-4503	246	9	5310-00-261-8278	69	31
5310-00-251-4503	246	14	5330-00-262-8195	BULK	48
5310-00-251-4503	247	13	5325-00-263-6648	496	3
5310-00-251-4503	247	24	5325-00-263-6651	79	25
5310-00-251-4503	248	9	4730-00-266-0535	334	19
5310-00-251-4503	249	6	4730-00-266-0536	366	12
5310-00-251-4503	249	21	4730-00-266-0538	366	14
5310-00-251-4503	249	33	4730-00-267-8945	401	2
5310-00-251-4503	251	15	5305-00-267-8952	387	7
5310-00-251-4503	251	25	5305-00-267-8953	279	2
5310-00-251-4503	251	39	5305-00-267-8953	388	12
5310-00-251-4503	252	17	5305-00-267-8953	43B	32
5310-00-251-4503	252	39	5305-00-267-8953	485	15
5310-00-251-4503	253	5	5305-00-267-8954	228	15
5310-00-251-4503	253	12	6680-00-269-0335	488	22
5310-00-251-4503	253	15	5305-00-269-2811	392	8
5310-00-251-4503	253	32	5305-00-269-3233	363	10
5310-00-251-4503	253	39	5305-00-269-3241	392	7
5310-00-251-4503	254	4	5305-00-269-3242	390	15
5310-00-251-4503	254	12	5310-00-269-3466	225	13
5310-00-251-4503	255	4	4730-00-270-4606	335	16
5310-00-251-4503	255	12	4730-00-270-4606	336	13
5310-00-251-4503	255	14	5510-00-270-6031	BULK	71
5310-00-251-4503	255	25	5325-00-270-8891	380	5
5310-00-251-4503	255	29	5310-00-274-8710	343	5
5310-00-251-4503	260	16	5310-00-274-8710	497	2

SECTION IV TM9-2320-280-24P C01

CROSS REFERENCE INDEXES

NATIONAL STOCK NUMBER INDEX

STOCK NUMBER	FIGURE NO	ITEM NO	STOCK NUMBER	FIGURE NO	ITEM NO
5310-00-274-8710	498	11	5325-00-291-9366	333	34
5310-00-274-8715	249	3	5325-00-291-9366	380	7
5310-00-274-8715	315	11	5306-00-292-8253	277	37
5310-00-274-9364	238	3	5120-00-293-0045	507	1
5320-00-275-8344	337	7	5330-00-297-6329	47	14
5325-00-276-6051	96	21	5310-00-297-8836	66A	7
5325-00-276-6090	496	21	5331-00-297-9990	72	24
5325-00-276-6091	369	4	5315-00-298-1481	437	11
5325-00-276-6228	489	6	6220-00-299-7425	63	3
5325-00-276-6343	70	14	6220-00-299-7425	63	9
5325-00-276-6343	473	9	6220-00-299-7426	63	3
5325-00-276-6343	481	10	6220-00-299-7426	63	9
4730-00-277-6347	366	24	2920-00-302-6342	KITS	102
4730-00-277-6347	384	15	5930-00-307-8856	46	19
5365-00-277-7341	142	5	5325-00-331-4774	43B	11
5365-00-277-7341	143	5	5310-00-333-7341	345	24
5365-00-277-7341	151	5	5310-00-333-7341	348	3
4730-00-278-2523	331	30	5935-00-333-9414	80	10
4730-00-278-2523	385	4	5935-00-333-9414	84	28
4730-00-278-2523	388	5	5935-00-333-9414	85	10
4730-00-278-3721	178	23	5935-00-333-9414	86	6
4730-00-278-3828	17	5	2805-00-336-1716	4	15
5325-00-281-8642	316	3	5935-00-338-2822	282	13
5325-00-281-8643	316	5	2530-00-340-1405	157	9
5325-00-281-8643	319	2	2530-00-340-1405	158	5
5325-00-281-8643	322	2	5930-00-345-5455	345	10
5325-00-282-1830	205	29	5930-00-345-5455	348	14
5340-00-282-7509	84	15	5325-00-349-8518	121	15
5340-00-282-7517	327	25	5310-00-350-2655	62	6
5340-00-282-7537	246	22	2520-00-352-2168	169	29
5340-00-282-7537	366	28	5330-00-353-0959	63	10
5340-00-282-7537	386	10	5305-00-353-0969	62	18
5940-00-283-5280	367	11	5310-00-355-5645	44	7
5310-00-285-7037	269	28	4730-00-359-9487	12	4
5315-00-285-7161	365	8	4730-00-366-3011	178	19
5315-00-285-7161	368	12	5305-00-366-3337	34	1
5325-00-285-8363	424	6	5325-00-371-8108	224	12
5340-00-286-2494	327	46	5325-00-371-8108	316	2
4730-00-287-1706	159	6	5325-00-371-8108	316	6
4730-00-287-1706	160	20	5325-00-371-8108	319	3
4730-00-287-1706	161	14	5325-00-371-8108	322	3
4730-00-287-1706	162	22	5325-00-371-8108	324	3
4730-00-287-1706	487B	8	5325-00-371-8108	375	2
4820-00-287-5627	335	17	5325-00-371-8108	375	5
4820-00-287-5627	336	14	5325-00-371-8108	378	2
5325-00-290-1960	70	13	5325-00-371-8108	378	5
5325-00-291-9366	92	13	5315-00-376-9034	433	10

I-19

SECTION IV TM9-2320-280-24P C01

CROSS REFERENCE INDEXES

NATIONAL STOCK NUMBER INDEX

STOCK NUMBER	FIGURE NO	ITEM NO	STOCK NUMBER	FIGURE NO	ITEM NO
3030-00-379-2815	32	2	5310-00-407-9566	487	15
5340-00-385-8820	17	11	5310-00-407-9566	175A	13
5340-00-385-8820	27	12	5310-00-407-9566	176A	3
5305-00-390-0524	83	29	5340-00-408-2432	191	24
5305-00-390-0524	272	8	5310-00-409-3333	144	3
5310-00-393-6685	83	37	5310-00-409-3333	152	3
5310-00-393-6685	85	19	5310-00-409-3333	181	16
5310-00-393-6685	86	12	5310-00-409-3333	182	2
5940-00-399-6676	75	12	5310-00-409-3333	183	13
5310-00-401-5321	237	13	5310-00-409-3333	184	19
5310-00-401-5321	258	6	5310-00-409-3333	188	28
5310-00-401-5321	262	4	5310-00-409-3333	195	13
5310-00-402-2778	51	12	5310-00-409-3333	270	35
3110-00-403-1488	172	20	5310-00-409-3333	271	5
5305-00-403-5130	345	14	5310-00-409-3333	271	22
5305-00-403-5130	348	5	5310-00-409-3333	456	20
5305-00-403-5130	348	13	5310-00-409-3333	471	3
3110-00-406-9608	148	18	5310-00-409-3355	176A	2
5310-00-407-9566	19	19	5315-00-417-5223	238	13
5310-00-407-9566	35	15	5310-00-419-0876	36	7
5310-00-407-9566	59	4	5310-00-419-0876	38	7
5310-00-407-9566	74	10	5310-00-419-0876	39	10
5310-00-407-9566	136	15	5310-00-419-0876	476	9
5310-00-407-9566	137	2	6240-00-419-3185	49	2
5310-00-407-9566	178	15	5306-00-425-8569	69	40
5310-00-407-9566	180	12	5310-00-429-3135	39	43
5310-00-407-9566	205	32	5310-00-429-3156	39	44
5310-00-407-9566	206	19	5310-00-429-3453	39	42
5310-00-407-9566	211	30	5310-00-432-3760	441	10
5310-00-407-9566	212	30	5305-00-432-4165	49	11
5310-00-407-9566	225	31	5305-00-432-4170	403	12
5310-00-407-9566	226	16	5305-00-432-4170	433	4
5310-00-407-9566	231	13	5305-00-432-4201	485	5
5310-00-407-9566	268	5	5305-00-432-4205	208	9
5310-00-407-9566	269	15	5305-00-432-4205	280	34
5310-00-407-9566	287	19	5305-00-432-4205	281	34
5310-00-407-9566	328	7	5305-00-432-4205	331	29
5310-00-407-9566	329	7	5305-00-432-4253	395	1
5310-00-407-9566	340	10	5305-00-432-4253	396	2
5310-00-407-9566	373	9	5305-00-432-4254	446	32
5310-00-407-9566	430	13	3110-00-436-3248	128	50
5310-00-407-9566	432	3	5340-00-438-1833	313	4
5310-00-407-9566	43B	15	5340-00-438-1833	443	4
5310-00-407-9566	454	3	5340-00-438-1836	477	9
5310-00-407-9566	461	6	4730-00-439-6021	335	1
5310-00-407-9566	473	5	5325-00-442-5845	155	7
5310-00-407-9566	474	8	5325-00-442-5845	280	42

SECTION IV TM9-2320-280-24P C01

CROSS REFERENCE INDEXES

NATIONAL STOCK NUMBER INDEX

STOCK NUMBER	FIGURE NO	ITEM NO	STOCK NUMBER	FIGURE NO	ITEM NO
5325-00-442-5845	281	42	5935-00-462-6603	93	9
5325-00-442-5845	289	19	5935-00-462-6603	445	7
5305-00-442-7347	40	49	5331-00-463-0200	60	10
6220-00-443-0589	62	21	5310-00-463-0268	62	25
5930-00-445-9274	55	6	5305-00-470-3321	74	11
5930-00-445-9274	56	18	5305-00-470-3321	283	26
5310-00-447-8774	103	2	4730-00-471-3102	366	13
5310-00-447-8774	127	1	5310-00-471-9243	131	2
5310-00-449-2376	167	26	2920-00-472-1723	KITS	68
5310-00-449-2376	191	16	5310-00-472-2963	421	9
5310-00-449-2381	35	7	5310-00-472-2963	422	4
5325-00-449-3001	208	5	2590-00-473-6331	392	6
3120-00-450-1905	45	18	4730-00-476-5859	302	6
5331-00-451-0118	128	51	4730-00-476-5859	303	15
5331-00-451-0118	129	26	4730-00-476-5859	305	8
5975-00-451-5001	5	8	4730-00-476-5859	306	6
5975-00-451-5001	435	8	4730-00-476-5859	309	4
2590-00-454-3620	93	23	4730-00-476-5859	310	6
2590-00-454-3620	95	37	4730-00-476-7127	302	3
2590-00-454-3620	96	30	4730-00-476-7127	303	4
2590-00-454-3620	303	9	4730-00-476-7127	304	11
2590-00-454-3620	304	5	4730-00-476-7127	305	5
2590-00-454-3620	338	2	4730-00-476-7127	306	3
2590-00-454-3620	339	6	4730-00-476-7127	309	10
2590-00-454-3620	410	32	5310-00-477-6768	482	8
2590-00-454-3620	412	13	5305-00-478-0273	73	2
5305-00-456-2582	345	8	5305-00-482-6825	254	29
5305-00-456-2582	348	6	5305-00-482-6825	281	7
5305-00-456-2582	348	15	5310-00-482-9493	33	6
5310-00-460-4490	218	6	5310-00-483-8791	321	18
5310-00-460-4490	257	5	5310-00-483-8791	376	14
5331-00-462-0907	61	11	5310-00-483-8791	382	6
5935-00-462-6603	65	4	5310-00-483-8791	389	10
5935-00-462-6603	65	27	5310-00-483-8791	397	5
5935-00-462-6603	75	11	3120-00-485-1017	165	3
5935-00-462-6603	76	14	3120-00-485-1017	166	8
5935-00-462-6603	80	2	8145-00-485-8250	126	3
5935-00-462-6603	81	7	5310-00-488-3889	2	8
5935-00-462-6603	83	20	5310-00-488-3889	140	1
5935-00-462-6603	84	32	5310-00-488-3889	147	1
5935-00-462-6603	85	15	5310-00-488-3889	179	3
5935-00-462-6603	86	8	5310-00-488-3889	180	4
5935-00-462-6603	87	2	5310-00-488-3889	181	9
5935-00-462-6603	87	4	5310-00-488-3889	182	4
5935-00-462-6603	89	5	5310-00-488-3889	183	14
5935-00-462-6603	90	6	5310-00-488-3889	184	1
5935-00-462-6603	92	4	5310-00-488-3889	185	7

SECTION IV TM9-2320-280-24P C01

CROSS REFERENCE INDEXES

NATIONAL STOCK NUMBER INDEX

STOCK NUMBER	FIGURE NO	ITEM NO	STOCK NUMBER	FIGURE NO	ITEM NO
5310-00-488-3889	186	7	5320-00-526-2945	213	30
5310-00-488-3889	187	11	4730-00-529-1237	384	3
5310-00-488-3889	188	5	6145-00-538-8219	BULK	69
5310-00-488-3889	188	38	6145-00-538-8222	BULK	68
5310-00-488-3889	191	1	4730-00-541-0793	31	1
5310-00-488-3889	199	22	4710-00-541-4935	288	11
5310-00-488-3889	242	8	5305-00-543-2419	1	6
5310-00-488-3889	245	26	5305-00-543-2419	2	2
5310-00-488-3889	251	47	5305-00-543-2419	23	8
5310-00-488-3889	270	22	5305-00-543-2419	251	43
5310-00-488-3889	283	18	5305-00-543-2419	329	1
5310-00-488-3889	315	5	5305-00-543-2419	341	1
5310-00-488-3889	429	36	5305-00-543-2419	425	14
5310-00-488-3889	441	5	5305-00-543-2419	463	1
5310-00-488-3889	442	4	5305-00-543-2866	272	34
5310-00-488-3889	444	3	5305-00-543-2866	273	5
5310-00-488-3889	446	28	5305-00-543-2866	274	26
5310-00-488-3889	453	1	5305-00-543-2866	275	21
5310-00-488-3889	455	6	5305-00-543-2866	276	23
5310-00-488-3889	456	17	5305-00-543-2866	33A	17
5310-00-488-3889	463	9	5305-00-543-2866	33B	12
5310-00-488-3889	471	1	5325-00-543-2902	345	38
5310-00-488-3889	487A	6	5325-00-543-2902	348	37
5310-00-488-3889	487B	1	5305-00-543-4372	60	1
5305-00-489-0743	77	18	5305-00-543-4372	61	1
5305-00-489-0743	88	14	5305-00-543-4372	240	6
5305-00-489-0743	88	28	5305-00-543-4372	332	35
5305-00-489-0743	443	5	5305-00-543-4372	370	26
5305-00-489-0743	444	6	5305-00-543-4709	337	2
5305-00-489-0743	457	16	5310-00-543-5101	88	49
5340-00-489-5684	269	8	5305-00-543-5828	52	14
4720-00-491-0102	BULK	27	5940-00-549-6581	69	15
5340-00-492-2313	208	11	5940-00-549-6583	69	18
2910-00-493-2138	9	3	5310-00-550-1130	68	7
5330-00-493-3876	KITS	84	5310-00-550-1130	269	32
5320-00-493-4101	300	8	5310-00-550-1130	327	14
5330-00-497-4633	128	8	5310-00-550-1130	366	4
5310-00-498-6675	395	9	5310-00-550-1130	384	7
5305-00-499-7694	51	9	5310-00-550-1130	484	6
5940-00-503-6184	41	17	5305-00-550-1537	205	24
3030-00-504-8682	490	11	5310-00-550-3714	380	12
6105-00-512-9225	333	13	5310-00-550-4725	256	7
5325-00-514-1299	143	21	5975-00-553-6995	380	13
5325-00-514-1299	151	21	2640-00-555-2834	167	11
5310-00-515-8776	453	14	5640-00-555-3011	467	12
5940-00-520-2447	313	18	4720-00-555-8015	BULK	28
2540-00-525-7067	191	15	4730-00-555-8292	16	43

I-22

SECTION IV TM9-2320-280-24P C01

CROSS REFERENCE INDEXES

NATIONAL STOCK NUMBER INDEX

STOCK NUMBER	FIGURE NO	ITEM NO	STOCK NUMBER	FIGURE NO	ITEM NO
2520-00-557-6619	100	16	5310-00-582-5965	342	5
5310-00-559-0070	379	12	5310-00-582-5965	343	15
5315-00-559-7467	99	13	5310-00-582-5965	365	5
5340-00-565-5378	273	12	5310-00-582-5965	368	23
4240-00-565-6059	307	2	5310-00-582-5965	387	8
4240-00-565-6059	308	8	5310-00-582-5965	410	37
4730-00-567-1630	393	13	5310-00-582-5965	414	16
5310-00-568-6077	100	12	5310-00-582-5965	416	20
5975-00-570-9598	18	9	5310-00-582-5965	425	8
5975-00-570-9598	41	15	5310-00-582-5965	426	5
5975-00-570-9598	77	28	5310-00-582-5965	427	9
5975-00-570-9598	379	7	5310-00-582-5965	428	19
5975-00-570-9598	381	2	5310-00-582-5965	430	41
5935-00-572-9180	483	6	5310-00-582-5965	443	2
5315-00-576-0265	132	30	5310-00-582-5965	444	8
5305-00-576-2335	345	18	5310-00-582-5965	444	9
5305-00-576-2335	348	12	5310-00-582-5965	446	12
5310-00-576-5752	333	17	5310-00-582-5965	451	2
5310-00-576-5752	349	4	5310-00-582-5965	467	5
6220-00-577-3434	63	1	5310-00-582-5965	492	16
6210-00-578-7970	311	12	5310-00-582-5965	492	18
6210-00-578-7970	312	8	5310-00-582-5965	498	18
5310-00-579-5554	72	12	5310-00-582-5965	66A	6
3110-00-580-3843	128	38	5305-00-582-9501	37	8
5331-00-580-6586	177	8	5305-00-582-9501	75	24
5331-00-580-6586	178	10	5305-00-582-9501	77	5
5331-00-580-6586	470	3	5305-00-582-9501	83	10
5331-00-580-6586	176A	25	5305-00-582-9501	84	4
5310-00-582-5965	28	7	5305-00-582-9501	85	22
5310-00-582-5965	53	4	5305-00-582-9501	88	7
5310-00-582-5965	62	26	5305-00-582-9501	216	4
5310-00-582-5965	75	35	5305-00-582-9501	327	43
5310-00-582-5965	93	18	5305-00-582-9501	468	13
5310-00-582-5965	155	12	5305-00-582-9501	488	6
5310-00-582-5965	156	17	5310-00-584-5005	468	9
5310-00-582-5965	205	23	5310-00-584-5272	69	2
5310-00-582-5965	228	14	5310-00-584-5272	103	19
5310-00-582-5965	237	11	5310-00-584-5272	156	21
5310-00-582-5965	240	24	5310-00-584-5272	174	7
5310-00-582-5965	250	9	5310-00-584-5272	240	2
5310-00-582-5965	267	6	5310-00-584-5272	240	12
5310-00-582-5965	286	37	5310-00-584-5272	271	16
5310-00-582-5965	287	11	5310-00-584-5272	271	28
5310-00-582-5965	308	5	5310-00-584-5272	277	39
5310-00-582-5965	311	3	5310-00-584-5272	33A	2
5310-00-582-5965	313	2	5310-00-584-5272	363	19
5310-00-582-5965	327	39	5310-00-584-5272	369	1

SECTION IV TM9-2320-280-24P C01

CROSS REFERENCE INDEXES

NATIONAL STOCK NUMBER INDEX

STOCK NUMBER	FIGURE NO	ITEM NO	STOCK NUMBER	FIGURE NO	ITEM NO
5310-00-584-5272	410	35	5310-00-637-9541	24	25
5310-00-584-5272	412	26	5310-00-637-9541	34	14
5310-00-584-5272	424	7	5310-00-637-9541	35	3
5310-00-584-5272	434	3	5310-00-637-9541	35	20
5310-00-584-5272	435	7	5310-00-637-9541	36	15
5310-00-584-5272	453	2	5310-00-637-9541	37	3
4010-00-585-2108	BULK	2	5310-00-637-9541	38	15
4730-00-585-2653	169	25	5310-00-637-9541	43	48
5331-00-585-6663	21	7	5310-00-637-9541	44	3
4730-00-586-8463	27	8	5310-00-637-9541	52	24
4730-00-595-1078	399	9	5310-00-637-9541	62	7
5310-00-595-6057	42	3	5310-00-637-9541	64	13
5310-00-595-6057	488	15	5310-00-637-9541	68	3
5310-00-595-6612	39	9	5310-00-637-9541	70	10
5310-00-595-7237	412	6	5310-00-637-9541	72	2
5310-00-595-7237	482	13	5310-00-637-9541	75	43
5310-00-595-7486	183	3	5310-00-637-9541	77	2
5310-00-595-7486	185	13	5310-00-637-9541	79	19
5310-00-595-7486	187	17	5310-00-637-9541	83	33
5310-00-595-7486	188	15	5310-00-637-9541	88	38
5305-00-600-8993	324	38	5310-00-637-9541	89	7
5305-00-601-7729	39	7	5310-00-637-9541	124	15
5935-00-605-9322	85	11	5310-00-637-9541	125	15
5935-00-605-9322	86	16	5310-00-637-9541	164	3
5305-00-614-3423	46	12	5310-00-637-9541	169	2
5310-00-614-3505	35	8	5310-00-637-9541	174	14
5935-00-614-3959	80	7	5310-00-637-9541	175	8
5935-00-614-3959	85	7	5310-00-637-9541	205	35
5935-00-614-3959	86	3	5310-00-637-9541	206	22
5310-00-615-1556	43B	24	5310-00-637-9541	286	9
5930-00-615-7897	367	6	5310-00-637-9541	286	35
5930-00-615-9376	52	30	5310-00-637-9541	328	17
5315-00-616-4261	133	12	5310-00-637-9541	33A	16
5305-00-616-4831	43B	35	5310-00-637-9541	33B	2
5315-00-616-5501	476	8	5310-00-637-9541	340	3
5315-00-616-5514	43	4	5310-00-637-9541	352	2
5315-00-616-5514	175A	7	5310-00-637-9541	363	11
5315-00-616-5526	35	9	5310-00-637-9541	372	5
6685-00-618-1822	363	28	5310-00-637-9541	379	23
5120-00-618-6902	504	14	5310-00-637-9541	38A	3
5310-00-619-1148	482	12	5310-00-637-9541	401	12
5310-00-625-5756	43B	16	5310-00-637-9541	43B	31
5310-00-637-9541	1	4	5310-00-637-9541	443	12
5310-00-637-9541	2	5	5310-00-637-9541	443	13
5310-00-637-9541	5	13	5310-00-637-9541	457	6
5310-00-637-9541	16	23	5310-00-637-9541	459	2
5310-00-637-9541	23	17	5310-00-637-9541	468	8

SECTION IV TM9-2320-280-24P C01

CROSS REFERENCE INDEXES

NATIONAL STOCK NUMBER INDEX

STOCK NUMBER	FIGURE NO	ITEM NO	STOCK NUMBER	FIGURE NO	ITEM NO
5310-00-637-9541	472	15	5340-00-689-9514	92	20
5310-00-637-9541	475	23	5340-00-689-9514	95	24
5310-00-637-9541	476	2	5340-00-689-9514	488	7
5310-00-637-9541	481	7	5975-00-697-7769	85	12
5310-00-637-9541	490	9	5975-00-697-7769	86	17
5305-00-638-0714	35	17	4730-00-701-7737	345	35
5305-00-638-0714	43	55	4730-00-701-7737	348	33
5305-00-638-8920	341	9	5340-00-702-2848	41	14
5305-00-638-8920	426	14	5340-00-702-2848	473	4
5305-00-638-8920	427	13	5340-00-702-2848	474	2
4030-00-641-3921	275	11	5340-00-702-2848	475	30
5530-00-641-6069	BULK	47	5360-00-704-4253	191	22
5310-00-641-9464	251	5	6145-00-705-6674	BULK	70
5930-00-655-1514	367	5	5940-00-705-6702	86	35
5310-00-655-9370	438	18	5940-00-705-6703	75	17
5305-00-655-9711	73	12	5940-00-705-6703	76	17
5310-00-655-9860	80	8	5940-00-705-6703	93	10
5310-00-655-9860	85	8	5940-00-705-6708	75	7
5310-00-655-9860	86	4	5940-00-705-6708	76	10
5310-00-656-0114	476	13	5940-00-705-6708	80	6
5310-00-656-0114	476	25	5940-00-705-6708	83	4
5305-00-660-2832	456	15	5940-00-705-6708	85	21
5120-00-677-2259	500	1	5940-00-705-6708	86	14
5342-00-678-1753	160	18	5940-00-705-6708	87	3
5342-00-678-1753	162	18	5940-00-705-6709	76	5
5342-00-678-1753	487A	8	5940-00-705-6709	85	30
5930-00-679-5925	345	15	5940-00-705-6709	86	20
5930-00-679-5925	348	38	5940-00-705-6709	89	6
5930-00-681-4727	396	10	5940-00-705-6711	75	16
5310-00-682-0775	257	6	5940-00-705-6711	76	16
5325-00-682-1762	116	19	5940-00-705-6711	80	4
2590-00-683-0598	68	5	5940-00-705-6711	81	18
5120-00-683-8597	505	5	5940-00-705-6711	83	1
4720-00-684-4033	BULK	18	5940-00-705-6711	86	36
5310-00-685-3228	327	11	5940-00-705-6714	75	2
5935-00-686-2599	85	16	5940-00-705-6715	75	15
5935-00-686-2599	86	9	5940-00-705-6715	76	15
5305-00-688-2111	103	13	5940-00-705-6732	41	7
5305-00-688-2111	144	16	5305-00-709-8283	234	19
5305-00-688-2111	152	16	5305-00-716-7680	180	14
5305-00-688-2111	438	7	5305-00-719-5003	248	29
3110-00-689-4076	128	2	5305-00-719-5017	246	11
5305-00-689-7472	148	26	5305-00-719-5021	339	14
5340-00-689-9514	84	33	5305-00-719-5021	409	12
5340-00-689-9514	85	4	5305-00-719-5021	413	3
5340-00-689-9514	88	2	5305-00-719-5021	415	3
5340-00-689-9514	91	14	5305-00-719-5021	486	11

SECTION IV TM9-2320-280-24P C01

CROSS REFERENCE INDEXES

NATIONAL STOCK NUMBER INDEX

STOCK NUMBER	FIGURE NO	ITEM NO	STOCK NUMBER	FIGURE NO	ITEM NO
5305-00-719-5221	390	2	5305-00-724-7263	196	16
5305-00-719-5235	437	12	5305-00-724-7264	196	12
5305-00-719-5238	397	16	5305-00-724-7264	196	16
5305-00-719-5239	397	8	5305-00-724-7265	196	1
5305-00-719-5239	437	12	5305-00-725-2317	36	16
5305-00-719-5240	191	14	5305-00-725-2317	205	20
5305-00-719-5241	397	17	5305-00-725-2317	245	23
5305-00-719-5241	397	20	5305-00-725-2317	270	9
5325-00-720-8064	245	4	5305-00-725-2317	270	13
5305-00-721-5665	482	14	5305-00-725-2317	270	30
5310-00-721-7809	35	12	5305-00-725-2317	272	34
5310-00-721-7809	398	12	5305-00-725-2317	273	3
5320-00-721-9062	197	22	5305-00-725-2317	274	26
5320-00-721-9062	199	46	5305-00-725-2317	275	22
5320-00-721-9062	212	17	5305-00-725-2317	277	12
5320-00-721-9062	288	9	5305-00-725-2317	286	34
5320-00-721-9062	289	20	5305-00-725-2317	33A	20
2520-00-722-7074	KITS	13	5305-00-725-2317	38A	4
5970-00-724-1907	BULK	40	5305-00-725-2317	463	21
5970-00-724-1910	91	6	5305-00-725-2317	490	1
5970-00-724-1910	BULK	37	5340-00-725-5280	75	38
5305-00-724-5812	333	20	5340-00-725-5280	77	20
5305-00-724-6783	368	9	5340-00-725-5280	177	12
5340-00-724-7038	59	9	5340-00-725-5280	178	25
5340-00-724-7038	83	18	6220-00-726-1916	63	1
5340-00-724-7038	84	3	5305-00-727-2283	437	8
5340-00-724-7038	88	21	5340-00-727-4774	394	1
5340-00-724-7038	91	9	4730-00-728-2393	6	10
5340-00-724-7038	95	44	5930-00-728-4328	55	19
5340-00-724-7038	410	21	5930-00-728-4328	56	16
5340-00-724-7038	414	18	5305-00-728-6274	213	26
5340-00-724-7038	416	21	5305-00-728-6274	214	21
5340-00-724-7038	428	20	6220-00-729-9295	63	2
5305-00-724-7219	140	8	5340-00-731-9611	480	7
5305-00-724-7219	147	8	5305-00-732-0511	103	18
5305-00-724-7220	140	10	5305-00-732-0511	363	21
5305-00-724-7220	147	11	5310-00-732-0558	44	4
5305-00-724-7220	157	21	5310-00-732-0558	62	8
5305-00-724-7220	158	14	5310-00-732-0558	75	42
5305-00-724-7228	194	8	5310-00-732-0558	77	1
5305-00-724-7247	185	3	5310-00-732-0558	89	9
5305-00-724-7248	186	4	5310-00-732-0558	401	11
5305-00-724-7248	457	18	5310-00-732-0558	490	15
5305-00-724-7254	186	3	5360-00-735-1126	128	57
5305-00-724-7254	196	12	5310-00-735-5396	191	26
5305-00-724-7254	470	8	5940-00-735-5520	41	10
5305-00-724-7263	196	1	5940-00-735-5520	44	13

I-26

SECTION IV TM9-2320-280-24P C01

CROSS REFERENCE INDEXES

NATIONAL STOCK NUMBER INDEX

STOCK NUMBER	FIGURE NO	ITEM NO	STOCK NUMBER	FIGURE NO	ITEM NO
5940-00-735-5520	44	19	5340-00-764-7051	396	12
5940-00-735-5520	44	24	5340-00-764-7051	468	11
5940-00-735-5520	69	36	5340-00-764-7052	77	11
5940-00-735-5520	72	16	5315-00-765-2190	279	19
5940-00-735-5520	73	25	6240-00-765-8443	47	2
5940-00-735-5520	474	11	5310-00-768-0318	69	1
5940-00-735-5520	480	26	5310-00-768-0318	363	20
5930-00-736-3539	39	29	5310-00-768-0319	35	19
5930-00-736-3539	59	19	5310-00-768-0319	43	49
5315-00-737-0134	403	9	5310-00-768-0319	62	27
5340-00-738-5182	457	8	5310-00-768-0319	487	11
4820-00-752-9040	29	9	5331-00-770-0242	387	15
5315-00-753-3895	202	22	5310-00-770-0243	387	16
5306-00-753-6996	403	13	5975-00-771-6634	85	17
2530-00-753-7285	169	21	5975-00-771-6634	86	10
5330-00-753-9072	67	5	5310-00-772-0442	36	6
5315-00-754-0848	153	21	5310-00-772-0442	38	6
5320-00-754-0992	198	5	5310-00-772-0442	43B	3
5320-00-754-0992	238	6	5365-00-772-2322	80	9
5325-00-754-1155	428	8	5365-00-772-2322	84	27
5310-00-755-7283	199	5	5365-00-772-2322	85	9
3110-00-756-2022	128	44	5365-00-772-2322	86	5
5305-00-757-8122	75	44	5935-00-772-2344	85	13
5305-00-757-8122	488	12	5935-00-772-2344	86	18
5305-00-757-8287	379	5	5935-00-772-2353	84	26
5310-00-761-0654	15	23	5935-00-772-2353	86	23
5310-00-761-0654	16	27	5935-00-772-3307	85	14
5310-00-761-0654	39	11	5935-00-772-3307	86	19
5310-00-761-0654	205	34	5935-00-773-1428	83	28
5310-00-761-0654	206	23	5935-00-773-1428	88	45
5310-00-761-0654	286	8	5310-00-774-9073	390	4
5310-00-761-3706	283	10	5310-00-774-9073	397	7
5310-00-761-3706	285	10	5310-00-774-9073	397	22
5310-00-761-6882	64	12	5310-00-775-5139	39	37
5310-00-761-6882	252	9	5305-00-777-2284	39	57
5310-00-761-6882	494	10	5305-00-781-3926	1	5
5310-00-761-6882	495	14	5305-00-781-3926	272	23
5305-00-763-7829	225	1	5305-00-781-3926	272	35
5310-00-763-8906	476	11	5305-00-781-3926	272	38
5310-00-763-8919	165	6	5305-00-781-3926	273	1
5310-00-763-8919	179	12	5305-00-781-3926	273	4
5340-00-764-7051	83	19	5305-00-781-3926	274	25
5340-00-764-7051	85	28	5305-00-781-3926	275	3
5340-00-764-7051	88	24	5305-00-781-3926	277	12
5340-00-764-7051	92	21	5305-00-781-3926	286	32
5340-00-764-7051	95	46	5305-00-781-3927	277	31
5340-00-764-7051	335	4	5305-00-781-3927	288	10

I-27

SECTION IV TM9-2320-280-24P C01

CROSS REFERENCE INDEXES

NATIONAL STOCK NUMBER INDEX

STOCK NUMBER	FIGURE NO	ITEM NO	STOCK NUMBER	FIGURE NO	ITEM NO
5305-00-781-3927	33A	19	4730-00-808-5090	312	9
5305-00-781-3928	274	12	5340-00-809-1490	75	23
5305-00-781-3928	274	23	5340-00-809-1490	88	37
5305-00-781-3928	277	31	5340-00-809-1490	400	11
5305-00-781-3928	288	6	5340-00-809-1492	77	4
5305-00-781-3930	170	21	5340-00-809-1492	83	5
5305-00-781-3930	275	14	5340-00-809-1492	84	29
5305-00-782-9489	23	1	5340-00-809-1492	88	13
5305-00-782-9489	23	14	5340-00-809-1492	96	22
5305-00-782-9489	24	13	5340-00-809-1492	125	12
5305-00-782-9489	24	22	5340-00-809-1492	315	10
5305-00-782-9489	202	35	5340-00-809-1492	444	7
5305-00-782-9489	272	38	5340-00-809-1492	478	2
5305-00-782-9489	273	4	5340-00-809-1494	44	17
5305-00-782-9489	274	13	5340-00-809-1494	64	27
5305-00-782-9489	398	22	5340-00-809-1494	75	29
5305-00-782-9489	410	4	5340-00-809-1494	84	11
5305-00-782-9489	412	4	5340-00-809-1494	88	19
5305-00-782-9489	428	3	5340-00-809-1494	338	29
5305-00-782-9489	490	3	5340-00-809-1494	339	25
4910-00-785-6437	505	15	5340-00-809-1494	480	22
5310-00-789-0398	349	15	5340-00-809-1500	75	50
4730-00-789-0951	177	3	5340-00-809-1500	77	22
4730-00-789-0951	178	1	5310-00-809-3078	4	31
5945-00-789-3706	49	8	5310-00-809-3078	71	11
5935-00-790-4614	56	11	5310-00-809-3078	180	20
5320-00-801-1548	348	11	5310-00-809-3078	203	2
2590-00-801-2355	51	8	5310-00-809-3078	206	3
5310-00-802-4701	482	10	5310-00-809-3078	233	2
4730-00-803-6266	334	11	5310-00-809-3078	243	16
5325-00-803-7305	99	25	5310-00-809-3078	247	6
5325-00-803-7305	133	7	5310-00-809-3078	248	19
5935-00-806-4183	93	11	5310-00-809-3078	266	15
5310-00-807-1466	273	14	5310-00-809-3078	375	21
5310-00-807-1466	275	13	5310-00-809-3078	430	35
5310-00-807-1466	276	9	5310-00-809-3078	446	13
5310-00-807-1475	279	14	5310-00-809-3079	455	5
5310-00-807-1476	279	18	5310-00-809-4058	12	1
5310-00-807-1477	279	26	5310-00-809-4058	16	22
5935-00-807-4109	62	22	5310-00-809-4058	19	4
5935-00-807-4109	366	31	5310-00-809-4058	21	18
5935-00-807-4109	386	9	5310-00-809-4058	42	7
5935-00-807-4109	469	10	5310-00-809-4058	50	14
5935-00-807-4109	483	1	5310-00-809-4058	51	2
4240-00-807-6856	311	7	5310-00-809-4058	62	28
4240-00-807-6856	312	3	5310-00-809-4058	64	14
4730-00-808-5090	311	13	5310-00-809-4058	64	21

SECTION IV TM9-2320-280-24P C01

CROSS REFERENCE INDEXES

NATIONAL STOCK NUMBER INDEX

STOCK NUMBER	FIGURE NO	ITEM NO	STOCK NUMBER	FIGURE NO	ITEM NO
5310-00-809-4058	69	4	5310-00-809-4058	326	11
5310-00-809-4058	70	11	5310-00-809-4058	327	44
5310-00-809-4058	80	16	5310-00-809-4058	328	14
5310-00-809-4058	84	19	5310-00-809-4058	331	26
5310-00-809-4058	84	21	5310-00-809-4058	343	16
5310-00-809-4058	141	15	5310-00-809-4058	365	16
5310-00-809-4058	155	5	5310-00-809-4058	365	25
5310-00-809-4058	169	11	5310-00-809-4058	366	16
5310-00-809-4058	177	11	5310-00-809-4058	368	22
5310-00-809-4058	178	6	5310-00-809-4058	375	29
5310-00-809-4058	197	29	5310-00-809-4058	379	14
5310-00-809-4058	198	8	5310-00-809-4058	382	5
5310-00-809-4058	198	14	5310-00-809-4058	384	8
5310-00-809-4058	203	12	5310-00-809-4058	388	9
5310-00-809-4058	204	12	5310-00-809-4058	389	9
5310-00-809-4058	206	25	5310-00-809-4058	397	9
5310-00-809-4058	207	7	5310-00-809-4058	402	2
5310-00-809-4058	211	27	5310-00-809-4058	429	12
5310-00-809-4058	212	26	5310-00-809-4058	430	15
5310-00-809-4058	214	5	5310-00-809-4058	475	8
5310-00-809-4058	225	42	5310-00-809-4058	481	8
5310-00-809-4058	227	30	5310-00-809-4058	492	8
5310-00-809-4058	238	19	5310-00-809-4058	497	4
5310-00-809-4058	241	13	5310-00-809-4061	1	3
5310-00-809-4058	243	7	5310-00-809-4061	8	1
5310-00-809-4058	246	23	5310-00-809-4061	37	11
5310-00-809-4058	248	26	5310-00-809-4061	170	22
5310-00-809-4058	249	1	5310-00-809-4085	35	23
5310-00-809-4058	250	13	5310-00-809-4085	36	2
5310-00-809-4058	250	18	5310-00-809-4085	38	2
5310-00-809-4058	267	2	5310-00-809-4085	136	4
5310-00-809-4058	267	4	5310-00-809-4085	144	15
5310-00-809-4058	269	31	5310-00-809-4085	152	15
5310-00-809-4058	272	3	5310-00-809-4085	169	15
5310-00-809-4058	280	49	5310-00-809-4085	171	2
5310-00-809-4058	281	49	5310-00-809-4085	202	33
5310-00-809-4058	287	24	5310-00-809-4085	234	9
5310-00-809-4058	288	21	5310-00-809-4085	239	6
5310-00-809-4058	288	29	5310-00-809-4085	271	10
5310-00-809-4058	292	10	5310-00-809-4085	271	25
5310-00-809-4058	293	30	5310-00-809-4085	313	22
5310-00-809-4058	294	5	5310-00-809-4085	398	4
5310-00-809-4058	304	8	5310-00-809-4085	402	9
5310-00-809-4058	312	2	5310-00-809-4085	438	19
5310-00-809-4058	321	12	5310-00-809-4085	456	16
5310-00-809-4058	323	8	5310-00-809-4085	472	11
5310-00-809-4058	325	14	5310-00-809-4085	475	11

SECTION IV TM9-2320-280-24P C01

CROSS REFERENCE INDEXES

NATIONAL STOCK NUMBER INDEX

STOCK NUMBER	FIGURE NO	ITEM NO	STOCK NUMBER	FIGURE NO	ITEM NO
5310-00-809-5997	209	13	5310-00-814-0673	212	16
5310-00-809-5997	292	19	5310-00-814-0673	226	12
5310-00-809-5997	429	37	5310-00-814-0673	230	16
5310-00-809-5998	33	2	5310-00-814-0673	242	5
5310-00-809-5998	142	17	5310-00-814-0673	250	23
5310-00-809-5998	143	17	5310-00-814-0673	266	1
5310-00-809-5998	150	17	5310-00-814-0673	268	10
5310-00-809-5998	151	17	5310-00-814-0673	268	29
5310-00-809-5998	165	9	5310-00-814-0673	269	16
5310-00-809-5998	166	6	5310-00-814-0673	284	12
5310-00-809-5998	293	9	5310-00-814-0673	288	20
5310-00-809-5998	295	7	5310-00-814-0673	292	15
5310-00-809-5998	295	13	5310-00-814-0673	328	6
5310-00-809-5998	363	17	5310-00-814-0673	328	15
5310-00-809-5998	390	3	5310-00-814-0673	329	9
5310-00-809-5998	397	15	5310-00-814-0673	333	21
5310-00-809-5998	455	3	5310-00-814-0673	338	27
5310-00-809-8533	178	22	5310-00-814-0673	339	22
5310-00-809-8533	385	10	5310-00-814-0673	364	7
5310-00-809-8546	46	10	5310-00-814-0673	375	31
5310-00-809-8546	446	8	5310-00-814-0673	390	5
5310-00-809-8546	447	9	5310-00-814-0673	401	9
5310-00-809-8546	449	3	5310-00-814-0673	409	4
5310-00-811-0966	291	4	5310-00-814-0673	411	4
5310-00-811-0966	297	4	5310-00-814-0673	413	6
5310-00-811-3494	252	4	5310-00-814-0673	414	11
5310-00-811-3494	379	3	5310-00-814-0673	415	14
5305-00-813-2785	375	10	5310-00-814-0673	416	14
5305-00-813-2785	375	25	5310-00-814-0673	417	11
5310-00-814-0672	144	4	5310-00-814-0673	418	8
5310-00-814-0672	152	4	5310-00-814-0673	419	9
5310-00-814-0672	272	24	5310-00-814-0673	420	4
5310-00-814-0673	74	8	5310-00-814-0673	423	4
5310-00-814-0673	83	25	5310-00-814-0673	428	14
5310-00-814-0673	144	4	5310-00-814-0673	429	18
5310-00-814-0673	152	4	5310-00-814-0673	440	11
5310-00-814-0673	153	25	5310-00-814-0673	445	2
5310-00-814-0673	154	6	5310-00-814-0673	463	12
5310-00-814-0673	169	33	5310-00-814-0673	468	7
5310-00-814-0673	189	6	5310-00-814-0673	480	4
5310-00-814-0673	190	1	5315-00-814-3529	230	11
5310-00-814-0673	205	1	5315-00-815-8840	153	4
5310-00-814-0673	206	10	5315-00-815-8840	155	17
5310-00-814-0673	206	13	5315-00-816-1794	165	7
5310-00-814-0673	208	29	5315-00-816-1794	166	7
5310-00-814-0673	209	8	4730-00-816-2787	327	23
5310-00-814-0673	211	19	5310-00-817-4623	261	14

SECTION IV TM9-2320-280-24P C01

CROSS REFERENCE INDEXES

NATIONAL STOCK NUMBER INDEX

STOCK NUMBER	FIGURE NO	ITEM NO	STOCK NUMBER	FIGURE NO	ITEM NO
5310-00-820-6653	193	6	5340-00-833-8476	410	18
5305-00-821-3869	23	23	5340-00-833-8476	412	15
5305-00-821-3869	24	27	5935-00-833-8561	86	25
5305-00-821-3869	251	26	5935-00-833-8561	366	33
5305-00-821-3869	251	44	5935-00-833-8561	386	12
5305-00-821-3869	272	27	5935-00-833-8561	472	7
5305-00-821-3869	285	4	5935-00-833-8561	475	2
5305-00-821-3869	414	2	5935-00-833-8561	479	5
5305-00-821-3869	462	11	5935-00-833-8561	483	8
5305-00-821-3869	470	15	5970-00-833-8562	75	13
5305-00-821-3869	471	6	5970-00-833-8562	76	11
5305-00-821-3869	486	8	5310-00-834-7606	438	13
5325-00-823-5999	198	2	5310-00-834-8732	44	6
5325-00-823-5999	199	11	5310-00-838-2024	283	15
5325-00-823-5999	225	23	5315-00-839-2325	21	19
5325-00-823-5999	239	9	5315-00-839-2325	21	29
5325-00-823-5999	317	7	5315-00-839-2325	238	18
5325-00-823-5999	318	11	5315-00-839-2325	250	10
5325-00-823-5999	321	8	5315-00-839-2326	274	5
5325-00-823-5999	325	19	5315-00-839-2326	276	31
5325-00-823-5999	373	6	5315-00-839-5820	202	18
5325-00-823-5999	376	7	5315-00-839-5820	243	9
5325-00-823-5999	376	11	5315-00-839-5820	293	10
5310-00-823-8803	193	5	5315-00-839-5821	134	1
5310-00-823-8803	199	41	5315-00-839-5822	443	17
5310-00-823-8803	437	4	2540-00-840-0022	328	12
5310-00-823-8804	5	6	5310-00-840-6222	169	14
5310-00-823-8804	15	21	5315-00-840-9653	243	11
5310-00-823-8804	18	3	5305-00-841-2044	72	14
5310-00-823-8804	21	2	5315-00-842-3044	97	6
5310-00-823-8804	21	28	5315-00-842-3044	99	18
5310-00-823-8804	253	2	5315-00-842-3044	101	1
5310-00-823-8804	494	4	5315-00-842-3044	133	9
5310-00-823-8804	495	12	5315-00-842-3044	153	13
5310-00-823-8804	496	7	5315-00-842-3044	155	19
5310-00-823-8804	66A	4	5315-00-842-3044	202	39
5305-00-824-2004	382	13	5315-00-842-3044	205	21
5970-00-827-6566	44	14	5315-00-842-3044	243	23
5970-00-827-6566	313	14	5315-00-842-3044	244	9
5315-00-828-5487	4	13	5315-00-842-3044	390	14
5315-00-829-1480	368	8	5340-00-843-7825	65	20
5330-00-830-1745	9	4	5340-00-843-7825	88	30
5325-00-832-5650	62	16	5340-00-843-7825	89	14
5305-00-832-5743	62	15	5340-00-843-7825	95	28
5310-00-832-6852	46	14	5340-00-843-7825	96	20
5340-00-833-8476	66	5	5340-00-843-7825	327	28
5340-00-833-8476	79	21	5330-00-843-9235	172	12

I-31

SECTION IV TM9-2320-280-24P C01

CROSS REFERENCE INDEXES

NATIONAL STOCK NUMBER INDEX

STOCK NUMBER	FIGURE NO	ITEM NO	STOCK NUMBER	FIGURE NO	ITEM NO
5306-00-844-0036	438	5	5305-00-857-6886	274	10
5306-00-844-0036	444	5	5305-00-857-6886	276	12
5305-00-844-1507	445	13	5305-00-857-6886	277	1
5315-00-846-0126	191	10	5320-00-864-6203	199	48
5315-00-846-0126	442	13	5320-00-865-8994	498	15
5315-00-846-0126	453	15	4730-00-871-6729	10	8
5935-00-846-3884	83	35	5310-00-877-5796	328	13
5935-00-846-3884	86	37	5310-00-877-5797	368	36
5305-00-846-5703	268	17	5310-00-877-5797	379	49
5305-00-846-5703	272	27	5310-00-877-5797	388	3
5305-00-846-5703	274	12	5310-00-877-5797	393	15
5305-00-846-5703	277	48	5310-00-877-5797	396	11
2920-00-848-3292	380	15	5310-00-877-5975	398	9
5330-00-848-4439	176	11	4730-00-877-6298	178	9
2530-00-848-4581	167	12	4730-00-877-6298	436	5
5310-00-849-6882	191	11	4730-00-877-6298	477	4
5320-00-850-3282	368	44	6220-00-880-1624	60	4
5310-00-850-6868	188	34	5120-00-880-4268	503	4
5310-00-850-6993	186	9	5310-00-880-7744	71	2
5310-00-850-6993	188	18	5310-00-880-7744	209	5
5310-00-850-7004	144	10	5310-00-880-7744	226	21
5310-00-850-7004	144	18	5310-00-880-7744	230	5
5310-00-850-7004	152	10	5310-00-880-7744	235	8
5310-00-850-7004	152	18	5310-00-880-7744	236	6
5310-00-850-7004	170	3	5310-00-880-7744	295	10
5310-00-850-7004	170	9	5310-00-880-7746	137	1
5310-00-850-7004	170	13	5310-00-880-9344	216	5
5310-00-850-7004	183	5	5310-00-880-9344	221	5
5310-00-850-7004	185	14	5340-00-881-5303	11	7
5310-00-850-7004	187	18	5310-00-883-2237	45	1
5310-00-850-7004	188	14	5325-00-886-1067	247	8
5310-00-850-7004	437	10	5310-00-889-2528	327	13
5310-00-852-8593	498	10	5310-00-889-2543	55	10
3110-00-854-1504	148	6	5305-00-889-3000	43	34
5340-00-854-6729	93	17	5305-00-889-3001	43	39
5340-00-854-6729	96	23	5305-00-889-3001	57	1
5340-00-854-6729	412	21	5305-00-889-3002	45	51
5340-00-854-6729	425	17	5310-00-889-8782	140	2
5340-00-854-6729	426	23	5310-00-889-8782	147	2
5340-00-854-6729	427	23	5315-00-890-3461	390	11
5340-00-854-6730	416	16	3110-00-892-0896	148	19
5340-00-854-6730	428	16	9905-00-893-3570	41	4
5305-00-855-0957	225	9	9905-00-893-3570	41	9
5305-00-855-0957	277	40	9905-00-893-3570	41	18
5305-00-855-0957	294	2	9905-00-893-3570	44	15
5310-00-855-1102	52	16	9905-00-893-3570	44	21
5930-00-856-1654	482	2	9905-00-893-3570	44	26

I-32

SECTION IV TM9-2320-280-24P C01

CROSS REFERENCE INDEXES

NATIONAL STOCK NUMBER INDEX

STOCK NUMBER	FIGURE NO	ITEM NO	STOCK NUMBER	FIGURE NO	ITEM NO
9905-00-893-3570	69	13	5975-00-903-2284	98	6
9905-00-893-3570	69	22	5975-00-903-2284	99	3
9905-00-893-3570	69	27	5975-00-903-2288	98	6
9905-00-893-3570	69	38	5310-00-905-0762	213	11
9905-00-893-3570	72	6	5310-00-905-0762	240	23
9905-00-893-3570	72	18	5310-00-905-0762	311	14
9905-00-893-3570	73	20	5310-00-905-0762	313	1
9905-00-893-3570	73	24	5310-00-905-0762	315	12
9905-00-893-3570	81	6	5310-00-905-0762	327	40
9905-00-893-3570	81	12	5310-00-905-0762	425	9
9905-00-893-3570	94	3	5310-00-905-0762	426	6
9905-00-893-3570	474	13	5310-00-905-0762	427	10
9905-00-893-3570	480	19	5310-00-905-0762	443	1
9905-00-893-3570	480	28	5310-00-905-0762	444	9
9905-00-893-3570	480	33	5310-00-905-0762	457	5
5925-00-894-8015	50	10	5340-00-905-0790	485	6
5310-00-896-0903	170	23	5310-00-905-4600	59	3
5310-00-896-0903	194	7	5310-00-905-4600	136	14
5310-00-896-0903	195	9	4730-00-908-3193	301	4
5310-00-896-0903	202	30	4730-00-908-3193	303	6
5310-00-896-0903	425	10	4730-00-908-3193	304	2
5310-00-896-0903	426	12	4730-00-908-3193	364	3
5310-00-896-0903	427	11	4730-00-908-3193	365	10
4730-00-897-5497	178	20	4730-00-908-3193	368	14
4730-00-897-5497	334	12	4730-00-908-3193	379	39
4730-00-897-5497	366	10	4730-00-908-3193	496	15
4730-00-897-5497	385	11	4730-00-908-3194	268	32
5320-00-899-0981	213	29	4730-00-908-3194	310	7
5320-00-899-0981	214	19	4730-00-908-3194	379	9
5320-00-899-0981	270	16	4730-00-908-3194	176A	22
5320-00-899-0981	315	3	4730-00-908-6292	10	14
5320-00-899-0981	337	12	4730-00-908-6292	24	2
5975-00-899-4606	282	10	4730-00-908-6292	331	20
3030-00-899-4888	490	11	4730-00-908-6292	368	3
5305-00-900-2546	382	1	4730-00-908-6293	387	2
3110-00-900-2560	118	2	4730-00-908-6293	393	16
3110-00-900-2560	120	11	4730-00-908-6294	27	1
4730-00-900-3296	178	23	4730-00-908-6294	337	26
5935-00-900-6281	366	32	4730-00-908-6294	343	3
5935-00-900-6281	386	11	4730-00-908-6294	368	20
5935-00-900-6281	469	11	4730-00-908-6294	387	5
5935-00-900-6281	483	3	2920-00-909-2483	35	6
5305-00-901-2099	394	5	4730-00-909-8627	17	16
5305-00-901-2134	385	13	4730-00-909-8627	29	8
9905-00-901-2942	357	7	4730-00-909-8627	379	34
5305-00-901-7250	390	7	5306-00-911-5005	433	1
5305-00-902-6643	383	2	4730-00-911-5707	6	1

I-33

SECTION IV TM9-2320-280-24P C01

CROSS REFERENCE INDEXES

NATIONAL STOCK NUMBER INDEX

STOCK NUMBER	FIGURE NO	ITEM NO	STOCK NUMBER	FIGURE NO	ITEM NO
4730-00-911-5707	177	3	5310-00-933-4310	435	3
5310-00-913-5474	70	8	5310-00-933-8118	282	15
5310-00-913-5474	481	5	5310-00-933-8119	482	18
5305-00-915-8087	2	14	5310-00-934-9747	333	28
5305-00-915-8087	156	22	5310-00-934-9748	282	16
4030-00-916-2141	191	20	5310-00-934-9751	45	49
5310-00-922-2017	64	15	5310-00-934-9751	327	34
5310-00-922-2017	64	25	5310-00-934-9751	333	18
5310-00-922-2017	67	3	5310-00-934-9751	482	22
5310-00-922-2017	88	42	5310-00-934-9754	380	1
5310-00-922-2017	225	49	5310-00-934-9757	48	1
5310-00-922-2017	226	29	5310-00-934-9758	64	16
5310-00-922-2017	370	17	5310-00-934-9758	64	26
5310-00-922-2017	465	3	5310-00-934-9758	74	9
5310-00-922-2017	487	3	5310-00-934-9758	88	43
5340-00-922-6300	58	11	5310-00-934-9758	208	28
5340-00-922-6300	59	11	5310-00-934-9758	213	14
5340-00-922-6300	84	2	5310-00-934-9758	214	10
5340-00-922-6300	88	22	5310-00-934-9758	240	1
5340-00-922-6300	93	22	5310-00-934-9758	240	13
5340-00-922-6300	95	31	5310-00-934-9758	240	20
5340-00-922-6300	96	28	5310-00-934-9758	241	7
5340-00-922-6300	306	9	5310-00-934-9758	269	2
5340-00-922-6300	310	4	5310-00-934-9758	283	30
5340-00-922-6300	338	28	5310-00-934-9758	327	30
5340-00-922-6300	379	2	5310-00-934-9758	335	14
5340-00-922-6300	410	20	5310-00-934-9758	336	11
5340-00-922-6300	414	19	5310-00-934-9758	442	8
5340-00-922-6300	416	19	5310-00-934-9764	63	5
5340-00-922-6300	428	18	5310-00-934-9764	85	6
5340-00-926-5449	313	20	5310-00-934-9764	169	19
5325-00-928-6214	65	7	5310-00-934-9764	246	33
1640-00-929-0041	BULK	50	5310-00-934-9764	293	15
5340-00-929-1794	477	7	5310-00-935-9021	2	15
5310-00-930-8214	167	6	5310-00-935-9021	23	4
5310-00-930-8214	245	27	5310-00-935-9021	24	10
2910-00-930-9367	366	25	5310-00-935-9021	26	6
2910-00-930-9367	384	6	5310-00-935-9021	52	23
5310-00-931-8167	203	21	5310-00-935-9021	103	17
5310-00-931-8167	204	9	5310-00-935-9021	144	14
5310-00-931-8167	473	6	5310-00-935-9021	152	14
5310-00-931-8167	474	9	5310-00-935-9021	159	4
2510-00-933-2893	266	7	5310-00-935-9021	160	9
2510-00-933-2895	266	7	5310-00-935-9021	164	4
5310-00-933-4310	282	7	5310-00-935-9021	169	1
5310-00-933-4310	403	3	5310-00-935-9021	187	10
5310-00-933-4310	433	2	5310-00-935-9021	189	2

I-34

SECTION IV TM9-2320-280-24P C01

CROSS REFERENCE INDEXES

NATIONAL STOCK NUMBER INDEX

STOCK NUMBER	FIGURE NO	ITEM NO	STOCK NUMBER	FIGURE NO	ITEM NO
5310-00-935-9021	190	3	5310-00-935-9022	368	18
5310-00-935-9021	205	9	5310-00-935-9022	379	22
5310-00-935-9021	210	14	5310-00-935-9022	380	14
5310-00-935-9021	245	30	5310-00-935-9022	388	10
5310-00-935-9021	251	45	5310-00-935-9022	393	12
5310-00-935-9021	266	24	5310-00-935-9022	411	19
5310-00-935-9021	268	19	5310-00-935-9022	484	12
5310-00-935-9021	270	19	5310-00-935-9022	485	17
5310-00-935-9021	270	21	5331-00-935-9136	3	3
5310-00-935-9021	272	29	5331-00-935-9136	400	2
5310-00-935-9021	273	22	5331-00-935-9136	479	3
5310-00-935-9021	274	17	2010-00-937-5599	122	26
5310-00-935-9021	275	19	5310-00-941-6019	250	14
5310-00-935-9021	276	19	5310-00-941-6019	378	10
5310-00-935-9021	277	4	5310-00-941-6019	378	13
5310-00-935-9021	278	12	5310-00-941-6019	389	1
5310-00-935-9021	283	8	5305-00-942-2196	456	22
5310-00-935-9021	285	9	9905-00-945-7637	86	2
5310-00-935-9021	286	28	9905-00-945-7637	94	4
5310-00-935-9021	288	13	9905-00-945-7637	480	14
5310-00-935-9021	292	16	5305-00-947-4355	183	17
5310-00-935-9021	293	6	5305-00-947-4358	181	10
5310-00-935-9021	369	10	5305-00-947-4358	182	6
5310-00-935-9021	372	6	5305-00-947-4358	188	39
5310-00-935-9021	398	25	5305-00-947-4359	183	8
5310-00-935-9021	409	8	5305-00-947-4360	183	16
5310-00-935-9021	410	9	5305-00-947-4360	184	13
5310-00-935-9021	411	14	5305-00-947-4361	144	1
5310-00-935-9021	412	7	5305-00-947-4361	152	1
5310-00-935-9021	414	9	5305-00-947-4361	195	2
5310-00-935-9021	415	13	5305-00-947-4362	144	22
5310-00-935-9021	416	10	5305-00-947-4363	471	4
5310-00-935-9021	425	16	4030-00-948-7315	293	25
5310-00-935-9021	426	20	4730-00-949-8694	335	5
5310-00-935-9021	427	20	4730-00-949-8694	336	3
5310-00-935-9021	428	10	5310-00-950-1310	274	21
5310-00-935-9021	431	7	5310-00-952-3567	374	12
5310-00-935-9021	440	17	4730-00-954-1251	18	5
5310-00-935-9021	441	11	4730-00-954-1251	141	10
5310-00-935-9021	456	18	4730-00-954-1251	399	16
5310-00-935-9021	463	3	4730-00-954-1281	5	1
5310-00-935-9021	464	1	4730-00-954-1281	436	4
5310-00-935-9021	470	13	5305-00-954-3937	243	29
5310-00-935-9021	471	5	5340-00-954-6014	18	11
5310-00-935-9021	484	13	5340-00-954-6014	89	8
5310-00-935-9021	490	10	5340-00-954-6014	124	16
5310-00-935-9022	64	29	5340-00-954-6014	125	16

I-35

SECTION IV TM9-2320-280-24P C01

CROSS REFERENCE INDEXES

NATIONAL STOCK NUMBER INDEX

STOCK NUMBER	FIGURE NO	ITEM NO	STOCK NUMBER	FIGURE NO	ITEM NO
5340-00-954-6014	159	12	5310-00-984-3807	324	9
5340-00-954-6014	160	3	5310-00-984-3807	325	25
5340-00-954-6014	161	12	5310-00-984-3807	325	26
5320-00-957-2500	357	2	5305-00-984-4976	55	3
5320-00-957-2507	282	12	5305-00-984-4983	55	2
5305-00-957-2635	447	6	5305-00-984-4988	213	19
5310-00-957-2677	296	2	5305-00-984-4988	367	16
5305-00-958-4357	72	13	5305-00-984-4992	79	13
5305-00-958-5246	247	3	5305-00-984-5675	277	34
5305-00-958-5246	271	15	5305-00-984-5678	233	1
5305-00-958-5246	338	19	5305-00-984-6191	47	7
5305-00-958-5246	411	16	5305-00-984-6191	354	11
5305-00-958-5253	225	46	5305-00-984-6192	72	21
5305-00-958-5253	226	25	5305-00-984-6192	73	9
5305-00-958-5253	226	31	5305-00-984-6193	84	23
5305-00-958-5254	226	25	5305-00-984-6193	280	36
5305-00-958-5254	226	31	5305-00-984-6193	281	36
5305-00-958-5262	383	5	5305-00-984-6193	300	12
5305-00-958-5469	225	46	5305-00-984-6193	453	5
5305-00-958-5469	226	31	5305-00-984-6194	52	21
5305-00-958-5471	249	17	5305-00-984-6194	59	24
5340-00-958-8457	282	17	5305-00-984-6194	273	11
5340-00-958-8457	303	3	5305-00-984-6194	379	1
5340-00-958-8457	338	4	5305-00-984-6195	48	4
5340-00-958-8457	416	22	5305-00-984-6195	275	7
5340-00-958-8457	428	21	5305-00-984-6195	276	5
5305-00-959-2704	382	8	5305-00-984-6195	282	14
5340-00-960-9340	172	22	5305-00-984-6195	337	6
2530-00-960-9363	172	13	5305-00-984-6195	343	10
5340-00-964-5267	75	45	5305-00-984-6195	378	9
5340-00-964-5267	77	19	5305-00-984-6195	378	25
5340-00-964-5267	487D	4	5305-00-984-6195	380	3
5310-00-964-7092	215	4	5305-00-984-6196	280	10
6240-00-966-3831	62	17	5305-00-984-6196	281	10
5315-00-973-8637	245	11	5305-00-984-6197	50	9
9905-00-977-2727	330	4	5305-00-984-6199	213	33
5930-00-978-8805	55	17	5305-00-984-6199	214	26
5930-00-978-8805	56	12	5305-00-984-6210	213	20
5310-00-982-4908	128	55	5305-00-984-6210	214	17
5310-00-982-4908	129	22	5305-00-984-6210	299	1
5310-00-982-4939	313	23	5305-00-984-6210	430	7
5940-00-983-4067	79	7	5305-00-984-6211	63	7
5940-00-983-4067	485	1	5305-00-984-6212	35	13
5940-00-983-6105	52	17	5305-00-984-6212	88	41
5315-00-983-7402	268	8	5305-00-984-6212	202	3
5331-00-984-3750	314	35	5305-00-984-6212	284	13
5310-00-984-3807	199	9	5305-00-984-6212	293	11

SECTION IV TM9-2320-280-24P C01

CROSS REFERENCE INDEXES

NATIONAL STOCK NUMBER INDEX

STOCK NUMBER	FIGURE NO	ITEM NO	STOCK NUMBER	FIGURE NO	ITEM NO
5305-00-984-6212	439	6	5975-00-985-6630	336	12
5305-00-984-6213	203	32	5975-00-985-6630	338	18
5305-00-984-6213	222	12	5975-00-985-6630	436	3
5305-00-984-6213	347	2	5975-00-985-6630	457	15
5305-00-984-6214	272	19	5975-00-985-6630	483	2
5305-00-984-6214	281	16	5975-00-985-6630	484	1
5975-00-984-6582	64	28	5340-00-988-1162	80	14
5975-00-984-6582	81	11	5305-00-988-1170	267	17
5975-00-984-6582	160	2	5305-00-988-1171	237	12
5975-00-984-6582	469	2	5305-00-988-1724	97	2
5975-00-984-6582	470	7	5305-00-988-1724	237	7
5975-00-984-6582	487B	2	5305-00-988-1724	330	3
5310-00-984-7042	73	3	5305-00-988-1725	49	7
5310-00-984-7042	482	9	5305-00-988-1725	51	1
5305-00-984-7341	377	5	5305-00-988-1725	59	25
5305-00-984-7341	447	7	5305-00-988-1725	75	37
5305-00-984-7341	448	10	5305-00-988-1725	83	29
5305-00-984-7341	449	7	5305-00-988-1725	84	18
5305-00-984-7341	450	4	5305-00-988-1725	88	31
5305-00-984-7342	251	17	5305-00-988-1725	88	36
5305-00-984-7342	251	42	5305-00-988-1725	213	6
5305-00-984-7342	252	20	5305-00-988-1725	214	2
5305-00-984-7342	446	19	5305-00-988-1725	267	1
5305-00-984-7342	448	1	5305-00-988-1725	280	14
5305-00-984-7343	323	14	5305-00-988-1725	281	14
5305-00-984-7343	429	15	5305-00-988-1725	313	5
5305-00-984-7343	449	1	5305-00-988-1725	316	15
5305-00-984-7352	446	6	5305-00-988-1725	320	14
5305-00-984-7363	412	23	5305-00-988-1725	320	26
5340-00-984-8540	19	13	5305-00-988-1725	321	28
5340-00-984-8540	88	11	5305-00-988-1725	378	20
5340-00-984-8540	89	10	5305-00-988-1727	198	18
5340-00-984-8540	96	26	5305-00-988-1727	213	8
5340-00-984-8540	338	15	5305-00-988-1727	254	21
5340-00-984-8540	339	24	5305-00-988-1727	376	17
5340-00-984-8540	414	23	5340-00-988-3186	365	21
5340-00-984-8541	44	11	5305-00-988-7614	315	9
4730-00-985-4804	505	16	5305-00-988-9106	248	10
5975-00-985-6630	75	21	5315-00-989-0534	244	11
5975-00-985-6630	83	38	5305-00-989-0830	292	8
5975-00-985-6630	85	31	5305-00-989-0830	293	19
5975-00-985-6630	89	12	5305-00-989-0830	440	12
5975-00-985-6630	94	10	5340-00-989-1771	77	3
5975-00-985-6630	95	27	5340-00-989-1771	141	12
5975-00-985-6630	96	32	5340-00-989-1771	338	13
5975-00-985-6630	178	21	5340-00-989-1771	484	2
5975-00-985-6630	335	15	5340-00-989-1771	487B	5

SECTION IV TM9-2320-280-24P C01

CROSS REFERENCE INDEXES

NATIONAL STOCK NUMBER INDEX

STOCK NUMBER	FIGURE NO	ITEM NO	STOCK NUMBER	FIGURE NO	ITEM NO
5340-00-989-1771	487D	2	5305-00-992-6143	178	16
5305-00-989-6265	39	32	5305-00-992-6143	436	11
5305-00-989-6265	393	11	5305-00-993-0190	43	37
5305-00-989-7434	45	43	5305-00-993-0190	297	7
5305-00-989-7434	52	29	5305-00-993-1848	52	3
5305-00-989-7434	54	6	5305-00-993-1848	54	11
5305-00-989-7434	55	20	5305-00-993-1848	55	22
5305-00-989-7434	56	7	5305-00-993-1848	59	18
5305-00-989-7434	59	23	5305-00-993-1848	79	24
5305-00-989-7434	282	2	5305-00-993-1848	96	19
5305-00-989-7434	317	5	5305-00-993-1848	246	2
5305-00-989-7434	338	10	5305-00-993-1848	251	31
5305-00-989-7434	396	3	5305-00-993-1848	252	34
5305-00-989-7434	398	15	5305-00-993-1848	255	33
5305-00-989-7434	416	28	5305-00-993-1848	325	9
5305-00-989-7435	52	28	5305-00-993-1848	446	20
5305-00-989-7435	95	25	5305-00-993-1848	448	4
5305-00-989-7435	207	10	5305-00-993-1851	65	23
5305-00-989-7435	255	6	5305-00-993-1851	260	6
5305-00-989-7435	258	5	5305-00-993-1851	289	27
5305-00-989-7435	260	8	5305-00-993-2457	376	20
5305-00-989-7435	337	19	5305-00-993-2457	382	4
5305-00-989-7435	338	9	5305-00-993-2457	389	7
5305-00-989-7435	379	53	5305-00-993-2460	391	1
5305-00-989-7435	447	3	5305-00-993-2460	397	1
5305-00-989-7435	449	6	5305-00-993-2738	397	2
5340-00-989-9222	83	9	6620-00-993-5546	66	10
5340-00-989-9222	84	9	5340-00-993-6207	77	14
5340-00-989-9222	88	20	5340-00-993-6207	89	15
5340-00-989-9222	95	33	5340-00-993-6207	334	15
5340-00-989-9222	282	5	5340-00-993-6207	366	20
5340-00-989-9222	385	2	5340-00-993-6207	368	42
5340-00-989-9222	414	15	5340-00-993-6207	379	46
5340-00-989-9222	445	5	5340-00-993-6207	386	6
5310-00-990-1361	487	5	5305-00-995-3440	46	18
5310-00-990-1361	487	6	5305-00-995-3442	446	3
5305-00-990-6444	43	36	5305-00-995-3444	58	7
5305-00-990-6444	45	30	5305-00-995-3444	63	7
5305-00-990-6444	45	47	5305-00-995-3444	247	10
5305-00-990-6444	52	27	5305-00-995-3444	256	2
5305-00-990-6444	396	6	5305-00-995-3444	257	3
5305-00-990-6444	485	26	5305-00-995-3444	260	14
5340-00-990-7610	332	24	5305-00-995-3444	262	9
5305-00-990-8632	273	1	5305-00-995-3444	263	9
5305-00-990-8632	274	25	5305-00-995-3444	291	2
5305-00-990-8632	276	22	5305-00-995-3444	297	6
5305-00-990-8632	277	31	5310-00-995-4130	72	8

SECTION IV TM9-2320-280-24P C01

CROSS REFERENCE INDEXES

NATIONAL STOCK NUMBER INDEX

STOCK NUMBER	FIGURE NO	ITEM NO	STOCK NUMBER	FIGURE NO	ITEM NO
5305-00-995-6311	368	27	9905-01-013-8723	96	3
5310-00-997-1888	88	39	4720-01-014-4915	BULK	62
5310-00-997-1888	324	28	5305-01-016-4344	453	18
5310-00-997-1888	375	19	5305-01-016-5469	71	12
5310-00-997-1888	375	27	5310-01-016-9348	283	22
5310-00-997-1888	378	27	5340-01-017-4630	13	3
6220-00-998-6142	62	19	5340-01-017-4630	14	3
4730-01-003-5105	123	4	9905-01-017-4748	348	10
4730-01-003-5105	141	23	5310-01-018-5332	324	16
4730-01-003-5105	149	12	5320-01-019-5694	355	3
4730-01-003-5105	313	8	5320-01-019-5694	370	21
4730-01-003-5105	399	6	5320-01-019-5694	371	2
4730-01-003-5105	443	10	5310-01-021-9027	167	10
4730-01-003-5105	457	2	5320-01-023-2529	28	4
4720-01-003-6706	BULK	20	5320-01-023-2529	65	18
4730-01-004-8346	141	20	5320-01-023-2529	198	1
4820-01-005-2994	363	3	5320-01-023-2529	199	10
5305-01-005-6715	243	27	5320-01-023-2529	225	22
5305-01-006-2052	252	37	5320-01-023-2529	229	2
5305-01-006-2053	208	25	5320-01-023-2529	316	8
5305-01-006-2053	279	29	5320-01-023-2529	318	10
5305-01-006-2053	429	11	5320-01-023-2529	320	6
5305-01-006-5736	74	5	5320-01-023-2529	320	29
5305-01-006-5736	303	11	5320-01-023-2529	324	15
2540-01-008-1501	348	26	5320-01-023-2529	324	37
5310-01-009-9785	337	13	5320-01-023-2529	325	30
5310-01-012-8962	24	9	5320-01-023-2529	370	11
5310-01-012-8962	393	4	2540-01-023-5116	191	12
5310-01-012-8962	395	21	2520-01-024-1273	12	31
9905-01-013-8723	65	6	5330-01-025-4212	121	22
9905-01-013-8723	65	14	5310-01-025-6444	383	8
9905-01-013-8723	75	3	5120-01-026-1666	511	9
9905-01-013-8723	76	3	4240-01-026-3112	307	1
9905-01-013-8723	79	8	4240-01-026-3112	308	7
9905-01-013-8723	80	3	5340-01-026-3250	203	10
9905-01-013-8723	81	4	5340-01-026-3250	378	12
9905-01-013-8723	81	10	3110-01-027-4475	142	9
9905-01-013-8723	81	16	3110-01-027-4475	143	9
9905-01-013-8723	83	2	3110-01-027-4475	150	9
9905-01-013-8723	84	30	3110-01-027-4475	151	9
9905-01-013-8723	85	2	5310-01-027-9392	KITS	101
9905-01-013-8723	89	3	5320-01-029-8205	239	8
9905-01-013-8723	90	3	5790-01-031-0736	BULK	39
9905-01-013-8723	91	4	5305-01-032-4165	40	16
9905-01-013-8723	92	5	5305-01-032-4165	43B	5
9905-01-013-8723	93	7	5310-01-032-4169	40	47
9905-01-013-8723	95	3	5325-01-032-4222	40	13

SECTION IV TM9-2320-280-24P C01

CROSS REFERENCE INDEXES

NATIONAL STOCK NUMBER INDEX

STOCK NUMBER	FIGURE NO	ITEM NO	STOCK NUMBER	FIGURE NO	ITEM NO
5310-01-032-4827	40	48	5330-01-043-5572	100	2
9905-01-032-7002	394	4	5330-01-043-5572	100	5
5355-01-032-7430	293	23	5330-01-044-0703	KITS	83
2530-01-033-1855	172	9	5130-01-044-7196	504	12
2530-01-033-4237	172	11	5320-01-044-7545	199	28
5320-01-033-8637	208	23	5130-01-045-3507	504	13
5320-01-033-8638	208	39	5310-01-046-0186	40	37
5320-01-033-8643	205	19	5310-01-046-5371	487	6
5320-01-033-9126	208	35	5310-01-046-5382	340	9
2530-01-034-1715	172	21	5325-01-047-3201	323	11
5320-01-034-1884	321	24	5325-01-047-3201	326	10
5320-01-034-1884	376	22	4730-01-048-7874	17	6
5325-01-034-2757	40	15	5970-01-048-9378	485	4
5975-01-034-5871	98	7	4730-01-048-9769	384	4
5975-01-034-5871	332	8	4730-01-049-1559	17	7
5975-01-034-5871	369	2	5340-01-049-9564	13	23
5975-01-034-5871	371	3	5340-01-049-9564	14	10
5975-01-034-5871	412	22	5970-01-049-9948	BULK	38
5975-01-034-5871	414	24	5895-01-050-2072	274	20
5975-01-034-5871	416	27	5310-01-050-6565	40	6
5975-01-034-5871	468	1	5310-01-050-6565	43B	39
5975-01-034-5871	470	4	5305-01-052-4438	280	7
5975-01-034-5871	472	6	5340-01-053-7128	224	2
5975-01-034-5871	473	7	5310-01-057-3098	351	3
5975-01-034-5871	474	1	4730-01-058-0900	149	11
5975-01-034-5871	475	3	5310-01-058-3353	382	10
5975-01-034-5871	479	4	5310-01-058-3353	390	9
5975-01-034-5871	480	1	5310-01-058-3353	392	11
5975-01-034-5871	492	14	4720-01-058-7213	443	11
5940-01-035-4212	79	9	4720-01-058-7213	BULK	61
5940-01-035-4212	485	2	5340-01-059-0114	72	9
5935-01-035-5139	367	9	5331-01-061-3000	467	9
5305-01-035-8766	74	14	5331-01-061-3000	493	7
5342-01-036-0649	237	3	5331-01-061-3000	494	12
5330-01-037-0663	62	9	5331-01-061-3000	496	16
3110-01-037-4661	40	14	5310-01-063-8522	351	2
5340-01-038-1493	74	16	4730-01-066-1278	6	9
5310-01-038-8500	172	1	5305-01-066-3431	350	12
5310-01-038-8501	172	10	5310-01-066-6759	19	18
5310-01-038-9579	375	11	5310-01-066-6759	233	4
5310-01-038-9579	375	14	5310-01-066-6759	237	6
5305-01-039-6633	44	28	5310-01-066-6759	248	14
3110-01-040-6541	172	7	5310-01-066-6759	249	12
5330-01-041-9721	BULK	5	5310-01-066-6759	249	34
4730-01-042-5266	505	17	5310-01-066-6759	252	9
5310-01-042-8391	40	50	5310-01-066-6759	266	16
5310-01-042-8391	43B	45	5310-01-066-6759	389	12

I-40

SECTION IV TM9-2320-280-24P C01

CROSS REFERENCE INDEXES

NATIONAL STOCK NUMBER INDEX

STOCK NUMBER	FIGURE NO	ITEM NO	STOCK NUMBER	FIGURE NO	ITEM NO
5306-01-068-4592	409	5	5320-01-084-9235	224	15
5306-01-068-4592	411	11	5320-01-084-9235	287	3
5306-01-068-4592	415	10	5920-01-085-0825	52	13
3010-01-068-6996	45	12	5920-01-085-0825	59	14
2920-01-068-7182	KITS	103	5365-01-085-0910	121	3
2530-01-069-0906	45	27	5310-01-085-1208	377	11
5310-01-069-5243	74	4	5320-01-085-1755	211	23
5310-01-069-5243	282	16	9320-01-085-2889	494	21
5310-01-069-5243	435	5	5320-01-085-9995	197	19
5310-01-069-5243	453	7	5320-01-085-9995	199	20
4730-01-070-7680	12	40	5320-01-085-9995	287	28
4730-01-070-7680	149	6	5320-01-085-9995	293	32
4730-01-070-7680	399	3	5320-01-086-1143	197	10
5340-01-071-2047	18	12	5320-01-086-1143	199	16
5340-01-071-2047	124	12	5320-01-086-1144	223	5
1640-01-072-1167	433	5	5320-01-086-1147	287	16
4730-01-074-0060	327	24	5320-01-086-1148	219	3
5320-01-075-0367	197	20	5320-01-086-1148	287	13
5305-01-075-0957	243	20	5320-01-086-1148	290	3
5305-01-075-0957	308	4	5330-01-086-5457	KITS	21A
5306-01-075-8519	209	12	5305-01-087-4478	5	9
5306-01-075-8519	430	12	5340-01-087-4612	230	23
5305-01-076-3150	50	5	5340-01-087-4612	231	28
5330-01-076-6172	62	10	5340-01-087-4612	235	19
5305-01-076-6308	43	45	5310-01-088-2490	340	2
5310-01-078-5996	261	1	5325-01-088-7468	327	17
5120-01-079-8023	505	1	4730-01-088-7798	7	6
5330-01-080-3253	104	15	4730-01-088-7798	333	3
5310-01-080-5746	487	13	5310-01-088-7962	254	14
5120-01-082-6436	503	6	5310-01-088-7962	492	17
5120-01-082-6447	503	7	4730-01-091-2809	15	13
5305-01-083-1591	465	2	4730-01-091-2809	16	12
4820-01-083-7993	176	10	4730-01-092-1904	331	2
5305-01-084-6067	124	17	4730-01-092-1904	332	9
5305-01-084-6067	125	17	5310-01-093-3493	492	17
5305-01-084-6067	366	3	6635-01-093-3710	499	1
5320-01-084-6101	197	26	6220-01-094-1440	62	1
5320-01-084-6101	199	37	5315-01-094-9025	403	11
5320-01-084-6101	290	10	5330-01-096-7699	78	2
5320-01-084-9234	211	24	5330-01-096-9650	172	19
5320-01-084-9234	212	23	2530-01-097-7659	172	3
5320-01-084-9234	258	9	5305-01-097-8178	350	10
5320-01-084-9234	259	1	2640-01-098-2029	167	21
5320-01-084-9234	260	17	5310-01-100-2067	62	5
5320-01-084-9235	197	24	5360-01-101-2143	245	13
5320-01-084-9235	199	38	5305-01-101-3312	443	7
5320-01-084-9235	212	8	5305-01-101-3312	457	4

SECTION IV TM9-2320-280-24P C01

CROSS REFERENCE INDEXES

NATIONAL STOCK NUMBER INDEX

STOCK NUMBER	FIGURE NO	ITEM NO	STOCK NUMBER	FIGURE NO	ITEM NO
5310-01-101-6046	340	8	5310-01-102-3170	272	9
5340-01-101-8262	18	2	5310-01-102-3170	274	4
5310-01-102-3170	13	17	5310-01-102-3170	274	9
5310-01-102-3170	14	23	5310-01-102-3170	275	4
5310-01-102-3170	15	2	5310-01-102-3170	276	2
5310-01-102-3170	16	2	5310-01-102-3170	276	30
5310-01-102-3170	17	10	5310-01-102-3170	277	22
5310-01-102-3170	27	13	5310-01-102-3170	285	2
5310-01-102-3170	29	4	5310-01-102-3170	286	38
5310-01-102-3170	43	47	5310-01-102-3170	331	10
5310-01-102-3170	97	3	5310-01-102-3170	332	3
5310-01-102-3170	98	14	5310-01-102-3170	333	8
5310-01-102-3170	99	7	5310-01-102-3170	398	17
5310-01-102-3170	124	11	5310-01-102-3170	399	14
5310-01-102-3170	125	11	5310-01-102-3170	400	12
5310-01-102-3170	153	2	5310-01-102-3170	409	10
5310-01-102-3170	155	13	5310-01-102-3170	411	2
5310-01-102-3170	169	28	5310-01-102-3170	411	9
5310-01-102-3170	198	21	5310-01-102-3170	413	4
5310-01-102-3170	202	25	5310-01-102-3170	415	16
5310-01-102-3170	203	6	5310-01-102-3170	430	23
5310-01-102-3170	204	2	5310-01-102-3170	430	37
5310-01-102-3170	233	5	5310-01-102-3170	440	9
5310-01-102-3170	241	4	5310-01-102-3170	446	15
5310-01-102-3170	243	14	5310-01-102-3170	453	9
5310-01-102-3170	244	8	5310-01-102-3170	484	7
5310-01-102-3170	245	10	5310-01-102-3170	486	6
5310-01-102-3170	245	17	5310-01-102-7356	154	1
5310-01-102-3170	245	33	5310-01-103-6042	176A	20
5310-01-102-3170	246	17	5305-01-104-4846	187	8
5310-01-102-3170	247	2	5305-01-104-4846	441	7
5310-01-102-3170	247	14	5130-01-104-5370	504	11
5310-01-102-3170	248	6	5342-01-104-7700	245	15
5310-01-102-3170	249	11	5310-01-105-9398	293	8
5310-01-102-3170	249	26	5310-01-106-1144	331	38
5310-01-102-3170	249	35	6220-01-107-2613	62	2
5310-01-102-3170	249	48	5331-01-107-4950	6	2
5310-01-102-3170	250	6	5305-01-109-0587	323	6
5310-01-102-3170	251	2	5310-01-110-1145	395	16
5310-01-102-3170	254	2	5340-01-110-4036	245	28
5310-01-102-3170	254	20	5340-01-110-4036	478	5
5310-01-102-3170	255	2	5305-01-111-2774	187	2
5310-01-102-3170	267	14	5305-01-111-2774	441	2
5310-01-102-3170	269	12	5306-01-114-0963	37	1
5310-01-102-3170	269	27	5306-01-114-0963	475	29
5310-01-102-3170	270	2	5306-01-114-0963	476	4
5310-01-102-3170	271	8	5340-01-114-7712	402	11

I-42

SECTION IV TM9-2320-280-24P C01

CROSS REFERENCE INDEXES

NATIONAL STOCK NUMBER INDEX

STOCK NUMBER	FIGURE NO	ITEM NO	STOCK NUMBER	FIGURE NO	ITEM NO
8315-01-115-7617	BULK	4	5310-01-119-1024	416	6
5305-01-117-3396	321	10	5310-01-119-1024	417	5
5305-01-117-3396	325	12	5310-01-119-1024	418	5
5305-01-117-3396	376	9	5310-01-119-1024	419	6
5310-01-117-3446	243	15	5310-01-119-1024	420	2
5360-01-118-5907	274	3	5310-01-119-1024	421	2
5360-01-118-5907	276	29	5310-01-119-1024	422	2
4730-01-118-8278	13	6	5310-01-119-1024	423	2
4730-01-118-8278	14	6	5310-01-119-1024	428	6
4730-01-118-8278	19	11	5310-01-119-1024	439	2
4730-01-118-8278	27	4	5310-01-119-1024	440	2
4730-01-118-8278	124	9	5310-01-119-1024	445	1
4730-01-118-8278	177	2	5310-01-119-1024	463	11
4730-01-118-8278	334	16	5310-01-119-3668	231	7
4730-01-118-8278	386	7	5310-01-119-3668	439	5
5310-01-119-1024	83	24	5310-01-119-3675	7	2
5310-01-119-1024	144	5	4730-01-119-6895	15	22
5310-01-119-1024	152	5	4730-01-119-6895	16	17
5310-01-119-1024	153	20	5305-01-120-4363	388	8
5310-01-119-1024	169	32	5310-01-121-1703	2	9
5310-01-119-1024	205	26	5310-01-121-1703	26	11
5310-01-119-1024	209	9	5310-01-121-1703	174	6
5310-01-119-1024	225	3	5310-01-121-1703	183	10
5310-01-119-1024	226	9	5310-01-121-1703	184	2
5310-01-119-1024	231	8	5310-01-121-1703	185	6
5310-01-119-1024	231	15	5310-01-121-1703	186	2
5310-01-119-1024	234	5	5310-01-121-1703	186	6
5310-01-119-1024	242	3	5310-01-121-1703	187	12
5310-01-119-1024	250	21	5310-01-121-1703	188	23
5310-01-119-1024	266	2	5310-01-121-1703	242	7
5310-01-119-1024	268	11	5310-01-121-1703	271	17
5310-01-119-1024	268	28	5310-01-121-1703	271	29
5310-01-119-1024	277	6	5310-01-121-1703	285	7
5310-01-119-1024	279	33	5310-01-121-1703	315	6
5310-01-119-1024	284	2	5310-01-121-1703	437	7
5310-01-119-1024	292	14	5310-01-121-1703	438	3
5310-01-119-1024	328	9	5310-01-121-1703	441	6
5310-01-119-1024	333	22	5310-01-121-1703	442	2
5310-01-119-1024	338	26	5310-01-121-1703	444	2
5310-01-119-1024	339	19	5310-01-121-1703	453	11
5310-01-119-1024	374	8	5310-01-121-1703	456	4
5310-01-119-1024	403	4	4330-01-121-6350	KITS	15
5310-01-119-1024	409	2	5305-01-122-0253	121	16
5310-01-119-1024	410	7	2540-01-123-1218	368	16
5310-01-119-1024	413	2	3040-01-123-4942	172	8
5310-01-119-1024	414	6	2530-01-123-8787	KITS	82
5310-01-119-1024	415	2	4030-01-124-8201	353	5

I-43

SECTION IV TM9-2320-280-24P C01

CROSS REFERENCE INDEXES

NATIONAL STOCK NUMBER INDEX

STOCK NUMBER	FIGURE NO	ITEM NO	STOCK NUMBER	FIGURE NO	ITEM NO
5340-01-125-1682	353	7	5325-01-136-4880	43B	21
2540-01-125-6154	388	16	5310-01-136-4888	40	17
5310-01-126-9404	329	12	5310-01-136-4888	43B	6
5310-01-127-2456	100	7	5325-01-136-7662	40	12
5310-01-127-2456	173	1	5325-01-136-7662	43B	19
2520-01-127-3969	116	17	5310-01-137-6801	350	8
6220-01-128-0087	62	4	5310-01-137-6801	351	8
5365-01-129-0399	271	13	5310-01-138-2605	314	23
5306-01-130-0457	463	8	5320-01-138-4239	219	10
5310-01-130-4274	196	7	5320-01-138-4239	235	4
5120-01-130-8865	500	4	5320-01-138-4239	235	18
5310-01-130-9065	471	9	5330-01-138-5190	KITS	86
5310-01-132-8275	12	12	5305-01-138-9540	46	4
5325-01-135-4290	268	35	5305-01-138-9540	97	11
5310-01-135-4730	131	1	5305-01-138-9540	367	1
5320-01-135-7319	197	11	5307-01-140-6594	268	38
5320-01-135-7319	199	15	5305-01-140-9118	175	7
5320-01-135-7319	211	3	5305-01-140-9118	272	24
5320-01-135-7319	213	37	4030-01-142-0456	276	4
5320-01-135-7319	214	30	5970-01-142-2282	96	17
5320-01-135-7319	218	1	5310-01-143-0512	12	20
5320-01-135-7319	219	1	5310-01-143-0512	23	22
5320-01-135-7319	227	39	5310-01-143-0512	127	14
5320-01-135-7319	237	17	5310-01-143-1679	39	45
5320-01-135-7319	278	1	5310-01-143-1679	39	76
5320-01-135-7319	291	6	5305-01-143-2328	327	38
5320-01-135-7319	292	5	5305-01-143-2328	477	8
5320-01-135-7319	365	1	5305-01-143-2328	480	2
5320-01-135-7319	366	26	5320-01-143-5075	197	27
5320-01-135-7319	368	5	5320-01-143-5075	199	35
5320-01-135-7319	369	8	5320-01-143-5075	208	13
5320-01-135-7319	379	21	5320-01-143-5075	213	36
5320-01-135-7319	384	14	5320-01-143-5075	214	29
4730-01-135-8310	148	30	5320-01-143-5075	220	6
5320-01-136-1782	64	2	5320-01-143-5075	300	15
5320-01-136-1782	197	8	5320-01-143-5079	208	10
5320-01-136-1782	199	17	5320-01-143-5079	220	1
5320-01-136-1782	212	9	5320-01-143-5079	230	24
5320-01-136-1782	229	5	5320-01-143-5079	231	29
5320-01-136-1782	238	11	5320-01-143-5079	369	15
5320-01-136-1782	287	1	5977-01-143-6996	39	68
5320-01-136-1782	289	6	5310-01-144-2779	172	18
5320-01-136-1785	219	4	5120-01-144-4484	499	3
5320-01-136-1785	287	14	5325-01-144-4871	68	8
5320-01-136-1785	290	8	5330-01-145-5376	39	69
5320-01-136-1787	288	3	5330-01-146-6053	111	3
5320-01-136-1787	290	5	6620-01-146-8006	30	1

I-44

SECTION IV TM9-2320-280-24P C01

CROSS REFERENCE INDEXES

NATIONAL STOCK NUMBER INDEX

STOCK NUMBER	FIGURE NO	ITEM NO	STOCK NUMBER	FIGURE NO	ITEM NO
5935-01-147-0148	95	21	5305-01-149-1938	31	8
5935-01-147-0148	96	2	2520-01-149-3809	113	16
5310-01-147-4052	144	2	5310-01-149-4407	26	13
5310-01-147-4052	152	2	5340-01-149-4434	399	8
5310-01-147-4052	165	2	2520-01-149-4993	128	61
5310-01-147-4052	166	2	2520-01-149-4993	129	33
5310-01-147-4052	181	11	3020-01-149-5049	111	14
5310-01-147-4052	182	1	4710-01-149-5075	11	30
5310-01-147-4052	183	7	4710-01-149-5076	11	11
5310-01-147-4052	184	14	4710-01-149-5077	11	24
5310-01-147-4052	188	17	5935-01-149-5165	76	2
5310-01-147-4052	188	29	5935-01-149-5165	79	6
5310-01-147-4052	195	1	5935-01-149-5165	86	27
5310-01-147-4052	270	34	5306-01-149-6280	4	4
5310-01-147-4052	470	10	5306-01-149-6280	8	5
2930-01-147-4198	30	3	5306-01-149-6280	30	16
2590-01-147-4285	4	2	5360-01-149-6308	176	9
4820-01-147-4294	100	3	3040-01-149-6706	111	1
5305-01-147-5864	39	52	3040-01-149-6759	116	2
5998-01-147-7895	39	74	5310-01-149-7793	121	11
5310-01-147-8743	490	5	5340-01-149-7811	116	21
2990-01-147-9284	7	3	2520-01-149-7861	106	11
5330-01-147-9808	31	4	2520-01-149-7861	KITS	28
5310-01-148-0240	68	4	4730-01-149-7935	118	5
4710-01-148-2659	11	1	3020-01-149-7938	106	23
5310-01-148-2687	129	35	3020-01-149-7941	106	10
4730-01-148-2758	31	2	3020-01-149-7941	107	8
4720-01-148-2768	BULK	14	5305-01-149-9673	31	11
5306-01-148-3666	30	10	3040-01-150-0407	106	7
5306-01-148-3667	11	5	4730-01-150-0879	6	6
5305-01-148-5915	8	4	4710-01-150-0971	11	16
5330-01-148-7492	104	7	5306-01-150-1190	31	12
5340-01-148-7528	10	5	5306-01-150-1190	122	33
5340-01-148-7529	11	31	5307-01-150-1227	8	2
5340-01-148-8349	10	4	5307-01-150-1227	30	5
6680-01-148-8875	66	4	5305-01-150-1521	11	14
3020-01-148-9548	128	41	5307-01-150-1549	31	13
4710-01-148-9580	11	10	2520-01-150-2279	111	10
4710-01-148-9581	11	28	2520-01-150-2280	113	7
4710-01-148-9582	11	17	2520-01-150-3931	112	15
5330-01-149-0874	30	12	2520-01-150-3932	109	3
2520-01-149-1221	105	2	2520-01-150-3932	110	3
2520-01-149-1221	106	2	2520-01-150-3932	111	16
4320-01-149-1866	121	1	2520-01-150-3932	112	16
5340-01-149-1867	121	4	5310-01-150-4003	127	9
2520-01-149-1868	109	14	5310-01-150-4003	136	3
2520-01-149-1868	114	3	5310-01-150-4003	144	14

I-45

SECTION IV TM9-2320-280-24P C01

CROSS REFERENCE INDEXES

NATIONAL STOCK NUMBER INDEX

STOCK NUMBER	FIGURE NO	ITEM NO	STOCK NUMBER	FIGURE NO	ITEM NO
5310-01-150-4003	152	14	5340-01-151-4964	116	5
5310-01-150-4003	271	12	3040-01-151-5663	116	9
5310-01-150-4003	271	27	3040-01-151-5663	117	16
5310-01-150-4003	398	8	5330-01-151-6106	118	1
5310-01-150-4003	464	8	5325-01-151-6117	4	3
5340-01-150-4106	11	20	5310-01-151-7347	140	7
5306-01-150-4835	104	4	5310-01-151-7347	147	7
4820-01-150-4964	118	11	5310-01-151-7347	170	20
5325-01-150-4982	106	15	5310-01-151-7347	185	4
5325-01-150-4982	107	23	5310-01-151-7347	187	3
5310-01-150-5919	116	24	5310-01-151-7347	195	11
5310-01-150-5919	117	10	5310-01-151-7347	196	2
5310-01-150-5921	121	20	5310-01-151-7347	441	3
5330-01-150-5928	116	20	5310-01-151-7347	457	19
5330-01-150-5944	10	1	5310-01-151-7347	470	11
5307-01-150-5991	8	3	5330-01-151-8364	118	3
5307-01-150-5992	365	17	5340-01-151-9956	11	4
5340-01-150-6026	11	15	5310-01-152-0598	51	4
5360-01-150-6086	116	6	5310-01-152-0598	69	34
5360-01-150-6086	117	5	5310-01-152-0598	77	13
5360-01-150-6087	116	8	5310-01-152-0598	83	15
5360-01-150-6091	109	7	5310-01-152-0598	83	31
5360-01-150-6091	111	7	5310-01-152-0598	88	17
5325-01-150-6092	109	13	5310-01-152-0598	88	29
5330-01-150-6239	116	10	5310-01-152-0598	89	11
5330-01-150-6239	117	15	5310-01-152-0598	95	40
5340-01-150-6275	11	21	5310-01-152-0598	155	6
5935-01-150-6319	100	4	5310-01-152-0598	214	9
2520-01-150-7609	113	3	5310-01-152-0598	304	7
5330-01-150-7744	4	28	5310-01-152-0598	307	6
5340-01-150-7774	11	29	5310-01-152-0598	308	15
5360-01-150-7829	116	23	5310-01-152-0598	313	11
5360-01-150-7829	117	9	5310-01-152-0598	320	18
5325-01-150-7830	113	1	5310-01-152-0598	320	31
5325-01-150-7830	114	10	5310-01-152-0598	339	29
5935-01-150-8322	86	26	5310-01-152-0598	378	19
5306-01-150-8713	118	10	5310-01-152-0598	410	33
5999-01-150-8808	86	7	5310-01-152-0598	412	14
5306-01-150-9497	104	13	5310-01-152-0598	414	22
5307-01-150-9538	11	6	5310-01-152-0598	416	25
5307-01-150-9538	12	17	5310-01-152-0598	428	24
5305-01-150-9781	4	25	5310-01-152-0598	443	14
4710-01-151-3663	118	6	5310-01-152-0598	453	3
2520-01-151-3857	116	15	5310-01-152-0598	457	13
5310-01-151-4137	116	3	5310-01-152-0598	480	23
5315-01-151-4180	9	5	3120-01-152-2613	128	73
5306-01-151-4925	9	6	6680-01-152-2845	176	1

I-46

SECTION IV TM9-2320-280-24P C01

CROSS REFERENCE INDEXES

NATIONAL STOCK NUMBER INDEX

STOCK NUMBER	FIGURE NO	ITEM NO	STOCK NUMBER	FIGURE NO	ITEM NO
5310-01-152-4229	121	12	3110-01-155-2600	128	67
5306-01-152-4693	22	1	3120-01-155-3509	128	42
5306-01-152-4696	128	21	5330-01-155-4383	113	13
5306-01-152-4696	130	21	5330-01-155-4393	128	70
4710-01-152-5798	104	14	3110-01-155-4438	109	15
5330-01-152-5941	121	19	3110-01-155-4440	128	1
5330-01-152-5942	118	13	3110-01-155-4440	130	38
5315-01-152-9029	100	15	3120-01-155-4462	121	17
5315-01-152-9031	116	4	3120-01-155-4463	121	17
5315-01-152-9038	117	7	3120-01-155-4464	121	17
5315-01-152-9039	116	7	3120-01-155-4465	121	17
5315-01-152-9039	117	7	3120-01-155-4466	121	17
5315-01-152-9040	116	7	3120-01-155-4467	121	17
5315-01-152-9040	117	7	3120-01-155-4468	113	9
5365-01-153-0872	104	5	3120-01-155-4468	114	18
5360-01-153-0933	113	8	3120-01-155-4470	128	40
4710-01-153-1636	11	18	4320-01-155-5145	176	6
4730-01-153-1871	345	40	4320-01-155-5153	176	13
4730-01-153-1871	348	32	4720-01-155-7784	BULK	26
5930-01-153-8215	68	10	3120-01-155-8713	128	43
5935-01-153-9320	76	13	3110-01-155-9862	128	37
5340-01-153-9425	177	14	4720-01-156-0549	BULK	12
2520-01-154-1185	118	14	4720-01-156-0550	BULK	13
5905-01-154-2354	498	6	5306-01-156-3730	445	13
3120-01-154-4369	106	30	5330-01-156-5147	4	7
3120-01-154-4369	107	25	3120-01-156-5189	106	8
5935-01-154-6233	79	5	3120-01-156-5189	121	5
3120-01-154-6272	487	12	5305-01-156-8692	157	18
5340-01-154-6559	116	22	5305-01-156-8692	158	17
5340-01-154-6559	117	8	3120-01-156-8763	111	15
5330-01-154-7159	4	10	3120-01-156-8763	112	14
3120-01-154-7174	106	14	5330-01-157-0856	8	6
3120-01-154-7174	106	28	5331-01-157-1884	176	14
3120-01-154-7174	107	15	5325-01-157-6764	128	25
3120-01-154-7174	107	24	5325-01-157-6764	128	47
3120-01-154-8516	106	30	5325-01-157-6764	130	41
3120-01-154-8516	107	25	5340-01-158-0098	79	4
3120-01-154-8517	106	30	5325-01-158-2182	113	6
3120-01-154-8517	107	25	5325-01-158-2182	114	4
5325-01-154-8561	106	3	5306-01-158-6224	7	1
5325-01-154-8561	107	3	5310-01-158-6257	75	36
5325-01-154-8562	107	1	5310-01-158-6257	77	17
9320-01-155-2369	BULK	58	3120-01-158-6304	106	18
5310-01-155-2503	7	5	9905-01-158-7981	357	4
5310-01-155-2503	10	6	5306-01-158-9917	12	22
5310-01-155-2503	75	27	5305-01-159-0065	175A	11
5310-01-155-2503	101	4	5340-01-159-1788	121	7

I-47

SECTION IV TM9-2320-280-24P C01

CROSS REFERENCE INDEXES

NATIONAL STOCK NUMBER INDEX

STOCK NUMBER	FIGURE NO	ITEM NO	STOCK NUMBER	FIGURE NO	ITEM NO
2530-01-159-2732	106	27	5340-01-163-4773	151	40
5330-01-159-2811	128	52	2520-01-163-4999	126	2
5330-01-159-2811	129	25	5930-01-163-6256	55	7
3120-01-159-5773	106	30	4730-01-163-7163	105	3
3120-01-159-5773	107	25	3040-01-163-7208	4	8
4720-01-159-5796	BULK	25	2520-01-163-7213	KITS	91
5340-01-159-6626	11	25	4720-01-163-7833	BULK	11
2520-01-159-7757	128	80	2520-01-163-7866	109	11
2520-01-159-7757	129	38	2520-01-163-7866	111	6
2590-01-159-8757	132	2	5935-01-163-8981	348	2
2590-01-159-8757	132	15	5935-01-163-8987	348	8
5342-01-160-4397	4	5	5310-01-164-0745	351	5
5325-01-160-4618	11	12	5310-01-164-1023	350	7
5315-01-160-4642	128	10	5310-01-164-1023	351	7
5315-01-160-4642	130	23	2540-01-164-6170	348	9
5325-01-160-4693	106	32	4730-01-164-7028	8	8
5325-01-160-4693	113	5	2520-01-164-7157	116	11
5325-01-160-4693	114	5	2520-01-164-7157	KITS	97
2530-01-160-9569	167	1	2520-01-164-7158	100	9
2520-01-160-9570	118	7	2520-01-164-7234	102	1
5331-01-160-9857	467	7	5306-01-164-8486	118	9
5331-01-160-9857	493	1	2540-01-165-0465	345	19
5331-01-160-9857	496	11	2540-01-165-0465	348	19
2640-01-161-2114	167	5	5340-01-165-0745	348	30
2520-01-161-2136	104	1	2540-01-165-0814	350	2
3120-01-161-4033	106	30	2540-01-165-0814	351	1
3120-01-161-4033	107	25	5310-01-165-1312	40	8
4820-01-161-6435	496	10	5310-01-165-1312	43B	28
5970-01-161-6796	BULK	31	5360-01-165-1563	4	16
5310-01-161-7308	54	13	5930-01-165-1657	55	18
2540-01-162-4339	348	35	5310-01-165-3331	4	30
3120-01-162-5787	106	30	5310-01-165-3331	7	8
3120-01-162-5787	107	25	5310-01-165-3331	11	26
5340-01-162-5883	108	10	5306-01-165-4283	128	23
5305-01-162-8512	12	2	5306-01-165-4283	130	19
2520-01-162-8985	132	3	5330-01-165-4333	113	12
2520-01-162-8985	132	7	5330-01-165-4333	121	18
2520-01-162-8985	132	14	6150-01-165-4667	345	26
2520-01-162-8985	132	21	6150-01-165-4667	348	24
5305-01-162-9713	176	12	5306-01-165-5582	4	12
5970-01-163-1103	BULK	35	2540-01-165-8175	345	31
5330-01-163-2614	116	1	2540-01-165-8175	348	25
5330-01-163-2614	117	3	2540-01-165-8176	345	29
4710-01-163-2805	345	33	2540-01-165-8176	348	31
4710-01-163-2805	348	28	2520-01-165-9563	100	13
4730-01-163-3544	11	2	2520-01-165-9563	118	8
5340-01-163-4773	150	34	5342-01-166-1534	11	22

I-48

SECTION IV TM9-2320-280-24P C01

CROSS REFERENCE INDEXES

NATIONAL STOCK NUMBER INDEX

STOCK NUMBER	FIGURE NO	ITEM NO	STOCK NUMBER	FIGURE NO	ITEM NO
5340-01-166-1672	25	2	5120-01-170-3278	501	3
5340-01-166-1672	103	5	5940-01-170-4956	41	10
5331-01-166-1712	493	3	5940-01-170-4956	41	19
4730-01-166-2244	104	3	5940-01-170-4956	313	15
4730-01-166-2244	105	11	5940-01-170-4956	474	11
3120-01-166-3677	106	29	5310-01-170-8765	411	18
5325-01-166-6324	128	74	2520-01-170-9826	111	12
3110-01-167-2443	KITS	88	5325-01-171-3392	109	5
3040-01-167-2836	106	31	5325-01-171-3392	110	10
5355-01-167-4114	54	7	5325-01-171-3392	111	9
3120-01-167-4172	KITS	87	5325-01-171-3392	112	9
6685-01-167-4298	67	1	2610-01-171-4746	168	1
5306-01-167-4346	128	64	4130-01-171-5997	494	13
5340-01-167-7794	10	7	2520-01-171-8258	137	4
5340-01-167-9314	394	2	5310-01-173-0941	163	5
2520-01-168-1983	106	16	5315-01-173-3397	277	19
2520-01-168-2060	113	4	5315-01-173-3397	277	46
5330-01-168-3870	128	66	4820-01-173-6883	14	27
5330-01-168-3870	129	3	5935-01-173-7654	75	18
2920-01-168-4127	35	18	2510-01-173-9316	266	10
2920-01-168-4127	43	54	2520-01-173-9673	128	58
2920-01-168-4128	KITS	100	5340-01-174-2271	142	24
2920-01-168-4129	KITS	104	5340-01-174-2271	143	32
2920-01-168-4130	KITS	105	5340-01-174-2271	150	24
2920-01-168-4131	43	27	5340-01-174-2271	151	32
2920-01-168-4132	43	33	5935-01-174-3669	75	19
5325-01-168-5729	106	12	2530-01-174-4174	142	31
2930-01-168-7870	34	2	2530-01-174-4174	150	31
2815-01-168-7871	4	9	5340-01-174-4894	370	23
4730-01-168-7872	6	5	5340-01-174-4894	394	3
2520-01-168-7878	128	60	2520-01-174-5849	148	15
2920-01-168-7891	44	2	2520-01-174-5919	132	11
2815-01-168-7892	1	2	3020-01-174-5984	142	29
2910-01-168-7905	9	1	3020-01-174-5984	150	29
2815-01-168-7909	8	7	3020-01-174-5985	142	10
2930-01-168-7911	5	2	3020-01-174-5985	150	10
2815-01-168-7912	4	22	2530-01-174-7441	156	1
2920-01-168-7914	35	5	2540-01-174-7696	206	14
2815-01-168-7917	8	10	3020-01-174-7810	128	79
2815-01-168-7918	8	11	5330-01-174-8090	116	12
2540-01-168-9481	351	11	5310-01-174-8091	128	20
2540-01-168-9482	350	9	5310-01-174-8091	129	36
2540-01-168-9482	351	9	3110-01-174-8136	109	16
3110-01-169-0734	KITS	89	5330-01-174-8145	148	14
5120-01-169-4876	501	2	5330-01-174-8146	148	7
2520-01-169-7674	111	2	2520-01-174-8153	KITS	94
4720-01-169-9891	BULK	19	5365-01-174-8154	128	28

I-49

SECTION IV TM9-2320-280-24P C01

CROSS REFERENCE INDEXES

NATIONAL STOCK NUMBER INDEX

STOCK NUMBER	FIGURE NO	ITEM NO	STOCK NUMBER	FIGURE NO	ITEM NO
3120-01-174-8155	128	33	2520-01-174-9554	128	18
3120-01-174-8156	128	78	2520-01-174-9554	129	32
5310-01-174-8158	128	11	2520-01-174-9554	129	34
5310-01-174-8158	130	22	2520-01-174-9580	148	12
5365-01-174-8174	KITS	45	5310-01-175-0617	148	10
5365-01-174-8184	KITS	74	3020-01-175-1962	128	75
5365-01-174-8185	128	29	3110-01-175-3538	KITS	75
5365-01-174-8186	142	4	5365-01-175-3593	KITS	44
5365-01-174-8186	143	4	2520-01-175-6401	128	13
5365-01-174-8186	150	4	3020-01-175-6477	128	32
5365-01-174-8186	151	4	2520-01-175-6492	113	2
5306-01-174-8492	128	56	2520-01-175-6494	132	1
5340-01-174-8493	142	13	5340-01-175-7208	140	9
5340-01-174-8493	143	13	2530-01-175-7210	152	6
5340-01-174-8493	150	13	2530-01-175-7211	144	6
5340-01-174-8493	151	13	2530-01-175-7212	144	20
5310-01-174-8607	142	21	2530-01-175-7212	152	20
5310-01-174-8607	143	28	2530-01-175-7213	144	20
5310-01-174-8607	150	21	2530-01-175-7213	152	20
5310-01-174-8607	151	28	2920-01-175-7214	66	2
5360-01-174-8613	132	5	2510-01-175-7216	182	11
5360-01-174-8614	132	10	2510-01-175-7217	183	12
5331-01-174-8618	128	77	2510-01-175-7218	183	22
5305-01-174-8625	148	24	2510-01-175-7219	182	17
5325-01-174-8626	106	19	2520-01-175-7220	103	1
5325-01-174-8626	107	21	2510-01-175-7221	179	11
3120-01-174-8630	128	35	9520-01-175-7222	184	5
3120-01-174-8631	128	34	2510-01-175-7224	181	17
5310-01-174-8632	142	8	5945-01-175-7318	118	4
5310-01-174-8632	150	8	5325-01-175-7442	387	13
5307-01-174-8640	128	68	5306-01-175-7577	167	3
5307-01-174-8640	129	7	3010-01-176-0557	128	39
5315-01-174-8644	116	7	5330-01-176-0825	142	12
5315-01-174-8644	117	7	5330-01-176-0825	143	12
5315-01-174-8645	116	7	5330-01-176-0825	150	12
5315-01-174-8645	117	7	5330-01-176-0825	151	12
5315-01-174-8646	116	7	5310-01-176-0839	142	25
5315-01-174-8646	117	7	5310-01-176-0839	143	33
5315-01-174-8647	116	7	5310-01-176-0839	150	25
5315-01-174-8647	117	7	5310-01-176-0839	151	33
3040-01-174-9130	148	1	5331-01-176-0923	167	4
2520-01-174-9287	128	26	2940-01-176-1427	489	1
2520-01-174-9288	128	31	5120-01-176-1845	500	5
2520-01-174-9291	128	5	3040-01-176-2835	132	9
5970-01-174-9449	73	19	2520-01-176-2840	132	12
5970-01-174-9449	73	23	2530-01-176-4649	152	6
5970-01-174-9449	480	27	2530-01-176-4650	144	6

I-50

SECTION IV TM9-2320-280-24P C01

CROSS REFERENCE INDEXES

NATIONAL STOCK NUMBER INDEX

STOCK NUMBER	FIGURE NO	ITEM NO	STOCK NUMBER	FIGURE NO	ITEM NO
2540-01-176-6521	205	12	5365-01-181-5085	150	23
3040-01-176-9584	128	76	9905-01-181-9456	357	3
3040-01-176-9615	128	53	5340-01-182-1074	22	2
3040-01-177-2427	128	46	4730-01-182-6565	30	15
3040-01-177-2428	128	36	4730-01-182-6565	31	7
3020-01-177-5441	128	24	3120-01-182-8417	121	17
3020-01-177-5441	128	48	5331-01-183-0971	6	8
5365-01-177-5720	142	26	5930-01-183-6637	66	9
5365-01-177-5720	143	34	5930-01-183-6757	67	7
5365-01-177-5720	150	26	5340-01-183-7233	60	3
5365-01-177-5720	151	34	5340-01-183-7233	61	3
4730-01-177-7501	104	23	5999-01-183-9530	327	18
4730-01-177-7501	105	5	5905-01-183-9636	333	1
4910-01-178-0722	500	3	6160-01-184-0643	71	1
4910-01-178-0724	510	1	6160-01-184-0728	71	5
4910-01-178-6551	511	7	2920-01-184-1054	45	21
5307-01-178-7445	31	10	2910-01-184-2159	15	10
4910-01-178-8865	499	5	5995-01-184-2228	79	1
2520-01-178-9768	128	72	2540-01-184-4389	165	4
5120-01-179-1032	503	8	2590-01-184-4475	184	3
4910-01-179-2517	503	5	4730-01-184-4760	141	25
5360-01-179-4106	192	1	5330-01-184-5421	157	12
5360-01-179-4107	192	1	5330-01-184-5421	158	8
4710-01-179-4346	400	5	2510-01-184-5497	331	12
6680-01-179-4349	400	4	2910-01-184-5498	331	5
6680-01-179-4350	3	4	2910-01-184-5499	331	1
4820-01-179-4869	511	15	2540-01-184-5500	328	10
4910-01-179-5530	502	3	2520-01-184-5501	97	1
2530-01-179-7511	KITS	14	2540-01-184-5502	99	4
2530-01-179-7589	163	1	2540-01-184-5503	327	21
2530-01-179-7590	164	1	5995-01-184-5544	80	1
2590-01-179-7602	313	12	5995-01-184-5545	75	1
2590-01-179-7602	443	15	5310-01-184-5784	194	2
5365-01-180-2585	148	5	5935-01-184-6417	85	24
5940-01-180-3655	69	5	5935-01-184-6417	86	15
5940-01-180-3655	481	1	5330-01-184-6492	103	10
6625-01-180-6542	47	10	5330-01-184-6492	400	6
2920-01-180-8666	43	23	5330-01-184-6500	6	3
6685-01-180-9037	47	9	2540-01-184-6602	230	3
5330-01-180-9099	142	11	4730-01-184-6971	159	8
5330-01-180-9099	143	11	4730-01-184-6971	160	6
5330-01-180-9099	150	11	4730-01-184-6971	161	9
5330-01-180-9099	151	11	4730-01-184-6971	162	17
5970-01-180-9776	75	8	4730-01-184-6971	487A	10
6620-01-181-1757	47	11	4730-01-184-6971	487D	6
2530-01-181-3907	100	14	2520-01-184-7036	99	22
5365-01-181-5085	142	23	6680-01-184-9214	103	8

SECTION IV TM9-2320-280-24P C01

CROSS REFERENCE INDEXES

NATIONAL STOCK NUMBER INDEX

STOCK NUMBER	FIGURE NO	ITEM NO	STOCK NUMBER	FIGURE NO	ITEM NO
2530-01-184-9821	156	19	9905-01-185-3207	358	3
5305-01-185-0114	213	24	7690-01-185-3208	358	1
5305-01-185-0114	225	15	2540-01-185-3214	402	1
5305-01-185-0114	234	14	2540-01-185-3216	402	KIT
5305-01-185-0114	246	6	2540-01-185-3216	452	2
5305-01-185-0114	254	16	9340-01-185-3757	225	55
5305-01-185-0114	260	20	2530-01-185-3879	158	12
5305-01-185-0114	260	25	2540-01-185-4387	235	9
5305-01-185-0114	287	7	3040-01-185-4388	99	21
5305-01-185-0114	321	22	3040-01-185-4388	133	11
5305-01-185-0114	326	5	2590-01-185-4390	240	16
5342-01-185-0116	225	56	2590-01-185-4390	429	26
5342-01-185-0116	226	37	2590-01-185-4391	240	16
5340-01-185-0386	205	3	5305-01-185-4628	156	2
5340-01-185-0386	206	8	5310-01-185-4672	137	16
5340-01-185-0401	225	45	5340-01-185-4959	225	5
5340-01-185-0401	226	32	5340-01-185-4959	226	40
5340-01-185-0404	225	43	4730-01-185-5348	12	41
5330-01-185-0587	21	24	4730-01-185-5348	135	3
6680-01-185-1264	67	4	4730-01-185-5348	141	26
5305-01-185-2211	216	8	4730-01-185-5348	149	14
5305-01-185-2211	221	4	4730-01-185-5348	399	4
2510-01-185-3107	205	37	2510-01-185-6112	2	4
5365-01-185-3108	204	4	2510-01-185-6113	2	10
9515-01-185-3109	267	15	2510-01-185-6114	2	12
9515-01-185-3110	267	19	2510-01-185-6115	2	1
9515-01-185-3111	267	8	2540-01-185-6117	46	8
2590-01-185-3112	277	11	2510-01-185-6118	205	22
2540-01-185-3113	277	24	2540-01-185-6119	240	8
3110-01-185-3114	277	9	4710-01-185-6179	10	3
2590-01-185-3115	314	32	2510-01-185-6647	182	19
2590-01-185-3116	314	24	2510-01-185-6648	182	19
9515-01-185-3118	267	12	4720-01-185-6673	BULK	22
9515-01-185-3119	267	20	2540-01-185-6710	185	9
9905-01-185-3127	358	2	2590-01-185-6711	182	12
9905-01-185-3129	354	12	2590-01-185-6711	183	20
9905-01-185-3131	354	12	2530-01-185-6712	157	23
9905-01-185-3132	354	12	2530-01-185-6712	158	13
9905-01-185-3134	354	12	2530-01-185-6713	157	20
9905-01-185-3138	354	12	2530-01-185-6713	158	15
9905-01-185-3139	354	12	2530-01-185-6714	158	9
9905-01-185-3143	355	1	2510-01-185-6715	202	29
2510-01-185-3155	268	3	5340-01-185-6997	225	10
6105-01-185-3187	314	27	5340-01-185-6998	225	27
2540-01-185-3197	187	1	5340-01-185-7024	225	18
3040-01-185-3201	314	8	5306-01-185-7048	1	6
6680-01-185-3202	400	3	5306-01-185-7048	103	13

I-52

SECTION IV TM9-2320-280-24P C01

CROSS REFERENCE INDEXES

NATIONAL STOCK NUMBER INDEX

STOCK NUMBER	FIGURE NO	ITEM NO	STOCK NUMBER	FIGURE NO	ITEM NO
5306-01-185-7048	138	18	2530-01-185-7998	158	10
5306-01-185-7048	139	16	3040-01-185-8011	314	9
5306-01-185-7048	145	18	2540-01-185-8015	403	14
5306-01-185-7048	146	16	2540-01-185-8016	402	1
5306-01-185-7048	175	10	5120-01-185-8024	501	10
5306-01-185-7051	66A	13	4720-01-185-8152	12	8
5306-01-185-7052	137	15	6150-01-185-8240	327	16
5306-01-185-7052	239	11	4710-01-185-8328	161	13
5306-01-185-7071	71	3	4710-01-185-8329	161	8
5305-01-185-7122	179	7	5340-01-185-8619	203	8
5305-01-185-7122	180	8	5340-01-185-8620	179	14
5305-01-185-7122	184	11	5305-01-185-8647	199	4
5310-01-185-7188	69	6	5305-01-185-8647	225	19
5310-01-185-7188	70	5	5305-01-185-8647	232	4
5310-01-185-7208	69	8	5305-01-185-8647	462	9
5310-01-185-7208	70	3	5305-01-185-8657	157	17
5310-01-185-7214	12	11	5305-01-185-8657	158	1
5310-01-185-7214	364	1	5305-01-185-8673	156	15
5310-01-185-7214	436	7	5305-01-185-8674	240	4
5310-01-185-7218	138	19	5305-01-185-8674	240	18
5310-01-185-7218	139	17	5305-01-185-8674	240	22
5310-01-185-7218	145	19	5305-01-185-8674	269	6
5310-01-185-7218	146	17	5305-01-185-8674	442	5
5315-01-185-7421	21	27	5310-01-185-8689	194	9
2510-01-185-7945	204	11	5310-01-185-8690	157	19
2510-01-185-7946	206	17	5310-01-185-8690	158	16
2590-01-185-7947	314	30	5310-01-185-8691	156	12
3020-01-185-7948	314	7	5310-01-185-8692	328	4
2590-01-185-7949	314	6	5330-01-185-8709	331	17
2510-01-185-7950	204	11	5330-01-185-8714	206	11
2540-01-185-7951	403	2	5330-01-185-8715	205	40
5120-01-185-7955	501	5	5340-01-185-8723	230	17
5120-01-185-7956	501	8	5340-01-185-8724	266	5
5120-01-185-7957	502	9	5315-01-185-8781	165	1
5340-01-185-7959	398	18	5315-01-185-8781	166	1
5120-01-185-7960	501	21	5360-01-185-8809	156	7
5120-01-185-7962	502	8	5340-01-185-8820	240	14
5120-01-185-7963	503	2	5340-01-185-8820	451	1
5120-01-185-7964	502	2	5340-01-185-8821	202	27
5120-01-185-7965	503	9	4730-01-185-9453	19	22
4910-01-185-7966	504	2	4730-01-185-9453	27	10
9905-01-185-7970	355	4	2540-01-185-9530	202	17
9905-01-185-7972	354	12	2530-01-185-9651	170	6
9905-01-185-7973	354	12	2540-01-185-9656	21	26
9905-01-185-7975	354	12	4710-01-185-9665	161	6
9905-01-185-7977	354	20	4710-01-185-9666	159	1
2530-01-185-7998	157	14	4710-01-185-9667	159	5

SECTION IV TM9-2320-280-24P C01

CROSS REFERENCE INDEXES

NATIONAL STOCK NUMBER INDEX

STOCK NUMBER	FIGURE NO	ITEM NO	STOCK NUMBER	FIGURE NO	ITEM NO
4710-01-185-9667	160	8	5340-01-186-5520	205	41
4710-01-185-9668	159	7	3120-01-186-5527	144	7
2540-01-186-0501	185	5	3120-01-186-5527	144	21
2540-01-186-0501	187	5	3120-01-186-5527	152	7
4820-01-186-0822	159	2	3120-01-186-5527	152	21
2590-01-186-0863	179	10	5315-01-186-5633	202	37
2590-01-186-0863	180	16	4730-01-186-5994	19	15
3040-01-186-0967	165	10	4720-01-186-6018	27	5
2540-01-186-0969	327	8	2910-01-186-6038	15	16
4710-01-186-1009	13	1	2540-01-186-6039	191	6
4710-01-186-1010	13	5	5310-01-186-6930	69	9
4710-01-186-1016	159	10	5310-01-186-7066	69	7
5310-01-186-1245	197	34	5310-01-186-7066	70	4
5310-01-186-1254	1	3	5310-01-186-7066	473	8
5340-01-186-1280	266	6	5306-01-186-7128	181	14
5340-01-186-1280	284	4	5306-01-186-7129	205	25
5340-01-186-1281	266	6	5340-01-186-7173	2	13
5340-01-186-1281	284	4	3120-01-186-7181	158	4
5365-01-186-1294	196	3	5340-01-186-7229	205	28
5365-01-186-1294	196	8	5365-01-186-7233	158	3
5340-01-186-1388	202	32	5342-01-186-7236	196	5
5340-01-186-1817	157	10	5342-01-186-7236	196	15
5340-01-186-1817	158	6	5342-01-186-7382	225	35
4720-01-186-2358	BULK	23	5342-01-186-7382	226	23
2510-01-186-3106	197	28	5340-01-186-7593	225	25
2510-01-186-3107	199	34	5340-01-186-7594	225	11
9515-01-186-3111	267	21	5340-01-186-7658	202	8
5180-01-186-3114	499	8	5340-01-186-7664	183	6
9905-01-186-3253	354	12	5310-01-186-7702	79	15
9905-01-186-3255	354	12	3120-01-186-7714	21	23
9905-01-186-3259	354	12	3120-01-186-7715	205	27
2510-01-186-3311	204	8	5330-01-186-7757	15	30
5310-01-186-3483	222	13	5330-01-186-7757	16	42
2590-01-186-3614	99	1	5365-01-186-7764	193	1
3040-01-186-3718	194	10	5980-01-186-8319	46	2
5360-01-186-3737	21	25	2590-01-186-8704	331	33
2530-01-186-3740	195	10	5340-01-186-8768	19	17
5360-01-186-4844	99	26	5340-01-186-9496	202	38
5360-01-186-4845	165	12	5340-01-186-9606	15	32
5360-01-186-4845	166	3	5340-01-186-9606	16	19
5310-01-186-5237	128	19	4710-01-186-9617	15	9
5310-01-186-5237	128	63	4710-01-186-9617	16	8
5306-01-186-5369	188	27	4710-01-186-9620	161	5
5306-01-186-5369	442	18	4710-01-186-9620	162	5
5305-01-186-5381	103	11	5307-01-187-0519	4	17
5305-01-186-5381	105	23	5310-01-187-0678	374	15
5310-01-186-5420	156	13	5980-01-187-0791	46	3

SECTION IV TM9-2320-280-24P C01

CROSS REFERENCE INDEXES

NATIONAL STOCK NUMBER INDEX

STOCK NUMBER	FIGURE NO	ITEM NO	STOCK NUMBER	FIGURE NO	ITEM NO
5340-01-187-0892	11	27	5340-01-188-1018	226	5
5340-01-187-0892	13	2	5310-01-188-1093	169	4
5340-01-187-0892	14	2	5307-01-188-1229	194	11
5340-01-187-0892	141	21	4720-01-188-1370	29	12
2920-01-187-1310	45	8	4720-01-188-3190	BULK	21
5325-01-187-1604	23	19	2940-01-188-3224	12	5
5325-01-187-1604	24	18	2590-01-188-3225	225	51
5325-01-187-1604	26	2	2590-01-188-3225	226	34
2510-01-187-2238	46	1	2510-01-188-3228	203	18
2510-01-187-2239	194	3	2540-01-188-3229	235	3
4720-01-187-3386	29	7	5342-01-188-3233	99	27
4720-01-187-3387	7	7	3040-01-188-3234	99	28
5310-01-187-3485	193	2	2590-01-188-3237	240	5
2510-01-187-3606	278	4	4710-01-188-3238	103	9
5120-01-187-3607	503	1	2540-01-188-3239	181	15
5120-01-187-3659	502	4	2590-01-188-3267	17	2
5120-01-187-3660	502	1	5995-01-188-3269	81	1
5330-01-187-5148	15	27	2990-01-188-3281	23	7
5330-01-187-5148	16	36	4730-01-188-3296	104	16
4710-01-187-6759	13	9	2940-01-188-3387	12	28
4710-01-187-6759	13	16	4710-01-188-3515	124	7
4710-01-187-6759	14	15	4710-01-188-3515	125	6
4710-01-187-6759	14	22	4710-01-188-3516	124	6
4720-01-187-6911	5	10	4710-01-188-3516	125	7
4730-01-187-6929	6	4	2530-01-188-3520	153	7
5342-01-187-7029	12	36	3040-01-188-3607	145	9
2990-01-187-7030	23	10	2540-01-188-3675	235	9
2510-01-187-7031	182	7	2930-01-188-3682	30	18
2510-01-187-7033	183	25	3040-01-188-3684	144	17
2540-01-187-7034	181	15	3040-01-188-3684	152	17
2510-01-187-7037	193	8	2530-01-188-3685	144	8
2590-01-187-7039	156	20	2530-01-188-3685	152	8
2530-01-187-7161	156	18	2940-01-188-3776	12	25
2990-01-187-7168	23	5	2920-01-188-3863	20	1
5305-01-187-8757	46	9	3020-01-188-3885	34	21
5305-01-187-8757	331	13	4720-01-188-3909	29	2
5305-01-187-8757	332	27	5315-01-188-4490	225	37
9905-01-187-9468	355	2	5315-01-188-4490	226	20
5305-01-187-9555	73	7	5340-01-188-5072	99	5
4710-01-188-0028	29	10	2590-01-188-5079	245	14
4710-01-188-0028	364	8	2590-01-188-5080	245	38
5310-01-188-0745	169	3	2590-01-188-5081	245	1
5365-01-188-0782	4	23	3120-01-188-5082	275	15
5330-01-188-0911	12	38	3120-01-188-5082	276	14
5340-01-188-1017	136	17	6140-01-188-5113	247	4
5340-01-188-1017	137	10	2940-01-188-5117	12	23
5340-01-188-1018	225	21	2520-01-188-5131	148	11

I-55

SECTION IV TM9-2320-280-24P C01

CROSS REFERENCE INDEXES

NATIONAL STOCK NUMBER INDEX

STOCK NUMBER	FIGURE NO	ITEM NO	STOCK NUMBER	FIGURE NO	ITEM NO
5305-01-188-5133	102	2	4720-01-189-2218	333	38
7690-01-188-5144	356	5	2510-01-189-3456	209	7
4710-01-188-6050	15	8	2540-01-189-3457	319	1
4710-01-188-6050	16	7	2510-01-189-3459	320	5
5310-01-188-6861	51	13	2510-01-189-3459	324	10
5310-01-188-6861	165	11	2510-01-189-3460	320	5
2510-01-188-7381	280	40	2510-01-189-3460	324	10
2510-01-188-7381	281	40	2540-01-189-3706	320	1
5340-01-188-7382	249	29	2540-01-189-3707	322	1
5340-01-188-7383	249	4	7690-01-189-3736	361	6
2590-01-188-7384	246	5	7690-01-189-3737	361	2
2590-01-188-7386	276	10	7690-01-189-3738	361	1
2590-01-188-7394	270	6	7690-01-189-3739	361	5
2520-01-188-8269	99	19	7690-01-189-3740	359	2
3040-01-188-8292	100	8	7690-01-189-3740	361	3
2530-01-188-8446	153	8	5305-01-189-3769	34	20
4710-01-188-8780	124	3	4730-01-189-4600	489	2
5305-01-188-9707	KITS	106	2910-01-189-4770	15	1
2520-01-189-0594	137	9	5977-01-189-4988	KITS	8
4710-01-189-0705	124	2	5342-01-189-5452	196	4
4710-01-189-0705	125	2	5342-01-189-5452	196	14
4720-01-189-0853	178	18	2520-01-189-6726	133	10
4730-01-189-0871	12	7	5340-01-189-6748	21	14
4330-01-189-0889	19	6	5340-01-189-7554	331	37
2510-01-189-0890	184	6	5340-01-189-7554	332	26
2510-01-189-0891	182	7	5340-01-189-7558	68	2
2530-01-189-0897	68	9	5340-01-189-7640	161	2
4330-01-189-1007	19	1	5340-01-189-7640	162	2
2520-01-189-1064	99	16	5340-01-189-7640	487A	9
2590-01-189-1067	225	53	5340-01-189-7640	487C	1
2590-01-189-1067	226	35	5340-01-189-7640	487D	5
2540-01-189-1074	185	1	5340-01-189-7739	266	19
4710-01-189-1080	124	4	5310-01-189-8468	34	4
3040-01-189-1637	138	7	5310-01-189-8476	159	9
3040-01-189-1637	145	7	5310-01-189-8476	160	5
2930-01-189-1744	34	5	5310-01-189-8476	161	10
2530-01-189-1745	184	10	5310-01-189-8476	162	16
2940-01-189-1809	12	3	5310-01-189-8476	487A	11
2510-01-189-1832	193	4	5310-01-189-8476	487D	7
4820-01-189-2107	333	37	5310-01-189-8485	165	8
3040-01-189-2134	99	24	2930-01-189-8643	34	18
2520-01-189-2135	KITS	43	2910-01-189-8851	4	11
3040-01-189-2172	133	14	4730-01-189-8854	15	12
2540-01-189-2193	191	8	4730-01-189-8854	16	11
2530-01-189-2195	170	11	2510-01-189-9717	207	1
4710-01-189-2208	161	11	2510-01-189-9720	204	11
4710-01-189-2209	161	3	9515-01-189-9721	267	12

I-56

SECTION IV TM9-2320-280-24P C01

CROSS REFERENCE INDEXES

NATIONAL STOCK NUMBER INDEX

STOCK NUMBER	FIGURE NO	ITEM NO	STOCK NUMBER	FIGURE NO	ITEM NO
2510-01-189-9724	316	7	4330-01-190-3579	KITS	5
2510-01-189-9724	320	19	2510-01-190-3862	193	4
2510-01-189-9724	324	24	5305-01-190-4073	19	2
2510-01-189-9725	316	7	5306-01-190-4553	19	3
2510-01-189-9725	320	19	5310-01-190-4607	34	15
2510-01-189-9725	324	24	5340-01-190-5100	15	28
4710-01-189-9727	273	23	5340-01-190-5100	16	38
9515-01-189-9728	205	37	5306-01-190-5760	136	2
2540-01-189-9729	204	11	5306-01-190-5760	137	14
2510-01-189-9730	204	3	5307-01-190-5854	34	12
9515-01-189-9731	267	15	5315-01-190-5908	266	20
9515-01-189-9732	267	8	5360-01-190-6214	34	11
9515-01-189-9733	267	20	5360-01-190-6218	99	15
2590-01-189-9734	273	6	2540-01-190-7079	333	11
2590-01-189-9736	274	24	5360-01-190-7234	KITS	9
2590-01-189-9736	276	21	5330-01-190-7509	KITS	42
5330-01-189-9738	23	3	5330-01-190-7510	KITS	41
5330-01-189-9738	24	15	5355-01-190-7635	99	2
2510-01-189-9741	268	16	5307-01-190-7650	103	12
5365-01-189-9743	204	4	5342-01-190-7735	103	15
2510-01-189-9744	206	17	9905-01-190-8425	354	12
9515-01-189-9745	267	19	9905-01-190-8427	354	12
9515-01-189-9746	267	19	2510-01-190-8458	204	3
2590-01-189-9747	274	14	7690-01-190-8501	359	5
2510-01-189-9748	280	40	7690-01-190-8501	360	3
2510-01-189-9748	281	40	5325-01-191-0595	194	4
2520-01-189-9750	128	4	5340-01-191-0746	15	26
2510-01-189-9828	274	1	5340-01-191-0746	16	37
2510-01-189-9828	276	25	5340-01-191-0748	15	25
5340-01-189-9979	202	15	5340-01-191-0748	16	33
3110-01-189-9980	21	22	2510-01-191-0971	204	8
5342-01-189-9982	2	7	2540-01-191-0973	452	5
5340-01-190-0333	179	18	7690-01-191-1312	359	6
5340-01-190-0333	180	28	7690-01-191-1313	359	1
5340-01-190-0402	12	29	7690-01-191-1314	359	10
5340-01-190-0807	136	1	7690-01-191-1315	359	9
5340-01-190-0807	137	13	7690-01-191-1316	359	8
5340-01-190-0810	194	5	5315-01-191-2611	133	4
5340-01-190-0815	244	19	5340-01-191-3222	203	17
2990-01-190-1089	23	15	3120-01-191-3232	99	23
2990-01-190-1089	24	24	3120-01-191-3232	133	8
5930-01-190-1231	54	9	5340-01-191-3359	206	12
5306-01-190-2193	71	4	5340-01-191-3526	15	33
5340-01-190-2248	141	18	4820-01-191-4262	172	6
5340-01-190-2248	162	1	3120-01-191-4637	KITS	109
5340-01-190-2248	181	13	7690-01-191-6467	359	7
3040-01-190-3574	138	9	2920-01-191-6534	KITS	110

I-57

SECTION IV TM9-2320-280-24P C01

CROSS REFERENCE INDEXES

NATIONAL STOCK NUMBER INDEX

STOCK NUMBER	FIGURE NO	ITEM NO	STOCK NUMBER	FIGURE NO	ITEM NO
5325-01-191-7555	208	22	2540-01-192-9711	403	5
5340-01-191-8161	170	16	2540-01-192-9712	403	1
5340-01-191-8161	195	5	2540-01-192-9712	403	6
2530-01-191-8740	KITS	71	2540-01-192-9713	403	1
2530-01-191-8741	KITS	36	2540-01-192-9713	403	6
3020-01-191-8784	148	27	2540-01-192-9714	403	21
7690-01-191-8793	359	4	2540-01-192-9715	403	21
2520-01-192-1260	109	9	2540-01-192-9716	267	7
3040-01-192-1632	230	15	2540-01-192-9718	277	21
5306-01-192-2207	169	26	4710-01-192-9720	277	15
5340-01-192-2256	443	8	2540-01-192-9721	269	10
5340-01-192-2256	457	7	2510-01-192-9723	278	4
2920-01-192-2956	KITS	107	5340-01-193-0251	15	29
2920-01-192-2959	KITS	99	5340-01-193-0251	16	41
2590-01-192-3021	314	22	2510-01-193-1807	207	1
5340-01-192-3414	182	5	6220-01-193-1970	62	14
3040-01-192-3673	488	19	5330-01-193-2338	266	9
2590-01-192-4425	399	11	2540-01-193-2711	403	22
2920-01-192-4473	KITS	108	5365-01-193-2951	169	18
2920-01-192-4474	KITS	11	9905-01-193-4065	354	5
2540-01-192-4500	199	30	9905-01-193-4065	370	20
2540-01-192-4501	403	15	5365-01-193-4479	403	20
2540-01-192-4502	399	12	5342-01-193-7088	488	14
2540-01-192-4504	269	10	5945-01-193-7175	51	5
2590-01-192-4525	277	8	4730-01-193-7390	345	12
4810-01-192-5817	177	5	4730-01-193-7390	348	17
2540-01-192-5937	277	41	2930-01-193-7802	31	3
2510-01-192-5947	199	32	5977-01-193-9931	68	11
2540-01-192-5948	267	10	2540-01-194-0197	277	26
2540-01-192-5949	277	28	2590-01-194-0309	243	6
2540-01-192-5949	278	9	2590-01-194-0309	244	5
2540-01-192-5949	440	5	2590-01-194-0309	244	13
2510-01-192-5950	270	15	5330-01-194-0472	8	9
4520-01-192-6073	343	9	5330-01-194-0473	15	11
4710-01-192-6113	277	42	5330-01-194-0473	16	10
7690-01-192-6369	361	10	5310-01-194-0481	34	3
7690-01-192-6370	361	9	5325-01-194-0562	17	3
7690-01-192-6371	359	3	5365-01-194-0761	267	16
6625-01-192-7498	69	28	5340-01-194-0887	270	7
4820-01-192-7678	399	18	5340-01-194-0887	340	5
4710-01-192-7965	268	31	5340-01-194-0888	270	27
4820-01-192-8030	399	2	2590-01-194-2048	250	22
4730-01-192-8086	141	19	2530-01-194-2049	170	10
2540-01-192-8283	270	15	5342-01-194-3128	197	13
5331-01-192-8892	34	6	5342-01-194-3128	199	54
5330-01-192-8916	203	16	2540-01-194-3323	345	1
4710-01-192-9546	268	36	2540-01-194-3323	368	19

SECTION IV TM9-2320-280-24P C01

CROSS REFERENCE INDEXES

NATIONAL STOCK NUMBER INDEX

STOCK NUMBER	FIGURE NO	ITEM NO	STOCK NUMBER	FIGURE NO	ITEM NO
5365-01-194-5093	270	18	2590-01-196-1290	325	11
5340-01-194-5294	4	14	2590-01-196-1290	376	6
4720-01-194-5334	398	32	2540-01-196-1291	321	13
5342-01-194-5806	269	30	2540-01-196-1291	376	10
5360-01-194-6070	271	4	3040-01-196-1493	281	33
5360-01-194-6070	271	21	4720-01-196-1636	27	9
5306-01-194-6433	97	5	4710-01-196-1642	3	2
5310-01-194-6459	277	36	5340-01-196-2573	403	19
6620-01-194-6586	66	11	5365-01-196-2699	243	18
2590-01-194-6979	249	37	5355-01-196-2770	133	1
2590-01-194-6980	249	30	5340-01-196-3297	267	19
2590-01-194-6981	249	5	5307-01-196-4717	4	26
2590-01-194-6990	333	35	5340-01-196-4720	188	3
5310-01-194-7066	199	56	2540-01-196-4721	188	16
5310-01-194-7066	208	7	2540-01-196-4722	188	16
5310-01-194-7066	331	34	2540-01-196-4723	188	21
4010-01-194-8546	266	8	2530-01-196-4724	208	1
4010-01-194-8546	284	5	2540-01-196-4726	12	33
5331-01-194-8966	34	10	2540-01-196-4920	188	7
5306-01-195-0277	269	17	5307-01-196-4937	142	32
5331-01-195-1500	34	9	5307-01-196-4937	143	24
5340-01-195-1575	69	3	5307-01-196-4937	150	32
5995-01-195-1918	282	1	5307-01-196-4937	151	24
6680-01-195-2146	488	11	2590-01-196-5228	4	1
5120-01-195-2721	501	4	2510-01-196-5311	203	4
4730-01-195-3803	400	7	5342-01-196-5312	209	11
3040-01-195-4173	281	33	2540-01-196-5313	225	8
5120-01-195-4551	501	11	2540-01-196-5313	325	2
5365-01-195-4949	277	5	2590-01-196-5314	321	7
5340-01-195-5397	243	19	2590-01-196-5314	325	11
5305-01-195-5818	8	14	2590-01-196-5314	376	6
5360-01-195-6200	243	25	2510-01-196-5316	280	15
5325-01-195-8002	71	6	2510-01-196-5316	281	15
5342-01-195-8039	280	15	2540-01-196-5317	280	22
5342-01-195-8039	281	15	2540-01-196-5317	281	22
5340-01-195-8057	315	2	2510-01-196-5506	280	40
5340-01-195-8837	269	22	2510-01-196-5506	281	40
5331-01-195-8889	488	18	5325-01-196-5631	245	8
5330-01-195-9049	270	24	2590-01-196-7281	BULK	43
5330-01-195-9083	268	2	5340-01-196-7733	246	26
5340-01-195-9107	280	18	5330-01-196-7968	280	26
5340-01-195-9107	281	18	5330-01-196-7969	281	26
5180-01-195-9777	500	2	5340-01-196-8028	403	17
5120-01-196-0084	501	1	5340-01-196-8158	269	23
5315-01-196-0277	277	25	5310-01-197-1100	270	36
5340-01-196-0572	277	38	5310-01-197-1161	209	1
2590-01-196-1290	321	7	5340-01-197-1238	269	5

I-59

SECTION IV TM9-2320-280-24P C01

CROSS REFERENCE INDEXES

NATIONAL STOCK NUMBER INDEX

STOCK NUMBER	FIGURE NO	ITEM NO	STOCK NUMBER	FIGURE NO	ITEM NO
5340-01-197-1294	220	2	5340-01-197-5477	198	17
5340-01-197-1294	230	20	5340-01-197-5477	321	20
5340-01-197-1294	231	25	5340-01-197-5477	325	23
5340-01-197-1294	235	16	2540-01-197-5478	324	1
5340-01-197-1444	268	12	7690-01-197-5500	356	4
5306-01-197-1492	108	1	7690-01-197-5500	357	1
5315-01-197-1494	323	9	2530-01-197-5502	490	14
5315-01-197-1494	326	8	5975-01-197-5505	409	3
5305-01-197-1594	271	2	5975-01-197-5505	411	3
5305-01-197-1594	271	19	5975-01-197-5505	430	48
2530-01-197-2160	170	17	3040-01-197-5510	368	41
2530-01-197-2160	195	6	2540-01-197-5525	325	5
5340-01-197-2382	398	28	2540-01-197-5526	326	1
5340-01-197-2438	62	24	2540-01-197-5528	375	4
5340-01-197-2642	270	25	2510-01-197-5546	375	12
5340-01-197-2716	270	29	2510-01-197-5546	378	11
5340-01-197-2717	268	7	2510-01-197-5547	375	12
5340-01-197-2736	269	18	2510-01-197-5547	378	11
5315-01-197-3122	245	5	5340-01-197-5548	37	5
5306-01-197-3274	150	27	5995-01-197-5554	82	1
5306-01-197-3277	153	10	5310-01-197-6253	254	26
5325-01-197-3460	153	22	5310-01-197-6253	287	6
5325-01-197-3460	332	21	5340-01-197-6714	281	29
5340-01-197-3461	26	9	5340-01-197-6715	245	2
5340-01-197-3478	245	19	5340-01-197-6716	245	6
5330-01-197-3620	208	31	5340-01-197-6753	268	39
2590-01-197-4898	317	6	5340-01-197-6755	277	49
2590-01-197-4898	373	5	5305-01-197-7547	68	12
2590-01-197-4899	272	37	5315-01-197-7563	275	8
2540-01-197-4900	280	22	2540-01-197-8071	374	16
2540-01-197-4900	281	22	5340-01-197-8239	269	25
5340-01-197-4901	280	9	5330-01-197-8298	268	15
2590-01-197-5446	199	49	5330-01-197-8298	295	6
2540-01-197-5447	197	7	2510-01-197-8572	246	20
2540-01-197-5448	197	2	2510-01-197-8572	253	24
2540-01-197-5449	365	6	2510-01-197-8572	254	19
2540-01-197-5450	368	4	5340-01-197-8673	216	3
9330-01-197-5458	374	13	5340-01-197-8673	252	19
2540-01-197-5460	365	23	5340-01-197-8673	253	36
2540-01-197-5463	374	20	5340-01-197-8673	266	12
5999-01-197-5465	410	KIT	5340-01-197-8673	317	10
5895-01-197-5467	412	KIT	5340-01-197-8673	321	23
5340-01-197-5469	286	4	5340-01-197-8673	323	15
5340-01-197-5470	416	5	5340-01-197-8673	326	6
2590-01-197-5474	286	39	5340-01-197-8673	377	6
2590-01-197-5475	286	39	5340-01-197-8674	228	13
2540-01-197-5476	325	15	5340-01-197-8674	248	16

SECTION IV TM9-2320-280-24P C01

CROSS REFERENCE INDEXES

NATIONAL STOCK NUMBER INDEX

STOCK NUMBER	FIGURE NO	ITEM NO	STOCK NUMBER	FIGURE NO	ITEM NO
5340-01-197-8674	429	14	5340-01-198-2248	247	17
5340-01-197-8674	440	22	5340-01-198-2248	249	39
5340-01-197-8674	446	31	5340-01-198-2249	246	16
5340-01-197-8675	280	6	5340-01-198-2253	271	30
5340-01-197-8675	281	6	5340-01-198-2253	274	16
3040-01-197-8727	272	26	5340-01-198-2254	249	24
2540-01-197-8735	323	1	5340-01-198-2255	246	10
5365-01-197-9383	280	21	5340-01-198-2255	247	20
5365-01-197-9383	281	21	5340-01-198-2255	248	11
5305-01-197-9417	463	14	5340-01-198-2255	249	14
2510-01-198-0333	398	2	5340-01-198-2256	249	43
5340-01-198-0480	75	49	5340-01-198-2257	249	45
5340-01-198-0480	77	27	2590-01-198-2895	317	6
3020-01-198-0633	35	32	2590-01-198-2895	373	5
5340-01-198-0686	375	9	5315-01-198-3346	245	12
5340-01-198-0686	375	13	5305-01-198-3385	323	2
5340-01-198-0686	378	8	5305-01-198-3385	323	7
5340-01-198-0686	378	14	5305-01-198-3385	324	22
5305-01-198-1619	324	11	5305-01-198-3385	324	34
5305-01-198-1619	324	25	5305-01-198-3385	326	2
5305-01-198-1619	375	16	5365-01-198-3399	248	12
5305-01-198-1621	272	14	5365-01-198-3400	248	21
5310-01-198-1722	53	5	5365-01-198-3401	248	18
5310-01-198-1722	277	20	5310-01-198-3428	280	27
5310-01-198-1723	225	20	5310-01-198-3428	281	27
5310-01-198-1723	246	15	5305-01-198-3440	463	7
5310-01-198-1723	251	35	5340-01-198-3457	248	13
5310-01-198-1723	252	32	5365-01-198-3463	248	17
5310-01-198-1723	253	27	5310-01-198-3487	280	13
5310-01-198-1723	254	18	5310-01-198-3487	281	13
5310-01-198-1723	255	30	5365-01-198-3505	248	22
5310-01-198-1723	263	3	5330-01-198-3521	429	27
5310-01-198-1723	264	6	5330-01-198-3521	BULK	51
5310-01-198-1723	266	11	5360-01-198-3525	280	43
5310-01-198-1723	318	6	5360-01-198-3525	281	43
5310-01-198-1723	333	27	5340-01-198-5165	246	4
5310-01-198-1723	429	16	5340-01-198-5165	450	7
5310-01-198-1723	446	21	5340-01-198-5455	269	19
5310-01-198-1723	448	6	3040-01-198-5699	99	14
5310-01-198-1723	450	6	2510-01-198-7413	225	34
5310-01-198-1724	197	4	2510-01-198-7413	226	24
5310-01-198-1724	417	8	3040-01-198-7473	273	10
5310-01-198-1724	430	18	5340-01-198-7580	280	29
5340-01-198-2240	280	8	2540-01-198-7581	281	29
5340-01-198-2240	281	8	5120-01-198-7583	499	4
5340-01-198-2241	280	8	5310-01-198-7585	167	7
5340-01-198-2241	281	8	5310-01-198-7585	167	20

SECTION IV TM9-2320-280-24P C01

CROSS REFERENCE INDEXES

NATIONAL STOCK NUMBER INDEX

STOCK NUMBER	FIGURE NO	ITEM NO	STOCK NUMBER	FIGURE NO	ITEM NO
5330-01-198-8789	280	26	5340-01-199-9679	326	9
5306-01-199-1314	270	3	5330-01-200-0466	23	13
2510-01-199-1498	280	9	5330-01-200-0466	24	21
2510-01-199-1499	280	32	5330-01-200-0466	398	24
2510-01-199-1499	281	32	1015-01-200-0869	250	17
5340-01-199-1662	268	14	1015-01-200-0869	252	6
5340-01-199-1663	246	3	6220-01-200-0897	197	17
5325-01-199-2254	322	6	6220-01-200-0897	199	51
5325-01-199-2254	324	6	3040-01-200-1015	280	33
5310-01-199-2293	227	32	2540-01-200-1994	316	9
5310-01-199-2293	377	2	2540-01-200-1994	320	9
2910-01-199-2355	10	2	2540-01-200-1994	320	21
9905-01-199-2371	367	2	2540-01-200-1994	324	12
2540-01-199-2390	378	KITS	2540-01-200-1994	324	26
2540-01-199-2396	366	2	2540-01-200-1994	375	17
5305-01-199-3371	266	13	2540-01-200-1994	378	17
5305-01-199-3371	317	9	2540-01-200-1995	316	14
5305-01-199-3371	318	5	2540-01-200-1995	320	12
5310-01-199-3440	209	4	2540-01-200-1995	320	24
5310-01-199-3440	225	40	2540-01-200-1995	324	19
5340-01-199-3495	245	31	2540-01-200-1995	324	31
5340-01-199-3498	280	6	2540-01-200-1995	375	22
5340-01-199-3498	281	6	2540-01-200-1995	378	22
5340-01-199-3499	272	21	2520-01-200-3096	136	16
5340-01-199-3510	375	26	2520-01-200-3097	136	6
5340-01-199-4989	272	30	3120-01-200-4105	268	34
2590-01-199-5423	21	15	4730-01-200-4277	141	9
2590-01-199-5425	285	1	4730-01-200-4277	399	20
4720-01-199-5717	BULK	7	2540-01-200-6611	318	1
2510-01-199-5748	321	17	2540-01-200-8393	316	4
2510-01-199-5748	446	5	2540-01-201-2356	281	9
2510-01-199-5748	446	22	2540-01-201-2357	281	9
2510-01-199-5748	448	5	5315-01-201-3592	169	5
2540-01-199-5812	280	29	5995-01-201-4128	410	3
5340-01-199-6103	269	19	5995-01-201-4128	412	3
2540-01-199-6759	317	1	5995-01-201-4129	410	2
2540-01-199-6759	321	2	5995-01-201-4129	412	2
2540-01-199-6759	376	1	5365-01-201-4749	280	5
2540-01-199-6760	318	2	5365-01-201-4749	281	5
2540-01-199-6760	389	6	5180-01-201-4978	504	10
2540-01-199-6761	318	3	5995-01-201-7495	410	1
2540-01-199-7778	327	5	5995-01-201-7495	412	1
2590-01-199-7977	248	24	5120-01-201-7857	501	20
5340-01-199-8258	247	19	5306-01-201-7969	133	6
5930-01-199-8853	66	3	1640-01-201-9062	311	8
5930-01-199-8853	66A	5	1640-01-201-9062	312	4
5340-01-199-9679	323	10	5340-01-201-9076	280	6

SECTION IV TM9-2320-280-24P C01

CROSS REFERENCE INDEXES

NATIONAL STOCK NUMBER INDEX

STOCK NUMBER	FIGURE NO	ITEM NO	STOCK NUMBER	FIGURE NO	ITEM NO
5340-01-201-9076	281	6	5340-01-203-2708	446	23
5305-01-202-2676	155	16	5340-01-203-2708	449	5
5340-01-202-7445	323	3	5340-01-203-2821	331	36
5340-01-202-7445	326	3	5342-01-203-2822	198	23
4710-01-202-8245	400	1	5342-01-203-2822	280	48
5330-01-202-8360	273	20	5342-01-203-2822	281	48
2540-01-202-8545	409	11	5342-01-203-2822	321	27
2540-01-202-8545	411	15	5342-01-203-2822	377	10
2540-01-202-8545	415	17	5310-01-203-3217	280	20
2590-01-202-8546	409	6	5310-01-203-3217	281	20
2590-01-202-8546	411	12	5310-01-203-3217	429	3
2590-01-202-8546	415	11	5310-01-203-3217	430	3
2540-01-202-8547	208	16	5310-01-203-3217	430	22
2540-01-202-8548	411	8	5310-01-203-3217	430	39
4931-01-202-8692	213	13	5310-01-203-3230	273	9
4931-01-202-8692	214	7	5340-01-203-3273	209	2
4931-01-202-8692	252	24	5340-01-203-4646	247	27
4931-01-202-8692	254	23	5340-01-203-5629	245	3
4931-01-202-8692	267	11	5340-01-203-5630	245	7
4931-01-202-8692	342	6	5340-01-203-5634	243	10
4931-01-202-8692	401	18	5340-01-203-5634	270	38
4931-01-202-8692	446	24	5342-01-203-5661	174	3
4931-01-202-8692	498	19	2530-01-203-5662	142	18
7690-01-203-0161	361	4	2530-01-203-5662	150	18
2540-01-203-0183	230	18	2530-01-203-5663	142	33
2540-01-203-0183	235	10	2530-01-203-5663	150	33
5340-01-203-0324	280	18	2540-01-203-5664	184	18
5340-01-203-0324	281	18	2530-01-203-5746	142	1
4010-01-203-0411	243	17	2530-01-203-5746	150	1
4010-01-203-0412	270	4	5340-01-203-6538	249	28
4710-01-203-0607	334	14	5340-01-203-6539	248	1
4710-01-203-0607	366	19	5340-01-203-6542	319	6
4710-01-203-0607	386	5	5340-01-203-6542	320	4
4710-01-203-0608	334	18	5340-01-203-6542	322	5
4710-01-203-0608	366	23	5340-01-203-6542	323	4
4710-01-203-0608	384	1	5340-01-203-6542	324	5
4710-01-203-0615	334	13	5340-01-203-6542	326	4
4710-01-203-0615	386	4	5340-01-203-6542	375	7
4730-01-203-1025	385	8	5330-01-203-6551	142	2
4710-01-203-1304	368	28	5330-01-203-6551	143	2
5340-01-203-1980	197	18	5330-01-203-6551	150	2
5340-01-203-1980	199	61	5330-01-203-6551	151	2
5340-01-203-1980	227	8	2540-01-203-7721	316	9
6210-01-203-2101	367	4	2540-01-203-7721	320	9
5306-01-203-2637	313	21	2540-01-203-7721	320	21
5340-01-203-2707	248	8	2540-01-203-7721	324	12
5340-01-203-2708	247	28	2540-01-203-7721	324	26

SECTION IV TM9-2320-280-24P C01

CROSS REFERENCE INDEXES

NATIONAL STOCK NUMBER INDEX

STOCK NUMBER	FIGURE NO	ITEM NO	STOCK NUMBER	FIGURE NO	ITEM NO
2540-01-203-7721	375	17	5340-01-204-3862	321	4
2540-01-203-7721	378	17	5340-01-204-3862	325	7
4720-01-203-7789	12	8	5340-01-204-3862	373	2
5305-01-203-8346	317	4	5340-01-204-3862	376	3
5305-01-203-8346	321	5	5340-01-204-3903	334	3
5305-01-203-8346	325	8	5340-01-204-3903	366	27
5315-01-203-8543	280	41	5340-01-204-3903	384	12
5315-01-203-8543	281	41	5340-01-204-3904	368	21
5340-01-203-8608	225	4	5305-01-204-4190	275	17
5340-01-203-8608	226	38	5305-01-204-4190	276	16
5340-01-203-8674	274	29	4730-01-204-5457	368	29
5340-01-203-8675	272	20	5305-01-204-6502	25	5
5340-01-203-8675	280	17	5305-01-204-6502	51	6
5340-01-203-8675	281	17	5305-01-204-6502	83	6
5330-01-203-9187	368	1	5305-01-204-6502	84	1
5340-01-203-9391	278	15	5305-01-204-6502	85	25
3130-01-203-9870	136	12	5305-01-204-6502	335	6
2520-01-203-9871	131	4	5310-01-204-6745	490	12
2540-01-203-9872	278	6	2520-01-204-7699	97	1
2510-01-203-9873	278	14	9905-01-204-7776	354	9
2590-01-203-9874	283	28	7690-01-204-7785	360	1
2590-01-203-9875	283	17	7690-01-204-7847	360	2
2510-01-203-9883	184	15	7690-01-204-7849	361	7
9905-01-203-9992	354	10	7690-01-204-7850	361	8
9905-01-203-9994	354	4	3020-01-204-8132	472	22
9905-01-203-9994	370	19	5940-01-204-8830	412	8
9905-01-203-9995	354	14	5342-01-204-9610	321	26
9905-01-203-9996	354	17	5342-01-204-9610	325	24
7690-01-204-0076	354	18	5342-01-204-9610	376	16
7690-01-204-0077	354	1	2590-01-205-2506	278	13
2520-01-204-0325	132	6	2510-01-205-2507	278	5
4210-01-204-0913	241	10	5342-01-205-2519	314	26
5310-01-204-1039	73	4	5315-01-205-2988	225	36
5310-01-204-1039	434	2	5315-01-205-2988	226	19
5310-01-204-1039	435	2	5325-01-205-3203	410	36
5306-01-204-2139	280	28	5325-01-205-3203	412	12
5306-01-204-2139	281	28	5340-01-205-5118	209	2
5315-01-204-2328	280	39	5340-01-205-5377	188	20
5315-01-204-2328	281	39	5325-01-205-5378	339	5
5340-01-204-2458	280	6	5325-01-205-5378	410	31
5340-01-204-2458	281	6	5325-01-205-5378	412	24
5340-01-204-2543	272	36	5975-01-205-5379	313	7
2530-01-204-2583	158	2	5975-01-205-5379	333	32
5340-01-204-2584	183	11	5975-01-205-5379	339	13
5310-01-204-3344	213	25	5975-01-205-5379	410	28
5310-01-204-3344	214	20	5975-01-205-5379	443	6
5340-01-204-3862	317	3	5975-01-205-5379	485	9

I-64

SECTION IV TM9-2320-280-24P C01

CROSS REFERENCE INDEXES

NATIONAL STOCK NUMBER INDEX

STOCK NUMBER	FIGURE NO	ITEM NO	STOCK NUMBER	FIGURE NO	ITEM NO
5342-01-205-5381	77	26	5305-01-206-7217	324	33
5325-01-205-5966	129	11	5305-01-206-7217	375	20
2990-01-205-5981	45	2	5305-01-206-7219	13	8
5305-01-205-6048	167	9	5305-01-206-7219	14	14
2510-01-205-6089	195	3	5305-01-206-7219	17	15
2920-01-205-8605	KITS	1	5305-01-206-7219	88	35
2590-01-205-8607	273	21	5310-01-206-7306	1	4
9905-01-205-8635	354	15	5310-01-206-7306	2	5
9905-01-205-8635	399	1	5310-01-206-7306	12	19
2920-01-205-8653	43	38	5310-01-206-7306	30	8
5340-01-205-9021	316	17	5310-01-206-7306	33	11
5340-01-205-9021	320	13	5310-01-206-7306	35	26
5340-01-205-9021	320	25	5310-01-206-7306	36	19
5340-01-205-9021	324	23	5310-01-206-7306	37	14
5340-01-205-9021	324	35	5310-01-206-7306	38	19
5340-01-205-9021	375	24	5310-01-206-7306	41	12
5340-01-205-9021	378	24	5310-01-206-7306	42	10
5340-01-205-9022	316	17	5310-01-206-7306	66	7
5340-01-205-9022	320	13	5310-01-206-7306	75	31
5340-01-205-9022	320	25	5310-01-206-7306	77	9
5340-01-205-9022	324	23	5310-01-206-7306	174	2
5340-01-205-9022	324	35	5310-01-206-7306	33A	9
5340-01-205-9022	375	24	5310-01-206-7306	364	6
5340-01-205-9022	378	24	5310-01-206-7306	379	37
5310-01-205-9056	316	13	5310-01-206-7306	472	2
5310-01-205-9056	320	11	5310-01-206-7306	473	10
5310-01-205-9056	320	23	5310-01-206-7306	474	4
5310-01-205-9056	324	18	5310-01-206-7306	475	14
5310-01-205-9056	324	30	5310-01-206-7306	476	16
5310-01-205-9056	378	21	5310-01-206-7306	480	8
2530-01-206-3874	KITS	4	5310-01-206-7306	490	7
2520-01-206-3875	128	69	5310-01-206-7306	490A	6
3040-01-206-3876	128	15	5365-01-206-7486	72	7
2520-01-206-3877	128	9	5340-01-206-7780	241	1
2540-01-206-4115	317	2	5305-01-206-8401	75	28
2540-01-206-4115	321	3	5305-01-206-8401	77	12
2540-01-206-4115	373	1	5305-01-206-8401	84	16
2540-01-206-4115	376	2	5305-01-206-8401	88	12
5340-01-206-5040	240	21	5305-01-206-8401	477	1
5340-01-206-5040	243	22	5340-01-207-0379	199	24
5310-01-206-5479	365	18	5340-01-207-0610	241	2
5340-01-206-5540	75	41	5340-01-207-0717	198	24
5315-01-206-7135	293	34	5340-01-207-0717	321	33
5306-01-206-7172	66	8	5340-01-207-0717	325	28
5306-01-206-7172	77	24	5340-01-207-0717	373	10
5306-01-206-7172	79	18	5975-01-207-1706	75	14
5305-01-206-7217	324	21	2590-01-207-3696	173	2

I-65

SECTION IV TM9-2320-280-24P C01

CROSS REFERENCE INDEXES

NATIONAL STOCK NUMBER INDEX

STOCK NUMBER	FIGURE NO	ITEM NO	STOCK NUMBER	FIGURE NO	ITEM NO
2530-01-207-6256	158	2	5330-01-209-5997	225	54
9905-01-207-6304	133	2	2590-01-209-6041	226	34
5306-01-207-7487	374	7	2590-01-209-6042	226	35
6150-01-207-8120	416	9	2590-01-209-6043	226	26
6150-01-207-8120	426	16	2540-01-209-6049	271	3
6150-01-207-8120	427	16	2540-01-209-6049	271	20
6150-01-207-8120	428	9	4710-01-209-6746	398	14
5307-01-207-9004	31	9	2540-01-209-6838	315	1
5305-01-208-0274	325	10	5340-01-209-6874	315	4
6210-01-208-4790	47	1	2540-01-209-6875	284	16
5310-01-208-5252	374	23	2540-01-209-6875	439	9
5325-01-208-5424	410	10	4030-01-209-7047	314	20
5325-01-208-5424	414	8	5340-01-209-7386	411	17
5325-01-208-5424	416	8	5340-01-209-7387	75	34
5310-01-208-7576	410	15	5340-01-209-7387	77	15
5310-01-208-7576	414	10	5305-01-209-7667	128	12
5310-01-208-7576	416	13	5305-01-209-7668	128	17
5310-01-208-7576	425	11	5305-01-209-7668	130	26
5310-01-208-7576	426	11	5330-01-209-7669	45	3
5310-01-208-7576	427	12	5310-01-209-7702	225	39
5310-01-208-7576	428	13	5310-01-209-7702	231	21
5340-01-208-7670	321	26	5365-01-209-7713	128	71
5340-01-208-7670	325	24	5360-01-209-7715	121	13
5340-01-208-7670	376	16	5330-01-209-7723	169	24
5340-01-208-8688	12	32	5331-01-209-7726	131	3
5975-01-208-9618	41	8	5340-01-209-7761	250	3
5975-01-208-9618	44	22	5340-01-209-7767	149	10
5975-01-208-9618	44	27	5340-01-209-7767	183	26
5975-01-208-9618	474	15	5340-01-209-7767	184	12
5975-01-208-9618	480	18	5340-01-209-7794	51	7
5975-01-208-9618	480	30	5340-01-209-7795	159	13
2510-01-209-0502	202	21	5340-01-209-7796	161	1
4910-01-209-0729	500	6	5340-01-209-7799	85	27
4730-01-209-0845	135	4	5340-01-209-7799	88	8
7690-01-209-0864	356	1	5340-01-209-7802	196	9
2510-01-209-1881	227	12	5340-01-209-7802	196	14
3040-01-209-3376	314	28	5340-01-209-7807	153	19
3040-01-209-3409	314	36	5340-01-209-7808	149	13
2510-01-209-3450	280	40	5330-01-209-7817	331	11
2510-01-209-3450	281	40	5340-01-209-7830	121	14
2540-01-209-4590	202	24	5365-01-209-7831	181	5
3020-01-209-4603	314	4	5365-01-209-7831	184	9
3020-01-209-4606	314	5	5365-01-209-7832	181	5
4710-01-209-4628	178	8	5365-01-209-7832	184	9
5340-01-209-4870	280	11	5340-01-209-7834	203	28
5340-01-209-4870	281	11	5330-01-209-7842	331	16
5935-01-209-5594	79	3	5325-01-209-7843	333	25

I-66

SECTION IV TM9-2320-280-24P C01

CROSS REFERENCE INDEXES

NATIONAL STOCK NUMBER INDEX

STOCK NUMBER	FIGURE NO	ITEM NO	STOCK NUMBER	FIGURE NO	ITEM NO
5325-01-209-7843	372	4	5365-01-210-4903	44	1
2590-01-210-0176	250	2	5305-01-210-4967	314	25
2990-01-210-0427	25	4	4730-01-210-5785	12	35
4720-01-210-0484	7	9	2910-01-210-5872	18	1
4030-01-210-0691	314	16	2590-01-210-6202	268	30
5340-01-210-0692	17	18	5305-01-210-6248	314	21
5310-01-210-0819	39	55	5305-01-210-6249	321	1
5945-01-210-1299	45	20	5305-01-210-6249	325	1
4910-01-210-1318	499	6	5305-01-210-6251	15	34
2530-01-210-1324	144	11	2910-01-210-6856	13	20
2530-01-210-1324	152	11	2510-01-210-6939	28	1
2540-01-210-1325	331	14	2930-01-210-6940	26	3
3020-01-210-1365	314	33	2510-01-210-6941	183	15
3020-01-210-1391	174	11	2510-01-210-6942	184	16
5365-01-210-1584	206	26	2990-01-210-6982	25	3
5310-01-210-1587	37	13	5310-01-210-7935	224	6
5310-01-210-1587	472	3	5310-01-210-7938	202	40
5310-01-210-1587	473	11	5340-01-210-7966	181	12
5310-01-210-1587	474	3	5120-01-210-8793	499	2
5310-01-210-1587	475	16	5310-01-211-0691	284	17
5310-01-210-1587	476	22	5310-01-211-0691	439	10
5340-01-210-1606	240	17	5310-01-211-0698	283	2
5365-01-210-1628	184	8	5310-01-211-0698	438	2
5365-01-210-1629	203	7	5330-01-211-0717	314	3
5365-01-210-1630	203	27	5340-01-211-0838	277	29
5325-01-210-1633	84	6	5340-01-211-0838	278	8
2990-01-210-2176	398	19	5340-01-211-0838	440	4
5305-01-210-2309	313	19	4730-01-211-0896	18	7
5305-01-210-2309	458	42	5330-01-211-1343	156	9
2520-01-210-2624	148	21	5305-01-211-1405	198	13
5340-01-210-2747	5	15	5305-01-211-1439	47	3
2510-01-210-2748	331	15	5305-01-211-1450	13	4
2540-01-210-2749	268	13	5305-01-211-1450	14	4
2540-01-210-2787	208	33	5305-01-211-1450	16	24
5360-01-210-2852	22	3	5305-01-211-1470	208	34
5325-01-210-2870	15	5	5340-01-211-1645	280	52
5325-01-210-2870	16	5	4720-01-211-1998	364	4
4820-01-210-3488	15	4	4720-01-211-1998	379	40
4820-01-210-3488	16	4	5342-01-211-2897	7	4
4710-01-210-3503	333	7	5305-01-211-3100	314	14
4710-01-210-3504	333	5	5305-01-211-3101	314	1
2520-01-210-3506	99	19	5340-01-211-3132	202	14
2530-01-210-3567	182	9	5340-01-211-3137	23	20
5995-01-210-3594	81	2	5340-01-211-3137	24	17
6150-01-210-3596	327	16	5975-01-211-3142	17	13
2990-01-210-4276	28	1	5975-01-211-3142	153	18
5340-01-210-4751	240	9	5975-01-211-3142	379	48

I-67

SECTION IV TM9-2320-280-24P C01

CROSS REFERENCE INDEXES

NATIONAL STOCK NUMBER INDEX

STOCK NUMBER	FIGURE NO	ITEM NO	STOCK NUMBER	FIGURE NO	ITEM NO
5360-01-211-3163	314	39	5340-01-212-4716	77	10
2510-01-211-3229	280	33	5365-01-212-4736	202	19
5305-01-211-3786	95	34	5340-01-212-4887	23	9
5310-01-211-3811	24	19	5340-01-212-4887	24	6
5310-01-211-3811	25	6	2590-01-212-4955	153	15
5310-01-211-3811	175	5	2590-01-212-4956	333	36
5340-01-211-4025	23	12	2540-01-212-4959	327	2
2920-01-211-4935	45	79	2920-01-212-4976	45	67
5310-01-211-5037	12	26	2540-01-212-5814	202	20
5365-01-211-5063	62	12	2540-01-212-5815	331	27
5340-01-211-5083	274	31	2920-01-212-5816	43	50
5365-01-211-5094	177	15	2920-01-212-5817	43	52
5342-01-211-5327	45	60	2920-01-212-5817	43	53
4530-01-211-5328	250	4	2590-01-212-5868	488	1
5340-01-211-5593	327	7	4720-01-212-6403	333	26
5330-01-211-5856	47	5	4820-01-212-6763	121	10
5305-01-211-6049	49	3	2520-01-212-7634	104	9
5365-01-211-6083	246	25	5340-01-212-7711	182	10
5340-01-211-6288	272	13	4720-01-212-8269	17	17
6105-01-211-6635	333	12	5935-01-212-9631	84	7
5355-01-211-6991	314	37	5330-01-213-1312	BULK	54
5355-01-211-6991	458	6	5310-01-213-1333	269	21
6210-01-211-7024	47	13	4720-01-213-1574	178	11
5305-01-211-7415	357	5	2590-01-213-1628	283	25
5340-01-211-7436	203	13	2590-01-213-1628	438	21
5305-01-211-7478	170	15	2590-01-213-1629	283	19
5305-01-211-7478	195	4	2590-01-213-1629	438	15
5340-01-211-7562	153	23	5305-01-213-4149	171	5
4140-01-211-8403	32	1	5310-01-213-4174	314	34
2990-01-211-8587	23	16	5310-01-213-4185	142	20
2990-01-211-8587	24	20	5310-01-213-4185	143	27
5365-01-212-0144	272	32	5310-01-213-4185	150	20
2590-01-212-1057	268	30	5310-01-213-4185	151	27
5310-01-212-2213	170	19	5310-01-213-4192	205	15
5310-01-212-2213	195	8	5310-01-213-4192	250	15
5310-01-212-2303	277	27	5365-01-213-4239	225	24
5365-01-212-2325	202	31	5365-01-213-4239	226	11
5325-01-212-2403	138	6	5340-01-213-4267	240	17
5325-01-212-2403	145	6	5340-01-213-4662	80	15
5340-01-212-2407	83	14	5340-01-213-4662	203	15
5930-01-212-3373	67	6	5340-01-213-4662	204	13
4720-01-212-3531	177	9	5340-01-213-4665	83	12
5340-01-212-3553	23	6	5340-01-213-4665	88	34
5340-01-212-3553	24	23	5340-01-213-4667	75	48
5340-01-212-3709	28	6	5340-01-213-4667	77	25
5340-01-212-4714	5	14	5305-01-213-5024	332	17
5340-01-212-4716	75	30	5315-01-213-5545	202	23

SECTION IV TM9-2320-280-24P C01

CROSS REFERENCE INDEXES

NATIONAL STOCK NUMBER INDEX

STOCK NUMBER	FIGURE NO	ITEM NO	STOCK NUMBER	FIGURE NO	ITEM NO
5340-01-213-5600	153	9	5985-01-215-9404	410	6
5340-01-213-5735	202	2	5985-01-215-9404	414	5
5340-01-213-5735	218	7	5985-01-215-9404	428	5
5340-01-213-5735	253	35	5330-01-216-0288	314	11
5365-01-213-5739	398	20	5365-01-216-0364	277	30
2540-01-213-8057	283	1	5365-01-216-0364	278	7
2540-01-213-8057	438	1	5365-01-216-0364	440	3
5365-01-214-0006	246	30	5365-01-216-1166	314	10
2590-01-214-1566	315	7	6220-01-216-5285	64	5
2590-01-214-1566	444	4	2540-01-216-6320	226	35
2540-01-214-1568	327	6	5310-01-216-7390	44	5
5340-01-214-1778	247	25	5310-01-216-7390	209	14
5340-01-214-2089	181	1	5310-01-216-7390	443	13
2520-01-214-2565	169	17	5310-01-216-7390	457	11
5340-01-214-3615	333	6	5331-01-216-7392	488	3
5360-01-214-3695	153	14	5331-01-216-7393	488	4
5340-01-214-3712	83	7	5180-01-216-8643	502	7
5940-01-214-4563	79	10	6695-01-216-8687	64	9
5340-01-214-4675	283	9	5935-01-216-8763	76	9
5340-01-214-4675	438	10	6220-01-216-9337	64	6
5310-01-214-4955	45	50	5310-01-217-0715	279	34
5365-01-214-4987	196	11	5340-01-217-2576	286	26
5340-01-214-4992	271	6	2590-01-217-2662	64	24
5340-01-214-4992	271	23	2590-01-217-2662	81	14
5340-01-214-4992	296	6	2510-01-217-2663	286	33
5325-01-214-5007	327	19	2510-01-217-2664	286	33
4720-01-214-5757	12	39	2590-01-217-2665	286	23
5310-01-214-7784	45	44	2540-01-217-2666	286	20
5310-01-214-7785	43	30	5340-01-217-2667	286	26
5970-01-214-7806	43	42	2540-01-217-2669	21	20
5342-01-214-7809	284	15	5970-01-217-4566	43	41
5342-01-214-7809	439	8	5310-01-217-5205	85	5
5325-01-214-7817	80	11	5310-01-217-5205	88	3
5305-01-215-3985	43	32	5310-01-217-5205	203	30
5305-01-215-3990	83	13	5310-01-217-5205	272	5
2590-01-215-4325	181	6	5310-01-217-5205	486	2
5305-01-215-5174	3	1	2590-01-217-5743	286	1
5305-01-215-5174	75	40	4010-01-217-7086	284	3
5305-01-215-5174	77	21	4010-01-217-7086	439	3
5305-01-215-5174	479	1	4710-01-217-9959	286	22
5310-01-215-5356	199	31	5320-01-218-0721	31	6
5310-01-215-5356	331	18	5315-01-218-1114	245	40
5310-01-215-5356	332	25	5330-01-218-1196	64	7
5310-01-215-5356	347	7	5305-01-218-1243	64	10
5340-01-215-5478	333	30	5360-01-218-1608	286	3
4010-01-215-7466	284	3	5910-01-218-1675	43	46
4010-01-215-7466	439	3	5310-01-218-3190	202	9

I-69

SECTION IV TM9-2320-280-24P C01

CROSS REFERENCE INDEXES

NATIONAL STOCK NUMBER INDEX

STOCK NUMBER	FIGURE NO	ITEM NO	STOCK NUMBER	FIGURE NO	ITEM NO
5310-01-218-3190	252	29	5310-01-232-1361	40	30
2910-01-218-4489	KITS	3	5310-01-232-1362	40	40
4010-01-218-4639	284	7	5310-01-232-1363	43B	9
4010-01-218-4640	284	9	5310-01-232-6617	43B	8
5320-01-218-5802	286	30	2590-01-232-8210	286	42
5310-01-218-7137	457	10	3020-01-232-9629	37	20
2930-01-218-8103	26	3	3020-01-232-9629	475	28
5120-01-218-8235	502	5	2510-01-233-1141	225	12
5365-01-218-9848	285	12	5340-01-233-1461	240	14
5365-01-218-9914	285	8	2510-01-233-7767	226	4
5365-01-218-9914	285	11	5310-01-234-9415	88	16
5365-01-219-3528	191	3	5310-01-234-9416	35	14
5977-01-219-7042	43	40	5310-01-234-9416	37	7
5320-01-219-7261	286	6	5310-01-234-9416	52	11
5320-01-219-7261	286	15	5310-01-234-9416	56	8
5340-01-219-7541	64	3	5310-01-234-9416	59	16
5340-01-220-3084	205	10	5310-01-234-9416	234	16
5310-01-220-3099	286	11	5310-01-234-9416	246	32
5320-01-221-4231	274	30	5310-01-234-9416	247	23
5305-01-221-7734	286	2	5310-01-234-9416	249	44
5820-01-222-1106	416	KIT	5310-01-234-9416	260	7
5820-01-222-1107	414	KIT	5310-01-234-9416	269	3
2920-01-222-2183	KITS	69	5310-01-234-9416	284	14
2540-01-223-3749	240	8	5310-01-234-9416	298	8
5306-01-223-4345	40	39	5310-01-234-9416	307	12
5320-01-223-7728	298	7	5310-01-234-9416	321	16
5935-01-223-9420	74	12	5310-01-234-9416	327	36
5305-01-224-1449	286	5	5310-01-234-9416	327	42
9905-01-224-5860	354	8	2590-01-235-8661	243	5
5310-01-225-0701	40	3	2590-01-235-8661	244	2
5365-01-225-0844	40	35	2590-01-235-8661	244	14
5120-01-227-1680	501	13	2530-01-235-8688	167	12
5120-01-227-3159	505	3	5305-01-236-4349	123	2
5310-01-228-3299	286	21	5305-01-236-4349	124	14
5120-01-229-0842	503	3	5305-01-236-4349	125	14
5340-01-229-3632	5	5	5340-01-236-5101	345	32
5305-01-229-7855	327	3	5340-01-236-5101	348	23
5306-01-230-3354	4	21	2510-01-236-6488	278	4
5310-01-231-0596	142	6	2540-01-236-6489	278	4
5310-01-231-0596	143	6	5930-01-237-7322	66	1
5310-01-231-0596	150	6	5970-01-238-0146	43	28
5310-01-231-0596	151	6	5935-01-238-0543	84	13
4910-01-231-1709	504	1	5935-01-238-0543	86	21
4910-01-231-3671	499	7	4910-01-238-2551	505	14
8145-01-231-3747	363	1	5330-01-238-3217	275	24
4010-01-231-5075	314	17	5340-01-238-9543	398	5
5310-01-231-7455	398	6	4730-01-239-5251	301	3

I-70

SECTION IV TM9-2320-280-24P C01

CROSS REFERENCE INDEXES

NATIONAL STOCK NUMBER INDEX

STOCK NUMBER	FIGURE NO	ITEM NO	STOCK NUMBER	FIGURE NO	ITEM NO
4730-01-239-5251	303	12	2590-01-247-7923	255	8
5340-01-242-6245	235	14	2590-01-247-7924	255	22
5325-01-242-7083	333	10	2510-01-247-7968	226	4
5325-01-242-7083	367	10	5330-01-247-8438	34	7
5305-01-243-3759	476	17	9905-01-248-1111	355	4
5306-01-243-4650	175	2	9905-01-248-1112	355	4
3040-01-244-3646	488	17	9905-01-248-1113	355	4
2920-01-244-4993	43	1	9905-01-248-1114	355	4
5340-01-244-5765	15	15	9905-01-248-1115	355	4
5340-01-244-5765	180	2	2510-01-248-1340	208	19
5306-01-244-7882	440	1	2920-01-248-2509	43	5
2520-01-244-9863	100	1	2590-01-248-2531	251	12
5342-01-246-1120	17	1	5340-01-248-2878	59	15
5330-01-246-1822	12	14	2530-01-248-4873	184	18
5330-01-246-1822	364	2	2590-01-248-4874	251	11
5365-01-246-8281	205	18	7690-01-248-4907	430	28
5340-01-246-8282	197	31	7690-01-248-4908	430	43
2540-01-246-8283	242	1	7690-01-248-4909	430	45
2540-01-246-8284	243	4	5340-01-248-7621	243	31
2540-01-246-8285	244	1	5340-01-248-7621	244	15
2540-01-246-8286	244	4	2590-01-248-7622	251	33
2510-01-246-8287	452	3	2590-01-248-7623	251	34
2590-01-246-8288	226	3	5340-01-248-7646	199	43
2540-01-246-8289	226	1	5340-01-248-7649	430	36
5340-01-246-8290	246	34	9905-01-248-7656	370	13
2590-01-246-8291	249	13	9905-01-248-7657	354	7
2540-01-246-8292	251	22	7690-01-248-7679	429	28
2540-01-246-8292	254	3	7690-01-248-7680	429	33
5340-01-246-8293	251	37	7690-01-248-7681	429	39
2590-01-246-8294	252	36	7690-01-248-7682	430	24
2590-01-246-8294	253	3	7690-01-248-7683	430	46
2590-01-246-8294	255	3	2510-01-248-8875	47	6
5355-01-247-3593	331	28	2590-01-248-9531	286	40
5120-01-247-6629	501	7	2590-01-248-9536	286	40
4730-01-247-6763	150	5	9905-01-248-9543	355	4
2590-01-247-7910	199	7	9905-01-248-9544	355	4
5340-01-247-7911	242	9	9905-01-248-9545	355	4
5340-01-247-7912	226	10	9905-01-248-9546	355	4
2540-01-247-7914	208	38	9905-01-248-9547	355	4
2540-01-247-7915	208	38	9905-01-248-9548	355	4
1005-01-247-7916	251	6	9905-01-248-9549	355	4
1010-01-247-7917	251	8	9905-01-248-9550	356	2
5340-01-247-7918	252	25	9905-01-248-9551	356	6
5340-01-247-7919	252	30	9905-01-248-9552	355	4
2590-01-247-7920	255	34	2510-01-248-9557	268	3
5340-01-247-7921	251	13	5210-01-249-0369	510	22
5340-01-247-7922	252	15	5315-01-249-0555	437	2

SECTION IV TM9-2320-280-24P C01

CROSS REFERENCE INDEXES

NATIONAL STOCK NUMBER INDEX

STOCK NUMBER	FIGURE NO	ITEM NO	STOCK NUMBER	FIGURE NO	ITEM NO
5310-01-249-0899	398	1	2510-01-250-7593	227	26
5310-01-249-0904	461	5	2510-01-250-7594	227	26
2520-01-249-1434	97	8	2510-01-250-7595	227	25
2520-01-249-1434	101	10	2510-01-250-7596	227	2
3040-01-249-1435	99	9	2510-01-250-7597	227	1
2520-01-249-1506	137	17	5310-01-250-7679	4	20
5340-01-249-1575	179	16	2510-01-250-7786	268	3
5340-01-249-1576	179	5	5310-01-250-7835	26	10
2530-01-249-1577	170	2	5310-01-250-7844	45	29
2590-01-249-1578	244	12	5310-01-251-0711	49	9
2590-01-249-1579	255	13	5310-01-251-0711	333	2
2510-01-249-1580	268	20	5340-01-251-0724	28	3
2510-01-249-1581	268	21	5305-01-251-0728	249	23
2510-01-249-1582	268	21	5305-01-251-0728	253	25
2510-01-249-1583	268	22	5305-01-251-0728	256	6
2540-01-249-1584	279	1	5305-01-251-0728	257	9
2510-01-249-1585	272	1	5305-01-251-0728	259	10
2510-01-249-1586	272	1	5305-01-251-0728	262	3
5340-01-249-1587	275	25	5305-01-251-0728	263	6
2590-01-249-1588	286	18	5305-01-251-0728	264	2
2540-01-249-1589	413	7	5305-01-251-0728	265	2
5340-01-249-1590	415	7	3120-01-251-0735	133	13
9515-01-249-1591	429	7	5310-01-251-0748	58	8
2540-01-249-1592	429	8	5310-01-251-0748	99	8
2540-01-249-1593	429	19	5310-01-251-0748	208	15
5340-01-249-1594	429	21	5310-01-251-0748	252	2
2540-01-249-1595	429	30	5310-01-251-0748	334	4
2590-01-249-1596	430	1	5310-01-251-0748	338	20
2540-01-249-1597	430	4	5310-01-251-0748	339	15
2540-01-249-1598	430	30	5310-01-251-0748	417	2
2510-01-249-1599	179	2	5310-01-251-0748	418	2
9905-01-249-1612	354	19	5310-01-251-0748	419	2
2920-01-249-3492	99	20	5310-01-251-0748	420	7
5342-01-249-3502	208	32	5310-01-251-0748	421	4
5342-01-249-5389	208	32	5310-01-251-0748	422	6
5342-01-250-1098	10	13	5310-01-251-0748	423	10
7690-01-250-1101	430	47	5310-01-251-0748	425	2
5305-01-250-6582	57	8	5310-01-251-0748	426	4
5305-01-250-6583	46	22	5310-01-251-0748	427	3
5305-01-250-6584	55	4	5340-01-251-0754	84	20
5305-01-250-6584	56	2	5310-01-251-0760	62	13
5310-01-250-6587	59	10	5330-01-251-1607	100	10
5310-01-250-6587	99	12	5330-01-251-1607	108	7
2540-01-250-7589	429	1	5306-01-251-1616	142	27
2540-01-250-7590	415	9	2590-01-251-3068	227	20
2540-01-250-7591	430	KIT	2590-01-251-3069	227	7
2540-01-250-7592	430	31	2590-01-251-3069	227	17

SECTION IV TM9-2320-280-24P C01

CROSS REFERENCE INDEXES

NATIONAL STOCK NUMBER INDEX

STOCK NUMBER	FIGURE NO	ITEM NO	STOCK NUMBER	FIGURE NO	ITEM NO
2590-01-251-3070	227	17	5340-01-252-8537	208	24
2590-01-251-3071	227	6	5305-01-252-9108	199	36
4710-01-251-3073	227	5	5365-01-252-9214	205	11
4710-01-251-3075	227	22	5310-01-253-0058	199	58
4710-01-251-3076	268	27	5310-01-253-0060	205	8
2510-01-251-5316	193	4	5310-01-253-0060	206	28
4710-01-251-6103	227	4	5306-01-253-1611	150	36
5995-01-251-6785	416	1	5306-01-253-1612	202	34
5995-01-251-6785	428	2	5310-01-253-1615	410	16
2510-01-251-8548	227	19	5310-01-253-1618	52	31
5306-01-251-9190	75	47	5310-01-253-1618	208	27
5306-01-251-9190	174	4	5310-01-253-1618	252	5
5305-01-251-9222	104	17	5310-01-253-1618	260	11
5995-01-251-9316	414	1	5310-01-253-1618	273	13
5995-01-251-9316	416	2	5310-01-253-1618	275	12
3040-01-251-9383	230	10	5310-01-253-1618	276	8
2510-01-251-9995	227	21	5310-01-253-1618	300	13
5330-01-252-0461	34	13	5310-01-253-1618	327	29
5310-01-252-0513	241	6	4720-01-253-2809	504	3
5310-01-252-0513	249	25	4720-01-253-2810	504	5
5310-01-252-0513	272	17	2530-01-253-2825	157	16
5340-01-252-0741	245	34	2530-01-253-2825	158	11
9330-01-252-1239	487A	1	2510-01-253-2839	272	10
9330-01-252-1239	BULK	66	5305-01-253-2989	240	10
5310-01-252-2999	59	7	5305-01-253-2993	205	33
5310-01-252-2999	75	32	5305-01-253-2993	206	20
5310-01-252-2999	84	10	5305-01-253-2998	241	11
5310-01-252-2999	153	6	5985-01-253-3514	414	4
5310-01-252-2999	330	1	5985-01-253-3514	425	13
2540-01-252-3386	270	33	5340-01-253-3757	197	23
2510-01-252-4466	227	10	5340-01-253-3757	199	40
2510-01-252-4756	243	21	2590-01-253-3905	235	5
5330-01-252-5377	197	32	2510-01-253-3907	280	1
5330-01-252-5378	199	60	5306-01-253-4426	179	8
4820-01-252-6169	504	4	5340-01-253-5479	246	19
2510-01-252-6240	277	3	5340-01-253-5479	253	34
1005-01-252-6299	251	3	5340-01-253-5479	254	25
1010-01-252-6300	251	9	5340-01-253-5479	255	27
5355-01-252-6504	280	38	2540-01-253-6112	349	16
5355-01-252-6504	281	38	5310-01-253-6439	203	14
5310-01-252-7285	196	6	6680-01-253-6782	504	9
2510-01-252-7903	284	8	5306-01-253-7073	4	19
2510-01-252-7904	284	10	5340-01-253-8377	272	12
5365-01-252-8026	208	30	5310-01-253-8437	127	10
2530-01-252-8362	170	7	5310-01-253-8437	230	13
2540-01-252-8460	270	33	5310-01-253-8437	234	11
2510-01-252-8461	284	6	5310-01-253-8437	235	12

I-73

SECTION IV TM9-2320-280-24P C01

CROSS REFERENCE INDEXES

NATIONAL STOCK NUMBER INDEX

STOCK NUMBER	FIGURE NO	ITEM NO	STOCK NUMBER	FIGURE NO	ITEM NO
5310-01-253-8437	239	12	5320-01-254-2283	332	40
5310-01-253-8438	286	27	5306-01-254-2357	174	10
5310-01-253-8440	410	34	5306-01-254-2358	148	3
5310-01-253-8440	412	25	5306-01-254-2359	177	16
5310-01-253-8928	331	23	5306-01-254-2360	279	37
5340-01-253-8933	277	14	5305-01-254-2451	249	40
5310-01-253-8948	179	4	5305-01-254-2452	248	2
5310-01-253-8948	180	5	5305-01-254-2452	272	22
5310-01-253-8948	181	2	5305-01-254-2453	269	26
5310-01-253-8948	182	3	5305-01-254-2459	268	1
5310-01-253-8948	187	12	5305-01-254-2459	270	40
5310-01-253-8948	188	2	5305-01-254-2460	92	17
5310-01-253-8948	191	2	5305-01-254-2460	199	25
5310-01-253-8948	194	1	5305-01-254-2460	497	1
5310-01-253-8948	283	3	5305-01-254-2461	398	13
5310-01-253-8948	417	15	5307-01-254-2504	278	11
5310-01-253-8949	248	15	5305-01-254-2521	174	9
5310-01-253-8949	280	4	5305-01-254-2521	66A	1
5310-01-253-8949	280	53	5305-01-254-2522	230	12
5310-01-253-8949	281	4	5305-01-254-2522	238	1
5310-01-253-8949	281	53	5310-01-254-4284	280	3
5310-01-253-8949	286	17	5310-01-254-4284	281	3
5310-01-253-8952	280	12	5365-01-254-4394	277	13
5310-01-253-8952	280	37	5310-01-254-5352	46	20
5310-01-253-8952	281	12	5310-01-254-5352	208	3
5310-01-253-8952	281	37	4730-01-254-6023	504	6
5310-01-253-8953	280	19	5306-01-254-6356	171	4
5310-01-253-8953	280	35	2540-01-254-6511	370	9
5310-01-253-8953	281	19	5340-01-254-6543	140	3
5310-01-253-8953	281	35	5340-01-254-7189	316	19
5310-01-253-8953	335	12	5340-01-254-7189	320	16
5310-01-253-8953	336	9	5340-01-254-7189	320	30
5310-01-253-8957	226	22	5340-01-254-7189	324	20
5340-01-253-8971	280	2	5340-01-254-7189	324	32
5340-01-253-8971	281	2	5340-01-254-7189	378	26
5365-01-253-8975	272	11	5340-01-254-7190	251	14
5365-01-253-8980	280	51	5340-01-254-7190	430	44
5365-01-253-8980	281	51	5340-01-254-7191	253	37
2510-01-254-1056	272	16	5340-01-254-7191	429	29
2510-01-254-1482	280	1	5340-01-254-7191	430	19
2510-01-254-1483	281	1	4710-01-254-7272	227	13
2510-01-254-1484	208	18	5305-01-255-0094	185	2
5360-01-254-1492	192	1	5330-01-255-0207	202	5
5342-01-254-1498	366	1	5330-01-255-0208	202	11
2510-01-254-1500	280	1	4730-01-255-0925	331	32
2510-01-254-1501	281	1	5365-01-255-0965	153	26
5320-01-254-2283	208	37	5365-01-255-0965	154	8

SECTION IV TM9-2320-280-24P C01

CROSS REFERENCE INDEXES

NATIONAL STOCK NUMBER INDEX

STOCK NUMBER	FIGURE NO	ITEM NO	STOCK NUMBER	FIGURE NO	ITEM NO
5365-01-255-0965	379	36	5340-01-255-9510	198	9
5305-01-255-1016	247	21	5340-01-255-9510	321	14
5305-01-255-1016	247	26	5340-01-255-9510	323	12
5305-01-255-1016	248	23	5340-01-255-9510	325	16
5305-01-255-1016	249	8	5340-01-255-9510	402	13
5305-01-255-1016	249	32	5340-01-255-9514	415	8
5305-01-255-1017	249	19	5340-01-255-9515	430	2
5340-01-255-1109	413	9	5360-01-255-9899	205	30
5340-01-255-1109	421	10	5340-01-255-9918	280	2
5340-01-255-1109	422	10	5340-01-255-9918	281	2
5325-01-255-1126	235	6	5340-01-255-9919	316	20
5340-01-255-1657	203	24	5340-01-255-9919	320	17
5306-01-255-2661	193	7	5340-01-255-9919	320	32
5306-01-255-2662	415	6	5340-01-255-9919	324	17
5305-01-255-2670	243	1	5340-01-255-9919	324	29
5305-01-255-2675	246	27	5305-01-256-0405	244	20
5310-01-255-2695	58	10	5305-01-256-0406	215	6
5310-01-255-2704	164	2	5305-01-256-0406	430	26
5340-01-255-2729	429	23	5310-01-256-0416	328	16
5315-01-255-2763	243	34	4030-01-256-0471	314	19
5315-01-255-2764	243	26	4030-01-256-0471	458	40
5360-01-255-2808	243	33	5340-01-256-0942	430	11
5340-01-255-2812	430	9	6685-01-256-3406	504	8
5340-01-255-2812	446	18	3030-01-256-3616	32	2
5340-01-255-2812	449	10	3030-01-256-3616	475	22
3020-01-255-2906	504	7	2590-01-256-3628	429	25
2590-01-255-2957	251	36	5310-01-256-4491	243	12
2590-01-255-2958	251	48	5310-01-256-4491	244	6
5305-01-255-3588	331	8	5340-01-256-4515	243	30
5305-01-255-3588	332	43	5340-01-256-4516	244	3
5365-01-255-3610	285	15	5340-01-256-4517	251	27
4030-01-255-3634	284	11	5340-01-256-4517	252	27
5305-01-255-4606	222	9	5340-01-256-4517	253	22
5305-01-255-4611	327	20	5340-01-256-4517	255	21
5365-01-255-4633	277	5	5365-01-256-4526	246	28
5365-01-255-4634	277	5	5340-01-256-4655	254	15
5340-01-255-4655	208	26	5340-01-256-4656	272	18
5305-01-255-4950	274	22	5340-01-256-4657	246	36
5330-01-255-4986	488	20	5340-01-256-4659	243	3
5365-01-255-4990	277	5	7690-01-256-4908	354	21
5365-01-255-4991	277	5	7690-01-256-4909	354	22
5325-01-255-5062	414	21	2530-01-256-4914	152	20
5325-01-255-5063	416	24	2930-01-256-5350	27	3
5325-01-255-5063	428	23	5305-01-256-6870	429	13
5320-01-255-6608	202	4	5305-01-256-6870	430	27
5340-01-255-9380	240	3	5305-01-256-6870	430	34
5310-01-255-9452	413	10	5340-01-256-9318	316	20

I-75

SECTION IV TM9-2320-280-24P C01

CROSS REFERENCE INDEXES

NATIONAL STOCK NUMBER INDEX

STOCK NUMBER	FIGURE NO	ITEM NO	STOCK NUMBER	FIGURE NO	ITEM NO
5340-01-256-9318	320	17	5340-01-258-4866	252	23
5340-01-256-9318	320	32	5340-01-258-6157	246	12
5340-01-256-9318	324	17	5340-01-258-6157	251	18
5340-01-256-9318	324	29	5340-01-258-6157	251	32
5365-01-256-9568	245	29	5340-01-258-6157	253	23
5325-01-257-0801	268	23	5340-01-258-6157	255	32
5340-01-257-0908	429	20	5340-01-258-6158	252	40
2510-01-257-1572	277	17	5340-01-258-6158	253	4
2510-01-257-1572	277	45	5340-01-258-6158	255	7
5340-01-257-2644	430	10	5340-01-258-6158	255	10
4710-01-257-2649	364	5	5340-01-258-6158	260	23
4710-01-257-2649	379	42	5340-01-258-8609	252	31
4720-01-257-2655	364	9	2590-01-258-9491	272	7
4720-01-257-2655	379	43	5340-01-259-0327	251	49
4730-01-257-3348	148	23	5340-01-259-0328	252	21
4720-01-257-3705	5	11	5340-01-259-0328	252	35
2540-01-257-3742	370	1	5340-01-259-0329	253	9
2510-01-257-3747	274	2	5340-01-259-0329	253	17
2510-01-257-3747	276	28	5340-01-259-0329	255	16
5340-01-257-3778	430	25	5340-01-259-0330	254	9
2530-01-257-3864	140	6	5340-01-259-0330	255	26
2510-01-257-3876	281	1	4140-01-259-2175	368	6
2540-01-257-3877	188	4	5340-01-259-3956	246	37
6220-01-257-3878	83	32	4210-01-259-5634	241	9
2510-01-257-3903	280	1	4210-01-259-5634	241	14
2510-01-257-3904	281	1	5320-01-259-6155	224	1
2530-01-257-3905	152	20	5320-01-259-6155	237	16
4730-01-257-4905	12	30	5320-01-259-6155	263	5
5342-01-257-7706	35	27	5305-01-259-6322	58	2
5310-01-257-7719	169	32	5305-01-259-6322	59	26
5310-01-257-7719	205	7	5305-01-259-6322	79	16
5310-01-257-7719	206	27	5305-01-259-6322	83	17
2590-01-257-8784	227	53	5305-01-259-6322	88	5
2590-01-257-8787	369	13	5305-01-259-6322	88	26
5310-01-258-1536	26	4	5305-01-259-6322	227	3
5310-01-258-1536	174	12	5305-01-259-6322	286	41
5310-01-258-1536	398	26	5305-01-259-6322	303	14
5340-01-258-1547	279	36	5305-01-259-6322	310	5
5320-01-258-2576	208	12	5305-01-259-6322	332	29
5340-01-258-4529	251	23	5305-01-259-6322	335	3
5340-01-258-4529	251	38	5305-01-259-6322	338	17
5340-01-258-4529	253	19	5305-01-259-6322	339	10
5340-01-258-4529	254	6	5305-01-259-6322	368	7
5340-01-258-4529	255	18	5305-01-259-6322	412	17
5340-01-258-4529	448	3	5305-01-259-6322	424	5
5340-01-258-4529	450	5	5305-01-259-6322	425	19
5330-01-258-4664	272	6	5305-01-259-6322	426	24

SECTION IV TM9-2320-280-24P C01

CROSS REFERENCE INDEXES

NATIONAL STOCK NUMBER INDEX

STOCK NUMBER	FIGURE NO	ITEM NO	STOCK NUMBER	FIGURE NO	ITEM NO
5305-01-259-6322	427	24	2510-01-262-6007	198	12
5305-01-259-6322	468	10	2510-01-262-6008	200	5
5320-01-259-7423	226	2	2510-01-262-6009	200	6
5320-01-259-7423	237	5	2510-01-262-6010	198	11
5340-01-259-7523	247	16	2510-01-262-6011	198	26
5310-01-259-7554	328	11	2590-01-262-7706	227	29
5340-01-259-8579	266	17	2590-01-262-7707	227	11
5340-01-259-8600	269	24	6670-01-262-8646	167	12
2590-01-260-0214	279	6	6670-01-262-8647	167	12
5342-01-260-0215	279	10	2590-01-262-9513	253	28
2590-01-260-0216	279	15	2590-01-262-9514	253	8
2590-01-260-0217	279	20	2590-01-262-9515	253	14
5340-01-260-0218	279	21	2540-01-262-9516	253	33
5360-01-260-0700	153	24	2540-01-262-9516	254	28
5360-01-260-0700	154	9	2510-01-262-9520	268	3
5340-01-260-4881	268	24	6670-01-263-2268	167	12
5310-01-260-5784	208	17	2590-01-263-3254	227	29
5340-01-260-5792	266	3	2590-01-263-3254	368	33
5342-01-260-7853	430	29	5306-01-263-6142	279	32
6150-01-260-8000	366	29	5306-01-263-8889	35	28
6150-01-260-8000	384	10	5306-01-263-8889	36	12
5340-01-260-9026	429	24	5306-01-263-8889	37	17
5340-01-260-9027	207	9	5306-01-263-8889	38	12
5340-01-260-9027	318	8	5306-01-263-8889	38A	8
5340-01-260-9027	325	21	5306-01-263-8889	472	14
5970-01-260-9132	39	75	5306-01-263-8889	475	21
5340-01-260-9874	266	3	5306-01-263-8889	476	19
5340-01-260-9940	286	13	5305-01-264-0923	245	9
5340-01-260-9942	217	2	5325-01-264-1522	207	3
2590-01-261-0523	279	22	5306-01-264-3531	174	1
5340-01-261-1564	251	41	5306-01-264-3531	472	16
2590-01-261-2636	202	6	5306-01-264-3531	475	24
2590-01-261-2637	202	12	5306-01-264-3531	476	20
2590-01-261-5484	202	13	5305-01-264-3555	230	14
2590-01-261-5485	230	22	5305-01-264-3602	183	1
2590-01-261-5485	231	27	5305-01-264-3602	185	11
6670-01-261-6844	167	12	5305-01-264-3602	187	15
6670-01-261-6845	167	12	5305-01-264-3602	188	11
6670-01-261-6846	167	12	5365-01-264-4023	196	13
6670-01-261-6847	167	12	5305-01-264-5809	155	10
2590-01-261-6851	227	24	5305-01-264-5809	226	33
2590-01-261-6851	369	12	5305-01-264-5809	227	36
6670-01-261-8011	167	12	5340-01-264-5833	234	13
6670-01-261-8012	167	12	5340-01-264-5833	252	16
6670-01-261-8013	167	12	5340-01-264-5833	253	38
2590-01-262-4980	279	4	5305-01-264-5864	46	7
2590-01-262-4981	279	7	5305-01-264-5874	199	50

I-77

SECTION IV TM9-2320-280-24P C01

CROSS REFERENCE INDEXES

NATIONAL STOCK NUMBER INDEX

STOCK NUMBER	FIGURE NO	ITEM NO	STOCK NUMBER	FIGURE NO	ITEM NO
5305-01-264-5874	343	19	2530-01-265-3153	157	4
5305-01-264-5874	495	9	2510-01-265-3155	293	17
5320-01-264-5978	199	14	4820-01-265-3156	492	10
5320-01-264-5978	217	3	2510-01-265-3157	293	17
5320-01-264-5978	219	6	4820-01-265-3159	492	12
5320-01-264-5978	245	35	2540-01-265-3160	292	7
5320-01-264-5978	264	1	4730-01-265-3166	301	6
5320-01-264-5978	265	1	4730-01-265-3166	302	10
5320-01-264-5978	295	9	4730-01-265-3166	303	1
5320-01-264-5978	297	2	4730-01-265-3166	304	1
5320-01-264-5978	300	5	4730-01-265-3166	305	3
5320-01-264-5978	307	7	4730-01-265-3166	306	1
5320-01-264-5978	308	13	4730-01-265-3166	309	2
5320-01-264-5978	342	8	4730-01-265-3166	310	2
5340-01-264-6104	197	5	2590-01-265-3185	155	3
5340-01-264-6108	430	14	2510-01-265-3194	293	22
2590-01-264-6531	251	29	4720-01-265-3213	336	2
5342-01-264-6532	211	2	4720-01-265-3222	301	2
5342-01-264-6533	212	2	4720-01-265-3223	492	5
7690-01-264-6536	358	4	4720-01-265-3224	492	13
7690-01-264-6536	443	18	4710-01-265-3228	335	2
5330-01-264-6537	363	5	4710-01-265-3228	335	8
5340-01-264-6540	363	8	4710-01-265-3228	336	5
5342-01-264-6543	363	22	4710-01-265-3229	335	7
5340-01-264-6544	363	6	4710-01-265-3229	336	4
2510-01-264-6546	211	13	4710-01-265-3230	493	8
2510-01-264-6546	212	14	4710-01-265-3231	334	8
3040-01-264-9554	279	8	4710-01-265-3232	492	2
2510-01-265-1126	213	21	4710-01-265-3233	492	1
2510-01-265-1127	213	31	2540-01-265-3234	54	1
5342-01-265-1129	293	27	4720-01-265-3237	301	1
2510-01-265-1130	214	28	4720-01-265-3238	343	12
9905-01-265-1131	59	22	4720-01-265-3239	302	13
9905-01-265-1132	59	22	4720-01-265-3239	305	2
7690-01-265-1133	362	3	2540-01-265-3266	343	21
7690-01-265-1134	362	4	4710-01-265-3275	334	10
7690-01-265-1135	362	5	4710-01-265-3276	493	4
2540-01-265-1137	224	11	4710-01-265-3277	493	2
2510-01-265-1138	211	1	2510-01-265-3281	297	3
2510-01-265-1139	211	1	2510-01-265-3281	297	10
2510-01-265-1140	212	1	2510-01-265-3282	291	7
2510-01-265-1141	212	1	2510-01-265-3282	297	10
2530-01-265-3148	157	4	2510-01-265-3283	297	9
4820-01-265-3149	494	7	4730-01-265-3284	302	5
4820-01-265-3149	496	31	4730-01-265-3284	305	7
4140-01-265-3151	342	1	4730-01-265-3285	304	3
4710-01-265-3152	496	17	4730-01-265-3285	306	5

SECTION IV TM9-2320-280-24P C01

CROSS REFERENCE INDEXES

NATIONAL STOCK NUMBER INDEX

STOCK NUMBER	FIGURE NO	ITEM NO	STOCK NUMBER	FIGURE NO	ITEM NO
4730-01-265-3285	309	6	5365-01-265-8937	283	20
2990-01-265-3297	346	2	5365-01-265-8937	438	16
2990-01-265-3298	347	1	5365-01-265-8938	283	20
5342-01-265-3672	212	2	5365-01-265-8938	438	16
5340-01-265-3676	363	18	5365-01-265-8939	283	20
5120-01-265-4872	501	6	5365-01-265-8939	438	16
2510-01-265-6953	214	16	5330-01-265-8946	223	3
6105-01-265-8634	494	19	5330-01-265-8947	223	1
6105-01-265-8634	498	14	5330-01-265-8948	213	1
4130-01-265-8691	344	1	5360-01-265-9116	156	4
6680-01-265-8698	496	19	5330-01-265-9231	293	33
4130-01-265-8726	496	1	6220-01-266-1651	65	24
5306-01-265-8767	262	5	5930-01-266-1809	494	8
5315-01-265-8771	39	23	5945-01-266-1810	59	5
5360-01-265-8798	157	3	5930-01-266-1811	58	14
5360-01-265-8799	157	3	3040-01-266-2122	109	17
5330-01-265-8803	293	1	5340-01-266-2995	202	7
5330-01-265-8804	214	1	4010-01-266-3842	292	18
5330-01-265-8805	214	25	5340-01-266-3843	346	4
5330-01-265-8806	223	3	5340-01-266-3844	287	18
5330-01-265-8807	223	1	5340-01-266-3845	293	14
5330-01-265-8808	213	32	5340-01-266-3846	287	9
5331-01-265-8809	492	11	5920-01-266-3908	39	62
5340-01-265-8858	289	23	9905-01-266-3913	362	2
5340-01-265-8859	292	12	5930-01-266-3917	496	30
5340-01-265-8883	157	2	5930-01-266-3919	58	4
5340-01-265-8895	262	6	5930-01-266-3919	396	14
5306-01-265-8897	287	31	5961-01-266-3984	59	6
5306-01-265-8897	294	7	2510-01-266-4003	215	3
5340-01-265-8905	283	24	5930-01-266-4592	55	11
5340-01-265-8905	438	20	5930-01-266-4592	56	17
5340-01-265-8911	211	25	5945-01-266-4593	52	18
5340-01-265-8911	212	28	5945-01-266-4593	56	9
5340-01-265-8911	213	27	5342-01-266-5689	211	2
5340-01-265-8911	214	22	2510-01-266-5690	218	3
5340-01-265-8921	262	8	5342-01-266-5691	217	1
5340-01-265-8922	238	2	2540-01-266-5702	229	3
5340-01-265-8923	258	7	7690-01-266-5706	59	20
5340-01-265-8923	259	3	5340-01-266-5708	224	14
5340-01-265-8923	260	3	2510-01-266-5711	216	10
5340-01-265-8923	262	11	5310-01-266-6392	488	16
5340-01-265-8923	287	27	5340-01-266-6398	227	15
5340-01-265-8923	447	4	5342-01-266-7559	289	8
5340-01-265-8924	216	6	2510-01-266-8013	277	18
5340-01-265-8925	256	5	2510-01-266-8013	277	44
5340-01-265-8925	257	10	5306-01-266-8274	202	28
5340-01-265-8925	289	1	5325-01-266-8414	207	2

SECTION IV TM9-2320-280-24P C01

CROSS REFERENCE INDEXES

NATIONAL STOCK NUMBER INDEX

STOCK NUMBER	FIGURE NO	ITEM NO	STOCK NUMBER	FIGURE NO	ITEM NO
5325-01-266-8414	208	20	9330-01-269-7114	287	17
5340-01-266-9173	313	3	6115-01-269-7911	39	21
5340-01-266-9173	443	3	6230-01-269-8054	65	22
5340-01-266-9173	457	21	6230-01-269-8054	289	28
5180-01-267-2907	509	1	5985-01-269-8271	341	2
5365-01-267-6769	279	13	5985-01-269-8272	340	6
2520-01-267-7371	148	11	5985-01-269-8274	341	4
5315-01-267-7570	403	8	5970-01-269-8384	39	14
5342-01-268-0921	275	23	5940-01-269-8396	95	19
2520-01-268-1051	363	2	5315-01-269-8504	213	17
5305-01-268-5680	333	40	5315-01-269-8504	214	13
2540-01-268-7203	299	3	5365-01-269-8671	211	8
5640-01-268-7204	222	7	5365-01-269-8671	212	5
5640-01-268-7204	495	3	5365-01-269-8672	211	29
6150-01-269-0067	89	1	5365-01-269-8672	212	31
5310-01-269-0786	26	8	5935-01-269-8992	52	19
5940-01-269-1816	89	4	5935-01-269-8992	56	6
6150-01-269-1839	339	21	5310-01-269-9245	411	6
6150-01-269-1840	58	13	5995-01-269-9525	338	14
6150-01-269-1840	90	1	5995-01-269-9525	339	11
6150-01-269-1848	338	6	4730-01-269-9530	336	1
6150-01-269-1848	339	7	5920-01-270-0090	52	10
6150-01-269-1849	338	1	5920-01-270-0090	59	13
6150-01-269-1854	39	65	5905-01-270-0966	342	3
6150-01-269-1856	39	40	5330-01-270-1353	274	7
6150-01-269-1857	39	46	5330-01-270-1353	276	26
6150-01-269-1858	39	64	6220-01-270-1454	64	19
6150-01-269-1859	56	5	5970-01-270-1465	39	27
6150-01-269-1861	56	19	5977-01-270-1466	39	66
6150-01-269-1862	56	20	5970-01-270-1467	39	60
6150-01-269-1863	39	51	5977-01-270-1468	39	72
6150-01-269-1897	93	1	4720-01-270-2226	304	4
5306-01-269-4319	121	21	2540-01-270-2250	213	5
4130-01-269-4921	496	27	2540-01-270-2250	214	33
6220-01-269-5263	65	3	3120-01-270-2645	39	18
6150-01-269-5275	91	1	3120-01-270-2646	496	4
6150-01-269-5298	95	1	5310-01-270-2661	494	2
6150-01-269-5299	94	1	5310-01-270-2661	495	13
6150-01-269-5300	92	1	5310-01-270-2661	496	8
5935-01-269-5994	52	4	5340-01-270-2681	55	9
5935-01-269-5994	57	3	5330-01-270-3668	39	26
5306-01-269-6254	279	9	5940-01-270-3700	92	2
2990-01-269-6625	337	1	5940-01-270-3700	93	6
6220-01-269-7066	65	12	5940-01-270-3700	95	7
6220-01-269-7067	55	8	5940-01-270-3700	96	11
6220-01-269-7067	56	10	5940-01-270-3701	93	4
4140-01-269-7101	494	18	5940-01-270-3702	95	15

I-80

SECTION IV TM9-2320-280-24P C01

CROSS REFERENCE INDEXES

NATIONAL STOCK NUMBER INDEX

STOCK NUMBER	FIGURE NO	ITEM NO	STOCK NUMBER	FIGURE NO	ITEM NO
5940-01-270-3702	96	4	5330-01-270-8315	426	15
2540-01-270-3776	337	21	5330-01-270-8315	427	15
2540-01-270-4353	298	6	5330-01-270-8316	39	67
5340-01-270-4415	298	1	5330-01-270-8317	343	7
5340-01-270-4415	346	7	5340-01-270-8370	211	18
5975-01-270-4587	58	3	5340-01-270-8370	212	15
2540-01-270-5065	238	5	5310-01-270-8377	39	13
4520-01-270-5080	298	4	5310-01-270-8415	39	22
5310-01-270-5394	343	22	5340-01-270-8431	212	25
5305-01-270-5418	211	21	5340-01-270-8431	258	10
5305-01-270-5418	212	19	5340-01-270-8431	259	13
5305-01-270-5419	342	2	5340-01-270-8431	260	24
5305-01-270-5419	496	26	5340-01-270-8432	212	22
5305-01-270-5419	498	7	5340-01-270-8432	258	8
5305-01-270-5435	335	10	5340-01-270-8432	259	2
5305-01-270-5435	336	7	5340-01-270-8432	260	18
5306-01-270-5448	35	29	5340-01-270-8433	211	20
5306-01-270-5448	36	13	5340-01-270-8433	212	18
5306-01-270-5448	38	13	5340-01-270-8435	213	28
5306-01-270-5448	33A	7	5340-01-270-8435	214	18
5306-01-270-5448	38A	9	5340-01-270-8455	264	7
2540-01-270-6458	290	4	5340-01-270-8455	265	6
2540-01-270-6459	219	5	5340-01-270-8483	213	12
2540-01-270-6459	290	7	5340-01-270-8483	214	6
2540-01-270-6460	290	9	5340-01-270-8484	39	49
2540-01-270-6537	236	3	5340-01-270-8489	338	23
2540-01-270-6538	236	4	5340-01-270-8489	339	18
2540-01-270-6539	236	5	5340-01-270-8489	486	10
2540-01-270-6540	237	2	5975-01-270-8546	59	2
2540-01-270-6541	237	2	5975-01-270-8552	294	1
5305-01-270-7225	227	23	2540-01-270-9344	229	1
5305-01-270-7225	228	2	2590-01-270-9446	295	5
5305-01-270-7225	484	17	2510-01-270-9447	295	4
3040-01-270-7917	238	10	3040-01-270-9466	37	6
2510-01-270-7919	347	5	4710-01-270-9507	288	8
5342-01-270-7921	307	13	4720-01-270-9616	303	7
5342-01-270-7921	308	1	4710-01-270-9617	310	3
2510-01-270-7922	222	23	4720-01-270-9618	310	8
2510-01-270-7925	296	10	4730-01-270-9652	311	11
4730-01-270-7931	301	8	4730-01-270-9652	312	7
3040-01-270-7961	238	15	2510-01-270-9707	300	1
5310-01-270-8189	65	21	2540-01-270-9711	238	4
5310-01-270-8189	91	8	5305-01-271-1806	191	13
5310-01-270-8189	93	20	5320-01-271-1834	211	6
5310-01-270-8189	216	2	5320-01-271-1834	212	3
5330-01-270-8315	340	4	5320-01-271-1834	296	18
5330-01-270-8315	341	3	5320-01-271-1834	346	8

I-81

SECTION IV TM9-2320-280-24P C01

CROSS REFERENCE INDEXES

NATIONAL STOCK NUMBER INDEX

STOCK NUMBER	FIGURE NO	ITEM NO	STOCK NUMBER	FIGURE NO	ITEM NO
5310-01-271-1841	39	61	5320-01-271-6357	261	8
5340-01-271-1846	211	22	5320-01-271-6357	289	22
5340-01-271-1846	212	20	5320-01-271-6357	291	1
2540-01-271-1946	288	27	5320-01-271-6357	293	3
2540-01-271-1954	311	6	5320-01-271-6357	295	1
2540-01-271-1955	311	18	5320-01-271-6357	296	14
2540-01-271-1991	293	20	5320-01-271-6357	297	1
5365-01-271-2491	266	26	5320-01-271-6357	298	3
5365-01-271-2491	439	4	5320-01-271-6357	299	4
1640-01-271-2598	279	16	5320-01-271-6357	300	7
2540-01-271-2832	290	2	5320-01-271-6357	307	8
2540-01-271-2839	347	9	5320-01-271-6357	308	10
2540-01-271-2840	289	3	5320-01-271-6357	334	2
2540-01-271-2841	289	3	5320-01-271-6357	343	1
4520-01-271-2949	300	11	5320-01-271-6357	347	10
5340-01-271-3059	363	7	5320-01-271-6357	365	14
2540-01-271-3769	300	9	5320-01-271-6357	369	7
2590-01-271-3770	289	26	5320-01-271-6357	417	9
2540-01-271-3771	289	12	5340-01-271-6361	215	5
2540-01-271-3772	288	5	3120-01-271-6432	39	59
2540-01-271-3773	288	28	5340-01-271-6455	494	20
3040-01-271-3823	39	41	5340-01-271-6461	296	11
3040-01-271-3824	238	17	3130-01-271-6464	39	3
2540-01-271-3828	236	2	5340-01-271-6465	494	14
5340-01-271-4281	279	25	5340-01-271-6466	494	15
5340-01-271-4282	279	30	5340-01-271-6467	496	28
2540-01-271-5071	263	1	4710-01-271-6957	302	12
2540-01-271-5072	237	8	4720-01-271-6985	331	24
5340-01-271-5866	211	5	2510-01-271-7085	228	10
5340-01-271-5866	291	3	4710-01-271-7937	287	30
5340-01-271-5866	297	5	2540-01-271-8009	238	9
4720-01-271-6151	300	6	2540-01-271-8010	263	4
3110-01-271-6355	39	20	5340-01-271-9573	220	4
5320-01-271-6357	71	9	5340-01-271-9574	220	4
5320-01-271-6357	197	6	2540-01-271-9806	289	15
5320-01-271-6357	199	12	2590-01-271-9807	289	26
5320-01-271-6357	211	10	5999-01-272-0018	338	12
5320-01-271-6357	213	2	5999-01-272-0018	339	9
5320-01-271-6357	214	23	5999-01-272-0018	410	12
5320-01-271-6357	215	1	4710-01-272-0513	302	11
5320-01-271-6357	216	11	2510-01-272-0536	224	3
5320-01-271-6357	217	7	2510-01-272-0537	222	6
5320-01-271-6357	220	3	2510-01-272-0538	288	14
5320-01-271-6357	230	21	5310-01-272-1356	56	4
5320-01-271-6357	231	26	5330-01-272-2601	222	4
5320-01-271-6357	235	17	5340-01-272-2635	289	7
5320-01-271-6357	245	36	5365-01-272-2663	39	73

SECTION IV TM9-2320-280-24P C01

CROSS REFERENCE INDEXES

NATIONAL STOCK NUMBER INDEX

STOCK NUMBER	FIGURE NO	ITEM NO	STOCK NUMBER	FIGURE NO	ITEM NO
5340-01-272-2677	39	34	5365-01-272-7482	300	14
5340-01-272-2732	224	17	5340-01-272-7493	288	31
5340-01-272-2801	218	4	5340-01-272-7495	287	2
5365-01-272-2806	289	18	5365-01-272-7502	39	15
5310-01-272-3311	494	9	5365-01-272-7504	335	11
5310-01-272-3311	496	6	5365-01-272-7504	336	8
5340-01-272-3353	213	3	5307-01-272-7539	39	54
5340-01-272-3361	219	7	4710-01-272-7565	302	2
5340-01-272-3389	58	6	4030-01-272-7573	311	9
5340-01-272-3402	287	10	4030-01-272-7573	312	5
5340-01-272-3403	257	7	5340-01-272-7712	311	1
2510-01-272-3951	288	12	5330-01-272-7727	223	2
2510-01-272-3952	288	14	5340-01-272-7814	218	8
2510-01-272-3953	288	16	5340-01-272-7815	260	9
5306-01-272-4798	39	24	5305-01-272-8302	39	78
5310-01-272-4823	292	9	5305-01-272-8322	292	22
5310-01-272-4831	39	48	5305-01-272-8323	292	20
5365-01-272-4866	237	4	5330-01-272-8337	39	71
5365-01-272-4867	237	10	5365-01-272-8376	298	5
5365-01-272-4877	39	50	5342-01-272-8391	492	4
5340-01-272-4935	289	24	5340-01-272-8397	293	4
5340-01-272-4942	217	4	5340-01-272-8405	289	17
5340-01-272-4962	289	21	5340-01-272-8409	215	2
5310-01-272-5464	300	10	5340-01-272-8410	214	31
5310-01-272-5470	238	14	5340-01-272-8431	259	8
5330-01-272-5474	39	2	5340-01-272-8431	264	4
5340-01-272-5516	39	5	5365-01-272-8435	93	24
5340-01-272-5517	219	7	5365-01-272-8435	95	41
5340-01-272-5594	287	29	5305-01-272-9025	39	47
5340-01-272-5595	58	1	5305-01-272-9034	339	23
5340-01-272-5611	213	38	5305-01-272-9042	56	25
5340-01-272-5652	289	21	5305-01-272-9042	57	2
5940-01-272-5800	52	26	5305-01-272-9042	65	2
5355-01-272-5824	56	24	5305-01-272-9042	92	11
5340-01-272-6520	264	5	5305-01-272-9042	95	29
5340-01-272-6520	265	5	5305-01-272-9042	283	27
5310-01-272-6546	287	32	5305-01-272-9042	304	9
5365-01-272-6574	298	2	5305-01-272-9042	339	1
5340-01-272-6578	219	2	5330-01-272-9071	39	17
5340-01-272-6633	238	12	5365-01-272-9082	308	6
5340-01-272-6634	490	13	5340-01-272-9103	237	8
5310-01-272-7449	238	23	5340-01-272-9105	288	18
5310-01-272-7449	287	26	5340-01-272-9109	294	3
5330-01-272-7454	39	4	5340-01-272-9145	222	16
5330-01-272-7471	347	8	5365-01-272-9275	96	31
5330-01-272-7472	347	6	5305-01-272-9959	238	7
5330-01-272-7473	295	3	5310-01-272-9981	142	35

SECTION IV TM9-2320-280-24P C01

CROSS REFERENCE INDEXES

NATIONAL STOCK NUMBER INDEX

STOCK NUMBER	FIGURE NO	ITEM NO	STOCK NUMBER	FIGURE NO	ITEM NO
5310-01-272-9981	143	41	5320-01-275-1998	64	2
5310-01-272-9985	289	25	5320-01-275-1998	197	14
5330-01-272-9995	219	8	5320-01-275-1998	199	53
5365-01-273-0010	298	9	5320-01-275-1998	212	21
5320-01-273-0020	287	15	5320-01-275-1998	332	12
5315-01-273-0096	287	22	5320-01-275-1998	379	27
5340-01-273-0136	260	5	5975-01-275-2695	229	4
6220-01-273-0500	47	4	5330-01-275-3381	296	8
2510-01-273-0572	343	8	5360-01-275-3512	218	2
5310-01-273-1635	64	20	2540-01-275-6202	287	21
5340-01-273-1680	289	7	2540-01-275-6203	288	15
5340-01-273-1684	214	24	2510-01-275-8026	211	9
5340-01-273-1686	290	6	5920-01-275-8096	39	58
5340-01-273-1721	492	3	5340-01-275-8567	253	29
5340-01-273-2342	213	10	5340-01-275-8567	279	31
5340-01-273-2342	214	8	5340-01-275-8568	256	1
5340-01-273-2368	490	6	5340-01-275-8568	257	4
5340-01-273-2380	59	21	5306-01-276-0848	35	21
5340-01-273-2381	215	8	2540-01-276-1451	224	5
5320-01-273-4601	224	10	2540-01-276-1451	237	15
5320-01-273-4601	294	8	9330-01-276-1470	287	8
5340-01-273-4667	287	25	5306-01-276-1621	283	11
5340-01-273-4668	57	9	5306-01-276-1621	438	12
5365-01-273-4690	338	3	5340-01-276-3615	220	7
5306-01-273-6333	154	2	5340-01-276-3615	237	1
4730-01-273-7660	343	11	5305-01-276-4855	157	1
2510-01-273-8376	213	35	4420-01-276-5050	495	15
5340-01-274-3021	52	15	2510-01-276-5058	54	8
5340-01-274-3184	237	14	2510-01-276-5058	55	1
5340-01-274-3287	260	12	2590-01-276-5071	54	3
5340-01-274-3295	293	13	5340-01-276-5850	290	1
5325-01-274-3317	95	30	2540-01-276-5928	289	11
5340-01-274-3498	216	1	2540-01-276-5935	224	7
5365-01-274-3573	339	26	5340-01-276-6673	211	12
2510-01-274-4234	455	1	5340-01-276-6673	212	6
2510-01-274-4234	455	KIT	5340-01-276-6738	496	18
5365-01-274-4674	338	5	5340-01-276-7415	213	7
5365-01-274-4675	95	38	5340-01-276-7415	214	3
2510-01-274-4901	222	24	2590-01-276-7633	65	1
5330-01-274-6106	296	5	6150-01-276-7698	54	4
5342-01-274-6210	290	1	5970-01-276-7753	BULK	41
5340-01-274-7337	258	1	2540-01-277-2313	52	1
5340-01-274-7337	259	9	5365-01-277-2387	283	13
5342-01-274-9884	155	8	5365-01-277-2387	438	14
5340-01-274-9899	238	16	5340-01-277-4426	219	9
5342-01-275-0532	217	5	5340-01-277-4427	155	4
5310-01-275-1189	222	14	5340-01-277-4613	215	7

SECTION IV TM9-2320-280-24P C01

CROSS REFERENCE INDEXES

NATIONAL STOCK NUMBER INDEX

STOCK NUMBER	FIGURE NO	ITEM NO	STOCK NUMBER	FIGURE NO	ITEM NO
5340-01-277-5129	258	3	5340-01-282-2670	265	4
5340-01-277-5129	259	6	5340-01-282-2670	289	16
5340-01-277-5129	260	19	5305-01-282-3386	224	9
5340-01-277-5129	262	14	5342-01-282-4853	71	8
5340-01-277-8125	213	40	3030-01-282-6968	35	1
5342-01-277-9101	147	6	3030-01-282-6968	472	19
5342-01-277-9102	147	3	5935-01-282-7833	52	6
4710-01-278-1053	12	10	5935-01-282-7833	57	5
5315-01-278-2123	292	21	2540-01-282-8562	300	16
5305-01-278-7118	494	17	5360-01-282-9316	280	23
5340-01-278-8588	213	39	5360-01-282-9316	281	23
5340-01-278-8588	214	32	9390-01-282-9949	296	7
5310-01-278-9555	208	21	5340-01-283-2458	247	9
5310-01-278-9555	283	23	5305-01-283-7776	293	21
5360-01-278-9723	238	20	9905-01-283-7937	345	17
4130-01-278-9837	495	10	7690-01-283-7938	345	4
4720-01-279-5149	331	31	5310-01-283-8482	383	7
2540-01-280-4159	293	18	6220-01-284-2709	61	8
2510-01-280-4240	293	2	2540-01-284-7091	345	2
5310-01-280-5796	379	25	4520-01-284-7099	345	22
5365-01-280-5875	337	25	4520-01-284-7099	KITS	7
5365-01-280-5875	338	25	5315-01-284-9812	21	13
5340-01-280-5985	311	16	2940-01-285-2942	343	20
5315-01-280-6185	230	9	6150-01-285-3901	96	1
5315-01-280-6185	287	23	6110-01-285-3902	349	9
5342-01-280-6264	288	25	5305-01-285-4923	44	10
5310-01-280-6751	39	56	4710-01-285-5123	496	20
4330-01-280-8417	19	7	2540-01-285-6066	498	4
5975-01-280-8922	57	10	2540-01-285-6087	56	1
5330-01-281-0911	293	29	2530-01-285-8381	169	23
2510-01-281-1032	228	6	5307-01-286-6007	279	24
5340-01-281-3429	288	18	5310-01-286-6077	44	9
5330-01-281-6523	BULK	52	5310-01-286-6077	234	3
5340-01-281-7795	496	5	5310-01-286-6077	239	4
5340-01-281-7798	498	12	5340-01-286-6191	62	11
4720-01-282-0143	496	22	5306-01-286-7182	203	25
4720-01-282-0160	496	24	5306-01-286-7182	204	10
5340-01-282-0469	498	2	5905-01-287-4255	349	3
5330-01-282-0913	341	5	5905-01-287-4256	349	11
5340-01-282-1115	311	15	5340-01-287-5825	227	34
4130-01-282-1895	496	2	5340-01-287-5825	227	35
5330-01-282-2208	498	13	5340-01-287-6352	262	15
5330-01-282-2213	277	10	5315-01-287-6520	491	4
5330-01-282-2214	496	12	5310-01-287-6543	345	27
5330-01-282-2214	498	1	5310-01-287-6557	349	12
2930-01-282-2524	498	9	5930-01-287-6698	58	5
5305-01-282-2667	39	31	5365-01-287-8645	491	9

I-85

SECTION IV TM9-2320-280-24P C01

CROSS REFERENCE INDEXES

NATIONAL STOCK NUMBER INDEX

STOCK NUMBER	FIGURE NO	ITEM NO	STOCK NUMBER	FIGURE NO	ITEM NO
5365-01-287-8701	491	17	5342-01-289-7708	295	12
5310-01-287-8726	345	13	5340-01-289-7793	494	16
5340-01-287-8761	260	15	5365-01-289-7852	35	11
5820-01-287-9612	428	KIT	5325-01-289-7859	74	7
6150-01-287-9917	345	43	5325-01-289-7859	91	13
6150-01-287-9918	345	9	5325-01-289-7859	493	6
6150-01-287-9923	349	10	2510-01-289-8258	226	36
6150-01-287-9924	349	2	2510-01-289-8259	226	36
6150-01-287-9925	349	1	2510-01-289-8260	226	36
6150-01-287-9926	349	13	2540-01-289-8328	345	28
5331-01-288-1062	467	4	2540-01-289-8329	345	39
5331-01-288-1062	491	10	2540-01-289-8330	KITS	79
5305-01-288-1129	349	7	5330-01-289-9231	425	12
5305-01-288-1130	349	5	2520-01-289-9663	491	1
5340-01-288-1242	491	24	2540-01-290-0715	345	36
5310-01-288-1257	213	34	4320-01-290-0748	491	26
5310-01-288-1257	214	27	5325-01-290-1695	91	12
5330-01-288-1307	KITS	85	5995-01-290-1719	426	1
5340-01-288-2127	292	6	5995-01-290-1719	427	1
2540-01-288-5240	KITS	47	5340-01-290-2263	337	14
5640-01-288-5547	493	5	5340-01-290-2265	295	2
5340-01-288-6550	307	10	5340-01-290-2279	417	6
5340-01-288-6550	308	11	5330-01-290-2709	337	9
5340-01-288-6551	55	15	5340-01-290-2761	337	11
5975-01-288-6594	345	11	5315-01-290-3327	156	10
2540-01-288-8567	KITS	80	2815-01-290-3705	491	23
2510-01-289-2233	226	36	2520-01-290-3921	KITS	112
5340-01-289-3228	498	16	5340-01-290-4887	425	4
5340-01-289-3231	337	10	5340-01-290-4887	427	5
5340-01-289-3243	429	2	5340-01-290-4888	426	7
5340-01-289-3244	342	7	2940-01-290-5014	337	8
5340-01-289-3245	295	11	2510-01-290-5675	423	5
2520-01-289-3617	KITS	111	5340-01-290-6234	128	65
5365-01-289-4434	35	10	5340-01-290-6234	129	37
5340-01-289-4474	307	11	5305-01-290-6290	345	5
5340-01-289-4475	53	2	5340-01-290-6331	345	3
6150-01-289-4761	424	1	5340-01-290-6342	417	14
5325-01-289-5038	92	19	5340-01-290-6343	426	3
5340-01-289-5055	308	12	3040-01-290-6757	337	22
5310-01-289-5455	154	4	3040-01-290-6758	337	16
5940-01-289-5955	96	14	3040-01-290-6760	337	23
5950-01-289-6007	491	8	5305-01-290-8206	21	10
5365-01-289-6169	92	18	5340-01-290-8360	21	9
5325-01-289-6196	95	45	5340-01-290-8390	417	17
5315-01-289-6212	271	11	5340-01-290-8473	423	6
5315-01-289-6212	271	24	2920-01-290-9245	45	17
5340-01-289-6216	53	1	2590-01-291-1033	417	13

SECTION IV TM9-2320-280-24P C01

CROSS REFERENCE INDEXES

NATIONAL STOCK NUMBER INDEX

STOCK NUMBER	FIGURE NO	ITEM NO	STOCK NUMBER	FIGURE NO	ITEM NO
2540-01-291-1043	343	2	5340-01-296-2814	294	6
5340-01-291-2301	307	5	5305-01-296-7762	343	6
7690-01-291-2974	354	6	5305-01-296-7762	494	1
2520-01-291-2975	140	4	5306-01-296-8600	491	22
2520-01-291-2975	147	4	4730-01-296-9318	337	4
5340-01-291-4537	417	12	4730-01-296-9318	398	31
2530-01-291-4597	154	3	2530-01-296-9333	138	8
5330-01-291-4615	BULK	46	2530-01-296-9333	139	8
6150-01-291-4809	345	41	2530-01-296-9333	146	8
5340-01-291-5186	425	7	2540-01-296-9358	343	13
5340-01-291-5186	427	8	4720-01-297-0255	177	7
5340-01-291-5711	417	10	5340-01-297-1549	296	13
5995-01-291-6377	425	1	5340-01-297-2800	21	5
5995-01-291-6377	426	2	6220-01-297-3217	61	9
5995-01-291-6377	427	2	5340-01-297-4127	64	1
5995-01-291-6384	428	1	4730-01-297-5111	337	5
5340-01-291-8915	418	3	2520-01-297-5200	148	8
5340-01-291-8915	419	3	6680-01-298-0498	47	12
5340-01-291-8915	420	8	5305-01-298-2436	332	45
5340-01-291-8915	421	5	4730-01-298-4363	401	3
5340-01-291-8915	422	7	5310-01-298-4686	279	5
5340-01-291-8915	423	11	5330-01-298-4713	211	28
5307-01-291-9018	30	5	5330-01-298-4713	212	24
5307-01-291-9018	490	4	5330-01-298-6009	296	9
5310-01-292-5354	42	12	5310-01-298-7770	398	21
5340-01-292-5402	42	9	5365-01-298-7780	293	28
5340-01-292-5402	475	7	5330-01-298-8127	340	7
3010-01-292-6428	148	9	5355-01-298-8571	280	44
5940-01-292-6907	345	6	5355-01-298-8571	281	44
5325-01-292-7558	485	24	5340-01-298-9691	417	14
5340-01-292-8404	37	12	5340-01-299-2963	252	28
5340-01-292-8404	38	3	5310-01-299-6460	327	4
5340-01-292-8404	42	1	4010-01-299-7699	293	24
5340-01-292-8404	475	12	5975-01-299-7781	52	7
5340-01-293-0125	35	24	5975-01-299-7781	57	6
5340-01-293-0125	42	1	4720-01-299-8469	337	20
6150-01-293-4074	349	6	2990-01-299-8820	401	13
5330-01-293-5345	163	4	5315-01-300-3012	280	24
5340-01-293-5558	163	2	5315-01-300-3012	281	24
2540-01-293-6926	368	2	5340-01-300-3715	141	13
5340-01-294-3230	300	3	5340-01-300-3715	149	7
9905-01-294-3356	52	8	4130-01-300-6350	494	3
9905-01-294-3356	57	7	2540-01-300-8744	KITS	73
5340-01-294-4415	300	2	2540-01-300-8745	KITS	72
5340-01-295-1088	431	5	5340-01-301-7929	147	10
5340-01-295-8169	431	1	2540-01-302-2595	432	KIT
5340-01-296-1736	494	6	4820-01-303-3982	172	15

SECTION IV TM9-2320-280-24P C01

CROSS REFERENCE INDEXES

NATIONAL STOCK NUMBER INDEX

STOCK NUMBER	FIGURE NO	ITEM NO	STOCK NUMBER	FIGURE NO	ITEM NO
5310-01-303-4701	43	31	4730-01-311-4294	17	4
5310-01-303-4701	425	6	5310-01-312-4777	203	22
5310-01-303-4701	426	9	5975-01-312-5469	417	1
5310-01-303-4701	427	7	2590-01-312-7355	BULK	44
2510-01-303-5769	47	8	2540-01-313-0678	230	6
5310-01-304-8733	43B	12	3020-01-313-0682	34	21
5310-01-304-8733	43B	30	5310-01-313-3562	329	16
5365-01-305-7067	128	22			
5365-01-305-7067	130	25			
5961-01-305-8848	60	8	5310-01-313-3562	461	4
3120-01-306-3577	314	31	4910-01-313-8839	353	1
5340-01-306-3606	300	4	4710-01-313-9340	379	31
3010-01-306-4113	458	24	2540-01-314-0142	397	KIT
2590-01-306-5881	227	9	2930-01-314-0145	379	33
4820-01-306-6838	172	14	2990-01-314-0151	379	30
5330-01-306-7887	458	23	2540-01-314-1121	284	16
5365-01-306-9955	458	9	2540-01-314-1121	439	9
2510-01-307-0152	331	7	2530-01-314-1129	157	22
2510-01-307-0152	332	44	2530-01-314-1129	158	12
3040-01-307-6157	458	3	2540-01-314-1130	379	26
2590-01-307-9303	468	14	2540-01-314-1188	329	2
5310-01-308-0131	199	8	2540-01-314-1189	329	3
5340-01-308-2168	482	20	2540-01-314-1190	329	3
5340-01-308-3851	484	3	5340-01-314-1955	329	17
1640-01-308-5097	BULK	50	5340-01-314-1956	329	15
4010-01-308-9168	482	6	5935-01-314-2084	380	9
5305-01-309-1521	458	1	2540-01-314-2101	329	5
5995-01-309-2953	82	1	5340-01-314-2445	379	35
5640-01-309-3866	200	4	2540-01-314-2782	320	2
5640-01-309-3867	200	1	2540-01-314-2786	316	1
5640-01-309-3868	200	2	4710-01-314-4483	11	10
5640-01-309-3870	200	3	4710-01-314-4484	11	17
5330-01-309-4340	482	16	4710-01-314-4485	11	16
2540-01-309-4459	23	11	4710-01-314-4486	11	24
2540-01-309-4459	24	12	4710-01-314-4487	11	18
6130-01-309-6458	39	53	4710-01-314-4488	11	28
5325-01-309-7164	379	16	4710-01-314-4489	11	30
5340-01-309-7900	296	16	2510-01-314-4892	397	18
5340-01-310-1189	213	16	5340-01-314-4951	83	26
5340-01-310-1658	214	12	5340-01-314-4951	445	11
2510-01-310-2324	274	6	6150-01-314-5241	381	4
5340-01-310-5321	224	4	5365-01-314-5544	379	4
5330-01-310-6780	4	27	5365-01-314-5544	379	6
2920-01-310-9973	39	28	5340-01-314-5567	379	11
2920-01-310-9978	39	16	5340-01-314-5602	104	19
5340-01-311-1633	104	18	5340-01-314-5602	105	27
5340-01-311-3033	138	4	6150-01-314-5643	381	1
5340-01-311-3033	145	4	5340-01-314-5957	210	15
5930-01-311-3610	54	10	5340-01-314-5957	213	23

I-88

SECTION IV TM9-2320-280-24P C01

CROSS REFERENCE INDEXES

NATIONAL STOCK NUMBER INDEX

STOCK NUMBER	FIGURE NO	ITEM NO	STOCK NUMBER	FIGURE NO	ITEM NO
5340-01-314-5957	225	14	5340-01-314-5957	289	2
5340-01-314-5957	231	3	5340-01-314-5957	292	2
5340-01-314-5957	234	15	5340-01-314-5957	318	4
5340-01-314-5957	240	11	5340-01-314-5957	374	3
5340-01-314-5957	246	1	5340-01-314-5957	429	9
5340-01-314-5957	246	7	5340-01-314-5957	430	5
5340-01-314-5957	247	11	5340-01-314-5957	430	21
5340-01-314-5957	247	22	5340-01-314-5957	430	32
5340-01-314-5957	248	3	5340-01-314-5957	446	5
5340-01-314-5957	249	7	5340-01-314-5957	447	2
5340-01-314-5957	249	15	5340-01-314-5957	448	2
5340-01-314-5957	249	22	5340-01-314-5957	449	4
5340-01-314-5957	249	31	5340-01-314-5957	450	3
5340-01-314-5957	251	16	5330-01-314-6781	104	21
5340-01-314-5957	251	24	5340-01-314-6838	379	51
5340-01-314-5957	251	30	2540-01-314-7834	230	1
5340-01-314-5957	251	40	2540-01-314-7834	235	1
5340-01-314-5957	252	18	2540-01-314-7835	230	7
5340-01-314-5957	252	33	2815-01-314-7940	1	1
5340-01-314-5957	252	38	3990-01-314-8393	199	19
5340-01-314-5957	253	6	3990-01-314-8393	251	20
5340-01-314-5957	253	11	3990-01-314-8393	270	12
5340-01-314-5957	253	16	3990-01-314-8393	296	3
5340-01-314-5957	253	26	2540-01-314-9320	381	KIT
5340-01-314-5957	253	31	2540-01-314-9378	230	2
5340-01-314-5957	254	5	2540-01-314-9378	235	2
5340-01-314-5957	254	11	2540-01-314-9379	432	1
5340-01-314-5957	254	17	2540-01-314-9380	432	1
5340-01-314-5957	255	5	6150-01-315-1148	381	3
5340-01-314-5957	255	11	5310-01-315-1535	379	38
5340-01-314-5957	255	15	5330-01-315-1609	104	20
5340-01-314-5957	255	24	5330-01-315-1609	105	28
5340-01-314-5957	255	31	2540-01-315-3143	234	2
5340-01-314-5957	256	3	2540-01-315-3143	239	1
5340-01-314-5957	257	2	4730-01-315-3242	385	9
5340-01-314-5957	258	4	2540-01-315-3358	234	2
5340-01-314-5957	259	4	5310-01-315-3403	234	10
5340-01-314-5957	260	1	5310-01-315-3403	235	13
5340-01-314-5957	262	2	5310-01-315-3403	239	7
5340-01-314-5957	262	10	5310-01-315-3403	402	10
5340-01-314-5957	263	2	5365-01-315-3595	329	4
5340-01-314-5957	264	3	5307-01-315-3597	8	12
5340-01-314-5957	265	3	5307-01-315-3597	36	18
5340-01-314-5957	268	40	5307-01-315-3597	38	18
5340-01-314-5957	270	39	5307-01-315-3597	42	2
5340-01-314-5957	286	14	5307-01-315-3597	475	15
5340-01-314-5957	287	5	5340-01-315-3611	329	10

I-89

SECTION IV TM9-2320-280-24P C01

CROSS REFERENCE INDEXES

NATIONAL STOCK NUMBER INDEX

STOCK NUMBER	FIGURE NO	ITEM NO	STOCK NUMBER	FIGURE NO	ITEM NO
2540-01-315-3762	395	6	5306-01-316-2433	234	1
5340-01-315-4120	379	47	5306-01-316-2433	239	10
4010-01-315-4179	314	15	4720-01-316-2538	366	22
5340-01-315-4955	379	15	4720-01-316-2538	368	39
5305-01-315-7066	239	2	4720-01-316-2538	386	8
5305-01-315-7066	402	4	2590-01-316-2581	319	4
5306-01-315-7087	239	3	4520-01-316-2585	387	9
5306-01-315-7088	234	7	2540-01-316-2597	387	6
5360-01-315-7211	316	11	2520-01-316-2630	178	13
5360-01-315-7211	320	10	2520-01-316-2630	436	6
5360-01-315-7211	320	22	2540-01-316-2697	396	9
5360-01-315-7211	324	13	5340-01-316-2959	329	15
5360-01-315-7211	324	27	5307-01-316-2986	329	14
5360-01-315-7211	375	18	2540-01-316-6624	321	KIT
5360-01-315-7211	378	18	5365-01-316-8966	382	2
5360-01-315-7212	316	11	5365-01-316-8985	392	5
5360-01-315-7212	320	10	5340-01-317-0147	181	17
5360-01-315-7212	320	22	5305-01-317-0409	393	6
5360-01-315-7212	324	13	5305-01-317-0409	395	20
5360-01-315-7212	324	27	2590-01-317-0534	389	13
5360-01-315-7212	375	18	2590-01-317-0535	389	13
5360-01-315-7212	378	18	2540-01-317-0728	372	KIT
5340-01-315-7223	72	23	2540-01-317-0820	390	1
5340-01-315-7223	73	10	4520-01-317-0929	387	3
7690-01-315-8539	354	3	2540-01-317-1024	390	10
7690-01-315-8540	380	11	4710-01-317-1044	387	14
5305-01-315-8649	69	33	4710-01-317-1045	385	7
5305-01-315-8649	434	4	4710-01-317-1046	384	2
5305-01-315-8649	435	9	5340-01-317-1470	181	4
2590-01-315-9128	188	8	5340-01-317-2675	388	17
5305-01-315-9512	392	2	2540-01-317-3309	379	28
5305-01-315-9524	395	15	2540-01-317-3350	396	5
5305-01-315-9683	397	4	2590-01-317-3952	395	2
5305-01-315-9882	395	24	2590-01-317-3953	383	4
5305-01-315-9882	397	12	5310-01-317-4022	391	6
2590-01-316-0084	185	9	2540-01-317-4289	393	10
2590-01-316-0084	186	14	2540-01-317-4289	395	10
2590-01-316-0084	187	13	2590-01-317-4854	393	8
2510-01-316-0216	389	8	2590-01-317-4855	395	18
4710-01-316-0222	11	11	2590-01-317-4856	383	3
2540-01-316-0892	452	4	2930-01-317-5358	28	5
5306-01-316-1456	234	12	5330-01-317-5392	393	5
5340-01-316-1507	329	17	5330-01-317-5392	395	22
4030-01-316-1551	183	2	5330-01-317-5393	395	4
4030-01-316-1551	185	12	6150-01-317-5853	72	25
4030-01-316-1551	187	16	6150-01-317-5853	434	1
4030-01-316-1551	188	12	6150-01-317-5853	435	1

I-90

SECTION IV TM9-2320-280-24P C01

CROSS REFERENCE INDEXES

NATIONAL STOCK NUMBER INDEX

STOCK NUMBER	FIGURE NO	ITEM NO	STOCK NUMBER	FIGURE NO	ITEM NO
5340-01-317-7501	389	5	2910-01-323-0123	368	35
2590-01-317-7561	395	23	6220-01-323-0431	396	13
9905-01-317-7986	380	10	2990-01-323-2562	365	7
9905-01-317-7987	380	6	2590-01-323-5153	155	9
5310-01-317-8164	395	8	5306-01-323-5534	198	22
2510-01-317-8258	396	1	5306-01-323-5534	331	4
2590-01-317-8292	395	3	5306-01-323-5535	198	22
5340-01-317-9086	391	7	5306-01-323-5535	207	8
5340-01-318-0212	433	11	5306-01-323-5535	280	50
5330-01-318-1998	395	5	5306-01-323-5535	281	50
5340-01-318-2040	391	2	9905-01-324-0886	396	7
5340-01-318-2041	392	10	5330-01-324-0906	121	2
5365-01-318-2066	388	4	5330-01-324-0906	122	6
5340-01-318-2180	397	3	5340-01-324-1076	431	6
5930-01-318-2809	380	8	5977-01-324-6323	45	45
5340-01-318-3366	433	3	6150-01-324-6355	387	18
5340-01-318-4893	397	6	6150-01-324-6356	383	9
5310-01-318-5237	389	4	5330-01-324-8260	172	19
2540-01-318-9229	230	8	4720-01-325-0204	399	17
5330-01-318-9780	395	7	5340-01-325-2891	433	6
2815-01-319-1433	1	2	4720-01-325-6985	370	4
2510-01-319-5952	397	10	2510-01-325-8741	375	8
5330-01-319-7302	387	10	2510-01-325-8741	375	15
2510-01-319-9384	395	17	5365-01-326-1153	368	25
5340-01-319-9426	391	9	5342-01-326-2583	365	19
5340-01-320-1079	379	45	2910-01-326-9221	10	2
5340-01-320-2060	393	2	5310-01-327-0387	337	15
5340-01-320-2060	395	13	3040-01-327-1426	431	4
2910-01-320-6645	366	37	2590-01-328-2904	437	KIT
2910-01-320-6645	384	11	2540-01-329-8074	323	KIT
5340-01-320-8662	387	11	2540-01-329-8649	378	1
2540-01-320-8731	401	KIT	4730-01-329-9151	29	11
2540-01-320-8918	388	13	3950-01-329-9890	314	29
3030-01-321-4482	37	19	2590-01-329-9910	314	2
3030-01-321-4482	475	27	2510-01-330-2249	378	11
2530-01-321-4497	436	KIT	2510-01-330-2250	378	11
1005-01-321-6774	431	KIT	2590-01-330-6102	314	13
5330-01-321-8622	BULK	53	2540-01-330-6169	319	1
5975-01-322-6373	396	8	2540-01-330-6170	322	1
4730-01-322-9160	302	4	2540-01-330-6171	316	4
4730-01-322-9160	305	6	2540-01-330-6172	320	2
4730-01-322-9160	306	4	2540-01-330-6173	316	1
4730-01-322-9160	309	5	2510-01-330-6174	316	7
4730-01-322-9871	368	15	2510-01-330-6174	320	19
2990-01-322-9879	368	17	2510-01-330-6175	316	7
2990-01-322-9880	365	11	2510-01-330-6175	320	19
2990-01-322-9881	368	13	2510-01-330-6176	320	5

SECTION IV TM9-2320-280-24P C01

CROSS REFERENCE INDEXES

NATIONAL STOCK NUMBER INDEX

STOCK NUMBER	FIGURE NO	ITEM NO	STOCK NUMBER	FIGURE NO	ITEM NO
2540-01-330-6177	320	1	4710-01-335-7523	277	15
2510-01-330-6576	320	5	5331-01-335-7592	19	8
5330-01-330-9645	118	13	2510-01-335-7799	197	25
5305-01-331-6284	437	3	5331-01-335-8878	167	16
5330-01-331-9995	118	1	2540-01-336-0791	378	4
2510-01-332-0128	271	1	5340-01-336-3004	251	27
2510-01-332-0129	271	1	5340-01-336-3004	252	27
5340-01-332-7515	208	8	5340-01-336-3004	253	22
5340-01-332-7516	437	9	5340-01-336-3004	254	27
5340-01-332-7599	331	19	5340-01-336-3004	255	21
5340-01-332-7599	332	23	2530-01-336-3127	167	19
5340-01-332-7599	370	5	5331-01-336-3236	73	13
5340-01-332-7602	225	6	2530-01-336-5740	167	14
5340-01-332-7602	226	39	5306-01-336-7175	167	15
5310-01-333-0060	283	16	2540-01-337-0242	323	5
5340-01-333-0162	346	3	2590-01-337-4071	433	KIT
3030-01-333-2286	37	19	5310-01-337-7034	42	12
2590-01-333-2999	377	12	6220-01-338-2059	65	9
5310-01-333-5245	62	29	2530-01-338-2730	167	18
5310-01-333-5245	379	50	2540-01-338-8081	402	KIT
5310-01-333-5245	485	27	2510-01-338-8087	209	KIT
5320-01-333-5545	197	21	5995-01-339-0190	282	1
5320-01-333-5545	199	47	5995-01-339-0190	435	4
5340-01-333-5851	234	8	6250-01-339-6271	65	11
4710-01-333-5995	277	42	5330-01-340-5627	205	6
5340-01-333-6038	197	3	5940-01-341-7689	93	2
2530-01-333-6068	157	5	5330-01-341-8963	206	6
2610-01-333-7632	168	1	6145-01-341-9591	BULK	67
2530-01-333-8263	157	5	2590-01-342-4918	433	KIT
5330-01-334-2834	205	5	2510-01-344-4169	271	18
5330-01-334-2834	206	5	5310-01-344-8250	16	32
5340-01-334-2887	239	5	2510-01-345-4365	271	1
5340-01-334-4241	234	20	4730-01-346-1063	167	24
2540-01-334-4333	403	18	2520-01-346-1374	137	4
2640-01-334-9453	167	17	5331-01-346-3806	167	25
5120-01-335-1094	505	1	4730-01-347-7342	385	6
5120-01-335-1096	505	1	5930-01-347-9216	327	10
5120-01-335-1099	505	1	5999-01-348-0302	40	19
5120-01-335-1100	505	1	5920-01-348-0303	40	32
6150-01-335-3087	39	79	3120-01-348-3364	40	20
2510-01-335-4170	199	55	3110-01-348-4867	40	10
2510-01-335-4171	197	12	3110-01-348-4867	43B	33
2540-01-335-4482	199	3	4730-01-348-6231	388	6
2640-01-335-4583	167	23	2510-01-348-6670	271	1
5120-01-335-5847	505	6	5315-01-348-6880	40	45
2510-01-335-7343	283	17	5365-01-348-6971	40	34
2510-01-335-7363	227	10	5306-01-348-8310	40	24

I-92

SECTION IV TM9-2320-280-24P C01

CROSS REFERENCE INDEXES

NATIONAL STOCK NUMBER INDEX

STOCK NUMBER	FIGURE NO	ITEM NO	STOCK NUMBER	FIGURE NO	ITEM NO
5310-01-348-8313	40	38	2520-01-357-5056	148	25
5310-01-348-8314	40	28	4730-01-357-5651	388	2
5310-01-348-8360	40	44	4730-01-357-6523	385	5
5310-01-348-8360	43B	2	2510-01-357-8789	187	4
5310-01-348-8384	40	43	2510-01-357-8789	441	4
5310-01-348-8385	40	36	2510-01-357-8789	456	1
5310-01-348-8386	40	33	2910-01-357-8798	15	1
5310-01-348-8392	40	29	2540-01-357-8939	288	1
5310-01-348-8393	40	25	5120-01-357-9123	501	19
5310-01-348-8398	40	51	2530-01-357-9708	164	1
5310-01-348-8398	43B	13	2530-01-357-9776	163	7
6115-01-349-5320	40	5	4710-01-357-9968	161	6
4140-01-350-0839	40	41	5975-01-358-0021	40	27
5975-01-350-1987	40	21	2510-01-358-1176	187	7
5975-01-350-1987	43B	37	2510-01-358-1176	441	12
7690-01-350-2094	358	4	2510-01-358-1176	456	21
5310-01-350-4257	40	22	2510-01-358-1177	187	7
5310-01-350-4257	43B	46	2510-01-358-1177	441	12
2510-01-350-4949	203	3	2510-01-358-1177	456	21
5970-01-350-5646	40	31	2510-01-358-1178	187	4
5310-01-350-8549	40	23	2510-01-358-1178	441	4
5310-01-350-8549	43B	36	2510-01-358-1178	456	1
2540-01-350-9380	379	18	4710-01-358-1943	159	1
5306-01-351-7742	345	23	4710-01-358-1943	160	12
5306-01-351-7742	348	22	4710-01-358-2127	13	21
2930-01-353-5794	31	3	4710-01-358-2127	14	7
5355-01-353-6934	327	12	2520-01-358-3160	140	4
2510-01-354-0417	271	18	2520-01-358-3160	147	4
5320-01-354-2547	220	5	5365-01-358-4641	130	17
2540-01-354-4291	315	1	5365-01-358-4642	132	24
2540-01-354-4291	444	1	5360-01-358-5420	132	28
5330-01-355-3686	4	24	5325-01-358-5450	129	5
2520-01-356-9189	136	6	5310-01-358-5461	129	16
2520-01-356-9197	127	3	5340-01-358-5588	180	6
5325-01-357-1080	316	18	4720-01-358-6100	130	13
5325-01-357-1080	320	33	4710-01-358-6397	13	22
5325-01-357-1080	324	39	4710-01-358-6397	14	8
5360-01-357-2413	192	1	5360-01-358-6563	129	29
5120-01-357-3629	501	16	5360-01-358-6564	132	26
5120-01-357-3630	501	17	5340-01-358-6697	74	3
5120-01-357-3631	501	18	3040-01-358-7034	488	17
5120-01-357-3632	501	12	5340-01-358-7595	180	26
5120-01-357-3633	501	15	2590-01-358-7643	132	18
5340-01-357-3996	15	10	2590-01-358-7644	132	19
5340-01-357-3996	16	9	2520-01-358-7698	131	5
2520-01-357-5043	136	16	5340-01-358-7940	132	29
2520-01-357-5044	137	18	5325-01-358-7958	129	10

I-93

SECTION IV TM9-2320-280-24P C01

CROSS REFERENCE INDEXES

NATIONAL STOCK NUMBER INDEX

STOCK NUMBER	FIGURE NO	ITEM NO	STOCK NUMBER	FIGURE NO	ITEM NO
5325-01-358-7959	129	17	4730-01-359-2217	130	14
5325-01-358-7960	129	20	3020-01-359-2529	130	43
5325-01-358-7961	130	42	3020-01-359-2529	130	47
5325-01-358-7962	130	34	6220-01-359-2870	61	10
5310-01-358-7974	130	4	4710-01-359-2956	15	6
5310-01-358-7975	130	7	4710-01-359-2956	16	6
5310-01-358-7978	132	25	4710-01-359-2956	366	18
5340-01-358-8000	129	30	5306-01-359-4529	129	1
4710-01-358-8410	130	12	5306-01-359-4530	130	2
4730-01-358-8538	175A	1	5305-01-359-4586	129	27
3040-01-358-8596	132	17	3110-01-359-6669	130	5
3020-01-358-8601	130	48	3120-01-359-6760	129	6
4320-01-358-8608	130	40	3120-01-359-6764	132	27
3020-01-358-8656	130	9	5970-01-359-7836	98	12
3020-01-358-8657	130	15	5340-01-359-8819	446	4
2520-01-358-8658	130	10	3110-01-359-8858	130	45
3020-01-358-8659	130	33	5340-01-359-8933	440	20
3020-01-358-8661	130	8	5365-01-359-9406	440	19
3020-01-358-8662	130	16	5340-01-359-9474	440	8
3040-01-358-8690	129	13	4820-01-359-9489	159	2
3040-01-358-8691	130	44	4820-01-359-9489	160	11
3040-01-358-8692	130	18	5306-01-360-0926	438	4
2520-01-358-8700	132	20	3120-01-360-0977	130	36
2520-01-358-8701	132	13	3130-01-360-1050	129	15
3040-01-358-8708	129	24	3130-01-360-1051	129	21
2520-01-358-8878	130	3	3130-01-360-1051	130	36
2520-01-358-8879	130	1	5306-01-360-1123	283	21
2520-01-358-8880	129	2	5306-01-360-1123	438	17
2520-01-358-8881	129	9	5306-01-360-1124	446	11
2520-01-358-8883	132	22	5306-01-360-1125	441	13
5340-01-358-9301	153	12	5306-01-360-1125	442	10
7690-01-358-9391	361	11	5306-01-360-1125	453	19
5330-01-358-9532	128	59	5306-01-360-1126	442	14
5330-01-358-9532	129	31	5306-01-360-1127	442	17
5330-01-358-9533	130	31	5306-01-360-1128	188	37
5330-01-358-9540	130	30	5306-01-360-1128	442	1
5330-01-358-9541	130	11	5306-01-360-1128	453	12
5331-01-358-9545	129	28	5306-01-360-1129	440	7
5305-01-359-0167	74	13	5306-01-360-1130	440	16
3040-01-359-1116	129	8	5305-01-360-1136	33	15
3040-01-359-1117	129	19	5340-01-360-1188	446	34
3040-01-359-1118	130	37	5340-01-360-1189	445	6
3040-01-359-1119	130	20	3120-01-360-1202	129	12
3020-01-359-1136	130	6	5340-01-360-1727	446	2
3020-01-359-1211	129	18	5340-01-360-1784	446	35
5330-01-359-1292	130	35	5340-01-360-1886	440	23
2510-01-359-2076	198	26	5340-01-360-1887	446	29

I-94

SECTION IV TM9-2320-280-24P C01

CROSS REFERENCE INDEXES

NATIONAL STOCK NUMBER INDEX

STOCK NUMBER	FIGURE NO	ITEM NO	STOCK NUMBER	FIGURE NO	ITEM NO
5340-01-360-2020	442	12	6220-01-362-5211	97	9
5340-01-360-2021	440	15	6150-01-362-5229	81	8
5340-01-360-2053	440	14	5310-01-362-6171	43B	10
4720-01-360-2380	27	7	5320-01-362-6195	247	7
2590-01-360-2417	446	25	2590-01-363-2084	442	19
2540-01-360-2482	442	16	2920-01-363-5173	40	2
5340-01-360-2979	446	17	2590-01-364-1534	442	15
3040-01-360-4448	130	29	2510-01-364-3120	KITS	46
5340-01-360-4909	446	7	3020-01-364-3398	148	27
5340-01-360-4909	447	1	9905-01-364-7342	354	12
5330-01-360-5271	105	17	4240-01-365-0982	307	15
2590-01-360-5411	442	19	4240-01-365-0982	308	3
2590-01-360-5412	442	11	5340-01-365-8796	287	4
2590-01-360-5413	442	6	9905-01-365-8849	448	11
2590-01-360-5414	442	6	5140-01-366-0642	512	8
3120-01-360-5960	130	46	5120-01-366-0893	512	1
5340-01-360-9070	446	33	5120-01-366-0894	512	9
2590-01-360-9536	442	11	5120-01-366-0895	512	5
2590-01-360-9537	442	9	5120-01-366-2520	512	2
5310-01-361-1152	438	9	6150-01-366-2916	445	12
5310-01-361-1163	439	7	5310-01-366-3539	33A	10
5310-01-361-1163	462	8	5120-01-366-4713	512	6
5310-01-361-1163	465	1	5120-01-366-6260	512	7
5310-01-361-1163	485	25	2510-01-366-6968	271	18
5340-01-361-1212	440	18	2510-01-366-6969	271	18
5120-01-361-3101	501	14	5120-01-366-9468	512	3
6220-01-361-5084	67	8	5340-01-367-2070	266	14
5305-01-361-5353	225	52	5120-01-367-3536	510	8
5305-01-361-5353	306	8	5120-01-367-3582	505	2
5305-01-361-5353	346	5	5120-01-367-3585	505	2
5305-01-361-5353	410	23	2510-01-369-3230	227	27
5305-01-361-5353	445	4	5306-01-371-4685	198	15
5305-01-361-5353	451	3	2640-01-371-8332	KITS	49
5340-01-361-6658	184	3	6220-01-372-3883	61	4
5340-01-361-7879	448	9	5999-01-372-5601	39	80
5340-01-361-7879	449	9	5999-01-372-5601	41	20
2530-01-361-8012	KITS	53	5999-01-372-5601	475	26
5330-01-361-8013	KITS	50	6680-01-372-5748	506	3
5330-01-361-8014	KITS	51	2510-01-372-9734	228	4
5330-01-361-8015	KITS	52	2510-01-373-0131	228	3
2590-01-361-8087	442	3	5925-01-373-1034	367	12
2590-01-361-8088	442	3	2510-01-373-2783	228	1
6150-01-361-8129	74	15	2510-01-373-2784	228	5
2540-01-361-8206	187	1	5120-01-373-4981	507	5
2540-01-361-8206	441	1	5120-01-373-4982	506	1
2510-01-361-8213	199	39	5120-01-373-4984	508	5
9905-01-362-2014	355	4	5120-01-373-4985	508	2

I-95

SECTION IV TM9-2320-280-24P C01

CROSS REFERENCE INDEXES

NATIONAL STOCK NUMBER INDEX

STOCK NUMBER	FIGURE NO	ITEM NO	STOCK NUMBER	FIGURE NO	ITEM NO
5120-01-373-4987	507	3	5365-01-381-6110	148	20
5120-01-373-4988	507	4	5365-01-381-6112	148	20
5120-01-373-4989	506	2	5365-01-381-6113	148	20
5120-01-373-5983	508	3	5365-01-381-6120	148	20
5120-01-373-5984	508	4	5365-01-381-6122	148	20
5120-01-373-5985	506	4	5365-01-381-6145	148	20
2510-01-373-7877	228	7	5365-01-381-6151	148	20
2510-01-373-8445	228	8	5365-01-381-6154	148	20
5120-01-373-8836	507	2	5365-01-381-6158	148	20
5120-01-373-8840	508	1	5365-01-381-6164	148	20
2510-01-374-0000	228	9	5365-01-381-6165	148	20
5310-01-374-4512	261	3	5365-01-381-6174	148	20
2510-01-376-1092	368	31	5365-01-381-6175	148	20
3120-01-377-5220	314	12	5365-01-381-6188	148	20
2590-01-377-6819	250	7	5365-01-381-6197	148	20
5365-01-378-1755	104	12	5365-01-381-6210	148	20
5340-01-378-5201	69	16	5365-01-381-6211	148	20
2510-01-378-5615	266	10	5365-01-381-6223	148	20
5340-01-378-7525	69	17	5365-01-381-6226	148	20
2920-01-378-8035	434	KITS	5365-01-381-6229	148	20
5330-01-378-8572	4	32	5365-01-381-6240	148	20
5330-01-379-1139	122	4	5365-01-381-6250	148	20
5331-01-380-2118	176A	13	5365-01-381-6258	148	20
5305-01-380-3395	1	5	5365-01-381-6261	148	20
5305-01-380-3395	2	6	5365-01-381-6276	148	20
5340-01-380-4561	153	1	5365-01-381-6278	148	20
5365-01-380-5345	148	20	5365-01-381-6292	148	20
5365-01-380-7340	231	23	5365-01-381-6303	148	20
5305-01-380-9163	333	14	5365-01-381-6313	148	20
5340-01-381-1266	401	19	5365-01-381-6317	148	20
5365-01-381-1783	401	7	5365-01-381-6332	148	20
5340-01-381-1808	140	9	2540-01-381-8392	231	1
5330-01-381-1810	327	33	2540-01-381-8988	231	18
5340-01-381-2045	401	17	2540-01-381-9331	231	20
2540-01-381-2079	231	19	5365-01-381-9582	148	20
5365-01-381-2215	303	8	5365-01-381-9588	148	20
5340-01-381-2248	333	16	5365-01-382-0973	148	17
5305-01-381-2296	231	11	5365-01-382-0986	454	5
5305-01-381-2305	231	2	5365-01-382-1058	148	17
5365-01-381-4487	148	17	5365-01-382-1132	148	17
5365-01-381-5166	148	20	5365-01-382-1135	148	17
5365-01-381-5174	148	20	5365-01-382-1157	148	17
5310-01-381-5328	231	24	5365-01-382-1159	148	17
5310-01-381-5328	454	4	5365-01-382-1162	148	17
5365-01-381-6076	148	20	5365-01-382-1166	148	17
5365-01-381-6080	148	20	5365-01-382-1170	148	17
5365-01-381-6086	148	20	5365-01-382-1171	148	17

SECTION IV TM9-2320-280-24P C01

CROSS REFERENCE INDEXES

NATIONAL STOCK NUMBER INDEX

STOCK NUMBER	FIGURE NO	ITEM NO	STOCK NUMBER	FIGURE NO	ITEM NO
5365-01-382-1177	148	17	5340-01-385-0057	377	3
5365-01-382-1178	231	30	9905-01-385-2633	354	12
5365-01-382-1181	148	17	9905-01-385-2639	354	12
5365-01-382-1235	148	17	2540-01-385-7497	289	8
5365-01-382-1240	148	17	2990-01-385-8988	401	5
5365-01-382-1273	148	17	2540-01-385-9000	327	31
4730-01-382-3165	302	1	2540-01-385-9031	327	37
4730-01-382-3165	305	1	2930-01-385-9108	333	19
5365-01-382-3447	148	17	2520-01-386-3384	137	19
5365-01-382-3676	148	17	2530-01-386-5499	167	18
5365-01-382-3693	148	17	5340-01-386-6067	292	4
5365-01-382-3694	148	17	2540-01-386-8790	327	32
5365-01-382-3700	148	17	5305-01-386-9052	431	8
5365-01-382-3710	148	17	5340-01-387-0090	231	5
5365-01-382-3715	148	17	9905-01-387-1145	354	12
5365-01-382-3718	148	17	9905-01-387-1146	354	12
5365-01-382-3726	148	17	9905-01-387-2746	354	12
5310-01-382-5031	142	19	9905-01-387-2752	354	12
5310-01-382-5031	143	26	9905-01-387-2761	354	12
5310-01-382-5031	150	19	9905-01-387-2762	354	12
5310-01-382-5031	151	26	2540-01-387-5578	440	13
7690-01-382-8471	354	14	5306-01-387-7457	270	31
2520-01-382-8728	138	1	3120-01-388-1527	327	8
2990-01-382-8796	401	6	5325-01-388-2363	456	10
2520-01-382-8874	138	1	5365-01-388-4449	148	17
5340-01-383-0462	21	3	5310-01-388-6205	398	7
5365-01-383-1925	138	5	5310-01-388-7420	274	19
5365-01-383-1925	145	5	5340-01-388-9512	KITS	81
2520-01-383-2387	145	1	2540-01-389-0984	286	7
5340-01-383-2397	401	10	5120-01-389-9992	501	22
5365-01-383-3722	304	6	5305-01-390-0547	484	4
2540-01-383-5232	227	2	5310-01-390-5105	43B	40
2510-01-383-5444	74	6	2540-01-390-5711	370	3
2510-01-383-5444	283	28	5120-01-391-5131	501	9
2540-01-383-5660	231	4	5331-01-392-0830	73	15
2990-01-383-5689	401	16	5331-01-392-1793	73	11
2530-01-383-5740	169	9	9905-01-392-5794	355	4
2540-01-383-6470	286	24	9905-01-392-5795	355	4
6160-01-384-0325	231	32	9905-01-392-5796	355	4
2540-01-384-1012	227	26	9905-01-392-5797	355	4
2540-01-384-1028	227	27	9905-01-392-5798	355	4
4720-01-384-1212	301	5	9905-01-392-5799	355	4
5935-01-384-2610	73	1	9905-01-392-5800	355	4
6160-01-384-3922	231	31	2520-01-392-8435	110	11
9905-01-384-5311	354	12	2520-01-392-8435	112	17
2530-01-384-7154	169	13	9905-01-393-1830	355	4
5340-01-384-9987	4	18	9905-01-393-1833	355	4

SECTION IV TM9-2320-280-24P C01

CROSS REFERENCE INDEXES

NATIONAL STOCK NUMBER INDEX

STOCK NUMBER	FIGURE NO	ITEM NO	STOCK NUMBER	FIGURE NO	ITEM NO
9905-01-393-1834	355	4	5340-01-395-0812	369	3
9905-01-393-3794	354	8	5340-01-395-1244	365	24
9905-01-393-3795	355	4	2590-01-395-2228	140	3
9905-01-393-5622	354	12	2590-01-395-2229	140	6
9905-01-393-5623	354	12	2540-01-395-2230	372	7
5305-01-393-6311	135	7	5310-01-395-3549	213	22
5305-01-393-6311	160	22	5310-01-395-3549	262	16
5305-01-393-6311	162	10	2540-01-395-3979	372	3
5305-01-393-6311	487A	12	4710-01-395-3982	366	11
5305-01-393-6311	487C	2	2540-01-395-4202	370	6
5305-01-393-6311	487D	3	2540-01-395-7999	188	3
9905-01-393-7128	355	4	5310-01-395-8747	30	9
5305-01-393-7801	210	11	5310-01-395-8747	35	25
5330-01-393-9101	9	2	5310-01-395-8747	36	20
9905-01-393-9357	354	12	5310-01-395-8747	37	15
5340-01-393-9371	369	11	5310-01-395-8747	38	20
5365-01-394-0440	370	14	5310-01-395-8747	41	11
5365-01-394-2394	369	6	5310-01-395-8747	42	11
5340-01-394-2408	155	15	5310-01-395-8747	77	8
5340-01-394-2409	390	12	5310-01-395-8747	472	1
5305-01-394-3543	370	16	5310-01-395-8747	474	5
4720-01-394-3747	178	18	5310-01-395-8747	475	13
2530-01-394-3748	KITS	56	5310-01-395-8747	476	21
2990-01-394-3751	365	12	5310-01-395-8747	490	8
2540-01-394-4454	375	KIT	2540-01-395-8771	402	KIT
2540-01-394-4787	402	6	2540-01-395-8785	375	1
2540-01-394-4788	402	6	6150-01-396-0906	386	2
2540-01-394-4792	402	6	2590-01-396-1422	232	1
4720-01-394-6166	370	2	2590-01-396-1424	232	1
2540-01-394-6167	368	11	5340-01-396-3950	43B	7
2530-01-394-6168	KITS	37	5935-01-396-3991	73	14
4710-01-394-6169	366	15	5325-01-396-4320	43B	18
4720-01-394-6170	368	46	2590-01-396-5007	233	3
4720-01-394-6170	370	8	2540-01-396-7756	402	1
5340-01-394-7288	370	22	2540-01-396-7757	402	1
5340-01-394-7853	369	9	2540-01-396-7759	402	1
2590-01-394-8134	147	3	5340-01-396-8322	377	3
4730-01-394-8345	368	37	3130-01-396-8388	43B	20
5340-01-394-8496	155	20	5310-01-396-8392	270	23
4140-01-394-8583	43B	42	5935-01-397-0201	86	28
5310-01-394-9280	370	10	5330-01-397-0374	105	14
2990-01-394-9670	365	9	5340-01-397-1388	262	1
2590-01-394-9672	234	4	4140-01-397-2228	43B	38
2590-01-394-9673	233	3	9905-01-397-3196	370	12
2540-01-394-9679	402	6	5935-01-397-3813	73	5
2540-01-394-9682	402	6	9905-01-397-6974	354	12
2540-01-394-9683	402	6	5330-01-398-3724	KITS	22

SECTION IV TM9-2320-280-24P C01

CROSS REFERENCE INDEXES

NATIONAL STOCK NUMBER INDEX

STOCK NUMBER	FIGURE NO	ITEM NO	STOCK NUMBER	FIGURE NO	ITEM NO
2520-01-398-4589	105	15	5340-01-408-6459	210	12
4330-01-398-8484	6	7	5340-01-408-6460	329	15
4730-01-399-0241	178	17	5340-01-408-6460	461	7
2540-01-399-0785	402	1	5340-01-408-6462	210	5
2920-01-399-0794	43B	43	5180-01-408-7050	511	KIT
2990-01-399-1023	105	9	5120-01-408-7051	510	15
2520-01-399-4691	105	1	5120-01-408-7051	510	21
2930-01-405-9885	26	1	5315-01-408-7089	261	16
5315-01-406-5019	353	6	5340-01-408-7091	261	13
5365-01-406-6336	455	4	5310-01-408-7138	462	4
2815-01-406-6675	1	1	6220-01-408-7785	49	1
5342-01-406-6962	196	10	5325-01-408-7969	115	11
5342-01-406-6962	196	15	5325-01-408-7970	115	16
4010-01-406-6963	353	8	5325-01-408-7971	115	8
5340-01-406-6964	453	13	5325-01-408-7972	115	10
5340-01-406-6965	453	8	5325-01-408-7973	120	2
5340-01-406-7413	455	8	5360-01-408-7977	122	15
5340-01-406-7418	453	6	5315-01-408-8020	108	8
2990-01-406-8738	54	5	5120-01-408-8173	511	13
5310-01-406-9129	73	16	5365-01-408-8507	261	2
2590-01-406-9821	252	8	5340-01-408-8508	261	9
2920-01-407-0532	36	4	5340-01-408-8509	261	10
2920-01-407-0532	43B	1	5340-01-408-8512	261	12
5977-01-407-2847	39	63	5340-01-408-8522	113	11
2540-01-407-3296	369	5	5340-01-408-8522	114	16
2530-01-407-3977	KITS	55	5340-01-408-8522	115	25
2510-01-407-6036	316	7	5340-01-408-8523	188	16
2510-01-407-6036	320	19	5340-01-408-8525	KITS	30
2510-01-407-6037	316	7	5340-01-408-8526	455	7
2510-01-407-6037	320	19	5340-01-408-8529	188	16
5320-01-407-6239	197	16	5310-01-408-9593	127	6
5340-01-407-7192	453	17	5325-01-409-0211	127	8
5305-01-407-7713	318	9	5310-01-409-0897	462	14
5305-01-407-7713	485	23	4730-01-409-1204	125	18
5305-01-407-7713	487	2	4820-01-409-1218	119	22
7690-01-407-8248	370	15	2520-01-409-1602	134	3
2540-01-407-9241	316	4	5315-01-409-1662	488	24
5305-01-407-9487	199	6	5330-01-409-1664	114	15
5307-01-408-4029	33	20	5330-01-409-1665	122	30
5307-01-408-4029	37	18	2520-01-409-1751	115	9
5307-01-408-4029	475	25	2520-01-409-1758	115	18
5307-01-408-4029	476	23	2520-01-409-1767	115	19
2520-01-408-4785	102	1	5330-01-409-1981	BULK	57
5340-01-408-5851	16	30	4720-01-409-2499	125	8
5340-01-408-6456	329	17	3040-01-409-2500	101	9
5340-01-408-6456	461	2	3040-01-409-2501	134	4
5340-01-408-6458	210	6	3040-01-409-2502	101	3

I-99

SECTION IV TM9-2320-280-24P C01

CROSS REFERENCE INDEXES

NATIONAL STOCK NUMBER INDEX

STOCK NUMBER	FIGURE NO	ITEM NO	STOCK NUMBER	FIGURE NO	ITEM NO
4710-01-409-2504	125	4	5306-01-411-2338	464	6
4710-01-409-2508	125	3	5306-01-411-2342	101	7
2520-01-409-2509	120	1	2990-01-411-2728	24	1
2520-01-409-2512	127	3	2530-01-411-2729	169	6
2540-01-410-7034	454	KIT	2540-01-411-2733	188	7
2540-01-410-7035	454	KIT	2530-01-411-2735	169	8
5340-01-410-7036	453	KIT	2510-01-411-2736	199	26
2510-01-410-7169	199	27	2590-01-411-2737	254	8
5330-01-410-7229	KITS	20	2520-01-411-2749	110	12
3040-01-410-8054	101	5	4330-01-411-2786	119	17
2520-01-410-8067	KITS	93	5310-01-411-3422	49	5
3040-01-410-8068	107	13	5310-01-411-3422	352	1
3040-01-410-8071	107	22	5310-01-411-3422	459	1
2520-01-410-8072	110	13	5310-01-411-3422	481	9
2520-01-410-8077	KITS	95	5310-01-411-3422	484	11
2520-01-410-8079	115	5	2540-01-411-3946	166	10
6150-01-410-8215	511	12	2990-01-411-3947	24	14
5120-01-410-8216	510	4	2990-01-411-3954	24	5
2520-01-410-8241	115	27	2590-01-411-3958	210	9
5305-01-410-8386	210	16	2520-01-411-3959	115	21
5305-01-410-8386	269	9	2510-01-411-4175	210	8
5180-01-410-8467	511	KIT	3040-01-411-4191	66A	8
2540-01-410-8789	51	10	5305-01-411-4455	453	16
2590-01-410-8791	261	6	5305-01-411-4456	261	15
2540-01-410-8793	352	4	3120-01-411-5783	122	31
2540-01-410-8793	459	4	3110-01-411-5784	115	15
2540-01-410-8794	329	3	3120-01-411-5787	122	31
2540-01-410-8794	461	3	3990-01-411-6575	210	7
4010-01-410-9099	353	9	2540-01-411-7519	453	10
2815-01-410-9710	1	1	5307-01-411-8340	261	4
5330-01-410-9840	16	39	4710-01-411-8488	11	18
3040-01-410-9965	101	8	4710-01-411-8489	11	16
2590-01-410-9967	261	5	4710-01-411-8492	11	17
2590-01-410-9968	197	9	2910-01-412-0047	14	28
5325-01-411-0066	210	13	5340-01-412-0141	KITS	48
5325-01-411-0066	225	48	2990-01-412-0142	24	16
5325-01-411-0066	226	28	4710-01-412-0273	14	1
5325-01-411-0066	269	7	3110-01-412-0490	115	3
5325-01-411-0066	460	4	3010-01-412-0673	129	19
5330-01-411-0367	16	31	5340-01-412-0866	409	6
2510-01-411-0652	226	27	5340-01-412-0866	411	12
2510-01-411-0652	460	1	5340-01-412-0867	162	6
2510-01-411-0653	267	8	4710-01-412-1667	11	24
5306-01-411-1596	122	1	4710-01-412-1669	11	11
2910-01-411-2124	16	1	2540-01-412-2661	320	1
5307-01-411-2336	103	14	2590-01-412-2664	103	7
5306-01-411-2338	16	28	5930-01-412-2836	66A	2

I-100

SECTION IV TM9-2320-280-24P C01

CROSS REFERENCE INDEXES

NATIONAL STOCK NUMBER INDEX

STOCK NUMBER	FIGURE NO	ITEM NO	STOCK NUMBER	FIGURE NO	ITEM NO
6150-01-412-3192	76	1	5305-01-412-5994	152	12
5340-01-412-3563	105	26	5305-01-412-5995	16	29
2590-01-412-3862	186	5	5305-01-412-5995	464	5
5310-01-412-4013	15	24	6220-01-412-6420	97	10
5310-01-412-4013	16	26	2510-01-412-6761	320	5
5310-01-412-4013	23	21	5340-01-412-7514	188	10
5310-01-412-4013	30	4	2590-01-412-7564	186	10
5310-01-412-4013	35	4	6150-01-412-7774	511	11
5310-01-412-4013	36	14	2540-01-412-8610	186	1
5310-01-412-4013	38	14	2540-01-412-8610	456	3
5310-01-412-4013	99	17	4710-01-412-8611	11	10
5310-01-412-4013	144	5	4710-01-412-8615	11	28
5310-01-412-4013	144	15	5325-01-412-8806	115	12
5310-01-412-4013	152	5	5935-01-412-9146	76	7
5310-01-412-4013	152	15	2590-01-412-9560	186	10
5310-01-412-4013	174	15	2590-01-412-9570	186	5
5310-01-412-4013	202	36	2520-01-413-0080	136	16
5310-01-412-4013	203	19	5365-01-413-0275	266	22
5310-01-412-4013	270	20	5340-01-413-0282	186	12
5310-01-412-4013	276	11	5340-01-413-0282	188	31
5310-01-412-4013	278	10	6150-01-413-0853	81	19
5310-01-412-4013	283	7	5945-01-413-0886	50	12
5310-01-412-4013	33A	15	2540-01-413-1356	320	2
5310-01-412-4013	33B	3	4710-01-413-1360	14	5
5310-01-412-4013	341	7	2530-01-413-1365	144	6
5310-01-412-4013	38A	2	2530-01-413-1366	144	8
5310-01-412-4013	409	7	2530-01-413-1366	152	8
5310-01-412-4013	410	5	6150-01-413-1845	82	1
5310-01-412-4013	411	13	6150-01-413-1847	82	1
5310-01-412-4013	412	5	2520-01-413-1899	129	13
5310-01-412-4013	414	3	2520-01-413-1900	129	9
5310-01-412-4013	415	12	2520-01-413-1902	129	18
5310-01-412-4013	416	4	2520-01-413-1904	130	29
5310-01-412-4013	425	15	5310-01-413-2049	166	9
5310-01-412-4013	426	13	5330-01-413-2118	105	25
5310-01-412-4013	427	14	2520-01-413-2595	98	1
5310-01-412-4013	428	4	2520-01-413-2612	98	1
5310-01-412-4013	456	19	5340-01-413-2689	24	11
5310-01-412-4013	462	3	2520-01-413-2724	137	18
5310-01-412-4013	463	2	2510-01-413-3259	226	4
5310-01-412-4013	464	2	5310-01-413-3276	225	32
5310-01-412-4013	476	3	5310-01-413-3276	226	17
5310-01-412-4013	490A	5	5310-01-413-3276	461	1
2510-01-412-4969	180	15	2510-01-413-3360	225	12
4730-01-412-5213	14	26	2510-01-413-3618	267	8
4730-01-412-5216	14	12	2530-01-413-3653	143	1
5305-01-412-5994	144	12	2530-01-413-3653	151	1

I-101

SECTION IV TM9-2320-280-24P C01

CROSS REFERENCE INDEXES

NATIONAL STOCK NUMBER INDEX

STOCK NUMBER	FIGURE NO	ITEM NO	STOCK NUMBER	FIGURE NO	ITEM NO
5330-01-413-3713	128	16	6150-01-416-7899	475	4
5330-01-413-3713	129	31	5331-01-417-1043	167	16
3040-01-413-4022	130	15	5306-01-417-2467	167	15
5340-01-413-4486	105	24	5975-01-417-3267	435	KIT
2520-01-413-5540	110	8	6150-01-417-3411	69	35
3120-01-413-6106	40	42	2530-01-417-4908	167	2
3120-01-413-6106	43B	44	2530-01-417-4908	167	14
6150-01-413-7946	70	2	3120-01-417-8178	130	31
6150-01-413-7946	481	2	2530-01-417-8450	167	8
2510-01-413-8045	182	17	2530-01-417-8450	167	19
3110-01-413-8094	129	6	2590-01-418-0310	457	KIT
4710-01-413-8230	162	3	2590-01-418-2135	457	KIT
2510-01-413-8872	320	5	6220-01-418-4404	63	8
2530-01-413-8886	KITS	58	2640-01-419-6200	167	27
6220-01-413-9828	97	9	2640-01-419-6202	167	17
5340-01-414-0701	71	1	2640-01-419-6205	KITS	57
2910-01-414-1272	10	2	5360-01-420-0480	458	8
2540-01-414-1275	24	4	5330-01-420-0705	16	40
5340-01-414-1454	319	5	3040-01-420-1887	130	18
5340-01-414-1454	320	3	2530-01-420-3837	170	6
5340-01-414-1454	322	4	2530-01-420-3839	170	2
5340-01-414-1454	324	4	2590-01-420-4179	485	KIT
5120-01-414-1849	511	14	2530-01-420-5180	170	7
5365-01-414-2895	143	31	2530-01-420-7892	152	6
5365-01-414-2895	151	31	2530-01-420-7893	152	6
5365-01-414-2902	143	23	2920-01-420-7894	33	14
5365-01-414-2902	151	23	2530-01-420-7895	144	20
			2530-01-420-7895 152		20
2990-01-414-4072	24	28	2530-01-420-7904	157	23
5331-01-414-4161	115	22	2530-01-420-7904	158	13
5330-01-414-6607	122	2	2530-01-420-8025	KITS	10
2530-01-414-7844	144	6	2930-01-420-8622	32	4
2510-01-414-7846	226	4	2530-01-420-8634	144	20
			2530-01-420-8634 152		20
5340-01-414-8959	77	23	2920-01-420-9968	38	4
6150-01-415-0535	41	1	2920-01-420-9968	40	1
5935-01-415-4322	76	18	5340-01-421-1415	138	15
5935-01-415-4392	75	20	5340-01-421-1415	145	15
5935-01-415-4392	76	20	5340-01-421-1423	138	4
5310-01-415-5245	76	19	5340-01-421-1423	145	4
5340-01-415-8672	210	2	5340-01-421-1782	281	52
5330-01-415-9612	130	20	2530-01-421-2563	170	10
5330-01-415-9613	130	30	2520-01-421-4588	139	1
5305-01-416-1269	130	27	2520-01-421-4589	146	1
5310-01-416-3009	130	28	5935-01-421-4801	83	27
5310-01-416-5447	101	6	5935-01-421-4801	88	47
6150-01-416-7899	41	6	2510-01-421-8067	227	18
6150-01-416-7899	70	12	5365-01-421-8382	138	14
6150-01-416-7899	474	10	5365-01-421-8382	145	14

SECTION IV TM9-2320-280-24P C01

CROSS REFERENCE INDEXES

NATIONAL STOCK NUMBER INDEX

STOCK NUMBER	FIGURE NO	ITEM NO	STOCK NUMBER	FIGURE NO	ITEM NO
5975-01-421-9718	BULK	64	3010-01-429-5333	458	29
5180-01-422-0138	510	11	5305-01-429-9137	43	26
5120-01-422-0139	510	14	5305-01-429-9149	35	21
5120-01-422-0173	510	12	5305-01-429-9149	36	1
5120-01-422-0219	510	13	5305-01-429-9149	38	1
2540-01-422-0253	327	1	5305-01-429-9149	472	10
5120-01-422-0334	510	9	5305-01-429-9149	475	10
5120-01-422-1293	511	6	5340-01-429-9352	210	1
5180-01-422-1294	511	1	5340-01-430-0037	458	43
5120-01-422-1297	511	2	5962-01-430-0182	50	4
5120-01-422-1300	510	5	5962-01-430-0208	50	4
5120-01-422-1301	511	5	5925-01-430-2318	48	3
5120-01-422-1302	511	3	5925-01-430-2318	469	13
5120-01-422-1306	511	4	2815-01-430-2599	4	22
5120-01-422-1308	510	6	5940-01-430-2764	50	6
5120-01-422-1310	510	16	4820-01-430-4132	14	13
5120-01-422-1313	510	3	2520-01-430-5294	KITS	62
5120-01-422-1318	510	19	5340-01-430-9239	321	25
5120-01-422-1324	510	20	5340-01-430-9239	376	21
5120-01-422-1326	510	7	5340-01-430-9240	321	25
5120-01-422-1329	510	10	5340-01-430-9240	376	21
5120-01-422-5406	510	17	2540-01-431-1338	KITS	59
5120-01-422-5411	510	18	2510-01-431-1339	460	KIT
5306-01-422-8649	66A	12	4820-01-431-2499	19	10
3040-01-422-9390	144	17	5945-01-431-5195	458	38
3040-01-422-9390	152	17	5935-01-431-5656	458	37
5120-01-423-0032	510	2	7220-01-431-8340	210	10
2530-01-423-1796	171	1	2540-01-431-9182	459	KIT
2520-01-423-1947	139	1	3120-01-432-1474	327	7
2510-01-423-2877	195	3	3020-01-432-2553	143	10
2520-01-423-5120	136	6	3020-01-432-2553	151	10
3120-01-424-1028	43B	44	2590-01-432-2691	457	17
5340-01-424-1300	138	2	5340-01-432-4870	189	5
5340-01-424-1300	145	2	5340-01-432-4876	182	12
3120-01-424-1525	43B	4	5340-01-432-4877	182	15
2540-01-424-7363	461	KIT	3990-01-432-5370	199	19
2540-01-425-1617	293	7	3990-01-432-5370	270	12
5330-01-425-5069	4	29	5310-01-432-6727	153	3
5340-01-426-3850	307	14	5310-01-432-6727	155	14
5340-01-426-3850	308	2	5310-01-432-6727	198	10
2990-01-426-4425	66A	11	5310-01-432-6727	208	4
4010-01-426-4536	458	39	5310-01-432-6727	227	31
4030-01-426-4537	458	41	5310-01-432-6727	261	11
5365-01-426-6643	458	26	5310-01-432-6727	321	15
5305-01-428-6635	458	46	5310-01-432-6727	325	17
2590-01-429-3008	486	KIT	5310-01-432-6727	326	12
6220-01-429-4596	47	13	5310-01-432-6727	332	2

SECTION IV TM9-2320-280-24P C01

CROSS REFERENCE INDEXES

NATIONAL STOCK NUMBER INDEX

STOCK NUMBER	FIGURE NO	ITEM NO	STOCK NUMBER	FIGURE NO	ITEM NO
5310-01-432-6727	377	8	3120-01-435-2495	458	34
5310-01-432-6727	486	4	3120-01-435-2560	458	30
5310-01-432-6727	487	10	2540-01-436-4175	231	6
5306-01-432-7900	33	12	2540-01-436-4175	454	1
5340-01-432-8678	183	20	5940-01-436-4561	93	15
5340-01-432-8680	140	9	5120-01-437-0480	511	8
5340-01-432-8681	160	25	5120-01-437-3658	505	5
5340-01-432-8681	162	19	5325-01-437-4175	319	7
5340-01-432-8681	487A	5	5306-01-437-8056	234	18
2540-01-432-9894	189	8	5306-01-437-8056	239	13
2540-01-432-9894	190	8	5306-01-437-8058	239	14
2540-01-432-9899	189	11	5340-01-437-8748	183	20
5340-01-433-0936	484	16	5340-01-437-8751	182	12
5340-01-433-2721	190	5	5365-01-438-1556	148	20
5340-01-433-2743	183	21	5340-01-438-4497	189	10
2510-01-433-4421	183	12	5980-01-438-7452	47	2
5340-01-433-6262	198	16	5980-01-438-8939	KITS	60
2530-01-433-8003	143	35	5310-01-439-1154	193	3
2530-01-433-8003	151	19	4730-01-439-2814	496	23
9905-01-433-8554	354	6	3040-01-439-3662	127	7
2530-01-433-8702	143	19	2520-01-439-5265	112	1
2530-01-433-8702	151	35	2520-01-439-5265	115	20
5305-01-433-9248	366	6	5340-01-439-5636	497	3
5365-01-434-0186	458	27	2815-01-439-5882	1	2
2520-01-434-0822	126	2	2815-01-439-6664	1	1
2520-01-434-0822	127	2	2815-01-439-6665	1	1
6105-01-434-1729	458	47	2520-01-439-6830	KITS	65
5340-01-434-1809	260	21	2540-01-439-7308	21	21
5340-01-434-2832	189	7	4710-01-439-8165	125	4
5340-01-434-2864	272	31	4710-01-439-8167	125	3
4710-01-434-3037	160	21	2590-01-439-8268	332	33
4710-01-434-3037	487A	13	2590-01-439-8268	370	28
4710-01-434-3151	160	19	2590-01-439-8268	411	20
4710-01-434-3151	162	20	3120-01-440-0043	456	8
4710-01-434-3151	487A	7	5940-01-440-0097	93	14
4710-01-434-3157	160	24	5962-01-440-0368	50	4
4710-01-434-3157	162	23	5340-01-440-1435	270	37
4710-01-434-3157	487B	7	5340-01-440-9678	11	25
4710-01-434-3159	162	21	5340-01-442-0308	125	21
4710-01-434-3159	487C	4	5330-01-442-2874	KITS	26
6110-01-434-5562	458	36	5340-01-442-4876	86	30
5340-01-434-8109	160	17	5940-01-442-6298	86	32
5340-01-434-8109	162	24	3020-01-442-7278	34	22
5340-01-434-8109	487A	5	5935-01-442-8569	86	22
5330-01-434-8611	30	2	6680-01-442-9413	478	1
3950-01-435-1014	458	31	6680-01-442-9413	488	2
3950-01-435-1632	458	28	4720-01-442-9875	160	1

I-104

SECTION IV TM9-2320-280-24P C01

CROSS REFERENCE INDEXES

NATIONAL STOCK NUMBER INDEX

STOCK NUMBER	FIGURE NO	ITEM NO	STOCK NUMBER	FIGURE NO	ITEM NO
4720-01-442-9875	487A	3	5935-01-445-5599	86	31
4720-01-443-3033	160	16	4720-01-445-5690	332	10
4720-01-443-3033	487A	2	4720-01-445-5705	332	41
2530-01-443-3405	KITS	66	2540-01-445-6029	379	17
4720-01-443-3481	162	9	2910-01-445-8097	368	30
4720-01-443-3481	487A	15	5305-01-446-0767	38	16
5340-01-443-6908	83	11	5305-01-446-0767	275	18
5340-01-443-6908	88	33	3040-01-446-3684	458	15
4720-01-443-8487	162	25	3040-01-446-3689	458	5
4720-01-443-8487	487A	16	3020-01-446-3695	458	25
5980-01-443-9093	63	8	3020-01-446-4387	458	12
5935-01-443-9124	86	29	5330-01-446-4696	458	20
5940-01-443-9580	95	4	5330-01-446-4696	458	32
4730-01-444-0701	30	13	9905-01-446-5768	355	4
3020-01-444-1359	458	10	9905-01-446-5769	354	12
3040-01-444-1756	458	2	9905-01-446-5770	354	12
3020-01-444-2311	175	3	9905-01-446-5771	355	4
2520-01-444-2711	120	5	5307-01-446-5930	103	14
2920-01-444-2749	38	11	2590-01-446-5957	458	19
5310-01-444-3084	77	16	2590-01-446-5962	458	16
2590-01-444-3317	33	9	9905-01-446-6187	355	4
3040-01-444-3350	456	13	5365-01-446-9408	128	22
2510-01-444-3360	16	21	5365-01-446-9408	130	25
2590-01-444-3399	456	14	6140-01-446-9498	69	23
3010-01-444-4018	458	11	4710-01-446-9570	11	1
3010-01-444-4019	458	14	5315-01-447-0492	108	4
3110-01-444-4020	458	7	5310-01-447-0968	108	3
3110-01-444-4100	458	13	2540-01-447-2236	332	14
2590-01-444-4365	37	16	5306-01-447-2568	33	13
2590-01-444-4365	475	19	2540-01-447-3457	231	9
6150-01-444-4437	44	12	3040-01-447-3797	107	7
2540-01-444-5340	332	20	2910-01-447-3911	16	13
3020-01-444-5447	38	8	5330-01-447-4762	103	3
3040-01-444-5640	51	11	5340-01-447-5839	78	3
4720-01-444-6433	332	11	9905-01-447-7799	354	12
2540-01-444-6615	332	19	3040-01-447-7995	108	2
2540-01-444-6642	332	30	3120-01-447-8658	458	21
2930-01-444-6649	31	3	3120-01-447-8658	458	33
2540-01-444-6655	332	22	3120-01-447-8663	458	18
6685-01-444-9478	30	1	5305-01-447-9227	332	31
7690-01-445-0444	456	2	5355-01-447-9655	98	4
7690-01-445-0456	361	11	2590-01-448-1105	315	7
4730-01-445-2358	30	15	2590-01-448-1105	457	12
4730-01-445-2358	30	17	5340-01-448-1213	488	8
6220-01-445-5058	49	10	4720-01-448-1655	331	3
5935-01-445-5144	86	33	5310-01-448-3219	332	32
5331-01-445-5285	458	4	5342-01-448-4215	104	18

I-105

SECTION IV TM9-2320-280-24P C01

CROSS REFERENCE INDEXES

NATIONAL STOCK NUMBER INDEX

STOCK NUMBER	FIGURE NO	ITEM NO	STOCK NUMBER	FIGURE NO	ITEM NO
5340-01-448-4245	472	13	2540-01-450-7686	319	1
5340-01-448-4316	278	2	5340-01-450-8693	487	4
5305-01-448-5074	458	17	4720-01-451-0894	370	7
5305-01-448-5080	458	22	5340-01-451-5903	487	14
3120-01-448-6268	456	12	2540-01-451-9345	379	20
3120-01-448-6269	456	9	5330-01-452-2474	12	27
2590-01-448-6351	456	7	2590-01-452-5034	369	13
3040-01-448-9611	104	2	2520-01-452-7569	126	2
9905-01-448-9783	355	4	2520-01-452-7569	127	3
9905-01-448-9784	355	4	2520-01-452-8365	126	1
9905-01-448-9786	354	12	4130-01-452-8773	466	KIT
9905-01-449-0476	355	4	3040-01-452-9065	129	8
5306-01-449-0510	235	11	2815-01-453-7403	1	1
5310-01-449-0628	50	7	2815-01-453-7404	1	1
2540-01-449-1718	379	20	4730-01-455-1220	467	3
5305-01-449-1983	169	31	4730-01-455-1231	466	3
2530-01-449-2495	143	22	4730-01-455-1231	467	2
2530-01-449-2495	151	22	4730-01-455-1259	466	2
6150-01-449-5445	44	23	4730-01-455-1276	466	1
6150-01-449-5445	480	25	2920-01-455-1626	474	KIT
2815-01-449-5585	4	6	4820-01-455-5020	467	8
5340-01-449-6854	41	8	4820-01-455-5021	466	8
5340-01-449-6854	474	15			
5340-01-449-7352	332	16	2815-01-455-8424	66A	14
5307-01-450-3072	127	11	5340-01-455-8700	439	KIT
5307-01-450-3072	180	9	2590-01-455-9123	332	13
2540-01-450-4015	319	1	2540-01-455-9308	332	18
2540-01-450-4017	316	1	2530-01-455-9330	KITS	63
2540-01-450-4018	316	4	4820-01-456-0172	19	23
2540-01-450-4019	320	1	5340-01-456-1500	98	9
2540-01-450-4020	322	1	5365-01-456-2032	143	30
2540-01-450-4021	316	1	5365-01-456-2032	151	30
2540-01-450-4024	322	1	2510-01-456-2277	194	3
2540-01-450-5477	320	2	2520-01-456-2736	114	6
2510-01-450-5479	316	7	5315-01-456-2737	117	7
2510-01-450-5479	320	19	5315-01-456-2738	117	7
2510-01-450-5479	324	24	5315-01-456-2739	117	7
2510-01-450-5480	316	7	5315-01-456-2740	117	7
2510-01-450-5480	320	19	5340-01-456-2741	117	19
2510-01-450-5480	324	24	5365-01-456-2742	117	10
7690-01-450-5481	354	12	5315-01-456-2743	117	7
2510-01-450-5482	320	5	5330-01-456-2744	120	10
2510-01-450-5482	324	10	5360-01-456-2745	117	13
2510-01-450-5483	320	5	5340-01-456-2751	163	9
2510-01-450-5483	324	10	4720-01-456-2955	162	7
5306-01-450-5947	230	14	2520-01-456-5014	115	6
5306-01-450-5947	235	15	4730-01-456-5446	366	9
5340-01-450-7521	387	1	5315-01-456-6254	117	7

SECTION IV TM9-2320-280-24P C01

CROSS REFERENCE INDEXES

NATIONAL STOCK NUMBER INDEX

STOCK NUMBER	FIGURE NO	ITEM NO	STOCK NUMBER	FIGURE NO	ITEM NO
2520-01-456-6256	120	20	4720-01-458-3326	466	10
4820-01-456-6257	125	19	5330-01-458-3810	269	4
4820-01-456-6257	332	14	5340-01-458-6386	163	8
4320-01-456-6258	122	34	5340-01-458-7147	162	26
5305-01-456-6358	314	40	5340-01-458-7147	487D	1
4720-01-456-6674	466	5	5340-01-458-7156	162	12
4720-01-456-6729	466	9	5340-01-458-7156	487D	1
2590-01-456-7879	471	KIT	5340-01-458-8236	160	23
5930-01-456-7880	120	13	5340-01-458-8236	487B	6
3040-01-456-7881	117	16	7690-01-458-8254	354	12
5315-01-456-7882	117	7	2520-01-459-0050	137	18
2520-01-456-7883	114	8	2530-01-459-0367	KITS	54
5330-01-456-7886	119	5	5342-01-459-4992	487	1
2520-01-456-7888	120	9	2520-01-459-5454	KITS	19
2510-01-456-7889	268	3	2530-01-459-5890	163	10
5330-01-456-8823	143	20	5330-01-459-6477	KITS	61
5330-01-456-8823	151	20	2520-01-459-8532	120	12
5340-01-457-0459	205	16	2530-01-459-9492	KITS	16
4710-01-457-0590	467	10	2530-01-459-9493	KITS	38
4730-01-457-0727	160	7	2530-01-459-9494	KITS	39
4730-01-457-0727	161	7	2530-01-459-9495	KITS	40
4730-01-457-0727	162	8	2530-01-459-9497	139	10
4730-01-457-0727	487A	14	2530-01-459-9497	146	10
4730-01-457-0727	487C	3	5305-01-459-9564	190	6
5340-01-457-0804	465	4	5305-01-459-9564	463	17
2510-01-457-0949	183	22	2530-01-460-2439	174	5
2530-01-457-1337	KITS	64	2530-01-460-2439	175	1
5340-01-457-1778	273	18	5355-01-460-4586	98	3
5935-01-457-2965	282	8	2520-01-460-4961	114	11
5310-01-457-3171	35	30	5340-01-460-5002	270	41
5310-01-457-3171	36	9	5340-01-460-5291	199	24
5310-01-457-3171	38	9	5340-01-460-5306	141	16
5310-01-457-3171	472	20	5935-01-460-5508	458	45
5310-01-457-3171	475	18	4730-01-460-5520	108	12
5365-01-457-3364	277	13	2930-01-460-7507	30	18
5340-01-457-5571	277	11	5340-01-460-8207	105	20
5340-01-457-5734	44	8	4820-01-460-8288	122	24
5365-01-457-8947	273	24	2520-01-460-8301	KITS	27
3110-01-457-9711	277	9	3010-01-460-8309	115	14
2590-01-457-9922	277	32	2590-01-460-8316	108	14
2590-01-457-9922	473	3	2590-01-460-8320	233	7
3020-01-458-0176	33	5	2590-01-460-8320	277	32
3020-01-458-0176	33A	5	2520-01-460-8323	102	1
3020-01-458-0185	33	3	5325-01-460-8350	465	5
5305-01-458-1670	274	10	5340-01-460-8361	122	21
5305-01-458-1670	274	23	5340-01-460-8362	119	23
5305-01-458-1670	276	18	5340-01-460-8363	119	29

I-107

SECTION IV TM9-2320-280-24P C01

CROSS REFERENCE INDEXES

NATIONAL STOCK NUMBER INDEX

STOCK NUMBER	FIGURE NO	ITEM NO	STOCK NUMBER	FIGURE NO	ITEM NO
5340-01-460-8364	119	27	3020-01-462-1202	33	4
5330-01-460-8987	115	4	3040-01-462-1286	KITS	25
5330-01-460-8988	105	7	5310-01-462-2175	232	3
5360-01-460-9068	122	12	5310-01-462-4459	10	15
5360-01-460-9069	119	3	5331-01-462-7294	122	32
5360-01-460-9070	122	23	2530-01-462-8068	158	12
5360-01-460-9072	122	28	2530-01-462-8079	157	22
5360-01-460-9073	122	18	2530-01-462-8079	158	12
5315-01-460-9907	108	11	2540-01-463-0200	98	2
5325-01-460-9908	105	8	2540-01-463-3097	332	39
3040-01-460-9964	108	13	5315-01-464-0812	130	24
4810-01-460-9975	122	16	2520-01-464-1772	103	1
4820-01-460-9988	122	27	5306-01-465-1564	137	3
4820-01-460-9991	122	13	5306-01-465-2140	105	19
4820-01-461-0011	122	20	3040-01-465-3178	156	11
2520-01-461-0073	114	9	5310-01-465-4525	33	7
2520-01-461-0074	114	7	5310-01-465-4525	512	4
2520-01-461-0076	110	2	5307-01-465-5796	30	11
2520-01-461-0078	115	17	5307-01-465-5796	472	17
2520-01-461-0083	115	13	2510-01-465-7620	379	19
2520-01-461-0085	115	7	2590-01-465-7756	463	5
2520-01-461-0092	114	1	2520-01-465-8809	122	9
2520-01-461-0099	110	9	2540-01-465-9030	463	13
2520-01-461-0099	112	8	5310-01-465-9727	463	19
2520-01-461-0107	107	14	4010-01-466-0849	462	7
2920-01-461-1065	98	5	5365-01-466-0850	463	15
2540-01-461-1129	227	19	2920-01-466-1855	476	7
2540-01-461-1134	369	14	5340-01-466-1982	463	10
4820-01-461-1703	466	6	5340-01-466-1985	485	19
2520-01-461-2374	105	22	5340-01-466-2425	463	18
5306-01-461-3555	105	13	5340-01-466-2438	485	18
5306-01-461-3557	105	12	5340-01-466-2451	485	11
5305-01-461-4396	332	38	5340-01-466-2468	480	6
5360-01-461-4931	108	9	5340-01-466-2606	485	20
2520-01-461-5133	122	25	5340-01-466-2611	485	22
2520-01-461-5648	122	29	5340-01-466-2620	480	11
5315-01-461-5649	122	14	3020-01-466-3603	476	10
3120-01-461-6677	122	31	2510-01-466-3752	182	11
3120-01-461-6679	122	31	2540-01-466-3770	463	16
3120-01-461-6680	122	19	2590-01-466-3824	464	4
2520-01-461-7072	KITS	67	2540-01-466-3849	463	4
2520-01-461-7074	KITS	67A	2590-01-466-3866	464	7
2815-01-461-7078	1	1	5340-01-466-4086	463	20
5325-01-461-7157	115	28	2590-01-466-4397	468	2
5340-01-461-7851	147	6	7690-01-466-5217	468	3
5310-01-461-8043	227	48	5340-01-466-5243	374	4
5315-01-461-8374	122	10	2590-01-466-5250	468	6

I-108

SECTION IV TM9-2320-280-24P C01

CROSS REFERENCE INDEXES

NATIONAL STOCK NUMBER INDEX

STOCK NUMBER	FIGURE NO	ITEM NO	STOCK NUMBER	FIGURE NO	ITEM NO
3030-01-466-5387	476	26	5330-01-470-6543	120	4
4710-01-466-5447	479	2	6150-01-470-7014	480	12
2540-01-466-6114	462	1	5340-01-470-7135	198	4
5342-01-466-6276	476	5	5340-01-470-7160	198	6
5342-01-466-6280	476	14	2590-01-471-0236	12	16
5342-01-466-6283	476	24	2930-01-471-0622	28	5
6680-01-466-7238	103	8	5999-01-471-0710	482	7
6680-01-466-7238	400	4	5340-01-471-1040	485	12
2590-01-466-7250	273	6	4720-01-471-5063	487D	KIT
6150-01-466-7528	472	18	5305-01-471-7623	485	21
3030-01-466-9476	32	3	3120-01-471-7844	122	31
5340-01-467-1669	249	47	9540-01-472-5874	88	48
5340-01-467-2127	282	11	2815-01-472-6312	1	1
5305-01-467-2711	486	1	5310-01-472-7368	270	10
5340-01-467-4153	486	7	5340-01-472-8534	98	11
6150-01-467-6716	41	6	6150-01-473-1169	469	1
6150-01-467-6716	472	5	6150-01-473-1174	470	5
6150-01-467-6716	474	10	6150-01-473-1178	469	3
2540-01-467-8313	464	KIT	5315-01-473-1590	462	6
2540-01-467-8341	464	KIT	2920-01-473-1763	36	11
2910-01-467-9029	10	2	5306-01-473-1895	33	10
5330-01-468-3604	115	1	3020-01-473-2048	36	8
5340-01-469-2979	16	15	2510-01-473-2309	202	1
5975-01-469-5558	76	21	4710-01-473-2352	333	7
5975-01-469-5558	472	9	6150-01-473-2435	81	14
5975-01-469-5558	475	5	4730-01-473-3279	4	11
5975-01-469-5558	476	6	5310-01-473-3373	143	8
5340-01-469-5774	487	7	5310-01-473-3373	151	8
5365-01-469-6343	332	37	4820-01-473-3580	178	3
2920-01-469-6903	66	2	5315-01-473-3682	227	43
6150-01-469-7930	480	31	5340-01-473-4031	178	2
6150-01-469-7941	480	34	7690-01-473-4550	356	3
5360-01-469-8077	120	22	5330-01-473-4701	104	21
7690-01-469-9591	466	4	5340-01-473-4800	241	15
7690-01-469-9596	467	6	2520-01-473-5029	126	1
2520-01-469-9893	126	1	2540-01-473-6304	227	37
5306-01-469-9903	120	3	2540-01-473-7107	227	41
5320-01-470-1545	197	16	2520-01-473-7410	KITS	67B
5330-01-470-1922	116	14	2540-01-473-7521	227	42
6150-01-470-2009	78	1	2540-01-473-7544	227	45
6150-01-470-2011	78	1	2540-01-473-7553	227	52
6150-01-470-2016	78	1	2540-01-473-7959	227	54
5340-01-470-2069	332	28	2540-01-473-8050	227	55
4720-01-470-3076	178	26	2510-01-473-8204	227	47
6160-01-470-4172	198	3	2540-01-473-8310	227	40
5962-01-470-4619	50	4	5330-01-473-8584	120	8
5340-01-470-5892	198	16	5340-01-473-8626	66A	10

I-109

SECTION IV TM9-2320-280-24P C01

CROSS REFERENCE INDEXES

NATIONAL STOCK NUMBER INDEX

STOCK NUMBER	FIGURE NO	ITEM NO	STOCK NUMBER	FIGURE NO	ITEM NO
6150-01-473-8640	56	14	6150-01-475-8835	469	6
6150-01-473-8663	483	4	2520-01-475-9665	120	9
2540-01-473-9040	470	9	5325-01-475-9697	269	14
2540-01-473-9045	470	9	5340-01-476-0352	110	14
6680-01-474-0384	482	3	4730-01-476-0850	13	13
5340-01-474-0954	484	9	4730-01-476-0850	14	19
5365-01-474-1057	470	16	2510-01-476-2171	268	3
5365-01-474-1057	471	7	2540-01-476-2172	324	1
5340-01-474-2315	41	3	5360-01-476-2692	120	7
5935-01-474-3033	84	14	3020-01-476-2701	143	38
5999-01-474-3213	484	KIT	3020-01-476-2701	151	38
2540-01-474-3217	333	11	2920-01-476-2716	98	10
5340-01-474-3279	484	8	5340-01-476-2840	188	33
2590-01-474-3326	471	2	7690-01-476-2842	354	16
5340-01-474-4002	120	21	5340-01-476-2845	188	33
5340-01-474-4006	122	9	7690-01-476-2855	354	12
5340-01-474-4011	105	4	5340-01-476-2963	188	26
5340-01-474-4011	125	22	5340-01-476-2972	188	36
5940-01-474-5066	92	8	5306-01-476-3500	120	16
7690-01-474-5928	469	5	5306-01-476-3501	120	14
7690-01-474-5935	470	6	5330-01-476-3866	31	4
5306-01-474-6117	120	18	5340-01-476-4374	332	4
2520-01-474-7051	KITS	29	5340-01-476-4374	372	1
2540-01-474-8562	332	36	4720-01-476-4656	393	17
5340-01-474-8830	463	22	5340-01-476-5072	188	30
2520-01-474-8868	114	12	5340-01-476-5585	11	9
2520-01-474-8871	110	8	5340-01-476-5616	228	12
4820-01-474-8911	KITS	18	7690-01-476-6089	354	4
5306-01-474-8959	119	10	7690-01-476-6101	354	7
4720-01-474-9148	332	7	5340-01-476-6182	250	12
6150-01-474-9676	482	4	5340-01-476-6205	250	11
7690-01-475-1197	482	1	5340-01-476-6243	188	21
5330-01-475-1671	482	5	5360-01-476-6291	117	6
5305-01-475-3014	476	15	7690-01-476-6507	354	8
2540-01-475-3102	231	14	7690-01-476-6510	354	10
2540-01-475-3129	482	15	3020-01-476-6520	175	4
5305-01-475-3154	276	13	5340-01-476-6587	250	8
5305-01-475-3154	33B	1	2540-01-476-6609	188	25
2920-01-475-3180	40	19	5340-01-476-6612	282	11
5305-01-475-3364	276	17	5340-01-476-7263	433	KIT
5975-01-475-3442	227	46	5962-01-476-7772	50	4
5340-01-475-3650	469	8	2540-01-476-7827	397	13
4710-01-475-3753	469	7	5330-01-476-9295	489	4
4720-01-475-3757	470	1	5310-01-476-9321	196	6
4720-01-475-4641	470	2	2540-01-476-9353	454	1
2540-01-475-4708	227	51	5330-01-476-9732	180	22
3940-01-475-4983	469	9	2540-01-477-0162	454	2

SECTION IV TM9-2320-280-24P C01

CROSS REFERENCE INDEXES

NATIONAL STOCK NUMBER INDEX

STOCK NUMBER	FIGURE NO	ITEM NO	STOCK NUMBER	FIGURE NO	ITEM NO
2540-01-477-0184	454	2	5950-01-480-5377	119	11
2540-01-477-0242	397	14	5950-01-480-5380	119	13
5310-01-477-0828	156	3	5305-01-480-5408	33	16
5305-01-477-1867	180	25	2520-01-480-5436	115	9
5305-01-477-2824	225	38	5365-01-480-6812	105	20
5365-01-477-2843	227	33	3040-01-480-7297	107	7
5331-01-477-3657	15	17	3020-01-480-7299	107	8
5331-01-477-3657	16	16	3020-01-480-7302	107	8
4710-01-477-3662	105	6	3040-01-480-7309	107	13
2510-01-477-3968	179	15	3020-01-480-7311	107	17
2510-01-477-3968	180	24	3020-01-480-7321	107	17
5325-01-477-5348	95	6	5325-01-480-7328	115	10
5325-01-477-6608	122	22	5325-01-480-7331	117	17
5331-01-477-6762	122	32	3040-01-480-7536	107	14
5340-01-477-7052	122	11	2520-01-480-7556	110	15
4710-01-477-9032	120	15	2520-01-480-7563	107	9
5340-01-477-9397	463	5	2520-01-480-7565	114	9
3040-01-477-9645	456	11	2520-01-480-7568	120	9
1095-01-478-3115	275	6	3040-01-480-7591	107	22
5305-01-478-3387	469	12	3040-01-480-7593	107	16
5330-01-478-3900	114	14	3040-01-480-7595	110	15
3110-01-478-4266	KITS	31	3010-01-480-7597	KITS	33
2835-01-478-4268	115	2	3020-01-480-7630	107	17
5330-01-478-4797	117	18	3020-01-480-8698	33	17
5310-01-478-5620	175A	10	3020-01-480-8698	33A	3
5330-01-478-5993	120	10	5340-01-481-0558	33	21
5330-01-478-5994	114	13	2520-01-481-3230	KITS	35A
5360-01-478-6550	117	6	5310-01-481-4029	291	5
2520-01-478-6604	112	7	5310-01-481-4029	297	8
2520-01-478-6607	112	11	7690-01-481-4552	448	12
2590-01-478-6799	181	17	7690-01-481-4906	355	4
2520-01-478-7282	137	19	7690-01-481-4908	354	12
5945-01-478-7862	119	24	5355-01-481-5517	231	17
5340-01-478-9995	120	19	2520-01-481-6120	50	1
5340-01-479-0211	227	38	4730-01-481-6278	177	4
2520-01-479-0847	114	8	5305-01-481-7254	169	31
3020-01-479-2069	112	13	2520-01-481-7690	120	9
5340-01-479-2514	117	19	2520-01-481-8344	KITS	98
5340-01-479-2517	117	17	2520-01-481-8460	KITS	32
5331-01-480-0748	115	26	2520-01-481-8478	115	20
5330-01-480-0750	114	15	2520-01-481-8479	110	11
5330-01-480-0752	KITS	34	2520-01-481-8479	112	17
5330-01-480-2502	105	21	2540-01-481-9178	50	3
5310-01-480-4161	33	8	2540-01-481-9194	50	2
3110-01-480-5240	107	2	3020-01-482-1834	33	18
5962-01-480-5247	50	4	3020-01-482-1834	33A	4
5365-01-480-5282	33	19	2520-01-482-4280	KITS	21

I-111

SECTION IV TM9-2320-280-24P C01

CROSS REFERENCE INDEXES

NATIONAL STOCK NUMBER INDEX

STOCK NUMBER	FIGURE NO	ITEM NO	STOCK NUMBER	FIGURE NO	ITEM NO
6680-01-482-4995	176A	14	5305-01-491-2143	175A	9
2920-01-483-2291	40	46	5310-01-491-2148	33B	11
2920-01-483-2291	43B	34	6110-01-491-2158	49	4
5975-01-484-1150	79	26	5340-01-491-2205	33A	13
7690-01-484-4111	357	6	5340-01-491-2205	33B	5
5340-01-484-6930	240	25	3020-01-491-2659	33B	6
5340-01-484-7646	182	5	3020-01-491-2671	33B	9
5340-01-484-7760	24	3	2530-01-491-2681	175A	3
2530-01-484-9573	158	2	2510-01-491-6919	462	KIT
2530-01-484-9574	158	2	5340-01-491-7497	120	17
4730-01-485-0065	135	5	4730-01-491-8047	178	4
2520-01-485-1813	138	12	5306-01-491-9414	117	20
2520-01-485-1813	145	12	5306-01-492-7301	188	39
2590-01-485-5455	471	12	4710-01-493-1712	11	30
5340-01-485-6862	470	14	5340-01-493-2538	11	31
4320-01-485-7480	122	3	5340-01-493-2555	11	32
2520-01-485-9405	112	2	5340-01-493-2569	4	18
9905-01-486-0051	355	4	3040-01-493-4036	112	7
5340-01-486-1005	458	44	4730-01-493-4056	176A	24
2540-01-487-3616	329	3	2530-01-493-5859	167	13
2540-01-487-3626	329	3	5340-01-493-9134	120	17
5330-01-487-7129	148	16	5331-01-494-2431	119	32
7690-01-488-5520	471	8	5340-01-494-6794	462	5
5365-01-488-5594	33B	8	5306-01-495-1744	492	19
3030-01-488-5606	32	3	4710-01-495-4268	492	1
3020-01-488-5635	175A	12	4710-01-495-4302	492	2
3020-01-488-5638	34	22	5331-01-495-4801	493	7
5365-01-488-5640	33A	11	5331-01-495-4810	492	21
5365-01-488-5640	33B	7	5331-01-495-4810	493	1
5340-01-488-5643	38A	6	5331-01-495-4810	490A	1
5340-01-488-6073	38A	1	5331-01-495-4814	492	23
4710-01-488-6076	333	7	5331-01-495-4814	493	3
4710-01-488-6143	176A	1	5331-01-495-4814	490A	2
2530-01-488-6147	176A	5	5331-01-495-4820	492	22
4720-01-488-6156	175A	14	5365-01-495-5340	462	10
4730-01-488-6163	333	41	4720-01-495-6202	492	24
5340-01-488-6179	38A	7	5340-01-495-7546	462	12
5340-01-488-6181	33A	8	4820-01-495-7582	492	12
5340-01-488-6185	176A	4	4820-01-495-7582	490A	3
2540-01-488-6429	371	1	4820-01-495-7601	492	10
2520-01-488-7016	120	12	2510-01-495-8335	347	9
5340-01-488-8389	240	25	5310-01-496-0115	327	49
2520-01-489-0849	KITS	67C	2590-01-496-1569	240	5
5310-01-490-7461	33B	10	2510-01-496-1886	347	5
3020-01-491-0776	33A	12	2510-01-496-1927	347	11
2920-01-491-2011	33A	18	4940-01-496-2178	347	10
5310-01-491-2014	175A	8	4730-01-496-3012	492	20

SECTION IV TM9-2320-280-24P C01

CROSS REFERENCE INDEXES

NATIONAL STOCK NUMBER INDEX

STOCK NUMBER	FIGURE NO	ITEM NO	STOCK NUMBER	FIGURE NO	ITEM NO
4730-01-496-3012	490A	4	5306-01-502-3440	136	13
4720-01-496-3717	492	5	2520-01-502-6586	118	14
2990-01-496-3766	398	KIT	5310-01-504-5729	490A	9
5330-01-496-5547	223	4	5306-01-504-6221	490A	8
5330-01-496-5548	223	4	5306-01-504-6223	490A	7
4330-01-496-5720	KITS	35	3030-01-504-6408	490A	11
6110-01-496-9260	52	1	5340-01-505-4197	490A	10
5962-01-497-1611	50	4	4720-01-505-4702	176A	23
5340-01-497-2269	33A	6	3010-01-505-7205	490A	12
5306-01-497-4738	9	7	5330-01-505-7214	176A	10
5340-01-498-7964	191	4	5331-01-505-7218	176A	7
2510-01-498-8000	397	21	5340-01-505-9463	176A	8
5340-01-498-9619	368	45	5340-01-506-1892	176A	9
5340-01-498-9625	368	45	4330-01-506-2918	176A	12
5340-01-499-0014	36	3	4730-01-506-3179	176A	6
2610-01-500-4806	167	13	5340-01-506-7986	191	24
5305-01-500-8280	88	25			

SECTION IV TM9-2320-280-24P C01

CROSS REFERENCE INDEXES

PART NUMBER INDEX

CAGE	PART NUMBER	STOCK NUMBER	FIGURE NO	ITEM NO
41947	A-327	4730-00-949-8694	335	5
		4730-00-949-8694	336	3
58536	A-A-52406-4SA	5340-01-333-0162	346	3
17576	A-A-52506	4730-00-908-6294	368	20
58536	A-A-59432-08	4730-01-135-8310	148	30
27767	A01-09314	5305-01-185-8673	156	15
24975	A010095461	5977-01-270-1466	39	66
22075	A02-08052	5310-01-186-5420	156	13
35510	A022096282	2920-01-310-9973	39	28
22075	A03-002204	5310-01-477-0828	156	3
24975	A041095972	6115-01-269-7911	39	21
89326	A22506-4	5310-00-809-4058	42	7
		5310-00-809-4058	402	2
		5310-00-809-4058	481	8
78570	A22673	5305-00-984-4988	367	16
10988	A27365	5331-00-451-0118	128	51
		5331-00-451-0118	129	26
80063	A3046166	5985-01-215-9404	410	6
		5985-01-215-9404	414	5
		5985-01-215-9404	428	5
03481	A31-125	5310-00-952-3567	374	12
80063	A3157855	5340-01-467-4153	486	7
58536	A524-72-1	2910-01-445-8097	368	30
58536	A52425-1	5940-00-549-6581	69	15
58536	A52432-1	2540-00-840-0022	328	12
58536	A52463-2-10	6240-00-044-6914	60	7
		6240-00-044-6914	61	7
		6240-00-044-6914	62	3
		6240-00-044-6914	64	11
05657	A6	4730-00-816-2787	327	23
53964	A609014	5310-00-809-4058	19	4
53964	A609015	5306-01-190-4553	19	3
53964	A618017	5331-01-335-7592	19	8
53964	A630003	2910-01-218-4489	KITS	3
35510	A777100841	5977-01-407-2847	39	63
53964	A910044	4330-01-190-3579	KITS	5
33116	A98-09439	2530-01-206-3874	KITS	4
58536	AA55569/01-009	5920-01-085-0825	52	13
		5920-01-085-0825	59	14
58536	AA55571/01-001	5925-01-430-2318	48	3
		5925-01-430-2318	469	13
58536	AA55571/02-001	5925-00-894-8015	50	10
11815	AAC-32	5320-01-362-6195	247	7
11815	AAP-610	5320-01-223-7728	298	7
54402	AD42BS	5320-00-899-0981	213	29
		5320-00-899-0981	214	19
		5320-00-899-0981	270	16
		5320-00-899-0981	315	3
		5320-00-899-0981	337	12

SECTION IV TM9-2320-280-24P C01

CROSS REFERENCE INDEXES

PART NUMBER INDEX

CAGE	PART NUMBER	STOCK NUMBER	FIGURE NO		ITEM NO
54402	AD44BS	5320-01-023-2529		325	30
90030	AD45BS	5320-00-275-8344		337	7
07707	AD48BS	5320-00-850-3282		368	44
21868	AD778-08B79	4720-00-111-5096		BULK	8
11078	ADS43	5320-00-083-5009		52	9
		5320-00-083-5009	197	1	
		5320-00-083-5009	208	6	
		5320-00-083-5009	211	4	
		5320-00-083-5009		296	12
		5320-00-083-5009	311	5	
00624	AE208-10	9330-00-011-2000		BULK	65
12387	AEA9630A	4730-00-908-3194		176A	22
72794	AJ7-40-CBC	5325-01-264-1522		207	3
78276	ALS4-1024-130	5325-01-411-0066		210	13
		5325-01-411-0066	225	48	
		5325-01-411-0066	226	28	
		5325-01-411-0066	269	7	
		5325-01-411-0066	460	4	
78276	ALS4-1024-225	5325-01-475-9697		269	14
19728	AMA-74			43	3
19728	AMA-75			43	2
19728	AMA-90			43	11
68505	AMA-95	2920-01-212-5816		43	50
68505	AMA1096	5910-01-218-1675		43	46
0EDY1	AMA2004AS	2920-01-180-8666		43	23
88044	AN365-1024A	5310-00-208-1918		202	41
		5310-00-208-1918	205	38	
		5310-00-208-1918	206	16	
		5310-00-208-1918	243	28	
		5310-00-208-1918	244	17	
		5310-00-208-1918	248	5	
		5310-00-208-1918	249	18	
		5310-00-208-1918	249	42	
		5310-00-208-1918	279	28	
88044	AN5H10A	5306-00-182-2027		198	25
		5306-00-182-2027		199	44
		5306-00-182-2027		317	12
		5306-00-182-2027	376	19	
88044	AN5H41	5306-00-292-8253		277	37
88044	AN5H6A	5306-00-182-2025		199	45
88044	AN5H7	5306-00-180-2748		321	32
88044	AN6-16A	5306-00-208-3639		279	12
88044	AN8008D6	5342-00-200-7564		307	4
		5342-00-200-7564	308	16	
88044	AN8013-2	5310-00-167-0893		16	25
		5310-00-167-0893		307	3
		5310-00-167-0893		308	14
		5310-00-167-0893	401	15	
55719	AN8506-4	5120-00-184-8398		505	1

SECTION IV TM9-2320-280-24P C01

CROSS REFERENCE INDEXES

PART NUMBER INDEX

CAGE PART NUMBER STOCK NUMBER FIGURE NO ITEM NO

CAGE	PART NUMBER	STOCK NUMBER	FIGURE NO	ITEM NO
88044	AN960-416L	5310-00-167-0835	279	3
88044	AN960-516L	5310-00-167-0836	279	17
88044	AN960-616L	5310-00-167-0837	279	11
88044	AN960-816	5310-00-167-0823	199	21
		5310-00-167-0823	245	25
		5310-00-167-0823	251	46
		5310-00-167-0823	252	14
		5310-00-167-0823	429	35
		5310-00-167-0823	446	27
87585	ANA9-5	5325-00-276-6228	489	6
78276	ATS2-813	5325-01-409-0211	127	8
78225	AX2022-02	2590-01-196-5228	4	1
12204	B0013444	3030-00-504-8682	490	11
73821	B0625-022-S	5310-01-203-3230	273	9
80204	B107.19	5120-01-373-4981	507	5
80204	B18212HRCZ080	5310-01-478-5620	175A	10
80204	B1821BH025C050N	5305-00-071-2506	25	1
		5305-00-071-2506	28	8
		5305-00-071-2506	55	12
		5305-00-071-2506	103	4
		5305-00-071-2506	156	16
		5305-00-071-2506	365	13
		5305-00-071-2506	365	26
		5305-00-071-2506	368	24
		5305-00-071-2506	375	30
		5305-00-071-2506	496	13
80204	B1821BH025C063N	5305-00-068-7837	28	8
		5305-00-068-7837	46	25
		5305-00-068-7837	155	11
		5305-00-068-7837	177	10
		5305-00-068-7837	178	5
		5305-00-068-7837	197	30
		5305-00-068-7837	199	42
		5305-00-068-7837	210	3
		5305-00-068-7837	274	8
		5305-00-068-7837	276	1
		5305-00-068-7837	176A	17
80204	B1821BH025C075N	5305-00-068-0508	15	20
		5305-00-068-0508	17	12
		5305-00-068-0508	29	6
		5305-00-068-0508	88	50
		5305-00-068-0508	208	2
		5305-00-068-0508	233	6
		5305-00-068-0508	241	12
		5305-00-068-0508	243	13
		5305-00-068-0508	252	3
		5305-00-068-0508	269	33
		5305-00-068-0508	270	28
		5305-00-068-0508	275	5

SECTION IV TM9-2320-280-24P C01

CROSS REFERENCE INDEXES

PART NUMBER INDEX

CAGE PART NUMBER STOCK NUMBER FIGURE NO ITEM NO

CAGE	PART NUMBER	STOCK NUMBER	FIGURE NO	ITEM NO
		5305-00-068-0508	283	14
		5305-00-068-0508	287	12
		5305-00-068-0508	327	45
		5305-00-068-0508	331	35
		5305-00-068-0508	332	5
		5305-00-068-0508	365	4
		5305-00-068-0508	368	40
		5305-00-068-0508	398	16
		5305-00-068-0508	399	21
		5305-00-068-0508	400	14
		5305-00-068-0508	43B	41
		5305-00-068-0508	457	14
		5305-00-068-0508	468	5
		5305-00-068-0508	478	4
		5305-00-068-0508	481	6
		5305-00-068-0508	485	10
		5305-00-068-0508	492	9
		5305-00-068-0508	498	8
80204	B1821BH025C088N	5305-00-071-2505	80	13
		5305-00-071-2505	98	15
		5305-00-071-2505	99	6
		5305-00-071-2505	202	16
		5305-00-071-2505	209	6
		5305-00-071-2505	245	37
		5305-00-071-2505	247	15
		5305-00-071-2505	280	54
		5305-00-071-2505	281	54
		5305-00-071-2505	311	4
		5305-00-071-2505	325	13
		5305-00-071-2505	326	13
		5305-00-071-2505	419	1
80204	B1821BH025C100N	5305-00-225-3843	21	17
		5305-00-225-3843	42	6
		5305-00-225-3843	95	42
		5305-00-225-3843	153	5
		5305-00-225-3843	169	10
		5305-00-225-3843	169	30
		5305-00-225-3843	203	1
		5305-00-225-3843	203	26
		5305-00-225-3843	204	1
		5305-00-225-3843	227	28
		5305-00-225-3843	247	1
		5305-00-225-3843	249	2
		5305-00-225-3843	250	5
		5305-00-225-3843	252	12
		5305-00-225-3843	254	13
		5305-00-225-3843	266	21
		5305-00-225-3843	267	5
		5305-00-225-3843	277	23

SECTION IV TM9-2320-280-24P C01

CROSS REFERENCE INDEXES

PART NUMBER INDEX

CAGE PART NUMBER STOCK NUMBER FIGURE NO ITEM NO

CAGE	PART NUMBER	STOCK NUMBER	FIGURE NO	ITEM NO
		5305-00-225-3843	285	14
		5305-00-225-3843	286	36
		5305-00-225-3843	292	11
		5305-00-225-3843	312	1
		5305-00-225-3843	334	6
		5305-00-225-3843	343	18
		5305-00-225-3843	352	3
		5305-00-225-3843	366	17
		5305-00-225-3843	368	34
		5305-00-225-3843	417	3
		5305-00-225-3843	430	42
		5305-00-225-3843	459	3
		5305-00-225-3843	475	6
		5305-00-225-3843	484	14
		5305-00-225-3843	486	12
		5305-00-225-3843	492	15
		5305-00-225-3843	494	11
80204	B1821BH025C113N	5305-00-225-3842	5	16
		5305-00-225-3842	202	10
		5305-00-225-3842	203	5
		5305-00-225-3842	204	5
		5305-00-225-3842	308	9
		5305-00-225-3842	430	16
80204	B1821BH025C125N	5305-00-068-0509	49	6
		5305-00-068-0509	198	7
		5305-00-068-0509	251	28
		5305-00-068-0509	252	11
		5305-00-068-0509	270	8
		5305-00-068-0509	285	3
		5305-00-068-0509	288	24
		5305-00-068-0509	288	32
		5305-00-068-0509	321	11
		5305-00-068-0509	398	10
		5305-00-068-0509	402	3
		5305-00-068-0509	425	3
		5305-00-068-0509	426	10
		5305-00-068-0509	427	4
		5305-00-068-0509	487	9
80204	B1821BH025C150N	5305-00-071-2509	83	29
		5305-00-071-2509	88	46
		5305-00-071-2509	225	41
		5305-00-071-2509	246	35
		5305-00-071-2509	248	27
		5305-00-071-2509	249	20
		5305-00-071-2509	252	10
		5305-00-071-2509	253	13
		5305-00-071-2509	253	20
		5305-00-071-2509	255	19
		5305-00-071-2509	289	9

SECTION IV TM9-2320-280-24P C01

CROSS REFERENCE INDEXES

PART NUMBER INDEX

CAGE	PART NUMBER	STOCK NUMBER	FIGURE NO	ITEM NO		
		5305-00-071-2509	389	11		
80204	B1821BH025C175N	5305-00-071-2510			153	11
		5305-00-071-2510	211	11		
		5305-00-071-2510	212	7		
		5305-00-071-2510	246	29		
		5305-00-071-2510	246	38		
		5305-00-071-2510	249	9		
		5305-00-071-2510	484	10		
80204	B1821BH025C200N	5305-00-071-2511			248	20
80204	B1821BH025C225N	5305-00-071-2512			293	26
80204	B1821BH025C250N	5305-00-071-2513			248	25
		5305-00-071-2513	249	36		
80204	B1821BH025C275N	5305-00-071-2514			251	10
80204	B1821BH025C450N	5305-01-202-2676			155	16
80204	B1821BH025F050N	5305-00-267-8952			387	7
80204	B1821BH025F063N	5305-00-267-8953			279	2
		5305-00-267-8953			388	12
		5305-00-267-8953			43B	32
		5305-00-267-8953	485	15		
80204	B1821BH025F088N	5306-00-068-0514			379	13
80204	B1821BH025F100N	5305-00-068-0515			64	22
		5305-00-068-0515			379	52
		5305-00-068-0515			387	17
		5305-00-068-0515	393	9		
80204	B1821BH025F125N	5305-00-267-8954			228	15
80204	B1821BH031C050N	5306-00-226-4822			35	16
80204	B1821BH031C056N	5306-00-226-4823			143	36
		5306-00-226-4823	151	36		
80204	B1821BH031C063N	5306-00-226-4824			273	19
		5306-00-226-4824	43B	14		
		5306-00-226-4824	488	23		
80204	B1821BH031C075N	5306-00-226-4825			19	20
		5306-00-226-4825	100	17		
		5306-00-226-4825	116	18		
		5306-00-226-4825	230	26		
		5306-00-226-4825	234	6		
		5306-00-226-4825	480	3		
		5306-00-226-4825	487	16		
		5306-00-226-4825	487B	3		
80204	B1821BH031C088N	5306-00-226-4826			230	19
		5306-00-226-4826	231	12		
		5306-00-226-4826	273	7		
		5306-00-226-4826	277	7		
		5306-00-226-4826	328	8		
80204	B1821BH031C100N	5306-00-226-4827			83	30
		5306-00-226-4827	204	6		
		5306-00-226-4827	225	29		
		5306-00-226-4827	226	14		
		5306-00-226-4827	242	10		

SECTION IV TM9-2320-280-24P C01

CROSS REFERENCE INDEXES

PART NUMBER INDEX

CAGE	PART NUMBER	STOCK NUMBER	FIGURE NO	ITEM NO
		5306-00-226-4827	287	20
		5306-00-226-4827	329	18
		5306-00-226-4827	473	1
		5306-00-226-4827	474	7
80204	B1821BH031C113N	5306-00-226-4828	160	13
		5306-00-226-4828	162	14
		5306-00-226-4828	225	2
		5306-00-226-4828	226	13
		5306-00-226-4828	284	1
		5306-00-226-4828	288	26
		5306-00-226-4828	292	13
		5306-00-226-4828	340	1
		5306-00-226-4828	346	1
		5306-00-226-4828	363	23
		5306-00-226-4828	374	22
		5306-00-226-4828	417	4
		5306-00-226-4828	418	7
		5306-00-226-4828	419	8
		5306-00-226-4828	420	5
		5306-00-226-4828	421	7
		5306-00-226-4828	422	9
		5306-00-226-4828	423	3
		5306-00-226-4828	439	1
		5306-00-226-4828	487D	8
80204	B1821BH031C125N	5306-00-226-4829	144	12
		5306-00-226-4829	152	12
		5306-00-226-4829	211	31
		5306-00-226-4829	212	29
		5306-00-226-4829	225	26
		5306-00-226-4829	226	8
		5306-00-226-4829	328	2
		5306-00-226-4829	390	6
		5306-00-226-4829	417	7
		5306-00-226-4829	423	7
		5306-00-226-4829	432	2
		5306-00-226-4829	461	8
80204	B1821BH031C150N	5305-00-226-4831	121	9
		5305-00-226-4831	242	2
		5305-00-226-4831	269	20
		5305-00-226-4831	279	35
		5305-00-226-4831	288	23
80204	B1821BH031C175N	5306-00-226-4832	121	8
80204	B1821BH031C200N	5306-00-226-4833	211	15
		5306-00-226-4833	212	11
		5306-00-226-4833	242	4
		5306-00-226-4833	377	15
80204	B1821BH031C250N	5306-00-226-4835	209	16
		5306-00-226-4835	338	24
		5306-00-226-4835	429	5

I-120

SECTION IV TM9-2320-280-24P C01

CROSS REFERENCE INDEXES

PART NUMBER INDEX

CAGE	PART NUMBER	STOCK NUMBER	FIGURE NO		ITEM NO	
80204	B1821BH031C325N	5306-00-226-4838			189	3
80204	B1821BH031F075N	5306-00-050-1238			433	9
80204	B1821BH031F088N	5305-01-407-9487			199	6
80204	B1821BH031F100N	5305-00-051-4076			324	7
		5305-00-051-4076	378	30		
80204	B1821BH031F113N	5306-00-051-4077			266	4
		5306-00-051-4077	321	29		
		5306-00-051-4077	321	34		
		5306-00-051-4077	325	29		
		5306-00-051-4077	413	11		
		5306-00-051-4077	415	5		
80204	B1821BH038C075N	5305-00-543-4372			60	1
		5305-00-543-4372	61	1		
		5305-00-543-4372	240	6		
		5305-00-543-4372	332	35		
		5305-00-543-4372	370	26		
80204	B1821BH038C088N	5305-01-140-9118			175	7
		5305-01-140-9118	272	24		
80204	B1821BH038C100D	5305-00-942-2196			456	22
80204	B1821BH038C100L	5306-01-114-0963			37	1
		5306-01-114-0963	475	29		
		5306-01-114-0963	476	4		
80204	B1821BH038C100N	5305-00-068-0510			2	6
		5305-00-068-0510	15	31		
		5305-00-068-0510	16	20		
		5305-00-068-0510	23	18		
		5305-00-068-0510	24	26		
		5305-00-068-0510	26	7		
		5305-00-068-0510	35	2		
		5305-00-068-0510	37	2		
		5305-00-068-0510	142	15		
		5305-00-068-0510	143	15		
		5305-00-068-0510	150	15		
		5305-00-068-0510	151	15		
		5305-00-068-0510	174	13		
		5305-00-068-0510	175	6		
		5305-00-068-0510	245	21		
		5305-00-068-0510	274	27		
		5305-00-068-0510	275	1		
		5305-00-068-0510	276	24		
		5305-00-068-0510	277	47		
		5305-00-068-0510	286	29		
		5305-00-068-0510	38A	5		
		5305-00-068-0510	445	13		
		5305-00-068-0510	463	6		
		5305-00-068-0510	464	3		
		5305-00-068-0510	471	11		
		5305-00-068-0510	476	1		
80204	B1821BH038C113N	5305-00-543-2419			1	6

SECTION IV TM9-2320-280-24P C01

CROSS REFERENCE INDEXES

PART NUMBER INDEX

CAGE	PART NUMBER	STOCK NUMBER	FIGURE NO	ITEM NO
		5305-00-543-2419	2	2
		5305-00-543-2419	23	8
		5305-00-543-2419	251	43
		5305-00-543-2419	329	1
		5305-00-543-2419	341	1
		5305-00-543-2419	425	14
		5305-00-543-2419	463	1
80204	B1821BH038C125N	5305-00-068-0511	24	8
		5305-00-068-0511	138	16
		5305-00-068-0511	139	15
		5305-00-068-0511	145	16
		5305-00-068-0511	146	15
		5305-00-068-0511	189	9
		5305-00-068-0511	190	7
		5305-00-068-0511	266	18
		5305-00-068-0511	272	33
		5305-00-068-0511	283	6
		5305-00-068-0511	431	2
		5305-00-068-0511	462	2
80204	B1821BH038C138N	5305-00-688-2111	103	13
		5305-00-688-2111	144	16
		5305-00-688-2111	152	16
		5305-00-688-2111	438	7
80204	B1821BH038C150N	5305-00-725-2317	36	16
		5305-00-725-2317	205	20
		5305-00-725-2317	245	23
		5305-00-725-2317	270	9
		5305-00-725-2317	270	13
		5305-00-725-2317	270	30
		5305-00-725-2317	272	34
		5305-00-725-2317	273	3
		5305-00-725-2317	274	26
		5305-00-725-2317	275	22
		5305-00-725-2317	277	12
		5305-00-725-2317	286	34
		5305-00-725-2317	33A	20
		5305-00-725-2317	38A	4
		5305-00-725-2317	463	21
		5305-00-725-2317	490	1
80204	B1821BH038C175N	5305-00-821-3869	23	23
		5305-00-821-3869	24	27
		5305-00-821-3869	251	26
		5305-00-821-3869	251	44
		5305-00-821-3869	272	27
		5305-00-821-3869	285	4
		5305-00-821-3869	414	2
		5305-00-821-3869	462	11
		5305-00-821-3869	470	15
		5305-00-821-3869	471	6

SECTION IV TM9-2320-280-24P C01

CROSS REFERENCE INDEXES

PART NUMBER INDEX

CAGE	PART NUMBER	STOCK NUMBER	FIGURE NO	ITEM NO
		5305-00-821-3869	486	8
80204	B1821BH038C200N	5305-00-782-9489	23	1
		5305-00-782-9489	23	14
		5305-00-782-9489	24	13
		5305-00-782-9489	24	22
		5305-00-782-9489	202	35
		5305-00-782-9489	272	38
		5305-00-782-9489	273	4
		5305-00-782-9489	274	13
		5305-00-782-9489	398	22
		5305-00-782-9489	410	4
		5305-00-782-9489	412	4
		5305-00-782-9489	428	3
		5305-00-782-9489	490	3
80204	B1821BH038C225N	5305-00-638-8920	341	9
		5305-00-638-8920	426	14
		5305-00-638-8920	427	13
80204	B1821BH038C250N	5305-00-543-2866	272	34
		5305-00-543-2866	273	5
		5305-00-543-2866	274	26
		5305-00-543-2866	275	21
		5305-00-543-2866	276	23
		5305-00-543-2866	33A	17
		5305-00-543-2866	33B	12
80204	B1821BH038C275N	5305-00-781-3926	1	5
		5305-00-781-3926	272	23
		5305-00-781-3926	272	35
		5305-00-781-3926	272	38
		5305-00-781-3926	273	1
		5305-00-781-3926	273	4
		5305-00-781-3926	274	25
		5305-00-781-3926	275	3
		5305-00-781-3926	277	12
		5305-00-781-3926	286	32
80204	B1821BH038C300N	5305-00-846-5703	268	17
		5305-00-846-5703	272	27
		5305-00-846-5703	274	12
		5305-00-846-5703	277	48
80204	B1821BH038C350N	5305-00-781-3927	277	31
		5305-00-781-3927	288	10
		5305-00-781-3927	33A	19
80204	B1821BH038C375N	5305-00-990-8632	273	1
		5305-00-990-8632	274	25
		5305-00-990-8632	276	22
		5305-00-990-8632	277	31
80204	B1821BH038C400N	5305-00-781-3928	274	12
		5305-00-781-3928	274	23
		5305-00-781-3928	277	31
		5305-00-781-3928	288	6

SECTION IV TM9-2320-280-24P C01

CROSS REFERENCE INDEXES

PART NUMBER INDEX

CAGE	PART NUMBER	STOCK NUMBER	FIGURE NO	ITEM NO
80204	B1821BH038C450N	5305-00-857-6886		274	10
		5305-00-857-6886	276	12
		5305-00-857-6886	277	1
80204	B1821BH038C475N	5305-00-781-3930		170	21
		5305-00-781-3930	275	14
80204	B1821BH038C550N	5305-01-458-1670		274	10
		5305-01-458-1670	274	23
		5305-01-458-1670	276	18
80204	B1821BH038C575N	5305-01-446-0767		38	16
		5305-01-446-0767	275	18
80204	B1821BH038C600N	5305-01-386-9052		431	8
80204	B1821BH038C700N	5305-01-475-3154		276	13
		5305-01-475-3154	33B	1
80204	B1821BH038C750N	5305-01-475-3364		276	17
80204	B1821BH038F063N	5305-00-269-3233		363	10
80204	B1821BH038F175N	5305-00-269-3241		392	7
80204	B1821BH038F200N	5305-00-269-3242		390	15
80204	B1821BH044C100N	5305-00-071-1786		136	5
		5305-00-071-1786	151	42
80204	B1821BH044C125N	5305-00-071-1788		425	5
		5305-00-071-1788	426	8
		5305-00-071-1788	427	6
80204	B1821BH044C150N	5305-00-071-2055		35	31
		5305-00-071-2055	152	16
80204	B1821BH044C175N	5305-00-071-2056		389	3
		5305-00-071-2056	393	3
		5305-00-071-2056	395	14
80204	B1821BH044C200N	5305-00-071-2057		37	21
		5305-00-071-2057	42	5
		5305-00-071-2057	462	13
80204	B1821BH044C250N	5305-00-071-2059		37	9
		5305-00-071-2059	42	5
80204	B1821BH044F975N	5305-01-429-9149		35	21
		5305-01-429-9149	36	1
		5305-01-429-9149	38	1
		5305-01-429-9149	472	10
		5305-01-429-9149	475	10
80204	B1821BH050C100N	5305-00-071-2066		140	5
		5305-00-071-2066	140	11
		5305-00-071-2066	147	9
		5305-00-071-2066	182	13
		5305-00-071-2066	182	21
		5305-00-071-2066	183	19
		5305-00-071-2066	296	4
80204	B1821BH050C113N	5305-00-732-0511		103	18
		5305-00-732-0511	363	21
80204	B1821BH050C125D	5305-00-915-8087		2	14
		5305-00-915-8087	156	22
80204	B1821BH050C125N	5305-00-071-2067		140	11

SECTION IV TM9-2320-280-24P C01

CROSS REFERENCE INDEXES

PART NUMBER INDEX

CAGE	PART NUMBER	STOCK NUMBER	FIGURE NO	ITEM NO		
		5305-00-071-2067	142	16		
		5305-00-071-2067	143	16		
		5305-00-071-2067	147	5		
		5305-00-071-2067	150	16		
		5305-00-071-2067	151	16		
		5305-00-071-2067	179	17		
		5305-00-071-2067	180	27		
		5305-00-071-2067	182	20		
		5305-00-071-2067	183	19		
		5305-00-071-2067	184	7		
		5305-00-071-2067	237	9		
		5305-00-071-2067	429	34		
		5305-00-071-2067	456	6		
		5305-00-071-2067	487A	4		
80204	B1821BH050C150N	5305-00-071-2069			142	34
		5305-00-071-2069	143	40		
		5305-00-071-2069	174	8		
		5305-00-071-2069	182	13		
		5305-00-071-2069	183	9		
		5305-00-071-2069	183	24		
		5305-00-071-2069	187	14		
		5305-00-071-2069	209	15		
		5305-00-071-2069	315	8		
		5305-00-071-2069	363	27		
		5305-00-071-2069	456	23		
80204	B1821BH050C175N	5305-00-071-2070			127	4
		5305-00-071-2070	181	8		
		5305-00-071-2070	186	13		
		5305-00-071-2070	242	6		
		5305-00-071-2070	245	24		
		5305-00-071-2070	429	38		
		5305-00-071-2070	438	6		
		5305-00-071-2070	441	14		
		5305-00-071-2070	446	26		
		5305-00-071-2070	456	24		
80204	B1821BH050C200N	5305-00-071-2071			26	12
		5305-00-071-2071	185	10		
		5305-00-071-2071	252	13		
		5305-00-071-2071	285	6		
		5305-00-071-2071	295	8		
		5305-00-071-2071	455	2		
		5305-00-071-2071	456	5		
80204	B1821BH050C225N	5305-00-071-2072			283	4
80204	B1821BH050C250N	5305-00-071-2073			188	24
		5305-00-071-2073	191	7		
80204	B1821BH050C275N	5305-00-071-2074			33	1
		5305-00-071-2074	182	18		
		5305-00-071-2074	183	18		
80204	B1821BH050C300N	5305-00-071-2075			182	14

SECTION IV TM9-2320-280-24P C01

CROSS REFERENCE INDEXES

PART NUMBER INDEX

CAGE PART NUMBER STOCK NUMBER FIGURE NO ITEM NO

CAGE	PART NUMBER	STOCK NUMBER	FIGURE NO	ITEM NO
		5305-00-071-2075	183	23
		5305-00-071-2075 188		6
		5305-00-071-2075 191		7
		5305-00-071-2075 33A		1
80204	B1821BH050C325N	5305-00-071-2076	147	9
		5305-00-071-2076	182	8
		5305-00-071-2076	182	16
		5305-00-071-2076 188		9
80204	B1821BH050C350N	5305-00-071-2077	191	5
80204	B1821BH050C375N	5305-00-071-2078	181	7
		5305-00-071-2078 184		4
		5305-00-071-2078 188		1
80204	B1821BH050C400N	5305-00-071-2079	181	3
		5305-00-071-2079	182	8
		5305-00-071-2079	188	24
		5305-00-071-2079 191		7
80204	B1821BH050C425N	5305-00-071-1783	188	22
80204	B1821BH050C450N	5305-00-071-2081	2	11
		5305-00-071-2081	179	13
		5305-00-071-2081	180	18
		5305-00-071-2081	184	17
80204	B1821BH050C475N	5305-00-071-2082	179	6
		5305-00-071-2082 180		7
80204	B1821BH050C500N	5305-00-071-2083	363	16
80204	B1821BH050C775N	5305-01-480-5408	33	16
80204	B1821BH050F150N	5305-00-719-5221	390	2
80204	B1821BH050F200N	5305-00-719-5238	397	16
80204	B1821BH050F275N	5305-00-719-5241	397	20
80204	B1821BH056C150N	5305-00-716-7680	180	14
80204	B1821BH056F950N	5305-01-475-3014	476	15
80204	B1821BH063C125N	5305-00-724-7219	140	8
		5305-00-724-7219 147		8
80204	B1821BH063C150N	5305-00-724-7220	140	10
		5305-00-724-7220	147	11
		5305-00-724-7220	157	21
		5305-00-724-7220	158	14
80204	B1821BH063C300N	5305-00-724-7228	194	8
80204	B1821BH063C350N	5305-00-724-7247	185	3
80204	B1821BH063C375N	5305-00-724-7248	186	4
		5305-00-724-7248	457	18
80204	B1821BH063C400N	5305-00-724-7254	186	3
		5305-00-724-7254	196	12
		5305-00-724-7254 470		8
80204	B1821BH063C425N	5305-00-724-7263	196	1
		5305-00-724-7263	196	16
80204	B1821BH063C450N	5305-00-724-7264	196	12
		5305-00-724-7264	196	16
80204	B1821BH063C475N	5305-00-724-7265	196	1
80204	B1821BH075C325N	5305-00-947-4355	183	17

SECTION IV TM9-2320-280-24P C01

CROSS REFERENCE INDEXES

PART NUMBER INDEX

CAGE	PART NUMBER	STOCK NUMBER	FIGURE NO	ITEM NO
80204	B1821BH075C400N	5305-00-947-4358	181	10
		5305-00-947-4358 182 6		
		5305-00-947-4358	188	39
80204	B1821BH075C425N	5305-00-947-4359	183	8
80204	B1821BH075C450N	5305-00-947-4360	183	16
		5305-00-947-4360	184	13
80204	B1821BH075C475N	5305-00-947-4361	144	1
		5305-00-947-4361 152 1		
		5305-00-947-4361 195 2		
80204	B1821BH075C500N	5305-00-947-4362	144	22
80204	B1821BH075C550N	5305-00-947-4363	471	4
80204	B18231B10020NF	5305-01-390-0547	484	4
80204	B18231B10025NF	5305-01-380-3395	1	5
		5305-01-380-3395 2 6		
80204	B18235B10110N	5306-01-504-6221	490A	8
80204	B18235B10130N	5306-01-504-6223	490A	7
80204	B18241B100	5310-01-366-3539	33A	10
80204	B18241B100F	5310-01-395-8747	30	9
		5310-01-395-8747	35	25
		5310-01-395-8747	36	20
		5310-01-395-8747	37	15
		5310-01-395-8747	38	20
		5310-01-395-8747	41	11
		5310-01-395-8747	42	11
		5310-01-395-8747 77 8		
		5310-01-395-8747 472 1		
		5310-01-395-8747 474 5		
		5310-01-395-8747	475	13
		5310-01-395-8747	476	21
		5310-01-395-8747 490 8		
83298	B20521-4	5315-00-616-5514	175A	7
28527	B2099690G1	5331-00-185-0075	496	14
81349	B47287-II-1/2	9330-01-252-1239	487A	1
		9330-01-252-1239	BULK	66
81349	B47287-II-1/2-27		162	15
81349	B47287-II-1/2-3		13	12
			14	18
81361	B5-19-1676-1	4730-00-808-5090	311	13
		4730-00-808-5090 312 9		
81361	B5-19-1829	4240-00-007-9453	311	10
		4240-00-007-9453 312 6		
88907	B54-32780	4730-00-278-2523	385	4
		4730-00-278-2523 388 5		
72962	B79NM-50	5310-01-299-6460	327	4
81205	BACG20AF8H1	5325-00-276-6051	96	21
81205	BACT14A6	4030-00-641-3921	275	11
11815	BALM-6BP-14	5320-01-254-2283	208	37
		5320-01-254-2283	332	40
11815	BAPKTR-64	5320-01-275-1998	64	2

I-127

SECTION IV TM9-2320-280-24P C01

CROSS REFERENCE INDEXES

PART NUMBER INDEX

CAGE PART NUMBER STOCK NUMBER FIGURE NO ITEM NO

CAGE	PART NUMBER	STOCK NUMBER	FIGURE NO	ITEM NO
		5320-01-275-1998	197	14
		5320-01-275-1998	199	53
		5320-01-275-1998	212	21
		5320-01-275-1998	332	12
		5320-01-275-1998	379	27
98410	BB-823-06	5940-00-283-5280	367	11
98410	BB-837-08	5940-00-143-4774	367	15
72794	BJR7-100CBC	5325-01-191-7555	208	22
72794	BJR7-90	5325-01-255-1126	235	6
31272	BLC5LA13S	5315-00-765-2190	279	19
74159	BLM8S18	5940-00-114-1315	75	6
55509	BM2337	3110-00-100-5920	150	22
4A439	BN252	4730-01-357-6523	385	5
99017	BPF-3-1/2	5340-01-194-0887	270	7
		5340-01-194-0887	340	5
99017	BPF-3/8-3	5340-01-194-0888	270	27
13940	BS 78505	5325-00-371-8108	224	12
		5325-00-371-8108	316	2
		5325-00-371-8108	316	6
		5325-00-371-8108	319	3
		5325-00-371-8108	322	3
		5325-00-371-8108	324	3
		5325-00-371-8108	375	2
		5325-00-371-8108	375	5
		5325-00-371-8108	378	2
		5325-00-371-8108	378	5
33287	BT-33-73F	6635-01-093-3710	499	1
78553	C-14329-8A04	5310-01-273-1635	64	20
82484	C-1703	5340-01-211-5593	327	7
98349	C-205-TEZ-35	2530-01-188-3685	144	8
		2530-01-188-3685	152	8
80604	C-4171	5120-01-026-1666	511	9
78553	C-8036-832	5310-01-272-5464	300	10
98410	C-828-06	5940-00-204-8990	367	13
98410	C-828-08	5940-00-143-4775	367	14
98410	C-828-10	5940-00-143-4794	72	10
98410	C-840-38	5940-00-113-9826	86	34
70472	C0360-045-1250S	5360-01-118-5907	274	3
		5360-01-118-5907	276	29
73821	C0720-081-1500S	5360-01-218-1608	286	3
73821	C0850-081-2250-M	5360-01-255-9899	205	30
75272	C0V0709Z1	5340-01-071-2047	124	12
30327	C110-10	4720-00-241-4435	BULK	24
01976	C17	5310-00-261-7340	341	8
78553	C183-012-4	5310-01-213-1333	269	21
07860	C21452	2590-00-473-6331	392	6
15434	C2798	5315-00-012-0123	144	13
81349	C3030	5325-00-184-9846	89	16
		5325-00-184-9846	368	10

I-128

SECTION IV TM9-2320-280-24P C01

CROSS REFERENCE INDEXES

PART NUMBER INDEX

CAGE	PART NUMBER	STOCK NUMBER	FIGURE NO	ITEM NO
08627	C3254	5340-01-209-7808	149	13
16004	C43974	4010-00-585-2108	BULK	2
78553	C46021-1024A-1	5310-01-186-3483	222	13
57958	C5136336	5340-01-308-2168	482	20
57958	C5136338	5330-01-309-4340	482	16
57958	C5136341	4010-01-308-9168	482	6
57958	C5136342	5340-01-308-3851	484	3
57958	C5136359	2590-01-307-9303	468	14
78553	C596-1024-24	5310-01-275-1189	222	14
03487	C766	5305-00-993-0190	43	37
		5305-00-993-0190 297	7	
72800	C7685-10A	5310-01-194-7066	199	56
		5310-01-194-7066 208	7	
		5310-01-194-7066	331	34
78563	C7740-1420	5310-01-253-6439	203	14
78553	C8117-1024	5310-01-215-5356	199	31
		5310-01-215-5356	331	18
71468	CA121003-3	5935-01-460-5508	458	45
71468	CA3106E28-21S-B-F80	5935-01-397-0201	86	28
97403	CARX5-583027	5325-00-720-8064	245	4
28818	CF40BH52	2590-01-312-7355	BULK	44
99862	CL7AHK4T	5355-01-032-7430	293	23
81348	CMDX2-3PT573036	5340-00-809-1492	77	4
		5340-00-809-1492 83	5	
		5340-00-809-1492	84	29
		5340-00-809-1492	88	13
		5340-00-809-1492	96	22
		5340-00-809-1492	125	12
		5340-00-809-1492	315	10
		5340-00-809-1492 444	7	
		5340-00-809-1492 478	2	
75272	COV-0909	5340-00-385-8820	17	11
		5340-00-385-8820	27	12
75272	COV-1109Z1	5340-01-153-9425	177	14
75272	COV-1309	5340-01-101-8262	18	2
75272	COV-1609	5340-01-229-3632	5	5
75272	COV-3311	5340-01-210-0692	17	18
75272	COVO7O9Z1	5340-01-071-2047	18	12
78500	CPL6N8	2520-00-352-2168	169	29
19207	CPR104420-1	4720-01-058-7213	443	11
		4720-01-058-7213	BULK 61	
19207	CPR104420-1-17		135	1
19207	CPR104420-1-24		135	8
34623	CPR104420-1-26		12	6
19207	CPR104420-1-28		12	6
19207	CPR104420-1-4		141	4
			457	3
19207	CPR104420-1-40		443	16

I-129

SECTION IV TM9-2320-280-24P C01

CROSS REFERENCE INDEXES

PART NUMBER INDEX

CAGE	PART NUMBER	STOCK NUMBER	FIGURE NO	ITEM NO
19207	CPR104420-1-70		313	6
			457	1
19207	CPR104420-2	4720-01-014-4915	BULK	62
63769	CR 535094-60	5330-01-203-6551	142	2
		5330-01-203-6551	143	2
		5330-01-203-6551	150	2
		5330-01-203-6551	151	2
11815	CR-212-6-4	5320-01-136-1785	219	4
		5320-01-136-1785	287	14
		5320-01-136-1785	290	8
11815	CR-212-6-5	5320-01-273-0020	287	15
11815	CR-213-4-2	5320-01-258-2576	208	12
11815	CR-213-4-5	5320-01-259-7423	226	2
		5320-01-259-7423	237	5
11815	CR-213-6-2	5320-01-143-5075	213	36
		5320-01-143-5075	214	29
11815	CR-213-6-3	5320-01-135-7319	211	3
		5320-01-135-7319	213	37
		5320-01-135-7319	214	30
		5320-01-135-7319	218	1
		5320-01-135-7319	219	1
		5320-01-135-7319	237	17
		5320-01-135-7319	292	5
		5320-01-135-7319	365	1
		5320-01-135-7319	384	14
11815	CR-213-6-5	5320-01-084-9235	224	15
		5320-01-084-9235	287	3
11815	CR-213-6-8	5320-01-273-4601	224	10
		5320-01-273-4601	294	8
11815	CR-3213-6-5	5320-01-084-9235	212	8
11815	CR-3213-6-6	5320-01-084-9234	211	24
11815	CR3212-6-3	5320-01-084-6101	197	26
		5320-01-084-6101	199	37
		5320-01-084-6101	290	10
11815	CR3212-6-5	5320-01-136-1787	288	3
11815	CR3212-6-7	5320-01-086-1147	287	16
11815	CR3213-4-3	5320-01-138-4239	219	10
		5320-01-138-4239	235	4
		5320-01-138-4239	235	18
11815	CR3213-6-06	5320-01-084-9234	212	23
		5320-01-084-9234	260	17
11815	CR3213-6-1	5320-01-085-1755	211	23
11815	CR3213-6-2	5320-01-143-5075	197	27
		5320-01-143-5075	199	35
		5320-01-143-5075	220	6
		5320-01-143-5075	300	15
11815	CR3213-6-3	5320-01-135-7319	197	11
		5320-01-135-7319	199	15
		5320-01-135-7319	227	39

SECTION IV TM9-2320-280-24P C01

CROSS REFERENCE INDEXES

PART NUMBER INDEX

CAGE	PART NUMBER	STOCK NUMBER	FIGURE NO	ITEM NO
11815	CR3213-6-4	5320-01-136-1782		197	8
		5320-01-136-1782	199	17
		5320-01-136-1782	212	9
		5320-01-136-1782	287	1
		5320-01-136-1782	289	6
11815	CR3213-6-5	5320-01-084-9235		197	24
		5320-01-084-9235	199	38
11815	CR3213-6-7	5320-01-085-9995		197	19
		5320-01-085-9995	199	20
		5320-01-085-9995	287	28
11815	CR3213-6-8	5320-01-086-1144		223	5
11815	CR3213-6-9	5320-01-086-1143		197	10
		5320-01-086-1143	199	16
11815	CR3242-6-2	5320-01-033-8643		205	19
11815	CR3242-6-4	5320-01-033-9126		208	35
11815	CR3243-6-3	5320-01-033-8638		208	39
11815	CR3243-6-4	5320-01-033-8637		208	23
11815	CR3243-6-5	5320-01-034-1884		321	24
		5320-01-034-1884	376	22
06090	CRN1RED	5970-00-827-6566		44	14
		5970-00-827-6566	313	14
60380	CRS22CP	3110-00-162-3766		286	12
16236	CS-4520-SV-0705	4520-00-217-5782		345	30
		4520-00-217-5782	348	29
7Z588	CT250L	4730-01-329-9151		29	11
75272	CWV-0407	5340-01-215-5478		333	30
33116	D00-09102	2530-01-174-7441		156	1
64878	D100-MIL-1	5180-01-201-4978		504	10
24866	D12604	2540-01-184-5503		327	21
81346	D1785	4710-00-228-2177		BULK	45
07860	D21059-16	5310-00-080-6004		8	13
		5310-00-080-6004	12	21
		5310-00-080-6004	97	7
		5310-00-080-6004	133	5
		5310-00-080-6004	155	18
		5310-00-080-6004	251	5
		5310-00-080-6004	270	32
		5310-00-080-6004	274	11
		5310-00-080-6004	275	2
81361	D5-19-1754	4240-00-565-6059		307	2
		4240-00-565-6059	308	8
81361	D5-19-2353	4240-01-026-3112		307	1
		4240-01-026-3112	308	7
62161	D528235-H1	2640-01-419-6200		167	27
84760	DB2829-4523	2910-01-199-2355		10	2
84760	DB2829-4879	2910-01-326-9221		10	2
84760	DB2831-5149	2910-01-414-1272		10	2
84760	DB2831-5209	2910-01-467-9029		10	2
93061	DC603-2	4820-00-752-9040		29	9

SECTION IV TM9-2320-280-24P C01

CROSS REFERENCE INDEXES

PART NUMBER INDEX

CAGE	PART NUMBER	STOCK NUMBER	FIGURE NO	ITEM NO
5A910	DC8218	6220-01-107-2613	62	2
34904	DC8226	5330-01-076-6172	62	10
13873	DGSCV-91-157	6220-01-128-0087	62	4
1FH08	DIN933M10X20	5306-01-158-9917	12	22
60827	DP3-01-01	6680-01-482-4995	176A	14
02126	DR-T10X3/4	5305-01-006-5736	74	5
		5305-01-006-5736	303	11
68505	DRA-3002D	2920-01-187-1310	45	8
70308	E06879-00	2540-01-203-0183	230	18
72658	E06879-00	2540-01-203-0183	235	10
01568	E1000-105-3500M	5360-01-255-2808	243	33
30327	E14A	4730-00-254-6211	334	9
7X868	E19250-00	2540-01-315-3358	234	2
81361	E5-19-1782	4240-00-807-6856	311	7
		4240-00-807-6856	312	3
34623	EC-RCSK1834-3-1	5340-01-412-3563	105	26
34623	EC112338639-2B1	2510-01-189-9744	206	17
34623	EC12338083	5340-01-197-6755	277	49
34623	EC12339050	5306-01-465-1564	137	3
34623	EC12460241B1	5340-01-432-8681	160	25
		5340-01-432-8681	162	19
		5340-01-432-8681	487A	5
34623	EC12460245B1	5340-01-432-4877	182	15
34623	EC12460253B1	5340-01-437-8751	182	12
34623	EC12460284B1	2510-01-457-0949	183	22
34623	ECEX4939B1	5340-01-457-5734	44	8
02978	ERNA245	5310-00-584-5272	69	2
		5310-00-584-5272	103	19
		5310-00-584-5272	174	7
		5310-00-584-5272	240	2
		5310-00-584-5272	240	12
		5310-00-584-5272	271	16
		5310-00-584-5272	271	28
		5310-00-584-5272	277	39
		5310-00-584-5272	453	2
34623	ET029	4720-01-185-6673	BULK	22
34623	EX4228-2	5340-01-434-8109	160	17
		5340-01-434-8109	162	24
		5340-01-434-8109	487A	5
34623	EX4231-3	4710-01-434-3151	160	19
		4710-01-434-3151	162	20
		4710-01-434-3151	487A	7
34623	EX4231-4	4710-01-434-3157	160	24
		4710-01-434-3157	162	23
		4710-01-434-3157	487B	7
34623	EX4233	4710-01-434-3037	160	21
		4710-01-434-3037	487A	13
34623	EX4234	4710-01-434-3159	162	21
		4710-01-434-3159	487C	4

SECTION IV TM9-2320-280-24P C01

CROSS REFERENCE INDEXES

PART NUMBER INDEX

CAGE	PART NUMBER	STOCK NUMBER	FIGURE NO	ITEM NO
34623	EX4319	7690-01-473-4550	356	3
34623	EX4717-1	5340-01-458-7156	162	12
		5340-01-458-7156	487D	1
34623	EX4717-2	5340-01-458-7147	162	26
		5340-01-458-7147	487D	1
34623	EX4786-4	4720-01-456-2955	162	7
34623	EX5061	7690-01-476-2855	354	12
19207	EX5063		180	1
34623	EXCE4718	5340-01-458-8236	160	23
		5340-01-458-8236	487B	6
78940	F-6D9K090B	2940-01-189-1809	12	3
29944	F07011	2540-01-315-3143	234	2
		2540-01-315-3143	239	1
27182	F35SRBXXBZXX	5340-01-271-6461	296	11
57526	F6TZ9B076A-A	5325-01-210-2870	15	5
		5325-01-210-2870	16	5
55719	FAM7E	5120-01-437-3658	505	5
01276	FB9665-02	4720-01-257-3705	5	11
16717	FB9778-01	4720-01-213-1574	178	11
55719	FC18B	5120-01-335-1094	505	1
55719	FC22B	5120-01-335-1096	505	1
55719	FC28B	5120-01-335-1099	505	1
55719	FC30B	5120-01-335-1100	505	1
55719	FCOM14	5120-01-079-8023	505	1
01276	FF3323-0706-259	4730-01-493-4056	176A	24
01276	FF90154-18		175A	2
01276	FF9311-36	4730-00-909-8627	379	34
16632	FLX400132057600	4720-01-299-8469	337	20
95760	FP-161	5340-01-274-7337	258	1
		5340-01-274-7337	259	9
55719	FTX40E	5120-01-367-3536	510	8
62161	G 6793	2530-01-386-5499	167	18
34623	G-00271166	5310-01-251-0760	62	13
34623	G-00271482	5310-01-272-1356	56	4
34623	G-11500829	5306-01-254-2357	174	10
34623	G-11501940	5306-01-251-9190	75	47
		5306-01-251-9190	174	4
75535	G-408-3/8	4030-01-256-0471	314	19
		4030-01-256-0471	458	40
34623	G0045696	5305-01-272-9034	339	23
12204	G121224	5315-01-206-7135	293	34
34634	G167084	5305-01-210-6249	325	1
24161	G3336US-10	4720-00-006-9541	BULK	10
34623	G45637	5315-01-218-1114	245	40
57733	G488213	2540-00-039-8258	56	13
78385	G704177	2540-01-165-8175	345	31
		2540-01-165-8175	348	25
78385	G704183	5935-01-163-8981	348	2
78385	G704213	2540-01-164-6170	348	9

I-133

SECTION IV TM9-2320-280-24P C01

CROSS REFERENCE INDEXES

PART NUMBER INDEX

CAGE PART NUMBER STOCK NUMBER FIGURE NO ITEM NO

CAGE	PART NUMBER	STOCK NUMBER	FIGURE	NO	ITEM	NO
78385	G704232	2540-01-165-0465			345	19
		2540-01-165-0465	348	19		
78385	G704234	5935-01-163-8987			348	8
78385	G704284	2540-01-165-0814			350	2
		2540-01-165-0814	351	1		
78385	G704288-1	2540-01-165-8176			345	29
		2540-01-165-8176	348	31		
78385	G704293	5340-01-165-0745			348	30
78385	G704373	6150-01-165-4667			345	26
		6150-01-165-4667	348	24		
78385	G704554	2540-01-008-1501			348	26
78385	G706014	2540-01-289-8328			345	28
78385	G706016	2540-01-253-6112			349	16
78385	G706024				349	14
78385	G706026	6150-01-293-4074			349	6
78385	G706033	5340-01-290-6331			345	3
78385	G706034				345	16
78385	G706035	2540-01-289-8329			345	39
78385	G706037	2540-01-162-4339			348	35
78385	G706039				350	11
78385	G706052-13	6150-01-287-9925			349	1
78385	G706052-15	6150-01-287-9924			349	2
78385	G706052-17	6150-01-287-9926			349	13
78385	G706052-19	6150-01-287-9923			349	10
78385	G706052-21	6150-01-287-9917			345	43
78385	G706052-25	6150-01-287-9918			345	9
78385	G706052-9	6150-01-291-4809			345	41
78385	G706055	4520-01-284-7099			345	22
		4520-01-284-7099	KITS		7	
78385	G706057	6110-01-285-3902			349	9
78385	G706064	2540-01-284-7091			345	2
78385	G706133	2540-01-168-9481			351	11
09527	GG0490	6685-01-180-9037			47	9
09527	GG0552	6620-01-181-1757			47	11
01276	GG308-NP08-18	4730-01-358-8538			175A	1
72794	GH7	5325-01-266-8414			207	2
		5325-01-266-8414	208	20		
19728	GK-174A				45	39
7X677	GM8670757	5306-01-150-4835			104	4
97783	GR011RED	9905-00-205-2795			212	10
82484	GS-2176	2540-01-422-0253			327	1
76599	H12SS	4730-00-908-3194			268	32
81718	H2525M	5310-00-637-9541			89	7
		5310-00-637-9541	205	35		
		5310-00-637-9541	401	12		
50992	H38C-2619	4820-01-189-2107			333	37
76599	H68SS	4730-01-189-0871			12	7
11815	H749A-456	5130-01-104-5370			504	11
88818	H795220608	5305-00-889-3000			43	34

I-134

SECTION IV TM9-2320-280-24P C01

CROSS REFERENCE INDEXES

PART NUMBER INDEX

CAGE	PART NUMBER	STOCK NUMBER	FIGURE NO	ITEM NO
54979	HC-9.56	4010-01-203-0411	243	17
0HDW7	HC3213-4-04	5320-01-143-5079	230	24
K5667	HE11595		175A	6
K5667	HE1250079/100		175A	5
K5667	HE1290019	5305-01-491-2143	175A	9
K5667	HE206000	5310-01-491-2014	175A	8
K5667	HE260128		175A	4 35708
HS32	4730-00-204-3491	29	1	
		4730-00-204-3491	177	6
78947	HS49745	5340-01-195-8837	269	22
78947	HS49746	5340-01-196-8158	269	23
7Z588	HTM-1200	4730-01-210-5785	12	35
21102	HV032504	2540-01-282-8562	300	16
21002	HV033596	5905-01-270-0966	342	3
21102	HV034186	2540-01-276-5935	224	7
24825 I F-5		2510-01-250-7594	227	26
24825	IF-3	2510-01-250-7595	227	25
55883	IPD8S5128	2530-01-069-0906	45	27
25341	J-21362	4910-01-178-6551	511	7
33287	J-21370-10	5120-01-423-0032	510	2
25341	J-21370-A	4910-01-178-0722	500	3
33287	J-21552	5120-01-179-1032	503	8
33287	J-21777-43	5120-01-366-9468	512	3
25341	J-23327	4910-01-178-0724	510	1
60380	J-2416	3110-01-155-4440	130	38
33287	J-25018-A	5120-01-410-8216	510	4
25341	J-29162	4910-01-179-5530	502	3
25341	J-29167	5120-01-170-3278	501	3
25341	J-29168	5120-01-169-4876	501	2
25341	J-29642-C	5180-01-267-2907	509	1
25341	J-33043	4820-01-179-4869	511	15
33287	J-33149-A	5120-01-218-8235	502	5
33287	J-33826	5120-01-357-3633	501	15
33287	J-33829	5120-01-357-3629	501	16
33287	J-33831	5120-01-357-3632	501	12
33287	J-33832		511	10
33287	J-33833	5120-01-357-3630	501	17
33287	J-33839	5120-01-357-3631	501	18
33287	J-33843	5120-01-361-3101	501	14
33287	J-34175-4	5120-01-366-0894	512	9
33287	J-35199	5180-01-216-8643	502	7
33287	J-35209	4730-01-042-5266	505	17
33287	J-35307	5120-01-265-4872	501	6
33287	J-35506-2	5120-01-366-6260	512	7
33287	J-35506-3	5120-01-366-0895	512	5
33287	J-37789-A	5120-01-422-1308	510	6
33287	J-38655	5120-01-422-1326	510	7
33287	J-38731	5180-01-422-0138	510	11
33287	J-38731-1	5120-01-422-0173	510	12

SECTION IV TM9-2320-280-24P C01

CROSS REFERENCE INDEXES

PART NUMBER INDEX

CAGE PART NUMBER STOCK NUMBER FIGURE NO ITEM NO

CAGE	PART NUMBER	STOCK NUMBER	FIGURE NO	ITEM NO
33287	J-38731-2	5120-01-422-0219	510	13
33287	J-38731-3	5120-01-422-0139	510	14
33287	J-38732	5120-01-422-1301	511	5
33287	J-38732-1	5120-01-422-1293	511	6
33287	J-38734	5120-01-422-1329	510	10
33287	J-38736-1	5120-01-422-1310	510	16
33287	J-38736-2	5120-01-422-5406	510	17
33287	J-38736-3	5120-01-422-5411	510	18
33287	J-38736-4	5120-01-422-1318	510	19
33287	J-38736-5	5120-01-422-1324	510	20
33287	J-38736-6	5120-01-408-7051	510	15
		5120-01-408-7051	510	21
33287	J-38737	5120-01-422-1313	510	3
33287	J-38739	5180-01-422-1294	511	1
33287	J-38739-1	5120-01-422-1297	511	2
33287	J-38739-2	5120-01-422-1302	511	3
33287	J-38739-3	5120-01-422-1306	511	4
33287	J-38868-A	5120-01-422-1300	510	5
33287	J-39084	5120-01-422-0334	510	9
33287	J-39250	5120-01-335-5847	505	6
33287	J-39250-1		505	11
33287	J-39524-1	5120-01-366-2520	512	2
33287	J-39524-2	5120-01-366-0893	512	1
33287	J-39524-4	5140-01-366-0642	512	8
33287	J-39524-5	5120-01-366-4713	512	6
25341	J-6278	5120-01-082-6447	503	7
33287	J-9276-3	5120-01-357-9123	501	19
03787	J0082	9905-00-202-3639	330	2
27182	J105CCLXBZ02	5342-01-264-6532	211	2
25341	J21359-A	5120-01-176-1845	500	5
33287	J21465-01	5180-01-195-9777	500	2
33287	J21664	4910-01-210-1318	499	6
33287	J21786	5120-01-185-7957	502	9
33287	J21787	5120-01-187-3660	502	1
25341	J21795-02	4910-01-178-8865	499	5
33287	J22661	5120-01-227-1680	501	13
33287	J24396	4910-01-209-0729	500	6
33287	J25033-B	4910-01-179-2517	503	5
33287	J25323-1	3020-01-255-2906	504	7
33287	J25323-16	4820-01-252-6169	504	4
33287	J25323-17	4720-01-253-2809	504	3
33287	J25323-18	4720-01-253-2810	504	5
33287	J25323-2	4730-01-254-6023	504	6
33287	J25323-5	6685-01-256-3406	504	8
33287	J25323-7	6680-01-253-6782	504	9
33287	J26999-30	4910-01-238-2551	505	14
33287	J29163	5120-01-195-2721	501	4
33287	J29166	5120-01-196-0084	501	1
33287	J29169	5120-01-185-7955	501	5

I-136

SECTION IV TM9-2320-280-24P C01

CROSS REFERENCE INDEXES

PART NUMBER INDEX

CAGE	PART NUMBER	STOCK NUMBER	FIGURE NO	ITEM NO
33287	J29170	5120-01-195-4551	501	11
33287	J29174	5120-01-185-7960	501	21
33287	J29185	5120-01-185-7956	501	8
33287	J29185-2	5120-01-247-6629	501	7
33287	J29369	5120-01-201-7857	501	20
33287	J3289-20	5120-01-144-4484	499	3
33287	J33127	5180-01-186-3114	499	8
33287	J33141	4910-01-231-1709	504	1
33287	J33142	5120-01-187-3659	502	4
33287	J33143	5120-01-187-3607	503	1
33287	J33144	5120-01-185-7963	503	2
33287	J35184	5120-01-229-0842	503	3
33287	J38869	5120-01-414-1849	511	14
33287	J39636	5120-01-389-9992	501	22
27182	J4501XXBZ01	5306-01-265-8897	287	31
		5306-01-265-8897	294	7
27182	J450XXLXBZ01	2540-01-385-7497	289	8
33287	J4670-01	5120-01-210-8793	499	2
27182	J50062XXBZ01	5340-01-276-6673	211	12
		5340-01-276-6673	212	6
27182	J500IHXXBZ01	5340-01-265-8911	211	25
		5340-01-265-8911	212	28
		5340-01-265-8911	213	27
		5340-01-265-8911	214	22
81343	J512	4730-00-278-3828	17	5
33287	J6221	5120-01-185-7965	503	9
33287	J6471-2	5120-01-130-8865	500	4
33287	J6471-8	5120-01-391-5131	501	9
33287	J6692	4910-00-785-6437	505	15
25341	J7624	5120-01-082-6436	503	6
33287	J7818	5120-01-185-8024	501	10
33287	J8092	5120-00-677-2259	500	1
33287	J8608	5120-01-185-7962	502	8
33287	J8611-01	5120-01-185-7964	502	2
33287	J8763-B	5120-01-198-7583	499	4
27182	J8UBXXFXBZ01	5340-01-265-8859	292	12
27182	J95CCXRXBZ01	5340-01-310-1658	214	12
27182	J95CCXXXBZ02	5340-01-310-1189	213	16
81343	JAE J1508	4730-00-908-3193	365	10
81343	JAEJ1508	4730-00-908-3193	301	4
		4730-00-908-3193	303	6
		4730-00-908-3193	304	2
		4730-00-908-3193	368	14
59556	K-500-PC-0001	5315-00-842-3044	101	1
78189	K1224	5310-00-995-4130	72	8
78940	L-111D30	2940-01-188-5117	12	23
78940	L-111DR30-L		12	24
98349	L-18-CPS-5-A-10	5340-01-191-8161	170	16
		5340-01-191-8161	195	5

I-137

SECTION IV TM9-2320-280-24P C01

CROSS REFERENCE INDEXES

PART NUMBER INDEX

CAGE PART NUMBER STOCK NUMBER FIGURE NO ITEM NO

CAGE	PART NUMBER	STOCK NUMBER	FIGURE NO	ITEM NO
98349	L-18-SV-5150-C-12	2530-01-197-2160	170	17
		2530-01-197-2160	195	6
98349	L-18-SV-5151-C-12	2530-01-189-2195	170	11
98349	L-18-VT-5276-B-11	2530-01-194-2049	170	10
98349	L-18-VT-5277-B-11	2510-01-205-6089	195	3
98349	L-20-MA-132-A-11	2530-01-252-8362	170	7
98349	L-20-MA-133-A-11	2530-01-249-1577	170	2
78940	L-223C24	5340-01-190-0402	12	29
46717	L-3604-13	4730-00-877-6298	436	5
78940	L-563C170	2940-01-188-3387	12	28
78940	L-638C2GS	2520-01-024-1273	12	31
46717	L-6493-1	5310-00-045-4007	84	25
16563	LD-1501	4720-01-187-3387	7	7
60038	LM603049LM603012	3110-00-892-0896	148	19
84256	LT1504-C6-08	1640-01-072-1167	433	5
10177	M-200-60-EP	5325-01-187-1604	26	2
92830	M-C1100-096-0875	5360-01-190-6218	99	15
81349	M13486/1-11	6145-00-538-8219	BULK	69
81349	M13486/1-14	6145-00-705-6674	BULK	70
81349	M13486/1-4	6145-01-341-9591	BULK	67
81349	M13486/1-5	6145-00-152-6499	81	17
81349	M13486/1-9	6145-00-538-8222	BULK	68
81349	M13516/1-1	5925-00-026-4767	380	4
		5925-00-026-4767	385	12
61465	M2025103	5310-00-167-0821	101	2
		5310-00-167-0821	382	9
		5310-00-167-0821	390	8
61208	M2046S	5310-00-595-7237	482	13
81343	M23053/15-112-0	5970-01-276-7753	BULK	41
81349	M23053/15-112-0-.75		94	7
81349	M23053/15-112-2		BULK	42
81349	M23053/15-112-2-.75		94	2
81349	M23053/15-202-0	5970-01-359-7836	98	12
81349	M23053/4-103-0	5970-01-048-9378	485	4
81349	M23053/4-202-2		BULK	29
81349	M23053/4-202-2-2		95	23
81349	M23053/4-203-2		BULK	30
81349	M23053/4-203-2-3		95	13
81349	M23053/4-302-0	5970-01-161-6796	BULK	31
81349	M23053/4-302-0-.75		93	3
81349	M23053/4-302-2		BULK	32
81349	M23053/4-302-2-2		95	8

SECTION IV TM9-2320-280-24P C01

CROSS REFERENCE INDEXES

PART NUMBER INDEX

CAGE	PART NUMBER	STOCK NUMBER	FIGURE NO	ITEM NO
81349	M23053/4-303-0		BULK	33
81349	M23053/4-303-0-.75		93	5
81349	M23053/4-303-0-1		85	32
81349	M23053/4-303-0-3	5970-01-142-2282	96	17
81349	M23053/4-303-2		BULK	34
81349	M23053/4-303-2-3		95	18
			96	8
81349	M23053/4-304-0	5970-01-163-1103	BULK	35
81349	M23053/4-304-0-1		69	37
81349	M23053/4-304-2		BULK	36
81349	M23053/4-304-2-2		474	12
81349	M23053/4-304-2-4		95	11
81349	M23053/6-103-6	5970-00-724-1910	91	6
		5970-00-724-1910	BULK	37
81349	M23053/6-104-2	5970-01-049-9948	BULK	38
81349	M23053/6-104-2-4		91	3
81349	M23053/6-107-0	5790-01-031-0736	BULK	39
81349	M23053/6-107-0-7		90	4
			92	9
81349	M23053/6-107-1	5970-00-724-1907	BULK	40
81349	M23053/6-107-1-7		92	7
81349	M23469/2-30605	5320-01-333-5545	197	21
		5320-01-333-5545	199	47
81349	M23469/2-50809		199	29
81349	M24066-2-354	5340-01-282-1115	311	15
81349	M24243/1-A403	5320-00-083-5009	270	17
		5320-00-083-5009	311	17
		5320-00-083-5009	317	8
		5320-00-083-5009	321	9
		5320-00-083-5009	325	18
		5320-00-083-5009	332	15
		5320-00-083-5009	354	13
		5320-00-083-5009	355	3
		5320-00-083-5009	373	7
		5320-00-083-5009	376	8
		5320-00-083-5009	376	12
81349	M24243/1-A404	5320-01-023-2529	65	18
		5320-01-023-2529	198	1
		5320-01-023-2529	199	10
		5320-01-023-2529	225	22
		5320-01-023-2529	229	2
		5320-01-023-2529	316	8
		5320-01-023-2529	318	10
		5320-01-023-2529	324	15
		5320-01-023-2529	370	11
81349	M24243/1-A405	5320-01-029-8205	239	8
81349	M24243/1-B604	5320-00-493-4101	300	8
81349	M24243/1-D404	5320-00-865-8994	498	15

SECTION IV TM9-2320-280-24P C01

CROSS REFERENCE INDEXES

PART NUMBER INDEX

CAGE	PART NUMBER	STOCK NUMBER	FIGURE NO		ITEM NO	
81349	M24243/1A402	5320-01-019-5694			355	3
		5320-01-019-5694	370	21		
		5320-01-019-5694	371	2		
81349	M24243/1A404	5320-01-023-2529			28	4
		5320-01-023-2529			320	6
		5320-01-023-2529			320	29
		5320-01-023-2529	324	37		
02768	M36-0790-10	5325-01-257-0801			268	23
81349	M43436/1-3	9905-00-893-3570			41	4
		9905-00-893-3570	41	9		
		9905-00-893-3570	41	18		
		9905-00-893-3570	44	15		
		9905-00-893-3570	44	21		
		9905-00-893-3570	44	26		
		9905-00-893-3570	69	13		
		9905-00-893-3570	69	22		
		9905-00-893-3570	69	27		
		9905-00-893-3570	69	38		
		9905-00-893-3570	72	6		
		9905-00-893-3570	72	18		
		9905-00-893-3570	73	20		
		9905-00-893-3570	73	24		
		9905-00-893-3570	81	6		
		9905-00-893-3570	81	12		
		9905-00-893-3570	94	3		
		9905-00-893-3570	474	13		
		9905-00-893-3570	480	19		
		9905-00-893-3570	480	28		
		9905-00-893-3570	480	33		
81349	M43436/1-4	9905-00-945-7637			86	2
		9905-00-945-7637	94	4		
		9905-00-945-7637	480	14		
81349	M43436/3-1	9905-01-013-8723			65	6
		9905-01-013-8723	65	14		
		9905-01-013-8723	75	3		
		9905-01-013-8723	76	3		
		9905-01-013-8723	79	8		
		9905-01-013-8723	80	3		
		9905-01-013-8723	81	4		
		9905-01-013-8723	81	10		
		9905-01-013-8723	81	16		
		9905-01-013-8723	83	2		
		9905-01-013-8723	84	30		
		9905-01-013-8723	85	2		
		9905-01-013-8723	89	3		
		9905-01-013-8723	90	3		
		9905-01-013-8723	91	4		
		9905-01-013-8723	92	5		
		9905-01-013-8723	93	7		

SECTION IV TM9-2320-280-24P C01

CROSS REFERENCE INDEXES

PART NUMBER INDEX

CAGE PART NUMBER STOCK NUMBER FIGURE NO ITEM NO

CAGE	PART NUMBER	STOCK NUMBER	FIGURE NO	ITEM NO
		9905-01-013-8723	95	3
		9905-01-013-8723	96	3
96906	M45912/3-6FG8C	5310-00-814-0672	144	4
		5310-00-814-0672	152	4
		5310-00-814-0672	272	24
81349	M45913-4FG8C	5310-00-935-9022	393	12
81349	M45913/1-4CG5C	5310-00-088-1251	50	13
		5310-00-088-1251	169	12
81349	M45913/3-4FG8C	5310-00-935-9022	411	19
81349	M45913/3-5FG8C	5310-00-241-6658	199	1
		5310-00-241-6658	277	35
		5310-00-241-6658	321	30
		5310-00-241-6658	402	8
81349	M45913/3-6CG8P	5310-01-249-0904	461	5
34623	M46792 2-1	2540-01-194-3323	368	19
81349	M5501/11-F1	5340-01-174-4894	370	23
		5340-01-174-4894	394	3
81349	M5501/11-F5	5340-00-727-4774	394	1
81349	M5501/11-F7	5340-01-167-9314	394	2
81349	M5501/7-F2	5340-01-026-3250	203	10
		5340-01-026-3250	378	12
60038	M802048M802011	3110-00-406-9608	148	18
81349	M83413/8-A05BD	6150-00-242-4885	313	24
81349	M83461/1-020	5331-01-107-4950	6	2
81349	M83461/1-029	5331-01-336-3236	73	13
81349	M83461/1-030	5331-01-392-1793	73	11
81349	M83461/1-208	5331-01-392-0830	73	15
81349	M83461/1-236	5331-01-183-0971	6	8
81349	M85049/52-1-24W	5935-01-174-3669	75	19
43334	M88048M88010	3110-00-854-1504	148	6
34623	MA128-21182	5330-01-037-0663	62	9
19728	MAD-110A		45	68
19728	MAW-37		45	53
19728	MBD-233B		45	64
19728	MBD-2566		45	38
19728	MBD-438		45	55
19728	MBD-439		45	63
19728	MBD-442		45	56
19728	MBD-444		45	54
19728	MCS-47A		45	11
24975	MEL50	3120-00-450-1905	45	18
19728	MES-117A		45	13
19728	MES-61		45	16
19728	MES-71A		45	71
19728	MES-72		45	72
19728	MES-78A		45	77
19728	MES-91		45	75
19728	MES-93		45	74
19728	MES-94		45	73

SECTION IV TM9-2320-280-24P C01

CROSS REFERENCE INDEXES

PART NUMBER INDEX

CAGE	PART NUMBER	STOCK NUMBER	FIGURE NO	ITEM NO
68505	MES-95S	5330-00-138-0251	45	41
68505	MEU-2042B	3010-01-068-6996	45	12
19728	MFJ-13A		45	46
24975	MFJ-2012BS	5977-01-189-4988	KITS	8
19728	MFJ-28A		45	70
24975	MFY-1064B	5977-01-324-6323	45	45
68505	MFY-109	2920-01-211-4935	45	79
19728	MFY-117		45	69
19728	MFY-133		45	7
19728	MFY-19		45	48
68505	MFY-19S	5360-01-190-7234	KITS	9
19728	MFY-21		45	66
19728	MFY-2101SS		45	42
68505	MFY-2130	2990-01-205-5981	45	2
19728	MFY-25		45	6
19728	MFY-53		45	10
19728	MFY-63		45	15
68505	MFY-6701UT	2920-01-168-7891	44	2
68505	MFY1102S	2920-01-212-4976	45	67
9K475	MGLP-B6-4	5320-01-271-6357	71	9
		5320-01-271-6357	199	12
		5320-01-271-6357	211	10
		5320-01-271-6357	213	2
		5320-01-271-6357	214	23
		5320-01-271-6357	215	1
		5320-01-271-6357	216	11
		5320-01-271-6357	217	7
		5320-01-271-6357	220	3
		5320-01-271-6357	245	36
		5320-01-271-6357	261	8
		5320-01-271-6357	289	22
		5320-01-271-6357	291	1
		5320-01-271-6357	293	3
		5320-01-271-6357	295	1
		5320-01-271-6357	296	14
		5320-01-271-6357	297	1
		5320-01-271-6357	298	3
		5320-01-271-6357	299	4
		5320-01-271-6357	300	7
		5320-01-271-6357	307	8
		5320-01-271-6357	308	10
		5320-01-271-6357	334	2
		5320-01-271-6357	343	1
		5320-01-271-6357	347	10
81349	MIL-DTL-49513/3	5310-00-935-9022	64	29
		5310-00-935-9022	368	18
		5310-00-935-9022	379	22
		5310-00-935-9022	380	14
		5310-00-935-9022	484	12

SECTION IV TM9-2320-280-24P C01

CROSS REFERENCE INDEXES

PART NUMBER INDEX

CAGE	PART NUMBER	STOCK NUMBER	FIGURE NO	ITEM NO
81349	MIL-P-18066	5530-00-641-6069	BULK	47
81361	MIL-PRF-51193E	4240-01-365-0982	307	15
		4240-01-365-0982	308	3
81349	MIL-PRF-62550/3	2540-01-194-3323	345	1
81349	MIL-W-83420/14IN		275	10
			276	6
19728	ML-128A		45	40
19728	ML-129	5310-00-883-2237	45	1
76599	MMF4-SS	4730-01-118-8278	13	6
		4730-01-118-8278	14	6
		4730-01-118-8278	19	11
		4730-01-118-8278	27	4
		4730-01-118-8278	124	9
		4730-01-118-8278	177	2
		4730-01-118-8278	334	16
		4730-01-118-8278	386	7
34623	MS-00034206-221	5305-01-255-4950	274	22
96906	MS-35338-45	5310-00-407-9566	180	12
		5310-00-407-9566	454	3
		5310-00-407-9566	473	5
		5310-00-407-9566	474	8
		5310-00-407-9566	487	15
		5310-00-407-9566	175A	13
96906	MS122031	5310-00-285-7037	269	28
96906	MS122083	5325-00-331-4774	43B	11
96906	MS15001-1	4730-00-050-4203	136	9
		4730-00-050-4203	136	18
		4730-00-050-4203	136	22
		4730-00-050-4203	137	8
		4730-00-050-4203	137	11
		4730-00-050-4203	137	23
		4730-00-050-4203	144	9
		4730-00-050-4203	144	19
		4730-00-050-4203	152	9
		4730-00-050-4203	152	19
		4730-00-050-4203	170	8
96906	MS15001-3	4730-00-050-4205	170	12
		4730-00-050-4205	170	18
		4730-00-050-4205	195	7
96906	MS15001-4	4730-00-050-4207	191	25
96906	MS15002-1	4730-00-172-0010	169	16
96906	MS15795-808	5310-00-619-1148	482	12
80205	MS15795-812	5310-00-625-5756	43B	16
80205	MS15795-815	5310-00-595-6057	488	15
80205	MS15795-819	5310-00-656-0114	476	25
80205	MS15795-820	5310-00-614-3505	35	8
80205	MS15795-841	5310-00-225-5328	79	14
		5310-00-225-5328	482	17
80205	MS15795-852	5310-01-304-8733	43B	12

SECTION IV TM9-2320-280-24P C01

CROSS REFERENCE INDEXES

PART NUMBER INDEX

CAGE	PART NUMBER	STOCK NUMBER	FIGURE NO	ITEM NO
		5310-01-304-8733	43B	30
80205	MS15795-908	5310-00-045-5207	69	32
80205	MS15795-918	5310-00-045-5218	70	6
		5310-00-045-5218	481	4
80205	MS16562-20	5315-00-814-3529	230	11
80205	MS16562-256	5315-00-753-3895	202	22
80205	MS16562-28	5315-00-616-4261	133	12
96906	MS16624-1062	5325-00-803-7305	99	25
		5325-00-803-7305	133	7
96906	MS16624-1066	5325-00-282-1830	205	29
96906	MS16633-1031	5325-00-682-1762	116	19
96906	MS16633-1050	5325-00-442-5845	155	7
		5325-00-442-5845	280	42
		5325-00-442-5845	281	42
		5325-00-442-5845	289	19
96906	MS16843-5	4030-00-237-8741	314	18
80205	MS16995-50	5305-00-988-7614	315	9
80205	MS16995-63	5305-00-225-7211	5	3
80205	MS16995-72	5305-00-992-6143	178	16
		5305-00-992-6143	436	11
80205	MS16995-77	5305-00-051-8605	149	9
		5305-00-051-8605	159	11
		5305-00-051-8605	161	4
		5305-00-051-8605	162	4
96906	MS16995-77	5305-00-051-8605	141	11
80205	MS16995-78	5305-00-226-7237	12	37
80205	MS16996-10	5305-00-052-6456	34	17
96906	MS17131-55	3110-00-056-9376	128	7
96906	MS17829-4F	5310-00-483-8791	321	18
		5310-00-483-8791	376	14
		5310-00-483-8791	382	6
		5310-00-483-8791	389	10
		5310-00-483-8791	397	5
96906	MS17986-C1013	5315-01-249-0555	437	2
96906	MS17986-C423	5315-00-376-9034	433	10
96906	MS17986-C840	5315-00-150-2793	276	3
80205	MS17986C1009	5315-01-473-1590	462	6
96906	MS19059-2410	3110-00-100-6162	314	38
96906	MS19061-20007	3110-00-900-2560	118	2
		3110-00-900-2560	120	11
96906	MS20002-12	5310-00-595-6612	39	9
96906	MS20002-20	5310-00-515-8776	453	14
96906	MS20392-3C67	5315-00-840-9653	243	11
96906	MS20392-4C69	5315-00-983-7402	268	8
96906	MS20392-4C77	5315-00-240-0251	243	32
96906	MS20392-4C95	5315-00-989-0534	244	11
96906	MS20392-6C17	5315-00-890-3461	390	11
96906	MS20426-AD6-8	5320-00-117-7288	289	10
80205	MS20426AD10-16	5320-00-117-6910	288	17

SECTION IV TM9-2320-280-24P C01

CROSS REFERENCE INDEXES

PART NUMBER INDEX

CAGE	PART NUMBER	STOCK NUMBER	FIGURE NO	ITEM NO
80205	MS20426AD6-12	5320-00-117-7291	213	4
80205	MS20426AD6-4	5320-00-526-2945	213	30
80205	MS20426AD6-7	5320-00-117-7287	224	16
80205	MS20426AD6-9	5320-00-117-7289	288	4
80205	MS20470AD6-10	5320-00-721-9062	197	22
		5320-00-721-9062	199	46
		5320-00-721-9062	212	17
		5320-00-721-9062	288	9
		5320-00-721-9062	289	20
80205	MS20470AD6-12	5320-00-117-6856	197	15
80205	MS20470AD6-7	5320-00-754-0992	198	5
		5320-00-754-0992	238	6
80205	MS20470AD6-8	5320-00-117-6853	197	15
		5320-00-117-6853	197	36
		5320-00-117-6853	199	52
		5320-00-117-6853	289	13
80205	MS20470B4-4	5320-00-119-6823	498	3
96906	MS20604AD6W4	5320-00-957-2507	282	12
96906	MS20604B4W1	5320-00-957-2500	357	2
96906	MS20659-103	5940-00-113-3143	90	5
		5940-00-113-3143	93	12
		5940-00-113-3143	95	5
		5940-00-113-3143	96	12
96906	MS20659-104	5940-00-107-1481	75	5
		5940-00-107-1481	76	6
		5940-00-107-1481	92	10
		5940-00-107-1481	96	6
96906	MS20659-105	5940-00-114-1300	95	9
		5940-00-114-1300	96	10
96906	MS20659-106	5940-00-113-9819	95	10
		5940-00-113-9819	96	7
96906	MS20659-108	5940-00-115-2674	41	2
		5940-00-115-2674	95	12
		5940-00-115-2674	96	13
96906	MS20659-109	5940-00-114-1317	91	7
96906	MS20659-110	5940-00-115-4992	91	2
		5940-00-115-4992	94	6
96906	MS20659-111	5940-00-115-2678	41	7
		5940-00-115-2678	474	16
96906	MS20659-118	5940-00-115-2684	480	35
96906	MS20659-124	5940-00-115-5023	480	13
96906	MS20659-129	5940-00-114-1314	41	5
		5940-00-114-1314	95	16
		5940-00-114-1314	96	16
96906	MS20659-135	5940-00-115-5001	480	21
96906	MS20659-141	5940-00-113-9825	75	4
		5940-00-113-9825	95	20
		5940-00-113-9825	96	15
96906	MS20659-142	5940-00-114-1315	76	12

SECTION IV TM9-2320-280-24P C01

CROSS REFERENCE INDEXES

PART NUMBER INDEX

CAGE	PART NUMBER	STOCK NUMBER	FIGURE NO	ITEM NO
96906	MS20659-143	5940-00-115-2676	91	5
		5940-00-115-2676 94		5
96906	MS20659-151	5940-00-115-2683	480	32
96906	MS20659-152	5940-00-115-2685	69	39
96906	MS20659-157	5940-00-115-5012	480	15
96906	MS20659-163	5940-00-113-3145	95	17
		5940-00-113-3145 96		9
96906	MS20659-165	5940-00-113-0954	93	13
96906	MS20659-166	5940-00-113-9821	367	7
96906	MS20668-4	1640-00-132-9181	275	9
		1640-00-132-9181 276		7
96906	MS20913-6S	4730-00-221-2140	363	12
80205	MS21042-08	5310-00-807-1466	273	14
		5310-00-807-1466 275	13	
		5310-00-807-1466 276		9
80205	MS21042L4	5310-00-807-1475	279	14
80205	MS21042L5	5310-00-807-1476	279	18
80205	MS21042L6	5310-00-807-1477	279	26
96906	MS21044-C3	5310-00-208-9255	480	10
		5310-00-208-9255 482	11	
96906	MS21044-N08	5310-00-811-3494	252	4
		5310-00-811-3494 379		3
96906	MS21044-N4	5310-00-877-5796	328	13
96906	MS21044-N5	5310-00-088-0553	392	4
80205	MS21044N3	5310-00-877-5797	368	36
		5310-00-877-5797 379	49	
		5310-00-877-5797 388		3
		5310-00-877-5797 393	15	
		5310-00-877-5797 396	11	
96906	MS21045-04	5310-00-889-2543	55	10
96906	MS21045-3	5310-00-061-7326	365	22
96906	MS21045-6	5310-00-982-4908	128	55
		5310-00-982-4908 129	22	
96906	MS21045-7	5310-00-274-9364	238	3
96906	MS21083N08	5310-00-941-6019	250	14
		5310-00-941-6019	378	10
		5310-00-941-6019	378	13
		5310-00-941-6019 389		1
96906	MS21245-8	5310-00-449-2376	167	26
		5310-00-449-2376 191	16	
96906	MS21245-L10	5310-00-449-2381	35	7
96906	MS21245-L12	5310-00-419-0876	36	7
		5310-00-419-0876	38	7
		5310-00-419-0876	39	10
		5310-00-419-0876 476		9
96906	MS21266-2N	5325-00-074-3301	BULK	6
80205	MS21318-20	5305-00-253-5614	197	35
80205	MS21318-47	5305-00-253-5626	191	18
80205	MS21318-48	5305-00-253-5627	357	8

SECTION IV TM9-2320-280-24P C01

CROSS REFERENCE INDEXES

PART NUMBER INDEX

CAGE	PART NUMBER	STOCK NUMBER	FIGURE NO	ITEM NO		
96906	MS21333-101	5340-00-088-6655			88	40
		5340-00-088-6655	93	27		
		5340-00-088-6655	410	27		
96906	MS21333-102	5340-00-984-8540			19	13
		5340-00-984-8540	88	11		
		5340-00-984-8540	89	10		
		5340-00-984-8540	96	26		
		5340-00-984-8540	338	15		
		5340-00-984-8540	339	24		
		5340-00-984-8540	414	23		
96906	MS21333-103	5340-00-854-6729			93	17
		5340-00-854-6729	96	23		
		5340-00-854-6729	412	21		
		5340-00-854-6729	425	17		
		5340-00-854-6729	426	23		
		5340-00-854-6729	427	23		
96906	MS21333-104	5340-00-088-1254			75	33
		5340-00-088-1254	84	17		
		5340-00-088-1254	88	15		
		5340-00-088-1254	155	1		
		5340-00-088-1254	416	26		
		5340-00-088-1254	176A	18		
96906	MS21333-105	5340-00-809-1494			44	17
		5340-00-809-1494	64	27		
		5340-00-809-1494	75	29		
		5340-00-809-1494	84	11		
		5340-00-809-1494	88	19		
		5340-00-809-1494	338	29		
		5340-00-809-1494	339	25		
		5340-00-809-1494	480	22		
96906	MS21333-106	5340-00-984-8541			44	11
96906	MS21333-107	5340-00-809-1500			75	50
		5340-00-809-1500	77	22		
96906	MS21333-109	5340-00-067-3868			135	2
		5340-00-067-3868	141	22		
96906	MS21333-112	5340-00-057-3043			160	15
		5340-00-057-3043			162	11
		5340-00-057-3043			424	2
		5340-00-057-3043	426	26		
96906	MS21333-113	5340-00-988-1162			80	14
96906	MS21333-114	5340-00-057-3052			69	41
		5340-00-057-3052	282	6		
80205	MS21333-116	5340-00-764-7052			77	11
96906	MS21333-119	5340-00-050-9077			123	3
		5340-00-050-9077			135	6
		5340-00-050-9077			149	8
		5340-00-050-9077	414	12		
96906	MS21333-120	5340-00-964-5267			75	45
		5340-00-964-5267	77	19		

I-147

SECTION IV TM9-2320-280-24P C01

CROSS REFERENCE INDEXES

PART NUMBER INDEX

CAGE	PART NUMBER	STOCK NUMBER	FIGURE NO	ITEM NO
		5340-00-964-5267	487D	4
96906	MS21333-121	5340-00-954-6014	18	11
		5340-00-954-6014	89	8
		5340-00-954-6014	124	16
		5340-00-954-6014	125	16
		5340-00-954-6014	159	12
		5340-00-954-6014	160	3
		5340-00-954-6014	161	12
96906	MS21333-122	5340-00-833-8476	66	5
		5340-00-833-8476	79	21
		5340-00-833-8476	410	18
		5340-00-833-8476	412	15
96906	MS21333-123	5340-00-989-1771	77	3
		5340-00-989-1771	141	12
		5340-00-989-1771	338	13
		5340-00-989-1771	484	2
		5340-00-989-1771	487B	5
		5340-00-989-1771	487D	2
96906	MS21333-124	5340-00-854-6730	416	16
		5340-00-854-6730	428	16
96906	MS21333-125	5340-00-725-5280	75	38
		5340-00-725-5280	77	20
		5340-00-725-5280	177	12
		5340-00-725-5280	178	25
96906	MS21333-126	5340-00-053-8994	75	39
		5340-00-053-8994	472	4
		5340-00-053-8994	474	2
		5340-00-053-8994	480	24
96906	MS21333-128	5340-00-702-2848	41	14
		5340-00-702-2848	473	4
		5340-00-702-2848	474	2
		5340-00-702-2848	475	30
80205	MS21333-129	5340-00-151-9651	12	18
		5340-00-151-9651	30	6
		5340-00-151-9651	41	14
96906	MS21333-19	5340-00-988-3186	365	21
96906	MS21333-36	5340-00-286-2494	327	46
96906	MS21333-41	5340-00-282-7537	246	22
		5340-00-282-7537	366	28
		5340-00-282-7537	386	10
96906	MS21333-42	5340-00-282-7517	327	25
96906	MS21333-45	5340-00-881-5303	11	7
96906	MS21333-62	5340-00-282-7509	84	15
96906	MS21333-65	5340-00-905-0790	485	6
96906	MS21333-66	5340-00-990-7610	332	24
96906	MS21333-67	5340-00-079-7837	21	16
		5340-00-079-7837	398	11
		5340-00-079-7837	400	9
		5340-00-079-7837	425	18

SECTION IV TM9-2320-280-24P C01

CROSS REFERENCE INDEXES

PART NUMBER INDEX

CAGE	PART NUMBER	STOCK NUMBER	FIGURE NO	ITEM NO
96906	MS21333-68	5340-00-843-7825		65	20
		5340-00-843-7825	88	30
		5340-00-843-7825	89	14
		5340-00-843-7825	95	28
		5340-00-843-7825	96	20
		5340-00-843-7825	327	28
96906	MS21333-69	5340-00-764-7051		83	19
		5340-00-764-7051	85	28
		5340-00-764-7051	88	24
		5340-00-764-7051	92	21
		5340-00-764-7051	95	46
		5340-00-764-7051	335	4
		5340-00-764-7051	396	12
		5340-00-764-7051	468	11
96906	MS21333-70	5340-00-689-9514		84	33
		5340-00-689-9514	85	4
		5340-00-689-9514	88	2
		5340-00-689-9514	91	14
		5340-00-689-9514	92	20
		5340-00-689-9514	95	24
		5340-00-689-9514	488	7
96906	MS21333-71	5340-00-057-2904		75	25
		5340-00-057-2904	77	6
		5340-00-057-2904	79	22
		5340-00-057-2904	83	22
		5340-00-057-2904	85	29
		5340-00-057-2904	88	23
		5340-00-057-2904	92	15
		5340-00-057-2904	93	19
		5340-00-057-2904	95	26
		5340-00-057-2904	96	24
		5340-00-057-2904	282	3
		5340-00-057-2904	338	8
		5340-00-057-2904	339	2
		5340-00-057-2904	366	7
		5340-00-057-2904	410	24
		5340-00-057-2904	412	18
		5340-00-057-2904	414	14
96906	MS21333-72	5340-00-091-3790		55	16
		5340-00-091-3790	75	46
		5340-00-091-3790	79	17
		5340-00-091-3790	83	16
		5340-00-091-3790	85	26
		5340-00-091-3790	88	6
		5340-00-091-3790	93	26
		5340-00-091-3790	96	25
		5340-00-091-3790	366	5
		5340-00-091-3790	424	4
		5340-00-091-3790	426	25

I-149

SECTION IV TM9-2320-280-24P C01

CROSS REFERENCE INDEXES

PART NUMBER INDEX

CAGE	PART NUMBER	STOCK NUMBER	FIGURE NO	ITEM NO
		5340-00-091-3790	427	25
		5340-00-091-3790	445	9
		5340-00-091-3790 488	13	
96906	MS21333-73	5340-00-057-2906	88	32
		5340-00-057-2906	92	12
		5340-00-057-2906	95	35
		5340-00-057-2906	410	19
		5340-00-057-2906	414	17
		5340-00-057-2906	416	18
		5340-00-057-2906	428	17
		5340-00-057-2906 453	4	
96906	MS21333-74	5340-00-989-9222	83	9
		5340-00-989-9222 84	9	
		5340-00-989-9222	88	20
		5340-00-989-9222	95	33
		5340-00-989-9222 282	5	
		5340-00-989-9222 385	2	
		5340-00-989-9222	414	15
		5340-00-989-9222 445	5	
96906	MS21333-75	5340-00-050-2740	59	9
		5340-00-050-2740	83	21
		5340-00-050-2740	84	12
		5340-00-050-2740	88	18
		5340-00-050-2740	91	11
		5340-00-050-2740 94	9	
		5340-00-050-2740	96	27
		5340-00-050-2740 282	5	
		5340-00-050-2740	338	16
		5340-00-050-2740	339	12
		5340-00-050-2740	410	22
		5340-00-050-2740	412	16
		5340-00-050-2740	414	20
		5340-00-050-2740	416	23
		5340-00-050-2740	416	29
		5340-00-050-2740 424	3	
		5340-00-050-2740	428	22
96906	MS21333-76	5340-00-724-7038	59	9
		5340-00-724-7038	83	18
		5340-00-724-7038 84	3	
		5340-00-724-7038	88	21
		5340-00-724-7038 91	9	
		5340-00-724-7038	95	44
		5340-00-724-7038	410	21
		5340-00-724-7038	414	18
		5340-00-724-7038	416	21
		5340-00-724-7038	428	20
96906	MS21333-77	5340-00-922-6300	58	11
		5340-00-922-6300	59	11
		5340-00-922-6300	84	2

I-150

SECTION IV TM9-2320-280-24P C01

CROSS REFERENCE INDEXES

PART NUMBER INDEX

CAGE	PART NUMBER	STOCK NUMBER	FIGURE NO	ITEM NO		
		5340-00-922-6300	88	22		
		5340-00-922-6300	93	22		
		5340-00-922-6300	95	31		
		5340-00-922-6300	96	28		
		5340-00-922-6300	306	9		
		5340-00-922-6300	310	4		
		5340-00-922-6300	338	28		
		5340-00-922-6300	379	2		
		5340-00-922-6300	410	20		
		5340-00-922-6300	414	19		
		5340-00-922-6300	416	19		
		5340-00-922-6300	428	18		
96906	MS21333-78	5340-00-958-8457			282	17
		5340-00-958-8457	303	3		
		5340-00-958-8457	338	4		
		5340-00-958-8457	416	22		
		5340-00-958-8457	428	21		
96906	MS21333-86	5340-00-193-4111			370	18
96906	MS21333-96	5340-00-088-1255			485	7
96906	MS21333-98	5340-00-809-1490			75	23
		5340-00-809-1490	88	37		
		5340-00-809-1490	400	11		
96906	MS21333-99	5340-00-993-6207			77	14
		5340-00-993-6207	89	15		
		5340-00-993-6207	334	15		
		5340-00-993-6207	366	20		
		5340-00-993-6207	368	42		
		5340-00-993-6207	379	46		
		5340-00-993-6207	386	6		
96906	MS21334-25	5340-01-049-9564			13	23
		5340-01-049-9564	14	10		
96906	MS21334-28	5340-00-926-5449			313	20
96906	MS21334-31	5340-00-929-1794			477	7
96906	MS21334-34	5340-00-731-9611			480	7
96906	MS24521-7	4730-00-100-3918			379	44
96906	MS24523-21	5930-00-681-4727			396	10
96906	MS24539	6620-00-993-5546			66	10
80205	MS24625-42	5305-00-045-7603			224	8
		5305-00-045-7603	268	26		
80205	MS24629-46	5305-00-855-0957			225	9
		5305-00-855-0957	277	40		
		5305-00-855-0957	294	2		
80205	MS24629-57	5305-00-052-6921			28	2
80205	MS24629-73	5305-00-204-4850			327	15
80205	MS24630-13	5305-00-052-8881			469	4
80205	MS24630-66	5305-01-109-0587			323	6
96906	MS24659-23G	5930-00-856-1654			482	2
80205	MS24665-132	5315-00-839-2325			238	18
		5315-00-839-2325	250	10		

I-151

SECTION IV TM9-2320-280-24P C01

CROSS REFERENCE INDEXES

PART NUMBER INDEX

CAGE	PART NUMBER	STOCK NUMBER	FIGURE NO	ITEM NO
80205	MS24665-134	5315-00-839-5820		202	18
		5315-00-839-5820	243	9
		5315-00-839-5820	293	10
80205	MS24665-208	5315-00-829-1480		368	8
80205	MS24665-281	5315-00-839-2326		274	5
		5315-00-839-2326	276	31
80205	MS24665-283	5315-00-842-3044		155	19
		5315-00-842-3044		243	23
		5315-00-842-3044		244	9
		5315-00-842-3044	390	14
80205	MS24665-298	5315-00-234-1861		268	9
80205	MS24665-319	5315-01-267-7570		403	8
80205	MS24665-351	5315-00-839-5821		134	1
80205	MS24665-353	5315-00-839-5822		443	17
80205	MS24665-355	5315-00-012-0123		152	13
		5315-00-012-0123	188	13
		5315-00-012-0123	195	12
96906	MS24665-355	5315-00-012-0123		183	4
		5315-00-012-0123	185	15
		5315-00-012-0123	187	19
80205	MS24665-357	5315-00-298-1481		437	11
80205	MS24665-377	5315-00-285-7161		365	8
		5315-00-285-7161	368	12
80205	MS24665-423	5315-00-013-7228		156	14
80205	MS24665-513	5315-00-239-8032		186	8
		5315-00-239-8032	188	19
		5315-00-239-8032	188	35
80205	MS24665-628	5315-00-846-0126		191	10
		5315-00-846-0126	442	13
		5315-00-846-0126	453	15
96906	MS24665-640	5315-00-243-1169		191	21
96906	MS24667-52Z	5305-01-393-7801		210	11
96906	MS24667-77	5305-00-052-2232		199	18
		5305-00-052-2232		251	19
		5305-00-052-2232		270	11
		5305-00-052-2232	296	1
96906	MS24693-S350	5305-00-901-7250		390	7
96906	MS24694-S98	5305-00-043-2682		279	23
81343	MS25036-112	5940-00-143-4794		72	20
		5940-00-143-4794	73	6
96906	MS25036-133	5940-00-115-5006		44	16
		5940-00-115-5006	69	14
		5940-00-115-5006	69	20
		5940-00-115-5006	69	25
		5940-00-115-5006	72	4
		5940-00-115-5006	72	19
		5940-00-115-5006	73	18
		5940-00-115-5006	73	22
		5940-00-115-5006	474	16

SECTION IV TM9-2320-280-24P C01

CROSS REFERENCE INDEXES

PART NUMBER INDEX

CAGE PART NUMBER STOCK NUMBER FIGURE NO ITEM NO

CAGE	PART NUMBER	STOCK NUMBER	FIGURE NO	ITEM NO
		5940-00-115-5006	480	17
96906	MS25043-32DA	5935-01-223-9420	74	12
96906	MS25171-1S	5975-00-553-6995	380	13
96906	MS25231-1873	6240-00-419-3185	49	2
96906	MS25306-222	5930-00-728-4328	55	19
		5930-00-728-4328	56	16
96906	MS25307-312	5930-00-978-8805	55	17
		5930-00-978-8805	56	12
96906	MS27040-13	5310-00-982-4939	313	23
96906	MS27130-20	5310-01-009-9785	337	13
96906	MS27130-29	5310-00-074-2351	292	1
96906	MS27130-31	5310-01-093-3493	492	17
96906	MS27130-A26	5310-00-141-3062	256	8
		5310-00-141-3062	257	8
		5310-00-141-3062	259	11
		5310-00-141-3062	262	7
96906	MS27130-A27	5310-00-003-6779	260	13
96906	MS27130-A28	5310-01-395-3549	213	22
		5310-01-395-3549	262	16
96906	MS27130-A31	5310-01-088-7962	254	14
		5310-01-088-7962	492	17
96906	MS27130-A37	5310-00-126-5168	228	11
96906	MS27130-A43	5310-01-046-5382	340	9
96906	MS27130-A50	5310-01-101-6046	340	8
96906	MS27130-CR27	5310-01-046-5371	487	6
96906	MS27130-CR31	5310-01-283-8482	383	7
96906	MS27130-S25	5310-00-990-1361	487	5
		5310-00-990-1361	487	6
96906	MS27130-S26	5310-00-269-3466	225	13
96906	MS27130-S32	5310-00-006-8291	225	44
96906	MS27130-S37	5310-00-460-4490	218	6
		5310-00-460-4490	257	5
96906	MS27130-S43	5310-00-080-2411	432	5
		5310-00-080-2411	487	13
96906	MS27130-S44	5310-01-080-5746	487	13
96906	MS27130-S71	5310-01-085-1208	377	11
96906	MS27130A32	5310-00-401-5321	237	13
		5310-00-401-5321	258	6
		5310-00-401-5321	262	4
96906	MS27142-2	5935-00-462-6603	65	4
		5935-00-462-6603	65	27
		5935-00-462-6603	75	11
		5935-00-462-6603	76	14
		5935-00-462-6603	80	2
		5935-00-462-6603	81	7
		5935-00-462-6603	83	20
		5935-00-462-6603	84	32
		5935-00-462-6603	85	15
		5935-00-462-6603	86	8

SECTION IV TM9-2320-280-24P C01

CROSS REFERENCE INDEXES

PART NUMBER INDEX

CAGE	PART NUMBER	STOCK NUMBER	FIGURE NO	ITEM NO
		5935-00-462-6603	87	2
		5935-00-462-6603	87	4
		5935-00-462-6603	89	5
		5935-00-462-6603	90	6
		5935-00-462-6603	92	4
		5935-00-462-6603	93	9
		5935-00-462-6603	445	7
96906	MS27142-3	5935-00-115-2306	65	17
		5935-00-115-2306	81	13
96906	MS27144-1	5935-00-167-7775	65	5
		5935-00-167-7775	65	26
		5935-00-167-7775	75	10
		5935-00-167-7775	76	4
		5935-00-167-7775	80	5
		5935-00-167-7775	81	3
		5935-00-167-7775	81	15
		5935-00-167-7775	83	3
		5935-00-167-7775	84	31
		5935-00-167-7775	85	1
		5935-00-167-7775	86	1
		5935-00-167-7775	87	1
		5935-00-167-7775	87	5
		5935-00-167-7775	89	2
		5935-00-167-7775	90	2
		5935-00-167-7775	92	6
		5935-00-167-7775	93	8
		5935-00-167-7775	95	2
		5935-00-167-7775	96	5
		5935-00-167-7775	367	8
		5935-00-167-7775	445	8
96906	MS27144-2	5935-00-115-2307	65	13
		5935-00-115-2307	81	9
		5935-00-115-2307	483	5
96906	MS27144-3	5935-00-184-6707	75	9
		5935-00-184-6707	76	8
		5935-00-184-6707	81	5
		5935-00-184-6707	85	20
		5935-00-184-6707	86	13
96906	MS27147-1	5935-00-900-6281	366	32
		5935-00-900-6281	386	11
		5935-00-900-6281	469	11
		5935-00-900-6281	483	3
96906	MS27183-10	5310-00-809-4058	12	1
		5310-00-809-4058	16	22
		5310-00-809-4058	21	18
		5310-00-809-4058	50	14
		5310-00-809-4058	51	2
		5310-00-809-4058	62	28
		5310-00-809-4058	64	14

SECTION IV TM9-2320-280-24P C01

CROSS REFERENCE INDEXES

PART NUMBER INDEX

CAGE PART NUMBER STOCK NUMBER FIGURE NO ITEM NO

CAGE	PART NUMBER	STOCK NUMBER	FIGURE NO	ITEM NO
		5310-00-809-4058	64	21
		5310-00-809-4058	69	4
		5310-00-809-4058	70	11
		5310-00-809-4058	84	19
		5310-00-809-4058	84	21
		5310-00-809-4058	141	15
		5310-00-809-4058	155	5
		5310-00-809-4058	169	11
		5310-00-809-4058	177	11
		5310-00-809-4058	178	6
		5310-00-809-4058	197	29
		5310-00-809-4058	198	8
		5310-00-809-4058	198	14
		5310-00-809-4058	203	12
		5310-00-809-4058	204	12
		5310-00-809-4058	206	25
		5310-00-809-4058	207	7
		5310-00-809-4058	211	27
		5310-00-809-4058	212	26
		5310-00-809-4058	214	5
		5310-00-809-4058	225	42
		5310-00-809-4058	227	30
		5310-00-809-4058	238	19
		5310-00-809-4058	241	13
		5310-00-809-4058	243	7
		5310-00-809-4058	246	23
		5310-00-809-4058	248	26
		5310-00-809-4058	249	1
		5310-00-809-4058	250	13
		5310-00-809-4058	250	18
		5310-00-809-4058	267	2
		5310-00-809-4058	267	4
		5310-00-809-4058	269	31
		5310-00-809-4058	272	3
		5310-00-809-4058	280	49
		5310-00-809-4058	281	49
		5310-00-809-4058	287	24
		5310-00-809-4058	288	21
		5310-00-809-4058	288	29
		5310-00-809-4058	292	10
		5310-00-809-4058	293	30
		5310-00-809-4058	294	5
		5310-00-809-4058	304	8
		5310-00-809-4058	312	2
		5310-00-809-4058	321	12
		5310-00-809-4058	323	8
		5310-00-809-4058	325	14
		5310-00-809-4058	326	11
		5310-00-809-4058	327	44

SECTION IV TM9-2320-280-24P C01

CROSS REFERENCE INDEXES

PART NUMBER INDEX

CAGE PART NUMBER STOCK NUMBER FIGURE NO ITEM NO

CAGE	PART NUMBER	STOCK NUMBER	FIGURE NO	ITEM NO
		5310-00-809-4058	328	14
		5310-00-809-4058	331	26
		5310-00-809-4058	343	16
		5310-00-809-4058	365	16
		5310-00-809-4058	365	25
		5310-00-809-4058	366	16
		5310-00-809-4058	368	22
		5310-00-809-4058	375	29
		5310-00-809-4058	379	14
		5310-00-809-4058	382	5
		5310-00-809-4058	384	8
		5310-00-809-4058	388	9
		5310-00-809-4058	389	9
		5310-00-809-4058	397	9
		5310-00-809-4058	429	12
		5310-00-809-4058	430	15
		5310-00-809-4058	475	8
		5310-00-809-4058	492	8
		5310-00-809-4058	497	4
96906	MS27183-11	5310-00-809-3078	4	31
		5310-00-809-3078	71	11
		5310-00-809-3078	180	20
		5310-00-809-3078	203	2
		5310-00-809-3078	206	3
		5310-00-809-3078	233	2
		5310-00-809-3078	243	16
		5310-00-809-3078	247	6
		5310-00-809-3078	248	19
		5310-00-809-3078	266	15
		5310-00-809-3078	375	21
		5310-00-809-3078	430	35
		5310-00-809-3078	446	13
96906	MS27183-12	5310-00-081-4219	5	4
		5310-00-081-4219	74	2
		5310-00-081-4219	142	28
		5310-00-081-4219	143	37
		5310-00-081-4219	150	28
		5310-00-081-4219	151	37
		5310-00-081-4219	154	7
		5310-00-081-4219	160	14
		5310-00-081-4219	162	13
		5310-00-081-4219	178	14
		5310-00-081-4219	180	11
		5310-00-081-4219	189	4
		5310-00-081-4219	190	2
		5310-00-081-4219	198	20
		5310-00-081-4219	199	2
		5310-00-081-4219	204	7
		5310-00-081-4219	209	3

I-156

SECTION IV TM9-2320-280-24P C01

CROSS REFERENCE INDEXES

PART NUMBER INDEX

CAGE	PART NUMBER	STOCK NUMBER	FIGURE NO	ITEM NO
		5310-00-081-4219	211	14
		5310-00-081-4219	212	12
		5310-00-081-4219	225	30
		5310-00-081-4219	226	15
		5310-00-081-4219	230	4
		5310-00-081-4219	235	7
		5310-00-081-4219	236	7
		5310-00-081-4219	243	24
		5310-00-081-4219	244	10
		5310-00-081-4219	273	8
		5310-00-081-4219	273	17
		5310-00-081-4219	277	33
		5310-00-081-4219	280	46
		5310-00-081-4219	281	46
		5310-00-081-4219	288	19
		5310-00-081-4219	321	19
		5310-00-081-4219	321	31
		5310-00-081-4219	324	8
		5310-00-081-4219	325	27
		5310-00-081-4219	329	6
		5310-00-081-4219	333	23
		5310-00-081-4219	363	15
		5310-00-081-4219	376	15
		5310-00-081-4219	377	14
		5310-00-081-4219	378	29
		5310-00-081-4219	402	7
		5310-00-081-4219	429	6
		5310-00-081-4219	432	4
		5310-00-081-4219	433	8
		5310-00-081-4219	436	9
		5310-00-081-4219	473	2
		5310-00-081-4219	474	6
		5310-00-081-4219	480	5
		5310-00-081-4219	487B	4
		5310-00-081-4219	487D	9
96906	MS27183-13	5310-00-087-7493	37	4
		5310-00-087-7493	52	25
		5310-00-087-7493	60	2
		5310-00-087-7493	61	2
		5310-00-087-7493	103	16
		5310-00-087-7493	153	16
		5310-00-087-7493	159	3
		5310-00-087-7493	160	10
		5310-00-087-7493	205	2
		5310-00-087-7493	206	1
		5310-00-087-7493	206	9
		5310-00-087-7493	206	29
		5310-00-087-7493	238	22
		5310-00-087-7493	240	7

SECTION IV TM9-2320-280-24P C01

CROSS REFERENCE INDEXES

PART NUMBER INDEX

CAGE	PART NUMBER	STOCK NUMBER	FIGURE NO	ITEM NO
		5310-00-087-7493	268	18
		5310-00-087-7493	270	14
		5310-00-087-7493	285	5
		5310-00-087-7493	286	10
		5310-00-087-7493	288	7
		5310-00-087-7493	292	17
		5310-00-087-7493	293	5
		5310-00-087-7493	398	23
		5310-00-087-7493	429	17
		5310-00-087-7493	436	8
		5310-00-087-7493	490	2
96906	MS27183-14	5310-00-080-6004	2	3
		5310-00-080-6004	23	2
		5310-00-080-6004	24	7
		5310-00-080-6004	36	17
		5310-00-080-6004	66	6
		5310-00-080-6004	79	20
		5310-00-080-6004	128	54
		5310-00-080-6004	129	23
		5310-00-080-6004	134	2
		5310-00-080-6004	142	14
		5310-00-080-6004	143	14
		5310-00-080-6004	150	14
		5310-00-080-6004	151	14
		5310-00-080-6004	163	6
		5310-00-080-6004	164	2
		5310-00-080-6004	175	9
		5310-00-080-6004	187	9
		5310-00-080-6004	189	1
		5310-00-080-6004	190	4
		5310-00-080-6004	245	22
		5310-00-080-6004	266	25
		5310-00-080-6004	272	25
		5310-00-080-6004	273	2
		5310-00-080-6004	277	2
		5310-00-080-6004	280	25
		5310-00-080-6004	281	25
		5310-00-080-6004	328	18
		5310-00-080-6004	329	13
		5310-00-080-6004	332	34
		5310-00-080-6004	392	9
		5310-00-080-6004	431	3
		5310-00-080-6004	438	8
		5310-00-080-6004	440	6
		5310-00-080-6004	441	8
		5310-00-080-6004	461	9
		5310-00-080-6004	471	10
96906	MS27183-15	5310-00-809-4061	1	3
		5310-00-809-4061	8	1

I-158

SECTION IV TM9-2320-280-24P C01

CROSS REFERENCE INDEXES

PART NUMBER INDEX

CAGE	PART NUMBER	STOCK NUMBER	FIGURE NO		ITEM NO
		5310-00-809-4061	37	11	
		5310-00-809-4061	170	22	
96906	MS27183-16	5310-00-809-4085		36	2
		5310-00-809-4085	38	2	
		5310-00-809-4085	136	4	
		5310-00-809-4085	144	15	
		5310-00-809-4085	169	15	
		5310-00-809-4085	171	2	
		5310-00-809-4085	202	33	
		5310-00-809-4085	234	9	
		5310-00-809-4085	239	6	
		5310-00-809-4085	271	10	
		5310-00-809-4085	271	25	
		5310-00-809-4085	313	22	
		5310-00-809-4085	398	4	
		5310-00-809-4085	402	9	
		5310-00-809-4085	438	19	
		5310-00-809-4085	475	11	
96906	MS27183-17	5310-00-809-5997		209	13
		5310-00-809-5997	292	19	
		5310-00-809-5997	429	37	
96906	MS27183-18	5310-00-809-5998		142	17
		5310-00-809-5998	143	17	
		5310-00-809-5998	150	17	
		5310-00-809-5998	151	17	
		5310-00-809-5998	165	9	
		5310-00-809-5998	166	6	
		5310-00-809-5998	293	9	
		5310-00-809-5998	295	7	
		5310-00-809-5998	295	13	
		5310-00-809-5998	363	17	
		5310-00-809-5998	390	3	
		5310-00-809-5998	397	15	
		5310-00-809-5998	455	3	
96906	MS27183-19	5310-00-809-3079		455	5
96906	MS27183-20	5310-00-068-5285		127	12
		5310-00-068-5285	165	5	
		5310-00-068-5285	166	4	
		5310-00-068-5285	179	9	
		5310-00-068-5285	180	10	
96906	MS27183-21	5310-00-823-8803		193	5
		5310-00-823-8803	199	41	
		5310-00-823-8803	437	4	
96906	MS27183-23	5310-00-809-8533		178	22
		5310-00-809-8533	385	10	
96906	MS27183-4	5310-00-950-1310		274	21
96906	MS27183-42	5310-00-014-5850		63	6
		5310-00-014-5850		93	25
		5310-00-014-5850		95	39

I-159

SECTION IV TM9-2320-280-24P C01

CROSS REFERENCE INDEXES

PART NUMBER INDEX

CAGE	PART NUMBER	STOCK NUMBER	FIGURE NO	ITEM NO
		5310-00-014-5850	96	29
		5310-00-014-5850	169	20
		5310-00-014-5850	199	33
		5310-00-014-5850	199	57
		5310-00-014-5850	202	42
		5310-00-014-5850	203	31
		5310-00-014-5850	205	39
		5310-00-014-5850	206	15
		5310-00-014-5850	213	15
		5310-00-014-5850	214	11
		5310-00-014-5850	222	11
		5310-00-014-5850	227	49
		5310-00-014-5850	232	2
		5310-00-014-5850	241	8
		5310-00-014-5850	243	2
		5310-00-014-5850	244	18
		5310-00-014-5850	246	8
		5310-00-014-5850	246	13
		5310-00-014-5850	247	12
		5310-00-014-5850	248	4
		5310-00-014-5850	249	41
		5310-00-014-5850	268	25
		5310-00-014-5850	272	15
		5310-00-014-5850	293	12
		5310-00-014-5850	299	2
		5310-00-014-5850	303	10
		5310-00-014-5850	331	22
		5310-00-014-5850	333	15
		5310-00-014-5850	337	17
		5310-00-014-5850	339	27
		5310-00-014-5850	347	4
		5310-00-014-5850	368	26
		5310-00-014-5850	374	14
		5310-00-014-5850	411	7
		5310-00-014-5850	429	4
		5310-00-014-5850	430	8
		5310-00-014-5850	430	40
		5310-00-014-5850	446	10
		5310-00-014-5850	447	8
		5310-00-014-5850	448	7
		5310-00-014-5850	449	8
		5310-00-014-5850	450	2
		5310-00-014-5850	480	9
96906	MS27183-57	5310-01-280-5796	379	25
96906	MS27183-8	5310-00-809-8546	46	10
		5310-00-809-8546	446	8
		5310-00-809-8546	447	9
		5310-00-809-8546	449	3
96906	MS27183-9	5310-00-823-8804	5	6

SECTION IV TM9-2320-280-24P C01

CROSS REFERENCE INDEXES

PART NUMBER INDEX

CAGE PART NUMBER STOCK NUMBER FIGURE NO ITEM NO

CAGE	PART NUMBER	STOCK NUMBER	FIGURE NO	ITEM NO
		5310-00-823-8804	15	21
		5310-00-823-8804	18	3
		5310-00-823-8804	21	2
		5310-00-823-8804	21	28
		5310-00-823-8804	253	2
		5310-00-823-8804	494	4
		5310-00-823-8804	495	12
		5310-00-823-8804	496	7
		5310-00-823-8804	66A	4
96906	MS27407-4	5930-00-445-9274	55	6
		5930-00-445-9274	56	18
96906	MS28775-110	5331-00-585-6663	21	7
81343	MS28775-222	5331-00-297-9990	72	24
81343	MS28775-223	5331-00-171-6649	163	11
96906	MS3108R18-11S	5935-00-806-4183	93	11
96906	MS3217-2075	5325-01-388-2363	456	10
96906	MS3367-1-0	5975-00-984-6582	64	28
		5975-00-984-6582	81	11
		5975-00-984-6582	160	2
		5975-00-984-6582	469	2
		5975-00-984-6582	487B	2
81343	MS3367-1-9	5975-00-074-2072	345	7
96906	MS3367-1-9	5975-00-074-2072	12	9
		5975-00-074-2072	98	8
		5975-00-074-2072	485	8
96906	MS3367-2-0	5975-00-899-4606	282	10
96906	MS3367-2-9	5975-00-156-3253	488	5
96906	MS3367-3-0	5975-00-985-6630	75	21
		5975-00-985-6630	83	38
		5975-00-985-6630	85	31
		5975-00-985-6630	89	12
		5975-00-985-6630	94	10
		5975-00-985-6630	95	27
		5975-00-985-6630	96	32
		5975-00-985-6630	178	21
		5975-00-985-6630	335	15
		5975-00-985-6630	336	12
		5975-00-985-6630	338	18
		5975-00-985-6630	436	3
		5975-00-985-6630	457	15
		5975-00-985-6630	483	2
		5975-00-985-6630	484	1
96906	MS3367-3-9	5975-00-451-5001	5	8
		5975-00-451-5001	435	8
96906	MS3367-4-0	5975-00-903-2284	99	3
96906	MS3367-4-2	5975-00-903-2288	98	6
96906	MS3367-5-0	5975-00-133-8687	51	14
		5975-00-133-8687	386	3
		5975-00-133-8687	388	11

SECTION IV TM9-2320-280-24P C01

CROSS REFERENCE INDEXES

PART NUMBER INDEX

CAGE	PART NUMBER	STOCK NUMBER	FIGURE NO	ITEM NO
96906	MS3367-5-9	5975-00-111-3208	492	6
96906	MS3367-7	5975-01-205-5379	313	7
		5975-01-205-5379	333	32
		5975-01-205-5379	339	13
		5975-01-205-5379	410	28
		5975-01-205-5379	443	6
		5975-01-205-5379	485	9
96906	MS3367-7-0	5975-01-034-5871	98	7
		5975-01-034-5871	332	8
		5975-01-034-5871	369	2
		5975-01-034-5871	371	3
		5975-01-034-5871	412	22
		5975-01-034-5871	414	24
		5975-01-034-5871	416	27
		5975-01-034-5871	468	1
		5975-01-034-5871	470	4
		5975-01-034-5871	472	6
		5975-01-034-5871	473	7
		5975-01-034-5871	474	1
		5975-01-034-5871	475	3
		5975-01-034-5871	479	4
		5975-01-034-5871	480	1
		5975-01-034-5871	492	14
96906	MS3367-7-9	5975-00-570-9598	18	9
		5975-00-570-9598	41	15
		5975-00-570-9598	77	28
		5975-00-570-9598	379	7
		5975-00-570-9598	381	2
96906	MS3456W12-5S	5935-01-147-0148	95	21
		5935-01-147-0148	96	2
96906	MS3456W18-11S	5935-01-035-5139	367	9
96906	MS3456W24-11S	5935-01-173-7654	75	18
96906	MS35058-22	5930-00-655-1514	367	5
96906	MS35059-21	5930-00-615-9376	52	30
96906	MS35059-31	5930-00-615-7897	367	6
80205	MS35190-272	5305-00-088-8332	374	2
		5305-00-088-8332	429	10
		5305-00-088-8332	430	6
		5305-00-088-8332	430	20
		5305-00-088-8332	430	33
96906	MS35190-273	5305-00-958-5471	249	17
80205	MS35190-289	5305-00-958-5246	247	3
		5305-00-958-5246	271	15
		5305-00-958-5246	338	19
80205	MS35190-290	5305-00-954-3937	243	29
80205	MS35190-305	5305-00-958-5469	225	46
		5305-00-958-5469	226	31
80205	MS35190-306	5305-00-958-5253	225	46
		5305-00-958-5253	226	25

SECTION IV TM9-2320-280-24P C01

CROSS REFERENCE INDEXES

PART NUMBER INDEX

CAGE	PART NUMBER	STOCK NUMBER	FIGURE NO	ITEM NO
		5305-00-958-5253 226		31
80205	MS35190-307	5305-00-958-5254	226	25
		5305-00-958-5254 226		31
80205	MS35190-324	5305-00-958-5262	383	5
96906	MS35191-272	5305-00-984-7363	412	23
96906	MS35191-273	5305-00-984-7341	377	5
		5305-00-984-7341 447		7
		5305-00-984-7341 448		10
		5305-00-984-7341 449		7
		5305-00-984-7341 450		4
96906	MS35191-274	5305-00-984-7342	251	17
		5305-00-984-7342 251		42
		5305-00-984-7342 252		20
		5305-00-984-7342 446		19
		5305-00-984-7342 448		1
96906	MS35191-276	5305-00-984-7343	323	14
		5305-00-984-7343 429		15
		5305-00-984-7343 449		1
96906	MS35191-277	5305-00-957-2635	447	6
96906	MS35191-278	5305-00-984-7352	446	6
96906	MS35191-292	5305-00-988-9106	248	10
96906	MS35191-324	5305-00-959-2704	382	8
96906	MS35191-326	5305-00-824-2004	382	13
96906	MS35191-348	5305-01-315-9683	397	4
80205	MS35206-219	5305-00-984-4976	55	3
80205	MS35206-226	5305-00-984-4983	55	2
80205	MS35206-228	5305-00-984-4988	213	19
80205	MS35206-231	5305-00-889-3001	57	1
96906	MS35206-231	5305-00-889-3001	43	39
80205	MS35206-232	5305-00-984-4992	79	13
96906	MS35206-242	5305-00-889-3002	45	51
80205	MS35206-243	5305-00-984-6191	47	7
		5305-00-984-6191 354		11
80205	MS35206-244	5305-00-984-6192	72	21
		5305-00-984-6192 73		9
80205	MS35206-245	5305-00-984-6193	84	23
		5305-00-984-6193 280		36
		5305-00-984-6193 281		36
		5305-00-984-6193 300		12
		5305-00-984-6193 453		5
80205	MS35206-246	5305-00-984-6194	52	21
		5305-00-984-6194	59	24
		5305-00-984-6194	273	11
		5305-00-984-6194 379		1
96906	MS35206-247	5305-00-984-6195	48	4
		5305-00-984-6195 275		7
		5305-00-984-6195 276		5
		5305-00-984-6195 282		14
		5305-00-984-6195 337		6

SECTION IV TM9-2320-280-24P C01

CROSS REFERENCE INDEXES

PART NUMBER INDEX

CAGE	PART NUMBER	STOCK NUMBER	FIGURE NO	ITEM NO
		5305-00-984-6195	343	10
		5305-00-984-6195 378	9	
		5305-00-984-6195	378	25
		5305-00-984-6195 380	3	
80063	MS35206-248	5305-00-984-6196	281	10
80205	MS35206-248	5305-00-984-6196	280	10
96906	MS35206-249	5305-00-984-6197	50	9
96906	MS35206-251	5305-00-984-6199	213	33
		5305-00-984-6199	214	26
96906	MS35206-263	5305-00-984-6210	213	20
		5305-00-984-6210	214	17
96906	MS35206-264	5305-00-984-6211	63	7
96906	MS35206-265	5305-00-984-6212	35	13
		5305-00-984-6212	88	41
		5305-00-984-6212 202	3	
		5305-00-984-6212	284	13
		5305-00-984-6212	293	11
		5305-00-984-6212 439	6	
96906	MS35206-266	5305-00-984-6213	203	32
		5305-00-984-6213	222	12
		5305-00-984-6213 347	2	
96906	MS35206-267	5305-00-984-6214	272	19
		5305-00-984-6214	281	16
96906	MS35206-280	5305-00-988-1724	97	2
		5305-00-988-1724 237	7	
		5305-00-988-1724 330	3	
96906	MS35206-281	5305-00-988-1725	49	7
		5305-00-988-1725 51	1	
		5305-00-988-1725	59	25
		5305-00-988-1725	75	37
		5305-00-988-1725	83	29
		5305-00-988-1725	84	18
		5305-00-988-1725	88	36
		5305-00-988-1725 213	6	
		5305-00-988-1725 214	2	
		5305-00-988-1725 267	1	
		5305-00-988-1725	280	14
		5305-00-988-1725	281	14
		5305-00-988-1725 313	5	
		5305-00-988-1725	316	15
		5305-00-988-1725	320	14
		5305-00-988-1725	320	26
		5305-00-988-1725	321	28
		5305-00-988-1725	378	20
96906	MS35206-283	5305-00-988-1727	198	18
		5305-00-988-1727	213	8
		5305-00-988-1727	254	21
		5305-00-988-1727	376	17
96906	MS35206-284	5305-00-988-1170	267	17

I-164

SECTION IV TM9-2320-280-24P C01

CROSS REFERENCE INDEXES

PART NUMBER INDEX

CAGE	PART NUMBER	STOCK NUMBER	FIGURE NO	ITEM NO
96906	MS35206-285	5305-00-988-1171	237	12
96906	MS35206-295	5305-00-984-5675	277	34
96906	MS35206-298	5305-00-984-5678	233	1
96906	MS35206-329	5305-00-150-3408	56	21
96906	MS35206-345	5305-00-482-6825	254	29
		5305-00-482-6825	281	7
96906	MS35207-242	5305-00-958-4357	72	13
96906	MS35207-261	5305-00-990-6444	43	36
		5305-00-990-6444	45	30
		5305-00-990-6444	45	47
		5305-00-990-6444	52	27
		5305-00-990-6444	396	6
		5305-00-990-6444	485	26
96906	MS35207-262	5305-00-989-6265	39	32
		5305-00-989-6265	393	11
96906	MS35207-263	5305-00-989-7434	45	43
		5305-00-989-7434	52	29
		5305-00-989-7434	54	6
		5305-00-989-7434	55	20
		5305-00-989-7434	56	7
		5305-00-989-7434	59	23
		5305-00-989-7434	282	2
		5305-00-989-7434	317	5
		5305-00-989-7434	338	10
		5305-00-989-7434	396	3
		5305-00-989-7434	398	15
		5305-00-989-7434	416	28
96906	MS35207-264	5305-00-989-7435	52	28
		5305-00-989-7435	95	25
		5305-00-989-7435	207	10
		5305-00-989-7435	255	6
		5305-00-989-7435	258	5
		5305-00-989-7435	260	8
		5305-00-989-7435	337	19
		5305-00-989-7435	338	9
		5305-00-989-7435	379	53
		5305-00-989-7435	447	3
		5305-00-989-7435	449	6
96906	MS35207-265	5305-00-993-1848	52	3
		5305-00-993-1848	54	11
		5305-00-993-1848	55	22
		5305-00-993-1848	59	18
		5305-00-993-1848	79	24
		5305-00-993-1848	96	19
		5305-00-993-1848	246	2
		5305-00-993-1848	252	34
		5305-00-993-1848	255	33
		5305-00-993-1848	325	9
		5305-00-993-1848	446	20

SECTION IV TM9-2320-280-24P C01

CROSS REFERENCE INDEXES

PART NUMBER INDEX

CAGE	PART NUMBER	STOCK NUMBER	FIGURE NO	ITEM NO
		5305-00-993-1848	448	4
96906	MS35207-266	5305-00-995-3444	58	7
		5305-00-995-3444	63	7
		5305-00-995-3444	247	10
		5305-00-995-3444	256	2
		5305-00-995-3444	257	3
		5305-00-995-3444	260	14
		5305-00-995-3444	262	9
		5305-00-995-3444	263	9
		5305-00-995-3444	291	2
		5305-00-995-3444	297	6
80205	MS35207-267	5305-00-993-1851	65	23
96906	MS35207-268	5305-00-995-3442	446	3
96906	MS35207-270	5305-00-995-3440	46	18
96906	MS35207-271	5305-00-995-6311	368	27
96906	MS35207-280	5305-00-993-2738	397	2
96906	MS35207-282	5305-00-993-2460	391	1
		5305-00-993-2460	397	1
96906	MS35207-284	5305-00-993-2457	376	20
		5305-00-993-2457	382	4
		5305-00-993-2457	389	7
80205	MS35215-53	5305-00-206-2508	69	30
96906	MS35265-43	5305-00-614-3423	46	12
96906	MS35266-65	5305-00-616-4831	43B	35
96906	MS35291-41	5306-00-226-4834	211	16
		5306-00-226-4834	212	13
96906	MS35307-355	5305-00-478-0273	73	2
96906	MS35307-357	5305-00-655-9711	73	12
96906	MS35307-361	5305-00-721-5665	482	14
96906	MS35307-389	5305-00-660-2832	456	15
80205	MS35308-384	5305-00-689-7472	148	26
96906	MS35309-306	5305-00-543-5828	52	14
96906	MS35333-35	5310-00-579-5554	72	12
96906	MS35333-38	5310-00-559-0070	379	12
96906	MS35333-39	5310-00-576-5752	333	17
		5310-00-576-5752	349	4
96906	MS35333-40	5310-00-550-1130	68	7
		5310-00-550-1130	269	32
		5310-00-550-1130	327	14
		5310-00-550-1130	366	4
		5310-00-550-1130	384	7
		5310-00-550-1130	484	6
96906	MS35333-41	5310-00-167-0721	436	10
96906	MS35333-42	5310-00-595-7237	412	6
96906	MS35333-43	5310-00-685-3228	327	11
96906	MS35333-44	5310-00-194-1483	46	15
		5310-00-194-1483	388	14
96906	MS35333-47	5310-00-550-3714	380	12
96906	MS35335-33	5310-00-209-0786	55	13

SECTION IV TM9-2320-280-24P C01

CROSS REFERENCE INDEXES

PART NUMBER INDEX

CAGE	PART NUMBER	STOCK NUMBER	FIGURE NO	ITEM NO		
96906	MS35336-53	5310-00-957-2677			296	2
96906	MS35338-100	5310-00-261-8278			69	31
96906	MS35338-103	5310-00-184-8971			69	29
		5310-00-184-8971	434	5		
		5310-00-184-8971	435	10		
96906	MS35338-141	5310-00-984-7042			73	3
		5310-00-984-7042	482	9		
96906	MS35338-27	5310-00-543-5101			88	49
96906	MS35338-41	5310-00-045-4007			39	30
		5310-00-045-4007	56	22		
		5310-00-045-4007	213	18		
		5310-00-045-4007	214	14		
		5310-00-045-4007	333	29		
96906	MS35338-42	5310-00-045-3299			46	13
		5310-00-045-3299	48	2		
		5310-00-045-3299	52	22		
		5310-00-045-3299	64	18		
		5310-00-045-3299	380	2		
80205	MS35338-43	5310-00-045-3296			240	19
		5310-00-045-3296	283	29		
		5310-00-045-3296	327	35		
		5310-00-045-3296	410	13		
		5310-00-045-3296	410	17		
		5310-00-045-3296	412	10		
		5310-00-045-3296	43B	23		
		5310-00-045-3296	460	3		
96906	MS35338-43	5310-00-045-3296			39	33
		5310-00-045-3296	54	12		
		5310-00-045-3296	55	21		
		5310-00-045-3296	59	17		
		5310-00-045-3296	260	10		
		5310-00-045-3296	335	13		
		5310-00-045-3296	336	10		
		5310-00-045-3296	337	18		
		5310-00-045-3296	338	11		
		5310-00-045-3296	416	15		
		5310-00-045-3296	426	22		
		5310-00-045-3296	427	17		
		5310-00-045-3296	428	15		
		5310-00-045-3296	442	7		
		5310-00-045-3296	482	21		
		5310-00-045-3296	485	16		
96906	MS35338-44	5310-00-582-5965			28	7
		5310-00-582-5965	53	4		
		5310-00-582-5965	62	26		
		5310-00-582-5965	75	35		
		5310-00-582-5965	93	18		
		5310-00-582-5965	155	12		
		5310-00-582-5965	156	17		

SECTION IV TM9-2320-280-24P C01

CROSS REFERENCE INDEXES

PART NUMBER INDEX

CAGE	PART NUMBER	STOCK NUMBER	FIGURE NO	ITEM NO
		5310-00-582-5965	205	23
		5310-00-582-5965	228	14
		5310-00-582-5965	237	11
		5310-00-582-5965	240	24
		5310-00-582-5965	250	9
		5310-00-582-5965	267	6
		5310-00-582-5965	286	37
		5310-00-582-5965	287	11
		5310-00-582-5965	308	5
		5310-00-582-5965	311	3
		5310-00-582-5965	313	2
		5310-00-582-5965	327	39
		5310-00-582-5965	342	5
		5310-00-582-5965	343	15
		5310-00-582-5965	365	5
		5310-00-582-5965	368	23
		5310-00-582-5965	387	8
		5310-00-582-5965	410	37
		5310-00-582-5965	414	16
		5310-00-582-5965	416	20
		5310-00-582-5965	425	8
		5310-00-582-5965	426	5
		5310-00-582-5965	427	9
		5310-00-582-5965	428	19
		5310-00-582-5965	430	41
		5310-00-582-5965	443	2
		5310-00-582-5965	444	8
		5310-00-582-5965	444	9
		5310-00-582-5965	446	12
		5310-00-582-5965	451	2
		5310-00-582-5965	467	5
		5310-00-582-5965	492	16
		5310-00-582-5965	492	18
		5310-00-582-5965	498	18
		5310-00-582-5965	66A	6
96906	MS35338-45	5310-00-407-9566	19	19
		5310-00-407-9566	35	15
		5310-00-407-9566	59	4
		5310-00-407-9566	74	10
		5310-00-407-9566	136	15
		5310-00-407-9566	137	2
		5310-00-407-9566	178	15
		5310-00-407-9566	205	32
		5310-00-407-9566	206	19
		5310-00-407-9566	211	30
		5310-00-407-9566	212	30
		5310-00-407-9566	225	31
		5310-00-407-9566	226	16
		5310-00-407-9566	231	13

SECTION IV TM9-2320-280-24P C01

CROSS REFERENCE INDEXES

PART NUMBER INDEX

CAGE	PART NUMBER	STOCK NUMBER	FIGURE NO	ITEM NO
		5310-00-407-9566	268	5
		5310-00-407-9566	269	15
		5310-00-407-9566	287	19
		5310-00-407-9566	328	7
		5310-00-407-9566	329	7
		5310-00-407-9566	340	10
		5310-00-407-9566	373	9
		5310-00-407-9566	430	13
		5310-00-407-9566	432	3
		5310-00-407-9566	43B	15
		5310-00-407-9566	461	6
96906	MS35338-46	5310-00-637-9541	1	4
		5310-00-637-9541	2	5
		5310-00-637-9541	5	13
		5310-00-637-9541	16	23
		5310-00-637-9541	23	17
		5310-00-637-9541	24	25
		5310-00-637-9541	34	14
		5310-00-637-9541	35	3
		5310-00-637-9541	35	20
		5310-00-637-9541	36	15
		5310-00-637-9541	37	3
		5310-00-637-9541	38	15
		5310-00-637-9541	43	48
		5310-00-637-9541	44	3
		5310-00-637-9541	52	24
		5310-00-637-9541	62	7
		5310-00-637-9541	64	13
		5310-00-637-9541	68	3
		5310-00-637-9541	70	10
		5310-00-637-9541	72	2
		5310-00-637-9541	75	43
		5310-00-637-9541	77	2
		5310-00-637-9541	79	19
		5310-00-637-9541	83	33
		5310-00-637-9541	88	38
		5310-00-637-9541	124	15
		5310-00-637-9541	125	15
		5310-00-637-9541	164	3
		5310-00-637-9541	169	2
		5310-00-637-9541	174	14
		5310-00-637-9541	175	8
		5310-00-637-9541	286	9
		5310-00-637-9541	286	35
		5310-00-637-9541	328	17
		5310-00-637-9541	33A	16
		5310-00-637-9541	33B	2
		5310-00-637-9541	340	3
		5310-00-637-9541	352	2

SECTION IV TM9-2320-280-24P C01

CROSS REFERENCE INDEXES

PART NUMBER INDEX

CAGE PART NUMBER STOCK NUMBER FIGURE NO ITEM NO

CAGE	PART NUMBER	STOCK NUMBER	FIGURE NO	ITEM NO
		5310-00-637-9541	363	11
		5310-00-637-9541	372	5
		5310-00-637-9541	379	23
		5310-00-637-9541	38A	3
		5310-00-637-9541	43B	31
		5310-00-637-9541	443	12
		5310-00-637-9541	443	13
		5310-00-637-9541	457	6
		5310-00-637-9541	459	2
		5310-00-637-9541	468	8
		5310-00-637-9541	472	15
		5310-00-637-9541	475	23
		5310-00-637-9541	476	2
		5310-00-637-9541	481	7
		5310-00-637-9541	490	9
96906	MS35338-47	5310-00-209-0965	35	22
		5310-00-209-0965	36	10
		5310-00-209-0965	37	10
		5310-00-209-0965	38	10
		5310-00-209-0965	42	4
		5310-00-209-0965	150	35
		5310-00-209-0965	151	41
		5310-00-209-0965	171	3
		5310-00-209-0965	472	21
		5310-00-209-0965	475	17
96906	MS35338-48	5310-00-584-5272	156	21
		5310-00-584-5272	33A	2
		5310-00-584-5272	363	19
		5310-00-584-5272	369	1
		5310-00-584-5272	410	35
		5310-00-584-5272	412	26
		5310-00-584-5272	424	7
		5310-00-584-5272	434	3
		5310-00-584-5272	435	7
96906	MS35338-49	5310-00-167-0680	476	12
96906	MS35338-50	5310-00-820-6653	193	6
96906	MS35338-63	5310-00-274-8715	249	3
		5310-00-274-8715	315	11
96906	MS35338-65	5310-00-011-5093	138	17
		5310-00-011-5093	139	14
		5310-00-011-5093	145	17
		5310-00-011-5093	146	14
		5310-00-011-5093	164	3
96906	MS35338-67	5310-00-011-6121	70	7
96906	MS35340-43	5310-00-721-7809	35	12
		5310-00-721-7809	398	12
96906	MS35340-47	5310-00-655-9370	438	18
96906	MS35340-48	5310-00-834-7606	438	13
96906	MS35387-1	9905-00-205-2795	296	17

SECTION IV TM9-2320-280-24P C01

CROSS REFERENCE INDEXES

PART NUMBER INDEX

CAGE	PART NUMBER	STOCK NUMBER	FIGURE NO	ITEM NO
96906	MS35421-1	6220-00-299-7425	63	3
		6220-00-299-7425	63	9
96906	MS35421-2	6220-00-299-7426	63	3
		6220-00-299-7426	63	9
96906	MS35422-1	6220-00-729-9295	63	2
96906	MS35423-1	6220-00-577-3434	63	1
96906	MS35423-2	6220-00-726-1916	63	1
96906	MS35425-68	5310-01-106-1144	331	38
96906	MS35425-74	5310-01-088-2490	340	2
96906	MS35425-75	5310-01-078-5996	261	1
96906	MS35436-16	5940-00-020-0072	79	12
96906	MS35489-11	5325-00-291-9366	92	13
		5325-00-291-9366	333	34
		5325-00-291-9366	380	7
96906	MS35489-110	5325-00-202-4005	80	17
96906	MS35489-115	5325-00-928-6214	65	7
96906	MS35489-135	5325-00-263-6648	496	3
96906	MS35489-143	5325-01-292-7558	485	24
96906	MS35489-149	5325-01-309-7164	379	16
96906	MS35489-17	5325-01-242-7083	333	10
		5325-01-242-7083	367	10
96906	MS35489-18	5325-00-276-6090	496	21
96906	MS35489-19	5325-00-276-6091	369	4
96906	MS35489-21	5325-00-270-8891	380	5
96906	MS35489-23	5325-00-276-6343	473	9
		5325-00-276-6343	481	10
96906	MS35489-27	5325-00-290-1960	70	13
96906	MS35489-39X	5325-01-274-3317	95	30
96906	MS35489-40	5325-00-185-0004	96	18
96906	MS35489-41	5325-01-088-7468	327	17
96906	MS35489-45	5325-00-285-8363	424	6
96906	MS35489-48	5325-00-174-9332	80	12
		5325-00-174-9332	315	13
96906	MS35489-50	5325-00-754-1155	428	8
96906	MS35492-28	5305-00-900-2546	382	1
96906	MS35492-76	5305-00-902-6643	383	2
96906	MS35493-51	5305-00-901-2099	394	5
96906	MS35493-55	5305-00-901-2134	385	13
96906	MS35493-74	5305-00-056-7879	385	1
		5305-00-056-7879	395	11
		5305-00-056-7879	396	4
96906	MS35493-76	5305-00-014-9926	387	12
		5305-00-014-9926	392	1
96906	MS35495-124	5305-01-315-9512	392	2
96906	MS35495-126	5305-01-315-9882	395	24
		5305-01-315-9882	397	12
96906	MS35495-127	5305-01-317-0409	393	6
		5305-01-317-0409	395	20
96906	MS35649-103	5310-00-852-8593	498	10

SECTION IV TM9-2320-280-24P C01

CROSS REFERENCE INDEXES

PART NUMBER INDEX

CAGE	PART NUMBER	STOCK NUMBER	FIGURE NO	ITEM NO
96906	MS35649-202	5310-00-934-9758	64	16
		5310-00-934-9758 64	26	
		5310-00-934-9758 74	9	
		5310-00-934-9758	88	43
		5310-00-934-9758	208	28
		5310-00-934-9758	213	14
		5310-00-934-9758	214	10
		5310-00-934-9758 240	1	
		5310-00-934-9758	240	13
		5310-00-934-9758	240	20
		5310-00-934-9758 241	7	
		5310-00-934-9758 269	2	
		5310-00-934-9758	283	30
		5310-00-934-9758	327	30
		5310-00-934-9758	335	14
		5310-00-934-9758	336	11
		5310-00-934-9758 442	8	
96906	MS35649-205	5310-00-934-9764	63	5
		5310-00-934-9764 85	6	
		5310-00-934-9764	169	19
		5310-00-934-9764	246	33
		5310-00-934-9764	293	15
96906	MS35649-2252	5310-00-997-1888	88	39
		5310-00-997-1888	324	28
		5310-00-997-1888	375	19
		5310-00-997-1888	375	27
		5310-00-997-1888	378	27
96906	MS35649-2255	5310-00-855-1102	52	16
96906	MS35649-2384	5310-00-477-6768	482	8
96906	MS35649-262	5310-00-934-9747	333	28
96906	MS35649-282	5310-00-934-9757	48	1
96906	MS35650-302	5310-00-934-9751	45	49
		5310-00-934-9751	327	34
		5310-00-934-9751	333	18
		5310-00-934-9751 482	22	
96906	MS35650-382	5310-00-934-9754	380	1
96906	MS35690-824	5310-00-010-3028	437	6
96906	MS35691-33	5310-00-834-8732	44	6
96906	MS35692-53	5310-01-280-6751	39	56
96906	MS35692-54	5310-00-850-7004	144	10
		5310-00-850-7004	144	18
		5310-00-850-7004	152	10
		5310-00-850-7004	152	18
		5310-00-850-7004 170	3	
		5310-00-850-7004 170	9	
		5310-00-850-7004	170	13
		5310-00-850-7004 183	5	
		5310-00-850-7004	185	14
		5310-00-850-7004	187	18

SECTION IV TM9-2320-280-24P C01

CROSS REFERENCE INDEXES

PART NUMBER INDEX

CAGE	PART NUMBER	STOCK NUMBER	FIGURE NO	ITEM NO
		5310-00-850-7004	188	14
		5310-00-850-7004	437	10
96906	MS35692-62	5310-00-850-6993	186	9
		5310-00-850-6993	188	18
96906	MS35692-9	5310-00-850-6868	188	34
96906	MS35692-94	5310-00-849-6882	191	11
80205	MS35751-42	5306-00-911-5005	433	1
80205	MS35751-43	5306-00-753-6996	403	13
96906	MS35751-44	5306-00-012-0231	403	10
80205	MS35756-17	5315-00-012-4553	36	5
		5315-00-012-4553	38	5
		5315-00-012-4553	39	8
		5315-00-012-4553	43B	26
80205	MS35756-20	5315-00-616-5501	476	8
80205	MS35756-6	5315-00-616-5514	43	4
80205	MS35756-8	5315-00-616-5526	35	9
96906	MS35764-1303	5306-01-068-4592	409	5
		5306-01-068-4592	411	11
		5306-01-068-4592	415	10
96906	MS35764-848	5306-01-253-1611	150	36
96906	MS35764-850	5306-01-253-1612	202	34
96906	MS35764-851	5306-01-203-2637	313	21
96906	MS35764-853	5306-01-276-0848	35	21
96906	MS35764-861	5306-01-254-6356	171	4
96906	MS35764-887	5306-01-253-4426	179	8
96906	MS35769-9	5330-01-355-3686	4	24
96906	MS35810-1	5315-00-417-5223	238	13
96906	MS35810-3	5315-00-754-0848	153	21
96906	MS35810-4	5315-00-815-8840	153	4
		5315-00-815-8840	155	17
96906	MS35810-6	5315-00-140-1938	293	16
96906	MS35823-7A	5340-01-053-7128	224	2
96906	MS35842-11	4730-00-908-3194	310	7
		4730-00-908-3194	379	9
96906	MS35842-14	4730-00-908-6292	10	14
		4730-00-908-6292	331	20
		4730-00-908-6292	368	3
96906	MS39061-4	5930-00-226-6429	46	6
96906	MS45904-60	5310-00-080-9786	88	27
		5310-00-080-9786	215	9
		5310-00-080-9786	222	10
		5310-00-080-9786	339	8
		5310-00-080-9786	346	6
		5310-00-080-9786	347	3
		5310-00-080-9786	426	18
		5310-00-080-9786	427	21
		5310-00-080-9786	467	1
96906	MS45904-68	5310-00-889-2528	327	13
96906	MS49006-6	4730-00-045-9769	142	30

SECTION IV TM9-2320-280-24P C01

CROSS REFERENCE INDEXES

PART NUMBER INDEX

CAGE	PART NUMBER	STOCK NUMBER	FIGURE NO	ITEM NO
		4730-00-045-9769	143	39
		4730-00-045-9769	150	30
		4730-00-045-9769	151	39
96906	MS51085-1	2910-00-025-3493	335	9
		2910-00-025-3493	336	6
		2910-00-025-3493	388	7
96906	MS51108-139	5305-01-331-6284	437	3
96906	MS51113-1	5930-00-307-8856	46	19
96906	MS51335-1	2540-00-078-6633	191	9
		2540-00-078-6633	191	17
96906	MS51412-1	5310-01-303-4701	43	31
		5310-01-303-4701	425	6
		5310-01-303-4701	426	9
		5310-01-303-4701	427	7
96906	MS51412-18	5310-01-253-1618	52	31
		5310-01-253-1618	208	27
		5310-01-253-1618	252	5
		5310-01-253-1618	260	11
		5310-01-253-1618	273	13
		5310-01-253-1618	275	12
		5310-01-253-1618	276	8
		5310-01-253-1618	300	13
		5310-01-253-1618	327	29
96906	MS51412-2	5310-01-234-9416	35	14
		5310-01-234-9416	37	7
		5310-01-234-9416	52	11
		5310-01-234-9416	56	8
		5310-01-234-9416	59	16
		5310-01-234-9416	234	16
		5310-01-234-9416	246	32
		5310-01-234-9416	247	23
		5310-01-234-9416	249	44
		5310-01-234-9416	260	7
		5310-01-234-9416	269	3
		5310-01-234-9416	284	14
		5310-01-234-9416	298	8
		5310-01-234-9416	307	12
		5310-01-234-9416	321	16
		5310-01-234-9416	327	36
		5310-01-234-9416	327	42
96906	MS51412-23	5310-01-333-0060	283	16
96906	MS51412-36	5310-01-449-0628	50	7
96906	MS51412-5	5310-01-234-9415	88	16
96906	MS51415-7	5310-01-218-7137	457	10
96906	MS51415-9	5310-01-216-7390	44	5
		5310-01-216-7390	209	14
		5310-01-216-7390	443	13
		5310-01-216-7390	457	11
96906	MS51470-03	5310-01-249-0899	398	1

SECTION IV TM9-2320-280-24P C01

CROSS REFERENCE INDEXES

PART NUMBER INDEX

CAGE	PART NUMBER	STOCK NUMBER	FIGURE NO	ITEM NO
96906	MS51471-01	5310-01-272-9981	142	35
		5310-01-272-9981	143	41
96906	MS51480-05	5305-01-264-0923	245	9
96906	MS51500A10-6	4730-01-066-1278	6	9
96906	MS51815-4P	4730-00-985-4804	505	16
96906	MS51844-63	4030-01-142-0456	276	4
96906	MS51848-13	5310-01-016-9348	283	22
96906	MS51849-100	5305-01-075-0957	243	20
		5305-01-075-0957	308	4
96906	MS51849-11	5305-00-164-7082	65	8
		5305-00-164-7082	65	15
96906	MS51849-33	5305-00-227-1543	333	31
96906	MS51849-53	5305-00-115-9406	349	8
96906	MS51849-55	5305-00-115-9934	205	14
96906	MS51849-64	5305-00-180-4966	39	6
		5305-00-180-4966	259	12
		5305-00-180-4966	263	7
		5305-00-180-4966	289	4
		5305-00-180-4966	292	3
		5305-00-180-4966	365	20
96906	MS51849-65	5305-01-006-2052	252	37
96906	MS51849-74	5305-00-470-3321	74	11
		5305-00-470-3321	283	26
96906	MS51849-75	5305-01-006-2053	208	25
		5305-01-006-2053	279	29
		5305-01-006-2053	429	11
96906	MS51849-76	5305-00-240-0194	67	2
		5305-00-240-0194	88	10
		5305-00-240-0194	169	22
		5305-00-240-0194	197	33
		5305-00-240-0194	199	59
		5305-00-240-0194	205	4
		5305-00-240-0194	206	7
		5305-00-240-0194	225	50
		5305-00-240-0194	226	30
		5305-00-240-0194	248	28
		5305-00-240-0194	286	31
		5305-00-240-0194	327	26
		5305-00-240-0194	411	5
		5305-00-240-0194	460	2
96906	MS51849-77	5305-01-005-6715	243	27
96906	MS51849-78	5305-00-240-6668	85	3
		5305-00-240-6668	88	1
		5305-00-240-6668	327	48
96906	MS51849-79	5305-00-240-6541	227	50
96906	MS51849-95	5305-01-016-4344	453	18
96906	MS51849-98	5305-00-390-0524	83	29
		5305-00-390-0524	272	8
96906	MS51850-69	5305-01-315-9524	395	15

SECTION IV TM9-2320-280-24P C01

CROSS REFERENCE INDEXES

PART NUMBER INDEX

CAGE	PART NUMBER	STOCK NUMBER	FIGURE NO	ITEM NO
96906	MS51851-126	5305-01-101-3312	443	7
		5305-01-101-3312	457	4
96906	MS51851-34	5305-01-076-3150	50	5
96906	MS51851-43	5305-00-757-8287	379	5
96906	MS51851-64	5305-00-757-8122	75	44
		5305-00-757-8122	488	12
96906	MS51851-85	5305-00-191-3640	313	10
		5305-00-191-3640	443	9
		5305-00-191-3640	457	9
96906	MS51851-95	5305-01-087-4478	5	9 96906
MS51859-20	5310-01-313-3562	329	16	
		5310-01-313-3562	461	4
96906	MS51861-22C	5305-01-298-2436	332	45
96906	MS51861-26	5305-00-432-4165	49	11
96906	MS51861-34	5305-00-058-1082	12	34
		5305-00-058-1082	274	28
96906	MS51861-35	5305-00-432-4170	403	12
		5305-00-432-4170	433	4
96906	MS51861-45	5305-00-432-4201	485	5
96906	MS51861-49	5305-00-432-4205	208	9
		5305-00-432-4205	280	34
		5305-00-432-4205	281	34
		5305-00-432-4205	331	29
96906	MS51861-67	5305-00-432-4253	395	1
		5305-00-432-4253	396	2
96906	MS51861-69	5305-00-432-4254	446	32
96906	MS51863-22	5305-01-076-6308	43	45
96906	MS51863-33	5305-01-138-9540	46	4
		5305-01-138-9540	97	11
		5305-01-138-9540	367	1
96906	MS51869-23	5305-01-039-6633	44	28
96906	MS51871-1	5305-01-084-6067	124	17
		5305-01-084-6067	125	17
		5305-01-084-6067	366	3
96906	MS51873-32	4730-01-315-3242	385	9
96906	MS51941-10	5310-01-025-6444	383	8
96906	MS51943-3	5310-00-472-2963	421	9
		5310-00-472-2963	422	4
96906	MS51943-31	5310-00-061-4650	5	7
		5310-00-061-4650	13	18
		5310-00-061-4650	14	24
		5310-00-061-4650	15	3
		5310-00-061-4650	16	3
		5310-00-061-4650	17	9
		5310-00-061-4650	18	4
		5310-00-061-4650	18	10
		5310-00-061-4650	21	1
		5310-00-061-4650	25	7
		5310-00-061-4650	27	14

SECTION IV TM9-2320-280-24P C01

CROSS REFERENCE INDEXES

PART NUMBER INDEX

CAGE PART NUMBER STOCK NUMBER FIGURE NO ITEM NO

CAGE	PART NUMBER	STOCK NUMBER	FIGURE NO	ITEM NO
		5310-00-061-4650	29	3
		5310-00-061-4650	42	8
		5310-00-061-4650	46	21
		5310-00-061-4650	50	8
		5310-00-061-4650	71	10
		5310-00-061-4650	84	22
		5310-00-061-4650	97	4
		5310-00-061-4650	98	13
		5310-00-061-4650	99	11
		5310-00-061-4650	124	13
		5310-00-061-4650	125	13
		5310-00-061-4650	141	17
		5310-00-061-4650	153	17
		5310-00-061-4650	169	27
		5310-00-061-4650	169	33
		5310-00-061-4650	177	13
		5310-00-061-4650	178	24
		5310-00-061-4650	202	26
		5310-00-061-4650	203	9
		5310-00-061-4650	203	23
		5310-00-061-4650	204	14
		5310-00-061-4650	206	24
		5310-00-061-4650	210	4
		5310-00-061-4650	211	26
		5310-00-061-4650	212	27
		5310-00-061-4650	241	5
		5310-00-061-4650	243	8
		5310-00-061-4650	244	7
		5310-00-061-4650	245	16
		5310-00-061-4650	245	32
		5310-00-061-4650	246	18
		5310-00-061-4650	247	5
		5310-00-061-4650	247	18
		5310-00-061-4650	248	7
		5310-00-061-4650	249	27
		5310-00-061-4650	250	19
		5310-00-061-4650	251	1
		5310-00-061-4650	252	22
		5310-00-061-4650	253	21
		5310-00-061-4650	254	22
		5310-00-061-4650	255	20
		5310-00-061-4650	267	3
		5310-00-061-4650	267	13
		5310-00-061-4650	269	13
		5310-00-061-4650	270	1
		5310-00-061-4650	270	26
		5310-00-061-4650	271	9
		5310-00-061-4650	272	2
		5310-00-061-4650	274	18

SECTION IV TM9-2320-280-24P C01

CROSS REFERENCE INDEXES

PART NUMBER INDEX

CAGE PART NUMBER STOCK NUMBER FIGURE NO ITEM NO

CAGE	PART NUMBER	STOCK NUMBER	FIGURE NO	ITEM NO
		5310-00-061-4650	275	20
		5310-00-061-4650	276	20
		5310-00-061-4650	285	13
		5310-00-061-4650	286	16
		5310-00-061-4650	288	22
		5310-00-061-4650	288	30
		5310-00-061-4650	289	5
		5310-00-061-4650	293	31
		5310-00-061-4650	316	21
		5310-00-061-4650	323	13
		5310-00-061-4650	331	9
		5310-00-061-4650	332	42
		5310-00-061-4650	333	39
		5310-00-061-4650	334	5
		5310-00-061-4650	338	22
		5310-00-061-4650	339	17
		5310-00-061-4650	343	17
		5310-00-061-4650	366	30
		5310-00-061-4650	368	32
		5310-00-061-4650	375	28
		5310-00-061-4650	376	13
		5310-00-061-4650	384	9
		5310-00-061-4650	398	27
		5310-00-061-4650	399	15
		5310-00-061-4650	400	13
		5310-00-061-4650	402	12
		5310-00-061-4650	409	9
		5310-00-061-4650	411	10
		5310-00-061-4650	413	5
		5310-00-061-4650	415	15
		5310-00-061-4650	417	18
		5310-00-061-4650	418	9
		5310-00-061-4650	419	4
		5310-00-061-4650	420	6
		5310-00-061-4650	421	6
		5310-00-061-4650	422	8
		5310-00-061-4650	423	9
		5310-00-061-4650	429	22
		5310-00-061-4650	430	17
		5310-00-061-4650	430	38
		5310-00-061-4650	440	21
		5310-00-061-4650	446	16
		5310-00-061-4650	468	4
		5310-00-061-4650	475	9
		5310-00-061-4650	477	5
		5310-00-061-4650	478	3
		5310-00-061-4650	484	15
		5310-00-061-4650	485	14
		5310-00-061-4650	486	9

SECTION IV TM9-2320-280-24P C01

CROSS REFERENCE INDEXES

PART NUMBER INDEX

CAGE	PART NUMBER	STOCK NUMBER	FIGURE NO	ITEM NO
		5310-00-061-4650	492	7
		5310-00-061-4650	176A	21
96906	MS51943-32	5310-00-935-9022	388	10
		5310-00-935-9022	485	17
96906	MS51943-33	5310-00-814-0673	74	8
		5310-00-814-0673	83	25
		5310-00-814-0673	144	4
		5310-00-814-0673	152	4
		5310-00-814-0673	153	25
		5310-00-814-0673	154	6
		5310-00-814-0673	169	33
		5310-00-814-0673	189	6
		5310-00-814-0673	190	1
		5310-00-814-0673	205	1
		5310-00-814-0673	206	10
		5310-00-814-0673	206	13
		5310-00-814-0673	208	29
		5310-00-814-0673	209	8
		5310-00-814-0673	211	19
		5310-00-814-0673	212	16
		5310-00-814-0673	226	12
		5310-00-814-0673	230	16
		5310-00-814-0673	242	5
		5310-00-814-0673	250	23
		5310-00-814-0673	266	1
		5310-00-814-0673	268	10
		5310-00-814-0673	268	29
		5310-00-814-0673	269	16
		5310-00-814-0673	284	12
		5310-00-814-0673	288	20
		5310-00-814-0673	292	15
		5310-00-814-0673	328	6
		5310-00-814-0673	328	15
		5310-00-814-0673	329	9
		5310-00-814-0673	333	21
		5310-00-814-0673	338	27
		5310-00-814-0673	339	22
		5310-00-814-0673	364	7
		5310-00-814-0673	375	31
		5310-00-814-0673	390	5
		5310-00-814-0673	401	9
		5310-00-814-0673	409	4
		5310-00-814-0673	411	4
		5310-00-814-0673	413	6
		5310-00-814-0673	414	11
		5310-00-814-0673	415	14
		5310-00-814-0673	416	14
		5310-00-814-0673	417	11
		5310-00-814-0673	418	8

SECTION IV TM9-2320-280-24P C01

CROSS REFERENCE INDEXES

PART NUMBER INDEX

CAGE	PART NUMBER	STOCK NUMBER	FIGURE NO	ITEM NO
		5310-00-814-0673	419	9
		5310-00-814-0673	420	4
		5310-00-814-0673	423	4
		5310-00-814-0673	428	14
		5310-00-814-0673	429	18
		5310-00-814-0673	440	11
		5310-00-814-0673	445	2
		5310-00-814-0673	463	12
		5310-00-814-0673	468	7
		5310-00-814-0673	480	4
96906	MS51943-34	5310-00-241-6658	154	5
		5310-00-241-6658	198	19
		5310-00-241-6658	273	16
		5310-00-241-6658	280	47
		5310-00-241-6658	281	47
		5310-00-241-6658	317	11
		5310-00-241-6658	373	8
		5310-00-241-6658	376	18
		5310-00-241-6658	378	28
		5310-00-241-6658	389	14
		5310-00-241-6658	415	4
		5310-00-241-6658	433	7
96906	MS51943-35	5310-00-935-9021	2	15
		5310-00-935-9021	23	4
		5310-00-935-9021	24	10
		5310-00-935-9021	26	6
		5310-00-935-9021	52	23
		5310-00-935-9021	103	17
		5310-00-935-9021	144	14
		5310-00-935-9021	152	14
		5310-00-935-9021	159	4
		5310-00-935-9021	160	9
		5310-00-935-9021	164	4
		5310-00-935-9021	169	1
		5310-00-935-9021	187	10
		5310-00-935-9021	189	2
		5310-00-935-9021	190	3
		5310-00-935-9021	205	9
		5310-00-935-9021	210	14
		5310-00-935-9021	245	30
		5310-00-935-9021	251	45
		5310-00-935-9021	266	24
		5310-00-935-9021	268	19
		5310-00-935-9021	270	19
		5310-00-935-9021	270	21
		5310-00-935-9021	272	29
		5310-00-935-9021	273	22
		5310-00-935-9021	274	17
		5310-00-935-9021	275	19

SECTION IV TM9-2320-280-24P C01

CROSS REFERENCE INDEXES

PART NUMBER INDEX

CAGE PART NUMBER STOCK NUMBER FIGURE NO ITEM NO

CAGE	PART NUMBER	STOCK NUMBER	FIGURE NO	ITEM NO
		5310-00-935-9021	276	19
		5310-00-935-9021	277	4
		5310-00-935-9021	278	12
		5310-00-935-9021	283	8
		5310-00-935-9021	285	9
		5310-00-935-9021	286	28
		5310-00-935-9021	288	13
		5310-00-935-9021	292	16
		5310-00-935-9021	293	6
		5310-00-935-9021	369	10
		5310-00-935-9021	372	6
		5310-00-935-9021	398	25
		5310-00-935-9021	409	8
		5310-00-935-9021	410	9
		5310-00-935-9021	411	14
		5310-00-935-9021	412	7
		5310-00-935-9021	414	9
		5310-00-935-9021	415	13
		5310-00-935-9021	416	10
		5310-00-935-9021	425	16
		5310-00-935-9021	426	20
		5310-00-935-9021	427	20
		5310-00-935-9021	428	10
		5310-00-935-9021	431	7
		5310-00-935-9021	440	17
		5310-00-935-9021	441	11
		5310-00-935-9021	456	18
		5310-00-935-9021	463	3
		5310-00-935-9021	464	1
		5310-00-935-9021	470	13
		5310-00-935-9021	471	5
		5310-00-935-9021	484	13
		5310-00-935-9021	490	10
96906	MS51943-37	5310-00-241-6659	170	14
96906	MS51943-39	5310-00-488-3889	2	8
		5310-00-488-3889	140	1
		5310-00-488-3889	147	1
		5310-00-488-3889	179	3
		5310-00-488-3889	180	4
		5310-00-488-3889	181	9
		5310-00-488-3889	182	4
		5310-00-488-3889	183	14
		5310-00-488-3889	184	1
		5310-00-488-3889	185	7
		5310-00-488-3889	186	7
		5310-00-488-3889	187	11
		5310-00-488-3889	188	5
		5310-00-488-3889	188	38
		5310-00-488-3889	191	1

SECTION IV TM9-2320-280-24P C01

CROSS REFERENCE INDEXES

PART NUMBER INDEX

CAGE	PART NUMBER	STOCK NUMBER	FIGURE NO	ITEM NO
		5310-00-488-3889	199	22
		5310-00-488-3889	242	8
		5310-00-488-3889	245	26
		5310-00-488-3889	251	47
		5310-00-488-3889	270	22
		5310-00-488-3889	283	18
		5310-00-488-3889	315	5
		5310-00-488-3889	429	36
		5310-00-488-3889	441	5
		5310-00-488-3889	442	4
		5310-00-488-3889	444	3
		5310-00-488-3889	446	28
		5310-00-488-3889	453	1
		5310-00-488-3889	455	6
		5310-00-488-3889	456	17
		5310-00-488-3889	463	9
		5310-00-488-3889	471	1
		5310-00-488-3889	487A	6
		5310-00-488-3889	487B	1
96906	MS51943-45	5310-00-409-3333	144	3
		5310-00-409-3333	152	3
		5310-00-409-3333	181	16
		5310-00-409-3333	182	2
		5310-00-409-3333	183	13
		5310-00-409-3333	184	19
		5310-00-409-3333	188	28
		5310-00-409-3333	195	13
		5310-00-409-3333	270	35
		5310-00-409-3333	271	5
		5310-00-409-3333	271	22
		5310-00-409-3333	456	20
		5310-00-409-3333	471	3
96906	MS51943-6	5310-01-344-8250	16	32
96906	MS51943-71	5310-01-408-7138	462	4
96906	MS51957-131	5305-00-182-9304	255	35
96906	MS51957-134	5305-01-052-4438	280	7
96906	MS51957-28	5305-00-054-6652	214	15
96906	MS51957-45	5305-00-054-6670	482	19
96906	MS51957-47	5305-00-054-6672	64	17
		5305-00-054-6672	252	7
96906	MS51957-63	5305-00-050-9229	347	2
96906	MS51957-64B	5305-01-083-1591	465	2
96906	MS51957-67	5305-00-050-9233	280	16
96906	MS51957-77	5305-00-071-1313	373	4
		5305-00-071-1313	376	4
96906	MS51957-79	5305-00-071-1315	316	16
		5305-00-071-1315	320	15
		5305-00-071-1315	320	27
		5305-00-071-1315	375	23

SECTION IV TM9-2320-280-24P C01

CROSS REFERENCE INDEXES

PART NUMBER INDEX

CAGE	PART NUMBER	STOCK NUMBER	FIGURE NO	ITEM NO		
		5305-00-071-1315	378	23		
96906	MS51957-8	5305-00-054-5642			238	21
96906	MS51957-83	5305-00-071-1318			214	4
		5305-00-071-1318	246	24		
		5305-00-071-1318	251	7		
		5305-00-071-1318	252	26		
		5305-00-071-1318	267	18		
		5305-00-071-1318	296	15		
96906	MS51958-63	5305-00-059-3659			227	14
		5305-00-059-3659	321	6		
		5305-00-059-3659	339	4		
		5305-00-059-3659	373	3		
		5305-00-059-3659	376	5		
		5305-00-059-3659	410	14		
		5305-00-059-3659	412	9		
		5305-00-059-3659	416	12		
		5305-00-059-3659	426	21		
		5305-00-059-3659	427	18		
		5305-00-059-3659	428	12		
96906	MS51958-64	5305-00-059-3660			65	19
		5305-00-059-3660	92	14		
		5305-00-059-3660	95	43		
		5305-00-059-3660	218	5		
		5305-00-059-3660	236	1		
		5305-00-059-3660	253	7		
		5305-00-059-3660	253	10		
		5305-00-059-3660	253	18		
		5305-00-059-3660	253	30		
		5305-00-059-3660	255	9		
		5305-00-059-3660	255	17		
		5305-00-059-3660	255	23		
		5305-00-059-3660	259	5		
		5305-00-059-3660	260	2		
		5305-00-059-3660	262	13		
		5305-00-059-3660	325	22		
		5305-00-059-3660	426	17		
		5305-00-059-3660	427	22		
96906	MS51958-65	5305-00-059-3661			255	9
		5305-00-059-3661	255	17		
		5305-00-059-3661	255	23		
		5305-00-059-3661	255	28		
		5305-00-059-3661	410	25		
		5305-00-059-3661	412	19		
96906	MS51959-81	5305-00-719-5021			339	14
		5305-00-719-5021	409	12		
		5305-00-719-5021	413	3		
		5305-00-719-5021	415	3		
		5305-00-719-5021	486	11		
96906	MS51959-82	5305-00-719-5017			246	11

SECTION IV TM9-2320-280-24P C01

CROSS REFERENCE INDEXES

PART NUMBER INDEX

CAGE	PART NUMBER	STOCK NUMBER	FIGURE NO	ITEM NO
96906	MS51959-84	5305-00-719-5003	248	29
96906	MS51959-99	5305-00-763-7829	225	1
96906	MS51960-65	5305-00-071-1322	246	31
		5305-00-071-1322	410	29
96906	MS51960-70	5305-00-071-1327	316	10
		5305-00-071-1327	320	8
		5305-00-071-1327	320	20
		5305-00-071-1327	378	16
80205	MS51964-65	5305-00-724-5812	333	20
96906	MS51965-29	5305-00-724-6783	368	9
96906	MS51967-12	5310-00-896-0903	170	23
		5310-00-896-0903	194	7
		5310-00-896-0903	195	9
		5310-00-896-0903	202	30
		5310-00-896-0903	425	10
		5310-00-896-0903	426	12
		5310-00-896-0903	427	11
96906	MS51967-15	5310-00-761-3706	283	10
		5310-00-761-3706	285	10
96906	MS51967-18	5310-00-763-8919	165	6
		5310-00-763-8919	179	12
96906	MS51967-2	5310-00-761-6882	64	12
		5310-00-761-6882	252	9
		5310-00-761-6882	494	10
		5310-00-761-6882	495	14
96906	MS51967-3	5310-00-905-0762	213	11
		5310-00-905-0762	240	23
		5310-00-905-0762	311	14
		5310-00-905-0762	313	1
		5310-00-905-0762	315	12
		5310-00-905-0762	327	40
		5310-00-905-0762	425	9
		5310-00-905-0762	426	6
		5310-00-905-0762	427	10
		5310-00-905-0762	443	1
		5310-00-905-0762	444	9
		5310-00-905-0762	457	5
96906	MS51967-45	5310-00-432-3760	441	10
96906	MS51967-5	5310-00-880-7744	71	2
		5310-00-880-7744	209	5
		5310-00-880-7744	226	21
		5310-00-880-7744	230	5
		5310-00-880-7744	235	8
		5310-00-880-7744	236	6
		5310-00-880-7744	295	10
96906	MS51967-6	5310-00-931-8167	203	21
		5310-00-931-8167	204	9
		5310-00-931-8167	473	6
		5310-00-931-8167	474	9

SECTION IV TM9-2320-280-24P C01

CROSS REFERENCE INDEXES

PART NUMBER INDEX

CAGE	PART NUMBER	STOCK NUMBER	FIGURE NO		ITEM NO
96906	MS51967-8	5310-00-732-0558		44	4
		5310-00-732-0558 62	8		
		5310-00-732-0558	75	42	
		5310-00-732-0558 77	1		
		5310-00-732-0558 89	9		
		5310-00-732-0558	401	11	
		5310-00-732-0558	490	15	
96906	MS51967-9	5310-00-761-0654		15	23
		5310-00-761-0654	16	27	
		5310-00-761-0654	39	11	
		5310-00-761-0654	205	34	
		5310-00-761-0654	206	23	
		5310-00-761-0654 286	8		
96906	MS51968-18	5310-00-763-8906		476	11
96906	MS51968-2	5310-00-768-0319		35	19
		5310-00-768-0319		43	49
		5310-00-768-0319		62	27
		5310-00-768-0319	487	11	
96906	MS51968-5	5310-00-880-7746		137	1
96906	MS51968-6	5310-00-905-4600		59	3
		5310-00-905-4600	136	14	
96906	MS51969-5	5310-00-913-5474		70	8
		5310-00-913-5474 481	5		
80205	MS51973-50	5305-00-728-6274		213	26
		5305-00-728-6274	214	21	
96906	MS51987-117	5315-01-278-2123		292	21
96906	MS51988-7	5310-00-930-8214		167	6
		5310-00-930-8214	245	27	
96906	MS51988-8	5310-00-447-8774		103	2
		5310-00-447-8774 127	1		
96906	MS521301A203R	4720-00-491-0102		BULK	27
96906	MS521301A203R 58			379	10
96906	MS521301A203R 7			379	32
96906	MS521301A203R 74			379	8
96906	MS521302B206R	4720-00-555-8015		BULK	28
96906	MS53000-1	2590-00-801-2355		51	8
96906	MS75021-2	5935-00-846-3884		83	35
		5935-00-846-3884	86	37	
96906	MS87006-33	4030-00-948-7315		293	25
96906	MS87006-53	4030-00-916-2141		191	20
96906	MS90724-39	5310-01-110-1145		395	16
80205	MS90725-14	5305-00-071-2237		338	21
		5305-00-071-2237	339	16	
		5305-00-071-2237 393	1		
		5305-00-071-2237	395	12	
		5305-00-071-2237 413	8		
		5305-00-071-2237 486	5		
80205	MS90725-15	5305-00-071-2236		266	23
		5305-00-071-2236	401	14	

SECTION IV TM9-2320-280-24P C01

CROSS REFERENCE INDEXES

PART NUMBER INDEX

CAGE	PART NUMBER	STOCK NUMBER	FIGURE NO	ITEM NO
80205	MS90725-31	5306-00-225-8496	180	13
80205	MS90725-33	5306-00-225-8498	250	20
		5306-00-225-8498	418	4
		5306-00-225-8498	419	5
		5306-00-225-8498	420	3
96906	MS90725-33	5306-00-225-8498	213	9
		5306-00-225-8498	421	1
		5306-00-225-8498	422	1
		5306-00-225-8498	423	1
80205	MS90725-34	5306-00-225-8499	153	27
		5306-00-225-8499	203	20
		5306-00-225-8499	205	31
		5306-00-225-8499	206	2
		5306-00-225-8499	206	18
		5306-00-225-8499	333	24
96906	MS90725-34	5306-00-225-8499	74	1
		5306-00-225-8499	154	10
		5306-00-225-8499	268	4
		5306-00-225-8499	329	11
		5306-00-225-8499	339	20
		5306-00-225-8499	402	5
		5306-00-225-8499	409	1
		5306-00-225-8499	410	8
		5306-00-225-8499	411	1
		5306-00-225-8499	413	1
		5306-00-225-8499	414	7
		5306-00-225-8499	415	1
		5306-00-225-8499	416	7
		5306-00-225-8499	418	6
		5306-00-225-8499	419	7
		5306-00-225-8499	420	1
		5306-00-225-8499	421	8
		5306-00-225-8499	422	3
		5306-00-225-8499	423	8
		5306-00-225-8499	428	7
80205	MS90725-36	5306-01-075-8519	209	12
		5306-01-075-8519	430	12
80205	MS90725-47	5306-00-225-8511	12	13
80205	MS90725-5	5305-00-068-0501	18	13
		5305-00-068-0501	118	12
		5305-00-068-0501	286	19
		5305-00-068-0501	307	9
96906	MS90725-58	5305-00-115-9526	72	1
		5305-00-115-9526	251	4
96906	MS90725-6	5305-00-068-0502	19	16
		5305-00-068-0502	21	4
		5305-00-068-0502	50	11
		5305-00-068-0502	53	3
		5305-00-068-0502	58	9

I-186

SECTION IV TM9-2320-280-24P C01

CROSS REFERENCE INDEXES

PART NUMBER INDEX

CAGE PART NUMBER STOCK NUMBER FIGURE NO ITEM NO

	5305-00-068-0502 68	1	
	5305-00-068-0502 70	9	
	5305-00-068-0502	84	24
	5305-00-068-0502	93	16
	5305-00-068-0502	124	10
	5305-00-068-0502	125	10
	5305-00-068-0502	141	14
	5305-00-068-0502 155	2	
	5305-00-068-0502	180	19
	5305-00-068-0502	203	11
	5305-00-068-0502 206	4	
	5305-00-068-0502	208	14
	5305-00-068-0502 241	3	
	5305-00-068-0502	244	16
	5305-00-068-0502	245	20
	5305-00-068-0502	250	16
	5305-00-068-0502	251	21
	5305-00-068-0502 252	1	
	5305-00-068-0502 253	1	
	5305-00-068-0502 254	1	
	5305-00-068-0502 255	1	
	5305-00-068-0502 267	9	
	5305-00-068-0502	269	11
	5305-00-068-0502 270	5	
	5305-00-068-0502 294	4	
	5305-00-068-0502 311	2	
	5305-00-068-0502 313	9	
	5305-00-068-0502	331	25
	5305-00-068-0502 333	9	
	5305-00-068-0502	338	30
	5305-00-068-0502	339	28
	5305-00-068-0502 342	4	
	5305-00-068-0502	343	14
	5305-00-068-0502 377	1	
	5305-00-068-0502	384	13
	5305-00-068-0502 393	7	
	5305-00-068-0502	395	19
	5305-00-068-0502	399	13
	5305-00-068-0502	410	30
	5305-00-068-0502	412	11
	5305-00-068-0502 416	3	
	5305-00-068-0502	421	11
	5305-00-068-0502	422	11
	5305-00-068-0502	440	10
	5305-00-068-0502 484	5	
	5305-00-068-0502 492	9	
	5305-00-068-0502 494	5	
	5305-00-068-0502	495	11
	5305-00-068-0502 496	9	

I-187

SECTION IV TM9-2320-280-24P C01

CROSS REFERENCE INDEXES

PART NUMBER INDEX

CAGE	PART NUMBER	STOCK NUMBER	FIGURE NO	ITEM NO
		5305-00-068-0502	497	5
		5305-00-068-0502	498	17
80205	MS90725-94	5305-00-069-5578	398	3
96906	MS90726-162	5305-00-727-2283	437	8
80205	MS90726-34	5306-00-225-9089	280	45
		5306-00-225-9089	281	45
80205	MS90726-42	5306-00-225-9097	392	3
96906	MS90727-116	5305-00-719-5239	397	8
		5305-00-719-5239	437	12
96906	MS90727-84	5305-00-709-8283	234	19
80204	MS90728-7	5305-00-071-2505	418	1
		5305-00-071-2505	420	9
		5305-00-071-2505	421	3
		5305-00-071-2505	422	5
		5305-00-071-2505	423	12
28666	MT161A	5330-01-282-2213	277	10
55719	MT95	4910-01-231-3671	499	7
19728	MZ-294		45	57
68505	MZ-414S	5310-01-214-7784	45	44
02697	N1173 2-015	5331-01-495-4814	490A	2
76761	N1602-1	2920-01-466-1855	476	7
76761	N3135	2920-01-483-2291	40	46
		2920-01-483-2291	43B	34
03538	N405P42C12	5310-00-407-9566	176A	3
76761	N7309	6115-01-349-5320	40	5
76761	N7312	2920-01-363-5173	40	2
76761	N7346		40	9
76761	N7347		40	11
76761	N7348		40	18
76761	N7370	4140-01-350-0839	40	41
76761	N7398		43B	17
76761	N7399		43B	29
76761	N7400		43B	27
76761	N7401	2920-01-399-0794	43B	43
76761	N7402		43B	25
76761	N7433		43B	17
76761	N7434		40	18
09817	N80P21012Q90	5305-00-988-1725	88	31
76761	N9005	5305-01-032-4165	40	16
		5305-01-032-4165	43B	5
76761	N9008	5325-01-032-4222	40	13
76761	N9009	5325-01-034-2757	40	15
76761	N9010	3110-01-037-4661	40	14
76761	N9015	5310-01-046-0186	40	37
76761	N9016	5310-01-032-4169	40	47
76761	N9018	5310-01-032-4827	40	48
76761	N9062	5310-01-232-1363	43B	9
76761	N9063	5310-01-050-6565	40	6
		5310-01-050-6565	43B	39

SECTION IV TM9-2320-280-24P C01

CROSS REFERENCE INDEXES

PART NUMBER INDEX

CAGE	PART NUMBER	STOCK NUMBER	FIGURE NO	ITEM NO
76761	N9077	5306-01-223-4345	40	39
76761	N9092	5310-01-390-5105	43B	40
76761	N9098	5310-01-225-0701	40	3
76761	N9099	5310-01-165-1312	40	8
		5310-01-165-1312	43B	28
76761	N9256	5325-01-136-4880	43B	21
76761	N9260	5325-01-136-7662	40	12
		5325-01-136-7662	43B	19
76761	N9265	5310-01-136-4888	40	17
		5310-01-136-4888	43B	6
76761	N9269	5365-01-225-0844	40	35
76761	N9318	5999-01-348-0302	40	19
76761	N9331	5310-01-232-1361	40	30
76761	N9338	5310-01-232-1362	40	40
76761	N9343	5310-01-232-6617	43B	8
76761	N9385	3110-01-348-4867	40	10
		3110-01-348-4867	43B	33
76761	N9393	5310-01-348-8384	40	43
76761	N9406	5310-01-362-6171	43B	10
76761	N9408	5310-01-348-8386	40	33
76761	N9410	5310-01-348-8398	40	51
		5310-01-348-8398	43B	13
76761	N9414	5920-01-348-0303	40	32
76761	N9416	5310-01-348-8360	40	44
		5310-01-348-8360	43B	2
76761	N9417	3120-01-348-3364	40	20
76761	N9420	5970-01-350-5646	40	31
76761	N9426	5310-01-348-8313	40	38
76761	N9451	5315-01-348-6880	40	45
76761	N9452		40	4
76761	N9453	5310-01-348-8314	40	28
76761	N9455	5310-01-348-8385	40	36
76761	N9457	5306-01-348-8310	40	24
76761	N9459	5310-01-348-8393	40	25
76761	N9460	5365-01-348-6971	40	34
76761	N9461	5310-01-348-8392	40	29
76761	N9464	5975-01-350-1987	40	21
		5975-01-350-1987	43B	37
76761	N9465	5310-01-350-8549	40	23
		5310-01-350-8549	43B	36
76761	N9467	5310-01-350-4257	40	22
		5310-01-350-4257	43B	46
76761	N9486	3130-01-396-8388	43B	20
76761	N9491	4140-01-397-2228	43B	38
76761	N9492	3120-01-424-1525	43B	4
76761	N9493	5340-01-396-3950	43B	7
76761	N9494	5325-01-396-4320	43B	18
76761	N9500	5975-01-358-0021	40	27
76761	N9502	2920-01-475-3180	40	19

I-189

SECTION IV TM9-2320-280-24P C01

CROSS REFERENCE INDEXES

PART NUMBER INDEX

CAGE	PART NUMBER	STOCK NUMBER	FIGURE NO	ITEM NO
76761	N9545	3120-01-424-1028	43B	44
80205	NAS-1635-3LE12	5305-01-117-3396	376	9
80205	NAS1080R06	5320-01-075-0367	197	20
80205	NAS1080R08	5320-01-044-7545	199	28
80205	NAS1149F0863P	5310-01-396-8392	270	23
80205	NAS1329A3B80	5310-00-811-0966	291	4
		5310-00-811-0966	297	4
80205	NAS1329A3K280	5310-00-682-0775	257	6
80205	NAS1329A4B60	5310-01-481-4029	291	5
		5310-01-481-4029	297	8
80205	NAS1330A3-116	5310-00-964-7092	215	4
80205	NAS1387-4	5940-00-113-9420	485	3
81343	NAS1464-050-10T	5340-01-450-8693	487	4
80205	NAS1635-3-12P	5305-00-442-7347	40	49
80205	NAS1635-3LE12	5305-01-117-3396	321	10
		5305-01-117-3396	325	12
80205	NAS1831A6E28	5340-01-248-2878	59	15
80205	NAS1922-0150	4730-01-322-9160	302	4
80205	NAS1922-0150-1	4730-01-322-9160	305	6
		4730-01-322-9160	306	4
		4730-01-322-9160	309	5
80205	NAS387-832-12P	5305-01-035-8766	74	14
80205	NAS453-10K75	5310-00-550-4725	256	7
80205	NAS561C4-18	5315-00-559-7467	99	13
80205	NAS601-12P	5305-00-813-2785	375	10
		5305-00-813-2785	375	25
80205	NAS9301BNS-4-04	5320-01-143-5079	208	10
		5320-01-143-5079	220	1
		5320-01-143-5079	231	29
		5320-01-143-5079	369	15
80205	NAS9301BNS-6-02	5320-01-143-5075	208	13
80205	NAS9301BNS-6-03	5320-01-135-7319	278	1
		5320-01-135-7319	291	6
		5320-01-135-7319	366	26
		5320-01-135-7319	368	5
		5320-01-135-7319	369	8
		5320-01-135-7319	379	21
80205	NAS9301BNS-6-04	5320-01-136-1782	64	2
		5320-01-136-1782	229	5
		5320-01-136-1782	238	11
80205	NAS9301BNS-6-06	5320-01-084-9234	258	9
		5320-01-084-9234	259	1
80205	NAS9301BNS-6-07	5320-01-085-9995	293	32
80205	NAS9302BNS-6-05	5320-01-136-1787	290	5
80205	NAS9302BNS-6-06	5320-01-086-1148	219	3
		5320-01-086-1148	287	13
		5320-01-086-1148	290	3
80205	NAS9303B-6-03	5320-01-354-2547	220	5
80205	NAS9303B-6-04	5320-01-470-1545	197	16

SECTION IV TM9-2320-280-24P C01

CROSS REFERENCE INDEXES

PART NUMBER INDEX

CAGE	PART NUMBER	STOCK NUMBER	FIGURE NO	ITEM NO
80205	NAS9303B-6-06	5320-01-407-6239	197	16
81348	NN-P-530	5330-00-262-8195	BULK	48
60380	NTA2435	3110-00-580-3843	128	38
21335	NTA2840	3110-00-756-2022	128	44
06968	ORD-112	5330-01-197-8298	268	15
		5330-01-197-8298	295	6
73165	OS34400	5330-01-310-6780	4	27
9G287	P-100	2530-00-848-4581	167	12
9G287	P-450	6670-01-261-6847	167	12
9G287	P-500	6670-01-262-8647	167	12
9G287	P-550	6670-01-263-2268	167	12
21335	P207K	3110-00-155-6152	148	13
65035	P49866-11	5310-00-637-9541	206	22
61424	P4EUB-4	4730-01-298-4363	401	3
81755	P5006-4G	4020-00-246-0688	BULK	3
81755	P5006-4G-9		72	11
			72	22
			73	8
18631	PDG6120-1/8-1-1/4	5315-00-187-9384	458	48
70040	PF1218	4330-01-398-8484	6	7
61424	PFT-6B	4720-01-169-9891	BULK	19
61424	PFT-88-BLK-100	4720-01-003-6706	BULK	20
00198	PK379	5315-00-816-1794	165	7
		5315-00-816-1794	166	7
16941	PL25D02P12	5305-01-210-6249	321	1
16941	PL25D040P12	5305-01-198-3385	323	2
		5305-01-198-3385	323	7
		5305-01-198-3385	324	22
		5305-01-198-3385	324	34
		5305-01-198-3385	326	2
16941	PL25D040P8	5305-01-206-7217	324	21
		5305-01-206-7217	324	33
		5305-01-206-7217	375	20
16941	PL25J02P12	5305-01-199-3371	266	13
		5305-01-199-3371	317	9
		5305-01-199-3371	318	5
16941	PL25J02P20	5305-01-198-1619	324	11
		5305-01-198-1619	324	25
		5305-01-198-1619	375	16
18876	PMS90727001-06	5306-00-068-0513	66A	3
80063	PPL-10323	5820-01-222-1107	414	KIT
80063	PPL-10324	5820-01-222-1106	416	KIT
68505	PS-1475SS	2920-01-290-9245	45	17
82647	PSA-25	5925-01-373-1034	367	12
58727	PT1	5640-00-555-3011	467	12
77445	PWA10132-3	5310-00-225-6993	188	32
76700	Q186245	5340-01-506-1892	176A	9
76700	Q206141	5340-01-505-9463	176A	8

I-191

SECTION IV TM9-2320-280-24P C01

CROSS REFERENCE INDEXES

PART NUMBER INDEX

CAGE PART NUMBER STOCK NUMBER FIGURE NO ITEM NO

CAGE	PART NUMBER	STOCK NUMBER	FIGURE NO	ITEM NO
76700	Q348148	4730-01-506-3179	176A	6
76700	Q63330		176A	11
76700	Q79699	5330-01-505-7214	176A	10
76700	Q79869	5331-01-380-2118	176A	13
76700	Q79870	5331-01-505-7218	176A	7
21335	QAR29523	3110-00-689-4076	128	2
77252	QR816	5310-00-007-1607	283	12
09094	QRP3C0855	5315-01-197-7563	275	8
71843	R-2423-A	2510-01-264-6546	211	13
		2510-01-264-6546	212	14
26051	R-451-N	5330-01-041-9721	BULK	5
81343	R-495-T		BULK	56
35510	R027100150		39	35
35510	R055102143	5998-01-147-7895	39	74
83065	RB1450-1-4IDX1-20D	4720-00-684-4033	BULK	18
0ALB6	RC01A008-07-05	4730-01-455-1231	466	3
		4730-01-455-1231	467	2
0ALB6	RC01A009-07-04	4730-01-455-1259	466	2
0ALB6	RC01A010-07-08	4730-01-455-1276	466	1
01276	RC01C-008	4820-01-461-1703	466	6
0ALB6	RC01C-009	4820-01-455-5021	466	8
0ALB6	RC01D007-07-04	4730-01-455-1220	467	3
34623	RCSK-18685	3020-01-444-5447	38	8
34623	RCSK17028-1	2520-01-421-4589	146	1
34623	RCSK17028-2	2520-01-423-1947	139	1
34623	RCSK17028-3	2520-01-421-4588	139	1
34623	RCSK17089-1	2530-01-420-7892	152	6
34623	RCSK17089-2	2530-01-420-7893	152	6
34623	RCSK17123	5340-01-433-2743	183	21
34623	RCSK17176	2510-01-433-4421	183	12
34623	RCSK17215	2530-01-420-8025	KITS	10
34623	RCSK17250	2530-01-420-7895	144	20
			152	20
34623	RCSK17805		160	4
34623	RCSK18121	4710-01-473-2352	333	7
34623	RCSK18220	3020-01-473-2048	36	8
34623	RCSK18221	3020-01-442-7278	34	22
34623	RCSK18268	2920-01-444-2749	38	11
34623	RCSK18278-6	2590-01-444-3317	33	9
34623	RCSK18330	2530-01-460-2439	174	5
		2530-01-460-2439	175	1
34623	RCSK18375	2990-01-411-3954	24	5
34623	RCSK18434-2	4720-01-442-9875	160	1
		4720-01-442-9875	487A	3
34623	RCSK18434-3	4720-01-443-3033	160	16
		4720-01-443-3033	487A	2
34623	RCSK18435-2	4720-01-443-8487	162	25
		4720-01-443-8487	487A	16
34623	RCSK18435-3	4720-01-443-3481	162	9

SECTION IV TM9-2320-280-24P C01

CROSS REFERENCE INDEXES

PART NUMBER INDEX

CAGE PART NUMBER STOCK NUMBER FIGURE NO ITEM NO

CAGE	PART NUMBER	STOCK NUMBER	FIGURE NO	ITEM NO
		4720-01-443-3481	487A	15
34623	RCSK18531	5340-01-476-2963	188	26
34623	RCSK18532	5340-01-476-2972	188	36
34623	RCSK18596	5340-01-476-5072	188	30
34623	RCSK18626	3020-01-458-0176	33	5
		3020-01-458-0176	33A	5
34623	RCSK18627	3020-01-458-0185	33	3
34623	RCSK18628	3020-01-462-1202	33	4
34623	RCSK18639	5330-01-476-9732	180	22
34623	RCSK18683	5365-01-480-5282	33	19
34623	RCSK18686	3020-01-444-2311	175	3
34623	RCSK18747-1	5340-01-476-2845	188	33
34623	RCSK18747-2	5340-01-476-2840	188	33
34623	RCSK18809	2510-01-477-3968	180	24
34623	RCSK19027	5305-01-477-1867	180	25
34623	RCSK19361	5340-01-488-6185	176A	4
27182	RD-860066	5342-01-266-5689	211	2
01276	RE01-0001-0720	4720-01-458-3326	466	10
01276	RE01-0003-0720		466	7
01276	REO1-0002-0720	4720-01-456-6674	466	5
01276	REO1-0003-0360	4720-01-456-6729	466	9
14429	RESINITEEP69C	5970-00-221-5301	44	20
		5970-00-221-5301	69	21
		5970-00-221-5301	69	26
		5970-00-221-5301	72	5
		5970-00-221-5301	313	17
81348	RR-C-271-6		191	19
33287	RS-3400-135		505	10
33287	RS-5700-125		505	13
33287	RS-5701-795		505	12
53551	RV817	5130-01-045-3507	504	13
53551	RV818	5130-01-044-7196	504	12
82484	S-3076	3120-01-432-1474	327	7
82484	S-3087	3120-01-388-1527	327	8
03481	S10P196	5310-01-210-7935	224	6
25472	S10P320	5310-01-197-6253	254	26
		5310-01-197-6253	287	6
70318	S147-1	5310-00-933-8118	282	15
76385	S1608	5325-00-174-9038	70	1
0VK23	S25P296	5310-01-496-0115	327	49
03481	S31P280	5310-01-198-1724	197	4
		5310-01-198-1724	417	8
		5310-01-198-1724	430	18
27182	S4501XRXBZXX	5342-01-266-7559	289	8
27914	S45043-3405	5310-00-934-9748	282	16
27182	S450XXLXBZXX	5342-01-265-1129	293	27
25472	S4P90	5310-01-388-7420	274	19
72794	S7-250	5325-00-449-3001	208	5
53964	SA609010	5305-01-190-4073	19	2

SECTION IV TM9-2320-280-24P C01

CROSS REFERENCE INDEXES

PART NUMBER INDEX

CAGE PART NUMBER STOCK NUMBER FIGURE NO ITEM NO

CAGE	PART NUMBER	STOCK NUMBER	FIGURE NO	ITEM NO
9D117	SAEJ562		480	20
9D117	SAEJ562/3/4		474	14
19728	SAT-1004F		45	31
68505	SAT-1004FS	2920-01-192-4474	KITS	11
19728	SAT-1014B		45	35
19728	SAT-32	5330-00-193-0850	45	32
19728	SAT-40		45	37
19728	SAT-46		45	28
19728	SAT-47		45	26
19728	SAT-48		45	24
19728	SAT-52		45	25
68505	SAT4108UT	5945-01-210-1299	45	20
68505	SAV-1008A	2920-01-184-1054	45	21
28520	SB-750-625-2840	5975-01-211-3142	17	13
		5975-01-211-3142	153	18
		5975-01-211-3142	379	48
63728	SC202494	5330-01-393-9101	9	2
53964	SC220072	4330-01-280-8417	19	7
99587	SC300079		19	5
53964	SC300081		19	9
80063	SCD13617-8	5310-00-809-5998	33	2
27737	SCE-2012	3110-01-155-9862	128	37
27737	SCE-2416	3110-01-155-4440	128	1
2L480	SD1130	9330-00-065-8682	BULK	1
34623	SF-5583585		99	10
34623	SF-5590748	2510-01-203-9873	278	14
34623	SF-5595841	2510-01-252-4466	227	10
9C234	SF-5597234	4730-01-265-3166	302	10
34623	SF-5597692	5640-01-309-3866	200	4
34623	SF-5597693	5640-01-309-3870	200	3
34623	SF-5597695	2530-01-285-8381	169	23
34623	SF5575898		71	7
34623	SF5575925	5330-01-197-3620	208	31
34623	SF5581351	2510-01-189-3460	320	5
		2510-01-189-3460	324	10
34623	SF5581723	2510-01-197-5546	375	12
		2510-01-197-5546	378	11
34623	SF5581724	2510-01-197-5547	375	12
		2510-01-197-5547	378	11
34623	SF5585243		327	47
34623	SF5585258		327	22
34623	SF5589377	2540-01-276-5928	289	11
34623	SF5590706		225	17
			226	7
17168	SF5592171	2540-01-265-1137	224	11
34623	SF5597161	2510-01-281-1032	228	6
34623	SF5597214		304	10
			309	8
34623	SF5597216		305	10

SECTION IV TM9-2320-280-24P C01

CROSS REFERENCE INDEXES

PART NUMBER INDEX

CAGE	PART NUMBER	STOCK NUMBER	FIGURE NO	ITEM NO
34623	SF5597218		309	3
34623	SF5597219		309	7
34623	SF5597221		309	13
34623	SF5597226		310	1
34623	SF5597227		309	12
34623	SF5597229		306	2
34623	SF5597230		306	7
34623	SF5597231		306	10
34623	SF5597232	4710-01-270-9617	310	3
34623	SF5597426	2540-01-270-6537	236	3
34623	SF5597431	2540-01-270-6538	236	4
34623	SF5597606	2540-01-271-2832	290	2
34623	SF5597721	4730-01-270-7931	301	8
34623	SF5597723	4720-01-270-2226	304	4
34623	SF5597725	4710-01-272-7565	302	2
34623	SF5597744	4730-01-265-3284	302	5
		4730-01-265-3284	305	7
34623	SF5597958		302	9
			303	2
34623	SF5597970		310	9
34623	SF5598132	4710-01-272-0513	302	11
34623	SF5598133	4710-01-271-6957	302	12
34623	SF5598134		302	7
			305	9
34623	SF5934157	9330-01-276-1470	287	8
34623	SF5934158	2540-01-270-6459	219	5
		2540-01-270-6459	290	7
34623	SF5934173	5340-01-270-8432	260	18
34623	SF5934182	5330-01-265-8808	213	32
34623	SF5934196	5340-01-271-1846	211	22
		5340-01-271-1846	212	20
34623	SF5934202	4720-01-270-9616	303	7
66295	SIZE 83	4730-01-092-1904	332	9
66295	SIZE83	4730-01-092-1904	331	2
62915	SK-11002-6	5306-00-225-8503	329	8
00334	SK2155	6685-00-618-1822	363	28
53421	SNP-22	4730-01-239-5251	301	3
		4730-01-239-5251	303	12
7Z588	SP-311	2590-01-212-5868	488	1
08484	SP1548-1	5310-00-112-2233	39	12
01428	SPM-4	3120-01-154-6272	487	12
55719	SRES18	5120-01-367-3582	505	2
55719	SRES24	5120-01-367-3585	505	2
19728	SS-88A		45	36
08718	SS2R-120-1804-000	5935-01-149-5165	79	6
08718	SS4R-120-1806-000	5935-01-154-6233	79	5
8J942	SSLQ-68	5320-01-255-6608	202	4

I-195

SECTION IV TM9-2320-280-24P C01

CROSS REFERENCE INDEXES

PART NUMBER INDEX

CAGE	PART NUMBER	STOCK NUMBER	FIGURE NO	ITEM NO		
11815	SSPS-04-03	5320-01-221-4231			274	30
06383	SST1M-C0	5975-00-903-2284			98	6
17875	T-18	2640-00-555-2834			167	11
06481	T-218758	4730-00-908-6293			387	2
		4730-00-908-6293	393	16		
74410	T-60-A0S-L	2540-01-189-2193			191	8
83553	T048-270-250L	5360-01-101-2143			245	13
28666	T2H2-38P2	3110-01-457-9711			277	9
53421	T30MR-0	5975-01-299-7781			52	7
		5975-01-299-7781	57	6		
59875	TD97203	5325-00-263-6651			79	25
56501	TG11	5940-00-705-6711			75	16
		5940-00-705-6711	76	16		
		5940-00-705-6711	80	4		
		5940-00-705-6711	81	18		
		5940-00-705-6711	83	1		
		5940-00-705-6711	86	36		
56501	TG14	5940-00-705-6714			75	2
56501	TG15	5940-00-705-6715			75	15
		5940-00-705-6715	76	15		
56501	TG2	5940-00-705-6702			86	35
56501	TG3	5940-00-705-6703			75	17
		5940-00-705-6703	76	17		
		5940-00-705-6703	93	10		
56501	TG33	5940-00-735-5520			41	10
		5940-00-735-5520	44	13		
		5940-00-735-5520	44	19		
		5940-00-735-5520	44	24		
		5940-00-735-5520	72	16		
		5940-00-735-5520	73	25		
		5940-00-735-5520	474	11		
		5940-00-735-5520	480	26		
56501	TG8	5940-00-705-6708			75	7
		5940-00-705-6708	76	10		
		5940-00-705-6708	80	6		
		5940-00-705-6708	83	4		
		5940-00-705-6708	85	21		
		5940-00-705-6708	86	14		
		5940-00-705-6708	87	3		
56501	TG9	5940-00-705-6709			76	5
		5940-00-705-6709			85	30
		5940-00-705-6709			86	20
		5940-00-705-6709	89	6		
55719	TLE60	5120-01-227-3159			505	3
60380	TRA 2840	3110-00-055-2100			128	45
09527	TS6058	6620-01-194-6586			66	11
71124	TURB0LEX76B	5970-00-221-5301			69	12
		5970-00-221-5301	72	17		
96508	TW6B	5120-00-683-8597			505	5

I-196

SECTION IV TM9-2320-280-24P C01

CROSS REFERENCE INDEXES

PART NUMBER INDEX

CAGE	PART NUMBER	STOCK NUMBER	FIGURE NO		ITEM NO
56501	TY25M0	5975-00-984-6582	470		7
81348	TYIV/CL1/TRVC8	2640-01-098-2029	167		21
81348	TYV/CL2/TR C1	2640-00-050-1229	167		22
14557	U-6236	5340-01-197-8239	269		25
92830	U500-0113	5310-01-105-9398	293		8
59199	UM5008	4730-00-954-1251	141		10
		4730-00-954-1251	399	16	
62161	VF0010	2640-01-419-6202	167		17
68505	VSF-16	5310-01-214-7785	43		30
80604	W-129-B	4910-00-105-2823	502		6
78940	W-250D53	2940-01-188-3776	12		25
99017	W-4	5340-01-242-6245	235		14
99017	W-5	5340-01-255-1657	203		24
79470	W21204	4730-00-897-5497	178		20
		4730-00-897-5497	334		12
		4730-00-897-5497	366		10
		4730-00-897-5497	385	11	
27647	W6000HMV	2590-01-179-7602	313		12
		2590-01-179-7602	443	15	
08484	WA-1429-10	5305-00-984-6210	299		1
		5305-00-984-6210	430	7	
34623	WA12338230	2510-01-466-3752	182		11
81348	WW-C-440B TYPE E 1/2	4730-00-267-8945	401		2
81348	WW-P-471ACABCB	4730-00-954-1281	5		1
		4730-00-954-1281	436	4	
81349	WW-P-471ACABUA	4730-01-166-2244	105		11
81349	WW-P-471AQABUA	4730-01-166-2244	104		3
66295	WWD48-58H	4730-00-908-3193	364		3
19728	X-3508		45		33
19728	X-3925SC		45		62
19728	X-4059		43		19
19728	X-4070		43		12
80293	X-4071		43		17
19728	X-4328		43		21
81348	X/GP3/TYRA/CL0/0/37-12.50R16.50/	2610-01-333-7632	168		1
63829	X10		491		3
63829	X11		491		6
63829	X12		491		25
63829	X13		491		27
63829	X14		491		14
63829	X15		491		15
63829	X16		491		16
63829	X17		491		18
63829	X18		491		13
14351	X601	5315-00-842-3044	97		6
		5315-00-842-3044	99		18
		5315-00-842-3044	133		9

SECTION IV TM9-2320-280-24P C01

CROSS REFERENCE INDEXES

PART NUMBER INDEX

CAGE	PART NUMBER	STOCK NUMBER	FIGURE NO	ITEM NO
		5315-00-842-3044	153	13
		5315-00-842-3044	202	39
		5315-00-842-3044	205	21
41885	X91430	5306-01-417-2467	167	15
19728	XA-1316		43	7
			43	13
19728	XA-1317		43	16
68505	XA-1642	5330-01-209-7669	45	3
19728	XA-744AP		45	52
19728	XA-744AV		43	8
68505	XA-744AV	5331-00-078-0265	43	22
19728	XA-744AW		43	24
19728	XA-744AY		43	10
19728	XA-744Z	5331-00-137-3450	45	9
78553	XAN-262-H	5325-01-166-6324	128	74
13940	XB78323-05001	5325-00-823-5999	198	2
		5325-00-823-5999	199	11
		5325-00-823-5999	225	23
		5325-00-823-5999	239	9
		5325-00-823-5999	317	7
		5325-00-823-5999	318	11
		5325-00-823-5999	321	8
		5325-00-823-5999	325	19
		5325-00-823-5999	373	6
		5325-00-823-5999	376	7
		5325-00-823-5999	376	11
63829	XX1		491	2
63829	XX3		491	21
63829	XX4		491	20
63829	XX5		491	19
63829	XX6		491	5
63829	XX7		491	7
63829	XX8		491	12
63829	XX9		491	11
78462	Y797-ABIFE-1 1/2 CP60	4820-01-455-5020	467	8
35510	Z0982-75723	5310-01-143-1679	39	45
35510	Z106005179	5305-00-777-2284	39	57
35510	Z116 096169	5920-01-275-8096	39	58
35510	Z116095517	5305-01-147-5864	39	52
76385	ZB-1079M.I.	5340-01-203-8675	280	17
		5340-01-203-8675	281	17
76385	ZX-2024	5330-01-275-3381	296	8
76385	ZX-4597	9390-01-282-9949	296	7
77820	10-40817-010	5935-00-772-2344	85	13
		5935-00-772-2344	86	18
77820	10-42622-235	5935-00-614-3959	80	7
		5935-00-614-3959	85	7
81834	10-7005-06	6695-01-216-8687	64	9

SECTION IV TM9-2320-280-24P C01

CROSS REFERENCE INDEXES

PART NUMBER INDEX

CAGE	PART NUMBER	STOCK NUMBER	FIGURE NO	ITEM NO
35510	100204	6150-01-269-1857	39	46
35510	100209	6150-01-335-3087	39	79
35510	100210	5340-01-270-8484	39	49
35510	100212	5340-01-272-2677	39	34
24975	100276	5330-01-272-7454	39	4
7X677	10054241	5330-01-096-7699	78	2
98853	100828-HN37	5940-01-443-9580	95	4
54905	10084	5340-01-426-3850	307	14
		5340-01-426-3850	308	2
91094	100951-D02	5940-01-269-1816	89	4
70898	100951X031-012-028	5310-00-641-9464	251	5
7X677	10137486	5330-01-150-5944	10	1
7X677	10137488	5330-01-149-0874	30	12
7X677	10137492	5330-01-147-9808	31	4
28891	101450-S02	5940-01-214-4563	79	10
7X677	10149551	4710-01-314-4489	11	30
7X677	10149552	4710-01-314-4483	11	10
7X677	10149553	4710-01-314-4488	11	28
7X677	10149554	4710-01-316-0222	11	11
7X677	10149555	4710-01-314-4487	11	18
7X677	10149556	4710-01-314-4485	11	16
7X677	10149557	4710-01-314-4486	11	24
7X677	10149558	4710-01-314-4484	11	17
7X677	10154669	4710-01-493-1712	11	30
60602	10166	5315-01-284-9812	21	13
7X677	10183956	4730-01-445-2358	30	15
		4730-01-445-2358	30	17
7X677	10183996		11	23
18876	10189683	5315-00-255-5580	268	6
7X677	10191497	5340-01-493-2555	11	32
19207	10192299	5340-01-197-2717	268	7
19207	10192302	5340-01-197-1444	268	12
18876	10224818	5325-00-886-1067	247	8
7X677	10225180	5340-01-493-2538	11	31
7X677	10243629		30	14
7X677	103384	5315-01-191-2611	133	4
34281	103700-01	2530-01-204-2583	158	2
34281	103700-02	2530-01-207-6256	158	2
34281	103700-09	2530-01-484-9574	158	2
34281	103700-10	2530-01-484-9573	158	2
34281	103706-05		158	7
34281	103706-06		158	7
34281	103706-07		158	7
34281	103706-08		158	7
34281	10371009	2530-01-185-6714	158	9
34281	103712-01	2530-01-185-7998	157	14
		2530-01-185-7998	158	10
70655	10392	5315-01-280-6185	230	9

SECTION IV TM9-2320-280-24P C01

CROSS REFERENCE INDEXES

PART NUMBER INDEX

CAGE PART NUMBER STOCK NUMBER FIGURE NO ITEM NO

CAGE	PART NUMBER	STOCK NUMBER	FIGURE NO	ITEM NO
		5315-01-280-6185	287	23
22075	10453002	2530-01-314-1129	157	22
		2530-01-314-1129	158	12
34281	104534-01	2530-01-185-6713	157	20
		2530-01-185-6713	158	15
7X677	10478147	5950-01-480-5380	119	13
7X677	10478148	5950-01-480-5377	119	11
7X677	10483630	5331-01-494-2431	119	32
34281	104870-01	2530-01-179-7511	KITS	14
7X677	104918		119	7
56161	1051759	5305-01-159-0065	175A	11
78385	10530A24	2540-00-113-4180	348	1
7X677	105455	5330-01-155-4393	128	70
68505	105651	6220-00-880-1624	60	4
18876	10579085-1	4010-00-224-482	BULK	49
60602	10618		21	12
01989	1062X6X4	4730-01-184-4760	141	25
27767	106397-03	2530-01-253-2825	157	16
		2530-01-253-2825	158	11
33116	106397-04		157	16
			158	11
79470	1064X4	4730-01-003-5105	123	4
		4730-01-003-5105	141	23
		4730-01-003-5105	149	12
		4730-01-003-5105	313	8
		4730-01-003-5105	399	6
		4730-01-003-5105	443	10
		4730-01-003-5105	457	2
79470	1064X6X6X4	4730-01-476-0850	13	13
		4730-01-476-0850	14	19
79470	1065X4	4730-01-058-0900	149	11
01989	1065X6	4730-01-211-0896	18	7
01989	1068X4	4730-01-185-5348	12	41
		4730-01-185-5348	135	3
		4730-01-185-5348	141	26
		4730-01-185-5348	149	14
		4730-01-185-5348	399	4
34281	106980-02	2530-01-185-6712	157	23
		2530-01-185-6712	158	13
01989	1069X4	4730-01-070-7680	12	40
		4730-01-070-7680	399	3
01989	1069X4X1	4730-01-195-3803	400	7
79470	1069X6	4730-00-041-2526	30	19
01989	1077X4	4730-01-192-8086	141	19
34281	108037-01	5365-01-186-7233	158	3
34281	108039-01	3120-01-186-7181	158	4
58538	10812	4010-01-203-0412	270	4
7X677	108597	5306-01-164-8486	118	9
19207	10886344-3	4720-00-005-5008	385	3

I-200

SECTION IV TM9-2320-280-24P C01

CROSS REFERENCE INDEXES

PART NUMBER INDEX

CAGE	PART NUMBER	STOCK NUMBER	FIGURE NO	ITEM NO
19207	10896651	9905-01-032-7002	394	4
51377	1090-06404-01	2930-01-168-7870	34	2
15434	109155	5315-00-839-2325	21	19
		5315-00-839-2325	21	29
19207	10918603	5310-00-877-5975	398	9
19207	10929868	2920-00-909-2483	35	6
34281	109413-01	4820-01-186-0822	159	2
34821	109433-01		157	6
34281	109457-01	5305-01-276-4855	157	1
19207	10948233	5930-00-679-5925	345	15
		5930-00-679-5925	348	38
19207	10948235	5330-00-089-0978	345	20
96881	10L18F	3120-01-191-3232	99	23
		3120-01-191-3232	133	8
79470	110-03604	4730-00-789-0951	177	3
		4730-00-789-0951	178	1
24617	110200	4730-00-266-0536	366	12
81100	110401-8SS	5310-00-933-8119	482	18
72582	111625	4730-00-142-2010	125	20
17773	11176106-5	5340-00-150-1658	83	23
		5340-00-150-1658	488	10
62161	1118BAC1048	505	8	01989
112-06136	4730-01-184-6971	159	8	
		4730-01-184-6971	161	9
78189	112-1008800-003	5305-01-204-6502	25	5
		5305-01-204-6502	51	6
		5305-01-204-6502	83	6
		5305-01-204-6502	335	6
19207	112446825	6150-01-416-7899	41	6
		6150-01-416-7899	70	12
29510	1125218C1	4720-00-065-8682	BULK	17
52304	113324	5305-00-071-2069	283	5
		5305-00-071-2069	417	16
45152	114356A	5310-00-061-4650	103	6
18876	11459081	8315-01-115-7617	BULK	4
24617	11500161	5310-01-416-5447	101	6
7X677	11500177	5310-01-185-7218	138	19
		5310-01-185-7218	139	17
		5310-01-185-7218	145	19
		5310-01-185-7218	146	17
7X677	11500207	5310-01-206-7306	1	4
		5310-01-206-7306	2	5
		5310-01-206-7306	12	19
		5310-01-206-7306	30	8
		5310-01-206-7306	33	11
		5310-01-206-7306	35	26
		5310-01-206-7306	36	19
		5310-01-206-7306	37	14
		5310-01-206-7306	38	19

SECTION IV TM9-2320-280-24P C01

CROSS REFERENCE INDEXES

PART NUMBER INDEX

CAGE	PART NUMBER	STOCK NUMBER	FIGURE NO	ITEM NO
		5310-01-206-7306	41	12
		5310-01-206-7306	42	10
		5310-01-206-7306	66	7
		5310-01-206-7306	75	31
		5310-01-206-7306	77	9
		5310-01-206-7306	174	2
		5310-01-206-7306	33A	9
		5310-01-206-7306	364	6
		5310-01-206-7306	379	37
		5310-01-206-7306	472	2
		5310-01-206-7306	473	10
		5310-01-206-7306	474	4
		5310-01-206-7306	475	14
		5310-01-206-7306	476	16
		5310-01-206-7306	480	8
		5310-01-206-7306	490	7
		5310-01-206-7306	490A	6
7X677	11500256	5310-01-210-1587	37	13
		5310-01-210-1587	472	3
		5310-01-210-1587	473	11
		5310-01-210-1587	474	3
		5310-01-210-1587	475	16
		5310-01-210-1587	476	22
24617	11500324	5310-01-204-6745	490	12
24617	11500362	5310-01-394-9280	370	10
24617	11500749	5310-01-165-3331	4	30
		5310-01-165-3331	7	8
		5310-01-165-3331	11	26
7X677	11500815	5305-01-149-1938	31	8
7X677	11500827	5305-01-254-2521	174	9
		5305-01-254-2521	66A	1
24617	11500831	5306-01-165-4283	128	23
		5306-01-165-4283	130	19
7X677	11500921	5306-01-150-1190	31	12
		5306-01-150-1190	122	33
7X677	11501032	5310-01-186-1254	1	3
24617	11501086	5310-01-504-5729	490A	9
24617	11501619	5306-01-473-1895	33	10
7X677	11501940	5306-01-206-7172	66	8
		5306-01-206-7172	77	24
		5306-01-206-7172	79	18
24617	11503428	5306-01-152-4696	128	21
		5306-01-152-4696	130	21
7X677	11503617	5305-01-150-1521	11	14
7X677	11503639	5310-01-155-2503	10	6
7X677	11503739	5310-01-158-6257	75	36
		5310-01-158-6257	77	17
24617	11503870	5306-01-474-6117	120	18
24617	11503962	5310-01-444-3084	77	16

I-202

SECTION IV TM9-2320-280-24P C01

CROSS REFERENCE INDEXES

PART NUMBER INDEX

CAGE	PART NUMBER	STOCK NUMBER	FIGURE NO		ITEM NO
7X677	11504512	5306-01-148-3667		11	5
24617	11504594	5306-01-447-2568		33	13
7X677	11504596	5306-01-243-4650		175	2
24617	11504617	5306-01-432-7900		33	12
24617	11504726	5975-01-475-3442		227	46
24617	11504986	5306-01-152-4693		22	1
7X677	11504986	5306-01-185-7051		66A	13
7X677	11505074	5306-01-411-2342		101	7
73342	11505328	5305-01-360-1136		33	15
7X677	11505885	5325-01-196-5631		245	8
24617	11505922	5310-01-315-1535		379	38
7X677	11508353	5306-01-253-7073		4	19
7X677	11508534	5306-01-230-3354		4	21
7X677	11508600	5306-01-165-5582		4	12
24617	11508687	5306-01-197-1492		108	1
7X677	11508805			105	10
7X677	11509669	5306-01-151-4925		9	6
24617	1150967	5305-01-149-9673		31	11
7X677	11513606	5306-01-185-7048		1	6
		5306-01-185-7048	103	13	
		5306-01-185-7048	138	18	
		5306-01-185-7048	139	16	
		5306-01-185-7048	145	18	
		5306-01-185-7048	146	16	
		5306-01-185-7048	175	10	
7X677	11513697	5306-01-474-8959		119	10
7X677	11514578	5325-01-480-7331		117	17
7X677	11514603	5310-01-155-2503		7	5
		5310-01-155-2503	75	27	
		5310-01-155-2503	101	4	
7X677	11515262	5306-01-476-3500		120	16
7X677	11515756	5306-01-465-2140		105	19
7X677	11515763	5305-01-186-5381		103	11
		5305-01-186-5381	105	23	
7X677	11516075	5310-01-206-5479		365	18
7X677	11516230	4730-01-444-0701		30	13
19207	11595179	5120-00-880-4268		503	4
7X677	116000	5310-01-187-0678		374	15
34623	11602903	2540-01-317-4289		393	10
		2540-01-317-4289	395	10	
19207	11608950-13	4730-00-908-6292		24	2
19207	11608950-18	4730-01-296-9318		337	4
		4730-01-296-9318	398	31	
19207	11608950-6	4730-00-278-2523		331	30
19207	11609693-1	2540-00-104-2889		389	2
19207	11609942	5340-01-320-8662		387	11
29930	116103	5330-00-135-6383		105	21
19207	11613631	5945-00-789-3706		49	8
19207	11614131	5930-00-134-5036		46	11

I-203

SECTION IV TM9-2320-280-24P C01

CROSS REFERENCE INDEXES

PART NUMBER INDEX

CAGE PART NUMBER STOCK NUMBER FIGURE NO ITEM NO

CAGE	PART NUMBER	STOCK NUMBER	FIGURE NO	ITEM NO
19207	11614131-1		46	16
19207	11630585	9905-01-181-9456	357	3
19207	11639519-1	5331-00-463-0200	60	10
19207	11639519-2	5331-00-462-0907	61	11
19207	11639541		60	5
19207	11639546	6220-00-179-4325	60	9
19207	11643398	9905-00-198-2728	354	2
19207	11644801	7690-00-138-5788	362	1
19207	11655788-2	5120-00-240-8702	505	4
19207	11663000	4730-00-150-6118	11	8
19207	11663057	5930-00-345-5455	345	10
		5930-00-345-5455	348	14
19207	11663058	4810-00-248-1635	348	40
19207	11663061	5905-00-251-7145	348	4
19207	11663279	5930-00-207-9422	51	3
19207	11668591	5935-01-184-6417	85	24
		5935-01-184-6417	86	15
19207	11668932	6220-01-094-1440	62	1
19207	11668979	5310-01-100-2067	62	5
19207	11669531	5935-01-314-2084	380	9
19207	11669705	2540-01-125-6154	388	16
19207	11675004	5340-01-059-0114	72	9
19207	11682088-1	5340-01-114-7712	402	11
24617	117212	5310-00-568-6077	100	12
76760	11797	5306-01-359-4530	130	2
94222	12-11014-12	5310-00-880-9344	216	5
		5310-00-880-9344	221	5
94222	12-12-404-11	5305-01-185-2211	216	8
		5305-01-185-2211	221	4
68505	12-333	2920-01-168-4131	43	27
71468	120-1805-000	5935-01-153-9320	76	13
K5667	1200359	2530-01-491-2681	175A	3
77060	12004436	5940-01-341-7689	93	2
34281	120100-07		157	7
34281	120100-08		157	8
34281	120100-09		157	13
34281	120100-10		157	11
30379	120217	5310-00-922-2017	370	17
		5310-00-922-2017	465	3
		5310-00-922-2017	487	3
72582	120217	5310-00-922-2017	64	15
		5310-00-922-2017	64	25
		5310-00-922-2017	67	3
		5310-00-922-2017	88	42
		5310-00-922-2017	225	49
		5310-00-922-2017	226	29
24617	120367	5310-00-012-0367	370	24
12204	120375	5310-00-584-5005	468	9
24617	120376	5310-00-409-3355	176A	2

I-204

SECTION IV TM9-2320-280-24P C01

CROSS REFERENCE INDEXES

PART NUMBER INDEX

CAGE	PART NUMBER	STOCK NUMBER	FIGURE NO	ITEM NO
79410	120380	5310-00-838-2024	283	15
24617	120384	5310-00-482-9493	33	6
29510	120384	5310-00-011-6121	127	5
		5310-00-011-6121	472	12
		5310-00-011-6121	475	20
		5310-00-011-6121	476	18
		5310-00-011-6121	481	3
24617	120388	5310-00-012-0388	33A	14
		5310-00-012-0388	33B	4
24617	120392	5310-01-103-6042	176A	20
81306	120393	5310-01-308-0131	199	8
21450	120487	4730-00-043-3750	345	34
		4730-00-043-3750	348	34
24617	120528	5340-01-166-1672	25	2
		5340-01-166-1672	103	5
72915	120613	5310-00-297-8836	66A	7
77060	12078084	5935-01-443-9124	86	29
77060	12084912	5940-01-442-6298	86	32
7X677	12143072	6150-01-470-2009	78	1
34281	1215567-01	5340-01-265-8883	157	2
77060	12160490	5935-01-445-5599	86	31
77060	12160494	5935-01-445-5144	86	33
27767	12173601	2530-01-333-8263	157	5
27767	12173602	2530-01-333-6068	157	5
16764	121743	5310-00-012-1743	269	29
34271	121743-01		157	15
7X677	12181688	6150-01-470-2016	78	1
7X677	12181863	6150-01-470-2011	78	1
41024	12214139	5305-01-282-3386	224	9
19207	12255608	5315-01-094-9025	403	11
19207	12258931-2	6680-01-148-8875	66	4
34345	12258932-6	6685-01-167-4298	67	1
19207	12265594	5895-01-050-2072	274	20
19207	12267802	5330-01-080-3253	104	15
19207	12275161	2540-01-123-1218	368	16
19207	12296696-2	5305-01-271-1806	191	13
19207	12298661	5340-01-110-4036	245	28
		5340-01-110-4036	478	5
78500	1229F1514	5310-00-080-6004	30	7
		5310-00-080-6004	41	13
19207	12300715-1	5325-01-175-7442	387	13
19207	12301363-1	2540-01-200-1994	316	9
		2540-01-200-1994	378	17
19207	12301363-2	2540-01-203-7721	316	9
		2540-01-203-7721	320	9
		2540-01-203-7721	324	12
		2540-01-203-7721	375	17
		2540-01-203-7721	378	17
19207	12301365	2540-01-200-1995	316	14

SECTION IV TM9-2320-280-24P C01

CROSS REFERENCE INDEXES

PART NUMBER INDEX

CAGE	PART NUMBER	STOCK NUMBER	FIGURE NO	ITEM NO
		2540-01-200-1995	320	12
		2540-01-200-1995	320	24
		2540-01-200-1995	324	19
		2540-01-200-1995	324	31
		2540-01-200-1995	375	22
		2540-01-200-1995	378	22
7X677	123153	5310-01-185-4672	137	16
19207	12331924-6		231	10
7X677	12337210	4330-01-121-6350	KITS	15
7X677	12337891	4730-00-166-9178	30	13
7X677	12338062	5310-01-148-2687	129	35
19207	12338064	8145-01-231-3747	363	1
19207	12338065		363	9
19207	12338066		363	4
19207	12338067	5340-01-271-3059	363	7
19207	12338068	5340-01-264-6544	363	6
19207	12338070	5342-01-264-6543	363	22
19207	12338071	5340-01-265-3676	363	18
19207	12338072-1	7690-01-484-4111	357	6
19207	12338073	5330-01-264-6537	363	5
19207	12338074	5340-01-264-6540	363	8
19207	12338078	2520-01-268-1051	363	2
19207	12338084	5325-01-205-5378	339	5
		5325-01-205-5378	410	31
		5325-01-205-5378	412	24
19207	12338085	5325-01-205-3203	410	36
		5325-01-205-3203	412	12
19207	12338086-2	5995-01-269-9525	338	14
19207	12338086-3	5995-01-201-7495	410	1
19207	12338086-4	5995-01-251-9316	414	1
		5995-01-251-9316	416	2
19207	12338089-2	6150-01-269-1848	338	6
		6150-01-269-1848	339	7
19207	12338089-3	5995-01-201-4129	410	2
		5995-01-201-4129	412	2
19201	12338089-4	5995-01-251-6785	416	1
		5995-01-251-6785	428	2
19207	12338092	2590-00-454-3620	93	23
		2590-00-454-3620	96	30
		2590-00-454-3620	338	2
		2590-00-454-3620	339	6
		2590-00-454-3620	410	32
19207	12338095	5310-01-208-7576	410	15
		5310-01-208-7576	414	10
		5310-01-208-7576	416	13
		5310-01-208-7576	425	11
		5310-01-208-7576	426	11
		5310-01-208-7576	427	12
		5310-01-208-7576	428	13

SECTION IV TM9-2320-280-24P C01

CROSS REFERENCE INDEXES

PART NUMBER INDEX

CAGE	PART NUMBER	STOCK NUMBER	FIGURE NO	ITEM NO
19207	12338098	5325-01-208-5424	410	10
		5325-01-208-5424	414	8
		5325-01-208-5424	416	8
34623	12338099-1	6150-01-207-8120	416	9
		6150-01-207-8120	426	16
		6150-01-207-8120	427	16
		6150-01-207-8120	428	9
19207	12338100-1	6150-01-269-1839	339	21
19207	12338100-2	6150-01-269-1849	338	1
34623	12338103	2540-01-250-7591	430	KITS
19207	12338105	2540-01-185-3216	402	KIT
		2540-01-185-3216	452	2
19207	12338119	2510-01-246-8287	452	3
19207	12338124	5310-01-413-2049	166	9
19207	12338146	2590-01-185-6711	182	12
		2590-01-185-6711	183	20
19207	12338147	2510-01-175-7221	179	11
19207	12338148	2510-01-187-7033	183	25
19207	12338150-1	5340-01-254-6543	140	3
19207	12338150-2	2530-01-257-3864	140	6
34623	12338150-6	2590-01-395-2228	140	3
9C234	12338150-7	2590-01-395-2229	140	6
19207	12338151-1	5342-01-277-9101	147	6
19207	12338151-6	5340-01-461-7851	147	6
34623	12338151-7	2590-01-394-8134	147	3
19207	12338152	2510-01-203-9883	184	15
34623	12338155-1	2510-01-185-6647	182	19
34623	12338155-2	2510-01-185-6648	182	19
19207	12338159	5340-01-214-2089	181	1
19207	12338163	5340-01-186-7173	2	13
19207	12338164	5340-01-205-5377	188	20
19207	12338165	2530-01-210-3567	182	9
19207	12338166	5340-01-192-3414	182	5
19207	12338167-1	2590-01-184-4475	184	3
19207	12338167-2	5340-01-361-6658	184	3
19207	12338170	5340-01-190-2248	141	18
		5340-01-190-2248	162	1
		5340-01-190-2248	181	13
19207	12338171	5365-01-210-1628	184	8
19207	12338172	5340-01-209-7767	149	10
		5340-01-209-7767	183	26
		5340-01-209-7767	184	12
19207	12338173	5340-01-210-7966	181	12
34623	12338174	2510-01-210-6942	184	16
19207	12338178	5340-01-185-8620	179	14
19207	12338181	5340-01-186-7664	183	6
19207	12338183	2510-01-185-6114	2	12
19207	12338185	2590-01-186-0863	179	10
		2590-01-186-0863	180	16

I-207

SECTION IV TM9-2320-280-24P C01

CROSS REFERENCE INDEXES

PART NUMBER INDEX

CAGE	PART NUMBER	STOCK NUMBER	FIGURE NO	ITEM NO
19207	12338186-20	5310-01-292-5354	42	12
19207	12338186-52	5365-01-216-0364	277	30
		5365-01-216-0364	278	7
		5365-01-216-0364	440	3
19207	12338186-62	5310-01-337-7034	42	12
19207	12338186-69	5365-01-466-0850	463	15
19207	12338186-88	5365-01-488-5594	33B	8
19207	12338186-89	5365-01-495-5340	462	10
19207	12338187-1	5340-01-249-1576	179	5
19207	12338189	2540-01-186-6039	191	6
19207	12338190	2540-01-203-5664	184	18
19207	12338191-1	2540-01-187-7034	181	15
19207	12338191-2	2540-01-188-3239	181	15
19207	12338193	9520-01-175-7222	184	5
19207	12338194	2540-01-196-4723	188	21
19207	12338197	5340-01-204-2584	183	11
19207	12338199	2540-01-186-0501	185	5
		2540-01-186-0501	187	5
34623	12338202-1	2540-01-196-4721	188	16
19207	12338202-2	2540-01-196-4722	188	16
19207	12338205	2510-01-185-6113	2	10
19207	12338215	5340-01-175-7208	140	9
19207	12338217	2510-01-175-7217	183	12
19207	12338218	2540-01-196-4920	188	7
19207	12338221	2990-01-187-7030	23	10
19207	12338222	2510-01-175-7218	183	22
19207	12338223	2510-01-175-7219	182	17
19207	12338224	2510-01-210-6941	183	15
19207	12338225-1	5305-01-264-3602	183	1
		5305-01-264-3602	185	11
		5305-01-264-3602	187	15
		5305-01-264-3602	188	11
19207	12338226-2	5365-01-209-7832	181	5
		5365-01-209-7832	184	9
19207	12338230	2510-01-175-7216	182	11
19207	12338233-1		179	1
19207	12338233-2		179	1
19207	12338233-3		179	1
19207	12338233-4		179	1
19207	12338233-5		179	1
19207	12338233-6		179	1
19207	12338234		179	1
19207	12338234-1		179	1
19207	12338240	2530-01-189-1745	184	10
19207	12338245	2590-01-215-4325	181	6
19207	12338246	2510-01-249-1599	179	2
19207	12338268-2	2530-01-176-4650	144	6
19207	12338270	3120-01-186-5527	144	7
		3120-01-186-5527	144	21

SECTION IV TM9-2320-280-24P C01

CROSS REFERENCE INDEXES

PART NUMBER INDEX

CAGE	PART NUMBER	STOCK NUMBER	FIGURE NO	ITEM NO
		3120-01-186-5527	152	7
		3120-01-186-5527	152	21
19207	12338290	2530-01-187-7161	156	18
19207	12338291	2530-01-188-8446	153	8
19207	12338294	5306-01-197-3277	153	10
19207	12338295	2530-01-188-3520	153	7
34623	12338296	4710-01-185-9668	159	7
19207	12338301	5340-01-209-7795	159	13
19207	12338302	5340-01-209-7796	161	1
19207	12338304	4710-01-189-2208	161	11
19207	12338306	4710-01-185-9667	159	5
19207	12338308	4710-01-189-2209	161	3
19207	12338315	2510-01-187-7037	193	8
19207	12338316-1	5360-01-179-4107	192	1
34623	12338316-2	5360-01-179-4106	192	1
34623	12338316-3	5360-01-254-1492	192	1
19207	12338317	5325-01-191-0595	194	4
19207	12338319	5340-01-190-0810	194	5
19207	12338320	2530-01-210-1324	144	11
		2530-01-210-1324	152	11
19207	12338322	5310-01-185-8689	194	9
34623	12338323	3040-01-186-3718	194	10
19207	12338324	5340-01-163-4773	150	34
		5340-01-163-4773	151	40
19207	12338328	3040-01-188-3684	144	17
34623	12338328	3040-01-188-3684	152	17
19207	12338329	2510-01-187-2239	194	3
19207	12338330-1		125	1
19207	12338330-19.5		176A	19
34623	12338330-2	4720-01-409-2499	125	8
19207	12338330-22.5		176A	16
19207	12338330-6		178	12
			436	1
19207	12338330-7		178	7
			436	2
19207	12338330-8		477	3
19207	12338330-9		477	2
34623	12338333	4710-01-189-0705	124	2
		4710-01-189-0705	125	2
19207	12338338	5325-01-187-1604	23	19
		5325-01-187-1604	24	18
19207	12338339	5330-01-200-0466	23	13
		5330-01-200-0466	24	21
		5330-01-200-0466	398	24
34623	12338340	5340-01-211-3137	23	20
		5340-01-211-3137	24	17
19207	12338342	5330-01-189-9738	23	3
		5330-01-189-9738	24	15
19207	12338343-1	5340-01-212-3553	23	6

I-209

SECTION IV TM9-2320-280-24P C01

CROSS REFERENCE INDEXES

PART NUMBER INDEX

CAGE PART NUMBER STOCK NUMBER FIGURE NO ITEM NO

CAGE	PART NUMBER	STOCK NUMBER	FIGURE NO	ITEM NO
		5340-01-212-3553	24	23
19207	12338343-2	5340-01-484-7760	24	3
19207	12338343-3	5340-01-383-2397	401	10
19207	12338346	2990-01-211-8587	23	16
		2990-01-211-8587	24	20
19207	12338350	2990-01-190-1089	23	15
34623	12338350	2990-01-190-1089	24	24
19207	12338351	2990-01-187-7168	23	5
19207	12338356	3120-01-186-7714	21	23
19207	12338362	5340-01-383-0462	21	3
19207	12338365	2540-01-185-9656	21	26
34623	12338369	2540-01-217-2669	21	20
19207	12338373	2940-01-188-3224	12	5
19207	12338379	5330-01-188-0911	12	38
19207	12338380	4720-01-185-8152	12	8
19207	12338382	5330-01-246-1822	12	14
		5330-01-246-1822	364	2
19207	12338390	3040-01-186-0967	165	10
19207	12338394	2540-01-184-4389	165	4
19207	12338415-3	2520-01-291-2975	140	4
		2520-01-291-2975	147	4
19207	12338425	2540-01-185-6710	185	9
19207	12338426	2540-01-189-1074	185	1
19200	12338428-2	6680-01-442-9413	478	1
		6680-01-442-9413	488	2
19207	12338433	9905-01-207-6304	133	2
19207	12338434	2920-01-249-3492	99	20
19207	12338435	5930-01-212-3373	67	6
19207	12338439	2540-01-184-5502	99	4
19207	12338440-1	2520-01-184-5501	97	1
19207	12338440-2	2520-01-204-7699	97	1
19207	12338441	5342-01-188-3233	99	27
19207	12338442-1	3040-01-185-4388	99	21
19207	12338443	2520-01-189-1064	99	16
19207	12338448	2520-01-189-6726	133	10
34623	12338449	3040-01-188-3234	99	28
19207	12338451	3040-01-189-2134	99	24
19207	12338452	3040-01-198-5699	99	14
19207	12338454	2520-01-184-7036	99	22
19207	12338456	3040-01-249-1435	99	9
19207	12338457	5340-01-188-5072	99	5
19207	12338458-1	2520-01-188-8269	99	19
19207	12338459	3040-01-189-2172	133	14
19207	12338460	5306-01-201-7969	133	6
19207	12338462	5330-01-211-5856	47	5
19207	12338463	6680-01-195-2146	488	11
19207	12338465	6210-01-208-4790	47	1
19207	12338466	2540-01-185-6117	46	8
19207	12338467	2510-01-248-8875	47	6

SECTION IV TM9-2320-280-24P C01

CROSS REFERENCE INDEXES

PART NUMBER INDEX

CAGE	PART NUMBER	STOCK NUMBER	FIGURE	NO	ITEM NO
19207	12338468	5905-01-183-9636	333		1
19207	12338470	9905-01-203-9992	354		10
19207	12338475	9905-01-203-9996	354		17
34623	12338477	2510-01-187-2238	46		1
19207	12338481	5340-01-266-9173	313		3
		5340-01-266-9173	443	3	
		5340-01-266-9173	457	21	
19207	12338482	2540-01-185-3197	187		1
19207	12338489	2540-01-209-6838	315		1
19207	12338489-1	2540-01-354-4291	315		1
		2540-01-354-4291	444	1	
34623	12338495	2590-01-214-1566	315		7
		2590-01-214-1566	444	4	
19207	12338503-1	2540-01-318-9229	230		8
19207	12338521-1	5365-01-413-0275	266		22
34623	12338521-1	5365-01-198-3400	248		21
19207	12338521-14	5365-01-255-0965	379		36
19207	12338521-5	5365-01-218-9914	285		8
		5365-01-218-9914	285	11	
19207	12338528	2540-01-184-6602	230		3
19207	12338536	3040-01-192-1632	230		15
19207	12338539	3040-01-251-9383	230		10
19207	12338540	5340-01-185-8723	230		17
19207	12338548-2	4710-01-359-2956	15		6
		4710-01-359-2956	16	6	
		4710-01-359-2956	366	18	
19207	12338549	4710-01-186-9617	15		9
		4710-01-186-9617	16	8	
19207	12338551-2	4720-01-212-8269	17		17
9C234	12338553-1	4720-01-316-2538	366		22
		4720-01-316-2538	368	39	
		4720-01-316-2538	386	8	
19207	12338557-1	5330-01-186-7757	15		30
		5330-01-186-7757	16	42	
19207	12338557-2	5330-01-187-5148	15		27
		5330-01-187-5148	16	36	
19207	12338558-2	5342-01-254-1498	366		1
19207	12338559	5340-01-204-3903	334		3
		5340-01-204-3903	366	27	
		5340-01-204-3903	384	12	
19207	12338560-1		13		14
			14	20	
19207	12338560-3		14		11
19207	12338560-4		14		21
19207	12338561-2		14		25
19207	12338563	5340-01-186-8768	19		17
19207	12338564	4710-01-203-0608	366		23
19207	12338567	2910-01-184-2159	15		10
19207	12338568	5340-01-186-9606	15		32

I-211

SECTION IV TM9-2320-280-24P C01

CROSS REFERENCE INDEXES

PART NUMBER INDEX

CAGE PART NUMBER STOCK NUMBER FIGURE NO ITEM NO

CAGE	PART NUMBER	STOCK NUMBER	FIGURE NO	ITEM NO
		5340-01-186-9606	16	19
19207	12338571	5340-01-191-0748	15	25
		5340-01-191-0748	16	33
19207	12338574	5340-01-190-5100	15	28
		5340-01-190-5100	16	38
19207	12338575	5340-01-193-0251	15	29
		5340-01-193-0251	16	41
19207	12338577	5340-01-476-5585	11	9
19207	12338578	2910-01-210-6856	13	20
34623	12338579	2910-01-189-4770	15	1
34623	12338580	2590-01-188-3267	17	2
19207	12338585	5330-01-194-0473	15	11
		5330-01-194-0473	16	10
19207	12338591	4720-01-189-0853	178	18
19207	12338591-1	4720-01-394-3747	178	18
34623	12338593	4710-01-209-4628	178	8
19207	12338596	5930-01-183-6637	66	9
19207	12338599	5930-01-237-7322	66	1
19207	12338603	5945-01-193-7175	51	5
19207	12338605	5340-01-286-6191	62	11
19207	12338608	5340-01-183-7233	60	3
		5340-01-183-7233	61	3
19207	12338609	5340-01-189-7558	68	2
19207	12338611	6220-01-193-1970	62	14
19207	12338622	2530-01-185-9651	170	6
19207	12338629	5340-01-220-3084	205	10
19207	12338630	5340-01-185-0386	205	3
		5340-01-185-0386	206	8
19207	12338633	5340-01-251-0754	84	20
19207	12338634-1	5340-01-186-5520	205	41
19207	12338635	5365-01-210-1584	206	26
19207	12338637	5365-01-246-8281	205	18
34623	12338639-2	2510-01-185-7946	206	17
19207	12338640-2	9515-01-189-9728	205	37
19207	12338650	5342-01-203-2822	198	23
		5342-01-203-2822	280	48
		5342-01-203-2822	281	48
		5342-01-203-2822	321	27
		5342-01-203-2822	377	10
19207	12338663	5340-01-192-2256	443	8
		5340-01-192-2256	457	7
19207	12338667	5340-01-253-3757	197	23
19207	12338670	2510-01-411-2736	199	26
19207	12338676	5340-01-219-7541	64	3
19207	12338696-1	5340-01-314-4951	83	26
		5340-01-314-4951	445	11
19207	12338709	6220-01-200-0897	197	17
		6220-01-200-0897	199	51
19207	12338711	5342-01-194-3128	197	13

I-212

SECTION IV TM9-2320-280-24P C01

CROSS REFERENCE INDEXES

PART NUMBER INDEX

CAGE	PART NUMBER	STOCK NUMBER	FIGURE NO	ITEM NO
		5342-01-194-3128	199	54
19207	12338716	2510-01-186-3107	199	34
19207	12338719	2510-01-262-6009	200	6
19207	12338723	3020-01-174-5984	142	29
		3020-01-174-5984	150	29
19207	12338725	5340-01-203-2821	331	36
19207	12338726	5340-01-209-7794	51	7
19207	12338734	2510-01-186-3106	197	28
19207	12338736	2510-01-262-6008	200	5
19207	12338738	2530-01-248-4873	184	18
19207	12338740	2510-01-262-6007	198	12
19207	12338765	6160-01-470-4172	198	3
19207	12338770	2815-01-168-7892	1	2
19207	12338770-4	2815-01-319-1433	1	2
19207	12338772	2920-01-175-7214	66	2
19207	12338773	3020-01-210-1391	174	11
19207	12338774	2590-01-207-3696	173	2
19207	12338775	5342-01-203-5661	174	3
19207	12338777	5340-01-212-4714	5	14
19207	12338779	5340-01-210-2747	5	15
19207	12338780	4730-01-168-7872	6	5
19207	12338781	3020-01-188-3885	34	21
19207	12338782	3020-01-313-0682	34	21
34623	12338786-1	2590-01-444-4365	37	16
		2590-01-444-4365	475	19
19207	12338787	2815-01-168-7918	8	11
19207	12338788	2815-01-168-7917	8	10
19207	12338792	2920-01-168-7914	35	5
19207	12338803	2640-01-161-2114	167	5
19207	12338807-4		226	6
			261	7
19207	12338807-6	5330-01-298-4713	211	28
		5330-01-298-4713	212	24
19207	12338810	2540-01-246-8289	226	1
19207	12338813	5315-01-205-2988	225	36
		5315-01-205-2988	226	19
19207	12338814	5340-01-207-0379	199	24
19207	12338818	2590-01-246-8288	226	3
19207	12338819	2590-01-209-6043	226	26
19207	12338822-1	5342-01-186-7382	225	35
		5342-01-186-7382	226	23
19207	12338822-2	5342-01-185-0116	225	56
		5342-01-185-0116	226	37
19207	12338824	5340-01-185-0404	225	43
19207	12338828-1		328	1
19207	12338828-2		328	1
19207	12338829	5340-01-188-1018	225	21
		5340-01-188-1018	226	5
19207	12338830	2510-01-250-7597	227	1

SECTION IV TM9-2320-280-24P C01

CROSS REFERENCE INDEXES

PART NUMBER INDEX

CAGE	PART NUMBER	STOCK NUMBER	FIGURE NO	ITEM NO
19207	12338831	2540-01-214-1568	327	6
19207	12338833	5340-01-247-7912	226	10
19207	12338834-1	5340-01-185-6998	225	27
19207	12338834-2	5340-01-186-7594	225	11
19207	12338835	5340-01-185-0401	225	45
		5340-01-185-0401	226	32
19207	12338836-1	5340-01-186-7593	225	25
19207	12338836-2	5340-01-185-6997	225	10
19207	12338838-2	2590-01-263-3254	368	33
19207	12338839-1	5340-01-314-5957	210	15
		5340-01-314-5957	213	23
		5340-01-314-5957	225	14
		5340-01-314-5957	231	3
		5340-01-314-5957	234	15
		5340-01-314-5957	240	11
		5340-01-314-5957	246	1
		5340-01-314-5957	246	7
		5340-01-314-5957	247	11
		5340-01-314-5957	247	22
		5340-01-314-5957	248	3
		5340-01-314-5957	249	7
		5340-01-314-5957	249	15
		5340-01-314-5957	249	22
		5340-01-314-5957	249	31
		5340-01-314-5957	251	16
		5340-01-314-5957	251	24
		5340-01-314-5957	251	30
		5340-01-314-5957	251	40
		5340-01-314-5957	252	18
		5340-01-314-5957	252	33
		5340-01-314-5957	252	38
		5340-01-314-5957	253	6
		5340-01-314-5957	253	11
		5340-01-314-5957	253	16
		5340-01-314-5957	253	26
		5340-01-314-5957	253	31
		5340-01-314-5957	254	5
		5340-01-314-5957	254	11
		5340-01-314-5957	254	17
		5340-01-314-5957	255	5
		5340-01-314-5957	255	11
		5340-01-314-5957	255	15
		5340-01-314-5957	255	24
		5340-01-314-5957	255	31
		5340-01-314-5957	256	3
		5340-01-314-5957	257	2
		5340-01-314-5957	258	4
		5340-01-314-5957	259	4
		5340-01-314-5957	260	1

SECTION IV TM9-2320-280-24P C01

CROSS REFERENCE INDEXES

PART NUMBER INDEX

CAGE	PART NUMBER	STOCK NUMBER	FIGURE NO	ITEM NO
		5340-01-314-5957	262	2
		5340-01-314-5957	262	10
		5340-01-314-5957	263	2
		5340-01-314-5957	264	3
		5340-01-314-5957	265	3
		5340-01-314-5957	268	40
		5340-01-314-5957	270	39
		5340-01-314-5957	286	14
		5340-01-314-5957	287	5
		5340-01-314-5957	289	2
		5340-01-314-5957	292	2
		5340-01-314-5957	318	4
		5340-01-314-5957	374	3
		5340-01-314-5957	429	9
		5340-01-314-5957	430	5
		5340-01-314-5957	430	21
		5340-01-314-5957	430	32
		5340-01-314-5957	446	5
		5340-01-314-5957	447	2
		5340-01-314-5957	448	2
		5340-01-314-5957	449	4
		5340-01-314-5957	450	3
19207	12338839-3	5340-01-434-1809	260	21
19207	12338839-4	5340-01-197-8673	216	3
		5340-01-197-8673	252	19
		5340-01-197-8673	253	36
		5340-01-197-8673	266	12
		5340-01-197-8673	317	10
		5340-01-197-8673	321	23
		5340-01-197-8673	323	15
		5340-01-197-8673	326	6
		5340-01-197-8673	377	6
19207	12338839-6	5340-01-197-8674	228	13
		5340-01-197-8674	440	22
		5340-01-197-8674	446	31
34623	12338839-7	5340-01-214-1778	247	25
19207	12338839-8	2510-01-199-5748	321	17
		2510-01-199-5748	446	5
		2510-01-199-5748	446	22
		2510-01-199-5748	448	5
19207	12338841-1	2510-01-198-7413	225	34
		2510-01-198-7413	226	24
19207	12338842	5342-01-187-7029	12	36
19207	12338843	2590-01-188-3225	225	51
19207	12338845	5340-01-185-7024	225	18
19207	12338847		328	5
19207	12338848		328	3
19207	12338849	2540-01-192-4500	199	30
19207	12338851-1	2540-01-216-6320	226	35

I-215

SECTION IV TM9-2320-280-24P C01

CROSS REFERENCE INDEXES

PART NUMBER INDEX

CAGE	PART NUMBER	STOCK NUMBER	FIGURE NO	ITEM NO
19207	12338851-2	2590-01-209-6042	226	35
19207	12338852	2590-01-189-1067	225	53
		2590-01-189-1067	226	35
19207	12338853	5340-01-332-7515	208	8
19207	12338860	5330-01-209-5997	225	54
19207	12338862	2590-01-209-6041	226	34
19207	12338863	2540-01-196-4726	12	33
19207	12338867	5340-01-358-9301	153	12
19207	12338879	5365-01-212-2325	202	31
19207	12338880	2510-01-209-0502	202	21
19207	12338884	5340-01-211-7436	203	13
19207	12338885	5340-01-191-3222	203	17
19207	12338886	5365-01-212-4736	202	19
19207	12338889	2540-01-212-5814	202	20
19207	12338891	2540-01-209-4590	202	24
19207	12338892	5365-01-210-1630	203	27
19207	12338893	5315-01-186-5633	202	37
19207	12338894	5315-01-213-5545	202	23
19207	12338895	3120-01-186-7715	205	27
19207	12338899	5340-01-185-8619	203	8
19207	12338900	5340-01-186-1388	202	32
19207	12338902	2590-01-261-5484	202	13
19207	12338903	5365-01-210-1629	203	7
19207	12338904	2590-01-261-2636	202	6
19207	12338905	2510-01-185-6715	202	29
19207	12338906	5340-01-209-7834	203	28
19207	12338907	2590-01-261-2637	202	12
19207	12338909	2540-01-185-9530	202	17
19207	12338911	5330-01-192-8916	203	16
19207	12338915	2510-01-185-6118	205	22
19207	12338920-1	5365-01-185-3108	204	4
19207	12338920-2	5365-01-189-9743	204	4
19207	12338921	2510-01-190-8458	204	3
34623	12338925-1	2510-01-185-7950	204	11
34623	12338925-2	2510-01-185-7945	204	11
19207	12338926	2510-01-189-9730	204	3
19207	12338936	2510-01-191-0971	204	8
19207	12338937	2510-01-186-3311	204	8
19207	12338940-1	2510-01-473-2309	202	1
19207	12338945-1	2540-01-247-7915	208	38
19207	12338945-2	2540-01-247-7914	208	38
19207	12338950	2540-01-210-2787	208	33
34623	12338952-1	5342-01-249-3502	208	32
34623	12338952-2	5342-01-249-5389	208	32
19207	12338957	4720-01-212-3531	177	9
19207	12338972	5340-01-214-4675	283	9
		5340-01-214-4675	438	10
19207	12338974-2	5340-01-260-5792	266	3
19207	12338976	5342-01-214-7809	284	15

I-216

SECTION IV TM9-2320-280-24P C01

CROSS REFERENCE INDEXES

PART NUMBER INDEX

CAGE	PART NUMBER	STOCK NUMBER	FIGURE NO	ITEM NO
		5342-01-214-7809	439	8
19207	12338978-2	2540-01-209-6875	284	16
		2540-01-209-6875	439	9
19207	12338981-1	2510-01-173-9316	266	10
34623	12338981-2	2510-01-378-5615	266	10
19207	12338996	2590-01-186-3614	99	1
19207	12338997	5331-01-216-7393	488	4
19207	12338998	5331-01-216-7392	488	3
34623	12339002	5331-01-195-8889	488	18
19207	12339006	5340-01-197-5477	198	17
		5340-01-197-5477	321	20
		5340-01-197-5477	325	23
19207	12339008	6220-01-257-3878	83	32
19207	12339009	5340-01-255-9510	198	9
		5340-01-255-9510	321	14
		5340-01-255-9510	323	12
		5340-01-255-9510	325	16
		5340-01-255-9510	402	13
19207	12339013	5306-01-190-2193	71	4
19207	12339014-2	2590-01-261-6851	369	12
19207	12339015-1	2590-01-251-3070	227	17
19207	12339015-2	2590-01-251-3071	227	6
19207	12339015-3	2590-01-251-3069	227	7
		2590-01-251-3069	227	17
34623	12339015-4	2590-01-251-3068	227	20
19207	12339017	5342-01-282-4853	71	8
19207	12339018	4710-01-254-7272	227	13
19207	12339019	2510-01-250-7593	227	26
19207	12339021	2510-01-369-3230	227	27
19207	12339026	2540-01-197-5448	197	2
19207	12339027	2540-01-197-5447	197	7
19207	12339028-1	9515-01-185-3111	267	8
34623	12339029-1	2540-01-192-9716	267	7
34623	12339029-2	2540-01-192-5948	267	10
19207	12339031-1	5340-01-470-5892	198	16
19207	12339031-2	5340-01-433-6262	198	16
19207	12339034	6160-01-184-0643	71	1
19207	12339035	6160-01-184-0728	71	5
19207	12339036	2590-01-203-9874	283	28
19207	12339036-1	2510-01-383-5444	74	6
		2510-01-383-5444	283	28
19207	12339038		230	25
19207	12339039	2510-01-251-9995	227	21
19207	12339043-1	2590-01-203-9875	283	17
19207	12339043-3	2510-01-335-7343	283	17
19207	12339044	2510-01-251-8548	227	19
19207	12339047-1	2540-01-188-3675	235	9
19207	12339047-2	2540-01-185-4387	235	9
19207	12339048	2590-01-197-5446	199	49

I-217

SECTION IV TM9-2320-280-24P C01

CROSS REFERENCE INDEXES

PART NUMBER INDEX

CAGE	PART NUMBER	STOCK NUMBER	FIGURE NO	ITEM NO
19207	12339050	5306-01-502-3440	136	13
19207	12339052	5310-01-185-7214	12	11
		5310-01-185-7214	364	1
		5310-01-185-7214	436	7
19207	12339053	7690-01-204-0077	354	1
19207	12339055	9905-01-248-9550	356	2
19207	12339056-2	9905-01-248-9551	356	6
19207	12339057	7690-01-265-1133	362	3
19207	12339059	7690-01-265-1134	362	4
19207	12339060	7690-01-265-1135	362	5
19207	12339063	7690-01-204-0076	354	18
19207	12339064	7690-01-197-5500	356	4
		7690-01-197-5500	357	1
19207	12339065	7690-01-209-0864	356	1
19207	12339076	9905-01-185-3138	354	12
19207	12339077	9905-01-185-3132	354	12
19207	12339080	9905-01-224-5860	354	8
19207	12339082	9905-01-185-3134	354	12
19207	12339085	9905-01-185-3131	354	12
19207	12339087	9905-01-187-9468	355	2
19207	12339090	9905-01-186-3255	354	12
19207	12339091	9905-01-185-7973	354	12
19207	12339095	9905-01-185-7972	354	12
19207	12339096	9905-01-185-7975	354	12
19207	12339099	9905-01-185-3139	354	12
19207	12339100	9905-01-203-9994	354	4
		9905-01-203-9994	370	19
19207	12339101	9905-01-203-9995	354	14
19207	12339104	9905-01-186-3259	354	12
19207	12339105	9905-01-185-7977	354	20
19207	12339106	9905-01-193-4065	354	5
		9905-01-193-4065	370	20
19207	12339107	9905-01-185-3143	355	1
19207	12339108	9905-01-248-7656	370	13
19207	12339109	9905-01-205-8635	354	15
		9905-01-205-8635	399	1
19207	12339110	9905-01-186-3253	354	12
19207	12339111	9905-01-248-7657	354	7
19207	12339114	5315-01-185-8781	165	1
		5315-01-185-8781	166	1
19207	12339132	2510-01-185-6112	2	4
19207	12339133	2510-01-185-6115	2	1
34623	12339134	2520-01-175-7220	103	1
19207	12339137	5342-01-211-2897	7	4
19207	12339142	5340-01-182-1074	22	2
19207	12339146	2520-01-161-2136	104	1
19207	12339147	4710-01-188-3238	103	9
19207	12339147-2	4710-01-179-4346	400	5
19207	12339154-1	5340-01-251-0724	28	3

SECTION IV TM9-2320-280-24P C01

CROSS REFERENCE INDEXES

PART NUMBER INDEX

CAGE	PART NUMBER	STOCK NUMBER	FIGURE NO	ITEM NO
19207	12339155-1	4720-01-186-6018	27	5
19207	12339159	4720-01-188-1370	29	12
19207	12339160	4720-01-188-3909	29	2
19207	12339162	4720-01-187-3386	29	7
19207	12339163	4720-01-196-1636	27	9
19207	12339165-1	2930-01-210-6940	26	3
19207	12339165-2	2930-01-218-8103	26	3
19207	12339166-1	2510-01-210-6939	28	1
19207	12339166-2	2990-01-210-4276	28	1
19207	12339167	6680-01-184-9214	103	8
19207	12339168	6680-01-179-4350	3	4
19207	12339170	4710-01-196-1642	3	2
19207	12339174	2990-01-210-0427	25	4
19207	12339176	3040-01-188-8292	100	8
19207	12339178	5342-01-190-7735	103	15
19207	12339181	2990-01-210-6982	25	3
19207	12339186	2540-01-257-3877	188	4
19207	12339187	5340-01-207-0717	321	33
		5340-01-207-0717	325	28
		5340-01-207-0717	373	10
19207	12339188		199	13
19207	12339194	2590-01-247-7910	199	7
19207	12339196-1	2540-01-335-4482	199	3
34623	12339198	4330-01-189-1007	19	1
19207	12339201-1	2510-01-307-0152	332	44
19207	12339203	6210-01-211-7024	47	13
19207	12339203-1	6220-01-429-4596	47	13
34623	12339206	5310-01-439-1154	193	3
19207	12339207	5307-01-188-1229	194	11
19207	12339208	2910-01-210-5872	18	1
19207	12339222	2510-01-192-5947	199	32
19207	12339223	4010-01-194-8546	266	8
		4010-01-194-8546	284	5
19207	12339224-1	4010-01-218-4639	284	7
19207	12339224-2	4010-01-218-4640	284	9
19207	12339225-1	4030-01-255-3634	284	11
19207	12339225-2		284	11
19207	12339227-1	2510-00-933-2893	266	7
19207	12339227-2	2510-00-933-2895	266	7
19207	12339228-7	5330-01-209-7817	331	11
19207	12339228-9	5330-01-193-2338	266	9
34623	12339229-1	4010-01-217-7086	284	3
		4010-01-217-7086	439	3
34623	12339229-2	4010-01-215-7466	284	3
		4010-01-215-7466	439	3
19207	12339232-1	9515-01-185-3118	267	12
19207	12339232-2	9515-01-189-9721	267	12
34623	12339235	4730-01-184-6971	160	6
		4730-01-184-6971	162	17

I-219

SECTION IV TM9-2320-280-24P C01

CROSS REFERENCE INDEXES

PART NUMBER INDEX

CAGE	PART NUMBER	STOCK NUMBER	FIGURE NO	ITEM NO
		4730-01-184-6971	487A	10
		4730-01-184-6971 487D		6
19207	12339239-1	5305-01-195-5818	8	14
19207	12339240	5305-01-189-3769	34	20
19207	12339244-3	5340-01-209-7802	196	9
		5340-01-209-7802 196	14	
19207	12339247-1	5342-01-186-7236	196	5
		5342-01-186-7236 196	15	
19207	12339247-3	5342-01-406-6962	196	10
		5342-01-406-6962 196	15	
19207	12339248-2	5310-01-252-7285	196	6
19207	12339248-3	5310-01-476-9321	196	6
19207	12339249-1	5365-01-264-4023	196	13
19207	12339251	4720-01-189-2218	333	38
19207	12339252	2590-01-186-8704	331	33
34623	12339255	5330-01-185-8709	331	17
19207	12339261	5340-01-189-7554	331	37
		5340-01-189-7554	332	26
19207	12339263	4710-01-210-3504	333	5
19207	12339265-1	4720-01-448-1655	331	3
19207	12339265-10	4720-01-394-6170	368	46
		4720-01-394-6170 370	8	
19207	12339265-16	4720-01-451-0894	370	7
19207	12339265-2		331	6
34623	12339265-3	4720-01-279-5149	331	31
19207	12339265-6	4720-01-394-6166	370	2
19207	12339265-7	4720-01-325-6985	370	4
34623	12339265-8		331	21
19207	12339265-9	4720-01-271-6985	331	24
19207	12339267	5330-01-209-7842	331	16
19207	12339273	2510-01-210-2748	331	15
19207	12339281	2540-01-212-5815	331	27
19207	12339283	2540-01-210-1325	331	14
19207	12339289	2910-01-184-5499	331	1
34623	12339289-2	2540-01-257-3742	370	1
19207	12339291	2510-01-184-5497	331	12
19207	12339292	2910-01-184-5498	331	5
19207	12339293-1		333	4
34623	12339296	5325-01-194-0562	17	3
19207	12339308	5310-01-188-1093	169	4
19207	12339310	5315-01-201-3592	169	5
19207	12339313	5310-01-188-6861	51	13
		5310-01-188-6861 165	11	
19207	12339316		44	18
19207	12339316-1		44	18
19207	12339317	6150-01-449-5445	44	23
		6150-01-449-5445 480	25	
19207	12339317-1		BULK	72
19207	12339321		73	21

I-220

SECTION IV TM9-2320-280-24P C01

CROSS REFERENCE INDEXES

PART NUMBER INDEX

CAGE	PART NUMBER	STOCK NUMBER	FIGURE NO	ITEM NO
19207	12339322	73	17	19207
12339323	2590-01-217-2662	64	24	
		2590-01-217-2662	81	14
19207	12339324	6150-01-473-2435	81	14
19207	12339327	5999-01-183-9530	327	18
19207	12339331	5975-01-484-1150	79	26
19207	12339332	5935-01-421-4801	83	27
		5935-01-421-4801	88	47
19207	12339333	5340-01-209-7387	75	34
		5340-01-209-7387	77	15
19207	12339334	5975-01-207-1706	75	14
19207	12339335	5340-01-212-4716	75	30
		5340-01-212-4716	77	10
19207	12339336	5340-01-206-5540	75	41
19207	12339337	5365-01-206-7486	72	7
19207	12339338	5935-01-212-9631	84	7
19207	12339342	5995-01-210-3594	81	2
19207	12339343	5340-01-198-0480	75	49
		5340-01-198-0480	77	27
19207	12339345	5995-01-188-3269	81	1
19207	12339345-1	5935-01-474-3033	84	14
19207	12339346	5995-01-197-5554	82	1
19207	12339347-1	6150-01-185-8240	327	16
19207	12339347-2	6150-01-210-3596	327	16
19207	12339349	5995-01-184-5544	80	1
19207	12339350	5995-01-184-5545	75	1
19207	12339351-3		95	22
19207	12339353-1	4820-01-359-9489	159	2
34623	12339353-1	4820-01-359-9489	160	11
19207	12339355-1	5320-01-271-6357	197	6
		5320-01-271-6357	230	21
		5320-01-271-6357	231	26
		5320-01-271-6357	235	17
		5320-01-271-6357	365	14
		5320-01-271-6357	369	7
		5320-01-271-6357	417	9
19207	12339355-2	5320-01-264-5978	199	14
		5320-01-264-5978	217	3
		5320-01-264-5978	219	6
		5320-01-264-5978	245	35
		5320-01-264-5978	264	1
		5320-01-264-5978	265	1
		5320-01-264-5978	295	9
		5320-01-264-5978	297	2
		5320-01-264-5978	300	5
		5320-01-264-5978	307	7
		5320-01-264-5978	308	13
		5320-01-264-5978	342	8
34623	12339359-12	3030-01-282-6968	35	1

SECTION IV TM9-2320-280-24P C01

CROSS REFERENCE INDEXES

PART NUMBER INDEX

CAGE	PART NUMBER	STOCK NUMBER	FIGURE NO	ITEM NO
		3030-01-282-6968	472	19
19207	12339359-14	3030-01-256-3616	475	22
19207	12339359-18	3030-01-321-4482	37	19
		3030-01-321-4482	475	27
19207	12339359-26	3030-01-333-2286	37	19
19207	12339359-28	3030-01-466-5387	476	26
19207	12339369	5360-01-198-3525	280	43
		5360-01-198-3525	281	43
19207	12339374-1	2510-01-188-7381	280	40
		2510-01-188-7381	281	40
19207	12339374-2	2510-01-189-9748	280	40
		2510-01-189-9748	281	40
19207	12339375-1	2510-01-209-3450	280	40
		2510-01-209-3450	281	40
19207	12339375-2	2510-01-196-5506	280	40
		2510-01-196-5506	281	40
19207	12339376-1		BULK	59
19207	12339376-1-AR		231	16
19207	12339377-1	2540-01-201-2356	281	9
19207	12339377-2	2540-01-201-2357	281	9
19207	12339378-1	2510-01-199-1498	280	9
19207	12339378-2	5340-01-197-4901	280	9
19207	12339379	5365-01-253-8980	280	51
		5365-01-253-8980	281	51
19207	12339380	5355-01-252-6504	280	38
		5355-01-252-6504	281	38
19207	12339386	5340-01-421-1782	281	52
19207	12339392	3020-01-232-9629	37	20
		3020-01-232-9629	475	28
19207	12339395	3020-01-198-0633	35	32
19207	12339397-1	5310-00-755-7283	199	5
19207	12339397-3	5310-01-231-7455	398	6
19207	12339397-4	5340-01-213-4662	80	15
		5340-01-213-4662	203	15
19207	12339397-5	5310-01-312-4777	203	22
19207	12339397-6	5310-01-388-6205	398	7
19207	12339402	5325-01-209-7843	333	25
		5325-01-209-7843	372	4
19207	12339404-2	5340-01-195-9107	280	18
		5340-01-195-9107	281	18
19207	12339406-2	5307-01-291-9018	30	5
		5307-01-291-9018	490	4
19207	12339406-3	5307-01-315-3597	8	12
		5307-01-315-3597	36	18
		5307-01-315-3597	38	18
		5307-01-315-3597	42	2
		5307-01-315-3597	475	15
19207	12339406-4	5307-01-465-5796	30	11
		5307-01-465-5796	472	17

I-222

SECTION IV TM9-2320-280-24P C01

CROSS REFERENCE INDEXES

PART NUMBER INDEX

CAGE	PART NUMBER	STOCK NUMBER	FIGURE NO		ITEM NO
19207	12339410	4730-01-187-6929		6	4
34623	12339411-2	4720-01-187-6911		5	10
19207	12339412	3020-01-204-8132		472	22
34623	12339413	5305-01-285-4923		44	10
19207	12339416-1	2510-01-196-5316		280	15
		2510-01-196-5316	281		15
19207	12339416-2	5342-01-195-8039		280	15
		5342-01-195-8039	281		15
19207	12339420-1	2590-01-253-3905		235	5
19207	12339426	5310-01-448-3219		332	32
19207	12339428	5305-01-211-1470		208	34
19207	12339431	2510-01-248-1340		208	19
19207	12339433	5310-01-218-3190		202	9
		5310-01-218-3190	252		29
19207	12339438	5340-01-266-6398		227	15
19207	12339448-1	5340-01-199-3498		281	6
34623	12339448-1	5340-01-199-3498		280	6
34623	12339448-4	5340-01-201-9076		280	6
		5340-01-201-9076	281		6
19207	12339450	2510-01-199-1499		280	32
		2510-01-199-1499	281		32
34623	12339451-1	3040-01-195-4173		281	33
34623	12339451-2	3040-01-196-1493		281	33
19207	12339452-1	3040-01-200-1015		280	33
34623	12339454-1	2540-01-199-5812		280	29
19207	12339462	5330-01-196-7969		281	26
19207	12339463-1	5330-01-198-8789		280	26
19207	12339463-2	5330-01-196-7968		280	26
19207	12339469-1	2510-01-254-1482		280	1
19207	12339469-2	2510-01-253-3907		280	1
19207	12339473-1	2510-01-254-1501		281	1
19207	12339473-2	2510-01-257-3904		281	1
19207	12339477-1	2510-01-257-3876		281	1
34623	12339477-2	2510-01-254-1483		281	1
19207	12339480-1	2510-01-257-3903		280	1
19207	12339480-2	2510-01-254-1500		280	1
19207	12339491-1	5342-01-250-1098		10	13
19207	12339492	5360-01-210-2852		22	3
19207	12339494	5930-01-199-8853		66	3
		5930-01-199-8853	66A		5
19207	12339501	5310-01-198-7585		167	7
		5310-01-198-7585	167		20
19207	12339503	5970-01-180-9776		75	8
19207	12339506	2510-01-289-8258		226	36
19207	12339507	2510-01-289-8260		226	36
19207	12339508	2510-01-289-8259		226	36
19207	12339509	2510-01-289-2233		226	36
19207	12339512	2510-01-233-1141		225	12
19207	12339513	2510-01-233-7767		226	4

I-223

SECTION IV TM9-2320-280-24P C01

CROSS REFERENCE INDEXES

PART NUMBER INDEX

CAGE	PART NUMBER	STOCK NUMBER			FIGURE NO	ITEM NO
19207	12339514	9340-01-185-3757			225	55
19207	12339518	5307-01-446-5930			103	14
19207	12339544	5315-01-289-6212			271	11
		5315-01-289-6212	271	24		
19207	12339545	5340-01-214-4992			271	6
		5340-01-214-4992	271	23		
		5340-01-214-4992	296	6		
19207	12339546	2540-01-209-6049			271	3
		2540-01-209-6049	271	20		
19207	12339547	5360-01-194-6070			271	4
		5360-01-194-6070	271	21		
19207	12339548	5305-01-197-1594			271	2
		5305-01-197-1594	271	19		
19207	12339559	5307-01-254-2504			278	11
19207	12339561	2540-01-192-5949			277	28
		2540-01-192-5949	278	9		
		2540-01-192-5949	440	5		
19207	12339562	2540-01-194-0197			277	26
19207	12339569	5340-01-448-4316			278	2
19207	12339579	5342-01-194-5806			269	30
34623	12339595	5330-01-195-9083			268	2
19207	12339599	5310-01-197-1100			270	36
19207	12339601	5340-01-460-5002			270	41
19207	12339602	3120-01-200-4105			268	34
19207	12339605	5340-01-197-2642			270	25
19207	12339610-1	2540-01-252-8460			270	33
34623	12339610-2	2540-01-252-3386			270	33
19207	12339612	2510-01-189-9741			268	16
34623	12339623-3	2510-01-192-9723			278	4
19207	12339624-1	2510-01-205-2507			278	5
19207	12339624-2	2540-01-203-9872			278	6
19207	12339625	2510-01-249-1583			268	22
19207	12339626-1	2590-01-205-2506			278	13
19207	12339631-2	2540-01-236-6489			278	4
19207	12339634-3				271	7
19207	12339634-4				271	7
19207	12339635-3				271	26
19207	12339635-4				271	26
19207	12339639-3				271	
					271	26
19207	12339639-4				271	7
19207	12339640-3				271	26
19207	12339641-1	2510-01-250-7786			268	3
19207	12339641-2	2510-01-248-9557			268	3
19207	12339642	2590-01-188-7394			270	6
19207	12339644-5	2510-01-345-4365			271	1
19207	12339644-6	2510-01-344-4169			271	18
19207	12339644-8	2510-01-354-0417			271	18
19207	12339644-9	2510-01-332-0128			271	1

I-224

SECTION IV TM9-2320-280-24P C01

CROSS REFERENCE INDEXES

PART NUMBER INDEX

CAGE	PART NUMBER	STOCK NUMBER	FIGURE NO	ITEM NO
19207	12339646-1	2540-01-192-4504	269	10
19207	12339646-2	2540-01-192-9721	269	10
19207	12339649-1	2510-01-249-1582	268	21
19207	12339649-2	2510-01-249-1581	268	21
19207	12339650	2510-01-249-1580	268	20
34623	12339651-1	2510-01-262-9520	268	3
34623	12339651-2	2510-01-185-3155	268	3
19207	12339655-10	2510-01-332-0129	271	1
19207	12339655-6	2510-01-366-6968	271	18
19207	12339655-7	2510-01-348-6670	271	1
19207	12339655-8	2510-01-366-6969	271	18
19207	12339661-1	5340-01-332-7602	225	6
		5340-01-332-7602	226	39
19207	12339662	5307-01-190-7650	103	12
19207	12339663	5315-01-190-5908	266	20
19207	12339679-1	2590-01-189-9736	274	24
		2590-01-189-9736	276	21
19207	12339679-2	2590-01-188-5080	245	38
19207	12339680	2590-01-188-5079	245	14
19207	12339689		274	15
19207	12339690	2590-01-189-9747	274	14
19207	12339694	5310-01-215-5356	332	25
		5310-01-215-5356	347	7
19207	12339697	2510-01-189-3456	209	7
19207	12339698-3	2510-01-335-7799	197	25
34623	12339698-4	2510-01-361-8213	199	39
19207	12339701-1	2510-01-335-4171	197	12
19207	12339707-1	2510-01-335-4170	199	55
19207	12339724	5310-01-194-6459	277	36
19207	12339726	5315-01-203-8543	280	41
		5315-01-203-8543	281	41
19207	12339728-3	5310-01-198-3487	280	13
		5310-01-198-3487	281	13
19207	12339730	5310-01-254-4284	280	3
		5310-01-254-4284	281	3
19207	12339731	5340-01-198-2240	280	8
		5340-01-198-2240	281	8
19207	12339750	6140-01-188-5113	247	4
19207	12339751	5365-01-256-9568	245	29
19207	12339753	5340-01-196-7733	246	26
19207	12339754	5340-01-246-8290	246	34
19207	12339755-4	5340-01-258-4866	252	23
19207	12339758	5340-01-199-8258	247	19
19207	12339759	5340-01-252-0741	245	34
19207	12339760	5340-01-256-4659	243	3
19207	12339763	5340-01-256-4515	243	30
19207	12339764	7690-01-191-6467	359	7
19207	12339767	5365-01-196-2699	243	18
19207	12339778	2590-01-235-8661	243	5

SECTION IV TM9-2320-280-24P C01

CROSS REFERENCE INDEXES

PART NUMBER INDEX

CAGE	PART NUMBER	STOCK NUMBER	FIGURE NO	ITEM NO
		2590-01-235-8661	244	2
		2590-01-235-8661	244	14
19207	12339779	5340-01-256-4516	244	3
19207	12339784	2510-01-252-4756	243	21
19207	12339798	5340-01-199-3495	245	31
19207	12339802	2590-01-194-6979	249	37
19207	12339803	2590-01-194-6980	249	30
19207	12339805	5340-01-188-7382	249	29
19207	12339806	2540-01-246-8286	244	4
34623	12339807	2540-01-246-8285	244	1
19207	12339809	2590-01-199-7977	248	24
19207	12339811	2590-01-194-6981	249	5
19207	12339813	5340-01-188-7383	249	4
19207	12339814-2	5365-01-214-0006	246	30
19207	12339814-6	5365-01-256-4526	246	28
19207	12339816-3	5340-01-360-4909	446	7
		5340-01-360-4909	447	1
19207	12339817-4	5340-01-265-8923	287	27
		5340-01-265-8923	447	4
19207	12339818	2590-01-246-8291	249	13
19207	12339819	2540-01-246-8284	243	4
19207	12339820	2590-01-249-1578	244	12
19207	12339821	2540-01-246-8283	242	1
19207	12339832	5340-01-238-9543	398	5
34623	12339833	4710-01-209-6746	398	14
19207	12339843	2510-01-198-0333	398	2
19207	12339844	4710-01-202-8245	400	1
19207	12339845	2540-01-192-4502	399	12
19207	12339846	6680-01-185-3202	400	3
34623	12339847	6680-01-179-4349	400	4
19207	12339848	5340-01-185-7959	398	18
19207	12339849	2990-01-210-2176	398	19
19207	12339856	2590-01-213-1628	283	25
		2590-01-213-1628	438	21
34623	12339857	2590-01-213-1629	283	19
		2590-01-213-1629	438	15
19207	12339862-2	5310-01-211-0698	283	2
		5310-01-211-0698	438	2
19207	12339875	2540-01-213-8057	283	1
		2540-01-213-8057	438	1
19207	12339884-1	4730-00-024-3971	27	6
		4730-00-024-3971	125	9
		4730-00-024-3971	368	38
		4730-00-024-3971	176A	15
19207	12339888	4730-01-186-5994	19	15
19207	12339889	4730-01-185-9453	19	22
34623	12339889	4730-01-185-9453	27	10
19207	12339890	5340-01-187-0892	11	27
		5340-01-187-0892	13	2

I-226

SECTION IV TM9-2320-280-24P C01

CROSS REFERENCE INDEXES

PART NUMBER INDEX

CAGE	PART NUMBER	STOCK NUMBER	FIGURE NO	ITEM NO
		5340-01-187-0892	14	2
		5340-01-187-0892	141	21
19207	12339900	9905-01-248-9544	355	4
19207	12339901	9905-01-248-9546	355	4
19207	12339902-1	2590-01-196-7281	BULK	43
19207	12339902-13	2590-01-471-0236	12	16
19207	12339902-7	5342-01-272-8391	492	4
19207	12339904	5340-01-470-7135	198	4
19207	12339905	5340-01-470-7160	198	6
19207	12339906	5340-01-272-6634	490	13
19207	12339907	2510-01-250-7596	227	2
34623	12339908	2510-01-209-1881	227	12
34623	12339909	2510-01-376-1092	368	31
19207	12339912	4710-01-217-9959	286	22
19207	12339913	5365-01-255-3610	285	15
19207	12339916	5340-01-197-5469	286	4
19207	12339921	2590-01-217-2665	286	23
19207	12339924	5310-01-228-3299	286	21
19207	12339925	2540-01-217-2666	286	20
34623	12339942	2540-01-389-0984	286	7
19207	12339944	2590-01-232-8210	286	42
19207	12339945	5340-01-367-2070	266	14
19207	12339947-1	2510-01-217-2663	286	33
19207	12339947-2	2510-01-217-2664	286	33
19207	12339951	2540-01-383-6470	286	24
19207	12339952	2590-01-248-9536	286	40
19207	12339954-1	5340-01-217-2667	286	26
19207	12339954-2	5340-01-217-2576	286	26
19207	12339957	2590-01-217-5743	286	1
19207	12339958-1	2590-01-197-5474	286	39
19207	12339958-2	2590-01-197-5475	286	39
19207	12339959	2590-01-199-5425	285	1
19207	12339961-1	5340-01-196-8028	403	17
34623	12339964-3	4730-01-485-0065	135	5
34623	12339981-2	4720-01-325-0204	399	17
19207	12339981-5		399	19
			401	1
19207	12339981-6		18	6
			401	4
19207	12339981-8		141	7
19207	12339981-9		141	1
34623	12339982	4810-01-192-5817	177	5
19207	12339987	9905-01-248-9545	355	4
19207	12339988-1	5340-01-246-8282	197	31
19207	12339988-2	5340-01-248-7646	199	43
19207	12339996	5340-01-197-2382	398	28
19207	12339997	2590-01-192-4425	399	11
19207	12339998-13	4931-01-202-8692	213	13
		4931-01-202-8692	214	7

SECTION IV TM9-2320-280-24P C01

CROSS REFERENCE INDEXES

PART NUMBER INDEX

CAGE PART NUMBER STOCK NUMBER FIGURE NO ITEM NO

CAGE	PART NUMBER	STOCK NUMBER	FIGURE NO	ITEM NO
		4931-01-202-8692	252	24
		4931-01-202-8692	254	23
		4931-01-202-8692	267	11
		4931-01-202-8692	342	6
		4931-01-202-8692	401	18
		4931-01-202-8692	446	24
		4931-01-202-8692	498	19
19207	12339998-14	5310-01-198-1723	225	20
		5310-01-198-1723	246	15
		5310-01-198-1723	251	35
		5310-01-198-1723	252	32
		5310-01-198-1723	253	27
		5310-01-198-1723	254	18
		5310-01-198-1723	255	30
		5310-01-198-1723	263	3
		5310-01-198-1723	264	6
		5310-01-198-1723	266	11
		5310-01-198-1723	318	6
		5310-01-198-1723	333	27
		5310-01-198-1723	429	16
34623	12339999	5340-01-189-7739	266	19
19207	12340001-2	5306-01-199-1314	270	3
19207	12340019	5340-01-196-4720	188	3
19207	12340020	5340-01-214-3712	83	7
19207	12340024	4730-01-188-3296	104	16
19207	12340025	2540-01-184-5500	328	10
19207	12340027	9905-01-249-1612	354	19
34623	12340028-1L	2510-01-193-1807	207	1
34623	12340028-2L	2510-01-189-9717	207	1
34623	12340029	4720-01-212-6403	333	26
19207	12340031	2540-01-196-5313	325	2
19207	12340039-1	2540-01-197-4900	280	22
19207	12340045	5365-01-271-2491	266	26
		5365-01-271-2491	439	4
19207	12340046	4720-01-360-2380	27	7
19207	12340048		44	12
34623	12340051	2540-01-188-3229	235	3
19207	12340053	5315-01-197-3122	245	5
19207	12340057	5340-01-197-5548	37	5
19207	12340058	5310-01-253-8437	127	10
		5310-01-253-8437	230	13
		5310-01-253-8437	234	11
		5310-01-253-8437	235	12
		5310-01-253-8437	239	12
19207	12340061	2930-01-256-5350	27	3
19207	12340063	2590-01-194-0309	243	6
		2590-01-194-0309	244	5
		2590-01-194-0309	244	13
19207	12340074	2530-01-189-0897	68	9

SECTION IV TM9-2320-280-24P C01

CROSS REFERENCE INDEXES

PART NUMBER INDEX

CAGE	PART NUMBER	STOCK NUMBER	FIGURE NO	ITEM NO
19207	12340078-3	5330-01-476-9295	489	4
19207	12340078-5		489	5
19207	12340087	5340-01-249-1587	275	25
19207	12340090-1	2510-01-257-1572	277	17
		2510-01-257-1572	277	45
19207	12340090-2	2510-01-266-8013	277	18
		2510-01-266-8013	277	44
19207	12340091	2510-01-254-1056	272	16
19207	12340095	5330-01-238-3217	275	24
19207	12340097	5340-01-203-8674	274	29
19207	12340098	5365-01-253-8975	272	11
19207	12340099-1	5365-01-195-4949	277	5
19207	12340103-3	2590-01-460-8320	233	7
		2590-01-460-8320	277	32
19207	12340103-4	2590-01-457-9922	277	32
		2590-01-457-9922	473	3
19207	12340104	2510-01-257-3747	274	2
		2510-01-257-3747	276	28
19207	123401103-2	2590-01-205-8607	273	21
19207	12340111	3040-01-198-7473	273	10
19207	12340115-1	5365-01-254-4394	277	13
19207	12340115-2	5365-01-457-3364	277	13
19207	12340116	2510-01-252-6240	277	3
19207	12340123	2590-01-258-9491	272	7
19207	12340124	5340-01-253-8377	272	12
19207	12340126	2510-01-253-2839	272	10
19207	12340129	5340-01-198-2249	246	16
19207	12340130	5340-01-253-5479	246	19
		5340-01-253-5479	253	34
		5340-01-253-5479	254	25
		5340-01-253-5479	255	27
19207	12340131	2540-01-202-8545	409	11
		2540-01-202-8545	411	15
		2540-01-202-8545	415	17
19207	12340132	5365-01-198-3463	248	17
19207	12340134	2590-01-202-8546	409	6
		2590-01-202-8546	411	12
		2590-01-202-8546	415	11
19207	12340135	1010-01-247-7917	251	8
19207	12340142	2590-01-194-2048	250	22
19207	12340143	2590-01-264-6531	251	29
34623	12340147-1	2590-01-248-7622	251	33
34623	12340147-2	2590-01-248-7623	251	34
19207	12340152	1010-01-252-6300	251	9
19207	12340153	2540-01-202-8548	411	8
19207	12340154	5340-01-246-8293	251	37
19207	12340155	2510-01-197-8572	246	20
		2510-01-197-8572	253	24
		2510-01-197-8572	254	19

SECTION IV TM9-2320-280-24P C01

CROSS REFERENCE INDEXES

PART NUMBER INDEX

CAGE PART NUMBER STOCK NUMBER FIGURE NO ITEM NO

CAGE	PART NUMBER	STOCK NUMBER	FIGURE NO	ITEM NO
19207	12340156	1005-01-252-6299	251	3
19207	12340158-2	2590-01-247-7920	255	34
19207	12340158-3	2540-01-262-9516	253	33
		2540-01-262-9516	254	28
19207	12340160	2590-01-255-2958	251	48
19207	12340161	2590-01-255-2957	251	36
19207	12340162	1005-01-247-7916	251	6
19207	12340163	5340-01-247-7921	251	13
19207	12340164	5340-01-247-7918	252	25
19207	12340166	2590-01-246-8294	252	36
		2590-01-246-8294	253	3
		2590-01-246-8294	255	3
19207	12340167	5975-01-197-5505	409	3
		5975-01-197-5505	411	3
		5975-01-197-5505	430	48
19207	12340170	2540-01-246-8292	251	22
		2540-01-246-8292	254	3
19207	12340174	2590-01-188-7384	246	5
19207	12340175-1	2590-01-247-7923	255	8
34623	12340175-2	2590-01-262-9514	253	8
34623	12340175-3	2590-01-262-9513	253	28
19207	12340179	5340-01-247-7922	252	15
19207	12340180	2590-01-247-7924	255	22
19207	12340182-1	5340-01-208-7670	321	26
		5340-01-208-7670	325	24
		5340-01-208-7670	376	16
19207	12340182-2	5342-01-204-9610	321	26
		5342-01-204-9610	325	24
19207	12340193-1	5340-01-205-9021	316	17
		5340-01-205-9021	320	13
		5340-01-205-9021	320	25
		5340-01-205-9021	324	23
		5340-01-205-9021	324	35
		5340-01-205-9021	375	24
		5340-01-205-9021	378	24
19207	12340193-2	5340-01-205-9022	316	17
		5340-01-205-9022	320	13
		5340-01-205-9022	320	25
		5340-01-205-9022	324	23
		5340-01-205-9022	324	35
		5340-01-205-9022	375	24
		5340-01-205-9022	378	24
19207	12340196	5340-01-199-9679	326	9
19207	12340199	2540-01-197-5526	326	1
19207	12340201	5340-01-254-7189	316	19
		5340-01-254-7189	320	16
		5340-01-254-7189	320	30
		5340-01-254-7189	324	20
		5340-01-254-7189	324	32

SECTION IV TM9-2320-280-24P C01

CROSS REFERENCE INDEXES

PART NUMBER INDEX

CAGE	PART NUMBER	STOCK NUMBER	FIGURE NO	ITEM NO
		5340-01-254-7189	378	26
19207	12340212-3	5325-01-357-1080	316	18
		5325-01-357-1080	320	33
		5325-01-357-1080	324	39
19207	12340214-1	5340-01-256-9318	316	20
		5340-01-256-9318	320	17
		5340-01-256-9318	320	32
		5340-01-256-9318	324	29
19207	12340222-1	2590-01-196-1290	321	7
		2590-01-196-1290	325	11
		2590-01-196-1290	376	6
19207	12340222-2	2590-01-196-5314	321	7
		2590-01-196-5314	325	11
		2590-01-196-5314	376	6
19207	12340226	2540-01-197-5478	324	1
19207	12340226-2	2540-01-476-2172	324	1
19207	12340231-10	2510-01-450-5480	316	7
		2510-01-450-5480	320	19
		2510-01-450-5480	324	24
19207	12340231-11	2510-01-450-5479	316	7
		2510-01-450-5479	320	19
		2510-01-450-5479	324	24
19207	12340231-2	2510-01-189-9725	316	7
19207	12340231-3	2510-01-330-6174	316	7
		2510-01-330-6174	320	19
19207	12340231-4	2510-01-330-6175	316	7
		2510-01-330-6175	320	19
34623	12340231-5	2510-01-407-6037	316	7
		2510-01-407-6037	320	19
34623	12340231-6	2510-01-407-6036	316	7
		2510-01-407-6036	320	19
19207	12340239	5340-01-204-3862	317	3
		5340-01-204-3862	321	4
		5340-01-204-3862	325	7
		5340-01-204-3862	373	2
		5340-01-204-3862	376	3
19207	12340241	2540-01-197-5525	325	5
19207	12340245-3	2510-01-330-6176	320	5
19207	12340245-4	2510-01-330-6576	320	5
34623	12340245-5	2510-01-412-6761	320	5
34623	12340245-6	2510-01-413-8872	320	5
19207	12340245-7	2510-01-450-5482	320	5
		2510-01-450-5482	324	10
19207	12340245-8	2510-01-450-5483	320	5
		2510-01-450-5483	324	10
19207	12340252	2530-01-196-4724	208	1
19207	12340258	5340-01-332-7599	331	19
		5340-01-332-7599	332	23
		5340-01-332-7599	370	5

I-231

SECTION IV TM9-2320-280-24P C01

CROSS REFERENCE INDEXES

PART NUMBER INDEX

CAGE	PART NUMBER	STOCK NUMBER	FIGURE NO		ITEM NO
19207	12340259	5305-01-254-2522	230		12
		5305-01-254-2522	238	1	
19207	12340259-10	5306-01-450-5947	230		14
		5306-01-450-5947	235	15	
19207	12340259-11	5306-01-437-8056	234		18
		5306-01-437-8056	239	13	
19207	12340259-12	5306-01-449-0510	235		11
19207	12340259-13	5306-01-437-8058	239		14
19207	12340259-2	5305-01-264-3555	230		14
19207	12340259-4	5306-01-315-7088	234		7
19207	12340259-5	5306-01-315-7087	239		3
19207	12340259-6	5306-01-316-1456	234		12
19207	12340259-7	5306-01-316-2433	234		1
		5306-01-316-2433	239	10	
19207	12340263	5310-01-269-0786	26		8
19207	12340264	5340-01-197-3461	26		9
19207	12340271	5340-01-210-4751	240		9
19207	12340275-1	5340-01-210-1606	240		17
19207	12340275-3	5340-01-488-8389	240		25
19207	12340275-4	5340-01-484-6930	240		25
19207	12340286	2590-01-188-3237	240		5
19207	12340286-3	2590-01-496-1569	240		5
34623	12340288	2540-01-185-6119	240		8
34623	12340289	2540-01-223-3749	240		8
19207	12340293	2590-01-197-4899	272		37
19207	12340298	2590-01-192-4525	277		8
19207	12340301	2540-01-192-9718	277		21
19207	12340302	3110-01-185-3114	277		9
19207	12340308		276		27
19207	12340310	3120-01-188-5082	275		15
		3120-01-188-5082	276	14	
19207	12340310-1		275		16
			276	15	
19207	12340311	2510-01-189-9828	274		1
		2510-01-189-9828	276	25	
19207	12340312	2590-01-185-3112	277		11
19207	12340314	5340-01-247-7919	252		30
19207	12340318	9905-01-248-1113	355		4
19207	12340320	9905-01-248-1114	355		4
19207	12340321	9905-01-248-9543	355		4
19207	12340322	9905-01-248-1112	355		4
19207	12340323	9905-01-185-7970	355		4
19207	12340324	9905-01-248-9548	355		4
19207	12340325	9905-01-248-9549	355		4
19207	12340326	9905-01-248-9552	355		4
19207	12340327	9905-01-248-1115	355		4
34623	12340331	4730-01-394-8345	368		37
19207	12340334	2540-01-185-3113	277		24
19207	12340336		273		15

SECTION IV TM9-2320-280-24P C01

CROSS REFERENCE INDEXES

PART NUMBER INDEX

CAGE	PART NUMBER	STOCK NUMBER	FIGURE NO	ITEM NO
19207	12340337	4710-01-189-9727	273	23
19207	12340338	2540-01-192-5937	277	41
19207	12340340	2510-01-249-1585	272	1
19207	12340342-1	4710-01-192-6113	277	42
19207	12340342-2	4710-01-192-9720	277	15
19207	12340342-3		277	43
19207	12340342-4		277	16
19207	12340342-5	4710-01-333-5995	277	42
19207	12340342-6	4710-01-335-7523	277	15
19207	12340346	2590-01-189-9734	273	6
19207	12340347	2510-01-249-1586	272	1
19207	12340349	1095-01-478-3115	275	6
19207	12340352-1	5306-01-323-5535	207	8
		5306-01-323-5535	280	50
19207	12340352-2	5306-01-323-5534	198	22
19207	12340360	2540-01-174-7696	206	14
19207	12340362	5935-01-238-0543	84	13
		5935-01-238-0543	86	21
19207	12340364	5340-01-197-5470	416	5
19207	12340366	5340-01-255-2729	429	23
19207	12340372	7690-01-248-7680	429	33
19207	12340373	7690-01-248-4908	430	43
19207	12340374	7690-01-248-7682	430	24
19207	12340375	7690-01-248-4909	430	45
19207	12340376	7690-01-250-1101	430	47
19207	12340377	7690-01-248-7679	429	28
19207	12340378	7690-01-248-7683	430	46
19207	12340379	7690-01-248-7681	429	39
19207	12340382	5340-01-249-1594	429	21
19207	12340387	5340-01-257-3778	430	25
19207	12340390	5340-01-255-9515	430	2
34623	12340395	5331-00-580-6586	177	8
		5331-00-580-6586	178	10
		5331-00-580-6586	176A	25
19207	12340401	2540-01-249-1595	429	30
19207	12340409	5340-01-248-7649	430	36
19207	12340411	2540-01-249-1598	430	30
19207	12340413	2540-01-249-1592	429	8
19207	12340414	2590-01-249-1596	430	1
19207	12340419	2540-01-249-1593	429	19
19207	12340420	2540-01-250-7592	430	31
19207	12340421	2540-01-249-1597	430	4
19207	12340422	2540-01-250-7589	429	1
19207	12340423	7690-01-248-4907	430	28
34623	12340432	2540-01-293-6926	368	2
19207	12340434	2540-01-197-5450	368	4
19207	12340436	2540-01-199-2396	366	2
19207	12340442	9905-01-199-2371	367	2
19207	12340448	5340-01-204-3904	368	21

SECTION IV TM9-2320-280-24P C01

CROSS REFERENCE INDEXES

PART NUMBER INDEX

CAGE PART NUMBER STOCK NUMBER FIGURE NO ITEM NO

CAGE	PART NUMBER	STOCK NUMBER	FIGURE NO	ITEM NO
19207	12340448-1	5340-01-476-5616	228	12
19207	12340466	6210-01-203-2101	367	4
19207	12340470	2540-01-254-6511	370	9
19207	12340474	5340-01-395-1244	365	24
19207	12340477	4140-01-259-2175	368	6
19207	12340486	5365-01-211-5094	177	15
19207	12340487	1015-01-200-0869	250	17
		1015-01-200-0869	252	6
19207	12340489	6150-01-260-8000	366	29
19207	12340492	4720-01-211-1998	364	4
		4720-01-211-1998	379	40
19207	12340493	4720-01-257-2655	364	9
		4720-01-257-2655	379	43
19207	12340498	4710-01-257-2649	364	5
		4710-01-257-2649	379	42
19207	12340507-4	5340-01-258-4529	251	23
		5340-01-258-4529	251	38
		5340-01-258-4529	253	19
		5340-01-258-4529	254	6
		5340-01-258-4529	255	18
		5340-01-258-4529	448	3
		5340-01-258-4529	450	5
19207	12340514	5305-01-188-5133	102	2
19207	12340515	5305-01-215-5174	3	1
		5305-01-215-5174	75	40
		5305-01-215-5174	77	21
		5305-01-215-5174	479	1
19207	12340517-1	5340-01-414-1454	319	5
		5340-01-414-1454	320	3
		5340-01-414-1454	322	4
		5340-01-414-1454	324	4
33729	12340526	5325-01-437-4175	319	7
19207	12340531-3	2590-01-194-6990	333	35
19207	12340531-4	2590-01-212-4956	333	36
19207	12340533-3	5325-00-185-0961	207	11
19207	12340535-1	2540-01-192-9715	403	21
34623	12340542-1	2540-01-394-9682	402	6
34623	12340542-2	2540-01-394-9683	402	6
19207	12340542-3	2540-01-394-4788	402	6
19207	12340542-4	2540-01-394-4787	402	6
19207	12340542-5	2540-01-394-4792	402	6
34623	12340542-6	2540-01-394-9679	402	6
19207	12340544	2540-01-185-7951	403	2
34623	12340545-2	2540-01-192-9713	403	1
		2540-01-192-9713	403	6
19207	12340549	2540-01-192-4501	403	15
19207	12340553	2540-01-185-8015	403	14
19207	12340560-1	2540-01-185-8016	402	1
19207	12340560-2	2540-01-185-3214	402	1

I-234

SECTION IV TM9-2320-280-24P C01

CROSS REFERENCE INDEXES

PART NUMBER INDEX

CAGE	PART NUMBER	STOCK NUMBER	FIGURE NO	ITEM NO
19207	12340560-3	2540-01-396-7759	402	1
19207	12340560-4	2540-01-399-0785	402	1
19207	12340560-5	2540-01-396-7756	402	1
34623	12340560-6	2540-01-396-7757	402	1
19207	12340565	3120-01-251-0735	133	13
19207	12340569	5340-01-247-7911	242	9
19207	12340570	6625-01-192-7498	69	28
19207	12340571-2	5340-01-460-5306	141	16
19207	12340581	2540-01-250-7590	415	9
34623	12340587-3	2990-01-394-9670	365	9
19207	12340589-1	5340-01-498-9619	368	45
19207	12340589-2	5340-01-498-9625	368	45
19207	12340591	2590-01-257-8787	369	13
19207	12340591-1	2590-01-452-5034	369	13
19207	12340593-1	2540-01-461-1134	369	14
34623	12340593-1	2540-01-461-1134	369	14
19207	12340595-2	5340-01-467-1669	249	47
19207	12340604		378	7
19207	12340605		378	15
19207	12340608	2510-01-325-8741	375	8
		2510-01-325-8741	375	15
19207	12340635	5340-01-255-1109	413	9
		5340-01-255-1109	421	10
		5340-01-255-1109	422	10
19207	12340636	5340-01-255-9514	415	8
19207	12340637	5985-01-253-3514	414	4
		5985-01-253-3514	425	13
19207	12340638	2540-01-202-8547	208	16
19207	12340639	5340-01-249-1590	415	7
19207	12340641	2540-01-249-1589	413	7
31272	12340648	5340-01-434-2864	272	31
19207	12340652	7690-01-256-4908	354	21
19207	12340653	9905-01-248-1111	355	4
19207	12340654	5340-01-476-6612	282	11
19207	12340655	5995-01-195-1918	282	1
19207	12340655-1	5995-01-339-0190	282	1
		5995-01-339-0190	435	4
34623	12340658	2510-01-254-1484	208	18
19207	12340661	2530-01-197-5502	490	14
19207	12340663-1	5340-01-213-4667	77	25
19207	12340664	5342-01-205-5381	77	26
19207	12340665	5340-01-213-4665	83	12
		5340-01-213-4665	88	34
19207	12340666	5340-01-443-6908	83	11
		5340-01-443-6908	88	33
19207	12340676	2540-01-189-3706	320	1
19207	12340676-11	2540-01-330-6177	320	1
34623	12340676-21	2540-01-412-2661	320	1
19207	12340676-31	2540-01-450-4019	320	1

I-235

SECTION IV TM9-2320-280-24P C01

CROSS REFERENCE INDEXES

PART NUMBER INDEX

CAGE PART NUMBER STOCK NUMBER FIGURE NO ITEM NO

CAGE	PART NUMBER	STOCK NUMBER	FIGURE NO		ITEM NO
19207	12340679	2540-01-196-1291	321		13
		2540-01-196-1291	376	10	
19207	12340682	2540-01-199-6759	317		1
		2540-01-199-6759	321	2	
		2540-01-199-6759	376	1	
19207	12340688-1	5340-01-196-3297	267		19
19207	12340689-1	9515-01-185-3110	267		19
19207	12340689-2	9515-01-189-9746	267		19
19207	12340690-1	9515-01-185-3109	267		15
19207	12340690-2	9515-01-189-9731	267		15
19207	12340693-1	9515-01-186-3111	267		21
19207	12340693-2		267		21
19207	12340694-2	9515-01-189-9733	267		20
19207	12340695		365		15
34623	12340696	2910-01-320-6645	366		37
		2910-01-320-6645	384	11	
19207	12340700-1		374		17
19207	12340700-10		374		10
19207	12340700-11		374		11
			374	24	
19207	12340700-12		374		5
19207	12340700-14		377		4
19207	12340700-15		377		13
19207	12340700-16		377		9
34623	12340700-2		374		18
34623	12340700-3		374		18
19207	12340700-4		374		16
19207	12340700-6	2540-01-197-8071	374		16
19207	12340700-7		374		9
19207	12340700-8		374		21
19207	12340700-9		374		19
19207	12340702		366		8
19207	12340703		366		35
19207	12340704		366		36
19207	12340706		368		43
19207	12340708	5340-01-466-5243	374		4
19207	12340708-1		374		6
19207	12340709-1	5340-01-203-3273	209		2
19207	12340709-2	5340-01-205-5118	209		2
19207	12340710	5340-01-198-0686	375		9
		5340-01-198-0686	375		13
		5340-01-198-0686	378		8
		5340-01-198-0686	378	14	
19207	12340713	2540-01-206-4115	317		2
		2540-01-206-4115	321		3
		2540-01-206-4115	373		1
		2540-01-206-4115	376	2	
19207	12340720	5342-01-196-5312	209		11
19207	12340721	5330-01-198-3521	429		27

SECTION IV TM9-2320-280-24P C01

CROSS REFERENCE INDEXES

PART NUMBER INDEX

CAGE	PART NUMBER	STOCK NUMBER	FIGURE NO	ITEM NO
		5330-01-198-3521	BULK	51
19207	12340721-1.25		209	10
19207	12340722		365	2
19207	12340736-13	2540-01-450-4021	316	1
19207	12340736-18	2540-01-450-4017	316	1
19207	12340736-7	2540-01-314-2786	316	1
19207	12340736-8	2540-01-330-6173	316	1
19207	12340737	2540-01-200-8393	316	4
19207	12340737-2	2540-01-330-6171	316	4
34623	12340737-3	2540-01-407-9241	316	4
19207	12340737-6	2540-01-450-4018	316	4
19207	12340744	2540-01-191-0973	452	5
19207	12340747	2540-01-199-6760	318	2
19207	12340761	2540-01-189-3457	319	1
19207	12340761-2	2540-01-330-6169	319	1
19207	12340761-3	2540-01-450-7686	319	1
19207	12340761-9	2540-01-450-4015	319	1
19207	12340764-1	2540-01-199-6761	318	3
19207	12340764-2	2540-01-200-6611	318	1
19207	12340768	2540-01-197-8735	323	1
19207	12340777	2540-01-189-3707	322	1
19207	12340777-10	2540-01-450-4020	322	1
19207	12340777-15	2540-01-450-4024	322	1
19207	12340777-5	2540-01-330-6170	322	1
19207	12340779	2590-01-249-1588	286	18
19207	12340791	5340-01-395-0812	369	3
19207	12340792	5305-01-264-5874	199	50
		5305-01-264-5874	343	19
		5305-01-264-5874	495	9
19207	12340799	4720-01-214-5757	12	39
19207	12340814	7690-01-204-7785	360	1
34623	12340823	2510-01-421-8067	227	18
34623	12340824	2590-01-306-5881	227	9
19207	12340828	4710-01-251-3073	227	5
34623	12340829-1	3040-01-197-5510	368	41
19207	12340832-3	2590-01-212-1057	268	30
19207	12340832-4	2590-01-210-6202	268	30
19207	12340836	2510-01-247-7968	226	4
19207	12340842	4210-01-259-5634	241	9
		4210-01-259-5634	241	14
19207	12340843	5995-01-309-2953	82	1
19207	12340845-2	5306-01-263-8889	35	28
		5306-01-263-8889	36	12
		5306-01-263-8889	37	17
		5306-01-263-8889	38	12
		5306-01-263-8889	38A	8
		5306-01-263-8889	472	14
		5306-01-263-8889	475	21
		5306-01-263-8889	476	19

I-237

SECTION IV TM9-2320-280-24P C01

CROSS REFERENCE INDEXES

PART NUMBER INDEX

CAGE	PART NUMBER	STOCK NUMBER	FIGURE NO	ITEM NO
19207	12340845-3	5306-01-270-5448	35	29
		5306-01-270-5448	36	13
		5306-01-270-5448	38	13
		5306-01-270-5448	33A	7
		5306-01-270-5448	38A	9
19207	12340845-4	5306-01-264-3531	472	16
		5306-01-264-3531	475	24
		5306-01-264-3531	476	20
19207	12340845-5	5306-01-495-1744	492	19
19207	12340846-1	5340-01-185-8820	240	14
		5340-01-185-8820	451	1
19207	12340852	7690-01-204-7847	360	2
19207	12340860-1	2530-01-256-4914	152	20
19207	12340860-2	2530-01-257-3905	152	20
19207	12340867	9905-01-190-8425	354	12
19207	12340868	9905-01-185-3129	354	12
19207	12340869	5340-01-460-5291	199	24
34623	12340871	2510-01-373-0131	228	3
19207	12340872	2510-01-373-2783	228	1
34623	12340873	2510-01-373-7877	228	7
19207	12340874	2510-01-372-9734	228	4
19207	12340875	2510-01-374-0000	228	9
34623	12340876	2510-01-373-8445	228	8
19207	12340877	2510-01-373-2784	228	5
19207	12340879	5340-01-270-8489	338	23
		5340-01-270-8489	339	18
		5340-01-270-8489	486	10
19207	12340880	5310-01-472-7368	270	10
19207	12340890	7690-01-191-8793	359	4
19207	12340891	7690-01-189-3737	361	2
19207	12340892	7690-01-204-7850	361	8
19207	12340893	7690-01-189-3736	361	6
19207	12340894	7690-01-189-3738	361	1
19207	12340895	7690-01-204-7849	361	7
19207	12340896	7690-01-203-0161	361	4
19207	12340897	7690-01-189-3739	361	5
19207	12340898	7690-01-191-1313	359	1
19207	12340899	7690-01-189-3740	359	2
		7690-01-189-3740	361	3
19207	12340900	7690-01-191-1314	359	10
19207	12340901	7690-01-192-6371	359	3
19207	12340902	7690-01-191-1315	359	9
19207	12340903	7690-01-191-1316	359	8
19207	12340904	7690-01-190-8501	359	5
		7690-01-190-8501	360	3
19207	12340905	7690-01-192-6369	361	10
19207	12340906	7690-01-192-6370	361	9
19207	12340908	7690-01-191-1312	359	6
19207	12340909	7690-01-188-5144	356	5

SECTION IV TM9-2320-280-24P C01

CROSS REFERENCE INDEXES

PART NUMBER INDEX

CAGE	PART NUMBER	STOCK NUMBER	FIGURE NO	ITEM NO
19207	12340910	9905-01-204-7776	354	9
19207	12340917	7690-01-256-4909	354	22
19207	12340923	7690-01-264-6536	443	18
19207	12340926-3	5330-01-298-6009	296	9
19207	12340927	5930-01-266-1811	58	14
19207	12340935	5340-01-272-7814	218	8
19207	12340941	5340-01-272-6578	219	2
34623	12340946	2540-01-357-8939	288	1
19207	12340952-1	5340-01-272-2635	289	7
19207	12340952-2	5340-01-273-1680	289	7
19207	12340955	2540-01-271-3771	289	12
19207	12340958	3040-01-270-7917	238	10
19207	12340961	5365-01-272-7482	300	14
19207	12340968	5340-01-270-8370	211	18
		5340-01-270-8370	212	15
19207	12340979	2540-01-270-9344	229	1
19207	12340983	5340-01-274-3021	52	15
19207	12340984	5365-01-269-8672	211	29
		5365-01-269-8672	212	31
19207	12340990-1	5340-01-272-6520	264	5
		5340-01-272-6520	265	5
19207	12340990-2	5340-01-270-8455	264	7
		5340-01-270-8455	265	6
19207	12340994-1	3040-01-271-3824	238	17
19207	12340995-1	3040-01-270-7961	238	15
19207	12341000	6150-01-269-0067	89	1
19207	12341003	5975-01-270-8552	294	1
19207	12341004	5340-01-272-9109	294	3
19207	12341012	2540-01-271-3772	288	5
19207	12341020	2510-01-272-3951	288	12
34623	12341027	2990-01-383-5689	401	16
19207	12341045-1	5340-01-272-5652	289	21
19207	12341045-2	5340-01-272-4962	289	21
19207	12341049	4710-01-271-7937	287	30
19207	12341064	5340-01-272-8405	289	17
19207	12341073	2510-01-265-3155	293	17
34623	12341076	2540-01-275-6202	287	21
19207	12341081-2	2510-01-265-3283	297	9
19207	12341082-1	2510-01-265-3281	297	3
		2510-01-265-3281	297	10
19207	12341082-2	2510-01-265-3282	291	7
		2510-01-265-3282	297	10
19207	12341088	2540-01-271-2839	347	9
34623	12341092-1	2510-01-265-3194	293	22
19207	12341108	5985-01-269-8274	341	4
19207	12341109	5340-01-273-1684	214	24
34623	12341109-2	2510-01-265-1127	213	31
19207	12341111	5342-01-266-5691	217	1
19207	12341112	2510-01-272-3953	288	16

SECTION IV TM9-2320-280-24P C01

CROSS REFERENCE INDEXES

PART NUMBER INDEX

CAGE	PART NUMBER	STOCK NUMBER	FIGURE NO		ITEM NO	
19207	12341113	2540-01-276-1451			224	5
		2540-01-276-1451	237	15		
19207	12341119	5340-01-266-5708			224	14
19207	12341120	2510-01-272-0536			224	3
19207	12341125	6150-01-269-5300			92	1
19207	12341130	5342-01-275-0532			217	5
19207	12341132	2540-01-271-1946			288	27
19207	12341133	2540-01-270-5065			238	5
19207	12341135	2540-01-271-8009			238	9
19207	12341138				238	8
19207	12341139	2540-01-268-7203			299	3
19207	12341142	2510-01-266-5690			218	3
19207	12341143-1	2540-01-271-2840			289	3
19207	12341143-2	2540-01-271-2841			289	3
19207	12341145-1	2590-01-271-3770			289	26
19207	12341145-2	2590-01-271-9807			289	26
19207	12341151	5999-01-372-5601			39	80
		5999-01-372-5601	41	20		
		5999-01-372-5601	475	26		
19207	12341155	2540-01-266-5702			229	3
34623	12341168	2540-01-271-1991			293	20
19207	12341176	2540-01-265-3160			292	7
19207	12341205	2540-01-270-6541			237	2
34623	12341206	2540-01-270-6540			237	2
19207	12341207	6150-01-269-5298			95	1
19207	12341211-1	5330-01-265-8803			293	1
19207	12341211-2	5330-01-265-8807			223	1
19207	12341211-3	5330-01-265-8806			223	3
19207	12341211-4	5330-01-265-9231			293	33
19207	12341211-8	5330-01-265-8947			223	1
19207	12341211-9	5330-01-265-8946			223	3
19207	12341217	2510-01-266-4003			215	3
19207	12341218	2510-01-270-7922			222	23
19207	12341219	2510-01-272-0537			222	6
19207	12341227	2540-01-270-3776			337	21
19207	12341229	5340-01-266-3843			346	4
19207	12341231-1	5365-01-273-4690			338	3
19207	12341231-2	5365-01-272-8435			93	24
		5365-01-272-8435	95	41		
19207	12341231-3	5365-01-274-3573			339	26
19207	12341231-4	5365-01-272-9275			96	31
19207	12341231-5	5365-01-383-3722			304	6
19207	12341231-6	5365-01-381-2215			303	8
19207	12341239	6220-01-269-7066			65	12
34623	12341246	2540-01-270-9711			238	4
19207	12341250	2510-01-280-4240			293	2
19207	12341271	2990-01-265-3297			346	2
19207	12341276	5340-01-274-3295			293	13
19207	12341286	4130-01-265-8691			344	1

I-240

SECTION IV TM9-2320-280-24P C01

CROSS REFERENCE INDEXES

PART NUMBER INDEX

CAGE	PART NUMBER	STOCK NUMBER	FIGURE NO	ITEM NO
19207	12341293	6150-01-269-5299	94	1
19207	12341294	6150-01-269-5275	91	1
19207	12341320-2	2540-01-270-6458	290	4
19207	12341320-4	2540-01-271-9806	289	15
19207	12341320-6	9330-01-269-7114	287	17
19207	12341326	7690-01-266-5706	59	20
19207	12341348	5330-01-458-3810	269	4
19207	12341353-1	5310-01-272-7449	238	23
		5310-01-272-7449	287	26
19207	12341354	5315-01-273-0096	287	22
19207	12341382	5340-01-386-6067	292	4
19207	12341384	5340-01-365-8796	287	4
34623	12341387-2	4030-01-272-7573	311	9
		4030-01-272-7573	312	5
19207	12341388	5315-01-269-8504	213	17
		5315-01-269-8504	214	13
19207	12341399	5310-01-272-9985	289	25
19207	12341401	5365-01-272-9082	308	6
19207	12341403	2510-01-265-6953	214	16
19207	12341407	5365-01-269-8671	211	8
		5365-01-269-8671	212	5
19207	12341408	6150-01-269-1840	58	13
		6150-01-269-1840	90	1
19207	12341414	4730-01-270-9652	311	11
		4730-01-270-9652	312	7
19207	12341416	5945-01-266-1810	59	5
19207	12341418	2510-01-273-8376	213	35
34623	12341420	4710-01-265-3229	335	7
		4710-01-265-3229	336	4
19207	12341429	2510-01-274-4901	222	24
19207	12341440	4710-01-265-3275	334	10
19207	12341450	2540-01-271-3828	236	2
19207	12341458	6150-01-269-1897	93	1
19207	12341466	5985-01-269-8272	340	6
19207	12341467	5975-01-270-4587	58	3
19207	12341485	5340-01-272-3353	213	3
19207	12341486	5340-01-277-8125	213	40
19207	12341487	5330-01-272-7471	347	8
34623	12341489-2	5365-01-272-6574	298	2
19207	12341491	5340-01-273-2381	215	8
19207	12341492	5975-01-275-2695	229	4
19207	12341496	5340-01-272-7495	287	2
19207	12341499	2540-01-270-6460	290	9
19207	12341500-1	5340-01-270-8432	212	22
		5340-01-270-8432	258	8
19207	12341500-2	5340-01-270-8433	211	20
		5340-01-270-8433	212	18
19207	12341500-3	5340-01-270-8431	212	25
		5340-01-270-8431	258	10

I-241

SECTION IV TM9-2320-280-24P C01

CROSS REFERENCE INDEXES

PART NUMBER INDEX

CAGE PART NUMBER STOCK NUMBER FIGURE NO ITEM NO

CAGE	PART NUMBER	STOCK NUMBER	FIGURE NO	ITEM NO
		5340-01-270-8431	259	13
		5340-01-270-8431	260	24
19207	12341507	5340-01-273-1686	290	6
19207	12341509	5340-01-277-5129	258	3
		5340-01-277-5129	259	6
		5340-01-277-5129	260	19
		5340-01-277-5129	262	14
19207	12341514	5330-01-272-7472	347	6
19207	12341533	2510-01-270-7919	347	5
19207	12341535	2540-01-271-3769	300	9
19207	12341536	2510-01-266-5711	216	10
19207	12341544	6220-01-270-1454	64	19
19207	12341551	4520-01-270-5080	298	4
19207	12341552	5340-01-270-4415	298	1
		5340-01-270-4415	346	7
19207	12341553	4520-01-271-2949	300	11
19207	12341555	2540-01-270-6539	236	5
19207	12341556-1	2510-01-272-0538	288	14
19207	12341556-2	2510-01-272-3952	288	14
19207	12341564	6110-01-496-9260	52	1
19207	12341565	5340-01-265-8858	289	23
19207	12341575	5340-01-297-4127	64	1
19207	12341582-1	4710-00-541-4935	288	11
19207	12341582-2	4710-01-270-9507	288	8
19207	12341583	4730-01-265-3166	301	6
		4730-01-265-3166	303	1
		4730-01-265-3166	304	1
		4730-01-265-3166	305	3
		4730-01-265-3166	306	1
		4730-01-265-3166	309	2
		4730-01-265-3166	310	2
19207	12341584	4730-00-476-7127	302	3
		4730-00-476-7127	303	4
		4730-00-476-7127	304	11
		4730-00-476-7127	305	5
		4730-00-476-7127	306	3
		4730-00-476-7127	309	10
19207	12341585	4730-00-476-5859	302	6
		4730-00-476-5859	303	15
		4730-00-476-5859	305	8
		4730-00-476-5859	306	6
		4730-00-476-5859	309	4
		4730-00-476-5859	310	6
19207	12341586	4730-01-265-3285	304	3
		4730-01-265-3285	306	5
		4730-01-265-3285	309	6
19207	12341586-3	4720-01-384-1212	301	5
19207	12341587-10		302	8
			305	4

I-242

SECTION IV TM9-2320-280-24P C01

CROSS REFERENCE INDEXES

PART NUMBER INDEX

CAGE	PART NUMBER	STOCK NUMBER	FIGURE NO	ITEM NO
19207	12341587-15		309	1
19207	12341587-2		309	9
19207	12341587-20		301	7
			303	13
19207	12341587-28		303	16
19207	12341587-4		303	5
			309	11
19207	12341591	4720-01-271-6151	300	6
34623	12341606-2	4730-01-382-3165	302	1
		4730-01-382-3165	305	1
19207	12341610	4140-01-265-3151	342	1
19207	12341614	4720-01-270-9618	310	8
19207	12341615-1	6150-01-269-1861	56	19
19207	12341615-3	6150-01-269-1862	56	20
19207	12341616	6150-01-269-1859	56	5
19207	12341617	5330-01-272-2601	222	4
19207	12341621	9905-01-265-1132	59	22
19207	12341622	9905-01-265-1131	59	22
19207	12341625-2	5330-01-265-8948	213	1
19207	12341625-4	5330-01-265-8804	214	1
34623	12341627	5325-01-289-6196	95	45
19207	12341637	5340-01-265-8924	216	6
19207	12341639	5940-01-272-5800	52	26
19207	12341640	5365-01-477-2843	227	33
19207	12341641	5340-01-381-1266	401	19
19207	12341657	5330-01-290-2709	337	9
19207	12341661	5340-01-289-3244	342	7
34623	12341662	2940-01-290-5014	337	8
19207	12341670	5365-01-272-2806	289	18
19207	12341680	5640-01-268-7204	222	7
		5640-01-268-7204	495	3
19207	12341681		495	6
19207	12341682		495	2
19207	12341684		495	8
19207	12341685		495	5
19207	12341689	2510-01-271-7085	228	10
19207	12341704-10		221	1
19207	12341704-11		221	2
19207	12341704-12		222	25
19207	12341704-13		222	1
19207	12341704-14		222	2
19207	12341704-16		222	21
19207	12341704-17		222	3
19207	12341704-18		222	8
19207	12341704-19		222	5
19207	12341704-21		222	15
19207	12341704-22		222	19
19207	12341704-23		222	18
19207	12341704-24		222	17

SECTION IV TM9-2320-280-24P C01

CROSS REFERENCE INDEXES

PART NUMBER INDEX

CAGE	PART NUMBER	STOCK NUMBER	FIGURE NO	ITEM NO
19207	12341704-6		216	7
19207	12341704-7		216	9
19207	12341704-8		221	7
19207	12341704-9		221	6
19207	12341711-1	5342-01-264-6533	212	2
19207	12341711-2	5342-01-265-3672	212	2
19207	12341714	5342-01-270-7921	307	13
		5342-01-270-7921	308	1
34623	12341726	5920-01-270-0090	52	10
		5920-01-270-0090	59	13
19207	12341730	5340-01-273-1721	492	3
34623	12341746	5340-01-276-7415	213	7
		5340-01-276-7415	214	3
19207	12341749	5935-01-269-5994	52	4
		5935-01-269-5994	57	3
34623	12341751	5930-01-287-6698	58	5
19207	12341757	9905-01-266-3913	362	2
19207	12341778	2510-01-265-1126	213	21
19207	12341779	2510-01-265-1130	214	28
19207	12341782-1	2510-01-265-1140	212	1
19207	12341782-2	2510-01-265-1141	212	1
19207	12341791-1	5305-01-272-8322	292	22
19207	12341791-2	5305-01-272-8323	292	20
19207	12341793	4820-00-287-5627	335	17
19207	12341795-1	4720-01-265-3237	301	1
19207	12341795-2	4720-01-265-3222	301	2
19207	12341796	5330-01-272-7727	223	2
19207	12341798	5340-01-271-5866	211	5
		5340-01-271-5866	291	3
		5340-01-271-5866	297	5
19207	12341800	5310-01-327-0387	337	15
19207	12341802	5340-01-265-8922	238	2
19207	12341809	3040-01-270-9466	37	6
19207	12341812	9905-01-190-8427	354	12
19207	12341813	2540-01-271-1955	311	18
19207	12341814	2540-01-271-1954	311	6
19207	12341815	5305-01-255-2670	243	1
19207	12341822	5305-01-270-7225	227	23
		5305-01-270-7225	228	2
		5305-01-270-7225	484	17
19207	12341836	9905-01-248-9547	355	4
19207	12341838	6220-01-269-7067	55	8
		6220-01-269-7067	56	10
19207	12341840	5961-01-266-3984	59	6
19207	12341848	5365-01-274-4674	338	5
19207	12341856	5310-01-286-6077	234	3
		5310-01-286-6077	239	4
19207	12341862	4720-01-265-3239	302	13
		4720-01-265-3239	305	2

I-244

SECTION IV TM9-2320-280-24P C01

CROSS REFERENCE INDEXES

PART NUMBER INDEX

CAGE	PART NUMBER	STOCK NUMBER	FIGURE NO	ITEM NO
19207	12341876	5330-01-270-8315	340	4
		5330-01-270-8315	341	3
		5330-01-270-8315	426	15
		5330-01-270-8315	427	15
19207	12341883	4730-01-297-5111	337	5
34623	12341888	5640-01-309-3867	200	1
34623	12341889	5640-01-309-3868	200	2
19207	12341891	5340-01-287-5825	227	34
19207	12341897	5310-01-289-5455	154	4
19207	12341900	5340-01-301-7929	147	10
19207	12341910	2540-01-277-2313	52	1
19207	12341967-1	5365-01-265-8937	438	16
34623	12341967-2	5365-01-265-8938	283	20
		5365-01-265-8938	438	16
19207	12341967-3	5365-01-265-8939	283	20
		5365-01-265-8939	438	16
19207	12341968	5365-01-277-2387	438	14
34623	12341968	5365-01-277-2387	283	13
19207	12341969	5340-01-265-8905	283	24
		5340-01-265-8905	438	20
34623	12341976	4720-01-265-3224	492	13
19207	12341977	4720-01-265-3223	492	5
19207	12341987		495	7
19207	12341989-1		495	4
19207	12341989-3		495	1
19207	12341997-1	5330-01-496-5547	223	4
19207	12341997-2	5330-01-496-5548	223	4
19207	12342000-1	5340-01-294-3230	300	3
19207	12342000-2	5340-01-294-4415	300	2
19207	12342000-3	5340-01-306-3606	300	4
19207	12342002	5365-01-298-7780	293	28
19207	12342003	5340-01-272-8397	293	4
34623	12342005	5340-01-296-2814	294	6
34623	12342007	4010-01-299-7699	293	24
19207	12342024	7690-01-315-8540	380	11
19207	12342026	9905-01-317-7987	380	6
19207	12342027	9905-01-317-7986	380	10
19207	12342036	5340-01-448-4245	472	13
19207	12342038	2540-01-350-9380	379	18
9C234	12342040	1005-01-321-6774	431	KITS
19207	12342043	5340-01-324-1076	431	6
19207	12342044-1	5340-01-295-8169	431	1
19207	12342044-2	5340-01-295-1088	431	5
19207	12342045	3040-01-327-1426	431	4
19207	12342047	2540-01-313-0678	230	6
34623	12342061	2540-01-314-7835	230	7
34623	12342068	2540-01-314-7834	230	1
		2540-01-314-7834	235	1
19207	12342076	5340-01-292-5402	475	7

I-245

SECTION IV TM9-2320-280-24P C01

CROSS REFERENCE INDEXES

PART NUMBER INDEX

CAGE	PART NUMBER	STOCK NUMBER	FIGURE NO	ITEM NO
19207	12342077	3990-01-314-8393	199	19
		3990-01-314-8393	251	20
		3990-01-314-8393	270	12
		3990-01-314-8393	296	3
19207	12342082	2540-01-197-5449	365	6
19207	12342083	4730-01-322-9871	368	15
19207	12342085	5342-01-326-2583	365	19
19207	12342086	2910-01-323-0123	368	35
19207	12342087-1	2540-01-394-6167	368	11
19207	12342087-2	2990-01-322-9880	365	11
19207	12342088	2540-01-197-5460	365	23
19207	12342089	2990-01-394-3751	365	12
19207	12342091	2990-01-322-9881	368	13
19207	12342092	5365-01-326-1153	368	25
19207	12342093	2990-01-322-9879	368	17
19207	12342094	2990-01-323-2562	365	7
19207	12342095	4710-01-313-9340	379	31
19207	12342096	4730-01-311-4294	17	4
19207	12342097		353	2
19207	12342099		353	3
19207	12342102	4910-01-313-8839	353	1
19207	12342102-2		353	10
19207	12342102-3		353	4
19207	12342102-4	4010-01-410-9099	353	9
19207	12342103		365	3
19207	12342104	5330-01-314-6781	104	21
19207	12342104-1	5330-01-473-4701	104	21
19207	12342105	5330-01-315-1609	104	20
		5330-01-315-1609	105	28
19207	12342106		104	22
19207	12342106-3		104	22
19207	12342107	5340-01-314-5602	104	19
		5340-01-314-5602	105	27
19207	12342121	5315-01-406-5019	353	6
34623	12342129	2540-01-314-9380	432	1
34623	12342130	2540-01-314-9379	432	1
19207	12342134	5340-01-316-1507	329	17
19207	12342135	5340-01-314-1955	329	17
19207	12342136	5340-01-316-2959	329	15
19207	12342137	5340-01-314-1956	329	15
19207	12342141	4010-01-406-6963	353	8
34623	12342142	2520-01-316-2630	178	13
		2520-01-316-2630	436	6
19207	12342144	5340-01-311-1633	104	18
19207	12342144-1	5342-01-448-4215	104	18
80212	12342152	2590-01-333-2999	377	12
19207	12342153-1	2540-01-329-8649	378	1
19207	12342153-2	2540-01-336-0791	378	4
19207	12342155	2510-01-330-2250	378	11

I-246

SECTION IV TM9-2320-280-24P C01

CROSS REFERENCE INDEXES

PART NUMBER INDEX

CAGE	PART NUMBER	STOCK NUMBER	FIGURE NO	ITEM NO
19207	12342157	2510-01-330-2249	378	11
19207	12342165	7690-01-315-8539	354	3
19207	12342166	2510-01-316-0216	389	8
19207	12342170		382	11
19207	12342171-1		382	7
19207	12342171-2		382	7
19207	12342172		382	14
19207	12342173	2590-01-317-4854	393	8
34623	12342175	2590-01-317-0534	389	13
19207	12342177		383	6
19207	12342178		383	1
34623	12342179	2590-01-317-4856	383	3
19207	12342180		382	15
19207	12342181		382	3
19207	12342182		382	3
19207	12342183		382	14
19207	12342184		382	12
19207	12342185	2590-01-317-4855	395	18
19207	12342187	2540-01-315-3762	395	6
19207	12342189	2540-01-320-8918	388	13
19207	12342191	4520-01-316-2585	387	9
19207	12342193	6150-01-324-6355	387	18
19207	12342194	9905-01-324-0886	396	7
19207	12342195	2590-01-317-3952	395	2
19207	12342198	5975-01-322-6373	396	8
19207	12342199	2540-01-317-3350	396	5
34623	12342200	2540-01-316-2697	396	9
19207	12342201	5340-01-320-2060	393	2
		5340-01-320-2060	395	13
34623	12342203	2590-01-317-3953	383	4
19207	12342210	5365-01-316-8966	382	2
34623	12342213	2540-01-316-2597	387	6
19207	12342214	2510-01-317-8258	396	1
34623	12342215	2590-01-317-7561	395	23
19207	12342216	2590-01-317-8292	395	3
19207	12342218	5340-01-318-2040	391	2
19207	12342219	5340-01-318-2041	392	10
19207	12342220-1		391	4
19207	12342220-2		391	8
19207	12342221		391	5
19207	12342223-1		397	19
19207	12342223-2		397	11
19207	12342226		391	3
19207	12342234	5340-01-317-9086	391	7
19207	12342235	5340-01-319-9426	391	9
19207	12342237	5365-01-316-8985	392	5
19207	12342248	5330-01-317-5392	393	5
		5330-01-317-5392	395	22
19207	12342255	6150-01-324-6356	383	9

I-247

SECTION IV TM9-2320-280-24P C01

CROSS REFERENCE INDEXES

PART NUMBER INDEX

CAGE	PART NUMBER	STOCK NUMBER	FIGURE NO	ITEM NO
19207	12342256	4520-01-317-0929	387	3
19207	12342257		382	15
19207	12342258	2510-01-319-9384	395	17
19207	12342260	5340-01-317-2675	388	17
19207	12342261-1	4720-01-476-4656	393	17
34623	12342264	2590-01-317-0535	389	13
19207	12342265		382	11
19207	12342266	5340-01-317-7501	389	5
34623	12342270	4710-01-317-1045	385	7
34623	12342276	4710-01-317-1044	387	14
34623	12342277	4710-01-317-1046	384	2
19207	12342281	5340-01-325-2891	433	6
19207	12342284	5340-01-318-3366	433	3
19207	12342285	5340-01-318-0212	433	11
19207	12342287	4010-01-466-0849	462	7
19207	12342294	2930-01-314-0145	379	33
19207	12342295	5340-01-333-6038	197	3
19207	12342296		211	7
			212	4
34623	12342301	2540-01-317-1024	390	10
19207	12342303	6150-01-317-5853	72	25
		6150-01-317-5853	434	1
		6150-01-317-5853	435	1
19207	12342306	5340-01-315-7223	72	23
		5340-01-315-7223	73	10
19207	12342308		201	6
19207	12342309		201	3
19207	12342310		201	4
19207	12342311		201	5
19207	12342312		201	1
19207	12342313		201	2
19207	12342314	2540-01-337-0242	323	5
19207	12342317	2510-01-319-5952	397	10
19207	12342322	5340-01-318-4893	397	6
19207	12342324	5340-01-318-2180	397	3
19207	12342327	2510-01-498-8000	397	21
19207	12342330	2510-01-314-4892	397	18
34623	12342332	2540-01-317-0820	390	1
19207	12342338	5305-01-315-8649	69	33
		5305-01-315-8649	434	4
		5305-01-315-8649	435	9
19207	12342343	5330-01-318-1998	395	5
19207	12342344	5330-01-318-9780	395	7
19207	12342345	5330-01-317-5393	395	4
19207	12342354	4030-01-316-1551	183	2
		4030-01-316-1551	185	12
		4030-01-316-1551	187	16
		4030-01-316-1551	188	12
19207	12342355	5330-01-334-2834	205	5

SECTION IV TM9-2320-280-24P C01

CROSS REFERENCE INDEXES

PART NUMBER INDEX

CAGE	PART NUMBER	STOCK NUMBER	FIGURE NO	ITEM NO
		5330-01-334-2834	206	5
19207	12342357	5330-01-340-5627	205	6
19207	12342358	5330-01-341-8963	206	6
19207	12342361	5340-01-317-1470	181	4
19207	12342362	5340-01-317-0147	181	17
19207	12342372	5340-01-334-4241	234	20
34623	12342373	2590-01-315-9128	188	8
19207	12342374	5340-01-334-2887	239	5
19207	12342381	5340-01-333-5851	234	8
19207	12342383	5340-01-336-3004	251	27
		5340-01-336-3004	252	27
		5340-01-336-3004	253	22
		5340-01-336-3004	254	27
		5340-01-336-3004	255	21
19207	12342387	2510-01-358-1178	187	4
		2510-01-358-1178	441	4
		2510-01-358-1178	456	1
19207	12342388	2510-01-357-8789	187	4
		2510-01-357-8789	441	4
		2510-01-357-8789	456	1
19207	12342389	2510-01-358-1176	187	7
		2510-01-358-1176	441	12
		2510-01-358-1176	456	21
19207	12342390	2510-01-358-1177	187	7
		2510-01-358-1177	441	12
		2510-01-358-1177	456	21
19207	12342395	2540-01-361-8206	187	1
		2540-01-361-8206	441	1
19207	12342403	6220-01-323-0431	396	13
19207	12342404	2590-01-316-0084	185	9
34623	12342404	2590-01-316-0084	186	14
		2590-01-316-0084	187	13
34623	12342408	2510-01-350-4949	203	3
19207	12342412		437	1
19207	12342418	5340-01-332-7516	437	9
34623	12342435-1	2930-01-471-0622	28	5
19207	12342439	7690-01-350-2094	358	4
19207	12342440	5305-01-481-7254	169	31
34623	12342441	2540-01-395-8785	375	1
19207	12342442	2540-01-197-5528	375	4
19207	12342455	2510-01-335-7363	227	10
19207	12342475	2540-01-314-2782	320	2
19207	12342475-1	2540-01-330-6172	320	2
19207	12342475-2	2540-01-413-1356	320	2
19207	12342475-6	2540-01-450-5477	320	2
19207	12342495	5340-01-385-0057	377	3
19207	12342496	5340-01-396-8322	377	3
34623	12342499-1	5305-01-259-6322	58	2
		5305-01-259-6322	59	26

SECTION IV TM9-2320-280-24P C01

CROSS REFERENCE INDEXES

PART NUMBER INDEX

CAGE	PART NUMBER	STOCK NUMBER	FIGURE NO	ITEM NO
		5305-01-259-6322	79	16
		5305-01-259-6322	83	17
		5305-01-259-6322	88	5
		5305-01-259-6322	88	26
		5305-01-259-6322	227	3
		5305-01-259-6322	286	41
		5305-01-259-6322	303	14
		5305-01-259-6322	310	5
		5305-01-259-6322	332	29
		5305-01-259-6322	335	3
		5305-01-259-6322	338	17
		5305-01-259-6322	339	10
		5305-01-259-6322	368	7
		5305-01-259-6322	412	17
		5305-01-259-6322	424	5
		5305-01-259-6322	425	19
		5305-01-259-6322	426	24
		5305-01-259-6322	427	24
		5305-01-259-6322	468	10
19207	12342499-2	5305-01-264-5809	155	10
		5305-01-264-5809	226	33
		5305-01-264-5809	227	36
19207	12342501	2540-01-186-0969	327	8
19207	12342501-1		327	9
19207	12342503		179	1
19207	12342509	5340-01-358-5588	180	6
19207	12342510	5340-01-358-7595	180	26
19207	12342515	2540-01-360-2482	442	16
19207	12342516	2590-01-360-5414	442	6
19207	12342517	2590-01-360-5413	442	6
19207	12342518	2590-01-360-5411	442	19
19207	12342519	2590-01-363-2084	442	19
19207	12342520	2590-01-360-5412	442	11
19207	12342521	2590-01-360-9536	442	11
34623	12342522	2590-01-361-8087	442	3
19207	12342523	2590-01-364-1534	442	15
19207	12342539	2590-01-360-9537	442	9
19207	12342540	2590-01-361-8088	442	3
19207	12342545	5307-01-408-4029	33	20
		5307-01-408-4029	37	18
		5307-01-408-4029	475	25
		5307-01-408-4029	476	23
19207	12342557	5365-01-359-9406	440	19
19207	12342558	2540-01-387-5578	440	13
19207	12342559	5340-01-359-9474	440	8
19207	12342560	5340-01-359-8933	440	20
19207	12342561	5340-01-360-2053	440	14
19207	12342562	5340-01-360-1886	440	23
34623	12342565	2990-01-299-8820	401	13

SECTION IV TM9-2320-280-24P C01

CROSS REFERENCE INDEXES

PART NUMBER INDEX

CAGE	PART NUMBER	STOCK NUMBER	FIGURE NO	ITEM NO
19207	12342566	2990-01-382-8796	401	6
19207	12342567-2	2990-01-385-8988	401	5
19207	12342569	5365-01-381-1783	401	7
9C234	12342570	2540-01-395-4202	370	6
19207	12342571	5340-01-394-7853	369	9
34623	12342572	2540-01-407-3296	369	5
19207	12342573	5340-01-393-9371	369	11
16615	12342574	2590-01-360-2417	446	25
19207	12342580	5340-01-360-1784	446	35
19207	12342583	5340-01-360-1188	446	34
19207	12342584	5340-01-360-1727	446	2
19207	12342585	5340-01-359-8819	446	4
19207	12342586	5340-01-360-2979	446	17
19207	12342587	5340-01-360-9070	446	33
19207	12342594	5340-01-360-1887	446	29
19207	12342595		446	1
19207	12342596		446	30
34623	12342602	3040-01-358-7034	488	17
19207	12342605	4710-01-395-3982	366	11
19207	12342606	4710-01-394-6169	366	15
19207	12342616	5360-01-357-2413	192	1
34623	12342620	2520-01-356-9189	136	6
19207	12342623	5340-01-361-1212	440	18
19207	12342625	7690-01-358-9391	361	11
19207	12342633	5331-01-335-8878	167	16
19207	12342634	2640-01-335-4583	167	23
19207	12342638	2640-01-334-9453	167	17
34623	12342639	2530-01-338-2730	167	18
19207	12342640	2530-01-336-3127	167	19
19207	12342641	2610-01-500-4806	167	13
19207	12342642	2530-01-336-5740	167	14
19207	12342643-1	2520-01-452-7569	126	2
		2520-01-452-7569	127	3
19207	12342646	5340-01-378-5201	69	16
19207	12342647	5340-01-378-7525	69	17
19207	12342648	9905-01-397-3196	370	12
19207	12342652	9905-01-364-7342	354	12
19207	12342653	9905-01-362-2014	355	4
19207	12342654	5935-01-415-4322	76	18
34623	12342655	4730-01-439-2814	496	23
19207	12342658	5340-01-358-6697	74	3
19207	12342661	2540-01-411-7519	453	10
19207	12342662	5340-01-406-6964	453	13
19207	12342663	5340-01-406-6965	453	8
19207	12342664	5340-01-406-7418	453	6
19207	12342665	5340-01-407-7192	453	17
19207	12342666	5340-01-360-2021	440	15
19207	12342667	6150-01-366-2916	445	12
19207	12342668	5340-01-360-1189	445	6

I-251

SECTION IV TM9-2320-280-24P C01

CROSS REFERENCE INDEXES

PART NUMBER INDEX

CAGE	PART NUMBER	STOCK NUMBER	FIGURE NO	ITEM NO
19207	12342669	7690-01-481-4552	448	12
19207	12342670	9905-01-365-8849	448	11
19207	12342679	5340-01-380-4561	153	1
19207	12342680	5310-01-382-5031	142	19
		5310-01-382-5031	143	26
		5310-01-382-5031	150	19
		5310-01-382-5031	151	26
19207	12342681	5930-01-347-9216	327	10
19207	12342682	5355-01-353-6934	327	12
9C234	12342683	2590-01-394-9672	234	4
19207	12342711	3120-01-413-6106	40	42
		3120-01-413-6106	43B	44
19207	12342712	5935-01-415-4392	75	20
		5935-01-415-4392	76	20
34623	12342745	2910-01-357-8798	15	1
19207	12342746	5340-01-357-3996	15	10
		5340-01-357-3996	16	9
19207	12342747	4710-01-358-6397	13	22
		4710-01-358-6397	14	8
19207	12342748	4710-01-358-2127	13	21
		4710-01-358-2127	14	7
19207	12342750	5340-01-360-2020	442	12
19207	12342751	5365-01-394-0440	370	14
19207	12342752	5365-01-394-2394	369	6
19207	12342754	5340-01-394-7288	370	22
9C234	12342755-2	2540-01-395-3979	372	3
9C234	12342756	2540-01-395-2230	372	7
19207	12342784	2590-01-377-6819	250	7
19207	12342785	5340-01-476-6587	250	8
19207	12342786	5340-01-476-6205	250	11
19207	12342787	5340-01-476-6182	250	12
34623	12342788	6220-01-362-5211	97	9
19207	12342789	6150-01-362-5229	81	8
19207	12342794	5331-01-346-3806	167	25
19207	12342795	5340-01-381-2045	401	17
19207	12342801	5340-01-381-1808	140	9
19207	12342805		278	3
19207	12342806	7690-01-407-8248	370	15
19207	12342808	2930-01-385-9108	333	19
19207	12342809	5340-01-381-2248	333	16
19207	12342855	2510-01-274-4234	455	1
19207	12342859	5365-01-406-6336	455	4
19207	12342864	5310-01-415-5245	76	19
34623	12342872	5330-01-452-2474	12	27
34623	12342875	2530-01-383-5740	169	9
19207	12342876	2530-01-384-7154	169	13
34623	12342878	2540-01-449-1718	379	20
19207	12342878-2	2540-01-451-9345	379	20
19207	12342880	2540-01-445-6029	379	17

I-252

SECTION IV TM9-2320-280-24P C01

CROSS REFERENCE INDEXES

PART NUMBER INDEX

CAGE	PART NUMBER	STOCK NUMBER	FIGURE NO	ITEM NO
19207	12342881	2510-01-465-7620	379	19
19207	12342883	5340-01-406-7413	455	8
19207	12342884	5340-01-408-8526	455	7
19207	12342886	5330-01-381-1810	327	33
34623	12342898	2530-01-357-9776	163	7
19207	12342900	4710-01-358-1943	159	1
		4710-01-358-1943	160	12
19207	12342901	4710-01-357-9968	161	6
19207	12342902	6150-01-396-0906	386	2
19207	12342903	5340-01-394-2409	390	12
19207	12342917	5935-01-384-2610	73	1
19207	12342920	5935-01-397-3813	73	5
19207	12342924	5935-01-396-3991	73	14
19207	12342925	5310-01-406-9129	73	16
19207	12342930	4140-01-394-8583	43B	42
34623	12342933	5331-01-477-3657	15	17
		5331-01-477-3657	16	16
19207	12342935		15	18
19207	12342938		15	19
19207	12342943	4910-01-185-7966	504	2
19207	12342947	4730-01-399-0241	178	17
34623	12342949	4710-01-412-0273	14	1
34623	12342950	4710-01-413-1360	14	5
34623	12342951-1	4730-01-412-5216	14	12
34623	12342951-2	4730-01-412-5213	14	26
19207	12342953-2	4820-01-456-0172	19	23
9C234	12342954	2590-01-396-1422	232	1
9C234	12342955	2590-01-396-1424	232	1
34623	12342960	2540-01-395-7999	188	3
19207	12342965	5340-01-394-2408	155	15
19207	12342966	5340-01-394-8496	155	20
19207	12342972-1		179	1
19207	12342972-2		179	1
19207	12342980	9905-01-392-5795	355	4
19207	12342981	9905-01-384-5311	354	12
19207	12342986	9905-01-387-2761	354	12
19207	12342987	9905-01-393-3795	355	4
19207	12342991	9905-01-393-1834	355	4
19207	12342992	9905-01-397-6974	354	12
19207	12342993	9905-01-392-5796	355	4
19207	12342995	9905-01-387-2752	354	12
19207	12342997	9905-01-393-9357	354	12
19207	12342999	9905-01-393-1830	355	4
19207	12343001	9905-01-392-5798	355	4
19207	12343002	9905-01-393-5622	354	12
19207	12343010	9905-01-387-1145	354	12
19207	12343011	9905-01-392-5797	355	4
19207	12343018	9905-01-393-7128	355	4
19207	12343020	9905-01-385-2639	354	12

SECTION IV TM9-2320-280-24P C01

CROSS REFERENCE INDEXES

PART NUMBER INDEX

CAGE PART NUMBER STOCK NUMBER FIGURE NO ITEM NO

CAGE	PART NUMBER	STOCK NUMBER	FIGURE NO		ITEM NO
19207	12343023	9905-01-387-2746		354	12
19207	12343025	9905-01-392-5799		355	4
19207	12343040	9905-01-392-5800		355	4
19207	12343041	9905-01-385-2633		354	12
19207	12343043	9905-01-392-5794		355	4
19207	12343044	9905-01-387-2762		354	12
19207	12343046	9905-01-387-1146		354	12
19207	12343048	9905-01-393-1833		355	4
19207	12343049	9905-01-392-5796		355	4
19207	12343050	9905-01-393-5623		354	12
19207	12343053	2540-01-383-5660		231	4
19207	12343054	7690-01-382-8471		354	14
19207	12343055	5365-01-382-0986		454	5
19207	12343058	9905-01-393-3794		354	8
19207	12343059	6160-01-384-3922		231	31
19207	12343062	5340-01-408-6456		329	17
		5340-01-408-6456	461	2	
19207	12343063	5340-01-408-6460		329	15
		5340-01-408-6460	461	7	
19207	123490694	5340-01-199-4989		272	30
76760	12353	2520-01-163-4999		126	2
19207	12356703-1	5980-01-187-0791		46	3
19207	12356703-3	6220-01-273-0500		47	4
19207	12356704	5930-01-183-6757		67	7
19207	12356764-1	5360-01-315-7212		316	11
		5360-01-315-7212	320	10	
		5360-01-315-7212	320	22	
		5360-01-315-7212	324	13	
		5360-01-315-7212	324	27	
		5360-01-315-7212	375	18	
		5360-01-315-7212	378	18	
19207	12356764-2	5360-01-315-7211		316	11
		5360-01-315-7211	320	10	
		5360-01-315-7211	320	22	
		5360-01-315-7211	324	13	
		5360-01-315-7211	324	27	
		5360-01-315-7211	375	18	
		5360-01-315-7211	378	18	
19207	12356766	5930-01-318-2809		380	8
19207	12356789	5330-01-319-7302		387	10
19207	12356795	5365-01-289-6169		92	18
19207	12357116	5365-01-289-4434		35	10
19207	12357126	5365-01-289-7852		35	11
19207	12360850-1	6220-01-284-2709		61	8
19207	12360865	5961-01-305-8848		60	8
19207	12360870-2	6220-01-297-3217		61	9
19207	12368374	5340-01-314-2445		379	35
19207	12368375	5340-01-315-4955		379	15
19207	12368387	5340-01-320-1079		379	45

SECTION IV TM9-2320-280-24P C01

CROSS REFERENCE INDEXES

PART NUMBER INDEX

CAGE	PART NUMBER	STOCK NUMBER	FIGURE NO	ITEM NO
19207	12368390	5340-01-314-6838	379	51
19207	12368391	5340-01-315-4120	379	47
19207	12368392	5340-01-314-5567	379	11
19207	12368394	6150-01-315-1148	381	3
19207	12368395	6150-01-314-5241	381	4
19207	12368396	6150-01-314-5643	381	1
19207	12368397-2	5365-01-314-5544	379	4
		5365-01-314-5544	379	6
19207	12368398	2540-01-314-1130	379	26
34623	12368400	2540-01-317-3309	379	28
19207	12368443	2990-01-314-0151	379	30
19207	12375837	6220-01-372-3883	61	4
19207	12375838		61	5
19207	12375841	6220-01-359-2870	61	10
34623	12380007	3020-01-482-1834	33	18
		3020-01-482-1834	33A	4
34623	12380024	3020-01-480-8698	33	17
		3020-01-480-8698	33A	3
19207	12387349-43	5310-00-061-4651	185	8
		5310-00-061-4651	186	11
		5310-00-061-4651	187	6
		5310-00-061-4651	194	12
		5310-00-061-4651	441	9
		5310-00-061-4651	442	20
		5310-00-061-4651	457	20
		5310-00-061-4651	470	12
19207	12412123	5310-01-374-4512	261	3
19207	12420924-004	5975-01-421-9718	BULK	64
9C234	12446700	2540-01-383-5232	227	2
19207	12446701	2540-01-384-1012	227	26
19207	12446702	2540-01-384-1028	227	27
19207	12446706	5340-01-473-4800	241	15
34623	12446709	2540-01-447-3457	231	9
34623	12446714	2540-01-381-8988	231	18
19207	12446724	5340-01-387-0090	231	5
19207	12446727		273	15
19207	12446730	5340-01-457-1778	273	18
19207	12446734	7690-01-445-0456	361	11
19207	12446735	5365-01-380-7340	231	23
19207	12446736	5365-01-382-1178	231	30
19207	12446737	6160-01-384-0325	231	32
19207	12446739	2590-01-396-5007	233	3
9C234	12446741	2590-01-394-9673	233	3
34623	12446747	2540-01-381-2079	231	19
19207	12446751	4820-01-431-2499	19	10
34623	12446753	3020-01-466-3603	476	10
34623	12446754	4710-01-466-5447	479	2
19207	12446755	5342-01-466-6283	476	24
19207	12446758	5342-01-466-6276	476	5

I-255

SECTION IV TM9-2320-280-24P C01

CROSS REFERENCE INDEXES

PART NUMBER INDEX

CAGE	PART NUMBER	STOCK NUMBER	FIGURE NO		ITEM NO
19207	12446762	2590-01-406-9821	252		8
34623	12446763	2590-01-439-8268	332		33
		2590-01-439-8268	370	28	
		2590-01-439-8268	411	20	
19207	12446765	5340-01-476-4374	332		4
		5340-01-476-4374	372	1	
19207	12446766	5340-01-412-0866	409		6
		5340-01-412-0866	411	12	
19207	12446767	3040-01-197-8727	272		26
19207	12446770	5340-01-457-0459	205		16
19207	12446771		205		13
34623	12446772	2590-01-411-2737	254		8
34623	12446773	2590-01-411-3958	210		9
19207	12446774	5340-01-408-6459	210		12
19207	12446780	7690-01-458-8254	354		12
19207	12446782	9905-01-448-9783	355		4
19207	12446783	9905-01-446-5769	354		12
19207	12446785	9905-01-446-6187	355		4
19207	12446786	9905-01-447-7799	354		12
19207	12446788	9905-01-448-9784	355		4
19207	12446789	9905-01-448-9786	354		12
19207	12446791	9905-01-446-5768	355		4
19207	12446792	9905-01-446-5770	354		12
19207	12446794	9905-01-446-5771	355		4
19207	12446798	7690-01-450-5481	354		12
19207	12446801	9905-01-449-0476	355		4
34623	12446802	2530-01-411-2735	169		8
19207	12446803	2530-01-411-2729	169		6
19207	12446811	3990-01-411-6575	210		7
19207	12446812	5340-01-415-8672	210		2
19207	12446813	5340-01-429-9352	210		1
34623	12446814	2510-01-411-4175	210		8
34623	12446819	2540-01-410-8793	352		4
		2540-01-410-8793	459	4	
19207	12446821-1	41			16 19207
12446821-2	6150-01-467-6716	41			6
		6150-01-467-6716	472	5	
		6150-01-467-6716	474	10	
19207	12446821-3	6150-01-417-3411	69		35
19207	12446821-4	6150-01-444-4437	44		12
19207	12446821-5	6150-01-415-0535	41		1
19207	12446821-6	6150-01-466-7528	472		18
19207	12446824	5340-01-414-0701	71		1
19207	12446825	6150-01-416-7899	474		10
		6150-01-416-7899	475	4	
19207	12446828	6150-01-412-3192	76		1
19207	12446845-1	6220-01-418-4404	63		8
19207	12446845-2	5980-01-443-9093	63		8
19207	12446846	2540-01-473-6304	227		37

SECTION IV TM9-2320-280-24P C01

CROSS REFERENCE INDEXES

PART NUMBER INDEX

CAGE	PART NUMBER	STOCK NUMBER	FIGURE NO	ITEM NO
19207	12446847	2540-01-473-7521	227	42
19207	12446848	2540-01-473-7553	227	52
19207	12446849	2540-01-473-8050	227	55
19207	12446850	2540-01-473-7544	227	45
19207	12446851	2540-01-473-7959	227	54
19207	12446852	2540-01-475-4708	227	51
19207	12446853	2540-01-473-8310	227	40
19207	12446854	2540-01-473-7107	227	41
19207	12446856	2510-01-473-8204	227	47
19207	12446857	5315-01-473-3682	227	43
19207	12446858	5340-01-479-0211	227	38
19207	12446868	2510-01-262-6011	198	26
19207	12446871-10	5325-01-460-8350	465	5
34623	12446878	2510-01-414-7846	226	4
19207	12446879	2510-01-413-3360	225	12
34623	12446880	2510-01-413-3259	226	4
19207	12446899	5340-01-408-6462	210	5
19207	12446900	5340-01-408-6458	210	6
19207	12446907	2510-01-456-7889	268	3
19207	12446908		269	1
19207	12446911	2510-01-476-2171	268	3
19207	12446912		269	1
34623	12446913-1	2510-01-411-0653	267	8
19207	12446913-2	2510-01-413-3618	267	8
19207	12446914	7220-01-431-8340	210	10
34623	12446928	2590-01-410-9968	197	9
19207	12446933	9540-01-472-5874	88	48
34623	12446953	2540-01-176-6521	205	12
19207	12446961	2540-01-461-1129	227	19
19207	12446966	2540-01-477-0162	454	2
19207	12446967	2540-01-477-0184	454	2
19207	12446973	2540-01-411-2733	188	7
19207	12446978-1		179	1
19207	12446981	2540-01-477-0242	397	14
19207	12446982	5340-01-450-7521	387	1
19207	12446983	2540-01-476-7827	397	13
19207	12446985	2930-01-405-9885	26	1
19207	12446987	3990-01-432-5370	199	19
		3990-01-432-5370	270	12
19207	12446988	2510-01-410-7169	199	27
19207	12446989	5340-01-408-8523	188	16
19207	12446990	5340-01-408-8529	188	16
19207	12446991	5340-01-476-6243	188	21
34623	12446993	2590-01-410-9967	261	5
34623	12446998	2590-01-410-8791	261	6
19207	12446999	5340-01-408-8512	261	12
19207	12447000	5340-01-408-8509	261	10
19207	12447001	2590-01-478-6799	181	17
19207	12447003	7690-01-476-6510	354	10

SECTION IV TM9-2320-280-24P C01

CROSS REFERENCE INDEXES

PART NUMBER INDEX

CAGE	PART NUMBER	STOCK NUMBER	FIGURE NO	ITEM NO
19207	12447007	2540-01-474-3217	333	11
19207	12447010-1	4720-01-474-9148	332	7
34623	12447010-2	4720-01-444-6433	332	11
19207	12447010-3	4720-01-445-5690	332	10
19207	12447011	7690-01-476-6089	354	4
34623	12447012	2540-01-444-6642	332	30
19207	12447013	5340-01-470-2069	332	28
19207	12447027	5305-01-449-1983	169	31
19207	12447028	4720-01-445-5705	332	41
34623	12447029	2510-01-411-0652	226	27
		2510-01-411-0652	460	1
19207	12447035	5307-01-411-8340	261	4
19207	12447036	5315-01-408-7089	261	16
19207	12447037	5365-01-408-8507	261	2
34623	12447046	2540-01-447-2236	332	14
19207	12447047	2540-01-474-8562	332	36
34623	12447048	2540-01-444-6655	332	22
34623	12447050	2540-01-444-5340	332	20
19207	12447051	5340-01-484-7646	182	5
34623	12447053	2510-01-412-4969	180	15
19207	12447054	2540-01-488-6429	371	1
19207	12447057-1	3120-01-448-6269	456	9
19207	12447057-2	3120-01-448-6268	456	12
19207	12447058	3120-01-440-0043	456	8
19207	12447060-1	3040-01-477-9645	456	11
34623	12447060-2	3040-01-444-3350	456	13
19207	12447073		125	5
19207	12447076	4820-01-456-6257	125	19
		4820-01-456-6257	332	14
34623	12447077	4730-01-409-1204	125	18
19207	12447078	2540-01-411-3946	166	10
19207	12447083	6220-01-408-7785	49	1
19207	12447084	6220-01-445-5058	49	10
34623	12447088	2590-01-444-3399	456	14
19207	12447089-1	5340-01-413-0282	186	12
		5340-01-413-0282	188	31
19207	12447089-2	5340-01-412-7514	188	10
19207	12447091	2540-01-412-8610	186	1
		2540-01-412-8610	456	3
19207	12447092	2590-01-412-7564	186	10
34623	12447093	2590-01-412-9560	186	10
19207	12447100	2590-01-412-9570	186	5
19207	12447101	2590-01-412-3862	186	5
19207	12447102	7690-01-445-0444	456	2
19207	12447103	5340-01-414-8959	77	23
19207	12447104	2590-01-448-6351	456	7
19207	12447108-1		1	2
19207	12447109	2920-01-420-9968	38	4
		2920-01-420-9968	40	1

I-258

SECTION IV TM9-2320-280-24P C01

CROSS REFERENCE INDEXES

PART NUMBER INDEX

CAGE	PART NUMBER	STOCK NUMBER	FIGURE NO	ITEM NO
19207	12447110	2920-01-407-0532	36	4
		2920-01-407-0532	43B	1
19207	12447123-1	5340-01-430-9240	321	25
		5340-01-430-9240	376	21
19207	12447124-1	5340-01-430-9239	321	25
		5340-01-430-9239	376	21
34623	12447130	2540-01-444-6615	332	19
19207	12447132	5340-01-449-7352	332	16
34623	12447143	2540-01-465-9030	463	13
34623	12447144	2540-01-466-3770	463	16
19207	12447145	5340-01-466-2425	463	18
19207	12447146	5340-01-466-1982	463	10
19207	12447148	5340-01-466-4086	463	20
19207	12447149	5310-01-465-9727	463	19
34623	12447150	2590-01-466-3866	464	7
34623	12447151	2590-01-466-3824	464	4
34623	12447163-2	2590-01-465-7756	463	5
19207	12447165	5340-01-412-0867	162	6
34623	12447166	4710-01-413-8230	162	3
34623	12447171	2590-01-448-1105	315	7
		2590-01-448-1105	457	12
19207	12447172	5330-01-447-4762	103	3
34623	12447173	2590-01-412-2664	103	7
34623	12447175	2520-01-464-1772	103	1
19207	12447176	3040-01-409-2502	101	3
19207	12447177	5307-01-450-3072	127	11
		5307-01-450-3072	180	9
19207	12447180	3040-01-439-3662	127	7
19207	12460066		101	11
19207	12460067	3040-01-409-2501	134	4
19207	12460070	6150-01-413-7946	70	2
		6150-01-413-7946	481	2
19207	12460082	2540-01-414-1275	24	4
19207	12460083	5340-01-413-2689	24	11
34623	12460086	2990-01-414-4072	24	28
19207	12460087	2990-01-411-2728	24	1
34623	12460090	2990-01-411-3947	24	14
19207	12460091	2990-01-412-0142	24	16
19207	12460092-1	6220-01-413-9828	97	9
9C234	12460092-2	6220-01-412-6420	97	10
19207	12460093	5340-01-413-4486	105	24
19207	12460095	5330-01-413-2118	105	25
19207	12460097	6150-01-413-0853	81	19
19207	12460097-1	5935-01-442-8569	86	22
19207	12460098	6150-01-413-1847	82	1
19207	12460099	6150-01-413-1845	82	1
34623	12460102	2910-01-412-0047	14	28
34623	12460104	2910-01-411-2124	16	1
34623	12460105	2910-01-447-3911	16	13

I-259

SECTION IV TM9-2320-280-24P C01

CROSS REFERENCE INDEXES

PART NUMBER INDEX

CAGE	PART NUMBER	STOCK NUMBER	FIGURE NO	ITEM NO
19207	12460108-3	5330-01-409-1981	BULK	57
19207	12460108-3-AR		16	35
19207	12460108-5		16	18
34623	12460112	2520-01-413-2612	98	1
19207	12460116	5945-01-413-0886	50	12
19207	12460120	6150-01-412-7774	511	11
19207	12460123	5930-01-412-2836	66A	2
19207	12460124	3040-01-411-4191	66A	8
19207	12460125	5340-01-473-8626	66A	10
19207	12460130	5340-01-442-4876	86	30
19207	12460132	5935-01-412-9146	76	7
19207	12460136	5120-01-408-8173	511	13
19207	12460137	6150-01-410-8215	511	12
19207	12460140	5340-01-469-2979	16	15
34623	12460143	2510-01-444-3360	16	21
19207	12460145-1	5306-01-411-2338	16	28
		5306-01-411-2338	464	6
19207	12460145-2	5305-01-412-5995	16	29
		5305-01-412-5995	464	5
34623	12460146	3040-01-409-2500	101	9
19207	12460151	2530-01-413-1365	144	6
19207	12460156	7690-01-476-6507	354	8
19207	12460157	7690-01-476-6101	354	7
19207	12460159	5305-01-412-5994	144	12
		5305-01-412-5994	152	12
19207	12460160		66A	9
19207	12460163-1	2815-01-442-8077	1	2
19207	12460163-11		1	2
19207	12460163-3	2815-01-439-5882	1	2
19207	12460163-5		1	2
19207	12460163-7		1	2
19207	12460163-9		1	2
19207	12460167	2530-01-414-7844	144	6
19207	12460168	2510-01-187-3606	278	4
19207	12460176	2530-01-493-5859	167	13
19207	12460199	5340-01-448-1213	488	8
19207	12460205	9905-01-433-8554	354	6
34623	12460206	2590-01-188-7386	276	10
19207	12460208	5365-01-457-8947	273	24
19207	12460212	2540-01-475-3102	231	14
19207	12460218	5340-01-466-2611	485	22
19207	12460219	5340-01-466-2606	485	20
19207	12460220-1	5340-01-471-1040	485	12
19207	12460220-2	5340-01-466-1985	485	19
19207	12460221	5340-01-466-2451	485	11
19207	12460222	5340-01-466-2438	485	18
19207	12460223	5305-01-471-7623	485	21
34623	12460224	2540-01-439-7308	21	21
19207	12460228	5980-01-438-7452	47	2

SECTION IV TM9-2320-280-24P C01

CROSS REFERENCE INDEXES

PART NUMBER INDEX

CAGE PART NUMBER STOCK NUMBER FIGURE NO ITEM NO

CAGE	PART NUMBER	STOCK NUMBER	FIGURE NO	ITEM NO
34623	12460229	3040-01-444-5640	51	11
19207	12460230	2540-01-410-8789	51	10
19207	12460232	2530-01-423-1796	171	1
19207	12460233	2590-01-455-9123	332	13
19207	12460243	2540-01-432-9894	189	8
		2540-01-432-9894	190	8
34623	12460268	2540-01-432-9899	189	11
19207	12460270	5340-01-432-4870	189	5
19207	12460270-1	5340-01-485-6862	470	14
19207	12460285	2510-01-413-8045	182	17
19207	12460290	5340-01-457-5571	277	11
19207	12460305	4730-01-481-6278	177	4
19207	12460314	6680-01-466-7238	103	8
		6680-01-466-7238	400	4
19207	12460329	5340-01-451-5903	487	14
19207	12460330	5342-01-459-4992	487	1
19207	12460331	5340-01-469-5774	487	7
19207	12460374	2520-01-267-7371	148	11
19207	12460375	2520-01-297-5200	148	8
19207	12460385	5365-01-469-6343	332	37
34623	12460386	2540-01-463-3097	332	39
34623	12460399	2520-01-481-6120	50	1
19207	12460407-1	2540-01-481-9178	50	3
19207	12460407-2	2540-01-481-9194	50	2
19207	12460414	5340-01-477-9397	463	5
19207	12460434	2540-01-455-9308	332	18
19207	12460442	2920-01-473-1763	36	11
19207	12460455		64	23
19207	12460465	5340-01-499-0014	36	3
19207	12460490		1	2
19207	12460491	5340-01-442-0308	125	21
19207	12460492	4710-01-439-8165	125	4
19207	12460493	4710-01-439-8167	125	3
19207	12460495	5340-01-466-2620	480	11
19207	12460496	5340-01-466-2468	480	6
19207	12460498-1	6150-01-469-7941	480	34
19207	12460498-2	6150-01-469-7930	480	31
19207	12460498-3		480	16
19207	12460498-4	6150-01-470-7014	480	12
19207	12460502	5342-01-466-6280	476	14
19207	12460506	5340-01-457-0804	465	4
19207	12460507	5340-01-472-8534	98	11
19207	12460542-2	5340-01-433-0936	484	16
19207	12460543	5340-01-474-3279	484	8
19207	12460544	5340-01-474-0954	484	9
34623	12460547	2590-01-466-5250	468	6
19207	12460548	7690-01-466-5217	468	3
34623	12460550	2540-01-466-6114	462	1
19207	12463123	4820-01-495-7582	490A	3

I-261

SECTION IV TM9-2320-280-24P C01

CROSS REFERENCE INDEXES

PART NUMBER INDEX

CAGE	PART NUMBER	STOCK NUMBER	FIGURE NO	ITEM NO
19207	12469029	6150-01-473-8663	483	4
19207	12469032	7690-01-475-1197	482	1
19207	12469034	2540-01-475-3129	482	15
19207	12469035	5330-01-475-1671	482	5
19207	12469043	6150-01-474-9676	482	4
19207	12469044	7690-01-469-9591	466	4
19207	12469045	7690-01-469-9596	467	6
19207	12469046	4710-01-457-0590	467	10
19207	12469047	5999-01-471-0710	482	7
19207	12469048		1	2
19207	12469061	4720-01-470-3076	178	26
19207	12469064	6150-01-473-1178	469	3
19207	12469065-1	6150-01-473-1169	469	1
19207	12469065-2	6150-01-473-1174	470	5
19207	12469066	7690-01-474-5928	469	5
19207	12469067	7690-01-474-5935	470	6
19207	12469087		1	2
19207	12469100	7690-01-481-4906	355	4
19207	12469101	7690-01-481-4908	354	12
19207	12469106	3020-01-476-6520	175	4
19207	12469107	5340-01-481-0558	33	21
34623	12469156	5977-01-193-9931	68	11
19207	12469158-2	2920-01-469-6903	66	2
19207	12469160		1	2
19207	12469169	5340-01-498-7964	191	4
19207	12469170	5340-01-506-7986	191	24
34623	12469215	4730-01-491-8047	178	4
34623	12469216	4820-01-473-3580	178	3
34623	12469217	5340-01-473-4031	178	2
19207	12469306	2510-01-477-3968	179	15
34623	12469322	7690-01-476-2842	354	16
19207	12469323	9905-01-486-0051	355	4
34623	12469390	7690-01-488-5520	471	8
19207	12469434	2590-01-485-5455	471	12
19207	12469444		1	2
19207	12469471	3020-01-488-5635	175A	12
34623	12469472	3020-01-488-5638	34	22
34623	12469473	5365-01-488-5640	33A	11
		5365-01-488-5640	33B	7
34623	12469474	5340-01-488-5643	38A	6
34623	12469475	5340-01-488-6073	38A	1
19207	12469476	5340-01-497-2269	33A	6
34623	12469478	4710-01-488-6143	176A	1
34623	12469479	4710-01-488-6076	333	7
34623	12469489	4720-01-488-6156	175A	14
34623	12469491	4730-01-488-6163	333	41
34623	12469497	5340-01-488-6179	38A	7
34623	12469498	5340-01-488-6181	33A	8
19207	12469516	4710-01-495-4302	492	2

SECTION IV TM9-2320-280-24P C01

CROSS REFERENCE INDEXES

PART NUMBER INDEX

CAGE	PART NUMBER	STOCK NUMBER	FIGURE NO	ITEM NO
19207	12469517	4710-01-495-4268	492	1
19207	12469518	4730-01-496-3012	492	20
		4730-01-496-3012	490A	4
19207	12469519	4720-01-495-6202	492	24
19207	12469520	4720-01-496-3717	492	5
19207	12469521	2510-01-495-8335	347	9
19207	12469522	4940-01-496-2178	347	10
19207	12479192	6110-01-491-2158	49	4
19207	12480349	2520-01-150-2280	113	7
19207	12480516	2510-01-496-1886	347	5
19207	12480518	2510-01-496-1927	347	11
19207	12480523-1	5340-01-494-6794	462	5
19207	12480523-2	5340-01-495-7546	462	12
19207	12480551	3030-01-504-6408	490A	11
19207	12480553	5340-01-505-4197	490A	10
19207	12480559	3010-01-505-7205	490A	12
08805	1251	6240-00-019-0877	63	4
7X677	12534772	2930-01-444-6649	31	3
7X677	1254002	3110-01-413-8094	129	6
7X677	12550863	5310-01-462-4459	10	15
7X677	12551497	6685-01-444-9478	30	1
7X677	12551502	5330-00-830-1745	9	4
7X677	12551591	5330-01-434-8611	30	2
7X677	12551755	2815-01-430-2599	4	22
7X677	12553484		31	5
7X677	12553488	5330-01-476-3866	31	4
7X677	12554091	2930-01-460-7507	30	18
7X677	12554092	2815-01-455-8424	66A	14
7X677	12554488	2990-01-426-4425	66A	11
7X677	12554808		11	13
7X677	12554835	4710-01-446-9570	11	1
7X677	12556688	4730-01-473-3279	4	11
7X677	12559801	5340-01-493-2569	4	18
7X677	12567769		4	9
45152	126488A	5999-01-150-8808	86	7
93334	127951	4730-00-012-7951	143	18
		4730-00-012-7951	151	18
34281	12821501	2530-01-462-8079	157	22
		2530-01-462-8079	158	12
34281	12821801	2530-01-462-8068	158	12
34281	12835400	157	16	
			158	11
06178	1295	4730-00-266-0538	366	14
14892	129599	5310-01-173-0941	163	5
96881	12L18F	3120-00-485-1017	165	3
		3120-00-485-1017	166	8
19728	12X-195		43	25
68505	12X-196		43	35
			43	57

I-263

SECTION IV TM9-2320-280-24P C01

CROSS REFERENCE INDEXES

PART NUMBER INDEX

CAGE	PART NUMBER	STOCK NUMBER	FIGURE NO	ITEM NO
19728	12X-198		45	80
19728	12X-201		45	58
10001	12Z3035-442	5310-00-595-6057	42	3
10001	12Z329PC92	4730-00-555-8292	16	43
06968	1301	5330-01-202-8360	273	20
18876	13048915-1	5342-01-104-7700	245	15
70082	1305A	2540-01-487-3626	329	3
08806	1309	6240-00-060-4707	65	25
62607	1309	6240-00-060-4707	65	10
12204	131482	5310-00-013-1482	215	10
70412	13194	2540-01-212-4959	327	2
97403	13219E0079	5510-00-270-6031	BULK	71
80132	13414-077	8145-00-485-8250	126	3
11862	134530	5310-01-186-7702	79	15
18876	13480720	5935-01-457-2965	282	8
18876	13493198		282	9
18876	13495871		282	1
18876	13495916	5340-01-467-2127	282	11
70082	1351	2540-01-410-8794	329	3
		2540-01-410-8794 461 3		
62161	1351BAC10SL		505	9
24617	135927		273	11
27647	13698	9905-01-185-3207	358	3
24617	137195	5315-00-012-0123	170	1
21450	137197	5340-01-214-3615	333	6
24617	137397	4730-00-014-2432	10	12
76760	13766	2520-01-174-9554	128	18
		2520-01-174-9554 129 34		
24617	138242	5305-00-366-3337	34	1
27647	13826	3120-01-306-3577	314	31
27647	13839	3020-01-209-4603	314	4
27647	13848	5330-01-211-0717	314	3
24617	138482	5310-01-462-2175	232	3
27647	13850	5305-01-211-3101	314	1
27647	13874	3020-01-209-4606	314	5
76760	13877	5325-01-358-7961	130	42
76760	13903	5306-01-167-4346	128	64
76760	13955	5325-01-358-7962	130	34
35510	13981	5305-01-272-9025	39	47
19728	13X-5549		45	5
68505	14-273	2920-01-248-2509	43	5
62826	1400164	5340-01-424-1300	138	2
		5340-01-424-1300 145 2		
62826	1400165	5340-01-421-1423	138	4
		5340-01-421-1423 145 4		
62826	1400166	5340-01-421-1415	138	15
		5340-01-421-1415 145 15		
62826	1400269	2530-01-459-9497	139	10
		2530-01-459-9497 146 10		

I-264

SECTION IV TM9-2320-280-24P C01

CROSS REFERENCE INDEXES

PART NUMBER INDEX

CAGE PART NUMBER STOCK NUMBER FIGURE NO ITEM NO

CAGE	PART NUMBER	STOCK NUMBER	FIGURE NO		ITEM NO
62826	1400270		139		2
			146	2	
62826	1400271		139		4
			146		4
7X677	14022649	5330-01-156-5147	4		7
7X677	14022650	5342-01-160-4397	4		5
7X677	14022683	5330-01-150-7744	4	28	
7X677	14022699	3040-01-163-7208	4		8
7X677	14022700	4730-01-150-0879	6		6
7X677	14028918	2930-01-147-4198	30		3
7X677	14028942	5325-01-151-6117	4		3
06968	1403	5330-01-213-1312	BULK		54
7X677	14033824	5340-01-159-6626	11		25
7X677	14033893	5340-01-151-9956	11		4
7X677	14033896	5340-01-148-8349	10		4
7X677	14033911	4710-01-149-5075	11		30
7X677	14033912	4710-01-148-9580	11		10
7X677	14033913	4710-01-148-9581	11		28
7X677	14033914	4710-01-149-5076	11		11
7X677	14033915	4710-01-153-1636	11		18
7X677	14033916	4710-01-150-0971	11		16
7X677	14033917	4710-01-149-5077	11		24
7X677	14033918	4710-01-148-9582	11		17
7X677	14033921	5340-01-150-4106	11		20
7X677	14033922	5340-01-150-6275	11		21
7X677	14033946	5307-01-150-1227	8		2
		5307-01-150-1227	30	5	
7X677	14033953	5340-01-150-7774	11		29
7X677	14033955	5340-01-150-6026	11		15
7X677	14037948	5315-01-160-4642	128		10
		5315-01-160-4642	130	23	
7X677	14037958	5330-01-159-2811	128		52
		5330-01-159-2811	129	25	
7X677	14037981	5325-01-157-6764	128		25
		5325-01-157-6764	128	47	
		5325-01-157-6764	130	41	
7X677	14044937	3120-01-155-3509	128		42
7X677	14045263	5307-01-150-1549	31		13
7X677	14050425	5315-01-151-4180	9		5
7X677	14050444	4730-01-164-7028	8		8
7X677	14056165	5306-01-148-3666	30		10
7X677	14060130	5305-01-243-3759	476		17
7X677	14061569	4710-01-148-2659	11		1
11862	14063339		10		11
7X677	14063340	5340-01-167-7794	10		7
7X677	14066301	5342-01-166-1534	11		22
11862	14066305-11		11		9
7X677	14066307	5307-01-150-5992	365		17
7X677	14071059	2590-01-147-4285	4		2

SECTION IV TM9-2320-280-24P C01

CROSS REFERENCE INDEXES

PART NUMBER INDEX

CAGE	PART NUMBER	STOCK NUMBER	FIGURE NO	ITEM NO
7X677	14071080	5307-01-178-7445	31	10
7X677	14077122	6620-01-146-8006	30	1
7X677	14079550	5330-00-107-3925	104	11
7X677	14095677	3120-01-155-4470	128	40
7X677	14095678	3120-01-152-2613	128	73
7X677	14098632	4710-01-412-8611	11	10
7X677	14098633	4710-01-412-8615	11	28
7X677	14098634	4710-01-412-1669	11	11
7X677	14098635	4710-01-411-8488	11	18
7X677	14098636	4710-01-411-8489	11	16
7X677	14098637	4710-01-412-1667	11	24
34623	14098638	4710-01-411-8492	11	17
27647	14401	2590-01-185-7949	314	6
72452	1459-254	5310-00-171-1734	39	25
7X677	147500		119	7
27647	14916	3020-01-446-4387	458	12
34623	1494255		205	17
27647	14964	5330-01-306-7887	458	23
96881	14L14-F	2530-00-753-7285	169	21
06349	14L15-1-1XX	5340-01-195-8057	315	2
76760	15105	5340-01-290-6234	129	37
27647	15271	3120-01-435-2560	458	30
15434	153014	2920-00-848-3292	380	15
79470	1537	4730-01-004-8346	141	20
10138	15406-022	5305-00-054-5650	282	14
81300	15440	3030-00-899-4888	490	11
92867	15480200	2590-01-212-4955	153	15
92867	15541800	2590-01-265-3185	155	3
92867	15541802	2590-01-323-5153	155	9
76760	15556	2520-01-358-8883	132	22
76760	15574	5330-01-359-1292	130	35
27647	15597	3040-01-185-3201	314	8
27647	15598	5365-01-216-1166	314	10
27647	15603	5305-01-309-1521	458	1
27647	15604	5355-01-211-6991	314	37
		5355-01-211-6991 458 6		
27647	15605	5331-01-445-5285	458	4
81343	15618240		BULK	60
7X677	15620999	4730-01-163-3544	11	2
7X677	15633464	5330-01-157-0856	8	6
7X677	15633465	2930-01-193-7802	31	3
27647	15634	5360-01-211-3163	314	39
27647	15643	3010-01-444-4018	458	11
27647	15647	3010-01-306-4113	458	24
24617	156471	5305-01-477-2824	225	38
27647	15654	3020-01-185-7948	314	7
27647	15655	2590-01-185-3116	314	24
27647	15657	7690-01-185-3208	358	1
27647	15663	2590-01-185-7947	314	30

SECTION IV TM9-2320-280-24P C01

CROSS REFERENCE INDEXES

PART NUMBER INDEX

CAGE	PART NUMBER	STOCK NUMBER	FIGURE NO	ITEM NO
27647	15664	2590-01-330-6102	314	13
27647	15665	6105-01-185-3187	314	27
27647	15667	4010-01-315-4179	314	15
27647	15668	5330-01-216-0288	314	11
7X677	1567442	5310-01-021-9027	167	10
27647	15686	5365-01-306-9955	458	9
24617	156966	5305-01-410-8386	210	16
		5305-01-410-8386	269	9
7X677	156971	5305-01-198-1621	272	14
24617	157125	5305-01-381-2305	231	2
76760	15715	5331-01-358-9545	129	28
7X677	157166	5305-01-185-0114	213	24
		5305-01-185-0114	225	15
		5305-01-185-0114	234	14
		5305-01-185-0114	246	6
		5305-01-185-0114	254	16
		5305-01-185-0114	260	20
		5305-01-185-0114	260	25
		5305-01-185-0114	287	7
		5305-01-185-0114	321	22
		5305-01-185-0114	326	5
76760	15720	3120-01-360-0977	130	36
7X677	15728526	5305-01-148-5915	8	4
70281	157432	5305-00-958-5246	411	16
79470	1582	4730-01-200-4277	141	9
		4730-01-200-4277	399	20
27647	15847		313	13
27647	15848		313	16
27647	15849	2590-01-192-3021	314	22
27647	15873	9905-01-185-3127	358	2
76760	15896	5330-01-415-9613	130	30
7X677	15957413	4730-01-456-5446	366	9
24617	159612	5305-01-359-0167	74	13
7X677	160046	5305-01-208-0274	325	10
24617	160052	5305-01-407-7713	318	9
		5305-01-407-7713	485	23
		5305-01-407-7713	487	2
7X677	160057	5305-01-185-8647	199	4
		5305-01-185-8647	225	19
		5305-01-185-8647	232	4
		5305-01-185-8647	462	9
24617	160069	5305-01-467-2711	486	1
27647	16013	2590-01-329-9910	314	2
27647	16064	5305-01-211-3100	314	14
76760	16064	3040-01-206-3876	128	15
76760	16081	5330-01-358-9533	130	31
76760	16104	5330-01-358-9540	130	30
79470	1611X3	4730-01-153-1871	345	40
		4730-01-153-1871	348	32

I-267

SECTION IV TM9-2320-280-24P C01

CROSS REFERENCE INDEXES

PART NUMBER INDEX

CAGE	PART NUMBER	STOCK NUMBER	FIGURE NO	ITEM NO
76760	16167	3120-01-359-6760	129	6
76760	16188		129	4
16764	16196390	5940-01-430-2764	50	6
16764	16197350	5962-01-430-0182	50	4
76760	16214	5330-01-358-9541	130	11
76760	16218	4720-01-358-6100	130	13
76760	16221	3040-01-360-4448	130	29
76760	16223		130	32
76760	16243	5325-01-358-5450	129	5
76760	16244	2520-01-358-8880	129	2
16764	16251210	5962-01-440-0368	50	4
16764	16266408	5962-01-470-4619	50	4
76760	16275	3040-01-358-8691	130	44
76760	16276	3020-01-358-8659	130	33
76760	16286	5310-01-358-7975	130	7
76760	16290	3020-01-358-8661	130	8
76760	16297	5365-01-358-4641	130	17
76760	16310	2590-01-358-7643	132	18
7X677	1635490	5306-01-149-6280	4	4
		5306-01-149-6280	8	5
		5306-01-149-6280	30	16
19220	16386(75)	5340-01-197-2736	269	18
76760	16397	5325-01-358-7958	129	10
76760	16433	3020-01-358-8657	130	15
76760	16447	5340-01-358-7940	132	29
76760	16449	3040-01-358-8596	132	17
76760	16456	5310-01-358-7978	132	25
76760	16457	3120-01-359-6764	132	27
27647	16464	4030-01-210-0691	314	16
76760	16466	5325-01-358-7959	129	17
76760	16467	5310-01-358-5461	129	16
76760	16472	2520-01-358-8701	132	13
76760	16473		132	16
76760	16474	2590-01-358-7644	132	19
7X677	164830	5305-01-255-3588	331	8
		5305-01-255-3588	332	43
76760	16538		129	14
76760	16550	3040-01-359-1117	129	19
27647	16554	3040-01-209-3376	314	28
76760	16651	2520-01-358-8878	130	3
76760	16662	3020-01-359-1211	129	18
76760	16686	4320-01-358-8608	130	40
7X677	167038	5305-01-250-6584	55	4
		5305-01-250-6584	56	2
24617	167163	5305-01-211-3786	95	34
59150	1675-182	4730-01-189-4600	489	2
76760	16754	5340-01-358-8000	129	30
76760	16756	5305-01-359-4586	129	27
30076	16779	5305-00-269-2811	392	8

I-268

SECTION IV TM9-2320-280-24P C01

CROSS REFERENCE INDEXES

PART NUMBER INDEX

CAGE	PART NUMBER	STOCK NUMBER	FIGURE NO	ITEM NO
76599	16HSS	4730-00-586-8463	27	8
57733	170677	5305-00-456-2582	345	8
		5305-00-456-2582	348	6
		5305-00-456-2582	348	15
76760	17071	5360-01-358-6563	129	29
7X677	172482	5305-01-236-4349	123	2
		5305-01-236-4349	124	14
		5305-01-236-4349	125	14
76760	17292	3040-01-359-1116	129	8
24617	173397	5305-01-411-4456	261	15
76760	17381	3040-01-358-8690	129	13
76760	17387	3040-01-358-8708	129	24
24617	174697	5306-00-425-8569	69	40
76760	17475	3120-01-417-8178	130	31
76760	17510	4710-01-358-8410	130	12
59150	1755-724		489	3
80064	1755683	5935-00-605-9322	85	11
		5935-00-605-9322	86	16
76760	17560	2520-01-358-8881	129	9
76760	17562		132	23
76760	17564	5360-01-358-5420	132	28
76760	17565	3040-01-358-8692	130	18
76760	17566	3020-01-359-1136	130	6
76760	17567	3110-01-359-6669	130	5
76760	17568	5310-01-358-7974	130	4
76760	17570	3020-01-358-8656	130	9
76760	17571	3020-01-359-2529	130	47
76760	17573	2520-01-358-8658	130	10
76760	17575	3020-01-358-8662	130	16
76760	17577	3020-01-358-8601	130	48
76760	17578		130	39
76760	17584	3040-01-359-1119	130	20
76760	17585	3040-01-359-1118	130	37
76760	17586	2520-01-358-8700	132	20
76760	17588	2520-01-358-8879	130	1
55787	1758CA		268	37
55787	1759CW		268	33
55787	1761CC		268	37
55787	1762CY		268	33
59150	1775-826	4720-01-188-3190	BULK	21
24617	178362	5310-00-061-0004	345	25
		5310-00-061-0004	348	36
76760	17844	5305-01-416-1269	130	27
76760	17855		129	14
7X677	178654	4730-00-287-1706	159	6
		4730-00-287-1706	160	20
		4730-00-287-1706	161	14
		4730-00-287-1706	162	22
		4730-00-287-1706	487B	8

I-269

SECTION IV TM9-2320-280-24P C01

CROSS REFERENCE INDEXES

PART NUMBER INDEX

CAGE	PART NUMBER	STOCK NUMBER	FIGURE NO	ITEM NO
76760	17869	3010-01-412-0673	129	19
10001	1788116	5310-00-498-6675	395	9
7X677	179148	5340-01-189-7640	161	2
		5340-01-189-7640	162	2
		5340-01-189-7640	487A	9
		5340-01-189-7640	487C	1
		5340-01-189-7640	487D	5
24617	179802	5306-01-150-8713	118	10
73183	179818	5305-01-122-0253	121	16
3M915	17C836R	5306-01-360-0926	438	4
51387	180021	5305-01-211-1405	198	13
29510	180022	5306-01-371-4685	198	15
27647	18135	3020-01-210-1365	314	33
27647	18136	3950-01-329-9890	314	29
27647	18138	3040-01-185-8011	314	9
27646	1833	5360-01-420-0480	458	8
27647	1834	3110-01-444-4020	458	7
21877	1851863-4	5310-01-117-3446	243	15
27647	18545	3120-01-377-5220	314	12
72582	186676	5305-01-016-5469	71	12
11862	186677	5305-00-844-1507	445	13
72582	186678	5305-00-989-0830	292	8
		5305-00-989-0830	293	19
		5305-00-989-0830	440	12
35510	1869	5315-01-265-8771	39	23
7X677	186965	5305-01-264-5864	46	7
24617	187838	5305-00-018-7838	88	44
		5305-00-018-7838	254	7
		5305-00-018-7838	254	10
		5305-00-018-7838	445	10
		5305-00-018-7838	487	8
29372	18L01-1X1AA	5340-01-206-5040	240	21
		5340-01-206-5040	243	22
71286	18L13-1-1AA	5340-01-190-0815	244	19
71286	18L13-1-AD	5340-01-255-9380	240	3
76760	19016	5330-01-413-3713	128	16
		5330-01-413-3713	129	31
72582	190171	5310-00-774-9073	390	4
		5310-00-774-9073	397	7
		5310-00-774-9073	397	22
24617	190254	5310-00-208-1918	88	4
		5310-00-208-1918	203	29
		5310-00-208-1918	272	4
		5310-00-208-1918	286	25
7X677	190254	5310-01-269-9245	411	6
24617	190652	4730-01-269-9530	336	1
50022	19111	2930-01-168-7911	5	2
76760	19134		130	32
76760	19140	2520-01-413-1904	130	29

SECTION IV TM9-2320-280-24P C01

CROSS REFERENCE INDEXES

PART NUMBER INDEX

CAGE	PART NUMBER	STOCK NUMBER	FIGURE NO	ITEM NO
13899	191410	4730-00-529-1237	384	3
72582	192481	5310-01-058-3353	382	10
		5310-01-058-3353	390	9
		5310-01-058-3353	392	11
7X677	193620	5320-01-218-0721	31	6
19728	19X-3437		45	78
19728	19X-4632		45	65
07482	1C817P63	5305-00-068-0508	446	36
02697	2-011N117370	5331-01-495-4801	493	7
02697	2-013N117370	5331-01-495-4810	492	21
		5331-01-495-4810	493	1
		5331-01-495-4810	490A	1
02697	2-015C557-70	5331-01-166-1712	493	3
02697	2-015N117370	5331-01-495-4814	492	23
		5331-01-495-4814	493	3
02697	2-016 N674-7	5331-00-984-3750	314	35
02697	2-017N117370	5331-01-495-4820	492	22
81343	2-2 130137B	4730-00-278-3721	178	23
95019	2-2-1209	2520-01-249-1506	137	17
95019	2-3-10021X		136	19
72447	2-3-10831X	2520-01-478-7282	137	19
95019	2-3-11491KX		136	19
95019	2-3-11781X	2520-01-386-3384	137	19
95019	2-3-7961KX	2520-01-189-0594	137	9
19220	2-300-P5	5360-01-282-9316	280	23
		5360-01-282-9316	281	23
95019	2-5-668X		136	8
			136	21
95019	2-5-848X		136	8
			136	21
			137	7
			137	21
95019	2-6-1318X		136	7
			136	20
95019	2-6-288X		136	7
			136	20
			137	6
			137	20
95019	2-60-1081		136	11
95019	2-60-1082		136	24
95019	2-60-108A2		136	24
95019	2-60-550		137	24
95019	2-60-550A2		137	24
95019	2-60-607		136	24
95019	2-60-899-1528		137	12
95019	2-60-899-1623		137	12
95019	2-60-996-1524		136	11
95019	2-70-59	5340-01-190-0807	136	1
		5340-01-190-0807	137	13

I-271

SECTION IV TM9-2320-280-24P C01

CROSS REFERENCE INDEXES

PART NUMBER INDEX

CAGE	PART NUMBER	STOCK NUMBER	FIGURE NO		ITEM NO	
0AKY3	20-8021	3030-01-466-9476			32	3
0AKY3	20-9646	3030-01-488-5606			32	3
73680	200275-7	5330-01-180-9099			142	11
64648	2003-0015-3	5305-00-059-3659			43B	22
24234	200335	5305-01-270-5419			342	2
		5305-01-270-5419	496	26		
		5305-01-270-5419	498	7		
24234	201685	5305-01-278-7118			494	17
24234	201686	5305-01-296-7762			343	6
		5305-01-296-7762	494	1		
01276	2023-6-10S	4730-00-728-2393			6 10 24234	
202751	5310-00-274-8710	343			5	
		5310-00-274-8710	497	2		
		5310-00-274-8710	498	11		
24234	202755	5310-01-270-2661			494	2
		5310-01-270-2661	495	13		
		5310-01-270-2661	496	8		
0DPV2	203400	2540-01-385-9000			327	31
0DPV2	203500	2540-01-385-9031			327	37
0DPV2	203600	2540-01-386-8790			327	32
24234	204434	3120-01-270-2646			496	4
24234	204531				496	29
7X677	2065616	5310-01-187-3485			193	2
20768	207-180201-0101	5340-01-264-6104			197	5
52854	207-320401-00	5340-01-185-4959			225	5
		5340-01-185-4959	226	40		
62826	2080005	138			13 62826	
2080007	5365-01-421-8382	138			14	
		5365-01-421-8382	145	14		
62826	2080021				145	13
62826	2090003				139	6
					146	6
62826	2090004				139	5
					146	5
29372	20L01-1-1AF	5340-01-197-1294			220	2
		5340-01-197-1294			230	20
		5340-01-197-1294			231	25
		5340-01-197-1294	235	16		
71286	20L02-1-1AA	5340-01-087-4612			230	23
		5340-01-087-4612	231	28		
		5340-01-087-4612	235	19		
71286	20L56-1-1AA	5340-01-408-8508			261	9
19728	20X-4061				43	60
19728	20X-4063				43	58
19728	20X-4222				43	56
					45	14
68505	20X-4383	5305-01-429-9137			43	26
27647	21009	3040-01-307-6157			458	3
41024	21022013-150IN.				320	7

I-272

SECTION IV TM9-2320-280-24P C01

CROSS REFERENCE INDEXES

PART NUMBER INDEX

CAGE	PART NUMBER	STOCK NUMBER	FIGURE NO	ITEM NO
41024	21022103	5330-01-321-8622	BULK	53
41024	21022103-150	324	14	
			324	36
41024	21022103-150IN		316	12
41024	21022103-150IN.		320	28
76760	21065	3040-01-413-4022	130	15
76760	21068	2520-01-413-1899	129	13
95019	211121X	5340-01-188-1017	136	17
		5340-01-188-1017	137	10
95019	211415X	3130-01-203-9870	136	12
55719	211FT	5120-00-152-2284	505	18
23862	2136455	5306-00-207-2221	64	4
		5306-00-207-2221	83	34
27647	21451	4010-01-231-5075	314	17
76760	21727	5330-01-415-9612	130	20
27647	21781	4030-01-209-7047	314	20
62826	2180007	2520-01-485-1813	138	12
		2520-01-485-1813	145	12
27647	21880	3040-01-209-3409	314	36
76760	21908	2520-01-409-2512	127	3
76760	21909	3040-01-420-1887	130	18
72962	21NE-040	5310-01-066-6759	19	18
		5310-01-066-6759	233	4
		5310-01-066-6759	237	6
		5310-01-066-6759	248	14
		5310-01-066-6759	249	12
		5310-01-066-6759	249	34
		5310-01-066-6759	252	9
		5310-01-066-6759	266	16
		5310-01-066-6759	389	12
72962	21NE083	5310-00-020-0358	438	11
70411	220C-J	4820-01-192-8030	399	2
10988	222-652	4730-01-193-7390	345	12
		4730-01-193-7390	348	17
0GE52	2227346	5340-01-456-2751	163	9
0GE52	2227677	5340-01-458-6386	163	8
0GE52	2228633	2530-01-459-5890	163	10
73370	2229379	5330-01-293-5345	163	4
73370	2230648	5340-01-293-5558	163	2
14894	2232134	2530-01-179-7589	163	1
73370	2232146		163	3
27647	22350	3110-01-444-4100	458	13
24617	223533	5306-01-492-7301	188	39
76760	22398	5365-01-446-9408	128	22
		5365-01-446-9408	130	25
7Z043	224N2V14	5340-01-474-2315	41	3
7X677	22511422	5340-01-148-7528	10	5
15946	22515630	5307-01-150-9538	11	6
		5307-01-150-9538	12	17

I-273

SECTION IV TM9-2320-280-24P C01

CROSS REFERENCE INDEXES

PART NUMBER INDEX

CAGE	PART NUMBER	STOCK NUMBER	FIGURE NO	ITEM NO
91816	2269B-2MM-.5	4820-01-430-4132	14	13
91816	2269B-2MM-3	4820-01-173-6883	14	27
62826	2280024		138	7
62826	2290033		139	7
			146	7
81834	23-0118-27	5305-01-218-1243	64	10
00724	23-01191-059	5305-00-993-1851	260	6
		5305-00-993-1851	289	27
80813	230-7/16-14	5355-01-196-2770	133	1
72582	23011420	5365-01-188-0782	4	23
73342	23015732	2520-01-170-9826	111	12
73342	23015880	5330-01-146-6053	111	3
73342	23017556	5315-01-152-9029	100	15
73342	23017763	2520-00-008-9987	109	2
06032	2310-0143-001	5310-01-121-1703	2	9
		5310-01-121-1703	174	6
		5310-01-121-1703	183	10
		5310-01-121-1703	184	2
		5310-01-121-1703	186	6
		5310-01-121-1703	285	7
		5310-01-121-1703	437	7
		5310-01-121-1703	438	3
		5310-01-121-1703	441	6
		5310-01-121-1703	442	2
		5310-01-121-1703	444	2
		5310-01-121-1703	453	11
		5310-01-121-1703	456	4
95019	231401	5306-01-190-5760	136	2
		5306-01-190-5760	137	14
76700	23244A	2990-01-188-3281	23	7
35510	23304	5305-01-272-8302	39	78
7X677	23500006	5307-01-187-0519	4	17
7X677	23500008	2910-01-189-8851	4	11
7X677	23500012	2930-01-188-3682	30	18
72825	23500024		10	10
7X677	23500032	5307-01-207-9004	31	9
7X677	23500035	2815-01-168-7909	8	7
11862	23500084		30	14
7X677	23500085		31	5
7X677	23500801	5306-01-422-8649	66A	12
7X677	23500832	5307-01-150-5991	8	3
72582	23502073	2815-01-168-7871	4	9
7X677	23502587	5330-01-378-8572	4	32
7X677	23503111		11	19
7X677	23503124	5325-01-160-4618	11	12
72582	23521131	5310-01-250-7679	4	20
78514	23716	5210-01-249-0369	510	22
12745	23728-1-74	5305-00-889-3001	43	39
62826	2390036		139	11

SECTION IV TM9-2320-280-24P C01

CROSS REFERENCE INDEXES

PART NUMBER INDEX

CAGE PART NUMBER STOCK NUMBER FIGURE NO ITEM NO

CAGE	PART NUMBER	STOCK NUMBER	FIGURE NO	ITEM NO
			146	11
62826	2390074		139	13
62826	2390076		139	12
			146	12
62826	2390077	2530-01-459-9492	KITS	16
62826	2390079		146	13
28520	2400	5325-01-289-7859	74	7
		5325-01-289-7859	91	13
		5325-01-289-7859	493	6
7X677	24200128	2835-01-478-4268	115	2
7X677	24200173	3040-01-447-7995	108	2
7X677	24200224	4730-01-460-5520	108	12
7X677	24200347	2590-01-460-8316	108	14
7X677	24200357	4320-01-456-6258	122	34
6X677	24200374	4710-01-477-3662	105	6
7X677	24200418	4820-01-474-8911	KITS	18
7X677	24200436	5360-01-476-2692	120	7
7X677	24200625		120	23
7X677	24200732	2520-01-481-7690	120	9
7X677	24200789	2520-01-459-5454	KITS	19
7X677	24200999		108	6
7X677	24201001		108	5
7X677	24201115	5330-01-456-2744	120	10
7X677	24201116	2520-01-456-6256	120	20
7X677	24201386		119	2
7X677	24201388	5330-01-456-7886	119	5
7X677	24201389		119	8
7X677	24202092	2520-01-456-5014	115	6
7X677	24202111	5330-01-410-7229	KITS	20
7X677	24202229	3040-01-480-7593	107	16
7X677	24202282	5340-01-456-2741	117	19
72590	24202329	2520-01-399-4691	105	1
7X677	24202346		107	9
7X677	24202357	5331-01-480-0748	115	26
7X677	24202358	5331-01-414-4161	115	22
7X677	24202359		112	5
7X677	24202360	5330-01-478-5994	114	13
7X677	24202361	5330-01-478-3900	114	14
7X677	24202552	2520-01-474-8868	114	12
7X677	24202577	5315-01-456-2737	117	7
7X677	24202578	5315-01-456-2739	117	7
7X677	24202579	5315-01-456-6254	117	7
7X677	24202580	5315-01-456-2738	117	7
7X677	24202581	5315-01-456-2740	117	7
7X677	24202582	5315-01-456-7882	117	7
7X677	24202583	5315-01-456-2743	117	7
7X677	24202611	5360-01-461-4931	108	9
7X677	24202631	3040-01-480-7595	110	15
7X677	24202646	2520-01-150-3932	109	3

I-275

SECTION IV TM9-2320-280-24P C01

CROSS REFERENCE INDEXES

PART NUMBER INDEX

CAGE	PART NUMBER	STOCK NUMBER	FIGURE NO	ITEM NO
		2520-01-150-3932	110	3
		2520-01-150-3932	111	16
		2520-01-150-3932	112	16
7X677	24202681		122	17
7X677	24202928		117	12
7X677	24202943	2520-01-459-8532	120	12
7X677	24202957		119	2
11862	24202959		119	21
7X677	24202963	2520-01-456-7888	120	9
7X677	24202966	2520-01-456-7883	114	8
7X677	24202976	2520-01-460-4961	114	11
7X677	24202985	3040-01-456-7881	117	16
7X677	24202986	5365-01-456-2742	117	10
7X677	24202987	5360-01-456-2745	117	13
7X677	24203366		105	2
7X677	24203377		105	18
7X677	24203397	3020-01-480-7299	107	8
7X677	24203398	3020-01-480-7630	107	17
7X677	24203399	3020-01-480-7321	107	17
7X677	24203400	3040-01-480-7309	107	13
7X677	24203499		112	3
7X677	24203511		114	16
61928	24203876	2990-01-399-1023	105	9
7X677	24204253	5330-01-478-5993	120	10
7X677	24204266		119	2
7X677	24204268	5330-01-473-8584	120	8
7X677	24204270	2520-01-475-9665	120	9
7X677	24204275	4710-01-477-9032	120	15
7X677	24204276	5340-01-491-7497	120	17
7X677	24204278	2520-01-482-4280	KITS	21
7X677	24204283	2520-01-481-8478	115	20
7X677	24204287	2520-01-409-1767	115	19
7X677	24204291	3040-01-480-7591	107	22
7X677	24204303	2520-01-465-8809	122	9
7X677	24204304		122	17
7X677	24204440	5330-01-086-5457	KITS	21A
7X677	24204441	5330-01-398-3724	KITS	22
7X677	24204449		KITS	23
7X677	24204472	1005-00-049-9324	KITS	24
7X677	24204497	2520-01-456-2736	114	6
7X677	24204835	2520-01-460-8323	102	1
7X677	24204957	2520-01-478-6604	112	7
7X677	24204961	2520-01-474-8871	110	8
7X677	24204962		107	10
7X677	24205103	5340-01-474-4011	105	4
		5340-01-474-4011	125	22
72590	24205114		105	1
7X677	24205123	5330-01-480-2502	105	21
72590	24205225		105	1

I-276

SECTION IV TM9-2320-280-24P C01

CROSS REFERENCE INDEXES

PART NUMBER INDEX

CAGE PART NUMBER STOCK NUMBER FIGURE NO ITEM NO

CAGE	PART NUMBER	STOCK NUMBER	FIGURE NO		ITEM NO
7X677	24205250	3040-01-462-1286	KITS		25
24617	24205251	5330-01-442-2874	KITS		26
7X677	24205348	2520-01-460-8301	KITS		27
7X677	24205373	5935-01-150-6319	100		4
7X677	24205398	5340-01-474-4006	122		9
7X677	24205551	2520-01-149-7861	106		11
		2520-01-149-7861	KITS	28	
7X677	24205552	2520-01-474-7051	KITS		29
7X677	24205560	2520-01-481-8479	110		11
		2520-01-481-8479	112	17	
7X677	24205722	5330-01-480-0750	114		15
7X677	24205786		122		7
7X677	24205818	2520-01-480-5436	115		9
7X677	24205827	2520-01-480-7565	114		9
7X677	24205833	5331-01-477-6762	122		32
7X677	24206024		120		24
7X677	24206025		120		23
7X677	24206571	5325-01-480-7328	115		10
7X677	24206749	5340-01-408-8525	KITS		30
7X677	24207130		114		17
7X677	24207168		105		2
7X677	24207180		114		17
7X677	24207247	3040-01-480-7297	107		7
7X677	24207501		112		4
7X677	24208527		105		2
72590	24208605		105		1
7X677	24208668		105		2
7X677	24208848	3110-01-478-4266	KITS		31
7X677	24208849	2520-01-481-8460	KITS		32
7X677	24208852	4320-01-485-7480	122		3
7X677	24209112	2520-01-488-7016	120		12
24617	24209225		110		5
7X677	24209244	2520-01-410-8072	110		13
7X677	24209276	5945-01-478-7862	119		24
7X677	24209311	2520-01-411-2749	110		12
7X677	24209475	3020-01-480-7302	107		8
7X677	24210080	2520-01-480-7556	110		15
7X677	24210468		105		15
7X677	24210605	5330-01-324-0906	122		6
7X677	24210954	3010-01-480-7597	KITS		33
7X677	24210955	5330-01-480-0752	KITS		34
7X677	24210956	4330-01-496-5720	KITS		35
7X677	24211287	4320-01-485-7480	122		3
72590	24211292		105		1
7X677	24211719	2520-01-480-7568	120		9
7X677	24212051	2520-01-480-7563	107		9
7X677	24213991	5365-01-480-6812	105		20
72590	24214094		105		1
72590	24214664		105		1

I-277

SECTION IV TM9-2320-280-24P C01

CROSS REFERENCE INDEXES

PART NUMBER INDEX

CAGE PART NUMBER STOCK NUMBER FIGURE NO ITEM NO

CAGE	PART NUMBER	STOCK NUMBER	FIGURE NO	ITEM NO
7X677	24215063	5340-01-209-7830	121	14
7X677	24217454	2520-01-481-3230	KITS	35A
7X677	24222077	5930-01-456-7880	120	13
7X677	24222160	3010-01-460-8309	115	14
7X677	24222173	2520-01-485-9405	112	2
24234	243046	6150-01-276-7698	54	4
24234	243047	2590-01-276-5071	54	3
24234	243048	2990-01-406-8738	54	5
16004	24338	9905-00-205-2795	211	17
35510	2434	5310-00-775-5139	39	37
24617	2436161	5310-01-102-3170	13	17
		5310-01-102-3170	14	23
		5310-01-102-3170	15	2
		5310-01-102-3170	16	2
		5310-01-102-3170	17	10
		5310-01-102-3170	27	13
		5310-01-102-3170	29	4
		5310-01-102-3170	43	47
		5310-01-102-3170	97	3
		5310-01-102-3170	98	14
		5310-01-102-3170	99	7
		5310-01-102-3170	124	11
		5310-01-102-3170	125	11
		5310-01-102-3170	153	2
		5310-01-102-3170	155	13
		5310-01-102-3170	169	28
		5310-01-102-3170	198	21
		5310-01-102-3170	202	25
		5310-01-102-3170	203	6
		5310-01-102-3170	204	2
		5310-01-102-3170	233	5
		5310-01-102-3170	241	4
		5310-01-102-3170	243	14
		5310-01-102-3170	244	8
		5310-01-102-3170	245	10
		5310-01-102-3170	245	17
		5310-01-102-3170	245	33
		5310-01-102-3170	246	17
		5310-01-102-3170	247	2
		5310-01-102-3170	247	14
		5310-01-102-3170	248	6
		5310-01-102-3170	249	11
		5310-01-102-3170	249	26
		5310-01-102-3170	249	35
		5310-01-102-3170	249	48
		5310-01-102-3170	250	6
		5310-01-102-3170	251	2
		5310-01-102-3170	254	2
		5310-01-102-3170	254	20

SECTION IV TM9-2320-280-24P C01

CROSS REFERENCE INDEXES

PART NUMBER INDEX

CAGE	PART NUMBER	STOCK NUMBER	FIGURE NO	ITEM NO
		5310-01-102-3170	255	2
		5310-01-102-3170	267	14
		5310-01-102-3170	269	12
		5310-01-102-3170	269	27
		5310-01-102-3170	270	2
		5310-01-102-3170	271	8
		5310-01-102-3170	272	9
		5310-01-102-3170	274	4
		5310-01-102-3170	274	9
		5310-01-102-3170	275	4
		5310-01-102-3170	276	2
		5310-01-102-3170	276	30
		5310-01-102-3170	277	22
		5310-01-102-3170	285	2
		5310-01-102-3170	286	38
		5310-01-102-3170	331	10
		5310-01-102-3170	332	3
		5310-01-102-3170	333	8
		5310-01-102-3170	398	17
		5310-01-102-3170	399	14
		5310-01-102-3170	400	12
		5310-01-102-3170	409	10
		5310-01-102-3170	411	2
		5310-01-102-3170	411	9
		5310-01-102-3170	413	4
		5310-01-102-3170	415	16
		5310-01-102-3170	430	23
		5310-01-102-3170	430	37
		5310-01-102-3170	440	9
		5310-01-102-3170	446	15
		5310-01-102-3170	453	9
		5310-01-102-3170	484	7
		5310-01-102-3170	486	6
24617	2436162	5310-01-119-1024	83	24
		5310-01-119-1024	144	5
		5310-01-119-1024	152	5
		5310-01-119-1024	153	20
		5310-01-119-1024	169	32
		5310-01-119-1024	205	26
		5310-01-119-1024	209	9
		5310-01-119-1024	225	3
		5310-01-119-1024	226	9
		5310-01-119-1024	231	8
		5310-01-119-1024	231	15
		5310-01-119-1024	234	5
		5310-01-119-1024	242	3
		5310-01-119-1024	250	21
		5310-01-119-1024	266	2
		5310-01-119-1024	268	11

SECTION IV TM9-2320-280-24P C01

CROSS REFERENCE INDEXES

PART NUMBER INDEX

CAGE	PART NUMBER	STOCK NUMBER	FIGURE NO	ITEM NO
		5310-01-119-1024	268	28
		5310-01-119-1024	277	6
		5310-01-119-1024	279	33
		5310-01-119-1024	284	2
		5310-01-119-1024	292	14
		5310-01-119-1024	328	9
		5310-01-119-1024	333	22
		5310-01-119-1024	338	26
		5310-01-119-1024	339	19
		5310-01-119-1024	374	8
		5310-01-119-1024	403	4
		5310-01-119-1024	409	2
		5310-01-119-1024	410	7
		5310-01-119-1024	413	2
		5310-01-119-1024	414	6
		5310-01-119-1024	415	2
		5310-01-119-1024	416	6
		5310-01-119-1024	417	5
		5310-01-119-1024	418	5
		5310-01-119-1024	419	6
		5310-01-119-1024	420	2
		5310-01-119-1024	421	2
		5310-01-119-1024	422	2
		5310-01-119-1024	423	2
		5310-01-119-1024	428	6
		5310-01-119-1024	439	2
		5310-01-119-1024	440	2
		5310-01-119-1024	445	1
		5310-01-119-1024	463	11
24617	2436163	5310-01-412-4013	15	24
		5310-01-412-4013	16	26
		5310-01-412-4013	23	21
		5310-01-412-4013	30	4
		5310-01-412-4013	35	4
		5310-01-412-4013	36	14
		5310-01-412-4013	38	14
		5310-01-412-4013	99	17
		5310-01-412-4013	144	5
		5310-01-412-4013	144	15
		5310-01-412-4013	152	5
		5310-01-412-4013	152	15
		5310-01-412-4013	174	15
		5310-01-412-4013	202	36
		5310-01-412-4013	203	19
		5310-01-412-4013	270	20
		5310-01-412-4013	276	11
		5310-01-412-4013	278	10
		5310-01-412-4013	283	7
		5310-01-412-4013	33A	15

SECTION IV TM9-2320-280-24P C01

CROSS REFERENCE INDEXES

PART NUMBER INDEX

CAGE	PART NUMBER	STOCK NUMBER	FIGURE NO	ITEM NO
		5310-01-412-4013	33B	3
		5310-01-412-4013	341	7
		5310-01-412-4013	38A	2
		5310-01-412-4013	409	7
		5310-01-412-4013	410	5
		5310-01-412-4013	411	13
		5310-01-412-4013	412	5
		5310-01-412-4013	414	3
		5310-01-412-4013	415	12
		5310-01-412-4013	416	4
		5310-01-412-4013	425	15
		5310-01-412-4013	426	13
		5310-01-412-4013	427	14
		5310-01-412-4013	428	4
		5310-01-412-4013	456	19
		5310-01-412-4013	462	3
		5310-01-412-4013	463	2
		5310-01-412-4013	464	2
		5310-01-412-4013	476	3
		5310-01-412-4013	490A	5
11862	2436165	5310-01-121-1703	184	2
		5310-01-121-1703	188	23
		5310-01-121-1703	242	7
		5310-01-121-1703	271	17
		5310-01-121-1703	271	29
		5310-01-121-1703	315	6
24617	2436165	5310-01-121-1703	26	11
		5310-01-121-1703	185	6
		5310-01-121-1703	186	2
		5310-01-121-1703	187	12
24617	2436167	5310-01-151-7347	140	7
		5310-01-151-7347	147	7
		5310-01-151-7347	170	20
		5310-01-151-7347	185	4
		5310-01-151-7347	187	3
		5310-01-151-7347	195	11
		5310-01-151-7347	196	2
		5310-01-151-7347	441	3
		5310-01-151-7347	457	19
		5310-01-151-7347	470	11
23862	2436168	5310-01-147-4052	152	2
24617	2436168	5310-01-147-4052	166	2
		5310-01-147-4052	183	7
		5310-01-147-4052	188	17
		5310-01-147-4052	188	29
		5310-01-147-4052	470	10
24617	24504716	5305-01-459-9564	463	17
27647	24563	3010-01-444-4019	458	14
20969	246X5X4	4730-00-277-6347	366	24

SECTION IV TM9-2320-280-24P C01

CROSS REFERENCE INDEXES

PART NUMBER INDEX

CAGE	PART NUMBER	STOCK NUMBER	FIGURE NO	ITEM NO
		4730-00-277-6347	384	15
27647	24800	6110-01-434-5562	458	36
27647	24832	3020-01-444-1359	458	10
72983	248X4	4730-00-900-3296	178	23
19220	249	5340-01-209-4870	280	11
		5340-01-209-4870	281	11
7X677	25042462	2990-01-147-9284	7	3
70040	25043364	4820-01-192-7678	399	18
80463	25055210	4730-01-189-8854	15	12
		4730-01-189-8854	16	11
61928	251-591	2930-01-353-5794	31	3
93334	2538608	3040-01-192-3673	488	19
16236	2540V0751	2540-01-314-9378	230	2
		2540-01-314-9378	235	2
7X677	2551880	4730-00-871-6729	10	8
80138	256A		56	3
53711	2590174	2910-00-930-9367	366	25
		2910-00-930-9367	384	6
99167	25932	5310-00-167-0834	279	27
7X677	26002516	5330-01-324-8260	172	19
52788	26049931		172	5
24234	260790	2930-01-282-2524	498	9
28527	2616950G001	5310-00-615-1556	43B	24
35510	26351	5310-01-210-0819	39	55
24234	265330	4140-01-269-7101	494	18
27647	26781	2590-01-185-3115	314	32
28520	2683	5340-01-203-1980	197	18
		5340-01-203-1980	199	61
		5340-01-203-1980	227	8
24617	271153	5305-01-393-6311	135	7
		5305-01-393-6311	160	22
		5305-01-393-6311	162	10
		5305-01-393-6311	487A	12
		5305-01-393-6311	487C	2
		5305-01-393-6311	487D	3
72582	271163	5310-01-069-5243	74	4
		5310-01-069-5243	282	16
		5310-01-069-5243	435	5
		5310-01-069-5243	453	7
24617	271168		92	16
			339	3
11862	271169	5310-00-124-9265	325	20
24617	271169	5310-00-124-9265	85	23
		5310-00-124-9265	95	32
		5310-00-124-9265	207	6
		5310-00-124-9265	227	16
		5310-00-124-9265	282	4
		5310-00-124-9265	338	31
		5310-00-124-9265	416	30

SECTION IV TM9-2320-280-24P C01

CROSS REFERENCE INDEXES

PART NUMBER INDEX

CAGE	PART NUMBER	STOCK NUMBER	FIGURE NO	ITEM NO
72582	271169	5310-00-124-9265	19	14
		5310-00-124-9265	52	12
		5310-00-124-9265	56	15
		5310-00-124-9265	58	12
		5310-00-124-9265	59	8
		5310-00-124-9265	75	26
		5310-00-124-9265	77	7
		5310-00-124-9265	79	23
		5310-00-124-9265	83	8
		5310-00-124-9265	84	8
		5310-00-124-9265	88	9
		5310-00-124-9265	231	22
		5310-00-124-9265	254	24
		5310-00-124-9265	258	2
		5310-00-124-9265	259	7
		5310-00-124-9265	260	22
		5310-00-124-9265	262	12
		5310-00-124-9265	263	8
		5310-00-124-9265	289	14
		5310-00-124-9265	318	7
		5310-00-124-9265	321	21
		5310-00-124-9265	323	16
		5310-00-124-9265	326	7
		5310-00-124-9265	327	41
		5310-00-124-9265	338	7
		5310-00-124-9265	400	10
		5310-00-124-9265	410	11
		5310-00-124-9265	410	26
		5310-00-124-9265	412	20
		5310-00-124-9265	414	13
		5310-00-124-9265	416	17
		5310-00-124-9265	426	19
		5310-00-124-9265	427	19
		5310-00-124-9265	428	11
		5310-00-124-9265	435	6
		5310-00-124-9265	445	3
		5310-00-124-9265	462	15
		5310-00-124-9265	462	16
		5310-00-124-9265	468	12
		5310-00-124-9265	477	6
		5310-00-124-9265	485	13
		5310-00-124-9265	488	9
24617	271172	5310-01-152-0598	51	4
		5310-01-152-0598	69	34
		5310-01-152-0598	77	13
		5310-01-152-0598	83	15
		5310-01-152-0598	83	31
		5310-01-152-0598	88	17
		5310-01-152-0598	88	29

SECTION IV TM9-2320-280-24P C01

CROSS REFERENCE INDEXES

PART NUMBER INDEX

CAGE	PART NUMBER	STOCK NUMBER	FIGURE NO	ITEM NO
		5310-01-152-0598	89	11
		5310-01-152-0598	95	40
		5310-01-152-0598	155	6
		5310-01-152-0598	214	9
		5310-01-152-0598	304	7
		5310-01-152-0598	307	6
		5310-01-152-0598	308	15
		5310-01-152-0598	313	11
		5310-01-152-0598	320	18
		5310-01-152-0598	320	31
		5310-01-152-0598	339	29
		5310-01-152-0598	378	19
		5310-01-152-0598	410	33
		5310-01-152-0598	412	14
		5310-01-152-0598	414	22
		5310-01-152-0598	416	25
		5310-01-152-0598	428	24
		5310-01-152-0598	443	14
		5310-01-152-0598	453	3
		5310-01-152-0598	457	13
		5310-01-152-0598	480	23
24617	271175	5310-01-333-5245	62	29
		5310-01-333-5245	379	50
		5310-01-333-5245	485	27
24617	271184	5310-00-933-4310	282	7
		5310-00-933-4310	403	3
		5310-00-933-4310	433	2
		5310-00-933-4310	435	3
28520	2723	5340-01-273-2380	59	21
7X677	272474	5310-01-204-1039	73	4
		5310-01-204-1039	434	2
		5310-01-204-1039	435	2
24617	272739	5310-01-317-8164	395	8
93061	272NTA-8-6	4730-01-119-6895	15	22
		4730-01-119-6895	16	17
7X677	274233	5305-01-211-6049	49	3
11862	274244	5331-00-935-9136	33	
		5331-00-935-9136	400	2
24617	274244	5331-00-935-9136	479	3
73342	274267	2010-00-937-5599	122	26
24234	274278	6105-01-265-8634	494	19
		6105-01-265-8634	498	14
73342	274613	5325-00-349-8518	121	15
7X677	274707	5305-01-187-8757	46	9
		5305-01-187-8757	331	13
		5305-01-187-8757	332	27
27647	27569	4010-01-426-4536	458	39
70485	2758	5325-00-249-6345	333	33
62826	277-0027	3040-01-189-1637	145	7

I-284

SECTION IV TM9-2320-280-24P C01

CROSS REFERENCE INDEXES

PART NUMBER INDEX

CAGE	PART NUMBER	STOCK NUMBER	FIGURE NO		ITEM NO
14892	2771302	2530-01-179-7590	164		1
62826	2780035		138		9
62826	2780037		138		9
27647	27855	2590-01-432-2691	457		17
62826	2790024		139		9
62826	2790026		139		9
62826	2790028		146		9
93061	279NTA-8-6	4730-01-091-2809	15		13
		4730-01-091-2809	16	12	
62826	287-0006	2530-01-191-8741	KITS		36
28520	2874	5325-01-289-5038	92		19
35510	2879		39		36
62826	2880015		138		3
			145	3	
62826	2880016		138		10
			145		10
62826	2880019	2530-01-394-6168	KITS		37
62826	2890019	139	3		
			146	3	
86403	2892399	6680-00-269-0335	488		22
81834	29-1063-02	6220-01-216-9337	64		6
24234	290081		55		14
24234	290101	5340-01-282-0469	498		2
24234	290227	4730-00-230-2959	496		25
24234	290249	5340-01-276-6738	496		18
24234	290308	5340-01-296-1736	494		6
24234	290309	5340-01-289-7793	494		16
41387	2938-2	5331-00-580-6586	470		3
73342	29505789	5365-01-378-1755	104		12
27647	29519	4030-01-426-4537	458		41
73342	29520052	5310-01-143-0512	12		20
		5310-01-143-0512	23	22	
		5310-01-143-0512	127	14	
73342	29530330	2520-00-172-1947	109		4
		2520-00-172-1947	110	4	
		2520-00-172-1947	111	17	
77060	2965718	5940-01-270-3701	93		4
77060	2973850	5940-01-270-3700	92		2
		5940-01-270-3700	93		6
		5940-01-270-3700	95		7
		5940-01-270-3700	96	11	
62826	298-0105	2520-01-382-8874	138		1
62826	298-0106	2520-01-382-8728	138		1
62826	298-0107	2520-01-383-2387	145		1
61361	298-212-93	5305-00-719-5241	397		17
77060	2984172	5940-01-270-3702	95		15
77066	2984172	5940-01-270-3702	96		4
77066	2984576	5940-01-289-5955	96		14
62826	2990054	2530-01-459-9493	KITS		38

I-285

SECTION IV TM9-2320-280-24P C01

CROSS REFERENCE INDEXES

PART NUMBER INDEX

CAGE	PART NUMBER	STOCK NUMBER	FIGURE	NO	ITEM NO
62826	2990055	2530-01-459-9494		KITS	39
62826	2990056	2530-01-459-9495		KITS	40
73680	2994-2106	5330-01-180-9099		143	11
		5330-01-180-9099	150		11
		5330-01-180-9099	151		11
73680	29960-0511	5330-01-358-9532		128	59
		5330-01-358-9532	129		31
11083	2D1959	6240-00-019-3093		60	6
		6240-00-019-3093	61		6
1E045	3-10	5340-01-260-9942		217	2
73957	3-32X3-4LG	5315-00-828-5487		4	13
19220	3-400-L (25)	5340-01-198-5455		269	19
19220	3-400-R	5340-01-199-6103		269	19
7Z588	30024H	4730-00-908-3193		379	39
		4730-00-908-3193	496		15
92878	30040-01	6105-00-512-9225		333	13
24234	300442	5331-01-160-9857		467	7
		5331-01-160-9857	493		1
		5331-01-160-9857	496		11
24234	300448	5331-01-061-3000		467	9
		5331-01-061-3000		493	7
		5331-01-061-3000		494	12
		5331-01-061-3000	496		16
24234	300506-72.0	5640-01-288-5547		493	5
24234	300515	4720-01-282-0143		496	22
60602	30071			21	8
60602	30073	5305-01-290-8206		21	10
51377	3018-01265-01	5331-01-192-8892		34	6
51377	3018-01339-01	5330-01-247-8438		34	7
51377	3018-01425-01	5330-01-252-0461		34	13
92878	30250	2540-01-190-7079		333	11
27647	30260	3010-01-429-5333		458	29
92878	30266	6105-01-211-6635		333	12
27647	30274	3120-01-447-8658		458	21
		3120-01-447-8658	458		33
27647	30275	5330-01-446-4696		458	20
		5330-01-446-4696	458		32
27647	30277	3120-01-447-8663		458	18
51377	3029-01371-01	5310-01-194-0481		34	3
51377	3030-00364-01	5310-01-190-4607		34	15
27647	30328	5305-01-448-5080		458	22
7Z588	30448-D	5315-01-409-1662		488	24
51377	3051-01405-01	5307-01-190-5854		34	12
00303	308	6240-00-155-7790		65	16
04NP3	309-552-321	2610-01-171-4746		168	1
12603	30H-18	5340-01-017-4630		13	3
		5340-01-017-4630	14		3
96795	31-119918-210	5305-00-059-3660		59	23
27647	31262	3040-01-446-3689		458	5

SECTION IV TM9-2320-280-24P C01

CROSS REFERENCE INDEXES

PART NUMBER INDEX

CAGE	PART NUMBER	STOCK NUMBER	FIGURE NO	ITEM NO
27647	31293	5340-01-486-1005	458	44
27647	31585	5340-01-430-0037	458	43
27647	31597	5935-01-431-5656	458	37
27647	31604	5945-01-431-5195	458	38
27647	31625	5305-01-428-6635	458	46
24234	319032	4720-01-265-3238	343	12
82465	31WLF3816	5310-00-355-5645	44	7
27647	32062	3120-01-435-2495	458	34
27647	32208	5365-01-434-0186	458	27
27647	32209	3040-01-446-3684	458	15
27647	32231	2590-01-446-5962	458	16
27647	32235	6105-01-434-1729	458	47
27647	32236	3950-01-435-1014	458	31
27647	32237	2590-01-446-5957	458	19
27647	32239	3020-01-446-3695	458	25
81495	330 2000	5310-00-087-7493	401	8
01347	330-10-10R	4820-01-005-2994	363	3
10001	33G1724	5325-00-276-6343	70	14
19220	34-4974-51X	5340-01-259-8600	269	24
50380	34017-14-AB	5310-01-130-9065	471	9
62226	34068-02	5355-01-481-5517	231	17
27647	34384	3040-01-444-1756	458	2
46156	34403-35-1	5315-01-188-4490	225	37
		5315-01-188-4490	226	20
39428	3458T44	1640-01-308-5097	BULK	50
96152	346-786	4820-01-210-3488	15	4
		4820-01-210-3488	16	4
27647	34848	5365-01-426-6643	458	26
68505	35-762	2920-01-168-4127	35	18
		2920-01-168-4127	43	54
02768	354-280308-00-0078	5325-01-197-3460	153	22
		5325-01-197-3460	332	21
39428	3543T41	4030-01-124-8201	353	5
7X677	354501	4730-01-148-2758	31	2
24234	354674	5330-01-282-2208	498	13
24234	354944	5330-01-282-2214	496	12
		5330-01-282-2214	498	1
60602	35745-44	2590-01-199-5423	21	15
60602	35752-12		21	11
0AT62	35A2C5	5305-00-225-3843	446	14
24975	36-630	5330-01-190-7510	KITS	41
7X677	360582	5360-01-165-1563	4	16
98853	361130-EA02	5940-01-440-0097	93	14
98853	361230-D	5940-01-436-4561	93	15
60602	36162-23	2590-01-466-4397	468	2
45152	362AX1	5310-00-080-6004	38	17
35510	36895	5310-00-045-3299	39	77
35510	36912	5310-01-143-1679	39	76

SECTION IV TM9-2320-280-24P C01

CROSS REFERENCE INDEXES

PART NUMBER INDEX

CAGE PART NUMBER STOCK NUMBER FIGURE NO ITEM NO

CAGE	PART NUMBER	STOCK NUMBER	FIGURE NO	ITEM NO
27647	36974	5305-01-448-5074	458	17
76599	36HSS	4730-00-909-8627	17	16
		4730-00-909-8627	29	8
94222	37-10-071-10	5340-00-492-2313	208	11
68505	37-123	5330-01-190-7509	KITS	42
7X677	3702366	2805-00-336-1716	4	15
7X677	3704871	5340-01-194-5294	4	14
7X677	3719599	2910-00-493-2138	9	3
7X677	3739798	4730-01-182-6565	30	15
		4730-01-182-6565	31	7
7X677	3787240	2520-01-159-7757	128	80
		2520-01-159-7757	129	38
7X677	3790768	5310-01-147-8743	490	5
19728	37X-4914		45	76
15434	3802287	2920-01-420-7894	33	14
7X677	3816659	5340-01-149-4434	399	8
81795	38266	5310-00-802-4701	482	10
73342	3829139	5306-00-024-6580	104	8
39428	3933T22	5340-01-125-1682	353	7
55787	393835	5307-01-140-6594	268	38
73342	3947086	5330-00-001-4904	111	4
81349	39TB10F	5940-00-983-6105	52	17
11083	3M9449	5940-00-705-6732	41	7
81343	4-060110B	4730-00-011-4627	348	16
81343	4-4 130139C	4730-01-347-7342	385	6
81343	4-4 130239C(CAD)	4730-01-209-0845	135	4
81343	4-4-4 130438B	4730-00-471-3102	366	13
81343	4-4-4 140438C	4730-01-048-9769	384	4
61864	400	5325-00-281-8642	316	3
31272	40000-10	2540-01-271-8010	263	4
24234	400084	4130-01-265-8726	496	1
92871	4004616	5310-01-189-8485	165	8
25341	40210		507	1
51377	4026-38368-01	2930-01-189-8643	34	18
24234	403475	4720-01-265-3213	336	2
51377	4040-38442-01		34	8
24234	404179	5930-01-266-1809	494	8
24234	404180	5930-01-266-3917	496	30
24234	404187	4820-01-265-3156	492	10
24234	404188	4820-01-265-3159	492	12
51377	4043-38443-01	2930-01-189-1744	34	5
24234	405475	4820-01-265-3149	494	7
		4820-01-265-3149	496	31
24234	405565	6680-01-265-8698	496	19
24234	405680	4710-01-265-3152	496	17
24234	405681	4710-01-285-5123	496	20
24234	405682	4710-01-265-3230	493	8
24234	405683	4710-01-265-3276	493	4
24234	405684	4710-01-265-3277	493	2

I-288

SECTION IV TM9-2320-280-24P C01

CROSS REFERENCE INDEXES

PART NUMBER INDEX

CAGE PART NUMBER STOCK NUMBER FIGURE NO ITEM NO

CAGE	PART NUMBER	STOCK NUMBER	FIGURE NO	ITEM NO
24234	405695	4720-01-282-0160	496	24
08627	406CD	5340-00-050-9077	141	24
24234	407234	2540-01-265-3266	343	21
51377	4073-38424-01		34	16
51377	4079-38441-01		34	19
82386	410-63	5305-00-582-9501	37	8
		5305-00-582-9501	75	24
		5305-00-582-9501	77	5
		5305-00-582-9501	83	10
		5305-00-582-9501	84	4
		5305-00-582-9501	85	22
		5305-00-582-9501	88	7
		5305-00-582-9501	216	4
		5305-00-582-9501	327	43
		5305-00-582-9501	468	13
		5305-00-582-9501	488	6
24234	410909	4130-01-300-6350	494	3
15434	410912	2540-01-291-1043	343	2
24234	410913	2540-01-265-3234	54	1
24234	410914 ITEM 8	5930-01-311-3610	54	10
24234	410915	2510-01-276-5058	54	8
		2510-01-276-5058	55	1
24234	410917	4130-01-282-1895	496	2
24234	410918	5340-01-281-7795	496	5
24234	410920	4130-01-269-4921	496	27
24234	410921	2540-01-285-6066	498	4
24234	410922	2540-01-296-9358	343	13
24234	410925	4420-01-276-5050	495	15
25341	41102	6680-01-372-5748	506	3
95019	41292	5330-01-174-8146	148	7
82978	4167924	5310-01-174-8091	128	20
		5310-01-174-8091	129	36
0U276	41945	5342-01-246-1120	17	1
76760	42054	2520-01-358-7698	131	5
25341	42106	5120-01-373-8836	507	2
25341	42107	5120-01-373-5983	508	3
25341	42109	5120-01-373-4989	506	2
86403	4210973	5310-01-186-5237	128	19
		5310-01-186-5237	128	63
25341	42113	5120-01-373-5984	508	4
25341	42124	5120-01-373-4985	508	2
06968	42130156-7		29	5
24161	4219-3681	4720-01-163-7833	BULK	11
24617	423531	5305-00-403-5130	345	14
		5305-00-403-5130	348	5
		5305-00-403-5130	348	13
24617	425570	5306-00-042-5570	84	5
		5306-00-042-5570	88	46
21450	426687		375	3

I-289

SECTION IV TM9-2320-280-24P C01

CROSS REFERENCE INDEXES

PART NUMBER INDEX

CAGE PART NUMBER STOCK NUMBER FIGURE NO ITEM NO

CAGE	PART NUMBER	STOCK NUMBER	FIGURE NO	ITEM NO
			375	6
			378	3
			378	6
7X677	427566	5306-00-844-0036	438	5
		5306-00-844-0036	444	5
24617	427583	5306-01-360-1129	440	7
20796	43-5791	3030-00-379-2815	32	2
25341	43204	5120-01-373-5985	506	4
25341	43205	5120-01-373-4987	507	3
25341	43208	5120-01-373-4984	508	5
25341	43210	5120-01-373-8840	508	1
25341	43305	5120-01-373-4982	506	1
35510	4340	5310-00-429-3135	39	43
25341	43501	5120-01-373-4988	507	4
70655	4375	5342-01-289-7708	295	12
24161	4377	4720-00-139-3968	BULK	9
31272	43977-12	5340-01-271-4282	279	30
24617	442393	4730-01-457-0727	487A	14
		4730-01-457-0727	487C	3
72582	442393	4730-01-457-0727	160	7
		4730-01-457-0727	161	7
		4730-01-457-0727	162	8
24617	446188	5310-00-044-6188	370	25
24617	446212	5310-00-044-6212	370	27
31272	44647-10	5340-01-203-5634	243	10
		5340-01-203-5634	270	38
31272	44647-11	5340-01-206-7780	241	1
31272	44647-12	5340-01-198-2255	246	10
		5340-01-198-2255	247	20
		5340-01-198-2255	248	11
		5340-01-198-2255	249	14
31272	44647-18	5340-01-265-8923	258	7
		5340-01-265-8923	259	3
		5340-01-265-8923	260	3
		5340-01-265-8923	262	11
31272	44647-19	5340-01-272-8431	259	8
		5340-01-272-8431	264	4
31272	44647-21	5340-01-272-2801	218	4
31272	44647-22	5340-01-272-3403	257	7
31272	44647-23	5340-01-287-6352	262	15
31272	44648-13	5340-01-203-6539	248	1
31272	44648-14	5340-01-203-2707	248	8
31272	44648-15	5340-01-203-4646	247	27
31272	44648-17	5340-01-203-2708	247	28
		5340-01-203-2708	446	23
		5340-01-203-2708	449	5
31272	44648-19	5340-01-203-6538	249	28
31272	44648-20	5340-01-198-2254	249	24
31272	44648-21	5340-01-198-2257	249	45

SECTION IV TM9-2320-280-24P C01

CROSS REFERENCE INDEXES

PART NUMBER INDEX

CAGE	PART NUMBER	STOCK NUMBER	FIGURE NO		ITEM NO
31272	44648-22	5340-01-198-2256		249	43
31272	44648-24	5340-01-260-9940		286	13
31272	44648-25	5340-01-256-0942		430	11
31272	44648-26	5340-01-255-2812		430	9
		5340-01-255-2812	446		18
		5340-01-255-2812	449		10
35510	4465	5305-00-601-7729		39	7
31272	44672-10	2590-01-185-4391		240	16
31272	44672-11	2590-01-185-4390		240	16
		2590-01-185-4390	429		26
31272	44702-10	5340-01-203-6542		319	6
		5340-01-203-6542	320		4
		5340-01-203-6542	322		5
		5340-01-203-6542	323		4
		5340-01-203-6542	324		5
		5340-01-203-6542	326		4
		5340-01-203-6542	375		7
87541	4485(US 27)	5340-01-278-8588		213	39
		5340-01-278-8588	214		32
7X677	448655	5305-01-206-8401		77	12
		5305-01-206-8401	88		12
		5305-01-206-8401	477		1
87541	4487(US26D)	5310-01-288-1257		213	34
		5310-01-288-1257	214		27
31272	44877	5340-01-199-3510		375	26
31272	44884	5340-01-197-6753		268	39
31272	44931-11	2590-01-256-3628		429	25
31272	44943	5340-01-198-3457		248	13
24617	449617	5305-01-478-3387		469	12
31272	44996-10	5340-01-266-3844		287	18
31272	45053-10	5340-01-199-1662		268	14
31272	45053-11	5340-01-198-2253		271	30
		5340-01-198-2253	274		16
31272	45057-10	5340-01-256-4656		272	18
31272	45065-10	5340-01-209-7761		250	3
31272	45079-10	2540-01-271-5071		263	1
31272	45084-11	5340-01-265-8925		256	5
		5340-01-265-8925	257		10
		5340-01-265-8925	289		1
31272	45084-12	5340-01-287-8761		260	15
31272	45085-11	5340-01-266-3846		287	9
31272	45087-10	5340-01-282-2670		265	4
		5340-01-282-2670	289		16
31272	45094-10	5340-01-272-3402		287	10
31272	45096-10	5340-01-198-5165		246	4
		5340-01-198-5165	450		7
31272	45096-11	5340-01-199-1663		246	3
31272	45096-13	5340-01-257-2644		430	10
31272	45160-10	5340-01-207-0610		241	2

SECTION IV TM9-2320-280-24P C01

CROSS REFERENCE INDEXES

PART NUMBER INDEX

CAGE	PART NUMBER	STOCK NUMBER	FIGURE NO	ITEM NO
24617	451786	5340-01-162-5883	108	10
31272	45187		250	1
31272	45188-10	4530-01-211-5328	250	4
31272	45189-10	2590-01-210-0176	250	2
31272	45243-10	5342-01-280-6264	288	25
31272	45262-11	5340-01-255-9918	280	2
		5340-01-255-9918 281 2		
31272	45262-13	5340-01-253-8971	280	2
		5340-01-253-8971 281 2		
24617	453006	5365-00-277-7341	142	5
		5365-00-277-7341 143 5		
		5365-00-277-7341 151 5		
34623	45302-11	5340-01-259-3956	246	37
24617	453039	5305-01-459-9564	190	6
31272	45307-10	5340-01-204-2543	272	36
7X677	453349	5305-01-215-3990	83	13
7X677	453571	3110-01-412-0490	115	3
72582	454065	5305-01-143-2328	327	38
		5305-01-143-2328 477 8		
		5305-01-143-2328 480 2		
21450	454097	4730-00-911-5707	6	1
		4730-00-911-5707 177 3		
31272	45410-10	5340-01-254-7191	253	37
		5340-01-254-7191	429	29
		5340-01-254-7191	430	19
31272	45410-11	5340-01-254-7190	251	14
		5340-01-254-7190	430	44
31272	45410-12	5340-01-259-0328	252	21
		5340-01-259-0328	252	35
31272	45410-14	5340-01-259-0329	253	9
		5340-01-259-0329	253	17
		5340-01-259-0329	255	16
31272	45410-15	5340-01-361-7879	448	9
		5340-01-361-7879 449 9		
31272	45410-16	5340-01-258-6158	252	40
		5340-01-258-6158 253 4		
		5340-01-258-6158 255 7		
		5340-01-258-6158	255	10
		5340-01-258-6158	260	23
0VU83	45410-18	5340-01-259-0330	254	9
31272	45410-19	5340-01-275-8567	253	29
		5340-01-275-8567	279	31
31272	45410-20	5340-01-264-5833	234	13
		5340-01-264-5833	252	16
		5340-01-264-5833	253	38
31272	45410-21	5340-01-275-8568	256	1
		5340-01-275-8568 257 4		
31272	45410-22	5340-01-272-7815	260	9
24617	454253	5340-01-038-1493	74	16

SECTION IV TM9-2320-280-24P C01

CROSS REFERENCE INDEXES

PART NUMBER INDEX

CAGE	PART NUMBER	STOCK NUMBER	FIGURE NO		ITEM NO
24617	454542	5305-00-499-7694		51	9
31272	45460-10	5340-01-253-8933		277	14
24617	454748	5310-01-038-9579		375	11
		5310-01-038-9579	375	14	
72582	454749	5310-00-164-1790		377	7
31272	45481-01	2590-01-260-0214		279	6
31272	45481-11	2540-01-249-1584		279	1
24617	454838	5306-01-360-1124		446	11
29510	454869	5306-01-244-7882		440	1
7X677	454977	5306-01-192-2207		169	26
24617	455000	5306-01-360-1125		441	13
		5306-01-360-1125	442	10	
		5306-01-360-1125	453	19	
24617	455002	5306-01-360-1126		442	14
24617	455004	5306-01-360-1127		442	17
24617	455006	5306-01-360-1128		188	37
		5306-01-360-1128	442	1	
		5306-01-360-1128	453	12	
31272	45502-10	5340-01-260-9026		429	24
31272	45503-10	5340-01-257-0908		429	20
31272	45602-10	2590-01-262-4981		279	7
31272	45602-11	2590-01-262-4980		279	4
31272	45606-10	5340-01-261-1564		251	41
31272	45606-11	5340-01-258-6157		246	12
		5340-01-258-6157	251	18	
		5340-01-258-6157	251	32	
		5340-01-258-6157	253	23	
		5340-01-258-6157	255	32	
31272	45606-12	5340-01-256-4655		254	15
31272	45607-10	5340-01-258-8609		252	31
31272	45608-10	2590-01-261-0523		279	22
31272	45610-10	5307-01-286-6007		279	24
31272	45611-10	5310-01-298-4686		279	5
31273	45613-10	5365-01-267-6769		279	13
31272	45615-10	5340-01-260-0218		279	21
31272	45616-11	2590-01-260-0217		279	20
31272	45617-13	3040-01-264-9554		279	8
31272	45618-10	5340-01-271-4281		279	25
31272	45619-11	2590-01-260-0216		279	15
31272	45621-10	5306-01-269-6254		279	9
31272	45623-10	1640-01-271-2598		279	16
7X677	456369	5315-01-196-0277		277	25
31272	45685-10	5340-01-299-2963		252	28
24617	456931	5305-01-500-8280		88	25
24617	456958	5306-01-360-1130		440	16
31272	45721-10	5342-01-260-0215		279	10
31272	45938-10	5340-01-397-1388		262	1
63829	4605-0470	5306-01-296-8600		491	22
13445	46211-03	5940-01-180-3655		69	5

I-293

SECTION IV TM9-2320-280-24P C01

CROSS REFERENCE INDEXES

PART NUMBER INDEX

CAGE	PART NUMBER	STOCK NUMBER	FIGURE NO	ITEM NO
		5940-01-180-3655	481	1
81263	46827	5305-01-197-7547	68	12
51377	4735-38449-02	4140-01-211-8403	32	1
51377	4735-42599-103	2930-01-420-8622	32	4
59150	4755-725	2940-01-176-1427	489	1
78385	476624	4730-00-701-7737	345	35
		4730-00-701-7737	348	33
7X677	477249	5330-01-154-7159	4 10 76599	
4816SS-305	4730-01-088-7798	7	6	
		4730-01-088-7798	333	3
8K755	48417	4820-01-495-7582	492	12
8K755	48418	4820-01-495-7601	492	10
78385	484487	4730-00-567-1630	393	13
78385	487283	5310-00-333-7341	345	24
		5310-00-333-7341	348	3
57733	487357	5305-00-576-2335	345	18
		5305-00-576-2335	348	12
78385	487370	5310-00-789-0398	349	15
12204	4883434AA		488	21
78385	488558	5305-00-135-3032	348	7
57733	488755	5320-00-801-1548	348	11
78385	488993	5325-00-543-2902	345	38
		5325-00-543-2902	348	37
78385	489142	5305-00-018-8370	345	37
		5305-00-018-8370	348	27
60602	49005-1224	5340-01-290-8360	21	9
24617	4919976	5310-01-317-4022	391	6
45152	4965	9320-01-155-2369	BULK	58
80753	49RED	9905-00-977-2727	330	4
79470	49X4X4	4730-00-366-3011	178	19
79470	49X6X2	4730-00-254-6227	334	7
76301	4M116-04004	5310-01-204-3344	213	25
		5310-01-204-3344	214	20
98465	4M25-4	5310-00-817-4623	261	14
81343	5 010112B(N5)	4730-00-260-8285	366	15
		4730-00-260-8285	384	5
81337	5-11-966-41	5310-00-087-7493	205	36
		5310-00-087-7493	206	21
81343	5-2 010102B	4730-00-266-0535	334	19
81343	5-2 010203CA(ZINC)	4730-01-204-5457	368	29
81343	5-2 430160C	4730-01-357-5651	388	2
81343	5-2 430260	4730-01-348-6231	388	6
95019	5-213X	2520-01-189-2135	KITS	43
81343	5-4 010203CA(ZINC)	4730-01-203-1025	385	8
13499	500-1114-003	5310-01-042-8391	40	50
		5310-01-042-8391	43B	45
79136	5008-125MD	5325-00-514-1299	143	21

I-294

SECTION IV TM9-2320-280-24P C01

CROSS REFERENCE INDEXES

PART NUMBER INDEX

CAGE	PART NUMBER	STOCK NUMBER	FIGURE NO	ITEM NO
		5325-00-514-1299	151	21
70412	51-11	2540-01-199-7778	327	5
60602	51078-1	5340-01-456-1500	98	9
24234	513310	5340-01-270-2681	55	9
24234	513394	5340-01-271-6466	494	15
24234	513395	5340-01-271-6465	494	14
24234	513396	5340-01-271-6455	494	20
24234	513409		55	5
24234	513431	5340-01-281-7798	498	12
24234	513438	5340-01-271-6467	496	28
24234	513444	5340-01-288-6551	55	15
24234	513447	2940-01-285-2942	343	20
24234	513448		54	2
24234	513450		498	5
24234	513462	2990-01-265-3298	347	1
24234	514277	5340-01-289-3228	498	16
71286	51L3-1-1-AA	5342-01-036-0649	237	3
71286	51L3-1X1AB	5340-01-209-6874	315	4
86403	52068157	3040-01-244-3646	488	17
96906	521301A203R 5.50		379	41
38506	52252	2590-01-316-2581	319	4
19207	5226154	5325-00-174-9829	136	10
		5325-00-174-9829	136	23
		5325-00-174-9829	137	5
62161	528236-05		505	7
60602	52826-1	3040-01-410-9965	101	8
60602	52827-1	3040-01-410-8054	101	5
19207	5294507	5310-00-350-2655	62	6
35510	5295	5310-01-271-1841	39	61
79136	5304-25-H	5325-01-047-3201	323	11
		5325-01-047-3201	326	10
19207	5310615	5310-00-463-0268	62	25
36251	5311	4730-00-585-2653	169	25
06046	535K1-49383	6680-01-474-0384	482	3
19207	5381088	5930-00-130-5349	46	17
19207	5381233	5310-00-832-6852	46	14
39428	5416K25	4730-01-257-4905	12	30
7Z043	5501	5975-01-280-8922	57	10
34623	5550554	5310-01-144-2779	172	18
34623	5568211	5307-01-196-4937	142	32
		5307-01-196-4937	143	24
		5307-01-196-4937	150	32
		5307-01-196-4937	151	24
34623	5568223	5310-01-174-8632	142	8
		5310-01-174-8632	150	8
34623	5568226	3110-01-027-4475	142	9
		3110-01-027-4475	143	9
		3110-01-027-4475	150	9
		3110-01-027-4475	151	9

I-295

SECTION IV TM9-2320-280-24P C01

CROSS REFERENCE INDEXES

PART NUMBER INDEX

CAGE	PART NUMBER	STOCK NUMBER	FIGURE NO	ITEM NO
34623	5568227	3110-00-100-5920	142	22
		3110-00-100-5920	143	29
		3110-00-100-5920	151	29
34623	5568228	3110-00-100-5937	142	3
		3110-00-100-5937	143	3
		3110-00-100-5937	150	3
		3110-00-100-5937	151	3
34623	5568310	5342-01-189-9982	2	7
34623	5568666	6625-01-180-6542	47	10
34623	5568667	6680-01-298-0498	47	12
34623	5569268	5365-01-186-1294	196	3
		5365-01-186-1294	196	8
34623	5569270	5342-01-189-5452	196	4
		5342-01-189-5452	196	14
34623	5569460	5340-01-189-6748	21	14
34623	5572142	5945-01-175-7318	118	4
34623	5573534		17	14
34623	5573684	5310-01-174-8607	142	21
		5310-01-174-8607	143	28
		5310-01-174-8607	150	21
		5310-01-174-8607	151	28
34623	5573687	5310-01-176-0839	142	25
		5310-01-176-0839	143	33
		5310-01-176-0839	150	25
		5310-01-176-0839	151	33
34623	5573688	5310-01-231-0596	142	6
		5310-01-231-0596	143	6
		5310-01-231-0596	150	6
		5310-01-231-0596	151	6
34623	5573690	5365-01-177-5720	142	26
		5365-01-177-5720	143	34
		5365-01-177-5720	150	26
		5365-01-177-5720	151	34
34623	5573692	5330-01-176-0825	142	12
		5330-01-176-0825	143	12
		5330-01-176-0825	150	12
		5330-01-176-0825	151	12
34623	5573989	5365-01-174-8186	142	4
		5365-01-174-8186	143	4
		5365-01-174-8186	150	4
		5365-01-174-8186	151	4
34623	5574002	5340-01-174-8493	142	13
		5340-01-174-8493	143	13
		5340-01-174-8493	150	13
		5340-01-174-8493	151	13
34623	5574377		269	1
34623	5574856	5330-01-194-0472	8	9
34623	5574872	4710-01-188-6050	15	8
		4710-01-188-6050	16	7

SECTION IV TM9-2320-280-24P C01

CROSS REFERENCE INDEXES

PART NUMBER INDEX

CAGE	PART NUMBER	STOCK NUMBER	FIGURE NO	ITEM NO
34623	5574879	9515-01-185-3119	267	20
34623	5574921	3020-01-174-5985	142	10
		3020-01-174-5985	150	10
34623	5574924	5340-01-174-2271	142	24
		5340-01-174-2271	143	32
		5340-01-174-2271	150	24
		5340-01-174-2271	151	32
34623	5574995	6680-01-185-1264	67	4
34623	5575418	5340-01-253-3757	199	40
34623	5575466	5310-01-197-1161	209	1
34623	5575702	5340-01-186-1280	266	6
		5340-01-186-1280	284	4
34623	5575703	5340-01-186-1281	266	6
		5340-01-186-1281	284	4
34623	5575752	5340-01-185-8724	266	5
34623	5575936	5320-01-259-6155	224	1
		5320-01-259-6155	237	16
		5320-01-259-6155	263	5
34623	5575940	5320-01-271-1834	211	6
		5320-01-271-1834	212	3
		5320-01-271-1834	296	18
		5320-01-271-1834	346	8
34623	5575967		328	10
34623	5577458	5310-01-220-3099	286	11
34623	5577552		17	8
34623	5577592	5365-01-219-3528	191	3
96139	5577626	5995-01-184-2228	79	1
34623	5577659	2530-01-184-9821	156	19
34623	5577765	5995-01-201-4128	410	3
		5995-01-201-4128	412	3
34623	5577813	3040-01-185-4388	133	11
34623	5577933	5330-01-184-6500	6	3
34623	5578150	5306-01-175-7577	167	3
34623	5578197		240	15
34623	5578215		79	2
34623	5578306-B	5935-01-209-5594	79	3
34623	5578331	9515-01-189-9732	267	8
34623	5578395	2540-01-193-2711	403	22
34623	5578396	2540-01-192-9711	403	5
34623	5578414	5340-01-186-7229	205	28
34623	5578418	5306-01-186-7129	205	25
34623	5578466	2540-01-192-9712	403	1
		2540-01-192-9712	403	6
34623	5578522	5340-01-186-9496	202	38
34623	5578547	2510-01-185-3107	205	37
34623	5578604	4710-01-188-0028	29	10
		4710-01-188-0028	364	8
34623	5578635	3110-01-189-9980	21	22
34623	5578660-B	5330-01-185-0587	21	24

I-297

SECTION IV TM9-2320-280-24P C01

CROSS REFERENCE INDEXES

PART NUMBER INDEX

CAGE PART NUMBER STOCK NUMBER FIGURE NO ITEM NO

CAGE	PART NUMBER	STOCK NUMBER	FIGURE NO	ITEM NO
34623	5578667	5360-01-186-4845	165	12
		5360-01-186-4845 166	3	
34623	5578673	5355-01-190-7635	99	2
34623	5578675	5360-01-186-4844	99	26
34623	5578694	4710-01-203-0615	334	13
		4710-01-203-0615 386	4	
34623	5578742	5360-01-186-3737	21	25
34623	5578749	2510-01-196-5311	203	4
34623	5578764	5315-01-185-7421	21	27
34623	5578802	2530-00-340-1405	157	9
		2530-00-340-1405 158	5	
34623	5578809	2530-01-185-3879	158	12
34623	5578822	2530-01-176-4649	152	6
34623	5578823	2530-01-175-7210	152	6
34623	5578874		69	24
34623	5578875		69	10
34623	5578876		69	19
34623	5578877		72	15
34623	5578878		72	3
34623	5578887		246	21
34623	5578910		13	11
			14	17
34623	5578911		13	15
34623	5578916	4710-01-185-9666	159	1
34623	5578918	4710-01-185-9665	161	6
34623	5578943	4720-01-194-5334	398	32
34623	5578947	4730-00-908-6294	27	1
		4730-00-908-6294	337	26
		4730-00-908-6294	343	3
		4730-00-908-6294 387	5	
34623	5578999		27	15
34623	5579007		13	7
34623	5579008		13	19
34623	5579061	2530-01-174-4174	142	31
		2530-01-174-4174 150	31	
34623	5579257	5340-01-203-8675	272	20
34623	5579292	9515-01-189-9745	267	19
34623	5579297	5365-01-194-0761	267	16
34623	5579324	5330-01-184-6492	103	10
		5330-01-184-6492 400	6	
34623	5579386		245	39
34623	5579424	3040-01-174-9130	148	1
34623	5579426		148	4
34623	5579428		148	29
34623	5579435		148	28
34623	5579436		148	22
34623	5579438	5305-01-174-8625	148	24
34623	5579442	5310-01-175-0617	148	10
34623	5579445	5365-01-175-3593	KITS	44

I-298

SECTION IV TM9-2320-280-24P C01

CROSS REFERENCE INDEXES

PART NUMBER INDEX

CAGE	PART NUMBER	STOCK NUMBER	FIGURE NO	ITEM NO
34623	5579445-54	5365-01-381-4487	148	17
34623	5579445-55	5365-01-382-3676	148	17
34623	5579445-56	5365-01-382-0973	148	17
34623	5579445-57	5365-01-382-3710	148	17
34623	5579445-58	5365-01-382-1170	148	17
34623	5579445-59	5365-01-382-1235	148	17
34623	5579445-60	5365-01-382-1181	148	17
34623	5579445-61	5365-01-382-1135	148	17
34623	5579445-62	5365-01-382-1177	148	17
34623	5579445-63	5365-01-382-1162	148	17
34623	5579445-64	5365-01-382-3693	148	17
34623	5579445-65	5365-01-382-3726	148	17
34623	5579445-66	5365-01-382-1273	148	17
34623	5579445-67	5365-01-382-1166	148	17
34623	5579445-68	5365-01-382-1157	148	17
34623	5579445-69	5365-01-382-1240	148	17
34623	5579445-70	5365-01-382-3718	148	17
34623	5579445-71	5365-01-382-1132	148	17
34623	5579445-72	5365-01-382-1171	148	17
34623	5579445-73	5365-01-382-3715	148	17
34623	5579445-74	5365-01-388-4449	148	17
34623	5579445-75	5365-01-382-3694	148	17
34623	5579445-76	5365-01-382-1159	148	17
34623	5579445-77	5365-01-382-3447	148	17
34623	5579445-78	5365-01-382-3700	148	17
34623	5579445-79	5365-01-382-1058	148	17
34623	5579450	5365-01-180-2585	148	5
34623	5579453	5365-01-174-8174	KITS	45
34623	5579453-44	5365-01-381-5166	148	20
34623	5579453-45	5365-01-381-9582	148	20
34623	5579453-46	5365-01-381-5174	148	20
34623	5579453-47	5365-01-381-6332	148	20
34623	5579453-48	5365-01-381-6120	148	20
34623	5579453-49	5365-01-380-5345	148	20
34623	5579453-50	5365-01-381-6151	148	20
34623	5579453-51	5365-01-381-6276	148	20
34623	5579453-52	5365-01-381-6223	148	20
34623	5579453-53	5365-01-381-6080	148	20
34623	5579453-54	5365-01-381-6250	148	20
34623	5579453-55	5365-01-381-6145	148	20
34623	5579453-56	5365-01-381-6174	148	20
34623	5579453-57	5365-01-381-6210	148	20
34623	5579453-58	5365-01-381-6292	148	20
34623	5579453-59	5365-01-381-6110	148	20
34623	5579453-60	5365-01-381-6158	148	20
34623	5579453-61	5365-01-381-6313	148	20
34623	5579453-62	5365-01-381-6226	148	20
34623	5579453-63	5365-01-381-6076	148	20
34623	5579453-64	5365-01-381-6303	148	20

SECTION IV TM9-2320-280-24P C01

CROSS REFERENCE INDEXES

PART NUMBER INDEX

CAGE	PART NUMBER	STOCK NUMBER	FIGURE NO		ITEM NO
34623	5579453-65	5365-01-381-6164	148		20
34623	5579453-66	5365-01-381-6197	148		20
34623	5579453-67	5365-01-381-6229	148		20
34623	5579453-68	5365-01-381-6086	148		20
34623	5579453-69	5365-01-381-6122	148		20
34623	5579453-70	5365-01-381-6154	148		20
34623	5579453-71	5365-01-381-6188	148		20
34623	5579453-72	5365-01-381-6317	148		20
34623	5579453-73	5365-01-381-6240	148		20
34623	5579453-74	5365-01-381-6112	148		20
34623	5579453-75	5365-01-381-6258	148		20
34623	5579453-76	5365-01-381-6165	148		20
34623	5579453-77	5365-01-381-6211	148		20
34623	5579453-78	5365-01-381-9588	148		20
34623	5579453-79	5365-01-438-1556	148		20
34623	5579453-80	5365-01-381-6113	148		20
34623	5579453-81	5365-01-381-6261	148		20
34623	5579453-82	5365-01-381-6175	148		20
34623	5579453-83	5365-01-381-6278	148		20
34623	5579474	2530-01-175-7213	144		20
		2530-01-175-7213	152	20	
34623	5579475	2530-01-175-7212	144		20
		2530-01-175-7212	152	20	
34623	5579485	2510-01-188-3228	203		18
34623	5579501	3020-01-191-8784	148		27
34623	5579538	2510-01-189-0890	184		6
34623	5579620	2510-01-189-0891	182		7
34623	5579621	2510-01-187-7031	182		7
34623	5579658	5325-01-214-5007	327		19
34623	5579716	2510-01-189-9720	204		11
34623	5579717	2540-01-189-9729	204		11
34623	5579811	5365-01-211-6083	246		25
34623	5579818	5340-01-195-5397	243		19
34623	5579894	4710-01-192-9546	268		36
34623	5579895	4710-01-192-7965	268		31
34623	5579941	5365-01-211-5063	62		12
34623	5581162	5330-01-252-5377	197		32
34623	5581163	5330-01-252-5378	199		60
34623	5581211		46		5
34623	5581213		199		23
34623	5581260	2590-01-188-3225	226		34
34623	5581313		325		3
34623	5581315	5325-01-199-2254	322		6
		5325-01-199-2254	324	6	
34623	5581319	2540-01-203-7721	320		21
		2540-01-203-7721	324	26	
34623	5581320	2540-01-200-1994	320		9
		2540-01-200-1994	320		21
		2540-01-200-1994	324		12

SECTION IV TM9-2320-280-24P C01

CROSS REFERENCE INDEXES

PART NUMBER INDEX

CAGE	PART NUMBER	STOCK NUMBER	FIGURE NO	ITEM NO
		2540-01-200-1994	324	26
		2540-01-200-1994	375	17
34623	5581321	5310-01-205-9056	316	13
		5310-01-205-9056	320	11
		5310-01-205-9056	320	23
		5310-01-205-9056	324	18
		5310-01-205-9056	324	30
		5310-01-205-9056	378	21
34623	5581331	2540-01-199-6760	389	6
34623	5581347	2540-01-197-5476	325	15
34623	5581348	2510-01-189-9724	316	7
		2510-01-189-9724	320	19
		2510-01-189-9724	324	24
34623	5581349	2510-01-189-9725	320	19
		2510-01-189-9725	324	24
34623	5581362	5340-01-207-0717	198	24
34623	5581364	5342-01-204-9610	376	16
34623	5581366		225	7
			325	4
34623	5581367	2540-01-196-5313	225	8
34623	5581371	2590-01-198-2895	317	6
		2590-01-198-2895	373	5
34623	5581372	2590-01-197-4898	317	6
		2590-01-197-4898	373	5
34623	5581387	5340-01-199-9679	323	10
34623	5581392-B	5315-01-197-1494	323	9
		5315-01-197-1494	326	8
34623	5581396	325	6	34623
5581439	5305-01-203-8346	317	4	
		5305-01-203-8346	321	5
		5305-01-203-8346	325	8
34623	5581461	5310-01-018-5332	324	16
34623	5581464	5340-01-202-7445	323	3
		5340-01-202-7445	326	3
13445	5582	5930-00-736-3539	39	29
		5930-00-736-3539	59	19
34623	5582303	2520-01-210-2624	148	21
34623	5582366	5310-01-189-8476	159	9
		5310-01-189-8476	160	5
		5310-01-189-8476	161	10
		5310-01-189-8476	162	16
		5310-01-189-8476	487A	11
		5310-01-189-8476	487D	7
34623	5582405	5340-01-191-3526	15	33
34623	5582412	5340-01-191-0746	15	26
		5340-01-191-0746	16	37
34623	5582423	5340-01-198-2248	247	17
		5340-01-198-2248	249	39
34623	5582436	2510-01-310-2324	274	6

I-301

SECTION IV TM9-2320-280-24P C01

CROSS REFERENCE INDEXES

PART NUMBER INDEX

CAGE	PART NUMBER	STOCK NUMBER	FIGURE NO	ITEM NO
34623	5582439	5340-01-199-3499	272	21
34623	5582479		124	1
34623	5582481	5975-01-208-9618	41	8
		5975-01-208-9618	44	22
		5975-01-208-9618	44	27
		5975-01-208-9618	474	15
		5975-01-208-9618	480	18
		5975-01-208-9618	480	30
34623	5582605	5340-01-191-3359	206	12
34623	5582606		15	14
34623	5582642		123	6
34623	5582643		141	2
34623	5582644		399	10
			400	8
34623	5582649		141	7
34623	5582653	4730-01-070-7680	149	6
34623	5582677	5310-01-210-7938	202	40
34623	5582763	5340-01-190-0333	179	18
		5340-01-190-0333	180	28
34623	5582815	4710-01-189-1080	124	4
34623	5582817	4710-01-188-3516	124	6
		4710-01-188-3516	125	7
34623	5582819	4710-01-188-3515	124	7
		4710-01-188-3515	125	6
34623	5582851	5365-01-193-2951	169	18
34623	5582897	2540-01-192-9714	403	21
34623	5582902		249	10
34623	5582936	5330-01-203-9187	368	1
34623	5582968	3010-01-292-6428	148	9
34623	5583050	5305-01-221-7734	286	2
34623	5583062-B	5305-01-224-1449	286	5
34623	5583066	5365-01-218-9848	285	12
34623	5583120		245	18
34623	5583141	5306-01-195-0277	269	17
34623	5583156	5365-01-194-5093	270	18
34623	5583215	5365-01-214-4987	196	11
34623	5583219	5365-01-209-7831	181	5
		5365-01-209-7831	184	9
34623	5583231	5340-01-203-9391	278	15
34623	5583235	5340-01-213-4267	240	17
34623	5583291	2540-01-210-2749	268	13
34623	5583352	4710-01-203-1304	368	28
34623	5583395	5980-01-186-8319	46	2
34623	5583555	5340-01-259-7523	247	16
34623	5583556-B	5340-01-283-2458	247	9
34623	5583558	5365-01-198-3399	248	12
34623	5583568	5340-01-197-8674	248	16
		5340-01-197-8674	429	14
34623	5583571	5365-01-198-3505	248	22

I-302

SECTION IV TM9-2320-280-24P C01

CROSS REFERENCE INDEXES

PART NUMBER INDEX

CAGE	PART NUMBER	STOCK NUMBER	FIGURE NO	ITEM NO
34623	5583572	5365-01-198-3401	248	18
34623	5583576	5342-01-193-7088	488	14
34623	5583606	5306-01-194-6433	97	5
34623	5583650	4710-01-188-8780	124	3
34623	5583676	2590-01-188-5081	245	1
34623	5583717	5340-01-209-7799	85	27
		5340-01-209-7799	88	8
34623	5583758		148	2
34623	5583993	5340-01-197-3478	245	19
34623	5584019	4710-01-186-1009	13	1
34623	5584021	4710-01-186-1010	13	5
34623	5584024	4710-01-187-6759	13	9
		4710-01-187-6759	13	16
		4710-01-187-6759	14	15
		4710-01-187-6759	14	22
34623	5584032	5340-01-196-0572	277	38
34623	5584077	5340-01-212-7711	182	10
34623	5584104	5340-01-197-6715	245	2
34623	5584105	5340-01-197-6716	245	6
34623	5584107	5315-01-198-3346	245	12
34623	5584110	2520-01-210-3506	99	19
34623	5584135	5340-01-203-5629	245	3
34623	5584136	5340-01-203-5630	245	7
34623	5584137	5365-01-181-5085	142	23
		5365-01-181-5085	150	23
34623	5584138	5340-01-197-2716	270	29
34623	5584142	4710-01-186-1016	159	10
34623	5584146	4710-01-185-9667	160	8
34623	5584151	2510-01-192-5950	270	15
34623	5584152	2540-01-192-8283	270	15
34623	5584173		141	8
34623	5584259	5340-01-198-7580	280	29
34623	5584260	5340-01-197-6714	281	29
34623	5584261	2540-01-198-7581	281	29
34623	5584264	2510-01-211-3229	280	33
34623	5584278	5315-01-204-2328	280	39
		5315-01-204-2328	281	39
34623	5584294	5340-01-211-1645	280	52
34623	5584299	5365-01-197-9383	280	21
		5365-01-197-9383	281	21
34623	5584302	5340-01-198-2241	280	8
		5340-01-198-2241	281	8
34623	5584325	5340-01-204-2458	280	6
		5340-01-204-2458	281	6
34623	5584341	2910-01-186-6038	15	16
34623	5584358		269	1
34623	5584372		141	3
34623	5584373		399	5
34623	5584383		141	1

SECTION IV TM9-2320-280-24P C01

CROSS REFERENCE INDEXES

PART NUMBER INDEX

CAGE	PART NUMBER	STOCK NUMBER	FIGURE NO		ITEM NO
34623	5584384	5330-01-209-7723		169	24
34623	5584391	5325-01-210-1633		84	6
34623	5584421	5340-01-203-0324		280	18
		5340-01-203-0324	281	18	
34623	5584436	5365-01-201-4749		280	5
		5365-01-201-4749	281	5	
34623	5584462	5310-01-213-4185		142	20
		5310-01-213-4185		143	27
		5310-01-213-4185		150	20
		5310-01-213-4185	151	27	
34623	5584559	4710-01-278-1053		12	10
34623	5584575-B	5365-01-213-5739		398	20
34623	5584588	4720-01-203-7789		12	8
34623	5584707			280	31
				281	31
34623	5584710	5310-01-203-3217		280	20
		5310-01-203-3217	281	20	
		5310-01-203-3217	429	3	
		5310-01-203-3217	430	3	
		5310-01-203-3217	430	22	
		5310-01-203-3217	430	39	
34623	5584713	5310-01-198-3428		280	27
		5310-01-198-3428	281	27	
34623	5584765	5340-01-211-5083		274	31
34623	5584766	5340-01-211-6288		272	13
34623	5584769	2530-01-175-7211		144	6
34623	5584813			207	5
34623	5584836			207	4
34623	5584854	5360-01-195-6200		243	25
34623	5584867	5340-01-197-8675		280	6
		5340-01-197-8675	281	6	
34623	5584891	5340-01-211-0838		277	29
		5340-01-211-0838	278	8	
		5340-01-211-0838	440	4	
34623	5584900	4710-01-186-9620		161	5
		4710-01-186-9620	162	5	
34623	5584963	5340-01-211-3132		202	14
34623	5584977	5340-01-196-2573		403	19
34623	5584999	5365-01-193-4479		403	20
34623	5585094	5340-01-208-8688		12	32
34623	5585103	5310-01-188-0745		169	3
34623	5585111	2510-01-303-5769		47	8
34623	5585132-C	5340-01-252-8537		208	24
34623	5585133	5340-01-255-4655		208	26
34623	5585143	5365-01-252-8026		208	30
34623	5585152	5310-01-185-8692		328	4
34623	5585154	5340-01-203-8608		225	4
		5340-01-203-8608	226	38	
34623	5585178	5365-01-213-4239		225	24

I-304

SECTION IV TM9-2320-280-24P C01

CROSS REFERENCE INDEXES

PART NUMBER INDEX

CAGE	PART NUMBER	STOCK NUMBER	FIGURE NO	ITEM NO
		5365-01-213-4239	226	11
34623	5585245		327	27
34623	5585259	5306-01-323-5535	281	50
34623	5585259-B	5306-01-323-5535	198	22
34623	5585445	2590-01-270-9446	295	5
34623	5585446	2510-01-270-9447	295	4
34623	5588022	2815-01-168-7912	4	22
34623	5588032	5340-01-209-7807	153	19
34623	5588062	5306-01-286-7182	203	25
		5306-01-286-7182	204	10
34623	5588063		280	30
			281	30
34623	5588064	5305-01-204-6502	84	1
		5305-01-204-6502	85	25
34623	5588167	5340-01-185-8821	202	27
34623	5588172	5340-01-186-7658	202	8
34623	5588179	5340-01-189-9979	202	15
34623	5588193	5340-01-291-5711	417	10
34623	5588629	5340-01-212-2407	83	14
34623	5588647		403	16
34623	5588648		403	16
34623	5588656	2540-01-334-4333	403	18
34623	5588675	5340-01-212-4887	23	9
		5340-01-212-4887	24	6
34623	5588694	4710-01-203-0607	334	14
		4710-01-203-0607	366	19
		4710-01-203-0607	386	5
34623	5588696	4710-01-203-0608	334	18
		4710-01-203-0608	384	1
34623	5588698		334	17
34623	5588701	4720-01-210-0484	7	9
34623	5588702	2510-01-262-6010	198	11
34623	5588705	5325-01-214-7817	80	11
34623	5588719	5340-01-209-7386	411	17
34623	5588767		19	12
34623	5588773		388	1
34623	5588778	2540-01-309-4459	23	11
		2540-01-309-4459	24	12
34623	5588785	5340-01-213-4667	75	48
34623	5588872		288	2
34623	5588877	5340-01-281-3429	288	18
34623	5588878	2540-01-275-6203	288	15
34623	5588904	5306-01-265-8767	262	5
34623	5588927	5305-01-272-9959	238	7
34623	5588931	2510-01-270-7925	296	10
34623	5588959	5340-01-266-3845	293	14
34623	5589012-B	5995-01-269-9525	339	11
34623	5589067	5306-01-204-2139	280	28
		5306-01-204-2139	281	28

I-305

SECTION IV TM9-2320-280-24P C01

CROSS REFERENCE INDEXES

PART NUMBER INDEX

CAGE PART NUMBER STOCK NUMBER FIGURE NO ITEM NO

CAGE	PART NUMBER	STOCK NUMBER	FIGURE NO	ITEM NO
34623	5589086	9330-01-197-5458	374	13
34623	5589167	5995-01-201-7495	412	1
34623	5589275		374	1
34623	5589357	5340-01-272-4942	217	4
34623	5589425	5340-01-272-2732	224	17
34623	5589505	5360-01-275-3512	218	2
34623	5589581	5340-01-274-9899	238	16
34623	5589748-C	5340-01-272-4935	289	24
34623	5589763	2510-01-270-9707	300	1
34623	5589776	2540-01-280-4159	293	18
34623	5589785		288	2
34623	5589853-A	5340-01-309-7900	296	16
34623	5589864	5340-01-272-5594	287	29
34623	5589868	5340-01-273-4667	287	25
34623	5589918	5340-01-310-5321	224	4
34623	5589947	5330-01-272-7473	295	3
34623	5589961	5340-01-272-8410	214	31
34623	5589970	5330-01-282-0913	341	5
34623	5590023		177	1
34623	5590024	4720-01-297-0255	177	7
34623	5590025		178	7
34623	5590029	5325-01-255-5062	414	21
34623	5590149	5320-01-218-5802	286	30
34623	5590150	5320-01-219-7261	286	6
		5320-01-219-7261	286	15
34623	5590155	5325-01-255-5063	416	24
		5325-01-255-5063	428	23
34623	5590244	5999-01-197-5465	410	KIT
34623	5590284	4710-01-210-3503	333	7
34623	5590387	5340-01-211-4025	23	12
34623	5590530	2590-01-187-7039	156	20
34623	5590540	3040-01-465-3178	156	11
34623	5590544		367	3
34623	5590550	5340-01-266-2995	202	7
34623	5590551	2520-01-188-5131	148	11
34623	5590556	5310-01-208-5252	374	23
34623	5590602	5360-01-214-3695	153	14
34623	5590630	2540-01-197-5463	374	20
34623	5590643	4710-01-185-8329	161	8
34623	5590644	4710-01-185-8328	161	13
34623	5590669	2540-01-314-1121	284	16
		2540-01-314-1121	439	9
34623	5591114	5340-01-260-4881	268	24
34623	5591115	4710-01-251-3076	268	27
34623	5591157		149	4
34623	5591158		149	1
34623	5591161		149	5
34623	5591168		123	5
34623	5591169	5340-01-197-2438	62	24

I-306

SECTION IV TM9-2320-280-24P C01

CROSS REFERENCE INDEXES

PART NUMBER INDEX

CAGE	PART NUMBER	STOCK NUMBER	FIGURE NO	ITEM NO
34623	5591200	5940-01-204-8830	412	8
34623	5591210		398	30
34623	5591252	5340-01-212-3709	28	6
34623	5591253		123	1
			399	7
34623	5591279	2530-01-203-5663	142	33
		2530-01-203-5663	150	33
34623	5591280	2530-01-203-5662	142	18
		2530-01-203-5662	150	18
34623	5591286	5307-01-196-4717	4	26
34623	5591313	5330-01-255-0207	202	5
34623	5591315	5330-01-255-0208	202	11
34623	5591333		403	7
34623	5591334		403	7
34623	5591345	5331-01-176-0923	167	4
34623	5591482		141	6
34623	5591483		141	5
			149	3
34623	5591507	5310-01-212-2303	277	27
34623	5591514	5310-01-217-0715	279	34
34623	5591553	2590-01-261-5485	230	22
		2590-01-261-5485	231	27
34623	5591575	5365-01-212-0144	272	32
34623	5591577-B	5365-01-252-9214	205	11
34623	5591614	4710-01-265-3231	334	8
34623	5591648	5340-01-213-5735	202	2
		5340-01-213-5735	218	7
		5340-01-213-5735	253	35
34623	5591707		398	29
34623	5591712	2590-01-262-7706	227	29
34623	5591713	2590-01-263-3254	227	29
34623	5591781	5305-01-185-7122	179	7
		5305-01-185-7122	180	8
		5305-01-185-7122	184	11
34623	5591935	2540-01-390-5711	370	3
34623	5591937		124	5
34623	5591938		124	8
34623	5591957	5340-01-213-4662	204	13
34623	5592010	5340-01-272-6633	238	12
34623	5592011	5360-01-278-9723	238	20
34623	5592019	2540-01-271-3773	288	28
34623	5592020	5340-01-272-7493	288	31
34623	5592085		217	6
34623	5592234	5340-01-297-1549	296	13
34623	5592725	2510-01-252-8461	284	6
34623	5592726	2510-01-252-7903	284	8
34623	5592727	2510-01-252-7904	284	10
34623	5592737	2510-01-175-7224	181	17
34623	5592814	5305-01-253-2998	241	11

I-307

SECTION IV TM9-2320-280-24P C01

CROSS REFERENCE INDEXES

PART NUMBER INDEX

CAGE	PART NUMBER	STOCK NUMBER	FIGURE NO	ITEM NO
34623	5592817	5305-01-256-6870		429	13
		5305-01-256-6870	430	27
		5305-01-256-6870	430	34
34623	5592819	5305-01-250-6582		57	8
34623	5592823	5305-01-251-0728		249	23
		5305-01-251-0728	253	25
		5305-01-251-0728	256	6
		5305-01-251-0728	257	9
		5305-01-251-0728	259	10
		5305-01-251-0728	262	3
		5305-01-251-0728	263	6
		5305-01-251-0728	264	2
		5305-01-251-0728	265	2
34623	5592838	5305-01-253-2989		240	10
34623	5592840	5305-01-254-2460		92	17
		5305-01-254-2460	199	25
		5305-01-254-2460	497	1
34623	5592846	5305-01-255-1017		249	19
34623	5592861	5305-01-254-2453		269	26
34623	5592862	5305-01-254-2451		249	40
34623	5592878	5305-01-250-6583		46	22
34623	5592879	5306-01-254-2360		279	37
34623	5592884	5305-01-255-1016		247	21
		5305-01-255-1016	247	26
		5305-01-255-1016	248	23
		5305-01-255-1016	249	8
		5305-01-255-1016	249	32
34623	5592899	5305-01-254-2461		398	13
34623	5592900	5305-01-255-4606		222	9
34623	5592916	5310-01-252-0513		241	6
		5310-01-252-0513	249	25
		5310-01-252-0513	272	17
34623	5592926	5310-01-253-8949		248	15
		5310-01-253-8949	280	4
		5310-01-253-8949	280	53
		5310-01-253-8949	281	4
		5310-01-253-8949	281	53
		5310-01-253-8949	286	17
34623	5592927	5310-01-253-8957		226	22
34623	5592952	5305-00-253-5609		133	3
34623	5592954	5305-01-206-8401		75	28
		5305-01-206-8401	84	16
34623	5592958	5310-01-253-1615		410	16
34623	5592970	5305-01-255-2675		246	27
34623	5592994	5305-01-252-9108		199	36
34623	5592995	5310-01-253-8928		331	23
34623	5592998	5310-01-250-6587		59	10
		5310-01-250-6587	99	12
34623	5592999	5310-01-253-0058		199	58

SECTION IV TM9-2320-280-24P C01

CROSS REFERENCE INDEXES

PART NUMBER INDEX

CAGE	PART NUMBER	STOCK NUMBER	FIGURE NO	ITEM NO
34623	5593006	5305-01-256-0406	215	6
		5305-01-256-0406 430	26	
34623	5593011	5305-01-251-9222	104	17
34623	5593024	5306-01-254-2359	177	16
34623	5593027	5310-01-255-9452	413	10
34623	5593030	5305-01-268-5680	333	40
34623	5593033	5310-01-252-2999	59	7
		5310-01-252-2999 75	32	
		5310-01-252-2999 84	10	
		5310-01-252-2999 153	6	
		5310-01-252-2999 330	1	
34623	5593035	5310-01-255-2695	58	10
34623	5593038	5310-01-251-0711	49	9
		5310-01-251-0711 333	2	
34623	5593047	5310-01-266-6392	488	16
34623	5593048	5310-00-251-4503	63	5
34623	5593050		337	24
34623	5593182	5340-01-259-0330	255	26
34623	5593189	2590-01-249-1579	255	13
34623	5593200	5306-01-387-7457	270	31
34623	5593213	5310-00-080-6004	26	5
34623	5593218	5310-01-253-8438	286	27
0MAY0	5593221	5310-01-258-1536	26	4
		5310-01-258-1536 174	12	
		5310-01-258-1536 398	26	
34623	5593225	5305-00-550-1537	205	24
34623	5593235	5310-01-257-7719	169	32
		5310-01-257-7719 205	7	
		5310-01-257-7719 206	27	
34623	5593241	5310-01-253-8952	280	12
		5310-01-253-8952	280	37
		5310-01-253-8952	281	12
		5310-01-253-8952 281	37	
34623	5593242	5310-01-253-8953	280	19
		5310-01-253-8953 280	35	
		5310-01-253-8953 281	19	
		5310-01-253-8953 281	35	
		5310-01-253-8953 335	12	
		5310-01-253-8953 336	9	
34623	5593258	5305-01-253-2993	205	33
		5305-01-253-2993 206	20	
34623	5593261	5310-01-253-0060	205	8
		5310-01-253-0060 206	28	
34623	5593284	5305-01-255-4611	327	20
34623	5593286	5305-01-254-2452	248	2
		5305-01-254-2452 272	22	
34623	5593308	5310-01-254-5352	46	20
		5310-01-254-5352 208	3	
34623	5593309	4730-01-255-0925	331	32

I-309

SECTION IV TM9-2320-280-24P C01

CROSS REFERENCE INDEXES

PART NUMBER INDEX

CAGE	PART NUMBER	STOCK NUMBER	FIGURE NO	ITEM NO
34623	5593312	5310-01-147-4052	144	2
		5310-01-147-4052	165	2
		5310-01-147-4052	181	11
		5310-01-147-4052	182	1
		5310-01-147-4052	184	14
		5310-01-147-4052	195	1
		5310-01-147-4052	270	34
34623	5593313	5305-01-254-2459	268	1
		5305-01-254-2459	270	40
34623	5593318	5310-01-278-9555	208	21
		5310-01-278-9555	283	23
34623	5593320	5310-01-253-8948	179	4
		5310-01-253-8948	180	5
		5310-01-253-8948	181	2
		5310-01-253-8948	182	3
		5310-01-253-8948	187	12
		5310-01-253-8948	188	2
		5310-01-253-8948	191	2
		5310-01-253-8948	194	1
		5310-01-253-8948	283	3
		5310-01-253-8948	417	15
34623	5593345	5306-01-263-6142	279	32
34623	5593348	5306-01-251-1616	142	27
34623	5593359	5306-01-255-2662	415	6
34623	5593387	5310-01-272-6546	287	32
34623	5593409	5305-01-272-9042	56	25
		5305-01-272-9042	57	2
		5305-01-272-9042	65	2
		5305-01-272-9042	92	11
		5305-01-272-9042	95	29
		5305-01-272-9042	283	27
		5305-01-272-9042	304	9
		5305-01-272-9042	339	1
34623	5593413	5310-01-270-8189	65	21
		5310-01-270-8189	91	8
		5310-01-270-8189	93	20
		5310-01-270-8189	216	2
34623	5593416	5310-00-889-8782	140	2
		5310-00-889-8782	147	2
34623	5593421	5305-01-270-5435	335	10
		5305-01-270-5435	336	7
3M915	5593430	5305-01-270-5418	211	21
		5305-01-270-5418	212	19
34623	5593432	5305-01-283-7776	293	21
34623	5593479	6150-01-473-8640	56	14
34623	5593495	5340-01-474-8830	463	22
34623	5593535	5325-01-290-1695	91	12
34623	5593597	5310-01-272-5470	238	14
34623	5593599	5935-01-282-7833	52	6

SECTION IV TM9-2320-280-24P C01

CROSS REFERENCE INDEXES

PART NUMBER INDEX

CAGE	PART NUMBER	STOCK NUMBER	FIGURE NO	ITEM NO
		5935-01-282-7833	57	5
34623	5593615	2510-01-307-0152	331	7
34623	5593624	5340-01-248-7621	243	31
		5340-01-248-7621	244	15
34623	5593625	5340-01-260-9027	207	9
		5340-01-260-9027	318	8
		5340-01-260-9027	325	21
34623	5593634	5305-01-256-0405	244	20
34623	5593641	2590-01-248-2531	251	12
34623	5593642	2590-01-248-4874	251	11
34623	5593661	2520-01-249-1434	97	8
		2520-01-249-1434	101	10
34623	5593675	5340-01-259-0327	251	49
34623	5593691	5310-01-251-0748	58	8
		5310-01-251-0748	99	8
		5310-01-251-0748	208	15
		5310-01-251-0748	252	2
		5310-01-251-0748	334	4
		5310-01-251-0748	338	20
		5310-01-251-0748	339	15
		5310-01-251-0748	417	2
		5310-01-251-0748	418	2
		5310-01-251-0748	419	2
		5310-01-251-0748	420	7
		5310-01-251-0748	421	4
		5310-01-251-0748	422	6
		5310-01-251-0748	423	10
		5310-01-251-0748	425	2
		5310-01-251-0748	426	4
		5310-01-251-0748	427	3
34623	5593694	5365-01-210-4903	44	1
34623	5593807	5340-01-258-1547	279	36
34623	5593817	2520-01-174-5849	148	15
34623	5593847	5360-01-265-9116	156	4
34623	5593910	5330-01-185-8715	205	40
34623	5593912	5330-01-185-8714	206	11
34623	5593926	2530-01-235-8688	167	12
34623	5593927	5342-01-268-0921	275	23
34623	5593930	5330-01-195-9049	270	24
34623	5593947	4710-01-251-3075	227	22
34623	5593978	5360-01-260-0700	153	24
		5360-01-260-0700	154	9
34623	5593982		208	36
34623	5593985		249	16
34623	5593990	5342-01-260-7853	430	29
34623	5593993		142	7
			143	7
			150	7
			151	7

I-311

SECTION IV TM9-2320-280-24P C01

CROSS REFERENCE INDEXES

PART NUMBER INDEX

CAGE	PART NUMBER	STOCK NUMBER	FIGURE NO	ITEM NO
34623	5594002	4710-01-251-6103	227	4
34623	5594076	2590-01-262-7707	227	11
34623	5594077	2590-01-261-6851	227	24
34623	5594083		18	8
34623	5594101-C	5340-01-297-2800	21	5
34623	5594178	5306-01-323-5534	331	4
34623	5594249	5340-01-289-3243	429	2
34623	5594252	9515-01-249-1591	429	7
34623	5594278		429	32
34623	5594317		52	5
			57	4
34623	5594319	9905-01-294-3356	52	8
		9905-01-294-3356	57	7
34623	5594531-B	5365-01-414-2902	143	23
		5365-01-414-2902	151	23
34623	5594572	5340-01-249-1575	179	16
34623	5594676	2590-01-248-9531	286	40
34623	5594703	5340-01-256-4657	246	36
34623	5595214	4730-01-247-6763	150	5
34623	5595216	5340-01-256-9318	324	17
34623	5595217	5340-01-255-9919	316	20
		5340-01-255-9919	320	17
		5340-01-255-9919	320	32
		5340-01-255-9919	324	17
		5340-01-255-9919	324	29
34623	5595227		269	1
34623	5595228		269	1
34623	5595393	2510-01-236-6488	278	4
34623	5595454	5365-01-186-7764	193	1
34623	5595531	5330-01-258-4664	272	6
34623	5595551	2590-01-291-1033	417	13
34623	5595643	5330-01-255-4986	488	20
34623	5595662	5306-01-254-2358	148	3
34623	5595746	6150-01-260-8000	384	10
34623	5595964	5310-01-256-0416	328	16
34623	5595966	6670-01-261-6844	167	12
34623	5595968	6670-01-261-6845	167	12
34623	5595969	6670-01-261-6846	167	12
34623	5595970	6670-01-262-8646	167	12
34623	5595971	6670-01-261-8011	167	12
34623	5595972	6670-01-261-8012	167	12
34623	5595973	6670-01-261-8013	167	12
34623	5595974	6670-01-261-6847	167	12
34623	5595995		429	31
34623	5596009	2590-01-262-9515	253	14
34623	5596025	4210-01-204-0913	241	10
34623	5596028	5340-01-256-4517	251	27
		5340-01-256-4517	252	27
		5340-01-256-4517	253	22

SECTION IV TM9-2320-280-24P C01

CROSS REFERENCE INDEXES

PART NUMBER INDEX

CAGE	PART NUMBER	STOCK NUMBER	FIGURE NO	ITEM NO
		5340-01-256-4517	255	21
3M915	5596167	5310-01-298-7770	398	21
34623	5596171		225	16
34623	5596172		225	33
			226	18
34623	5596198	5340-01-440-1435	270	37
34623	5596249	5355-01-247-3593	331	28
34623	5596271	5306-01-185-7071	71	3
34623	5596306	5365-01-255-0965	153	26
		5365-01-255-0965	154	8
34623	5596320	7690-01-264-6536	358	4
34623	5596374	5340-01-260-9874	266	3
34623	5596459	5315-01-255-2763	243	34
34623	5596460	5315-01-255-2764	243	26
34623	5596566	5310-01-253-8440	410	34
		5310-01-253-8440	412	25
34623	5596580	5330-01-270-1353	274	7
		5330-01-270-1353	276	26
34623	5596642	5365-01-255-4990	277	5
34623	5596643	5365-01-255-4991	277	5
34623	5596644	5365-01-255-4633	277	5
34623	5596645	5365-01-255-4634	277	5
34623	5596736	5340-01-259-8579	266	17
34623	5596794	5340-01-290-2279	417	6
34623	5596795	5975-01-312-5469	417	1
34623	5596797	5340-01-290-8473	423	6
34623	5596803	5306-01-273-6333	154	2
34623	5596814	5365-01-272-7504	335	11
		5365-01-272-7504	336	8
34623	5596830	2590-01-257-8784	227	53
34623	5596843	5340-01-290-4887	425	4
		5340-01-290-4887	427	5
34623	5596946	5820-01-287-9612	428	KIT
34623	5597035	5365-01-280-5875	337	25
		5365-01-280-5875	338	25
34623	5597048	5310-01-260-5784	208	17
34623	5597050	4730-01-257-3348	148	23
34623	5597086	5340-01-290-8390	417	17
34623	5597090	5340-01-270-8483	213	12
		5340-01-270-8483	214	6
34623	5597096	5340-01-273-2342	213	10
		5340-01-273-2342	214	8
34623	5597128		52	2
34623	5597136-C	6150-01-289-4761	424	1
34623	5597137	5995-01-290-1719	426	1
		5995-01-290-1719	427	1
34623	5597138	5995-01-291-6377	425	1
		5995-01-291-6377	426	2
		5995-01-291-6377	427	2

SECTION IV TM9-2320-280-24P C01

CROSS REFERENCE INDEXES

PART NUMBER INDEX

CAGE	PART NUMBER	STOCK NUMBER	FIGURE NO	ITEM NO
34623	5597142	5342-01-277-9102	147	3
34623	5597150	5340-01-291-5186	425	7
		5340-01-291-5186 427		8
34623	5597151	5340-01-290-6343	426	3
34623	5597155	5340-01-290-4888	426	7
34623	5597172	5340-01-186-1817	158	6
34623	5597186	5340-01-264-6108	430	14
34623	5597197		59	1
34623	5597208		92	3
			95	14
34623	5597243	2590-01-276-7633	65	1
34623	5597248	5935-01-269-8992	52	19
		5935-01-269-8992 56		6
34623	5597249	5945-01-266-4593	52	18
		5945-01-266-4593 56		9
34623	5597252	5306-01-264-3531	174	1
34623	5597262		149	2
34623	5597267	5340-01-273-4668	57	9
34623	5597277	5325-01-477-5348	95	6
34623	5597286	5340-01-244-5765	15	15
		5340-01-244-5765 180		2
34623	5597334	2530-01-291-4597	154	3
34623	5597344	5342-01-257-7706	35	27
34623	5597347	5310-01-259-7554	328	11
34623	5597349	5306-01-276-1621	283	11
		5306-01-276-1621 438		12
34623	5597350	5340-01-290-6342	417	14
34623	5597351	5365-01-265-8937	283	20
34623	5597356	5340-01-291-4537	417	12
34623	5597357	5340-01-291-8915	418	3
		5340-01-291-8915 419		3
		5340-01-291-8915 420		8
		5340-01-291-8915 421		5
		5340-01-291-8915 422		7
		5340-01-291-8915	423	11
34623	5597360	5340-01-277-4427	155	4
34623	5597371	5330-01-289-9231	425	12
34623	5597372	5340-01-298-9691	417	14
34623	5597396	5340-01-272-3361	219	7
34623	5597397	5340-01-272-5517	219	7
34623	5597496	5995-01-291-6384	428	1
34623	5597571	5340-01-271-9573	220	4
34623	5597572	5340-01-271-9574	220	4
34623	5597579	2510-01-265-1138	211	1
34623	5597580	2510-01-265-1139	211	1
34623	5597593		52	20
34623	5597595	2510-01-290-5675	423	5
34623	5597600	6150-01-285-3901	96	1
34623	5597601		59	1

SECTION IV TM9-2320-280-24P C01

CROSS REFERENCE INDEXES

PART NUMBER INDEX

CAGE	PART NUMBER	STOCK NUMBER	FIGURE NO	ITEM NO
34623	5597602	5340-01-276-5850	290	1
34623	5597603	5342-01-274-6210	290	1
34623	5597687	5330-01-281-0911	293	29
34623	5597689	7690-01-291-2974	354	6
34623	5597710	5340-01-274-3498	216	1
34623	5597736		56	23
34623	5597737	2540-01-285-6087	56	1
34623	5597738	5930-01-266-4592	55	11
		5930-01-266-4592	56	17
34623	5597753	5330-01-272-9995	219	8
34623	5597754	5340-01-277-4426	219	9
34623	5597768	1640-01-201-9062	311	8
		1640-01-201-9062	312	4
34623	5597777	2520-01-358-3160	140	4
		2520-01-358-3160	147	4
34623	5597808	5340-01-272-9105	288	18
34623	5597820	5340-01-274-3287	260	12
34623	5597826	5365-01-272-4866	237	4
34623	5597827	5340-01-274-3184	237	14
34623	5597842	5340-01-272-8409	215	2
34623	5597870	2510-01-275-8026	211	9
34623	5597890	5342-01-274-9884	155	8
34623	5597895	2510-01-265-3157	293	17
34623	5597900		343	4
34623	5597916	5331-01-265-8809	492	11
9C234	5597929	4710-01-265-3232	492	2
34623	5597935	4710-01-265-3233	492	1
34623	5597961	5365-01-272-4867	237	10
34623	5597985	5330-01-298-8127	340	7
34623	5597988	5340-01-272-3389	58	6
34623	5598002	2990-01-269-6625	337	1
34623	5598011	5930-01-266-3919	58	4
		5930-01-266-3919	396	14
34623	5598032		222	22
34623	5598046	5340-01-272-5595	58	1
34623	5598052	5340-01-288-2127	292	6
34623	5598059	5340-01-272-9145	222	16
34623	5598061		222	20
34623	5598103	5340-01-287-5825	227	35
34623	5598104	2540-01-270-4353	298	6
34623	5598107	5365-01-273-0010	298	9
34623	5598113	5365-01-272-8376	298	5
34623	5598117	5355-01-272-5824	56	24
34623	5598121	5999-01-272-0018	338	12
		5999-01-272-0018	339	9
		5999-01-272-0018	410	12
34623	5598124	5340-01-280-5985	311	16
34623	5598130	5975-01-270-8546	59	2
34623	5598131	5340-01-272-7712	311	1

I-315

SECTION IV TM9-2320-280-24P C01

CROSS REFERENCE INDEXES

PART NUMBER INDEX

CAGE	PART NUMBER	STOCK NUMBER	FIGURE NO	ITEM NO
34623	5598145	5340-01-273-0136	260	5
34623	5598163	5340-01-290-2265	295	2
34623	5598208	5340-01-300-3715	141	13
		5340-01-300-3715	149	7
34623	5598209	2530-01-296-9333	138	8
		2530-01-296-9333	139	8
		2530-01-296-9333	146	8
34623	5598213	3030-01-256-3616	32	2
34623	5598218	5340-01-289-6216	53	1
34623	5598221	5340-01-289-4475	53	2
34623	5598247	3040-01-290-6758	337	16
34623	5598250-B	5340-01-290-2263	337	14
34623	5598253	5340-01-289-3231	337	10
34623	5598256-C	5340-01-290-2761	337	11
34623	5598266	4710-01-265-3228	335	2
		4710-01-265-3228	335	8
		4710-01-265-3228	336	5
34623	5598272	4820-00-287-5627	336	14
34623	5598278	5340-01-289-5055	308	12
34623	5598279	5340-01-288-6550	307	10
		5340-01-288-6550	308	11
34623	5598282		490	11
34623	5598285	5340-01-291-2301	307	5
34623	5598286	5340-01-289-4474	307	11
34623	5598293	5340-01-439-5636	497	3
34623	5598302	5340-01-273-2368	490	6
34623	5598307		497	6
34623	5598309	3040-01-290-6757	337	22
34623	5598310	3040-01-290-6760	337	23
34623	5598315		337	3
34623	5598627	5310-01-286-6077	44	9
34623	5598650	5985-01-269-8271	341	2
34623	5598767	2530-01-203-5746	142	1
		2530-01-203-5746	150	1
34623	5598864	2540-01-425-1617	293	7
34623	5598904	5340-01-289-3245	295	11
7X677	560613	5340-01-148-7529	11	31
86928	5610-140-62	5310-01-256-4491	243	12
		5310-01-256-4491	244	6
70040	5614017	2920-01-188-3863	20	1
99688	56480	5331-00-185-0075	467	11
19220	565698-4	5315-01-300-3012	280	24
		5315-01-300-3012	281	24
7X677	5667628	5310-00-166-8567	172	17
7X677	5688037	5360-01-149-6308	176	9
48018	5688049	5330-00-848-4439	176	11
52788	5692682		176	8
19207	5703776	2920-00-472-1723	KITS	68
19207	5703776-1	2920-01-222-2183	KITS	69

SECTION IV TM9-2320-280-24P C01

CROSS REFERENCE INDEXES

PART NUMBER INDEX

CAGE	PART NUMBER	STOCK NUMBER	FIGURE NO	ITEM NO
19207	5704052	2540-00-200-4249	KITS	70
34623	5705606	2530-01-191-8740	KITS	71
34623	5705611	2540-01-314-0142	397	KIT
19207	5705612	2530-01-321-4497	436	KITS
19207	5705613	2590-01-337-4071	433	KITS
19207	5705615	2540-01-314-9320	381	KIT
19207	5705618	2540-01-300-8745	KITS	72
19207	5705619	2540-01-300-8744	KITS	73
19207	5705623	2920-01-378-8035	434	KITS
19207	5705624	5975-01-417-3267	435	KITS
19207	5705628	2815-01-314-7940	1	1
19207	5705692	2540-01-199-2390	378	KITS
34623	5705698	2540-01-317-0728	372	KIT
19207	5705703	2540-01-320-8731	401	KIT
34623	5705704	2540-01-302-2595	432	KITS
34623	5714222	3040-01-452-9065	129	8
34623	5714229	2520-01-413-1902	129	18
34623	5714233	5310-01-416-3009	130	28
34623	5714499	3950-01-435-1632	458	28
34623	5714585	2920-01-461-1065	98	5
34623	5714588	2540-01-463-0200	98	2
34623	5714648	5962-01-430-0208	50	4
34623	5715160	2530-01-449-2495	143	22
		2530-01-449-2495	151	22
34623	5715668	5950-01-480-5377	119	11
34623	5715895	5962-01-480-5247	50	4
34623	5715925G1	2510-01-359-2076	198	26
34623	5715976	2590-01-466-7250	273	6
34623	5716028	5962-01-497-1611	50	4
34623	5716242	2540-01-476-6609	188	25
34623	5716414		KITS	75A
34623	5740008	5307-01-174-8640	128	68
		5307-01-174-8640	129	7
34623	5740013	2520-01-168-7878	128	60
34623	5740017	5330-01-168-3870	128	66
		5330-01-168-3870	129	3
34623	5740018	2520-01-189-9750	128	4
34623	5740090	2520-01-178-9768	128	72
34623	5740091	3040-01-176-9584	128	76
34623	5740092	3040-01-177-2427	128	46
34623	5740093	3040-01-177-2428	128	36
34623	5740099	5331-01-174-8618	128	77
34623	5740100	5365-01-174-8185	128	29
34623	5740104	3010-01-176-0557	128	39
34623	5740105	3020-01-148-9548	128	41
34623	5740108	3020-01-175-6477	128	32
34623	5740109	3120-01-155-8713	128	43
34623	5740114	2520-01-162-8985	132	3
		2520-01-162-8985	132	7

SECTION IV TM9-2320-280-24P C01

CROSS REFERENCE INDEXES

PART NUMBER INDEX

CAGE	PART NUMBER	STOCK NUMBER	FIGURE NO	ITEM NO
		2520-01-162-8985	132	14
		2520-01-162-8985	132	21
34623	5740115	5360-01-174-8613	132	5
34623	5740116		132	8
34623	5740117	3020-01-175-1962	128	75
34623	5740118	3120-01-174-8156	128	78
34623	5740119	2520-01-175-6401	128	13
34623	5740125	2520-01-174-5919	132	11
34623	5740126	5360-01-174-8614	132	10
34623	5740218	3020-01-174-7810	128	79
34623	5740219	5365-01-174-8154	128	28
34623	5740220	3040-01-176-2835	132	9
34623	5740221	2520-01-176-2840	132	12
34623	5740222	2520-01-174-9287	128	26
34623	5740223	2590-01-159-8757	132	2
		2590-01-159-8757	132	15
34623	5740224	5365-01-305-7067	128	22
		5365-01-305-7067	130	25
34623	5740225	2520-01-149-4993	128	61
		2520-01-149-4993	129	33
34623	5740226		132	4
34623	5740227	2520-01-175-6494	132	1
34623	5740228	5365-01-174-8184	KITS	74
34623	5740229	3110-01-175-3538	KITS	75
34623	5740230	5310-01-174-8158	128	11
		5310-01-174-8158	130	22
34623	5740231	3040-01-176-9615	128	53
34623	5740232	2520-01-173-9673	128	58
34623	5740233	2520-01-174-9291	128	5
34623	5740235	3120-01-174-8155	128	33
34623	5740236	3120-01-174-8631	128	34
34623	5740237	3120-01-174-8630	128	35
34623	5740253	5310-01-189-8468	34	4
34623	5740255		138	9
34623	5740256	3040-01-188-3607	145	9
34623	5740257		138	10
			145	10
34623	5740258		138	3
			145	3
34623	5740269	5306-01-497-4738	9	7
34623	5740278	5360-01-190-6214	34	11
34623	5740279	2520-01-174-9288	128	31
34623	5740321	2530-01-181-3907	100	14
34623	5740429		176	3
34623	5740430		176	2
34623	5740431		176	5
34623	5740432	4320-01-155-5145	176	6
34623	5740433	4820-01-083-7993	176	10
34623	5740436	5331-01-157-1884	176	14
34623	5740442		172	4
34623	5740444	5310-01-038-8500	172	1

SECTION IV TM9-2320-280-24P C01

CROSS REFERENCE INDEXES

PART NUMBER INDEX

CAGE	PART NUMBER	STOCK NUMBER	FIGURE NO	ITEM NO
34623	5740446		172	2
34623	5740448	2530-01-097-7659	172	3
34623	5740452	5330-01-096-9650	172	19
34623	5740453	3110-00-403-1488	172	20
34623	5740454		172	5
34623	5740456	5340-00-960-9340	172	22
34623	5740458	2530-01-034-1715	172	21
34623	5740462	3040-01-123-4942	172	8
34623	5740464		176	4
34623	5740465	2530-01-033-1855	172	9
34623	5740476	3110-01-040-6541	172	7
34623	5740495	4820-01-306-6838	172	14
34623	5740497	4820-01-303-3982	172	15
34623	5740498		172	16
34623	5740499	2530-00-960-9363	172	13
34623	5740503	2520-01-206-3875	128	69
34623	5740504	5360-00-735-1126	128	57
34623	5740505	5306-01-174-8492	128	56
34623	5740512	3020-01-177-5441	128	24
		3020-01-177-5441	128	48
34623	5740514	2520-01-206-3877	128	9
34623	5740515		128	62
34623	5740517		128	30
34623	5740519		128	3
34623	5740520		128	6
34623	5740521		128	6
34623	5740522		128	6
34623	5740523		128	6
34623	5740524		128	6
34623	5740525		128	6
34623	5740526		128	49
34623	5740549	5365-01-383-1925	138	5
		5365-01-383-1925	145	5
34623	5740550		138	11
			145	11
34623	5740551	3040-01-189-1637	138	7
34623	5740552	3040-01-190-3574	138	9
34623	5740702		10	9
34623	5740908		156	8
34623	5740909	5360-01-185-8809	156	7
34623	5740910		156	5
34623	5740911		156	6
34623	5740912	5305-01-185-4628	156	2
34623	5740913	5315-01-290-3327	156	10
34623	5740916	5310-01-185-8691	156	12
34623	5741060	2520-01-204-0325	132	6
34623	5741062	5331-01-209-7726	131	3
34623	5741063	2520-01-203-9871	131	4
34623	5741074	5305-01-209-7667	128	12

I-319

SECTION IV TM9-2320-280-24P C01

CROSS REFERENCE INDEXES

PART NUMBER INDEX

CAGE	PART NUMBER	STOCK NUMBER	FIGURE NO	ITEM NO
34623	5741075	5305-01-209-7668	128	17
		5305-01-209-7668	130	26
34623	5741076	3110-01-155-2600	128	67
34623	5741077	5365-01-209-7713	128	71
34623	5741083	9905-01-158-7981	357	4
34623	5741084	5305-01-211-7415	357	5
34623	5741098	5325-01-212-2403	138	6
		5325-01-212-2403	145	6
34623	5741120	5342-01-211-5327	45	60
34623	5741141	5330-01-211-1343	156	9
34623	5741145	5310-01-038-8501	172	10
34623	5741221	5330-01-291-4615	BULK	46
34623	5741427	5310-01-272-3311	494	9
		5310-01-272-3311	496	6
34623	5741471	5330-01-270-8317	343	7
34623	5741475	5310-01-270-5394	343	22
34623	5741525	2510-01-273-0572	343	8
34623	5741547		138	2
			145	2
34623	5741548	5340-01-311-3033	138	4
		5340-01-311-3033	145	4
34623	5741611	4710-01-185-6179	10	3
34623	5741613	5340-01-384-9987	4	18
34623	5741746		176	7
34623	5741748	2530-01-265-3153	157	4
34623	5741750	2530-01-265-3148	157	4
34623	5741892	5330-00-843-9235	172	12
34623	5741908	2540-01-289-8330	KITS	79
34623	5741994	5360-01-265-8799	157	3
34623	5741995	5360-01-265-8798	157	3
34623	5742041	2540-01-288-8567	KITS	80
34623	5743156	5355-01-460-4586	98	3
34623	5743159	2920-01-476-2716	98	10
34623	5743744	2520-01-413-1900	129	9
19207	57K0106	2590-01-328-2904	437	KITS
19207	57K0107	2510-01-364-3120	KITS	46
19207	57K0109	2540-01-288-5240	KITS	47
19207	57K0112	2540-01-316-0892	452	4
19207	57K0123	2540-01-316-6624	321	KIT
19207	57K0137	2540-01-394-4454	375	KIT
19207	57K0158	2540-01-329-8074	323	KIT
19207	57K0216	2540-01-338-8081	402	KIT
19207	57K0217	2510-01-338-8087	209	KIT
19207	57K0218	2590-01-342-4918	433	KITS
34623	57K0220	2540-01-395-8771	402	KIT
19207	57K0222	5340-01-476-7263	433	KIT
19207	57K0223	5340-01-412-0141	KITS	48
19207	57K0227	2510-01-274-4234	455	KIT
19207	57K0237	2640-01-371-8332	KITS	49

I-320

SECTION IV TM9-2320-280-24P C01

CROSS REFERENCE INDEXES

PART NUMBER INDEX

CAGE	PART NUMBER	STOCK NUMBER	FIGURE NO	ITEM NO
19207	57K0239	5330-01-361-8013	KITS	50
19207	57K0240	5330-01-361-8014	KITS	51
19207	57K0241	5330-01-361-8015	KITS	52
19207	57K0242	2530-01-361-8012	KITS	53
19207	57K0262	2530-01-459-0367	KITS	54
34623	57K0264	2530-01-407-3977	KITS	55
19207	57K0274	2530-01-394-3748	KITS	56
19207	57K0285	2590-01-418-0310	457	KIT
19207	57K0297	2640-01-419-6205	KITS	57
19207	57K1640	5340-01-410-7036	453	KITS
19207	57K1641		440	KITS
34623	57K1641		452	1
19207	57K3196	2540-01-410-7035	454	KIT
19207	57K3197	2540-01-410-7034	454	KITS
34623	57K3205	2530-01-413-8886	KITS	58
19207	57K3206	2510-01-431-1339	460	KIT
19207	57K3209	2540-01-431-9182	459	KIT
19207	57K3213	2540-01-424-7363	461	KIT
34623	57K3217	2590-01-418-2135	457	KIT
19207	57K3218	5180-01-408-7050	511	KIT
19207	57K3219	5180-01-410-8467	511	KIT
19207	57K3222	2540-01-431-1338	KITS	59
19207	57K3223	2815-01-410-9710	1	1
19207	57K3226	2815-01-406-6675	1	1
19207	57K3233	2590-01-420-4179	485	KIT
19207	57K3238	5980-01-438-8939	KITS	60
19207	57K3488	2590-01-429-3008	486	KIT
19207	57K3489	5330-01-459-6477	KITS	61
34623	57K3492	2540-01-467-8313	464	KIT
34623	57K3495	2540-01-467-8341	464	KIT
19207	57K3498	2520-01-434-0822	126	2
		2520-01-434-0822	127	2
19207	57K3502	2520-01-430-5294	KITS	62
19207	57K3503	4720-01-471-5063	487D	KIT
19207	57K3505	2520-01-469-9893	126	1
19207	57K3506	2520-01-473-5029	126	1
19207	57K3507	2520-01-452-8365	126	1
19207	57K3512	2530-01-455-9330	KITS	63
19207	57K3515	2530-01-457-1337	KITS	64
19207	57K3519	2920-01-455-1626	474	KIT
19207	57K3523	2520-01-439-6830	KITS	65
19207	57K3524	2815-01-439-6665	1	1
19207	57K3525	2815-01-439-6664	1	1
19207	57K3527	2530-01-443-3405	KITS	66
19207	57K3532	5999-01-474-3213	484	KIT
19207	57K3534	4130-01-452-8773	466	KIT
19207	57K3536	2815-01-453-7403	1	1
19207	57K3537	2815-01-453-7404	1	1
19207	57K3539	2520-01-461-7072	KITS	67

SECTION IV TM9-2320-280-24P C01

CROSS REFERENCE INDEXES

PART NUMBER INDEX

CAGE PART NUMBER STOCK NUMBER FIGURE NO ITEM NO

CAGE	PART NUMBER	STOCK NUMBER	FIGURE NO	ITEM NO
19207	57K3541	2590-01-456-7879	471	KIT
19207	57K3542	5340-01-455-8700	439	KIT
19207	57K3543	2815-01-461-7078	1	1
19207	57K3547	2815-01-472-6312	1	1
19207	57K3549	2520-01-461-7074	KITS	67A
19207	57K3553	2990-01-496-3766	398	KIT
19207	57K3555	2815-01-461-7078	1	1
19207	57K3556	2815-01-439-6664	1	1
19207	57K3558	2520-01-473-7410	KITS	67B
19207	57K3566	2815-01-461-7078	1	1
19207	57K3567	2815-01-439-6664	1	1
19207	57K3569	2520-01-489-0849	KITS	67C
19207	57K4387	2510-01-491-6919	462	KIT
99688	585810	4130-01-171-5997	494	13
99688	58583	4820-01-161-6435	496	10
99688	586230	4520-01-192-6073	343	9
99688	58668	5355-01-167-4114	54	7
99688	58701	9320-01-085-2889	494	21
34623	5903101	5935-01-216-8763	76	9
60602	59267-1	5355-01-447-9655	98	4
34623	5933947	5340-01-272-5611	213	38
34623	5933974	5310-01-272-4823	292	9
34623	5933977	5340-01-270-8432	259	2
34623	5934007	5310-00-177-7529	256	4
		5310-00-177-7529	257	1
		5310-00-177-7529	260	4
34623	5934065	4010-01-266-3842	292	18
34623	5934085	5330-01-274-6106	296	5
34623	5934184	5330-01-265-8805	214	25
34623	5934219	4130-01-278-9837	495	10
34623	5934310		341	6
34623	5934761	5310-01-480-4161	33	8
34623	5935091	5340-01-293-0125	35	24
		5340-01-293-0125	42	1
34623	5935092	5340-01-292-8404	37	12
		5340-01-292-8404	38	3
		5340-01-292-8404	42	1
		5340-01-292-8404	475	12
34623	5935093	5340-01-292-5402	42	9
34623	5935102		221	3
34623	5935356	2930-01-317-5358	28	5
34623	5935390	3020-01-364-3398	148	27
34623	5935391		148	28
34623	5935392		148	29
34623	5935394	2520-01-357-5043	136	16
34623	5935425	5340-01-388-9512	KITS	81
34623	5936107	2520-01-357-5056	148	25
34623	5937339	2520-01-357-5044	137	18
34623	5937387	2520-01-356-9197	127	3

I-322

SECTION IV TM9-2320-280-24P C01

CROSS REFERENCE INDEXES

PART NUMBER INDEX

CAGE	PART NUMBER	STOCK NUMBER	FIGURE NO	ITEM NO
34623	5938607	3130-01-360-1051	129	21
		3130-01-360-1051	130	36
34623	5938619	3120-01-360-1202	129	12
34623	5938621	3130-01-360-1050	129	15
34623	5938630	5365-01-358-4642	132	24
34623	5938633	5360-01-358-6564	132	26
34623	5938653	3120-01-360-5960	130	46
34623	5938654	3110-01-359-8858	130	45
34623	5938663	3020-01-359-2529	130	43
34623	5938665	5315-01-464-0812	130	24
34623	5938666	4730-01-359-2217	130	14
34623	5938685	2520-01-174-9554	129	32
34623	5938705	6220-01-361-5084	67	8
34623	5939396	2530-01-357-9708	164	1
34623	5939517	5330-01-487-7129	148	16
73331	5939841	5330-00-353-0959	63	10
34623	5982526	5310-01-198-1722	53	5
		5310-01-198-1722	277	20
34623	5982528	5310-01-198-1723	446	21
		5310-01-198-1723	448	6
		5310-01-198-1723	450	6
34623	5993089	5340-01-276-3615	220	7
		5340-01-276-3615	237	1
34623	5995301	5340-01-158-0098	79	4
34623	5995488	5365-01-274-4675	95	38
35510	59982S	5310-00-429-3156	39	44
81343	6-2 010102B	4730-00-439-6021	335	1
78189	6-321-4	5305-00-638-0714	35	17
		5305-00-638-0714	43	55
81343	6-4 010103B	4730-00-270-4606	335	16
		4730-00-270-4606	336	13
34623	6000300	5340-01-408-5851	16	30
34623	6000329	5330-01-420-0705	16	40
34623	6000331	5330-01-410-9840	16	39
34623	6000333	5330-01-411-0367	16	31
34623	6000334		16	34
34623	6001552	2540-01-473-9045	470	9
34623	6001553	2540-01-473-9040	470	9
34623	6001696	2520-01-409-1602	134	3
34623	6001784	5365-01-456-2032	143	30
		5365-01-456-2032	151	30
34623	6002218	2530-01-420-7904	157	23
		2530-01-420-7904	158	13
34623	6002219	5340-01-432-8680	140	9
34623	6002224	5340-01-432-4876	182	12
34623	6002226	5340-01-432-8678	183	20
34623	6002227	5340-01-437-8748	183	20
34623	6002343	5340-01-438-4497	189	10
34623	6002344	5340-01-434-2832	189	7

SECTION IV TM9-2320-280-24P C01

CROSS REFERENCE INDEXES

PART NUMBER INDEX

CAGE PART NUMBER STOCK NUMBER FIGURE NO ITEM NO

CAGE	PART NUMBER	STOCK NUMBER	FIGURE NO	ITEM NO		
34623	6002346	5340-01-433-2721			190	5
89944	60025077004	4720-00-006-0048			BULK	63
34623	6002607	2520-01-413-0080			136	16
34623	6002629	2520-01-423-5120			136	6
34623	6002771	4710-01-409-2504			125	4
34623	6002779	4710-01-409-2508			125	3
34623	6002942	5307-01-411-2336			103	14
34623	6003666	2520-01-413-2724			137	18
34623	6003868	2530-01-420-3837			170	6
34623	6004838	2530-01-420-5180			170	7
34623	6004839	2530-01-420-3839			170	2
34623	6004872	2510-01-456-2277			194	3
27647	6005	5342-01-205-2519			314	26
34623	6005120	2530-01-433-8003			143	35
		2530-01-433-8003	151	19		
34623	6005121	2530-01-433-8702			143	19
		2530-01-433-8702	151	35		
34623	6005193	5330-01-456-8823			143	20
		5330-01-456-8823	151	20		
34623	6005274-B1	2510-01-423-2877			195	3
34623	6005790-A	2520-01-413-2595			98	1
001Y9	600634	3020-01-491-2671			33B	9
001Y9	600635	3020-01-491-0776			33A	12
001Y9	600636	3020-01-491-2659			33B	6
001Y9	600637	2920-01-491-2011			33A	18
001Y9	600638	5340-01-491-2205			33A	13
		5340-01-491-2205	33B	5		
001Y9	600639	5310-01-490-7461			33B	10
001Y9	600640	5310-01-491-2148			33B	11
34623	6006645	3040-01-422-9390			144	17
		3040-01-422-9390	152	17		
34623	6007749	2520-01-459-0050			137	18
34623	6007755	2530-01-413-1366			144	8
		2530-01-413-1366	152	8		
34623	6008558	2540-01-466-3849			463	4
19204	6008993	5305-00-600-8993			324	38
34623	6009349				143	25
					151	25
34623	6009352	2530-01-413-3653			143	1
		2530-01-413-3653	151	1		
34623	6009452	3020-01-476-2701			143	38
		3020-01-476-2701	151	38		
34623	6009453	3020-01-432-2553			143	10
		3020-01-432-2553	151	10		
34623	6009454	5310-01-473-3373			143	8
		5310-01-473-3373	151	8		
34623	6009472	5330-01-174-8145			148	14
34623	6010942	2520-01-174-9580			148	12
34623	6013199	2530-01-420-8634			144	20
					152	20

I-324

SECTION IV TM9-2320-280-24P C01

CROSS REFERENCE INDEXES

PART NUMBER INDEX

CAGE PART NUMBER STOCK NUMBER FIGURE NO ITEM NO

CAGE	PART NUMBER	STOCK NUMBER	FIGURE NO		ITEM NO
00779	60187-2	5940-01-474-5066	92		8
93334	6023054	5310-00-471-9243	131		2
42280	61-0542-1	5310-00-809-4058	80		16
81834	61-2028-01	5330-01-218-1196	64		7
70082	6185	5307-01-316-2986	329		14
99688	62229	5310-01-161-7308	54		13
7X677	6259194	5325-01-205-5966	129		11
81646	62606	4730-00-024-3971	13		10
		4730-00-024-3971	14	16	
		4730-00-024-3971	19	21	
70412	62886	5305-01-229-7855	327		3
77060	6288634-L	5940-01-269-8396	95		19
7Z588	63060	4730-01-273-7660	343		11
63829	6320-9620	4320-01-290-0748	491		26
35510	6385	6150-01-269-1854	39		65
70082	6388	2540-01-314-1188	329		2
70082	6389	5340-01-315-3611	329		10
70082	6390	2540-01-314-2101	329		5
94222	64-10-201-90	5340-01-277-4613	215		7
19220	6411B	5340-00-489-5684	269		8
73342	6437741		104		6
70040	6472357	2910-01-168-7905	9		1
81300	6490610019	4720-01-186-2358	BULK		23
76760	6500318	5306-01-359-4529	129		1
99688	65139	5930-01-165-1657	55		18
99688	65140	5930-01-163-6256	55		7
99688	65158	5905-01-154-2354	498		6
99688	65171	5930-01-190-1231	54		9
12436	667201-011	5905-00-258-6788	59		12
24975	66X-3331	5310-01-250-7844	45		29
7Z588	6706	4730-00-024-3971	366		21
81646	6772-1	4730-00-359-9487	12		4
27647	6779	5310-01-213-4174	314		34
82942	68-412121-1	5330-01-281-6523	BULK		52
73342	6884730	5325-01-171-3392	109		5
		5325-01-171-3392	111	9	
80604	6888	5120-01-437-0480	511		8
60285	6893-2	5306-00-068-0513	68		6
		5306-00-068-0513	379		24
		5306-00-068-0513	379		29
		5306-00-068-0513	388	15	
70485	698W	5340-00-565-5378	273		12
17446	6LC-2R06	5320-00-864-6203	199		48
19207	6TLMFTYI	6140-01-446-9498	69		23
19728	6X-4336		45		23
12361	7-099-000154	5305-00-993-1848	251		31
19220	7-1910	2540-01-270-2250	213		5
		2540-01-270-2250	214	33	
19220	7-2414	5340-01-270-8435	213		28

I-325

SECTION IV TM9-2320-280-24P C01

CROSS REFERENCE INDEXES

PART NUMBER INDEX

CAGE	PART NUMBER	STOCK NUMBER	FIGURE NO	ITEM NO
		5340-01-270-8435	214	18
76445	70112	2510-01-189-1832	193	4
76445	70113	2510-01-190-3862	193	4
76445	70114	2510-01-251-5316	193	4
78385	702903	5330-00-089-0978	348	20
78385	703546	5340-01-236-5101	345	32
		5340-01-236-5101	348	23
78385	703547	5306-01-351-7742	345	23
		5306-01-351-7742	348	22
78385	703611	5305-01-288-1130	349	5
78385	704181	2540-01-168-9482	350	9
		2540-01-168-9482	351	9
78385	704190	5310-01-137-6801	350	8
		5310-01-137-6801	351	8
78385	704191	5310-01-164-0745	351	5
78385	704192		351	6
78385	704206	5305-01-097-8178	350	10
78385	704225	2540-00-216-5722	348	18
78385	704363	4710-01-163-2805	345	33
		4710-01-163-2805	348	28
19207	7044253	5360-00-704-4253	191	22
78385	704501	9905-01-017-4748	348	10
78385	704678	5310-01-164-1023	350	7
		5310-01-164-1023	351	7
78385	705032	5305-01-066-3431	350	12
78385	705068	5310-01-057-3098	351	3
78385	705117		351	10
78385	705211	5305-01-288-1129	349	7
78385	705237	5310-01-063-8522	351	2
78385	705267	5975-01-288-6594	345	11
78385	705587		350	1
19207	7056640	5970-01-174-9449	73	19
		5970-01-174-9449	73	23
		5970-01-174-9449	480	27
78385	705930	5310-01-287-8726	345	13
78385	705944		350	3
			351	4
78385	706015	9905-01-283-7937	345	17
78385	706025	7690-01-283-7938	345	4
78385	706041	2540-01-290-0715	345	36
78385	706045	5905-01-287-4256	349	11
78385	706049		350	5
78385	706050GV	5310-01-287-6557	349	12
78385	706053	5905-01-287-4255	349	3
78385	706062		350	4
78385	706063		350	6
78385	706070	5940-01-292-6907	345	6
78385	706074	5305-01-290-6290	345	5
78385	706131	5310-01-287-6543	345	27

SECTION IV TM9-2320-280-24P C01

CROSS REFERENCE INDEXES

PART NUMBER INDEX

CAGE	PART NUMBER	STOCK NUMBER	FIGURE NO	ITEM NO
19207	7063812	5310-00-126-3842	169	7
19207	7064829	5940-00-520-2447	313	18
77335	71-510	5120-00-618-6902	504	14
13008	7168	5340-01-233-1461	240	14
78385	718768-23	5331-00-089-0998	345	21
		5331-00-089-0998	348	21
73680	71X6284	5330-00-497-4633	128	8
70082	7243	5365-01-315-3595	329	4
19207	7261674	5935-00-338-2822	282	13
35510	7266	5310-01-272-4831	39	48
19207	7320533	5940-00-549-6583	69	18
19207	7331407	5342-00-678-1753	160	18
		5342-00-678-1753	162	18
		5342-00-678-1753	487A	8
19207	7346712	5306-00-182-2027	373	11
19207	7355392	5340-00-408-2432	191	24
19207	7355393	3040-00-127-5322	191	23
19207	7355396	5310-00-735-5396	191	26
19207	7355520	5940-00-735-5520	69	36
19207	7358626	5330-00-297-6329	47	14
19207	7370134	5315-00-737-0134	403	9
19207	7385182	5340-00-738-5182	457	8
19207	7397780	5340-00-438-1836	477	9
19207	7397934	5340-01-408-7091	261	13
35510	74107	5310-01-270-8415	39	22
7X677	7471009	3110-01-480-5240	107	2
30327	75167	4730-00-541-0793	31	1
19207	7524314	2540-00-525-7067	191	15
19207	7527645	5975-00-697-7769	85	12
		5975-00-697-7769	86	17
19207	7529300	5365-01-129-0399	271	13
19207	7538146	5305-00-353-0969	62	18
19207	7539072	5330-00-753-9072	67	5
35510	75451	5310-00-429-3453	39	42
96508	760-2601	5340-01-265-8895	262	6
19207	7700242	5331-00-770-0242	387	15
19207	7700243	5310-00-770-0243	387	16
19207	7716634	5975-00-771-6634	85	17
		5975-00-771-6634	86	10
19207	7720442	5310-00-772-0442	36	6
		5310-00-772-0442	38	6
		5310-00-772-0442	43B	3
19207	7721654	5340-00-438-1833	313	4
		5340-00-438-1833	443	4
19207	7722322	5365-00-772-2322	80	9
		5365-00-772-2322	84	27
		5365-00-772-2322	85	9
		5365-00-772-2322	86	5
19207	7722333	5365-00-090-5426	83	36

I-327

SECTION IV TM9-2320-280-24P C01

CROSS REFERENCE INDEXES

PART NUMBER INDEX

CAGE	PART NUMBER	STOCK NUMBER	FIGURE NO	ITEM NO
		5365-00-090-5426	85	18
		5365-00-090-5426	86	11
19207	7722353	5935-00-772-2353	84	26
		5935-00-772-2353	86	23
19204	7723307	5935-00-772-3307	85	14
		5935-00-772-3307	86	19
19207	7723308	5935-00-333-9414	80	10
		5935-00-333-9414	84	28
		5935-00-333-9414	85	10
		5935-00-333-9414	86	6
19207	7723309	5310-00-393-6685	83	37
		5310-00-393-6685	85	19
		5310-00-393-6685	86	12
19207	7725882	5310-00-167-0822	194	6
16528	7728777	5940-00-983-4067	79	7
		5940-00-983-4067	485	1
19207	7728780	5940-01-035-4212	79	9
		5940-01-035-4212	485	2
19207	7731428	5935-00-773-1428	83	28
		5935-00-773-1428	88	45
19207	7748814	2910-00-073-0165	334	1
03007	7792 LH-02	2540-01-271-5072	237	8
03007	7792RH02	5340-01-272-9103	237	8
35510	77960	3120-01-270-2645	39	18
52788	7817725	2530-01-033-4237	172	11
52788	7818445		172	4
7X677	7830239	5305-01-162-9713	176	12
7X677	7830913	4320-01-155-5153	176	13
7X677	7834183	6680-01-152-2845	176	1
7X677	7834284	4820-01-191-4262	172	6
7X677	7846626	2530-01-123-8787	KITS	82
52788	7848522	5330-01-044-0703	KITS	83
7X677	7849306	2815-01-449-5585	4	6
19207	7951057	2540-00-177-8108	393	14
19207	7951748	5935-00-790-4614	56	11
19207	7973326	9905-00-901-2942	357	7
19207	7982843	5310-01-255-2704	164	2
19207	7982907	5935-00-214-0904	40	26
		5935-00-214-0904	86	24
		5935-00-214-0904	366	34
		5935-00-214-0904	386	1
		5935-00-214-0904	472	8
		5935-00-214-0904	475	1
		5935-00-214-0904	479	6
		5935-00-214-0904	483	7
11083	7N7782	5935-01-150-8322	86	26
11083	7N9738	5935-01-149-5165	76	2
		5935-01-149-5165	86	27
81343	8 100110B	4730-01-048-7874	17	6

SECTION IV TM9-2320-280-24P C01

CROSS REFERENCE INDEXES

PART NUMBER INDEX

CAGE	PART NUMBER	STOCK NUMBER	FIGURE NO	ITEM NO
81343	8 100115B	4730-01-049-1559	17	7
83330	8-1930-0112-912	6210-00-578-7970	311	12
		6210-00-578-7970	312	8
62226	8.6046	2540-01-381-8392	231	1
62226	8.6272-01	2540-01-436-4175	231	6
		2540-01-436-4175	454	1
62226	8.6272-02	2540-01-476-9353	454	1
18876	8034668-159	5970-00-057-9987	44	25
19207	809223	2590-00-454-3620	95	37
		2590-00-454-3620	303	9
		2590-00-454-3620	304	5
		2590-00-454-3620	412	13
60703	81087-5	4730-01-074-0060	327	24
19207	8329899	2540-00-040-2129	271	14
13445	83320	5310-01-186-6930	69	9
19207	8338561	5935-00-833-8561	86	25
		5935-00-833-8561	366	33
		5935-00-833-8561	386	12
		5935-00-833-8561	472	7
		5935-00-833-8561	475	2
		5935-00-833-8561	479	5
		5935-00-833-8561	483	8
19207	8338562	5970-00-833-8562	75	13
		5970-00-833-8562	76	11
19207	8338564	5940-00-399-6676	75	12
19207	8338566	5935-00-572-9180	483	6
19207	8359764	5365-01-318-2066	388	4
63829	8363-6330	5340-01-288-1242	491	24
19207	8380196	2540-01-023-5116	191	12
63829	8385-0520	5331-01-288-1062	467	4
		5331-01-288-1062	491	10
63829	8385-0880	5365-01-287-8701	491	17
63829	8385-6410	5365-01-287-8645	491	9
63829	8385-9611	5330-01-288-1307	KITS	85
63829	8476-9631	2815-01-290-3705	491	23
18876	8486066	4730-00-877-6298	178	9
		4730-00-877-6298	477	4
13445	85031	5310-01-186-7066	69	7
		5310-01-186-7066	70	4
		5310-01-186-7066	473	8
63829	8527-6041	5950-01-289-6007	491	8
27647	8548	5305-01-210-2309	313	19
		5305-01-210-2309	458	42
13445	85764	5310-01-185-7188	69	6
		5310-01-185-7188	70	5
7X677	8623039	5310-01-152-4229	121	12
7X677	8623078	5315-00-038-3059	121	6
7X677	8623090	5360-01-209-7715	121	13
11862	8623101		109	19

I-329

SECTION IV TM9-2320-280-24P C01

CROSS REFERENCE INDEXES

PART NUMBER INDEX

CAGE PART NUMBER STOCK NUMBER FIGURE NO ITEM NO

CAGE	PART NUMBER	STOCK NUMBER	FIGURE NO	ITEM NO
11862	8623102		109	8
63005	8623102	2840-00-001-4903	111	5
11862	8623104	3110-00-005-0873	109	6
		3110-00-005-0873	111	8
7X677	8623105	5325-01-171-3392	110	10
		5325-01-171-3392	112	9
7X677	8623112	5325-00-007-3052	109	1
		5325-00-007-3052	110	1
		5325-00-007-3052	111	11
		5325-00-007-3052	112	10
7X677	8623116	3110-01-174-8136	109	16
7X677	8623120	5325-01-150-6092	109	13
7X677	8623121		109	18
11862	8623143		113	14
7X677	8623145	5360-01-153-0933	113	8
7X677	8623149	5325-01-150-7830	113	1
		5325-01-150-7830	114	10
7X677	8623151	2520-01-150-7609	113	3
7X677	8623152	2520-01-175-6492	113	2
7X677	8623153	5325-01-158-2182	113	6
		5325-01-158-2182	114	4
7X677	8623157	4710-01-151-3663	118	6
7X677	8623174	5330-01-150-5928	116	20
11862	8623183		106	6
7X677	8623183		107	4
7X677	8623194		106	24
			107	12
11862	8623196		106	26
7X677	8623196		107	10
7X677	8623202	3020-01-149-7938	106	23
11862	8623204		106	20
7X677	8623204		106	4
			107	6
			107	18
11862	8623206		106	22
7X677	8623206		107	20
7X677	8623211	3120-01-155-4468	113	9
		3120-01-155-4468	114	18
7X677	8623262	2520-01-127-3969	116	17
7X677	8623263	5330-00-001-1984	116	16
7X677	8623292	5340-01-159-1788	121	7
7X677	8623300	3120-01-155-4462	121	17
7X677	8623301	3120-01-155-4463	121	17
7X677	8623302	3120-01-182-8417	121	17
7X677	8623303	3120-01-155-4464	121	17
7X677	8623304	3120-01-155-4465	121	17
7X677	8623305	3120-01-155-4466	121	17
7X677	8623306	3120-01-155-4467	121	17
7X677	8623368	5310-01-149-7793	121	11

SECTION IV TM9-2320-280-24P C01

CROSS REFERENCE INDEXES

PART NUMBER INDEX

CAGE	PART NUMBER	STOCK NUMBER	FIGURE NO	ITEM NO
7X677	8623430	5330-01-150-6239	116	10
		5330-01-150-6239 117	15	
7X677	8623437	5325-01-160-4693	106	32
		5325-01-160-4693 113	5	
		5325-01-160-4693 114	5	
7X677	8623489	5360-01-150-6086	116	6
		5360-01-150-6086 117	5	
7X677	8623561	5330-01-152-5942	118	13
7X677	8623592	4820-01-150-4964	118	11
7X677	8623664	5340-01-154-6559	116	22
		5340-01-154-6559 117	8	
7X677	8623666	5360-01-150-7829	116	23
		5360-01-150-7829 117	9	
7X677	8623671	5330-01-174-8090	116	12
7X677	8623741	5330-01-151-8364	118	3
7X677	8623744	5310-01-150-5919	116	24
		5310-01-150-5919 117	10	
7X677	8623849	2520-01-150-3931	112	15
11862	8623917	5330-01-138-5190	KITS	86
7X677	8623920	3120-01-167-4172	KITS	87
7X677	8623921	3110-01-167-2443	KITS	88
7X677	8623922	3110-01-169-0734	KITS	89
7X677	8623941	3120-01-166-3677	106	29
7X677	8623944	3120-01-156-5189	106	8
		3120-01-156-5189 121	5	
7X677	8623978	5330-01-152-5941	121	19
7X677	8624101	5360-01-150-6091	109	7
		5360-01-150-6091 111	7	
7X677	8624136	5340-01-149-7811	116	21
7X677	8624138	3040-01-151-5663	116	9
		3040-01-151-5663 117	16	
7X677	8624139	5315-01-152-9038	117	7
7X677	8624140	5315-01-152-9039	116	7
		5315-01-152-9039 117	7	
7X677	8624141	5315-01-152-9040	116	7
		5315-01-152-9040 117	7	
7X677	8624196	3120-01-156-8763	111	15
		3120-01-156-8763 112	14	
7X677	8624209	5340-01-408-8522	113	11
		5340-01-408-8522 114	16	
		5340-01-408-8522 115	25	
7X677	8624256	5360-01-150-6087	116	8
7X677	8624391	5360-01-476-6291	117	6
7X677	8624781	3120-00-255-5697	111	13
		3120-00-255-5697 112	12	
11862	8624784		106	5
			106	21
7X677	8624784		107	5
			107	19

I-331

SECTION IV TM9-2320-280-24P C01

CROSS REFERENCE INDEXES

PART NUMBER INDEX

CAGE	PART NUMBER	STOCK NUMBER	FIGURE NO		ITEM NO
7X677	8624908	2520-01-169-7674		111	2
7X677	8625221	5325-01-174-8626		106	19
		5325-01-174-8626	107		21
7X677	8625401	3120-01-154-8516		106	30
		3120-01-154-8516	107		25
7X677	8625402	3120-01-159-5773		106	30
		3120-01-159-5773	107		25
7X677	8625403	3120-01-154-4369		106	30
		3120-01-154-4369	107		25
7X677	8625404	3120-01-161-4033		106	30
		3120-01-161-4033	107		25
7X677	8625405	3120-01-162-5787		106	30
		3120-01-162-5787	107		25
7X677	8625406	3120-01-154-8517		106	30
		3120-01-154-8517	107		25
7X677	8625717	3020-01-149-5049		111	14
7X677	8625736	3040-01-167-2836		106	31
7X677	8625773	2520-01-165-9563		100	13
		2520-01-165-9563	118		8
7X677	8625955	5340-01-149-1867		121	4
7X677	8625990			KITS	90
7X677	8626112	5325-01-168-5729		106	12
7X677	8626173	5325-01-150-4982		106	15
		5325-01-150-4982	107		23
11862	8626252			100	6
7X677	8626281	5310-01-150-5921		121	20
7X677	8626356	5330-01-165-4333		113	12
		5330-01-165-4333	121		18
7X677	8626372	3120-01-154-7174		106	14
		3120-01-154-7174		106	28
		3120-01-154-7174		107	15
		3120-01-154-7174	107		24
7X677	8626423	5330-01-163-2614		116	1
		5330-01-163-2614	117		3
11862	8626806		106	9	7X677
8626807	3020-01-149-7941		106		10
		3020-01-149-7941	107		8
7X677	8626809	3040-01-150-0407		106	7
11862	8626812			113	10
7X677	8626816	5325-01-154-8562		107	1
11862	8626817			106	25
7X677	8626817			107	11
7X677	8626879	5315-01-152-9031		116	4
7X677	8626881	5340-01-151-4964		116	5
7X677	8626884	5310-01-151-4137		116	3
24617	8626916	5330-01-025-4212		121	22
72590	8626982			100	11
7X677	8627153	5330-01-470-1922		116	14
7X677	8627192	5315-01-174-8644		116	7

I-332

SECTION IV TM9-2320-280-24P C01

CROSS REFERENCE INDEXES

PART NUMBER INDEX

CAGE	PART NUMBER	STOCK NUMBER	FIGURE NO	ITEM NO
		5315-01-174-8644	117	7
7X677	8627193	5315-01-174-8645	116	7
		5315-01-174-8645	117	7
7X677	8627194	5315-01-174-8646	116	7
		5315-01-174-8646	117	7
7X677	8627195	5315-01-174-8647	116	7
		5315-01-174-8647	117	7
7X677	8627334	2520-01-149-1868	109	14
		2520-01-149-1868	114	3
7X677	8627379		113	15
11862	8627385	2520-01-149-3809	113	16
11862	8627422		109	10
7X677	8627509	4730-01-149-7935	118	5
7X677	8627627	5330-01-155-4383	113	13
73342	8627650	2520-00-557-6619	100	16
7X677	8627657	5325-01-154-8561	106	3
		5325-01-154-8561	107	3
72590	8627892		106	13
7X677	8627989	2520-01-163-7213	KITS	91
7X677	8629487	3110-01-155-4438	109	15
7X677	8629523	5365-01-153-0872	104	5
7X677	8629526	4710-01-152-5798	104	14
7X677	8629796	2520-01-154-1185	118	14
7X677	8629961	5365-01-085-0910	121	3
7X677	8629991	3040-01-448-9611	104	2
11862	8633056		106	17
11862	8633059	2520-01-168-1983	106	16
11862	8633075		106	1
7X677	8633173	2520-01-461-0092	114	1
7X677	8633208	5306-01-150-9497	104	13
7X677	8637742	4730-01-163-7163	105	3
7X677	8639718	4820-01-212-6763	121	10
7X677	8639743	5305-01-150-9781	4	25
7X677	8648178	5310-01-447-0968	108	3
29930	865168R91	3110-00-436-3248	128	50
7X677	8653985	3120-01-158-6304	106	18
7X677	8654701		105	16
7X677	8654716	5330-01-324-0906	121	2
		5330-01-324-0906	122	6
7X677	8655027	2520-01-160-9570	118	7
72590	8655036	2520-01-212-7634	104	9
11862	8655037		104	10
7X677	8655186	4730-01-177-7501	104	23
		4730-01-177-7501	105	5
7X677	8655280	5306-01-269-4319	121	21
7X677	8655625	5330-01-148-7492	104	7
7X677	8655843	5360-01-478-6550	117	6
7X677	8656613	5330-01-397-0374	105	14
7X677	8656942	2520-01-164-7234	102	1

I-333

SECTION IV TM9-2320-280-24P C01

CROSS REFERENCE INDEXES

PART NUMBER INDEX

CAGE	PART NUMBER	STOCK NUMBER	FIGURE NO	ITEM NO
7X677	8657163	5330-01-251-1607	100	10
		5330-01-251-1607	108	7
73342	8658110	5330-01-043-5572	100	2
		5330-01-043-5572	100	5
7X677	8661271		115	23
7X677	8661568	5325-01-408-7971	115	8
7X677	8661571	2520-01-409-1758	115	18
7X677	8661572	5325-01-408-7970	115	16
7X677	8661577	2520-01-410-8241	115	27
7X677	8661578	5325-01-461-7157	115	28
7X677	8661579	2520-01-411-3959	115	21
7X677	8661582	2520-01-409-1751	115	9
7X677	8661598	3040-01-410-8071	107	22
7X677	8661602	5330-01-379-1139	122	4
7X677	8661613	3040-01-410-8068	107	13
7X677	8661623	2520-01-461-0083	115	13
7X677	8661629	2520-01-461-0085	115	7
7X677	8661639	5330-01-470-6543	120	4
7X677	8661647		120	6
7X677	8661658		114	12
7X677	8661682		122	7
7X677	8661683		122	8
7X677	8661692	5315-01-460-9907	108	11
7X677	8661693	3040-01-460-9964	108	13
7X677	8661708	3040-01-447-3797	107	7
7X677	8661709	4330-01-411-2786	119	17
7X677	8661760	5330-01-414-6607	122	2
7X677	8661762	5306-01-461-3557	105	12
7X677	8661764	2520-01-461-0099	110	9
		2520-01-461-0099	112	8
7X677	8661766	4810-01-460-9975	122	16
7X677	8661767	5360-01-408-7977	122	15
7X677	8661768	5315-01-461-5649	122	14
7X677	8661789	5331-01-462-7294	122	32
7X677	8661792	5306-01-411-1596	122	1
7X677	8661798	5360-01-460-9070	122	23
7X677	8661801		119	4
7X677	8661804		119	12
7X677	8661805		119	14
11862	8661807		119	21
7X677	8661821		119	16
7X677	8661822		119	19
7X677	8661829		120	25
7X677	8661834	2520-01-461-2374	105	22
7X677	8661837	2520-01-478-6607	112	11
7X677	8661838	5315-01-461-8374	122	10
7X677	8661867		119	18
7X677	8661871		119	25
7X677	8661873		119	15

SECTION IV TM9-2320-280-24P C01

CROSS REFERENCE INDEXES

PART NUMBER INDEX

CAGE	PART NUMBER	STOCK NUMBER	FIGURE NO	ITEM NO
7X677	8661890		117	14
7X677	8661891	5360-01-460-9069	119	3
7X677	8661893	5330-01-460-8987	115	4
24617	8661894	5330-01-468-3604	115	1
7X677	8665830		105	16
7X677	8670200	2520-01-163-7866	109	11
		2520-01-163-7866	111	6
7X677	8670201	2520-01-192-1260	109	9
7X677	8670393	5330-01-151-6106	118	1
7X677	8670447	5330-01-330-9645	118	13
7X677	8670452	2520-01-502-6586	118	14
7X677	8670458	5330-01-331-9995	118	1
7X677	8675508	3040-01-493-4036	112	7
61928	8675511	2520-01-413-5540	110	8
7X677	8675517	5330-01-460-8988	105	7
7X677	8675519	2520-01-461-0073	114	9
7X677	8675520		114	2
7X677	8675521	2520-01-479-0847	114	8
7X677	8675522	2520-01-461-0074	114	7
7X677	8675523	5325-01-408-7969	115	11
7X677	8675524	5325-01-460-9908	105	8
7X677	8675529	5340-01-460-8207	105	20
7X677	8675533	5306-01-469-9903	120	3
7X677	8675535		119	9
7X677	8675537	5340-01-493-9134	120	17
7X677	8675539	5340-01-447-5839	78	3
7X677	8675557	2520-01-461-0076	110	2
7X677	8675558	5340-01-476-0352	110	14
7X677	8675578		109	12
7X677	8675611	3040-01-480-7536	107	14
7X677	8675612	2520-01-461-0107	107	14
7X677	8675634	5360-01-469-8077	120	22
7X677	8675728	5330-01-478-4797	117	18
7X677	8675729	5330-01-409-1664	114	15
11862	8675732		119	6
7X677	8676000	5306-01-491-9414	117	20
7X677	8676074		119	20
7X677	8676222	5306-01-476-3501	120	14
7X677	8676418	5340-01-474-4002	120	21
7X677	8677091	3020-01-479-2069	112	13
24617	8677483		110	6
7X677	8677571	3120-01-461-6679	122	31
7X677	8677572	3120-01-471-7844	122	31
7X677	8677573	3120-01-411-5783	122	31
7X677	8677574	3120-01-411-5787	122	31
7X677	8677575	3120-01-461-6677	122	31
7X677	8677582		110	7
			112	4
7X677	8677583		112	6

SECTION IV TM9-2320-280-24P C01

CROSS REFERENCE INDEXES

PART NUMBER INDEX

CAGE	PART NUMBER	STOCK NUMBER	FIGURE NO	ITEM NO
7X677	8677743	5330-01-360-5271	105	17
7X677	8677782	5330-01-409-1665	122	30
7X677	8677887	5325-01-408-7972	115	10
7X677	8678122	5315-01-408-8020	108	8
7X677	8678257	4820-01-460-9991	122	13
7X677	8678258	5360-01-460-9068	122	12
7X677	8678292		117	12
7X677	8678294		117	11
7X677	8678337	4820-01-409-1218	119	22
7X677	8678347	5315-01-447-0492	108	4
7X677	8678523	5340-01-460-8364	119	27
7X677	8678537	3040-01-149-6759	116	2
7X677	8679439	5325-01-412-8806	115	12
7X677	8679465	5325-01-477-6608	122	22
7X677	8679940	2520-01-150-2279	111	10
7X677	8679943	4820-01-147-4294	100	3
7X677	8679950	3040-01-149-6706	111	1
7X677	8679960	3040-01-266-2122	109	17
7X677	8679975	2520-01-151-3857	116	15
72590	8679988	4320-01-149-1866	121	1
7X677	8679990	2520-01-164-7158	100	9
7X677	8680547	4820-01-460-8288	122	24
7X677	8680548	3120-01-461-6680	122	19
7X677	8680549	4820-01-461-0011	122	20
7X677	8680551	5360-01-460-9073	122	18
61928	8680816	2520-01-392-8435	110	11
		2520-01-392-8435	112	17
7X677	8680911	2520-01-410-8067	KITS	93
7X677	8680914	2520-01-174-8153	KITS	94
7X677	8680915	2520-01-410-8077	KITS	95
7X677	8680918	2520-01-410-8079	115	5
7X677	8680929	2520-01-164-7157	116	11
		2520-01-164-7157	KITS	97
72590	8680938	2520-01-168-2060	113	4
7X677	8681069	5360-01-460-9072	122	28
7X677	8681070	4820-01-460-9988	122	27
7X677	8681449	2520-01-461-0078	115	17
11862	8681620	2530-01-159-2732	106	27
7X677	8681899		119	26
11862	8682194		122	5
7X677	8682215	5340-01-479-2514	117	19
7X677	8682802	2520-01-439-5265	112	1
		2520-01-439-5265	115	20
7X677	8682855	2520-01-461-5133	122	25
7X677	8682856	5340-01-460-8361	122	21
7X677	8682857	2520-01-461-5648	122	29
7X677	8682858	5340-01-477-7052	122	11
7X677	8683075		114	17
61928	8683088	2520-01-444-2711	120	5

SECTION IV TM9-2320-280-24P C01

CROSS REFERENCE INDEXES

PART NUMBER INDEX

CAGE	PART NUMBER	STOCK NUMBER	FIGURE NO	ITEM NO
7X677	8683188		119	28
7X677	8683771	3020-01-480-7311	107	17
7X677	8684205		120	8
7X677	8684213		119	31
7X677	8684214		119	30
7X677	8684215	5340-01-460-8363	119	29
7X677	8684217	5340-01-460-8362	119	23
7X677	8684218		119	1
7X677	8684220	2520-01-409-2509	120	1
7X677	8684221	2520-01-398-4589	105	15
7X677	8685472		117	4
7X677	8685473		117	1
7X677	8685474		117	2
7X677	8685921	2520-01-481-8344	KITS	98
7X677	8686122		115	24
7X677	8686124	5306-01-461-3555	105	13
19207	8689220	5940-00-503-6184	41	17
19207	8689221	5940-01-170-4956	41	10
		5940-01-170-4956	41	19
		5940-01-170-4956	313	15
		5940-01-170-4956	474	11
19207	8701325	5310-00-655-9860	80	8
		5310-00-655-9860	85	8
		5310-00-655-9860	86	4
76700	870889A	4330-01-506-2918	176A	12
19207	8712289	5310-00-044-3340	40	7
19207	8712289-4	5310-00-840-6222	169	14
19204	8712289-5	5310-00-044-3342	127	13
		5310-00-044-3342	180	17
19207	8724258	5935-00-686-2599	85	16
		5935-00-686-2599	86	9
11862	8727200		116	13
19207	8741435	5310-00-241-6921	62	20
19207	8741437	5305-00-832-5743	62	15
19207	8741442	5325-00-088-6147	62	23
19207	8741446	5325-00-832-5650	62	16
19207	8741447	6220-00-998-6142	62	19
19207	8741461	6220-00-443-0589	62	21
19207	8741491	6240-00-966-3831	62	17
19207	8741492	5935-00-807-4109	62	22
		5935-00-807-4109	366	31
		5935-00-807-4109	386	9
		5935-00-807-4109	469	10
		5935-00-807-4109	483	1
19207	8754124	5325-01-144-4871	68	8
19207	8754125	5930-01-153-8215	68	10
56232	876105-012	5310-00-063-7360	345	42
		5310-00-063-7360	348	39
27647	8762	5310-01-138-2605	314	23

I-337

SECTION IV TM9-2320-280-24P C01

CROSS REFERENCE INDEXES

PART NUMBER INDEX

CAGE	PART NUMBER	STOCK NUMBER	FIGURE NO	ITEM NO
27647	8763	5305-01-210-6248	314	21
13445	88587	5310-01-185-7208	69	8
		5310-01-185-7208	70	3
72582	8924145	4730-00-803-6266	334	11
41885	89305	2530-01-160-9569	167	1
39428	8930T31	1640-00-929-0041	BULK	50
41885	89324	5305-01-205-6048	167	9
76760	89331A	5325-01-358-7960	129	20
19712	8MA-3010AS	2920-01-244-4993	43	1
11083	8S4680	5330-00-493-3876	KITS	84
19728	8X-1681		45	22
19728	8X-179		45	59
19728	8X-3439		45	81
19728	8X-349		45	34
19728	8X-4678		45	19
19728	8X-830		45	61
19220	9-300-U-L	2540-01-197-4900	281	22
19220	9-300-U-R	2540-01-196-5317	280	22
		2540-01-196-5317	281	22
68505	90-2187	2920-01-168-4128	KITS	100
68505	90-2188	5310-01-027-9392	KITS	101
68505	90-2206	2920-00-302-6342	KITS	102
68505	90-2225	2920-01-068-7182	KITS	103
68505	90-2531	2920-01-168-4129	KITS	104
68505	90-2532	2920-01-168-4130	KITS	105
68505	90-2836	5305-01-188-9707	KITS	106
68505	90-2837	2920-01-192-2956	KITS	107
68505	90-2838	2920-01-192-4473	KITS	108
19728	90-2840	3120-01-191-4637	KITS	109
68505	90-2841	2920-01-191-6534	KITS	110
68505	90-816	2920-01-192-2959	KITS	99
27182	90-SN	5340-01-271-6361	215	5
98624	9002-00181-48	5331-01-194-8966	34	10
51377	9002-00741-58	5331-01-195-1500	34	9
19207	900529	3110-00-108-9247	458	35
78500	902-04-48-244	2520-01-214-2565	169	17
81834	90221	6220-00-001-1514	64	8
41885	90297	5306-01-336-7175	167	15
09094	903C4-16	5315-00-973-8637	245	11
41885	90619	4730-01-346-1063	167	24
73342	907425	3110-00-155-6152	128	14
13940	91-BS-78403		224	13
			324	2
13940	91-BS-78403-1E	5325-00-281-8643	316	5
		5325-00-281-8643	319	2
		5325-00-281-8643	322	2
63829	9103-9800	2520-01-289-3617	KITS	111
39428	91248A721	5305-00-719-5240	191	14
95019	913359-1528	2520-01-171-8258	137	4

SECTION IV TM9-2320-280-24P C01

CROSS REFERENCE INDEXES

PART NUMBER INDEX

CAGE	PART NUMBER	STOCK NUMBER	FIGURE NO	ITEM NO
76760	91386	5315-00-576-0265	132	30
63829	9140-9810	2520-01-290-3921	KITS	112
95019	914201-2626	2520-01-200-3096	136	16
95019	914202-1524	2520-01-200-3097	136	6
41885	91427EBKBC	2530-01-417-4908	167	2
		2530-01-417-4908	167	14
41885	91428EBKBC	2530-01-417-8450	167	8
		2530-01-417-8450	167	19
K1076	9145-105-00B	5310-01-127-2456	100	7
		5310-01-127-2456	173	1
95019	915051-1624	2520-01-346-1374	137	4
73165	91598	5330-01-425-5069	4	29
41885	91610	5331-01-417-1043	167	16
63829	9162-0121	5315-01-287-6520	491	4
68505	92-2923	5310-01-214-4955	45	50
0VDA9	922-621-02	6220-01-269-5263	65	3
0VDA9	922-720-11	6220-01-266-1651	65	24
0VDA9	922-721-00	6230-01-269-8054	65	22
		6230-01-269-8054	289	28
0VDA9	922-925-10	6250-01-339-6271	65	11
0VDA9	922-929-00	6220-01-338-2059	65	9
7X677	930678	3110-01-411-5784	115	15
7X677	930679	2520-01-149-1221	105	2
		2520-01-149-1221	106	2
34623	934245	4730-00-954-1251	18	5
24617	9356973	5962-01-476-7772	50	4
63829	9383-6010	2520-01-289-9663	491	1
7X677	9411030	4730-00-050-4205	170	4
27387	941180	2530-01-186-3740	195	10
24617	9411807	5310-01-461-8043	227	48
24617	9411893	5310-00-251-4503	234	17
		5310-00-251-4503	246	9
		5310-00-251-4503	246	14
		5310-00-251-4503	247	13
		5310-00-251-4503	247	24
		5310-00-251-4503	248	9
		5310-00-251-4503	249	6
		5310-00-251-4503	249	21
		5310-00-251-4503	249	33
		5310-00-251-4503	251	15
		5310-00-251-4503	251	25
		5310-00-251-4503	251	39
		5310-00-251-4503	252	17
		5310-00-251-4503	252	39
		5310-00-251-4503	253	5
		5310-00-251-4503	253	12
		5310-00-251-4503	253	15
		5310-00-251-4503	253	32
		5310-00-251-4503	253	39

SECTION IV TM9-2320-280-24P C01

CROSS REFERENCE INDEXES

PART NUMBER INDEX

CAGE	PART NUMBER	STOCK NUMBER	FIGURE NO	ITEM NO
		5310-00-251-4503 254		4
		5310-00-251-4503	254	12
		5310-00-251-4503 255		4
		5310-00-251-4503	255	12
		5310-00-251-4503	255	14
		5310-00-251-4503	255	25
		5310-00-251-4503	255	29
		5310-00-251-4503	260	16
		5310-00-251-4503	416	11
		5310-00-251-4503 446		9
		5310-00-251-4503 447		5
		5310-00-251-4503 448		8
		5310-00-251-4503 449		2
		5310-00-251-4503 450		1
		5310-00-251-4503 486		3
34623	9412281	5315-01-173-3397	277	19
		5315-01-173-3397	277	46
24617	9413266	5310-01-361-1152	438	9
7X677	9413349	5325-01-408-7973	120	2
24617	9413509	5310-00-768-0318	69	1
7X677	9413523	5340-01-479-2517	117	17
24617	9413583	5310-01-381-5328	231	24
		5310-01-381-5328 454		4
7X677	9414224	5305-01-211-1439	47	3
24617	9414241	5305-01-361-5353	225	52
		5305-01-361-5353 306		8
		5305-01-361-5353 346		5
		5305-01-361-5353	410	23
		5305-01-361-5353 445		4
		5305-01-361-5353 451		3
24617	9414295	5305-01-433-9248	366	6
24617	9414313	5305-01-380-9163	333	14
86403	9414722	5305-01-213-5024	332	17
24617	9414726	5305-01-447-9227	332	31
24617	9415477	5305-00-489-0743	77	18
		5305-00-489-0743	88	14
		5305-00-489-0743	88	28
		5305-00-489-0743 443		5
		5305-00-489-0743 444		6
		5305-00-489-0743	457	16
7X677	9415560	5306-01-186-5369	188	27
		5306-01-186-5369	442	18
24617	9415778	5305-01-394-3543	370	16
54132	9415779	5305-01-120-4363	388	8
7X677	9415786	5305-01-187-9555	73	7
7X677	9415857	5305-01-211-1450	13	4
		5305-01-211-1450 14		4
		5305-01-211-1450	16	24
24617	9416128	5310-00-402-2778	51	12

SECTION IV TM9-2320-280-24P C01

CROSS REFERENCE INDEXES

PART NUMBER INDEX

CAGE	PART NUMBER	STOCK NUMBER	FIGURE NO	ITEM NO
24617	9416402	5305-01-456-6358	314	40
24617	9416918	5310-01-012-8962	24	9
		5310-01-012-8962 393	4	
		5310-01-012-8962	395	21
24617	9417028	5310-01-457-3171	35	30
		5310-01-457-3171 36	9	
		5310-01-457-3171 38	9	
		5310-01-457-3171	472	20
		5310-01-457-3171	475	18
24617	9417373	5310-01-361-1163	439	7
		5310-01-361-1163	462	8
		5310-01-361-1163	465	1
		5310-01-361-1163	485	25
1W358	9417497		227	44
24617	9417497		227	44
24617	9417714	5310-01-217-5205	85	5
		5310-01-217-5205 88	3	
		5310-01-217-5205	203	30
		5310-01-217-5205 272	5	
		5310-01-217-5205 486	2	
7X677	9417793	5310-01-211-3811	24	19
		5310-01-211-3811 25	6	
		5310-01-211-3811 175	5	
24617	9417901	4730-00-050-4203	170	5
24617	9418304	5310-01-119-3675	7	2
24617	9418346	5305-00-179-8946	89	13
		5305-00-179-8946	91	10
		5305-00-179-8946	93	21
		5305-00-179-8946 94	8	
		5305-00-179-8946	95	36
24617	9418753	5310-01-465-4525	33	7
		5310-01-465-4525 512	4	
16764	9418913	5305-00-841-2044	72	14
24617	9418924	5310-01-132-8275	12	12
34623	9419011	5306-01-266-8274	202	28
24617	9419032	5306-01-158-6224	7	1
7X677	9419079	5305-01-211-7478	170	15
		5305-01-211-7478 195	4	
7X677	9419143	5310-01-148-0240	68	4
24617	9419265	5310-01-170-8765	411	18
7X677	9419293	5306-01-185-7052	137	15
		5306-01-185-7052	239	11
7X677	9419343	5306-01-197-3274	150	27
24617	9419456	5310-01-318-5237	389	4
24617	9419471	5310-01-432-6727	153	3
		5310-01-432-6727	155	14
		5310-01-432-6727	198	10
		5310-01-432-6727 208	4	
		5310-01-432-6727	227	31

I-341

SECTION IV TM9-2320-280-24P C01

CROSS REFERENCE INDEXES

PART NUMBER INDEX

CAGE	PART NUMBER	STOCK NUMBER	FIGURE NO	ITEM NO
		5310-01-432-6727	261	11
		5310-01-432-6727	321	15
		5310-01-432-6727	325	17
		5310-01-432-6727	326	12
		5310-01-432-6727	332	2
		5310-01-432-6727	377	8
		5310-01-432-6727	486	4
		5310-01-432-6727	487	10
24617	9419476	5310-00-984-3807	199	9
		5310-00-984-3807	324	9
		5310-00-984-3807	325	25
		5310-00-984-3807	325	26
24617	9419482	5310-01-409-0897	462	14
24617	9419836	5305-01-215-3985	43	32
24617	9420022	5310-00-809-4085	35	23
		5310-00-809-4085	152	15
		5310-00-809-4085	456	16
		5310-00-809-4085	472	11
24617	9421073	5305-01-162-8512	12	2
24617	9421077	5310-01-102-7356	154	1
7X677	9421294	5305-01-185-8674	240	4
		5305-01-185-8674	240	18
		5305-01-185-8674	240	22
		5305-01-185-8674	269	6
		5305-01-185-8674	442	5
7X677	9421394	5310-01-199-2293	227	32
		5310-01-199-2293	377	2
24617	9421705	5305-01-381-2296	231	11
7X677	9422042	5306-01-207-7487	374	7
24617	9422277	5310-01-126-9404	329	12
24617	9422295	5310-01-119-3668	231	7
		5310-01-119-3668	439	5
24617	9422299	5310-01-150-4003	127	9
		5310-01-150-4003	136	3
		5310-01-150-4003	144	14
		5310-01-150-4003	152	14
		5310-01-150-4003	271	12
		5310-01-150-4003	271	27
		5310-01-150-4003	398	8
		5310-01-150-4003	464	8
7X677	9422302	5310-01-184-5784	194	2
11862	9422303	5310-00-044-3342	166	5
24617	9422305	5310-01-130-4274	196	7
7X677	9422306	5310-00-198-6691	437	5
7X677	9422771	5310-01-211-0691	284	17
		5310-01-211-0691	439	10
7X677	9423534	5310-01-213-4192	205	15
		5310-01-213-4192	250	15
24617	9423557	5306-01-360-1123	283	21

I-342

SECTION IV TM9-2320-280-24P C01

CROSS REFERENCE INDEXES

PART NUMBER INDEX

CAGE	PART NUMBER	STOCK NUMBER	FIGURE NO	ITEM NO
		5306-01-360-1123	438	17
7X677	9423869	5310-01-186-1245	197	34
24617	9423995	5305-00-719-5235	437	12
7X677	9424258	5310-01-199-3440	209	4
		5310-01-199-3440	225	40
7X677	9424760	5306-01-255-2661	193	7
7X677	9425078	5305-01-210-6251	15	34
3M915	9425170	5305-01-461-4396	332	38
76700	94252A	2530-01-488-6147	176A	5
24617	9425339	5305-01-315-7066	239	2
		5305-01-315-7066	402	4
24617	9426241	5305-01-206-7219	13	8
		5305-01-206-7219	14	14
		5305-01-206-7219	17	15
		5305-01-206-7219	88	35
7X677	9427162	5306-01-186-7128	181	14
24617	9427321	5305-01-104-4846	187	8
		5305-01-104-4846	441	7
7X677	9427812	5305-01-255-0094	185	2
24617	9428039	5305-01-111-2774	187	2
		5305-01-111-2774	441	2
7X677	9428308	5310-01-212-2213	170	19
		5310-01-212-2213	195	8
7X677	9428747	5305-01-204-4190	275	17
		5305-01-204-4190	276	16
7X677	9428795	5305-01-197-9417	463	14
7X677	9428839	5310-01-408-9593	127	6
7X677	9428851	5310-01-250-7835	26	10
7X677	9429048	5310-01-209-7702	225	39
		5310-01-209-7702	231	21
24617	9430516	5305-01-411-4455	453	16
24617	9430677	5306-01-130-0457	463	8
7X677	9430724	5305-01-198-3440	463	7
7X677	9430761	5305-01-213-4149	171	5
7X677	9434068	3110-00-198-0492	128	27
24617	9436356		487	6
24617	9436711		487	4
7X677	9437847	5306-01-156-3730	445	13
7X677	9437889	5340-01-195-1575	69	3
24617	9438218		13	24
			14	9
7X677	9438257	4720-01-155-7784	BULK	26
7X677	9438315	4720-01-148-2768	BULK	14
7X677	9438373	4720-01-159-5796	BULK	25
7X677	9438381	4720-01-156-0550	BULK	13
7X677	9439046	4720-01-156-0549	BULK	12
11862	9439363		11	3
7X677	9440949	4720-01-199-5717	BULK	7
72582	9442899	5310-01-135-4730	131	1

SECTION IV TM9-2320-280-24P C01

CROSS REFERENCE INDEXES

PART NUMBER INDEX

CAGE	PART NUMBER	STOCK NUMBER	FIGURE NO	ITEM NO
24617	9442900	5310-01-149-4407	26	13
24617	9442938	5310-01-315-3403	234	10
		5310-01-315-3403	235	13
		5310-01-315-3403	239	7
		5310-01-315-3403	402	10
92867	95001418	5340-01-213-5600	153	9
92867	95001419	5340-01-211-7562	153	23
35510	95467	5970-01-269-8384	39	14
35510	95469	5306-01-272-4798	39	24
35510	95483	5330-01-272-9071	39	17
35510	95484	5307-01-272-7539	39	54
35510	95498	5330-01-270-3668	39	26
35510	95504	5305-01-282-2667	39	31
35510	95510	3130-01-271-6464	39	3
35510	95513	5330-01-272-5474	39	2
24975	95515	5977-01-270-1468	39	72
35510	95516	5365-01-272-2663	39	73
35510	95552	5340-01-272-5516	39	5
35510	95605	5310-01-270-8377	39	13
35510	95965	5970-01-260-9132	39	75
20722	9609A838-002	5340-0-149-6854	474	15
40152	961458	6240-00-765-8443	47	2
35510	96154		39	38
35510	96156	6150-01-269-1858	39	64
35510	96160	6150-01-269-1856	39	40
35510	96162	6150-01-269-1863	39	51
35510	96171	3120-01-271-6432	39	59
35510	96172	5330-01-270-8316	39	67
35510	96180	5365-01-272-4877	39	50
35510	96187	5970-01-270-1467	39	60
35510	96188	5920-01-266-3908	39	62
35510	96247		39	39
35510	96259	5970-01-270-1465	39	27
70485	963	5325-01-195-8002	71	6
80252	9630	2590-00-683-0598	68	5
35510	96377	3040-01-271-3823	39	41
35510	96520	2920-01-310-9978	39	16
76760	96711-B		130	35
35510	96745	5365-01-272-7502	39	15
35510	96756	6130-01-309-6458	39	53
35510	96764	3110-01-271-6355	39	20
35510	96768	5330-01-272-8337	39	71
35510	96769		39	70
35510	96846		39	19
20722	96909A383-002	5340-01-449-6854	41	8
73002	9781210307	4720-01-505-4702	176A	23
27647	980074 5553	5305-01-210-4967	314	25
39428	98019A510	5310-00-656-0114	476	13
0GZB7	983-60-50085-A	3940-01-475-4983	469	9

SECTION IV TM9-2320-280-24P C01

CROSS REFERENCE INDEXES

PART NUMBER INDEX

CAGE	PART NUMBER	STOCK NUMBER	FIGURE NO	ITEM NO
0GZB7	983-60-50089	5340-01-475-3650	469	8
0GZB7	983-74-1001	6150-01-475-8835	469	6
0GZB7	983-75-50050CV	2590-01-474-3326	471	2
0GZB7	983-88-00C36	4720-01-475-3757	470	1
0GZB7	983-88-00C47	4720-01-475-4641	470	2
0GZB7	983-90-50201	4710-01-475-3753	469	7
0GZB7	983-92-52000	5365-01-474-1057	470	16
		5365-01-474-1057	471	7
35510	99479	5977-01-143-6996	39	68
35510	99514	5330-01-145-5376	39	69
41024	99890451	2510-01-189-3459	320	5
		2510-01-189-3459	324	10

I-345

By Order of the Secretary of the Army:

                                          ERIC K. SHINSEKI
General, United States Army
Chief of Staff

Official:

JOEL B. HUDSON
Administrative Assistant to the
Secretary of the Army
05696

By Order of the Secretary of the Air Force:

                                          RONALD R. FOGLEMAN
General, United States Air Force
Chief of Staff

Official:

HENRY VICCELLIO, JR.
General, United States Air Force
Commander, Air Force Materiel Command

Distribution:

    To be distributed in accordance with the initial distribution number (IDN) 381008, requirements for TM 9-2320-280-24P-2.

RECOMMENDED CHANGES TO PUBLICATIONS AND BLANK FORMS For use of this form, see AR 25-30; the proponentagency is ODISC4.	Use Part II (reverse) for Repair Parts and Special Tools Lists (RPSTL) and Supply Catalogs/Supply Manuals (SC/SM).	DATE:
TO: *(Forward to proponent of publication or form) (include ZIP code)* AMSTA-LC-CI Tech Pubs, TACOM-RI 1 Rock Island Arsenal Rock Island, IL 61299-7630	**FROM:** *(Activity and location) (include ZIP code)*	

PART I - ALL PUBLICATIONS (EXCEPT RPSTL AND SC/SM) AND BLANK FORMS

PUBLICATION/FORM NUMBER	DATE	TITLE

ITEM NO.	PAGE NO.	PARA-GRAPH	LINE NO.*	FIG-URE NO.	TABLE NO.	RECOMMENDED CHANGES AND REASON *(Provide exact wording of recommended changes, if possible).*
						SAMPLE

* *Reference to line numbers within the paragraph or subparagraph.*

TYPED NAME, GRADE, OR TITLE	TELEPHONE EXCHANGE/AUTOVON, PLUS EXTENSION	SIGNATURE

DA FORM 2028, FEB 74 REPLACES DA FORM 2028, 1 DEC 68, WHICH WILL BE USAPPC V3.00

TO: *(Forward direct to addressee listed in publication)* AMSTA-LC-CI Tech Pubs, TACOM-RI 1 Rock Island Arsenal Rock Island, IL 61299-7630	FROM: *(Activity and location) (include ZIP code)* Co. B, 1st BN, 2nd Brigade Ft. Hood, TX 76445	DATE: 14 Jul 2003

PART II - REPAIR PARTS AND SPECIAL TOOLS LISTS AND SUPPLY CATALOGS/SUPPLY MANUALS

PUBLICATION/FORM NUMBER	DATE	TITLE
TM 9-2320-280-24P-2	30 MARCH 2001	TECHNICAL MANUAL, UNIT, DIRECT SUPPORT AND GENERAL SUPPORT MAINTENANCE REPAIR PARTS AND SPECIAL TOOLS LIST

PAGE NO.	LINE NO.	NATIONAL STOCK NUMBER	REFERENCE NO.	FIGURE NO.	TOTAL NO. OF MAJOR ITEMS SUPPORTED	RECOMMENDED ACTION
420-1	5	6,7	N/A	420	6,7	1 Part numbers for items 6 and 7 are incorrectly listed; item No. 6 should be P/N 1834 and item No. 7 should be P/N 1833.

SAMPLE

PART III - REMARKS *(Any general remarks or recommendations, or suggestions for improvement of publications and blank forms. Additional blank sheets may be used if more space is needed.)*

TYPED NAME, GRADE, OR TITLE	TELEPHONE EXCHANGE/AUTOVON, PLUS EXTENSION	SIGNATURE
Pat Smith, ILT	AV272-4162	*Pat Smith, ILT*

USAPPC V3.00

RECOMMENDED CHANGES TO PUBLICATIONS AND BLANK FORMS For use of this form, see AR 25-30; the proponentagency is ODISC4.	Use Part II (reverse) for Repair Parts and Special Tools Lists (RPSTL) and Supply Catalogs/Supply Manuals (SC/SM).	DATE:
TO: *(Forward to proponent of publication or form) (include ZIP code)* **AMSTA-LC-CI Tech Pubs, TACOM-RI** 1 Rock Island Arsenal Rock Island, IL 61299-7630	**FROM:** *(Activity and location) (include ZIP code)*	

PART I - ALL PUBLICATIONS (EXCEPT RPSTL AND SC/SM) AND BLANK FORMS

PUBLICATION/FORM NUMBER	DATE	TITLE

ITEM NO.	PAGE NO.	PARA-GRAPH	LINE NO.*	FIG-URE NO.	TABLE NO.	RECOMMENDED CHANGES AND REASON *(Provide exact wording of recommended changes, if possible).*

* *Reference to line numbers within the paragraph or subparagraph.*

TYPED NAME, GRADE, OR TITLE	TELEPHONE EXCHANGE/AUTOVON, PLUS EXTENSION	SIGNATURE

DA FORM 2028, FEB 74   REPLACES DA FORM 2028, 1 DEC 68, WHICH WILL BE   USAPPC V3.00

TO: *(Forward direct to addressee listed in publication)* AMSTA-LC-CI Tech Pubs, TACOM-RI 1 Rock Island Arsenal Rock Island, IL 61299-7630	FROM: *(Activity and location)* *(include ZIP code)*	DATE:

PART II - REPAIR PARTS AND SPECIAL TOOLS LISTS AND SUPPLY CATALOGS/SUPPLY MANUALS

PUBLICATION/FORM NUMBER	DATE	TITLE
TM 9-2320-280-24P-2	30 MARCH 2001	TECHNICAL MANUAL, UNIT, DIRECT SUPPORT AND GENERAL SUPPORT MAINTENANCE REPAIR PARTS AND SPECIAL TOOLS LIST

PAGE NO.		LINE NO.	NATIONAL STOCK NUMBER	REFERENCE NO.	FIGURE NO.		TOTAL NO. OF MAJOR ITEMS SUPPORTED	RECOMMENDED ACTION

PART III - REMARKS *(Any general remarks or recommendations, or suggestions for improvement of publications and blank forms. Additional blank sheets may be used if more space is needed.)*

TYPED NAME, GRADE, OR TITLE	TELEPHONE EXCHANGE/AUTOVON, PLUS EXTENSION	SIGNATURE

USAPPC V3.00

RECOMMENDED CHANGES TO PUBLICATIONS AND BLANK FORMS For use of this form, see AR 25-30; the proponent agency is ODISC4.	Use Part II (reverse) for Repair Parts and Special Tools Lists (RPSTL) and Supply Catalogs/Supply Manuals (SC/SM).	DATE:
TO: *(Forward to proponent of publication or form) (include ZIP code)* **AMSTA-LC-CI Tech Pubs, TACOM-RI** 1 Rock Island Arsenal Rock Island, IL 61299-7630	**FROM:** *(Activity and location) (include ZIP code)*	

PART I - ALL PUBLICATIONS (EXCEPT RPSTL AND SC/SM) AND BLANK FORMS

PUBLICATION/FORM NUMBER	DATE	TITLE

ITEM NO.	PAGE NO.	PARA-GRAPH	LINE NO.*	FIG-URE NO.	TABLE NO.	RECOMMENDED CHANGES AND REASON *(Provide exact wording of recommended changes, if possible).*

* *Reference to line numbers within the paragraph or subparagraph.*

TYPED NAME, GRADE, OR TITLE	TELEPHONE EXCHANGE/AUTOVON, PLUS EXTENSION	SIGNATURE

DA FORM 2028, FEB 74   REPLACES DA FORM 2028, 1 DEC 68, WHICH WILL BE   USAPPC V3.00

TO: *(Forward direct to addressee listed in publication)* **AMSTA-LC-CI Tech Pubs, TACOM-RI** 1 Rock Island Arsenal Rock Island, IL 61299-7630	FROM: *(Activity and location)* *(include ZIP code)*	DATE:

| PART II - REPAIR PARTS AND SPECIAL TOOLS LISTS AND SUPPLY CATALOGS/SUPPLY MANUALS ||||
| PUBLICATION/FORM NUMBER
 TM 9-2320-280-24P-2 || DATE
 30 MARCH 2001 | TITLE **TECHNICAL MANUAL, UNIT, DIRECT SUPPORT AND GENERAL SUPPORT MAINTENANCE REPAIR PARTS AND SPECIAL TOOLS LIST** |

PAGE NO.	LINE NO.	NATIONAL STOCK NUMBER	REFERENCE NO.	FIGURE NO.	TOTAL NO. OF MAJOR ITEMS SUPPORTED	RECOMMENDED ACTION

PART III - REMARKS *(Any general remarks or recommendations, or suggestions for improvement of publications and blank forms. Additional blank sheets may be used if more space is needed.)*

TYPED NAME, GRADE, OR TITLE	TELEPHONE EXCHANGE/AUTOVON, PLUS EXTENSION	SIGNATURE

USAPPC V3.00

RECOMMENDED CHANGES TO PUBLICATIONS AND BLANK FORMS For use of this form, see AR 25-30; the proponent agency is ODISC4.	Use Part II (reverse) for Repair Parts and Special Tools Lists (RPSTL) and Supply Catalogs/Supply Manuals (SC/SM).	DATE:
TO: *(Forward to proponent of publication or form) (include ZIP code)* **AMSTA-LC-CI Tech Pubs, TACOM-RI** 1 Rock Island Arsenal Rock Island, IL 61299-7630	**FROM:** *(Activity and location) (include ZIP code)*	

PART I - ALL PUBLICATIONS (EXCEPT RPSTL AND SC/SM) AND BLANK FORMS

PUBLICATION/FORM NUMBER	DATE	TITLE

ITEM NO.	PAGE NO.	PARA-GRAPH	LINE NO.*	FIG-URE NO.	TABLE NO.	RECOMMENDED CHANGES AND REASON *(Provide exact wording of recommended changes, if possible).*

* *Reference to line numbers within the paragraph or subparagraph.*

TYPED NAME, GRADE, OR TITLE	TELEPHONE EXCHANGE/AUTOVON, PLUS EXTENSION	SIGNATURE

DA FORM 2028, FEB 74   REPLACES DA FORM 2028, 1 DEC 68, WHICH WILL BE   USAPPC V3.00

TO: *(Forward direct to addressee listed in publication)* AMSTA-LC-CI Tech Pubs, TACOM-RI 1 Rock Island Arsenal Rock Island, IL 61299-7630	FROM: *(Activity and location)* *(include ZIP code)*	DATE:

PART II - REPAIR PARTS AND SPECIAL TOOLS LISTS AND SUPPLY CATALOGS/SUPPLY MANUALS

PUBLICATION/FORM NUMBER	DATE	TITLE
TM 9-2320-280-24P-2	30 MARCH 2001	TECHNICAL MANUAL, UNIT, DIRECT SUPPORT AND GENERAL SUPPORT MAINTENANCE REPAIR PARTS AND SPECIAL TOOLS LIST

PAGE NO.		LINE NO.	NATIONAL STOCK NUMBER	REFERENCE NO.	FIGURE NO.		TOTAL NO. OF MAJOR ITEMS SUPPORTED	RECOMMENDED ACTION

PART III - REMARKS *(Any general remarks or recommendations, or suggestions for improvement of publications and blank forms. Additional blank sheets may be used if more space is needed.)*

TYPED NAME, GRADE, OR TITLE	TELEPHONE EXCHANGE/AUTOVON, PLUS EXTENSION	SIGNATURE

USAPPC V3.00

PIN: 074666-001

www.ingramcontent.com/pod-product-compliance
Lightning Source LLC
Chambersburg PA
CBHW081341070526
44578CB00005B/688